NUMERICAL METHODS IN GEOMECHANICS / INNSBRUC

PROCEEDINGS OF THE SIXTH INTERNATIONAL CONFERENCE ON NUMERICAL METHODS IN GEOMECHANICS / INNSBRUCK / 11-15 APRIL 1988

Numerical Methods in Geomechanics Innsbruck 1988

Edited by
G.SWOBODA
Institute of Structural Engineering, University of Innsbruck

VOLUME THREE:
10 *Tunnels and underground openings*
11 *Dynamic and earthquake engineering problems, blasting*
12 *Mining applications*
13 *Interpretation of field measurements, back analysis*
14 *Use of microcomputers*
15 *CAD, mesh generation, software*

Published on behalf of the International Committee for Numerical Methods in Geomechanics by
A.A.BALKEMA / ROTTERDAM / BROOKFIELD / 1988

ORGANIZING COMMITTEES

International Conference Committee

Prof. G.Swoboda (Chairman) University of Innsbruck, Austria
Prof. C.S.Desai (Co-Chairman) University of Arizona, USA
Prof. H.Duddeck (Co-Chairman) Technical University of Braunschweig, FR Germany
Prof. W.Wittke (Co-Chairman), Technical University of Aachen, FR Germany
Prof. T.Adachi, Kyoto University, Japan
Prof. D.Aubry, Ecole Centrale des Arts, France
Dr. G.Beer, CSIRO, Division Geomechanics, Australia
Prof. Y.K.Cheung, University of Hong Kong, Hong Kong
Prof. G.Clough, Virginia State University, USA
Dr. J.T.Christian, Stone & Webster Engineer Corp., USA
Prof. W.D.L.Finn, University of British Columbia, Canada
Prof. G.Gioda, Politecnico di Milano, Italy
Prof. G.Gudehus, Technical University of Karlsruhe, FR Germany
Prof. Y.Ichikawa, Nagoya University, Japan
Prof. K.Kovári, ETH-Hönggerberg, Switzerland
Prof. Y.M.Lin, Northeast University, People's Republic of China
Prof. S.Ukhov, Moscow Civil Engineering Institute, USSR
Prof. F.Medina, Universidad de Chile, Chile
Prof. Z.Mróz, Polish Academy of Science, Poland
Prof. J.Prevost, Princeton University, USA
Prof. J.Smith, University of Manchester, UK
Prof. C.Tanimoto, Kyoto University, Japan
Prof. S.Valliappan, University of New South Wales, Australia
Prof. A.Varadarajan, ITT, Delhi, India
Prof. S.J.Wang, Academia Sinica, People's Republic of China
Prof. N.E.Wiberg, Chalmers University of Technology, Sweden
Prof. O.C.Zienkiewicz, University of Wales, UK

International Committee for Numerical Methods in Geomechanics

Prof. T.Adachi, Japan
Prof. D.Aubry, France
Prof. A.S.Balasubramaniam, Thailand
Prof. J.R.Booker, Australia
Dr. C.A.Brebbia, UK
Prof. Y.K.Cheung, Hong Kong
Dr. J.T.Christian, USA
Dr. A.Cividini, Italy
Prof. C.S.Desai, USA (Chairman)
Prof. J.M.Duncan, USA
Prof. Z.Eisenstein, Canada
Prof. A.J.Ferrante, Brazil
Prof. W.D.L.Finn, Canada
Dr. J.Geertsma, Netherlands
Prof. J.Ghaboussi, USA
Prof. K.Höeg, Norway
Prof. K.Ishihara, Japan
Prof. T.Kawamoto, Japan
Prof. K.Kovári, Switzerland
Prof. S.Prakash, India
Prof. J.M.Roesset, USA
Prof. I.M.Smith, UK
Prof. V.I.Solomin, USSR
Prof. G.Swoboda, Austria
Prof. S.Valliappan, Australia
Prof. C.Viggiani, Italy
Prof. S.Wang, People's Republic of China
Prof. N.E.Wiberg, Sweden
Prof. W.Wittke, FR Germany
Prof. O.C.Zienkiewicz, UK

1)
624.1513'0184
NUM

The texts of the various papers in this volume were set individually by typists under the supervision of each of the authors concerned.

Published by
A.A.Balkema, P.O.Box 1675, 3000 BR Rotterdam, Netherlands
A.A.Balkema Publishers, Old Post Road, Brookfield, VT 05036, USA

For the complete set of three volumes ISBN 90 6191 809 X
For volume 1: ISBN 90 6191 810 3
For volume 2: ISBN 90 6191 811 1
For volume 3: ISBN 90 6191 812 X

© 1988 A.A.Balkema, Rotterdam

Printed in the Netherlands

Numerical Methods in Geomechanics (Innsbruck 1988), Swoboda (ed.)
© 1988 Balkema, Rotterdam. ISBN 90 6191 809 X

Contents

10 *Tunnels and underground openings*

13 *Interpretation of field measurements, back analysis*

14 *Use of microcomputers*

15 *CAD, mesh generation, software*

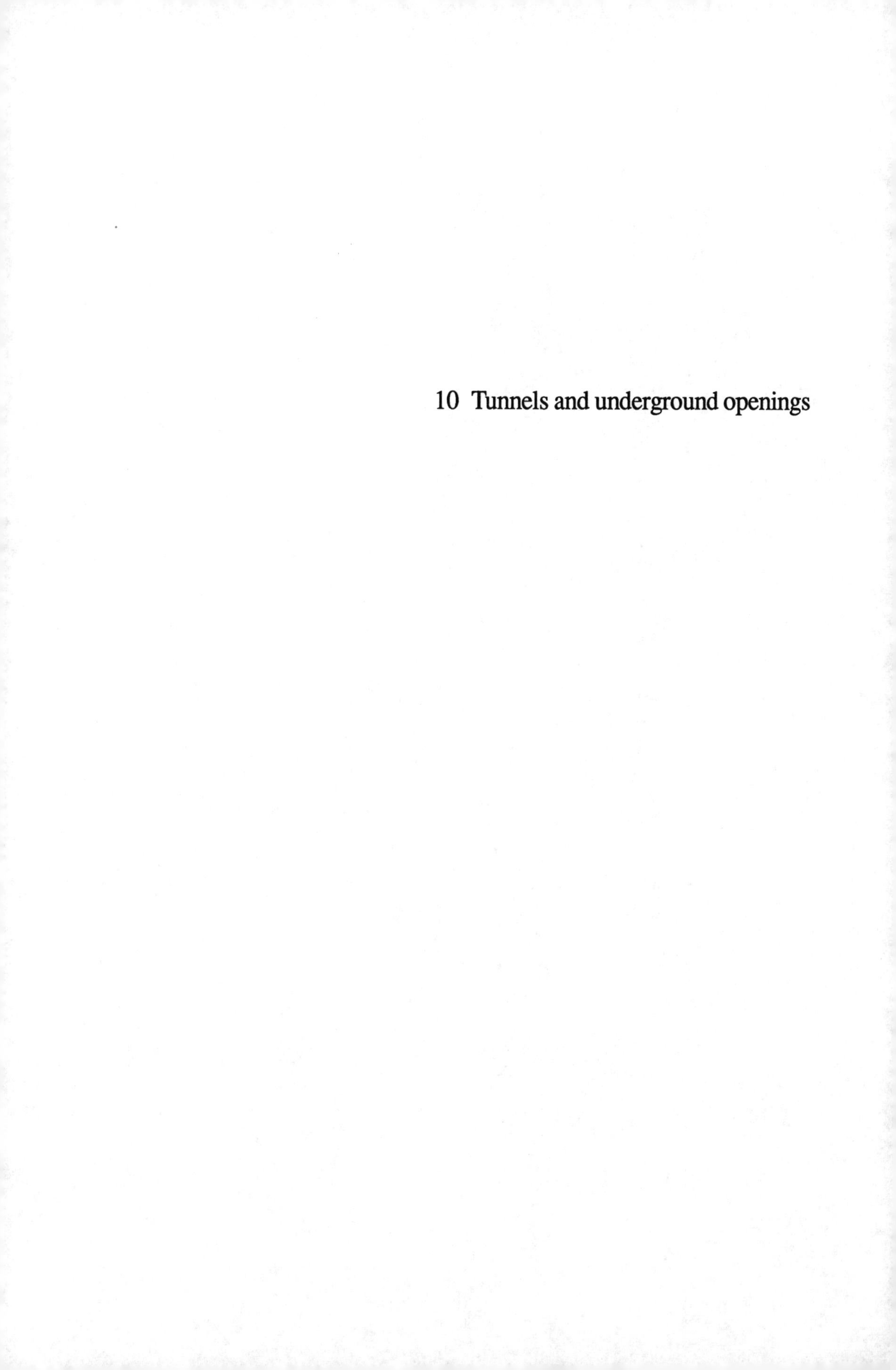

10 Tunnels and underground openings

Numerical Methods in Geomechanics (Innsbruck 1988), Swoboda (ed.)
© 1988 Balkema, Rotterdam. ISBN 90 6191 809 X

Numerical analysis and reality in tunnelling – Verification by measurements ?

Th.Baumann
Philipp Holzmann AG, FR Germany

ABSTRACT: Successfull tunnelling requires that excavation, support and other structural measures are planned in such a manner that the occurring forces,stresses and deformations can be mastered in a safe and economic way. Of particular importance is the activation of the load carrying capacity of the ground by controlled relaxation, and the corresponding design of the support, especially of the shotcrete lining. Which construction methods, constructional applications, principles and concepts are suited best for this task is the subject of many discussions, proceeding not seldomly without regard to quantitative facts and sliding off into philosophics. This study deals with the question of how far the dominating influences can be accounted for numerically by computational models and verified by the results of in-situ measurements. With examples for typical results of computations and measurements, the conditions are discussed, under which computational methods can provide reliable informations and diminish the dependence on subjective estimations and intuitive interpretations.

1 INTRODUCTION

Numerical analyses are to explain or predict the behaviour of structures on the basis of physical laws. In tunnelling there are additional difficulties, compared to other engineering structures:

- The acting loads are the consequence of the construction process and, thus, not the input data but the result of the design.
- The constitutive laws of rock resp. soil, the most important materials in tunnelling, are much less verified and realistic than those of e.g. steel or concrete.
- Concerning the computational models it is necessary to make far-reaching simplifications, because it is neither possible nor useful to simulate the construction progress and its effect on the structure in all its individual stages.

Therefore, the structural analyses of different consultants for the same project can differ considerably from each other [6]. It is often hard to decide wether such differences indicate the range of spreadings possible in nature, or if they are due to the shortcomings of the numerical methods. One goal of in situ-measurements is to provide informations about the real structural behaviour in typical situations. By that means, it is also possible to evaluate the results of the numerical analysis, and possibly to improve the underlying assumptions. Only after we are able to define a realistic computational model for a situation known by measurement, we can think about parameter studies to assess the inherent uncertainties of the underground, or to predict the effect of design modifications.

The use of numerical analyses is not to replace but to sustain and - as far as possible - to objectify the "synthetic, visually intuitive interpretation" [6] of measuring results and other observations predominating especially in the NATM.

2 IDEALIZED TUNNELLING PROCESS

For the conception of computational models and measurements as well as for intuitiv interpretations a qualitativ idea about the essential proceedings of tunnel construction is needed. Fig. 1 shows an idealized description of tunnel drivages with shotcrete construction methods. The corresponding computational steps of a 2D-calculation (plane strain) are shown in fig. 2 (for an immediate ring gap closure according to fig. 11). To simulate multiple stage excavation methods additional calculation steps are necessary.

The radial deformation of the contour of excavation (w) and the rock pressure (q)

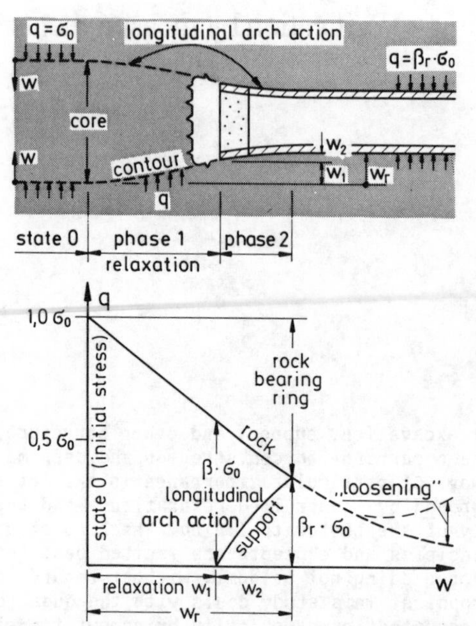

Fig.1. Conception of support model for tunnel drives with shotcrete construction methods

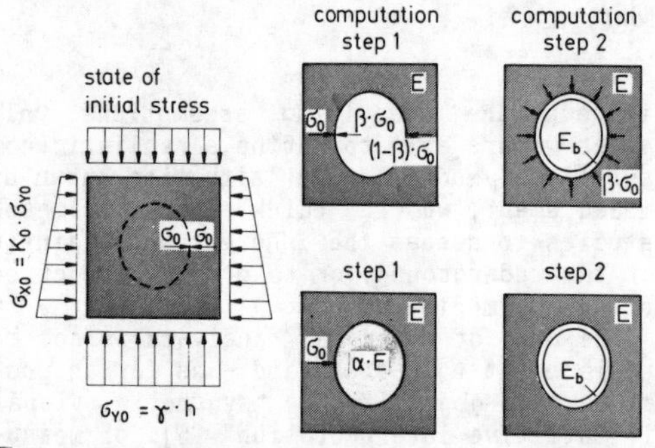

Fig.2. Simulation of typical construction states by a 2D computational model

acting on this contour resp. on the tunnel lining are considered as dominant factors. The supporting effect of the rock in circumferential direction is activated primarily by the deformations w_1 occurring in phase 1 ("relaxation").

In addition, an arch action in the rock is formed in longitudinal direction, spanning the unsupported space between the face and the closed tunnel lining. As the excavation is proceeding, the load due to this arch is carried by the compound system shotcrete lining/rock and induces the deformations w_2 (phase 2).

This process can be approximately described by the plane computational model according to fig. 2 as follows: Starting from the initial state of stress step 1 simulates the events of phase 1. The longitudinal arch action is accounted for by $\beta \cdot \sigma_0$. In step 2, $\beta \cdot \sigma_0$ acts as a load on the compound system lining/rock (according to phase 2 in fig. 1). Of course the stresses and deformations of the two subsequent computational systems are superimposing the initial state of stress.

In order to describe the longitudinal arch action, an elastic core with reduced stiffness $\alpha \cdot E$ supporting the excavated cross-section can be used instead of the supporting stress $\beta \cdot \sigma_0$. This core is introduced in step 1 and removed again in step 2 [3], [4].

For the safe support of tunnels driven in loose rock the shotcrete lining must sustain a well defined part of the initial rock pressure and has to be dimensioned for it. Additionally, in shallow tunnels in urban aereas, which are sensitive to settlements, the deformations of the ground must be limited to minimum values by an immediate ring closure of the lining. Computations according to fig. 2 are suited and necessary especially for this cases.

In the design of several deep tunnels, the shotcrete lining had not a bearing or statical function, but should only seal the rock in connection with bolts, thus preventing "loosening" although allowing simultaneously large deformations of the rock. In some cases, the shotcrete lining was even weakened intentionally by longitudinal "unloading-slits". The conflicting demands of such tunnelling methods - large deformations and relaxation and, at the same time, avoidance of "loosening" and destabilisation - cannot be achieved by the known constitutive laws, and thus it is hard to simulate this type of proceeding by computational models.

The results of calculations according to fig. 2 are controlled first by the supporting factor β (resp. the stiffness α of the supporting core), and in addition by the elastic and nonelastic material characteristics. The correct determination of these parameters requires measurements of deformations as well as forces, resp. stresses.

Measurements from inside the tunnel, i.e. after excavation, can provide only values corresponding to w_2 and $\beta_r \cdot \sigma_0$ (see fig. 1), i.e. u_1, u_2, r_1, r_2, N, M acc. fig. 3. Which fraction of w_2 is represented by such deformations measured from inside the tunnel is often hard to define, because this depends to a great extent on the instant, when the first measurement defining the reference point for the subsequent readings is done (e.g. immediatly after excavation or after shotcreting). Values such as f_1, f_2, s_o, s_u, ε_{yy} according fig. 3 associated with the relaxation (corresponding to w_1 or w_r) can only be measured from outside the tunnel.

Examples for such measuring results and their interpretation are discussed in section 4.

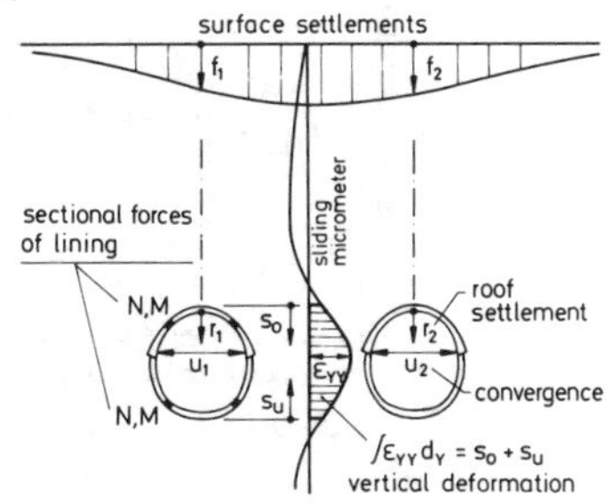

Fig.3. Measurement section of a shallow tunnel (2 tubes)

3 MEASUREMENT OF STRESSES AND STRAINS OF TUNNEL LININGS

The design and dimensioning of shotcrete linings - possibly in connection with tunnel arches and bolts - is not possible without an idea of the internal forces (due to the rock or soil pressure $\beta_r \cdot \sigma_o$ according to fig. 1). Therefore, a reliable measurement of these forces is of a fundamental importance, no matter if numerical analyses are made or not.

Direct stress measurements are only possible in gases and fluids. For solid materials indirect measuring methods must be applied, concluding from deformations to stresses. The measurement of shotcrete linings is complicated by the fact that the stresses build up mainly in the young concrete age. Thereby, the deformations due to creep, shrinkage and variations in temperature are considerably larger than the elastic deformations which are caused by the wanted stresses. For this reason, it is impossible to make a reliable statement about the stresses, when deformations are measured on the shotcrete or on reinforcing bars.

Stress gauges measure the fluid pressure which leads to a minimum extension of a flat tin cell (dimensions ≈ 150 x 150 x 5 mm), embedded directly in the concrete. Then this pressure is assumed to be identical with the concrete stress.

The application of stress gauges in prestressed concrete pressure vessels led to a careful and detailled investigation of the prerequisits for a successful application of this measuring method [1]. The result were special requirements for the formation of the contact surfaces between concrete and gauge, as well as the necessity to eliminate the lack of contact between concrete and gauge, which results from the dissipating heat of hydration cycle of the con-

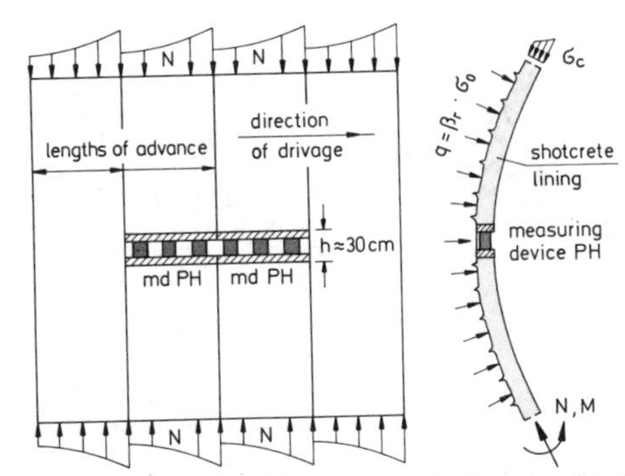

Fig.4. Forces acting on a shotcrete lining and device for measuring these forces

crete, by a so-called repressurizing. Only the changes of concrete stress occurring after this time could then be determined by this measuring method.

However, the cited requirements cannot be fulfilled in a shotcrete lining. In this case, the stresses arising during the hydration cycle are of prime interest. In addition, the contact surfaces cannot be formed in the required way when installed in a shotcrete lining. Finally, the stresses to be measured are distributed discontinuously. Therefore, it is problematic to conclude from local stress measurements to stress resultants like normal forces and moments. For these reasons, stress gauges embedded in shotcrete produce results which are spreading considerably and hard to be interpreted.

Since, on the other hand, reliable informations about the forces in tunnel linings are indispensible, a measuring method especially adapted for this purpose has been developped at the Philipp Holzmann AG [5].

With regard to the possible state of stress in tunnel linings the measuring device is designed as follows (fig. 4):

- It covers the entire cross-section in order to determine the non-uniform stress distribution caused by bending moments.
- To account for the discontinuous force distributions in longitudinal direction, several devices of a single length of 1 m are mounted behind each other in order to span at least one length of advance.
- The measuring device has approximately the same stiffness as the hardened shotcrete.
- Its height (ca. 30 cm) is small in comparison to the tunnel diameter. Thus, the difference between the forces in the measuring device and in the undisturbed tunnel lining is limited in size and extent to negligible small values even for the young shotcrete, which is less rigid than the measuring device and undergoes deformations

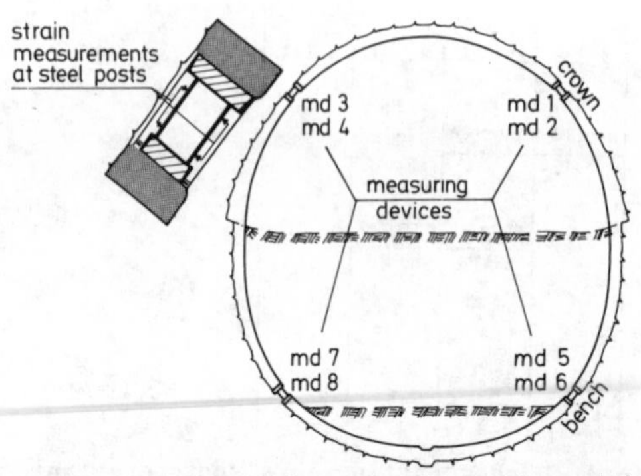

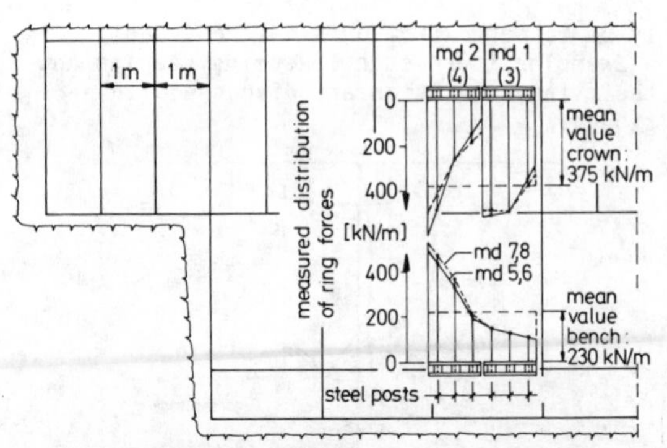

Fig.5. Equipment of a tunnel section with 8 measuring devices and measuring results

Fig.6. Measuring device (without covering)

Fig.7. Mounting of measuring device before shotcreting

due to creep and shrinkage.

The measuring device consists of two longitudinal girders being interconnected by 3 steel posts which transfer the compressive normal forces and the bending moments (fig. 4, 5, 6). Strain measurements at the front

Fig.8. Tunnel section with 8 measuring devices

and the back of the posts allow to determine the sectional forces. For these measurements, vibrating wire strain gauges are used. This guarantees the best long-term constancy of zero-setting even under the rough conditions of tunnelling (as e.g. humidity, renewing of damaged cables). Heavy tin coverings, which do not transfer normal forces, protect the vibrating wire strain gauges, and provide for approximately constant temperatures in the posts. In fig. 6 the tin covering is not yet mounted.

The measuring devices are installed together with the tunnel arches after excavation and before shotcreting (fig. 7). At this stage, the first measurement defining the reference point (M = N = 0) is made. A tunnel cross-section equipped with 8 measuring devices is shown in fig. 8.

Typical measuring results of a shallow tunnel are presented in fig. 5. The consistency of the results obtained in the right half of the lining (md 1, 2, 5, 6) and the ones in the left half (md 3, 4, 7, 8) is remarkably well. The mean forces (crown: 375 kN/m; bench: 230 kN/m) are useful for

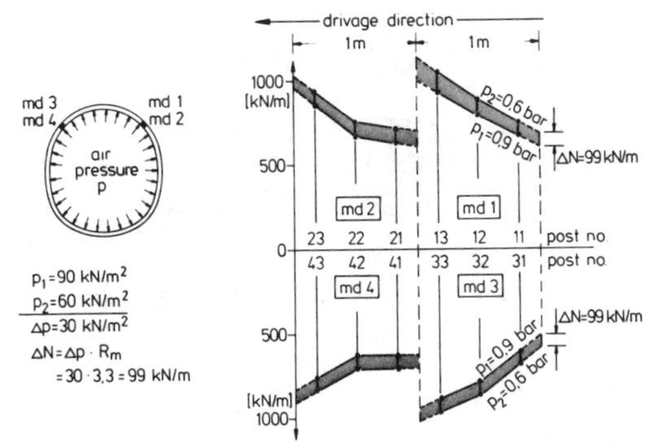

Fig.9. Variation of forces in a shotcrete lining due to a change in compressed air pressure

checking the results of 2D-calculations according fig. 2. The distributions in longitudinal direction enable us to check 3D-FEM-predictions of the stresses due to excavation [2].

Fig. 9 indicates the forces measured in the individual posts of 4 measuring devices before and after lowering the air pressure from 0,9 to 0,6 bar in a compressed air drivage. For each of the 12 posts, the measured change of force corresponds very well to the theoretical value ($\triangle N = \triangle p \cdot R_m = 30 \cdot 3,3 = 99$ kN/m). Further results are presented in fig's. 16, 18 and 19.

Thus, the described method allows the forces in shotcrete linings to be measured in a reliable way. It serves equally well to determine the forces in final linings of cast concrete (fig. 10).

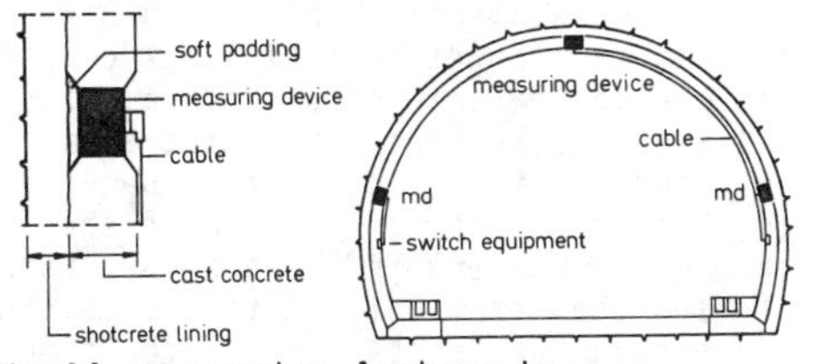

Fig.10. Measuring devices in a final tunnel lining of cast concrete

4 DISCUSSION OF RESULTS OF NUMERICAL ANALYSES AND MEASUREMENTS

4.1 Shallow tunnel in soft ground

20 m below ground level, two tunnel tubes were driven one after the other into the tertiary sands and marls under historical buildings very sensitive to settlements (excavation sequence according to fig. 11). The ground-water level in the section under consideration was about 4 m above the roof.

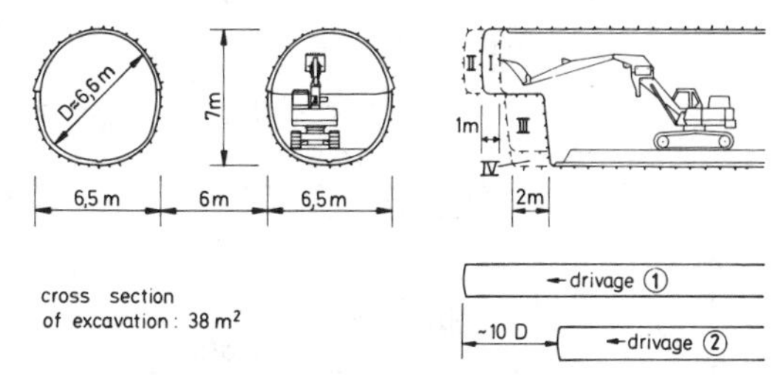

Fig.11. Excavation sequence of a shallow tunnel

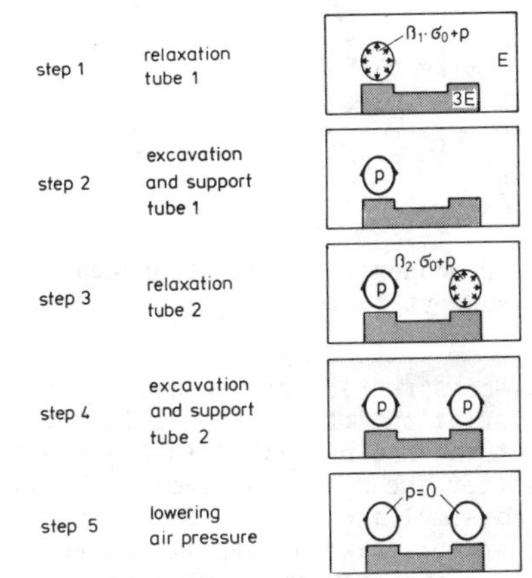

Fig.12. Simulation of tunnel drive by 5 computation steps

Since it should not be lowered, the drivage was executed under compressed air. The measurements made in a typical cross-section are shown in fig. 3.

The advancing of the two tubes was simulated in a plane FEM-model in 5 steps (fig. 12). Compared to a drivage under atmospheric pressure, the air pressure p increases the supporting effect of $\beta \cdot \sigma_o$ to $\beta \cdot \sigma_o + p$. The final lowering of the air pressure causes an additional load p on the compound system shotcrete/soil. This was simulated in step 5. For the soil an elastoplastic material was assumed (Mohr-Coulomb failure criterion, associated flow rule). As result of the parameter studies performed with the FEM-programm TUNNEL of RIB/RZB (Stuttgart), the decisive forces and deformations as a function of the supporting factor β are presented in fig. 13 [3], [4].

The following in situ-measurements turned out to be useful for evaluating the results of the numerical analyses:

- surface settlement f_1, f_2 (leveling)
- deformations ε_{yy}, s_o, s_u of the soil between the two tubes (sliding micrometer)
- normal forces and bending moments in the shotcrete lining (measuring devices PH)

Measuring results for the deformations in

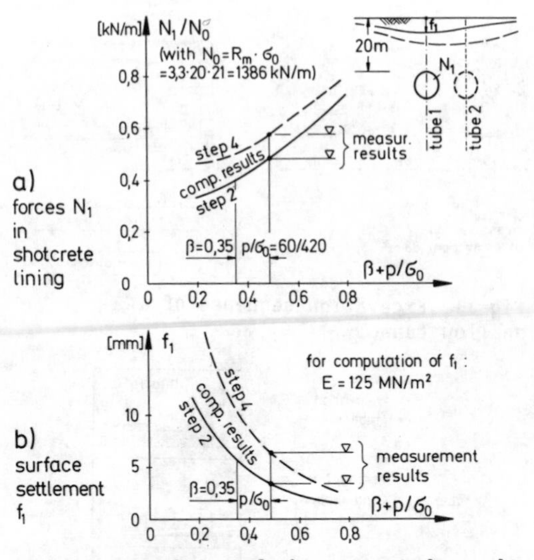

Fig.13. Dependence of the computed results on the support factor β

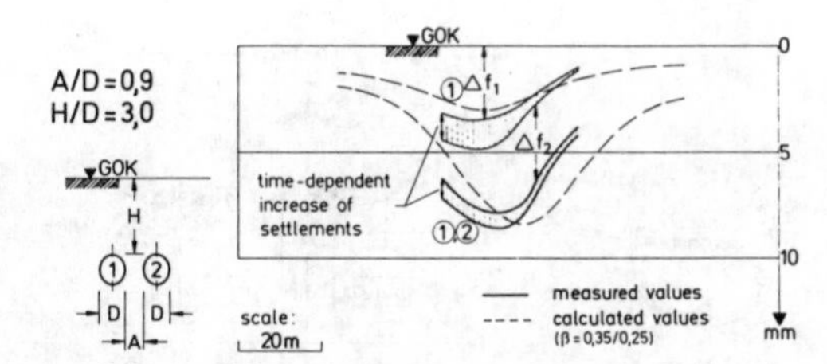

Fig.14. Surface settlements (in MS 7)

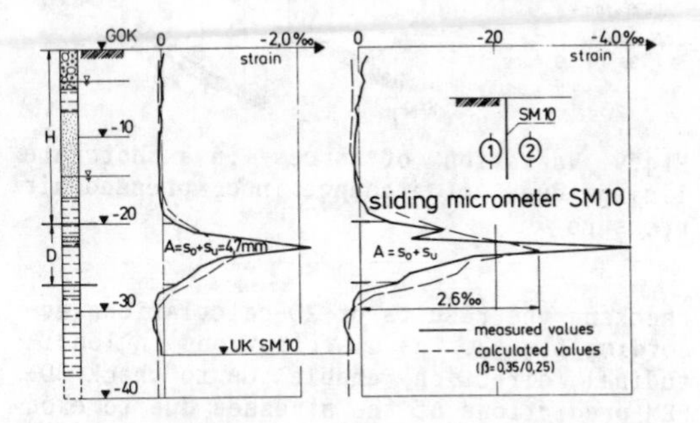

Fig.15. Vertical deformations of soil (MS 7)

typical construction states and for the temporal variation of the forces in the shotcrete lining are presented in fig's. 14, 15 and 16 (for the measuring section MS 7). The required simultaneous consideration of measuring and calculating results is facilitated by a comparison in tabular form as shown in fig. 17.

The horizontal convergence and the roof settlement were measured as well (u_1, u_2, s_1, s_2 in fig. 3). However, these values were not particularly helpful for the numerical evaluation, because they only represented an unknown fraction of the deformation associated with w_2.

The evaluation of computational and measuring results for the cross-section under consideration, which was advanced under an increased air pressure of 0,6 bar, was essentially conducted in the following steps [4]: Because of the relatively soft soil (according to laboratory tests: $E \approx 125$ MN/m²) the computational normal force in the tunnel lining is hardly affected by E, but depends mainly on the supporting factor β. The value of β leading to the measured normal force could, therefore, be determined directly from fig. 13a as $\beta + p/\sigma_0 = 0{,}35 + 60/420$.

When working with a uniform soil-modulus, the calculations according to fig. 2 resp. 12, delivered unrealistic deformations: Compared to the measured values, the computed elevations of the tunnel bottom (s_u) are too large and, at the same time, the settlements s_o and f_1 (line 2 in fig. 17) too small. The necessary increase of the stiffness of the soil under the tubes (at unloading due to the excavation) has intentionally not been achieved by a complex constitutive law, because there wasn't sufficient experimental assurance for it. Instead, the E-modulus of the soil under the tubes was increased by the factor 3. By that means, a satisfactory agreement between measured and calculated results was obtained (fig. 17, line 1 and 3).

Thus, a computational model was found which describes the measured forces and deformations consistently and which permits to investigate the effects of limited changes of the geological and structural situation like, for instance, variations of the air pressure in the tunnel or a non-staggered but synchronous drivage of the two tubes. For tunnelling in urban areas, where not only the stability of the tunnel but also a minimum of surface settlements must be guaranteed, such models verified by measurements are an important planning tool [7].

4.2 Excavation by blasting in destabilized rock

Fig. 18 shows the forces measured in the shotcrete lining of a tunnel lying about 14 m under a non-urban area. The maximum surface settlement was 11 cm. The measured forces amounted to about half the value to be expected when taking the full weight of 14 m overburden into account. For the cor-

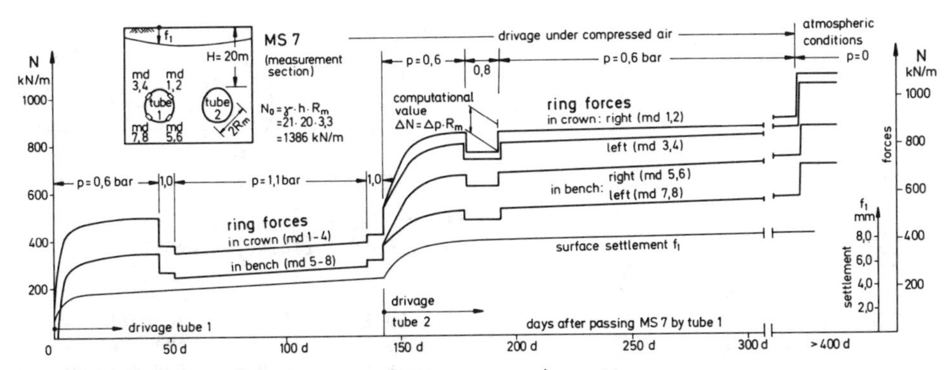

Fig.16. Time-dependent development of forces in the shotcrete lining of tube 1 (MS 7)

line			f_1	s_o	s_u	s_o+s_u	N_{crown}
			mm	mm	mm	mm	kN/m
1	measurement results		3,4	3,9	0,8	4,7	486
2	calculation results in step 2	E uniform (p=0,6 bar, E)	2,1	2,1	3,2	5,3	538
3		E increased below tubes (p=0,6 bar, E, 3E)	3,0	3,5	1,4	4,9	524

Fig.17. Effect of increasing the E-modulus of the soil below the tubes

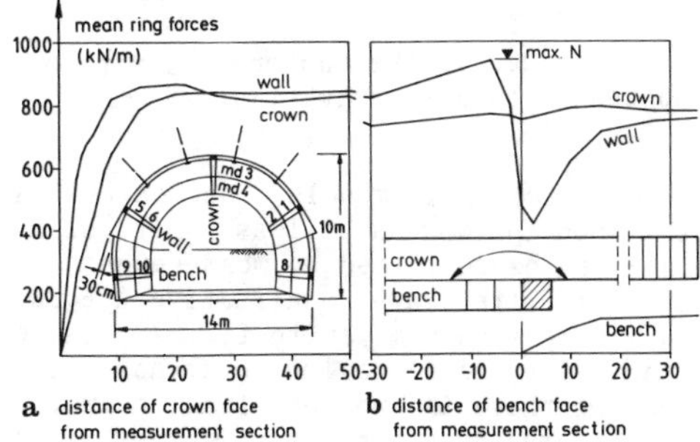

Fig.18. Forces in the shotcrete lining of a shallow tunnel (blast excavation)

responding value β ≈ 0,5 and an E-modulus of the rock determined by laboratory tests, only a fraction of the measured surface settlement of 11 cm was obtained, even with models according to fig. 2 and in consideration of an elastoplastic behaviour of the rock. This indicates that shear resp. failure planes might have formed in the rock, causing large displacements which could not be reproduced by the numerical model.

Fig. 19 shows the forces in the shotcrete lining of a tunnel lying about 120 m under ground level. They amounted to about 1 % of the values to be expected for the full weight of 120 m overburden. Surface settlements have not been observed. Total deformations in the tunnel area could not be measured because of the high overburden. Also in this case, computations according to fig. 2 could not be confirmed by measurements, and did not give additional insights.

For tunnel drivages according to fig. 18 and 19 with far-advanced crown, the maximum compressive force in the wall has occurred at the subsequent drivage of the bench. The digging out of the crown base leads to the vault effect of shotcrete lining indicated in fig. 18b, and increases the force in the wall for a short period. This increase grows proportional to the excavation length of the bench. As can be learned from failures, this is often the most critical state for the stability of a tunnel. The numerical investigation of this situation would require a 3D-model which also accounts for the shell-effect of the shotcrete lining. However, for the practical design of shotcrete linings it seems more convenient to evaluate this increase of the forces and its dependence on the bench excavation length empirically on the basis of reliable measurements.

5 CONCLUDING REMARKS

Success or failure in tunnelling depends on many influences concerning the rock as well as the structural procedures. Correspondingly, the experiences, and especially those resulting from critical situations and local failures, are very diverse [7], [8]. The ideas and models presented in the previous chapters describe only a small part of the reality staying behind these experiences. Nevertheless, it seems to be worthwhile to develop such simplified models and to verify them by measurements. In helping to understand the regular situations, they facilitate as well the understanding of unusual situations and the planning of means necessary to avoid risks.

One should strive for clarity, in how far these conceptions are physically meaningful, which means that they can be used as bases

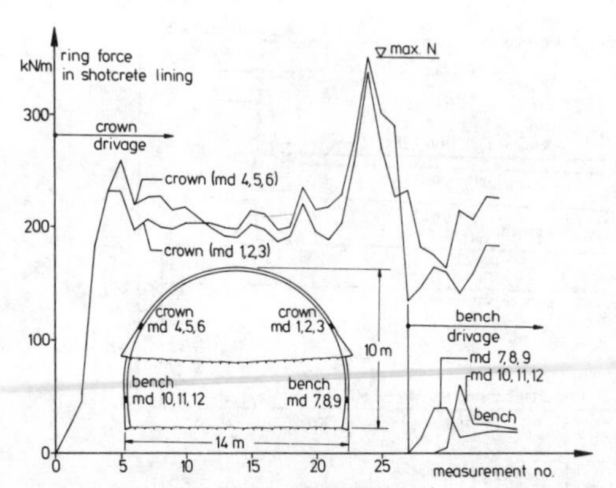

Fig.19. Forces in the shotcrete lining of a deep tunnel (blast excavation)

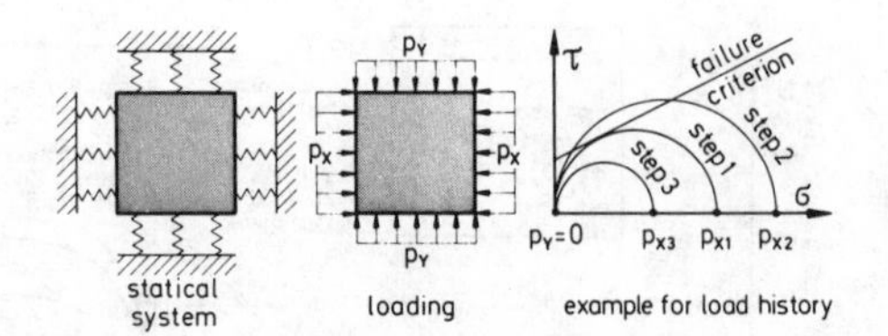

Fig.20. Test of the elastoplastic behaviour assumed in a finite element

for computational models, and in how far both - conceptions as well as numerical models - can be confirmed by measurements.

In the literature, qualitative diagrams of q-w-relations similar to fig. 1 are very common. They are called rock characteristics. If there is no rotational symmetry, an arbitrary number of such curves can be produced for the same cross-section by numerical models, since every point of the circumference has its own q- and w-curve.

Measurements, on the other hand, yield the displacement w_z (or a fraction of it) and the stress $\beta_r \cdot \sigma_o$ at some selected points of the tunnel circumference after the tunnel excavation, but, by no means, continuous curves as indicated in fig. 1. Such curves can - except for rotational symmetry - hardly be applied for a quantitative description of the rock behaviour basing on computations or measurements.

When qualitative conceptions are proved and quantified by numerical analyses, a verification by measurement seems possible. For this purpose, first of all, the dependence of measurable values on the parameters of the computation has to be described in a clear way, as for example shown in fig. 13. Next, measured values of forces as well as deformations must be available which can be assigned clearly to the construction phases determined in the computational model.

Results measured in one cross-section of a tunnel can verify numerical parameter studies acc. to fig. 13 in one point only. The "correct" shape of the curves going through these verified points can only be obtained by a "correct", that means physically approved computational model. The verification of computer programms is here of great importance. Among others, the aptitude of the underlying formulations and assumptions for displacements and the constitutive laws are to be evaluated. To this end, tests acc. fig. 20 for different load histories can be useful. In order to prove failure criterions and flow rules, the element is supported by elastic springs thus preventing numerical instability.

The requirement, that the measured deformations can clearly be assigned to the computation steps, can be fulfilled when total deformations relative to the initial stress situation are measured. This is, in general, only possible with measurements on or from the surface above the tunnel. For this reason and also because of the initial stress situation being harder to define with increasing tunnel depth, the verification of computation models by measuring results is getting more difficult for deep tunnels, and numerical analyses depend increasingly on intuitive interpretations.

Bibliography

[1] B. Grainger: Evaluation of the Glötzl stress gauge for use in concrete structures, C.E.R.L. Report R D/L/N 51/78, 1978

[2] K. Schikora: Plane and Spatial Finite Element Calculations in Tunnelling, Tunnel 3, 158-161 (1984)

[3] Th. Baumann, H.M. Hilber: Zur Berechnung von U-Bahn-Tunnels im Lockergestein, Fin. Elemente-Anwendungen in der Baupraxis, Ernst & Sohn, Berlin, 1985

[4] Th. Baumann, B. Sulke, Th. Trysna: Einsatz von Messung und Rechnung bei Spritzbetonbauweisen im Lockergestein, Bautechnik 62, 330-337, 368-374 (1985)

[5] Th. Baumann: Messung der Beanspruchung von Tunnelschalen, Bauingenieur 60, 449-454 (1985)

[6] H. Duddeck, A. Städing, Entwurfskonzept und Realität beim Standsicherheitsnachweis von Tunneln im deutschen Mittelgebirge, Forschung + Praxis, Band 30, Alba Verlag, Düsseldorf 1986

[7] J. Weber: Limits of Shotcrete Construction Methods in Urban Railway Tunnelling, Tunnel 3, 116-126 (1987)

[8] F. Schrewe: Findings in Conjunction with the German Federal Railway's New Routes, Tunnel 3, 102-116 (1987)

Numerical Methods in Geomechanics (Innsbruck 1988), Swoboda (ed.)
© 1988 Balkema, Rotterdam. ISBN 90 6191 809 X

Application of the 'decoupled finite element analysis' in tunnelling

G.Swoboda
University of Innsbruck, Austria

H.Ebner
ILF Consulting Engineers, Innsbruck, Austria

S.J.Wang & J.M.Zhang
Academia Sinica, Institute of Geology, Beijing, People's Republic of China

ABSTRACT: This paper deals with a way to simulate the failure of rock along element boundaries. In order to describe the failure process, we use a constraint element that is numerically very stable. The failure forms that occur are tensile failure and sliding. The generation of the complex meshes is performed with a special mesh generator. This simulation is put to practical use for tunnel excavation in jointed rock, whereby the typical failure mechanism for vertically jointed rock during tunnel driving is simulated. This model can also be used to check laboratory test results of jointed rock specimens.

1. INTRODUCTION

The finite element method has become an integral part of today's tunnel design work. However, the value of the finite element calculations depends to a great extent on the parameters and the model assumptions. Very often tunnels are driven in jointed or stratified rock and can only be described with difficulty as a continuum. The FEM is much easier to apply to plastic soils, as back analysis has shown [1].

Goodman's [2] works first presented an element that describes the rock jointing. This element can only be used to describe individual faults and is also numerically very instable. Improvements for this element have been proposed by many authors. In recent years, models have been developed to break down the discontinuum into blocks. In the rigid body model (RBM) by Cundall [6], the rock mass is broken down into rigid blocks. This is a model that does not take consideration of the elastic deformations and is certainly not suitable for calculating shallow tunnel structures. In the further developments by Asai, the elastic properties are simulated by springs in the nodes. In places where the block's displacement share is very large, a usable formulation of a model around the discontinuum definitely has to be described.

The approach taken here, namely that of the "decoupled finite element method", uses finite elements as elastic blocks. According to the faults, these are coupled by constraint elements [8]. These elements can give consideration to the opening and closing of the joint.

2. THEORY OF THE CONSTRAINT JOINT ELEMENT

If a continuum consists of two separate systems A and B (Fig. 1), the equation system (Fig. 2) breaks down into two independent blocks.

$$[K]\{a\} - \{F\} = 0 \qquad (1a)$$

with

$$\begin{bmatrix} K_A & 0 \\ 0 & K_B \end{bmatrix} \begin{Bmatrix} a_A \\ a_B \end{Bmatrix} - \begin{Bmatrix} F_A \\ F_B \end{Bmatrix} = 0 \qquad (1b)$$

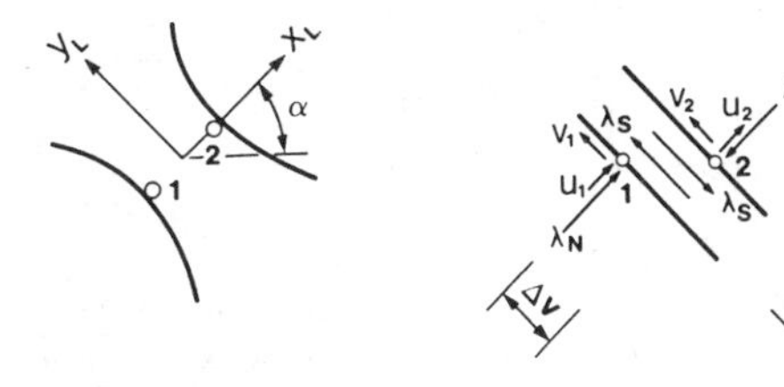

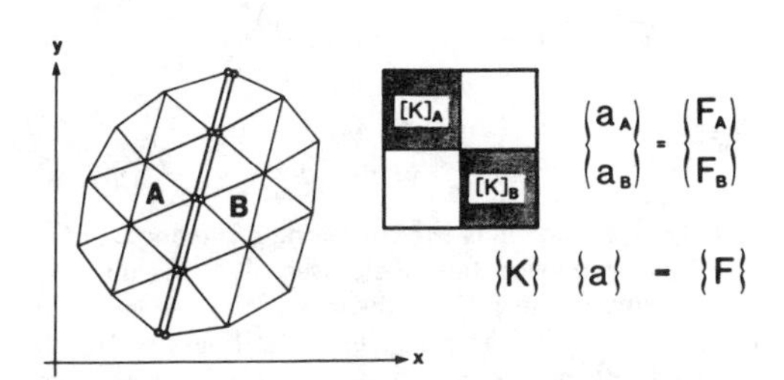

Fig.1 Coupled system

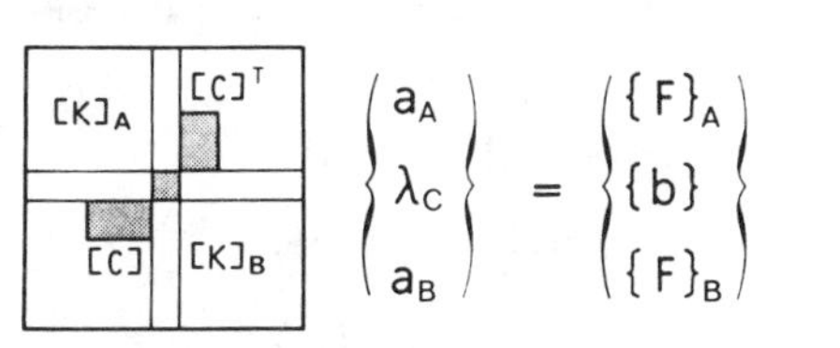

Fig.2 Decoupled constraint element

Coupling of the region A with the region B is done by means of two-node constraint elements (Fig. 3). For this purpose, pairs of nodes that have separate local displacements are defined along the joint

$$\{a_1^L\} = \begin{Bmatrix} u_1 \\ v_1 \end{Bmatrix} \qquad \{a_2^L\} = \begin{Bmatrix} u_2 \\ v_2 \end{Bmatrix} \qquad (2a)$$

or in a global system

$$\{a_1\} = [T]\{a_1^L\} \qquad \{a_2\} = [T]\{a_2^L\} \qquad (2b)$$

with

$$[T] = \begin{bmatrix} cos\phi & sin\phi \\ -sin\phi & cos\phi \end{bmatrix}$$

Further, coupling forces $\{\lambda\}$ are defined in the pairs of nodes in the local coordinate system

$$\{\lambda\} = \begin{Bmatrix} \lambda_N \\ \lambda_S \end{Bmatrix} \qquad (3)$$

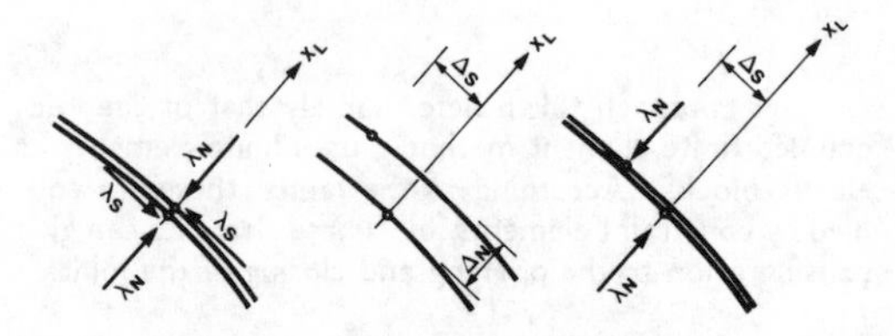

Fig.3 States of the constraint element

The virtual internal work of the different node displacements is:

$$A_{i,a} = \delta(\{a_2^L\} - \{a_1^L\})^T \{\lambda\} \qquad (4)$$

or, after transforming the displacements into the global system

$$\begin{aligned} A_{i,a} &= \delta((-[T]\,|\,[T])\{a\})^T \{\lambda\} \\ &= \delta([C]\{a\})^T \{\lambda\} \\ &= \delta\{a\}^T [C]^T \{\lambda\} \end{aligned} \qquad (5)$$

The corresponding external work is

$$A_{e,a} = \delta\{a\}^T \{F\} \qquad (6)$$

whereby $\{F\}$ is the nodal forces from the external load. Beyond this, the additional internal forces also perform virtual work

$$A_{i,\lambda} = \delta\{\lambda\}^T \{[C]\{a\}\} \qquad (7)$$

The corresponding external work reads

$$A_{e,\lambda} = \delta\{\lambda\}^T \{a_r\} \qquad (8)$$

whereby $\{a_r\}$ is the initial displacements of the nodes.

By equating internal and external work the following results:

$$\begin{aligned} A_i &= A_e \\ A_{i,a} + A_{i,\lambda} &= A_{e,a} + A_{e,\lambda} \end{aligned} \qquad (9)$$

and by inserting (5), (6), (7) and (8)

$$\delta\begin{Bmatrix} a \\ \lambda \end{Bmatrix}^T \begin{bmatrix} 0 & [C]^T \\ [C] & 0 \end{bmatrix} \begin{Bmatrix} a \\ \lambda \end{Bmatrix} = \delta\begin{Bmatrix} a \\ \lambda \end{Bmatrix}^T \begin{Bmatrix} F \\ a_r \end{Bmatrix} \qquad (10)$$

the stiffness matrix of the constraint joint element is received

$$\begin{bmatrix} 0 & [C]^T \\ [C] & 0 \end{bmatrix} \begin{Bmatrix} a \\ \lambda \end{Bmatrix} = \begin{Bmatrix} F \\ a_r \end{Bmatrix} \qquad (11)$$

From this it is seen that this element has no elastic stiffness; only additional constraints are introduced, by means of which coupling can be forced.

3. DECOUPLING

In order to be able to orient the local system of coordinates, two additional nodes, 3 and 4, are introduced, as shown in Fig. 3. These are also used to formulate the failure criterion for decoupling in stresses and not in forces.

The calculation is performed incrementally in load steps, whereby the fracture criteria are examined. This means that for each of these conditions (K) the criteria will be examined starting with the condition of the last iteration $(K-1)$. This gives the following forces and displacments: joint displacement

$$\{\Delta\}^K = \{\Delta\}^{K-1} + \{\Delta\}$$

coupling forces

$$\{\lambda\}^K = \{\lambda\}^{K-1} + \{\lambda\} \qquad (12)$$

loading

$$\{F\}^K = \{F\}^{K-1} + \{F\}$$

whereby the added values of the pertinent load steps are without indices.

The friction law developed by Coulomb is used as the failure criterion, which gives the maximum coupling forces: shear

$$F_t^K = (c - \sigma_N tan\varphi)\, L\, d \frac{\lambda_S^K}{|\lambda_S|^K} \qquad (13a)$$

tension

$$F_z^K = \sigma_z L\, d \qquad (13b)$$

In these equations, c is the cohesion, φ the angle of friction, σ_z the tension failure stresses, L is the influence of the element and d the thickness of the element. The decision matrix in Table 1 gives the stiffness matrix described in [8] for the conditions "fix state", "free state" and "slip state". The conditions are illustrated in Fig. 3.

Iteration n n - 1	Fix	Slip	Free
Fix	$\lambda_N^K \le F_z^K$ $\lambda_S^K \le F_t^K$	$\lambda_N^K \le F_z^K$ $\lambda_S^K > F_t^K$	$\lambda_N^K > F_z^K$
Slip	$\lambda_N^K \le F_z^K$ $\Delta_S F_t^K < 0$	$\lambda_N^K \le F_z^K$ $\Delta_S F_t^K \ge 0$	$\lambda_N^K > F_z^K$
Free	$\Delta_N^K < 0$		$\Delta_N^K \ge 0$

4. TEST EXAMPLE

In order to examine the element, a static system was chosen consisting of a cantilever beam with two cantilevers, one over the other. Only friction forces corresponding to Coulomb's Law can be transferred on their contact surfaces. Fig. 4 shows the static system and the finite element network. The constants are

module of elasticity
Poisson ratio
friction angle
cohesion
tension failure stress

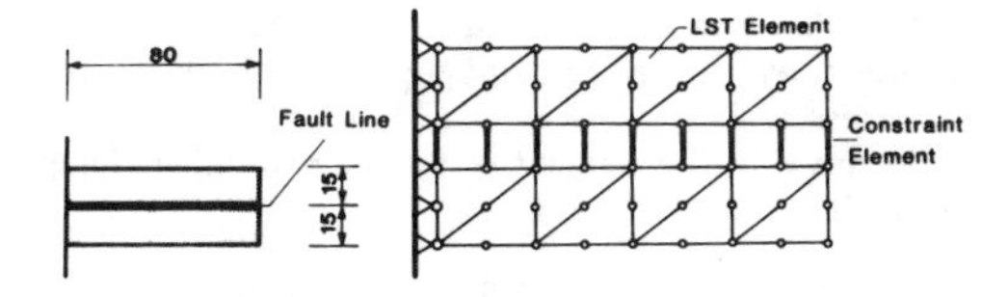

Fig.4 Test example: two-part cantilever

The lower cantilever is loaded with a concentrated load. The effect of the load is demonstrated in Fig. 5. Each loading case is considered as a loading step in this calculation and the loads applied as an incremental additional load. The changes in the system are caused by the constraint element's opening, closing and sliding. About ten iterations were needed for the iterative calculation of a loading condition.

Fig. 5 illustrates the loadings and their corresponding displacements and stresses. In loading 1, the low amount of transferable tensile stresses causes the contact surface to open. The resulting bending stresses are shown in the column on the right. Thus, the upper part of the cantilever remains without bending stresses; only the lower part is loaded.

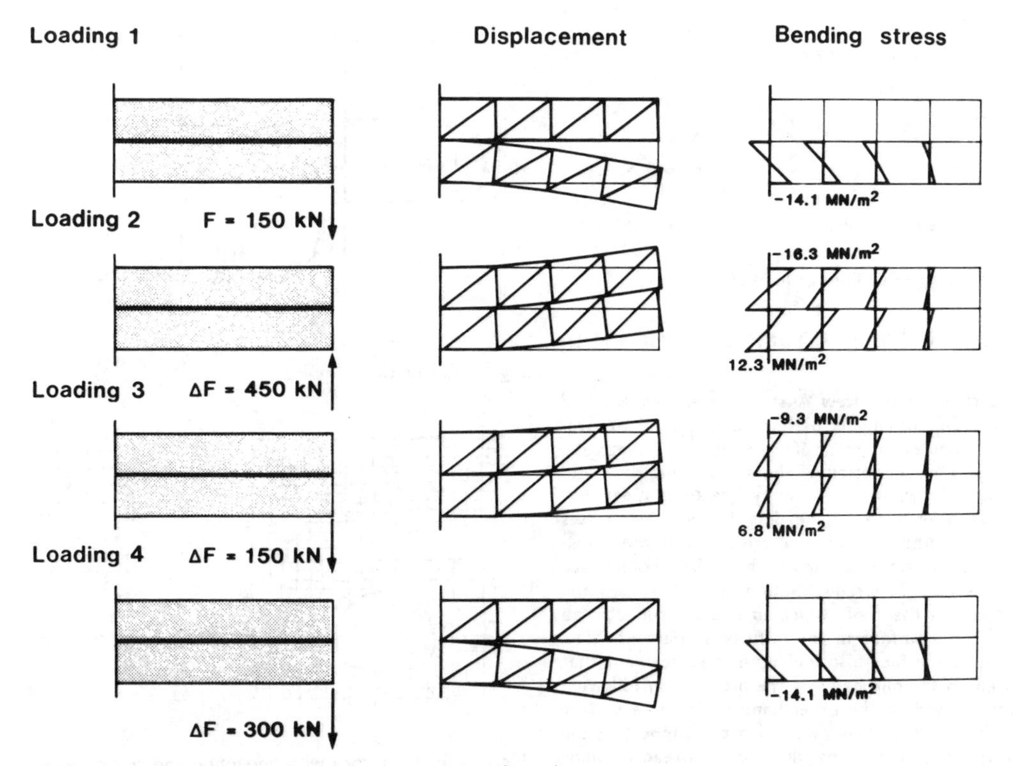

Fig.5 Incremental loading, displacement and bending stresses

Loading step 2 gives a total load of

$$F = 150 - 450 = -300\ KN$$

This causes the joint to close. Both cantilevers are loaded. Loading steps 3 and 4 return to the starting condition, which also corresponds to the displacements and stresses.

In this way, the element was examined under changing loads.

5. MESH GENERATION OF JOINTED ROCK

Jointed rock is described in the finite element mesh as rock along given possible fault lines, where constraint elements make it possible for the failure mechanisms described above to take place. In the framework of interactive mesh generation by means of a digitizer, the NETDIG program [9] includes the possibility of introducing individual fault lines after completion of the mesh topology. For this purpose, the node numbers on both sides of the fault line are reassigned, whereby one node retains its original number and the second, newly introduced node receives the highest node number available. The reference nodes in Fig. 3 have to be included in the generation.

Special problems arise in this algorithm when two fault lines intersect. In this case, appropriate constraint element combinations have to be generated, as shown in Fig. 6, in order to permit all elements to move along the given lines. A total of up to eight intersecting lines can be generated in the NETDIG program system.

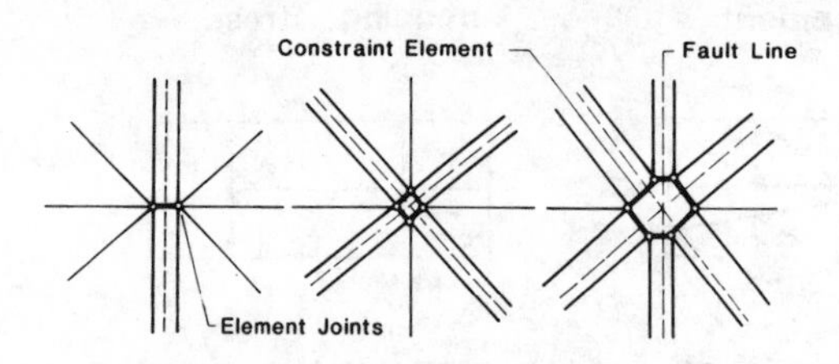

Fig.6 Crossing fault lines of jointed rock

6. CALCULATION OF A TUNNEL IN DISCONTINUUM

According to the New Austrian Tunneling Method (NATM) [10], driving is performed in partial excavation, with rapid securement of the cavity using shotcrete. This effects an activation of the surrounding rock and economic calculation of the cavity's securing measures. After excavation of the top heading, the primary stresses in the undisturbed soil or rock are redistributed transverse to the direction of driving and also via the face by arch action. This redistribution of loads causes displacements in the roof, which in shallow tunnels can extend to the surface in the form of settlement. The area behind the face also undergoes displacement that is caused on the one hand by the arch action following excavation, and on the other hand by the free surface of the face, which does away with the support action in the tunnel's longitudinal direction. Instead of plane displacement, plane stress is approximated.

Due to the extensive calculations entailed, the work is generally calculated on a plane model, whereby the driving is approximated by individual sections transverse to the tunnel axis. Literature makes reference to three simulation procedures that take consideration of the influences in the tunnel's longitudinal direction. For example, there is the possibility of variation of the module of elasticity of the shotcrete arch [12], or calculation according to the partial-load or support-load method [13], [14].

In the work at hand, simulation is performed using the stiffness reduction method [15], [16], that has been seen to be the simplest possibility. This method takes account of the displacements caused by strain relief by considering the future excavation area as a support core. The stresses here are applied on the excavation's periphery as nodal forces. The support core, which has a certain defined stiffness, acts as an elastic bed for the surrounding elements. If the support core has a theoretically infinitely large stiffness, preliminary relief of strain will preclude any displacements. Thus, the state of stress from the previous loads remains unchanged. Realistic values for the stiffness of the support core, that were confirmed in comparisons with calculations using in-situ measurements, are in a range of 0.2 to 3 times the rock's module of elasticity.

Fig. 7 shows the excavation procedure as well as the analyzed cross section of the individual construction steps.

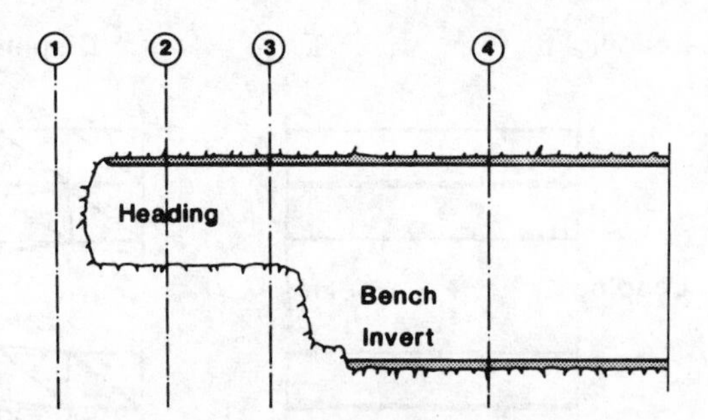

Fig.7 Excavation sequence

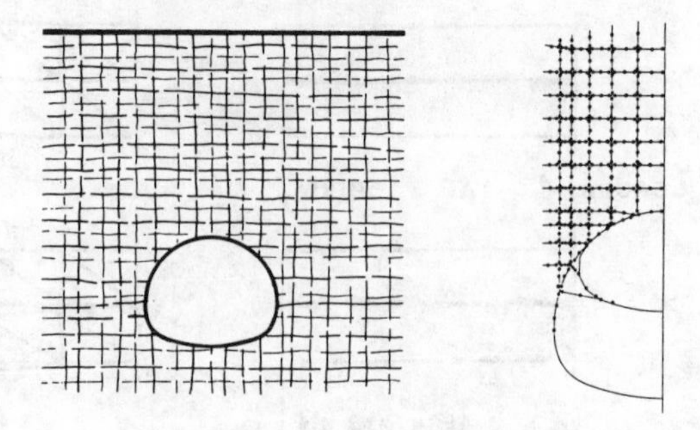

Fig.8 Tunnel in rock with horizontal and vertical jointing, arrangement of constraint elements.

Tunnel driving is subsequently examined in a discontinuous rock mass with uninterrupted fault system, as shown in Fig. 8. In Fig. 9, construction step 3, a chimney-like cave-in can be seen, a failure mechanism often observed in such jointed rock. The load applied by the failure body presses particularly on the foot of the top heading's shotcrete lining, causing the roof to heave slightly in construction step 4. Model tests by Feder [11], Fig. 10, give a similar picture. The principal stresses seen in Fig. 11 clearly show the roof block. The stresses have receded acutely. Only the undisturbed rock mass fully transfers the loads.

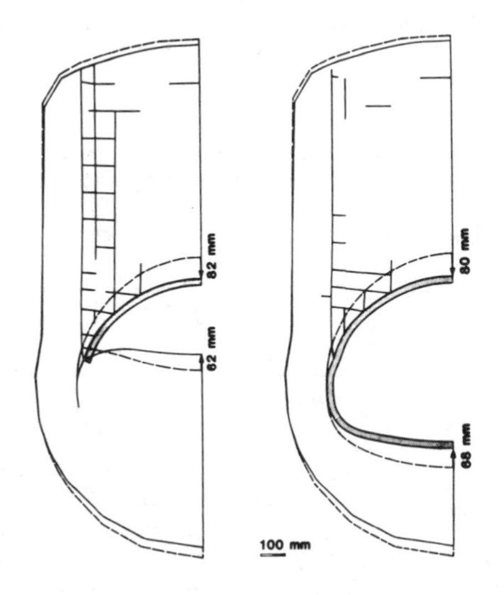

Fig.9 Fault system and displacements in construction steps 3 and 4

Fig.10 Model tests in jointed rock [11]

Also examined was the masonry-like fault system pictured in Fig. 12. Both the joint system in Fig. 13 and the principal stresses in Fig. 14 differ considerably from the fault system examined above.

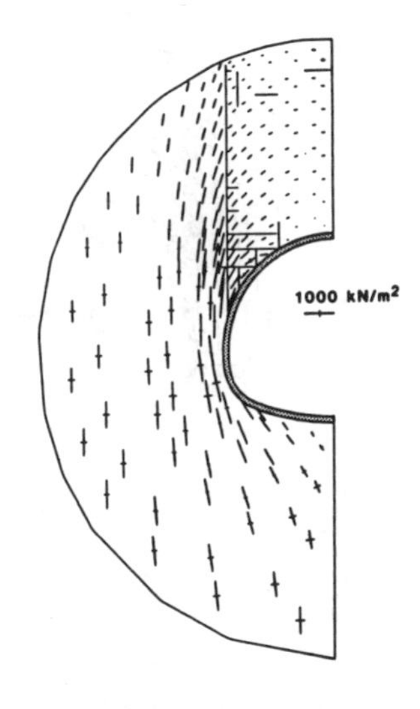

Fig.11 Stresses, construction step 4

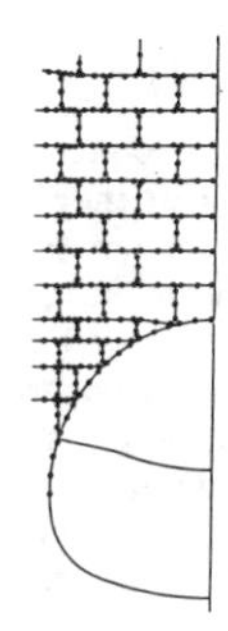

Fig.12 Masonry-like fault system, arrangement of constraint elements

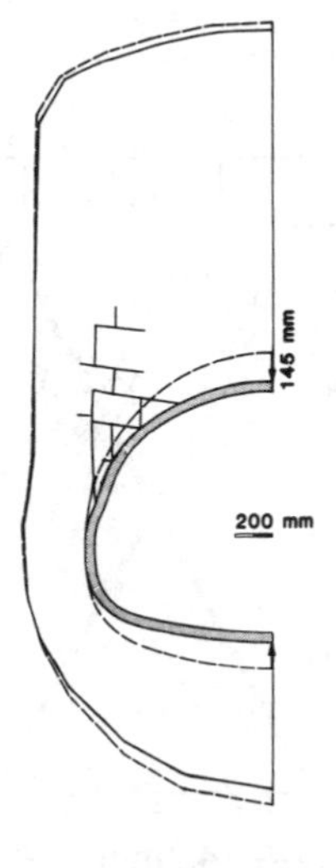

Fig.13 Fault system and displacement in construction step 4

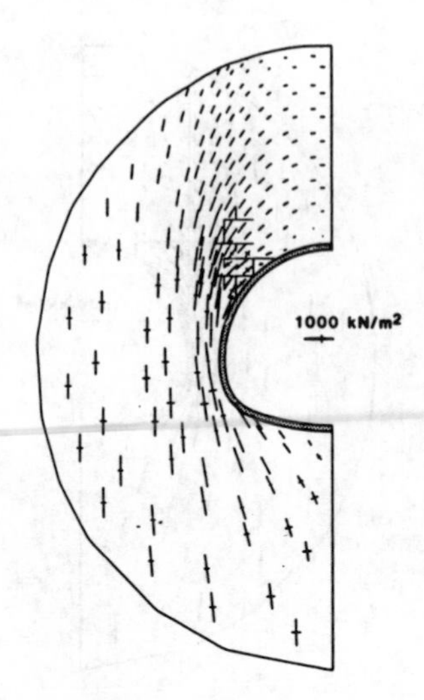

Fig.14 Stresses, construction step 4

7. ANALYSIS OF TEST SPECIMENS

Joints in rock are characterized by the fact that even when they have pronounced forms, the joint in reality does not take on a continuous form, but alternates between smooth joint surfaces and bridges. Test specimens were tested under biaxial conditions. An appropriate, comparative numerical model (Fig. 15) was set up in order to be able to make further statements on the characteristic values in the joint zone.

The characteristic values of the undisturbed rock were determined as follows:

$$E = 20000\ KN/m^2$$
$$\nu = 0,2$$

and those of the undisturbed slip surface with:

$$c = 0$$
$$\varphi = 12,4^\circ$$

whereby c is the cohesion and φ the angle of friction.

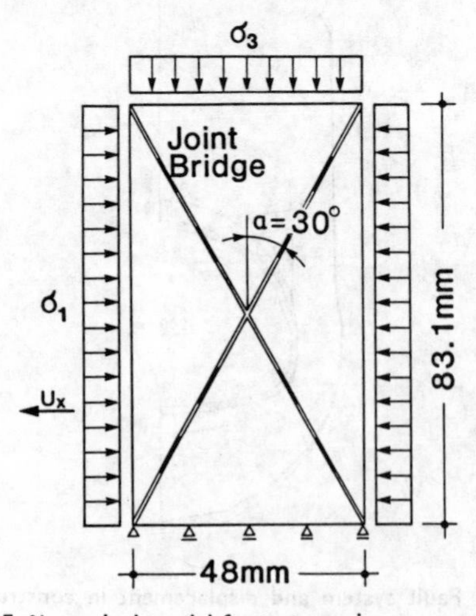

Fig.15 Numerical model for test specimens

The following characteristic values, calculated from the undisturbed specimen, were assumed for the rock bridges that are of crucial importance for the failure process:

rock bridge type 1:

$$c = 28,1\ KN/m^2$$
$$\varphi = 34,6^\circ$$

For **rock bridge type 2** it was assumed that after the failure the characteristic values drop to those of the undisturbed joint. The residual cohesion c_R and the angle of residual friction φ_R now have the values:

$$c_R = 0$$
$$\varphi_R = 12,4^\circ$$

The samples were tested at a constant ratio of horizontal stress σ_1 to vertical stress σ_3 :

$$n = \frac{\sigma_3}{\sigma_1} \tag{14}$$

with the results for $n = 5$ and $n = 10$ shown in Fig. 16, whereby $\sigma_1 = 5.0\ KN/m^2$ and $\sigma_1 = 10\ KN/m^2$.

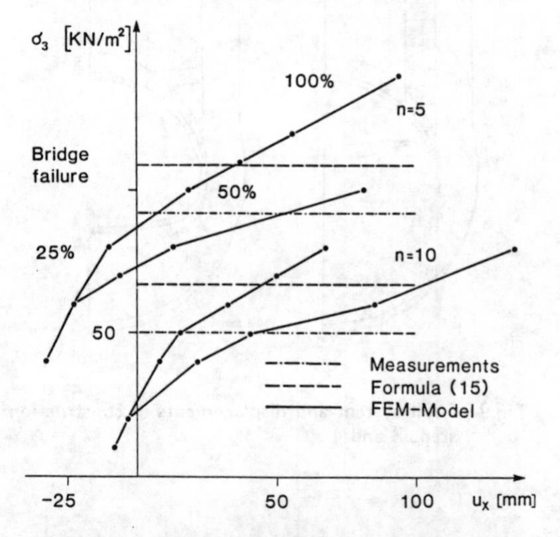

Fig.16 Comparison between vertical stress σ_3 and horizontal displacements u_x

The measurement readings were compared with a theoretically analyzed assumption taken from a paper published in [17]. This gives the following failure stress:

$$\sigma_3 = \frac{n\ c_R\ C_2\ C_1}{(n-1)C_1\ si\ co - tan\varphi_R\ C_2(co^2 + si^2)} \tag{15}$$

with

$$si = sin\alpha \qquad co = cos\alpha$$

$$C_1 = 1 - X_s + \frac{X_s}{m_1} \qquad m_1 = \frac{E_R}{E_J} = 10$$

$$C_2 = 1 - X_s + \frac{X_s}{m_2} \qquad m_2 = \frac{G_R}{E_J} = 10$$

and

X_s seperationfactor $\left(\frac{2}{3}\right)$

C_R cohesion fo the rock bridge
φ_R friction angle for the rock bridge
α orientation of the fault
E_R modul of elasticity of rock mass
E_J modul of elasticity of joint bridge
G_R shear modul of elasticity of the rock mass
G_J shear modul of elasticity of the joint mass

This is however, clearly lower than the maximum values calculated in the numerical model. At a ratio of $n = 10$, the measured failure is $\sigma_3 = 50KN/m^2$ and the theoretical stress $\sigma_3 = 68KN/m^2$. In the numerical model the measured value, under the assumption of rock bridge type 2, is reached at failure of 50% of the rock bridges. Thereafter, however, the model becomes clearly stiffer. This is probably due to additional cinematic stiffening of the model at the intersecting point of the slip surfaces. As shown in Fig. 17, the result is not a continuous slip surface, because that would require mechanical destruction of intersecting point A, something that cannot be simulated in the model. More realistic values for the failure stress could probably be obtained by coupling with a plastic analysis.

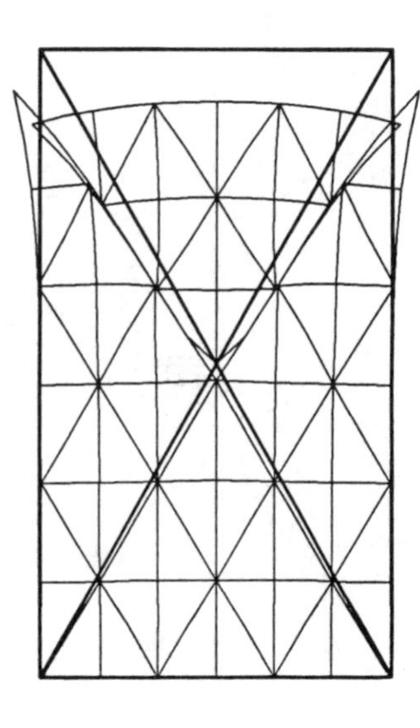

Fig.17 Displacement of numerical model

8. CONCLUSION

With the "decoupled finite element method" it was possible to show that the rock mass's discontinuum can be described well. In contrast to the rigid body method, this method permits the elastic properties to be taken fully into consideration, a fact that is of great significance for shallow tunnels. Furthermore, the method can be coupled with a classic plastic calculation. This means that the micro fractures in the given rock blocks can be described with the help of a failure body. In this way, a numerical model was created that solves the known problems in the calculation of shallow tunnels in jointed rock.

REFERENCES

[1] G.SWOBODA, F.LAABMAYR, I.MADER: Grundlagen und Entwicklung bei Entwurf und Berechnung im seichtliegenden Tunnel. Teil II: Verformungsrückrechnung. Felsbau 4 (1986) im Druck.

[2] R.E.GOODMAN, R.L.TAYLOR and T.BREKKE: A model for the mechanics of jointed rock. Journal of Soil Mechanics and Found. Div., ASCE 94, p. 637-658. (1968).

[3] C.S. DESAI, M.M. ZAMAN, J.G. LIGHTNER, H.J. SIRIWARDANE: Thin-layer element for interfaces and joints. Internal Journal for Numerical and Analytical Methods in Geomechanics 8, p. 19-43 (1984).

[4] W. WITTKE. Static analysis for underground openings in jointed rock. in Ch.S. Desai, J.T. Christian: Numerical Methods in Geotechnical Engineering. McGraw Hill, New York 1977.

[5] O.C. ZIENKIEWICZ, C. DULLAGE: Analysis of non-linear problems in rock mechanics with particular reference to jointed rock systems. Proc. 2nd International Congess on Rock Mechanics Sec., 8-14 (1970).

[6] T. MAINI, P. CUNDALL, J. MARTI, P. BERESFORD, N. LAST, M. ASGIAN: Computer modelling of jointed rock mass. Technical Reprint N-78-4. U.S. Army Engineers, Vicksburg, 1978.

[7] T. ASAI, M. NISHIMURA, T. SAITO, M. TERADA: Effects of rock bolting in discontinuous rock mass. 5th International Conference on Numerical Methods in Geomechanics, Nagoya, p. 1273-1280 (1985).

[8] M.G. KATONA: a simple contact-friction interface element with application to buried culverts. Internal Journal for Numerical and Analytical Methods in Geomechanics, 7, p. 371-384 (1983).

[9] NETDIG: Digitalisierung von Finite-Element-Netzen V 2.1 Manual University of Innsbruck 1985.

[10] L. MÜLLER. Der Felsbau I, Enke Verlag, Stuttgart 1963.

[11] G. FEDER: Einfluß von Bauverfahren, Anisotropie und Ausbruchsform auf die Konvergenz und den Stützmittelbedarf tiefliegender Hohlraumbauten. In: Strassenforschung, Bundesministerium für Bauten und Technik, Heft 124, 1979.

[12] R. PÖTTLER: Ideeller Elastizitätsmodul zur Abschätzung der Spritzbetonbeanspruchung bei Felshohlraumbauten. Felsbau, 3, p. 136-139 (1985).

[13] M. BAUDENDISTEL: Zum Entwurf von Tunneln mit großem Ausbruchsquerschnitt. Rock Mechanics, Supplement 8, p. 75-100 (1979).

[14] K. SCHIKORA, T. FINK: Berechnungsmethoden moderner bergmännischer Bauweisen beim UBahnbau. Bauingenieur, 57, p. 193-198 (1982).

[15] G. SWOBODA, F. LAABMAYR: Zusammenhang zwischen elektronischer Berechnung und Messung.

Stand und Entwicklung für seichtliegende Tunnel. Rock Mechanics, Supplement 8, p. 29-42 (1979).

[16] T. BAUMANN, H.M. HILBER: Zur Berechnung von U-Bahn Tunnels im Lockergestein. In: Finite Elemente: Anwendung in der Baupraxis. Ernst & Sohn, Berlin 1985.

[17] S. J. WANG: Rock mechanical testing (unpublished)

Numerical Methods in Geomechanics (Innsbruck 1988), Swoboda (ed.)
© 1988 Balkema, Rotterdam. ISBN 90 6191 809 X

A numerical method for the analysis of unlined pressure tunnels in jointed rock

Ming Lü
Institute of Water Conservancy and Hydroelectric Power Research, Beijing, People's Republic of China
E.T.Brown
University of Queensland, Australia

ABSTRACT: A joint element finite element method is developed for the numerical analysis of unlined pressure tunnels in jointed rock. A range of realistic hydro-mechanical responses of joints may be modelled and three-dimensional stress field effects are accounted for. Numerical analyses of generic cases simulate pressure tunnel failures by hydraulic jacking and hydraulic shearing of joints and by uplift of the ground surface.

1 INTRODUCTION

Hydraulic pressure tunnels are often constructed in strong, massive rock which inevitably contains some discontinuities. Although they are usually lined with steel or concrete, considerable economies can be achieved by omitting the lining in major sections of pressure tunnels. In this case, the pre-existing discontinuities play dominant roles in determining the most likely failure mode of a tunnel. Water under high pressure may penetrate the discontinuities, decreasing the effective normal stress across them and introducing the possibility of failure by hydraulic jacking, hydraulic shearing on the discontinuities or uplift of the ground surface. The empirical methods currently used to establish the required location and length of a lining consider mainly the effect of the overburden rock and do not address the problem adequately (Lu, 1987). A more comprehensive method of design analysis of such tunnels is required. This paper describes a finite element method based approach to the numerical analysis of such problems, which permits the calculation of the three dimensional stress distribution around a tunnel carrying internal water pressure for complete plane strain conditions (Brady and Bray, 1978) and the evaluation of the potential for hydraulic jacking, hydraulic shearing and uplift of the ground surface.

2 THE PROPOSED NUMERICAL METHOD

2.1 General features

A joint-element finite element method is used to calculate the stress distribution in the rock surrounding the pressure tunnel. Using an estimated initial stress field, excavation is simulated and the internal water pressure is applied progressively. The critical mode of failure in any case can be identified from the computational results.

The intact rock blocks are assumed to be linear elastic and may fail under induced tension according to a simple tensile cut-off criterion. Particular attention is paid to modelling the responses of discontinuities (which will be described generically as "joints" hereafter). Gerrard (1986) gives a useful review of the empirical relations that have been proposed for this purpose. Barton's empirical joint model (Barton et al 1985, Barton 1986) is used as the basis of the numerical model because it describes well the pre- and post- peak plasticity and associated dilation behaviour of joints and can be used in coupled mechanical and hydraulic analyses.

In order to avoid the ill-conditioning associated with high normal stiffnesses of joint elements (Ghaboussi et al 1973), the relative displacements of the two joint walls are taken to be the essential and independent unknowns. Special joint elements are used to model joint intersections. A complete numerical procedure is introduced for calculating the water pressures acting on the joint walls. Joint permeabilities are evaluated from Barton's empirical relations.

Pressure tunnels of the type being considered here are often constructed in mountainous terrain with their axes not parallel to the ground surface or to a principal stress direction. Under these conditions, a standard plane strain solution may be inadequate and a three-dimensional analysis may be required. Some aspects of the three-dimensional nature of the problem are allowed for in the present analysis by using the concept of complete plane strain introduced by Brady and Bray (1978). In this case, excavation-induced displacements in the x, y, z co-ordinate directions are functions of x, z only (where the y co-ordinate direction is parallel to the tunnel axis) and the strain components ε_x, ε_z, γ_{xy} and γ_{zx} are non-zero, in general. No co-ordinate direction need be parallel to a principal in situ stress direction.

2.2 Mechanical model for joint elements

Barton's empirical equations for peak shear strength, τ_p, and the associated dilation angle, $\Phi°_{dpk}$, are

$$\tau_p = \sigma_n' \tan\left[JRC_p \log_{10}\left(\frac{JCS}{\sigma_n'}\right) + \Phi_r'\right] \quad (1)$$

$$\text{and } \Phi°_{dpk} = \tfrac{1}{2} JRC_p \log_{10}\left(\frac{JCS}{\sigma_n'}\right) \quad (2)$$

where σ_n' = effective normal stress,
JRC_p = joint roughness coefficient at peak strength,
JCS = joint wall compressive strength
Φ_r' = basic residual friction angle

At very low normal stresses the dilation angle may approach infinity and the shear strength may become negative. In order to avoid this, a modification is made in the region $0 < \sigma_n'/JCS < 0.01$, such that the dilation angle can be never greater than 50°.

Barton et al (1985) introduced the concepts of mobilized joint roughness coefficient, JRC_m, mobilized shear strength, τ_m, and mobilized dilation angle, $\Phi°_{d\,mob}$, at any stage of shearing such that

$$\tau_m = \sigma_n' \tan\left[JRC_m \log_{10}\left(\frac{JCS}{\sigma_n'}\right) + \Phi_r'\right] \quad (3)$$

$$\Phi°_{dmob} = \tfrac{1}{2} JRC_m \log_{10}\left(\frac{JCS}{\sigma_n'}\right) \quad (4)$$

This enables the process of shearing at constant normal stress to be described in five stages as illustrated in Figure 1. The first stage is modelled by a linear relation between shear stress, τ, and shear displacement, δ_s. The region between point A and the peak at P is modelled as elasto-plastic with strain hardening and that from P to R as elasto-plastic with strain softening.

A yield function corresponding to the

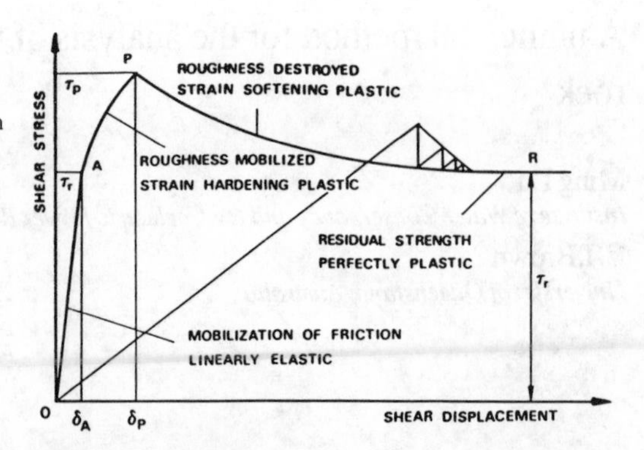

Figure 1. Elasto-plastic shear stress-shear displacement model illustrating the secant stiffness iteration method.

shear strength criterion, equation (3), can be written as

$$F = \tau - \sigma_n' \tan\left[JRC_m \log_{10}\left(\frac{JCS}{\sigma_n'}\right) + \Phi_r'\right] = 0 \quad (5)$$

where τ is the resultant shear stress on the joint plane and compression is taken as positive.

A non-associated flow rule is used to avoid the over-estimation of dilation given by the associated flow rule which was often assumed by early joint models (e.g. Pande and Xiong, 1982). Assume that the plastic potential function, G, is given by

$$G = \tau + f(\sigma_n') = \text{constant} \quad (6)$$

so that

$$\frac{\partial G}{\partial \tau} = 1, \quad \frac{\partial G}{\partial \sigma_n'} = \frac{\partial f}{\partial \sigma_n'} \quad (7)$$

The tangent of the dilation angle can be reasonably expressed as

$$\tan\Phi_d = \dot{\varepsilon}^p_n / \dot{\varepsilon}^p_s = \frac{\partial G}{\partial \sigma_n'} \Big/ \frac{\partial G}{\partial \tau} = \frac{\partial f}{\partial \sigma_n'} \quad (8)$$

where $\dot{\varepsilon}^p_n$, $\dot{\varepsilon}^p_s$ are the plastic components of incremental normal and shear strain.

From equation (8)

$$f(\sigma_n') = \int_0^{\sigma_n'} \tan\left[JRC_m \log_{10}\left(\frac{JCS}{\sigma_n'}\right)\right] d\sigma_n' \quad (9)$$

Substitution for $f(\sigma_n')$ into equation (6) gives an expression for G. This need not be evaluated explicitly because it is values of $\partial G/\partial \tau$, $\partial G/\partial \sigma_n'$, which are required in deriving the elasto-plastic stress-displacement matrix

$$[D^{ep}] = [D^e] - [D^p] \quad (10)$$

where $[D^e]$ is the elastic stress-displacement matrix, and $[D^p]$ is the plastic stress-displacement matrix

$$[D^p] = \frac{[D^e]\frac{\partial G}{\partial\{\sigma\}}\left[\frac{\partial F}{\partial\{\sigma\}}\right]^T[D^e]}{A + \left[\frac{\partial F}{\partial\{\sigma\}}\right]^T[D^e]\frac{\partial G}{\partial\{\sigma\}}} \qquad (11)$$

For complete plane strain, the elemental stress vector for a joint element is

$$\{\sigma\} = [\tau_{zx} \quad \tau_{yz} \; \sigma_n']^T \qquad (12)$$

The elemental displacement vector is

$$\{\delta\} = [\Delta u \quad \Delta v \quad \Delta w]^T \qquad (13)$$

and the elastic stress-displacement matrix is

$$[D^e] = \begin{bmatrix} K_{sx} & & \\ & K_{sy} & \\ & & K_n \end{bmatrix} \qquad (14)$$

The parameter A which defines plastic hardening or softening characteristics can be derived from the relation (Zienkiewicz 1977)

$$A = -\frac{\partial F}{\partial W^p}\{\sigma\}^T\frac{\partial G}{\partial\{\sigma\}} \qquad (15)$$

where W^p is the plastic work done during deformation.

The resulting expression for $[D^p]$ in terms of stress components and joint mechanical properties has been given by Lu(1987) and by Lu and Brown (1987). The expression includes a term $d(JRC_m)/d\delta_s$ which can be evaluated from the experimental or assumed $JRC_m - \delta_s$ relation.

Deformability in the direction normal to the joint plane is described by the hyperbola (Fig.2)

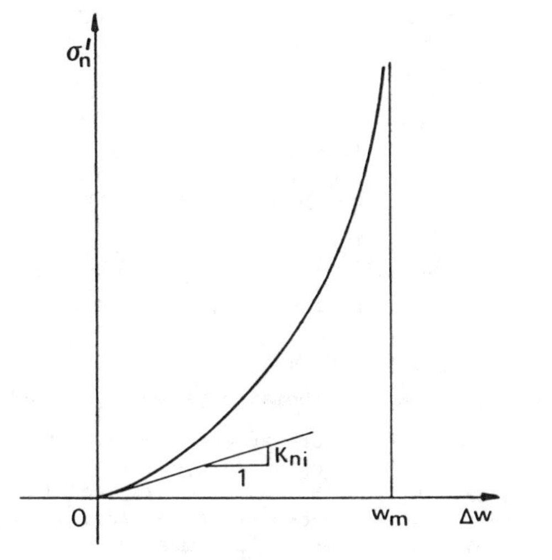

Figure 2. Hyperbolic effective normal stress (σ_n') - normal displacement (Δw) relation.

$$\sigma_n' = \frac{w_m \, \Delta w \, K_{ni}}{w_m - \Delta w} \qquad (20)$$

where w_m = maximum joint closure
and K_{ni} = initial normal stiffness coefficient

For simplicity, equation (20) is used for both joint closure and opening although experimental evidence (e.g. Barton et al 1985) indicates that hysteresis is involved with the opening curve being the steeper. The corresponding normal stiffness coefficient is given by

$$K_n = \frac{K_{ni}}{\left(1-\frac{\Delta w}{w_m}\right)^2} \qquad (21)$$

2.3 Non-linear computation method

In the plastic zone, an incremental secant stiffness iteration method (Fig.1) is used in the non-linear computation. This avoids the numerical difficulty associated with negative stiffness coefficients in the strain softening zone. Covergence is ensured providing that the equilibrium condition is satisfied. In this method, load increments are applied in a stepwise manner. A new global stiffness matrix is calculated at the beginning of each load step and remains constant throughout the step. An iteration procedure which, in some respects, is similar to the modified Newton-Raphson method (Zienkiewicz, 1977) is used to obtain the solution for each step.

2.4 Joint elements

For the ordinary four noded joint elements, the elemental stiffness matrix can be written as

$$[K^e] = \frac{l}{6}\begin{bmatrix}[K_{11}] & [K_{12}] \\ [K_{21}] & [K_{22}]\end{bmatrix} \qquad (22)$$

where l = length of the element,
$[K_{ij}]$ = submatrixes in which
$[K_{12}] = [K_{21}] = [D]$
and $[K_{11}] = [K_{22}] = 2[D]$

To be realistic, the joint elements must be able to simulate intersecting joints. Therefore, T shaped and cross joint elements are introduced. Figure 3(a) shows the T shaped joint element which can be regarded as three intersecting joints, J1, J2 and J3. At the intersection the relative displacements between points A and N and

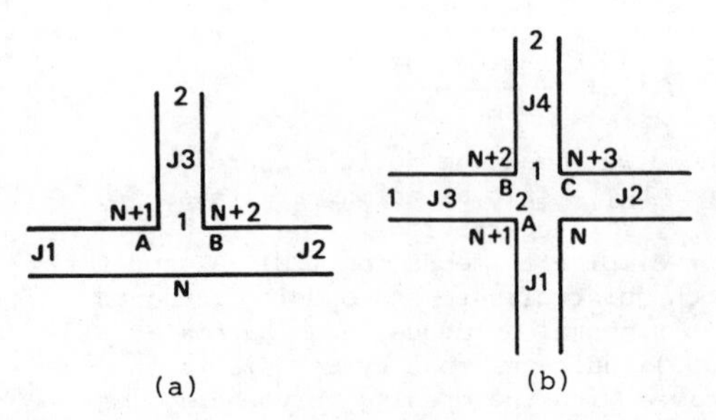

Figure 3(a). T shaped joint element, and (b) cross joint element

between points B and N are taken to be the independent unknowns, $\{\delta_{N+1}\}$ and $\{\delta_{N+2}\}$. Thus, J1 and J2 are ordinary joint elements. Only J3 has special features. For J3, the relative displacement between points A and B is $\{\delta_{N+1}\} - \{\delta_{N+2}\}$. We can consider J3 as having three 'nodes' N+1, N+2 and 2. Thus, its elemental stiffness matrix is

$$[K^e] = \begin{bmatrix} [K_{11}] & -[K_{11}] & [K_{12}] \\ -[K_{11}] & [K_{11}] & -[K_{12}] \\ [K_{21}] & -[K_{21}] & [K_{22}] \end{bmatrix} \quad (23)$$

Figure 3(b) shows the cross joint element. It can be regarded as four intersecting joint elements J1, J2, J3 and J4. The relative displacements between points A and N, B and N and C and N are taken to be the independent unknowns. Here, J1 and J2 are ordinary joint elements. For J3, the relative displacement between points C and B is $\{\delta_{N+2}\} - \{\delta_{N+1}\}$. If we define J3 with the three nodes 1, N+2 and N+1, its elemental stiffness matrix is

$$[K^e] = \begin{bmatrix} [K_{11}] & [K_{12}] & -[K_{12}] \\ [K_{12}] & [K_{22}] & -[K_{22}] \\ -[K_{12}] & [K_{22}] & [K_{22}] \end{bmatrix} \quad (24)$$

Joint element J4 is the same as J3 in the T shaped joint element.

2.5 Determination of water pressure forces acting on joint walls

The effect of water under pressure is accounted for as forces acting on the joint walls. Because of the assumption that the intact rock matrix is impermeable and water flow can only take place along joints, a finite element discontinuous medium model is used to determine the water pressure forces. The finite element discretization of the basic equation for joint flow (Zienkiewicz, 1977) can be written as

$$[H]\ \{\Phi\} = \{F_p\} \quad (25)$$

where $[H]$ = global seepage 'stiffness matrix',
$\{\Phi\}$ = flow potential vector where $\Phi = p_w/\gamma_w$,
p_w = water pressure acting on the joint wall,
γ_w = unit weight of water,
and F_p = seepage vector in terms of the flow rate at each node

For a two noded joint flow element, the elemental seepage stiffness matrix is

$$[H^e] = \frac{K_p}{l} \begin{bmatrix} 1 & 1 \\ -1 & 1 \end{bmatrix} \quad (26)$$

where l = length of the element,
and K_p = permeability coefficient of the joint

The permeability coefficient is evaluated from the joint aperture based on Barton's modified cubic flow law (Barton et al, 1985). The initial mechanical aperture E_o of a joint dpends upon its surface characteristics and may be estimated from the empirical equation

$$E_o = \frac{JRC_p}{5}\ (0.2\ \frac{\sigma_c}{JCS} - 0.1)\ \text{mm} \quad (27)$$

where σ_c is the unconfined compressive strength of the rock adjacent to the joint wall. At any stage, the residual mechanical aparture E is

$$E = E_o - \Delta w_j \quad (28)$$

where Δw_j is the joint closure or opening obtained from the finite element computation. The corresponding conducting aparture e is determined from Barton's modified cubic flow law as

$$e = \frac{E^2}{JRC_p^{2.5}}\ \mu m \quad (29)$$

For laminar flow, permeability k_p is calculated from e as

$$k_p = e^2/12 \quad (30)$$

and the permeability coefficient is given by

$$K_p = k_p\ g/\nu_m \quad (31)$$

where ν_m is the kinematic viscosity of water.

The water pressure at each node in the joint network is obtained by solving equation (25). The corresponding nodal water pressure forces are then readily calculated. However, care must be taken with the T shaped and the cross joint elements.

3 EXAMPLES OF NUMERICAL ANALYSIS OF UNLINED PRESSURE TUNNELS

The numerical method described herein has been validated for single joint behaviour (Lü and Brown 1987) and has been used by Lü (1987) in comprehensive studies of the failure modes of generic cases of unlined pressure tunnels. The examples presented here are chosen to illustrate some of the features of unlined pressure tunnel failures.

3.1 Generic case 1 - tunnel and ground surface connected by a single vertical joint

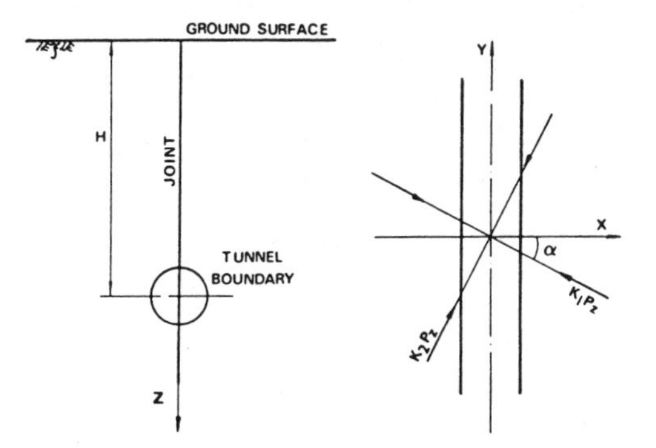

Figure 4. A simple case of an unlined pressure tunnel in complete plane strain (generic case 1).

Figure 4 illustrates the simple case of an unlined circular tunnel with a radius of 1m connected to the ground surface by a single vertical joint. The figure is drawn with the tunnel axis and the upper ground surface horizontal. This, and the subsequent figures showing illustrative examples, could equally well represent the important practical case in which the tunnel axis is parallel to a sloping upper ground surface and the plane of the section is not vertical.

A numerical solution has been obtained for this problem for the case in which

(a) the mechanical properties of the rock and joint are given by γ_r (unit weight) = 26.5 kN/m³, E_r (Young's modulus) = 20 GPa, ν_r(Poisson's ratio) = 0.25, w_m = 2.0mm, JRC_p = 10, JCS = 150 MPa, Φ_r' = 30°, L(a scaling parameter in Barton's model) = 1.0m and K_{ni} = 500 MPa/m;

(b) the in situ stress field is defined by $P_z = \gamma_r z$, K_1 = 1.5, K_2 = 1.0 and α = 0°;

(c) the internal water pressure is increased progressively in increments of 2m of hydraulic head which corresponds to a water pressure of 19.6 kPa.

The illustrative examples solved all involve shallow tunnels because clear failure modes can be developed in these cases with a minimum of computational effort. The results show that, as might be anticipated, hydraulic jacking of the joint is the unique failure mode for this problem geometry and stress field. At values of H/D = 3.5, 5.0 and 7.5, where D is the tunnel diameter and H is the thickness of superincumbent rock, the critical water pressures at which hydraulic jacking of the joint propagates to the ground surface are 176, 216 and 314 kPa, respectively.

Simple closed form solutions for deep tunnels based on the Kirsch equations (Hoek 1982, Lü 1987) show that the nature of the initial stress field plays an important role in determining failure conditions for pressure tunnels. Clearly, the initial horizontal stress K_1P_z will influence the pressure required to produce hydraulic jacking in generic case 1. For the case analysed above with K_2 = 1.0 α = 0° and H/D = 5.0, the critical water pressures for values of K_1 of 0.5, 1.0, 1.5 and 2.0 are 78, 157, 216 and 294 kPa, respectively.

A joint which intersects the tunnel and is subject to an initial shear stress may develop additional shear displacement and "fail" in shear as the effective normal stress is decreased following application of the internal water pressure. This phenomenon is called hydraulic shearing and may be explained by reference to Figure 5. Curve 1 is the shear stress-displacement plot for the original effective normal stress, σ'_{n1}. At a lower normal effective stress, σ'_{n2}, curve 2 applies. If the shear stress remains unchanged and the "correct" curve is to be followed, the shear displacement must

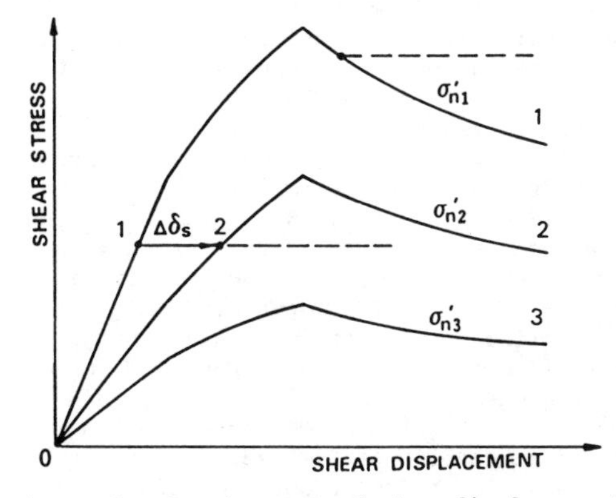

Figure 5. Development of shear displacement under reduced effective normal stress and constant shear stress.

increase from point 1 to point 2. If the normal stress decreases further to σ'_{n3}, the current shear stress cannot be sustained, slip will occur and the stresses will be redistributed.

Hydraulic shearing is a major failure mode of unlined pressure tunnels. It can occur in generic case 1 if the tunnel axis is not parallel to a principal stress direction and if the two lateral stress coefficients, K_1 and K_2, are not equal. Table 1 shows the results of calculations for the case previously considered with K_1 = 1.5, K_2 = 1.0, H/D = 5.0 and α varying from 0 to 45°. The results indicate that when the initial shear stresses are high, hydraulic shearing may become a more critical failure mode than hydraulic jacking.

Table 1. Failure modes and critical water pressures for generic case 1 with H/D = 5.0, K_1 = 1.5, K_2 = 1.0 and varying α.

α(°)	Failure mode	Critical water pressure (kPa)
0	Jacking	216
15	Jacking/shearing	216
30	Shearing	196
45	Shearing	157

3.2 Generic case 2- a joint parallel to the ground surface connected to the tunnel by a single vertical joint

An analysis of unlined pressure tunnel failure mechanisms (Lü 1987) showed the great danger associated with joints parallel and close to the ground surface connected to the tunnel by other joints. Water under pressure may penetrate the near surface joint causing hydraulic jacking and uplift of the ground surface.

For the generic case shown in Figure 6 results are presented for the two locations of the horizontal joint given by H_1 = 0.9m and H_1 = 1.7m. The same material properties as those used for generic case 1 are adopted with K_1 = 1.5, K_2 = 1.0 and α = 0°. The results show that as the internal water pressure increases, both the vertical and the horizontal joints open gradually with the horizontal joints opening at the greater rate (Figure 7). At water pressures of 49 kPa when H_1 = 1.7m and 29 kPa when H_1 = 0.9m the effective normal stress on the horizontal joint becomes zero and an abrupt increase in the joint opening occurs. This produces significant upwards displacement of the rock block above the horizontal joint and uplift of the ground surface. In this case, a lower tunnel water pressure is required to produce hydraulic jacking of the horizontal joint than that required to produce jacking of the vertical joint to the surface in generic case 1. Note that, although the vertical joint does not jack open in this case, it opens sufficiently to allow water under pressure to penetrate and jack open the horizontal joint.

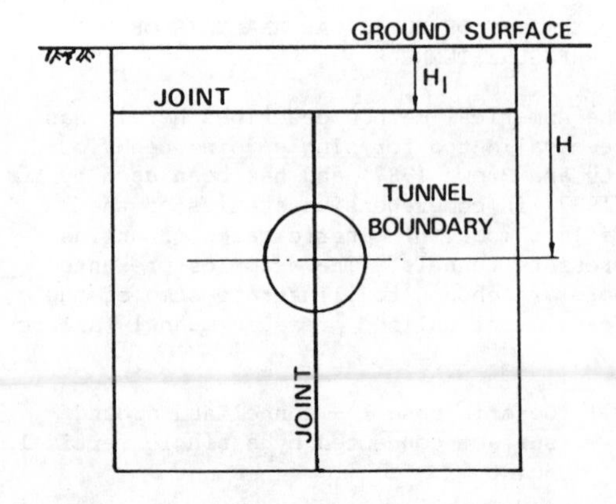

Figure 6. Unlined pressure tunnel in jointed rock (generic case 2).

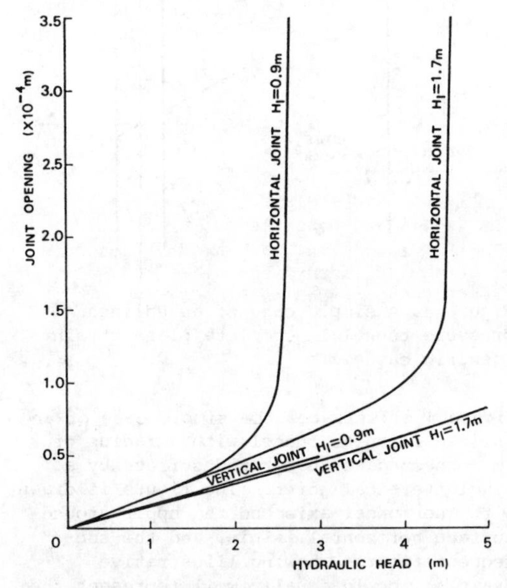

Figure 7. Development of joint opening in a numerical example of generic case 2.

3.3 Simulation of a hypothetical practical case

The results of a numerical analysis of a hypothetical practical case are presented to demonstrate how the method may be applied to real engineering problems. Figure 8 illustrates a case in which a

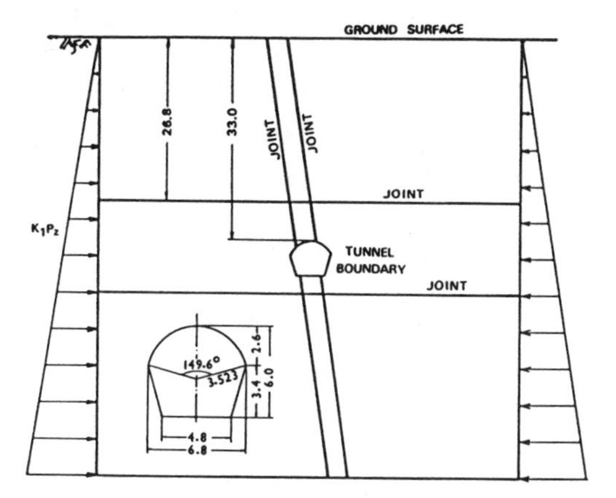

Figure 8. A hypothetical practical case

horseshoe shaped unlined pressure tunnel is excavated 33m below the ground surface in a rock mass containing two persistent horizontal and two persistent vertical discontinuities. Two possible failure modes are identified in this case - hydraulic shearing on the inclined discontinuities and uplift of the ground surface associated with hydraulic jacking of the upper horizontal joint. The numerical solution uses the parameters E_r = 10 GPa, ν_r = 0.25, γ_r = 26.5 kN/m³, w_m = 2.0mm, JRC_p = 10, JCS = 100 MPa, Φ'_r = 25°, L = 1.0m, K_{ni} = 2.0 GPa/m, K_1 = 1.5, K_2 = 1.0 and α = 30°.

For these parameters, the critical failure mode is hydraulic jacking of the upper horizontal joint which occurs at an internal hydraulic head of 90m of water or a water pressure of 883 kPa. Figure 9 shows the computed values of ground surface uplift. It so happens that this result is

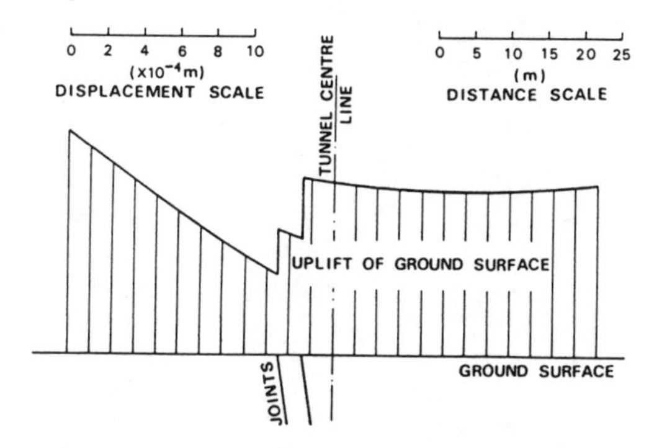

Figure 9. Ground surface uplift for the hypothetical practical case

almost identical with that obtained using the classical depth-of-cover calculation (Jaeger 1979). Such would not be the case under other conditions, particularly those involving low values of the lateral stress coefficient, K_1 (Lü 1987).

4 CONCLUSIONS

A numerical method has been developed which uses Barton's empirical joint model and the theory of elasto-plasticity with a non-associated flow rule to reproduce the experimentally observed pre- and post- peak plasticity behaviour and associated dilational characteristics of rock joints. The incremental secant stiffness iteration procedure developed ensures numerical convergence in cases of non-liner behaviour including strain softening.

Three major failure modes, hydraulic jacking, hydraulic shearing and uplift of the ground surface, have been identified for unlined pressure tunnels. The effects of in situ stresses, overburden thickness and the locations of joints parallel to the ground surface have been demonstrated. Low lateral in situ stresses, thin overburden rock and the existence of a joint parallel and near to the ground surface are adverse conditions for unlined pressure tunnels. The analysis of a hypothetical practical case further verified the capability of the proposed method in dealing with real engineering problems.

ACKNOWLEDGEMENTS

The work described herein was carried out in the Department of Mineral Resources Engineering, Imperial College of Science and Technology, London, U.K. The problems of unlined pressure tunnel design were drawn to the authors' attention by Dr. Evert Hoek.

REFERENCES

Barton, N.R. 1986. Deformation phenomena in jointed rock. Géotechnique 36: 147-167.

Barton, N., S. Bandis & K. Bakhtar 1985. Strength, deformation and conductivity coupling of rock joints. Int. J. Rock Mech. Min. Sci. & Geomech. Abstr. 22: 121-140.

Brady, B.H.G. & J.W. Bray 1978. The boundary element method for determining stresses and displacements around long openings in a triaxial stress field.

Int. J. Rock Mech. Min. Sci. & Geomech. Abstr. 15: 21-28.

Gerrard, C.M. 1986. Shear failure of rock joints: appropriate constraints for empirical relations. Int. J. Rock Mech. Min. Sci. & Geomech. Abstr. 23: 421-429.

Ghaboussi, J., E.L. Wilson & J. Isenberg 1973. Finite element for rock joints and interfaces. J. Soil Mech. Foundns Div., ASCE 94: 833-848.

Hoek, E. 1982. Personal communication.

Jaeger, C. 1979. Rock mechanics and engineering, 2nd edn. Cambridge: Cambridge University Press.

Lü, M. 1987. A numerical method for the analysis of unlined pressure tunnels in jointed rock. Ph.D. thesis, Univ. of London.

Lü, M. & E.T. Brown 1987. A joint model for use in the numerical analysis of unlined pressure tunnels in jointed rock. Proc. Conf. on Numerical Methods in Geomechanics, Vysoké Tatry, Czechoslovakia 2: 13-21.

Pande, G.N. & W. Xiong 1982. An improved multi-laminate model of jointed rock masses. In R. Dunger et al (eds), Numerical models in geomechanics, 218-226. Rotterdam: Balkema.

Zienkiewicz, O.C. 1977. The finite element method in engineering science, 3rd edn. London: McGraw-Hill.

Numerical Methods in Geomechanics (Innsbruck 1988), Swoboda (ed.)
© 1988 Balkema, Rotterdam. ISBN 90 6191 809 X

Three-dimensional simulation of an advancing tunnel supported with forepoles, shotcrete, steel ribs and rockbolts

Ö.Aydan, T.Kyoya, Y.Ichikawa & T.Kawamoto
Nagoya University, Japan
T.Ito
Toyota Technical College, Japan
Y.Shimizu
Meijo University, Nagoya, Japan

ASTRACT: A three dimensional simulation of the tunnel excavation by the finite element method has been carried out with particular emphasis on the reinforcement effect of forepoles as well as of shotcrete and rockbolts on the deformational behaviour and the stability of the tunnel. Forepoles were represented in the analysis by a rockbolt element proposed by the authors previously. The results of the three dimensional numerical analysis and in-situ measurements are presented and compared with each other. A detailed discussion is also given on the reinforcement effect of the forepoles.

1 INTRODUCTION

The excavation of underground openings with a shallow overburden through soft ground is a very difficult engineering task and requires great care for not only the stability of the opening but also its adverse effects on the surrounding.

The tunnel dealt in this paper has been excavated through a soft ground and overburden ranged inbetween 8 and 30 meters within the considered section of the tunnel. An asphalt plant was situated along the alignment where the overburden was about 8 meters. The primary tunnel support was consisted of forepoles, shotcrete, light steel sets and rockbolts. Because of the low overburden and poor ground conditions, an instrumentation program has been undertaken to investigate the performance of the tunnel and also the effects on the adjacent surface structures during and after the construction. Meanwhile, a three dimensional simulation of the tunnel excavation by the finite element method has been carried out with particular emphasis on the reinforcement effect of forepoles as well as of shotcrete and rockbolts on the deformational behaviour and the stability of the tunnel. Forepoles are represented in the analysis by a rockbolt element proposed by the authors previously (Aydan et al. 1985, 1986).

In this paper, results of the three dimensional numerical analysis and in-situ measurements are presented and compared with each other. A detailed discussion is also given on the reinforcement effect of the forepoles.

2 REPRESENTATION OF ROCKBOLTS AND FOREPOLES

Most of the available numerical models for the representation of forepoles and rockbolts in numerical analysis except that of John and Van Dillen (1983) are not capable of representing those truely as they do not take into account the interaction between ground and the steel bar and also the resistance offered by the bolts or forepoles against shearing

A realistic representation of rockbolts and forepoles in numerical analyses requires the consideration of stiffness and resistance of the steel bar against axial and shear loadings as well as the interaction between the steel bar and ground. The element proposed by the authors (1985, 1986) takes the above facts into the consideration. This element is briefly described herein with a particular emphasis on the constitutive modelling of the rockbolt system as most of the details can be found in the previous articles of the authors.

2.1 Constitutive modelling of rockbolt system

The constitutive relationships for the steel bar and grout annulus or interfaces between grout-rock or steel-grout are based upon the multi-response theory proposed by Ichikawa et al. (1985, 1988). These relationships are briefly outlined as follows.

Constitutive relationship for steel bar: The response functions of the steel bar against axial and shear stresses can be given as;

$$\begin{aligned} \gamma_b^e &= \Phi_b^e(\tau_b), \quad \gamma_b^p = \Phi_b^p(\tau_b), \\ \varepsilon_b^e &= \Psi_b^e(\sigma_b), \quad \varepsilon_b^p = \Psi_b^p(\sigma_b). \end{aligned} \tag{1}$$

where γ_b, τ_b are the shear strain and stress, and σ_b, ε_b are the axial strain and stress. It should be noted that the assumed form of the response functions implies that the axial and shear responses are independent of each other and the lateral expansion or compression of the steel bar during axial loading is omitted. As the behaviour of the steel bar is of strain hardening type, the specific forms of the response functions can be written by the use of the Laplace transformation concept (see Ichikawa et al. 1985). The incremental constitutive equation is obtained by differentiating the above response functions. For example, the incremental constitutive equation for the elastic behaviour takes the following matrix form

$$\left\{ \begin{array}{c} d\gamma_b^e \\ d\varepsilon_b^e \end{array} \right\} = \left[\begin{array}{cc} \frac{\partial \Phi_b^e}{\partial \tau_b} & 0 \\ 0 & \frac{\partial \Psi_b^e}{\partial \sigma_b} \end{array} \right] \left\{ \begin{array}{c} d\tau_b \\ d\sigma_b \end{array} \right\} \tag{2}$$

When a linear elastic behaviour is considered, one will have the followings:

$$\frac{\partial \Phi_b^e}{\partial \tau_b} = \frac{1}{G_b}, \quad \frac{\partial \Psi_b^e}{\partial \sigma_b} = \frac{1}{E_b}$$

Similarly, the incremental constitutive equation for the elasto-plastic behaviour is given in the matrix form as follows

$$\left\{ \begin{array}{c} d\gamma_b^{ep} \\ d\varepsilon_b^{ep} \end{array} \right\} = \left[\begin{array}{cc} \frac{\partial \Phi_b^e}{\partial \tau_b} + \frac{\partial \Phi_b^p}{\partial \tau_b} & 0 \\ 0 & \frac{\partial \Psi_b^e}{\partial \sigma_b} + \frac{\partial \Psi_b^p}{\partial \sigma_b} \end{array} \right] \left\{ \begin{array}{c} d\tau_b \\ d\sigma_b \end{array} \right\} \tag{3}$$

<u>Constitutive relationship for grout annulus and interfaces:</u> The grout annulus or interfaces are herein regarded as a plane with a finite thickness. The thickness of the plane is assumed to be associated with the thickness of shear bands experienced in tests or in nature and, if exists, the height of asperities of the plane. Assigning a thickness to such planes also makes the physical meaning of some parameters be clear and their determination from the tests easy. The response functions of the grout annulus or interfaces against axial and shear stresses can also be written as;

$$\begin{aligned} \gamma_{ga}^e &= \Phi_{ga}^e(\tau_{ga}), \quad \gamma_{ga}^p = \Phi_{ga}^p(\tau_{ga}, \sigma_{ga}), \\ \varepsilon_{ga}^e &= \Psi_b^e(\sigma_{ga}), \quad \varepsilon_b^p = \Psi_b^p(\tau_{ga}, \sigma_{ga}). \end{aligned} \tag{4}$$

where γ_{ga}, τ_{ga} are the shear strain and stress. $\varepsilon_{ga}, \sigma_{ga}$ are the normal strain and stress. Note that the response functions for the plastic behaviour are different from those for the steel bar. As the plastic strains are associated with the internal friction angle and the dilatancy, these are the functions of τ_{ga} and σ_{ga}.

The behaviour of the interfaces and grout against shearing is different in some respects from the behaviour of strain hardening materials. As the plastic behaviour of interfaces or grout annulus involves debonding during shearing the constitutive equation must be capable of expressing such a phenomena. Therefore, the specific forms of the response functions may be written by the use of the concept of Fourier transformation. An approximate form of the response functions is obtained by discretization. As an example, the response function $\Psi(\tau_{ga}, \sigma_{ga})$ is shown in Figure 1.

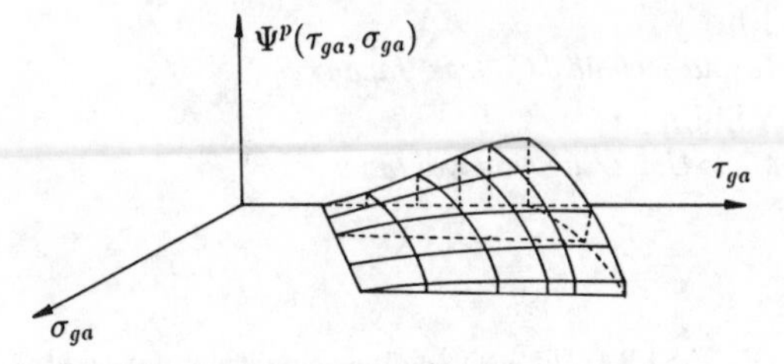

Fig.1 Schematic diagram of response function

The incremental constitutive equation of the grout annulus or interfaces is obtained by differentiating the response functions as shown previously. For example, the incremental constitutive equation for the elastic behaviour takes the following matrix form

$$\left\{ \begin{array}{c} d\gamma_{ga}^e \\ d\varepsilon_{ga}^e \end{array} \right\} = \left[\begin{array}{cc} \frac{\partial \Phi_{ga}^e}{\partial \tau_{ga}} & 0 \\ 0 & \frac{\partial \Psi_{ga}^e}{\partial \sigma_{ga}} \end{array} \right] \left\{ \begin{array}{c} d\tau_{ga} \\ d\sigma_{ga} \end{array} \right\} \tag{5}$$

For a linear elastic behaviour, we have the following:

$$\frac{\partial \Phi_{ga}^e}{\partial \tau_{ga}} = \frac{1}{G_{ga}}, \quad \frac{\partial \Psi_{ga}^e}{\partial \sigma_{ga}} = \frac{1}{E_{ga}} \tag{6}$$

Similarly the incremental constitutive equation for the elasto-plastic behaviour is given in the following matrix form:

$$\left\{ \begin{array}{c} d\gamma_{ga}^{ep} \\ d\varepsilon_{ga}^{ep} \end{array} \right\} = \left[\begin{array}{cc} \frac{\partial \Phi_{ga}^e}{\partial \tau_{ga}} + \frac{\partial \Phi_{ga}^p}{\partial \tau_{ga}} & \frac{\partial \Phi_{ga}^p}{\partial \sigma_{ga}} \\ \frac{\partial \Psi_{ga}^p}{\partial \tau_{ga}} & \frac{\partial \Psi_{ga}^e}{\partial \sigma_{ga}} + \frac{\partial \Psi_{ga}^p}{\partial \sigma_{ga}} \end{array} \right] \left\{ \begin{array}{c} d\tau_{ga} \\ d\sigma_{ga} \end{array} \right\} \tag{7}$$

The terms $\frac{\partial \Phi_{ga}^p}{\partial \sigma_{ga}}$ and $\frac{\partial \Psi_{ga}^p}{\partial \tau_{ga}}$ in the above matrix are associated with the internal friction coefficient and dilatancy factors respectively.

2.2 Rockbolt element

The rockbolt element developed by the authors (1985, 1986) takes into account the true behaviour of the steel bar as well as those of interfaces and grout annulus which are closely associated with the interaction phenomena between the bolts and ground. The derivation of the stiffness matrix for the bolt element will not be outlined herein as the derivation can be found elsewhere. However, the basic assumptions made in the mathematical modelling of the rockbolt system are summarised as follows; the steel bar has an axial and shear stiffness. The grout annulus is considered to be an axisymmetric body around the bar and its shear strain is evaluated in terms of relative displacements of the interfaces, dimensions of the bar and borehole depending upon the possible failure location. A three dimensional perspective view of the element is shown in Figure 2.

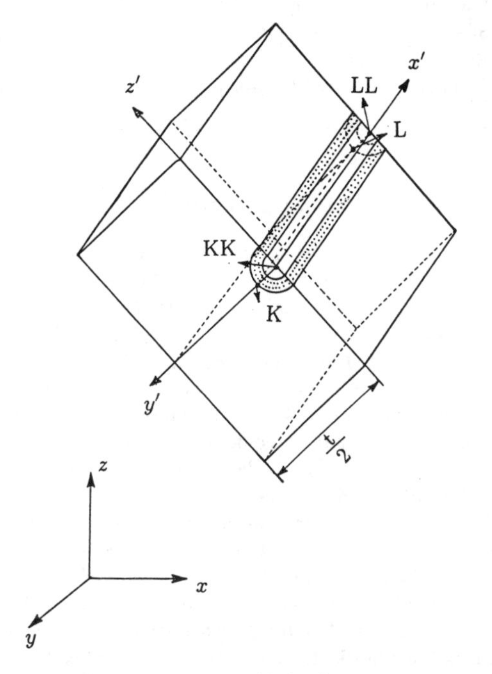

Fig.2 Perspective view of bolt element

The stiffness matrix of the bolt element in the explicit form for the case that the approximation functions are linear can be shown to be:

The better representation may also be obtained by increasing the order of the approximation functions.

3 HAE TUNNEL PROJECT

3.1 Outline of project

A project for a railway line in the mountainous western Japan was undertaken A major part of the line had to be passed through the mountains by tunnels of various length amounting the one third of the total length of the railway line. Of these tunnels, the tunnel under consideration is 1079 m long and is a single track railway tunnel of 6.5 m high and 6 m wide. The mountains be passed through are four and valleys are three. The overburden ranges inbetween 20 to 65 m under the mountain peaks and about 8 meters under the valley bottoms along the tunnel alignment. At one of the valley the tunnel had to be passed under an asphalt plant which was an sensitive structure to even a small amount of settlement.

3.2 Geology

The geology of the site consists of three different formations; aluvial deposits, shale and gabro. The aluvial deposit formation is consisted of clay, silt, gravel of 2-30 mm in size. The shale formation is highly weathered near the ground surface up to a depth of 20 m and bedding planes and discontinuities are almost unrecognisable. Thus, it is regarded as a soil-like material. Gabro is also highly weathered near the ground

$$\boldsymbol{K}^e = \begin{bmatrix} 2K_{ga} & 0 & 0 & K_{ga} & 0 & 0 & -2K_{ga} & 0 & 0 & -K_{ga} & 0 & 0 \\ 0 & K_s & 0 & 0 & -K_s & 0 & 0 & 0 & 0 & 0 & 0 & 0 \\ 0 & 0 & K_s & 0 & 0 & -K_s & 0 & 0 & 0 & 0 & 0 & 0 \\ K_{ga} & 0 & 0 & 2K_{ga} & 0 & 0 & -K_{ga} & 0 & 0 & -2K_{ga} & 0 & 0 \\ 0 & -K_s & 0 & 0 & K_s & 0 & 0 & 0 & 0 & 0 & 0 & 0 \\ 0 & 0 & -K_s & 0 & 0 & K_s & 0 & 0 & 0 & 0 & 0 & 0 \\ -2K_{ga} & 0 & 0 & -K_{ga} & 0 & 0 & K_b + 2K_{ga} & 0 & 0 & -K_b + K_{ga} & 0 & 0 \\ 0 & 0 & 0 & 0 & 0 & 0 & 0 & 0 & 0 & 0 & 0 & 0 \\ 0 & 0 & 0 & 0 & 0 & 0 & 0 & 0 & 0 & 0 & 0 & 0 \\ -K_{ga} & 0 & 0 & -2K_{ga} & 0 & 0 & -K_b + K_{ga} & 0 & 0 & K_b + 2K_{ga} & 0 & 0 \\ 0 & 0 & 0 & 0 & 0 & 0 & 0 & 0 & 0 & 0 & 0 & 0 \\ 0 & 0 & 0 & 0 & 0 & 0 & 0 & 0 & 0 & 0 & 0 & 0 \end{bmatrix} \quad (8)$$

where

$$K_{ga} = \pi G_{ga} \frac{x'_{LL} - x'_{KK}}{3\ln(r_h/r_b)} \quad K_b = \frac{E_b A}{x'_{LL} - x'_{KK}} \quad K_s = \frac{G_b A}{x'_{LL} - x'_{KK}} \quad A = \pi r_b^2$$

surface. The portion of the tunnel under the consideration is, however, only consists of aluvial deposit and weathered shale as illustrated in Fig.3.

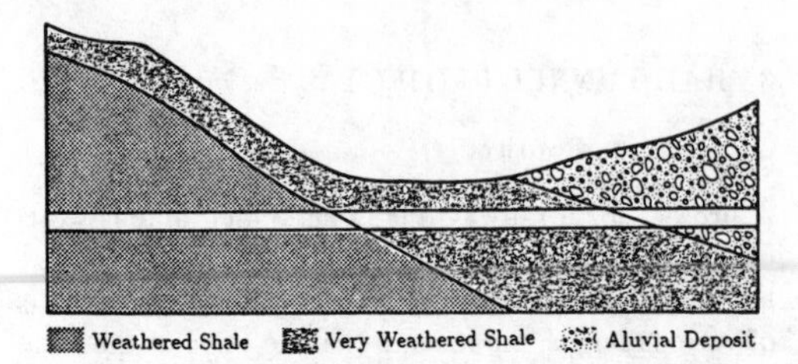

Fig.3 Geology of site

3.3 Material properties

To determine the material properties some laboratory and in-situ tests were undertaken. Laboratory tests involved the uniaxial and triaxial tests on ground samples, permeability tests. In addition, in-situ standart penetration tests (SPT), pressuremeter tests and permeability tests. The material properties obtained from these tests are given in Table 1.

Table.1 Material properties

Formations	E MPa	ν	C kPa	ϕ (^{0})
Aluvial Deposit	50	0.3	50	35
Very Weathered Shale	200	0.3	300	30
Weathered Shale	800	0.25	950	35

3.4 Instrumentation and monitoring

An instrumentation program was undertaken to observe the deformational behaviour of ground near the asphalt plant. The items of the instrumentation included the relative displacement measurements by extensometers, convergence measurements of tunnel, axial and bending stress measurements in forepoles, precise measurements of settlement of ground surface and rain fall and ground water level measurements.

3.5 Support and excavation

The initial tunnel support was consisted of shotcrete, forepoles, steel sets and rockbolts. The final support was a concrete lining cast with the purpose of reducing the friction against air flow, the water inflow into the tunnel and additional safety against any form of instability. The dimensions and properties of the support members are given in Table 2. The excavation of tunnel carried out by a short bench excavation method. the upper bench excavation was followed by the lower bench by approximately 6 m.

Table 2. Dimensions and material properties of support members

Support Member	Dimensions	E GPa	ν	σ_c or σ_t MPa
Shotcrete	t=10 mm	5	0.2	10
Steel sets	A=21 cm^2	210	0.3	300
Rockbolt or Forepoles	r_b =12.5 mm r_h =18.5 mm	210	0.3	300
Grout		5	0.25	8

4 FINITE ELEMENT METHOD

Two types of finite element analysis are carried out. The first type of the analyses was a three dimensional elastic finite element analysis. In this analysis, ground medium, shotcrete, steel ribs, forepoles and rockbolts were represented by 8 node isoparametric elements, 4 node membrane elements, truss elements, and rockbolt elements respectively. The finite element mesh used in the two and three dimensional analyses were generated by an specially developed auto-mesh code.

Initial ground stresses are assumed to be resulting from gravity. First, an analysis was carried out to determine the initial stresses in the medium and a full-face excavation procedure was simulated step by step. The support elements were assumed to be installed immediately after the excavation made. Three dimensional block diagram of analysed region is shown in Figure 4.

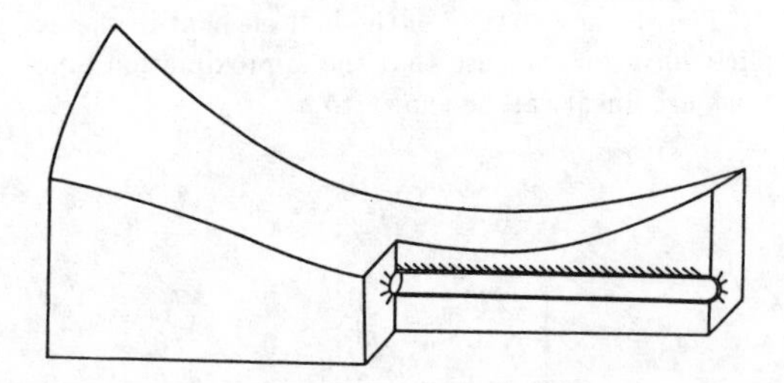

Fig.4 Block diagram of 3-D analysed region

As the three dimensional analysis was an elastic analysis, a two dimensional elasto-plastic analysis was carried out in order to see whether any yielding is likely to be appear in the vicinity of the ground about the tunnel section of interest.

5 RESULTS, COMPARISONS AND DISCUSSIONS

Results of the finite element analyses are herein reported and compared with the measured ones. In addition, the performance of support members with particular emphasis on the forepoles are discussed.

Two dimensional elasto-plastic finite element anaysis has indicated that the yielding in the ground was unlikely to occur about the tunnel for the given material properties. The results of the two dimensional analysis are also well comparable with those of the three dimensional analysis when the face effect ceases.

Figures 5 - 7 show the normalised settlement along the roof line, settlement of the ground surface at the measurement station 1, and displacement of ground above the roof and in sidewalls together with the measured results, respectively. As it is apparent from the figures the calculated results well agree with the measured ones.

Next, we discuss the performance of support members. Figures 8 and 9 show axial stress distributions in shotcrete and steel sets, respectively when the face effect disappears. As noted from the figures, the distributions are very similar to each other and are in a spine-like shape. In addition, the maximum axial stress locations are diagonal. This was thought to be resulting from the inclined ground surface. The maximum shotcrete stress is about 7.3 MPa which is less than the uniaxial strength of the shotcrete. The maximum axial stresses in the steel sets occur near the foot plates and are about 250 MPa which are also less than the yielding strength of the steel.

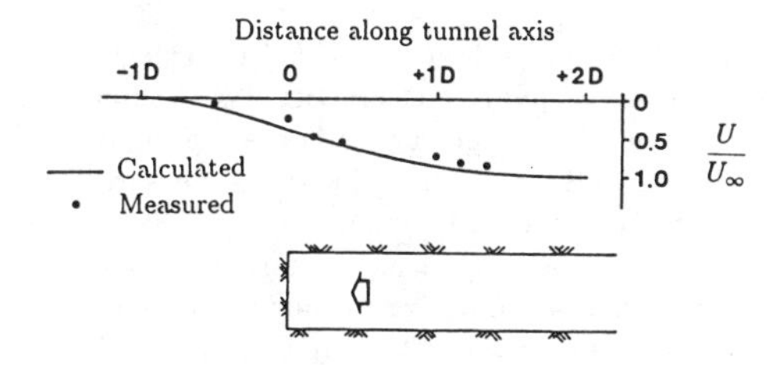

Fig.5 Settlement along roof line

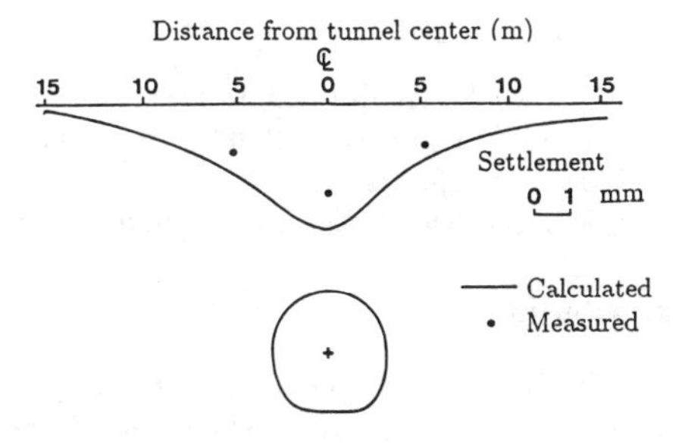

Fig.6 Ground surface settlement at Station 1

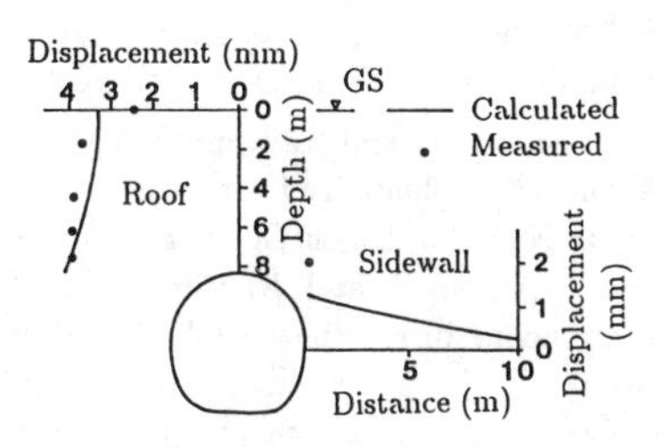

Fig.7 Displacement of ground in the roof and sidewalls

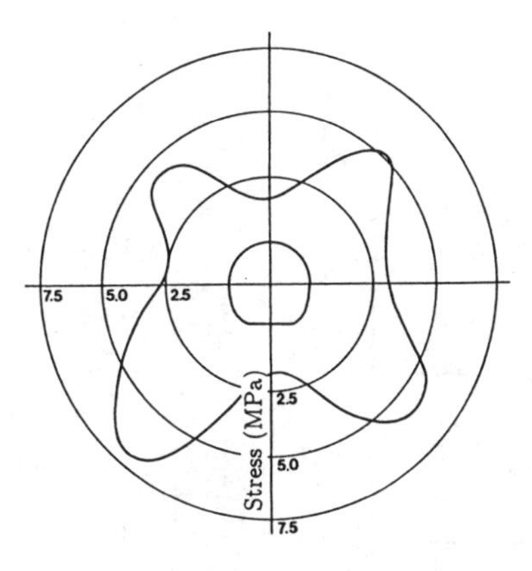

Fig.8 Axial stress distribution in shotcrete

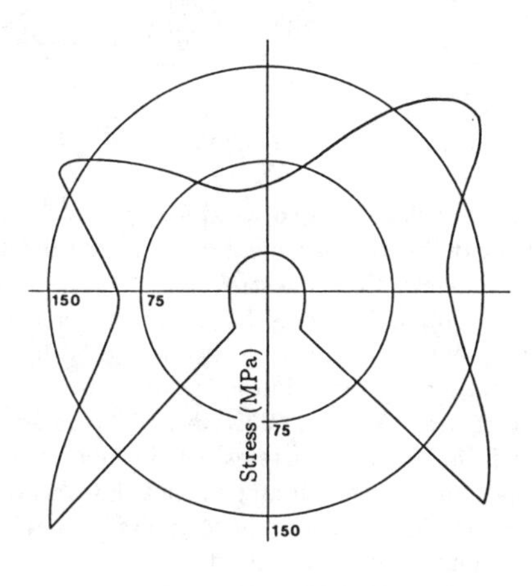

Fig.9 Axial stress distribution in steel ribs

Figure 10 shows the axial stress distributions in the bolts installed about the tunnel. Although it is unfortunate that no comparisons can be made between calculated results and measurements since no axial stress measurements were made on the rockbolts, the calculated results gives a clear picture of the likely performance of the bolts about the tunnel at the ultimate stage. The axial stress distributions in bolts are closely associated with the deformational behaviour of the ground. The axial stresses in bolts installed near the shoulders are much greater than those observed near the spring lines as expected.

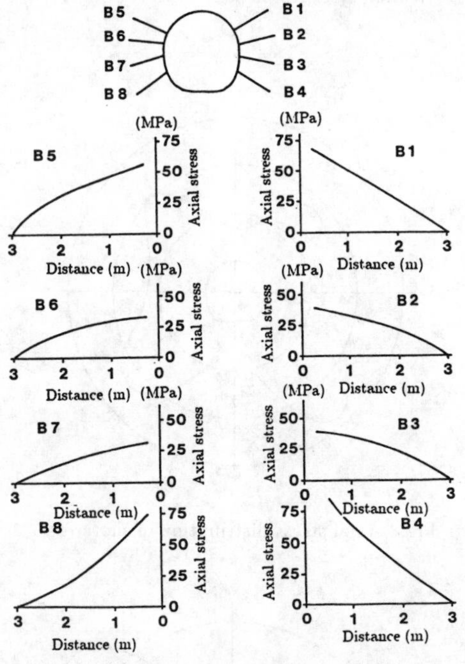

Fig.10 Axial stress distributions in bolts

Finally, the calculated axial stresses in forepoles are compared with those measured during tunnelling. Figure 11 shows the axial stress distributions in forepoles at various distances from the tunnel face together with the measured results. As noted from the figure, the axial stresses in the forepoles are in tensile character near the face. However, with the advance of the face, the axial stresses distributions tends to become compressive. Although it not shown here, the axial stress distributions in the forepoles becomes completely compressive when the tunnel face is far of a distance over the two times the tunnel diameter. This experimental fact is also verified by the results of the finite element analysis. The results also indicate that the reinforcement effect of the forepoles will disappear when the tunnel face is far of a distance of one times the tunnel diameter from the respective section.

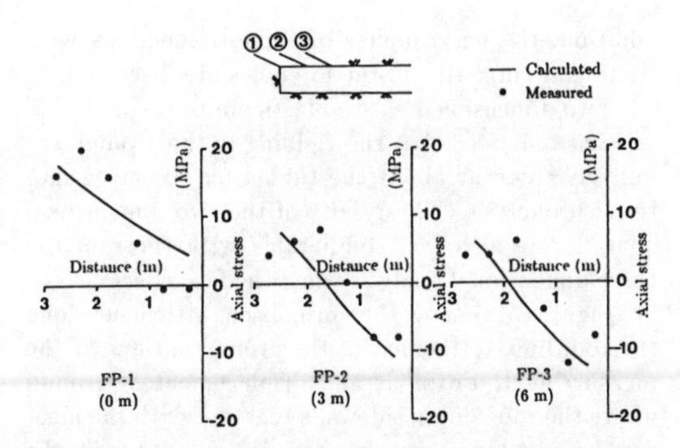

Fig.11 Axial stress distributions in forepoles

6 CONCLUSIONS

Results of the three dimensional finite element analysis of an advancing tunnel supported with forepoles, shotcrete, steel ribs and rockbolts are reported. The calculated results are compared with the measurements made at the respective tunnel and discussed. The three dimensional finite element analysis gives a clear picture of what is happening in the vicinity of ground and of the performance of the support members during the face advance. In addition, the calculated results are well comparable with measurements.

ACKNOWLEDGEMENTS

The authors would like to thank to Mr. A. Kashima, Japan Railway Construction Public Corp. and N. Takeda, Taisei Corp. for the assistance and information given on the field measurements and in-situ tests.

REFERENCES

Aydan, Ö., Ichikawa, Y. and Kawamoto, T. (1985): A finite Element for grouted rockbolts and their anchorage mechanism. Procs. 1st Nat. Symp. on Num. Meths. in Geotechnical Engineering, Tokyo.

Aydan, Ö., Ichikawa, Y. and Kawamoto, T. (1986): Reinforcement of geotechnical engineering structures by grouted rockbolts . Procs. of Int. Symp. on Eng. in Complex Rock Formations. Beijing.

Ichikawa, Y. Kyoya, T. and Kawamoto, T. (1985): Incremental theory of plasticity for rocks. Procs. of 5th Int. Conf. on Num Meths. in Geomechanics, Nagoya.

Ichikawa, Y. Kyoya, T. and Kawamoto, T. (1988): Incremental theory of elasticity and plasticity under cyclic loading. To be appeared in Procs. of 6th Int. Conf. on Num Meths. in Geomechanics, Innsbruck.

John, C.M. and Van Dillen, D.E. (1983): Rockbolts: A new representation and its application in tunnel design. Procs. of 24th U.S. Symp. on Rock Mechs.

Numerical Methods in Geomechanics (Innsbruck 1988), Swoboda (ed.)
© 1988 Balkema, Rotterdam. ISBN 90 6191 809 X

Analysis of a multiple tunnel interaction problem

S.L.Lee, K.W.Lo & L.K.Chang
National University of Singapore

ABSTRACT: Four closely interacting mass rapid transit tunnels have been constructed through generally very stiff Old Alluvium overlain by soft clayey or loose granular sediments of the Kallang formation and fill. The tunnel construction problem was modelled by an assemblage of isoparametric finite elements and the results of analysis compared with field measurements as well as empirical prediction. An elastic ideal was found to be a reasonable assumption for estimating intervening ground movements due to tunnel interaction. As a whole, predictions of bending moment and thrust distributions of linings also compared well with measurement results. A maximum radial deformation ratio of 0.1% decreasing linearly with pillar width between tunnels was found to be appropriate for determining interaction bending moments.

1 INTRODUCTION

Between February 1985 and September 1986, four tunnels of 5.85m extrados diameter each were constructed between Raffles Place and City Hall Stations of the Singapore mass rapid transit system. The tunnels formed an intricate pattern of interweaving alignments, thereby providing a unique opportunity to study an unusual case of multiple tunnel interaction. Accordingly, four lateral lines of instrumentation were installed to monitor ground and lining responses to tunnel excavations as shown in Fig 1, a typical sectional view of which is presented in Fig 2. Details of the installation, monitoring and results of instrumentation from which the following field data have been extracted may be referred to in earlier publications (Lo et al, 1987a, 1987b and 1987c).

A finite element model of the tunnelling problem was also developed as shown in Fig 3. The mixed assemblage of eight-node isoparametric quadrilateral and six-node triangular quadratic elements was adopted for accuracy of solution. In all, some 750 quadrilaterals and 50 triangles were employed to form a mesh with some 5000 degrees of freedom. To minimise storage, an out-of core method of solution by Gaussian elimination (Hinton et al, 1977) was employed. Computations were carried out on an IBM 3081 mainframe, requiring some 10 minutes' central processing time and 10 minutes' peripheral processing time in arriving at a solution. As indicated in Fig 3, a double layer of quadrilateral elements was necessary at tunnel linings to ensure that computed bending moments and thrusts kept within ± 10% of known solutions (Curtis, 1976; Einstein, 1979). To simulate ground stress relief prior to placement of tunnel linings, an initial peripheral take given by the difference between measured overbreak and thickness of lining grout annulus after setting was applied. Subsequent to installation, net unbalanced ground stresses were applied as reverse tractions on lining to simulate remaining ground unloading.

2 GROUND CONDITIONS AND METHOD OF CONSTRUCTION

Generally at the site a surface cover of fill was underlain by intercalating deposits of the Kallang formation, either consisting of soft silty or organic clay, or loose silty sands. These surface deposits in turn overlay very stiff clayey sands of the Old Alluvium formation. A high groundwater table was encountered in the vicinity of the interface between fill cover and Kallang deposits. Indicative

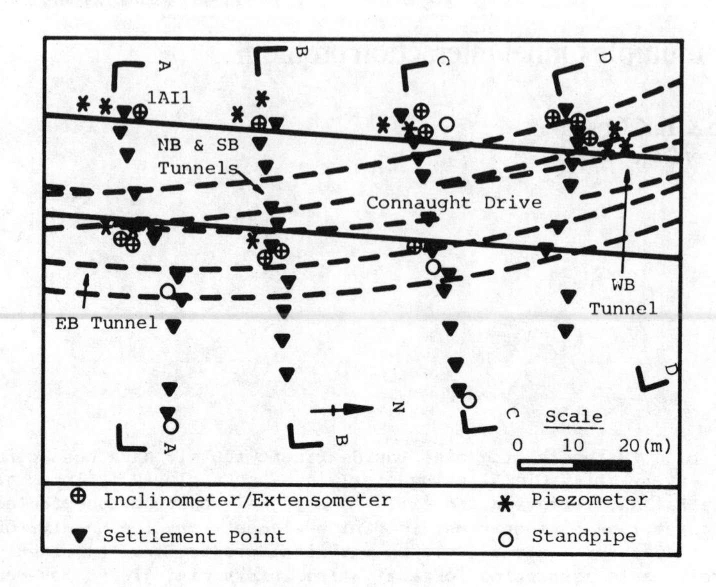

Figure 1. Instrumentation layout

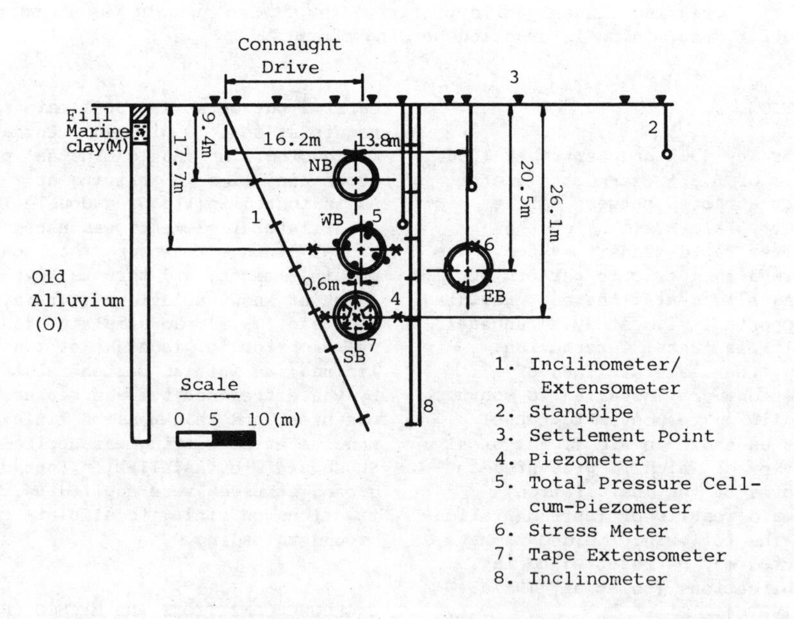

Figure 2. Section A-A

properties of various soil types obtained from both field and laboratory tests are included in Figs 3 and 4, the latter corresponding to ground conditions along northbound tunnel alignment.

Headings were advanced by conventional semi-mechanical shield tunnelling, face excavations being carried out by means of a backhoe. Generally, five standard precast concrete lining segments of 3.5m length and one key segment of 1m length, extrados, were bolted together to form a sectional ring of 1m width. Lining rings thus assembled were then longitudinally

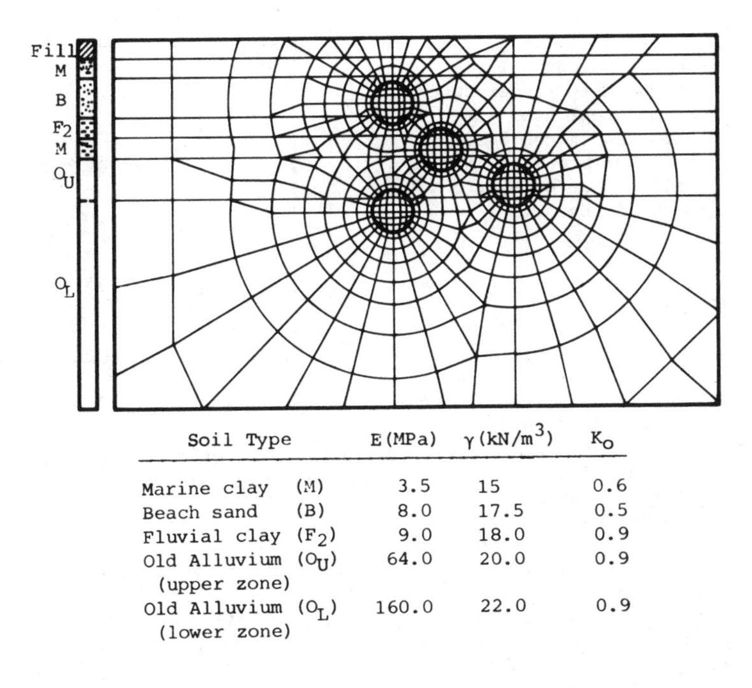

Soil Type		E(MPa)	γ(kN/m^3)	K_o
Marine clay	(M)	3.5	15	0.6
Beach sand	(B)	8.0	17.5	0.5
Fluvial clay	(F_2)	9.0	18.0	0.9
Old Alluvium (upper zone)	(O_U)	64.0	20.0	0.9
Old Alluvium (lower zone)	(O_L)	160.0	22.0	0.9

Figure 3. Finite element mesh for section C-C

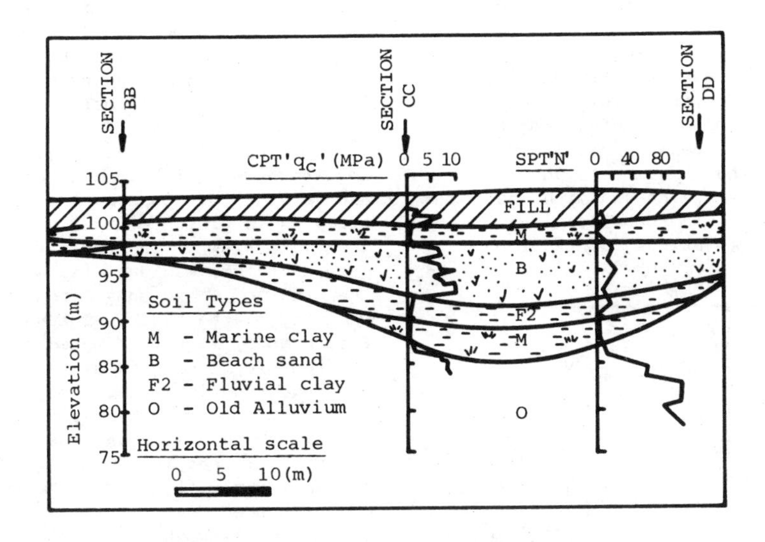

Figure 4. Longitudinal soil profile

bolted to installed lining in succession. To minimise peripheral take, a bentonite-cement mixture was pumped into the annular gap of some 75mm between excavated ground surface and lining extrados. After hardening of the primary grout, secondary grouting was carried out to fill persisting voids in the annulus. Furthermore, an insitu buried channel of waterlogged Kallang deposits was stabilised by cement followed by chemical jet grouting prior to excavations for the upper westbound, and latterly northbound tunnels passing through these ground conditions. In addition to soil improvement, tunnel face support was provided by compressed air pressure of up to 1.5 bars in conjunction with closely-packed timber breasting with occasional gaps to allow face excavations to take place by means of pneumatic spades.

3. TUNNEL INTERACTION

3.1 Ground movement

Results from the study have indicated that movements of the intervening ground between tunnels due to interaction may be reasonably modelled by an elastic soil idealisation. Fig 5, for instance, provides a comparison between analytical results and measurements of lateral ground movement along inclinometer tubing 1AI1 due to excavations for northbound tunnel, with southbound, eastbound and westbound tunnels completed according to stated sequence. Even with a maximum displacement of only some 3.5mm, differences in respective results were discernible although slight, generally within 1mm of each other. Significant ground movements were limited to within one diameter below tunnel invert level, that is to a greater depth than usual for single tunnel response. This may be attributed to westbound cavity effectively reducing the underlying ground stiffness. Fig 6, on the other hand, compares model predictions of vertical ground movements due to interactions of southbound with eastbound and hence westbound tunnels, with measurements at various points along

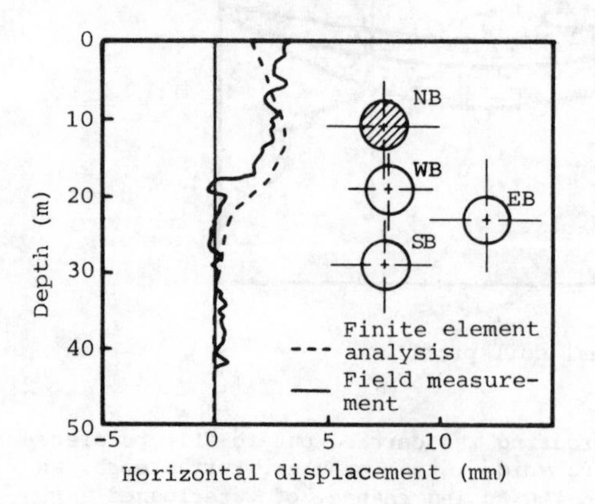

Figure 5. Horizontal displacement due to NB tunnel excavations

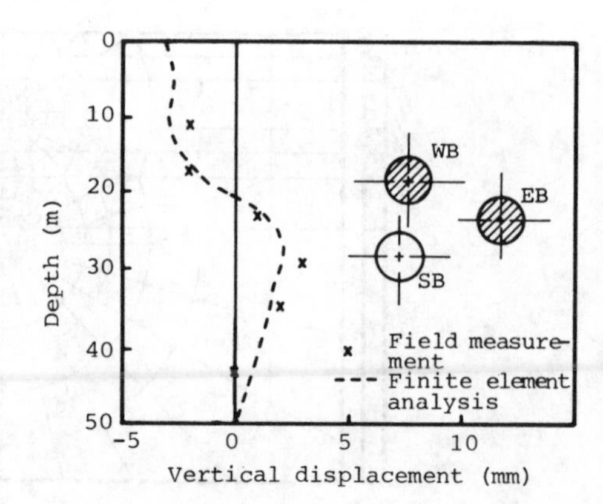

Figure 6. Vertical displacement due to EB and WB tunnel excavations

spider magnet extensometer 1AI1. These displacements were of similar magnitude as the above lateral movements, and generally good agreement was also obtained between prediction and measurement results. As in the previous case, significant ground movements occurred to greater depths below nearby westbound tunnel than would be expected for a corresponding single tunnel excavation. Also, ground stress relief due to initial southbound tunnel excavations led to lowering of the transition between downward to upward ground displacements to well below westbound springing level.

An example of cumulative general ground movements resulting from interactions amongst the four consecutive tunnel arrivals is presented in Fig 7. Accordingly, in view of the relatively stiff ground medium, maximum displacement was the order of 20 mm only, occuring in the vicinity of the eastbound tunnel crown. Furthermore, it may be noted that where two tunnels faced each other directly, a net reduction in movement of the intervening ground occurred compared to the single tunnel case due to counteracting stress relief by respective tunnels. This tendency was mitigated to some extent by the presence of an existing lining causing the adjacent ground to arch around it. The general increase in ground stiffness with depth also had a modifying influence on the above trends.

3.2 Lining loads

Figs 8 and 9 present typical comparisons between bending moment and thrust

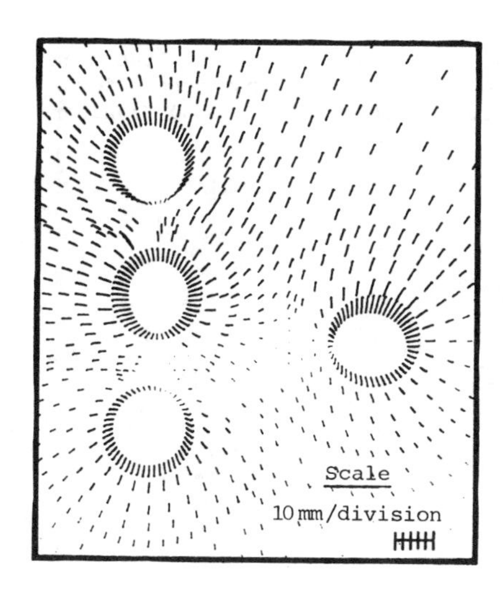

Figure 7. General ground displacements due to tunnel excavations

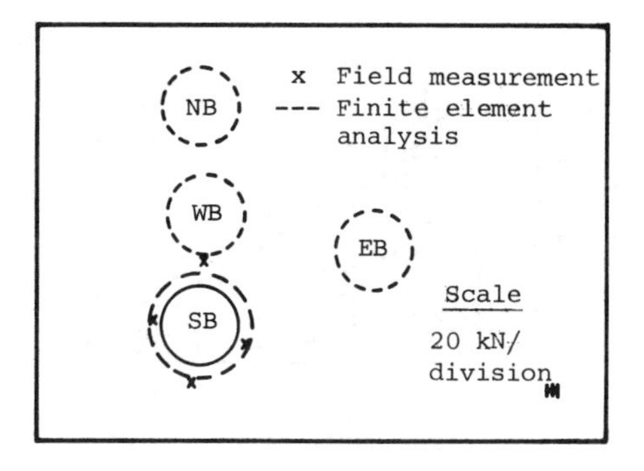

Figure 9. Axial thrust at SB after installation

distributions around a lining ring, alternatively determined via the finite element model and from stress meter readings of top and bottom reinforcement bars in lining segments. Generally, absolute values of bending moment and thrust by prediction varied within 50% and 30% respectively of those by measurement. Nevertheless, some extreme discrepancies in results also did occur, which could partly be attributed to the presence of nearby joints staggered in both longitudinal and transverse directions, at which significant local redistribution of bending moments could have taken place. Another possibility could be relative rotation between contact faces of adjacent lining segments resulting in eccentricity of thrust loading. A general tendency was for linings to distort towards excavations for subsequent tunnel arrivals. For example, southbound tunnel in Fig 8, which was initially squashed horizontally, protruded upwards due to excavations for westbound, and latterly northbound tunnels. This tendency also occurred in newly installed linings in the presence of existing tunnels, although to a lesser degree. Thus, when westbound tunnel was constructed, slight protrusions developed in its lining towards southbound and eastbound tunnels, less so the latter due to wider separation. On the other hand, the blind side of a lining from an excavation or existing opening was not significantly affected thus due to ground arching around it. As to the distribution of thrust, elongation towards an opening led to reduction in compression on opposite sides of a lining approximately parallel to the line between tunnel centres. Furthermore, maximum increase in bending moment and lowering of thrust occurred wherever an excavation was made directly above installed lining.

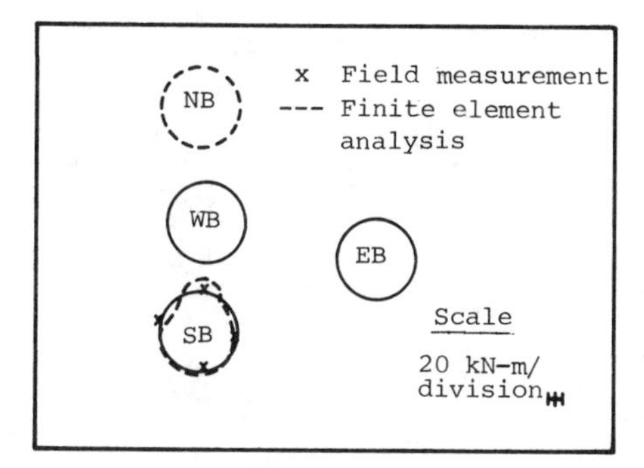

Figure 8. Cumulative bending moments at SB tunnel due to EB and WB tunnel excavations

Fig 10 contains plots of maximum percentage radial deformation, indicative of interaction moments at various tunnel lining sections, in relation to proximity of subsequent arrivals. Superimposed on these plots is the corresponding band of results reported by Peck (1969), as obtained from case studies. Accordingly, a linear variation from maximum of 0.1%, that is the lower practical limit of interaction, appears to be appropriate to the present project, as also found to be the most common case by Peck. In view of

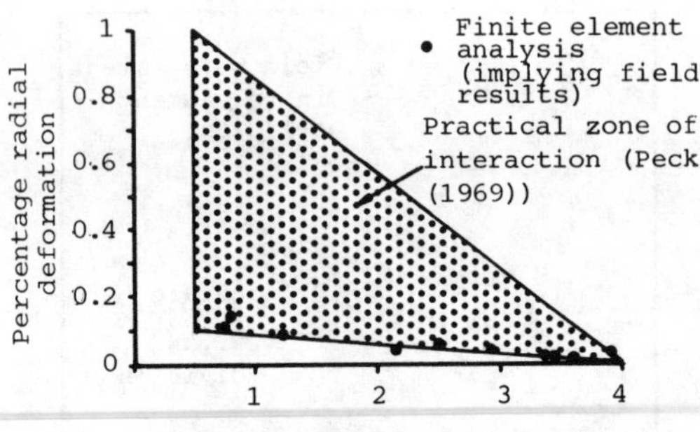

Figure 10. Comparison of finite element analyses with practical results of tunnel interaction

reasonably good agreement between measured and predicted bending moments discussed above, the same trend may also be taken as applying to field measurements at this site.

4 ACKNOWLEDGEMENT

The research upon which this paper is based was funded by the Science Council of Singapore under RDAS Grant No. C/81/04-06 which is gratefully acknowledged.

REFERENCES

D.J. Curtis, Discussion on the circular tunnel in elastic ground, Geotechnique 26, 231-237 (1976).

H.H. Einstein and C.W. Schwartz, Simplified analysis for tunnel supports, Journ. of Geotech. Engg. Div. ASCE, 105, GT4, 499-518 (1979).

E. Hinton and D.R.J. Owen, Finite element programming, Academic Press, London, 1977.

K.W. Lo, L.K. Chang, C.F. Leung, S.L. Lee, H. Makino and H. Tajima, Field instrumentation of a multiple tunnel interaction problem, Proc. 8th Asian Regional Conf., Kyoto, 1987a.

K.W. Lo, L.K. Chang, C.F. Leung, S.L. Lee, H. Makino and T. Mihara, Tunnels in close proximity, Proc. Singapore MRT Conf., Singapore, 1987b.

K.W. Lo, L.K. Chang, C.F. Leung, S.L. Lee, H. Makino and T. Mihara, Field measurements at a multiple tunnel interaction site, Proc. 2nd Int Symp. Field Measure. Geomech., Kobe, 1987c.

R.B. Peck, Design of tunnel liners and support systems, Final Report for Office of High Speed Ground Transportation, Washington D.C., 1969.

Numerical Methods in Geomechanics (Innsbruck 1988), Swoboda (ed.)
© 1988 Balkema, Rotterdam. ISBN 90 6191 809 X

Study of tunnel face in a gravel site

S.Chaffois & P.Laréal
Geotechnical Laboratory, INSA, Lyon, France
J.Monnet
IRIGM, Grenoble University, France
C.Chapeau
Centre d'Etudes des Tunnels, Bron, France

ABSTRACT: This study is related to the finite element computation of the tunnel face of the Lyon Subway, which was driven with a slurry pressure shield.

The theoretical model of soil behaviour uses only 5 constant parameters which have a physical significance. It allows dilatancy along shearing process. The 3-dimensional computation shows the relation between the slurry pressure at the tunnel face, its displacements, the plastic areas, and the settlement in the open.

The field measurements are conducted by inclinometer, tassometer, and by topography. The results are compared to the finite element ones.

1 INTRODUCTION

This study concerns the Line D of the Lyon Subway (France) for its subfluvial part. The selected technical option is a shallow tunnel excavated through immersed sands and gravels using a bentonite slurry shield.

The study is focused on the ground behaviour according to the slurry pressure applied to the face. The paper will first present the numerical study, then some results of experimental measurements which are compared with calculations.

2 PRESENT ANALYSIS ON THE FACE MODELIZATION

The Report of the U.S. Department of Transportation by Einstein and al.(1980) emphasizes some interesting questions :

1. To excavate a tunnel results in unloading, rotation of main stresses, elasto-plastic behaviour, creepage...
2. Ground parameters allowing to describe these phenomena are largely unknown.
3. It is impossible to use analytic methods to quantify these phenomena. Ranken and al.(1978) state that these methods can only be applied to deep excavated tunnels showing a neglictable stress variation with the depth. Therefore we have to apply complex numerical methods that can use several types of discretization.

Ranken,...(1978) and Schwartz, Einstein (1980) present axisymmetric modelizations using the Drucker-Prager law, modified or not. The results show that the plastication areas are restricted to the face and displacements oriented to the face while becoming quickly neglictable with a greater distance from it.

Also, a study by Romo, Diaz (1981) with two-dimensional deformation on an undrained cohesive ground, using an hyperbolic strain-deformation law, shows that the face unstability does not question the massive unstability. The authors can define a stability factor and a safety factor with respect to yielding.

Finally, one three-dimensional study was performed by Katzenbach and Breth (1981) applying the Duncan et al. law, showing that, during the excavation of a tunnel through clayey marls, the ground decompresses vertically beyond the face and compresses horizontally.

No modelization of a shallow tunnel face in a granular soil was carried out in order to simulate its "in-situ" behaviour. Consequently we worked out the modelization by choosing a three-dimensional discretization of ground with initial anisotropic stresses and a behaviour law specific to granular soils, for calculation by the Finite Element Method.

3 USED LAW OF BEHAVIOUR

The study of a tunnel face requires to determine a true triaxial state. Nowadays two types of law are available :

a) incremental laws

b) elastoplastic laws.

For strictly physical reasons, Chaffois and Monnet (1985) developed a three-dimensional law of elastoplastic type.As our ground is granular, the behaviour law required four constant parameters that can be determined from laboratory or field tests. This law consists of three stages :

a) Elastic stage

It is determined by the generalized Hooke law, and requires to know the value of E and ν.

b) Work-hardening stage

This stage is characterized by a ground dilatancy with a non standard plastic flow.

In the stress space, the plastic deformation is given for the direction by the normality to the function

$$f_1(\sigma) = (\sigma_1 + \sigma_2 + \sigma_3)^3 - L_1 \sigma_1 \sigma_2 \sigma_3$$

and the yield function used the incremental energetic formulation

$$\frac{\tau_{oct}}{\sigma_{oct}} = tg\,\varphi_\mu + \frac{d\,\varepsilon^p_{oct}}{d\,\varepsilon'^p_{oct} \cos(\gamma - \gamma')}$$

The intensity of the plastic deformation is obtained from the yield function :

$$f_2(\sigma) = \tau_{oct} + \sigma_{oct}\left(tg\,\varphi_\mu + \frac{\varepsilon^p_{oct}}{\varepsilon'^p_{oct}} \right)$$

We notice the emergence of the parameter φ_μ (interparticulate angle of friction) which is directly connected to L1.

c) Perfect plastic flow stage

The yield function that defines the limit state is of the following form :

$$f_3(\sigma) = (\sigma_1 + \sigma_2 + \sigma_3)^3 - L_2 \sigma_1 \sigma_2 \sigma_3$$

The parameter L2 is deduced from the ground internal friction angle

The direction of plastic deformation is obtained in the same way as work-hardening.

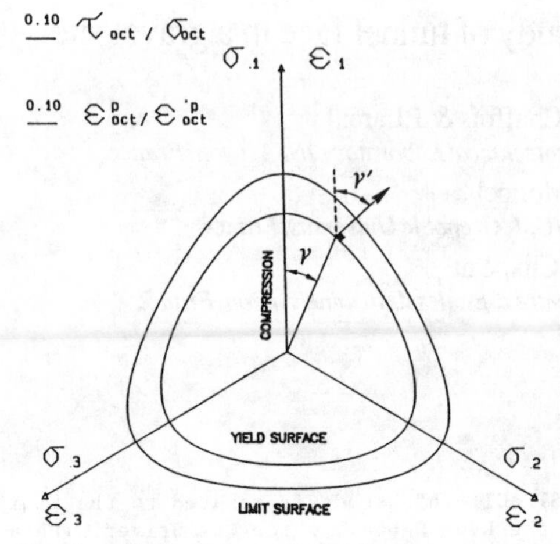

Fig.1 Definition of failure surfaces

4 MODELIZATION OF SHIELD TUNNELLING

4.1 The shield

The fundamental scheme is given on Fig.2; four major functions can be noticed :

1) Ground excavation

2) Temporary support

a) of the face by bentonite slurry pressure, controlled via a compressed air tank

b) of the tunnel behing the face by a steel sleeve forming the shield.

3) Hydraulic mucking

4) Lining with pressure pumped concrete.

It can be noticed that the tunnel boring machine shows a cylindrical steel sidewall that can be considered as perfectly rigid, thus preventing any displacement. The small inclination of the cutting wheel is not modelized.

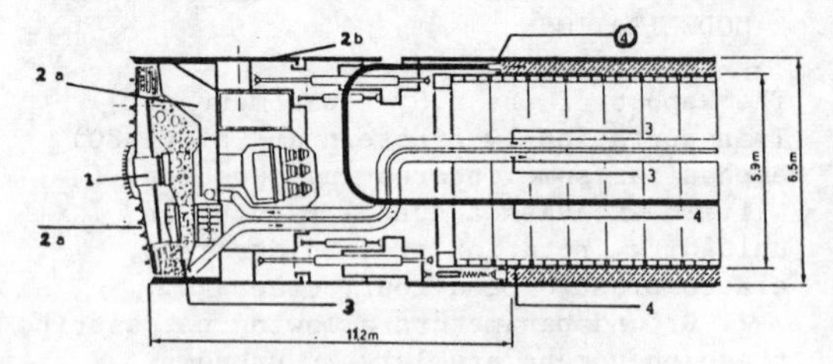

Fig.2 Bentonite shield scheme

4.2 The geotechnical context

The tunnel is excavated through fewly cohesive sand-gravel alluvia , with dilating shearing behaviour, lying under an unlowered watertable. The geological section and the mechanical ground characteristics are given in Fig.3.

The Young modulus E is obtained for load-unload cycles in the pressiometric test. φ is deduced from field tests with a direct shearing box. φ_μ and ν are given by conventional triaxial tests.

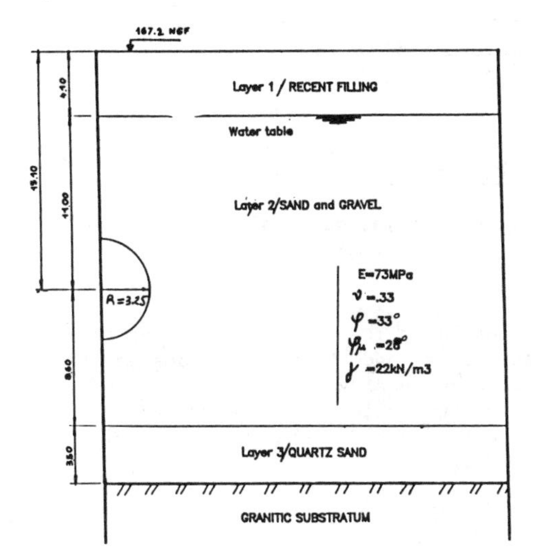

Fig.3 Geotechnical profile

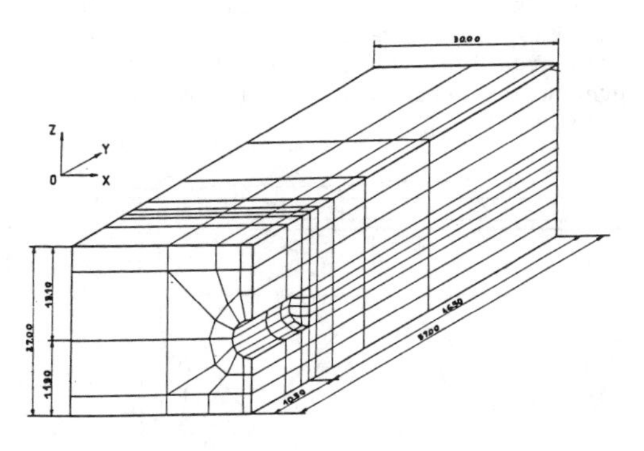

Fig.4 Mesh

4.3 Mesh design

From the bibliographical analysis and the sliding lines that could develop in the ground, we have defined the mesh given in Fig.4.

The parallelipiped is discretized into 272 20-node brick elements; we obtain a 1493-node mesh with three freedom levels. The rigidity matrix used 4,500 Ko storage with a 630 band width.

4.4 The calculation stages

They are :

1) Application of the ground specific gravity in order to obtain the initial stresses with a rest coefficient of 0.43.

2) The ground is suppressed inside the tunnel and the nodes blocked to simulate the support.

3) Application of stresses to the face after cancellation of displacements. The first stress variation is obtained from the following formula :

$$\Delta P = \sigma_y + P_w - (\sigma_b + P)$$

σ_y = horizontal stress due to earth weight
P_w = water pressure
σ_b = slurry pressure
P = overpressure applied to the face $\Big\} P_b$

4) The increments of the following calculation consist in decreasing pressure P on the face level, obtained by an a-priori determined decompression, applied at each node of the face.

5 RESULTS

5.1 Characteristic curve of the face

We call "characteristic curve of the face" the relation between the displacement of a point on the <u>face</u> (centre) and the pressure applied to the <u>face</u> (Fig.5).

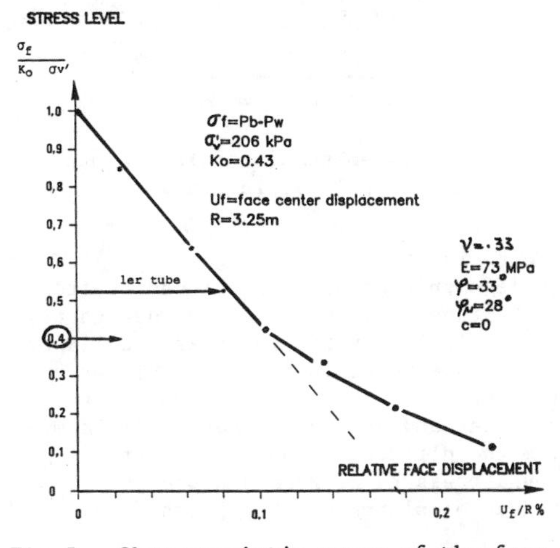

Fig.5 Characteristic curve of the face

We note that this curve is non linear. For a pressure P applied to the slurry higher than 0.13 MPa, i.e. at an effective stress level $(P_b - P_w)/k_o\,\sigma'_v$ = 40%, the ground decompression is linear. Beyond this level of horizontal stress, work-hardening, then plasticity develops before the face.

The calculation also demonstrates that the face movements remain low (5 to 8 mm) and that the increased displacement resulting from the plastic phase is most restricted with SL = 0.1.

5.2 Displacement of face along the vertical axis

For each increment of decrease in the pressure applied to the slurry the horizontal displacements of the vertical face axis are drawn (Fig.6).

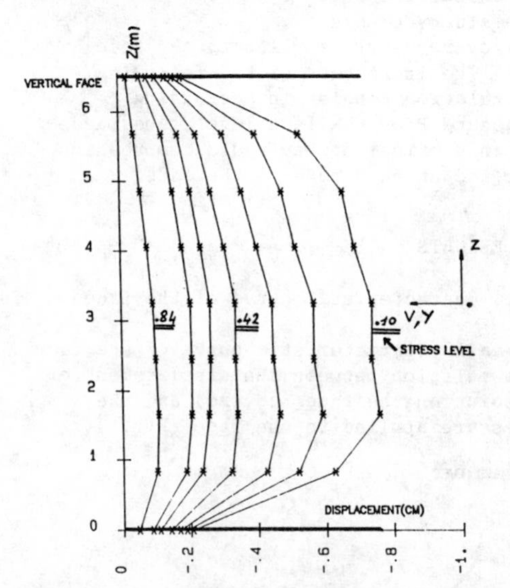

Fig.6 Displacement of the face along the tunnel vertical axis

We notice that the strongest displacements occur in the lowest third of the tunnel. Thus we can suppose that failure is likely to happen as a collapse of the face foot. In fact, in this area, the slurry can oppose less strongly to the water and ground pressure. This comes from the different slopes of water and ground pressure diagram and pressure diagram of slurry and top pressure P.

5.3 Displacement inside the ground

Fig.7 gives the displacements in a plane parallel to the tunnel axis with a SL = 0.1.

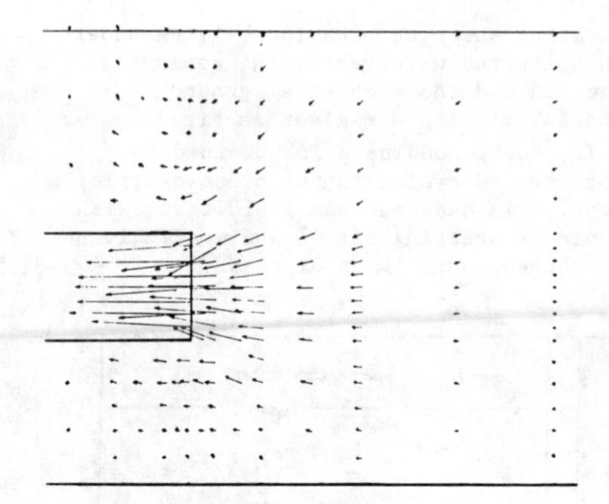

Fig.7 Displacements along the tunnel axis (ZOY plane)

These displacements are directed to the face centre and their amplitudes decrease with a decreasing depth.

Also, we notice that the ground decompresses under the future tunnel invert.

5.4 Plasticity area

Since the program provided the stress state of every mesh node, we could define the yielding area developing near the face. This area is given in fig.8 for a low pressure P applied to the slurry (SL = 0.1).

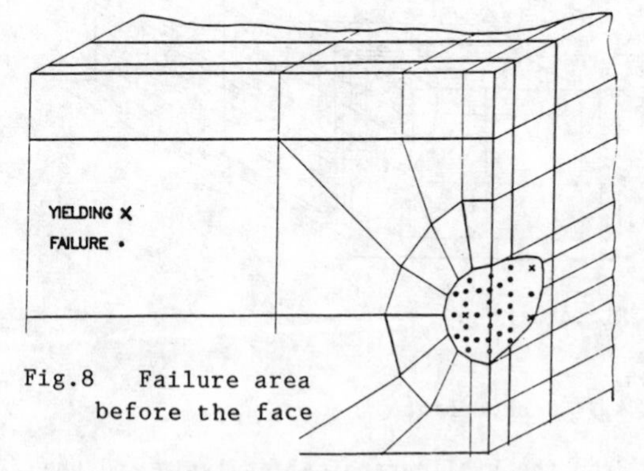

Fig.8 Failure area before the face

The yielding area is most precisely located as in bibliographical results, mainly for cohesive soils; here, for a granular soil, the yielding area is practically located in a semisphere centered in the middle of the face with a radius equal to that of the tunnel. Also an arching effect is developing around this yielding area.

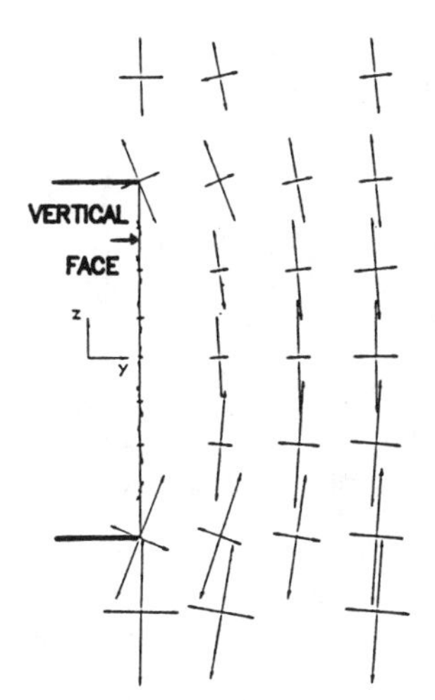

Fig.9 Direction of main stresses at face in a vertical plane

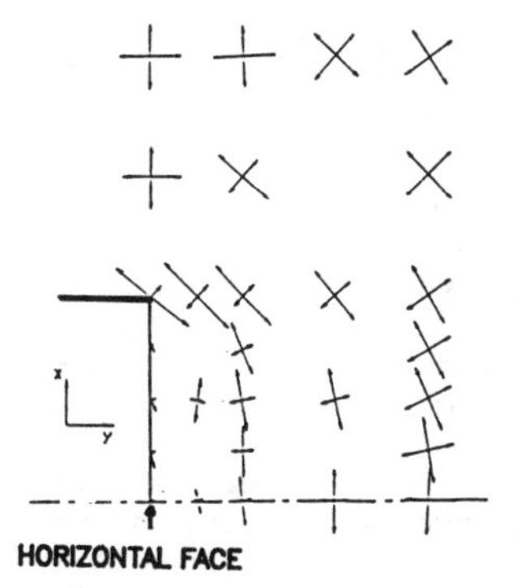

Fig.10 Direction of main stresses at face in a horizontal plane

5.5 Direction of the main stresses

In Fig.9 and 10 the directions of the main stresses near the face are drawn.

The face in itself modifies the main stresses. The minor main stresses are directed to the face. The major and intermediate main stresses lie on arcs of a circle approximately centered to the face axis. This confirms that an arching effect is developing both in horizontal and vertical planes.

5.6 Displacement on surface

It was also interesting to know the settlements on surface, since the tunnel has to cross under built areas. We see in Fig.11 the curves of vertical isodisplacements open.

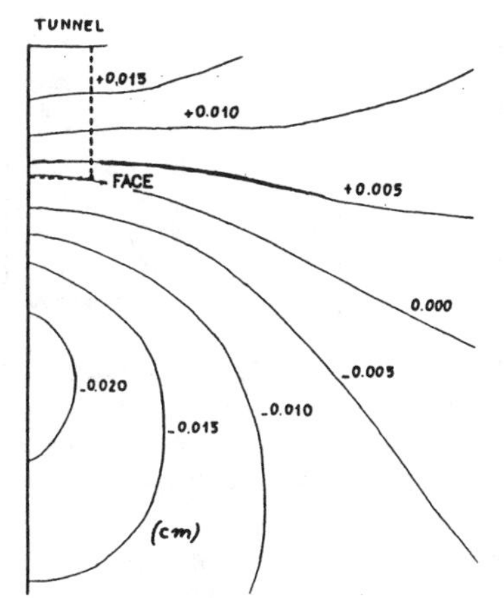

Fig.11 Isodisplacement curves in the open

We notice that these displacements are quite small (0.22 mm). The maximal settlement occurs at a distance of about 1.5 times the tunnel diameter.

6 MEASURED DISPLACEMENTS

Within a concerted research action on the Lyon Subway, a measuring section was installed in the Presqu'Ile, between the two rivers Saône and Rhône, on the Place Bellecour. The results are reported on by Chapeau and Kastner (1987). We present now the results, principally for the face, concerning :

- the vertical displacements measured by topometric means and Gloetz extensometers,
- the horizontal displacements measured by inclinometry.

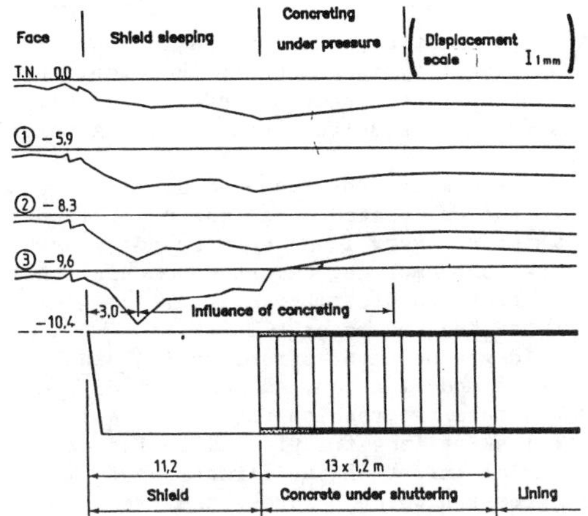

Fig.12 Vertical displacements in overburden - Longitudinal section -

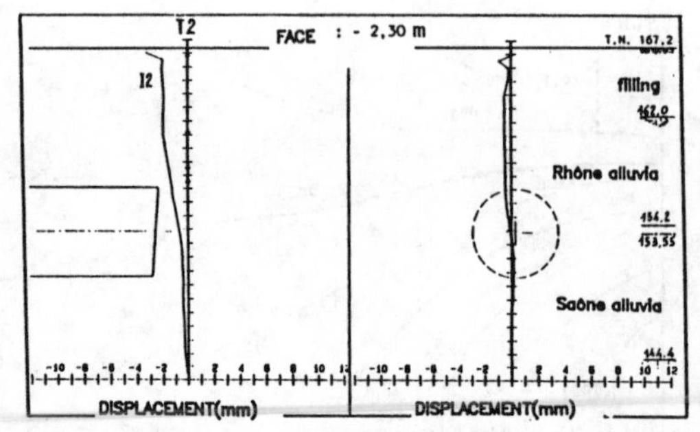

Fig.13 Horizontal displacements before the face on the tunnel axis

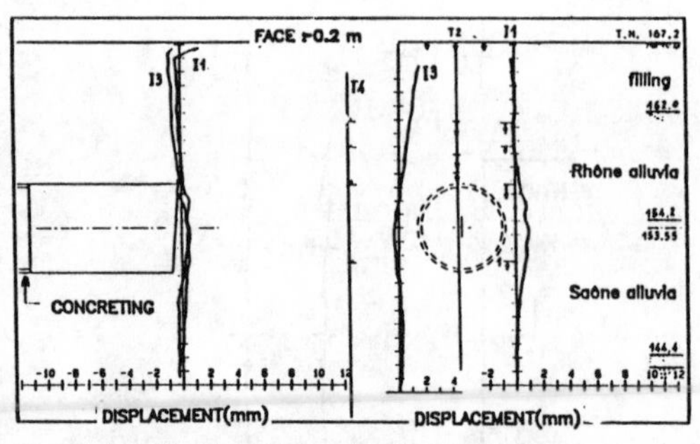

Fig.14 Horizontal displacements before the face

Fig.12 shows the evolution of vertical displacements for various depths along the vertical tunnel axis as the shield proceeds. If we examine the ground just near the shield - measurement No.3 - we establish a millimetric settlement at face crossing, still increased to 3 mm with the passage of the shield skirt.

Then a ground rising occurs resulting from the combined effect of dilatancy of shearing ground and pressure concreting (0.2 to 0.4 MPa).

In the overburden ground the same phenomenon, however reduced, can be established. The maximal settlement also amounts to 3 mm. Concerning the horizontal displacements before the face (Fig.13) the displacements of the overburden ground to the tunnel have a millimetric amplitude.

On the tunnel sides the face crossing induces transverse centrifugal displacements of lateral expansion and longitudinal displacements in the direction of the shield progression (Fig.14).

7 CONCLUSIONS

The modelization helped to decide on measurement types and their location; it was used as reference to interprete the displacement results measured at the shield face.

If we compare the measured displacements with the calculated ones, their amplitudes - which are very low - correspond; however there is no more agreement concerning their direction and distribution, this emphasizing the major role of the shield-ground interaction, unsimulated by this modelization. As a matter of fact, the shield progression results in a lateral steel skirt/face friction. This induces an axial thrust in the direction of progression which balances the face depression generated by the excavation process and reinforces the arching effect demonstrated by this calculation in granular soil with a 3D specific law of behaviour.

The face "characteristic curve" determined by this calculation allowed assessing the horizontal stress level to be applied to the face (Fig.5) during the works to prevent the development of plastic areas and ensure safe works.

REFERENCES

Chaffois, S. & J.Monnet 1985. Model of sand behaviour towards shearing in 3-dimensional conditions of stress and deformation. Revue Française de Géotechnique No.32, p.59-69.

Chapeau, C. & R.Kastner 1987. Displacement measurements associated with shield tunnelling, ENPC Paris, Interaction Sols-Structures, p. 313-320.

Einstein, H.H., C.W.Schwartz, W.Steiner, M.M.Baligh & R.E.Levitt 1980. Improved design for tunnel supports: Analysis method and ground structure behaviour. -A-Review Vol.II May 1980, US Department of Transportation Research.

Katzenbach & H.Breth 1981. Nonlinear 3-D analysis for NATM in Frankfurt clay. Proceedings of the 10th International Conference on Soil Mechanics and Foundation engineering, Stockholm 1981, Tome 1 p. 315-318

Ranken, R.E., J.Ghaboussi & A.J.Hendron 1978. Analysis of ground liner interaction for tunnels. US Department of Transportation.

Schwartz, C.W. & H.H.Einstein 1980. Improved design of tunnel supports: Simplified analysis for ground structures interaction in tunnelling. Vol.I, June 1980.

Romo, M.P. & C.Diaz 1981. Face stability and ground settlement in shield tunnelling. Proceedings of the 10th International Conference on Soil Mechanics and Foundation Engineering, Stockholm 1981, Tome 1

Numerical Methods in Geomechanics (Innsbruck 1988), Swoboda (ed.)
© 1988 Balkema, Rotterdam. ISBN 90 6191 809 X

Two-dimensional calculation model in tunneling – Verification by measurement results and by spatial calculation

K.Schikora & B.Ostermeier
Technische Universität München, FR Germany

ABSTRACT: The drivage of tunnels creates in soil a three-dimensional, time dependent state of stress. A highly simplified two dimensional calculation model has been developed which includes the mean characteristics of the soil-tunnel supporting system. Results obtained are compared with measurement and spatial calculations.

1 INTRODUCTION

The drivage of tunnels in soil causes very complex three-dimensional bearing systems of soil and tunnel-shell whose geometric proportions are varying with each drivage phase (Fig.1). During an excavating phase the soil surrounding the excavated area can deform into tunnel section and creates around the excavated

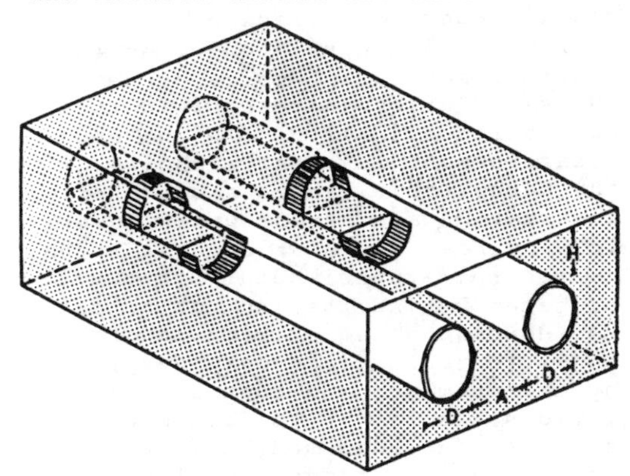

FIG.1 STATES OF DRIVAGE OF A DOUBLE TUNNEL

area a local three-dimensional state of stress which causes loadings of the adjacent shotcrete shell (Fig.3). If one considers a special tunnel section it receives an additional loading by every excavating phase, which decreases the more the greater the distance of drivage becomes. In sufficient distance of the excavated area (greater than one tunnel diameter) every tunnel-section has got nearly the same loading (Fig.2). Considered idealized there appear only supporting effects perpendicular to the tunnel axis. The longitudinal supporting effects have been repealed absolutely by superposition of the additional supporting effects of each excavating phase. Such a state of stress is called plane state of strain. This case of load can be evaluated by means of disk calculations. A further drivage phase - for instance the excavation of bench following the excavation of crown - causes new longitudinal supporting effects, which pass with increasing distance to the drivage by superposition to a state of plane strain. If excavating phases follow close togehter the state of plane strain occurs finally after termination of drivage.

2 CALCULATION MODEL

Exactly simulated calculations of tunnel drivage should be done in a three dimensional way by response of rheological aspects (Fig.2). Nowadays this procedure is only partly possible with available finite element programs and makes economically little sense. Highly simplifiing it is tried to determine tunnel drivage by means of disk calculations (Fig.3, /1/, /2/, /3/). The soil-tunnel system is divided into disks by means of sections perpendicular to the tunnel axis. The thickness of a disk corresponds to one length of advance. The division is carried out in a way that disk 1 covers the excavated sector. Disk 0 located in front of it covers the face. The contour of excavation is secured by shotcrete at disk 2. Similar considerations can be applied to the bench in the area of disks i - 1 to i + 1. If one considers disk 1 seperated from the main bearing system soil-tunnel, it is weakened by the excavation of the crown and is tending to deform into the excavated area. Such deformations are partly reduced by shear stresses which affect the intersection planes to the more rigid neighbouring disks 0 and 2. Highly idealized the supporting shear stresses are concentrated along the contour of excavation and can be substituted by supporting forces. The distribution of these supporting forces is unfortunately unknown, however a comparable supporting effect can be simulated by means of a weightless soft core. The unsecured area of drivage corresponds to a disk of which the weight of the soil is removed at the crown and the rigidity in this area is reduced by the value α. The parameter α has decisive influence on deformations and body forces in

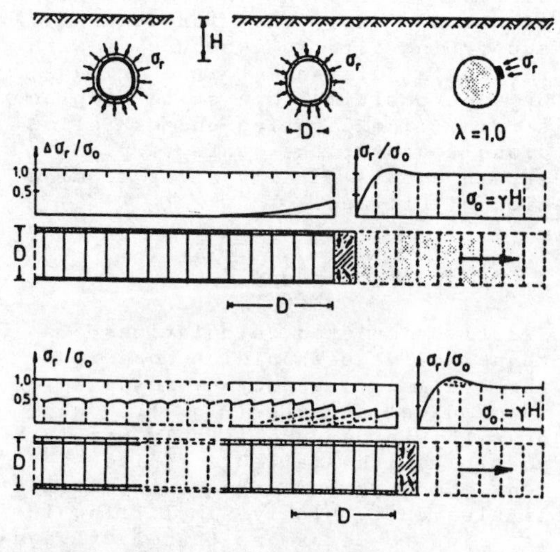

FIG.2 STRESS CONDITION BETWEEN SHOTCRETE SHELL AND SOIL ($\Delta\sigma$: ALTERATION OF STRESS PER ADVANCE)

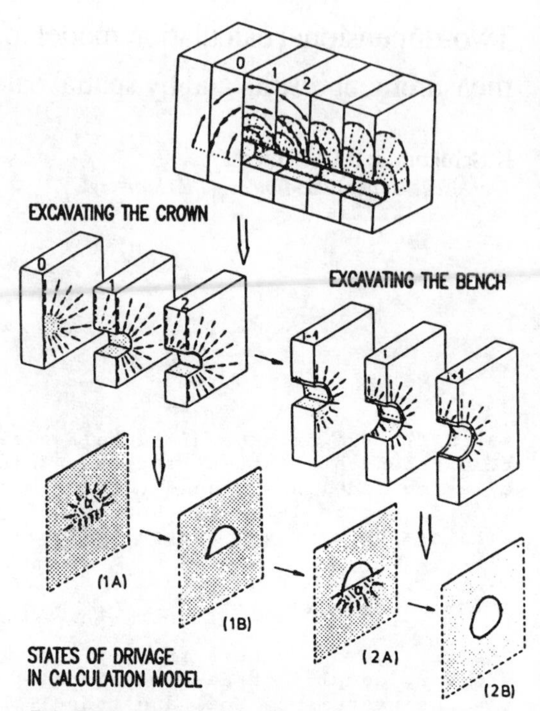

FIG.3 BEARING SYSTEM SOIL - TUNNEL

the shotcrete shell. In the following calculation step the weightless soft core representing the longitudinal supporting effect of the soil is removed and completed by an edge reinforcement which corresponds to the securing by shotcrete. In this considered disk with hole and edge reinforcement a suitable choise of the parameter α causes a state of deformation and stress which describes the real state of stress of the system soil-tunnel about one tunnel-diameter behind the excavation phase. How much the parameter α corresponds to reality can only be proved by measurements and spatial calculations. As far as the bench excavation is concerned similar consideration can be applied. The longitudinal supporting effect in soil is simulated by a soft core. The longitudinal supporting effect of the shotcrete shell in the crown could be described by means of tangential springs in the contour of the crown of shotcrete shell. As the springs dominate in the crown base sector, for the sake of simplicity the spring effect is concentrated in the crown base sector and is replaced by a flexible elastic rod which is located in the system line of the future shotcrete shell of the bench sector. The strain rigidity of the rod can be related to the rigidity of the soil. As considerable plastifications of soil under the crown base appear without this rod, considerable longitudinal supporting effects can be achieved inspite of the rod's little strain rigidity. By means of the decribed idealisations and simplifications it is possible to simulate the divage of a double tunnel by means of eight disk calculations (Fig.4).

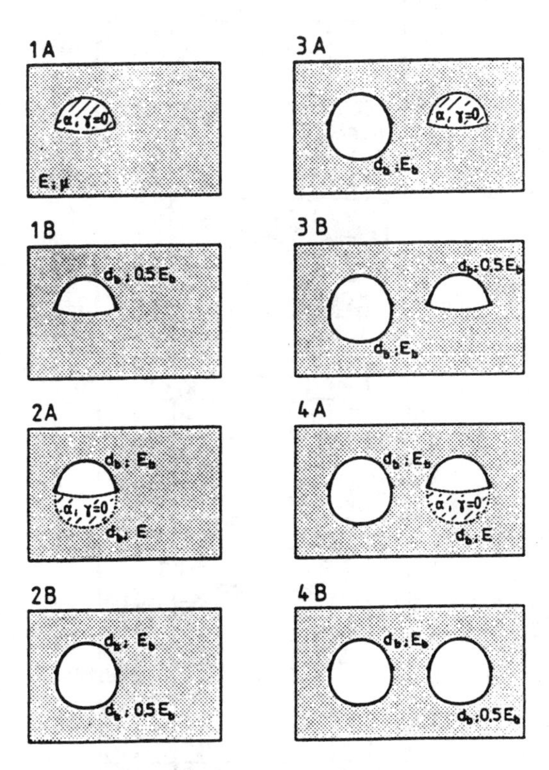

FIG.4 CALCULATION MODEL FOR A DOUBLE TUNNEL

Drivage phase 1A: The drivage phase-breastwork and excavation of the crown-corresponds to a disk deformed by its dead weight in the case of which the dead weight in the crown sector is removed and the rigidity is reduced by a factor α.

Drivage phase 1B: The drivage state of crown secured by shotcrete is obtained by means of a disk with hole and edge reinforcement. The load along the contour of the hole corresponds to the supporting forces which the softened crown core had accepted in state 1A.

Drivage phase 2A: The drivage phase - bench excavation - is described by a disk which is constructed in the bench section like phase 1B. In the bench section the dead weight is removed, the rigidity reduced by a faktor α and the bench contour is described by an elastic rod. The rod possesses the dimension of the future shotcrete shell and an E-module in the magnitude of the E-modul of soil.

Drivage phase 2B: This phase describes the state of the completely excavated and secured first tunnel. The calculation is carried out on a disk with hole and edge reinforcement. The load corresponds to the state of stress in soil after phase 2A additional the supporting forces which act upon the soft bench core or rather the elastic rod.

Four further drivage phases follow for the second tunnel, which are constructed in accordance with the described drivage phases of the first.

3 COMPARING CALCULATIONS-MEASUREMENT

The efficiency of the described calculation method is demonstrated in the following sample of a double tunnel in Munich underground.

Data of calculation: Mean diameter of tunnel D = 6,6 m, soil overburden H = 6,o m, gap between tunnels A/D = = o,25, soil parameters of quarternary gravel with loose layers γ = = 22 kN/m³, ρ = 37,5°, c = 4 kN/m², E = 1oo kN/m². The deformations were measured by levelling and extensometers, the axial forces in shotcrete were obtained by stress-strain measurements. The starting parameter α has decisive influence on the deformations and body forces. For α = = o,5 to α = o,15 the roof deformations differ up to 5o%. Similar effects go for the body forces. The Fig.5 shows the measured and calculated deformation curves behave differently in the individual measurement cross-sections; therefore a dispersal range was determined from maximum and minimum values, which covers all measurement results. In addition these figures contain the calculation results for short and large advance lengths or for crown bases with and without reinforcement. In case of short advance lengths the longitudinal supporting effect is strong, reduction factors of α = o,5 are recommended. If large advance lengths arise the longitudinal supporting is less, reduction factors α = o,3 or in extreme cases α = o,15 are required. In the case of measurement cross-section VI during drivage of tunnel 2 the deformations on surface do not differ much from the calculation values however the roof deformations grow 9 mm larger than the calculation values for good supporting effect. During drivage loosening must have taken place directly above the roof of tunnel 1; during the drivage of tunnel 2 it must have led to increased plastification and softening above both tunnels. Such extreme deformations can only be described by considerable small reduction factors α. Measurements of body-forces show similar results. The axial forces of the shotcrete shells were determined for crown bases with and without reinforcement. A comparison of axial forces in the base of crown obtained by calculation and measurement coincides with one another in every drivage phase. In the lower inner wall of the tunnel the calculated axial forces are lower than the measured forces, while by calculation the supporting force of the inner crown of the second tunnel is greater than the measured one. This can be explained by the fact that beneath the crown base of the second tunnel a stronger plastification has taken place in situ than in the calculation and thus a lower crown supporting force is produced. For reasons of equilibrium a stronger load arises on the lower inner wall of tunnel 1.

The discribed calculationsmethod can be used in extreme cases too. In the tunnel Hohe Wart of German federal railway's new route there were performed deformation and body stress measurements /4/. The overburden of the tunnel amounts to 14 m, the full section is 125 m². The rock around the tunnel mainly consists of shell limestone with strongly cutted layers ($E = 25o$ MN/m², $c = o,1o$ MN/m², $\rho = 24°$, $\gamma = 22$ kN/m³). At the roof of the tunnel deformations of 12 cm were ascertained. A calculation with the given rock

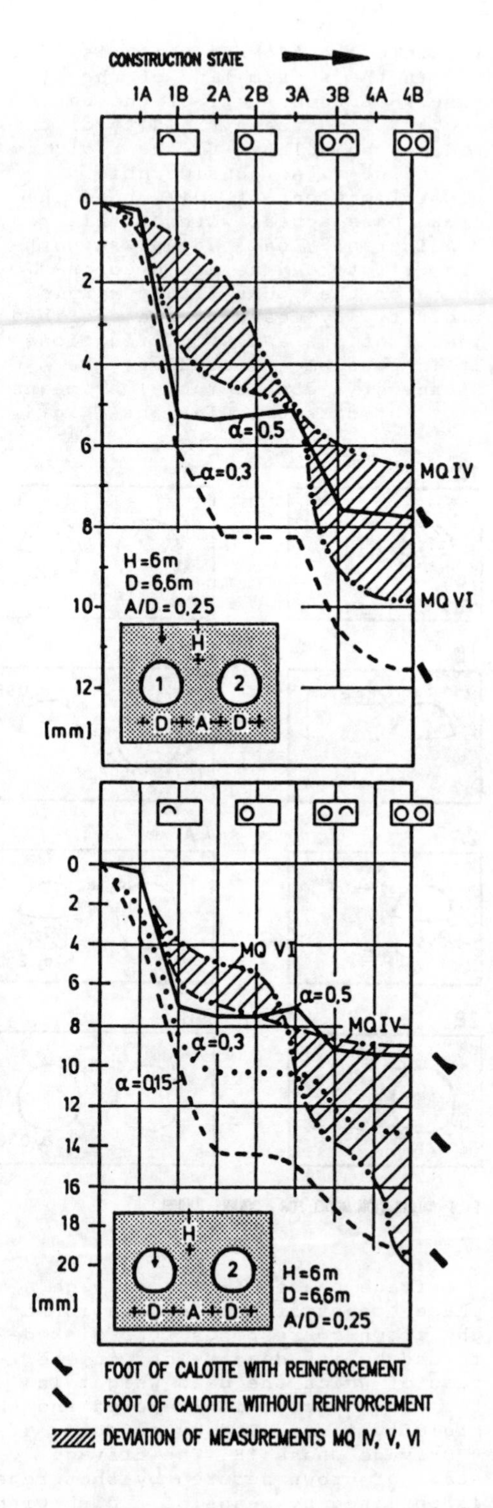

FIG.5 DEFORMATIONS IN THE SURFACE OF THE GROUND AND IN THE ROOF AREA ABOVE TUNNEL 1 OF A DOUBLE TUNNEL

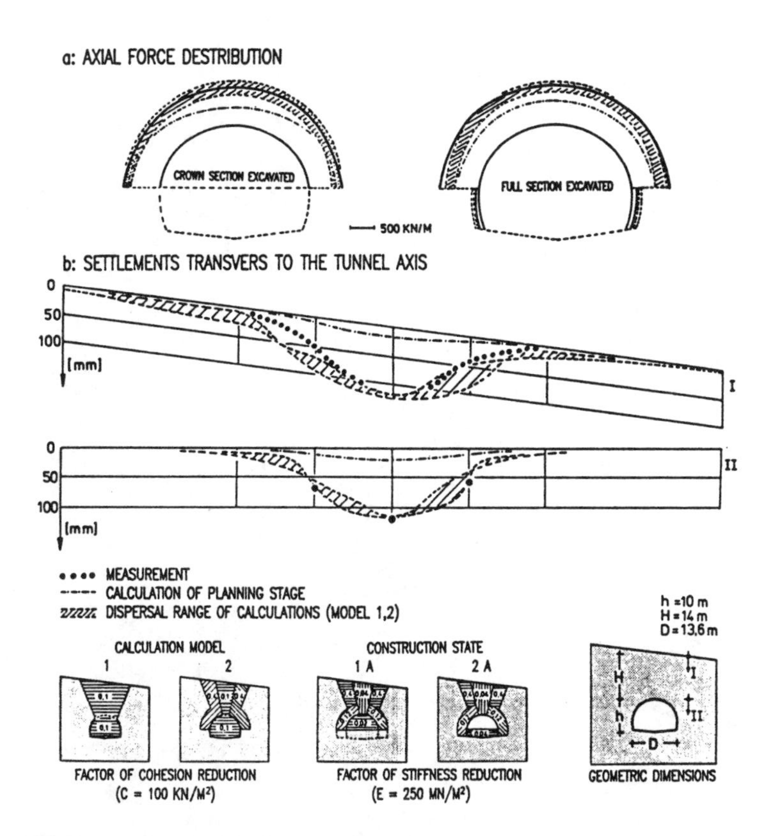

FIG.6 AXIAL FORCE DESTRIBUTION AND SETTLEMENTS TRANSVERS TO THE TUNNEL AXIS

parameters would have obtained roof deformations in a scale of 2 cm and axial forces in the bench area of 7oo to 8oo kN/m. This discrepancy between measurement and calculation caused a detailed consideration of rock conditions which showed strong faults in the area over the tunnel. On ground surface the deformation hollow was extreme narrow in size of one tunnel diameter. By extreme reduction of rock parameters (E-module and cohesion) in the area above the tunnel roof plastifications and large displacements can be produced having an similar effect upon the state of deformation as slipping in cracks. A comparsion of the calculation and measurement shows statisfactory results.

4 CONCLUSIONS

The conclusions and comparisons with measurements have shown, that tunnel calculations can be done by means of disk calculations instead of three-dimensional calculations, if a proper principal parameter α is chosen.

5 REFERENCES

/1/ Schikora, K., Fink, Th.: Berechnungsmethoden moderner bergmännischer Bauweisen beim U-Bahn-Bau. Bauingenieur 57 (1982) (1982), 193-198

/2/ Schikora, K.: Berechnungsmodell und Meßergebnisse eines seichtliegenden Doppeltunnels im quartären Kies. Tunnel 3/1982, 153-161

/3/ Schikora, K.: Doppelröhrentunnel der Münchner U-Bahn. Tunnel 2/1983, 71-79

/4/ Messungen an der Neubaustrecke der Deutschen Bundesbahn Tunnel Hohe Wart, Meßquerschnitt bei Station 745, Philipp Holzmann AG, Hauptniederl. München

Numerical Methods in Geomechanics (Innsbruck 1988), Swoboda (ed.)
© 1988 Balkema, Rotterdam. ISBN 90 6191 809 X

FE analysis of underground openings in creeping rock salt

W.Berwanger
Büro für Planung und Ingenieurtechnik GmbH, Grenzach-Wyhlen, FR Germany

ABSTRACT: The first basic examinations as to the convergence behaviour of gas-storing caverns in rock salt, with different removal and filling rhythms, will be presented. Using a model cavern, several given operating pressure cycles will be calculated, on the basis of the FEM. These calculations will be supplemented by thoughts on the effects of the time of lowering the pressure and the pressure level there within the cavern. On a long-term scale the purpose of this will be to help develop an optimizing method for gas withdrawal in cavern fields.

1 INTRODUCTION

Over the last years, many underground storage caverns were created worldwide to store petroleum products and gases. On the one hand they serve to store national energy reserves and on the other to balance out seasonal fluctuations in energy consumption. Such storage caverns are created preferably in salt dome formations, where a sufficiently long-term geological stability can be expected and where the stringent requirements as to imperviousness of the storage room as well the insolubility of the rock considering the stored material are fulfilled.

The salt dome formations which are usually formed in great depths, often make it necessary for the caverns to be created in depths of more than 1000 m. Besides problems with the stability of the cavern, this brings considerable disadvantages in connection with long-term economical operating of such a cavern. These caverns are subjected to a permanent volume loss as the result of the high creep rate of the salt rock under high tensions and temperatures, which can only be kept at a justifiable level by a sufficiently high operating level in the caverns. This support pressure acting as an inside pressure on the cavity edge decreases the tension difference to the cavity edge and thus the creep-influencing, deviatoric rock tension parts.

The fundamental interest of the cavern operator in reducing the cavity losses conflicts with the necessity of exchanging the stored product, with removal and filling cycles often shortly after each other. A minimization of the cavity losses can be achieved, in principle, by maintaining the maximum permissible pressure level, however the pressure reductions connected with removals of the stored product cause the average pressure to sink inevitably below the maximum pressure. Occasionally, extensive pressure reductions in connection with repair and maintenance work cannot be avoided. The aim must therefore be to control the removals of the stored product from the caverns in such a way, that not only the volume losses of single caverns, but also the total losses of the entire cavern field are minimized.

In connection with the development of effective optimizing methods, a number of problems must be solved that result mainly from the fact that the affects of pressure decreases on the volume loss rates not only depend on the actual operating pressure, but also on cavern-specific factors, as e.g. the depth, geometry, rock behaviour in the surroundings of the cavern, as well as on the cavern history. Due to the therefore extremely different behaviour pattern of each cavern, optimization of the total behaviour can only succeed, if the behaviour pattern of each cavern of the field, under different pressure levels and at different times, is known respectively can be estimated sufficiently.

2 A COMPARISON OF OPERATING PRESSURE CYCLES

The extent to which the cavern history is able to influence the loss rates will be shown first with a comparison of several operating pressure cycles that were calculated with a calculation model according to Fig. 1, using axisymmetric finite elements. Taking into consideration rock tension increasing linearly with the depth as well as increasing temperatures, the convergence development of 5 pressure levels were compared, using the elastic-viscoelastic material law LUBBY 2 /1/

$$\dot{\underline{\varepsilon}}^{v} = \left[\frac{1}{\eta_m} + \frac{1}{\eta_k}\left(1 - \frac{\varepsilon^{v}_{ef,p}}{\sigma_{ef}}\,\bar{G}_k\right)\right]\frac{3}{2}\cdot\underline{M}_2\cdot\underline{\sigma}$$

This material law allows, in salt rock, the comprehension of transient creep behaviour, especially noticeable in the case of load increases. In comparison, only an attempt has been made to investigate (/2/, /3/) into the creep behaviour in the course of decreasing deviatoric states of tension, as they occur during increases of operating pressure in parts of the surrounding rock, so that the creep calculations in the pressure building area can only be said to have a qualitative character. Furthermore the dependence on time of the sinking phases remains unconsidered.

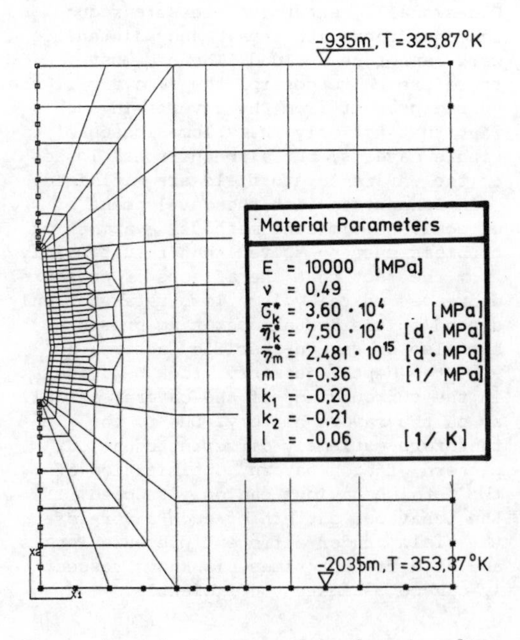

Figure 1. Calculation model of rock salt cavern

In the following calculations the pressure conditions were changed to a great extent, with no consideration of any stability problems, in order to point out clearly the influence of sinking. In all cases the sinking phases are adapted to the maximum pressures in such a way, that the product of sinking duration and pressure difference corresponds to the product of the following excess pressure respectively recovery duration. Thus in all cases the average pressure level of 200 bars was maintained over a time period of more than 3 years. Whereas in a first calculation a constant inner pressure of 200 bars is maintained for more than approx. 3 years (Fig.2, curve 1), the other cases are distinguished by strong, short-term pressure decreases to a minimum pressure of 50 bars, with different sinking rhythmus.

In total, pressure sinking as per Fig. 2 leads to marked volume losses with, at the same time, highly increased convergence speeds ("Convergence Shock" as per Hentschel /4/). The volume loss over a sinking period of 60 days (curve 2) thus amounts to more than 10 % and is therefore above the total convergence, that had levelled out at an order of magnitude of K = 7.4 % at the time of sinking. Evidently the volume loss cannot be compensated by the subsequent pressure increase. The pressure increases cause at first, due to an elastic deformation value in the material law formulation as well as the fictively presumed time-independent pressure sinking, a slight "widening" of the cavern. The strong convergence development is thus stopped directly, the loss rates come closer, on a long-term basis, to the order of magnitude corresponding to the actual operating pressure.

If compared to one decrease of the operating pressure the removal of the stored product is divided into a total of 4 decreases, each lasting 15 days, the convergence losses continue to increase. Even after only a few decreases the volume losses exceed those of the 2nd operating pressure cycle (curve 2). Dependence on the sinking rhythm is apparent here. Decreases quickly following each other (curve 3) cause in total slighter volumn losses than decreases at large time intervals (curve 4). The total loss increases in total roughly 1 resp. 3 % as compared with one decrease.

If the decreases are limited to a minimum pressure of 100 bars (curve 5), the volume losses decrease understandably however still reach, after 100 days, order of magnitude of more than 5 % as compared with a constantly maintained pressure level. This effect can be attributed mainly to the

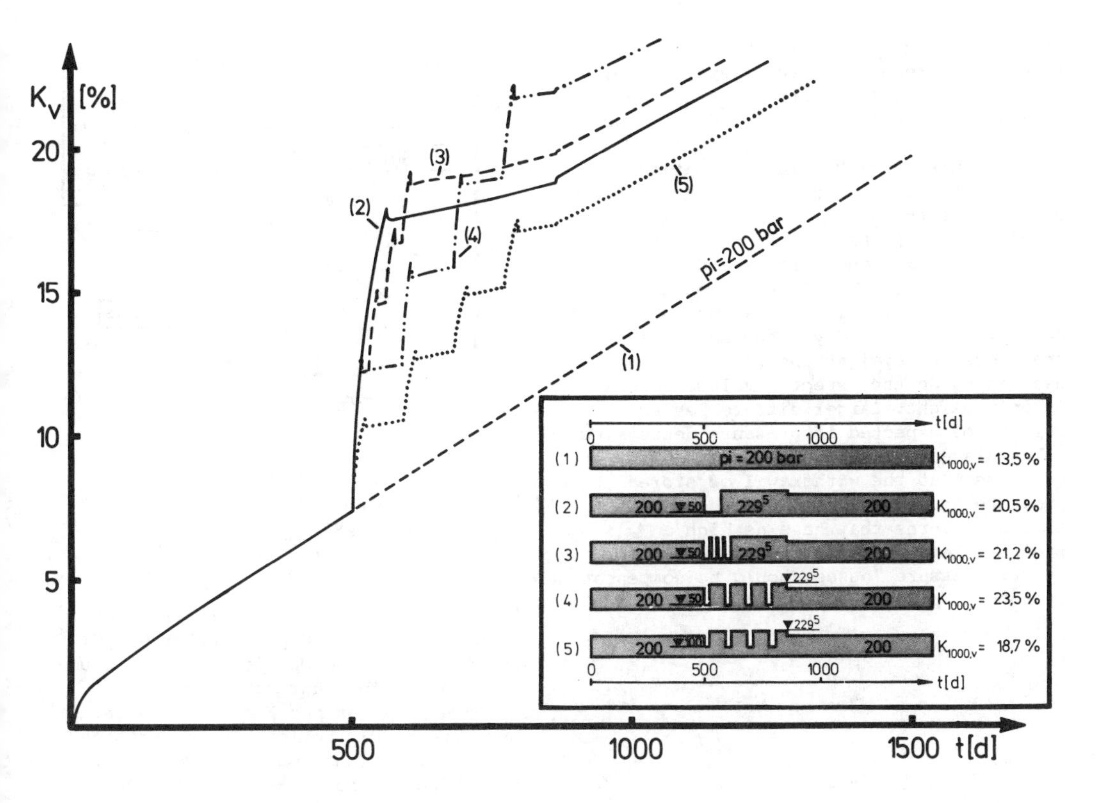

Figure 2. Convergences rates dependent on some operating pressure cycles

fact that the transient creep range distinguished by a high distortion speed is passed through several times. One pressure decrease over a long period of time should therefore be considered as more advantageous than several ones shortly after each other.

3 DECREASES OF DIFFERENT PRESSURE LEVELS

Numerical calculations taking the cavern history into consideration can certainly only be carried out in exceptional cases. On the one hand the calculations are extremely time-consuming and on the other hand the cavern history must be known, so that only post-calculations are possible.

Therefore the aim must be to develop a method with which, within a short period, optimum removal rates may be obtained, not only from a purely operational, but also from a rock-mechanics viewpoint. From this follows that the effects of a removal of the stored product for each cavern is to be judged before the removal. Usually, the expected long-term convergence rates in the secondary creep range are used for a qualitative evaluation, however greatly underestimating the loss rates in the case of short-term decreases. Quantitative statements can only be made, when those convergence rates are taken into consideration, that cofrespond the most to the actual state, i.e. such convergence rates that reflect the actual pressure level, the decrease time and the planned pressure difference. The decrease time must be taken into consideration, as it can only be assumed in exceptional cases, that the decreases take place from a stationary state.

In Fig. 3 therefore, convergence progressions during pressure decreases around Pi = 20, 50 and 100 bars are shown, assuming different pressure levels (150, 200, 250 bars). The decreases take place in a stationary state after 500 days and are each followed over a period of 250 days. To make a better interpretation of the results, the dependence on the pressure decrease of the calculated volume losses is also shown in Fig. 4. It can be seen that the sensitiveness as regards pressure decreases, taking low inital pressures, increases greatly. If decreasing is carried out for only a few days, the influence of the pressure level at the time of decreasing on the expected volume losses is only slight. Larger differences are only to be expected if pressure decreasing takes place over a longer period. This confirms that the withdrawal of stored products should, if possible, be limited to such caverns that have as high a pressure level as possible or that, at least, pressure losses should be compensated as soon as possible.

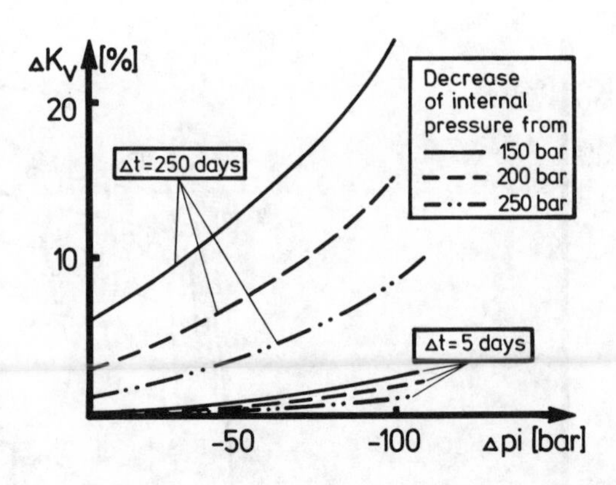

Figure 4. Volume losses dependent on pressure decreases after 5 and 250 days.

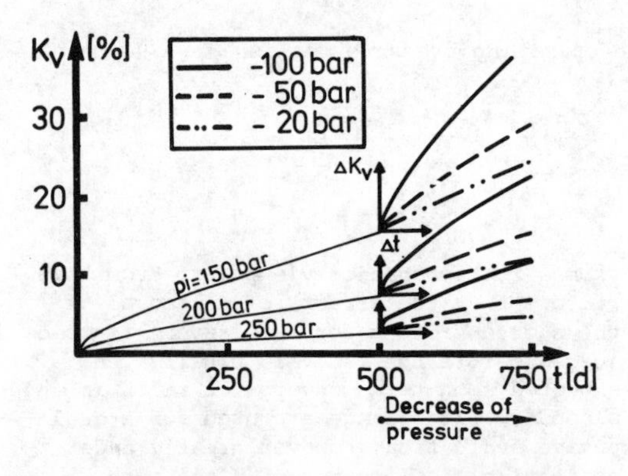

Figure 3. Convergence progressions during pressure decreases

4 DECREASES AT DIFFERENT TIMES

It remains to be investigated, if the decrease time also has a similarly marked influence on the convergence behaviour of a cavern. To this purpose, taking a stationary state below pi = 250 bars, the pressure was reduced first by 50 bars after 500 days, as per Fig. 5. Further reductions of 50 bars each took place after 5, 10 and 50 days. The comparison of the volume losses determined during the second decrease phase as per Fig. 6 shows that there are hardly noticeable differences in the cavern behaviour. However there is the tendency, that an early decrease time may be connected with slightly increased volume losses. Should further tests confirm the convergence behaviour to be largly independent of the decrease time, the cavern history could be neglected when making an approximate calculation for the development of an optimizing method.

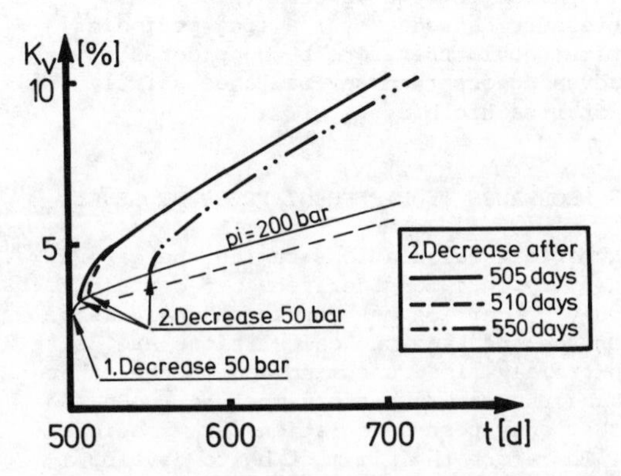

Figure 5. Convergence progressions dependent on pressure decreases at different times

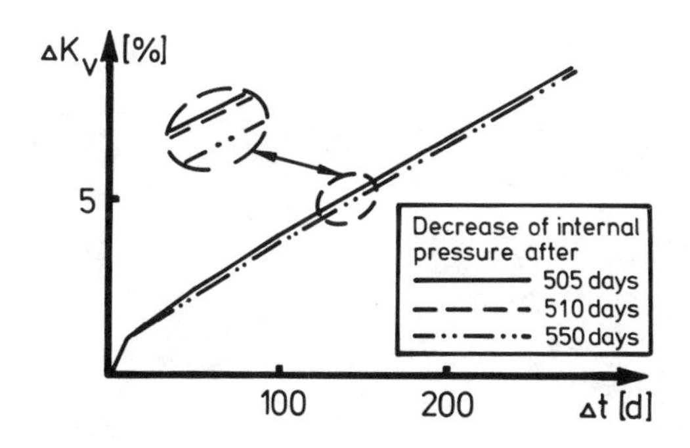

Figure 6. Volume losses determined during the second decrease phase

5 SUMMARY AND EVALUATION

The existing calculations deal, in the beginnings, with the cavern behaviour during the unavoidable pressure reduction during cavern operation. It can be shown that the decrease rhythms have an not unimportant influence on the long-term volume losses of a cavern. They should therefore be taken into consideration in the development of optimizing methods, whereas pressure increases can be largely neglected in this connection.

The calculations show that, in general, the initial pressure state, in connection with the size of the pressure decrease, and not so much the decrease time, influence the volume loss rates.

With systematic examinations, it should be possible to describe the behaviour of a rock salt cavern as exactly as possible, so that the effects of pressure decreases can be estimated quantitatively. If the behaviour of the various caverns within a cavern field is dealt with in this way, it should be possible, on this basis, to develop a method in future, that makes removal optimization possible, under the given conditions (operating period, upper/lower pressure limit, maximum/minimum removal rates etc.).

REFERENCES

/1/ K.H. Lux and S. Heusermann, Creep Tests on Rock Salt with Changing Load as a Basis for the Verification of Theoretical Materials Laws, Proc. 6th Int. Symp. on Salt, Toronto, 1983.

/2/ W. Herrmann and H.S. Lauson, Analysis of Stress Drop in Multistage Creep Experiments, Predictions of Exponential Creep Law, Sandia National Laboratories, SAND-81-1612, 1981.

/3/ U. Hunsche, Results and Interpretation of Creep Experiments on Rock Salt, Proc. 1st Conf. on the Mechanical Behaviour of Salt, Penn State University 1981.

/4/ J. Hentschel, Kavernenprojekt Etzel. Vortrag im Kolloquium Angewandte Gebirgsmechanik im Tunnel- und Kavernenbau, Institut für Unterirdisches Bauen, Universität Hannover, SS 1982.

Numerical Methods in Geomechanics (Innsbruck 1988), Swoboda (ed.)
© 1988 Balkema, Rotterdam. ISBN 90 6191 809 X

Tunnel design using the 'trapdoor' problem

N.C.Koutsabeloulis
BP Research Centre, Sunbury-on-Thames, Middlesex, UK

ABSTRACT: The 'trapdoor' problem has been used in conjunction with the finite element method to provide an understanding of the stress distribution around underground cavities. Using the principles of continuum mechanics, the Ground Reaction Curves (GRC) above the cavity roof have been computed assuming elasto-plastic soil behaviour. Parametric studies have provided a sufficient number of GRC which were statistically processed to provide an equation, which could provide an indication of the long term stability of the unlined cavity.

1 INTRODUCTION

Terzaghi (1936), considered the knowledge of the pressure distribution in the soil above a 'trapdoor' to be a pre-requisite for the clearer understanding of the stress distribution around tunnels.

The 'trapdoor' model, as conceived by Terzaghi (1936) has two modes of displacement. The 'passive' and the 'active' modes.

The 'passive' mode could be used to compute the pull-out forces of plate anchors or other structures embedded in the ground. Numerical modelling of this problem has been performed by Koutsabeloulis (1985), and Koutsabeloulis and Griffiths (1986), which yielded results very similar to those obtained experimentally by Rowe and Davis (1982).

Using the principles of continuum mechanics and assuming an elasto-perfectly plastic soil behaviour, an equation was proposed which could compute the pull-out force of an embedded structure, correlating both material and geometrical data.

The 'active' mode can be used to predict the Ground Reaction Curve (GRC) above the tunnel roof which could provide information regarding the support system required and its dimensioning.

The tunnelling aspects relate to the evaluation of the loosening pressure of sands and/or rocks after excavation takes place during the construction of tunnels. From the engineering point of view, an underground structure can be considered to represent a foreign inclusion inside a mass having definite rheological properties, and subjected to gravity forces. If the rheological properties of the inclusion are different from those of the surrounding mass, a perturbation in the original stress field will occur around the inclusion, disappearing rapidly with distance according to St-Venant's principle. The contact forces between the ground and the inclusion, which are of main interest in the design of underground structures, can therefore, in principle, be determined by the methods of continuum mechanics, if the rheological properties of both the inclusion and the ground are known, or if the displacements of the former are prescribed. In many cases it is common for the exact rheological properties not to be known, particularly for those soils that are mixed with oil and gas. For those cases, Harris et al (1979) suggested that preliminary stability analyses should be performed using the 'trapdoor' problem. Einstein and Schwartz (1975) reported than in cases of tunnel openings supported after the load corresponding to the free-field stresses has been applied, the simple assumption of external loading instead of excavation unloading may lead to support forces that are 50%-100% too conservative. The use of the 'trapdoor' problem in an analysis was fully supported by Szechy (1973). For the problem of a 'trapdoor' displacing downwards, limited analytical solutions and experimental data exist.

2 NUMERICAL SOLUTION TECHNIQUE

The numerical approach used with the finite element method was that of the 'initial stress' (Zienkiewicz et al., 1969, Smith 1982) which has been shown (Koutsabeloulis and Griffiths 1986) to be an efficient and versatile way of solving plasticity problems in geomechanics. This approach iterates using equivalent elastic solutions until any stresses that originally violated yield have returned to the surface of any previously specified yield criterion within quite strict tolerances.

The convergence criterion was implemented by observing the change in the 'body forces' from one iteration to the next. In the present work, convergence was said to have occurred when the change in 'body forces', non-dimensionalised with respect to the largest absolute value, nowhere exceeded 1%. Equilibrium and continuity were also satisfied in the usual way using a displacement finite element formulation.

Fifteen-node triangular, isoparametric elements (Nagtegaal et al 1974, Sloan and Randolph 1982) were used throughout the analysis using 12 integration points under plane strain conditions.

3 SOIL CONSTITUTIVE EQUATIONS

The solutions presented here consider that the soil behaves as an elastic, perfectly plastic material. Plane strain conditions are assumed using non-associated flow rules as they derived by the Mohr-Coulomb and/or Von-Mise's yield criteria.

Although in heavily jointed rock the principles of continuum mechanics are not directly applicable, Singh (1973) and Kulhawy and Goodman (1980) have shown that equivalent elastic properties could be derived based on the properties of the rock mass, the joints and the joint spacing. Using those equivalent elastic properties, the continuum mechanics principles could be adopted.

As the aim of the calculations was to predict estimates of collapse loads necessary to cause general shear failure rather than the settlements before failure, the stress-strain behaviour assumed to act within the failure surface was considered relatively unimportant.

The properties assigned to the soil/rock within the failure surface were chosen arbitrarily to be E = 130 MPa, γ = 20 kN/m^3, and K_o = 1.0. The soil/rock was taken to be mainly cohesionless and a range of friction angles was considered.

Rocks usually exhibit significant shear strength due to the high cohesion term, but possess limited tensile strength. Thus, to model zero tension conditions, the cohesion term within the Mohr-Coulomb yield criterion was set equal to zero.

4 NUMERICAL MODELLING OF THE 'TRAPDOOR' PROBLEM

The 'trapdoor' problem as defined by Terzaghi (1936) is shown in Figure 1. Rowe and Davis (1982) showed by using the 'passive' mode of the problem, that for L/D = 5, the modelling effects were insignificant. Thus the L/D ratio for the present analysis was fixed to L/D = 5. The height was fixed to 6m and to obtain different H/D ratios, the 'trapdoor' width was altered, keeping H and L/D constant.

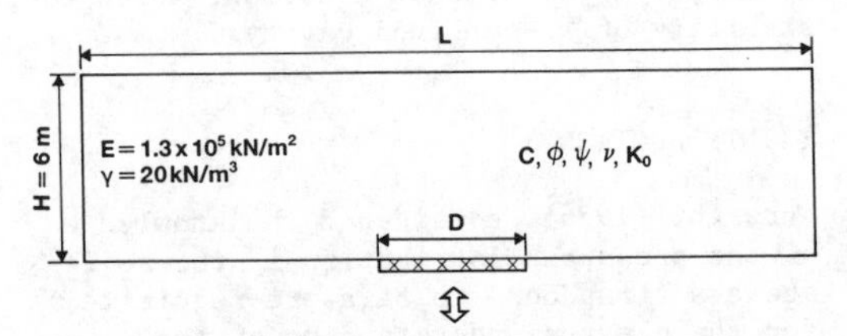

Figure 1. Definition of the trap door problem.

The finite element discretisation of the problem is shown in Figure 2, in a similar fashion to that shown by De Borst and Vermeer (1984). To avoid singularities, the lower boundary of the first row of elements beside the 'trapdoor' was given a linear displacement distribution, with the leftmost mode remaining fixed. The 'trapdoor' itself received a uniform set of prescribed displacements.

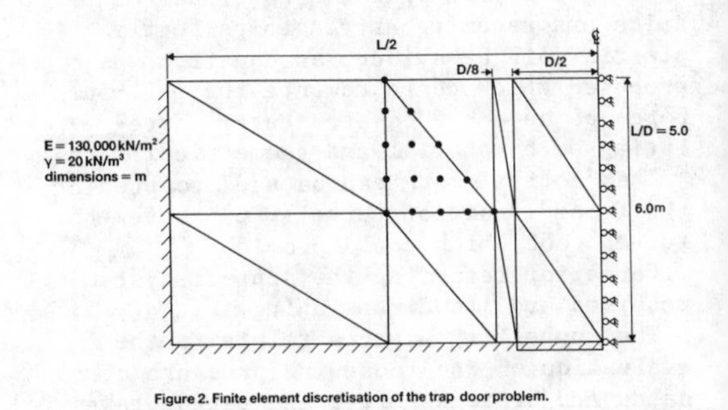

Figure 2. Finite element discretisation of the trap door problem.

The rational for using displacement rather than load control was explained by

Griffiths (1982). Failure under displacement control is indicated by levelling out of the averaged stresses above the 'trapdoor' which, having reached the ultimate capacity, remain at that value.

To benchmark, this present numerical modelling, the problem for which Davis (1968) proposed an upper bound solution was analysed. The soil was assumed to be frictionless, but with self-weight. Thus, the soil was said to obey the Von Mise's yield criterion. According to Davis (1968), the residual stress above the 'trapdoor' for H/D = 1 is given by

$$(\gamma H - P)/C = 2 \qquad (1)$$

where γ is the soil unit weight,
C its cohesion, and
P is the residual stress above the trapdoor

Setting C = 50 KPa, γ = 20 kN/m^3 and H = 20m, equation 1 becomes

$$P = 0.75\gamma H \qquad (2)$$

The numerical solutions obtained using the mesh of Figure 2 is shown in Figuree 3. As can be observed, the numerical result yields to the Davis (1968) solution.

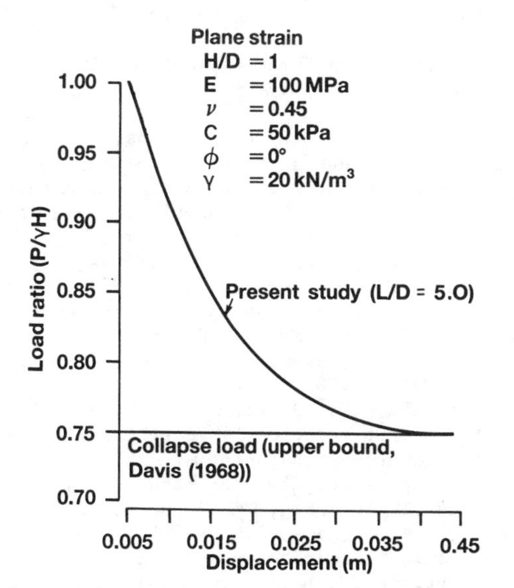

Figure 3. Load ratio versus imposed displacement.

Superimposing on the GRC, such as that of Figure 3, the stiffness of the lining, an equilibrium point can be computed in the sense of Brown et al (1981). This equilibrium point will define the soil-structure interaction load. If this is not done, support forces that are 50%-100% too conservative may be obtained as pointed out by Einstein and Schwartz (1979).

Figure 4 shows the computed failure loads for different H/D ratios considering ϕ = 20°, ψ = 0° and K_o = 1. It was observed that increasing the H/D ratio resulted in a lower failure load ratio P/γH. However, for this particular friction angle the load ratios did not become zero, which implies that for a tunnel design there may exist an ultimate load ratio which must be carried by its lining.

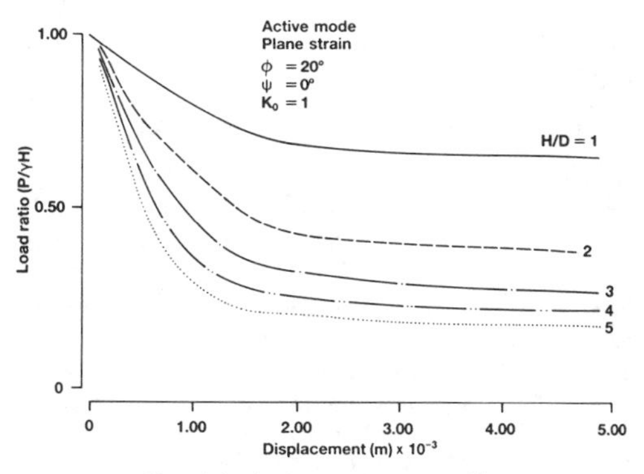

Figure 4. Load ratio versus displacement (ϕ constant).

This load ratio tended to zero as the friction angle increased as shown in Figure 5. This implies that the tunnel lining at high displacements may take no loading from the soil as can happen for tunnel constructed in 'rock' material.

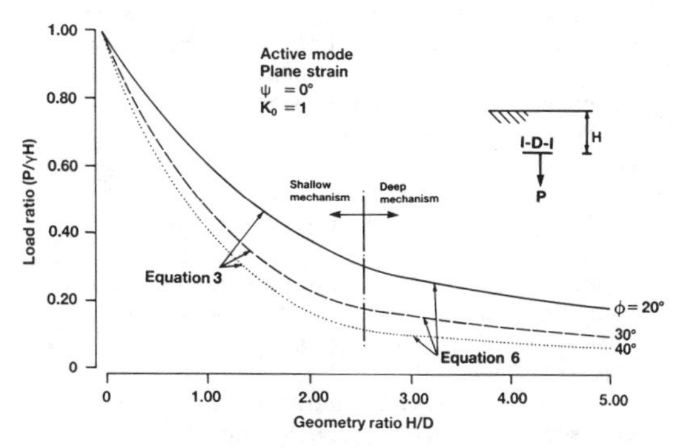

Figure 5. Ultimate load ratio versus H/D.

The authors found that the soil behaviour of Figure 5 can be close approximated by the consideration of two types of failure mechanism, a) a 'shallow' failure mechanism and b) a 'deep' failure mechanism. The change from one mechanism to another occurs at a depth ratio of around H/D = 2.5.

For 'shallow' failure mechanisms, ie H/D<2.5 the following expression is proposed, which gives good agreement with the computed results as shown in Figure 5.

$$P/\gamma H = {\frac{\sin\phi}{\sin\phi + \cos\phi}}^{(GH/D)\tan\phi} \quad (3)$$

where:-

$$G = 1 + 10*\tan(b) \text{ if } 20° < \phi < 30° \text{ and} \quad (4)$$

$$G = 1 + \tan \frac{3\phi - 2b}{6} \text{ if } \phi \quad 30°$$

If H/D   2.5 then defining:-

$$G_1 = \tan^2 (45 - \phi/2) \quad (5)$$

and

$$G_2 = \sin \phi/2)/(\sin \phi/2) + \cos \phi/2))$$

$$G_3 = 4C_2C_1$$

$$\text{Then } P/\gamma H = G_1 G_2^{(G_3H/D) \tan\phi} \quad (6)$$

where, b of the Equation 4 was taken to vary linearly with the friction angle. At $\phi = 20°$, b was set equal to zero, but at $\phi = 40°$, b was set equal to 6.5°. For any ϕ angle, b could be computed from a linear extrapolation of the above b values at $\phi = 20°$ and $\phi = 40°$. This angle is introduced to account for an arching shear band forming above the 'trapdoor'.

5 CONCLUSIONS

The finite element method has been implemented to model the active mode of displacement of the 'trapdoor' problem. Results obtained have been presented as expressions obtained empirically using curve fitting methods. These expressions take account of the geometry, the material properties and the type of failure mechanism developed above the 'trapdoor'.

The soil/rock was assume to possess zero tensile strength and obeyed either the Mohr-Coulomb or the Von-Mise's criterion.

The GRC have been obtained, which using the stiffness of the loading can provide us with the working equilibrium point of the soil/rock-structure system.

The 'trapdoor' model as presented in this study can not provide the full solution to the open cavity problem, but it can certainly provide fast preliminary solutions to what it is expected to be a very complicated design problem.

REFERENCES

Brown, E.T., Bray, I.N., Labanyi, B. & Hoek, E. (1981). Ground response curves for rock tunnels. ASCE, Journal of the Geotechnical Engineering, Vol.109, No.1, January, pp.15-39.

Davis, E.H. (1968). Theories of plasticity and the failure of soil masses. In: Soil Mechanics Selected Topics, (ed) I.K. Lee, p.341, Butterworths, 1968.

De Borst R. & Vermeer, P.A. (1984). Possibilities and Limitations of Finite elements for limit analysis. Geotechnique, 34, No.2, pp.199-210.

Einstein & Schwartz (1975). Simplified Analysis for Tunnel Supports. ASCE, Journal of the Geotechnical Engineering Div, Vol.109, No.1, January, pp.15-39.

Griffiths, D.V. (1982). Computation of Bearing Capacity Factors Using Finite Elements. Geotechnique 32, No.3, pp.195-202.

Harris, M.C. Poppen, S. & Morgenstern, N.R. (1979). Tunnels in Oil Sands. Journal of Canadian Petroleum Technology, 18, 4, pp.34-40.

Koutsabeloulis N.C. (1985). Numerical modelling of soil plasticity under static and dynamic loading. Ph.D. thesis, University of Manchester, Dept of Civil Engineering.

Koutsabeloulis, N.C. & Griffiths, D.V. (1986). Numerical evaluation of uplift forces in sands using the trapdoor problem. Proceedings of European Conference on Numerical Methods in Geomechanics, University of Stuttgart, 16-18 Sept.

Koutsabeloulis, N.C. & Griffiths, D.V. (1986). Accelerators for non-linear problems using the finite element method 3rd International Conference on Numerical Methods for Non-linear Problems, 15-18 Sept, Dubrovnik, Yugoslavia,

Kulhawy, F.H. & Goodman, R.E. (1980). Design of Foundations on Discontinuous Rock. International conference on structural foundations on rock, Sydney, 7-9 May, pp.209-220.

Nagtegaal, J.C., Parks, D.M. & Rise, J.R. (1974). On numerically modelling accurate finite element solutions in the fully plastic range. Computer meth. appl. mech. engineering, Vol.4, pp.153-177.

Rowe, R.K. & Davis, E.H. (1982). The behaviour of anchor plates in sand. Geotechnique 32, No.12, pp.15-41.

Singh, B. (1973). Continuum characterisation of jointed rock masses. Part II, significance of low shear modulus, Int. J. Rock Mech Min Sci and Geomech; Abstr., Vol.10, pp.337-340.

Sloan, S.W. & Randolph M.F. (1982). Numerical prediction of collapse loads using finite element methods. IJNMAG,

Vol.6, pp.47-76.
Smith I.M. (1982). Programming the finite element method, with application to geomechanics. John Wiley & Sons, U.K.
Szechy, K. (1973). The art of tunnelling, 2nd Edition pub. Adademiai Kiado, Budapest.
Terzaghi, K. (1936). Stress distribution in dry and saturated sand above a yielding trapdoor. Proceedings of international conference in soil mechanics, Cambridge, Mass., Vol.1, 22-26, June, pp.307-311.
Zienkiewicz, O.C., Valliappan, S. & King, I.P. (1969). Elasto-plastic solutions of engineering problems, initial stress, finite element approach. IJMME, Vol.1, pp.75-100.

Numerical Methods in Geomechanics (Innsbruck 1988), Swoboda (ed.)
© 1988 Balkema, Rotterdam. ISBN 90 6191 809 X

Rock mass modelling for large underground powerhouses

K.Hönisch
Lahmeyer International GmbH, Geotechnics Department, Frankfurt, FR Germany

ABSTRACT: Results of stability analyses by means of FEM are reported for four major hydropower projects. The jointed rock mass was idealized by the multi-laminate material model. A sequence of required parameter variations for similar projects is derived to determine appropriate support measures.

INTRODUCTION

Design and stability analyses for underground power plants of 20 to 30 m width and 35 to 50 m height require most reliable and proper modelling of the rock mass behaviour to grant a sufficient margin of safety as well as economic construction. Compared to former and actual other reasonable design tools the finite element method has proved its superior but not exclusive capability in this regard.

GENERAL MODEL

Minimum requirements on the rock quality must exist in order to enable a competitive underground powerhouse solution. In this case the rock mass is described as a homogeneous material of relatively high strength (UCS ≥ 4 MPa, C ≥ 1 MPa, ϕ ≥ 35°, Tens. ≥ 0.5 MPa) which is separated by mostly three sets of discontinuities. In most cases, these have lower friction angles and frequently no cohesion. A tensile strength is not taken into consideration.

Generally, without knowledge of the actual location of each discontinuity in the rock mass, surrounding of the later excavation, it must be assumed that these joints are existing at every point of the rock mass which is affected by the construction.

Suitable models were developed to introduce this idea into numerical analyses culminating in the multi-laminate rock mass model [6]. With this concept and its earlier versions, experience is at hand for about ten years with private consultants and twenty years with research institutions [5]. The principal advantage of the mentioned viscoplastic model is the permission of incompatible stresses during the iterative redistribution.

The orientations of discontinuities are represented by most probable values or reasonably unfavourable deviations extrapolated from investigation results. Other properties of discontinuities like roughness, undulation and opening width are sufficiently introduced by means of adjusted shear strength parameters before and after failure. The option of a stress-dependancy of shear strength is advisable for larger stress ranges [3].

In the computations, the joint spacing is not regarded as a parameter directly, although every analyst is fully aware of its decisive character for the overall rock mass stability. The spacing can be introduced to a limited effect by lowering the intact rock strength parameters. Keeping in mind the usual refinement of the FE-grid the actual state of stress is tested every 2.5 to 3 m along the excavation line with regard to failure. This is perfectly sufficient due to the smooth stress changes at the contour even for a typical joint spacing of 0.3 to 1 m. Due to the comparably rapid stress changes perpendicular to the contour line a finer gradation of elements in this direction is applied.

This described method as well as FEM itself is not capable to simulate intercalations of thin layers in the rock mass with respect to deformability and strength in other ways than by averageing.

REQUIREMENTS FOR THE ANALYSIS

On the basis of the mentioned rock mass model for the actual geotechnical parameters and powerhouse design, the following subjects have to be investigated:

i) how far away from the excavation line is the possible overstressing of rock existing,

ii) is a stable redistribution of incompatible stress states possible,

iii) are possible plastic deformation increments compatible with strength of rock mass and supporting structures,

iv) which are the most effective support measures for the actual situation ?

Based on experience with analyses in jointed hard rock for projects mentioned later, the following priority list is derived which illustrates the necessary capabilities for a useful analysis tool:

i) different rock types and related rock mechanical properties within the same analyzed structure

ii) elastic and non-elastic deformations at different excavation stages (a neglecting of these deformations is acceptable only at feasibility level)

iii) existing of parallel openings of comparable size at appropriate distance (for a pillar width less than 75% of cavern height this subject requires 2nd priority)

iv) anisotropy of deformability which at least is induced by intercalations other than horizontal ones

v) unusual variations of the primary state of stress (e.g. a cavern situated close to a valley slope requires 1st priority)

vi) unusual load cases, weight of heavy support structures

vii) different support concepts (sometimes dispensable due to preliminary support analyses)

viii) effects of seepage to cavern.

While the most requirements for geotechnical models and computer codes are self-understandable it must be stated that for fulfillment of tasks nos. ii & vii the choice of a shear and tensile strength along joints must be dependable on the excavation step and the distinction whether the joint is at failure state or not.

Geotechnical model and computer code must be truly three-dimensional to prevent restriction to the analysis of joints striking parallel to the cavern axis. This extremely unfavourable, special situation is avoided as a rule because of support requirements rising significantly.

After establishing a reference case from sensitivity studies, the final proof of validity for the rock mass model and its numerical representation is the comparison of predicted and monitored deformations.

A fictious, homogeneous and isotropic rock mass model with decreased material strength parameters regarding joint orientation, spacing and strength is neither reliable nor competitive to support requirements for rock caverns. It is useful for feasibility stages or cross checks and it may be recommendable for rock qualities which seem now unacceptable for caverns (rock mass rating ≤ 40, quality index ≤ 1).

DESIGN METHODS AND TOOLS

Complementary design methods for the finite element method are limit equilibrium wedge analyses, analytical stress calculations with shear failure check and the use of other experience with similar projects. The comparibility of other projects can be found out by similar joint orientation, spacing and shear strength or preferably by means of

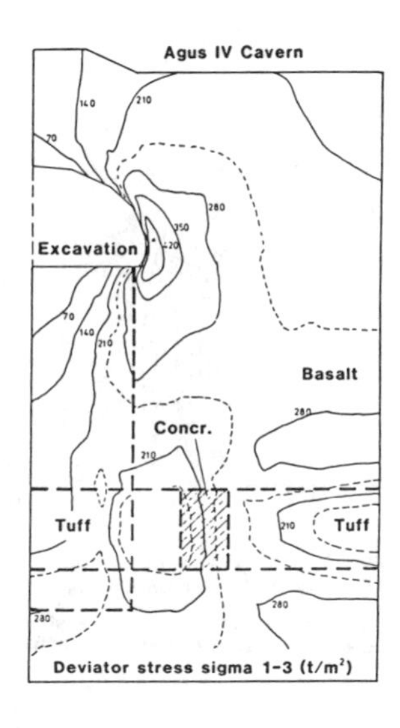

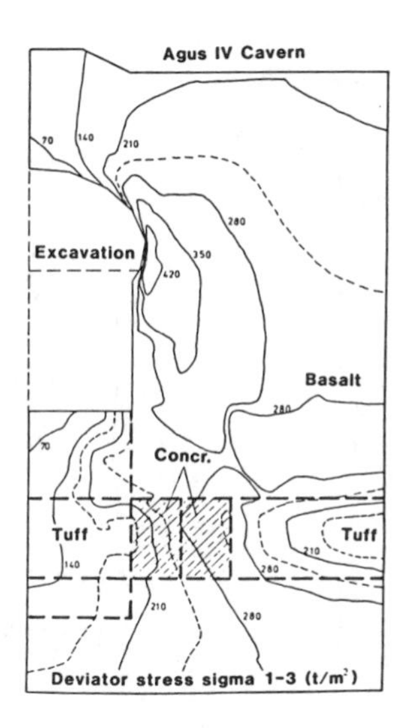

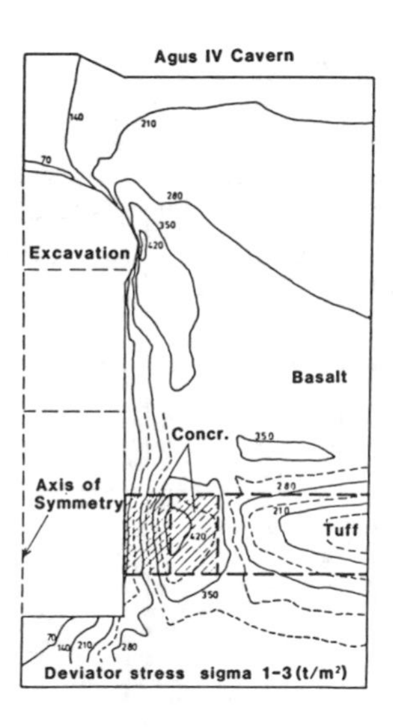

Figure 1a-c Rock mass loading at different excavation stages of a cavern (symmetry assumed)

rock mass classification. This rock description method is no tool for determination of support pressures and only a second source for geotechnical parameters. It is a method of cross-checking and division of rock mass into homogeneous rock units. A possible extended use of classification methods will be investigated in the future.

EXAMPLES

1) The powerhouse cavern Agus IV, Philippines, with dimensions of 19x 37.5x80 m was excavated in highly jointed Basalt [1]. The overburden is 100 m. Bedding and joint sets are arranged horizontally and vertically. A horizontal tuff layer of 7 m thickness located in the lower part was decisive for the excavation stability. It was replaced by concrete up to 8 m distance from the cavern walls. Required width, concrete quality, construction sequence and its completion in advance to cavern excavation were determined by FEM-calculations. In figure 1a-c, the influence of the excavation process and the neighbourhood of three different materials is illustrated by example of the principal stress difference σ_1-σ_3. It can be seen that

- the tuff layer is least loaded
- load is transferred to the concrete plug (outer part)
- close excavation of cavern and part of plug will cause critical stress disturbances
- final loading of cavern wall is as favourable as possible

Alternative solutions for support of the tuff layer did not yield the same suitable final stress distribution. Additionally, placed rock tendons 3, 10 and 20 m above the plug are not introduced in this analysis as well as the extensive grouting measures [1].

2) For the feasibility study Bakun, Sarawak/Malaysia, a powerhouse cavern with dimensions 32x50x300 m was designed to be situated in regularly jointed greywacke and intercalated shales. The overburden is 140 m. Bedding and longitudinal joints are dipping with 60° and 30°, respectively and transversal joints with 75° [8].

An important aspect of the stability analyses was a moderate anisotropy of deformability ($n=E_{\parallel}/E_{\perp}=2$). If this was considered in addition to the comparably low bedding shear strength, the results changed as follows (see figures 2a,b):

- maximum total horizontal displacement rose by 40%
- maximum plastic horizontal displacements and extent of initially overstressed zones was doubled

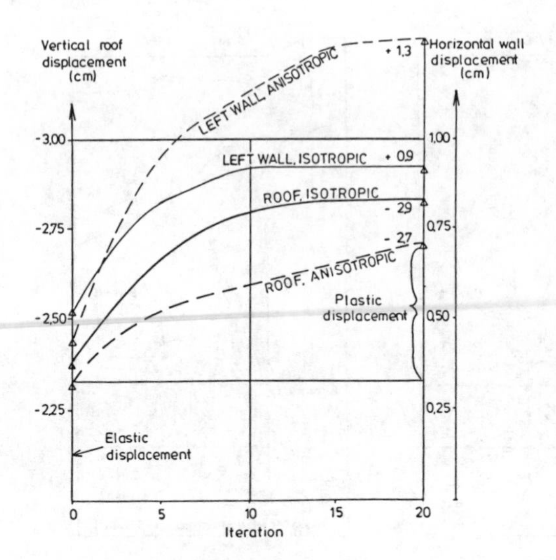

Figure 2a Deformation increments at a cavern

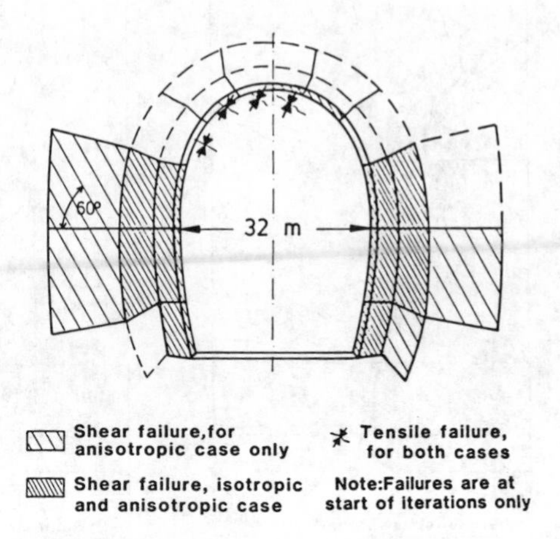

Figure 2b Overstressed zones around a cavern

- maximum tensile stress increased by 20%
- sufficient numerical stability was observed only after 20 instead of 10 iterations.

It must be noted that such results were obtained only for an alternative parallel strike of bedding and cavern walls, which could only have occurred at the front walls. The influence would be considerably smaller for moderate dip angles, too.

3) For replacement of the nearly 60 years old pumped storage plant Koepchenwerk, Germany, a powerhouse shaft with dimensions 19x55x45 m was designed in moderately jointed shale with sandstone intercalations. The plant is under construction now. Bedding and two joint sets are dipping moderately (10° to 20°) and steeply (70° to 90°), respectively [7]. The unusual shape of the shaft with its long parallel walls, an adjacent cut slope, the moderate anisotropy of deformability ($n=E_{\parallel}/E_{\perp}=1.5$) and possible additional horizontal stresses made FEM-analyses indispensable.

A comparison of predicted and observed horizontal wall deformations indicates the existence of additional horizontal stresses in the magnitude of the stresses in accordance with elastic parameters (see figure 3). This is in agreement with in-situ stress results.

Consideration of such additional horizontal stresses had favourable influence on the stability of the walls.

The totally different results at the mouth of the shaft are due to significantly increased tensile and shear failure without additional horizontal stresses and subsequent horizontal irreversible displacements.

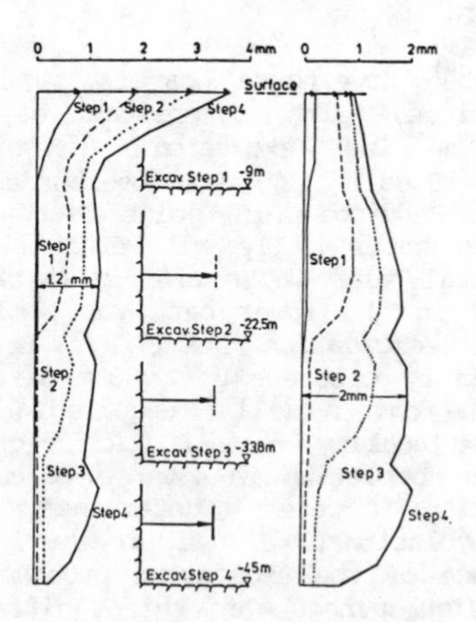

Figure 3 Horizontal displacements of a shaft (transverse section, symmetry assumed)

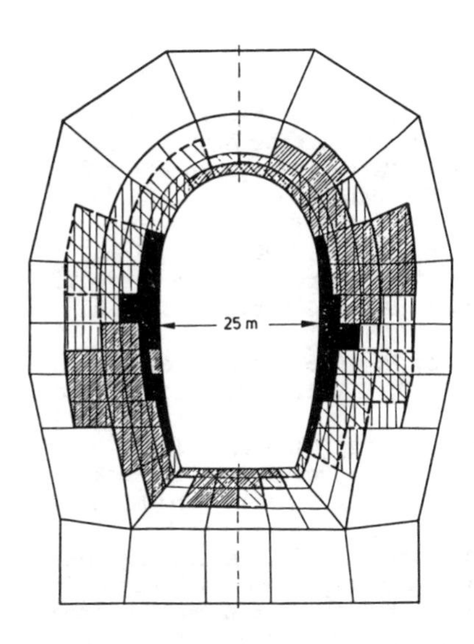

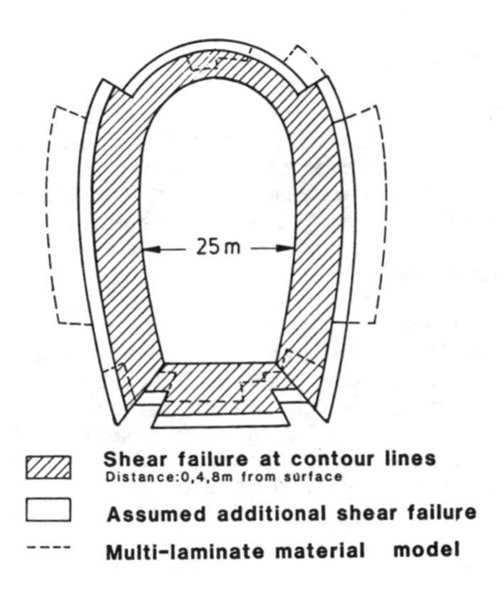

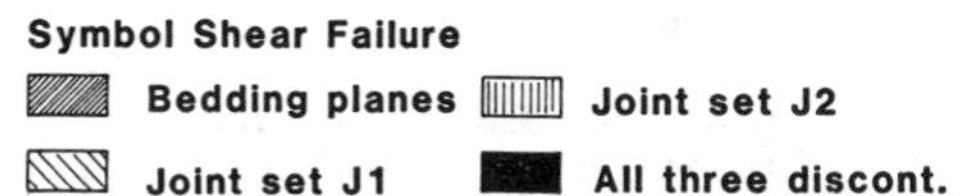

Figure 4a Failure zones due to bedding and two joint sets

In fact, the zones with tensile stresses were extended to 8 instead of 4 m depth and the deviator stress was doubled. Due to provision of a concrete gallery retaining max. 8 m of alluvial deposits, this result is irrelevant.

Figure 3 shows also that the idealization of four instead of e.g. 20 excavation steps causes irregularities in the distribution of wall displacements mainly due to calculatory shear failure at the temporary floors.

4 For the Siah Bishe pumped storage project, Iran, a powerhouse cavern with dimensions of 25x40x110 m was designed. The plant is situated in a sequence of shales, quartzitic sandstones and volcanic rocks [2]. Bedding planes and longitudinal joints are striking with 60° deviation from the cavern axis. The dip angles are approx. 60° opposite to each other. A second, vertical joint set is striking parallel to the cavern axis. The max. overburden is 280 m.

Significant analysis features are considerably high lateral in-situ stresses, different bedding & joint shear strength, anisotropy of deformability (n=2), temporary excavation stages and the transformer cavern parallel to the power cavern.

Figure 4b Failure zones for a fictious homogeneous material

Reported are in figure 4a,b the overstressed zones according to the multi-laminate joint representation and an elastic BEM stress check with fictious material strength. The FEM-results show clearly the asymmetric bedding failure zone high up at the right wall and down at the left and for joints J1 vice versa. Asymmetry is more pronounced for less cohesion on bedding. The zones with failure due to joints J2 are symmetrical.

With derived parameters m=0.5 and s=0.002 according to [4] representing the same rock quality and rating ($Q \approx 3$, $RMR \approx 55$ for shale), we get somewhat less extended failure in both walls and more failure in the floor by a shear failure check. The agreement is acceptable for three joint sets existing simultaneously with similar strength.

The possible differences are evident if the limited continuation of joints J1,2 is introduced in the analysis. Then only black and densely hatched zones would remain. In this case, failure mode and most effective type of support cannot be detected by this approach, even with the use of viscoplastic FEM-analyis.

CONCLUSION

The multi-laminate rock mass model is effective and reliable for the analysis of the influence of all significant rock mechanical parameters, joint behaviours, anisotropy and primary stresses as well as the construction process. A viscoplastic approach used within MISES 3 since 1984 [3] is providing considerable advantages for rock mass modelling compared to the perfectly plastic approach used earlier. These are dilatancy, smooth stress redistribution and reasonable strain increments. The future use of a critical strain concept for rock is enabled in this context.

Annex 1: Rock Mass Parameters

Anisotropy is noted with "/"

Site	V[GPa];	ν[-]	ø[°]	c[MPa]
Agus IV	2.5	0.20	45	10
	1.8	0.25	40	5
	1.0	0.30	30	2.5
Bakun	21/14	0.25	50	10
	8/4	0.25	40	6
Koepchen-werk	3/2	0.25	38	1
	5/3	0.25	42	3.5
Siah Bishe	15.0	0.25	50	18
	7.5	0.30	40	12
	10/5	0.30	40	12

Annex 2: Joint Strength Parameters

Site	Bedding ø [°]; c	Joint sets ø [°]; c
Agus IV	35;0	40;1
	30;0	30;1
	30;0	30;1
Bakun	40;1	45;1
	32;0.5	40;1
Koepchen-werk	20;0	30-35;0
	25;0	35-40;0.05-0.1
Siah Bishe	20;0.05	30;0.1
	20;0.05	30;0.05
	25;0.0	30;0.05

REFERENCES

[1] Almero, R.A. et al. 1984. Design and construction of the Agus IV Power Cavern, Parts 1-3. Water Power 36; Nos. 6, 7 and 9

[2] Franzmann, G. 1985. Planning Iran's Siah Bishe pumped storage project. Intern. Power Generation, No. 11

[3] Haas, W. 1986. User's manual MISES 3 Rev. 8.0, 4 Vol., TDV Pircher, Graz

[4] Hoek, E. & E.T. Brown 1980. Underground Excavations in Rock. London: Instit. Mining and Metallurgy

[5] Malina, H. 1969. Berechnung von Spannungsumlagerungen in Fels und Boden mit Hilfe der Elementenmethode. Publ. Inst. Soil Mech. Rock Mech. Univ. Karlsruhe, Vol. 40

[6] Pande, G.N. & W. Xiang 1982. An improved multi-laminate model of jointed rock masses. Int. Symp. num. meth. Geom. Zürich.Balkema

[7] Schenk, V. et al. 1986. Geotechnical investigations for Koepchenwerk. Water Power 38; No. 7.

[8] Schenk, V. & S.S. Lee 1987. Geotechnical aspects of the Bakun dam in Sarawak/Malaysia. Proc. 9th Southeast Asian Geot. Conf. Bangkok

Numerical Methods in Geomechanics (Innsbruck 1988), Swoboda (ed.)
© 1988 Balkema, Rotterdam. ISBN 90 6191 809 X

Behaviour of concrete tunnels in viscoelastic time-dependent rocks

B.B.Budkowska & Q.Fu
Concordia University, Montreal, Canada
K.Ghavami
Pontificia Universidade Catolica do Rio de Janeiro, Brazil

An analytical approach using two dimensional finite element method for the study of a concrete tunnel which has been cracked after the construction is presented in this paper. In the analysis the surrounding soft rock is considered as a viscoelastic time dependent material. The relation between volumetric parts of stress and strain tensor is assumed as an elastic while their deviatoric components of stress and strain tensor are related to viscoelastic relation. The static load time dependent is obtained by the application of Heaviside functions. The tunnel is considered as an elastic media.

1. INTRODUCTION

In the recent past, there has been an increasing interest in modeling the rock mass not as an elastic-plastic-brittle material but rather as an elastic viscous mass. For such a mass, several theoretical solutions for ground pressures on lining tunnel, valid for linear viscoelastic rock behaviour are available. Kaiser and Morgenstein (1983), Ladanyi and Gill (1984), Lombardi (1977), Panet and Guenot (1982), Lo and Yuew (1981). Some researchers have tried to look into the mechanism of time dependent stress effect by proposing and testing phenomonological models, others have tried to observe the behaviour and fit an equation which could be extrapolated.

To show the state of stress in a cracked concrete tunnel, subjected to internal water pressure and constructed in heavily disturbed rock, the viscoelasticity theory is applied. The viscoelastic rock is described by a special constitutive model, in which volumetric part of stress-strain is assumed to follow the propotionality law, while their deviators are related to viscoelasticity relationship. A general formulation in integral form is applied with respect of time. Considering the numerical approach according to time variable, the integral recurrent method is applied. With respect to spatial variables, finite element procedure is implemented. The concrete tunnel was analysed for different types of the load, which is described by the ratio of the horizontal pressure to the vertical one (denoted as coefficient K). For specific cases, the solution in terms of the displacement field is obtained for instantaneous and infinite time. The increase of stresses with time for a nonhomogeneous problem is analysed for different values of K and various viscoelastic properties of the soft rock.

2. ANALYTICAL FORMULATION OF THE PROBLEM

To consider time dependence behaviour of soft rock surrounding concrete tunnel, subjected to internal water pressure the analysis was concentrated on the effect of rheological properties of soft rocks located around the elastic type tunnel. The problem was considered as plane strain. For the constitutive model of soft rocks within isotropy assumption, it is assumed, that volumetric components of stress σ_{kk} and strain ε_{kk} tensors are related by proportionality law, while the deviatoric components of stress σ_{ij} and strain ε_{ij} tensors are related by viscoelasticity law.

The chosen viscoelastic model is described by two basic material functions, i.e. creep and relaxation curve as shown in Fig. 1 and Fig. 2.

The viscoelasticity problem can be formulated by differential operators or by functionals which are

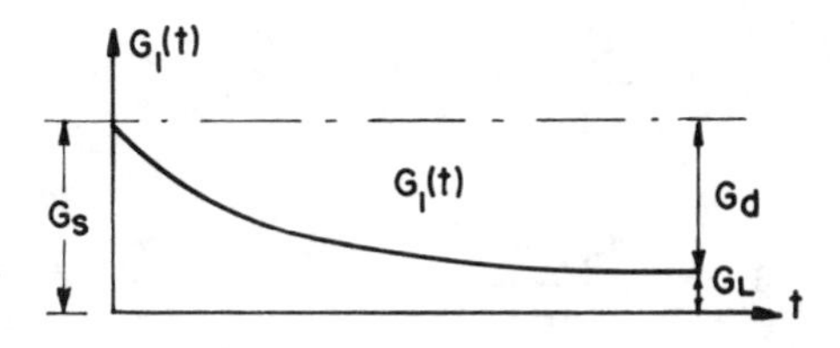

Fig. 1 Relaxation function for the viscoelastic standard model

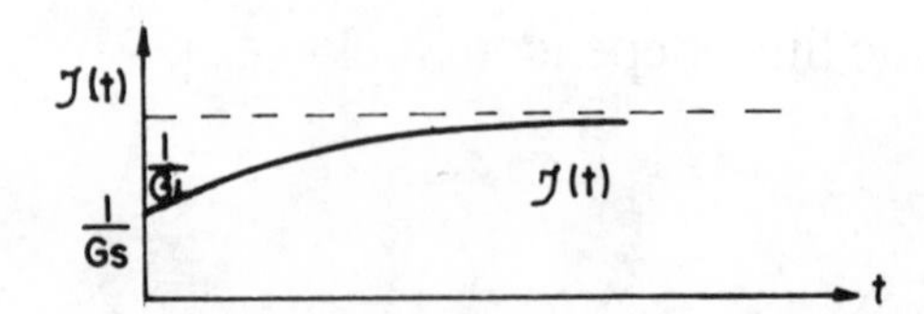

Fig. 2 Creep function for the viscoelastic standard model

of integral type. The latter approach is more general and has certain advantages as it gives better results with regards to stability of solution.

The formulation starts from the postulate of the existence of viscoelastic functional which makes it possible to express the current value of stress in terms of the history of deformation and its current value.

Thus the constitutive relation for stress-strain viscoelastic material is:

$$\sigma_{ij} = \int_{-\infty}^{t} G_{ijkl}\ (t-\tau)\frac{d\varepsilon_{kl}(\tau)}{d\tau}d\tau \qquad (1)$$

where G_{ijkl} is the fourth order relaxation tensor.

The equation (1) can be inverted so that for specified state of stress, the strain can be expressed as:

$$\varepsilon_{ij} = (t) = \int_{-\infty}^{t} J_{ijkl}\,(t-\tau)\frac{d\sigma_{kl}(\tau)}{d\tau}\,(d\tau) \qquad (2)$$

Where J_{ijkl} is the fourth order creep tensor.

The assumption of material isotropy makes it possible to separate the state of stress and deformation into two independent parts, defined as deviatoric and volumetric. Thereby equation (1) can be written independently for deviators of stress σ_{ij} and strain ε_{ij}, and then for volumetric components of stress σ_{kk} and strain ε_{kk}, as:

$$\bar{\sigma}_{ij} = \int_{-\infty}^{t} G_{ijkl}\ (t-\tau)\frac{d\bar{\varepsilon}^{a}_{kl}(\tau)}{d\tau}d\tau \qquad (3a)$$

$$\sigma_{kk} = \int_{-\infty}^{t} G_2\,(t-\tau)\frac{d\varepsilon_{kk}(\tau)}{d\tau}d\tau \qquad (3b)$$

where $G_1(t)$ and G_2 (t) are two independent functions connected with deviatoric and volumetric states respectively.

Similarly, the isotropy assumption applied to equation (2), leads to the following equations:

$$\bar{\varepsilon}_{ij} = \int_{-\infty}^{t} J_1\,(t-\tau)\frac{d\bar{\sigma}_{ij}(\tau)}{d\tau}d\tau \qquad (4a)$$

$$\varepsilon_{kk} = \int_{-\infty}^{t} J_2\,(t-\tau)\frac{d\sigma_{kk}(\tau)}{d\tau}d\tau \qquad (4b)$$

where J_1 (t) and J_2 (t) are independent creep functions connected with deviatoric and volumetric states respectively.

For an isotropic viscoelastic material the deviatoric relation is described by time dependent relaxation function which follows the transition from viscoelastic to elastic type material for two particular time instants i.e. $t = +0$ and $t = \infty$. At these two moments, the material behaves elastically which implies, that the relationship between deviators of stress and strain is described by shear modulus which has different values as $G_o \neq G_\infty$ for $t = 0$ and $t = \infty$. This implies that the material is characterized by bulk modulus K, which appears in the volumetric stress-strain relation ship and two bounded values of shear modulus G_o and G_∞.

The existence $G_o \neq G_\infty$ implies that there exist two values of the Poisson's ratio i.e. ν_o for $t = +0$ and ν_∞ for $t = \infty$.

For the standard viscoelastic model, the relaxation function is given as:

$$2G_1(t) = 2G_L + (2G_s - 2G_L)e^{-\beta t} \qquad (5)$$

where the definition of material constants is indicated on the Fig. 1. The creep function associated to relaxation curve, (see Fig. 2) for the considered model is written as:

$$J(t) = \frac{1}{2}\left[\frac{1}{G_L} + \left(\frac{1}{G_s} - \frac{1}{G_L}\right)e^{-\beta\frac{G_L}{G_s}t}\right] \qquad (6)$$

Finally the concise form of the constitutive equation applied in the analysis of viscoelastic medium can be written as:

$$\sigma_{ij} = K\,\varepsilon_{kk} + 2\,G(t)\,\bar{\varepsilon}_{ij} \qquad (7)$$

where the first term represents the law applied to volumetric part, while the second term stands for relationship described by equation (3a).

3. NUMERICAL FORMULATION OF THE VISCOELASTIC PROBLEM

As is has been mentioned, viscoelastic constitutive relations can be formulated by differential operators which leads in numerical implementation to finite differences method with respect to time variables. However, this method has very serious drawback, since it is very sensitive to the length of the time step and is conditionally stable. The latter means that the relationship between physical properties of the medium and the length of the time increment, affects the stability of the solution. If this condition is not satisfied the solution has divergent and oscillating character.

In the integral numerical solution with respect to time this negative characteristic does not exist. In turn, this relationship between deviators of stress and strain tensor should be defined for arbitrary time instant t_n. From a basic formulation, this relationship can be written as:

$$\bar{\sigma}_{ij}(t_n) = 2(G_L \bar{\varepsilon}_{ij}(t_n) + G_D e^{-\beta t_n} \bar{\varepsilon}_{ij}(0) + \int_0^{t_n} G_D e^{-\beta(t_n-\tau)} \dot{\bar{\varepsilon}}_{ij} \, d\tau) \tag{8}$$

where dot over ε_{ij} means the differentiation with respect to time.

The first term on the right side of equation (8) describes the current dependence, while the other two terms represent the history of the deformation in time.

To define the computational procedure, the relationship for the following time instant should be formulated, to form the dependence between the current time instant and the previous time step. This is obtained formally by application of recurrent procedure, i.e.

$$\bar{\sigma}_{ij}(t_n+1) = 2[G_L \Delta\bar{\varepsilon}_{ij}(t_n+1) + G_L \bar{\varepsilon}_{ij}(t_n) + e^{-\beta\Delta t}(G_D e^{-\beta t_n} \bar{\varepsilon}_{ij}(0) + \int_0^{t_n} G_D e^{-\beta(t_n-\tau)} \dot{\bar{\varepsilon}}_{ij}(\tau)\, d\tau) + \int_{t_n}^{t_n+1} G_D e^{-\beta(t_{n+1}-\tau)} \dot{\bar{\varepsilon}}_{ij}(\tau)\, d\tau] \tag{9}$$

The analysis of equation (9) reveals, that the term which is multiplied by $e^{-\beta\Delta t}$ considers the history of the deformation, while the last integral represents as the increment of history of deformation which after integration has the finite form as:

$$\int_{t_n}^{t_{n+1}} G_D e^{-\beta(t_{n+1}-\tau)} \dot{\bar{\varepsilon}}_{ij}(\tau)\, d\tau = G_D \frac{1-e^{-\beta\Delta t}}{\beta\Delta t} \Delta\bar{\varepsilon}_{ij}(t_{n+1}) \tag{10}$$

Denoting the term connected with history of deformation by I_n, the final expression of the numerical formulation with respect to time is:

$$\bar{\sigma}_{ij}(t_{n+1}) = 2[G_L \bar{\varepsilon}_{ij}(t_n) + (G_L + G_D \frac{1-e^{-\beta\Delta t}}{\beta\Delta t}) \cdot \Delta\bar{\varepsilon}_{ij}(t_{n+1}) + e^{-\beta\Delta t} I_n] \tag{11}$$

The term representing history of deformation I_n, after each step of computations must be updated according to the following formula:

$$I_{n+1} = e^{-\beta\Delta t} I_n + \Delta I_n \tag{12}$$

where ΔI_n is the increment of the history of deformation.

Implementation of the above described procedure to finite element formulation, with respect to spatial variables, gives the following relationship:

$$([K]^E + [K]^V (G_L + G_D \frac{1-e^{-\beta\Delta t}}{\beta\Delta t})) \cdot \{V_{n+1}\} = \{P_{n+1}\} - [K]^V \cdot (e^{-\beta\Delta t} I_n - G_D \frac{1-e^{-\beta\Delta t}}{\beta\Delta t} \{V\}) \tag{13}$$

where,
$[K]^E$ - is the stiffness matrix of the elastic material,
$[K]^V$ - is the stiffness matrix of the viscous material,
$\{V_{n+1}\}$, $\{V_n\}$ - are displacement vectors connected at t_{n+1} and t_n
$\{P_{n+1}\}$ - the the load vector for t_{n+1}

Having obtained the solution in terms of the displacement field, the stress components are computed according to the following formula:

$$\sigma_{ij}(t_{n+1}) = [D]^E [B] \{V_{n+1}\} + [D]^V [B] \cdot (G_L \{V_{n+1}\} + e^{-\beta\Delta t} I_n + G_D \frac{1-e^{-\beta\Delta t}}{\beta\Delta t} \cdot (\{V_{n+1}\} - \{V_n\})) \tag{14}$$

where,
$[D]^E$ - is the matrix of elastic properties of material,
$[D]^V$ - is the matrix of viscous properties
[B] - is the deformation matrix.

4. NUMERICAL EXAMPLE AND THE ANALYSIS OF THE RESULTS

To analyse the state of the stress distribution in a 30 cm thick concrete tunnel of 20 km in length and 3 m in diameter, which has been constructed in a very complex geological formation a computer program was developed based on the above theoretical concept. Part of the tunnel lining which goes through a heavily disturbed rock region showed relatively large horizontal cracks (See Fig.3) due time, while at the stretches where the surrounding rock is rigid and has got very low deformability no appreciable cracking has been observed, Ghavami 1986, Budkowska, Ghavami, Fu, 1987. Hence a limited parametric study of the load- stress-deformation states in the concrete tunnel for the possible set of geological conditions which can be ascribed to the disturb rocks has been carried out for better understanding of the monitoring results.

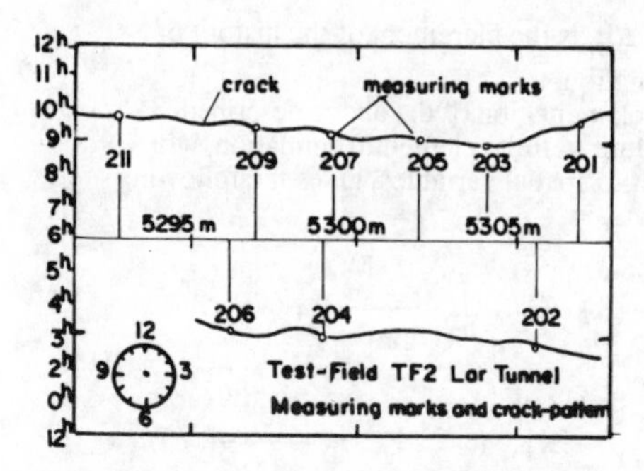

Fig. 3 Horizontal cracks which appeared in tunnel

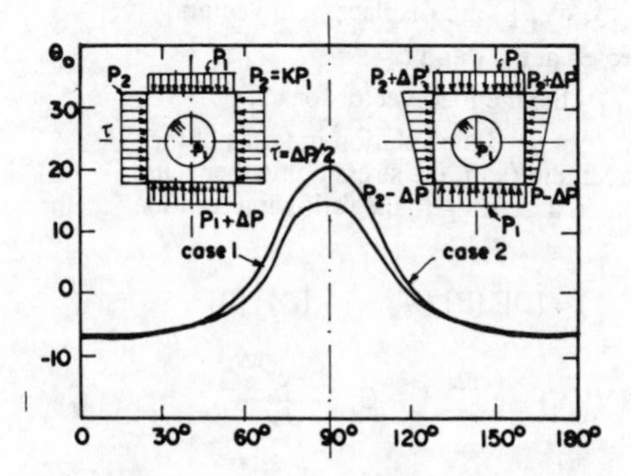

Fig. 4 Circumferential normal stresses in tunnel for two types of load conditions

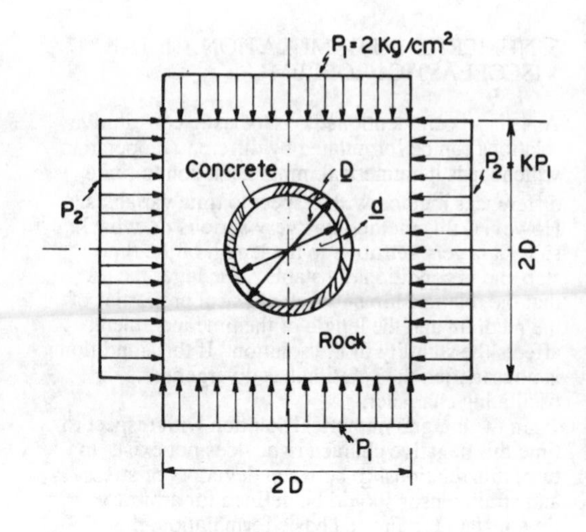

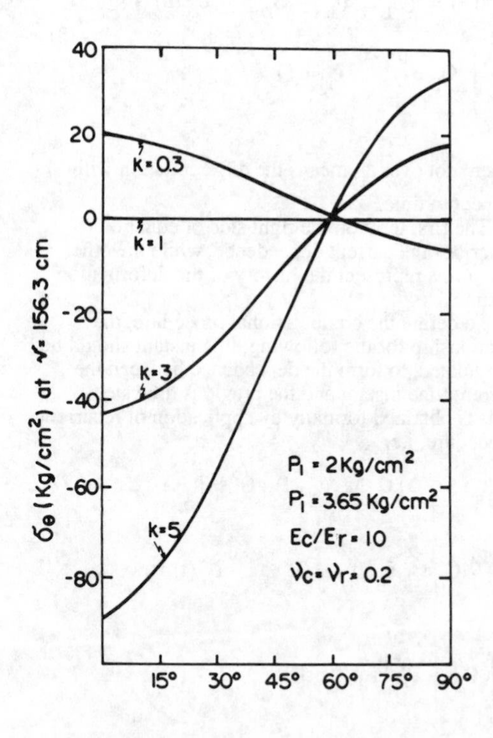

Fig. 5 Distribution of circumferential normal stresses σ_θ for different values of coefficient K.
E_c, E_r - Young modulus for concrete and rock respectively
ν_c, ν_r - The Poisson's ratio for concrete and rock respectively

Special attention, in the analysis was given tot he circumferential stress, σ_θ, because of the crack type.

The distribution of the circumferential normal stresses "σ_θ" obtained for two different types of loading is shown in Fig. 4. The first case takes into consideration the influence of shear forces, of the magnitude of $\Delta P = P1/2$ for $K = 3$.

The loading conditions correspond to Case 2 assume, that the lateral pressures are proportional to the lateral deflections which results in a bigger magnitude of horizonatal pressure in upper part and smaller in lower part.

The results for two loading cases in terms of circumferential normal stresses σ_θ along the circumferential of the tunnel are also shown in Fig. 4. However, the analysis of bisymmetrical load conditions presented in Fig. 5, for different values of coefficient K, gave more unfavorable results compared with the previously described cases.

The distribution of the principle stresses, σ_1, for last case is shown in Fig. 6.

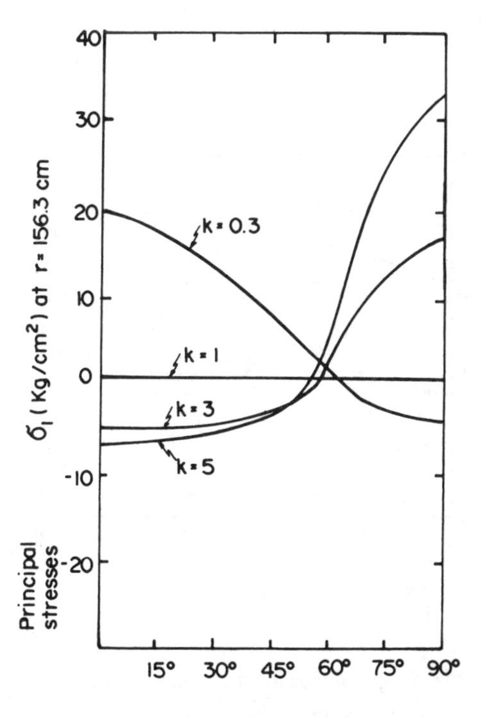

Fig. 6 Distribution of principal stresses σ_1 for different values of ceofficient K.

The discretization of the spatial domain into finite element mesh as well as the boundary conditions and type of finite element (element number 28) for this problem are shown in Fig. 7.

On the basis of the qualitative analysis, the chosen mesh can be considered as satisfactory.

To check the correctness of the method a simple case as a homogeneous medium subjected to constant load was analysed. In this case, only displacement field, which is the function of time is obtained. Moreover, the introduced viscoelastic standard model makes it possible to obtain final form of the displacement field on the basis of final value theorem. The results for different values of coefficient K for the analysed examples are shown in Fig. 8 and Fig. 9.

Finally, the presence of the tunnel in the soft rock creates nonhomogeneous medium. The tunnel, in contrast to the soft rock is assumed to be of elastic type material. This situation induces changes in stress distribution with time in the tunnel lining.

The changes in the circumferential normal stresses, for different values of the coefficient K are given in Fig. 10. Curves A - denotes elastic solution for $t = +0$, which is obtained from viscoelastic formulation by application of initial value theorem. Curves B give the solution for the case when $t = \infty$. Curves C and D represent the influence of viscoelastic properties of the disturbed rocks on the stress distribution of the tunnel with time. The time scale in the analysis is expressed in the parameter of the relaxation time of the viscoelastic model. It can be noted in Fig. 10, that shorter the values of relaxation time the faster are the changes in the stress (see Curve D), while for the higher values of relaxation times the process of stress increase is well decelerated.

From this follows that rheological properties of the surrounding medium has an important influence on the changes of the stress distribution in the lining tunnel.

5. CONCLUSIONS AND FINAL REMARKS

Based on the application of the viscoelasticity theory for soft rock, the influence of rocks pressure with time in the stress distribution of tunnel lining, which has been cracked after construction, and was subjected to internal water pressure, was fund to be a relatively significant factor. In the analysis, the influence of different types of rock pressure on the tunnel lining was studied. It was established that the loading case expressed in terms of coefficient "K" which is the ratio of the horizontal rock pressure to vertical one is the appropriate type of loading to describe the case under study with time.

A special type of constitutive model in which volumetric part of stress and strain was assumed to follow the proportionality law, while the deviators were related to viscoelasticity law was used. Regarding the rheological model a general formulation in which the constitutive relationship

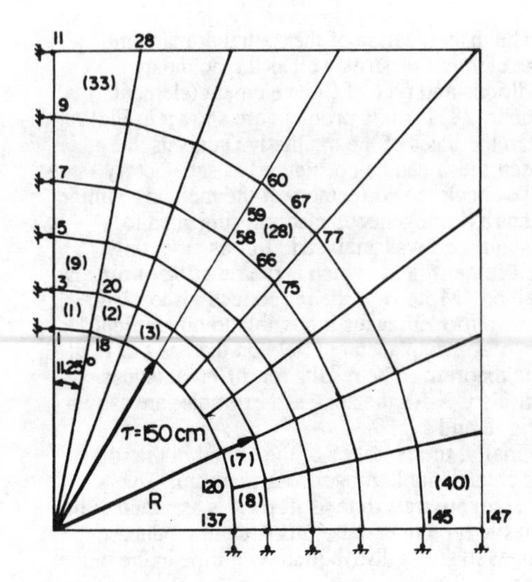

Fig. 7 Geometry, finite element mesh, boundary conditions, and the type of finite element for analysed problem

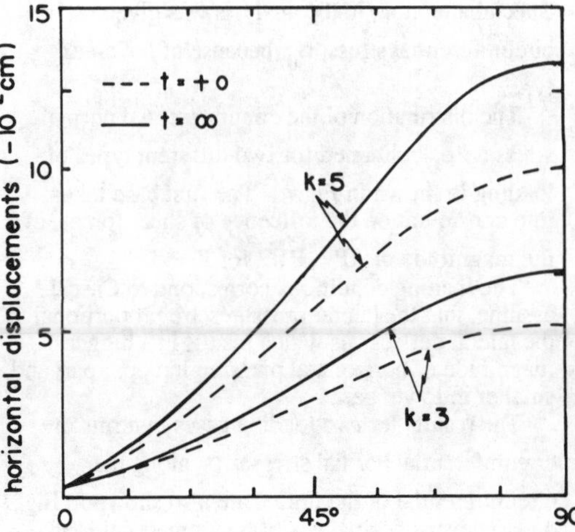

Fig. 9 Distribution of horizontal displacements along the circumference of the opening in homogeneous medium

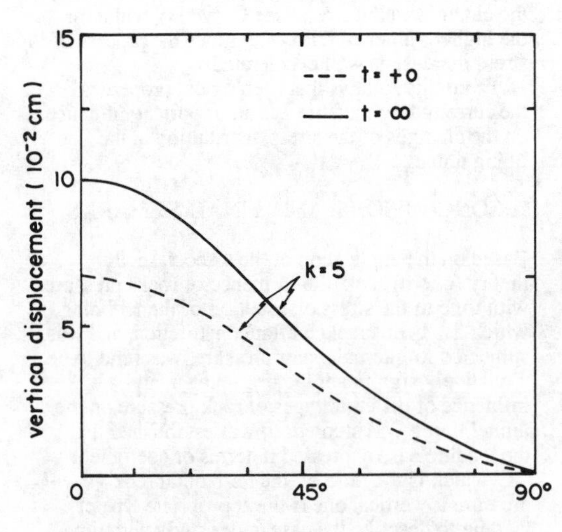

Fig. 8 Distribution of vertical displacements along the circumference of the opening in homogeneous medium

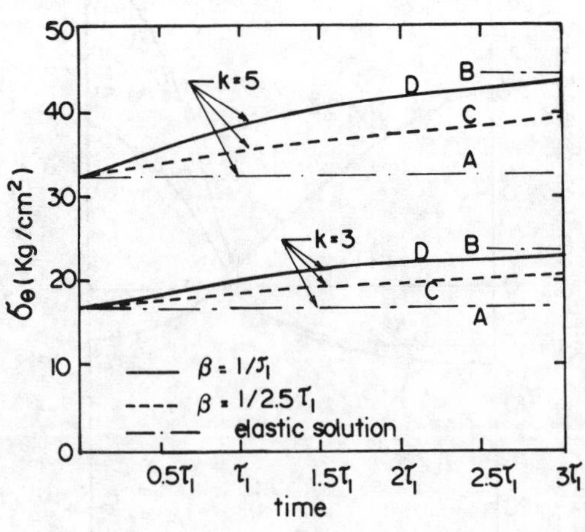

Fig. 10 Changes of the circumferential mormal stresses σ_θ in time for different values of coefficient K.

was expressed in integral form was applied. For the numerical formulation the integral recurrent procedure with respect to time which provided a better numerical stability in comparison to finite difference method was implemented. With respect to spatial variables, the finite element method as applied.

In the analysis attention was given on rheological properties of the surrounding soft rocks, which affected the increase of stresses, in time, in the elastic type tunnel. In this manner the formulated problem was treated as a nonhomogeneous, since different types of materials according to their constitutive models were used. This nonhomogeneity in the analysis implied redistribution of stresses in time.

REFERENCES

Kaiser, P.K., and Morgenstein, N.R., "Time Dependent Behaviour of Tunnels in Highly Stressed Rock" Proc. of the 5th Int. Congress on Rock Mech. Melbourne, Australia, 1983.

Ladanyi, B. and Gill, D.E., "Tunnel Lining, Design in Creeping Rock" Proc. ISRM Symposium on Design and Performance of Underground Excavations, Cambridge, U.K. 1984.

Lombardi, G., "Long-Term Measurements in Underground Opening and Their Interpretation With Special Consideration to the Rheological Behaviour of Rock" Field Measurements in Rock. Mech. Vol. 2 AA, Balkam, Rotterdam, Holland, 1977 pp. 839-858.

Panet, M. and Guenot, A., "Analysis of Convergence Behind the Face of a Tunnel" Tunnelling 82, The Inst. of Mining and Metallurgy, 1982. pp. 197-204.

Lo, K.Y. and Yuew, C.M.K., "Design of Tunnel Lining in Rock for Long Term Time Effects" Canadian Geotechnical Journal, Vol. 18, No. 1, 1981. pp. 24-39.

Ghavami, K. "Causes of Cracks in Kalan Tunnel Lining" Report on the Multipurpose Lar Dam and Related Problems to Tehran Water Board. Prepared by Filho, P., Ghavami, K. and Totis, E., Civil Eng. Dept. PUC-RJ, Report RJ 01/1986. pp. 18-25.

Budkowska, B.B., Ghavami, K. and Fu., Q., "Parametric study of tunnel in Soft Rocks Considering the Time Effect" Report 1, Civil Engineering Department, Concordia University, Montreal, Canada 1987 (in preparation).

ACKNOWLEDGEMENT

Financial support granted by Natural Sciences and Engineering Research Council of Canada under Grant A4686 is gratefully acknowledged.

Numerical Methods in Geomechanics (Innsbruck 1988), Swoboda (ed.)
© 1988 Balkema, Rotterdam. ISBN 90 6191 809 X

Geonumerical computations for the determination of critical deformations in shallow tunnelling

H.Wagner & A.Schulter
Mayreder Consult Ges.m.b.H., Linz, Austria

ABSTRACT: It is an essential part when planning inner urban tunnels to indicate expected deformations. Deformations on the surface and deformations within the tunnels do have the same importance.

Detailed investigations do usually include considerations of several loading conditions. Each loading condition does correspond with the specific deformation and contributes partially to the total deformation.

There is a serie of parameters influencing the development of deformations with respect to tendency and absolute value.
This are among others soil parameters material parameters and specific assumptions with respect to the construction concept applied. Because of the results calculated, it would become necessary to change the construction method to be used even during planning stage.

Before such changes lead to satisfying calculation results, in many cases they are considered to be far away from reality. But in fact calculation is the only possibility in connection with experiences to give the designer a tool thus judging the structural design and the construction concept to be used.

When calculating tunnels usually numerical methods,in tunnelling mostly FE-methods and more seldom analytical methods are used. Within this paper the very typical project of Metro section E-5 of Washington D.C. is analized using the two dimensional model.
It has shown that it is possible to define new criterias, based on geonumerical calculations, for critical deformations and more economical solutions, while the importance of geomechanical measurements during construction is unchanged.

1. INTRODUCTION

In the course of planning shallow tunnels in soft ground the question of what are the "critical deformations" usually arises both from the client side, the designers side and the contractors side. There is a special wish, coming from the structural point of view, that absolute values have to be available, where all participants at the development of an underground structure have to orient themselves during construction. On the other hand, observational methods, like the "NEW AUSTRIAN TUNNELLING METHODS", take much more than absolute values into consideration, when considering time and geomechanical measurements.

When judging critical deformations no defined criterias and no special size of deformation is considered to be critical. In many cases stresses within the structure or within the soils are considered to be relevant, instead of deformations. Only plastified zones are registered to be interesting.

During design stage the question for critical deformation is skillfully delegated to the construction at the site. It is the responsibility of the site manager, to correlate the results of the geomechanical in-situ measurements with the safety risks of the miners in the tunnel.

It has shown in the past, that simple rules of thumb are on the safe side. This rules give rough but unsatisfying and mostly uneconomical estimations of the deformation. Because many and significant deviations have been observed, it seems to be necessary to develop new and more differentiated algorithms for critical deformations. Depending on the geometry of the excavated tunnel special consideration should be given to geology, overburden and construction method used.

Geomechanical in-situ measurements are an integrated component of that methods, thus observing the real development of interaction between freshly exposed underground respective lining and the surrounding soil, which is considered to be part of the lining. Time is the other, much more difficult parameter; it can be measured in-situ, mostly in relation to the location of the tunnel heading, but it cannot be simulated in a computer model.

Therefore the question has to be answered, whether it is possible in general to find such values, where they have to be related to, how they should cover the time depending changes and what safety level requirements have to be met. An attempt will be made towards a simple relation between overburden and idealized tunnel diameter to define critical areas of deformations, to when the support measures of whatever design can remain unchanged, and only in the case that these deformations are exceeding, additional measures to strengthen the lining have to be taken.

2. PRESENT DESIGN PRACTICE

There is no doubt, FEM-calculations in usual soft-ground tunnelling design practice are the only tool to simulate complicated three-dimensional loading conditions, whether a three-dimensional calculation nor a two-dimensional one is used. But it is also clear, that every such calculation is worth only as much as the experience of the designer behind the calculation.

It is the usual case to analyse loading conditions as expected during construction. As a result of FEM-calculation there is a stress and strain condition related to each loading condition.

Basically two cases are the most important ones:
After the primary stress condition has been analysed the stress relieve condition shows deformations, which take place just before excavation works are done in the very cross section.
The second important condition is the one, when the initial shotcrete lining has developed its final strength.

The problem in practice at the construction site now starts with the question, where the one or the other, ot both deformations are critical or not, respectively which one the critical deformation, measured in-situ, at the construction site will be.

The common answer for this question in the past has been usually indifferent and therefore not satisfying. NATM as a construction concept was usually called a semi-empirical method with not only measurements but also surprises as an integrated part of the concept.
Absolute values of deformations have been unpopular. The unpredictable characteristic of the development of deformations during different stages of construction was observed to be essential. Practical experience of tunnel engineers is necessary to be permanently related with the indication of the respective instrument. The aura of "foggy mystery" was awarded in many cases to the development of a project instead of practical and economic solutions.

3. REQUIREMENTS DURING CONSTRUCTION

It can be stated as a general rule, that resident engineers as well as contractor site managers have to have an indication, of from where on measured deformations are critical and in consequence of that, what the proper measures to stop this development will be.

Modern geomechanical measurement methods have been developed, to measure very accurate deformations, caused by stress relieve as well as by excavated and initially lined tunnels.

It is important to know, whether the calculated deformations do correspond with these measured deformations, or not. Especially it would be worth to know this before the final lining is installed. Therefore, the calculated stress relieve deformation has to be compared with the measured one to check whether the lining corresponds satisfactorily with the construction, or not.

4. ANALYTICAL CONSIDERATIONS OF A TYPICAL EXAMPLE

Single track tunnel of the Subway section E-5 of Washington Metro has proven to be suitable for the following geonumerical analysis, being a typical example of a shallow tunnel under soft ground conditions.

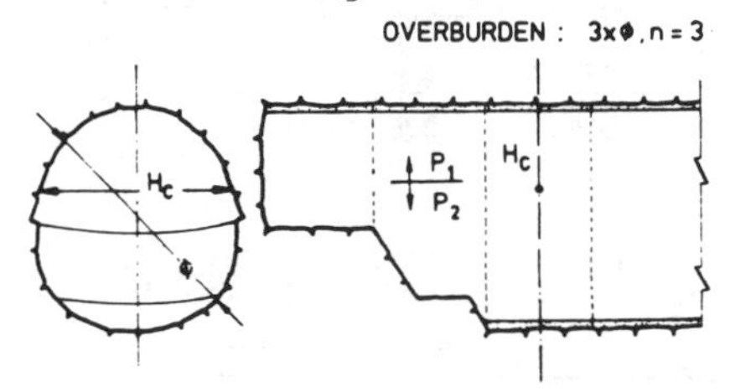

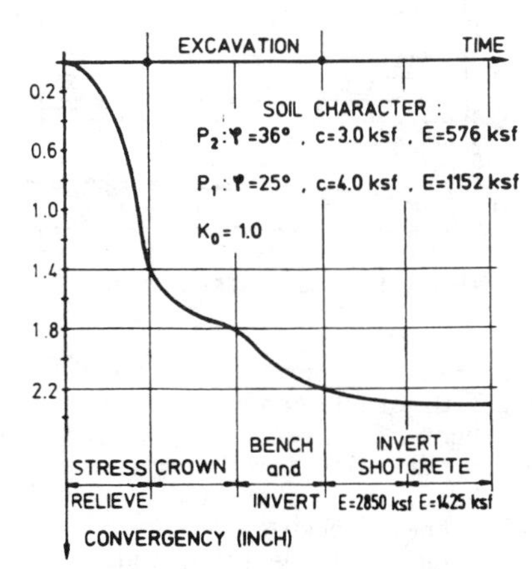

FIG. 1 FEM-DEFORMATION DEVELOPMENT OF METRO WASHINGTON SECTION E5

In Figure 1 the calculated development of the horizontal convergency is shown. Different construction stages are included, starting with the stress relieve in front of the tunnel face, followed by crown, bench and invert excavation and support. Finally the development of the modulus of elasticity of the shotcrete and its influence on deformation development is also shown.

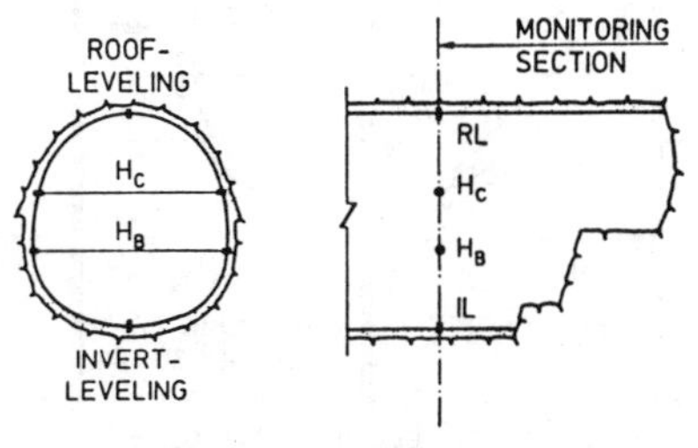

FIG. 2 RELEVANT DEFORMATION INDICATIONS

Usually several horizontal convergencies and roof and invert levelling are included in a minimum deformation program of a typical single track subway tunnel as shown in Fig.2 above.

FIG. 3A PRIMARY STRESS CONDITION

FIG. 3B FULL FACE STRESS RELIEV

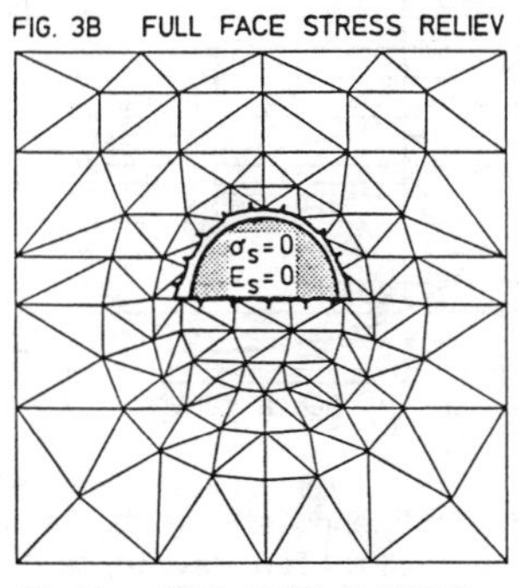

FIG. 3C INITIAL LINING IN CROWN

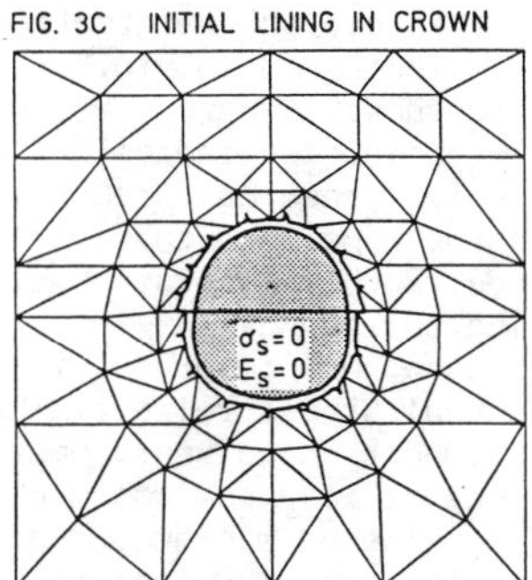

FIG.3D CLOSED RING INITIAL LINING

The results of deformation development as shown in Fig.1 are based on the two-dimensional FINITE-ELEMENT-ANALYSIS done with Program Final, as developed at INNSBRUCK UNIVERSITY by G.Swoboda.

In Fig.3 the basic load conditions investigated are shown. With respect to critical deformations, this loading cases do correspond not only with the minimum requirement, but also with the practical feasibility of measurement possibilities at the face.

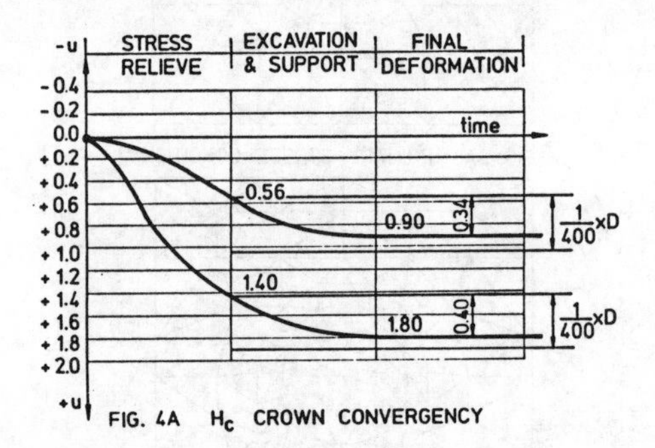

FIG. 4A H_C CROWN CONVERGENCY

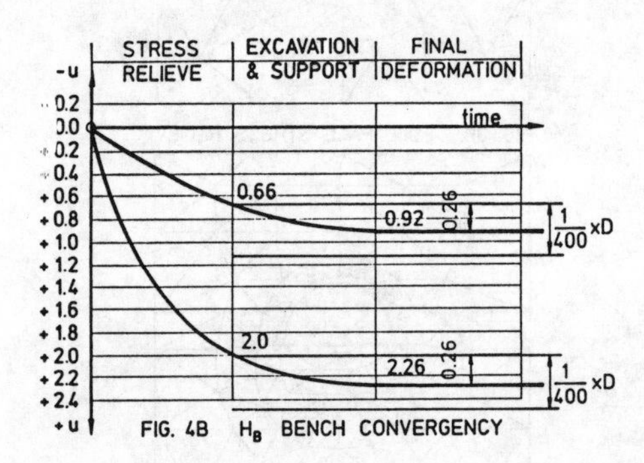

FIG. 4B H_B BENCH CONVERGENCY

Fig.4 and Fig.5 show the most important calculated deformation developments, related to the relevant deformation distances as indicated in Fig.2.
Every figure shows 2 curves, as the geomechanic investigation indicated two different soil mechanical properties, describing the upper and the lower limit of tolerance. Both values are basis for dimensioning the tunnels.
All curves show, how important it is, to know the actual stress relieve. Therefore it is very important to know, what the actual stress relieve is, which could be done e.g. by sliding micrometer measurements. Also the actual deformation will be considered critical, if the value for

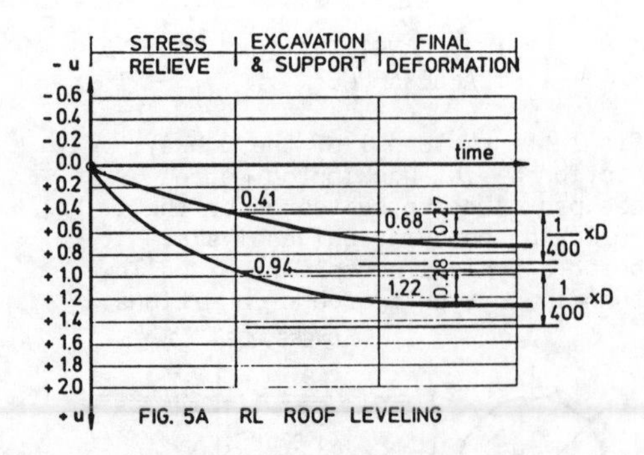

FIG. 5A RL ROOF LEVELING

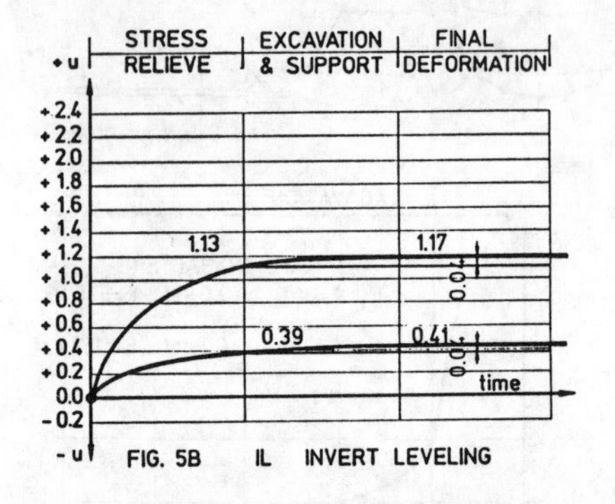

FIG. 5B IL INVERT LEVELING

the critical deformation will exceed 1/400 x D, starting from the end of the amount of stress relieve. This statement is not valid for deformations of the invert, which are usually significantly lower, because invert installation means closure of the ring, thus reducing and ending the deformation development.

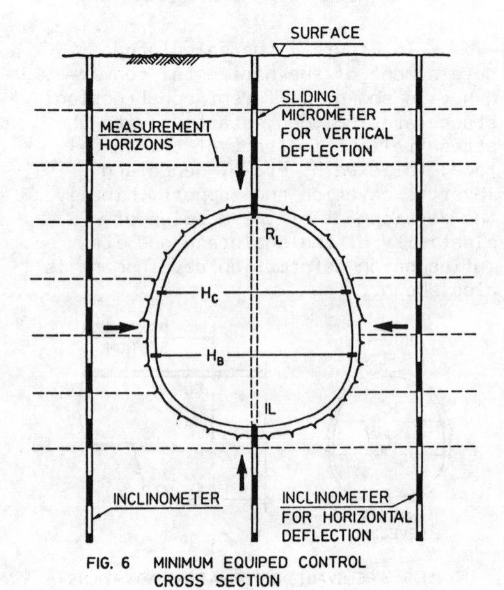

FIG. 6 MINIMUM EQUIPED CONTROL CROSS SECTION

In order to be able to verify curves of deformations, and to follow the calculated system, a measurement program as shown in Figure 6 has to be carried out on the basis of minimum equipped cross sections. In consequence of the requirement, to define critical deformations, a closed system is proposed, consisting of on the one hand FEM-calculations and design of linings, based on upper and lower limits of soil parameters, and on the other hand of relevant instruments to measure those values to be compared with the calculation.

5. RECOMMENDATIONS FOR SHALLOW SOFT GROUND TUNNELS

After analysing several construction projects under shallow soft ground conditions following recommendations are made. It has to be assumed that this recommendations will give the user the practical necessary indication. This indication is also connected with the hope, that more scientific work will be done on this subject which would also be related closely to underground construction works.

From the many possibilities to differ in shallow soft ground tunnelling 2 specific cases with respect to lining stiffness have been selected to give general recommendations for the limits of critical deformations.

In Figure 7A it is assumed, that as a result of the structural calculation the relation between thickness (lining stiffness) and tunnel diameter is in the range of 1/100. The thickness of lining does take into consideration soil mechanic parameters. The recommended value of critical deformation U_{crit} is now found by introducing overburden and tunnel diameter in the diagram.

Based on also structural calculations in this case the relation between lining thickness and tunnel diameter is in the range of 1/25. This two values, 1/100 and 1/25, represent usual upper and lower limits of shallow tunnels under soft ground conditions. It is therefore possible, to interpolate every other different relation between lining stiffness and diameter between this two limits.

6. CRITICAL REMARKS

Statements regarding critical deformations as made before are strictly related to shallow tunnels under soft ground or mixed face conditions. It is assumed that whatever construction concept will be used the ring closer time is short and in about one diameter distance from the face. NATM is able to fullfill this requirement as well as shield driven tunnels with in-situ concrete lining, while precast concrete lining is excluded.

Nowadays highly computerized tunnel computations allow considerations of almost every number of parameters. In spite of that fact experienced tunnel engineers are of unchanged importance, especially when considering the small range of lining dimensions of shallow tunnels.

The defined criterias for critical deformations seem to reduce the freedom of making decisions depending on actual behaviour of lining/unterground interaction.

On the other hand it satisfies the need of all design and construction engineers to have a simple useable definition where safety is ending and unsafety is starting. This means that all deformations developing within the proposed limits of critical deformations do not have any impact on design and construction. The principal ideas as presented in this paper when applied on the wide variety of soil/structure interactions could be extended systematically and developed in table works.

7. CONSLUSION

Critical deformations as defined above give a principal indication of safety or unsafety when calculating and measuring tunnel deformations. They are of special value when it has to be decided whether the experiences of a specialized tunnel consultant should be added, or not.

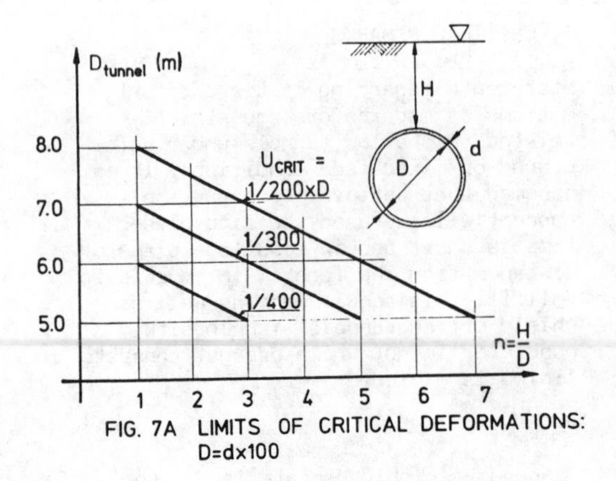

FIG. 7A LIMITS OF CRITICAL DEFORMATIONS: D=dx100

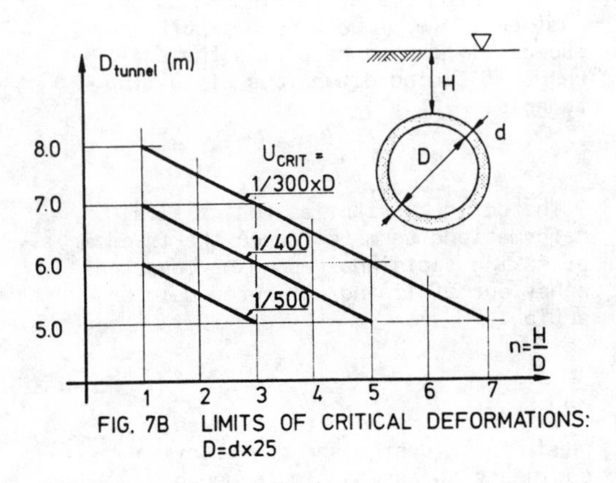

FIG. 7B LIMITS OF CRITICAL DEFORMATIONS: D=dx25

8. ACKNOWLEDGEMENT

The authors feel bound to express their thanks to Mr.VERNON GARRETT JR., Director of Engineering and Architecture of WASHINGTON METROPOLITAN AREA TRANSIT AUTHORITY (WMATA), Mr.MOHAMED IRSHAD, Chief Structural Engineer of DE LEUW CARTER COMPANY, Mr.ROBERT J.JENNY and Mr.PRAKASH M. DONDE, President and Vice President of JENNY ENGINEERING CORPORATION, for their impetus and activ cooperation within the work on designing subway section E-5 of WASHINGTON METRO GREENBELT ROUTE.

9. LITERATURE

SWOBODA, G., 1984. Final, Finite Element Analyse linearer und nicht linearer Strukturen (Finite Element Analysis of linear and non linear structures) Version 4-7. Universität Innsbruck.

HOFFMANN, H., 1975. STUVA-Tagung, Heft 19. Prognose und Kontrolle der Verformungen und Spannungen im Tunnelbau (Prognosis and control of stresses and strains in tunnelling) S.94-101

JENNY, R.J. et al., 1987. NATM design for soft ground Washington Metro Tunnels. Transportation Research Board, NRC, Washington/D.C.

WAGNER, H./HINKEL, W., Vienna, 1987. The New Austrian Tunnelling Method. ASCE, Tunnel/Underground Construction Seminar, New York.

WEBER, J./LAABMAYER, F., 1973. Messungen, Auswertungen und ihre Bedeutung (Measurements, evaluation and their relavance). Moderner Tunnelbau bei der Münchner U-Bahn (Modern tunnelling of Munich Subway) Springer Verlag (73-96)

WAGNER, H., 1981. Reportes de Asesoramiento Tecnico y Transferencia de Tecnologia del Nuevo Metodo Austriaco de Tuneles (Reports of technical assistance and transfer of technology of NATM) Unpublished.

Numerical Methods in Geomechanics (Innsbruck 1988), Swoboda (ed.)
© 1988 Balkema, Rotterdam. ISBN 90 6191 809 X

Finite element analysis of underground openings using space elasto-plastic theory

Y.R.Zheng
Air Force College of Engineering, Xian, People's Republic of China
Z.H.Xu & J.Chu
PLA Logistics Institute of Engineering and Technology, Chongqing, People's Republic of China
W.Wu
Shanxi Mechanical Institute, Xian, People's Republic of China

ABSTRACT: Based on strain space formulated elasto-plastic theory, a finite element method for underground openings has been proposed. The corresponding numerical program has also been compiled. From the results of the examples, it is shown that the proposed method are both simple and reliable and can be used in solving geotechnical problems.

1 INTRODUCTION

The problems involved in the geotechnical engineering are usually solved based on stress space plasticity theory. Because of the unstable property of rocks and soils, the strain space formulated plasticity turned out to be a more suitable theory in dealing with geotechnical problems than that in stress space. The studies (Chen and Zheng 1985, Chu 1985, Zheng et al 1986) show that the plasticity theory formulated in strain space has the following advantages over one in stress space:

1). The plastic theory in strain space has an extensive unity. It is valid for the full range of elastic-plastic deformation including, as a special case, the theory of elastic-perfectly plastic materials.

2). In finite element analyses for non-linear problems, the use of displacement method is consistent with the description in strain space. Thus, the numerical procedures may be simpler and more direct.

3). Hardening (or softening) parameters are easier to evaluate because of most of them are functions of strain components in plastic constitutive equations.

4). The plane strain problems may be more conveniently solved using the strain space formulated plasticity.

5). This theory has provided a theoritical basis for observational approaches which is often used in underground engineering by monitoring displacements. Moreover, the direct measurement in experiments is strain or displacement, therefore, the expression in strain space may avoid certain difficults in data-processing.

2 THE STRAIN SPACE FORMULATED PLASTIC THEORY

2.1 Comparisons of hardening or softening rules between two spaces

The yield surface in strain space may be expressed as follows by adopted the plastic internal variable formulation:

$$g(\epsilon, \epsilon^p, H_\alpha) = 0 \qquad (2-1)$$

where: ϵ, ϵ^p— tensors of strain and plastic strain respectively.
H_α— the hardening or softening parameters related to p.i.v.

In the same way, the yield surface in stress space may be given by:

$$f(\sigma, \sigma^p, H_\alpha) = 0 \qquad (2-2)$$

where: σ, σ^p— tensors of stress and plastic stress respectively.

The relationship between ϵ^p and σ^p is as follows:

$$\sigma^p = \mathbf{D}\epsilon^p$$

where: $\mathbf{D}$ — the elastic matrix.

Generally speaking, the yield surface in strain space and stress space can be transformed from one to another:

$$
\begin{aligned}
f(\sigma,\sigma^p,H_\alpha) &= f(\mathbf{D}\epsilon^e,\mathbf{D}\epsilon^p,H_\alpha)\\
&= f(\mathbf{D}(\epsilon-\epsilon^p),\mathbf{D}\epsilon^p,H_\alpha) \qquad (2-3)\\
&= g(\epsilon,\epsilon^p,H_\alpha)
\end{aligned}
$$

For geotechnical materials, formula (2-2) may be commonly specified as:

$$\bar{f}(\sigma-\alpha\sigma^p)-H(H_\alpha)=0 \qquad (2-4)$$

where:

α— a parameter related to p.i.v., $\alpha\sigma^p$ is a variable describing the kinematic hardening (or softening) rule.

$H(H_\alpha)$— the isotropic hardening (or softening) function.

The corresponding formulation of equation (2-4) in strain space is given by:

$$\bar{g}(\mathbf{D}\epsilon-(1+\alpha)\mathbf{D}\epsilon^p)-H(H_\alpha)=0 \qquad (2-5)$$

Now, using the equations (2-4) and (2-5), the moving patterns of the subsequent yield surface may be discussed as follows:

1). If $\alpha=0$, $H_\alpha=0$, then formulae (2-4) and (2-5) will change into:

$$\bar{f}(\sigma)=0$$
$$\bar{g}(\mathbf{D}\epsilon-\mathbf{D}\epsilon^p)=0$$

The above formulae imply that in this condition, both the size and position of yield surface in stress space are changed. Therefore a perfectly plastic material has been described. On the other hand, although the size of a yield surface in strain space does not change, the position of it changes with the developing of ϵ^p, that is, kinematic hardening happens.

2). If $\alpha=0$, $H_\alpha\neq0$, so, $H(H_\alpha)\neq0$, then

$$\bar{f}(\sigma)-H(H_\alpha)=0$$
$$\bar{g}(\mathbf{D}\epsilon-\mathbf{D}\epsilon^p)-H(H_\alpha)=0$$

In this case, the yield surface in stress space is in isotropic hardening with $H(H_\alpha)$. However, the yield surface in strain space is not only in isotropic hardening but also in kinematic hardening.

3). If $\alpha\neq0$, $H_\alpha=0$, then

$$\bar{f}(\sigma-\alpha\sigma^p)=0$$
$$\bar{g}(\mathbf{D}\epsilon-(1+\alpha)\mathbf{D}\epsilon^p)=0$$

The yield surfaces in both spaces are in kinematic hardening but the moving distances of yield surfaces are different in different spaces.

Some useful conclusions can be drawn from the above discussions. The sizes of yield surfaces in both surfaces are similar and the rule of isotropic hardening is the same but the rule of kinematic hardening is different. The yield surface in strain space is always in kinematic hardening as the ϵ^p developed. But the yield surface in stress space may be motionless or only in istropic hardening. In transforming a yield surface from stress space into strain space, the kinematic hardening features should be reflected in the yield function.

2.2 The loading-unloading criteria in strain space

The loading-unloading criteria in strain space can be unified written as (Qu and Yin 1981):

$$g=0,\ \hat{g}=\left(\frac{\partial g}{\partial\epsilon}\right)\dot{\epsilon}\begin{cases}<0, & \text{unloading;}\\ =0, & \text{neutral loading;}\\ >0, & \text{loading.}\end{cases} \qquad (2-6)$$

The above formula is suitable for hardening, softening and perfectly-plastic materials. In the loading condition, $g=0$, $\hat{g}>0$, corresponds to three material responses in stress space. In order to recognize whether the response is hardening, softening or perfectly-plastic, the following criteria have been proposed by the authors (Chu et al 1985):

$$\frac{1}{A_1}\left(\frac{\partial g}{\partial\epsilon}\right)\left(\frac{\partial g}{\partial\epsilon}\right)^t\begin{cases}<1, & \text{hardening;}\\ =1, & \text{perfectly-plastic;}\\ >1, & \text{softening.}\end{cases} \qquad (2-7)$$

where A_1 will be introduced later.

2.3 The constitutive equation in strain space

From generalized Hooke's law, the constitutive equation in strain space can be deduced as (Chu 1986):

$$\{d\sigma\}=[D_{ep}]\{d\epsilon\} \qquad (2-8a)$$

$$[D_{ep}]=[D]-\frac{[D]}{A_1}\left(\frac{\partial g}{\partial\epsilon}\right)\left(\frac{\partial g}{\partial\epsilon}\right)^t \qquad (2-8b)$$

where:

$$A_1=-\left[\left(\frac{\partial g}{\partial\epsilon^p}\right)^t\left(\frac{\partial g}{\partial\epsilon}\right)+A\right] \qquad (2-9)$$

A is a function determined by hardening parameters. For different kinds of hardening parameters H_α, the corresponding function A is given in Table 1.

Table 1. Relations between H_α and A

H_α	A
$W^p = \int \sigma^t d\epsilon^p = \int (d\epsilon^e)^t D d\epsilon^p$	$\{\frac{\partial g}{\partial W^p}\}^t \{d\epsilon^e\}^t [D] \{\frac{\partial g}{\partial \epsilon}\}$
$\epsilon_v^p = \epsilon_{ij}^p \delta_{ij}$	$\{\frac{\partial g}{\partial \epsilon_v^p}\}^t \{\delta_{ij}\} \{\frac{\partial g}{\partial \epsilon}\}$
$\bar{\gamma}^p = \int \sqrt{d\epsilon_{ij}^p d\epsilon_{ij}^p}$	$\{\frac{\partial g}{\partial \gamma^p}\}^t [\{\frac{\partial g}{\partial \epsilon}\}^t \{\frac{\partial g}{\partial \epsilon}\}]^{\frac{1}{2}} \{\frac{\partial g}{\partial \epsilon}\}$
ϵ_{ij}^p	$\{\frac{\partial g}{\partial \epsilon^p}\}^t \{\frac{\partial g}{\partial \epsilon}\}$
$(\epsilon_v^p, \bar{\gamma}^p)$ (For conventional tri-axial tests)	$\{\frac{\partial g}{\partial \epsilon_v^p}\}^t \{\delta_{ij}\} + \{\frac{\partial g}{\partial \bar{\gamma}^p}\}^t [\{\frac{\partial g}{\partial \epsilon}\}^t \{\frac{\partial g}{\partial \epsilon}\}]^{\frac{1}{2}}$
$(\epsilon_v^p, \bar{\gamma}^p)$ (For truely tri-axial tests)	$\{\frac{\partial g}{\partial \epsilon_v^p}\}^t \{\frac{\partial g}{\partial \epsilon_m}\} + \{\frac{\partial g}{\partial \bar{\gamma}^p}\}^t [(\frac{\partial g}{\partial \epsilon})^2 + (\frac{1}{q_\epsilon} \frac{\partial g}{\partial \theta_\epsilon^e})^2]^{\frac{1}{2}}$ or $\{\frac{\partial g}{\partial \epsilon_v^p}\}^t \{\frac{\partial g}{\partial \epsilon_m}\} + \{\frac{\partial g}{\partial \bar{\gamma}^p}\}^t \{\frac{\partial g}{\partial q_\epsilon}\} / \cos^2(\theta_\epsilon^p - \theta_\epsilon^e)$

3 NUMERICAL IMPLEMENTATION OF STRAIN SPACE FORMULATED PLASTICITY

3.1 The determination of yield function in strain space

Yield functions in strain space can be determined by experiments similar to locating the yield points in stress space. An alternative method is by transforming the yield function from stress space into strain space. Such transformation requires the relation between stress and elastic strain. By generalized Hooke's law, we have:

$$I_1 = 3KJ_{1e} \quad or \quad I_1 = 3K\epsilon_m^e \qquad (3-1a)$$
$$I_2' = (2G)^2 J_{2e}' \qquad (3-1b)$$
$$I_3' = (2G)^3 J_{3e}' \qquad (3-1c)$$
$$\theta_\sigma = \theta_\epsilon^e \qquad (3-1d)$$

where:

K, G— elastic bulk modulus and elastic shear modulus respectively.

ϵ_m^e— the mean elastic strain.

I_1, I_2', I_3'— the first stress, second and third deviatoric stress invariant respectively.

J_{1e}, J_{2e}', J_{3e}'— the first invariant of elastic strain, second and third invariant of deviatoric elastic strain respectively.

$\theta_\sigma, \theta_\epsilon^e$— the stress Lode angle and elastic strain Lode angle respectively.

Through formulae (3-1) the generalized yield function (Zheng and Chen 1984) in stress space can be transformed into strain space as follows. The generalized yield function in stress space is:

$$f = \bar{f} - H(H_\alpha) = 0 \qquad (3-2a)$$
$$\bar{f} = \beta\sigma_m^2 + \alpha_1\sigma_m - k + \bar{\sigma}_0^n \qquad (3-2b)$$

Transforming into strain space:

$$g = \bar{g} - H(H_\alpha) = 0 \qquad (3-3a)$$
$$\bar{g} = \beta'\epsilon_m^{e^2} + \alpha_1'\epsilon_m^e - k' + \bar{\epsilon}_0^n \qquad (3-3b)$$

where: $\sigma_m = (1/3)I_1$ is the mean stress, and

$$\bar{\sigma}_0 = \frac{\sqrt{I_2'}}{g_1(\theta_\sigma)}, \qquad \bar{\epsilon}_0 = \frac{\sqrt{J_{2e}'}}{g_2(\theta_\epsilon^e)}$$

$\alpha, \alpha', \beta, \beta', k, k', n$ are parameters to be determined. The relations between them are as follows:

$$g_1(\theta_\epsilon^e) = g_2(\theta_\sigma) \qquad (3-4a)$$

$$\beta' = \frac{(3K)^2}{(2G)^n}\beta \qquad (3-4b)$$

$$\alpha_1' = \frac{3K}{(2G)^n}\alpha_1 \qquad (3-4c)$$

$$k' = \frac{k}{(2G)^n} \qquad (3-4d)$$

Where the values of $\beta, \alpha, k, n, g_1(\theta)$ are related to the yield criteria that one selects. The relations between the coefficients and the selected yield functions are reported in Zheng and Chen (1984).

$g_1(\theta_\sigma), g_2(\theta_\epsilon^e)$ represent the variations of a yield surface in π-plane with θ_σ and θ_ϵ^e. In order to be more accurate, the following function is suggested by the authors (Zheng et al 1986):

$$g_1(\theta_\sigma) = \frac{2K}{(1+K) - (1-K)\sin 3\theta_\sigma + \alpha\cos^2 3\theta_\sigma}$$

the value of α is between $0.2 \sim 0.4$.

3.2 The calculation of elasto-plastic matrix

The steps to determine the elasto-plastic matrix of a given material are: (1) to select the hardening parameter H_α according to the material properties; (2) to specify the yield surface g based on experiments; and (3) to calculate the corresponding function A. Then the elasto-plastic matrix can be readily obtained.

The generalized yield function in strain space (3-3) can be more generally expressed as:

$$g = \bar{g} - H(H_\alpha) \qquad (3-5a)$$

$$\bar{g} = \bar{g}(\epsilon_m^e, \sqrt{J_{2e}'}, J_{3e}') \qquad (3-5b)$$

From the above equations, we have:

$$\left\{\frac{\partial g}{\partial \epsilon}\right\} = \left\{\frac{\partial \bar{g}}{\partial \epsilon}\right\} - \left\{\frac{\partial H}{\partial \epsilon}\right\}$$

$$\left\{\frac{\partial \bar{g}}{\partial \epsilon}\right\} = \beta_1\left\{\frac{\partial \epsilon_m^e}{\partial \epsilon}\right\} + \beta_2\left\{\frac{\partial \sqrt{J_{2e}'}}{\partial \epsilon}\right\} + \beta_3\left\{\frac{\partial J_{3e}'}{\partial \epsilon}\right\}$$

$$\left\{\frac{\partial H}{\partial \epsilon}\right\} = \left(\frac{\partial H}{\partial H_\alpha}\right)^t \left(\frac{\partial H_\alpha}{\partial \epsilon}\right) = \left(\frac{\partial H}{\partial H_\alpha}\right)^t h_1$$

where:

$$\beta_1 = \frac{\partial \bar{g}}{\partial \epsilon_m^e}, \qquad \beta_2 = \frac{\partial \bar{g}}{\partial \sqrt{J_{2e}'}}, \qquad \beta_3 = \frac{\partial \bar{g}}{\partial J_{3e}'},$$

and

$$h_1 = \left\{\frac{\partial H_\alpha}{\partial \epsilon}\right\}$$

β_i $(i = 1,2,3)$ and h_1 may be calculated according to the specifed $\bar{g}$ and H_α. For different yield function $\bar{g}$, the expression of β_i is given by Chen and Zheng (1985). From the definitions of J_{1e}, J_{2e}' and J_{3e}', we can get the following terms:

$$\left\{\frac{\partial \epsilon_m^e}{\partial \epsilon_m}\right\} = (1/3)[1 \quad 1 \quad 1 \quad 0 \quad 0 \quad 0]^t$$

$$\left\{\frac{\partial \sqrt{J_{2e}'}}{\partial \epsilon}\right\} = \frac{1}{2\sqrt{J_{2e}'}}[e_x^e \quad e_y^e \quad e_z^e \quad 2e_{xy}^e \quad 2e_{yz}^e \quad 2e_{xz}^e]^t$$

$$\left\{\frac{\partial J_{3e}'}{\partial \epsilon}\right\} = \Big[e_y^e e_z^e - e_{yz}^{e^2} \quad e_z^e e_x^e - e_{zx}^{e^2} \quad e_x^e e_y^e - e_{xy}^{e^2} \quad 2(e_{xy}^e e_{xz}^e - e_x^e e_{yz}^e) \quad 2(e_{yz}^e e_{yz}^e - e_y^e e_{xz}^e) \quad 2(e_{xz}^e e_{yz}^e - e_z^e e_{xy}^e)\Big]^t + (1/3)J_{2e}'[1 \quad 1 \quad 1 \quad 0 \quad 0 \quad 0]^t$$

and

$$\left\{\frac{\partial g}{\partial \epsilon^p}\right\} = \left\{\frac{\partial \bar{g}}{\partial \epsilon^p}\right\} - \left\{\frac{\partial H}{\partial \epsilon^p}\right\} \qquad (3-7)$$

Because ϵ_m^e, $\sqrt{J_{2e}'}$ and J_{3e}' are the functions of $\{\epsilon - \epsilon^e\}$, so:

$$\left\{\frac{\partial \bar{g}}{\epsilon^p}\right\} = -\left\{\frac{\partial \bar{g}}{\partial \epsilon}\right\}$$

Hence:

$$\left\{\frac{\partial g}{\partial \epsilon^p}\right\} = -\left\{\frac{\partial \bar{g}}{\partial \epsilon}\right\} - \left(\frac{\partial H}{\partial H_\alpha}\right)^t h_2$$

Where:

$$h_2 = \left\{\frac{\partial H_\alpha}{\partial \epsilon^p}\right\}$$

After getting $\left\{\frac{\partial g}{\partial \epsilon}\right\}$ and $\left\{\frac{\partial g}{\partial \epsilon^p}\right\}$, the constitutive equation can be obtained through equation (2-8).

3.3 The initial stress method in strain space

The initial stress method is widely used in solving elasto-plastic problems. Applying this method in strain space appears more convenient than in stress space.

The process of initial stress method in strain space is that suppose a strain increment $\{\Delta\epsilon_i\}$ at a certain loading level has been obtained, then the plastic stress increment can be calculated as $[D_p]\{\delta\epsilon_i\}$, which will be taken as the modified loading term and go on for the next iteration.

The modified loading term is :

$$\{\Delta P_0\} = \sum \int_v [B]^t[D_p]\{\Delta\epsilon_i\}dV$$

Where $[B]$ is the geometrical matrix.

The iterative formula is:

$$[K_0]\{\Delta U_i\}_0 = \{\Delta P_i\} + \{\Delta P_0\}_0 = \{\Delta P_i\}$$

Where $[K_0]$ is the initial stiffness matrix and $\{\Delta U_i\}_0$ is the displacement increment matrix.

In the proposed method, the plastic strain can be used as the criterion of convergence:

$$\{\Delta\epsilon_i^p\} = [D][D_p]\{\Delta\epsilon_i\}$$

The feature of the discussed method is that owing to the avoidence of stress increment in the calculation, this method is simpler and more convenient than that of stress space.

4 NUMERICAL EXAMPLES

According to the theory and numerical method presented in this paper, a finite element program is coded. Two examples are presented in the following.

The first example is a semi-circle-shapped underground opening in a soil mass. The discrete meshes of the crossed section is shown in Fig. 1. Four yield functions are used in the calculation. The results of Drucker-Prager yield function are illustrated in Fig. 2 and Fig. 3, in which the comparisons are made with the results from stress space program. From the figures it can be seen that the calculated displacement in the boundary of the opening is almost the same and the difference between the predicted plastic zone is relatively small.

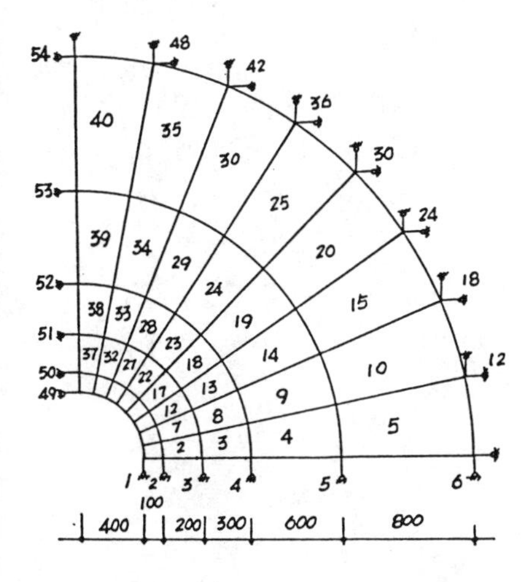

Figure 1. F.E. discrete meshes in cross section

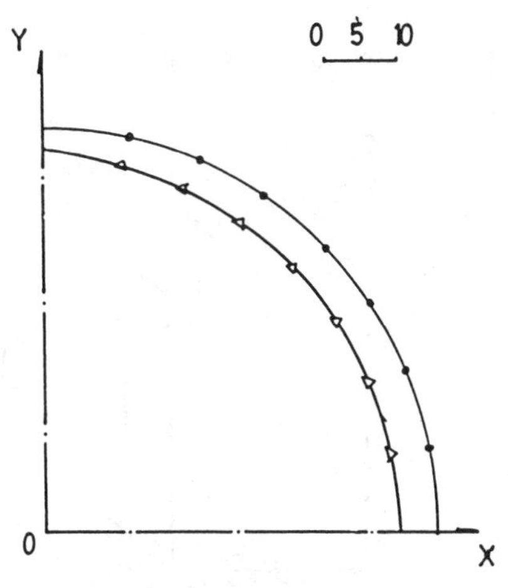

Figure 2. The radial displacement in boundary

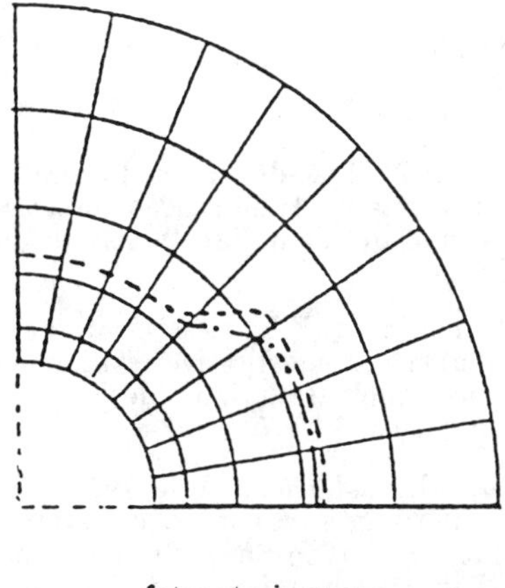

Figure 3. The plastic zone calculated

The second example is a rock cavern whose shape is depicted in Fig. 4. Five pairs of displacements in the side of the cavern have been measured after excavation. The vertical initial earth stress is obtained by back-analysis as $\sigma_y = 1.772kMa$, which is approximately equal to the embedded stress of the cavern in the depth of $45.0M$. The calculated parameter of lateral pressure is 0.79, nearly equal to the in-suit measured value.

Besides, the other works (Chang 1985, Yoder and Wihirley 1984) also indicate that a more accurate results will be expected if method in strain space is used instead of stress space.

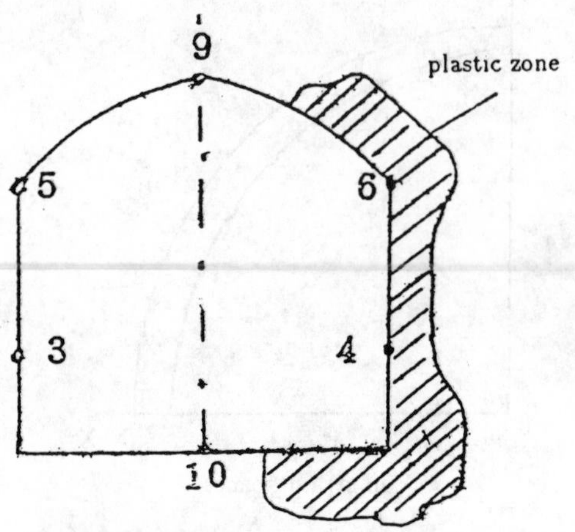

Figure 4. Illustration for example 2

REFERENCES

Chang, Z.X. 1985. Three-dimensional charateristics of concrete in strain space and stress distribution of arch dam. PhD thesis. Qinghua Univ. China.

Chen, C.A. and Zheng, Y.R. 1985. Geotechnical yield criterion and constitutive relations in strain space. Appl. Maths. and Mech. English Edition in China. 6:No.7.

Chu, J., Xu, Z.H. and Zheng, Y.R. 1985. The loading criteria in stress space and strain space. Proc. Symp. on Shear Strength and Consti. Relationships of Soils. China. (in Chinese).

Chu, J. 1986. The strain space formulated plasticity and its application in geomechanics. Master thesis. The PLA Logistics Institute of Engg. and Tech. China. (in Chinese).

Qu, S.N. and Yin, Y.Q. 1981. On the postulates of Drucker and Ily'ushin in plasticity. Chinese J. Mech. No.5. (in Chinese).

Yoder, P.J. and Whirley, R.G. 1984. On the numerical implementation of elasto-plastic models. J. Appl. Mech. 51:283-288.

Zheng, Y.R. and Chen, C.A. 1984. Yield criterion and constitutive relations for perfect-plastic rock-soils. Chinese J. of Geot. Engg. No.5, (in Chinese).

Zheng, R.Y., Chu, J. and Xu, Z.H. 1986. The strain space formulated elasto-plastic theory and its finite element implementation. Computers and Geomechanics. 2:373-388.

Numerical Methods in Geomechanics (Innsbruck 1988), Swoboda (ed.)
© 1988 Balkema, Rotterdam. ISBN 90 6191 809 X

Zürich Mass Rapid Transit line – Soil mechanical analysis for soft ground tunneling by application of the freezing method

Klaus Mettier
Electrowatt Engineering Services Ltd, Zurich, Switzerland

ABSTRACT: On the example of a completed tunnel construction it is shown how with the aid of numerical analysis data for the design and the implementation of a project can be obtained.

Preliminary calculations performed with a simple Spring-model yielded general indications for the selection of a suitable profile.

An FE-analysis furnished basic data for the design. Thereby, the three-dimensionally and chronologically variable interaction between tunnel structure and surrounding ground was simulated through the introduction of individual steps of load or stress changes - corresponding to the individual construction phases.

Deviations between the modelled and the real conditions may occur. Therefore, the results of the numerical analysis must be interpreted, i.e. must be subjected to critical evaluation. This is shown here on the example of the problem "Surface settlements".

1. INTRODUCTION

Numerical analysis are an important tool for the design and the implementation of subsurface structures. Under investigation is a model recreating the reality. On one hand basic aspects can be cleared and on the other hand important basic data for the design are generated.

In the application of numerical models for subsurface construction projects care must be taken of the complicated three-dimensionally and chronologically variable interactions between the tunnel structure and its surrounding ground. Through sensible application, this may be achieved without problems with a two-dimensional model.

Further, it shall be noted that for the model the often complex geotechnical reality must be idealized.

To obtain relevant design data, basic requirements are experience in the tunnelling process and profound knowledge of the behaviour of the particular ground.

On the following example of the "Rämistrasse-Tunnel" it is attempted to explain the individual aspects.

2. THE RAEMISTRASSE-TUNNEL PROJECT

2.1 General Overview

Within the framework of the implementation of a new MRT railway system, several tunnels in downtown Zurich are being constructed. One of the most interesting sectors is the so called Rämistrasse Tunnel.

For a distance of 65 meters it passes underneath the very busy Rämistrasse and a following park area.

The overburden in the Rämistrasse measures a scant 2.5 to 3.5 m; it reaches 8 to 12 m in the park area. The width of the excavation varies between 16 and 21 m and the height between 11 and 13.5 m. The cross section of 150 m2 below Rämistrasse increases to 220 m2 at the eastern end of the tunnel (Fig. 1 to Fig. 4).

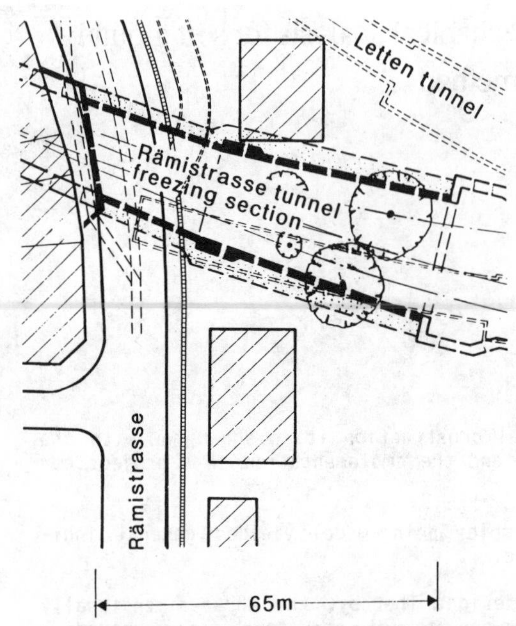

Fig. 1: Situation of Rämistrasse tunnel section

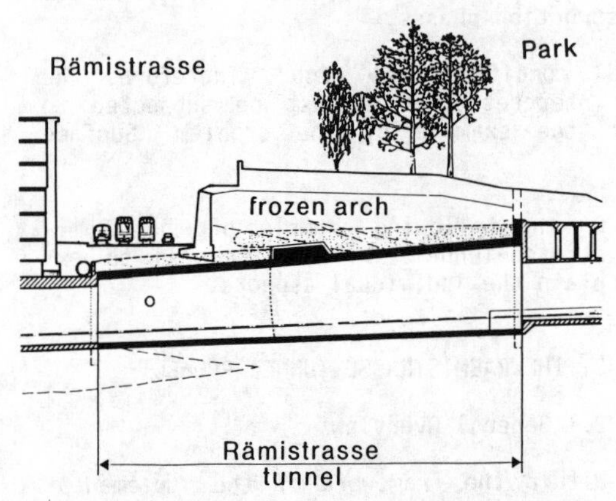

Fig. 2: Longitudinal section Rämistrasse-tunnel section

2.2 Geology

To a large degree the subsoil consists of glacially prestressed moraine material and lake deposits. The silts and sands are poor in clay and have inclusions and layers of course sand and gravel. The strata may be level or inclined due to glacial pressure. Boulders and erratic blocks are found in the mainly fine grained soil. Standing groundwater is present in the lower half of the profile and poses a risk of loss of soil stability. In the street area artificial (man made) fill material is also encountered.

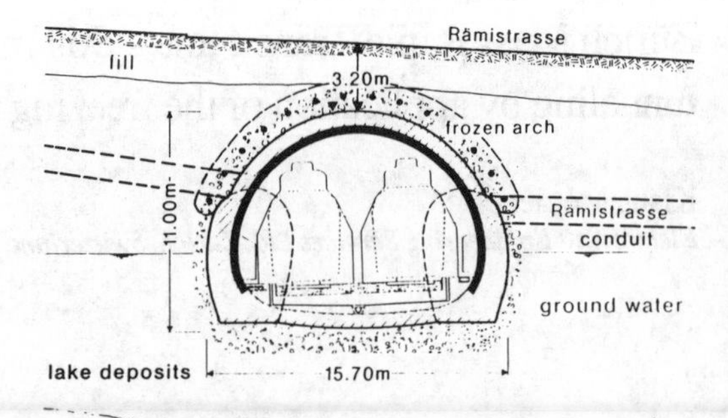

Fig. 3: Cross section in Rämistrasse

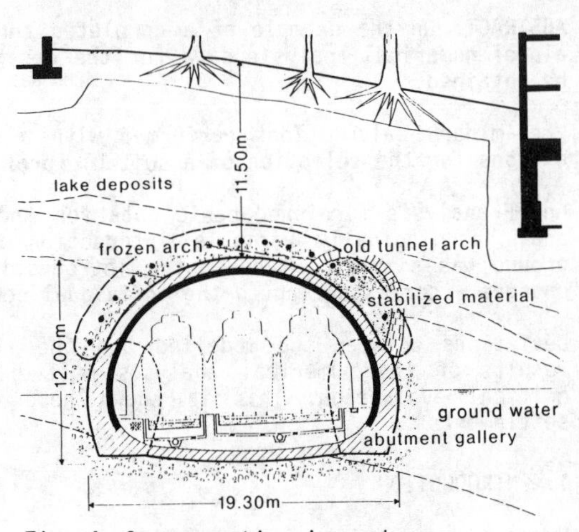

Fig. 4: Cross section in park area

2.3 Further Parameters

Additional parameters and interferences that were to be taken into account:

- an active railway tunnel in close proximity
- a disused railway tunnel constructed 100 years ago intersects the new tunnel arch; this old tunnel was filled with cement-stabilized material before commencement of the work
- two tall old buildings standing next to the tunnel
- a 10 m high retaining wall bordering the Rämistrasse must be passed under, immediately underneath the entrance to an underground car parking facility
- a sewer main crosses the tunnel profile in the Rämistrasse

2.4 Construction Method and Procedure

Considering the tunnel features, such as:

- the exceptionally large cross section
- the shallow depth of the cover
- the underpassing of a well travelled thoroughfare with heavy road and tramway traffic
- the complexity of the ambient soil conditions
- the parameters and adverse conditions mentioned above

led to the choice of the construction method and procedure that is based on the employment of the freezing method. In short, the construction concept can be summarized as follows:

- as a first step, two abutment galleries are excavated, lined and then concreted
- formation of a frozen arch around the future cavity of the heading (temporary strengthening of the ground area)
- the heading is excavated under the protection of this frozen arch. The lining of thicknesses of between 40 cm and 60 cm is placed immediately afterwards
- excavating the bench and placing the invert arch in 4 meter stages.

3. NUMERICAL ANALYSIS

3.1 General Overview

The complex reality as described was to be modelled such, that the chosen model yielded reliable, i.e. workable results.

Besides the structural and stability analysis, a forecast concerning the tunnel's possible adverse effects on the environment, e.g. settlings on the surface, especially with respects to the buildings and the influence on the existing railway tunnel remaining in operation.

The numerical investigations were divided into two parts:

(a) Preliminary calculations employing the Spring-Model; diverse profile shapes where investigated and first indications as to lining thicknesses were gained.

(b) FE-analysis for the definitive construction design, under consideration of the individual steps of the chosen construction method.

3.2 Preliminary Calculations

The questions to be answered were:

- most suitable shape of the tunnel profile under the given circumstances and first indications concerning the lining thicknesses required.

Considering the design status at the time, the simple Spring-model was used to investigate two possible profile shapes:

- Model I : Egg-shaped profile without abutment foundation
- Model II: Horseshoe profile with abutment fundation.

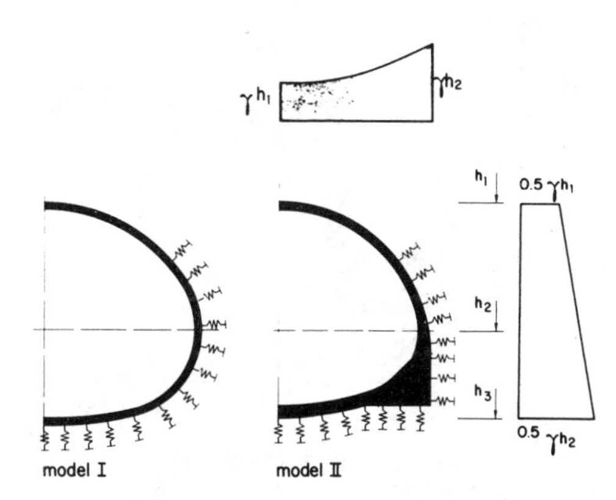

Fig. 5: Models for preliminary calculations

The calculations made evident that in Model I the supporting effect of the ground of the lateral areas must contribute to a great extent to the equilibrium, therefore, the practical results would depend heavily on the quality of the ground. Further, considerable deformations and as a consequence large section forces were to be expected.

On the other hand, the results of Model II showed that the supporting function of the ground need to be less, i.e. the quality of the ground is of less importance. The deformations and the section forces are correspondingly less.

Since a variety of ground material qualities had to be taken into account, a profile according to Model II was pursued for the development of the design. Of course, constructional considerations were also instrumental for this decision.

Some of the results of these investigations are shown in Fig. 6.

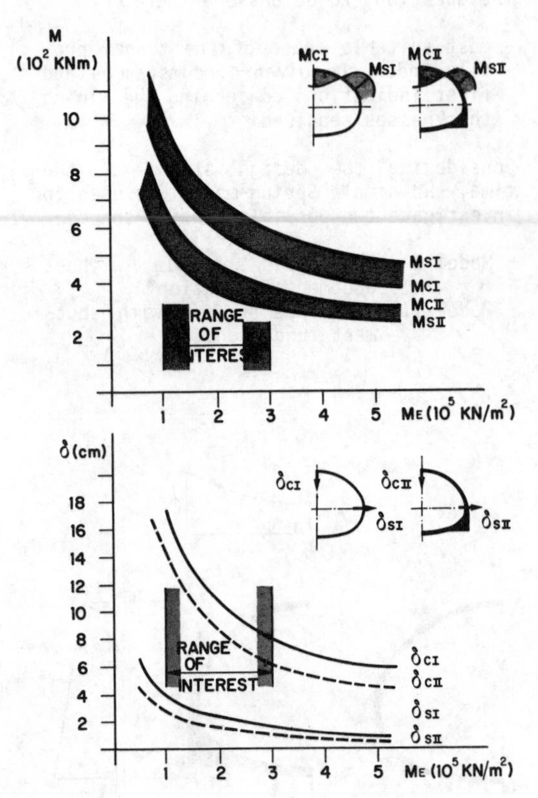

Fig. 6: Results of the preliminary calculations

3.3 FE-Analysis for the Detail Design

Computations were carried out with the FE-program, system Staub (1) to obtain a quantitative prediction of the behaviour of the soil and the frozen arch during the tunnelling and to provide a basis for:

- the structural design of the frozen mass and the tunnel arch
- stability analysis of the soil in the different excavation stages
- estimation of the influence on the active rail tunnel nearby (Letten tunnel)
- prognoses of settlings on the surface.

A two-dimensional system comprising tunnel, frozen arch and soil was examined.

For the modelling of the tunnel surroundings, the partially very complicated reality was simplified; for example, the filled-in old tunnel was disregarded in the model.

Observations during construction and after its completion confirmed the validity of the simplifications introduced.

For the obtainment of most reliable data the question posed was: how can the very complex three-dimensional and chronological construction progress and the accompanying change in ground stress be realistically derived from a twodimensional model?

The ultimately selected procedure is based on the experiences gained from the construction of previous tunnels with similar ground conditions. The basic idea can be summarized thus:

- in practice, three-dimensionally as well as chronologically, the construction progress is a fluid process from the condition of the undisturbed ground (primary condition, t = 0) up to the condition of the completed tunnel structure inclusive all after effects ($t = \infty$);
- for calculating, the work sequence was subdivided into 6 phases, whereby the end of each phase was numerically analyzed;
- steps in load or stress changes (estimated, i.e. derived from experience) were introduced accordingly.

The calculation process is summarized in Fig. 7. "Run 1", that is not shown stands for the calculation of the primary condition (undisturbed ground).

At this point be it mentioned that - independent of the FE-model employed - the determination of the steps for these load or stress changes has a decisive influence on the results. Realistic, i.e. valid results are, therefore, only achieved if at this point the pertinent experience is applied.

run 2: abutment galleries, pre-relief of the soil	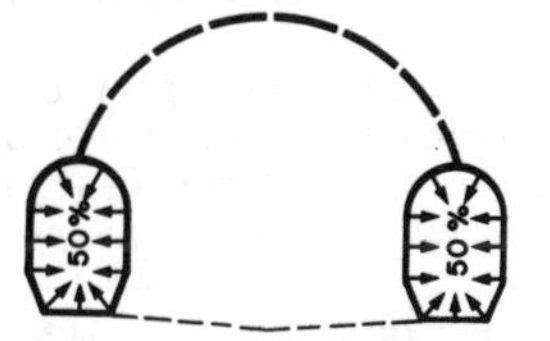 50 % 50 %	- excavation of the abutment galleries . the boundary of the cavities is loaded with 50 % of the stresses (nodal forces or "loads") of run 1.
run 3: abutment galleries, main loading of the galleries	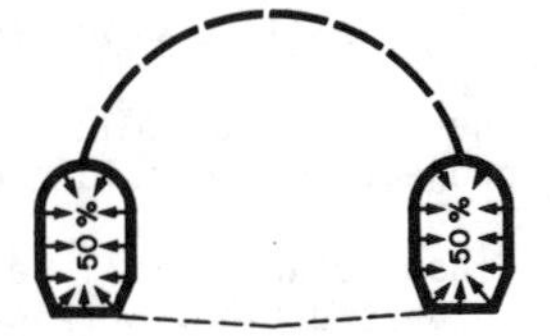 50 % 50 %	- lining of the abutment galleries . loading the boundary of the galleries with the remaining 50 % of the "loads" from run 1
run 4: heading, pre-relief of the soil	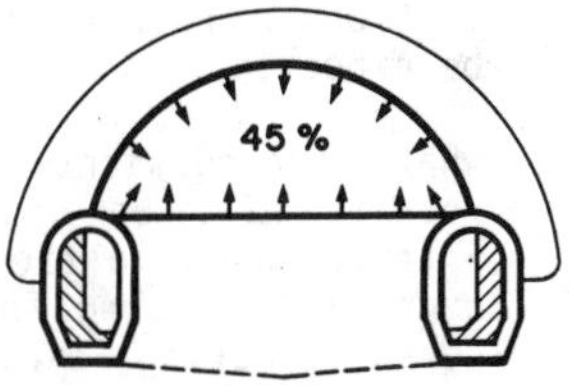 45 %	- concreting of the abutments - formation of the frozen arch - excavation of the heading . loading the boundary of the heading with 45 % of the "loads" from run 3
run 5: heading, main loading	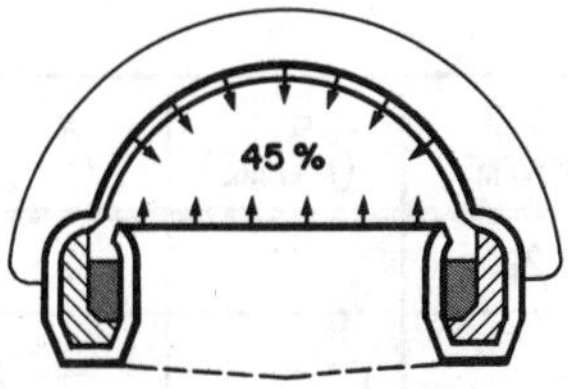 45 %	- lining of the heading - cut the lining of the galleries - fill the galleries with excavated material . loading the boundary of the heading with 45 % of the "loads" from run 3
run 6: heading, thawing of the frozen arch	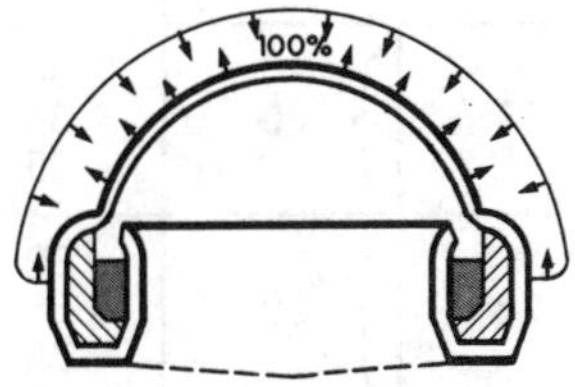 100%	- thawing of the frozen arch (change of ground parameters) . loading the boundary of the frozen arch with 100 % of the "loads" from run 5
run 7: final stage	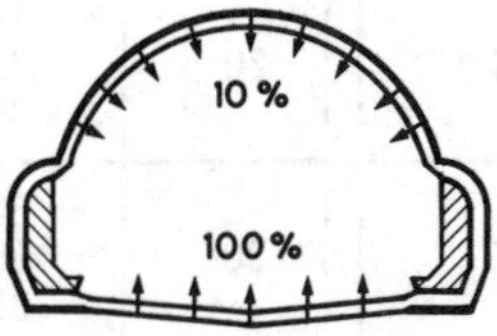 10 % 100 %	- excavation of the bench - placing of the invert arch . loading the boundary of the floor with 100 % of the "loads" from run 6, and the boundary of the heading with 10 % of run 3

Fig. 7: Calculation process of FE-analysis

In the case of the "Rämistrasse-Tunnel" two cross sections were evaluated by calculations; one cross section in the Rämistrasse and one cross section in the park area, near the eastern portal.

The model for the latter is shown in Fig. 8.

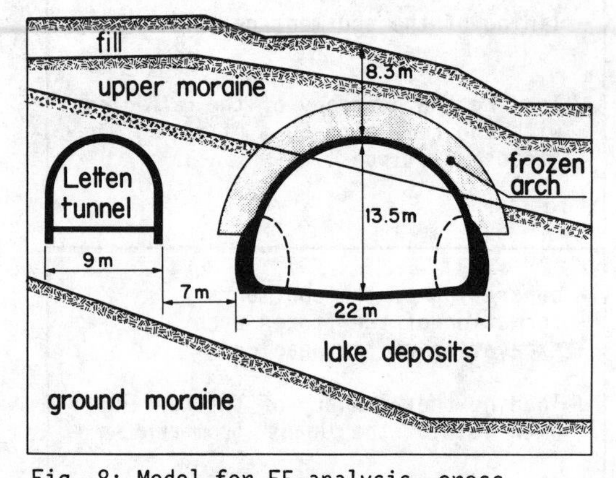

Fig. 8: Model for FE-analysis, cross section in park area

Several combinations of ground parameters were examined (Table 1).

As mentioned previously, the numerical investigations yield basic data and indications for the design as well as for its execution.

However, how are the terms "basic data" and "indications" to be interpreted? For sure not the uncritical application of the results. Rather, the individual results shall be weighted and analyzed.

The following points for the Rämistrasse project were considered:

- how accurate will reality (execution of construction, work phases, etc.) be reflected in the idealized numerical model, e.g. in which direction must deviations caused by this idealization to be expected?

- what repercussions on the individual results will be caused by additional interferences in the ground structure, such as freezing and thawing procedures, grouting, etc.?

Tab. 1: Ground parameters

	Calculation No.	ME (KN/m2)	γ (KN/m3)	c (KN/m2)	Φ (°)
Fill	1 ÷ 3	20'000	20	0	25
Upper Moraine	1 2 3	100'000 75'000 100'000	23	20	32 35 35
Lake Deposits	1 2 3	150'000 75'000 150'000	23	20	30 35 35
Ground Moraine	1 2 3	200'000	23	20	30 35 35
Frozen Soil	1 ÷ 3	250'000	23	750	0
Thawed Soil, Moraine	1 2 3	50'000 37'500 50'000	23	20	32 35 35
Thawed Soil, Lake Deposits	1 2 3	75'000 37'500 75'000	23	20	30 35 35
Pressure λo, All Types of Soil	1 2 3	0.6 0.6 0.4			

- what influence will be exerted by the disregarded disturbance factors or irregularities in the ground structure (old filled-in tunnel; old utility lines, etc.)?

Only the most carefully weighted answers to these questions, e.g. the facits derived therefrom, will lead to useful, relevant values for the project.

The example "Surface settlements" - it would go too far if all the voluminous data be reproduced here - shall demonstrate how the results from the FE-analysis have been applied in the project.

Before begin of construction it was important to furnish a prediction as to surface settlements. The endangering of nearby structures (buildings, retaining wall, tramway tracks) was to be assessed.

Basis for this prognosis were the results of the FE-calculations. The settlements to be expected were determined based on the weighted evaluation of the model (Fig. 9):

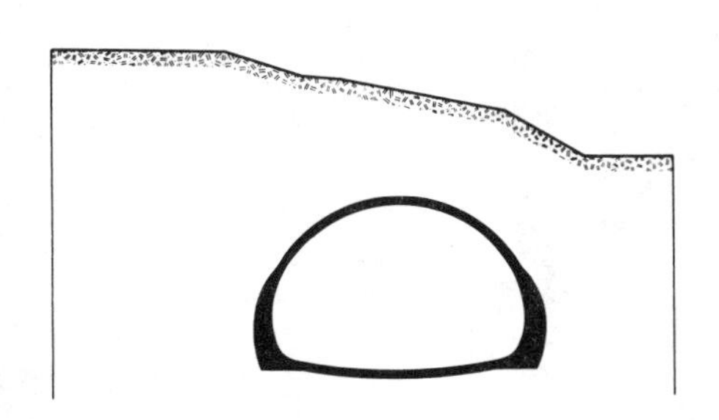

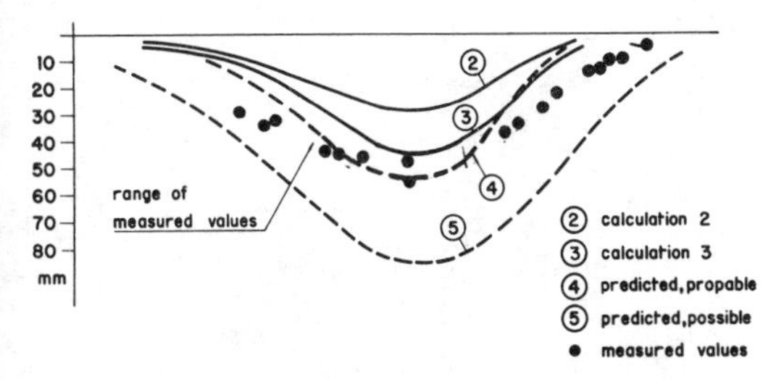

Fig. 9: Surface settlements: calculated, predicted and measured values

- the probable settlements (200 % of the settlements from calculation 2).

- and as an extreme case, the possible settlements (200 % of the settlements from calculation 3, whereby a widening of the settling through by 50 % was considered).

  Comparing the values with the actual measurements of the settlements showed that the chosen method (or the evaluation of the model) was very reasonable.

4. CONCLUDING REMARKS

The numerical analysis shall, however, always be performed on an idealized model in which the reality will be more or less streamlined. Therefore, the results must undergo a critical assessment before they can be introduced into the design.

A realistic integration of the steps of load or stress changes into the model is of equal importance.

Measurements and field observations during and after construction are important complementary elements. These allow constant comparison of the original assumptions and the expected consequences with actual data gained. The frequent checks also reveal correspondence with or deviations from the modelled conditions. As a further benefit a data base will be available for future projects having similar geotechnical characteristics.

P.S. The construction of this demanding tunnel section was successfully carried out from 1985 to 1986.

REFERENCES

(1) K. Kovari, H. Hagedorn, P. Fritz: Parametric studies as aid in tunnelling, paragraph 1.1: the program system STAUB/STAUPP. Proc. of second int. conf. on numerical methods in geomechanics, volume II, Virginia, 1976.

Numerical Methods in Geomechanics (Innsbruck 1988), Swoboda (ed.)
© 1988 Balkema, Rotterdam. ISBN 90 6191 809 X

Influence of the heading face and a two-dimensional calculation model of tunnel linings

V.H.Vassilev & T.N.Hristov
Institute of Water Problems, Sofia, Bulgaria

ABSTRACT: The paper describes a two-dimensional model based on FEM by means of which stresses in tunnel linings can be calculated taking into account the influence of the heading face. The model belongs to the so called "stiffness reduction models". At a previous calculation stage, the realization of the preliminary elastic deformations in the area of the heading face is imitated by reducing the value of Young's modulusof the excavated body (tunnel) from the rock massif. In order to calculate this lovered value, we should know only Young's modulus and Poisson's ratios of the rock massif. Some grafhs and a formula have been obtained for the practical application of the model. A comparison has been made with the three-dimensional FEM analysis.

1 INTRODUCTION

The construction of tunnels is a process of carrying out two successive basic operations, i.e excavating part of the rock massif and builging up the tunnel lining. As the tunnel lining is built some time after the excavation, at a certain distance from the tunnel face, it is obvious that it will take up only some loading of the above lying earth layers. This fact cannot be accounted for directly in the two-dimensional FEM analysis. It is for that reason that the spatial bearing effect of the rock massif and the preliminary deformations in the area of the tunnel face are often neglected in calculating the stresses in tunnel linings.

The best possible solution to the problem can be reached by using three-dimensional analysis taking into consideration the successive building of the structure. This, however, is not justified from economic point of view and is difficult to carry out in practice alonh the whole tunnel. A number of two-dimensional models were therefore proposed during the past few years (3,5,6,1).

Two are the basic ways to express spatial-temporary processes in the area of the tunnel face in these models:
a) presenting the loading on the lining as decreasing function in time (load reduction models); b) presenting the stiffness of the excavated body from the rock massif as decreasing function in time (stiffness reduction models).

A model has been worked out in this paper using the second approach to solve the problem.

2 CALCULATION MODEL

The experimental (8) and numerical three-dimensional analyses (2,4,7) have shown that the augmentation of the stresses in the tunnel lining rapidly attenuate with the advance of excavation when moving away from the heading face. The stresses reach a certain value at that, which is smaller compared to that obtained by two-dimensional FEM analysis.

A model is presented here, the basic points of which have been shown elsewhere (1,6). The model allows an easy calculation of these stresses using the two-dimensional FEM analysis.

Figure 1 shows the succession of making the calculations.

At the first stage the excavation area is covered with finite elements with module of elasticity Eo (figure 1b). The value of Eo should de such as to allow given vertical displacement Ua=Uo of point A. At the second calculation stage (figure 1c), the lining is introduced with its actual geometrical dimensions and stiffness, the stresses in it being in proportion to displacement U2 (figure 1a).

For this purpose, the following two problems must be solved:

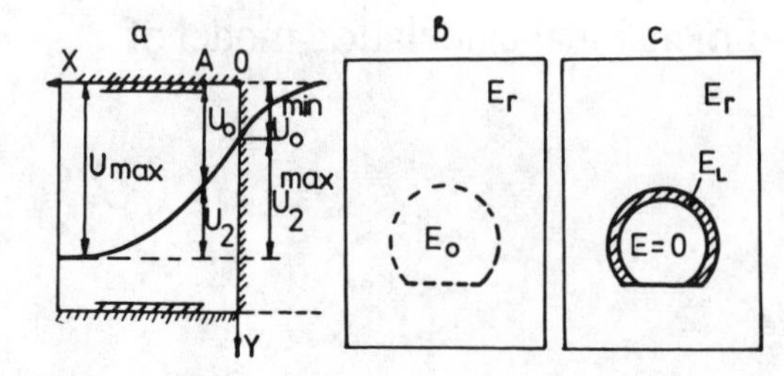

Figure 1 Stages of calculation

a) to determine the preliminary deformations in the area of the heading face,
b) to determine the function Uo/Umax= f (Eo/Er).

2.1 Calculating the preliminary deformations in the area of the heading face

As a suitable measure here for the value of the preliminary deformations, the ratio Uo/Umax has been accepted, where Uo=U (X=Xa), Umax=U(X=X∞).This ratio can be calculated using natural measurements or the three-dimensional analysis. In this case, the three-dimensional FEM analysis has been used, the ratio m=Uo/Umax being calculated for four shapes (figure 2) of the cross section.

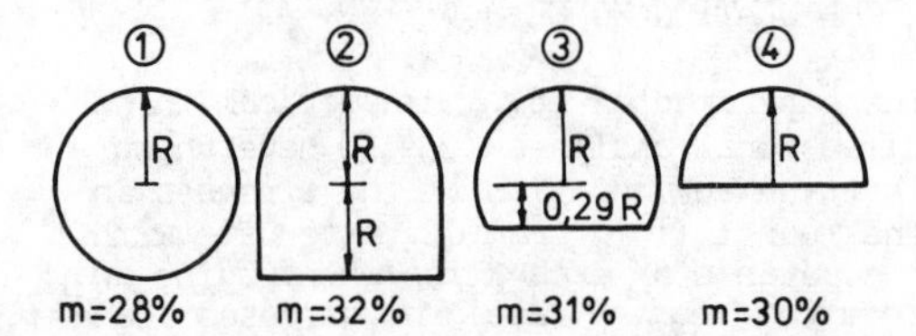

Figure 2 Cross-section's shapes of tunnel

The value of m shown in figure 2 refer to the cross-section vhich is in close proximity to the tunnel face, that is X=0.It should de noted that the influence of the shape of cross-section has no effect whatever at a distance smaller than 1R.

2.2 Calculating of Uo/Umax=f (Eo/Er,μ_r)

The ratio Uo/Umax as function of Eo module has been calculated for the four shapes of the cross-section shown in figure 2. This has been done by using two-dimensional FEM analysis for value of Poisson's ratio 0÷.45. The grafhs obtained to calculate Uo/Umax= f (Eo/Er) are given in figures 3,4,5 and 6.

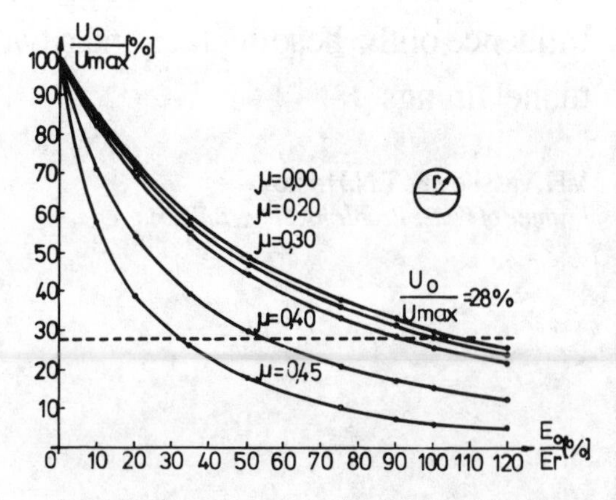

Figure 3 Function Uo/Umax=f (Eo/Er)

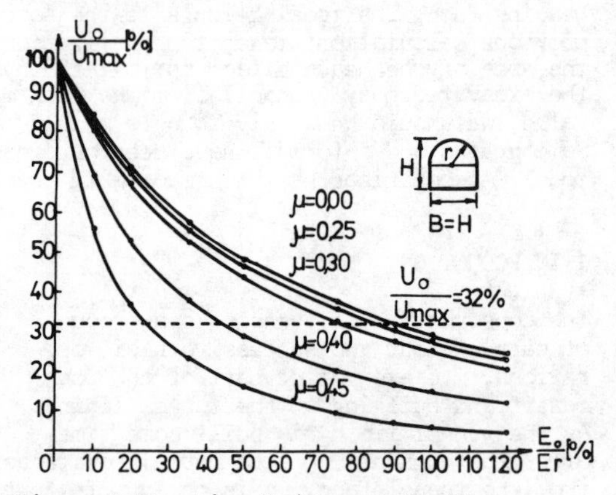

Figure 4 Function Uo/Umax=f (Eo/Er)

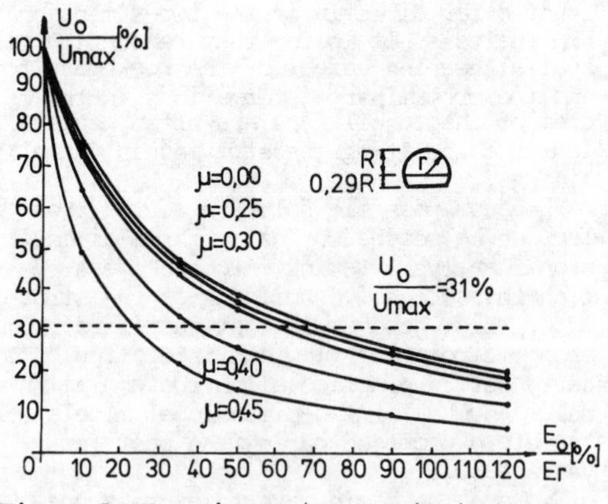

Figure 5 Function Uo/Umax=f (Eo/Er)

If the lining is built close to the heading face,stresses will appear in it,corresponding to U2 max deformations,i.2. stresses with maximal values $\{\sigma\}$ max.Practically

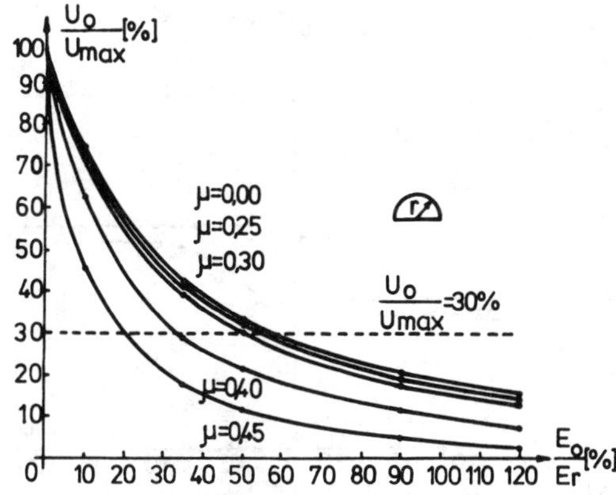

Figure 6 Function Uo/Umax=f (Eo/Er)

the lining is built at some distance from the tunnel face, so that the stresses $\{\sigma\}$ in it will be smaller than $\{\sigma\}$ max.

Let us accept (for safety's sake) that the lining will be built at a distance X= Xa=0 from the heading face. Then:

$$Uo=Umin \tag{1}$$

$$Uo/Umax=mmin=const. \tag{2}$$

If we now enter the value of mmin=const. for each shape of the cross-section on the ordinate axis of each of the figures 3,4, 5 and 6 for a given value of Poisson's ratio of the rock massif, we can calculate the value n=Eo/Er, vhere Er is Young's modulus of the rock massif. We thus obtain the function n=f (μ_r) which can be approximated to a polynomial raised to the fourt power with sufficcient accuracy:

$$n=a_0+a_1\mu_r+a_2\mu_r^2+a_3\mu_r^3+a_4\mu_r^4 \tag{3}$$

The coefficients ao ÷ a4 have been calculated using the least sguares method, and their values for the investigated shapes of the cross-section are given in figure 7.

The preliminary work in applying the two-dimensional model can thus reduosced to following: on the basis of the well-known Poisson's ratios μ_r and Young's modulus Er of the rock massif, n is determined from the graphs in fig. 3,4,5,6 or from formula 3. Then the value of Eo is calculated:

$$Eo = nEr. \tag{4}$$

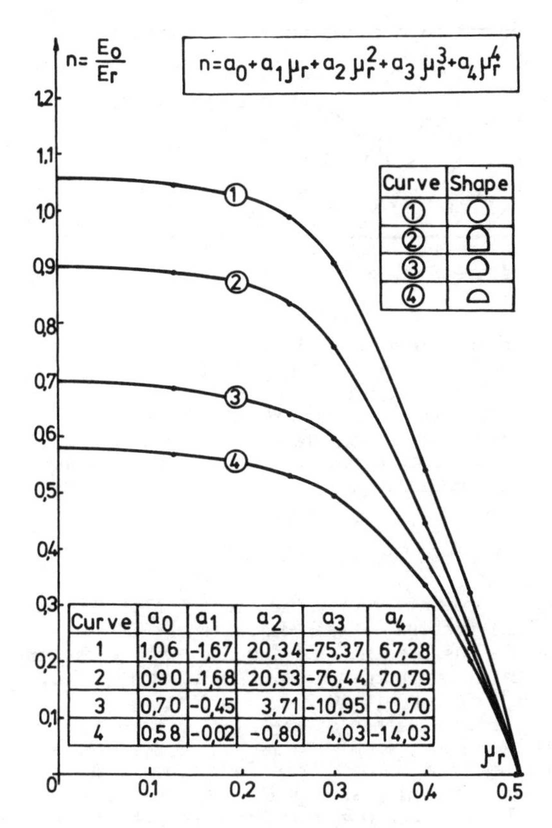

Curve	a_0	a_1	a_2	a_3	a_4
1	1,06	-1,67	20,34	-75,37	67,28
2	0,90	-1,68	20,53	-76,44	70,79
3	0,70	-0,45	3,71	-10,95	-0,70
4	0,58	-0,02	-0,80	4,03	-14,03

Figure 7 Function n=Eo/Er=f (μ_r)

After that the stresses in the tunnel lining are calculated taking into account the preliminary elastic deformations in the area of the tunnel according to the schemes in figures 1b, 1c using two-dimensional FEM analysis.

The model presented above has been created provided the rock massif is an isotropic medium. It also allows, however, to approximately calculate the anisotropic stratification of the rock massif, parallel to the longitudinal axis of the tunnel. For this purpose, it is necessary to calculate the preliminary elastic deformations in the area of the tunnel face into cosideration the anisotropy. This is done when the finite elements covering the light aperture of the tunnel are accepted to be transversal-isotropic medium with two modulus of elasticity Er1 and Er2, and two Poisson's ratios μ_{r1} and μ_{r2}, the plane of isotropy having the same angle of fall as in the surrounding rock massif (figure 8).

The values of Eo1 and Eo2 are calculated using formula 4:

$$Eo1 = n1.Er1 \tag{5}$$

$$Eo2 = n2.Er2. \tag{6}$$

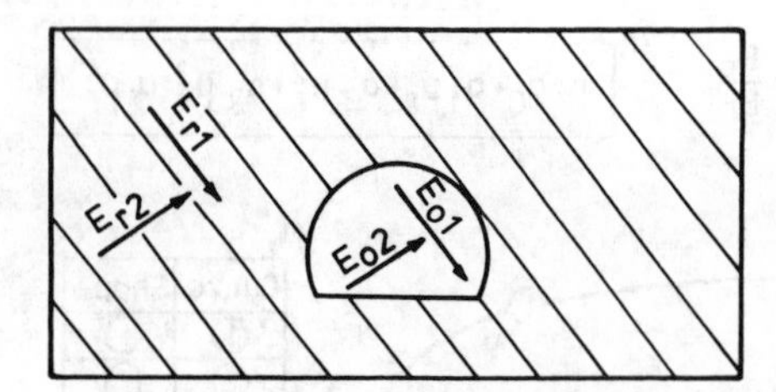

Figure 8 Case of transversal isotropy

On the other hand, n1 and n2 are obtained when in formula 3 instead of μ_r, μ_{r1} and μ_{r2} are replaced in succession.

Thus the stresses in the lining at the next calculation stage are determined after taking into account the anisotropy of the rock massif.

3 VERIFICATION OF THE MODEL

In order to verify the precision of the model proposed, three-dimensional FEM analysis has been made of a tunnel with diameter D=6m, overburden H=50m, Er=700 Mpa, μr=0.3, E_L=25000 Mpa, μ_L=0.20, d_L= 0.50m.

Five advances of the heading fase have been made, each with length of 2.0 m. After the fourt advance, the increased stresses in the lining are egual to 0. The stresses σ_z have been compared for 6 points of the lining (figure 9).

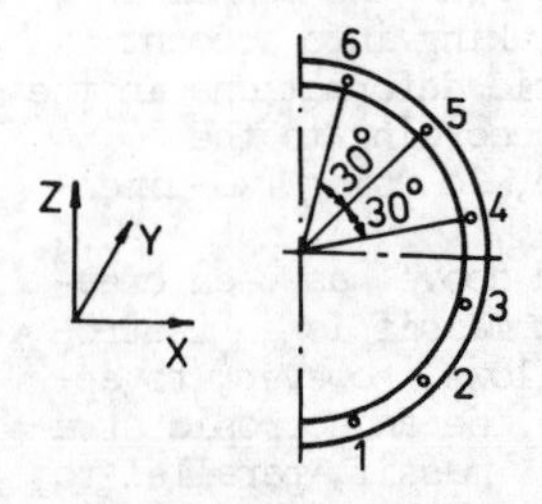

Fig. 9 Location of the points of comparison

The data obtained from the two-mensional and three-dimensional FEM analysis are shown in fig. 10. It is obvious that the coincidence of the results obtained is very good. The values of σ_z obtained using the two-dimensional model is slightly higher compared to that from the three-dimensional analysis because the value of Eo has been calculated for Uo=Umin, while in fact the length of each ring from the lining is 2m long, and Eo should be respectively accepted on an averaged value of Uo>Umin.

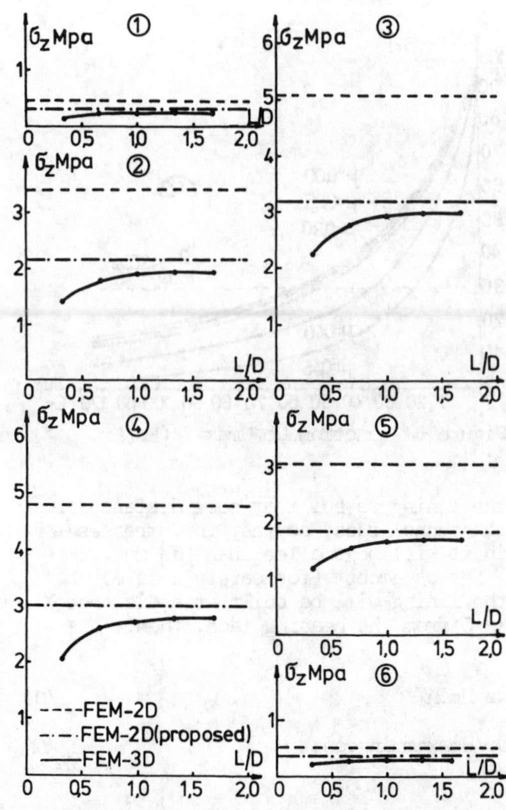

Figure 10 Results from 2D and 3D FEM analysis

4 CONCLUSIONS

A two-dimensional model, based on FEM, to calculate stresses in tunnel linings is presented in this paper, taking into account the influence of the tunnel face. The following basic conclusions can be drawn:

- graphs and formula have been drawn for the practical application of the model in four shapes of tunnel cross section
- in order to use the graphs and the formula, one should only know Young's modulus and Poisson's ratios of the rock massif
- the comparison of the results obtained by the 2D and 3D FEM analysis has shown good coincidence. It also makes it possible to take into account the influence of the tunnel face using calculation resources.

REFERENCES

[1] T.Hristov and V. Vassilev, On the problem of the two-dimensional calculation of the streses in the tunnel lining taking into account the influ-

ence of the heading face by finite element method, Geotechik special issu, 105-107 (1985)

[2] N.A.Jdankin and S.B.Kolokolov, Three-dimensional analysis of undergroung opening with limited lengt (in russian), Applied mechanics, tom 19, No 4 (1983)

[3] F.Laabmayer and G.Svoboda, Zusammenhand zwiscen electronicher berechnung und messung. Stand der entwicklung fur seichtliegende tunnel, Rock Mechanics 8, 29-42 (1979)

[4] S.Semprich, Berechnung der spanungen und verformungen im bereich der ortsbrust von tunnelbauverken im fels, Veroffentlichungen des Institutes fur grundbau, bodenmechanik, felsmechanik und verkehrswasserbau der RWTH Aachen, Helft 8, (1980)

[5] G.Svoboda, Finite element analysis of the new Austrian tunneling method. Proc. of the Third intern. conf. on numerical methods in geomechanics,Aachen (1979)

[6] V.Vassilev, Influence of the heading face on the streses in the tunnel linings (in bulgarian). Water problems 21, Sofia (1984)

[7] V.Vassilev, Programme package for three-dimensional investigations of underground structures by FEM (in bulgarian), Technical mind 5, Sofia (1986)

[8] W.H.Ward, Groundsupport for tunnels in weak rocks, Geotechnigue 28,No 2, 133-172 (1978)

Numerical Methods in Geomechanics (Innsbruck 1988), Swoboda (ed.)
© 1988 Balkema, Rotterdam. ISBN 90 6191 809 X

Finite element analysis of tunnel linings with emphasis on the nonlinearity of concrete

E.Haugeneder & M.Mehl
Technodat GmbH, Vienna, Austria

ABSTRACT: The conventional analyses of tunnels where a linear elastic material is assumed for the concrete lining, is not satisfactory and sometimes contradictory. This is not surprising because the deformation of the soil causes restraints in the lining.

In the present work a material model for reinforced concrete is developed which incorporates nonlinearity and cracking. With this model finite element analyses of a shallow cylindrical tunnel are performed. Some disadvantages of the conventional analyses are discussed and a modern safety concept based on EUROCODE 2 is introduced.

The results show that the ultimate load and the structural safety of the tunnel can be calculated satisfactory.

1) Introduction

Tunnel construction is a challenging work whose success depends mainly on the quality of the tunnelling team. Nevertheless, numerical calculations became a valuable tool to estimate the settlement of the surface and the stresses in the shotcrete lining. Very often the calculations are performed by the use of the Finite Element Method (FEM) including nonlinear material behaviour for the soil [1,2,3,4]. In general two-dimensional analyses are performed in order to limit the effort to a feasable amount. Excavation sequences are simulated by adequate reduction of the element stiffnesses inside the tunnel cross-section.

Comparisons of numerical results and measured quantities show a good agreement of the settlement but less coincidence of stress resultants in the tunnel lining. From practical experience we know that the stress resultants are often overestimated. This uncertainty about the real forces and the safety concept taken from usual civil engineering structures can result in very conservative dimensions of the tunnel lining [5].

An improvement of the calculations has to start from the fundamentals of a tunnel construction: The basic concept of the New Austrian Tunneling Method (NATM) is the interaction of the soil or rock mass and the outer tunnel lining [6]. Before the shotcrete lining is produced the soil is allowed to deform so its bearing capacity can be activated. Afterwards, soil and tunnel lining interact with each other and a part of the load is carried by the shotcrete. Therefore, considerations about the safety against ultimate load have to comprise the whole structure. The calculations should include the essential nonlinearities of soil and concrete.

In the following emphasis is given to develope a proper material model for the concrete tunnel lining. Further, results of a number of calculations are presented using the new model. The calculations have been carried out on a circular tunnel embedded in a homogenous and isotropic soil, where only the thickness of the tunnel lining and the material parameters are varied. The lining is assumed to be reinforced shotcrete. Anchors and supporting arches are not considered.

2) Numerical model for reinforced concrete

Reinforced concrete is a compound material consisting of concrete and steel. Concrete is able to carry the pressure load whereas the reinforcing steel carries tensile forces in those areas where the concrete cracks. Because of this very inhomogenous structure and the special type of load carrying behaviour a great number of material models are available in the literature [7,8,9]. As far as is known by the authors, in all models the two aggregates concrete and steel are considered separately.

In the concrete models the stress situations are grouped as follows (referring to a state of plane stress): compression-compression, compression-tension and tension-tension. For each stress state its own material law is defined. For pure compression we have:

a) Plasticity models, which are based on the mathematical theory of Plasticity, and are therefore defined by one or more yield surfaces, a flow rule and a hardening or softening function or by combinations of both. They are in general three-dimensional.

b) Non-linear elastic models, which in general follow very closely a selected number of test results. Most of the models are two-dimensional, because only a relatively few number of experiments are available for three dimensions.

c) Other models, e.g. based on the endochronoc theory or on the continuous damage theory.

For pure tension most of the models consider smeared cracks and include a tension stiffening behaviour. The latter should simulate the fact that concrete cracks (tensile cracks) do not propagate like cracks in a brittle material. The parameters that control the tension stiffening effect depend mostly on the energy release rate. Other models consider each individual crack, which consequently requires a modification of the finite element mesh.

The mixed stress states including both compression and tension at the same point are treated with a more or less continuous transition from one to the other model description.

There are also different models for the reinforcing steel in the literature. Depending on the size of the structure the reinforcement is either smeared or each bar is modelled individually. In each case an elastic - perfectly plastic behaviour is the proper model.

Separating concrete and steel in the numerical model of reinforced concrete has the disadvantage that the model parameters have to be evaluated from experiments on plain concrete specimen and on steel bars. This disadvantage is avoided, if reinforced concrete is treated as *one* material since practical requirements on a concrete structure, such as maximum crack spacing and maximum crack width, guarantee a minimum percentage of main reinforcement, a web reinforcement in panels and slabs, and stirrups in beams and columns. Therefore, a new model for reinforced concrete based on the compound material will be developed in the following, where reinforcement and cracks are considered to be smeared. Emphasis is taken on an easy to use model that does not require test data other than already available.

Reinforced Concrete in pure compression

A plasticity yield function is used in a pure compression state. Parameters are compressive and tensile strength. For a bi-axial stress state the failure surface follows closely the Kupfer-Gerstle curve [10].

$$F(\sigma) = 3J_2 + \sigma_P I_1 + \frac{I_1^2}{5} - \frac{\sigma_P^2}{5} \qquad (1)$$

Eq.(1) is a modification of the fracture criterion of Buyukozturk [11]. The function is developed such that the quadratic total stress - total strain relationship in uniaxial compression according to the Austrian standards (ÖNORM) is also matched. After reaching the ultimate compressive stress the model behaves perfectly plastic. The effect of the steel is neglected, since after the concrete has crushed also the reinforcement looses its strength.

Reinforced Concrete in pure tension

The experimental background for this relationship comes from the works of Falkner [12] and of Hartl [13]. Falkner tested a great number of columns in uniaxial tension to investigate the stresses due to restraint conditions and the crack width and crack spacing. One of the essential results is: The first crack appears if the average stress reaches the tensile strength. This is independent from the amount of reinforcement and from the number of bars and their diameters in the cross-section. If the load is further applied additional cracks appear until a minimum crack width is reached. Whereas the location of the first cracks is random the final picture shoes a uniform crack distribution. Only then the load can be increased until the steel bars begin to yield.

It is well known from biaxial experiments that cracks in reinforced concrete appear in directions normal to principal tensile stresses. This is shown very clearly in Collins' test results [14]. Also, there is negligible influence on the tensile strength from the other principal stresses.

Therefore, the numerical model first includes a transformation into the system of principal stresses. Parameters are the tensile strength, the percentage of reinforcement in each direction, the quotient of elastic steel modulus/elastic concrete modulus, and the yield strain of steel. Fig. 1 depicts the relationship between stresses and strains for reinforced concrete in tension.

The term "tension stiffening"

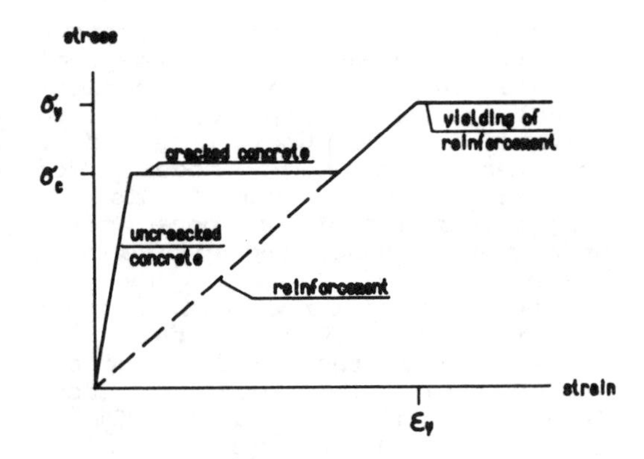

Fig.1 - Model for reinforced concrete in tension

does not appear as a parameter since this phenomenon is inherent in the present numerical model for reinforced concrete. For the same reason any considerations about energy release rates and crack propagations are obsolete.

Reinforced Concrete in compression/tension stress states

The compressive strength is reduced if tension appears in one of the other principal directions. To get a continuous surface between pure tension and pure compression a linear dependency is assumed.

A number of calculations have been done to prove the material model in restraint conditions and in dimensioning problems, to analyse the redestributions of the inner forces [15] and to compare with test results

3) Examples of a cylindrical tunnel

Today it is state-of-the-art, that the design of an NATM tunnel includes a finite element analysis with nonlinear material models for soil and rock. On the other hand the tunnel lining is generally modelled by a linear elastic material. Based on the resulting stresses from such an analysis and on the relevant standards (ÖNORM) the required amount of reinforcement is calculated. The aim of this part is to show that this technique does not take into account the specific physical situation and gives therefore unrealistic results. Numerical calculations of a cylindrical tunnel serve as a vehicle to demonstrate the usefulness of a nonlinear material model for concrete. The tunnel is a simplification of a one- track underground tunnel and with a high overburden.

The analysed tunnel has an inner diameter of 6.40m, but the thicknesses of the outer lining are varied from 10cm, 20cm, to 25cm. The tunnel is covered by 20m of homogenous and isotropic soil. For all analyses the same mesh layout has been used. The finite element mesh takes advantage of the symmetry. The tunnel lining is idealyzed by four rows of plane elements to give accurate results of the stresses and strains. The material parameters for concrete as well as for soil are shown in fig.2.

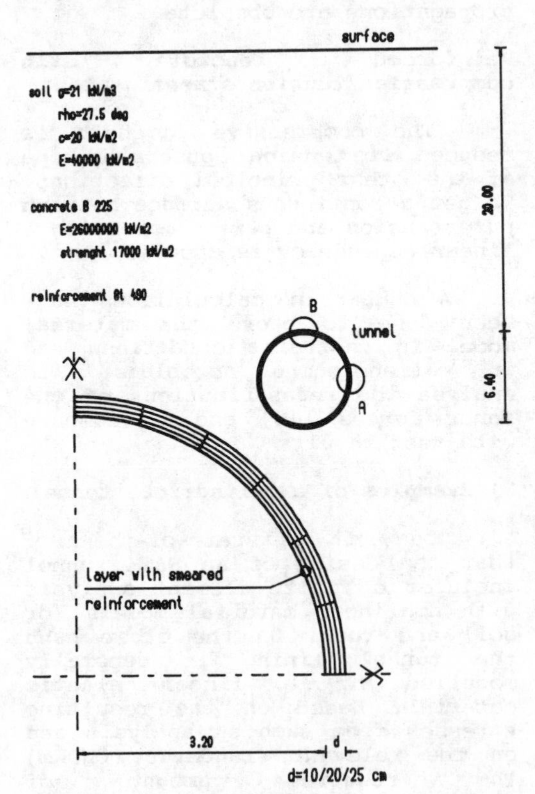

Fig.2 - Analysed tunnel

The excavation process according to the NATM is simulated by the following steps:

0) Primary state of stresses
1) Pre-relaxation
2) Complete excavation (1 - 3 load increments)

In step 2 the stiffness of the outer tunnel lining is activated. The finite element analysis is carried out by the use of TPS10.

Fig.5 shows the results of the calculations. Normal forces and bending moments at the points A and B (see fig.2) versus the shell thickness are drawn. Regarding the results for linear elastic shell material at first, we recognize almost constant normal forces but strongly increasing bending moments with increasing thickness. This is not surprising because we know that the load onto a tunnel is rather a prescribed deformation than a constant force.

Therefore, from a conventional calculation we get: A thicker shell needs more reinforcement and a higher quality of concrete than a thinner one (see fig.3). This figure shows the required amount of reinforcement and quality of concrete at the points A and B. While the 10cm shell has no tension at all at point B the 25cm shell needs 9.1cm^2 reinforcement.

From fig.4 we are confirmed in dealing with a prescribed deformation. This figure shows the change of curvature of the tunnel lining calculated from the bending moment and the linear elastic stiffness. The essence is that the change of curvature is almost independent of the shell thickness; the bending moment is therefore linearly dependent on the shell stiffness.

Introducing a nonlinear material model for concrete as described in section 2 those shells which are thicker than 10cm show a moment reduction due to the nonlinear compression state and the cracked regions (fig.5). The reduction is most significant for the 25cm shell.

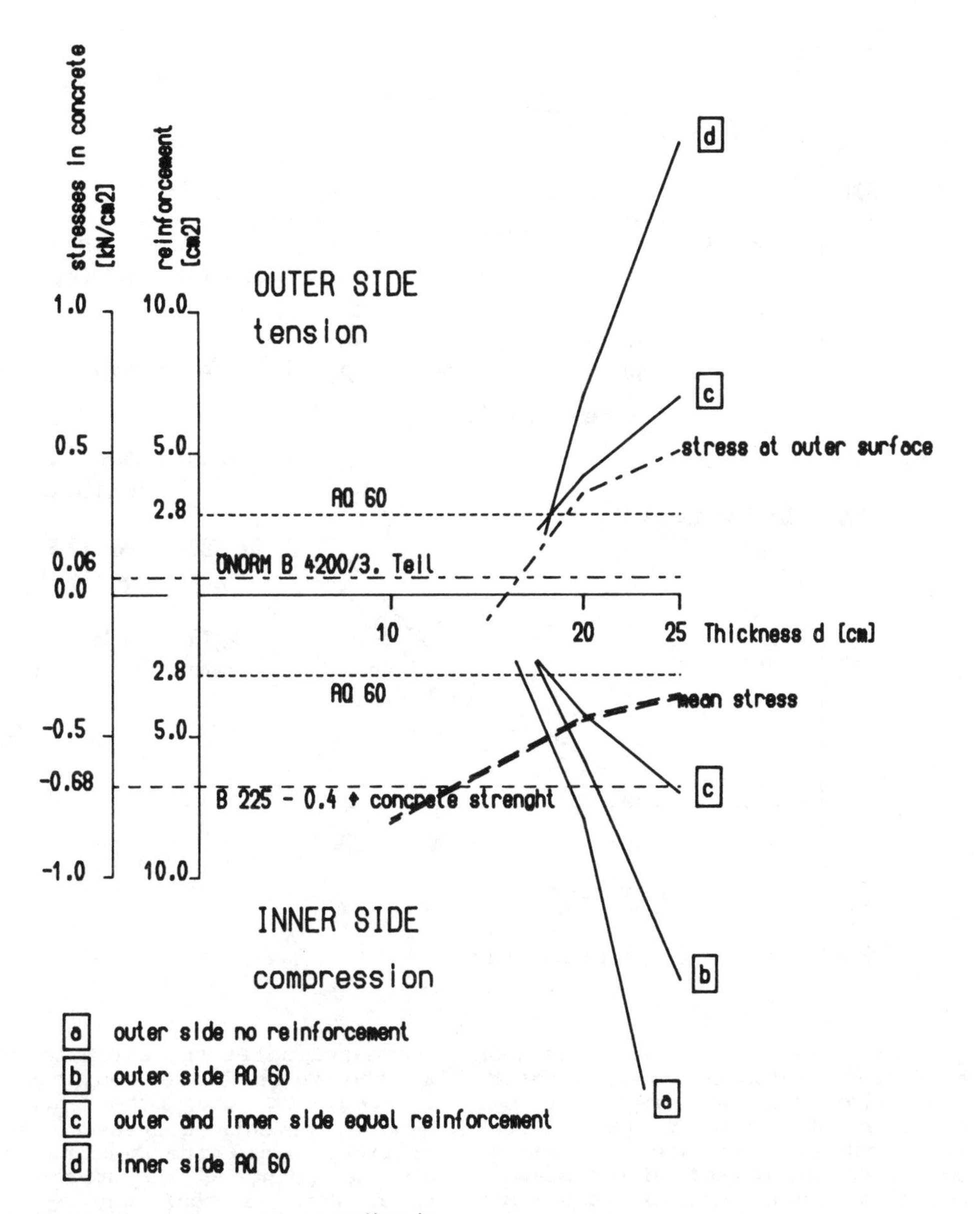

Fig.3 - Required reinforcement according to ÖNORM B 4200/9. Teil and stresses in concrete at the side (point A) in a conventional analysis

d [cm]	10	20	25
side (A)	0.67	0.65	0.58
top (B)	0.37	0.53	0.49

Fig.4 - Change of curvature [kN/cm2]

4) Structural safety

The previously mentioned calculations are based on the actually existing loads and material parameters. The resulting stresses in the lining are multiplied by safety factors and are used for dimensioning as if they were results of an ultimate load analysis. Strictly speaking this procedure is not correct.

To get deeper insight in the behaviour of the structure an

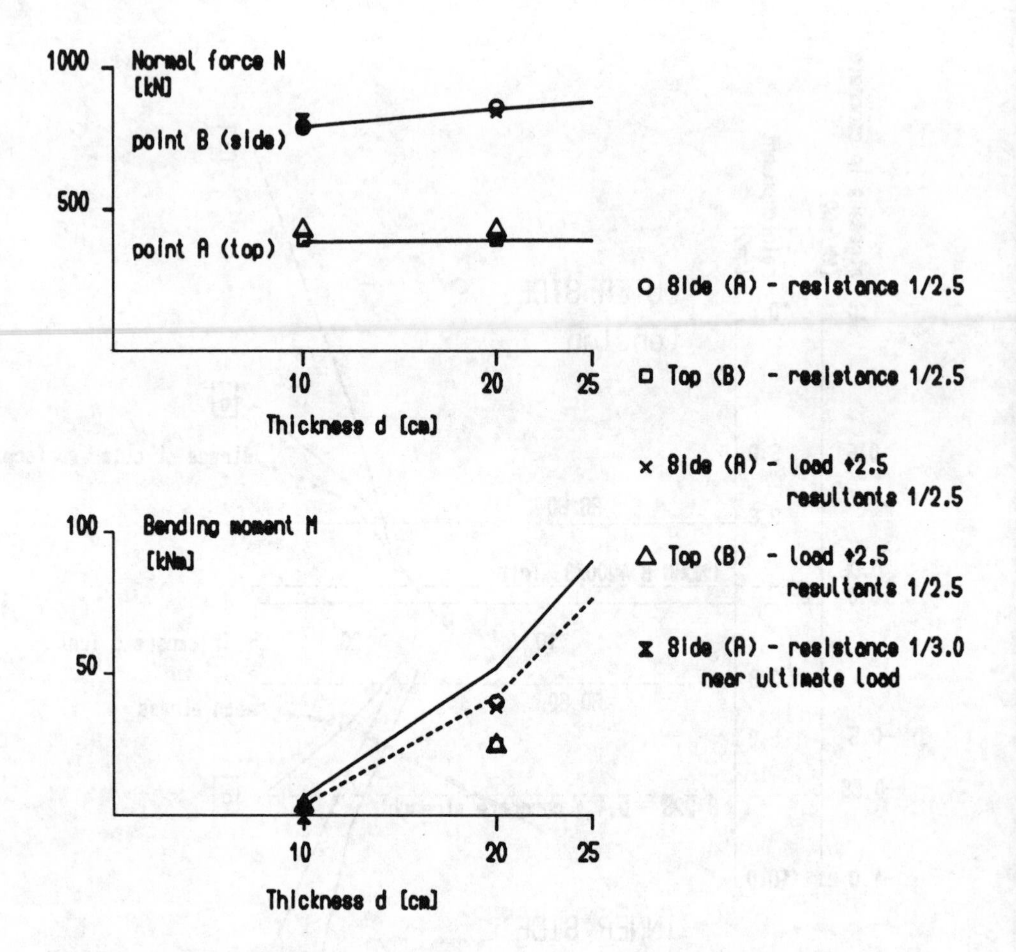

Fig.5 - Results of the calculations

ultimate load analysis has been performed. Following modern safety considerations, see e.g. the new Austrian standards or the EUROCODE 2, we have to increase the load or to reduce the structure resistance. For test reasons and to yield limit cases the two steps have been applied individually leaving the soil parameters out of consideration.

The load has been increased by applying a surface load equivalent to an increased overburden. Three different calculations for the 10cm lining have been done with load factors of 2.5, 3.0, and 3.4. After load factor 3.0 a critical state was reached where structural failure must be assumed. As one could foresee the normal forces increased until the ultimate stress at the tunnel side (point A) was reached. On the other hand the bending moments decreased. At the critical state the bending moment at the point A was approximately zero. This is confirmed by fig.6. This figure shows the strains at the surfaces of the lining. The strains at the side (point A) exceed 0.2% compressive strain, which indicates also damage of the structure.

In a second series of calculations the resistance of the lining has been reduced by factors of 0.4 and 0.33, but the load was held constant. These factors are equivalent to the load factors above (2.5 and 3.0, respectively) in the sense of ÖNORM B4040.

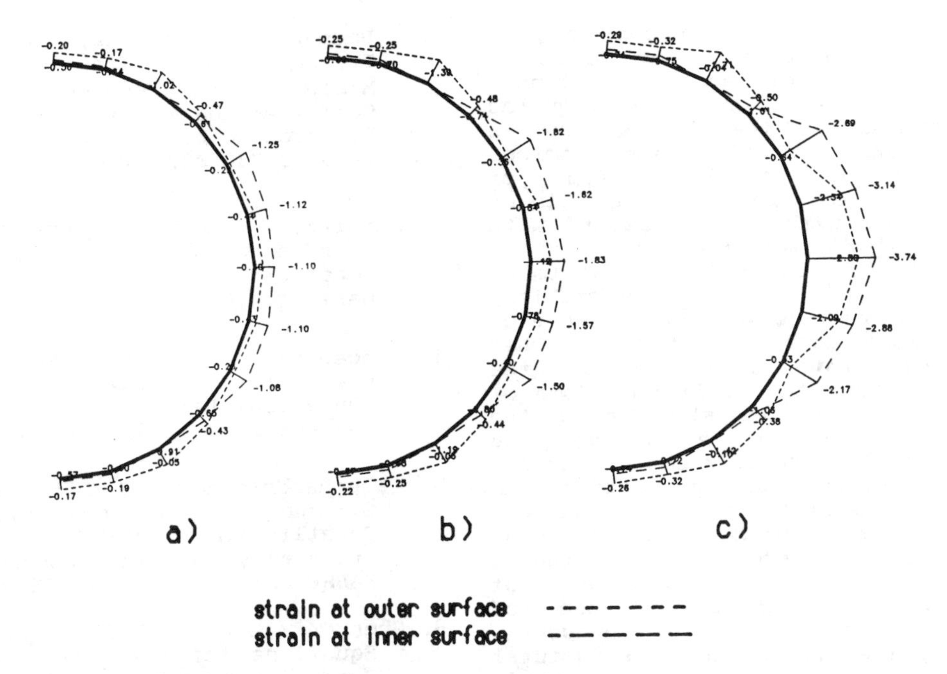

Fig.6 - Strains at the surfaces of the 10cm lining
at the side (point A)
a) load *2.5 b) load *3.0 c) load *3.4

Similar to the previous results a critical state has been reached at a reduction factor of 0.33 and the bending moment droped to zero.

From these calculations we see that structural failure occurs if the mean normal stress reaches the ultimate stress. Considering the mean normal stresses in fig.3 (service load) we find them approximately equal to the stresses at ultimate load for reduced resistance. We would therefore have a safety of 3.0 against failure. On the other hand the normal forces due to service load times a safety factor of 3.0 are approximately the normal forces at load factor 3.0 (ultimate load). In each of the failure cases the bending moments are very small and are not relevant for failure.

In another calculation the concept of ÖNORM B4040 and EUROCODE 2 has been applied. Since the soil is also part of the structure but safety factors based on probabilistic methods are not known, estinated factors have been introduced. The actual values are: load *1.5, ultimate concrete stress /1.5, yield strain of steel /1.15, coefficient of friction /1.15, cohesion /2. The numerical procedure converged, i.e. the tunnel was not damaged. A comparison of the normal forces and strains with the previous analyses concerning the increased loads shows an overall state of stress in the lining of little more than under 2.5 * service load.

5) Conclusions

A number of analyses have been performed on a one-track underground tunnel. It was assumed that the surrounding soil is able to carry load, and the overburden is hig enough to allow force redestribution. Groundwater pressure was not considered. The results show that under these assumptions the bending moments in an outer tunnel lining decrease rapidly if the ultimate load is approached. Therefore we find a nearly constant state of stress across the shell thickness at the limit load level. In our case it seems to be possible to determine the safety of an outer tunnel lining or to get the required thickness of the lining from a calculation of the normal forces only. This calculated type of failure of the tunnel lining coincides with experience and with reports of cases of damages.

A thickness of 10cm seems to be just enough in the sample problem, but in practice we must be aware of the fact that a number of other effects have influence on the design of a tunnel lining. Some of them are the low strength of the shotcrete in its early age, the practical requirements during construction, and imperfections and variations of the shell thickness.

Further work should be done including different types of tunnel geometries and heights of overburden, different types of soil, and groundwater pressure. Also, the indication of structural failure is currently unsatisfactory. A softening behaviour for reinforced concrete as well as for soil might help [16].

The authors would like to thank the Austrian Ministery of Commerce "Straßenforschung" grant no.667, for the financial support.

Literature

1. Baumann, Th.; Hilber, H.M.: Zur Berechnung von U-Bahn Tunneln im Lockergestein. Finite Elemente - Anwendung in der Baupraxis. Ernst u. Sohn, Berlin 1985.

2. Swoboda, G.; Beer, G.: Städtischer Tunnelbau - Rechnemodelle und Resultatinterpretation als Grundlage für Planung und Bauausführung. Finite Elemente - Anwendung in der Baupraxis. Ernst u. Sohn, Berlin 1985.

3. Haberl, G.; Haugeneder, E.: Nichtlineare Finite-Element-Berechnung von Tunnelbauwerken. Finite Elemente - Anwendung in der Baupraxis. Ernst u. Sohn, Berlin 1985.

4. Baumann, Th.;Sulke, B.-M.; Trysna, Th.: Einsatz von Messung und Rechnung bei Spritzbetonbauweisen im Lockergestein. Bautechnik 1985, S.330-337, 368-374.

5. Schultz, H.: Berechnung oberflächennaher Tunnel. Konstruktiver Ing. Bau - Berichte 40.

6. Rabcewicz, L.v.; Sattler, K.: Die Neue Österreichische Tunnelbauweise. Der Bauingenieur 1965, S.289-301.

7. Eberhardsteiner, J.; Meschke, G.; Mang, H.A.: Comparison of Constitutive Models for Triaxially Loaded Concrete. IABSE Colloquium Delft 1987.

8. Chen, W.F.: Constitutive Equations for Concrete. IABSE Symposium Kopenhagen 1979.

9. Milford, R.V.; Schnobrich, W.C.: Numerical Model for Cracked Reinforced Concrete. CAA - CAE of Concrete Structures, Split 1984.

10. Kupfer, H.: Das Verhalten des Betons unter mehrachsiger Kurzzeitbelastung unter besonderer Berücksichtigung der zweiachsigen Beanspruchung. DASt, Heft 229, 1973.

11. Shareef, S.S.; Buyukozturk, O.: Constitutive Modelling of Concrete in Finite Element Analysis. MIT Report Nr. R83-16, 1983.

12. Falkner, H.: Zur Frage der Rißbildung durch Eigen- und Zwängsspannungen infolge Temperatur in Stahlbetonbauteilen. DASt, Heft 208, 1969.

13. Hartl, G.: Die Arbeitslinie "eingebetteter Stäbe" bei Erst- und Kurzzeitbelastung. Diss. Universität Innsbruck, 1977.

14. Vecchio, F; Collins, M.P.: The Response of Reinforced Concrete to In-Plane Shear and Normal Stresses. University of Toronto, Publ. Nr.82-03, 1982.

15. Mehl, M.; Haugeneder, E.: Redestribution of Inner Forces in Hyperstatic Reinforced Concrete Structures. IABSE Colloquium Delft 1987.

16. De Borst, R.: Stability and Uniqueness in Numerical Modelling of Concrete Structures. IABSE Colloquium Delft 1987.

Numerical Methods in Geomechanics (Innsbruck 1988), Swoboda (ed.)
© 1988 Balkema, Rotterdam. ISBN 90 6191 809 X

Flexible steel tunnel lining response to unloading-reloading cycle

Mohammad Irshad
De Leuw, Cather & Company, Washington, D.C., USA
Jerome S.B.Iffland
Iffland Kavanagh Waterbury, P.C., New York, USA
Royce A.Drake, Jr.
Washington Metropolitan Area Transit Authority, Washington, D.C., USA

ABSTRACT: Three case histories of successful use of planar beam-spring model, used to simulate the behavior of flexible steel tunnel lining under symmetrical and asymmetrical loading conditions, are presented. The results of the linear-elastic computer analyses are compared with the findings of in-situ monitoring programs, and good agreement between the computed and actual lining displacements is reported. The flexible lining, used in the construction of many mass-transit tunnels in downtown Washington, D.C., represents a complex three-dimensional structural system that employs thin-walled bolted steel segments. The paper discusses how the force-analysis results, facilitated by the discrete-model, were used to verify that the liner distress, including the possibility of buckling due to disturbances caused by proposed adjacent construction of high-rise buildings, would be precluded. The relative ease of use, versatility and the practical value of prudently prepared beam-spring models for forecasting the flexible lining response in soft ground tunnels, is highlighted. A brief discussion of some of the closed-form and computer-based analytical methods is included to serve as a general reference.

1 INTRODUCTION

Mass-transit subway construction in a downtown area is usually followed by a spate of building construction along the newly opened lines. Some of the new buildings might be sited in close proximity of, or even directly above, the existing subway system. In such situations a complete analysis of the impact of the proposed construction on the existing tunnels becomes necessary.

Customarily, bored-tunnel lining is designed for symmetrical short term and long term loading conditions. The new adjacent construction, performed after several years of subway construction, disturbs the prevailing state-of-equilibrium in the ground mass around the tunnels. The basement excavation, for example, results in temporary unloading of the lining and is usually not symmetrical. The subsequent construction of the new building adds permanent superimposed loads on the tunnel that do not restore the original state-of-stress within and around the lining. Throughout the adjacent construction process, the lining is exposed to various loading conditions, dependent on the procedure adopted for performing excavation, and the manner in which new construction loads are superimposed on the tunnel lining.

An evaluation of the impact of adjacent construction on flexible steel lining needs careful consideration because of the following factors:

- Thin-walled, shell type, construction implies that premature lining distress due to buckling could occur if the lining is subjected to excessive symmetrical and asymmetrical cross-sectional distortion. Because of the segmental design, beneficial effects of structural continuity - such as are found in monolithic cast-in-place concrete lining - are not present. The bolted joints are not moment connections, and their integrity throughout, and after the adjacent construction process must be ensured.
- The pronounced flexible nature of the lining implies that timely corrective action may be difficult to take in case excessive distortion of the lining cross-section takes place during the

adjacent construction process. Modelling of flexible liner response, under actual adjacent construction - related conditions is complex, and does not represent every-day design office activity.

The three case histories of use of beam-spring computer analysis described later in this paper relate to the work performed between 1985 and 1987 on the Washington D.C. Metro Project. Generally, the complete analysis of segmental steel tunnel lining for a given adjacent construction condition includes the following:

(i) Planar analysis of lining for various excavations sequence (beam-spring model).
(ii) Lateral shift of the tunnel lining; horizontal beam action due to asymmetrical loading induced by excavation (computer model or closed-form solution).
(iii) Superpositioning of (i) and (ii) for a given excavation condition to produce worst effects on the lining.
(iv) Checking of transverse bolted connections between the 30 or 48 inch wide segments forming a tunnel ring. Also checking the joint opening associated with the cross section distortion.
(v) Checking of longitudinal bolted connections between the 30 or 48 inch wide lining rings.
(vi) Checking for buckling.
(vii) Checking for post-construction loading conditions.

The thrust of this paper is to discuss only Item (i) that deals with planar analysis. Experience indicates that cross-sectional distortions, rather than the horizontal beam action due to sidesway conditions, usually dominate the lining behavior.

2 SELECTION OF ANALYTICAL METHOD

The lining analyses, such as the ones described herein, are usually performed under tight time constraints in a Consulting Engineers' design office, necessitating selection of an optimal analytical tool that must meet the following criteria:

- Good accuracy
- Ease of use
- Versatility
- Quick turn-around time
- Some prior familiarity
- Economy

While considerable literature has been published on the design of linings, usually for symmetrical loading, relatively little information is available on analyzing the in-situ response of as-built lining under diverse, unloading and reloading, conditions. Closed-form solutions have inherent deficiency even in terms of their applications to design, because they are incapable of dealing with an asymmetrical loading regime. The computational simplifications of some well known closed-form formulations such as Peck, et al, 1972, and Muir-Wood, 1975, are dependent on the assumption of symmetrical loading, rendering them unsuitable for analyzing adjacent-construction impact on existing tunnels which involves complex asymmetrical loading.

Elasto-plastic finite element method (FEM) computer programs such as Mises-3 and FINAL, have increasingly been used in Europe and more recently in the U.S.A. to design shotcrete and concrete tunnel linings and other underground structures. These FEM programs represent powerful analytical tools in the field of geomechanics, capable of handling complex loading and boundary conditions. However, an FEM analysis, whether elastic or elasto-plastic, makes considerable demands in terms of technical competence of engineers using it, is more time consuming than a routine planar frame analysis routine, and is much more expensive to use than the simpler beam-spring method. Experience indicates that analysis of one tunnel cross-section using FEM technique could run into several thousand dollars. Based on the cost-benefit ratio, and on sheer technical attributes, the beam-spring method can probably match the FEM analysis when the following points are taken into consideration:

- Both FEM and the beam-spring methods use planar analysis. (Three-dimensional FEM analysis involving a large model usually cannot be justified in a design office).
- Both FEM and the beam-spring method model the liner as a linear-elastic structure. (Elasto-plastic elements in the FEM analysis are more commonly used to model the ground continuum, not

the lining).

The main-advantage of FEM method lies in its ability to simulate the ground movement and surface settlements. However, when dealing with the flexible steel tunnel lining, it is known beforehand that lining deformation must be controlled within a narrow range to avoid structural instability. Hence, no significant surface settlements are associated with the permissible distortion and displacement of lining surface. Thus, there is no justification to resort to more complex FEM analysis.

3 TESTING OF SYMMETRICALLY LOADED MODEL

In order to test the reliability of the beam-spring model, the results of an analysis were compared to the results of analyses using a number of other models.

In all cases, the loading was symmetrical. The tunnel liner investigated was that used on Section D4a of the Washington Metropolitan Area Transit Authority Metro system. This particular design was modified by the subcontractor (Commercial Shearing, Inc.) and the result of their analysis is also included in the comparison. The several tunnel liner design models studied were as follows:

I Commercial Shearing - A model based on correction of at-rest moments and forces by moments and forces resulting from applied liner deflections to simulate soil-structure interaction.

II Muir-Woods, 1975 - An elastic closed form model based on liner response within a stressed ground mass. (Also denoted as an excavation loading model).

III Peck, Hendron and Mohraz, 1972 - An elastic closed formed model based on liner response in a ground mass subjected to an externally applied pressure. (Also denoted as an excavation pressure model).

IV Ranken, Ghaboussi and Hendron, 1978 - An elastic closed form model based on liner response within a stressed ground mass. (Also denoted as an excavation loading model). This is the same as the Muir-Woods model except that Poisson's ratios for the liner and the surrounding ground are included in the analysis).

V Stress Model A - An elastic spring supported model with no springs in tension utilizing WMATA long term loading conditions which utilizes uniform horizontal pressure.

VI Stress Model B - An elastic spring supported model with no springs in tension utilizing WMATA long term loading but with horizontal pressures increasing linearly from the top of the tunnel (where the value equals the WMATA horizontal loading) to the bottom of the tunnel based on the increase in lateral pressure with height.

METHOD	THRUST (kips)	MAXIMUM MOMENT (Foot Kips)
I	54.9	10.0
II	52.8	5.4
III	52.8	14.2
IV	57.7	6.6
V	55.4	3.1
VI	58.7	11.4

Fig.1 Comparison of thrusts and moments in tunnel liners for different analysis models

The results of these analyses are shown in Figure 1. As can be seen the STRESS Model B provides a conservative but reasonable estimate of the maximum thrust in the liner. The largest variation is in the order of 11% while the average variation is approximately 7%. The various methods show a wider range in variations in maximum moments. Considering the spread, the STRESS Model B also gives a conservative and reasonable estimate of the maximum moment.

4 BEAM-SPRING MODEL DETAILS

The beam-spring model is well known in tunnel lining design. Therefore, only a brief description of this method, as adapted for the flexible liner analysis, is presented.

The circular steel tunnel lining shown in Fig. 2 is discretized using a series of straight interconnected beam elements. Sufficiently large number of beam elements is used to reasonably accurately replace the actual configuration of the lining in the mathematical model.

The beam-spring analytical model is utilized to simulate the planar soil-structure interaction. Discrete, linear-elastic, springs replace the elasto-plastic soil continuum surrounding the tunnel lining. The springs are attached to the lining at the model points. Both radial and a combination of radial-and-tangential springs may be used, dependent on site-specific conditions, and on the investigator's judgement (Fig. 3 & 4). Generally, where tangential springs were used, a spring contact of 1/4 that for radial spring has been used. A tributary area concept is utilized in assigning stiffness to springs. Modulus of subgrade reaction (K) rather than the Young's modulus is used in computing the spring stiffness.

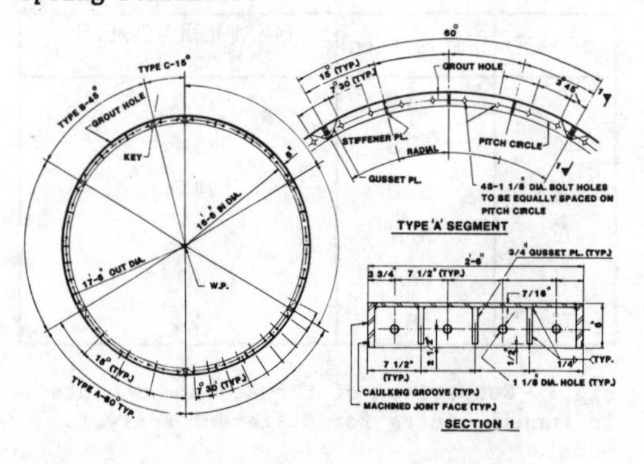

Fig.2 Segmental steel tunnel lining-typical details

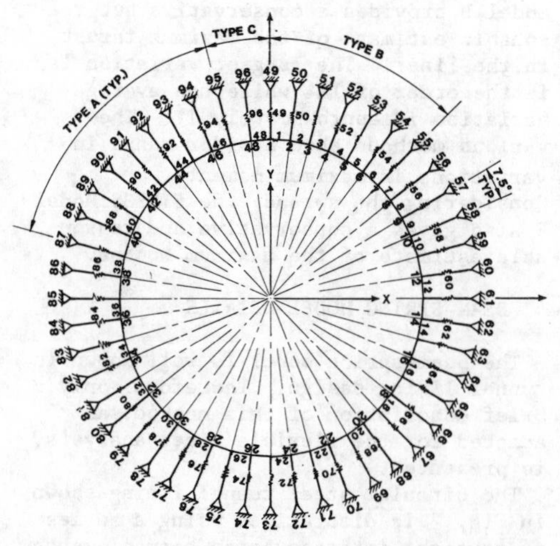

Fig.3 Radial spring model for 1201 New York Avenue

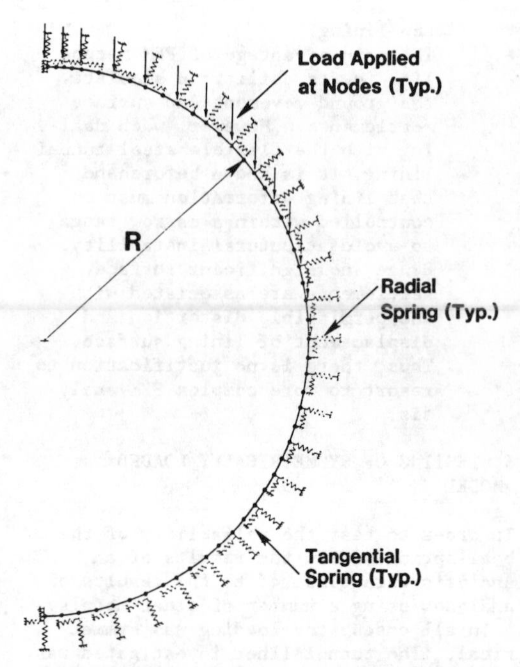

Fig.4 Structural model of liner support system showing radial and tangential springs

Loads are assigned to each nodal joint using the tributary areas of the vertical and horizontal loading diagrams. The familiar, stiffness-method based frame analysis computer program STRESS is then used to analyze the lining response for various loading conditions. After each run, tension springs are removed and the analysis re-run until all remaining springs show compression.

The analysis results provides lining displacements and forces. The lining displacement data is utilized to establish that portion of the lining that represents unsupported condition - lining moving away from the ground. This information is used to check whether or not potential buckling of the lining is of any concern. Lining forces such as bending moment and axial thrust are examined in each beam-element and liner stresses are checked for compliance with the permissible values. The beam-spring model offers many advantages. The properties of the tunnel lining, properties of the surrounding soil and loads can vary from node to node. The shape of the lining can also be varied to correspond to as-built conditions. It is this flexibility as well as the simplicity of the model that enhances its practical use in a design office.

5 BEAM-SPRING MODEL ANALYSES AND FIELD MONITORING DATA - CASE HISTORIES

In all, three WMATA tunnels were investigated using the beam-spring method described earlier. Two cases of unloading of tunnels were investigated by Iffland Kavanagh Waterbury (IKW). One case of unloading and subsequent reloading (post-construction condition) was investigated by De Leuw Cather & Co. (DCCO). DCCO also reduced and analyzed the field monitoring data for all three tunnels and periodically reported the finding to WMATA, including comments on the safety of the ongoing construction process. Construction of two buildings, Aerospace Center and 1201 New York Avenue, over, and adjacent to, the tunnels have been completed and the monitoring program discontinued. Construction of the third building, Crowne Plaza Hotel, is under way and so is the field monitoring program.

Brief details of each case history are provided below.

5.1 Aerospace Center

The Aerospace Center is a building with 10 floors above grade and 3 floors below grade. The construction parallels WMATA Metro tunnels with the southwest corner of the building projecting over the inbound Metro tunnel. The excavation required for the construction unloads the subway tunnel and the required cofferdam introduces additional vertical loads over the tunnel, both resulting in deformation of the tunnel liner. These construction loading conditions are shown in Figure 5. The pre-construction loading condition is the one described as Model VI in Section 3.

While the tunnel monitoring during construction consisted of monitoring deflections at 10 different cross-sections in the tunnel over a period of several months, the significant measurements are those at the critical location at the end of the foundation excavation process and after all bracing and shoring were in place. Referring to Figure 6, the theoretical horizontal and vertical deflections and the measured horizontal and vertical deflections for this critical section at the end of the excavation process were:

Horizontal deflection x:
- Theoretical: 0.047 inches
- Measured: 0.061 inches

Vertical deflection y:
- Theoretical: 0.063 inches
- Measured: 0.065 inches

Considering the number of assumptions that must be made in any analysis of the type and understanding that geotechnical calculations are a long ways from being an exact science, the above results are remarkably reassuring for the method of analysis utilized. Additional data is shown in FIG 11.

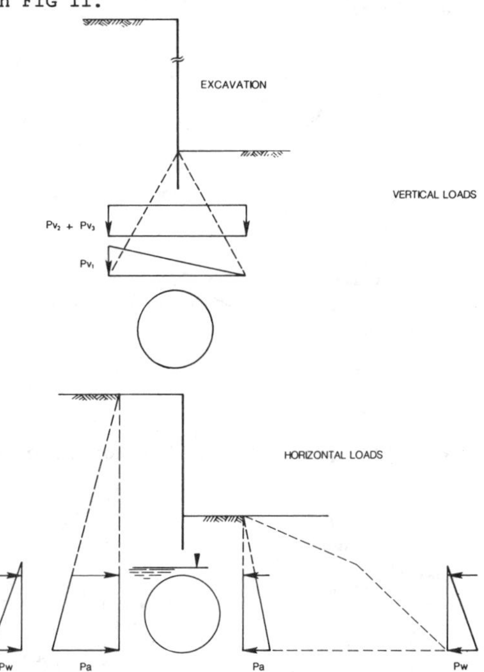

Fig.5 Construction loads for excavation of the Aerospace Center, Washington, D.C.

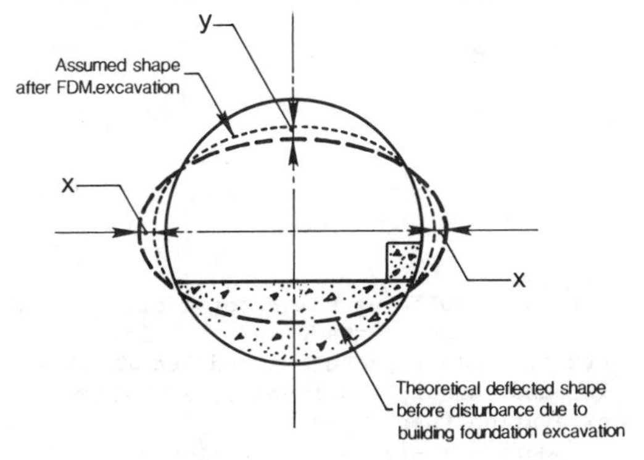

Fig.6 Aerospace Center - tunnel movements after excavation

5.2 1201 New York Avenue

The general layout of this building with respect to WMATA tunnels is shown in Fig. 7. The complex is composed of two structures known as Phase 1 and Phase 2 buildings.

Phase 1 has a total of 15 floors, 12 above grade and 3 below grade. The south-west corner of the Phase 1 building overlies WMATA's outbound tunnel. Phase 1 involved deep excavation, creating substantial asymmetrical loading on the tunnels. The extent of excavation is shown in Fig. 8.

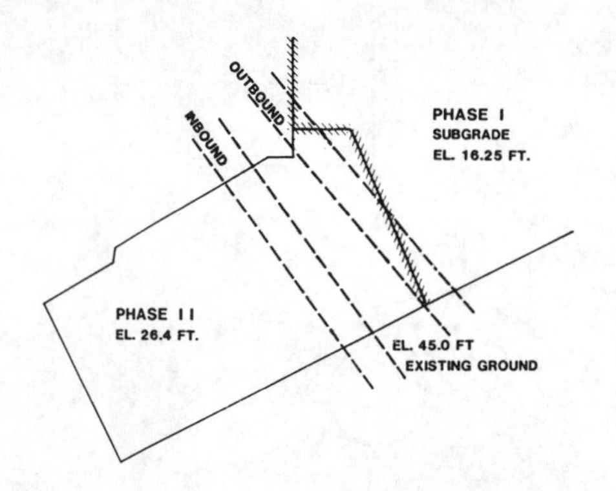

Fig.7 Site plan for 1201 New York Avenue

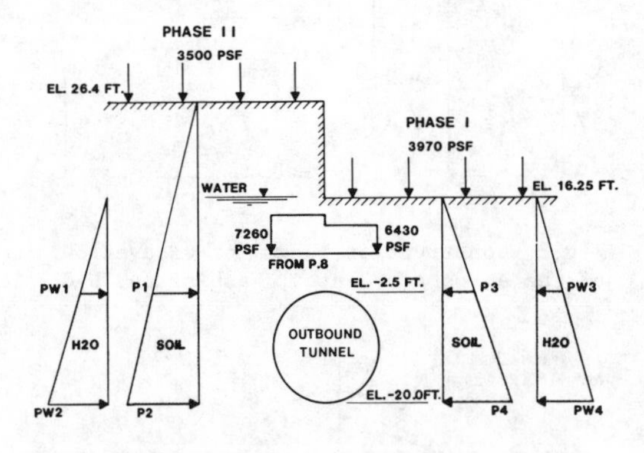

Fig.8 An asymmetrical loading condition for 1201 New York Avenue

Phase 2 building has a total of 17 floors, 12 above grade and 5 below grade. This structure straddles WMATA's inbound tunnel, and involves a shallower excavation than Phase 1.

Numerous analyses were performed to assess the worst-possible impact on the tunnels during the excavation process. It was established that with six feet of earth removed first over both tunnels, to mitigate the adverse effect of asymmetrical loading, the proposed excavation would be feasible insofar as the tunnel integrity was concerned.

The analysis used a modulus of subgrade reaction value of 250 kips per square foot. Gross cross-sectional properties of the steel tunnel liner were used in the beam-spring model. Liner stresses and deflections under all asymmetrical loadings were found acceptable. Maximum tunnel deflection was computed at 3/8-inch in element 47 (Fig. 3). An additional 1/8-inch maximum deflection due to lateral shift was computed.

A post-construction analysis was also performed to assess the effects of new building loads on the tunnel. Two conditions were considered:

Condition 1: Totally completed construction of Phase 1 and Phase 2 buildings.

Condition 2: Phase 1 completed, with only Phase 2 foundations and one floor level in place.

The analysis results indicated that Condition 1 governed. In this case 95% of the total liner strength is utilized. The maximum liner deflection occurs in the crown, and its magnitude is 1/2-inch.

A comparison between the computed and measured horizontal displacements of the lining at springline is shown in Fig. 9.

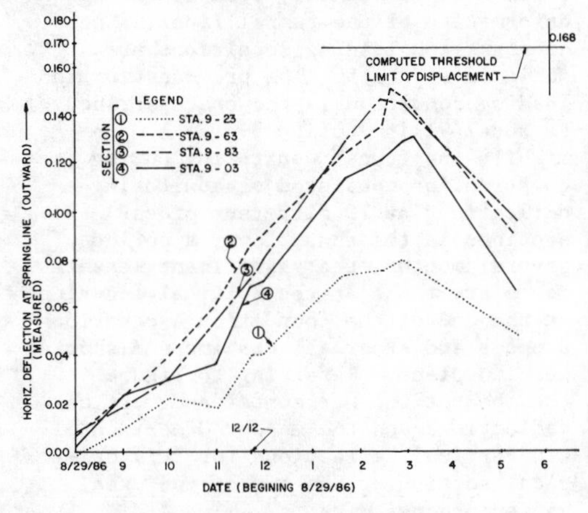

Fig.9 Horizontal displacements at springline for 1201 New York Avenue

It can be seen that the upperbound value of theoretical displacements compares well with the maximum displacement measured in-situ. Similar conservative results were obtained for the vertical displacement at the lining crown, indicating that the beam spring model can predict the flexible lining response with sufficient margin of safety.

5.3 Crowne Plaza Hotel Building

The hotel construction extended below grade to approximately the depth of WMATA's Metro Tunnels and to within a one foot horizontal distance of the nearest tunnel. The required excavation unloads the subway tunnels with resulting deformation in the tunnel liners. In addition, the soldier pile and sheeting, required for the excavation, which is braced by a combination of cross-lot struts, rakers and tie-backs, introduces further tunnel distortions resulting from the installation process. The construction loading conditions are shown in Figure 10. The preconstruction loading condition is similar to the one described as Model VI in Section 3.

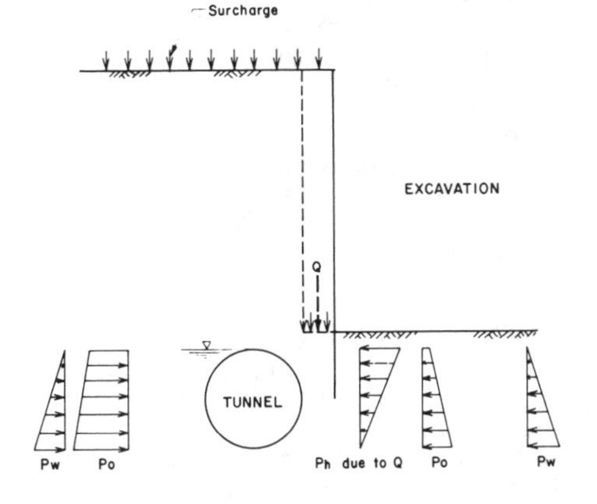

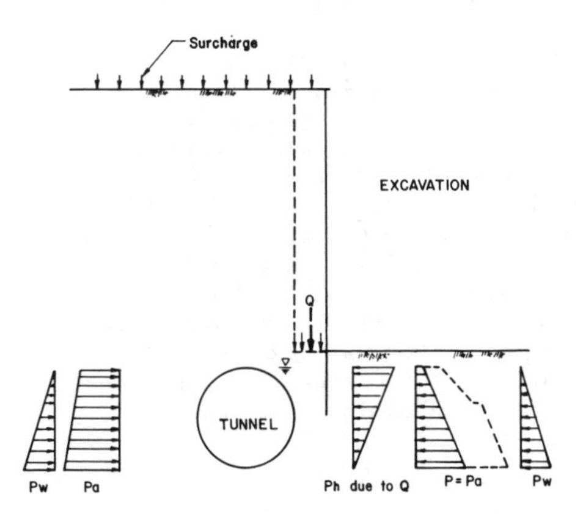

Fig.10 Construction loads for excavation of Crowne Plaza Hotel; Washington, D.C.

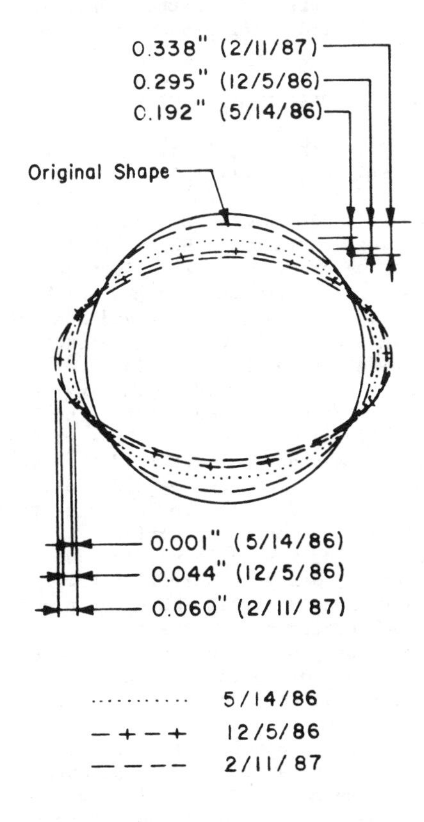

Fig.11 Aerospace Center - tunnel movement at the critical section.

Unlike the Aerospace Center project, the tunnel liner adjacent to the excavation of the Crowne Plaza Hotel exhibited an initial deflection during installation of the soldier piles. This occurred when excavation had only progressed downward 15 feet from the surface. It was a result of not following specified procedures which required use of a slurry while pre-augering the holes for the soldier piles and also use of a casing following the drilling auger no farther than 15 feet from the auger tip. Neither of these procedures were followed and holes for soldier piles quite close to the tunnel apparently collapsed with resulting tunnel liner movement. After this happened, it was necessary to measure, as well as predict tunnel liner movement, during excavation from an initial residual deflection. In addition, a prediction was also required on the total allowable movement the tunnel liners would be permitted to move without overstressing the liners. The latter analysis was also performed using a beam-spring model.

6 CONCLUSIONS AND RECOMMENDATIONS

Beam-spring analytical models can be efficiently and confidently used to described flexible steel tunnel lining response under complex unloading and reloading conditions. However, field monitoring programs must be instituted, not only to verify the results of the analysis, but also to safeguard against improper construction procedures not accounted for in the analysis assumptions. The field monitoring program must be continued until the tunnel lining displacements have stabilized, or a good indication of the stabilization trend has been demonstrated by the displacement plots at the completion of the construction.

Beam-spring methods have been used in the past to design tunnel linings. However, some designers have not placed too much confidence in the accuracy of deflections provided by this analysis. The work discussed in this paper suggests that this necessarily need not be so.

For future projects, liner strain measurements are recommended to augment the in-situ tape and optical survey measurements. This data could then be used to compare the theoretical and actual stresses in the liner, along with a comparison of liner displacements which are discussed in this paper.

REFERENCES

De Leuw, Cather & Company, 1986, "Adjacent Construction Review for 1201 New York Avenue (D-1) and Aerospace Center (D-4a). Computer Modeling of Effect of Building Excavation". Report submitted to WMATA

De Leuw, Cather & Company, 1986, "1201 New York Avenue Post - Construction Analysis D-1 Adjacent Construction" - Report submitted to WMATA

De Leuw, Cather & Company, 1987, "Adjacent Construction Analysis of Tunnel Liner Monitoring Data - 1201 New York Avenue Section D-1" - Report submitted to WMATA

De Leuw, Cather & Company, 1987, "Aerospace Center Building, Section D-4a, Analysis of Tunnel Liner Monitoring Data" - Report submitted to WMATA

Iffland Kavanagh Waterbury, P.C., 1985, "Investigation of the Effect on WMATA Tunnels by Construction of Proposed Aerospace Center", June, New York, N.Y.

Iffland Kavanagh Waterbury, P.C., 1986, "Investigation of the Effect on WMATA Tunnels by Construction of Proposed Crowne Plaza Hotel", November, New York, N.Y.

Muir-Wood, A.M., 1975, "The Circular Tunnel in Elastic Ground, "Geotechnique, Vol. 25, No. 1, March, pp. 115-127.

O'Rourke, T.D., 1984, editor, "Guidelines for Tunnel Lining Design", ASCE, pp. 1-82

Peck, R. B., Hendron Jr., A.J., and Mohraz, B., 1972, "State of The Art of Soft Ground Tunneling", Proceedings, Rapid Excavation and Tunneling Conference, Chicago, Illinois, Vol. 1, pp. 259-286.

Ranken, R.E., Ghaboussi, J., and Hendron Jr., A.J., 1978, "Analysis of Ground Liner Interaction for Tunnels", Report No. UMTA-IL-06-0043-78-3, U. S. Department of Transportation, Office of Secretary and Urban Mass Transit Administration, October, pp. 1-441.

Numerical Methods in Geomechanics (Innsbruck 1988), Swoboda (ed.)
© 1988 Balkema, Rotterdam. ISBN 90 6191 809 X

Numerical analysis of tunnel behaviour in creep-susceptible clays

Robert Y.K.Liang
University of Akron, Ohio, USA

ABSTRACT: Considerable effort has been devoted in the past decade toward the development of numerical procedures for tunnels in saturated, two-phase cohesive media, but little has been done to study the combined effects of hydrodynamic consolidation and creep. Recently, a constitutive model that considers the time-dependent behavior of soft clays has been developed and evaluated through a joint research effort by Stanford University and University of California, Berkeley. A versatile 2-dimensional finite element program has been developed as well. This numerical procedure provides an ideal tool for investigating the effects of creep on tunnel behavior in clays, particularly in terms of time-dependent variations of deformation field, pore pressure field, and liner pressures. In this paper, the abstract of the constitutive model as well as the implementation of coupled elasto-plastic consolidation program is briefly described. A set of results of parametric study on the effect of creep and existence of gaps between liners and surrounding soils is given. It is shown that ground surface deformation is greatest when gaps and creep are present. Inclusion of creep consideration results in higher pore pressure response and retarded pore pressure dissipation. Finally, for the stiff liner system, such as concrete liners, liner pressures increase with time. The increase of liner pressures is more pronounced when both hydrodynamic consolidation and creep effect are considered.

1 INTRODUCTION

For highly plastic, or recently deposited "young" clays, creep can be an important contributor to the time-dependent deformation when subjected to boundary loads. To completely describe the stress-strain-volume change-time behavior of saturated soft clays under general stress and strain conditions, a constitutive model considering both the hydrodynamic lag and creep phenomena has been developed and verified. A two-dimensional plane-strain and axi-symmetrical finite element program with implementations of such a constitutive model has been developed. The validity of the numerical procedure has been established through a comprehensive evaluation process which includes (i) single element type comparisons with laboratory test performed with various stress paths and drainage conditions, (ii) centrifuge model test results, and (iii) field case study. Based on these extensive evaluations, the numerical procedure seems to provide adequate computational capability for 2-D boundary value problems involving time-dependent deformation and pore pressure responses of soft, creep-susceptible clays.

Numerical analyses of tunneling in two-phase clay media have been developed in the past. However, very little effort was devoted to study the possible effects of creep on the tunnel behavior. As observed from results of various laboratory test, such as undrained creep tests and drained creep tests as well as from small-scale model tests of some kinds of prototype structures, the consideration of creep behavior sometimes can be important. The excavation of underground opening and subsequent installation of supporting structures such as concrete liners are a complex boundary value problem. The effects of creep on the system performance therefore can only be studied by either model tests or numerical simulations. Model tests require consideration of scaling effect. Also, study of creep effects in model tests can be time-consuming. As an alternative, numerical simulation technique using a well verified finite element program can provide much needed insights on such a problem. The purpose of this paper is to present the formulation and implementation of a creep-inclusive constitutive model in a 2-D coupled finite element program and also to present parametric study results of tunnel behavior in creep-susceptible clays.

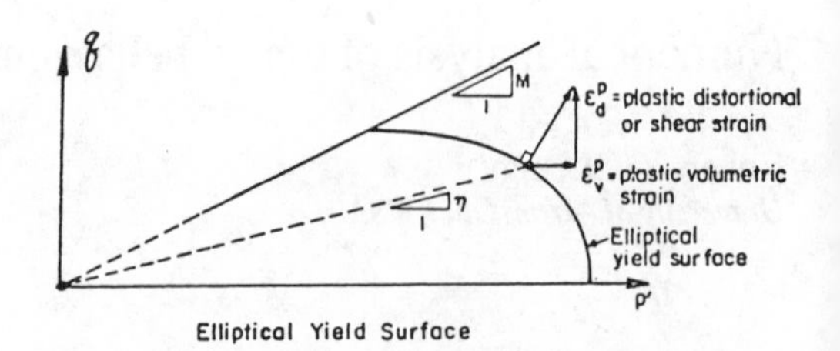

Fig. 2 Cap model + scaling procedure

2 CONSTITUTIVE EQUATIONS

The stress-strain-volume change-time relationship of fully saturated soft clays was formulated on the basis of phenomenologically observed behavior and the classical critical state theory originally forwarded by Roscoe and Burland (1968). The total deformation was considered to consist of two components: immediate and delayed deformations, according to Bjerrum's concept (1967). The immediate deformation refers to those caused by the change of effective stress. Due to hydrodynamic lag in cohesive soils, this change of effective stress is a time-dependent process and it is usually described by a diffusion theory. The delayed deformation refers to those caused by viscous nature of cohesive media which would include undrained or drained creep. Fig. 1 shows the concept of the immediate/delayed deformation and its relationship to the more conventional description of primary/secondary compression.

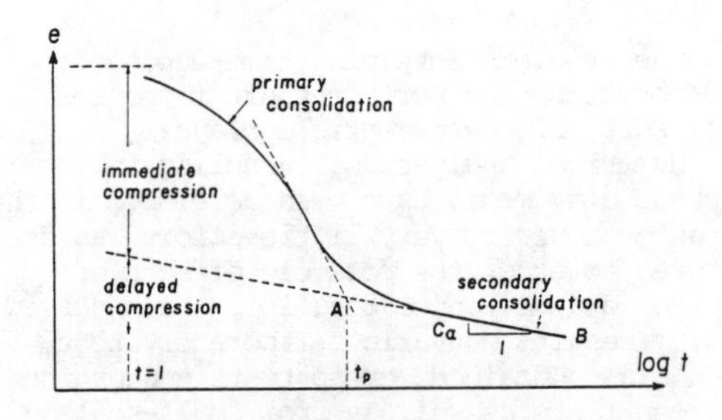

Fig. 1 Immediate/delayed vs. primary/secondary

Because some adopted phenomenological stress-strain relations in the constitutive equation are described in terms of deviatoric and volumetric stress. The constitutive model further separates the stress tensor into deviatoric and volumetric components. The relationships between deviatoric stress and deviatoric strain, however, is dependent upon the volumetric stress-volumetric strain relations.

In brief, the immediate strains are described by the modified Cam-clay formulation. The elliptic cap and the Mohr-coulomb failure line in the p-q diagram (see Fig. 2) defines the yield surface (Eq.1). The hardening law is provided by the isotropic consolidation curve in which the pre-consolidation pressure (or equivalent pre-consolidation pressure in the case of overconsolidated clays) governs the evolution of the yield surface. By evoking the consistency requirement on the yield surface and by assuming the normality rule applies, both immediate volumetric and immediate deviatoric strains can be calculated from a given effective stress increment, which in fact is controlled by the excess pore pressure diffusion process.

$$\frac{q^2}{M^2} + P(P-P_c) = 0 \qquad (1)$$

$$P = \sigma_{oct} , \; q = \frac{3}{\sqrt{2}} \tau_{oct}$$

The delayed strains occur with passage of time, even under constant effective stress condition. The phenomenologically derived (but verified from the thermodynamics rate process theory) Singh-Mitchell creep equation (Singh and Mitchell, 1968), is adopted for calculating the deviatoric part of delayed strains. By further assuming that normality rule applies to the delayed strain increments on the equivalent yield surface, the corresponding delayed volumetric strain increments can be calculated. The Singh-Mitchell equation can be written as Eq. 2.

$$\dot{\varepsilon}_a = A \exp(\bar{\alpha} \bar{D}) \left(\frac{t_i}{t}\right)^m \qquad (2)$$

where m, A, and $\bar{\alpha}$ are material constants to be determined from either triaxial undrained or drained creep tests.

The development of quasi-preconsolidation due to aging (drained creep under constant effective stress) is considered in the constitutive model by using two state variables --volumetric time and deviatoric time. These two variables are indicative of differences between the current state (e,p,q) with respect that it would be without creep effect. For example, the volumetric time state variable is calculated as in Eq. 3. Similarly, the deviatoric time state variable can be calculated from Eq. 4.

$$t_v = (t_v)_i \exp\left(\frac{e_2 - e_1}{\psi}\right) \tag{3}$$

$$t_d = \left[\frac{(\gamma_1-\gamma_2)(1-m)}{A \exp(\bar{\alpha}\,\bar{D})\,(t_d)_i^m}\right]^{\frac{1}{1-m}}, \text{ if } m=1 \tag{4}$$

$$= (t_d)_i \exp\left(\frac{\gamma_1-\gamma_2}{A \exp(\bar{\alpha}\,\bar{D})}\right) \text{ if } m=1$$

As pointed out by Leonards and Ramiah (1968), the development of quasi-preconsolidation pressure is to cause the material to behave stiffer upon subsequent loading. In order to account for this aging-induced strain hardening behavior, another set of hardening law is provided as in Eq. 5.

$$\frac{\partial P_c}{\partial t_v} = \frac{\psi}{\lambda - k}\;\frac{P_c}{t_v} \tag{5}$$

In summary, the constitutive equation developed is an elasto-plastic plasticity model but with inclusions of creep as well as quasi-preconsolidation effects. The model requires a total of 10 material constants which can be determined from standard triaxial testing apparatus. Preferably, a series of isotropically consolidated undrained tests and isotropically consolidated undrained creep tests should be performed to determine these material constants. Table 1 summarizes the required material constants and methods of obtaining them.

3 FINITE ELEMENT PROCEDURE

One of the advantages of formulating a creep-inclusive constitutive model using the scheme illustrated above is that the conventional numerial solution procedures developed for elasto-plastic consolidation problems can be readily adopted. This is because the general stress-strain equation can be reduced to Eq. 6, in which the stress relaxation term, $\dot{\sigma}^t_{ij}$, can be calculated at the beginning of each time increment, thus allowing for it to be brought to "the right hand side of equation".

Table 1 Material constants and numerical values used in the case study

MODEL PARAMETERS		
PARAMETER	SYMBOL	VALUE
Virgin compression index[1]	λ	0.147
Recompression index[1]	κ	0.06
Secondary compression coefficient[1]	ψ	0.0019
Hyperbolic stress-strain parameters[2]	a b R_f	0.0062 2.728 0.9
Singh-Mitchell creep parameters[3]	A $\bar{a}$ m	0.000144 2.475 0.642
Permeability[1]	k_h, k_v	0.00113, 0.00054
Slope of critical state line	M	1.4
Void ratio[1] at P_c = 1 KPa	E_A	3.56

[1] From 'triaxial' IC or convential consolidation test

[2] From ICU test with pore pressure measurement

[3] From ICU-creep test

$$\dot{\sigma}_{ij} = C^{ep}_{ijkl}\,\dot{\varepsilon}_{kl} - \dot{\sigma}^t_{ij} \tag{6}$$

In the development of the finite element formulation, four governing equations were evoked:

1. Equilibrium between the incremental stresses and the incremental body forces (Eq. 7)

$$\frac{\partial \dot{\sigma}_{ij}}{\partial x_j} - \dot{F}_i = 0 \tag{7}$$

2. Stress-strain relation between the incremental effective stresses and the incremental strains (Eq. 8)

$$\dot{\sigma}'_{ij} = C^{ep}_{ijkl}\,\dot{\varepsilon}_{kl} + \dot{\sigma}^t_{ij} \tag{8}$$

3. Darcy's law for the rate of flow of water through soil (Eq. 9)

$$\dot{\nu}_i = \frac{k_{ij}}{\partial_w}\,\frac{\partial \dot{P}}{\partial x_j} \tag{9}$$

4. Relationship between the rate of volume decrease of the element and the rate at which water is expelled, assuming the pore water is incompressible relative to the soil skeleton (Eq.10)

$$\frac{\partial \dot{\nu}_i}{\partial x_j} = \frac{\partial \dot{\varepsilon}_v}{\partial t} \tag{10}$$

The coupled equations for an element can be written in matrix form (Carter, Booker and Small, 1979):

$$\begin{bmatrix} K & -L^T \\ -L & -\beta dt\Phi \end{bmatrix} \begin{bmatrix} \Delta\delta \\ \Delta q \end{bmatrix} = \begin{bmatrix} f \\ ndt+\Phi q_o dt \end{bmatrix} \quad (11)$$

where the element stiffness matrix [K] is given by

$$[K] = \int_V [B]^T [D] \; [B] \; dv \quad (12)$$

where [B] is the strain-displacement matrix.

The matrix [L] couples the stiffness equation with those of fluid flow and is given by

$$[L]^T = \int_V \{B_V\}^T \{N_P\}^T dv \quad (13)$$

where $\{B_V\}$ is volumetric strain tensor, $\{N_P\}$ is pore pressure shape function.

The fluid flow matrix is given by

$$[\Phi] = \frac{1}{\gamma_w} \int_V [E]^T [K]^T [E] \; dv \quad (14)$$

where
$[E]^T = [\{\frac{\partial N_P}{\partial x}\}, \{\frac{\partial N_P}{\partial y}\}, \{\frac{\partial N_P}{\partial z}\}]$, [K] is the permeability matrix.

The existence of a hydrostatic pore pressure distribution is accounted for by adding the following term to the load vector:

$$\{n\} = \int_V [E]^T \; [K]^T \; [i_g] \; dv \quad (15)$$

The detailed description of the constitutive model formulation, its implementation in a coupled elasto-plastic consolidation scheme, and its verification against test results can be found in Borja and Kavazanjian (1985), Mitchell and Liang (1987).

4 PARAMETRIC STUDY

Several FEM analyses were conducted to study the effects of soil creep and gaps on the deformation, pore pressure response and liner pressures. Because of space limitations, only a set of analysis results will be presented herein.

The problem analyzed is presented by the FEM mesh shown in Fig. 3. A circular opening of 11 meter in diameter is excavated with an overburden cover of approximately 8 meter. A 1.75 meter thick of concrete liner is installed in place. The excavation is assumed to take place in one day in the numerical analyses, although a more realistic excavation process can be used in real situation. The mesh consists of a total of 237 nodal points. A Q8P4 element (8 displacement nodal points and 4 pore pressure nodal points) is used as shown in the mesh. To simulate the gap condition, an artificial soft material with Young's modulus in the neighborhood of 10 Kpa is used. A total of 110 time incremental steps were used for a time period of 300 days.

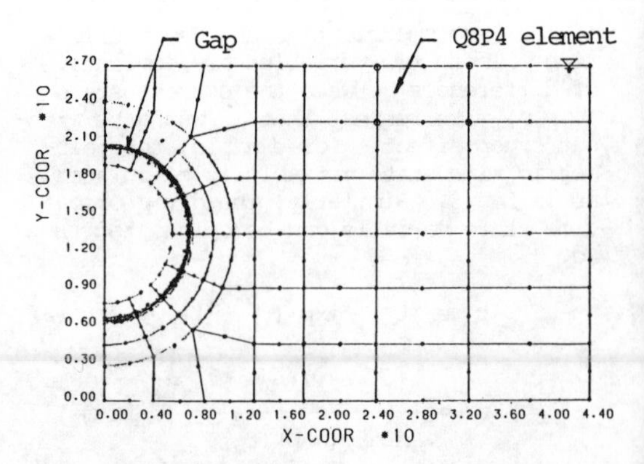

Fig. 3 FEM mesh

Four case study results will be shown in this paper. These include the situation where no gaps exist between the soil and the liner and the case where a gap of 0.02 meters exists. For both cases, two sets of analyses involving consideration of creep and no-creep were performed. Thus, the total number of cases studied is four.

4.1 Ground surface settlement

The ground surface settlement as a function of time for the four cases studied is shown in Fig. 4(a) through Fig. 4(d). It can be seen that the cases including creep have larger vertical deformation directly above the opening. However, some heaves occur in the far field away from the opening. This phenomenon was not predicted by non-creep analyses.

An examination of differences between the cases where gap exists and where gap does not exist, it can be concluded that gaps contribute to the immediate deformation, but they do not affect the long term deformation. This is certainly not surprising as the gaps are simulated in the numerical analysis as a very soft material, they deform rather readily under the stress changes caused by excavation.

4.2 Deformation field

To gain a better appreciation about the deformation pattern of the soil mass surrounding the opening, the displacements of several key nodal points are plotted as a function of time in Fig. 5 through Fig. 7.

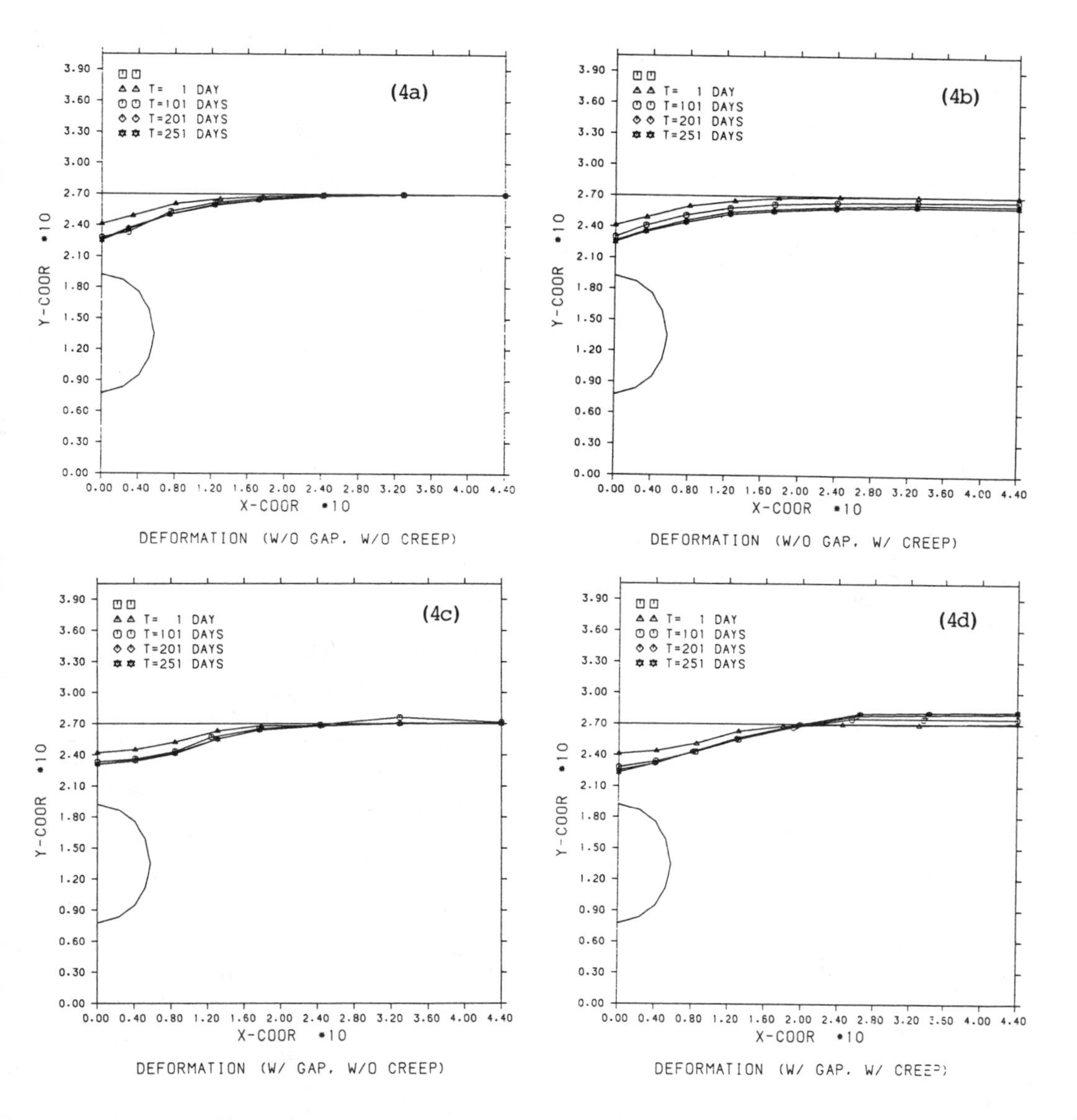

Fig. 4 (a - d) Ground surface settlements

Fig. 5 shows that the soil mass directly underneath the opening experiences expansion (or heave) immediately after tunnel excavation. However, with passage of time and the associated pore pressure dissipation, the soil mass compresses, resulting in downward movement with time. The effect of gaps is to allow the soil to expand stronger than it would be where gaps do not exist. Also, notice that creep seems to retard the recovery of heave. Similar observation can be made for nodal point 121 (see Fig. 6) which is in the crown position of the tunnel. The gaps allow the soil to deform more. Also, creep seems to accelerate the downward movement. The displacements of nodal point 65, which is at the spring line of the tunnel is shown vectorially in Fig. 7. It can be seen that because of the complex stress changes taking place, the effect of creep and gaps can be very profound. As a result of the presence of gaps, the point moves inward significantly at the beginning of excavation. The point moves outward during subsequent consolidation and creep. A less amount of movements was observed for no-gap cases.

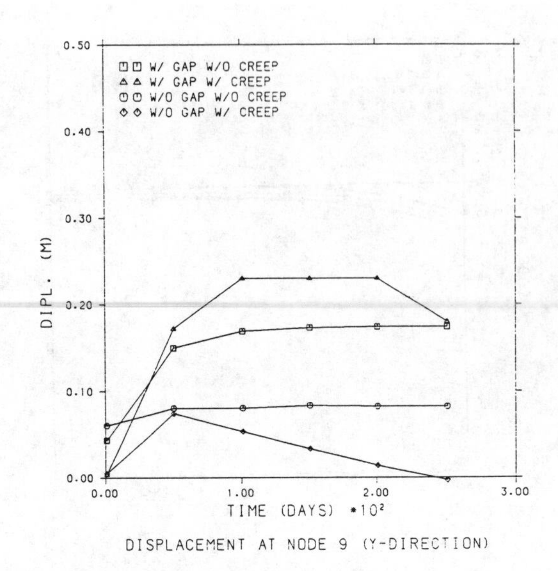

Fig. 5 Nodal heave

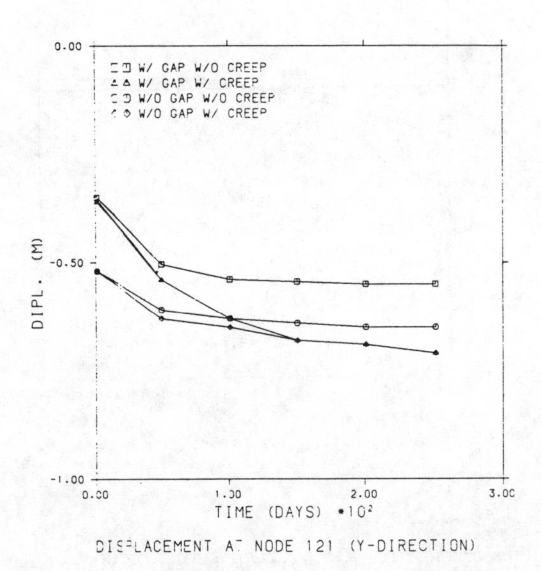

Fig. 6 Nodal 121

4.3 Pore pressure distribution

Pore pressure variations as a function of time in several representative elements are shonw from Fig. 8(a) to Fig. 8(e). Fig. 8(a) shows the pore pressure changes in element 1. As a result of expansion, negative pore pressure developed which then dissipated with time due to the diffusion process. Different pore pressure dissipa-

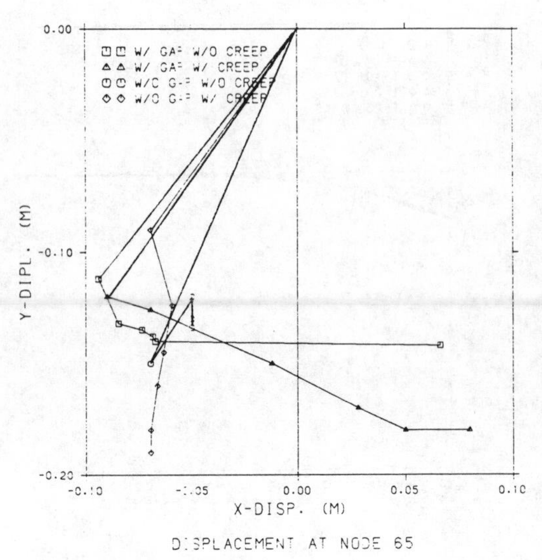

Fig. 7 Nodal 65

tion rates were observed for creep and no-creep cases.

Fig. 8(b) shows pore pressure variations of element 5, which is close to the spring line of the tunnel. A consistent positive pore pressure developed. Because the nature of coupling between deformation and pore pressure in the numerical formulation, it is expected that larger deformation in the soil mass would induce larger pore pressure. This is apparently the case.

Fig. 8(c) shows the pore pressure for element 8 which is directly above the crown of the tunnel. Both positive or negative pore pressures were possible, depending

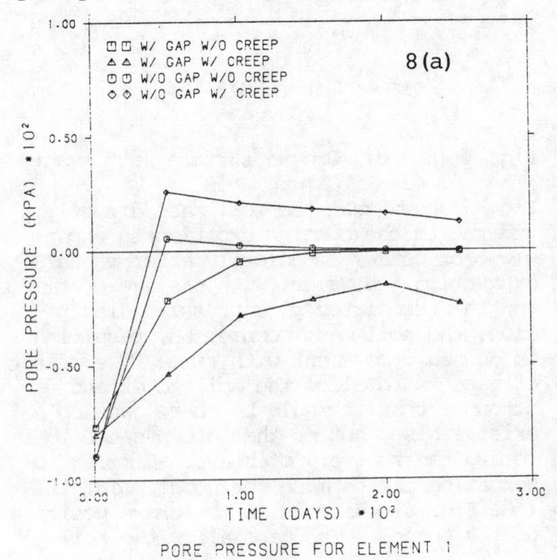

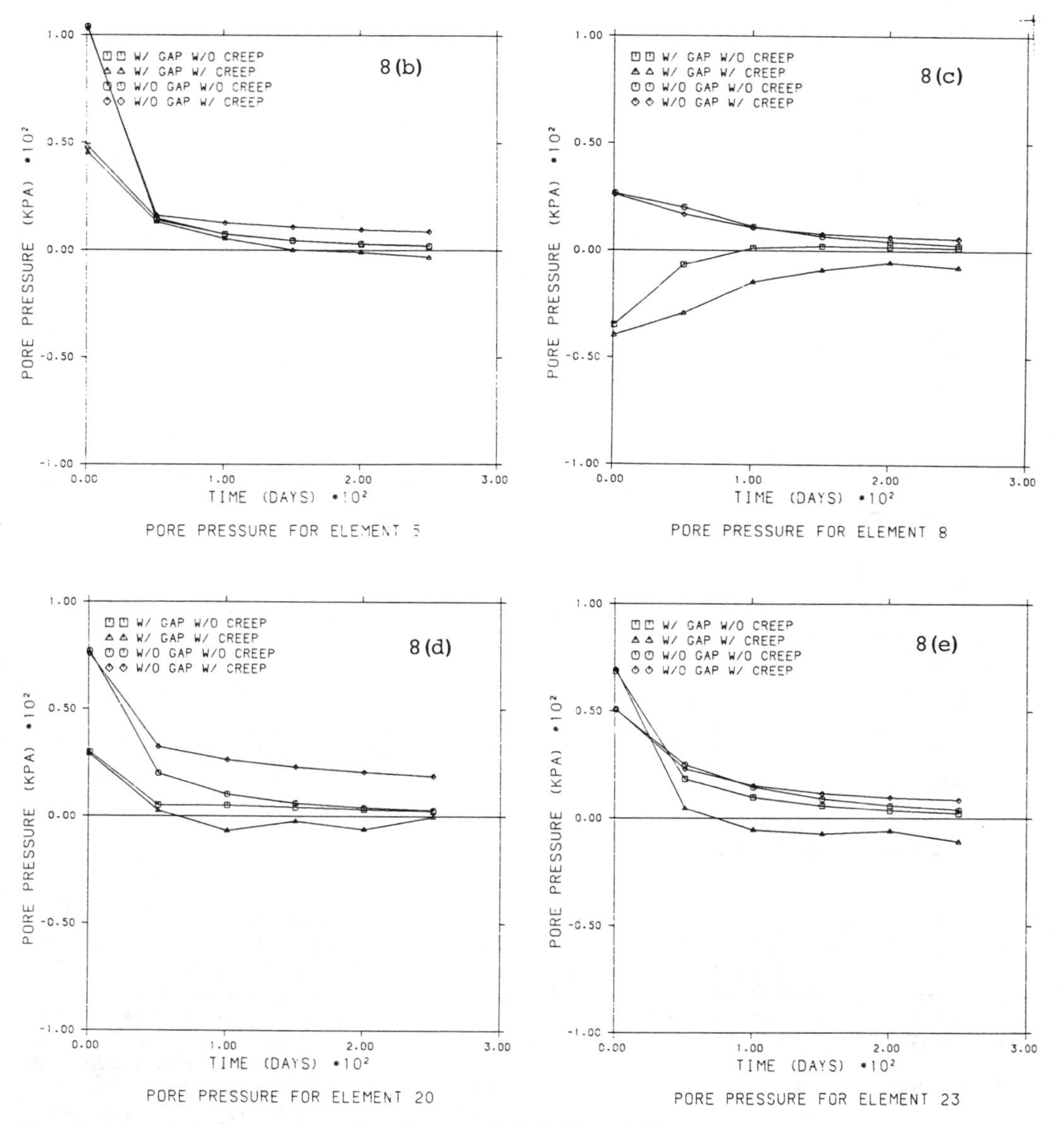

Fig. 8 (a - e) Pore pressure in elements 1,5,8,20 and 23

upon the amount of expansion experience by the soil element.

The pore pressure variations for soil elements in the far field, such as elements 20 and 23, are shown in Fig. 8(d) and Fig. 8(e), respectively. Again, the magnitudes and rates of dissipation of pore pressures are significantly affected by creep as well as existence of gaps.

4.4 Liner pressure distributions

One of the major design considerations of stiff liners (concrete liners) in the soft clay media is the strength against soil' pressure. As the concrete liner is a fairly non-yielding material, it is expected that earth pressure on the liner will gradually build up with time, particularly in creep-susceptible clays. The numerical procedures provide quantitative predictions of possible ranges of earth pressure increases. As shown in Fig. 9(a) through Fig. 9(d), the liner pressures increase and sometimes redistribute with passage of time. Since this study is to understand the qual-

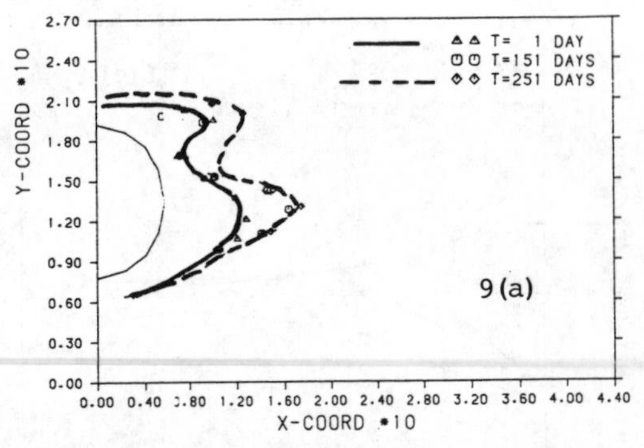

STRESS DISTRIBUTION (W/O GAP, W/O CREEP)

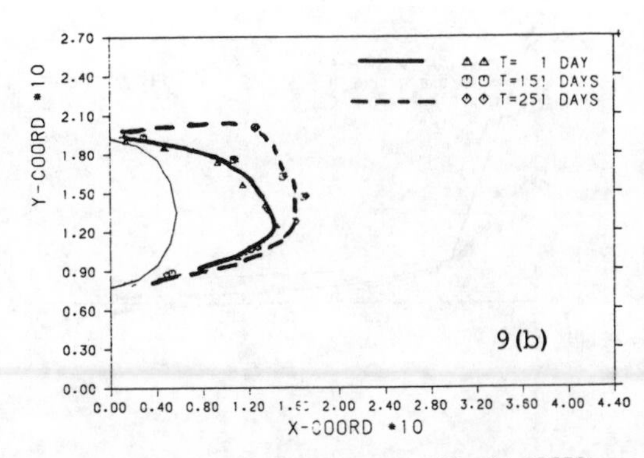

STRESS DISTRIBUTION (W/O GAP, W/ CREEP)

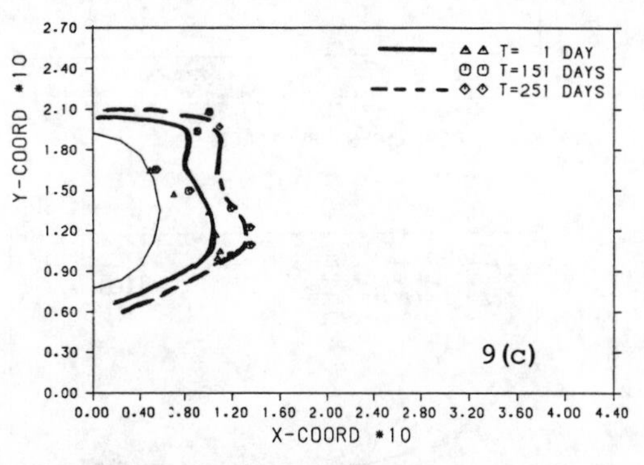

STRESS DISTRIBUTION (W/ GAP, W/ CREEP)

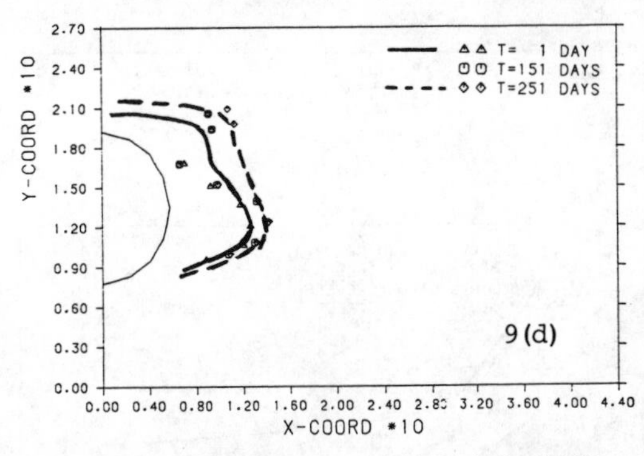

STRESS DISTRIBUTION (W/ GAP, W/O CREEP)

Fig. 9 (a - d) Liner pressure distribution

itative behavioral pattern, therefore, no numerical values were indicated in the figures. Changes of geometry conditions, soil properties, material properties would certainly alter the numerical values of the computed liner pressures.

5 CONCLUSION REMARKS

Numerical analysis procedures for tunneling behavior in soft clays have been in existence in the past; however, this paper presents a numerical procedure which accounts for the combined influence of hydrodynamic lags and creep. As the creep-inclusive constitutive model and its numerical implementations have been validated extensively through comparisons with laboratory test results as well as centrifuge model studies, the results of parametric study of tunnel problem as presented herein should have adequate reliability -- at least to qualitatively demonstrate the significance of creep effects, particularly for those highly plastic, young clay deposits.

Based on the case studies presented herein, it can be concluded that the time-dependent variations of deformation field, pore pressure field, and liner pressure will vary significantly as a consequence of creep and gap effects. Ground surface settlements are greatest for the case where both gap and creep are present. The pore pressure response, in general, is higher if creep is considered. The pore pressure dissipation may be retarded as a result of viscous creep effect. Liner pressures in general increase with time in the stiff liner system; a larger increase of liner pressure is expected in the creep-susceptible clay media.

REFERENCES

Bjerrum, L. 1967. Engineering geology of normally consolidated marine clays as related to settlements of buildings. Geotechnique, 17:2:82-118.

Borja, R.I. and Kavazanjian, Jr., E. 1985. A constitutive model for the stress-

strain-time behavior of wet clays. Geotechnique 35:3:283-298.
Carter, C., Booker, J.R., and Small, J.C. 1979. The analysis of finie elasto-plastic consolidation. Int. J. for Num. and Anal. Meths. in Geomechanics. 3:107-129.
Leonards, G.A. and Ramiah, B.K. 1960. Time effects in the consolidation of clays. ASTM SPT 254:116-130.
Liang, R.Y.K. and Mitchell, J.K. 1987. Centrifuge evaluation of a numerical model for clay. Accepted for ASCE, J. of Geot. Eng.
Roscoe, K.H. and Burland, J.B. 1968. On the generalized stress-strain behavior of 'wet' clay. Engineering Plasticity, Cambridge University Press, 535-609.
Singh, A. and Mitchell, J.K. 1968. Generalized stress-strain-time function for soils. J. Soil Mech. Found. Eng., ASCE 94:SM1:21-46.

Numerical Methods in Geomechanics (Innsbruck 1988), Swoboda (ed.)
© 1988 Balkema, Rotterdam. ISBN 90 6191 809 X

Prediction of tunnelling effect on groundwater condition by the water balance block method

K.Daito & K.Ueshita
School of Geotechnical Engineering, Nagoya University, Japan

ABSTRACT: It is reported that a number of tunnelling works have influenced water environment and caused troubles to people in surrounding area. Therefore we must predict this problem and take steps to find a solution for this problem. In this paper, a method using water balance block model and tank model is proposed to predict the change of tunnel discharge, groundwater level and river discharge. As a case study, this method was used to predict the tunnel discharge and the change of groundwater level in a mountainous area of a newly constructed highway tunnel and proved to be useful.

1 INTRODUCTION

Recently, tunnel construction works in mountainous areas for railways and highways have increased in Japan. The daily life of the people in the surrounding areas of constructed tunnels have been affected by drying up of wells and springs.

If these problems were not predicted at the early stage of the tunnel construction and the prevention works were not taken, when these problems occurred, lawsuits regarding compensation were filed which eventually led to the delay or stop of the construction works. In order to avoid these problems and smoothen the process of construction, we have to find out how the groundwater condition will be affected during and after the tunnel construction and have to take steps to solve these problems.

In this paper, the water balance block simulation method is practically presented based on Oshima's study [1] showing a case study on the tunnel construction in a mountainous area in Japan.

2 PARAMETERS RELATING TO WATER BALANCE SIMULATION

2.1 Fundamental equation of water balance

A water balance model is shown in Figure 1. In this model, the relationship of each parameter is given by

$$R = E + D + G + Q + SS \qquad (1)$$

where

- R : Precipitation (rainfall and snow),
- E : Total evaporation from ponds, ground surface and plants,
- D : Surface runoff from the higher to the lower regions,
- G : Groundwater flow from the higher to the lower head,
- Q : Tunnel discharge,
- SS : Surface storage and storage in soil.

The parameters, which are used in the water balance simulation, are briefly explained below.

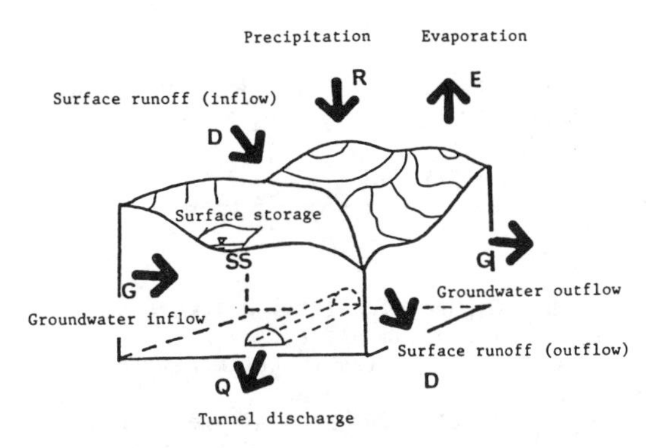

Fig.1 The water balance model

2.2 Precipitation

In order to know the precipitation of a region, measured data is necessary. In this research, the quantity of the melted snow and ice is also considered as precipitation.

2.3 Evaporation

In this research, the Hamon equation [2] is used to estimate the evaporation, which is given by

$$E_p = 1.40\ D_0{}^2 \cdot p_t \qquad (2)$$

where
- E_p : Total evaporation (mm/day),
- D_0 : Sun shine time (12 hr/day),
- p_t : Absolute saturated humidity related with monthly average temperature (g/m³).

2.4 Surface runoff

In this research, surface runoff, subsurface runoff and groundwater runoff are calculated by using the tank model which was presented for flood analysis [3] as shown in Figure 2. But when the depth between the ground surface and groundwater table is small, only surface runoff and groundwater runoff are considered, and subsurface runoff is not necessary to be considered.

2.5 Groundwater flow

Groundwater flow (outflow and inflow) is calculated with Darcy's law among the adjoining water balance blocks described later.

2.6 Tunnel discharge

In this research, the tunnel discharge q is estimated using the following equation [4].

$$q = \frac{2\pi k\ (\ H - h_{x=x_1}\ )}{\ln\ (\ 2x_1\ /\ r_0\ )} \qquad (3)$$

where
- H : Groundwater level measured from the datum plane,
- x_1 : Depth of the tunnel measured from the datum plane,
- $h_{x=x_1}$: Total water head at the tunnel measured from the datum plane,
- k : coefficient of permeability,
- r_0 : Radius of the tunnel.

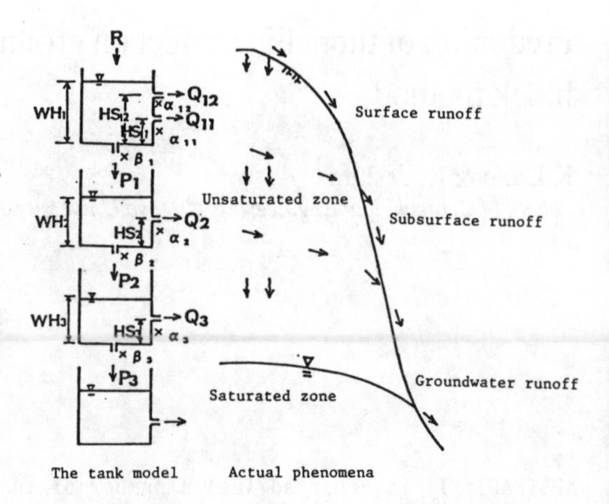

Fig.2 Concept of the tank model

The tunnel discharge which is used in the water balance simulation, is obtained from multiplying equation (3) by the tunnel length in the block.

3 WATER BALANCE RELATIONSHIP BETWEEN SIMULATION BLOCKS

3.1 Concept of water balance block

In the water balance block model when there is not a tunnel, groundwater level and surface runoff are affected only due to the change of weather, for example precipitation, temperature, wind velocity, etc.. When a tunnel exsists, it is affected by the following:

1. Tunnel discharge calculated by equation (3) occures during tunnelling.
2. The groundwater level in the block will become lower by the tunnel discharge. Therefore the tunnel block will influence the adjoining six blocks.
3. In the block which is separated far from the tunnel, the change in the groundwater level is small.
4. The change in the goundwater level will influence the river discharge.

3.2 Piezometric change in water balance block due to tunnel discharge

When the coefficient of permeability, groundwater head, tunnel diameter, tunnel length and tunnel location in the mountainous region are known, the tunnel discharge can be calculated by using equation (3). The water head in the block will become less and the average decrease in water head (ΔH) is given by

$$\Delta H = Q / P \qquad (4)$$

where

Q : Tunnel discharge,

P : Effective porosity.

As shown in Figure 3, block C is the center of adjoining six bloc and the water head in block C is related with the water heads in adjoining blocks. If the groundwater head, the coefficient of permeability and effective porosity are known in each block, the change of water head in block C during a short time is calculated by considering Darcy's flow between block C and blocks F, R, B, L, O and U.

Water head (H), coefficient of permeability (k) and effective porosity (P) of block C are expressed by H_c, k_c and P_c using suffix of the block. If the width of blocks are ℓ and the height of the block is m, average coefficient of permeability between blocks C and F (k_{cf}) is given by

$$k_{cf} = 2k_c k_f / (k_c + k_f) \qquad (5)$$

Therefore the inflow per unit time from block F to block C (Q_{cf}) is given by

$$Q_{cf} = \{ k_{cf} (H_f - H_c) / \ell \} \ell m \qquad (6)$$

The water head in block F is decreased by Q_{cf}/P and becomes $H_f - Q_{cf}/P$. On the otherside, the water head in block C is increased by Q_{cf}/P and becomes $H_c + Q_{cf}/P$. Thus, the difference in water head between both blocks becomes smaller.

Using the same method as above, the inflow per unit time from the blocks B, L, R, O and U to block C can be calculated. If Q_{cb}, Q_{cl}, Q_{cr}, Q_{cp} and Q_{cu} are calculated, the rise of head in block C (ΔH_c) can be shown by

$$\Delta H_c = (Q_{cf} + Q_{cb} + Q_{cl} + Q_{cr} + Q_{co} + Q_{cu}) / P_c \qquad (7)$$

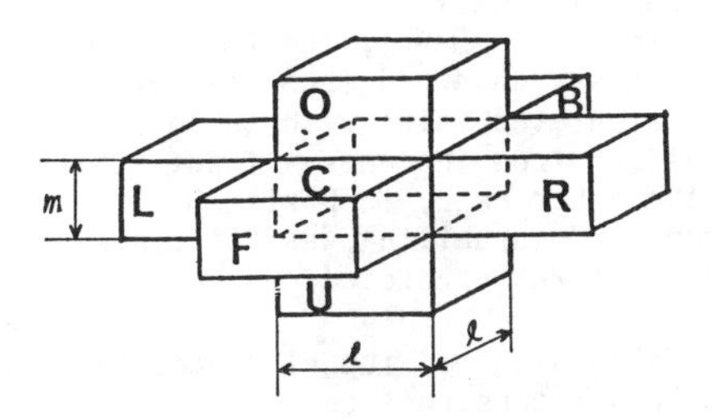

Fig.3 Block C and adjoining six blocks in the water balance block model

For the above calculation the following parameters are needed.

1. In relation to the tunnel : The tunnel diameter and position of tunnel face with time.

2. In relation to block : Coefficient of permeability, effective porosity and groundwater head.

4 WATER BALANCE SIMULATION IN THE CASE OF SHIOJIRI TUNNEL CONSTRUCTION AREA

4.1 Topography, geology and hydrological condition of the considered area

In this chapter, Shiojiri tunnel (about 1.8 km length) on the Nagano route of the Chuo expressway constructed by the Japanese Highway Corporation is used as the case study to predict the change of groundwater level, river and tunnel discharge during and after tunnelling.

The most of the route of the Chuo expressway passes through the area of the Tertiary or the Quarternary deposits. There are several faults in this area which should be considered for study on groundwater condition.

In this region, the surface water is being used as drinking water and irrigation water. Therefore the influence for the hydrological environment during and after tunnel construction has to be predicted, and neccessary steps should be taken to avoid troubles of water.

4.2 Analytical procedure of water balance simulation

The investigation committee on groundwater condition before and after tunnel construction in the Chubu branch of Japanese Society of Civil Engineers surveyed the surrounding area of the Siojiri tunnel. The east-west and north-south geological cross sections of the area were drawn by every 200 m intereval. Coefficient of permeability of each layer was determined by the pumping test in a bored hole.

The analyzing area is shown in Figure 4. The length of east-west is 3600 m and the length of north-south is 2200 m, where the south-west part is bordered by fault line. So the considered area is 7.09 km^2.

With reference to the geological cross sections, this analyzing area is modeled with 10635 blocks as shown in Figure 5. Size of each block is 100 m $\times$ 100 m $\times$ 50 m. Figure 6 shows the geological property of a modeled layer at M.S.L. +850 $\sim$ 900 m as an example. The coefficient of permeability and effective porosity are shown in Table 1.

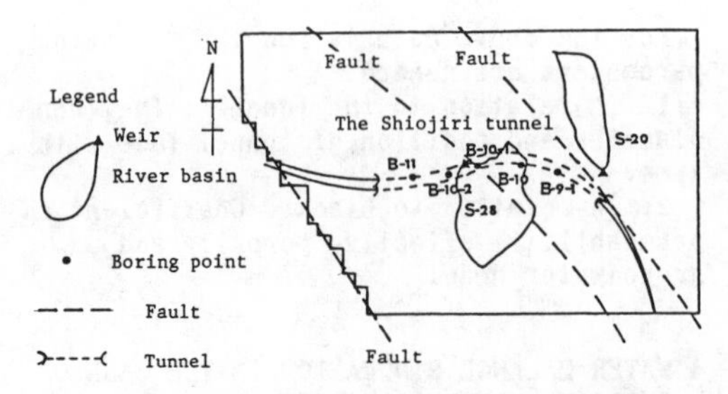

Fig.4 The route of the Shiojiri tunnel and the analyzing area

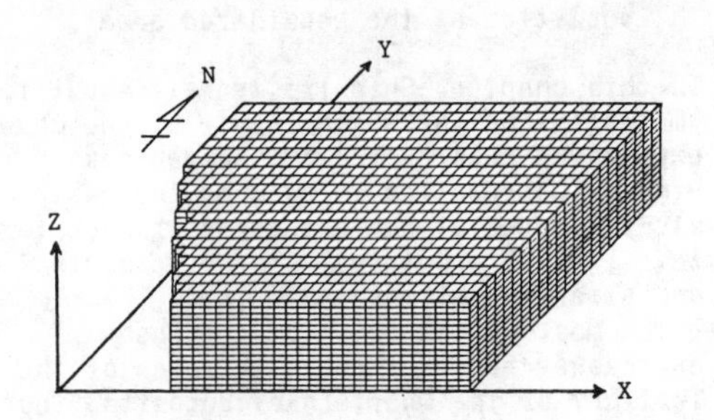

Fig.5 Blocks of the considered area for the water balance block model

Initial groundwater level can be obtained from the geological cross sections which shows investigated groundwater table.

The river discharge of basins are measured by weirs at the lower reaches. For the block where the depth from the ground surface to the groundwater level is less than 30 m, a one-tank model is used to analogize downflow of groundwater through unsaturated zone, and for the depth more than 30 m, a three-tank model is used.

The boundary conditions of the model are described below:

1. At the all outside boundaries, groundwater level is assumed to be constant.
2. At the bottom of the model which is about 300 m lower than the tunnel, impermeable condition is assumed.

4.3 Calculation of groudwater level before and during tunnel construction

Figures 7a and 7b show changes in observed groundwater level and calculated values for B-9-1 and B-11 observation wells before and during tunnel construction period (from April 1, 1984 to November 30, 1985). Computed groundwater level without tunnel construction is shown in the same figure for comparison.

From these figures, the following are known:

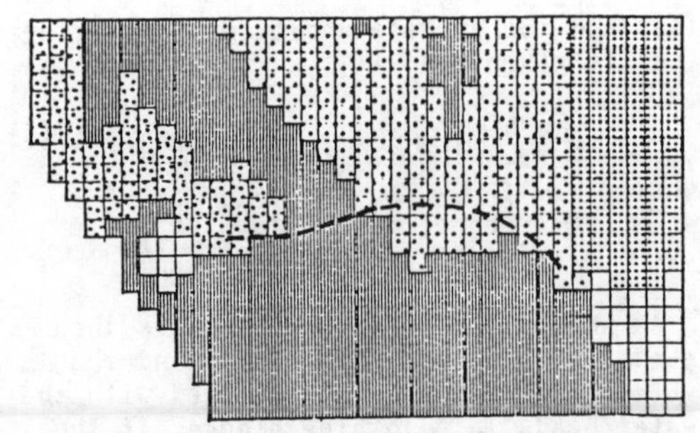

M.S.L.+850～900 m (The layer including the tunnel)

Legend

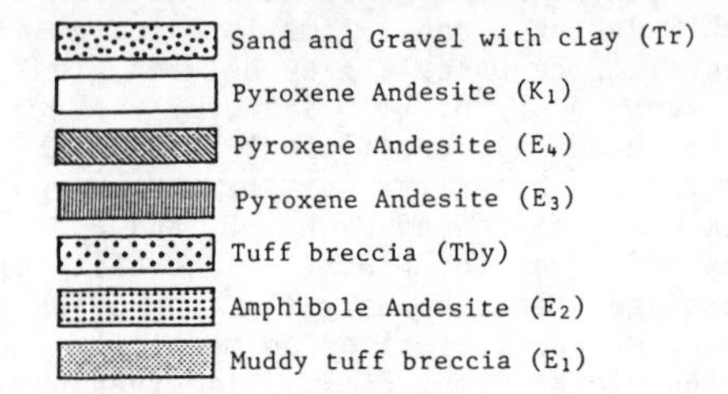

Fig.6 Example of the geological feature of the ground model

Table 1 Coefficient of permeability and effective porosity used in the water balance model

Symbol of geological feature	Coefficient of permeability (cm/s)	Effective porosity
Tr	2.40×10^{-5}	0.20
Any(K_1)	4.77×10^{-4}	0.20
Any(E_4)	1.00×10^{-4}	0.20
Any(E_3)	5.33×10^{-4}	0.15
Tby	1.10×10^{-5}	0.07
Anh(E_2)	2.40×10^{-5}	0.05
Tbm(E_1)	1.10×10^{-5}	0.12

1. In figures 7a and 7b, the observed change and the calculated ones show almost the same tendency.
2. In boring B-9-1, the influence of tunnelling was observed from the period when the calculated value with tunnelling began to differ from the ones without tunnelling. On the other hand, in boring B-11, the influence of tunnelling was not almost observed as it was predicted.

4.4 Calculation of tunnel discharge during tunnel construction

Figure 8 shows the observed and calculated

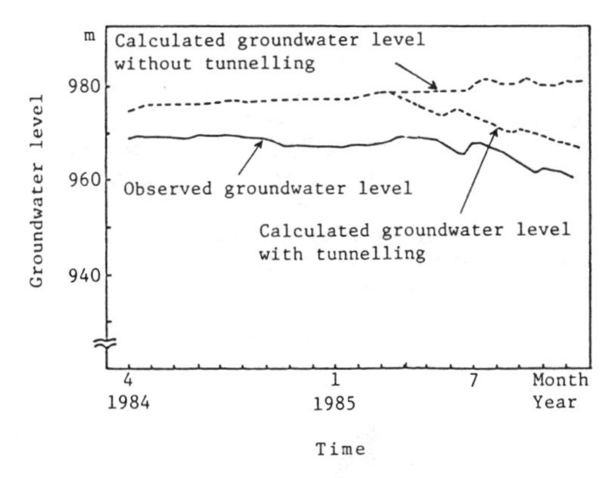

Fig.7a The change of groundwater level in B-9-1 observation well

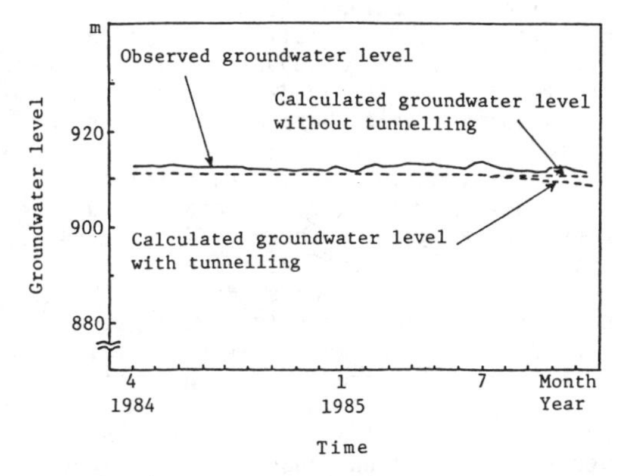

Fig.7b The change of groundwater level in B-11 observation well

values of tunnel discharge at the Okaya entrance and at the Shiojiri entrance.

From this figure, the following are known:

1. At the Okaya entrance, Observed value of tunnel discharge at the middle of July, 1985 increased rapidly. From calculation of the discharge, at the start of tunnelling it is found that continuous discharge is related to tunnel length. Because of this, for the Okaya entrance side, it is thought that due to difference in topography and the ground model which is obtained from calculation, the calculated groundwater level is higher than the actual groundwater level.

The observed entrance discharge became peak after July 10, 1985 and also showed a tendency to decrease, but the peak was not clearly calculated. This was maybe caused by the assumption of the homogeneous block for the complicated area.

2. At the Shiojiri entrance, the tunnel

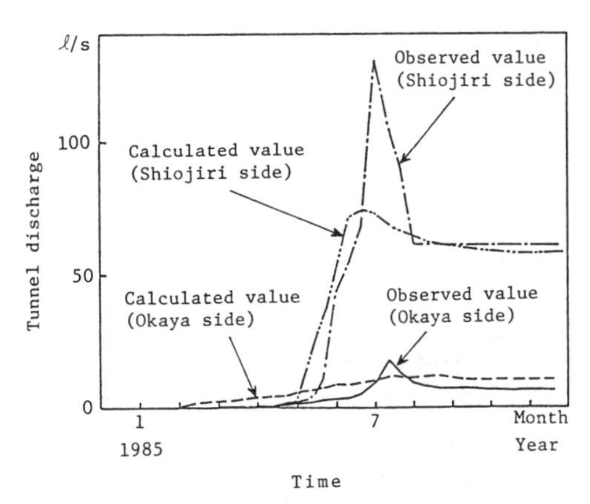

Fig.8 Tunnel discharge at each entrance side

discharge is quite large. Both the observed value of discharge and the calculated value at the entrance show that peak discharge occured in the middle of July, 1985. But the calculated discharge is somewhat different from observed one because the influence of the fault could not be accurately reflected in this simulation.

4.5 Prediction of hydrological environment after tunnel construction

The change in hydrological environment is calculated by assuming that weather will not change extraordinarily for five years from January, 1986. The predicted change of the tunnel discharge, groundwater level and the river discharge are shown in Figures 9a, 9b and 9c, respectively.

From these figures, the following are known:

1. After the completion of tunnel, the tunnel discharge decreases for one year, but after April, 1987, the tunnel discharge becomes almost constant of 53 ℓ/s.

2. The lowering of groundwater level in B-9-1 and B-10-2 observation wells caused by tunnelling can be seen clearly, and the goundwater level will become constant after about three years from the completion of tunnelling. The groundwater level in B-11 observation well decreases continuously due to the time lag of the effect of tunnelling.

3. The discharge quantity of the river basin S-20 changes with the season of year only, but it does not show the effect due to the tunnel construction.

The contour lines of lowering of groundwater level from December 2, 1984 to January 1, 1990 is shown in Figure 10.

From this figure, it is known that the north-east part of the area where is an

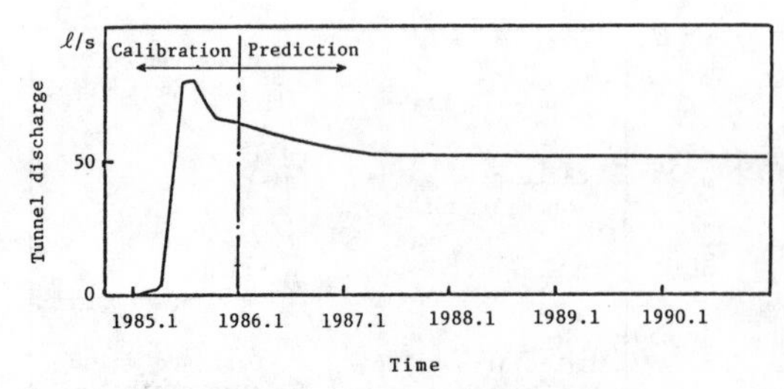

Fig.9a Predicted change of the tunnel discharge

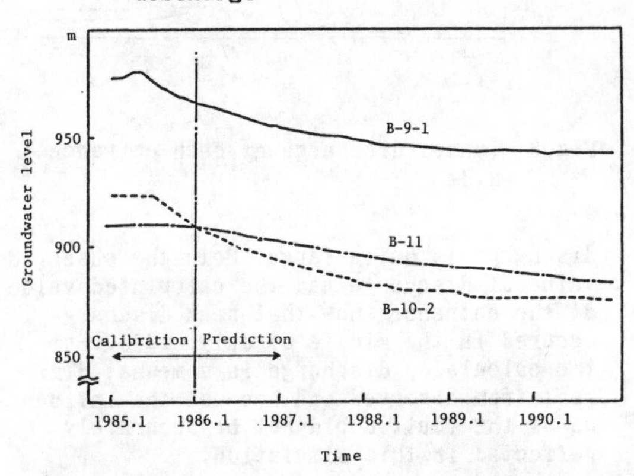

Fig.9b Predicted change of the groundwater level

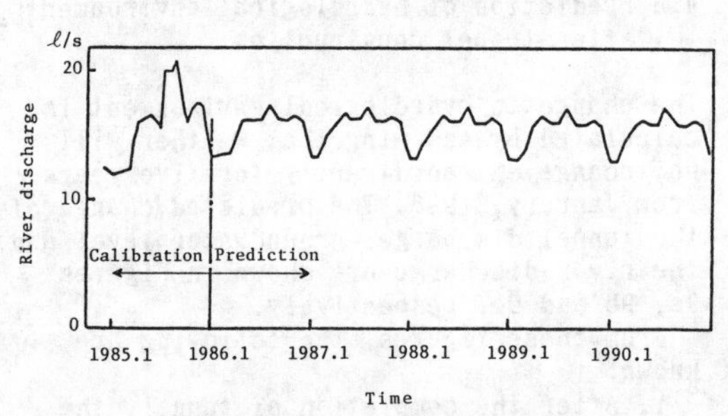

Fig.9c Predicted change of the river discharge in the basin S-20

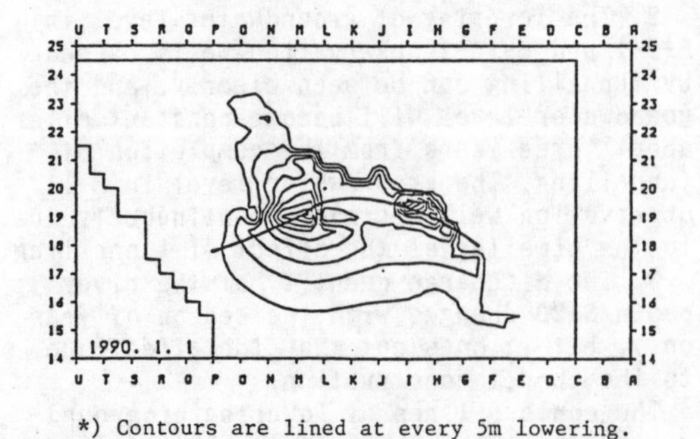

*) Contours are lined at every 5m lowering.

Fig.10 Predicted lowering of groundwater level by tunnelling from December 1984 to January 1990

impermeable fault, will not be affected by the construction of the tunnel.

Therefore, as conclusion of the above, it is found that due to the effect of the tunnel construction, the groundwater condition will be unsteady for about three years after the completion of tunnel. After that period, the groundwater condition of the region where a tunnel exsists will be in equilibrium state.

5 CONCLUSIONS

In this paper, the tank model and the block model are used to know the effect of tunnel construction on groundwater condition. Conclusions of this paper are summarized as follows:

1. The water balance block method used here is an easy computational method predicting the tunnel discharge of groundwater and the change of groundwater condition in a mountainous area where a tunnel is constructed.
2. Geological condition (coefficient of permeability and effective porosity) of the ground model gives a large difference on the result of the analysis. Therefore the geological condition of studied area has to be thoroughly and properly investigated.
3. If the ground model can be modified correctly during the tunnel construction, future hydrological environment can be predicted fitly. Therefore the method used here is found to be adequate for the case of tunnel construction.
4. The calculation time of the water balance block method is 1.6 min. for one year phenomena of groundwater hydrology with Facom M-382 of Nagoya University Computer Center and is much shorter than the analysis of FEM [5].

REFERENCES

[1] H.Oshima: Hydro-geological study on the tunnel discharge and the change of water balance by tunnelling, Railway Technical Research Institute Report, No.1228, 1983 (in Japanese)
[2] W.R.Hamon: Estimating Potential Evapotranspiration, ASCE, HY3, 2817 (1961)
[3] M.Sugawara: On the Analysis of Runoff Structure about Several Japanese Rivers, Japanese Journal of Geophysics, Vol2, No.4, 1-76 (1961)
[4] M.Muskat: The flow of homogeneous fluids through porous media, 1946
[5] K.Ueshita, et al.: Prediction of tunnelling effect on groundwater condition, Proc. 5th Int. Conf. on Numer. Meth. in Geomech, Vol.2, 1215-1219 (1985)

Numerical Methods in Geomechanics (Innsbruck 1988), Swoboda (ed.)
© 1988 Balkema, Rotterdam. ISBN 90 6191 809 X

Deformation behaviour of the tunnel under the excavation of crossing tunnel

S.Tsuchiyama & M.Hayakawa
The Chubu Electric Power Co., Inc., Nagoya, Japan
T.Shinokawa & H.Konno
Sato Kogyo Co., Ltd, Tokyo, Japan

ABSTRACT: The influence on the existing main tunnel when a new access tunnel is excavated to cross obliquely by the angle 45 degrees to the existing tunnel is discussed through a numerical analysis. The analysis is 3-dimensional finite element analysis in consideration of face progression due to excavation. Two types of the excavation patterns are examined; the access tunnel is excavated from and to the main tunnel, respectively. It is found that the influence area along the main tunnel is the order of tunnel diameter in the obtuse angle side and about 3 times to the tunnel diameter in the acute angle side from the point of intersection. The change of displacement appears according to the excavation to the main tunnel and it disappears according to the excavation from the main tunnel when face progresses 2 or 3 times to the tunnel diameter from the intersection of the tunnels.

1 INTRODUCTION

Circumstance with underground structures such as tunnels are more and more diversified. Sometimes they should be constructed not only under complicated geological conditions but also under complicated configuration. For example, tunnels which can not be considered as a single tunnel, like a twin tunnel or a crossing tunnel, need to be constructed.

When two tunnels cross to each other the tunnel intersection has the 3-dimensional structural configuration. Moreover, the ground behavior around the tunnel intersection is different from that around the single tunnel because secondary state of the stress of the ground under the tunnel excavation interferes with each other. High stress concentration and the occurrence and expansion of the unstable area, for example, are expected there. On the design and the construction of the tunnel intersection, it is important to grasp the influence area and magnitude and to examine the stability.

Some studies have already been presented on the tunnel intersection. Hocking analyzed cross-junction, T-junction and Y-junction by the boundary integral method and presented the extent and the magnitude of the three dimensional stress concentrations [1]. Takino et al. reported the measurements and the analysis carried out in the Enasan Tunnel phase 2 [2], [3]. Shinokawa et al. examined the stability of the pillar using the 2-dimensional boundary element analysis when two tunnels cross, in which the distance between the tunnels was changed so as to recognize the 3-dimensional behavior of the tunnels and the pillar [4].

In this paper, the ground behavior around the tunnel intersection is examined by using the 3-dimensional finite element analysis. It is supposed that a new access tunnel is excavated to cross with 45 degrees to the existing main tunnel.

2 OUTLINE OF NUMERICAL ANALYSIS

A 3-dimensional finite element elastic analysis program is used to analyze the behavior. The program can take into account excavation steps by reducing elements and nodes.

A main tunnel and an access tunnel cross at an angle of 45 degrees. The shapes of the main tunnel and the access tunnel are circular, the diameters of which are indicated as D. The modelled region has the dimensions with 15D width, 15D depth and 10D height, but only the upper half is analyzed because of the symmetry condition. Analytical model is shown in

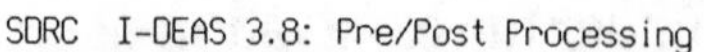
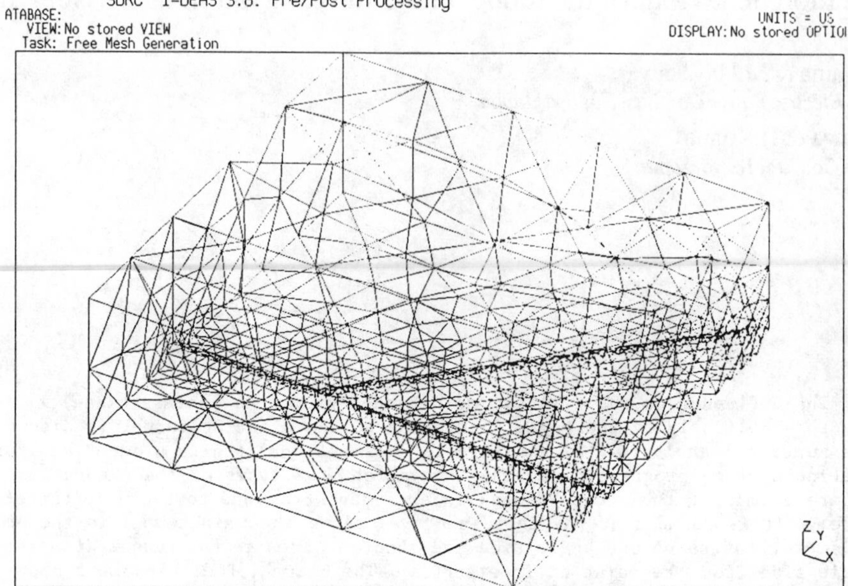

Figure 1. Analytical model

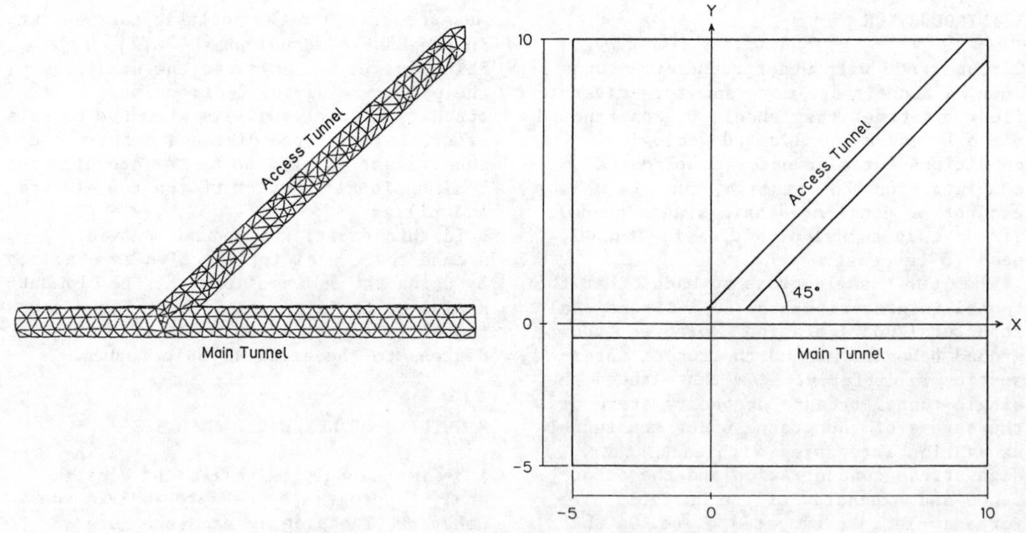

Figure 2. Analitical model (tunnels)

Figs. 1 and 2. For simplicity, the geometrical configuration and the displacement are represented by the dimensionless quantities by dividing the tunnel diameter, i.e., the tunnel diameter is represented, for example, 1.0. In the analysis, lateral displacements on the lateral planes are fixed and vertical displacement on the lower plane is fixed (symmetry condition); the other are free to move. Linear tetrahedrons are used for elements; total number of elements is 8725 and total number of nodes is 1913.

The following two cases are selected as parameters.

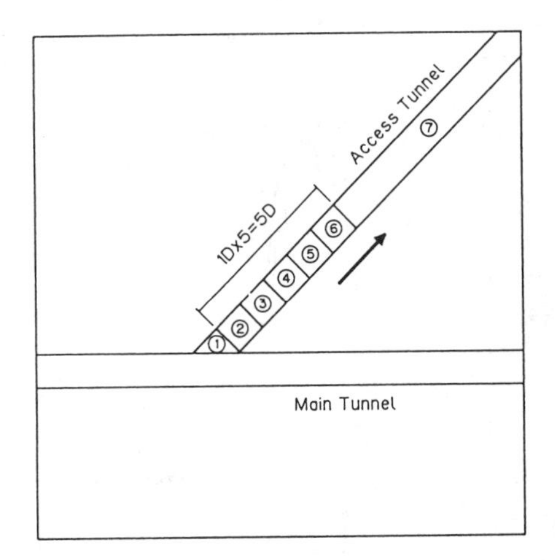

Figure 3. Analytical steps (excavation steps, CASE-1)

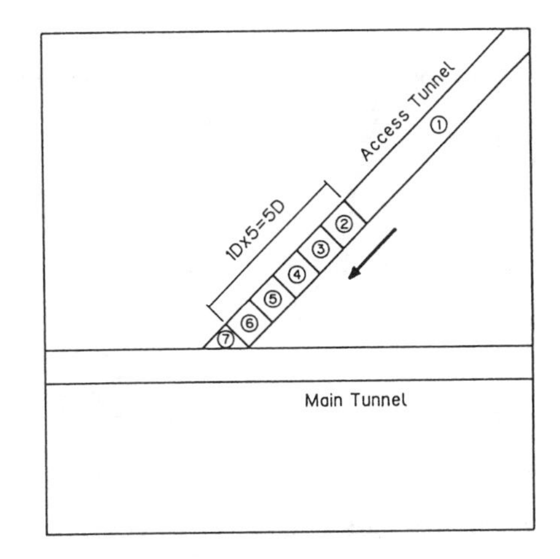

Figure 4. Analytical steps (excavation steps, CASE-2)

CASE-1 : An access tunnel is excavated from the main tunnel
CASE-2 : An access tunnel is excavated to the main tunnel

The analysis is done under the following steps.

(1) Set initial state
(2) Excavate access tunnel.

Initial state is obtained by applying unit distributed force (Pz=-1.0) in the vertical direction on the top surface after the excavation of the main tunnel. Excavation steps of the analysis are shown in Figs. 3 and 4. The circled numbers in the figures indicate excavation step. Constants used in the analysis are shown in Table 1. It is noted that Poisson's ratio used in the initial state analysis is 0.47 and that used in the excavation steps analysis is 0.3.

Mesh generation is done by the software GEOMOD and SUPERTAB which are subsystems of the CAE application software SDRC-I-DEAS.

Table 1. Analytical coefficients

Modulus of Elasticity	1000.0
Poisson's ratio	0.47 initial state
	0.3 excavation steps

3 INFLUENCE ON MAIN TUNNEL

Much attention are usually paid on the stress state when examining the ground behavior around the tunnel intersection because high stresses and stress concentrations are expected. However, it is difficult to measure stresses in the construction site; some other index may be more convenient to recognize the behavior of the tunnels under excavation. The convergence measurement is most popular and is used frequently. Then, in this paper, the behavior of the tunnel intersection is examined with respect to deformations. Especially, according to the excavation of a new access tunnel, how a existing main tunnel is influenced is discussed.

The face progression is one of the important issues when the ground behavior at the tunnel excavation stage is discussed. However, the analysis considering the face progression was hardly done in such a problem considered here, 3-dimensional tunnel intersection problem. In the following analysis, the excavation analysis taking into account the face progression is performed.

3.1 Results of CASE-1

The first case (CASE-1) is the case when the access tunnel is excavated from the main tunnel. Vertical and horizontal displacement at the crown of the main tunnel are shown in Figs. 5 and 6, and horizontal displacements at the spring line of the main tunnel are shown in Figs.

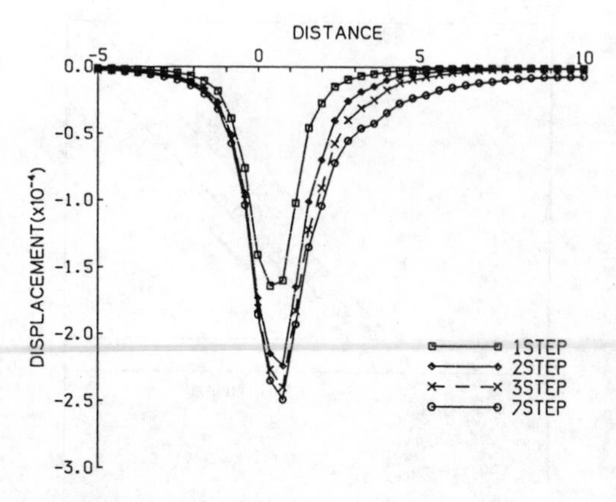

Figure 5. Vertical displacement at the crown of main tunnel (CASE-1)

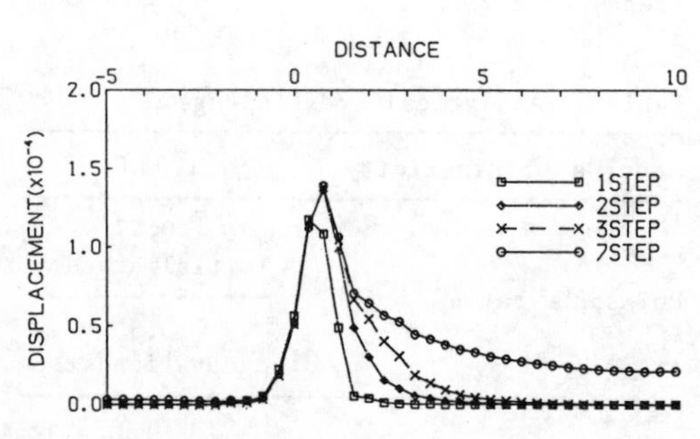

Figure 6. Horizontal displacement (Y-direction) at the crown of main tunnel (CASE-1)

7 and 8.

It is seen from Fig. 5 that influence area with respect to vertical displacement at the crown by the excavation of the access tunnel is about 2D in the obtuse angle side of the access tunnel and about 5D in the acute angle side of the access tunnel from the point of intersection. It is seen from Fig. 6 that the influence area with respect to horizontal displacement at the crown is about 1D in the obtuse angle side and about 3D in the acute angle side.

It is obvious from Fig. 7 that horizontal displacement at the spring line of the main tunnel on the side of the access tunnel is remarkably influenced by the excavation of the access tunnel. It is also regarded that influence area is about 1D in the obtuse angle side and 5D in the acute angle side. However, the effect on the horizontal displacement at the spring line of the main tunnel on the opposite

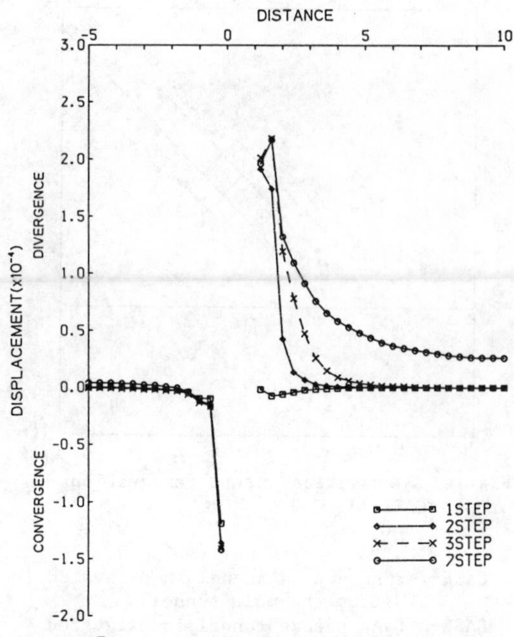

Figure 7. Horizontal displacement (Y-direction) at the sprong line on the side of access tunnel (CASE-1)

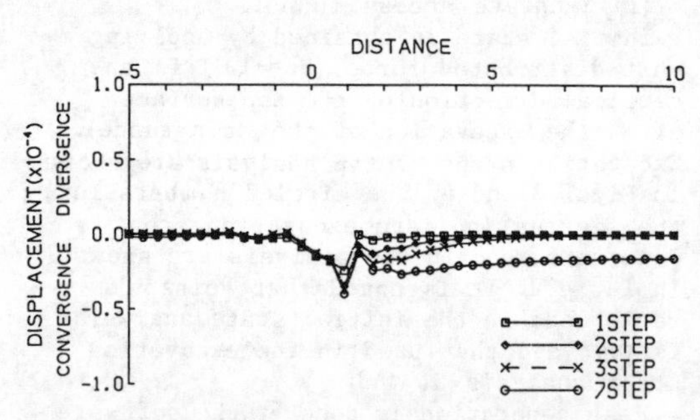

Figure 8. Horizontal displacement (Y-direction) at the spring line on the opposite side of access tunnel (CASE-1)

side of the access tunnel is not observed clearly except near the point of intersection where slight disturbance is observed (Fig. 8).

Next, the effect of the face progression is investigated. As can be seen from the results of CASE-1, Figs. 5-8, the deformation by the excavation of the access tunnel almost converges at the third step at which face distance from the point of intersection is about 3 times of the tunnel diameter.

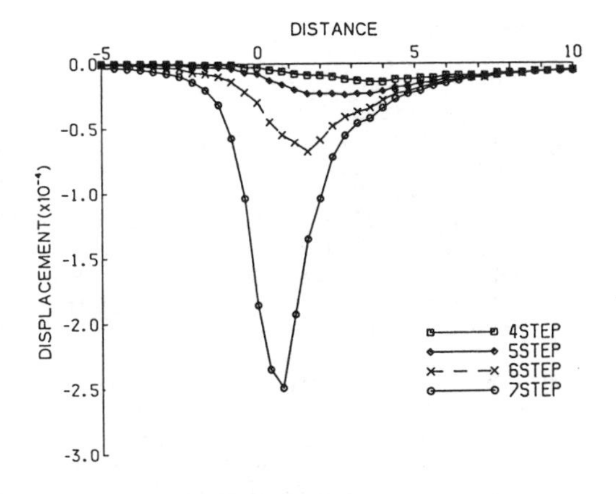

Figure 9. Vertical displacement at the crown of main tunnel (CASE-2)

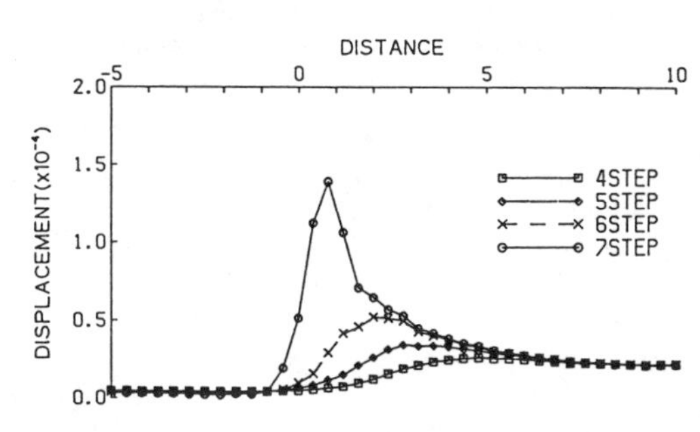

Figure 10. Horizontal displacement (Y-direction) at the crown of main tunnel (CASE-2)

3.2 Results of CASE-2

The next case (CASE-2) is the case that the access tunnel is excavated to the main tunnel. Vertical and horizontal displacement at the crown of the main tunnel are shown in Figs. 9 and 10, respectively, and horizontal displacement at the spring line of the main tunnel are shown in Figs. 11 and 12.

Because of the nature of the elastic analysis, the results of CASE-1 and CASE-2 should be the same at the final state; this is confirmed in Figs. 9-12.

Regarding to the effect of the face progression, the effect of the excavation appears at the fifth excavation step at which the distance to the face is twice of the tunnel diameter.

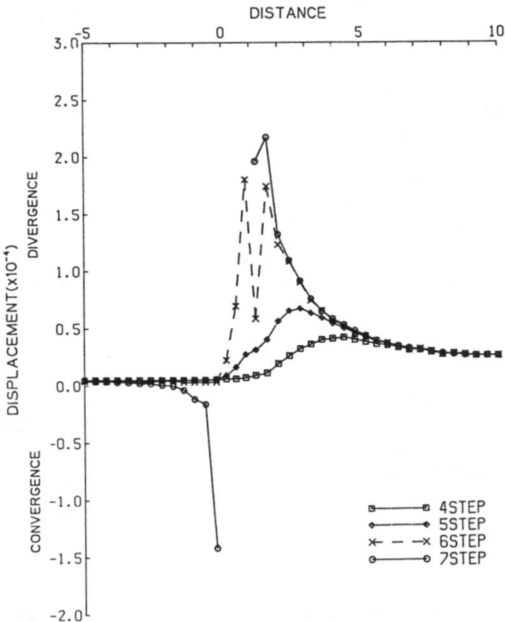

Figure 11. Horizontal displacement (Y-direction) at the spring line on the side of access tunnel (CASE-2)

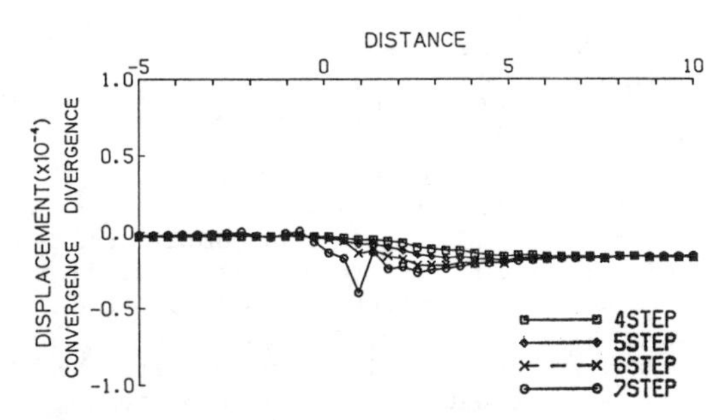

Figure 12. Horizontal displacement (Y-direction) at the spring line on the opposite side of access tunnel (CASE-2)

3.3 Discussion

The influence zone around the main tunnel by the excavation of the access tunnel is 1D in the obtuse angle side of the access tunnel and 3D in the acute angle side of the access tunnel; the effect is clearly observed on the displacements at Figs. 5-7 and Figs. 9-11. Horizontal displacement at the spring line of the main tunnel on the opposite side of the access tunnel, however, is hardly influenced by the excavation of the access tunnel. Moreover, the displacements of the main tunnel gradually decreases along the axis of the tunnel in the direction of the acute angle, whereas that decreases rapidly in the direction of the obtuse angle and the effect almost disappears beyond the tunnel diameter.

Regarding to the time effect due to excavation, the change of the displacement appears according to the excavation to the main tunnel and disappears according to the excavation from the main tunnel when face passes the section 2-3D far from the intersection of the tunnels.

4 CONCLUDING REMARKS

The ground behavior around the tunnel intersection was analyzed by the 3-dimensional finite element analysis. The influence on the tunnel intersection was examined with respect to deformations which are measured frequently at the tunnel excavation. The influence area on the main tunnel when the access tunnel was excavated was grasped. Moreover, the influence time on the main tunnel according to the progression of the access tunnel was extracted.

In this paper, however, only the case that cross angle is 45 degrees was treated. In the actual construction, however, the case of more acute cross angles, especially the angles less than 30 degrees, may sometimes be required. Therefore the studies with these cases should also be performed.

Furthermore, it is usually complicated and troublesome work to prepare input data for the finite element analysis with the large size although convenient mesh generator can be used. To improve it, the boundary element analysis is to be more convenient because it has an advantageous in reducing input data. Therefore a future subject seems to be the one to apply the 3-dimensional boundary element analysis to the tunnel intersection.

ACKNOWLEDGMENTS

The analytical program used in this paper is developed by Dr. Y.Ichikawa of Nagoya University. The authors wish to thank him for access to the program.

The mesh generation to the analysis is made by the CAE application software, SDRC-I-DEAS. The authors also wish to thank Mr. T.Moriya of C. Itoh Techno-Science Co., Ltd. for help to mesh generation.

REFERENCES

[1] G.Hocking, Stresses around tunnel intersections, Computer Methods in Tunnel Design, Ed. A.Burt, The institution of civil engineers, London, 41-60, 1978

[2] K.Takino, H.Kimura, K.Kamemura and T,Kawamoto, 3-Dimensional Ground Behavior at Tunnel Intersection, Proc. Int. Symp. on Field Measurements in Geomechanics, Balkema, Rotterdom, 1237-1246, 1983

[3] K.Takino, H.Kimura, N.Takeda and F.Ito, Three-dimensional behaviour of tunnel intersection, Proc. 5th Int. Conf. on Num. Methods in Geomechanics, Balkema, Rotterdam, 1185-1192, 1985

[4] T.Shinokawa and H.Konno, Stability of Pillar around Tunnel Intersection, Proc. 19th Symp. on Rock Mechanics, Committee on Rock Mechanics J.S.C.E., Tokyo, 271-275, 1987 (in Japanese)

Numerical Methods in Geomechanics (Innsbruck 1988), Swoboda (ed.)
© 1988 Balkema, Rotterdam. ISBN 90 6191 809 X

Analysis of stresses in the pillar zone of twin circular interacting tunnels

R.K.Srivastava
M.N.R. Engineering College, Allahabad, India
K.G.Sharma & A.Varadarajan
Indian Institute of Technology, Delhi

ABSTRACT : Underground openings are extremely complex structures. The complication is more in case of multiple openings where interaction effects are considerable. Analysis of stresses induced in the pillar zone of interacting openings is very important in understanding the stability of multiple underground excavations. Elastic and elasto-plastic finite element analysis of two interacting circular underground openings have been carried out for plane strain condition using Hoek-Brown yield criterion. Three pillar width (W) to tunnel diameter (D) ratios and three insitu stress conditions for each W/D ratio have been considered. The tunnels have been considered to be excavated simultaneously. The results have been compared with the excavation of that of a single tunnel and the effect of tunnel interaction on the stresses developed in the pillar zone has been brought out.

1 INTRODUCTION

Underground openings are being increasingly used for a variety of purposes e.g. hydro-electric projects, mining, railway, highway and sewage disposal networks, oil and gas storage and nuclear waste disposal. This has considerably increased the interest and concern of Geotechnical engineers to understand its behaviour.

For the design of underground openings, briefly, the requirement is to know the insitu stress field, induced stress field and related deformations when an opening is made, the characteristics of support system and interaction behaviour of support and ground. Analysis and understanding of stresses induced in the pillar zone is important and of special interest and use in case of multiple underground openings. An accurate interpretation of geological conditions is also essential for any rational design.

The methods of analysis and design generally, used by practising engineers in case of underground openings are empirical formulae, standardised codes and closed form solutions. These methods as such are not adequate to deal with such complicated situations. The only realistic approach is to resort to numerical methods such as finite element method (Desai and Abel, 1972; Zienkiewicz, 1977).

In case of analysis of interacting tunnels, very limited literature is available. A detailed review of the available literature has been presented by Srivastava (1985). It has been observed that in case of interacting tunnels generally elastic analysis has been carried out and in case of elasto-plastic analysis, Drucker-Prager yield criterion has been used. Recently, Srivastava, Sharma and Varadarajan (1986) have reported a finite element analysis of tunnels using Drucker-Prager, Mohr-Coulomb and Hoek-Brown yield criteria and have advocated the use of the yield criterion proposed by Hoek and Brown (1980) in the elasto-plastic finite element analysis of underground openings.

In present study elasto-plastic finite element method has been used for the analysis of stresses induced in the pillar zone of two deep circular interacting tunnels. Hoek-Brown yield criterion has been used for the elasto-plastic analysis. Three pillar width (W) to tunnel diameter (D) ratios of 0.3, 0.6 and 1.2 have been chosen.

Three insitu stress ratios (Ko) viz. 0.5, 1.0 and 1.5 have been considered to simulate possible field conditions for each W/D ratio. The tunnels have been considered to be excavated simultaneously. Both elastic and elasto-plastic analysis have been carried out. The results have been compared with those of a single opening. Variation of stresses induced in the pillar zone at springing level have been plotted. Also presented are the variations in principal stresses and deviatoric stresses with W/D for various cases analysed.

2 ELASTO-VISCOPLASTICITY

In the present study elasto-viscoplastic theory developed by Zienkiewicz and Cormeau (1974) has been adopted and used as an artifice to obtain elasto-plastic solution. The theory is briefly presented in this section. The state of stress at a point is represented by the stress vector $\{\sigma\}$,

$(\{\sigma\}^T = [\sigma_x \ \sigma_y \ \sigma_z \ t_{xy} \ t_{yz} \ t_{zx}])$

and total strain by the vector $\{\epsilon\}$,

$(\{\epsilon\}^T = [\epsilon_x \ \epsilon_y \ \epsilon_z \ \gamma_{xy} \ \gamma_{yz} \ \gamma_{zx}])$.

The total strain at a point is the sum of elastic and viscoplastic strains,

$$\text{i.e.} \quad \{\epsilon\} = \{\epsilon^e\} + \{\epsilon^{vp}\} \quad (1)$$

where $\{\epsilon^e\}$ and $\{\epsilon^{vp}\}$ are the elastic and viscoplastic strain vectors, respectively.

The stresses are related to the total and viscoplastic strains through the relation

$$\{\sigma\} = [D]\,(\{\epsilon\} - \{\epsilon^{vp}\}) \quad (2)$$

Where [D] is the elasticity matrix. The elements of [D] are functions of Young's modulus and Poisson's ratio. The yield function in general can be written in the form

$$F = F(\{\sigma\}, \{\epsilon^{vp}\}) = 0 \quad (3)$$

Equation 3 represents in general the conditions of plasticity, hardening/ softening at a point. The viscoplastic strain rates are given by the flow rule

$$\{\dot\epsilon^{vp}\} = \mu \langle F \rangle \frac{\partial F}{\partial \{\sigma\}} \quad (4)$$

In which $\{\dot\epsilon^{vp}\}$ is the viscoplastic strain vector, μ is the fluidity parameter (taken as 1 for elasto-plastic analysis), $\langle \rangle$ is used to indicate that if $F \le 0$, $\langle F \rangle = 0$, and $F > 0$, $\langle F \rangle = F$

3 YIELD OR FAILURE CRITERION

The empirical failure criterion as proposed by Hoek and Brown (1980) is given by

$$\sigma_1 = \sigma_3 + (m\,\sigma_c\,\sigma_3 + s\sigma_c^2)^{\frac{1}{2}} \quad (5)$$

where σ_1 and σ_3 are major and minor principal stresses respectively, σ_c is the uniaxial compressive strength of the intact rock material and m and s are dimensionless empirical constants.

Equation 5 can be written in the form

$$F = \sigma_1 - \sigma_3 - (m\sigma_c\,\sigma_3 + s\sigma_c^2)^{\frac{1}{2}} = 0 \quad (6)$$

which is applicable to perfectly plastic material. Till $F \le 0$, original rock mass is linear elastic, when $F > 0$, the rock mass starts yielding.

4 COMPUTER PROGRAM

A computer program has been developed on ICL 2960 for elasto-viscoplastic finite element analysis for plane strain condition. Eight noded isoparametric elements and 2x2 Gauss point integration have been used.

The procedure adopted to simulate excavation is that proposed by Chandrasekaran and King (1974). For the present study in case of simulataneous excavation, both tunnels are excavated (full face) in single stage and for comparison with the case of a single tunnel, only one of the two tunnels is excavated (fullface).

5 CASES ANALYSED

A schematic diagram shown in fig.1 indicates the tunnel diameter, material

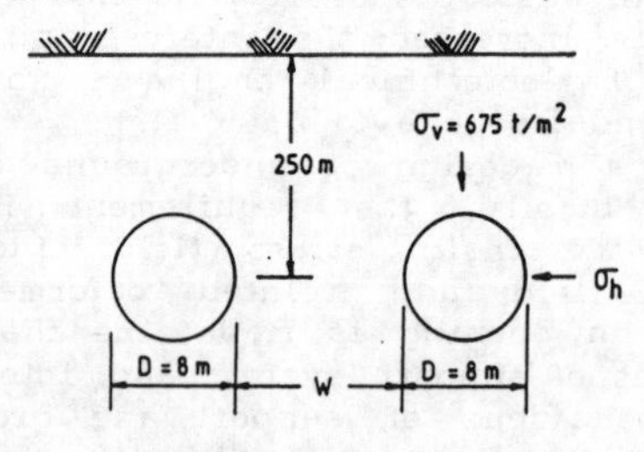

Fig.1 Schematic diagram-interacting tunnels

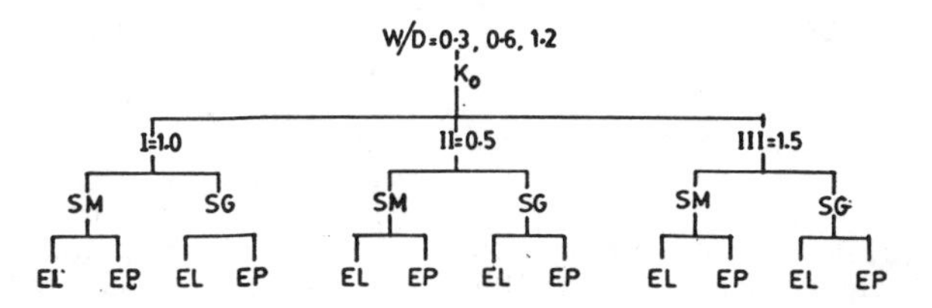

SM = Simultaneous Excavation
SG = Single Tunnel

EL = Elastic Analysis
EP = Elasto-plastic Analysis

Fig. 2 Schematic Diagram - Indicating Cases Analysed

properties and the W/D and insitu stress ratios chosen for the analysis. In fig.2 various cases analysed are shown and explained schematically.

6 ANALYSIS

For the three W/D ratios, three finite element discretizations have been used by increasing suitably the number of elements in the pillar zone with increasing width of the pillar. For W/D ratios of 0.3, 0.6 and 1.2, the mesh has 296, 324 and 344 elements and 949, 1037 and 1098 nodes respectively.

A typical mesh (for W/D ratio of 0.6) is shown is fig.3. The extent of rock medium taken on the crown, abutment and invert sides of the tunnel has been kept eight times the tunnel radius. The outer boundary of the mesh is assumed to be rigid and rough. Close to the tunnel boundary and in the pillar zone, the mesh is comparatively finer and farther away larger size elements have been used. The result of elastic and elasto-plastic analysis are discussed in the following section.

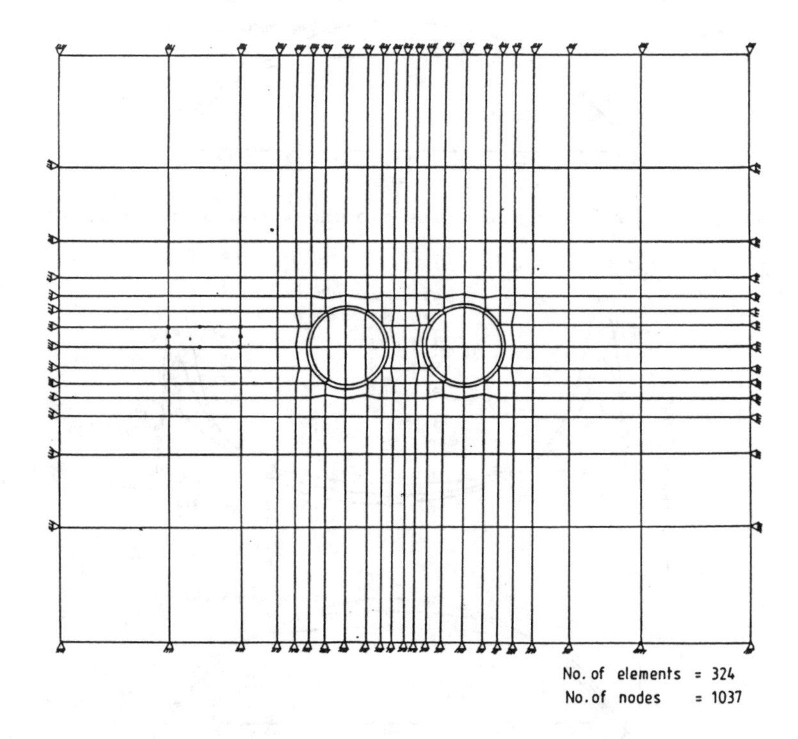

Fig. 3 Finite element discretization - interacting tunnels (W/D = 0.6)

7 RESULTS AND DISCUSSION

7.1 Variation of Principal stresses along pillar width

For W/D = 1.2 figures 4 and 5 show the variation of principal stresses in the pillar at springing level for the elastic and elasto-plastic analysis for the three insitu stress ratios. It is observed that the major principal stresses are highest in the case of Ko = 0.5 and least in the case of Ko = 1.5 at and near the tunnel boundary for elastic and elasto-plastic analysis respectively. But, at the centre of the pillar zone reverse trend is observed. In the case of minor principal stresses, highest value is observed for Ko = 1.5 and least for Ko = 0.5 at the centre of the pillar zone, for both elastic and elasto-plastic analysis. At the tunnel boundary the difference for the three insitu stress ratios is negligible.

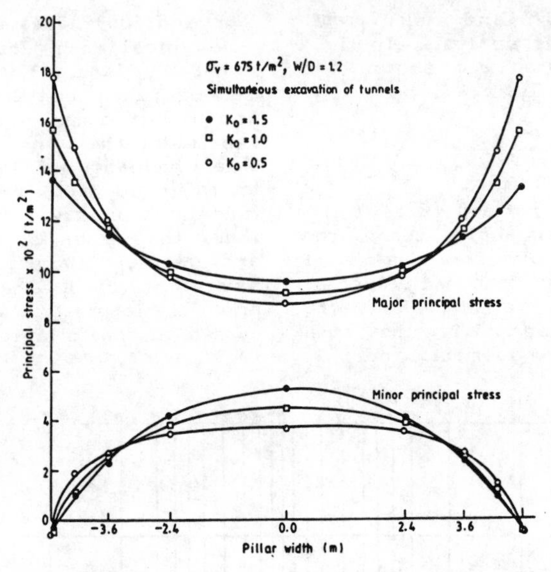

Fig. 4 Variation of principal stresses along pillar width - elastic analysis

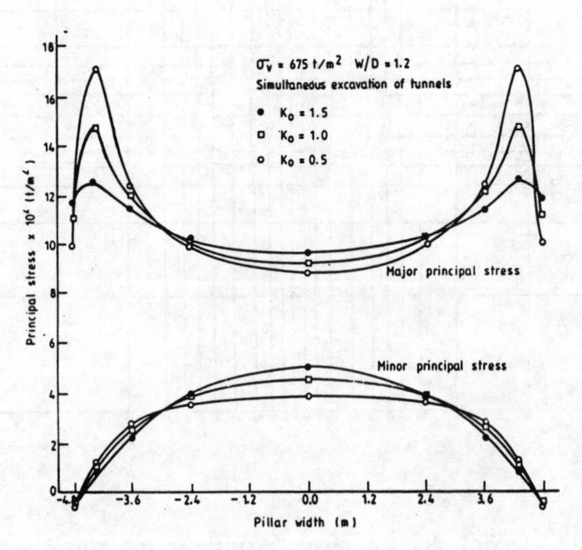

Fig. 5 Variation of principal stresses along pillar width -elasto-plastic analysis

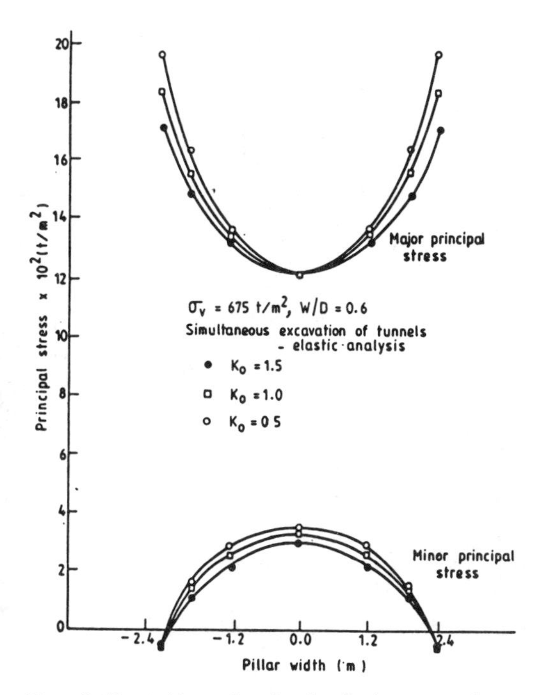

Fig. 6 Variation of principal stresses along pillar width - elastic analysis

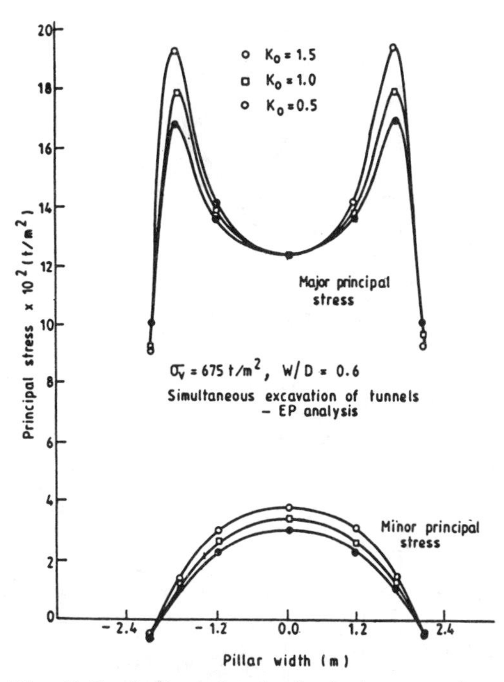

Fig. 7 Variation of principal stresses along pillar width - elasto plastic analysis

Figure 6 and 7 show the variation of principal stresses for W/D ratio of 0.6 and Figures 8 & 9 show the variation of principal stresses for W/D ratio of 0.3, at springing level for the three insitu stress ratios for elastic and elasto-plastic analyses. For these two W/D ratios, it is observed that, Ko = 0.5 gives the highest major principal stresses in the pillar and for Ko = 1.5 least stresses are observed in the pillar for elastic and elasto-plastic analyses. In the case of mincr principal stresses, highest value is observed for Ko = 0.5 and least for Ko = 1.5 at the centre of the pillar zone, for both elastic and elasto-plastic analyses. At the tunnel boundary the difference for the three insitu stress ratios is negligible for these W/D ratios also.

7.2 Comparison of principal stresses between interacting and single tunnels (elastic analysis).

Figure 10 shows the variation of percentage difference in principal stresses with W/D ratio at the tunnel boundary and the centre of the pillar zone for major principal stress and at the centre of the pillar for minor principal stress, between interacting and single tunnels for the three insitu stress ratios.

At the tunnel boundary (Fig.10a), the percentage difference in major principal stress decreases with increase in W/D ratio. The percentage difference is highest for Ko = 1.5 and lowest for Ko = 0.5. The rate of decrease is more in the case of Ko = 1.5 as compared to other two stress ratios. With increasing W/D ratio the percentage difference shows a tendency to reach zero value which is expected also, because at large W/D ratio, there will be no interaction effect and the tunnels would behave as a single tunnel.

The variation of percentage difference in major principal stresses with W/D ratio is shown in fig.10b at the centre of the pillar zone. The difference in percentage for the three insitu stress ratios at a particular W/D ratio is small. At smaller W/D ratio, the percentage difference is high and it decreases with increasing W/D ratio and tends towards zero for all the three stress ratios as expected.

The variation of percentage difference in minor principal stresses with

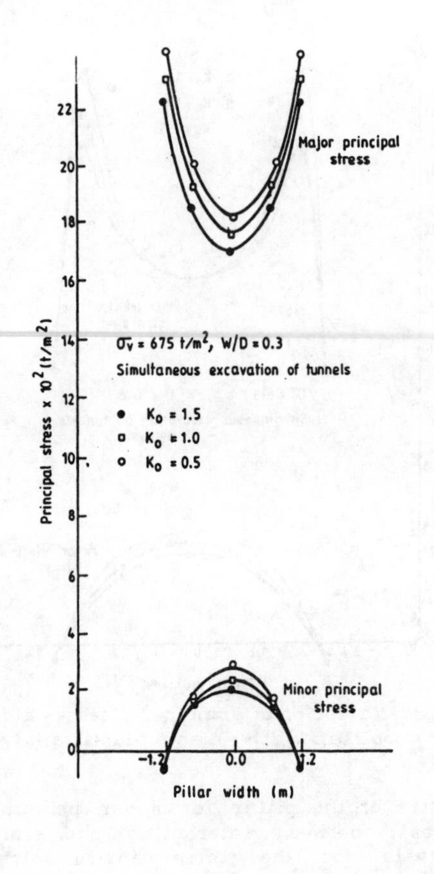

Fig. 8 Variation of principal stresses along pillar width - elastic analysis

Fig. 9 Variation of principal stresses along pillar width - elasto plastic analysis

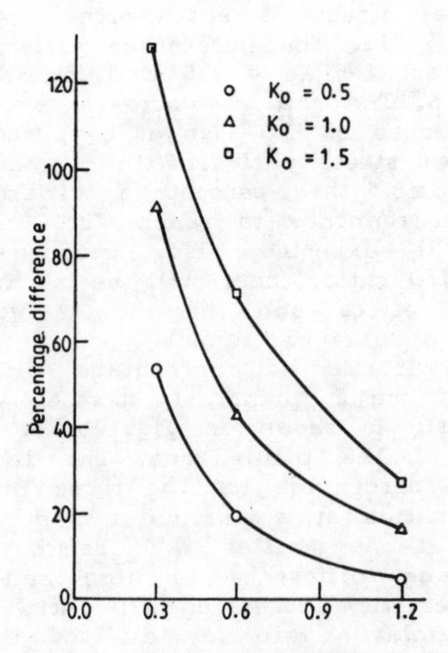

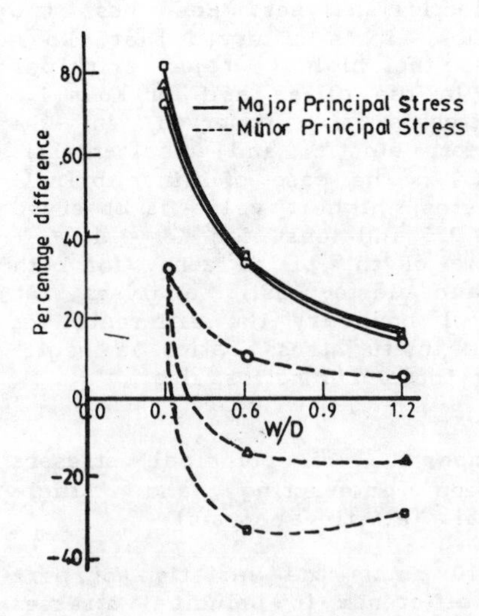

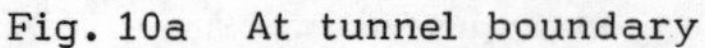

Fig. 10a At tunnel boundary

Fig. 10b At centre of pillar zone

Fig. 10 Variation of percentage difference in principal stresses between interacting and single tunnels with W/D (elastic analysis)

W/D ratio is also shown in Fig.10b at the centre of pillar zone. It is seen that for Ko = 0.5, the percentage difference is positive and decreases with increase in W/D ratio. In the case of stress ratio Ko = 1.0, the percentage difference is a small positive value at W/D = 0.3, which becomes zero and then negative with increase in W/D ratio and again shows a return towards zero percentage difference with further increase in W/D ratio. The insitu stress ratio Ko = 1.5 also shows a similar trend. In this case, the percentage difference is a negative value even for small W/D ratio of 0.3. It decreases further and is expected to become zero again with further increase in W/D ratio.

The variation of ($\sigma_1 - \sigma_3$) at the tunnel boundary and the centre of pillar zone with W/D ratio for the three insitu stress ratios is shown in Fig.11.

For the three insitu stress ratios, the difference in principal stresses, i.e. ($\sigma_1 - \sigma_3$) is high at smaller W/D ratio of 0.3 and it decreases with increase in W/D ratio towards the level of difference indicated for single tunnels. The difference in ($\sigma_1 - \sigma_3$) of interacting and single tunnel is largest for Ko = 1.5 at tunnel boundary and of about the same order at the centre of pillar zone for the three insitu stress ratios. It is observed that the W/D ratio required to reduce the difference between interacting and single tunnels is expected to be largest for Ko = 1.5.

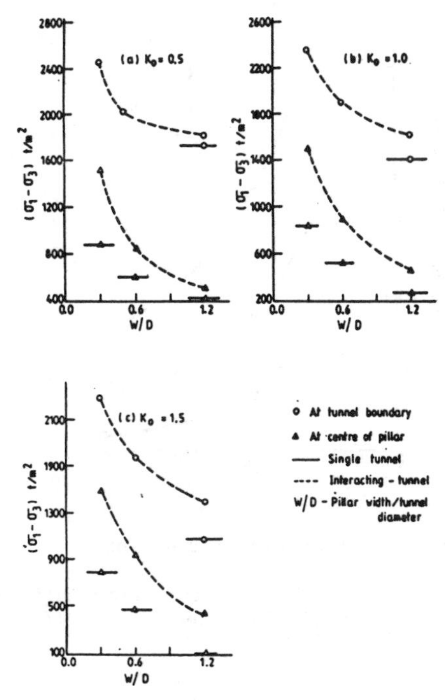

Fig. 11 Variation of deviatoric stress with W/D for single and interacting tunnels - elastic analysis

7.3 Comparision of principal stresses between interacting and single tunnels (elasto-plastic analysis).

Figure 12 shows the variation of percentage difference in principal stresses with W/D ratio near the tunnel boundary (where maximum major principal stress is observed) and at the centre of the pillar zone, for the three insitu stress ratios.

Near the tunnel boundary (Fig.12a), maximum percentage difference is indicated in the case of smaller W/D ratio of 0.3 and this decreases as W/D ratio is increased. Among the three insitu stress ratios, the percentage difference is highest for Ko = 1.5, and lowest for Ko = 0.5. The rate of increase in percentage difference with decrease in W/D ratio is largest for Ko = 1.5, and comparatively smaller for Ko = 0.5. Ko = 1.0 case remains in between. With increase in W/D ratio, the percentage difference shows a tendency to reach zero level at large W/D ratio when interaction effects would die down.

The variation of percentage difference in major principal stresses with W/D ratio is shown in fig.12b at the centre of the pillar zone. The difference in percentage difference for the three insitu stress ratios is rather small. In general, the percentage differences are high at smaller W/D ratios and they decreas with increasing W/D ratio and tend towards zero as expected.

The percentage difference in minor principal stresses at the centre of the pillar zone is also shown in Fig.12b. In this case, at W/D = 0.3, the percentage differences are positive and for Ko = 0.5 the difference decreases from higher positive value at smaller W/D ratio towards zero as W/D ratio increases. For Ko = 1.0 and 1.5, it is interesting to note that at W/D = 0.3, the percentage difference is positive but with increase in W/D ratio it becomes zero and then negative. With further increase in W/D ratio, it is expected to

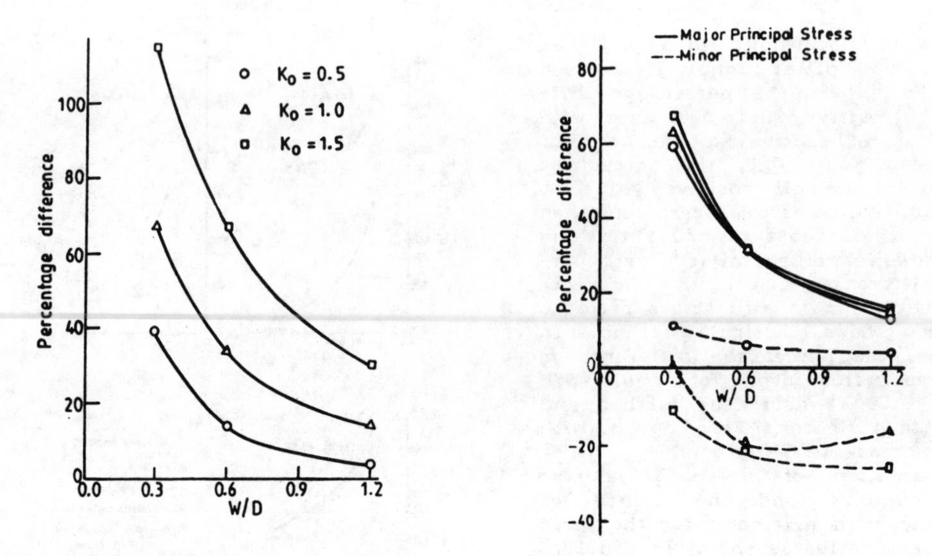

Fig. 12a Near tunnel boundary

Fig. 12b At centre of pillar zone

Fig. 12 Variation of percentage difference in principal stresses between interacting and single tunnels with W/D (elasto-plastic analysis)

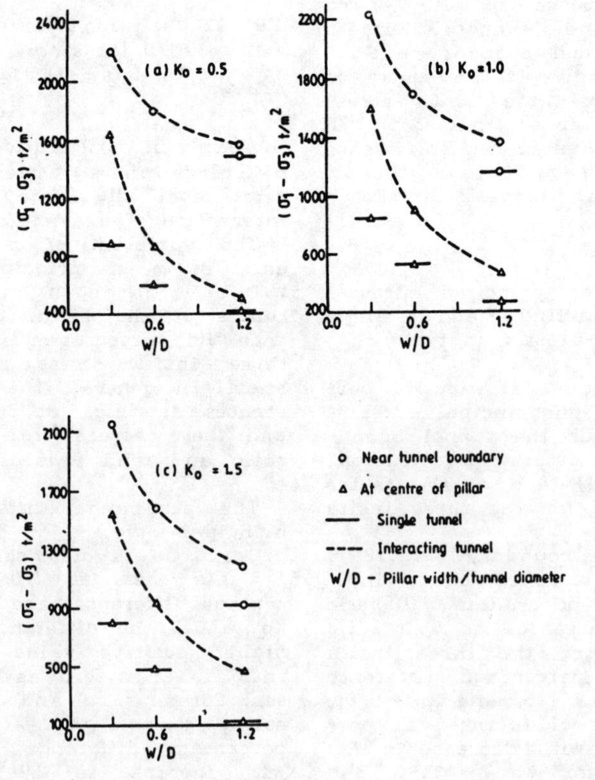

Fig. 13 Variation of deviatoric stress with W/D for single and interacting tunnels (elasto - plastic analysis)

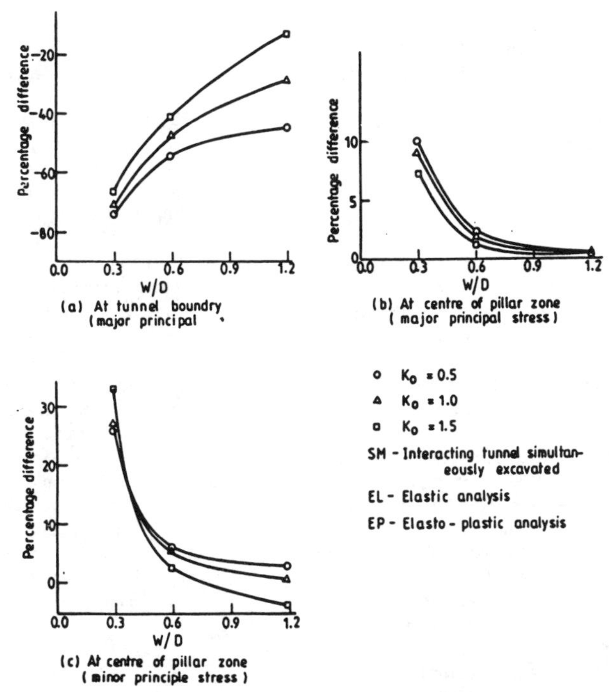

Fig. 14 Variation of percentage difference in principal stresses between SMEP and SMEL with W/D

show a tendency to reach zero percentage differene level.

The variation of (σ_1 - σ_3) near the tunnel boundary (where maximum principal stress is observed) and at the centre of the pillar zone with W/D ratio for the three insitu stress ratios is shown in Fig.13. The level of difference in the case of single tunnel is also shown.

In this case also, for the three insitu stress ratios, the difference in principal stresses is high at smaller W/D ratio of 0.3. It is observed that in this case also (σ_1 - σ_3) decreases with increase in W/D ratio and tends to reach the level of difference indicated by that of a single tunnel. From the difference shown between (σ_1 - σ_3) of the interacting and single tunnels at W/D = 1.2, it is expected that for higher value of insitu stress ratio of 1.5, the W/D ratio required to reduce the interaction effects would be larger as compared to smaller insitu stress ratios.

7.4 Comparison of principal stresses in interacting tunnels-elasto-plastic and elastic analyses.

In Fig. 14 are shown the variation of percentage differences in principal stress between the elasto-plastic and elastic analyses of simultaneously excavated tunnels for the three insitu stress ratios.

At the tunnel boundry (Fig.14a) the percentage difference in major principal stress is a large negative value (indicating that the stresses are higher for elastic analysis) for W/D=0.3 which decreases with increase in W/D ratio (smaller negative value) and tends to become constant and reach the level of percentage difference indicated for the case of single tunnels.

At the centre of pillar zone (Fig. 14b), the difference in the percentage difference is small and it starts decreasing from a comparatively higher value to a constant value as W/D increases.

The variation of percentage difference in the case of minor principal stress at the centre of pillar zone is shown in fig.14c. At W/D ratio of 0.3, the percentage difference is comparatively higher and positive and it decreases with increase in W/D ratio at faster rate. In this case also, the percentage difference tends to become constant at the level of percentage difference for single tunnels.

8 CONCLUSION

From the elastic and elasto-plastic analysis carried out for the case of two deep circular interacting tunnels having three different spacings giving a W/D ratio of 0.3, 0.6 and 1.2 and three insitu stress conditions (0.5, 1.0 and 1.5), following conclusions are drawn

(i) In the pillar zone for the smaller W/D ratios of 0.6 and 0.3 (where interaction affects are expected to be more pronounced, the principal stresses developed have highest value for smaller insitu stress ratio (Ko=0.5).

(ii) Considering the material behaviour as elastic, a comparison of principal stresses and deviatoric stresses ($\sigma_1 - \sigma_3$)in interacting and single tunnel at the boundary and the centre of the pillar zone has indicated that for higher insitu stress ratio of 1.5and smaller W/D ratio of 0.3,the difference is highest and it decreases with decrease in insitu stress ratio and increase in W/D ratio. The difference in principal stresses is also highest for Ko=1.5 and W/D ratio of 0.3.

(iii) In case the material behaviour is considered as elasto-plastic, similar trends (as in the case of elastic analysis) in percentage differences in principal stresses for interacting and single tunnels is observed.

(iv) From a study of elasto-plastic and elastic analyses of interacting tunnels, it has been concluded that, the difference in major principal stress is higher at tunnel boundary for smaller W/D ratio of 0.3 and this tends to become constant as W/D ratio increases.The difference in minor principal stresses is marginal.

9 REFERENCES

[1] C.S.Desai and J.F.Abel, Introduction to the finite element method, Van Nostrand Reinhold Co., New York, 1972

[2] E.Hoek and E.T.Brown, Underground excavation in Rock, Institution of Mining and Metallurgy, London, 1980

[3] O.C.Zienkiewicz, The finite element method, McGraw Hill, Berkshire, 1977

[4] O.C.Zienkiewicz and I.C.Cormeau: Visco-plasticity, plasticity and Creep in elastic solids: A unified numerical solution approach, Int. J. Num. Meth. Engg., 8, 821-845, 1974

[5] R.K.Srivastava, Elasto-plastic finite element analyses of single and Interacting tunnels, Ph.D. Thesis submitted to IIT, Delhi, India, 1985

[6] R.K.Srivastava, K.G.Sharma and A.Varadarajan, Finite element analysis of tunnels using different yield criteria, 2nd Int. Symp. on Numerical models in Geomechanics, Ghent, Belgium, 381-389, 1986

[7] V.S. Chandrasekaran and G.J.W. King, simulation of excavation using finite elements, J.Geotech. Engg. Div. Proc. ASCE, 100, 1087--1089, 1974

Numerical Methods in Geomechanics (Innsbruck 1988), Swoboda (ed.)
© 1988 Balkema, Rotterdam. ISBN 90 6191 809 X

Applying the theory of plastic intensification to calculate the stress in shotcrete and rockbolt lining of hydraulic pressure tunnel and its surrounding rock

Younian Lu & Xiaohong Cai
Jian Prefectural Bureau of Water Conservancy and Hydroelectric Power, Jiangxi Province, People's Republic of China

ABSTRACT: In this paper the shear stress $\tau(\gamma)$ produced by the bond between anchor bar and surrounding rock is considered as a body force. Analytic expressions for the stress in shotcrete and rockbolt lining of hydraulic pressure tunnel and its surrounding rock are established by using the theory of plastic intensification. Computational equations for the stress of anchor bar and its length are also derived. Therefore this paper provides a computational method for the engineering design of hydraulic pressure tunnels.

1 INTRODUCTION

Two papers on calculating the stress in shotcrete and rockbolt lining of hydraulic pressure tunnel and its surroundding rock were described by using the ideal elastoplastic theory and the Mohr-Coulomb yield criterion Refer to Refs. (1) and (2). Owing to diversities in the properties of physical mechanics of surrounding rock, either of the foregoing methods is unsuitable for calculating the stress of the surrounding rock which has plastic intensification properties and of the tunnel of shotcrete and rockbolt lining. In order to make good use of the property of plastic intensification of surrounding rock and reduce the construction cost of tunnels, a method for calculating the stress in shotcrete and rockbolt lining of hydraulic pressure tunnel and its surrounding rock must be established by using the theory of plastic intensification.

Based on the deformation curves of rock mass tests, they are simplified to broken line forms (Fig.1). According to the similarity of stress-deformation curves, analytic expressions for calculating the stress in shotcrete and rockbolt lining of hydraulic pressure tunnel and its surrounding rock are established in terms of the linear intensification model of plastic theory (Fig.2).

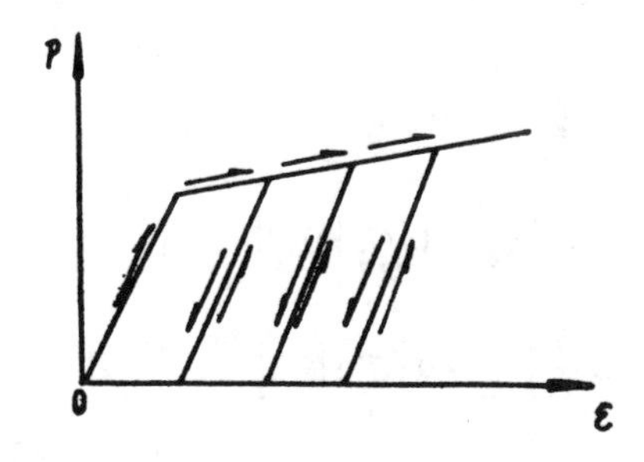

Fig.1 Rock mass stress-strain curves

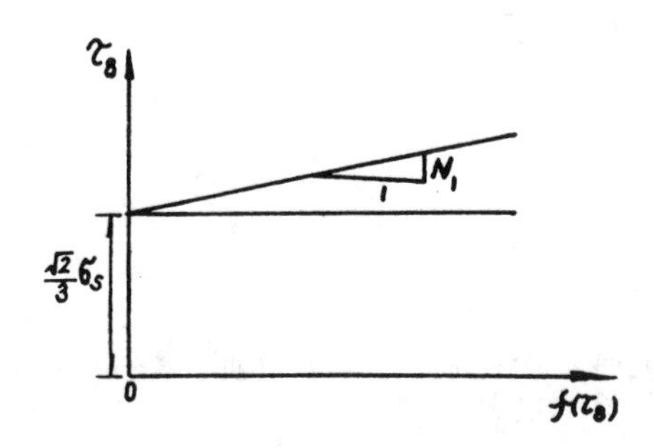

Fig.2 The linear intensification curve of rock masses

2 FUNDAMENTAL EQUATIONS

2.1 The Yield Criterion

Assuming that rock masses obey the Mises yield criterion, using the octahedral stress τ_8 may obtain

$$k=\frac{\sqrt{6}}{2}\tau_8 \qquad (1)$$

$$\tau_8=\frac{1}{3}\sqrt{(\sigma_r-\sigma_\theta)^2+(\sigma_\theta-\sigma_z)^2+(\sigma_z-\sigma_r)^2} \qquad (2)$$

where k - the yield property parameter of rock masses, it can be determined by simple tensile yield test.

2.2 The Constitutive Relation (using cylindrical coordinates)

1. The Elastic Stage

$$\left.\begin{aligned}
\varepsilon_r&=\frac{1}{E}\left[\sigma_r-\mu(\sigma_\theta+\sigma_z)\right]\\
\sigma_\theta&=\frac{1}{E}\left[\sigma_\theta-\mu(\sigma_z+\sigma_r)\right]\\
\sigma_z&=\frac{1}{E}\left[\sigma_z-\mu(\sigma_r+\sigma_\theta)\right]\\
\varepsilon_{r\theta}&=\frac{1}{2G}\tau_{r\theta}\\
\varepsilon_{\theta z}&=\frac{1}{2G}\tau_{\theta z}\\
\varepsilon_{zr}&=\frac{1}{2G}\tau_{zr}
\end{aligned}\right\} \qquad (3)$$

2. The Plastic Stage (The incremental theory)

$$\left.\begin{aligned}
d\varepsilon_r^p&=\frac{F(\tau_8)}{3\tau_8}\left[\sigma_r-\frac{1}{2}(\sigma_\theta+\sigma_z)\right]d\tau_8\\
d\varepsilon_\theta^p&=\frac{F(\tau_8)}{3\tau_8}\left[\sigma_\theta-\frac{1}{2}(\sigma_z+\sigma_r)\right]d\tau_8\\
d\varepsilon_z^p&=\frac{F(\tau_8)}{3\tau_8}\left[\sigma_z-\frac{1}{2}(\sigma_r+\sigma_\theta)\right]d\tau_8\\
d\varepsilon_{r\theta}^p&=\frac{F(\tau_8)}{2\tau_8}\tau_{r\theta}\,d\tau_8\\
d\varepsilon_{\theta z}^p&=\frac{F(\tau_8)}{2\tau_8}\tau_{\theta z}\,d\tau_8\\
d\varepsilon_{zr}^p&=\frac{F(\tau_8)}{2\tau_8}\tau_{zr}\,d\tau_8\\
F(\tau_8)&=f'(\tau_8)
\end{aligned}\right\} \qquad (4)$$

2.3 The Stress-Strain Relation of Linear Intensification

$$f(\tau_8)=\frac{\tau_8-\frac{\sqrt{2}}{3}\sigma_s}{N_1}\ ,\quad \left(\tau_8\geqslant\frac{\sqrt{2}}{3}\sigma_s\right) \qquad (5)$$

where σ_s - the yield stress of simple tension test of rock masses

N_1 - the characteristic parameter of rock mass deformation

2.4 Basic Method

The shotcrete and rockbolt lining in hydraulic pressure tunnel is a kind of flexible lining with low rigidity and high strength, and is capable of transmitting a higher internal water pressure to the surrounding rock. Under the action of a certain load, a plastic zone can be formed in the surrounding rock. A computation sketch of the stress and displacement in the shotcrete and rockbolt lining with reinforcing fabric and its surrounding rock is shown in Fig.3.

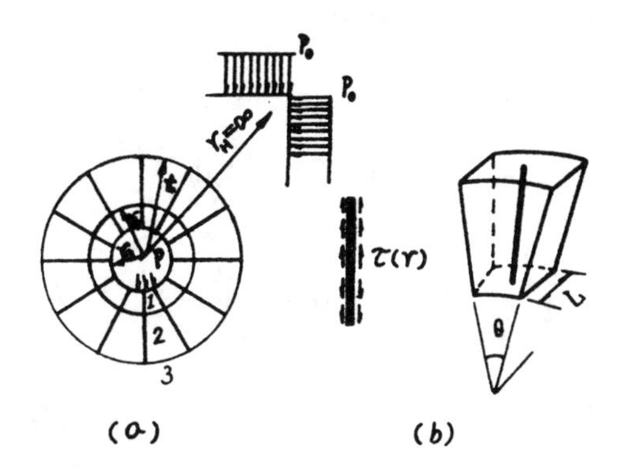

Fig. 3 The computation sketch of the stress and displacement of the three-layered cylinder

The first layer shown in Fig.3 is the elastic zone in the shotcrete a and rockbolt lining with reinforcing fabric (r_b - r_a); the second - the plastic zone of the surrounding rock (t - r_b); the third - the elastic zone of the surrounding rock with inside radius "t" and outside radius "infinite".

The resultant $\tau(r)\pi D dr$ of forces applied to the lateral surface of a micro-section dr of anchor bar is uniformly distributed to its surrounding medium and the transformed into a body force.

$$H(r)=\frac{\pi D \tau(r)}{L\theta r} \qquad (6)$$

where D - the diameter of anchor bar

L - the axial interval of anchor bars

θ - the circumferential angle between adjacent anchor bars

3 THE STRESS AND DISPLACEMENT IN THE INFINITE ELASTIC ZONE OF SURROUNDING ROCK

As shown in Fig,3 the second layer is considered as a thick-walled cylinder with an inside radius t and an outside radius $r_H=\infty$, and is subjected to a uniformly distributed pressure P (initial stress). Since there are no anchor bars there, then H(r)=0.

Let the stress expression be

$$\left.\begin{aligned} \sigma_r &= A' - \frac{B'}{r^2} \\ \sigma_\theta &= A' + \frac{B'}{r^2} \end{aligned}\right\} \qquad (7)$$

For contact surfaces of the elastic and plastic zones, we have the following equations:

$$\sigma_\theta\Big|_{r=t} - \sigma_r\Big|_{r=t} = \frac{2\sqrt{3}}{3}\sigma_s \qquad (8)$$

Its boundary condition is

$$\sigma_r\Big|_{r=\infty} = -P_0 \qquad (9)$$

Using Eqs. (8) and (9), two constants may be determined.

$$\left.\begin{aligned} A' &= -P_0 \\ B' &= \frac{\sqrt{3}\,t^2}{3}\sigma_s \end{aligned}\right\} \quad (10)$$

Substituting Eq. (10) into Eq.(7), we get the following stress expressions:

$$\left.\begin{aligned} \sigma_r &= -P_0 - \frac{\sqrt{3}\,t^2}{3r^2}\sigma_s \\ \sigma_\theta &= -P_0 + \frac{\sqrt{3}\,t^2}{3r^2}\sigma_s \end{aligned}\right\} \quad (11)$$

Based on Hooke's law and geometric equation, a displacement expression for the peripheral elastic zone can be expressed as

$$u = r\varepsilon_\theta = \frac{(1+\mu_3)r}{E_3}\left[\frac{\sqrt{3}\,t^2\sigma_s}{3r^2} - (1-2\mu_3)P_0\right] \quad (12)$$

where E_3 and μ_3 are the elastic modulus and Poisson's ratioin the infinite elastic zone of plastic zone of surrounding rock, respectively.

4THE STRESS AND DISPLACEMENT IN THE PLASTIC ZONE OF SURROUNDING ROCK

4.1 The Stress in Plastic Zone of Surrounding Rock

For elastoplastic plane strain problems, we have

$$\sigma_z = \frac{1}{2}(\sigma_r + \sigma_\theta) \quad (13)$$

Inserting Eq. (13) into Eq. (2) yields

$$\tau_8 = \frac{1}{\sqrt{6}}(\sigma_\theta - \sigma_r) \quad (14)$$

Elastic strains in the plastic zone can be written as

$$\left.\begin{aligned} \varepsilon_r^e &= \frac{1}{E}\left[(1-2\mu_2)\sigma_r - \frac{3\sqrt{6}}{2}\mu_2\tau_8\right] \\ \varepsilon_\theta^e &= \frac{1}{E}\left[(1-2\mu_2)\sigma_r + \frac{(2-\mu_2)\sqrt{6}}{2}\tau_8\right] \end{aligned}\right\} \quad (15)$$

Plastic strains in the plastic zone become

$$\left.\begin{aligned} \varepsilon_r^p &= -\frac{\sqrt{6}}{4}f(\tau_8) \\ \varepsilon_\theta^p &= \frac{\sqrt{6}}{4}f(\tau_8) \end{aligned}\right\} \quad (16)$$

Total atrain in the plastic zone can be expressed as

$$\left.\begin{aligned} \frac{du}{dr} &= \varepsilon_r = \varepsilon_r^e + \varepsilon_r^p = \\ &= \frac{1}{E_2}\left[(1-2\mu_2)\sigma_r - \frac{3\sqrt{6}}{2}\mu_2\tau_8\right] - \frac{\sqrt{6}}{4}f(\tau_8) \\ \frac{u}{r} &= \varepsilon_\theta = \varepsilon_\theta^e + \varepsilon_\theta^p = \\ &= \frac{1}{E_2}\left[(1-2\mu_2)\sigma_r + \frac{(2-\mu_2)\sqrt{6}}{2}\tau_8\right] + \frac{\sqrt{6}}{4}f(\tau_8) \end{aligned}\right\} \quad (17)$$

Using Eq.(17) into the harmonic equation of deformation,

$$\frac{d\varepsilon_\theta}{dr} + \frac{\varepsilon_\theta - \varepsilon_r}{r} = 0 \qquad (18)$$

we can write

$$\frac{1}{E_2}\left[(1-2\mu_2) r\frac{d\sigma_r}{dr} + \frac{(2-\mu_2)\sqrt{6}}{2} r\frac{d\tau_8}{dr}\right] + \\ + \frac{(1-\mu_2)\sqrt{6}}{E_2}\tau_8 + \frac{\sqrt{6}}{4}\left[2f(\tau_8) + r\frac{df(\tau_8)}{dr}\right] = 0 \qquad (19)$$

There are anchor bars in the plastic zone. The body force H(r) may be considered. We can obtain the following equilibrium equation

$$\frac{d\sigma_r}{dr} + \frac{\sigma_r - \sigma_\theta}{r} + H(r) = 0 \qquad (20)$$

Applying Eq.(14)into the above equation gives

$$\frac{d\sigma_r}{dr} = \frac{\sqrt{6}\,\tau_8}{r} - H(r) \qquad (21)$$

Deriving with respect to Eq.(5),we get

$$\frac{df(\tau_8)}{dr} = \frac{1}{N_1}\frac{d\tau_8}{dr} \qquad (22)$$

Putting Eqs.(5), (21)and(22)into Eq.(19) obtains

$$\frac{d\tau_8}{dr} + \frac{2}{r}\tau_8 = \alpha H(r) + \frac{2\beta}{r} \qquad (23)$$

where

$$\alpha = \frac{4N_1(1-2\mu_2)}{\sqrt{6}\left[2(2-\mu_2)N_1 + E_2\right]}$$

$$\beta = \frac{\sqrt{2}\,E_2\sigma_s}{6(2-\mu_2)N_1 + 3E_2}$$

Eq.(23)is a linear differential equation of first order. Its general solution is

$$\tau_8 = \frac{D}{r^2} + \frac{\alpha}{r^2}\int H(r)r^2dr + \beta \qquad (24)$$

Placing Eq.(24)into Eq.(21)and integrating, and then using Eq.(14), we find

$$\left.\begin{aligned}\sigma_r &= -\frac{\sqrt{6}D}{2r^2} + \sqrt{6}\beta\ln r + \\ &+ \sqrt{6}\alpha\int\frac{1}{r^3}\int H(r)r^2drdr - \\ &- \int H(r)dr + C \\ \sigma_\theta &= \frac{\sqrt{6}D}{2r^2} + \sqrt{6}\beta(1+\ln r) + \\ &+ \sqrt{6}\alpha\left[\int\frac{1}{r^3}\int H(r)r^2drdr + \frac{1}{r^2}\int H(r)r^2dr\right] - \\ &- \int H(r)dr + C\end{aligned}\right\} \qquad (25)$$

where the integration constants C and D may be determined by the following two contact conditions, i.e.,

$$\left.\begin{aligned}\sigma_r^p\big|_{r=t} &= \sigma_r^e\big|_{r=t} \\ \sigma_\theta^p\big|_{r=t} &= \sigma_\theta^e\big|_{r=t}\end{aligned}\right\} \qquad (26)$$

After plugging Eqs. (11)and (25) into Eq.(26)and performing mathematical calculation, we have

$$C = -P_0 - \sqrt{6}\beta\left(\frac{1}{2} + \ln t\right) - \\ -\sqrt{6}\alpha\int\frac{1}{r^3}\int H(r)r^2drdr\Big|_{r=t} - \frac{\sqrt{6}\alpha}{2t^2}\int H(r)r^2dr\Big|_{r=t} - \\ - \int H(r)dr\Big|_{r=t} \qquad (27)$$

$$D = t^2\left(\frac{\sqrt{2}}{3}\sigma_s - \beta\right) - \alpha\int H(r)r^2dr\Big|_{r=t} \qquad (28)$$

Entering constants C and D into Eq.(25), we get the stress expressions in the plastic zone.

$$\left.\begin{aligned}
\sigma_r = & -\sqrt{6}\left[\left(\frac{\sqrt{2}}{3}\sigma_s-\beta\right)\frac{t^2}{2r^2}-\beta\left(\frac{1}{2}+\ln\frac{t}{r}\right)\right]-P_0- \\
& -\sqrt{6}\alpha\int_r^t\frac{1}{r^3}\int r^2H(r)\,dr\,dr+ \\
& +\int_r^t H(r)\,dr+ \\
& +\frac{\sqrt{6}\alpha}{2}\left(\frac{1}{r^2}-\frac{1}{t^2}\right)\int r^2H(r)\,dr\Big|_{r=t} \\
\sigma_\theta = & \sqrt{6}\left[\left(\frac{\sqrt{2}}{3}\sigma_s-\beta\right)\frac{t^2}{2r^2}+\beta\left(\frac{1}{2}-\ln\frac{t}{r}\right)\right]-P_0- \\
& -\sqrt{6}\alpha\int_r^t\frac{1}{r^3}\int r^2H(r)\,dr\,dr+\int_r^t H(r)\,dr- \\
& -\frac{\sqrt{6}\alpha}{2r^2}\int_r^t r^2H(r)\,dr- \\
& -\left[\frac{\sqrt{6}\alpha}{2r^2}\int r^2H(r)\,dr\right]\Big|_{r=t}
\end{aligned}\right\} \tag{29}$$

4.2 The Displacement in the Plastic Zone

According to the incompressible condition, we can write

$$\varepsilon_r+\varepsilon_\theta+\varepsilon_z=0 \tag{30}$$

After substituting the geometric geomtric equation into the above equation and considering the plane strain problem, i.e., $\varepsilon_z = 0$, we give

$$\frac{du}{dr}+\frac{u}{r}=0 \tag{31}$$

The solution of Eq.(31) become

$$u=\frac{C'}{r} \tag{32}$$

which C' is the integration constant which is determined by boundary and contact conditions. The conditions are that the displacement in the plastic zone, where r = t, is equal to one in the peripheral elastic zone, where r = t, of the plastic zone, i.e.,

$$u^p\Big|_{r=t}=u^e\Big|_{r=t} \tag{33}$$

Substituting Eqs. (12)and (32) into Eq.(33) obtains

$$C'=\frac{(1+\mu_3)t^2}{E_3}\left[\frac{\sqrt{3}}{3}\sigma_s-(1-2\mu_3)P_0\right] \tag{34}$$

Inserting Eq. (34) into Eq. (32) yields the displacement expression in the plastic zone

$$u=\frac{(1+\mu_3)t^2}{E_3 r}\left[\frac{\sqrt{3}}{3}\sigma_s-(1-2\mu_3)P_0\right] \tag{35}$$

where E_3, μ_3 and the foregoing E_2 and μ_2 are all of the same medium. It should be equal in numerical values, i.e., $E_3 = E_2$; $\mu_3=\mu_2$; the different subscripts marked here are all for the convenience of equation derivation.

5 THE STRESS AND DISPLACEMENT IN THE SHOTCRETE AND ROCKBOLT LINING

The shotcrete and rockbolt lining is an elastic zone and there are anchor bars there. The body force H(r) must be considered. Its stress should also satisfy the equilibrium equation (20), namely,

$$\frac{d\sigma_r}{dr}+\frac{\sigma_r-\sigma_\theta}{r}+H(r)=0 \tag{20}$$

The geometric equations are

$$\left.\begin{aligned}
\varepsilon_r&=\frac{du}{dr}\\
\varepsilon_\theta&=\frac{u}{r}
\end{aligned}\right\} \tag{36}$$

The physical equations for plane strain problems are

$$\left.\begin{aligned}
\sigma_r&=\frac{E_1(1-\mu_1)}{(1+\mu_1)(1-2\mu_1)}\left(\varepsilon_r+\frac{\mu_1}{1-\mu_1}\varepsilon_\theta\right)\\
\sigma_\theta&=\frac{E_1(1-\mu_1)}{(1+\mu_1)(1-2\mu_1)}\left(\frac{\mu_1}{1-\mu_1}\varepsilon_r+\varepsilon_\theta\right)
\end{aligned}\right\} \tag{37}$$

in which E and are the elastic modulus and Poisson's ratio of shotcrete and rockbolt lining, respectively. Using Eq.(36) into Eq. (37) becomes

$$\left.\begin{aligned} \sigma_r &= \frac{E_1(1-\mu_1)}{(1+\mu_1)(1-2\mu_1)}\left[\frac{du}{dr} + \frac{\mu_1}{(1-\mu_1)}\frac{u}{r}\right] \\ \sigma_\theta &= \frac{E_1(1-\mu_1)}{(1+\mu_1)(1-2\mu_1)}\left[\frac{\mu_1}{(1-\mu_1)}\frac{du}{dr} + \frac{u}{r}\right] \end{aligned}\right\} \quad (38)$$

Applying Eq. (38) into Eq.(20), one can get the equilibrium equation which is represented by displacement

$$\frac{d^2u}{dr^2} + \frac{1}{r}\frac{du}{dr} - \frac{u}{r^2} = K_1(r) \quad (39)$$

where

$$K_1(r) = -\frac{(1+\mu_1)(1-2\mu_1)}{E_1(1-\mu_1)}H(r) \quad (40)$$

The general solution to Eq. (39) is

$$u(r) = \frac{r}{2}\int K_1(r)dr - \frac{1}{2r}\int r^2K_1(r) + $$

$$+ Ar + \frac{B}{r} \quad (41)$$

After putting Eq. (41) into Eq, (38), the general solution for stress component becomes

$$\left.\begin{aligned} \sigma_r &= \frac{E_1}{(1+\mu_1)(1-2\mu_1)}\left[\frac{1}{2}\int K_1(r)dr + \right. \\ &\left. + \frac{(1-2\mu_1)}{2r^2}\int r^2K_1(r)dr + A - (1-2\mu_1)\frac{B}{r^2}\right] \\ \sigma_\theta &= \frac{E_1}{(1+\mu_1)(1-2\mu_1)}\left[\frac{1}{2}\int K_1(r)dr - \right. \\ &\left. - \frac{(1-2\mu_1)}{2r^2}\int r^2K_1(r)dr + A + (1-2\mu_1)\frac{B}{r^2}\right] \end{aligned}\right\}$$

(42)

in which the integration constants A and B can be determined by the following boundary and contact conditions.

$$\sigma_r\Big|_{r=r_a} = -P \quad (43)$$

$$\sigma_r\Big|_{r=r_b} = -\sqrt{6}\Big[\Big(\frac{\sqrt{2}}{3}\sigma_s - \beta\Big)\frac{t^2}{2r_b^2} - $$

$$-\beta\Big(\frac{1}{2} + \ln\frac{t}{r_b}\Big) - P_0 - \sqrt{6}\alpha\int_{r_t}^{t}\frac{1}{r^3}\int r^2H(r)drdr + $$

$$+\int_r^t H(r)dr + \frac{\sqrt{6}\alpha}{2}\Big(\frac{1}{r_b^2} - \frac{1}{t^2}\Big)\int r^2H(r)dr\Big|_{r=t} \quad (44)$$

In Eq.(44), the stress in the plastic zone of surrounding rock can be obtained from the first formula in the Eq. (29) (where $r = r_b$).

Entering Eq.(42) into Eq. (43) and Eq.(44),we can determine integration constants A and B. substituting A and B into Eqs.(41) and (42), the expressions for the stress and displacement in the shotcrete and rockbolt lining can be written as:

$$\left.\begin{aligned} \sigma_r = \frac{E_1}{(1+\mu_1)(1-2\mu_1)}\Big\{& f(r) + \\ &+ \frac{(1+\mu_1)(1-2\mu_1)}{E_1}\Big[\frac{r_a^2}{r_b^2-r_a^2}\Big(1-\frac{r_b^2}{r^2}\Big)P - \\ &- \frac{r_b^2}{r_b^2-r_a^2}\Big(1-\frac{r_a^2}{r^2}\Big)P_0\Big] - (1-2\mu_1)(\eta+\xi)\Big(1-\frac{r_a^2}{r^2}\Big) + \\ &+ \frac{f(r_a)r_a^2}{r_b^2-r_a^2}\Big(1-\frac{r_b^2}{r^2}\Big) - \frac{f(r_b)r_b^2}{r_b^2-r_a^2}\Big(1-\frac{r_a^2}{r^2}\Big)\Big\} \\ \sigma_\theta = \frac{E_1}{(1+\mu_1)(1-2\mu_1)}\Big\{& \overline{f(r)} + \\ &+ \frac{(1+\mu_1)(1-2\mu_1)}{E_1}\Big[\frac{r_a^2}{r_b^2-r_a^2}\Big(1-\frac{r_b^2}{r^2}\Big)P - \\ &- \frac{r_b^2}{r_b^2-r_a^2}\Big(1+\frac{r_a^2}{r^2}\Big)P_0\Big] - (1-2\mu_1)(\eta+\xi)\Big(1+\frac{r_a^2}{r^2}\Big) + \\ &+ \frac{f(r_a)r_a^2}{r_b^2-r_a^2}\Big(1+\frac{r_b^2}{r^2}\Big) - \frac{f(r_b)r_b^2}{r_b^2-r_a^2}\Big(1+\frac{r_a^2}{r^2}\Big)\Big\} \end{aligned}\right\}$$

$$u(r)=\zeta(r)+\frac{(1+\mu_1)}{E_1}\left\{\frac{r_a^2}{r_b^2-r_a^2}\left[(1-2\mu_1)r+\frac{r_b^2}{r}\right]P-\right.$$
$$\left.-\frac{r_b^2}{r_b^2-r_a^2}\left[(1-2\mu_1)r+\frac{r_a^2}{r}\right]P_0\right\}-$$
$$-(\eta+\xi)\left[(1-2\mu_1)r+\frac{r_a^2}{r}\right]+$$
$$+\frac{f(r_a)r_a^2}{r_b^2-r_a^2}\left[r+\frac{r_b^2}{r(1-2\mu_1)}\right]-$$
$$-\frac{f(r_b)r_b^2}{r_b^2-r_a^2}\left[r+\frac{r_a^2}{r(1-2\mu_1)}\right] \qquad (45)$$

where

$$f(r)=\frac{1}{2}\int K_1(r)dr+\frac{1-2\mu_1}{2r^2}\int r^2K_1(r)dr$$

$$\overline{f(r)}=\frac{1}{2}\int K_1(r)dr-\frac{1-2\mu_1}{2r^2}\int r^2K_1(r)dr$$

$$\zeta(r)=\frac{r}{2}\int K_1(r)dr-\frac{1}{2r}\int r^2K_1(r)dr$$

$$\eta=\frac{(1+\mu_1)r_b^2}{E_1(r_b^2-r_a^2)}\left\{\sqrt{6}\left[\left(\frac{\sqrt{2}}{3}\sigma_s-\beta\right)\frac{t^2}{2r_b^2}-\right.\right.$$
$$\left.\left.-\beta\left(\frac{1}{2}+\ln\frac{t}{r_b}\right)-\int_{r_b}^{t}H(r)dr\right]\right\}$$

$$\xi=\frac{(1+\mu_1)r_b^2}{E_1(r_b^2-r_a^2)}\cdot\frac{\sqrt{6}\alpha}{2}\left[2\int_{r_b}^{t}\frac{1}{r^3}\int r^2H(r)drdr-\right.$$
$$\left.-\left(\frac{1}{r_b^2}-\frac{1}{t^2}\right)\int r^2H(r)dr\Big|_{r=t}\right]$$

$$f(r_a)=f(r)\Big|_{r=r_a}$$

$$f(r_b)=f(r)\Big|_{r=r_b}$$

6 THE DETERMINATION OF RADIUS IN THE PLSDTIC ZONE AND THE LENGTH OF ANCHOR BAR

Since the displacement in the shotcrete and rockbolt lining and in the plastic zone of surrounding rock, where $r = r_b$, should satisfy continuity condition, a relational expression between the internal water pressure P of tunnel and the radius t of the plastic zone of surrounding rock can be derived. The continuity condition of displacement is

$$u^{l}\Big|_{r=r_b}=u^{p}\Big|_{r=r_b} \qquad (46)$$

where u^l - the radial displacement in the shotcrete and rockbolt lining, calculated in accordance with the third formula of Eq.(45),

u^p -the radial displacement in the plastic zone of surrounding rock, calculated in accordance with Eq.(35). Inserting the third formula of Eq. (45) and Eq.(35) into Eq. (46), after rearrangement, yields

$$P=\frac{(r_b^2-r_a^2)E_1}{2(1-\mu_1^2)r_a^2r_b}\left\{-\zeta(r_b)+\right.$$
$$+\left\{\frac{(1+\mu_1)r_b}{E_1(r_b^2-r_a^2)}\left[(1-2\mu_1)r_b^2+r_a^2\right]-\right.$$
$$\left.-\frac{(1+\mu_3)(1-2\mu_3)t^2}{E_3r_b}\right\}P_0+$$
$$+\frac{(\eta+\xi)}{r_b}\left[(1-2\mu_1)r_b^2+r_a^2\right]-$$
$$-\frac{r_b}{(1-2\mu_1)(r_b^2-r_a^2)}\left\{2(1-\mu_1)r_a^2f(r_a)-\right.$$
$$\left.\left.-\left[(1-2\mu_1)r_b^2+r_a^2\right]f(r_b)\right\}+\frac{\sqrt{3}(1+\mu_3)t^2}{3E_3r_b}\sigma_s\right\} \qquad (47)$$

where

$$\zeta(r_b)=\zeta(r)\Big|_{r=r_b}$$

Eq. (47) indicates that, when a plastic zone emerges in the surrounding rock of hydraulic pressure tunnel with shotcrete and rockbolt lining, the above relation must exist between the internal water pressure P and the characteristic parameter of tunnel.

Given internal water pressure P, we can obtain radius t of the plastic zone ofsurrounding rock by using

Eq.(47) and numerical method, and then can determinethe computational length of anchor bar. As shown in Fig. 3, the length of anchor bar must satisfy the following formula:

$$l \geq (t - r_a) \qquad (48)$$

For practical purposes of engineering design, the above formula can be written as the following equation

$$l = (t - r_a) + e \qquad (49)$$

where e - the structural length of anchor bar

7 THE DETERMINATION OF SHEAR STRESS $\tau(r)$

According to Eq.(45), we find the stress and displacement in the shotcrete and rockbolt lining. In accordance with Eqs.(29) and (35), we get the stress and displacement in the plastic zone of surrounding rock. In the light of Eqs. (11) and (12), we yield the stress and displacement in the peripheral elastic zone of plastic zone of surrounding rock.

However, the interaction force $\tau(r)$ between the anchor bar and the medium of surrounding rock are unknown. Thus, the body force H(r), $k_1(r)$ value and its integral quantity cannot be computed actually.

On the basis of the model experiment of anchor bar, the bond shear stress distribution between the anchor bar, and the surrounding rock is a quadratic curve which has opposite-sign shear stress values in the ends. Thus theshear stress can be given by the following equation

$$\tau(r) = a_0 + a_1 r + a_2 r^2 \qquad (50)$$

Using the expression Eq. (50) of $\tau(r)$ into Eq. (6) becomes

$$H(r) = \frac{\pi D \tau(r)}{L \theta r} = \frac{\pi D}{L \theta r}(a_0 + a_1 r + a_2 r^2) \qquad (51)$$

Applying Eq. (41)into Eq. (40) obtains

$$K_1(r) = -\frac{(1+\mu_1)(1-2\mu_1)}{E_1(1-\mu_1)} \cdot \frac{\pi D}{L \theta r}(a_0 + a_1 r + a_2 r^2) \qquad (52)$$

in which a_0, a_1 and a_2 are undetermined coefficients.

Putting Eqs.(51)and (52) into the third equation in Eq. (45)and Eq.(35), we find the radial displacement ui (i=1,2) of the points in the shotcrete and rockbolt lining and in the plastic zone of surrounding rock. It is only dependent upon the radial coordinates r. The expression of $u_i(r)$ also contain undetermined coefficients a_0, a_1 and a_2. These coefficients are determined by the deformation coordination condition of the surrounding rock, the shotcrete and rockbolt lining and the anchor bar.

When the anchor bar is acted upon by $\tau(r)$, the stress $\sigma_0(r)$ in the cross section of the anchor bar at the radial coordinates r becomes

$$\sigma_0(r) = \frac{4}{D}\left[a_0(r - r_a) + \frac{a_1}{2}(r^2 - r_a^2) + \frac{a_2}{3}(r^3 - r_a^3)\right] \qquad (53)$$

where the stress $\sigma_0(r)$ of the anchor bar is epually that the tension is positive and the compression is negative. Let the elastic modulus of anchor bar be E_0, then the strain of anchor bar the tunnel radius r yields

$$\varepsilon_0(r) = \frac{4}{DE_0}\left[a_0(r - r_a) + \frac{a_1}{2}(r^2 - r_a^2) + \frac{a_2}{3}(r^3 - r_a^3)\right] \qquad (54)$$

Therefore, the displacement of anchor bar at the tunnel radius r should satisfy the following relational expression

$$u(r) = u_0(r_a) + \frac{4}{DE_0}\left[a_0\left(\frac{r^2}{2} - r_a r + \frac{r_a^2}{2}\right) + a_1\left(\frac{r^3}{6} - \frac{r_a^2 r}{2} + \frac{r_a^3}{3}\right) + a_2\left(\frac{r^4}{12} - \frac{r_a^3 r}{3} + \frac{r_a^4}{4}\right)\right] \qquad (55)$$

where $u_0(r)$ is the radial displacement of anchor bar at $r = r_a$, According to the deformation coordination condition, we have

$$u_0(r_a) = u(r_a) \qquad (56)$$

in which $u(r_a)$ is the radial displacement in the shotcrete and rockbolt lining at $r = r_a$. It may be computed by the third equation in Eq. (45).

Undetermined coefficients a_0, a_1 and a_2 may be determined by the following conditions:

7.1 The anchor bar must satisfy the static equilibrium candition under the action of the shear stress $\tau(r)$

$$\int_{r_a}^{t} \tau(r)\pi D\,dr = 0$$

namely

$$a_0(t-r_a) + \frac{a_1}{2}(t^2-r_a^2) + \frac{a_2}{3}(t^3-r_a^3) = 0 \tag{57}$$

7.2 In accordance with the deformation coordination condition, the displacement $u_0(r)$ of the anchor bar should be equal to one $u(r)$ of the medium (the surrounding rock and the shotcrete and rockbolt lining), that is

$$u_0(r) = u(r) \tag{58}$$

Since the longitudinal distribution of the shear stress $\tau(r)$ along the anchor bar is a quadratic curve, there are three coefficients a_0, a_1 and a_2. To determine the three coefficients, three equations must be established. One equation (57) has been established as above. Therefore, provided that we establish two equations in terms of the coordination condition of deformation, we can find the coefficients. When $r = r_b$ and $r = t$, the following equality can be established.

$$u_0(r_b) = u(r_b) \tag{59}$$

$$u_0(t) = u(t) \tag{60}$$

By solving Eqs.(58),(59) and (60) simultaneously, the undetermined coefficients a_0, a_1 and a_2 can be found.

Having found the coefficients, we can determine $\tau(r)$, $H(r)$ and $K_1(r)$. Substituting the determined $H(r)$ and $K_1(r)$ into expressions for the stress and displacement, one can yield the stress and displacement in the shotcrete and rockbolt lining, in the plastic zone of surrounding rock and the peripheral elastic zone. The displacement, stress and strain in cross sections of anchor bar can be computed by Eqs.(55), (54) and (53), respectively.

ACKNOWLEDGMENT

The authors are very grateful to Wang Zenxing, engineer at the Jiangxi Wire Communication Factory, for his help.

REFERENCES

1 Younian Lu and Xiaohong Cai, Applying Mohr-Coulomb Yield Criterion to Calculate Elastoplastic Stress in Adhesive Shotcrete and Rockbolt Lining of Hydraulic Pressure Tunnel and its Surrounding Rock, Proceedings of the International Symposium on Engineering in Complex Rock Formations, 1986, Beijing, China.

2 Xiaohong Cai and Younian Lu, Calculating Formula for Rock Resistance Coefficient K of Cylindrical Pressure Tunnel under Coulomb Yield Criterion, Proceedings of the First Annual Meeting of the Mechanics Institute of Jiangxi Province, 1984, Nanchang, China.

3 Xu Zhigang and Wang Mingshu, Model Experimental Reseach of Full-length Anchored Pin Timbering, Proceedings of the Inaugural Meeting of the Chinese Rock Mechanics and Engineering Institute and Academic Conference, 1985, Beijing, China.

Numerical Methods in Geomechanics (Innsbruck 1988), Swoboda (ed.)
© 1988 Balkema, Rotterdam. ISBN 90 6191 809 X

Numerical-analytical technique of optimizing the layout of parallel tunnels in seismic regions

Nina N.Fotieva & A.N.Kozlov
Tula Polytechnical Institute, USSR

ABSTRACT: The technique of establishing optimum distances between an arbitrary number of different size parallel tunnels erected in seismic regions is offered in the report presented. Conditions were summed up. Average stresses from static loads seismic effects of earthquakes in each pillar do not exceed ultimate limits of rock upon a uniaxial compression and a uniaxial strain is assumed to be the criterion of optimization.

Modern large hydraulic power systems are specified as a rule by the presence of different size parallel tunnels mutually influencing each other. To obtain the required strength and seismic stability of such structures and a decrease in the cost of their erection the problems of optimizing the layout of tunnels, i.e. evaluating distances are up-to-date. The distances taking into account their cost are to be minimum but are to secure at the same time the preservation of pillars with the safety coefficients given both in normal conditions and at seismic earthquake effects.

From the mathematical point of view the problem of establishing an optimum variant of the **N** layout of different size parallel workings with centres situated on one straight line are expressed as follows: such sizes $S_{i,i+1}$ of pillars between the i-th. and the i+1-st. workings which would secure the preservation of pillars with the ultimate limit coefficients given $C_{i,i+1}$ and bring to a minimum the general extent of the workings' complex in the cross-section plane have to be determined. Therefore the function of the aim must satisfy the condition

$$f(S_{1,2};S_{2,3};\ldots;S_{N-1,N})=\sum_{i=1}^{N-1} S_{i,i+1} \qquad (1)$$

at layed on restrictions

$$K_{i,i+1}(S_{1,2};S_{2,3};\ldots;S_{N-1,N})\geq C_{i,i+1} \quad (i=1,2,\ldots,N-1) \qquad (2)$$

where $K_{i,i+1}$ is the safety factor coefficient of the pillar between the i-th. and the i+1-st. workings being estimated by the formula

$$K_{i,i+1}=\min\left[\frac{[\sigma_c]_{i,i+1}}{\sigma^{(1)}_{i,i+1}};\frac{[\sigma_s]_{i,i+1}}{\sigma^{(2)}_{i,i+1}}\right] \qquad (3)$$

$[\sigma_c]_{i,i+1}$, $[\sigma_s]_{i,i+1}$ are the ultimate limits of the pillar rocks upon a uniaxial compression and a uniaxial strain correspondingly; $\sigma^{(1)}_{i,i+1}$, $\sigma^{(2)}_{i,i+1}$ are the summed maximum average compressive and tensional (if they are obtained as a result of the design) vertical stresses in the pillar from static stresses (weight of the rock itself, initial tectonic stresses, internal head) and seismic earthquake effects correspondingly.

By average stresses in pillars we have in view the relation of the integral derived from vertical stresses in pillars to its weight. The solution of the elasticity theory flat problems for medium weakened by a finite number of non-equal circular holes with centres distributed arbitrarily upon a horizontal straight line at corresponding boundary conditions are applied for determining the pillars' stressed state derived from static loads and from long seismic waves spread in the massif during earthquakes. For determing the strength of the pillars with seismic effects being taken into consideration maximum values of average stresses that may appear in pillars at different combinations of long waves of compression-stress (the longitudinal ones) and heaves (the transverse ones) of any direction in cross-section plane of tunnels coming simultaneously (the worst possible case) are applied. The approach described was suggested by Fotieva (1980).

The solutions of two elasticity theory flat quasi-statical problems are applied for evaluating the pillar stresses state upon the arbitrary directed action of long longitudinal and transverse waves. The designed schemes are given in fig. 1a,b.

Stresses upon infinity are assigned by the familiar formulae (Napetvaridze, 1959)

$$P=\frac{1}{2\pi}K_c\gamma C_1T_o\ ,\ \xi=\frac{\nu_o}{1-\nu_o}\ ,\ Q=\frac{1}{2\pi}K_c\gamma C_2T_o \quad (4)$$

where K_c is the seismicity coefficient, characterising the earthquake strength; T_o is the prevailing period of the oscillating rock particles; C_1, C_2 is the speed of spreading long longitudinal and transverse waves correspondingly; γ is the rock specific weight; ν_o is the Poisson Coefficient of rock.

External loads upon outlines of holes L_m (m=1,2,...,N) are not present, i.e.

$$\sigma_r^{(m)}=0\ ,\ \tau_{r\theta}^{(m)}=0\ . \quad (5)$$

The elasticity theory problems formulated are solved by the D.I. Sherman Method (1951) with the application of the theory of the complex variables analytical function theory, complex series and the property of the Cauchy Type Integrals (Muskhelishvili, 1966).

Average vertical stresses in pillare $\sigma_{i,i+1}^{(p)}$ from the action of the long longitudinal wave falling at an arbitrary angle α of the transverse wave are determined from the solution of the first problem (fig. 1a):

$$\sigma_{i,i+1}^{(p)}=\frac{1}{x_{i+1}-x_i-R_{i+1}-R_i}\int_{x_i+R_i}^{x_{i+1}-R_{i+1}}\sigma_y^{(p)}\big|_{y=0}dx\ , \quad (6)$$

where $\sigma_y^{(p)}|_{y=0}$ are the vertical stresses in pillars from longitudinal waves.

Average stresses in pillars $\sigma_{i,i+1}^{(s)}$ from the actions directed at the angle α of transverse wave are simularly evaluated with the second problem being solved (fig. 1b).

The sum and difference of the expressions received for average stresses, specifying the pillars stressed state in totallity of actions of longitudinal and transverse waves directed arbitrarily and coming simultaneously are investigated in each pillar upon extremum an the angle of the falling waves α by solving equation

$$\frac{\partial}{\partial\alpha}\left[\sigma_{i,i+1}^{(p)}\pm\sigma_{i,i+1}^{(s)}\right]=0\ ,\ (i=1,2,\dots,N-1) \quad (7)$$

and the angle of fall and combination of waves are determined for every pillar, the average stresses in the given pillar being maximum at the absolute value. After that values of the maximum average stresses in each pillar from seismic effects are evaluated

$$\sigma_{i,i+1}^{(c)}=max\left|\sigma_{i,i+1}^{(p)}\pm\sigma_{i,i+1}^{(s)}\right| \quad (8)$$

Average stresses in pillars from the action of tectonic forces in the massif $\sigma_{i,i+1}^{(N_1)}$ when the main axes of initial stresses may be inclined with respect to the vertical and the horizontal are evaluated from the solution of the first problem (fig. 1a) in a quotient case

$$P=N_1\ ,\ \xi=\tilde{\lambda}\ ,\ \alpha=\tilde{\alpha}\ , \quad (9)$$

where N_1 is the value of the bigger main initial stress, $\tilde{\lambda}=N_2/N_1$ is the relation of the main stresses $\tilde{\alpha}$ is the angle of the slope N_1 to the horizontal.

The stressed state of the pillars from the action of weight of the rock itself $\sigma_{i,i+1}^{(\gamma H)}$ is evaluated from the solution of the very problem in the quotient case.

$$P=\gamma H\ ,\ \xi=\lambda\ ,\ \alpha=\frac{\pi}{2} \quad (10)$$

where λ is the coefficient of the lateral rock pressure in an intact

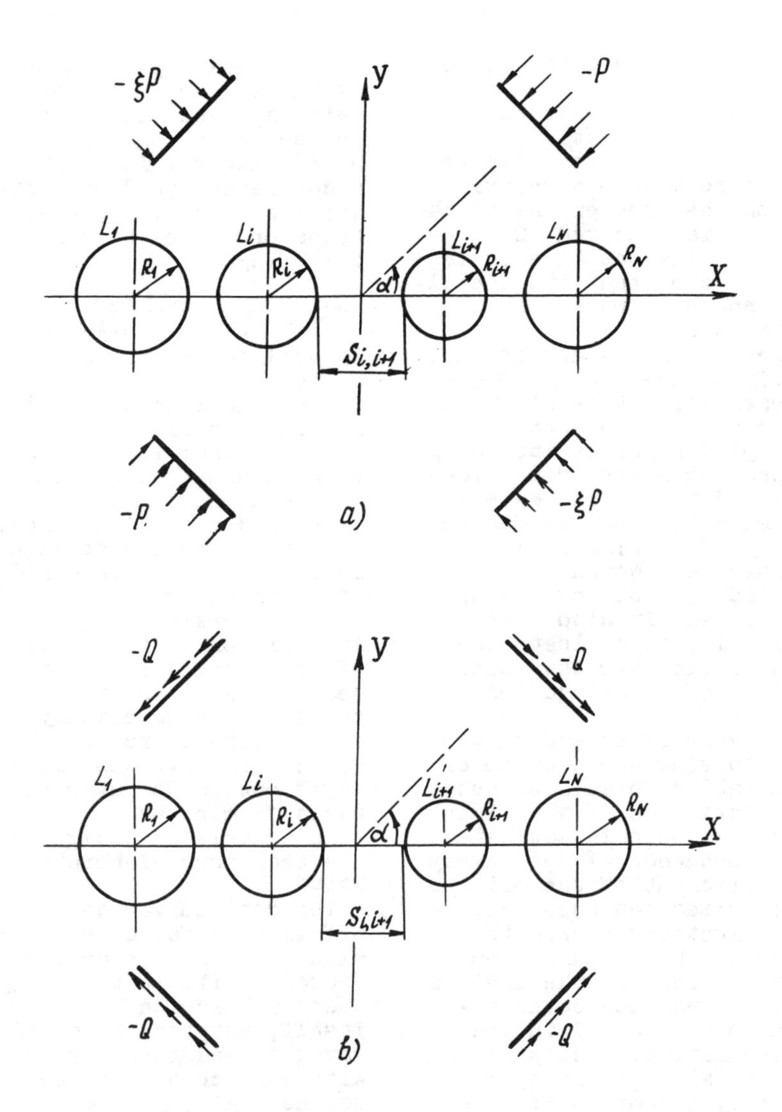

Fig. 1. Designs schemes for determining the pillars' stressed state from arbitrary directed long waves: a - longitudinal, b - transverse

massif.

The elasticity theory flat problem for the sphere shown in fig. 1 during the action upon outlines of holes L_m ($m = 1,2,...,N$) of uniform pressure - p_m ($m = 1,2,..,N$) has been solved for determining average stresses in pillars from the internal head action $\sigma_{i,i+1}^{(w)}$.

As the weight of the rocks themselves or the tectonic forces action in the massif calls forth compressive (negative) vertical stresses in pillars, the internal head calls forth tensile (positive) stresses and seismic effects call forth stresses of different signs, the two most dangerous stress combinations are investigated for determining the pillar safety factor coefficient

$$\sigma_{i,i+1}^{(1)} = \sigma_{i,i+1}^{(\gamma H)} - \sigma_{i,i+1}^{(c)}, \quad \sigma_{i,i+1}^{(2)} = \sigma_{i,i+1}^{(\gamma H)} + \sigma_{i,i+1}^{(c)} + \sigma_{i,i+1}^{(w)} \quad (11).$$

In case of tectonic forces acting in the massif, stresses $\sigma_{i,i+1}^{(\gamma H)}$

in formulae (11) are substituted by $\sigma^{(N)}_{i,i+1}$.

If both of the values received $\sigma^{(1)}_{i,i+1}$ and $\sigma^{(2)}_{i,i+1}$ are negative, then for comparison with the ultimate strength of rock upon a uniaxial compression the greater one in absolute value is selected. On the value $\sigma^{(2)}_{i,i+1}$ being positive it is additionally compared with the ultimate strength of rock upon the uniaxial strain.

A complete design algorithm has been received. The algorithm offers a possibility to evaluate all components of the stresses tensor in pillar points including those appearing apart from a longitudinal and transverse wave of an assigned direction besides the possibility of finding maximum average stresses, which are further applied for solving the optimized problem. It also offers the possibility to evaluate the normal tangential stresses upon the outlines of cross-section of workings.

With the view of selecting the method of solving the problem of the optimization formulated above the characteristics of the limit functions have been investigated, i.e. the dependence of the safety factor coefficient of the pillars from their sizes has been evaluated. Investigations showed that the dependence has a monotonous non-linear character, the minimum of the limit function being attained on condition (2) in the form of equalities, i.e. pillars are of equal strength at optimum sizes (safety factor coefficient is taken into account).

Applying the principle of equal strength one can formulate the problem set as an inverse one, it is necessary to find such sizes of pillars $S^*_{1,2}$, $S^*_{2,3}$, . . . , $S^*_{N-1,N}$, as to fulfill the conditions

$$K_{i,i+1}(S^*_{1,2}; S^*_{2,3}; \dots; S^*_{N-1,N}) = C_{i,i+1} \qquad (12)$$

The task set was reduced to solving the system of non-linear algebriac equations at the numerical solution. The main criterion in selecting the method of solving the system of non-linear equations was the condition of its quick convergence. For the solution the Stephen Generalised Method built on the iteration process consisting of the simple iteration being twice applied and the method of the secant being once applied has been selected. The method possesses a quadratic speed of convergence and its program can easily be transferred to the computer. The following

$$S^{(0)}_{i,i+1} = \frac{R_i + R_{i+1}}{[\sigma_c]_{i,i+1} \cdot C_{i,i+1} / \gamma H - 1} \qquad (13)$$

is applied as its initial approximation.

The algorithm of solving optimization problems has been elaborated and a program in the FORTRAN algorithm language has been made taking into consideration the approach mentioned above. The program allows the stressed state of the pillars to be determined and the strength at the distance given between workings to be evaluated in addition to the optimum variant of the tunnels layout being scanned. The designs stated were shown to be operating quickly, for example the search of the optimum variant of five tunnels being layed out on Computer ES - 1033 takes 15 minutes.

An example of design illustrating the technique elaborated is given below.

The optimum variant of the layed off five different size tunnels with the aim of securing safety factor coefficient is $C_{i,i+1} = 1.5$ under the action of the rock weight itself, internal head and seismic waves called forth by an earthquake with a force of 9 in number is scanned.

The initial data for the design are $R_1 = R_3 = 3$m, $R_2 = 1$m, $R_4 = R_5 = 2$m, $H = 80$m, $[\sigma_c]_{i,i+1} = 8$MP, $[\sigma_s]_{i,i+1} = 2$MPa, $\lambda = 0.33$, $\gamma = 27$kN/m, $E_0 = 800$MPa, $\nu_0 = 0.25$, $K_c = 0.1$, $T_0 = 0.5$s , $P_1 = P_2 = P_3 = 0$, $P_4 = P_5 = 0.6$MPa,

The design made showed that for securing the necessary safety factor of the pillars it is necessary to give the following sizes $S_{1,2} = 3.6$m, $S_{2,3} = 3.75$m, $S_{3,4} = 4.47$m, $S_{4,5} = 3.66$m, the basic part of the maximum average stresses in pillars (up to 73.4%) being called forth by the weight of the rock lying above, the contribution of seismic effects being 18.4%, and that of the internal being 8.2%. The epures of distributing stresses received vitness that the pillars between tunnels are in a stable condition as the summed up stresses from the acting loads at their most unfavourable combinations surpass the rock ultimate limit only in separate points of the first and second pillars and not more than 13%.

The technique developed of the optimization being based upon the solutions of the elasticity theory flat problems for the medium weakened by round holes, the question concerning the possibility of the technique being applied for designing tunnels of non-circular cross-section is under observation in the report presented. With this aim the comparison of the results of solving the quotient problems known and also the data of experiments upon patterns from optically active materials for several holes on non-circular form (square, oval, in forms of slots and vaults) with the results obtained applying the technique developed at the radii of round holes equal to half of the width of non-circular ones has been conducted. Comparison of the results showed that in spite of substantial differences in nature of stresses distribution, average vertical stresses in abutements between holes of non-circular forms at the ratio of their width $h/d \leq 1$ differ little (not more than 10%) from average stresses in abutements between corresponding round holes. At the error $h/d > 1$ they greatly depend upon distances between holes. On the basis of the analysis conducted the field of the application technique developed (with an error not exceeding 10) at the design of two similar workings of non-circular cross-section depending upon their sizes (ratio of their height to their span h/d) and relative distances between them S/d has been determined (fig. 2).

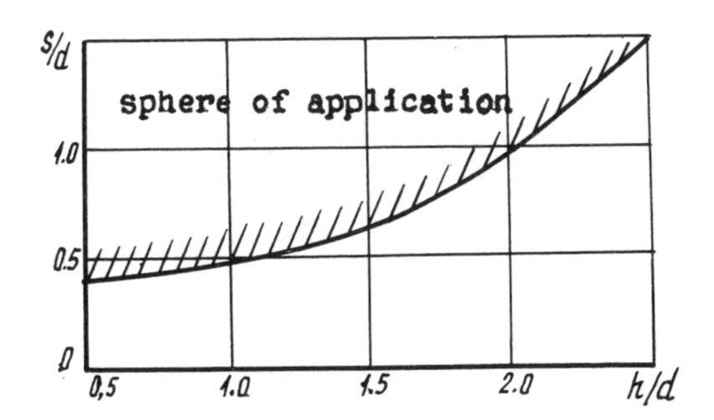

Fig. 2. The field of applying optimization technique developed in designing parallel non-circular cross-section tunnels

At present the technique offered is generalised for each case when centres of tunnels cross-section are not situated in one straight line.

REFERENCES

[1] N.N. Fotieva, Design of underground structure lining in seismic region, Nedra, Moscow, 1980.

[2] S.G. Napetvoridze, Seismic stability of hydro-engineering structures, Gosstroyizdat, Moscow, 1959.

[3] N.I. Muskhelishvili, Some basic problems of Mathematical Elasticity theory, Nauka, Moscow, 1966.

[4] D.I. Sherman, On stresses in plane heavy medium with two similar circular holes symmetry distributed. Applied Mathematics, vol. 155, 751-761, 6 (1951).

Numerical Methods in Geomechanics (Innsbruck 1988), Swoboda (ed.)
© 1988 Balkema, Rotterdam. ISBN 90 6191 809 X

Simulation of a tunnel behaviour from data provided by a reconnaissance gallery

M.Deffayet & A.Robert
Centre d'Etudes des Tunnels, Bron, France

ABSTRACT : Fundamentally the numerical methods applied to the study of underground structures allow the simulation of the behaviour of the tunnel surrounding material and supporting structures, and the calculation of stresses ans strains. In practice however, the major difficulty often emerges when determining values to be selected for the distinctive parameters introduced in the model. In the example presented here, the parameter values are deduced from the analysis of convergence measurements performed in a reconnaissance gallery excavated in sandy clays. This analysis based on a viscoelastic model also allowed the assessment of the impact of the various parameters. As these values were determined with the assumption of a gallery with circular cross section, the passage to a gallery of any cross section requires a finite element calculation. This calculation, used with plane deformation, defines the short term behaviour, i.e. during excavation, taking account of the face progression and works steps (upper half-section, sidewalls, lower half-section). The calculation of delayed effects allows the assessment of the longterm behaviour. In addition to a more accurate description of the ground, this process provides - based on the reconnaissance gallery - predominent data for the calculation of the final tunnel cross section, whatever its shape and excavation method may be.

1 INTRODUCTION

The excavation of a large section tunnel in grounds of poor mechanical characteristics sets the problem of the general stability of the structure. The laboratory tests performed within the feasibility study often show a large dispersion that reflects the non-homogeneity of the encountered materials and does not allow approaching the global behaviour of the surrounding ground.

In the case described here a small section exploratory adit has been excavated to complement the conventional geotechnical tests. It allowed on the one part to observe in the field the mechanical holding of materials, on the other part the accurate follow-up of the excavation behaviour with time. The essential objective was to obtain basic parameters for the most accurate definition of the ground mechanical behaviour, especially those which are determinative in the definite case of a tunnel excavation.

This behaviour was mainly quantified by the relative convergences of the gallery walls. These depend on the material, its rheological characteristics, and the installed support. Various authors, e.g. Sulem,Panet and Guenot /1/, /2/, have proposed a theoretical formulation considering the elastoplastic deformation resulting from the face and distinguishing between the effects of time and the average stress.

In this contribution, we take account of these both effects, before considering the face progression-time effect connection ; the main reason is that the viscous character of the deformation results in a dependency towards the whole stress development. Therefore the integration is based on the whole stress variations.

2 DESCRIPTION OF THE EXPLORATORY ADIT

The crossing of the A30 Motorway through the Butte du Bois des Chênes in Hayange requires to excavate two one-way tubes of 130 m² in section each. After completion of the first investigations and considering the dispersion of results, it was decided to bore an exploratory adit of small diameter, in order to get a more precise evaluation on the feasibility and cost of final works.

2.1 Geological conditions

The gallery is located in Toarcian grounds made of sandy clays with intercalated sandstone banks of 0.2-0.5 m in thickness. Above them is the ferriferous Aalenian formation of about 20 m in height.

2.2 Gallery geometry

The gallery is rectilinear and 150 m long; it is located between the two final tubes of the tunnel. The cross section is 9.6 m²; the cross profile is given in Fig.1. The maximal overburden is 36 m.

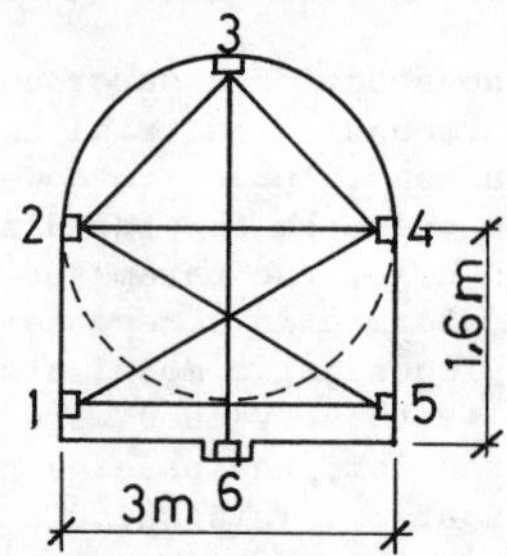

Fig.1 : Cross section profile

2.3 Installed supports

Three types of support were installed :
- TH21 ribs and lagging
- TH21 ribs, bolting and shotcrete
- Bolting and shotcrete.

The type of support was changed according to the characteristics of the encountered grounds.

In the interpretation of the displacements, the rigidity of support - and especially of bolts and ribs - depends on the type of sealing or wedging. AFTES /3/ have proposed various formulations. For our part, we express the support at every section by a global parameter - ks - based on the field measurements ; thus its value is only an "averaged" representation of the support, a plain calculation medium ; the major objective remains the surrounding ground properly said.

2.4 Control of gallery displacements

The excavation proceeded using a Westfalia roadheader, with an average progression of 7.50 m/week. It was essential to measure the displacements of the gallery as soon as the face is passing through, in order to distinguish the role of the face that acts for the first metres and the role of the delayed strains. These measurements were performed by means of devices determining the relative convergences.

Every 10 m, the cross profile of the gallery was equipped with 5 or 6 measuring shifts sealed into the ground as nearest as possible to the face. Fig. 1 shows the location of these shifts. Then, the measurements proceeded periodically by means of a telescopic stick of aluminium alloy, giving the length of each of the seven chords with a precision of 1/20th mm. Every measurement was recorded with the precise time of measurement and the face position at that time.

2.5 Results of the convergence measurements

For each instrumented section and each chord in this section, a curve defines the trend of convergence according to time on the one part and distance to the face on the other part (Fig. 2).

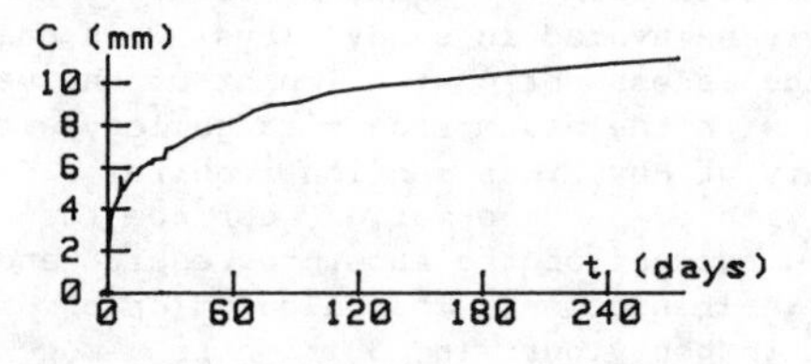

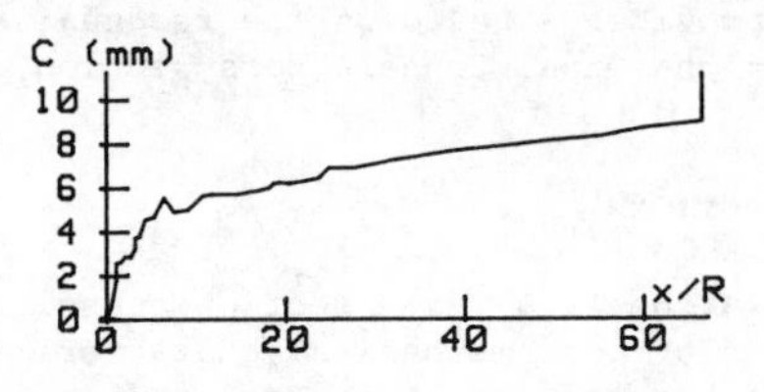

Fig. 2 : Trend of convergence curves according to time and distance x to the face

These two curves can be deducted from each other when the relation $x = f(t)$ defining the instantaneous progression is known. The theoretical analysis of these convergence curves allows approaching the determinative parameters of the behaviour.

3 USED THEORICAL MODEL

3.1 Displacement components

A number of rheological models were presented to simulate the deformation of a deep-located gallery. Panet /4/ and Nguyen Minh /5/ proposed solutions with viscous deformations. In the following text we distinguish two components of the displacement :

a) the face effect : this effect is most

important in the vicinity of the face, which opposes to the gallery closing. We use the Sakuraï formulation /6/ that proposes in the case of a circular gallery with initial isotropic stress σ_0 a decrease of the radial stress equal to :

$$\Delta\sigma = \sigma_0 \left(\lambda_0 + (1-\lambda_0)\left(1 - e^{-\frac{x}{X}}\right)\right)$$

This formula allows considering the three-dimensional aspect in a calculation restricted to the plane deformation.

λ_0 is a coefficient of about 0.3
x is the distance to the face
X is the influential distance of the face (about the gallery diameter).

Assuming $\lambda = \lambda_0 + (1-\lambda_0)(1 - e^{-\frac{x}{X}})$ we find again the coefficient λ utilized in the convergence-confinement method (AFTES /3/). Thus a law of evolution of this factor is defined.

b) the effect resulting from the delayed behaviour : the model representing the ground is of Kelvin-Voigt type (Fig. 3) ; i.e., for a constant stress variation, the radial displacement of the wall in a circular tunnel with initial isotropic stress is written as follows :

$$u_r = \sigma R \left(\frac{1}{2G_0} + \frac{1}{2G_1} e^{-\frac{t}{T_1}} \right) \quad \text{with} \quad T_1 = \frac{\eta_1}{G_1}$$

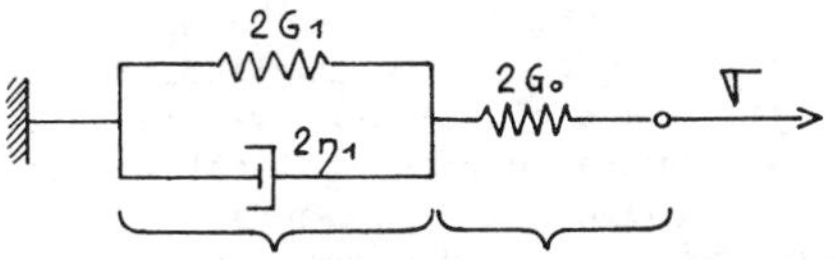

2G1 2G0 σ 2η1

Viscous part Elastic part

Fig. 3 : Used model

Go and G1 are respectively the instantaneous and delayed shear modula.

At very long term $u_r = \frac{\sigma R}{2G_\infty}$ with $\frac{1}{G_\infty} = \frac{1}{G_0} + \frac{1}{G_1}$

R is the radius

3.2 Hypothesis and connection

Considering the circular excavation, the initial isotropic stress condition and the homogeneous medium, the connection of the two previous effects reflects in a varying stress $\Delta\sigma$ calculated according to Sakuraï and applied to the Kelvin-Voigt model (Fig. 4) :

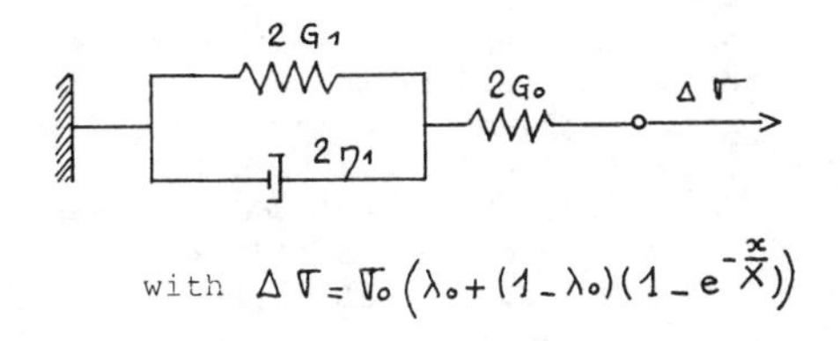

with $\Delta\sigma = \sigma_0\left(\lambda_0 + (1-\lambda_0)(1 - e^{-\frac{x}{X}})\right)$

Fig. 4 : Connection

The displacement integration and calculation in such a configuration require to define a relation between the distance to the face x and time t. In a first stage, we suppose that these two values are connected by $x = V_a t$: linear relation with V_a: average progression speed.

3.3 Consideration of the support

The previous calculation leads to determining the convergences in an unsupported ground. In fact, soon after the excavation, a support is installed, thus modifying the subsequent convergences.

Let us consider t_s as the time between the moment when the face passed through and the moment when the support becomes efficient. The support is supposed elastic, of rigidity k_s , so that $p = k_s \frac{u_r}{r}$ (AFTES /3/).

Finally, the model used is that of Fig. 5 : (Nedez /7/).

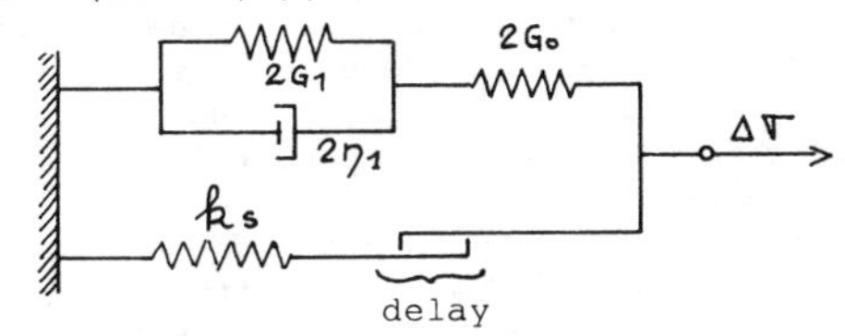

Fig. 5 : Scheme with support

The essential point in this representation is the delay in the support installation.

3.4 Developing the calculation

The theoretical solution of this rheological model supplies for each time t refering to t = 0 which is the moment when the face passes through the investigated section, the value of radial displacement $u_r(t)$. This provides :

$$u_r(t) = \mathcal{A}\left(1 - e^{-\frac{t-t_s}{T_a}}\right) + \mathcal{B}\left(1 - e^{-\frac{t-t_s}{T}}\right) + \mathcal{C}$$

for any time t over ts.

The values $\mathcal{A}, \mathcal{B}, \mathcal{C}$, Ta and T can be explicited according to the model parameters G_0, G_1, η_1, ks and λ_0, ts.

We have :

$$T_a = \frac{X}{V_a}$$

$$T_1 = \frac{\eta_1}{G_1} \qquad \theta_1 = \frac{T_1}{T_a}$$

$$T = \frac{\eta_1 \, (2G_0 + k_s)}{2G_1G_0 + k_sG_0 + k_sG_1} \qquad \theta = \frac{T}{T_a}$$

$$P = \frac{2G_0 + k_s}{2} \qquad Q = \frac{2G_0G_1 + k_sG_0 + k_sG_1}{2(G_0 + G_1)}$$

$$\mathcal{A}=\frac{\sigma_0 R}{2P}(1-\lambda_0)\left(\frac{1}{\theta_{-1}}\right)\left(\theta-\frac{P}{Q}\right)e^{-\frac{t_s}{T_a}}$$

$$\mathcal{B}=\frac{\sigma_0 R}{2P}\left(\frac{P}{Q}-\frac{\theta}{\theta_1}\right)\left(\frac{\theta_1-\lambda_0}{\theta_1-1}\right)e^{-\frac{t_s}{T_1}} + \frac{\sigma_0 R}{2P}(1-\lambda_0)\left(\theta-\frac{P}{Q}\right)\left(\frac{1}{\theta_1-1}-\frac{1}{\theta_{-1}}\right)e^{-\frac{t_s}{T_a}}$$

$$\mathcal{C}=u_r(t_s)=\lambda_0\frac{\sigma_0 R}{2G_0} + \frac{\sigma_0 R}{2G_0}(1-\lambda_0)\left(\frac{1}{\theta_1-1}\right)\left(\theta_1-\frac{G_0}{G_\infty}\right)\left(1-e^{-\frac{t_s}{T_a}}\right) + \frac{\sigma_0 R}{2G_0}\left(1+\frac{1-\lambda_0}{\theta_1-1}\right)\left(\frac{G_0}{G_\infty}-1\right)\left(1-e^{-\frac{t_s}{T_1}}\right)$$

When expressing the displacement, the first exponential is much more important at short term, since it is a function of Ta which illustrates the face effect.

The second appears only for longer times; T, by its expression, depends on the model viscosity. In addition, it can be established that T/Ta > 10, this confirming this hypothesis.

For an easier reading and a better representation of these two components we write:

$$u_2(t)=\underbrace{\mathcal{A}\left(1-e^{-\frac{x-x_s}{X}}\right)}_{\text{face effect}}+\underbrace{\mathcal{B}\left(1-e^{-\frac{t-t_s}{T}}\right)}_{\text{delayed effect}}+\mathcal{C}$$

$\mathcal{C}$ is displacement at t = ts

3.5 Obtaining mechanical parameters

The experimental convergence curves give the trend of chord shortening from a reference time tc (of the first measurement) to the present time t.

Considering a diametral chord, we have :

$$C(t)=2\left(u_r(t)-u_r(t_c)\right)$$

This relation gives the connection between the theory and measurements. A quite simple calculation program (Dubigny /8/) allows to pass from the experimental curve to a theoretical curve best ajusted by the least error squares method, thus defining the values $\mathcal{A}$, $\mathcal{B}$, $\mathcal{C}$ and T. This method utilizes the convergences when the displacements are stabilited and convergences, at the moment of the support installation and the relative influences of the two exponentials with time.

If $\mathcal{A}$, $\mathcal{B}$, $\mathcal{C}$, X, T, and also ts, Ta, θ, σ_0, R are known, it is possible to solve the equations connecting the coefficients and to go back to Go, G1, η_1, and ks in a second stage.

4 RESULTS AND APPLICATIONS

4.1 Mechanical characteristics of encountered grounds

The analysis of modula E_0 and E_∞ (corresponding to Go and G_∞) at the scale of each chord in each section allows to bring face to face the obtained values and the relevant geological profile. A plain explanation can therefore be given for too high or too small modula.

More generally, the whole results allow to distinguish two categories of sections :

	Eo (MPa)	E_∞ (MPa)	η_1 (MPa.j)
category 1	201	107	10 140
category 2	840	600	103 300

The higher values must be connected with a better rigidity of the section due to sandstone banks that give a better stability to the whole surrounding system.

However it must be emphasized the possible variations within the same section, according to chords, due to the non-homogeneity of grounds.

It is more difficult to interprete the results obtained for ks, rigidity of the support. It is difficult to associate this general modulus with each of the support components that have a different way of action, with or without a longitudinal effect, and a horseshoe shape which prevents any approach by radial displacement. Gaudin et al. /9/ have attempted to compare the respective actions of supports.

It should be noticed that we have to try to get much more closer to the circular gallery and axial symmetry with respect to supports, in view to have an easier analysis of the measurements.

4.2 Application to a large section

The parameters obtained in the gallery can be used only if the geological profiles of the final tube are most close to those of the gallery ; due to the dispersion of the values even inside the same massive, this imposes a quite near gallery parallel to the final tube.

In this precise case, the most pessimistic values reached in galleries - both with respect to short-term and long-term modula and viscosity coefficient - have been introduced in the CESAR Finite Elements Code developed by LCPC for a viscoelastic analysis.

Thus, the time-schedule of the final tube excavation with the support and lining stages are combined with the evolution of the ground characteristics, constantly.

The interest of this process is to take into account any excavation stage and a direct representation of support elements. Fig. 6 shows the meshing of the final tube and the works schedule including : excavation of /1/, shotcreting of /2/, excavation of /3/, shotcreting of /4/, concreting of /5/, excavation of /6/ and concreting of /7/

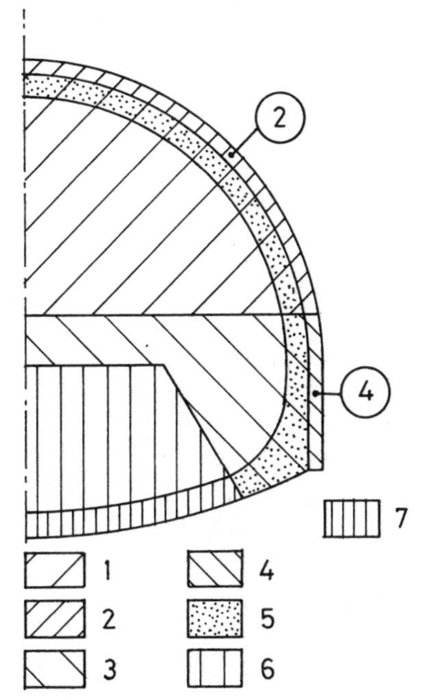

Fig. 6 : Works schedule

To get accurate results in this type of calculation, it is necessary to have available sufficiently long-lasting measurements to consider practically all delayed effects and have the best approach of the long-term ground characteristics.

Then the works stages equate more precisely with a precise state in the ground evolution.

5. CONCLUSION

The essential objective of this study was to approach the major characteristics of the ground in order to design a tunnel. To this purpose, the analysis of the convergence curves of a test gallery can - from a plain model considering the two main effects, which are deconfinement and delayed displacements - provide modula that can be directly introduced in a calculation code of visco-elastic type. The precision will be all the better as the hypothesis used to determine the characteristics will be better validated. Especially it is necessary to :

- get as close as possible to a circular exploratory adit,
- install supports that are not singular points (ribs are difficult to approach theoretically) or do not call for complex soil-structure interactions,
- give as much as possible an axial symmetry to the structure in order to be free of orthoradial phenomena.

Works are under progress and it is not yet possible to give a correct comparison of calculation results and field measurements for the large section tubes.

We still have to pass from the global value representative of the support rigidity to an analytic approach of the role of these support components. The major difficulty remains the interval between the theory and the practical requirements of the worksite and construction process.

REFERENCES

/1/ J.Sulem, M.Panet and A.Guenot : An analytical solution for time-dependent displacements in a circular tunnel, Int. Journal of Rock Mech. and Min. Sci., Vol 24, p 155-164 (1987).

/2/ J.Sulem, M.Panet and A.Guenot : Closure analysis in deep tunnels. Int. Journal of Rock Mech. and Min. Sci., Vol 24, p 145-154 (1987).

/3/ AFTES : Emploi de la méthode convergence-confinement. Tunnels et ouvrages souterrains, Numéro Spécial Novembre 84, p 149-169 (1984).

/4/ M.Panet : Time-dependent deformations in underground works. Proc. 4th Congr. Int. Soc. Rock Mech., Vol 3, p 291-301, Montreux (1979).

/5/ D.Nguyen Minh : modèles rhéologiques pour l'analyse du comportement différé des galeries profondes, Congr. Int. Grands ouvrages souterrains, Florence, Vol. 2, p 659-666 (1986).

/6/ S.Sakuraï : Approximate time-dependent analysis of tunnel support structure considering progress of tunnel face, Int. J. for Num. and Analy. methods in Geom., Vol 2, p 159-175 (1978).

/7/ V.Nedez : Dimensionnement du soutènement à partir de mesures de convergence relative effectuées dans une galerie de reconnaissance, CETu, Bron (1985).

/8/ P.Dubigny : Réalisation d'un outil informatique d'exploitation des mesures de convergence, CETu, Bron (1987).

/9/ B.Gaudin, J.P.Folacci, M.Panet and L. Salva : Soutènement d'une galerie dans les marnes du Cénomanien, 10th Int. Congr. of soil Mech. and found Eng., Stockholm (1981).

Numerical Methods in Geomechanics (Innsbruck 1988), Swoboda (ed.)
© 1988 Balkema, Rotterdam. ISBN 90 6191 809 X

Theoretical investigation of a yielding precast reinforced concrete lining in oligocene clay

M.Müller
Technical University, Budapest, Hungary

ABSTRACT: A tunnel lining has been constructed for supporting an entry tunnel in a weak rock being in a depth of 400 m under the floor level. Its support resistance is considerably higher than that of the constructions applied before, but it is yielding as for normal forces and moments. By applying this new construction smaller tunnel wall displacements, rock loosening and plastic fields in the rock are necessary for stabilizing the surrounding rocks.

The presented roadway support is advantageous, as after completion an equilibrium condition free from displacements will develop and the cavity surrounding is supported jointly by the wall and the rock zone.

The paper introduces the construction, the theoretical investigations made by Numerical Method having been necessary for making the construction, and the results of field measurements.

Cross section of large-diameter entry tunnels built in oligocene clay, supported on TH vault and lining boards soon much decreased, and in their surrounding, extensive elastically failed plastic rock zones arose.

In the following the dimensioning of a roadway support of increased load bearing capacity will be presented, permitting a rock deformation and loosening not greater than needed to reduce the initial rock stresses in the void edge surrounding to match the wall support resistance.

Structural member of the roadway support is a tunnel wall consisting of seven precast B 400 r.c. units d = 20 cm thick, 1 m in length along the tunnel axis, in-situ assembled to circular lining. /Fig. 1./

Wall units within the ring are hinged, and yield in annular direction due to special rubber joints.

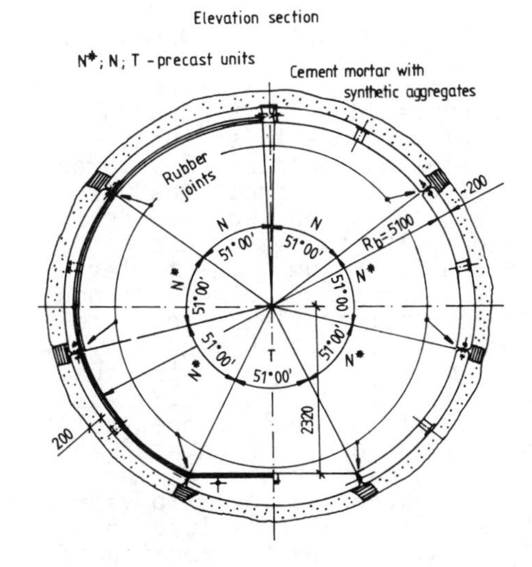

Fig. 1. Precast r.c. tunnel-lining

/with characteristic curves seen in Fig. 2./

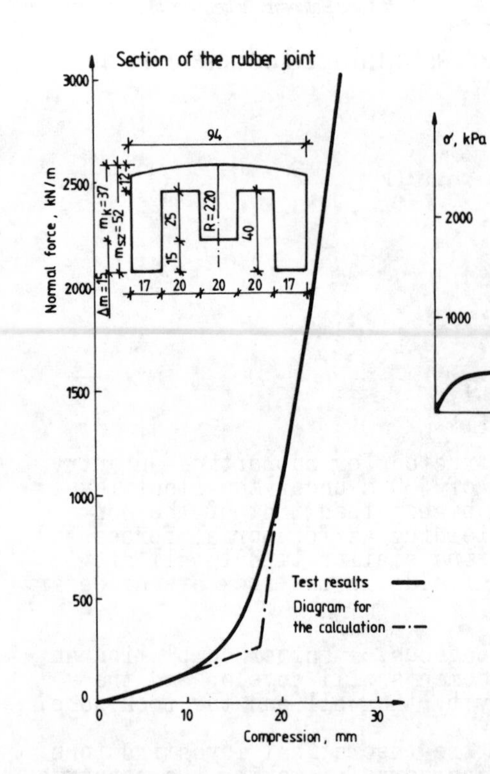

Fig.2. Deformation diagram of rubber joint unit

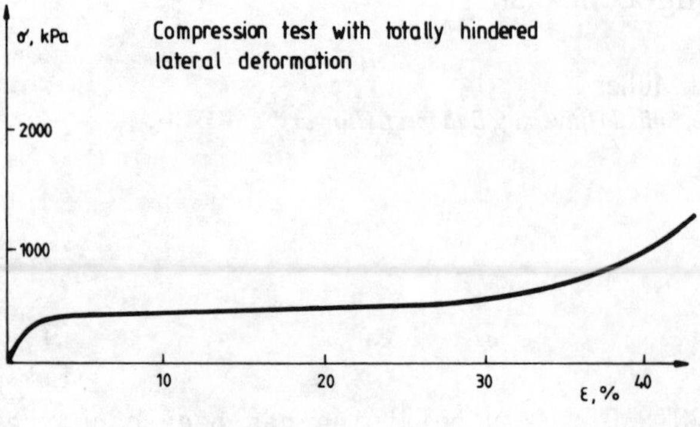

Compression test with totally hindered lateral deformation

σ, kPa

ε, %

Fig. 3. Deformation diagram of polystyrene added cement mortar

There is no structural connection between rings 1 m wide. The back void between tunnel wall and rough-surfaced rock is grouted by a cement mortar with synthetic aggregates and special admixtures, such as not to transmit load from the rock to the wall exceeding its load capacity but gets compressed to a smaller volume. The mortar with the actually designed composition is compressible up to 30 to 35 per cent of its thickness without significantly increasing the transmitted load. /Fig. 3./ Radial yielding of the wall, hence radial displacement of the excavated rock edge, that is, decrease of the initial, excessive loads to values not exceeding the wall load capacity is due to the interaction of rubber joints and back void grouting.

Variation of the stress/strain tensor space of this roadway support in interaction with the surrounding rock, calculation of the necessary displacement, deformations and stresses in the lining and rock has been done with finite element computer program making use of material laws corresponding test results gained with both grouting material and rock. Elastic failure has been reckoned with in terms of the Mohr-Coulomb failure condition.

At the elastic failure of a rock finite element, plastic yield starts at a level below ultimate stress Q $/Q^x = nQ/$, and after failure, the rock behaves as a quasi-elastic material, its stress is obtained from:

$$p' = nQ + /\bar{p} - Q/ \frac{E_r}{E_o}$$

Stress $\bar{p}$ is a physically impossible auxiliary quantity of calculation. /Fig. 4./

Back void grouting has been simulated, in conformity with its characteristic curve obtained in the tests, by a material low involving a linear-elastic section discribed by constants E_o and $0 < \nu < 0{,}5$, then, after elastic failure, a quasi-elastic section where $E_r = \frac{E_o}{100}$ and $\nu = 0$, that is the modulus of compressibility is also reduced to one hundredth of its original magnitude $/M_r = \frac{M_o}{100}/$.

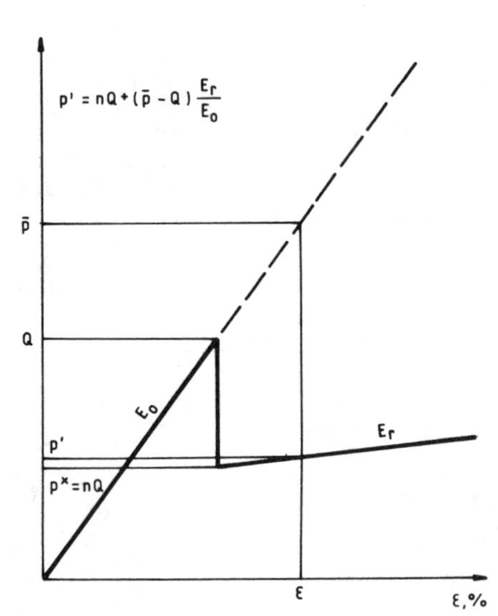

Fig. 4. Constitution law of rock

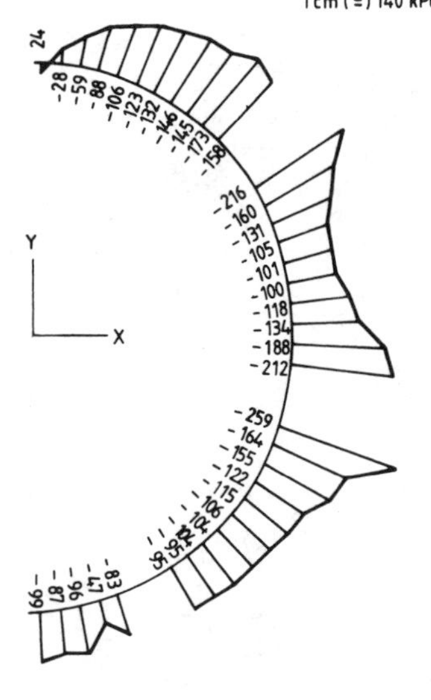

Fig.5. Stresses transmitted to the lining

Thus, it is a material with no lateral displacement in compression after failure. Elastic failure is where the deviator stress $/\sigma_1 - \sigma_3/$ exceeds the ultimate stress of the material, actually 600 kPa.

Computation and field measurement results have led to the following conclusions:

- According to computations, the grout layer designed to 200 mm has a maximum compression of 65 mm, and this void edge displacement reduces initial stresses to 130 kPa as a mean /Fig. 5/.

Field measurements showed a compression of 80 mm for the grout layer 25 cm thick /Fig. 6/, fairly agreeing with computations, but the initial geostatic stresses decreased by much less at the averaged edge than computed. Namely, stresses transmitted to the lining 807.4 kPa rather than 130 kPa as computed /Fig. 7./

According to checking computations, this fact may be attributed to a slight loss of the grout deformability $/\alpha/$ near $\alpha = 0.001$, thus much increasing the transferred stresses /Fig. 8./

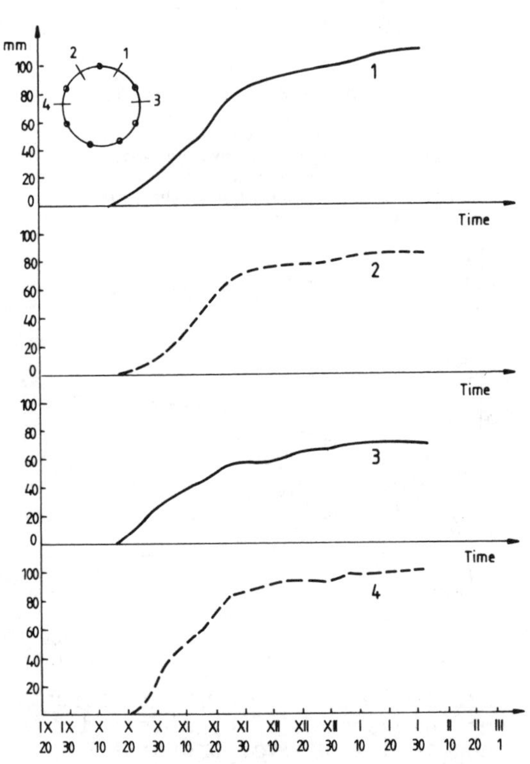

Fig. 6. Compression of polystyrene added cement mortar back void grouting

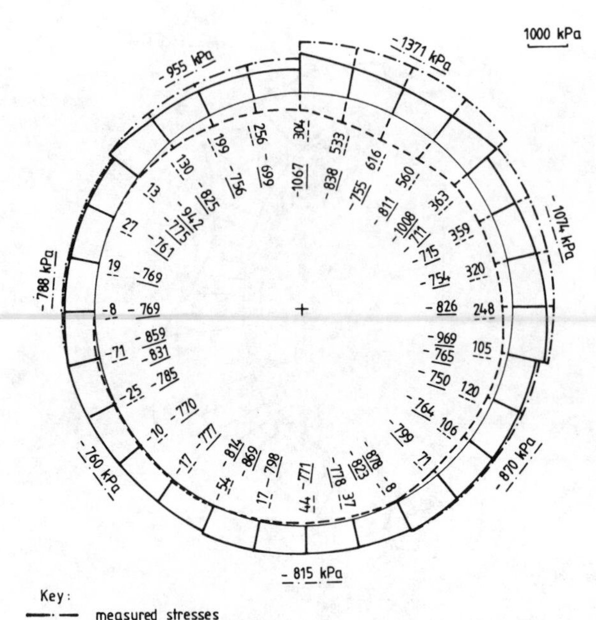

Fig. 7. Measured stresses transmitted to the lining

$M_t = \alpha M_0$ where M_0 - elastic modulus of compressibility
M_1 - modulus of compressibility after failure
α = 0,01 - material having the property of big volume decrement after failure
α = 1 - material with constant volume after failure

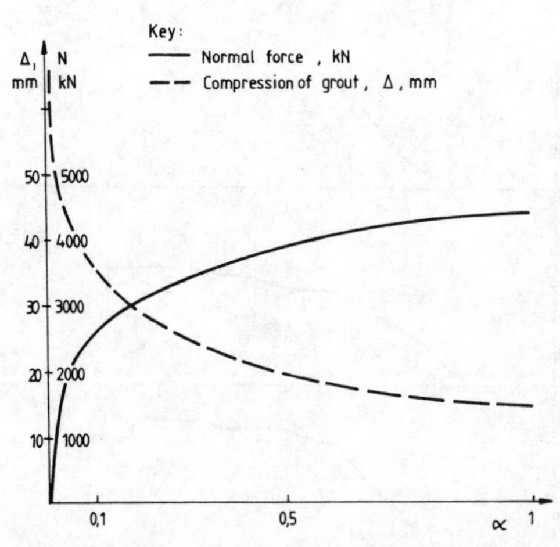

Fig. 8. Variation of normal force and compression of plastic added back-void grout vs. property of grout concerning volume decrement /α/

- According to computations, maximum normal force in the lining N_{max} = 606 kN/m, moment M_{max} = 55 to 56 kNm/m, and eccentricity e_{max} = 8 to 15 cm. Normal forces obtained in field tests were higher than computed but more uniform, while moments, and thereby eccentricities, are less significant /Fig. 9/.

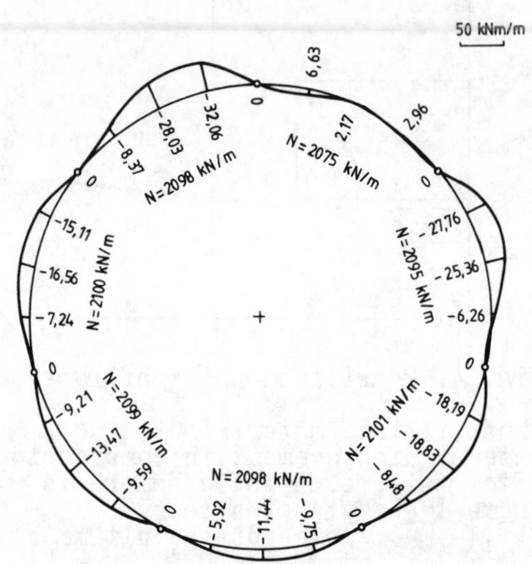

Fig. 9. Moments and normal forces calculated from the measured stresses

- Until November 15, pressures acting on the wall symmetrically grew, thereafter they became asymmetric. Thereupon the shear force in the top hinge increased, and after having exceeded its shear strength, key stones failed in shear.

Displacement possible in the top hinge resulted in the redistribution of eccentric loads and reduced the eccentricity /Fig. 10/.

This phenomenon may be attributed to the construction of another tunnel of a diameter D_k = 6 m at a distance of about 33 m, and parallel to the axis of the test object, constructed by blasting and with TH arch support. All these lead to the conclusion

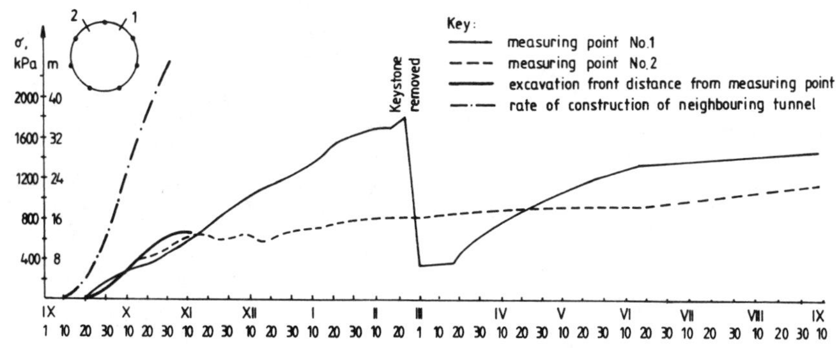

Fig. 10. Measured stresses acting on the lining

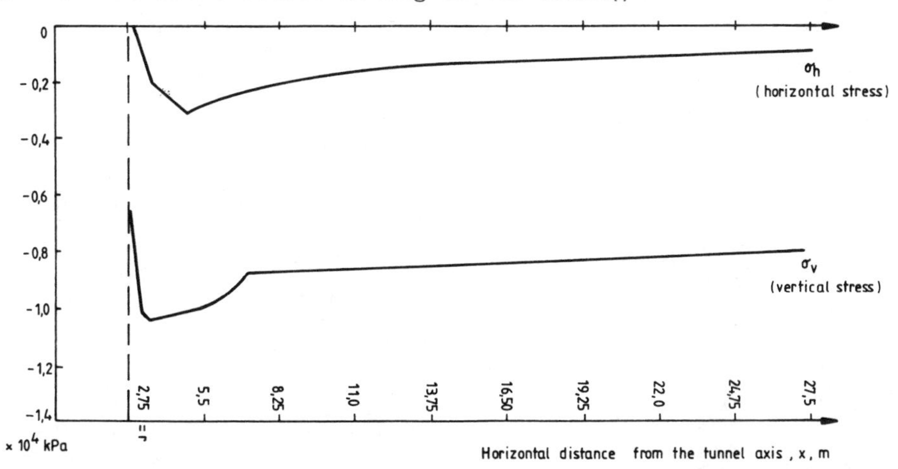

Fig. 11. Rock stresses at the level of horizontal diameter

that the lining system can bear significant asymmetric loads until the shear load capacity of hinges is exhausted. Thus, ultimate shear strengthes of hinges are advisably determined for a load asymmetric to the proportion 1:2 with a mean value such that its action on the wall produces exactly the ultimate normal force in the wall lining.

Evaluation of the stresses in the tunnel surrounding the rock showed an arching, that is, a rock ring of increased stress develop around the lining /Fig. 11/, and plastic hinges arise only where and to a degree that is necessary to the development of arching.

According to computations and field measurements, the presented support, yieldings to bending and normal loads, is advantageous, since it helps to develop an equilibrium condition with no displacement in the tunnel surroundings. Loads are borne in common by the lining and the rock, and the lining is subject to high normal forces and low moments.

Numerical Methods in Geomechanics (Innsbruck 1988), Swoboda (ed.)
© 1988 Balkema, Rotterdam. ISBN 90 6191 809 X

A method for the analysis and design of flexible retaining structures – Application to a strutted excavation

Giovanni Barla
Turin University of Technology, Italy
Bruno Becci
Center of Structural Analysis, Milan, Italy
Adolfo Colombo
Metropolitana Milanese, Milan, Italy
Roberto Nova
Milan University of Technology, Italy
Renato Peduzzi
Milan, Italy

ABSTRACT: A simple and economic method for the analysis of flexible retaining structures is presented. Soil is modelled as a bed of elastic-plastic strainhardening springs. The procedure to determine the constitutive parameters on the basis of the results of geotechnical investigation is outlined. The method is applied to the analysis of a strutted excavation and computed results are compared with measured displacements and forces in the struts. It is shown that, despite the simplicity of the method and the scanty geotechnical information available, the experimental trend is correctly matched. A similar analysis with the finite element method would require a computation time two orders of magnitude longer.

1 INTRODUCTION

The analysis of an anchored retaining structure in a stratified soil is usually performed by means of the finite element method. This procedure is certainly the most effective and versatile available today, but from the point of view of practical applications it is not totally free from shortcomings.

Indeed, because of the cost of a single run, the finite element method is normally employed for the analysis of a structure which has been designed by means of other, traditional, methods. It is then used more as a verification tool than as an aid to design.

Moreover, the constitutive law of soil behaviour usually adopted is elastic-perfectly plastic, although this assumption is clearly far from reality. The use of more realistic constitutive laws would infact largely increase the cost. As a consequence, finite element results can not be considered as fully representative of the actual soil-structure behaviour.

Finally, it is sometime necessary to have reliable information not only on the working conditions of the structure, but even on the collapse situation. A typical example is that of an excavation which should be deepened with respect to the original level for which the diaphragm wall has been designed. The finite element method is not very effective in studying failure conditions and, in such cases, special care is necessary in the choice of the element type and of the mesh.

To bypass the aforementioned shortcomings an alternative method is presented in this paper. It allows to analyse anchored retaining structures in complex soil conditions and is very simple and economic. The procedure can be considered as a modified Winkler method, in which the behaviour of the springs that schematize soil is represented as elastic-plastic strainhardening. In this way emphasis is put more on the constitutive law than on the respect of continuity of deformation. Indeed, while soil behaviour is largely stress path dependent and consequently a correct modelling of the constitutive law may be vital, continuity of deformation is ensured by the retaining structure.

The method proposed is very fast so that many analyses may be performed in the time necessary for a single finite element run, thus allowing for a rational design optimization. Simplicity has however a non-negligible cost in that much of the information retrievable with a finite element analysis can not be obtained with the alternative method. Surface settlements, for instance, can be predicted only by means of a semiempirical procedure and no information can be derived on the state of stress and strain within the soil mass. Also, the method can be employed only in cases in which the geometry of the problem is simple and regular.

To show the capabilities of the method, a

practical application will be presented. It will be shown that the model gives reasonable agreement between calculated and observed displacements of a strutted excavation, even if all the geotechnical information necessary for calibrating the constitutive parameters only comes from the results of conventional SPT tests.

2 SOIL MODELLING

Soil is schematized as a bed of springs of stiffness K_i such that there exists a linear relationship between the increment of the force in the spring DS_i and the corresponding incremental displacement Ds_i

$$DS_i = K_i \; Ds_i \qquad (1)$$

The spring stiffness is assumed to be given by

$$K_i = \frac{a \; E_{s_i} \; t_i}{L} \qquad (2)$$

The essential novelty of the method consists in the choice of the soil stiffness E_s. In fact, at variance with the conventional Winkler method, E_s is assumed to vary not only with the state of stress but also with the direction of the force increment. The behaviour of a virgin soil is in fact non linear and irreversible and the soil stiffness is markedly different for virgin loading or unloading-reloading conditions.

The soil stiffness can be then equal either to the virgin stiffness E_{ep} or to the unloading-reloading stiffness E_{ur}. The soil is charachterized by the virgin stiffness only when the state of stress is equal to the maximum past stress and the stress is increasing, as shown in fig. 1.

Assume that the soil-structure interaction is a plane strain problem and that the vertical and horizontal stresses are principal stresses. Thus the relevant part of the state of stress is fully identified by two parameters, σ'_v and σ'_h. If σ'_{vp} and σ'_{hp} are the maximum past vertical and horizontal stresses, it is possible to define, in the plane of the assumed principal stresses, an elastic domain which is limited by the Coulomb failure lines and the lines $\sigma'_v=\sigma'_{vp}$, $\sigma'_h=\sigma'_{hp}$.

Assume also, for sake of simplicity, that soil is granular. Stress increments such as AB, CD, EF are all characterized by the unloading reloading stiffness E_{ur}, while GI is linked to E_{ep}. Since for

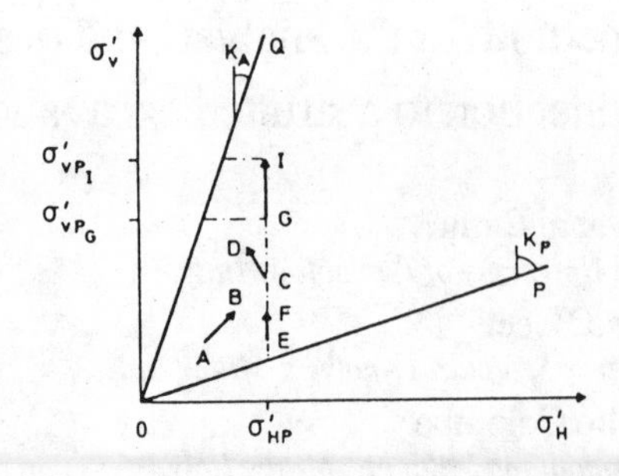

Fig.1 Failure condition, elastic domain and anisotropic hardening for granular soil.

this latter increment σ'_v becomes equal to σ'_{vI}, an expansion of the elastic domain occurs. Note however that, since σ'_h does not change, σ'_{hp} remains constant, that is hardening is anisotropic.

The two failure conditions are given by:

$$\sigma'_h = K_a \; \sigma'_v \qquad (3)$$

$$\sigma'_h = K_p \; \sigma'_v \qquad (4)$$

where K_a and K_p are the active and passive earth pressure coefficients. In the following, K_a will be equal to the Rankine value, while K_p will be chosen in such a way to implicitly take into account friction between soil and wall.

Note that E_{ur} and E_{ep} need not to be constant. They can be related to the value of the mean pressure p', for instance, in the following way

$$E_{ur} = R_{ur} \; (p/pa)^n \qquad (5)$$

$$E_{ep} = R_{ep} \; (p/pa)^n \qquad (6)$$

where p_a is a reference pressure and R_{ur}, R_{ep}, n are experimental constants.

Of the other parameters which characterize the spring stiffness, t_i is simply the influence area of the i-th spring, a is a non-dimensional constant and L is a geometric term which is characteristic of the problem. The latter should take into account the width of the soil involved in the movement of the wall. It will be then different for the active and passive zones, and will depend on the depth of excavation.

It will be assumed for the springs at the back of the wall:

$$L_M = 2/3 \; H' \tan (45° - \phi/2) \qquad (7)$$

while in front of it

$$L_V = 2/3\ (H'-D)\ \tan\ (45° + \phi/2) \quad (8)$$

where ϕ is the effective friction angle of the soil, D is the depth of the excavation, and H' is the effective length of the wall, that is

$$H' = \min\ (2D,H) \quad (9)$$

where H is the total structure length. The assumed value for H' is arbitrary, but is corroborated by numerical results on sample problems (Nova,Becci (1987)). Finally the non-dimensional parameter a can also be currently set equal to 1.

3 MODELLING OF THE EXCAVATION

The state of stress in the soil after the diaphragm wall insertion is assumed to be equal to the stress state at rest, which is taken as known. Each step of the excavation is divided into two phases. In the first one, the wall is assumed to be fixed while the excavation proceeds.

There is only a variation of the stresses in the springs in front of the wall since the vertical stress is reduced and the horizontal one will vary according to the law

$$\sigma'_h = \sigma'_v\ K_o^{NC} (OCR)^m \quad (10)$$

In eq. (10) K_o^{NC} is the coefficient of earth pressure at rest for the soil in the virgin condition, while m is an empirical coefficient, ranging between 0 and 1, which is generally close to 0.5. In the following, m will be taken equal to one, assuming hence σ'_h=const. Note, however, that the ratio between σ'_h and σ'_v should always be limited by K_p. Near the base of the excavation σ'_v is very small so that Eq.10 is no more valid and

$$\sigma'_h = K_p\ \sigma'_v \quad (11)$$

Let now the wall be free to move. Since the state of stress in the springs is no more self-equilibrated, the wall will depart from its initial position until a new equilibrium configuration is achieved. A first guess on the stiffness of the nonlinear springs is necessary. However the computation of the new equilibrium condition is carried on by checking at each step the respect of the material constitutive law (conformity). By means of a conventional iterative scheme (e.g. see Bathe (1982)), at the end of the excavation phase, an equilibrated and conform solution is obtained.

When convergence is achieved, a new excavation step can be performed, as well as the insertion of prestressed anchors, or the casting of provisional struts. These latter are simply made by adding new stiffness contributions to the total stiffness of the structure. Initial stresses in anchors or struts are easily simulated by means of initial strain states in elastic elements. The insertion of the current soil model in the framework of a general finite element procedure allows in fact to study general conditions of geometry and loading.

The effects of cohesion, surcharges, overconsolidation, water pressures can be easily taken into account. Settlements of the soil surface can be evaluated following a procedure suggested by Bransby and Milligan (1975) once the dilatancy at failure is known.

The method has been successfully tested so far against experimental data obtained with reduced scale models (Nova and Becci (1987)).

4 AN APPLICATION TO A STRUTTED EXCAVATION

The method presented has been used for the analysis of the diaphragm walls executed in connection with the Lamarmora Station of the line no.3 of the Milan Underground. The site plan and a typical cross section of the excavation are shown in Fig.2. Following the completion of the openwork structure, the neighboring tunnel is to be excavated in stages.

One of the diaphragm walls is anchored at two levels, whereas the other is linked to it by 13 m long provisional struts (Fig.2).

In the final design condition, the adjacent tunnel is to be separated by the diaphragm wall by a thin wall of jetgrouted soil.

In order to be able to analyze the behaviour of the soil and of the structures during excavation, a monitoring program was adopted including measurements of convergences of the diaphragm walls and around the tunnel, stresses in the structural elements such as anchors and steel struts, and strains in the soil (Barla et al., 1986). In particular, the anchors were instrumented by load cells, whereas the axial loads in the struts were monitored by thin flat jacks.

According to the results of measurements during excavation of the openwork station, the anchored wall experienced virtually no movement, with the loads in the anchors remaining nearly equal to the prestressing level. For the purpose of the numerical

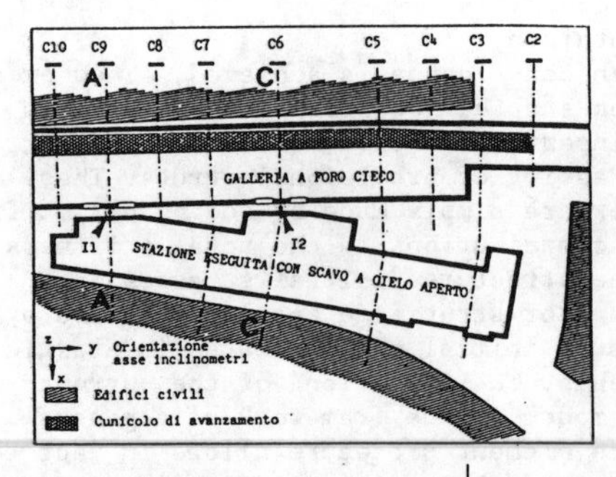

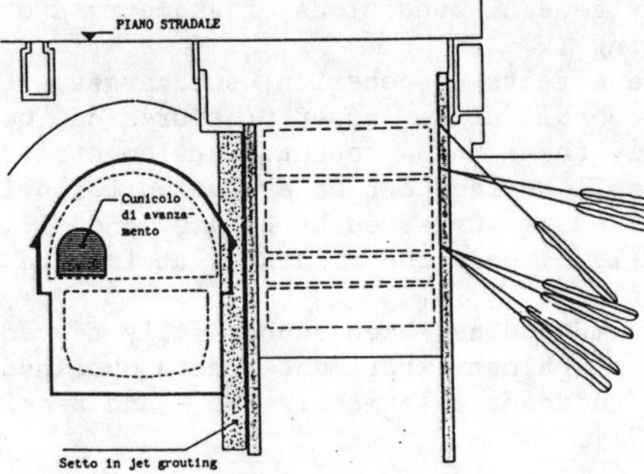

Fig.2 Milan Underground Lamarmora Station (MM - linea 3): site plan and typical cross section (from Barla et al., (1986)).

analyses discussed in the following, the measured relative displacements between the two diaphragm walls will be assumed to be the absolute displacements for the wall at the tunnel side.

The tunnel excavation started after the completion of the openwork excavation: it is then possible to analyze the behaviour of the strutted diaphragm walls by using the method presented in this paper.

The soil at the site is a sandy gravel, with silt. A typical simplified soil profile is shown in Fig.3 together with the measured SPT blow count. By means of the empirical relationships between N and friction angle ϕ proposed by Bazaraa (1967), that takes the influence of the overburden pressure into account, and assuming that the density of the sand is 1.8 Mg/m^3, it is possible to derive that the angle of friction is nearly constant with depth and is equal to 35.8° . The coefficient of earth pressure at rest is then assumed to be equal to 0.415 from Jaky simplified relationship, while the active and passive coefficients K_a=.262 and K_p=11. are determined taking into account the influence of friction between soil and wall.

The virgin stiffness E_{ep} is derived by means of the following empirical relationship (D'Appolonia et. al. (1970)):

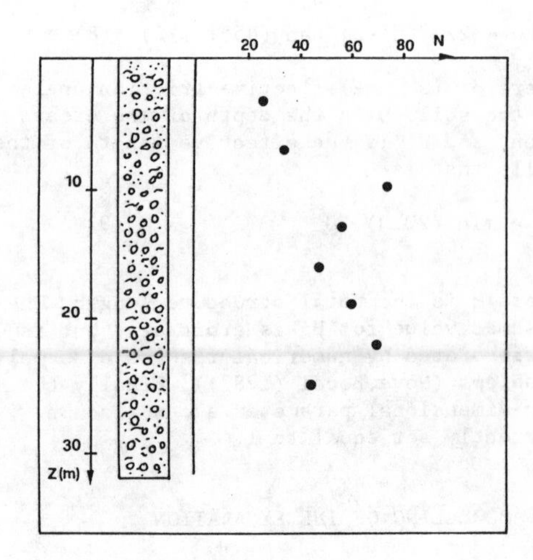

Fig.3 Soil profile, SPT blow count vs. depth.

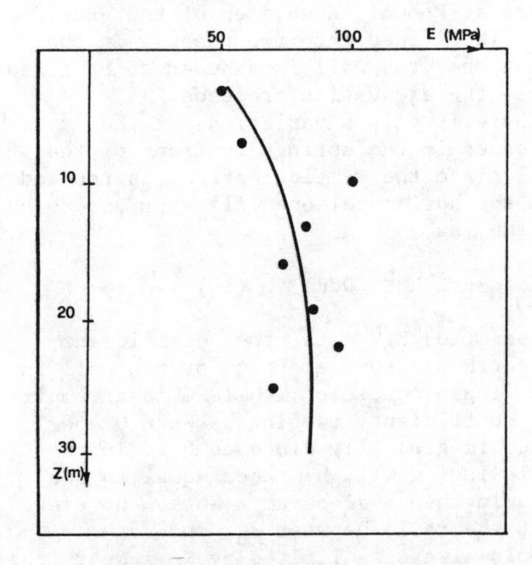

Fig.4 E_{ep} moduli from eq. (12) (dots); assumed interpolation (solid curve) according to eq. (6).

$$E_{ep} = 21.6 + 1.06\ N \quad (12)$$

where N is the blow count and E_{ep} is in Mpa.

Soil stiffness varies with depth and, consequently, with consolidation pressure. In the case considered, a single curve of the type of Eq.6 gives an interpolation of data which is accurate enough for the present purposes, as shown in fig.4. It is found that R_{ep}=63.4 Mpa while n=0.26. This can be considered a fortunate condition due to the homogeneity of the soil deposit. In more complex cases, however, the determination of such parameters could

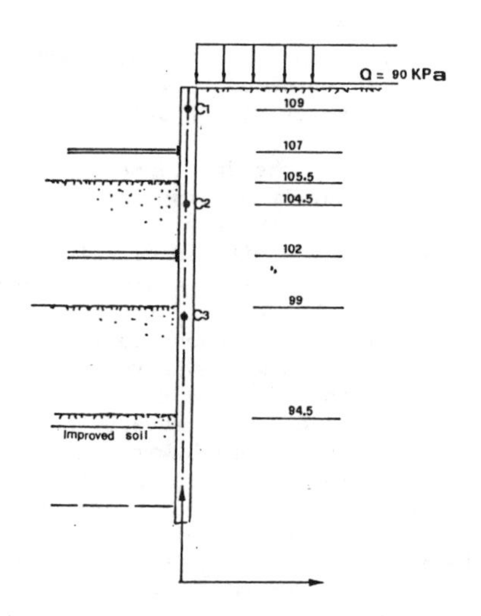

Fig.5 Main levels for mathematical model of the diaphragm wall excavation problem.

be more difficult and more than one set of parameters should be employed for the description of soil characteristics. Finally on the basis of previous experience the remaining parameter R_{ur} has been taken as 1.6 R_{ep}.

The stiffness of the diaphragm wall has been increased to take account of the effect of the jet-grouted soil. To calculate such stiffness the Young modulus of the improved soil has been taken equal to 10000 MPa and no shear has been assumed between the wall and the grouted soil.

The excavation procedure has been divided in the following steps:

1) excavation from surface (110. m over sea level) to 105.5 m o.s.l.

2) the first strut is forced into position (107 m o.s.l.). Readings of the convergences at the 107 m and 104.5 m levels start at this point, in conjunction with monitoring of the load in the first strut.

3) excavation proceeds down to 99.5 m o.s.l.

4) second strut is forced into position (103 m o.s.l.). Convergence and load in the strut at the same level start to be monitored.

5) 3 m of soil under the final excavation depth are consolidated by means of a water-cement mixture. From experimental data the Young modulus of such a soil, considered to be indefinitely elastic, has been taken equal to 3 times the initial value of the untreated soil at the same depth. Then excavation is deepened to the final depth (94.5 m o.s.l.).

The excavation steps have been further subdivided in substeps to take account of stiffness non-linearity. Fig.5 shows a schematic picture of the diaphragm wall with evidence of measuring points and main excavation steps.

Fig. 6 shows the comparison between calculated (dashed line) and measured (full line) displacements (fig. 6a) and compressive forces in the struts (fig. 6b). The trend of the experimental results is reasonably well matched and the calculated results are of the same order of magnitude of the experimental ones. Calculated bending moments are shown in fig. 7.

It is worth mentioning that the entire analysis required no more than two minutes on a HP 9000/550 computer equipped with 3 CPUs and 1.2 Gbyte RAM.

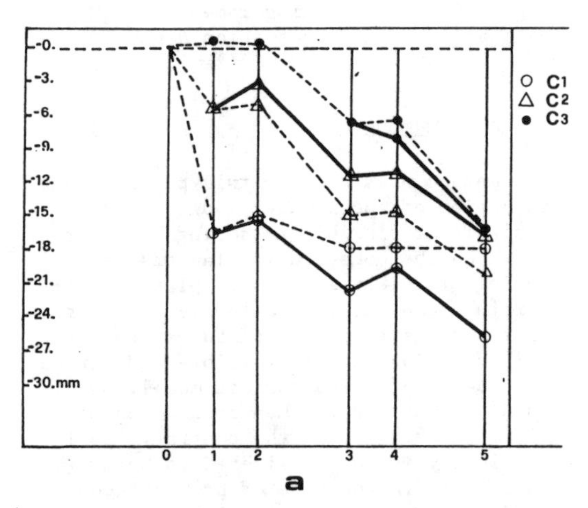

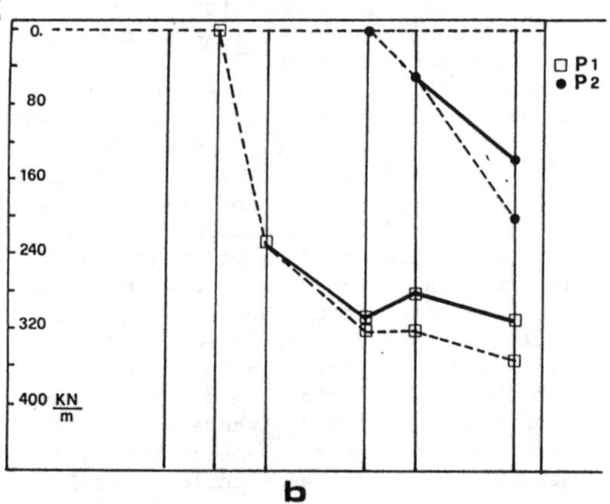

Fig.6 Comparison between experimental data (solid lines) and computed results (dashed lines): a) lateral deflections (convergences) at measurements levels; b) incremental axial forces in struts.

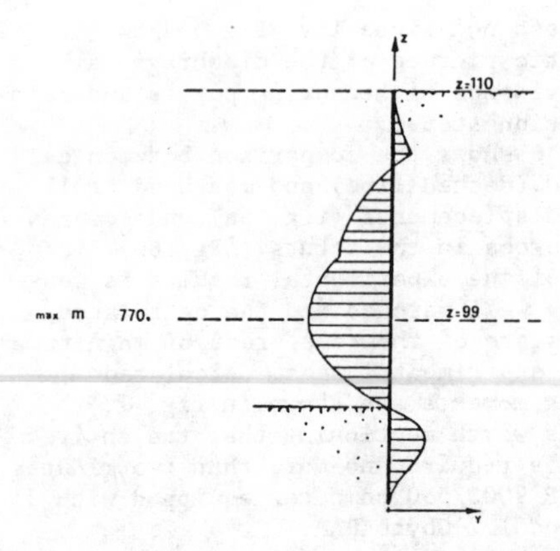

Fig.7 Computed bending moments in the wall at the final excavation depth.

5 CONCLUSIONS

The method presented in this paper is an extension of the Winkler model for the analysis of flexible retaining structures in which the constitutive law of the springs is taken as elastic-plastic strainhardening. The rationale for this choice comes from the consideration that soil behaviour is irreversible and stress path dependent. The correct modelling of such a behaviour is then considered of vital importance. On the contrary, soil continuity is not considered to be as important, in the type of problems considered, since the continuity of soil deformation is ensured by the stiffness of the retaining structure.

The parameters used in the constitutive law are all traditional soil constants such as friction angle, virgin and unloading-reloading stiffness, coefficients of earth pressure. They can be then determined as for other traditional methods. For the application presented, a strutted excavation in an alluvional soil, all the parameters have been determined starting from the results of an SPT test via accepted empirical or semiempirical relationships.

In the cases examined in a previous paper (Nova and Becci (1987)) where simple reduced scale models have been analyzed, the soil conditions were uniform, soil properties well studied and plane strain conditions were ensured. In all the cases considered the agreement between calculated and experimental results was quite satisfactory. In the case presented here, despite the scanty experimental information on soil characteristics, and the complexity of the excavation procedure, the method is able to match the observed trend of displacements and forces in the struts. Quantitative agreement is not fully satisfactory but the deviation between observed and calculated results can be still considered acceptable.

The reason for the development of this model was primarily a reason of cost. A finite element analysis with a strainhardening law or even with an elastic perfectly plastic law of a similar problem could be more then two orders of magnitude more expensive then the one performed. Such a situation prevents, nowadays, the use of finite elements as a tool for optimum design. On the contrary several dozens of analyses can be performed with the model presented in the time necessary for a simple finite element run. Moreover the amount of input data is dramatically reduced. The strainhardening spring method can therefore be used not only for verification but even for a modern computer aided design of flexible retaining structures.

6 ACKNOWLEDGEMENTS

Authors are indebted to Metropolitana Milanese for permission of publishing the data. G. Barla and R. Nova are also indebted with MPI for financial support.

7 REFERENCES

Barla G., Colombo A., Pavan F. 1986. Controlli e misure in corso d'opera nella stazione Lamarmora della Metropolitana Milanese, Proc. Symp. 'Large Underground Openings', Florence 1:627-633.

Bathe K.J. 1982. Finite Elements Procedures in Engineering Analysis. Prentice Hall, Englewood Cliffs, N.J.

Bazaraa A.R. 1967. Use of the standard penetration test for estimating settlements of shallow foundations on sand. Ph.D. Th. University of Illinois, Urbana.

Bransby P.L, Milligan G.W.E, (1975). Soil deformations near cantilever sheetpile walls. Geotech. 25, 2, 175-195.

D'Appolonia D.J., D'Appolonia E., Brisette R.F., 1970. Discussion on settlements of spread footings on sand. ASCE J. SMFD 96.

Nova R., Becci B., 1987. A method for the analysis and design of flexible retaining structures. Proc. Sym. 'Soil Structures Interactions', Paris, 657-664.

Numerical Methods in Geomechanics (Innsbruck 1988), Swoboda (ed.)
© 1988 Balkema, Rotterdam. ISBN 90 6191 809 X

An application of a strainhardening model to the design of tunnels in sand

E.Botti
Metropolitana Milanese, Milan, Italy

G.Canetta
Center of Structural Analysis, Milan, Italy

R.Nova
Milan University of Technology, Italy

R.Peduzzi
Milan, Italy

ABSTRACT: The paper describes the finite element analysis of an urban railway tunnel in a sandy soil improved by grouting. The constitutive model employed to describe the behaviour of natural sand is elastic plastic strainhardening. To allow an easy determination of constitutive parameters from conventional in situ tests, the model has been conceived to be characterized by four parameters, two elastic and two plastic. However, the choice of the yield function is such that the dependence of the behaviour on the Lode angle is correctly matched. The model for grouted sand is elastic perfectly plastic. The grouting pressures induced in the soil are modelled as a field of selfequilibrated stresses. The results obtained are compared with the results of previous analyses performed with elastic perfectly plastic models. It is shown that the analysis with the strainhardening model gives closer prediction of the measured tunnel behaviour.

1. INTRODUCTION

In the last decade a considerable number of constitutive models of soil behaviour have been formulated. Many of them are able to reproduce the experimental results with reasonable accuracy, not only qualitatively but also quantitatively. A long way has been run since the appearance of the original Cam-Clay model (Schofield and Wroth (1968)).

Despite the evident success in describing the observed soil behaviour, however, these models are seldom used for the analysis of engineering problems. The reason is that soil behaviour is overwhelmingly complex and, accordingly, models able to reproduce all its aspects enjoy a complicated theoretical structure and are characterized by quite a number of constitutive parameters. Often, only large computers may cope with theoretical complexity and it is virtually impossible to determine the appropriate values of the constants from routine tests, especially when a non-cohesive soil is involved in the problem to be analysed.

Because of such difficulties, practicing engineers tend to ignore the most advanced theories and make use of simple elastic-perfectly plastic laws which are characterized by few parameters and may be determined by means of simple in-situ tests. In this way, all the major advances in the understanding of soil behaviour of the last twenty years have little or no feedback on geotechnical engineering.

For practical purposes, however, the most advanced models are possibly unduly complicated. If loading is quasistatic and there are no unloading-reloading cycles, even much simpler elastic-plastic strainhardening models give reasonable predictions. Since the theoretical structure of such models is simple and the number of consitutive parameters is limited, they may be successfully used in practice even if large computers are not available and the information on soil properties is scanty and comes form SPT and CPT tests only.

The aim of this paper is to present a simplified model for a granular, non-cohesive material, and to show the results of a practical application. The model is elastic-plastic strainhardening and is characterized by four consitutive parameters which may be related to SPT blows or other results of in-situ tests by means of empirical relations. In fact the parameters are linked to traditional geotechnical constants such as friction angle or elastic moduli for which such relations already exist.

The application concerns the excavation of an urban tunnel in sand. Soil properties have been improved by grouting. The results obtained are compared with those derived by means of an elastic-perfectly plastic analysis and both are compared with experimental data whenever possible.

It is shown that the strainhardening model

reproduces better than the elastic-plastic one the observed displacement pattern.

2. THE STRAINHARDENING MODEL 'LAMBER'

The constitutive model that will be employed for the sake of convenience will be referred to as 'Lamber', from the name of a small river which touches Milan. The model is elastic plastic strainhardening (or softening) and is conceptually similar to Cam Clay. The main difference lies in the choice of the expression for the yield function that, at variance with Cam Clay, is function of the three stress invariants. The yield function expands or contracts depending on the value of the hardening modulus which may be either positive or negative. Admissible stress states are bounded by a limiting surface which coincides with the Matsuoka-Nakai (1974) failure condition. This latter criterion may be written in the following way:

$$3/2(\xi-1)J_{2\eta} - \xi J_{3\eta} - 3(\xi-3) = 0 \qquad (1)$$

where ξ is a constitutive parameter, while $J_{2\eta}$ and $J_{3\eta}$ are the second and third invariants of the tensor η_{ij} defined as

$$\eta_{ij} = s_{ij}/p' \qquad (2)$$

The tensor s_{ij} is the stress deviator while p' is the hydrostatic effective stress. The second and third invariants are defined as:

$$J_{2\eta} = \eta_{ij}\eta_{ij} \qquad (3)$$

$$J_{3\eta} = \eta_{ij}\eta_{jk}\eta_{ki} \qquad (4)$$

The yield function is given by:

$$f = 3/2(\xi-1)J_{2\eta} - \xi J_{3\eta} + 3(\xi-3)\ln\frac{p'}{p_c} \qquad (5)$$

where p_c is the maximum hydrostatic preconsolidation pressure, which depends on the history of the soil element considered. The yield function is defined only for values of p' larger than p_c/e. When $p'=p_c/e$ Eq. (1) and Eq. (5) coincide. Stress states for which p' is less than p_c/e cannot lie on the yield surface. The complete picture of the yield surface and of the limiting surface is given in fig. 1.

It is assumed that p_c depends only on the plastic volumetric strain experienced:

$$p_c = p_{c0} \exp(v^p/\chi) \qquad (6)$$

p_{c0} is a dummy reference pressure while it is easy to recognize that χ is a plastic logarithmic volumetric compliance. In a purely hydrostatic test χ would be the slope of the plot plastic volumetric strain, natural logarithm of hydrostatic pressure, fig. 2.

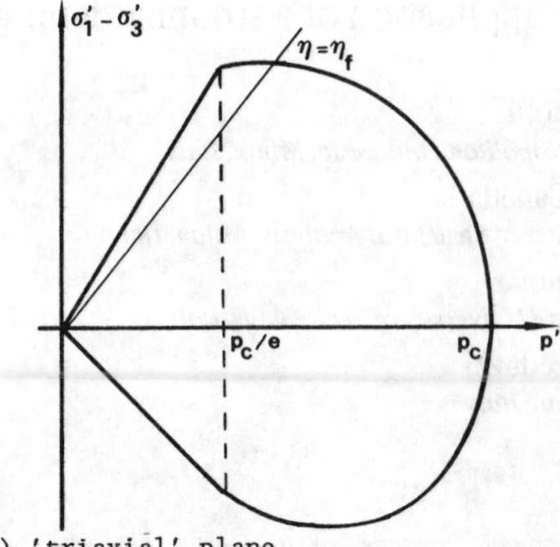

a) 'triaxial' plane

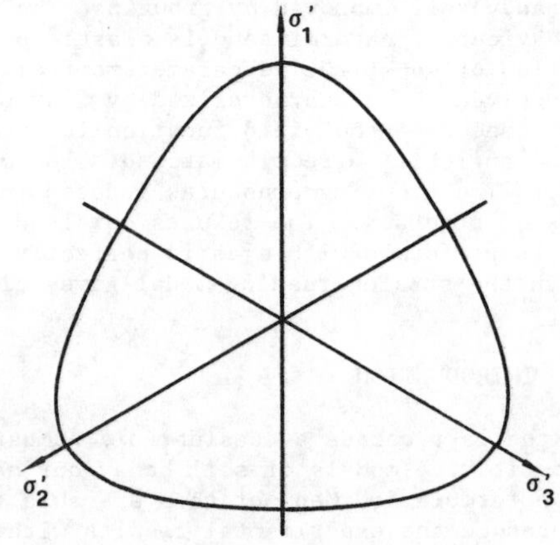

b) deviatoric plane $\sigma_2=\sigma_3$

Fig. 1. Yield and limiting surface.

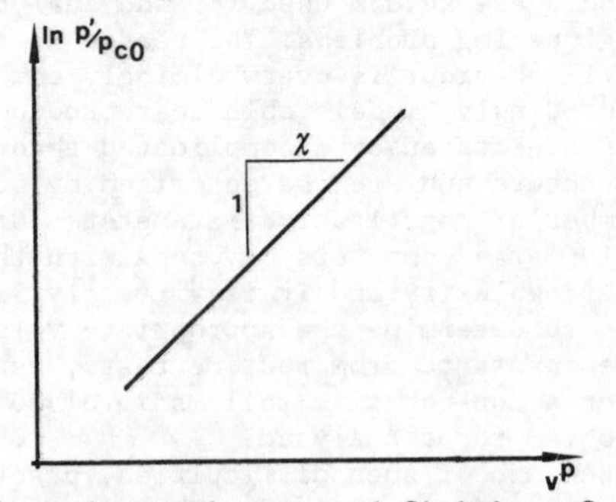

Fig. 2. Hydrostatic test: definition of χ.

The plastic potential is assumed to coincide with the yield function. Although this has been demonstrated to be far from experimental reality, the assumption of the validity of the normality rule is taken for limiting the number of the constitutive parameters and to ensure the symmetry of the stiffness matrix of the soil element, what allows well known computational advantages.

Plastic strain rates can be then obtained as

$$\dot{\epsilon}^{p}_{ij} = \frac{1}{H} \frac{\partial f}{\partial \sigma'_{ij}} \frac{\partial f}{\partial \sigma'_{hk}} \dot{\sigma}'_{hk} \qquad (7)$$

where the hardening modulus H is derived via the Prager's consistency rule:

$$H = \frac{9(\xi-3)}{\chi p'} \{\xi J_{3\eta} - (\xi-1)J_{2\eta} + \xi-3\} \qquad (8)$$

For a virgin soil, i.e. normally consolidated, failure will occur when H=0. In axisymmetric conditions ($\sigma_2=\sigma_3$) the stress ratio η, defined as

$$\eta \equiv (\sigma'_1-\sigma'_3)/p' \qquad (9)$$

takes the value η_f that is linked to ξ via Eq. (8). It is easy to show that

$$\xi = \frac{3 - 2/3\eta_f^2}{2/9\eta_f^3 - 2/3\ \eta_f^2 + 1} \qquad (10)$$

Since η_f is linked to the apparent friction angle ϕ' by the relation

$$\eta_f = \frac{6\sin\phi'}{3-\sin\phi'} \qquad (11)$$

also the constitutive parameter ξ may be linked to ϕ' by a one to one correspondence.

Elastic strains are linked to effective stresses by means of a nonlinear elastic law. It is assumed that the bulk modulus K linearly varies with p', while the shear modulus G is taken as constant.

Four parameters and the dummy constant p_{c0} fully characterize the model. This latter constant has little practical relevance since the initial state of stress due to selfweight is normally associated with zero strains. In each point p_{c0} may be simply taken as the initial maximum hydrostatic stress ever experienced. The parameter ξ may be easily determined from the assumed value of the friction angle. This in turn may be determined by means of empirical relationships with the SPT blow count or the Dutch cone penetration resistance.

The shear modulus may be chosen as for a purely elastic analysis, while the determination of the bulk modulus and of the plastic compliance is more difficult. In fact several empirical relations exist that allow to determine what is normally taken as an 'elastic modulus'. Since the actual soil behaviour is not elastic, however, the 'elastic modulus' determined by means of the empirical relations is in fact a sort of weighted average between the proper elastic stiffness, which is associated with conditions of unloading / reloading, and the elastic plastic stiffness associated to virgin loading. Moreover, since the elastic plastic stiffness is stress-path dependent, it is clear that an unambiguous recipe for determinig χ and K from SPT or CPT tests does not yet exist. In the following, therefore, the parameters χ and K will be chosen on the basis of previous experience matured in modelling soil behaviour in fully controlled laboratory test. From results in isotropic loading tests, elastic, i.e. reversible,volumetric strains may vary between 1/3 and 2/3 of total volumetric strains.

On the other hand χ may be linked to friction angle by another empirical relation such as:

$$\chi = (1.9-\eta_f)/62 \qquad \eta_f<1.9\ (\phi'<46.2°) \qquad (12)$$

Eq. (12) has been derived on the basis of few experimental data shown in fig. 3 and should be used cautiously. It should be

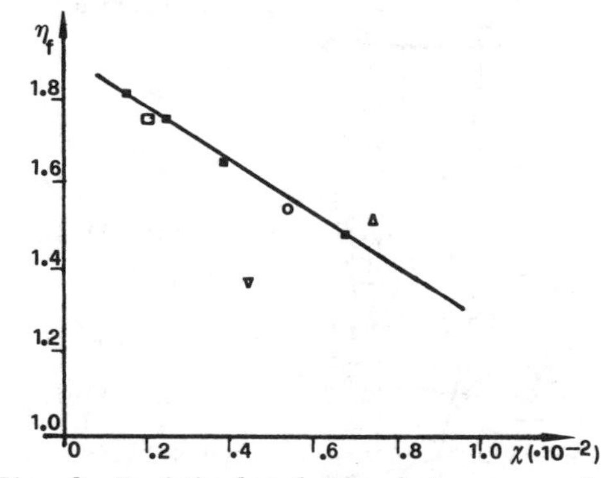

Fig. 3. Empirical relation between χ and η_f.

then considered as a very rough first approximation. More adequate relations will be proposed when enough experience will be gained on actual behaviour of real structures.

3. THE MODELLING OF A TUNNEL EXCAVATION IN SAND

The constitutive law presented has been employed for the finite element analysis of the excavation of an urban tunnel of the new line of the Milan Underground. To fit the narrow streets of the historic part of the town, a special profile with superposed rails has been chosen. A typical section of the tunnel is shown in fig. 4, where the 550 isoparametric 8-noded element mesh is also plotted.

The soil at site is alluvional sand and gravel. A typical stratigraphic profile together with the measured SPT blow counts is shown in fig. 5.

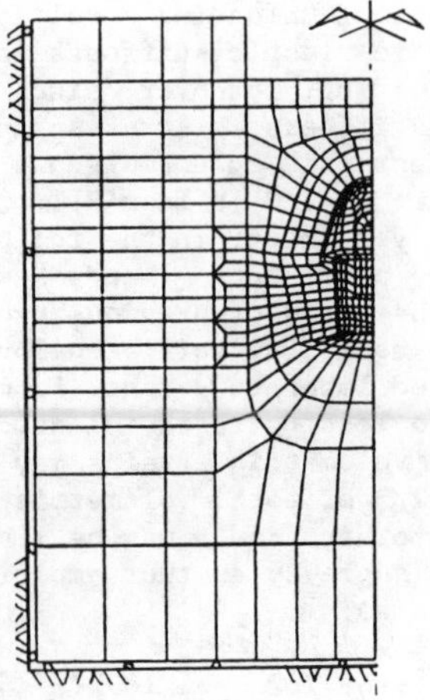

a) Finite element mesh

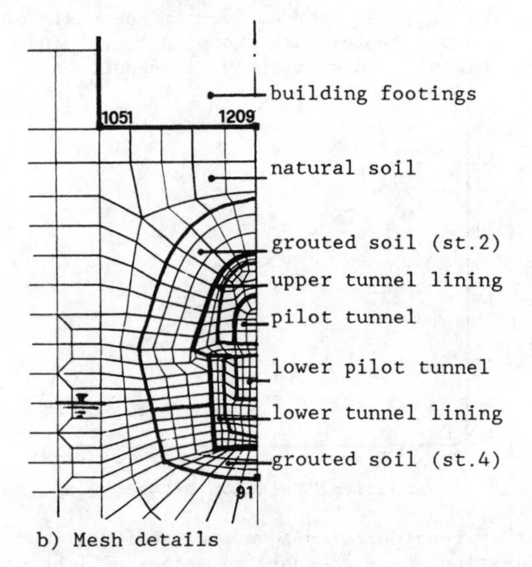

b) Mesh details

Fig. 4. Section of tunnel studied and finite element mesh.

The tunnel excavation has been carried out in the following steps:

1. Excavation of a 3.3 m wide pilot tunnel with shotcrete lining.
2. Grouting of a 3.5 m thick layer of soil surrounding the excavation profile of the upper tunnel and the shoulders of the lower one; grouting is not extended below the invert; injections are performed from the pilot tunnel; cement grouting is complemented with silicate additives where necessary (in sand and sandy gravel).
3. Excavation of the upper tunnel, with provisional steel ribs and shotcrete lining;
4. Grouting of a 2 m thick layer of soil below the profile of the invert of the lower tunnel; injections are performed from the floor of the upper tunnel; after the completion of the step, grouted soil forms a continuous ring surrounding the excavation profile.
5. Casting of the upper tunnel floor slab and crown arch lining.
6. Excavation of a 3 m wide pilot tunnel below the upper part floor slab; the bottom of this tunnel coincides with water level; walls have no lining and their stability relies upon apparent cohesion.
7. Lowering and shoulder excavation of the lower tunnel; walls are provisionally lined with steel struts and shotcrete.
8. Casting of the lining of invert arch and walls of the lower tunnel.

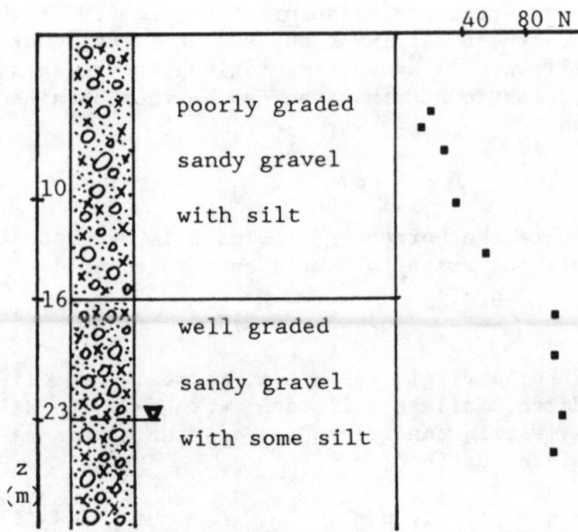

Fig. 5. Typical simplified soil profile.

The excavation procedure has been simulated by means of 16 loading steps on a plane strain model. Starting from imposed geostatic conditions, excavation, grouting and concrete casting are modelled with 'death' and 'birth' of finite elements and removal and application of respective weights.

Grouting improves mechanical performances of soil and introduces a self equilibrated force system as a consequence of injection pressures. This is simulated by means of enhanced material parameters and introducing an anelastic isotropic strain to prestress grouted soil arches. The numerical value of the imposed strain is determined in a semi-empirical way to obtain node displacements in agreement with measured surface heave.

Because of the plane strain approach adopted, in the crown excavation phase the crown provisional lining is supposed to begin acting simultaneously with crown excavation; in this way the settlement is neglected which takes place ahead from excavation front, as a consequence of front wedge decompression. To evaluate this settlement a single step analysis is performed in which crown excavation takes place with no lining; a fraction of the node displacements calculated in this way is then added to the node settlements of the crown excavation phase as calculated in the main analysis. According to experimental observations in similar conditions (e.g. Peduzzi et al. (1986)) this fraction has been taken equal to one third.

To see the influence of the constitutive model adopted on the results, various analyses with two different constitutive models have been conducted. The former is the one described in this paper, while the latter is a conventional elastic- perfectly plastic model with associate flow rule and Drucker-Prager yield condition.

Indeed several similar tunnels of the Milan Underground had previously been studied with the latter model. The calculated results were in substantial agreement with the measured settlements; however, during the excavation steps, the calculated heave was far larger than the observed, as a consequence of the low elastic modulus adopted to represent both virgin compression and unloading / reloading behaviour. The new material model has been conceived in order to avoid this discrepancy.

Different procedures have been followed to determine the constitutive parameters from available information on soil properties.

The elastic-perfectly plastic model is in this case governed by three parameters, the Young's modulus E, the Poisson's ratio ν and the effective friction angle ϕ'. The Young's modulus has been derived by means of the following empirical relationship (D'Appolonia et al. (1970)):

$$E = 21.6 + 1.06N \quad (13)$$

where N is the SPT blow count and E is in MPa. The Poisson's ratio has been taken equal to .25 which is a common value for the alluvional soil considered, while the friction angle has been derived from the empirical relation between ϕ' and N proposed by Bazaraa (1967), that takes the influence of overburden pressure into account. As far as strength is concerned, the soil can be conveniently subdivided in an upper layer 16 m thick with ϕ'=35° and a lower layer with ϕ'=38.5°.

The parameters for the Lamber model were calculated as follows. From the values of the friction angles calculated as above, the values of η_f have been determined from Eq. (11). Then the values of ξ and χ have been derived form Eq.s (10) and (12) respectively. The elastic bulk modulus has been taken in such a way to be three times the current elastic-plastic bulk modulus. Finally the elastic shear modulus was chosen in such a way that the initial apparent Poisson's ratio be .25.

For all the analyses performed the grouted soil in supposed to be elastic-perfectly plastic. The elastic modulus E is taken to be 210MPa. Although experimental results show much higher values, this value was taken as an average to account for macroscopic discontinuity in soil injection. The other parameters were taken as ν=0.25, ϕ'=35° while the cohesion is .2 MPa (average value as for E).

Fig. 6a,6b and 6c show the calculated vertical displacement of three critical points versus in-situ measured settlements in the eight construction phases; fig.s 7 and 8 represent typical stress and displacement conditions.

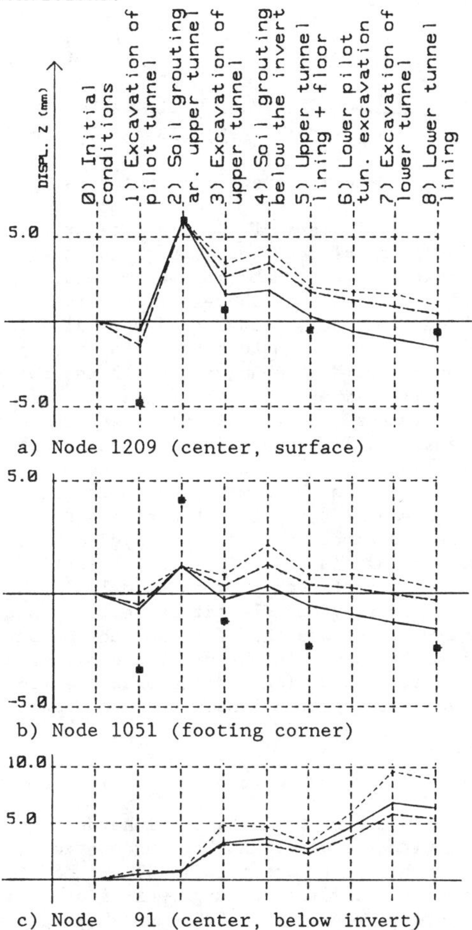

a) Node 1209 (center, surface)

b) Node 1051 (footing corner)

c) Node 91 (center, below invert)

Fig. 6. Comparison of dispacements.

As it can be seen from fig.s 6a,6b, which relate to surface points, the settlement predictions obtained with Lamber (solid line) are closer to experimental results (single square symbols) than those of the elastic-perfectly plastic model (short dash line); fig. 6c, in which the heave of a point below the excavation profile is plotted, shows the unrealistic higher swelling of the elastic-perfectly plastic model with respect to that predicted with Lamber. This behaviour is emphasized in fig. 7, which represents the displaced patterns in the crown excavation phase for the two models (amplif. 200 times).

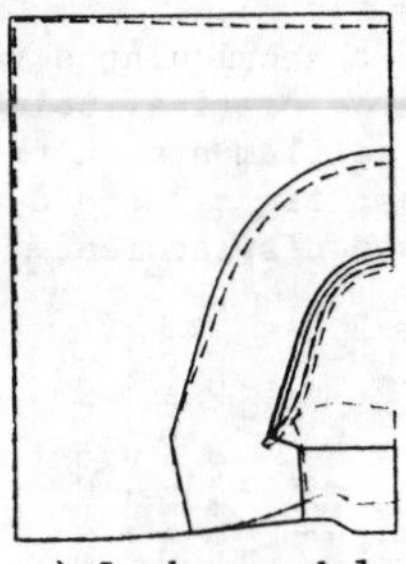

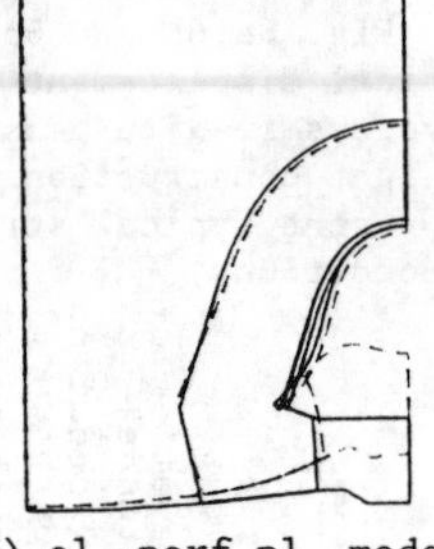

a) Lamber model   b) el.-perf.pl. model

Fig. 7. Deformed shape comparison (step 3)

The predictions of the elastic-perfectly plastic model could be improved with a different choice of material parameters. A third analysis has therefore been run, in which the elastic modulus of natural soil has been chosen in such a way to be equivalent to the bulk modulus used in the analysis with Lamber; also the friction angle has been reduced so that, in plane strain conditions, the limiting state is close to that of Lamber. The results are plotted with long dash lines on the same fig.s 6a,6b and 6c. The heave of the point at depth is now very close to Lamber's (the deep layer's behaviour being elastic). The surface settlements do not depart much from those calculated with the previous elastic-perfectly plastic analysis and are thus far from the measured ones. At least, in the problem considered, the strainhardening model proves then to give better predictions.

4. CONCLUSIONS

In this paper, a simple strainhardening constitutive model has been presented, which has been conceived to allow the numerical analysis of engineering problems. In order to do that, the complexity of the mathematical structure of the model and the number of constitutive parameters has been kept to the minimum. In the application presented it is shown that the constitutive parameters can be derived by combining the scanty information coming from SPT test, some empirical relationships between the blow count and the other geotechnical parameters and previous experience gained

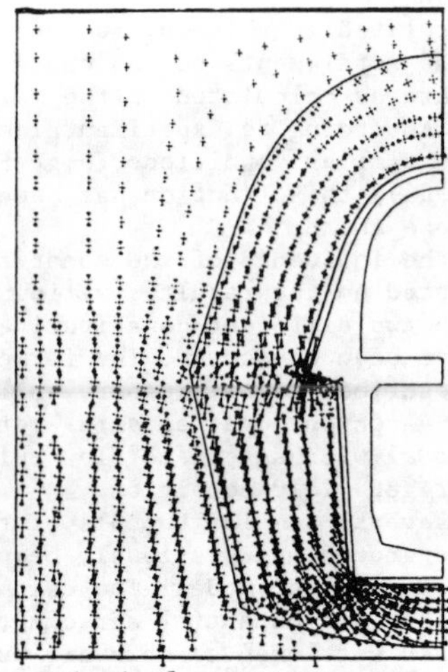

Fig. 8. Principal stresses.

in modelling laboratory soil behaviour.

The results obtained in the analysis of a tunnel with superposed rails are compared with the results obtained by means of a more classical analysis utilizing an elastic-perfectly plastic constitutive model. The former analysis gives predictions which are closer to observed data. The model presented constitutes then a promising tool for the solution of boundary value problems of practical interest.

5. ACKNOWLEDGEMENTS

Authors are indebted to Metropolitana Milanese for the contribution in performing the in situ measurements. R.Nova is also grateful to MPI for financial support.

REFERENCES

Bazaraa A.R. 1967. Use of the standard penetration test for estimating settlements of shallow foundations on sand. Ph.D. Th. University of Illinois, Urbana.

D'Appolonia D.J., D'Appolonia E., Brisette R.F. 1970. Discussion on settlement of spread footings on sand. ASCE J.SMFD 96.

Matsuoka H. and Nakai T. 1984. Stress deformation and strength characteristics under three different principal stresses. Proc. ISCE 232: 59-70.

Peduzzi R., Becci B. and Canetta G. 1986. Design and construction of blind-hole tunnels for line 3 of the Milan subway (underground). Problems and contibutions of the numerical calculation with finite element models. Proc. I86 1: 866-874.

Schofield A.N. and Wroth C.P. 1968. Critical state soil mechanics. Wiley.

11 Dynamic and earthquake engineering problems, blasting

Numerical Methods in Geomechanics (Innsbruck 1988), Swoboda (ed.)
© 1988 Balkema, Rotterdam. ISBN 90 6191 809 X

Continuum-boundary element and experimental models of soil-foundations interaction

L.Gaul, P.Klein & M.Plenge
University of the German Armed Forces, Hamburg, FR Germany

ABSTRACT: Dynamic responses of isolated rigid footings and flexible foundations, interaction between adjacent structures along with propagation and attenuation (by excavation) of surface waves in corresponding viscoelastic soil foundations are summarized. Three parameter constitutive models of soil improve the properties of Kelvin-Voigt and constant hysteretic model. Substructuring via continuum and/or boundary element method (BEM) for relaxed and perfectly bonded contact is verified by experimentation on a sand box (small-scaled laboratory foundation model).

1 INTRODUCTION

Design of foundations supporting vibrating machineries as well as the subsequent modification of dynamic responses after construction demand proper accounting of the interaction phenomena between those footings and the associated subsoil regions (Kehl, Bauer, Simon 1986, Gaul 1986, Aboul-Ella, Novak 1980). Interaction effects (through the underlying soil) between adjacent structures increase, when their natural frequencies become closer, when their masses increase and as the separation distance decreases (Roesset, Gonzalez 1977, Gaul 1980, Ottenstreuer 1981, Weber 1987).

The prediction of surface wave propagation becomes important since buildings and vibration-sensitive installations are effected by them. Herein the plane wave idealization (Haupt 1980) is extended to three-dimensional (3-d) cases.

Most frequently the interaction between the structure, the base plate and the subsoil is analyzed by the substructure technique. Structures are usually discretized by finite elements (FE), by multibody systems (Kehl, Bauer, Simon 1986) or treated by analytical dynamic stiffness matrices (Gaul 1980). Figures 1a and 1b depict, respectively, interactions of adjacent structures and the case of a shallow footing, as substructures.

Besides simplified soil models such as the truncated cone one (Gaul 1987), the substructure soil is specifically discretized by finite elements (Waas 1972, Kausel 1974, Werkle 1986), surface elements formulated by the continuum approach for the elastic (Holzlöhner 1969) or viscoelastic soil (Gaul 1980) or boundary elements (BE) (Dominguez 1978, Huh, Schmid and Ottenstreuer

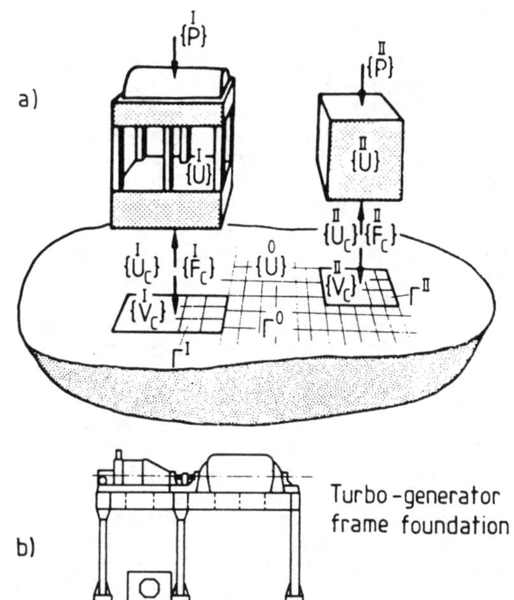

Fig. 1 Substructures of soil-structure interaction

1983, Huh, Willms 1984). Unlike the FE or finite domain (FD) methods the BEM requires only the discretization of the surface. Thus the BEM is ideally suited for three-dimensional linear interaction problems. Furthermore, the radiation condition is automatically taken into account, thereby the need of non-reflecting boundaries is eliminated. Employing the frequency domain BEM, the dynamic response of rigid or flexible three-dimensional structures with embedded or surface-foundations of arbitrary shape resting on viscoelastic soil can be studied. The prediction of interaction between adjacent structures is improved by replacing relaxed boundary conditions (Ottenstreuer 1981) by the perfectly bonded contact at the foundation interfaces.
The study of more general constitutive equations of viscoelastic soil initiated with a simplified soil model (Gaul 1987) is extended to the 3-d problem and matches with measured viscoelastic soil data. Numerical refinements and adapted discretizations allow for calculations of 3-d surface wave propagation in the nearfield of adjacent foundations including attenuation effects by an excavation in the soil.

The solution of dual integral equations of SSI by truncated series expansions (Triantafyllidis 1986) avoids discretization errors and satisfies the radiation condition as well as the interface boundary conditions continuously. But the approach is limited to the kinematic interaction with rigid surface foundations of simple shape. Arbitrary shapes and flexible slabs can be described by the continuum approach based on surface elements, where pointwise relaxed boundary conditions at element nodes is implemented (Gaul 1980).
This approach has been recently extended for the boundary conditions of perfectly bonded contact as well (Mesquita, Gaul 1987). The restriction of surface foundations is circumvented by substructure deletion utilizing a combination of the above mentioned continuum and discrete solution technique to describe embedment effects (Dasgupta 1980, Gaul 1986).
Errors by the truncation of surface discretization when adopting the BEM are found to decrease very rapidly with increasing number of elements (Weber 1986).
Irregular near-field can be coupled with the excavated homogeneous semi infinite domain described by BEM or substructure deletion.

Theoretical investigations of the present paper are accompanied by experimental research on SSI. Dynamic response of small scale frame foundations and driving point impedances of rigid bases on a sand box were measured (Gaul 1986, Kleinwort 1987) The detection of surface wave fields by laser interferometer is in progress.

2 SUBSTRUCTURE TECHNIQUE

The frequency dependent complex valued dynamic stiffness matrices $[K(i\omega)]$ describe the substructure behaviour in the frequency domain, where ω is the circular frequency and $i := \sqrt{-1}$, refer to Figure 1a. The substructure matrices of soil $[\overset{S}{K}]$ and both structures $[\overset{I}{K}]$, $[\overset{II}{K}]$ are coupled by compatibility requirements of generalized displacements $\{U_C\}$ and forces $\{F_C\}$ at the contact nodes of the interface I and II leading to

$$\begin{bmatrix} [\overset{I}{K}] & & \\ & [\overset{S}{K}] & \\ & & [\overset{II}{K}] \end{bmatrix} \begin{bmatrix} \{\overset{I}{U}\} \\ \{\overset{I}{U}_C\} \\ \{\overset{II}{U}_C\} \\ \{\overset{II}{U}\} \end{bmatrix} = \begin{bmatrix} \{\overset{I}{P}\} \\ \{0\} \\ \{0\} \\ \{\overset{II}{P}\} \end{bmatrix} + \begin{bmatrix} & \\ & [\overset{S}{K}] \\ & \end{bmatrix} \begin{bmatrix} \{0\} \\ \{\overset{I}{V}_C\} \\ \{\overset{II}{V}_C\} \\ \{0\} \end{bmatrix} \quad (1)$$

with generalized forces of active excitation $\{P\}$ and displacements $\{V_C\}$ of seismic excitation at the unloaded interfaces generated by incoming waves. The solution of equation (1) leads to complex amplitudes $\{U\} = \{U_R\} + i\{U_I\}$ corresponding to real displacements $\{u(t)\} = \{U_R\}\cos\omega t - \{U_I\}\sin\omega t$ for harmonic excitation.
The transient response in the time domain is obtained by inverse transformation, e.g. using Fast Fourier Transform (FFT).

3 CONSTITUTIVE EQUATIONS OF VISCOELASTIC SOIL

A better prediction of resonant magnifications is obtained with hysteretic losses taken into account primarily if the radiation damping is of same order of magnitude. However this is true for rocking and torsional vibrations of rigid foundations (Gaul 1987), while foundation flexibility reduces internal damping.
The correspondence principle replaces the elastic constitutive equation, relating the deviatoric states of stress and strain

$$s_{ij}(t) = 2Ge_{ij}(t) \quad (2)$$

by the viscoelastic law of differential operator type

$$\sum_{k=0}^{N} p_k \left(\frac{d}{dt}\right)^k s_{ij} = 2 \sum_{k=0}^{M} q_k \left(\frac{d}{dt}\right)^k e_{ij} \quad (3)$$

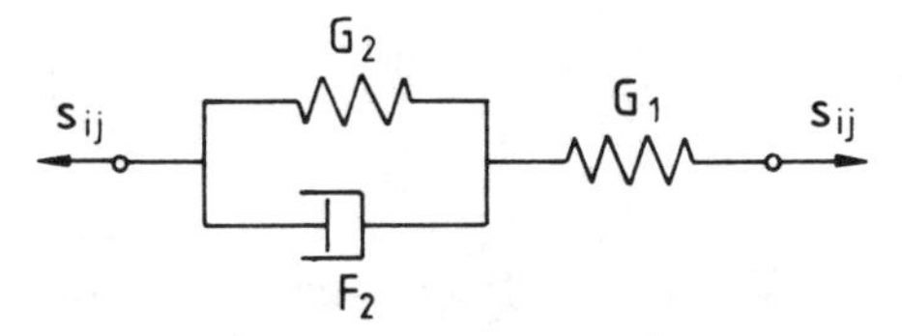

Fig. 2 Three-parameter viscoelastic model

or by the hereditary integral type

$$s_{ij}(t) = 2\int_{-\infty}^{t} G(t-\tau)\,\frac{de_{ij}(\tau)}{d\tau}\,d\tau \quad , \quad (4)$$

where the shear modulus G is replaced by the relaxation function G(t). The same holds for the hydrostatic state. Steady state harmonic functions of time such as $e_{ij} = \bar{e}_{ij}\exp(i\omega t)$ yield

$$\bar{s}_{ij} = 2G^{*}(i\omega)\,\bar{e}_{ij} \quad (5)$$

from equations (3,4).
Storage modulus and loss modulus $G'(\omega)$ and $G''(\omega)$ respectively as real and imaginary part of complex modulus G^{*}, written alternatively by introducing the damping factor $\eta(\omega) = G''(\omega)/G'(\omega)$,

$$G^{*}(i\omega) = 2\,\frac{\sum_{k=0}^{M} q_k (i\omega)^k}{\sum_{k=0}^{N} p_k (i\omega)^k} = G'(\omega) + iG''(\omega) = G'(\omega)\left[1+i\eta(\omega)\right] \quad (6)$$

are related to the relaxation function by

$$G'(\omega) = \overset{o}{G} + \omega\int_{o}^{\infty} \hat{G}(u)\sin(\omega u)\,du, \qquad G''(\omega) = \omega\int_{o}^{\infty} \hat{G}(u)\cos(\omega u)\,du \quad (7)$$

where $G(t) = \overset{o}{G} + \hat{G}(t)$, $\hat{G}(t) \rightarrow 0$ for $t \rightarrow \infty$.

The rheological three-parameter model of Figure 2 describes the standard test of relaxation and creep with instantaneous elasticity and asymptotic elastic behaviour. The corresponding parameters of the constitutive equations (3,4) are given by

$$p_1 = \frac{F_2}{G_1+G_2} \;,\; q_o = \frac{G_1G_2}{G_1+G_2} \;,\; q_1 = \frac{G_1F_2}{G_1+G_2} \;,\; q_1 > p_1q_o$$

$$G(t) = q_o\left[1-\left(1-\frac{q_1}{q_op_1}\right)\exp\left(-\frac{t}{p_1}\right)\right] \quad . \quad (8)$$

Assuming the two spring constants to be equal $G_1 = G_2 = 2G$ and $F_2 = 2G\xi a_o/\omega$ with the frequency parameter $a_o = \omega b/c_s$ yields the complex modulus

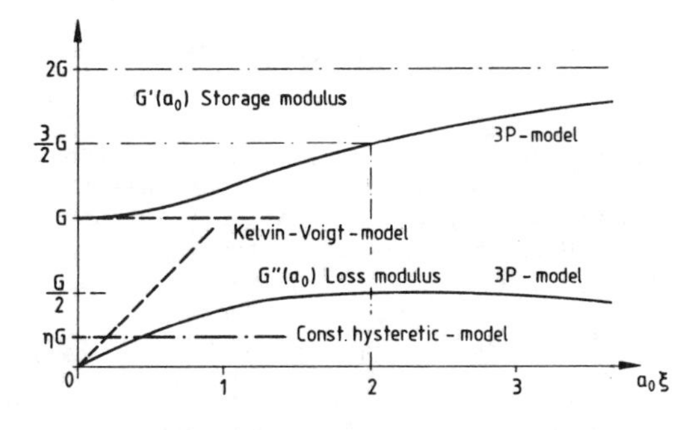

Fig. 3 Storage and loss moduli of material models

$$G^{*}(ia_o) = G\,\frac{(1+ia_o\xi)}{(1+ia_o\xi/2)} \quad . \quad (9)$$

If the spring G_1 is replaced by a rigid connection we arrive at the two-parameter Kelvin-Voigt model which does not exhibit instantaneous elasticity

$$G^{*}(ia_o) = G(1+ia_o\xi) \quad . \quad (10)$$

In a limited frequency band the model of constant hysteretic damping $\eta(\omega) \equiv \eta$ is supported by experimental results (Crandall, Kurzweil, Nigam 1971)

$$G^{*}(ia_o) = G(1+i\eta) \quad . \quad (11)$$

This model violates causality (Gaul, Bohlen, Kempfle 1985) in a weak sence if transient motions are described and it does not correspond to a rheological model strictly. A so called static hysteresis exists at vanishing frequency.
Storage and loss moduli of the material models are compared in Figure 3.

4 SUBSTRUCTURE SOIL

The substructure behaviour of soil in equation (1) (Figure 1a) is derived from the equation of motion of the viscoelastic continuum assuming zero body forces in terms of displacements $u_i(x_j,t)$

$$\int_{-\infty}^{t} E_D(t-\tau)\frac{\partial u}{\partial \tau}{}_{j,ji}\,d\tau - e_{ijk}e_{klm}\int_{-\infty}^{t} G(t-\tau)\frac{\partial u}{\partial \tau}{}_{m,lj}\,d\tau = \rho\,\frac{\partial^2 u_i}{\partial t^2}. \quad (12)$$

Steady-state elastodynamics is represented by taking the Fourier-transform of equation (12). With the transformed variables $U_i(\vec{x},\omega) = \mathcal{F}[u_i(\vec{x},t)]$ this yields an elliptical problem for the equations of the domain

$$c_D^2 U_{j,ji} - e_{ijk} e_{klm} c_S^2 U_{m,lj} + \omega^2 U_i = 0 \quad (13)$$

and transformed boundary conditions, where no initial conditions enter. The relaxation functions for plane dilatation $E_D(t)$ and shear $G(t)$ are replaced by the complex moduli

$$E_D^* = \lambda^* + 2G^* = E_D(1+i\eta_D),\ G^* = G(1+i\eta_S) \quad (14)$$

in the complex propagation velocities of dilatational and distortional waves

$$c_D^2 = E_D^*/\rho,\ c_S^2 = G^*/\rho\ .$$

4.1 Numerical treatment by BEM

Using the integral representation of equation (13), the displacement at the boundary point $\vec{\xi}$ may be written

$$c_{ij}(\vec{\xi})U_j(\vec{\xi}) = \int_\Gamma [U^*_{ij}(\vec{x},\vec{\xi})t_j(\vec{x}) - T^*_{ij}(\vec{x},\vec{\xi})U_j(\vec{x})]\,d\Gamma \quad (15)$$

where $\vec{x}$ are the boundary points to which the integral extends, $U_i(\vec{x})$ and $t_i(\vec{x})$ are displacement and traction components at $\vec{x}$, $U^*_{ij}(\vec{x},\vec{\xi},\omega)$ and $T^*_{ij}(\vec{x},\vec{\xi},\omega)$ stand for displacement and traction components of the fundamental solution at the field point $\vec{x}$ when the unit point load is applied at the load point $\vec{\xi}$ following the i direction, and c_{ij} is a coefficient that depends on the geometry of the boundary at $\vec{\xi}$ ($c_{ij} = \delta_{ij}/2$ for smooth boundary). One can solve the integral equation for a sufficient number of ω and numerically invert $U_i(\vec{x},\omega)$ to obtain the time-dependent displacement. Viscoelastic material properties enter the fundamental solutions (Cruse, Rizzo 1968), e.g.

$$U^*_{ij} = \frac{1}{4\pi G^*}[\psi(r,c_S,c_D) - \chi(r,c_S,c_D)r_{,i}r_{,j}]$$

where the potentials ψ and χ depend on the distance $r = (r_i r_i)^{1/2}$, $r_i = x_i - \xi_i$ and complex wave velocities.
The integrals in equation (15) are discretized into the sum of integrals extending over the boundary elements in Figure 1a. This yields for a boundary node α

$$c^\alpha_{ij} U^\alpha_j = U^{\alpha\beta}_{ij} t^\beta_j - T^{\alpha\beta}_{ij} U^\beta_j \quad (16)$$

where
$$U^{\alpha\beta}_{ij} = \int_{\Gamma_{el}} U^{*\alpha}_{ij}\,\Omega^\beta d\Gamma$$
$$T^{\alpha\beta}_{ij} = \int_{\Gamma_{el}} T^{*\alpha}_{ij}\,\Omega^\beta d\Gamma$$

contain the shape functions $\Omega^\beta(\vec{x})$ for the boundary elements ($\Omega^\beta(\vec{x}) = 1$ for constant elements), U^β_j and t^β_j are the displacement and traction nodal values. After numerical integration, the matrix formulation of equation (16) for all nodes and smooth boundary yields

$$\underline{\hat{T}}\underline{u} = \underline{U}\underline{t}\ ,\ \underline{\hat{T}} = \frac{1}{2}\ \underline{E}+\underline{T}\ , \quad (17)$$

where $\underline{u} = \{\underline{U}^1\underline{U}^2..\underline{U}^\alpha..\underline{U}^n\}^T, \underline{U}^\alpha = \{U^\alpha_1 U^\alpha_2 U^\alpha_3\}^T$

$$\underline{t} = \{\underline{t}^1\underline{t}^2..\underline{t}^\alpha..\underline{t}^n\}^T, \underline{t}^\alpha = \{t^\alpha_1 t^\alpha_2 t^\alpha_3\}^T. \quad (18)$$

4.2 Dynamic stiffness matrix of soil for flexible foundations

The dynamic stiffness matrix of substructure soil in equation (1) is obtained from equation (17) by partitioning the nodal displacements according to Figure 1a

$$\underline{U}^{-1}\underline{\hat{T}}\begin{bmatrix}\underline{U}^I\\ \underline{U}^{II}\\ \underline{U}^O\end{bmatrix} = \begin{bmatrix}\underline{t}^I\\ \underline{t}^{II}\\ \underline{t}^O\end{bmatrix}\ . \quad (19)$$

The discretization of the surface surrounding the foundation interfaces is truncated according to error studies of Weber (1986). The corresponding displacements $\underline{U}^O$ are eliminated by the boundary condition of vanishing tractions $\underline{t}^O = \underline{O}$. This yields

$$\underline{Q}\begin{bmatrix}\underline{U}^I\\ \underline{U}^{II}\end{bmatrix} = \begin{bmatrix}\underline{t}^I\\ \underline{t}^{II}\end{bmatrix} \quad (20)$$

and the dynamic stiffness matrix of soil is obtained by converting the tractions to nodal forces (multiplication with element areas for constant elements).
No decoupling assumptions of normal and shear stresses (relaxed boundary conditions, Ottenstreuer 1981) are introduced.
Embedment of foundations and the interaction of foundations with soil excavations in the near field are also considered.

4.3 Numerical treatment by continuum approach

The solution of equation (13) by the continuum approach is based on the discretization of foundation interfaces Γ^I and Γ^{II} (Figure 1a). The elements are loaded by the unknown interface stresses. The assumption of constant stresses per element define stress boundary value problems, which are solved by superposition of dilatational and distortional wave fields utilizing Fourier's integral theorem. The vertical contact traction t_3 acting on one surface element $|x_1|\leq a, |x_2|\leq b$ generates vertical and horizontal surface displacement fields, e.g.

$$U_3(x_1,x_2,0) = \frac{1}{2\pi}\iint_{-\infty}^{\infty} \bar{H}_{33}(\beta,\gamma,\omega)\,\bar{t}_3(\beta,\gamma)\exp[i(\beta x+\gamma y)]\,d\beta d\gamma \qquad (21)$$

with the harmonic wave compliance $\bar{H}_{33}$ and the 2-d Fourier transform of the exciting traction

$$\bar{t}_3(\beta,\gamma) = \frac{2t_3}{\pi}\;\frac{\sin\beta a\,\sin\gamma b}{\beta\gamma}\;. \qquad (22)$$

Integration (Gaul 1980) for nodal coordinates x_1^α, x_2^α and loaded element β with center coordinates x_1^β, x_2^β leads to the compliance $h_{33}^{\alpha\beta}(x_1^\alpha - x_1^\beta,\ x_2^\alpha - x_2^\beta,\ \omega)$. Superposition of displacements generated by all the element tractions on Γ^I and Γ^{II} yields at node α

$$U_3^\alpha = h_{31}^{\alpha\beta}t_1^\beta + h_{32}^{\alpha\beta}t_2^\beta + h_{33}^{\alpha\beta}t_3^\beta\;. \qquad (23)$$

Summarizing the displacements U_1^α, U_2^α, U_3^α according to equation (18), partitioning the variables with respect to the interfaces Γ^I and Γ^{II} and inverting the compliance matrix leads to the dynamic stiffness matrix of soil for flexible foundations obtained in Equation (20) by BEM.

The important difference is that no truncation of surface discretization has been necessary, because the stress boundary value problem is formulated for a traction free surface around the loaded elements. The compliance elements for the 2-d problem are discussed by Mesquita, Gaul (1987).

4.4 Substructure deletion by a combined continuum-BEM approach

The described continuum approach is restricted to foundations on a plane surface. Enormous mathematical complications are encountered in a continuum formulation, when satisfying the boundary conditions at the soil interface contour of an embedded foundation. When applying the BEM, the surface discretization has to be truncated for nonrelaxed boundary conditions implying a discretization error.
These difficulties are circumvented by substructure deletion (Figure 4) utilizing the available continuum and discrete solution technique.

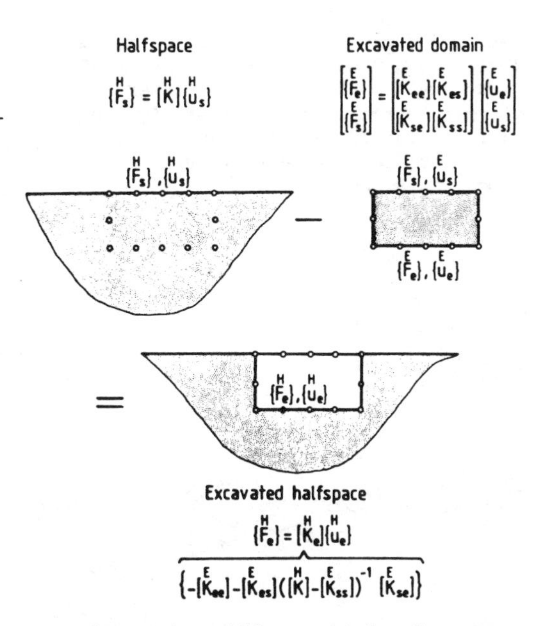

Fig. 4 Dynamic stiffness matrix of excavated soil for an embedded structure

The dynamic stiffness matrix of the excavated halfspace $[\overset{H}{K_e}]$ is calculated from the known dynamic stiffness matrix of the halfspace obtained by the continuum approach $[\overset{H}{K}]$ and the dynamic stiffness matrix $[\overset{E}{K}]$ of the excavated domain. Finite element discretization of the excavated domain (Dasgupta 1980) requires to condense the internal nodal degrees of freedom out and leads to $[\overset{E}{K}] = [K] - \omega^2 [M]$ by static condensation.
In the present paper surface discretization is applied by the BEM, where condensation drops out. As simultaneous prescription of nodal forces and displacements on the common boundary points is required, Dasgupta (1979) demonstrated that well posedness can be guaranteed if and only if the discrete models can reproduce those results which are the counterparts of Almansi's triviality (Almansi 1907).
These requirements and the known dynamic stiffness matrices of halfspace and excavated domain lead to the dynamic stiffness matrix of the excavated halfspace $[\overset{H}{K_e}]$ (Figure 4). Interaction with an embedded base plate can be calculated by coupling the dynamic stiffness matrices.

4.5 Dynamic stiffness matrix of soil for two adjacent rigid massless bases

Basic interaction effects between two adjacent foundations are described first by analyzing the dynamic stiffness matrix of soil loaded by two rigid massless bases (Figure 5). The kinematic restrictions of rigid body motion are introduced in equation (20) and the tractions are integrated to yield the resulting forces and moments.

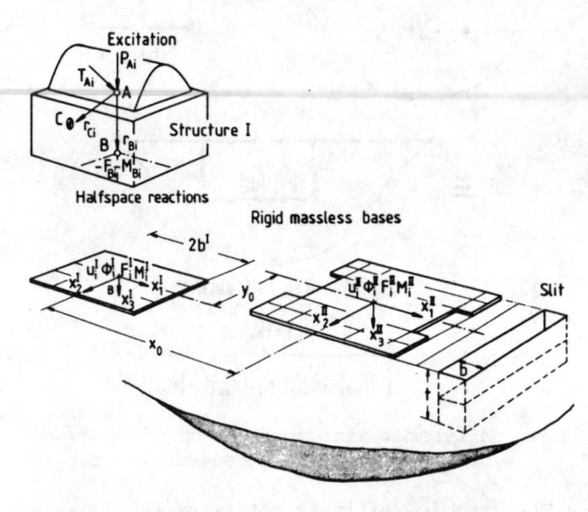

Fig. 5 Mixed boundary value problem for substructure soil

The displacement vector U_i^k of boundary element i at the interface of base k = I,II is related to the rigid body displacements U^k and the vector of small rotation angles ϕ^k by

$$U_i^k = U^k + (\tilde{r}_i^k)^T \phi^k \qquad k = I,\ II \qquad i = 1,..,n_k \tag{24}$$

with skew-symmetric matrix $\tilde{r}_i$ measuring the location of node number $(r_3^k \equiv 0)$. The hypermatrix formulation of all equations (24) leads to

$$\begin{bmatrix} U^I \\ U^{II} \end{bmatrix} = \overbrace{\begin{bmatrix} a^I & 0 \\ 0 & a^{II} \end{bmatrix}}^{a} \begin{bmatrix} U_B^I \\ U_B^{II} \end{bmatrix}, \tag{25}$$

where $U_B^k = [u_1^k u_2^k u_3^k\ \phi_1^k \phi_2^k \phi_3^k]^T \qquad k = I,II$

and $a^k = \begin{bmatrix} E & E & \dots & E \\ (r_1^k)^T & (r_2^k)^T & \dots & (r_{nk}^k)^T \end{bmatrix}$.

The traction vectors t_i^k multiplied with the diagonal matrix A of boundary element areas of constant shape functions are related to the resulting forces and moments in the vectors

$$F_B^k = [F_1^k F_2^k F_3^k M_1^k M_2^k M_3^k]^T \qquad k = I,II$$

by

$$\begin{bmatrix} F_B^I \\ F_B^{II} \end{bmatrix} = a^T A \begin{bmatrix} t^I \\ t^{II} \end{bmatrix} \tag{26}$$

The symmetric complex dynamic 12 x 12 stiffness matrix of soil K_B follows from equations (20, 25, 26) as

$$K_B = K^R + iK^I = a^T S Q a \tag{27}$$

and can be partitioned according to the influence of the bases I and II

$$\begin{bmatrix} F_B^I \\ F_B^{II} \end{bmatrix} = \begin{bmatrix} K_B^{I,I} & K_B^{I,II} \\ (K_B^{I,II})^T & K_B^{II,II} \end{bmatrix} \begin{bmatrix} U_B^I \\ U_B^{II} \end{bmatrix}. \tag{28}$$

By introducing the non-dimensional variables in equation (28)

$$\bar{u}_i^k = \frac{u_i^k}{b},\ \bar{\phi}_i = \phi_i,\ \bar{F}_i^k = \frac{F_i^k}{G(b^I)^2},\ \bar{M}_i^k = \frac{M_i^k}{G(b^I)^3}$$

the stiffness matrix K_B is made dimensionless with the frequency parameter

$a_o = \dfrac{\omega b^I}{\sqrt{G/\rho}}$, and is represented by

$$\bar{K} = c + i a_o d \quad . \tag{29}$$

The elements of c and d will be partitioned according to equation (28). They present lumped parameters of substructure soil:

- The equivalent spring coefficients c_{ij} describe the influence of elastic restoring and inertia forces
- The equivalent damping coefficients d_{ij} describe the radiation of wave energy, known as geometrical damping

In the present paper c_{ij} and d_{ij} depend additionally on the energy dissipation in soil governed by the viscoelastic constitutive equations.

5 SOIL STRUCTURE INTERACTION

5.1 Interaction between two rigid structures and soil

As an example of two sensitive structures interacting with soil two adjacent block foundations (Figure 5) are now considered. The Newton-Euler equations for each structure I, II with respect to body fixed reference point A

$$m(\ddot{u}_A + \dot{\tilde{\Omega}} r_c + \tilde{\Omega}\tilde{\Omega} r_c) = -F_A + P_A \tag{30}$$
$$I_A \dot{\Omega} + \tilde{\Omega} I_A \Omega + m\tilde{r}_c \ddot{u}_A = -M_A + T_A$$

are linearized for small angular displacements. The angular velocity Ω is approximated by the rotation vector $\dot{\underline{\phi}}$, $\underline{\Omega} = \dot{\underline{\phi}}$ and quadratic terms of Ω are neglected. The halfspace reactions acting in the interface at B are reduced to reference point A

$$\underline{F}_A = \underline{F}_B \; , \; \underline{M}_A = \underline{M}_B + \tilde{\underline{r}}_B \underline{F}_B \tag{31}$$

and the substructure soil (Equation 28) is coupled by reducing the displacement vector in equation (25) to reference point A

$$\underline{u}_B = \underline{u}_A + \tilde{\underline{r}}_B^T \underline{\phi} \; , \; \underline{\phi}_B = \underline{\phi}_A = \underline{\phi} \; . \tag{32}$$

Equations (27) to (32) for two rigid structures are summarized leading to the matrix equation of motion which is consistent due to frequency dependent parameters

$$\Bigg(-\omega^2 \begin{bmatrix} \underline{M}^I & \\ & \underline{M}^{II} \end{bmatrix} + \underbrace{\begin{bmatrix} \underline{A}^I & \\ & \underline{A}^{II} \end{bmatrix}^T \overbrace{\begin{bmatrix} \underline{K}_B^{I,I} & \underline{K}_B^{I,II} \\ \underline{K}_B^{II,I} & \underline{K}_B^{II,II} \end{bmatrix}}^{\underline{K}_B} \begin{bmatrix} \underline{A}^I & \\ & \underline{A}^{II} \end{bmatrix}}_{\underline{K}_A}\Bigg) \begin{bmatrix} \underline{u}_A^I \\ \underline{u}_A^{II} \end{bmatrix} = \begin{bmatrix} \underline{P}_A \\ \underline{T}_A \end{bmatrix} \tag{33}$$

The submatrices are defined for structure $k = I,II$ by

$$\underline{M}^k = \begin{bmatrix} m\underline{E} & -m\tilde{\underline{r}}_c^k \\ m\tilde{\underline{r}}_c^k & \underline{I}_A \end{bmatrix} \; \underline{A}^k = \begin{bmatrix} \underline{E} & -\tilde{\underline{r}}_c^k \\ \underline{O} & \underline{E} \end{bmatrix} \; . \tag{34}$$

New variables are introduced according to those which led to equation (29) for the forces and moments of active excitation $\underline{P}_A$, $\underline{T}_A$ Equation (33) can now be represented in the well known dimensionless form

$$[-a_o^2 \underline{B} + \underline{c}_A(a_o) + i a_o \underline{d}_A(a_o)]\underline{\bar{U}} = \underline{\bar{P}} \; . \tag{35}$$

Results of the frequency response are henceforth presented in terms of the following mass ratios. The elements of the 12 x 12 matrix $\underline{B}$ for proper i,j are

$$B_{ij} = \frac{m^k}{\rho (b^I)^3} \; ; \; \frac{m^k r_{ci}^k}{\rho (b^I)^4} \; ; \; \frac{I_{Aij}^k}{\rho (b^I)^5} \; . \tag{36}$$

Inversion of the frequency dependent coefficient matrix in equation (35) leads to the 12x12 complex elements $\bar{V}_{ij}(a_o)$ which belong to the frequency response matrix in

$$\underline{\bar{U}} = \underline{\bar{V}}\,(a_o, B_{ij})\; \underline{\bar{P}} \; . \tag{37}$$

5.3 Interaction with turbomachinery frame foundations

In the interest of brevity the approaches for describing the interaction between frame foundations and soil (Gaul 1986) are omitted here.

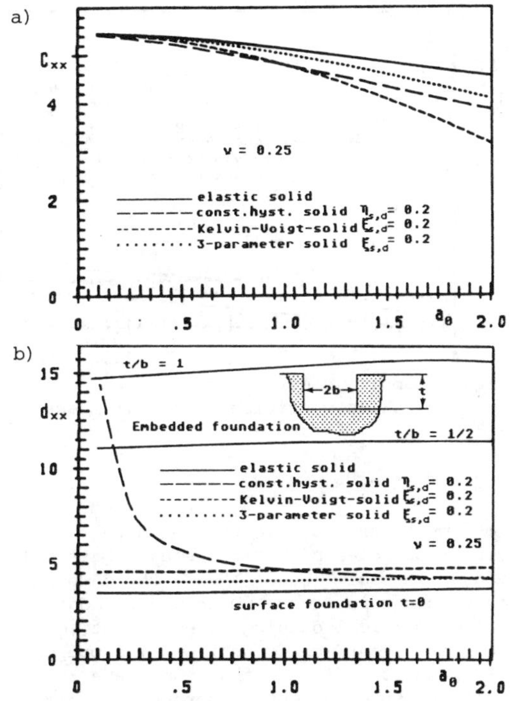

Fig. 6 Lumped parameters of soil
Horizontal vibration of a rigid base

6 NUMERICAL RESULTS

6.1 Lumped parameters of viscoelastic soil for one rigid surface foundation

Spring and damping coefficients of the lumped soil model defined in equation (29), as plotted in Figure 6, correspond to horizontal vibration of a rigid massless square base. The influence of three viscoelastic models is compared with the elastic soil model, which gives rise to only the geometrical damping. The increase of damping coefficients is caused by material damping. It should be mentioned here that the static hysteresis of the constant hysteretic solid leads to a singularity of the damping coefficient as $a_o \rightarrow 0$.

The linear frequency dependent increase of damping associated with the Kelvin-Voigt solid is not supported by measurements. Both shortcomings are removed by the three-parameter standard viscoelastic solid. Whereas geometrical damping predominates material damping for vertical and horizontal vibration modes, the damping factors for the rocking mode in Figure 7 indicate that both influences are of same order of magnitude in the practically relevant low frequency range. The same holds for the torsional mode with low geometrical damping (Gaul 1987, Weber 1987).

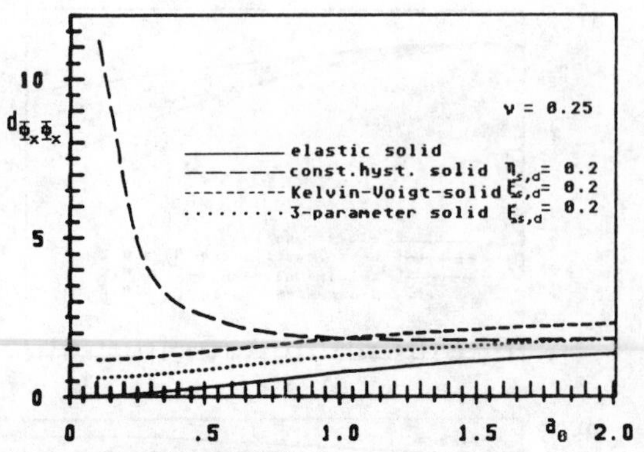

Fig. 7 Damping coefficient of soil
Rocking vibration of a rigid base

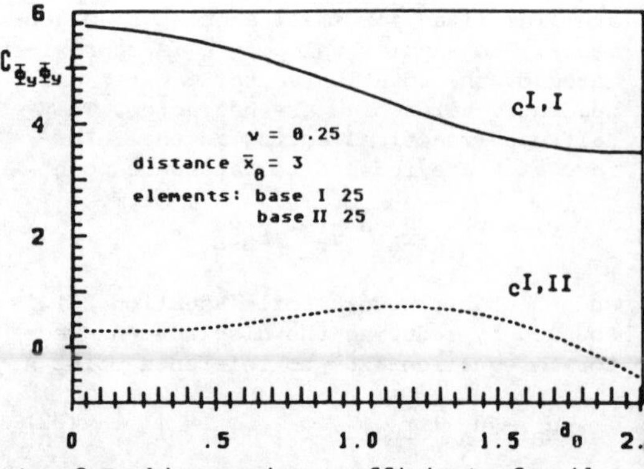

Fig. 8 Rocking spring coefficient of soil
Interaction between two rigid bases

6.2 Effect of embedment

Embedment effects have been studied for all vibration modes of a rigid square base welded to the soil (Weber 1986). The damping coefficient in Figure 6b shows the additional geometrical damping by embedment compared with a surface foundation. The present 3-d results indicate that amplitude reduction due to embedment is more pronounced in rocking and torsion modes than in vertical and horizontal modes. This is underlined by 2-d results (v. Estorff, Schmid 1984).

6.3 Effect of interaction between two rigid massless surface foundations

Interaction effects through the underlying viscoelastic soil between two rigid square bases separated by a distance $\bar{x}_o = x_o/b^I$, $\bar{y}_o = y_o/b^I = 0$ (Figure 5) are now discussed. The load vector $\underline{F}_B^I$ given in equation (26) acts harmonically in time on base I, while base II is not loaded.
The rocking interaction for example is described in Figure 8 where the spring coefficient of the load base $C^{I,I}_{\phi y\phi y}$ is compared with $C^{I,II}_{\phi y\phi y}$ of the second base (Equation 33).

For low frequencies when the bases are close to each other ($x_o/b^I = 3$) a positiv rocking angle of base I is associated with small rocking angle of base II in the reverse direction. The influence of distance between two rigid bases is discussed next. Again the lumped parameters of soil are partitioned according to equation (33). The horizontal spring coefficient $C^{I,I}_{xx}$ in Figure 9a is affected by interaction until the distance exceeds $\bar{x}_o = 5$ and then approaches the value that pertains to a single base.

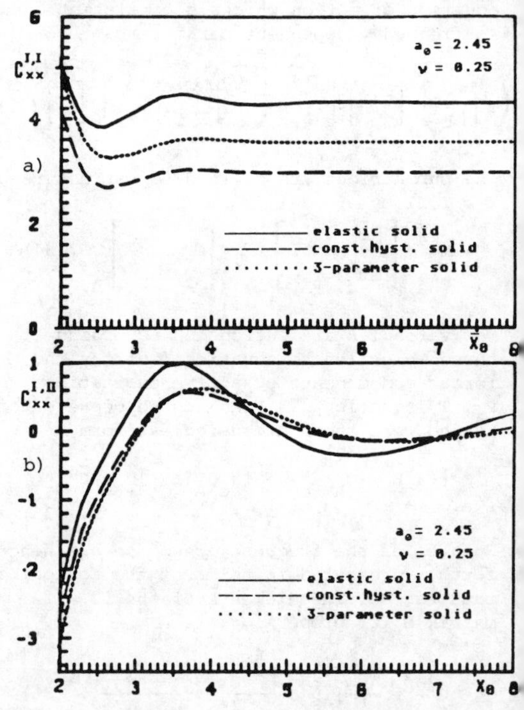

Fig. 9 Horizontal spring coefficient of soil
Interaction between two rigid bases

The same holds for the damping coefficient. The corresponding interaction spring coefficient $C^{I,II}_{xx}$ in Figure 9b dies out with increasing distance. The rocking spring coefficients $C^{I,I}_{\phi y\phi y}$ in Figure 10 a and $C^{I,II}_{\phi y\phi y}$ in Figure 10 b behave the same way but the distance at which interaction practically vanishes is significantly increased. Comparison between parameters that belong to the elastic soil with those of the viscoelastic soil (Figure 10 b) emphazises the pronounced effect of material damping on the interaction.

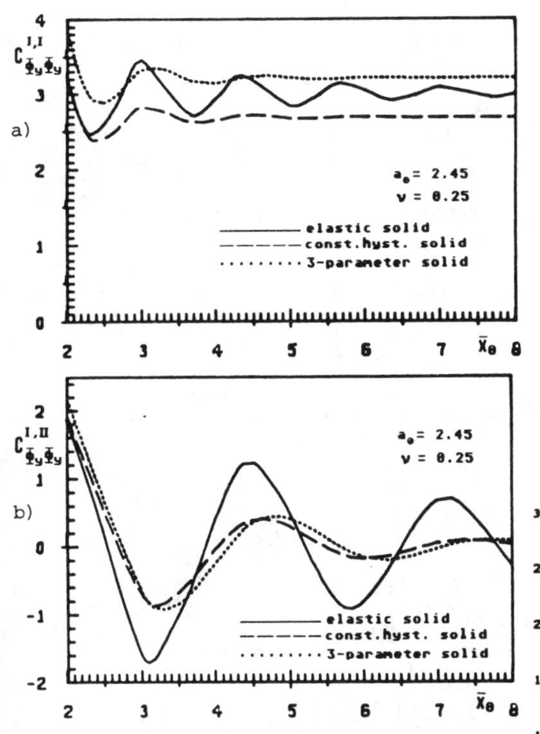

Fig. 10 Rocking spring coefficient of soil Interaction between two rigid bases

6.4 Structure-soil-structure interaction

The interaction between two block foundations with the same square base area and the same mass ratios is considered. They are centered along the x-axis at a distance $x_o/b^I = 3$ apart. Structure I is excited by a rotating unbalanced mass m_u which acts with horizontal force

$$P_x^I = m_u\, e\, \omega^2 \cos\omega t$$

in a height of $Z_B = (1/2)b^I$.

In Figure 11 the amplitudes of horizontal displacements at the center of structures I and II versus the frequency parameter are presented.

The coupled rocking and sliding mode of vibration gives rise to two resonant magnifications. The presence of a second non-excited block foundation increases the resonant magnification when compared with the response of a single block one.

Figure 12compares the response of a vertically excited block foundation I with the response of the non-excited foundation II at a distance $x_o = 3b^I$. The effects of an increasing mass ratio B_{xx}^I are to decrease

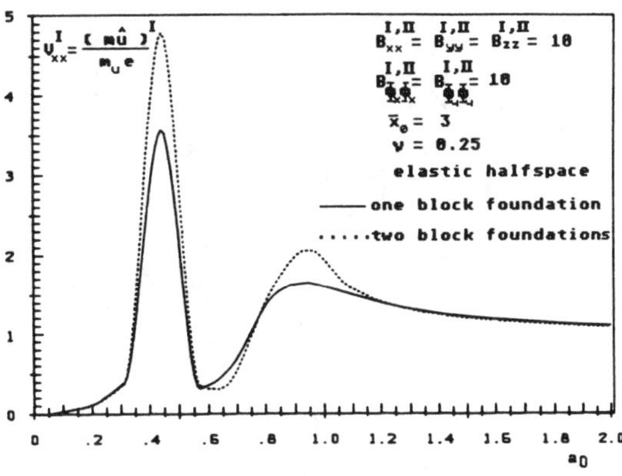

Fig. 11 Horizontal amplitude response of a single and two interacting block foundations

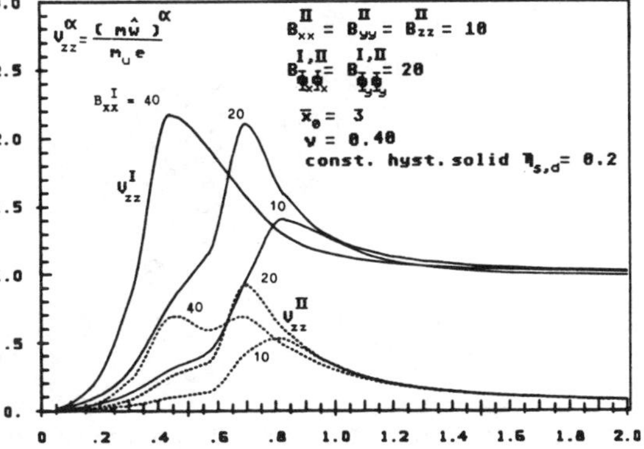

Fig. 12 Vertical amplitude response of two interacting block foundations

the resonant frequencies and the geometrical damping. One of the effects of an adjoining foundation is the excitation of modes of vibration which would not appear if the structure were alone. When only one foundation is excited by an external force, these effects are due to a feedback from the passive foundation and they decay with increasing distance (Weber 1987). When the two foundations are excited,they become more significant and their rate of decay with distance is much slower. An important effect is a change in the natural frequencies of the combined soil-structure system as illustrated in Figure 12.

6.5 Dynamic contact stresses

The dynamic stress distribution in the interface between rigid bases and soil follows

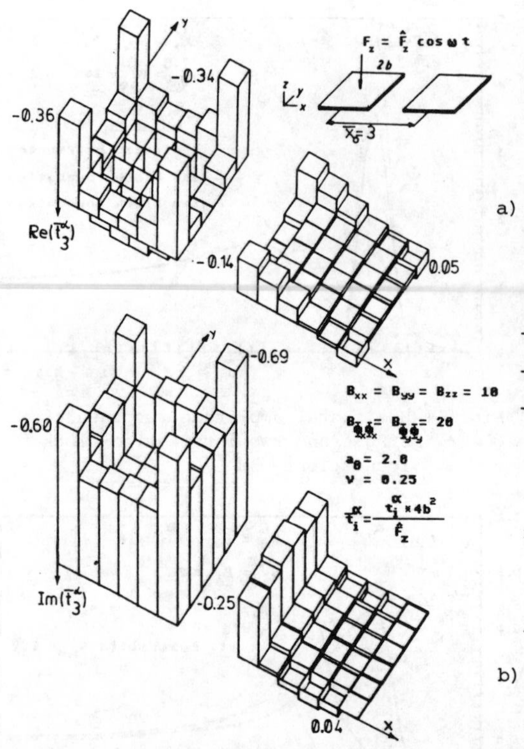

Fig. 13 Dynamic stress distribution at foundation interfaces

from the calculated rigid body motion of the foundation in equation (32). Equation (25) yields the boundary element displacements U_i^α which govern the corresponding surface tractions and stresses $t_i^\alpha = \sigma_{3i}^\alpha$ by equation (20).

The stress distribution in Figure 13 corresponds to vertical excitation $\hat{F}_z^I \exp(i\omega t)$ of base I and passive base II.

Nondimensionalized real and imaginary parts of the complex tractions $t_3^\alpha = \sigma_{33}^\alpha$ are plotted in Figure 13. They govern the time dependent stress by

$$t_3^\alpha(t) = (\mathrm{Re}\,\bar{t}_3^\alpha \cos\omega t - \mathrm{Im}\,\bar{t}_3^\alpha \sin\omega t)\,t_m$$

with mean stress $t_m = \hat{F}_z/(4b^{I2})$.

The phase shift of the stress distribution indicates that the excitation frequency is higher than the vertical resonance frequency.

The well known stress singularities at the interface boundaries show up. The feedback of the passive foundation causes a slightly unsymmetry in the stress distribution. This is due to dominant rocking motion of the passive base.

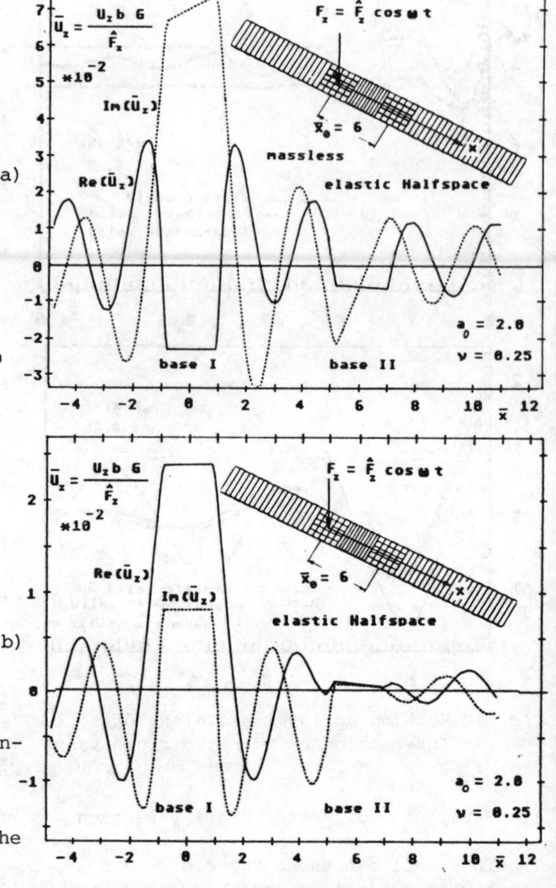

Fig. 14 Vertikal surface displacement fields Interaction between two bases
a) massless bases
b) bases with mass

6.6 Propagation of surface waves

The surface displacements at the unloaded outer boundary Γ_o (Figure 1a) are determined by equation (19) from the known displacements and tractions at the interfaces of two bases. The surface displacements along the center lines of both bases in Figure 14a are excited by vertically loaded massless base I and massless passive base II. The plotted real and imaginary parts govern the displacement field by

$$u_i^\alpha(t) = \mathrm{Re}U_i^\alpha \cos\omega t - \mathrm{Im}U_i^\alpha \sin\omega t.$$

With inertia properties of the bases included, the displacement field in Figure 14b indicates a phase shift of the overcritically excited base I associated with considerable reduction of rocking of both bases.

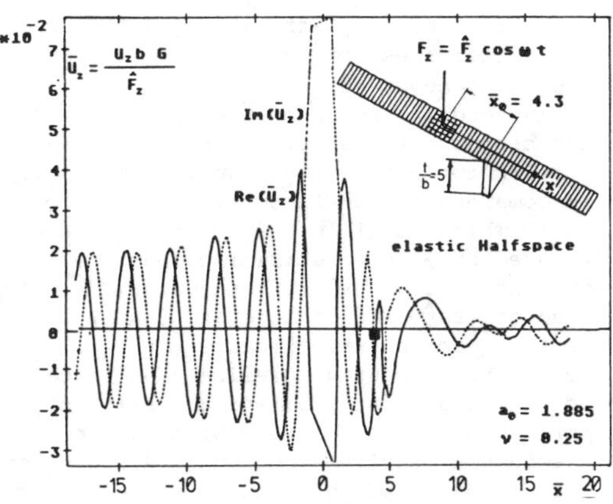

Fig. 15 Vertical surface displacement field Interaction between one base and a slit

An important problem area of reflections and transmissions of wave fields by soil excavations or obstacles is now briefly adressed.

The unsymmetric surface displacement field in Figure 15 is influenced by a deep slit on one side of the vertically excited base. Similar results of a 2-d problem (Haupt 1985) have been obtained by FE discretization utilising an influence matrix boundary condition at the vertical boundaries of the domain.

7 EXPERIMENTAL INVESTIGATION

Experimental investigations on SSI are carried out at the Universität der Bundeswehr Hamburg by the authors. The setup is shown in Figure 16. One or two rigid model footings interact with a model halfspace consisting of sand mixed with gravel in a box. The sand is compressed in a spherical domain underneath the footings while the zone with loose sand proves to be an effective energy absorber in a suitable frequency domain. Deviations of the halfspace assumptions by trapping the energy in the finite box domain are analysed by BE calculations.

Static and dynamic footing tests are performed. The dynamic tests include free vibrations excited by hammer impuls as well as forced vibrations by sine sweep and random excitation of vertical, horizontal, rocking and torsional modes of vibration. Lumped parameters of soil are evaluated from measured response impedances. The detection of surface wave fields by laser interferometer is in progress.

An experimental program has been started to measure the interaction phenomena that have been calculated in this constribution (Kleinwort 1987) as well as the influence of inhomogeneity in soil such as layering. Fig. 17a shows the transfer function of coupled rocking and sliding motion of a rigid circular footing. The horizontal acceleration of free vibrations excited by hammer impuls is measured and transformed in the frequency domain by FFT.

According to the response of the square foundation with different geometrie and inertia properties in Figure 11 two resonant magnifications occur.

Fig. 16 Model halfspace and measuring setup for experimental static and dynamic SSI investigations

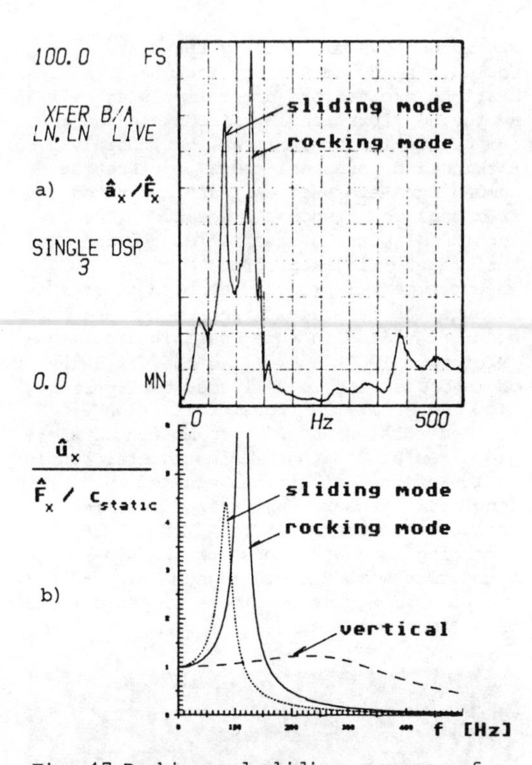

Fig. 17 Rocking and sliding response of a rigid circular footing
a) measured coupled response
b) calculated uncoupled response

Figure 17b shows the calculated transfer functions of uncoupled rocking and sliding response. The prediction of resonance frequencies is satisfactory.

8 CONCLUSIONS

The 3-d response of flexible and rigid structures interacting with soil is formulated by substructuring.
Single embedded and surface footings and the interaction between adjacent footings are considered. The main effects are worked out by means of BE discretization, continuum approach and a combination by substructure deletion for viscoelastic soil and rigid body description of footings. The feedback between an excited footing and a passive footing depends on the pronounced effect of material damping together with geometrical damping and on the distance and inertia properties of the footings. The simplification of relaxed boundary conditions is replaced by perfectly bonded contact. The combination of an inhouse BE program and a multibody description of structures proves to be efficient for 3-d calculations of structural response, dynamic stresses at foundation interfaces and surface wave fields of the homogeneous halfspace.
In the ongoing research the inhomogeneity has been incorporated. Layered viscoelastic soil is implemented by coupling layers of BE domains with different material properties. Thick flexible base plates are also treated by BE discretization. Surface irregularities e.g. by slits and obstacles indeed generate reflections and transmissions of the incoming wave field.
Both, the presented numerical/analytical methods and first promissing results of experimental investigations with footings on a sand box provide a good understanding of the essential SSI effects. For the purpose of analytical understanding,design calculation and eventual construction the approach of continuum/BE method validated by experimental observation is strongly recommended.

ACKNOWLEDGEMENTS

The assistance of Dr. G. Dasgupta, Associate Professor at Columbia University in the City of New York, currently a visiting Humbold Fellow at the University of the German Armed Forces Hamburg is gratefully acknowledged.

REFERENCES

Aboul-Ella, F.A. and Novak, M. 1980. Dynamic response of pile-supported turbine-generator foundations. Journal of the Engineering Mechanics Division, Proc. ASCE, Vol 106, EMG: 1215-1232.
Almansi, E., 1907. Un Teorema Sulle Deformazioni Elastiche dei Solidi Isotropi, Atti della reale accademia dei nazionale Lincei, Vol. 16: 865-868.
Crandall, S.H., Kurzweil, L.G., Nigam, A.K. 1971. On the measurement of Poisson's ratio for modeling clay. Experimental Mechanics, 11: 402-407.
Cruse, T.A., Rizzo, F.J. 1968. A direct formulation of the general transient elastodynamic problem I. Journal of Mathematical Analysis and Applications, 22: 244-259.
Dasgupta, G., 1979. Wellposedness of substructure Deletion Formulations. Proceedings Sixteenth Midwestern Mechanics Conference Manhatten, Kansas, Vol. 10.
Dasgupta, G., 1980. Foundation Impedance Matrices by Substructure Deletion. Journal of the Engineering Mechanics Division, ASCE, Vol. 106, EM3:517-523.
Dominguez, J. 1978. Dynamic stiffness of rectangular foundations. Publ. No. R78-20 Dept. of Civil Engg. M.I.T.

Estorff, O. von, Schmid, G. 1984. Application of the boundary element method to the analysis of the vibration behaviour of strip foundations on a soil layer. Proceedings Int.Symp. on Dynamic Soil-Structure Interaction, Minneapolis:11-18

Gaul, L. 1980. Dynamics of frame foundations interacting with soil. Journal of Mechanical Design, Vol. 102, No. 2: 303-310.

Gaul, L, 1986. Machine-foundation-soil interaction; combined continuum and boundary element approach. Revista Brasileira de Ciências Mecânicas Rio de Janeiro Vol. VIII, no. 3:169-198.

Gaul, L. 1987. Baugrund- und Fundamentdämpfung. VDI Berichte 627:201-230

Gaul, L., Bohlen, S. and Kempfle, S. 1985. Transient and forced oscillations of systems with constant hysteretic damping. Mechanics Research Communications, Vol. 12, 4:187-201.

Haupt, W. 1980. Abschirmung von Gebäuden gegen Erschütterungen im Boden. Deutsche Gesellschaft für Erd- und Grundbau, Grundbautagung in Mainz:117-139.

Haupt, W. 1985. Erschütterungsabschirmung in gefrorenem Boden. Veröffentlichungen des Grundbauinstitutes der Landesgewerbeanstalt Bayern, Nürnberg, Heft 43

Holzlöhner, U. 1969. Schwingungen des elastischen Halbraumes bei Erregung auf einer Rechteckfläche. Ingenieur-Archiv, 18:370-379.

Huh, Y., Schmid, G. and Ottenstreuer, M. 1983. Evaluation of kinematic interaction of soil-foundation by boundary element method. SMIRT7 Chicago, Paper K 814

Huh, Y., Willms, G. 1984. Ermittlung des Schwingungsverhaltens starrer Gründungskörper mit der Randelementmethode. SFB 151 Tragwerksdynamik Ruhr-Universität Bochum, Bericht Nr. 2.

Kausel, E. 1974. Forced vibrations of circular foundations on layered media. Research Report R74-11, Department of Civil Engineering, M.I.T.

Kehl, K., Bauer, J., Simon, G. 1986. Untersuchung von Turbine/Fundament-Modellen bei unterschiedlicher Sohlfugenausführung. VDI Berichte 603:175-191

Kleinwort, A. 1987. Analytische und experimentelle Untersuchung zur Fundament-Baugrund-Wechselwirkung. Diplomarbeit am Institut für Mechanik, UniBw Hamburg (not published)

Mesquita, E., Gaul, L. 1987. Wechselwirkung von Fundamenten mit dem Baugrund unter Berücksichtigung verschweißten Kontaktes. Dynamische Probleme - Modellierung und Wirklichkeit - Universität Hannover, to be published.

Ottenstreuer, M. 1981. Ein Beitrag zur Darstellung der Wechselwirkung zwischen Bauwerk und Baugrund unter Verwendung des Verfahrens der Randelemente. Institut für Konstruktiven Ingenieurbau, Ruhr-Universität Bochum, Mitteilung Nr. 81-6.

Roesset, J.M. and Gonzalez, J.J. 1978. Dynamic interaction between adjacent structures. Proc. DMSR Karlsruhe, Vol. 1:127-166.

Triantafyllidis, Th. 1986. Dynamic stiffness of rigid rectangular foundations on the half-space. Earthquake Engineering and Structural Dynamics, Vol. 14:391-411.

Waas, G. 1972. Linear two-dimensional analysis of soil dynamics problems in semi-infinite layered media, Ph.D. Thesis, University of California, Berkeley.

Weber, Th. 1986. Zur Berechnung der Wechselwirkung eines starren Maschinenfundamentes mit dem viskoelastischen Halbraum. Studienarbeit am Institut für Mechanik, UniBw Hamburg (not published)

Weber, Th. 1987. Berechnung der Nahfeldwellenausbreitung infolge schwingender Fundamente mit der Boundary Element Methode. Diplomarbeit am Institut für Mechanik, UniBw Hamburg (not published).

Werkle, H. 1986. Dynamic finite element analysis of three-dimensional soil models with a transmitting element. Earthquake Engineering and Structural Dynamics, Vol. 14:41-60.

Numerical Methods in Geomechanics (Innsbruck 1988), Swoboda (ed.)
© 1988 Balkema, Rotterdam. ISBN 90 6191 809 X

Non-linear response of a soil with a harmonic loading: A numerical-analytical method

J.F.Heitz & G.Bonnet
Institut de Mécanique de Grenoble, France
T.Avril
EDF-REAL, Chambéry, France

ABSTRACT : From a generalization of classic linear equivalent model (1), one interpretation of a surface harmonic seismic test is propounded. This one rests on a nearfield approximation of elastodynamic law and on a perturbation method. This semi-analytical computation allows to evaluate the non-linear adjustment and hence the characteristic parameter of non-linear behaviour. This paper presents first results of the direct problem especially concerning the focusing of results depending on the chosen discretization parameters of the non-linear domain. Moreover, the efficiency of the interpretation method is proved.

1 INTRODUCTION

The elaboration of more and more sophisticated behaviour law for soils let a large field of investigation for the in-situ determination of characteristic parameters introduced in those laws.

Some dynamic tests showed that the shear modulus G decays as the distorsion strain level increases in soils. This non-linear behaviour induces important effects on temporal and spectral soil response.

That is the reason why it is important to take this particular behaviour into account by introducing a characteristic parameter.

The purpose of our work is to interpret at best a surface harmonic seismic test. Then, the determination of a characteristic parameter of the non-linear behaviour can be used to simulation programs of foundations behaviour under seismic sollicitation like GEFDYN, FLUSH (2) ...

2. MODELING OF NON-LINEAR BEHAVIOUR IN SOILS

From the SEED's mean curve (3) set on figure 1, different analytical expressions have been suggested to characterize the dependence of stiffness with the shear strain level in soils.

Among these approaches, the hyperbolic model of HARDIN and DRNEVICH (4) :

$$\frac{G}{G_o}(\gamma) = \frac{1}{1 + \alpha_1 . \gamma} \qquad (2.1)$$

fits at best to SEED's experimental mean curve if $\alpha_1 = 5.32\ 10^3 (\pm 1\%)$ and $\gamma \leqslant 10^{-4}$ (see table 1)

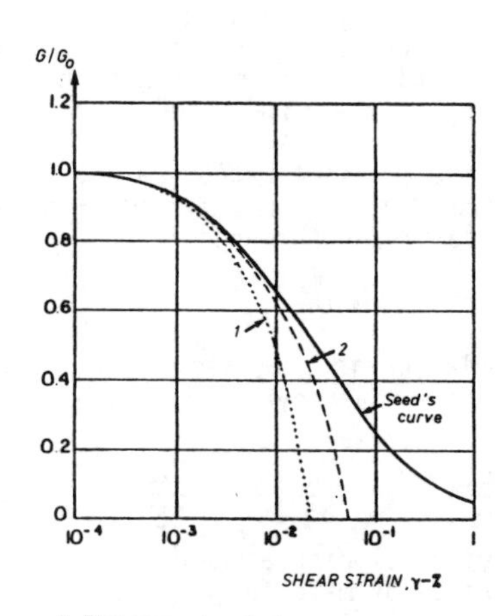

Figure 1 Dependence of shear modulus with shear strain level. SEED's and modeling approaches.

Table 1. Variation of α_1 with shear strain

Shear strain, γ	G/Go	α_1 ($\times 10^{-3}$)
10^{-5}	0.95	5.26
10^{-4}	0.65	5.38
10^{-3}	0.25	3.00

For shear strains below or equal to 10^{-4}, it is possible to develop in series (2.1) as :

$$\frac{G}{Go}(\gamma) = 1 - \alpha_1\gamma + \alpha_2\gamma^2 - \alpha_3\gamma^3 + \cdots \quad (2.2)$$

To take into account the symmetry of non-linear behaviour in tension and traction, it is necessary to rewrite (2.2) :

$$\frac{G}{Go}(\gamma) = 1 - \alpha_1|\gamma| + \alpha_2\gamma^2 - \alpha_3|\gamma|^3 + \cdots \quad (2.3)$$

3 A ONE-DIMENSIONAL EXAMPLE

The last expression is directly applicable to a simple scheme of one-dimensional shear wave propagation in a soil layer. A behaviour simulation of a soil sample submitted to a shear strain harmonic excitation at the base (see Figure 2) is computed.

The spectral acceleration ratios between the third (or fifth) harmonic and the fundamental is established below.

3.1 Constitutive law

The simulation is computed by introducing (2.3) in the elastodynamic law :

$$\frac{\partial}{\partial z}\left[G(\gamma)\cdot\frac{\partial u}{\partial z}\right] = \rho\frac{\partial^2 u}{\partial t^2} \quad (3.1)$$

u : horizontal displacement
ρ : density of material

This substitution leads to generate a second member F, depending on shear strain.

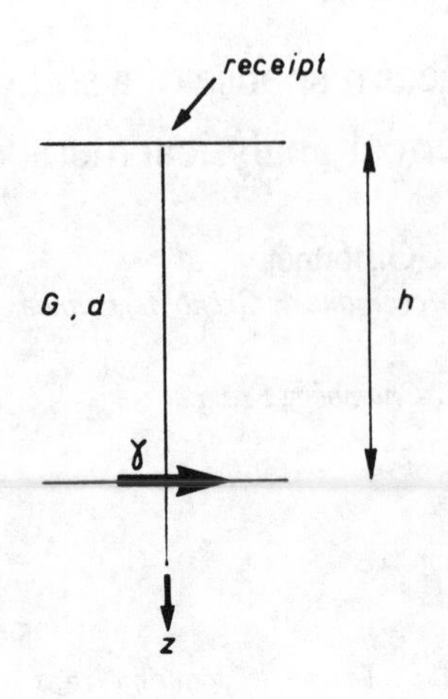

Figure 2 Simulation of SH-wave propagation in a soil sample

For the truncated development :

$$G(\gamma) = Go - G_1|\gamma| + G_2\gamma^2 - G_3|\gamma|^3 \quad (3.2)$$

with Go, linear shear modulus and Gi ($i \neq o$), non linear shear moduli, then :

$$F(\gamma) = G_1\left[|\gamma|\frac{\partial\gamma}{\partial z} + \gamma\frac{\partial|\gamma|}{\partial z}\right] - 3G_2\gamma^2\frac{\partial\gamma}{\partial z} + G_3\left[3\gamma^3\frac{\partial|\gamma|}{\partial z} + |\gamma|^3\frac{\partial\gamma}{\partial z}\right] \quad (3.3)$$

3.2 Odd harmonics and non-linear behaviour

Let us consider that $u = \bar{u}$ coswt and then, $\gamma = \bar{\gamma}$ coswt (w, fundamental frequency), the expression (3.3) becomes :

$$F(\bar{\gamma},t) = a_1(\bar{\gamma})\cdot|\cos\omega t|\cos\omega t + a_2(\bar{\gamma})\cos^3\omega t + a_3(\bar{\gamma})\cdot|\cos\omega t|\cos^3\omega t \quad (3.4)$$

With the help of a development in Fourier's serie (classical harmonic analysis), (3.4) is rewritten :

$$F(\bar{\gamma},t) = \sum_j b_j(\bar{\gamma})\cdot\cos j\omega t \quad (3.5)$$

The computation shows that :

$$b_{2j}(\bar{\gamma}) = 0 \quad \text{and} \quad b_{2j+1}(\bar{\gamma}) \neq 0$$

Finally, we find :

$$b_1(\bar{\gamma}) = 0.85\, a_1(\bar{\gamma}) + 0.75\, a_2(\bar{\gamma}) + 0.70\, a_3(\bar{\gamma})$$

$$b_3(\bar{\gamma}) = 0.17\, a_1(\bar{\gamma}) + 0.25\, a_2(\bar{\gamma}) + 0.28\, a_3(\bar{\gamma})$$

$$b_5(\bar{\gamma}) = -0.02\, a_1(\bar{\gamma}) - 0.05\, a_3(\bar{\gamma})$$

From the value α_1 (see § 2) and the assumption of α_i - dependence such as :

$$\alpha_i = \alpha_1^{\,i} / i!$$

a non-linear response spectrum of accelerations has been simulated for different shear strain level (see Figure 3).

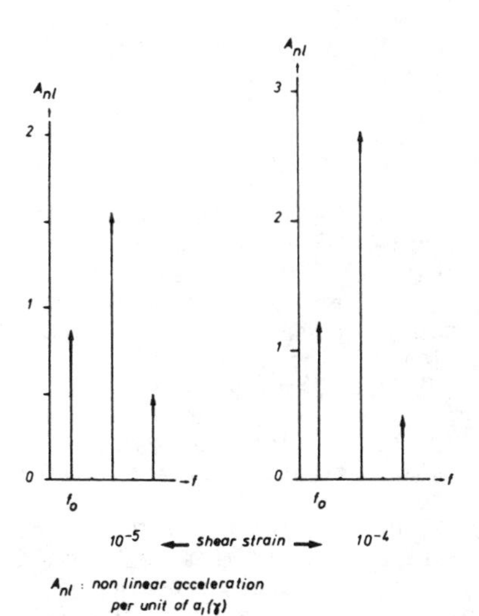

Figure 3 Aspect of non-linear acceleration spectra depending on shear strain

The figure 3 emphasizes the importance of the non-linear acceleration on the third harmonic compared to the fundamental.

3.3 Spectral ratio of non linear and linear accelerations.

A first approximation of behaviour law is to be considered for simulation :

$$\frac{G}{G_0} = 1 - \alpha_1 |\gamma| \qquad (3.6)$$

Moreover, two assumptions are used to the computation. Firstly, we assume that the material is homogeneous and secondly, the damping factor is constant in the whole material for a given shear strain level generated at the base of the soil sample.

Then, the computation of non-linear one (Anl) and linear acceleration (Al) leads us to constitute the ratios R_3 (h/λ) and R_5 (h/λ) defined by :

$$R_3(h/\lambda) = \frac{Anl\,(\omega_0 = 3\omega)}{(Al + Anl)\,(\omega_0 = \omega)} \qquad (3.7.a)$$

$$R_5(h/\lambda) = \frac{Anl\,(\omega_0 = 5\omega)}{(Al + Anl)\,(\omega_0 = \omega)} \qquad (3.7.b)$$

where h is the thickness of the soil sample and λ, the wavelength connected to SH-wave velocity and fundamental frequency.

Figure 4 presents these two ratios and shows that the main result is due to the importance of R_3 for discrete values of h/λ .

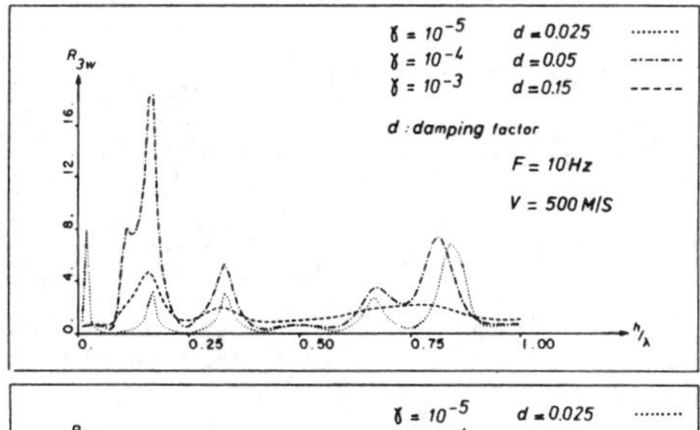

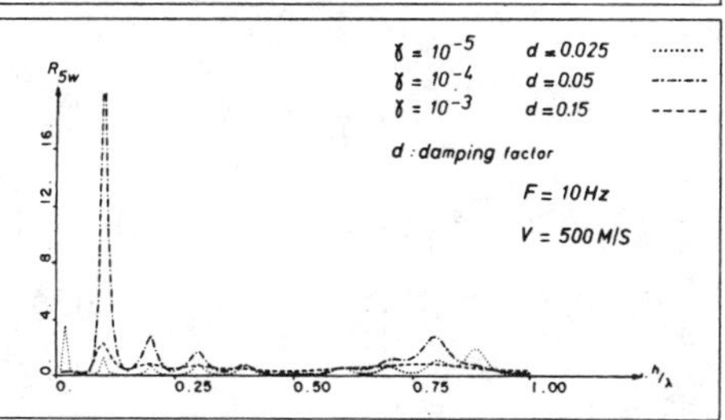

Figure 4 Spectral acceleration ratios depending on h/λ and shear strain level

Finally, we can conclude that the case for which the third harmonic acceleration is equal to the fundamental one, taking non-linear behaviour of soil into account is equivalent to consider a displacement of response spectrum.

4 GENERALIZATION OF APPROXIMATE NON-LINEAR BEHAVIOUR LAW TO THE THREE-DIMENSIONAL CASE

For the three-dimensional case, we must generalize (3.6). But the shear strain depends on the location in the soil and it is necessary to carry a writing independent of the chosen referential, the second invariant of the deviatoric part of strain tensor is used :

$$I_2 = \sum_{\substack{i,j \\ i \neq j}} \left[\left(\varepsilon_{ii} - \frac{\varepsilon_{kk}}{3}\right)\left(\varepsilon_{jj} - \frac{\varepsilon_{kk}}{3}\right) - \varepsilon_{ij}^2 \right] \quad (4.1)$$

Then, for a given stress level :

$$\frac{G}{G_0}(I_2) = 1 - \alpha_1 |I_2|^{1/2} \quad (4.2)$$

with α_1 , characteristic parameter of approximate non-linear behaviour.(4.2) is the first approach on Figure 1.

For shear strain level below or equal to 10^{-4}, $\alpha_1 |I_2|^{1/2} = \varepsilon \ll 1$. More precisely, for $\gamma \leq 5.10^{-5}$, ε appears as an adjustment of the linear case. Then, a perturbation method is used to compute the non-linear displacement.

5 COMPUTATION PRINCIPLE

Including (4.2) in the elastodynamic law, we have :

$$(\lambda_0 + G_0) u_{j,ji} + G_0 u_{i,jj} - \rho \ddot{u}_i = F_i \quad (5.1)$$

with, G_0 and λ_0 , LAME's elastic parameters and Fi, source term of non linearities such as :

$$F_i = \alpha_1 |I_2|^{1/2} \rho \ddot{u}_i + \frac{1}{2} \operatorname{sgn}(I_2) . |I_2|^{-1/2} \times$$

$$\left[\lambda_1 (I_{2,i} . \varepsilon_{jj}) + 2 G_1 (I_{2,j} . \varepsilon_{ij}) \right] \quad (5.2)$$

if we consider that Poisson's ratio is constant and then, the variation of λ is similar of G.

The equation (5.1) is like :

$$L(u) = F(u) = \varepsilon F^*(u) \quad (5.3)$$

L is a linear differential operator and F, a source function. As ε is small, it is logical to expect that the displacement u is next to the displacement uo obtained from the same equation except the second member. uo is called generator solution (5) which is used to accede to the sought non-linear solution Ul.

We express the difference between these two solutions as a truncated development in series with respect to ε , such as :

$$U - U_0 = U_1 = \varepsilon U_1^* \quad (5.4)$$

Replacing u by the above development in (5.3), we obtain :

$$L(u_0 + u_1) = F(u_0 + u_1)$$

But : $L(u_0) = 0$ and $u_0 \gg u_1$, the problem to solve becomes :

$$L(u_1) = \varepsilon F^*(u_0) \quad (5.5)$$

The chosen computation method is typically a perturbation method. It is divided in three steps :

- Seek the elastic linear solution Uo
- Compute the source of non linearities F from Uo
- Seek the non-linear solution Ul by solving (5.5)

6 THE SURFACE HARMONIC SEISMIC TEST

This test is the stationnary RAYLEIGH-wave test fit to measurement of the non-linear acceleration. It consists of the generation, at a point of the soil surface, of a harmonic loading and the reception of the wave field at another point(s). It requires a dynamic source (vibrating piston) and accelerometers. The input sollicitation is monitored by a SOLARTRON generator (spectral analyser) connected to an amplifier furnishing the required dynamic input load (see fig. 5).

The experimental measure consists in shifts and amplitudes recording got at the receipt.

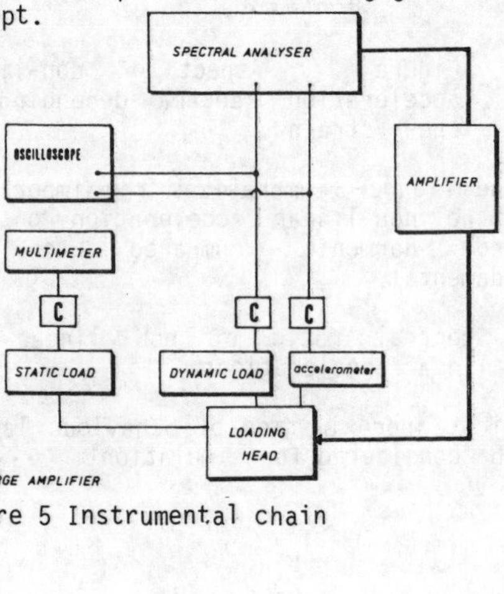

SPECTRAL ANALYSER
OSCILLOSCOPE
AMPLIFIER
MULTIMETER
C
C
C
STATIC LOAD
DYNAMIC LOAD
accelerometer
LOADING HEAD
C : CHARGE AMPLIFIER

Figure 5 Instrumental chain

7 COMPUTATION OF WAVEFIELD GENERATED DURING TESTS

Nearby the loading source, we must admit that soil response is in phase with the sollicitation. Then, the computation becomes quasi-static except an uniform modulation of the response, and the equations (5.1) and (5.2) are simplified. We obtain :

$$(\lambda o + Go)\, u_{j,ji} + Go\, u_{i,jj} = F_i \quad (7.1)$$

with :

$$F_i = sgn(I_2) . |I_2|^{-1/2} \left\{ I_{2,i} . \varepsilon_{jj} + I_{2,j} . \varepsilon_{ij} \right\} G_1 \quad (7.2)$$

The linear elastic solution is obtained from the elastic solutions given by GERRARD and HARRISON (6). The evaluation of the non-linear adjustment requires the computation of non-linear sources and the expression of GREEN's tensor in axisymmetric outline (7) (8) to apply a superposition principle such as :

$$u_{i\,NL}(x) = \int_{\Omega_\xi} u^*_{ij}(x;\xi)\, F_j(\xi)\, d\Omega_\xi$$

u^*_{ij} is the GREEN's displacement. It is corresponding to the elementary displacement created at a point x in the direction i and due to an unit load acting at a point ξ in the direction j. $u_{i\,NL}$ is the displacement at the point x in the direction i consequent upon the compound of the displacements created by a bulk load Fj acting within the non-linear domain Ω_ξ . Note that the GREEN's components are expressed in function of the first, second and third complete elliptic integrals (9).

8 COMPUTATION PROGRAM

The computation program executes a numerical integration. The discretization of Ω_ξ is classical because it consists in the generation of a mesh including N quadrilateral elements (see figure 6).

A Gaussian quadrature is applied to every element. It is important to note that a singular integration is needed near by the receipt point (if this one is inside Ω_ξ). The applied method is based to

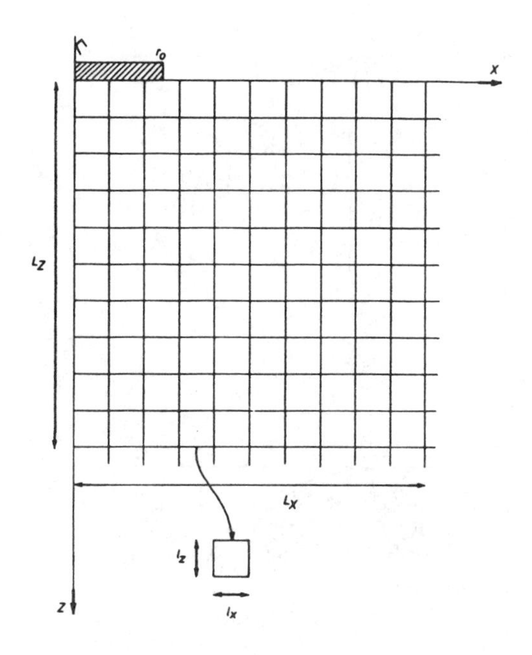

Figure 6 Mesh and Parameters of discretization

MANG's transform (10), that equalizes the singular function (in $1/_R$) by a geometrical transform.

An originality of the program is the computation of Fj terms by a subroutine written from REDUCE Language. This last one executes a formal computation of heavy expressions. It permits to express I_2 and $\vec{grad}\, I_2$ in an analytical form.

First, we deal with a direct problem :

$$U_{z\,NL}(x) = G_1 . \left[\int_{\Omega_\xi} u^*_{zj}(x;\xi)\, F_j(\xi)\, d\Omega_\xi \right] \quad (8.1)$$

by introducing G1 obtained from α_1(see § 2) and an experimental value of Go.

This computation permitted to test the stability of results by the variation of discretization parameters and its sensitivity with respect to the variation of experimental parameters. A correlation between the direct result and the estimation of the effective non-linear acceleration measured in situ is presented in the following part.

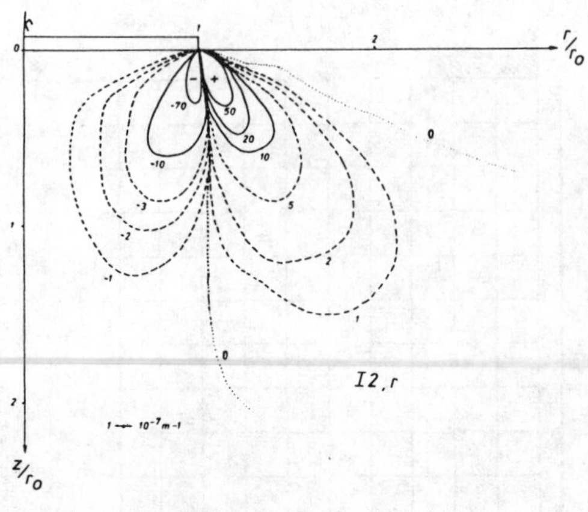

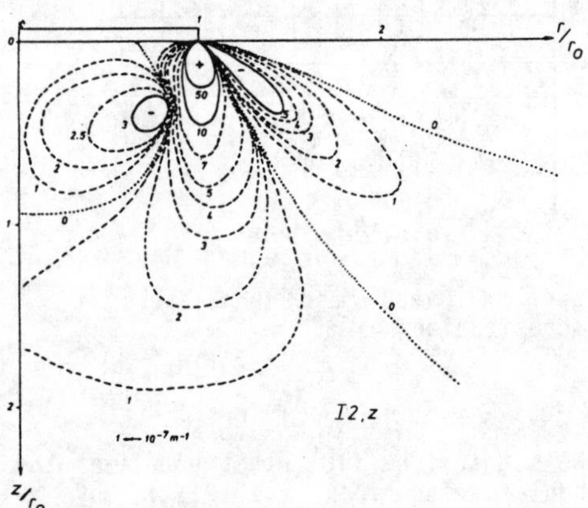

Figure 7 ($\vec{grad}$ I2) values within the domain

9 NUMERICAL RESULTS

The variation of integration domain extent L and that one of the element size l allow to access to an optimal discretization. This last one is a compromise between the accuracy of results and the computation cost if we tolerate an error below 15 %.

It is logical to think that the best result is obtained from Lmax and lmin and consequently, a maximal elements number. Indeed, $\vec{grad}$ I2 included in the function which is integrated (mapped figure 7) often presents polarity changes. That is why it is forbidden to too minimize L and increase l. We must consider :

$$L \geqslant 5 \quad ro$$
$$l \leqslant 0.5 \; ro$$

with ro, piston radius used as unit for discretization.

Different cases submitted to computation give the results presented in the following tables. (L and l expressed per unit of ro), and on the figure 8.

Table 2 Variation of L extent (lx = 0.4 and lz = 1.0)

Lx	Lz	U_z NL (m)	error
4	10	$5.04 \; 10^{-6}$	___
8	10	$4.86 \; 10^{-6}$	3.6 %
4	20	$4.97 \; 10^{-6}$	1.4 %

Table 3 Variation of lz extent (Lx = 4.0, lx = 0.4 and Lz = 10.0)

lz	U_z NL (m)	error
2	$4.35 \; 10^{-6}$	16.1 %
1	$5.04 \; 10^{-6}$	30.7 %
0.5	$3.85 \; 10^{-6}$	4.0 %
0.4	$3.70 \; 10^{-6}$	___

Table 4 Variation of L extent and lz with lx = 0.4

Lz	lz	Lx	U_z NL (m)	error
8.0	0.5	4.0	$3.97 \; 10^{-6}$	5.7 %
8.0	0.5	4.8	$3.88 \; 10^{-6}$	3.4 %
8.0	0.4	4.8	$3.75 \; 10^{-6}$	___
10.0	0.5	4.0	$3.85 \; 10^{-6}$	2.6 %
10.0	0.5	4.8	$3.68 \; 10^{-6}$	1.9 %

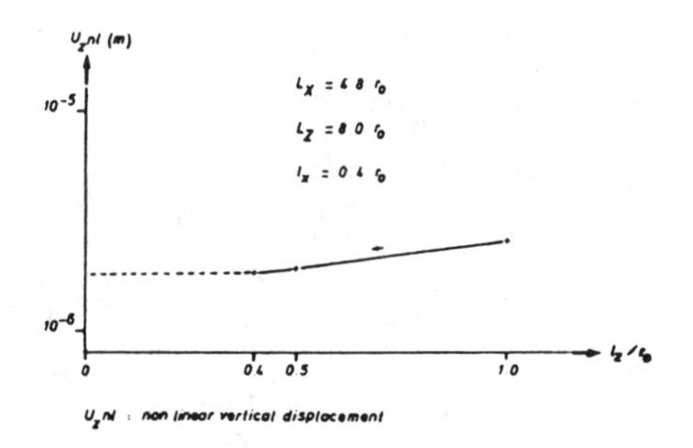

Figure 8 Focusing of results

If we tolerate an error less than 15 %, we can reduce the computation cost by a factor equal to 3.

However, we hold the more accurate result for the following part of this work such as :

$$U_z\ NL = 3.75\ 10^{-6}\ m$$

With respect to the weighting coefficient on harmonic acceleration (see § 3.2), we finally obtain :

$$R_w = \frac{A_z NL\ (\omega_0 = w)}{A_z L\ (\omega_0 = w)} = 3.9\ \%$$

$$R_{3w} = \frac{A_z NL\ (\omega_0 = 3w)}{(A_z L + A_z NL)(\omega_0 = w)} = 6.8\ \%$$

The idea is next to consider the following behaviour law (second approach on Figure 1)

$$\frac{G}{G_0}(I_2) = 1 - \alpha_1 |I_2|^{1/2} + \alpha_2 . I_2$$

with : $\alpha_2 = \alpha_1^2/2$

Then, the previous result becomes :

$$U_z\ NL\ (wo = w) = 2.67\ 10^{-5}\ m$$

$$U_z\ NL\ (wo = 3w) = 8.40\ 10^{-5} m$$

and consequently :

$$R_w = 32.9\ \%\ ;\ R_{3w} = 77.8\ \%$$

This last result can explain the experimental spectra got in situ, agreed with the fact that the α_1-value does not correspond with the effective value of the non-linear parameter of the experimental soil. The figures 9 and 10 allow to compare the computation and experimental results.

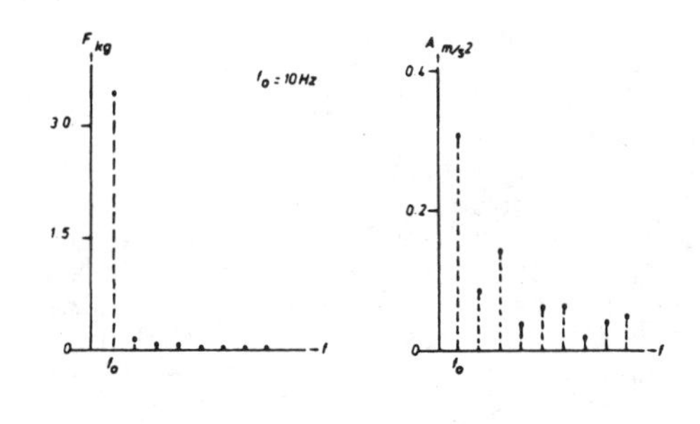

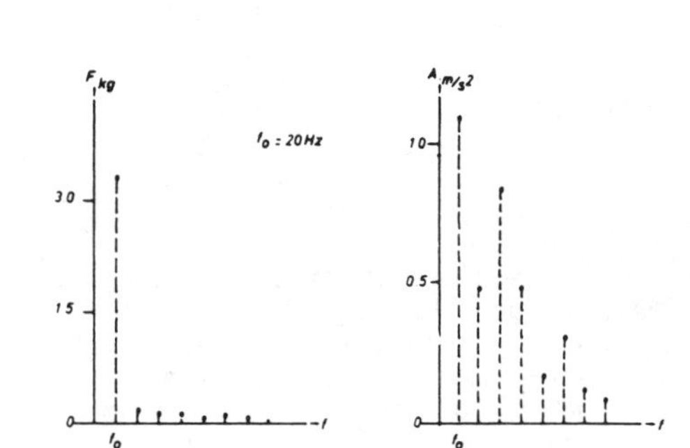

Figure 9 Experimental spectral accelerations

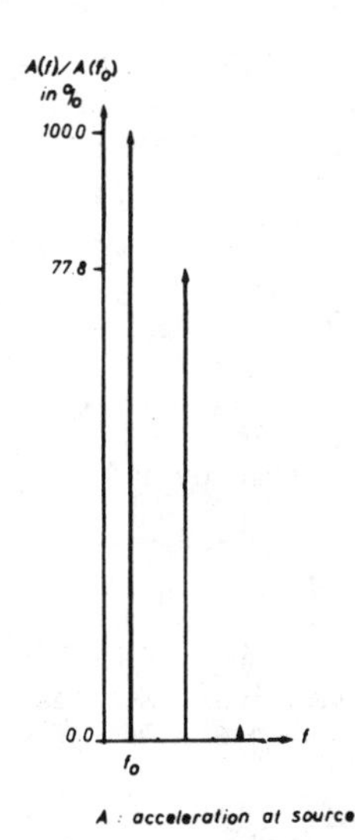

Figure 10 Numerical spectral accelerations

We note common aspects between these two figures, especially concerning the ratio between the third harmonic acceleration and the fundamental one.

At the present time, we cannot deal with the inverse problem. We are waiting more experimental results for this purpose and we hope to make the behaviour law more accurate. That is we have done for the last result with respect to the fact that taking a longer development in series of hyperbolic model leads to a more accurate result and an extent of the modeling validity.

10 CONCLUSION

The development in truncated series of hyperbolic model allows to study the spectral effects of non-linear soil behaviour. The interest of a such work is obvious. It permits to modelize properly an acceleration response spectrum.

The experimental test and its interpretation define a characteristic parameter of non-linear behaviour from a semi-analytical computation. This one minimizes the classical errors of numerical methods. By a restricted computation timing and a certain stability, this interpretation shows its efficiency. Consequently, its adjustment to another tests (as Cross-hole test) is possible and must give good results.

REFERENCES

(1) H.B. Seed and I.M. Idriss, Soil moduli and damping factors for dynamic response analysis, Earthq. Eng. Res. Cent., Rep. N° 70-10, Univ. of California, Berkeley (1970).

(2) H. Modaressi, D. Aubry and P. Mouroux, Wave propagation in a saturated porous media, Proceedings of the 8th European Conference on Earthquake Engineering, Section Analytical Methods in Soil Dynamics, Lisbonne (1986).

(3) H.B. Seed, K.L. Lee ; I.M. Idriss and F. Makdisi, Analysis of the slides in the San Francisco Dams during the Earthquake of Feb. 9, 1971, EERC Rep. N° 73-2, Univ. of California, Berkeley, (1973).

(4) B.O Hardin and V.P. Drnevich, shear modulus and damping in soils I. Measurement and Parameter Effects, II. Design equations and curves, Techn. Rep. Univ. Kentaky UKY, N° 26 and 27-70 - CE2, Soil Mech. series (1970).

(5) A. Blaquière, Analyse des systèmes non-linéaires, Chap. 2, pp. 88-94, Bibl. des Sciences et Techniques Nucléaires - INSTN (Saclay), P.U.F., Paris, 1966.

(6) C.M. Gerrard and W.J. Harrison, Circular loads applied to a cross-anisotropic half-space, in Elastic Solutions for Soil and Rock Mechanics by G. Davis and E. Poulos, Wiley and sons Ed., 1974.

(7) C.A. Brebbia, J.C.F. Telles and L.C. Wrobel, Boundary Element Techniques. Theory and applications in Engineering, Springer Verlag, Berlin, 1984.

(8) M. Abramowitz and I.A. Stegun, Handbook of mathematical functions, Dover Publications, 1968.

(9) P.F. Byrd and M.D. Friedmann, Handbook of elliptic integrals for Engineers and Physicists, Springer Verlag, Berlin, 1954.

(10) H.A. Mang, Hong Bao Li, Guo-Ming Han, A new method for evaluating singular integrals in stress analysis of solids by the direct boundary element technique, Int. J. for num. meth. in Eng. vol 21, pp. 2071-2098 (1985).

Numerical Methods in Geomechanics (Innsbruck 1988), Swoboda (ed.)
© 1988 Balkema, Rotterdam. ISBN 90 6191 809 X

Nonlinear pile driving problems calculated in the Laplace domain

P.Hillmer & G.Schmid
Ruhr-Universität, Bochum, FR Germany

ABSTRACT: In this paper a method is presented which allows to consider effects of wave propagation in the soil as well as local nonlinearities. Using a step by step method in the Laplace domain the calculation will be expanded for nonlinear pile driving problems. The nonlinear calculation will be divided in a number of linear steps, where every linear step satisfies obviously the superposition condition. The nonlinearity will be dealt with in the time domain and the results of every step will be transformed between Laplace and time domain and vice versa. Because of the large number of transformations a very fast and precisely calculating transform algorithm is needed. Therefore the Laplace transform is reduced to the Fourier transform and a FFT algorithm is used, which is undoubtedly very efficient.

1 INTRODUCTION

Interaction problems are most common solved in frequency domain, which allows to use the substructure method in dynamic analysis. That means that the substructures can be calculated separately utilizing different methods. Because of the superposition principle involved linearity has to be presumed. In case of nonlinear effects the general linear interaction technique is no longer valid. Therefore most nonlinear problems as e.g. pile driving are dealt with in time domain, but than the frequency dependent geometric damping effects of the soil cannot be considered any more.

A lot of papers have been published considering the nonlinear pile driving problem. Analytical solutions of this problem can only be found for simple cases (see [1], [2]). Following the original work by Smith [3] different numerical methods and computer programs have been produced [4],[5], [6]. Due to the nonlinearities involved in the problem nearly all of them work in the time domain. The influence of the surrounding soil is modeled by discrete spring-damper elements.
In this paper a new method for the calculation of the nonlinear pile driving effects is presented. It utilizes the Laplace transform. - In contrast to the Fourier transform the Laplace transform has the advantage of easily handling arbitrary amounts of damping. - The nonlinear effects will be dealt with in the time domain while all other calculations will take place in the Laplace domain.

2 PILE ELEMENT

Application of the Laplace transform with respect to time enables one to use the exact deformation of a bar element with continuously distributed masses as shape function in the calculation. This is in contrast to the conventional finite element method, which is based on the lumped or consistent mass representation and therefore leads to approximate solutions of the problem.
Considering a uniform elastic bar as shown in figure 1 the following partial differential equation describes the axial motions (see [10]):

$$EA\frac{\partial^2 u}{\partial x^2} = m\frac{\partial^2 u}{\partial t^2} + b\frac{\partial u}{\partial t} + cu + n(x,t) \, , \qquad (1)$$

where A is the area of the cross section, E is the modulus of elasticity, m is the mass per unit length, b is the external viscous damping of the soil, c is the external soil friction and n(x,t) is the external load distribution.

The terms $b \cdot \partial u/\partial t$ and $c \cdot u$ may model the effects of the surrounding soil.
The transformation of equation (1) into the Laplace domain will give an ordinary differential equation which is easy to solve:

$$EA\frac{\partial^2 \tilde{u}}{\partial x^2} - (ms^2+bs+c)\cdot\tilde{u} = \tilde{n}_1 + (\tilde{n}_r - \tilde{n}_1)\cdot x/L, \quad (2)$$

where $s=\delta+i\omega$ is the Laplace transform parameter and $(\tilde{\ })$ indicates variables transformed into the Laplace domain. Considering the boundary conditions and the positive directions as indicated in figure 1 one will get the following nodal force-displacement relations:

$$\begin{bmatrix} \tilde{U}_1 \\ \tilde{U}_r \end{bmatrix} = \begin{bmatrix} \tilde{F}_1 & \tilde{F}_2 \\ \tilde{F}_2 & \tilde{F}_1 \end{bmatrix} \begin{bmatrix} \tilde{u}_1 \\ \tilde{u}_r \end{bmatrix} + \begin{bmatrix} \tilde{G}_1 \\ \tilde{G}_2 \end{bmatrix}, \quad (3)$$

where:

$$\tilde{F}_1 = EA\cdot Z\cdot\frac{e^{-ZL}+e^{ZL}}{e^{ZL}-e^{-ZL}},$$

$$\tilde{F}_2 = EA\cdot Z\cdot\frac{-2}{e^{ZL}-e^{-ZL}},$$

$$\tilde{G}_1 = \frac{\tilde{n}_1\cdot(e^{ZL}+e^{-ZL})-2\tilde{n}_r}{Z\cdot(e^{ZL}-e^{-ZL})} - \frac{\tilde{n}_r-\tilde{n}_1}{L\cdot Z^2},$$

$$\tilde{G}_2 = \frac{\tilde{n}_r\cdot(e^{ZL}+e^{-ZL})-2\tilde{n}_1}{Z\cdot(e^{ZL}-e^{-ZL})} - \frac{\tilde{n}_r-\tilde{n}_1}{L\cdot Z^2},$$

$$Z^2 = (ms^2+bs+c)/EA \quad .$$

Equation (3) is valid for linear elastic material behavior which is described by HOOKE's law:

$$\sigma = E \cdot \varepsilon \quad , \quad (4)$$

where σ is the stress and ε the strain. Transforming equation (4) into the Laplace domain will give:

$$\tilde{\sigma} = \tilde{E} \cdot \tilde{\varepsilon} \quad . \quad (5)$$

With the help of the correspondence principle [13] it is possible to incorporate general viscoelastic models by replacing the real constant E by a corresponding complex value $\tilde{E}$:

$$\tilde{E} = \frac{q_o+q_1 s+q_2 s^2+\ldots}{1+p_1 s+p_2 s^2+p_3 s^3+\ldots} \quad , \quad (6)$$

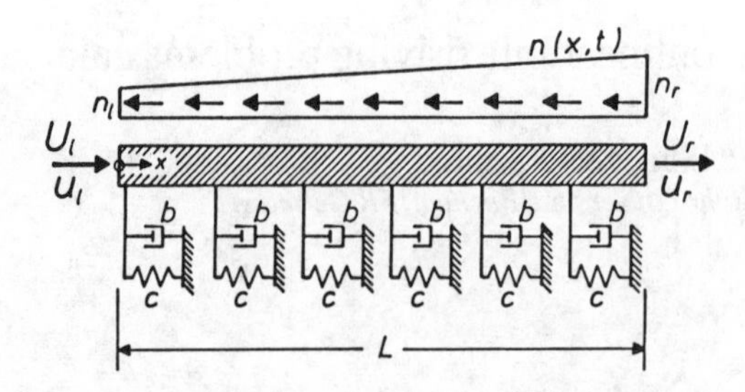

Figure 1. Uniform elastic bar element with positive nodal displacements and forces

where $q_o,\ldots,q_n$ and $p_1,\ldots\ldots,p_m$ are real constant values.

3 SYSTEMS WITH LOCAL NONLINEARITIES

Due to the superposition principle involved it seems at first sight impossible to analyse nonlinear problems with the integral transform method. But we can, as most time integration methods do, formulate the nonlinear behavior in an incremental way. For every increment the linearity is satisfied. Within the increment the integral transform method can be applied. The instant of time, when the system will change its linear characteristic, has to be determined in the time domain. The calculation of nonlinear structures in the Laplace domain can be performed by a single step or a total step method [7]. In this paper the total step method shall be used, which has been proved to be advantageous for these problems [8]. Two incremental steps of this method shall now demonstrate this technique:

-First step in the Laplace domain:

Transform of the dynamic loads $\{f(t)\}$ will give $\{\tilde{f}\}_o$.

$$[\tilde{A}]_o \; \{\tilde{x}\}_o = \{\tilde{f}\}_o \quad (7)$$

Solution of (7) will give $\{\tilde{x}\}_o$.

Transformation will give $\{x(t)\}_o$.

-Second step in the Laplace domain:
System changes its characteristic at t_k
(The following is valid only for $t>t_k$)

$$[\tilde{A}]_1 \; \{\tilde{x}\}_1 = \{\tilde{f}\}_1 \quad (8)$$

Using the increments: $\{\Delta\tilde{f}\}_o = \{\tilde{f}\}_1 - \{\tilde{f}\}_o$,

$\{\Delta\tilde{x}\}_o = \{\tilde{x}\}_1 - \{\tilde{x}\}_o$,

$[\Delta\tilde{A}]_o = [\tilde{A}]_1 - [\tilde{A}]_o$,

and taking (7) into account will give:

$$[\tilde{A}]_1 \{\Delta\tilde{x}\}_o = \{\Delta\tilde{f}\}_o - [\Delta\tilde{A}]_o \{\tilde{x}\}_o \quad (9)$$

Solution of (9) will give $\{\Delta\tilde{x}\}_o$.

Transformation will give $\{\Delta x(t)\}_o$.

-Superposition to the total solution after two steps:

$$\{x(t)\}_1 = \{x(t)\}_o + \{\Delta x(t)\}_o \quad (10)$$

In figure 2 the algorithm is presented in general form in a flow chart.

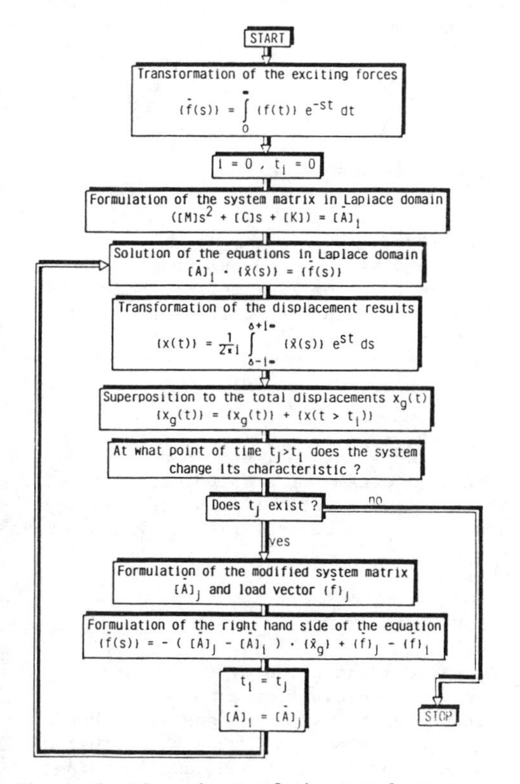

Figure 2. Flow chart of the total step method

4 NONLINEARITIES DUE TO PILE DRIVING

4.1 Soil modeled as distributed spring-damper system

In connection with the nonlinearities, which will appear at the pile skin during pile driving, the following assumptions will be made: The pile itself will stay in the linear range only. The surrounding soil shall be modeled as a continuously distributed spring-damper system (see figure 1). The damper is linearly viscous. The spring represents the external friction effects which follow the bilinear model of figure 3. The same assumptions are usually made in the time integration methods [5], [9].

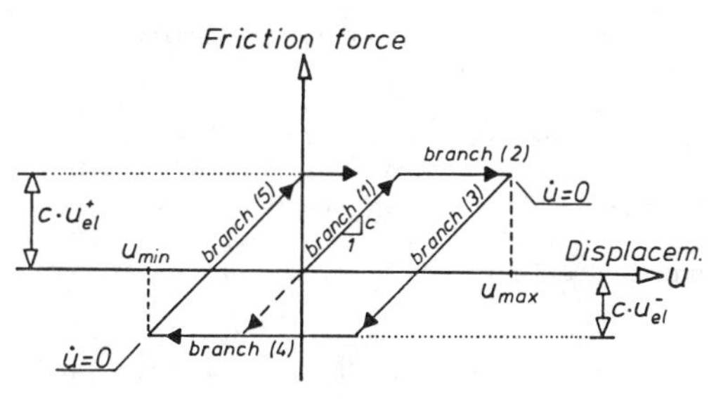

Figure 3. Bilinear model for the skin friction

The inclusion of the bilinear model (figure 3) into equation (2) results for every branch in a different differential equation:

$$\frac{\partial^2\tilde{u}}{\partial x^2} - \frac{ms^2}{EA}\cdot\tilde{u} - \frac{bs}{EA}\cdot\tilde{u} - \tilde{R}$$

$$= \frac{1}{EA}\cdot\{\tilde{n}_l+(\tilde{n}_r-\tilde{n}_l)\cdot x/L\}, \quad (11)$$

where:

in branch (1): $\tilde{R} = \frac{c}{EA}\cdot\tilde{u}$

in branch (2): $\tilde{R} = \frac{c}{EAs}\cdot u^+_{el}$

in branch (3): $\tilde{R} = \frac{c}{EA}\cdot\tilde{u} + \frac{c}{EAs}\cdot(-u_{max}+u^+_{el})$

in branch (4): $\tilde{R} = \frac{c}{EAs}\cdot u^-_{el}$

in branch (5): $\tilde{R} = \frac{c}{EA}\cdot\tilde{u} + \frac{c}{EAs}\cdot(-u_{min}+u^-_{el})$

where u^+_{el} and u^-_{el} are the maximal elastic displacements in positive and negative direction, respectively; u_{max} and u_{min} are the maximal and minimal displacements, respectively.

In contrast to the real situation the change of characteristics in our model (change of differential equation (11)) can only take place for the whole pile element at once. To determine the instant of time, when the system changes its characteristics, a mean value calculated from the displacements at the element ends will be used:

$$\bar{u}(t) = 0.5 \cdot (u_l(t) + u_r(t)) \quad (12)$$

For the transition e.g. from branch (1) to branch (2) it will be checked:

$$\bar{u}(t) = 0.5 \cdot (u_l(t)+u_r(t)) > u^+_{el} \quad (13)$$

The conditions for the other transitions can be derived easily using figure 3.

4.2 Soil modeled as boundary element substructure

The Laplace transform technique allows to use the substructure method for dynamic problems. When we use the boundary element method to model the soil, the geometric damping effects are implicitly considered.

The considered pile-soil system is assumed to be axi-symmetric. The unknown displacements are at the ends of every pile element. The soil will be modeled with constant boundary elements. Pile elements and boundary elements have the same length. The dynamic equations of the discrete soil model are:

$$[\tilde{K}^B] \{\tilde{u}^B\} = \{\tilde{P}^B\} \quad , \quad (14)$$

where $[\tilde{K}^B]$ is the dynamic stiffness matrix of the soil in the Laplace domain, $\{\tilde{u}^B\}$ are the displacements at the soil surface and $\{\tilde{P}^B\}$ are the forces at the soil surface.
We assume that only vertical degrees of freedom result from equation (14). For the coupling of the pile and the soil model we need identical degrees of freedom. The transformation of end nodes (pile model) to mid nodes (soil model) is given by:

$$\{\tilde{u}^B\} = [T] \{\tilde{u}^P\} \quad , \quad (15)$$

where $\{\tilde{u}^B\}$ is the vector of displacements at the element centres, $\{\tilde{u}^P\}$ is the vector of displacements at the element ends and [T] is the transformation matrix, which may e.g. be given using the assumption that the value at the element centre is the mean value of the ones at the respective element ends. Employing the principle of virtual work and coupling of the two models will result in:

$$([\tilde{K}^S]+[T]^T [\tilde{K}^B] [T]) \{\tilde{u}^P\} = \{\tilde{P}\} \quad , \quad (16)$$

where $[\tilde{K}^S]$ is the stiffness matrix of the pile and $\{\tilde{P}\}$ is the vector of external loads at the element end nodes.
With this method compatibility between soil and pile is satisfied only at the node points. For the nonlinear effects a similar model as shown in figure 3 can be used (for more details see [8]).

5 NUMERICAL LAPLACE TRANSFORM

A lot of papers have been published comparing different numerical methods (e.g. [10], [11]). Nearly all of them concluded that those methods will give the best results which work with complex data.
In our paper the method which refers the Laplace transform to the Fourier transform will be used:

$$L(f(t)) = \tilde{f}(s) = \int_{t=0}^{\infty} f^*(t)\, e^{-i\omega t}\, dt \quad , \quad (17)$$

where $f^*(t) = f(t)\, e^{-\delta t}$ when $t \geq 0$
$= 0$ when $t<0$

(see [12]).
To reduce the existing numerical errors the time function will be smoothed by multiplying the Fourier coefficients with the following factors:

$$\sigma_k = (\sin k2\pi/N)/(k2\pi/N) \quad , \quad (18)$$

where $k = -N,\ldots\ldots,N$.
For the constant real part δ of the Laplace transform parameter a value of about $\delta=8/T$, where T is the total time space, has been proved to produce best results (for details see [8]).

6 NUMERICAL EXAMPLE

For the demonstration of the presented method an example from a paper by Klingmüller [5] will be considered here. The pile is shown in figure 4 together with the existing friction and damping values. The load function is bell shaped and defined by the given formula. The pile model consists of four pile elements and one additional element modeling the toe resistance. The surrounding soil is modeled with continuously distributed spring-damper elements as shown in figure 1. For the numerical calculation the following values

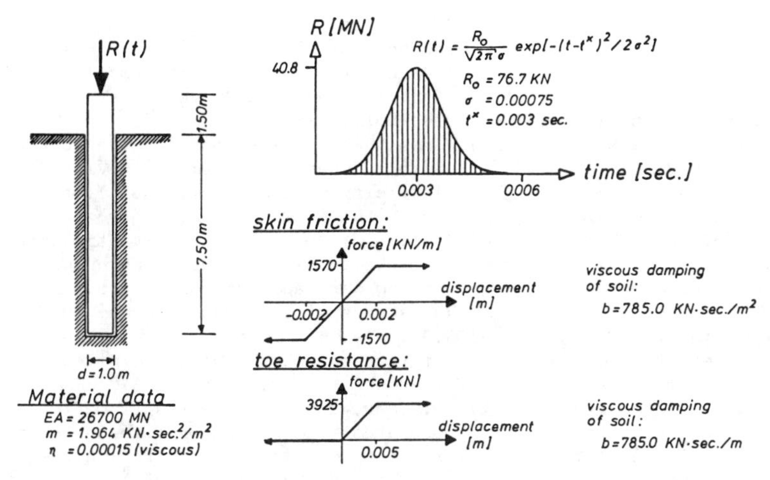

Figure 4. Nonlinear pile driving problem

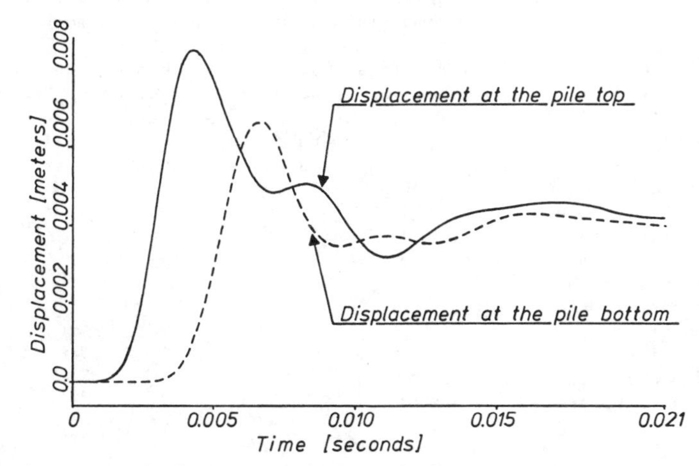

Figure 5. Displacement function of pile top and toe

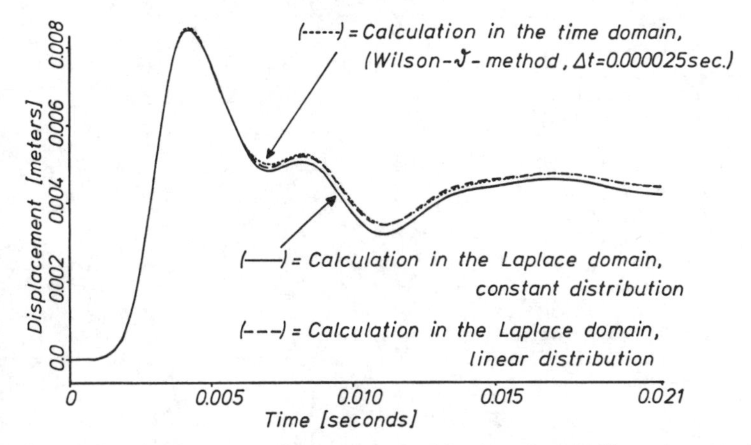

Figure 6. Displacement function of pile top for different methods

have been chosen:

$s = \delta + i \cdot k \cdot \Delta\omega$ (k=0,1,....,128)
$\delta = 8 / T = 300$
$\Delta\omega = 245.44$ rad/sec (frequency step size)
$T = 0.0256$ sec (total time space)
$\Delta t = 0.0001$ sec (time step size)

Results of the numerical calculation are presented in the figures 5 and 6. Figure 5 shows the displacement function of the pile top and toe. In figure 6 the displacement function of the pile top is presented calculated with different methods. The results of our method are presented for the assumption of constant ($n_l = n_r$) and also of linear distributed external loads over the element length. For the calculation with the Wilson-θ method 21 pile elements have been used. The good agreement of the results demonstrates the applicability of our method.

7 CONCLUSIONS

The presented method allows to use the Laplace transform method for nonlinear pile driving problems. Together with the substructure method the boundary element method can be chosen to model the soil to represent the radiation damping effects of the ground. The exact deformation of the longitudinal vibrating bar can be incorporated into the calculation. Therefore, in contrast to the time domain methods, the length of the pile element has not to be subdivided to increase accuracy. Different viscoelastic structural damping models may be considered with the help of the correspondence principle.

REFERENCES

[1] R.W. Clough and J. Penzien, Dynamics of structures, McGraw Hill, Tokio, 1975
[2] B. Hansen and H. Denver: Wave equation analysis of a pile - an analytic model, Seminar on the Application on Stress-Wave-Theory on piles, Stockholm, 1980
[3] E.A.L. Smith: Pile-Driving Analysis by the Wave Equation, Journal of the Soil Mech. and Found. Div., 86, 35-61 (1960)
[4] J.E. Bowles, Analytical and Computer Methods in Foundation Engineering, McGraw Hill, Tokio, 1974
[5] O. Klingmüller: Computational Tools for Dynamic Pile Testing, Sec. Int. Conf. on the Appl. of Stress-Wave-Theory on Piles, Stockholm, 1984
[6] G.G. Goble, F. Rausche, G.E. Likins: The analysis of pile driving - A state-of-the-art, Seminar on the Application of Stress-Wave Theory on Piles, Stockholm, 1980
[7] W. Matthees: A strategy for the solution of soil dynamic problems involving plasticity by transform, Int. J. for Num. Meth. in Eng., 18, 1601-1611 (1982)
[8] P.Hillmer, Berechnung von Stabtragwerken mit lokalen Nichtlinearitäten unter Verwendung der Laplacetransformation, Diss. Ruhr-Universität Bochum, Bochum 1987
[9] G.G.Goble, K.Fricke, G.E.Likins: Driving stresses in concrete piles, J. Prestr. Concr. Inst. 21, 70-88 (1976)
[10] G.V.Narayanan, D.E.Beskos: Numerical operational methods for time-dependent linear problems, Int. J. for Num. Meth. in Eng. 18, 1829-1854 (1982)
[11] B.Davies, B.Martin: Numerical inversion of the Laplace transform: a survey and comparison of methods, J. of Comp. Phy. 33, 1-32 (1979)
[12] W.Krings, H.Waller: Numerisches Berechnen von Schwingungs- und Kiechvorgängen mit der Laplace-Transformation, Ingenieur-Archiv 44, 335-346 (1975)
[13] W.Flügge, Viscoelasticity, Springer Verlag, Berlin, 1975

Numerical Methods in Geomechanics (Innsbruck 1988), Swoboda (ed.)
© 1988 Balkema, Rotterdam. ISBN 90 6191 809 X

Static and dynamic analysis of moderately thick plates on Pasternak foundation using classical plate theory

Hans Irschik & Rudolf Heuer
Technical University of Vienna, Austria

ABSTRACT: Forced vibrations of moderately thick plates on two-parameter, Pasternak-type foundations are considered. Influence of plate shear and rotatory inertia are taken into account according to Mindlin. Formulation is in the frequency domain. An analogy to thin plates without foundations is given. This analogy to classical plate theory is complete in the case of polygonal plan-forms and hinged support conditions. A further reduction to membrane-type solutions is shown, and numerical results for trapezoidal plates are derived using an advanced BEM.

1 BASIC EQUATIONS AND ANALOGY

This paper is concerned with linear elastic plates of moderate thickness. The influence of shear and rotatory inertia is taken into account according to Mindlin's theory /1/. (Analogous theories, see Reissner /2/, can be used by changing coefficients in the governing differential equations.)

With respect to Soil-Mechanics, we consider a foundation of the plate domain according to Pasternak's two-parameter model, which contains the widely used Winkler's model as a special case. (Following Vlasov, Leont'ev /3/, Pasternak's model corresponds to a consistent Ritz-type discretization of a linear elastic soil strip.)

Furthermore, we consider time-harmonic lateral loadings of the form $p.\exp(i\omega t)$ with amplitude p and forcing frequency ω. Considering steady-state vibrations, we calculate the undamped frequency response functions for the plate deflection, where the static case is the limit $\omega \to 0$. (The damped complex frequency response functions are derived by subsequent implementation of light damping, and they enter the FFT-algorithm for the transient response. This numerically advantageous frequency domain strategy is described in Ziegler /4/, Irschik, Heuer, Ziegler /5/, and is not covered in this paper.)

Our investigation is started by reducing Mindlin's sixth-order system of differential equations to a fourth-order equation for the deflection, see /1/ and Irschik /6/, /7/ for special cases. Replacing, e.g., the lateral loading in /1/ by p and the reaction forces of the foundation, and omitting the term $\exp(i\omega t)$, renders:

$$\bar{K}\Delta\Delta w - \bar{n}\Delta w - \bar{\mu} w = \bar{q} - \bar{K}(1+\nu)\Delta\bar{\kappa} \quad (1)$$

with

$$\bar{K} = K(1+es),\ K = Gh^3/6(1-\nu), \quad (2)$$

$$\bar{n} = e(1-sr\omega^2) - K(\rho h\omega^2 - d)s - r\omega^2, \quad (3)$$

$$\bar{\mu} = (\rho h\omega^2 - d)(1-sr\omega^2), \quad (4)$$

$$\bar{q} = p(1-sr\omega^2), \quad (5)$$

$$\bar{\kappa} = ps / [(1+es)(1+\nu)], \quad (6)$$

where:

- w... amplitude of deflection,
- h... constant plate thickness,
- G,ν ... elastic moduli of the plate,
- d, e ... elastic moduli of the (homogeneous) foundation, d denotes Winkler's modulus,
- $s = 1/\kappa^2 Gh$ ... influence of plate shear,
- ρ... (effective) mass density of the plate,
- $r = \rho h^3/12$... influence of rotatory inertia,
- Δ... Laplace's operator.

Note that $\bar{K}$, $\bar{n}$ and $\bar{\mu}w$ in equ. (1) correspond to bending stiffness, hydrostatic (tensile) in-plane force and inertia forces of a fictitious Lagrange-Kirchhoff plate without foundation of the plate domain. Furthermore, the loading side corresponds to a lateral loading $\overline{q}$, equ. (5), accompanied by a fictitious imposed thermal curvature of amplitude $\overline{\kappa}$, Eq. (6), of this Lagrange-Kirchhoff plate. (In the classical Lagrange-Kirchhoff theory of thin plates the influence of rotatory inertia and shear is neglected, see Ziegler /8/, 279-285.)

2 A COMPLETE ANALOGY FOR POLYGONAL PLATES WITH HINGED SUPPORTS

The above analogy between the Mindlin-Pasternak problem and the Lagrange-Kirchhoff-type field equation (1) gives a convenient tool for derivation of particular solutions using classical plate theory. The analogy is complete, however, in the case of plates with polygonal planforms and hinged support conditions. This has been shown in /6/ for the special case $\omega=0$, d=0, e=0, and is extended to forced steady state vibrations of plates on a Pasternak foundation in the following:

Within the theory of shear-deformable plates, hinged supports conveniently may be modelled in the form, see /1/:

$$C : w=0, \psi_s=0, m_n=K(\psi_{n,n}+\nu\psi_{s,s})=0 \quad (7)$$

with rotations ψ, bending moments m and a local (n,s)-coordinate system at boundary C with normal (n). Restricting to polygonal contours C, the second of equs. (7) can be replaced by $\psi_{s,s}=0$ and, thus, the third one by $\psi_{n,n}=0$.

The equation of momentum for a differential plate element leads to

$$q_{n,n}+q_{s,s}=(-p+dw-e\Delta w)-\rho h\omega^2 w, \quad (8)$$

where the shearing forces are given by

$$q_n=(\psi_n+w_{,n})\kappa^2 Gh, \text{ etc.} \quad (9)$$

Evaluating equ. (8) at the boundary C finally leads to two boundary conditions in w:

$$C : w=0, \Delta w=-\overline{\kappa}(1+\nu) \quad (10)$$

Equ. (1) together with boundary conditions (10) form a complete boundary value problem. Moreover, equs. (10) correspond to the boundary conditions of a polygonal Lagrange-Kirchhoff plate with hinged supports and imposed thermal curvature $\overline{\kappa}$, compare Irschik /9/. This, however, had to be shown for completeness.

3 USE OF MEMBRANE-TYPE SOLUTIONS

The complete analogy between the Mindlin-Pasternak problem of sec. 2 and the corresponding Lagrange-Kirchhoff plate with imposed loadings $\overline{q}$, $\overline{\kappa}$ and $\overline{n}$ enables a calculation of w using standard Langrange-Kirchhoff-type routines. (Analogous considerations concerning bending moments and shearing forces are left to an additional investigation; see /6/ for the special case $\omega=0$, d=0, e=0.)

Subsequently, however, we apply a classical method in the theory of thin plates, where the total deflection is decomposed into a set of two second order problems corresponding to deflections of prestressed linear elastic membranes. For the static case, this tradition dates back to Marcus /10/. For the eigenvalue problem associated with equ. (1), p=0, a decomposition using Helmholtz-equations was given by Federhofer /11/. This type of decomposition has been used in /7/ for an investigation of various eigenvalue problems of initially stressed Mindlin plates with a Pasternak foundation of the plate domain. In the following, we give an extension to the case of steady-state forced motions, $p \neq 0$. (The case of imposed support motions, but p=0, d=0, e=0, has been treated in /5/ analogously. Leaving the plate edges fixed, this case of support motion is not incorporated in the present paper for the sake of brevity.)

In extension of the homogeneous cases treated in /7/, we seek a solution of the boundary value problem eqs. (1), (10) in the form

$$w=w_1+w_2, \quad (11)$$

where the w_j are solutions of the inhomogeneous Helmholtz-Klein-Gordon equations

$$\Delta w_j+\alpha_j w_j=-\alpha_j\vartheta_j p \quad (12)$$

with

$$\alpha_j^2=-[\overline{n}\mp(\overline{n}^2+4\overline{K}\,\overline{\mu})^{\frac{1}{2}}]/2\overline{K},\ j=1,2. \quad (13)$$

Note that

$$\alpha_j^2+(\overline{n}\alpha_j-\overline{\mu})/\overline{K}=0, \quad (14)$$

$$\alpha_1+\alpha_2=-\overline{n}/\overline{K}. \quad (15)$$

Inserting equs. (11), (12) into equ. (1), it is seen that the coefficients of w_1, w_2 cancel identically due to equs. (14), while the vanishing of the coefficients of p and Δp lead to two equations for the unknown ϑ_j:

$$\alpha_1^2\vartheta_1+\alpha_2^2\vartheta_2+\overline{n}[\alpha_1\vartheta_1+\alpha_2\vartheta_2]/K=\delta, \quad (16)$$

$$\alpha_1\vartheta_1+\alpha_2\vartheta_2=\varepsilon, \quad (17)$$

with

$$\delta = (1-sr\,\omega^2)/\bar{K}, \quad \varepsilon = s/(1+es). \qquad (18)$$

Inserting equs. (11), (12) into the boundary conditions (10) gives

$$C: w_1 + w_2 = 0, \qquad (19)$$

$$-\alpha_1 w_1 - \alpha_2 w_2 - (\alpha_1 \vartheta_1 + \alpha_2 \vartheta_2)p = -\varepsilon p. \qquad (20)$$

Equ. (20) coincides with equ. (17) if both w_1 and w_2 vanish at C:

$$C: w_j = 0,\ j=1,2. \qquad (21)$$

Hence, the Mindlin-Pasternak problem (1) with (inhomogeneous) boundary conditions (10) is governed by the two second-order Helmholtz-type equations (12) with Dirichlet's conditions (21), where the coefficients ϑ_j follow from equs. (16), (17):

$$\alpha_j \vartheta_j = \pm(\delta + \alpha_j \varepsilon)/(\alpha_1 - \alpha_2). \qquad (22).$$

4 NONDIMENSIONAL FORMULATION AND NUMERICAL IMPLEMENTATION USING AN ADVANCED BEM

A numerical implementation with a minimum number of inputs, is achieved by scaling w by a characteristic span a. Hence it is seen that plates of geometrically similar shape result in the same ratio $\tilde{w}/\tilde{p}$, if the following set of non-dimensional inputs is the same:

$\lambda = \omega a/c$ ... nondimensional frequency (with wave speed $c=(G/\rho)^{1/2}$),

ν ... Poisson's ratio,

κ^2... shear factor ($\kappa^2 = \pi^2/12$ in /1/),

$\tilde{h}=h/a$... thickness-to-span ratio,

$\tilde{p}_o=p_o/G$ load factor (with characteristic itensity p_o) corresponding to a linear scaling of w/a,

$\tilde{d}=da/G$, $\tilde{e}=e/Ga$... nondimensional soil parameters.

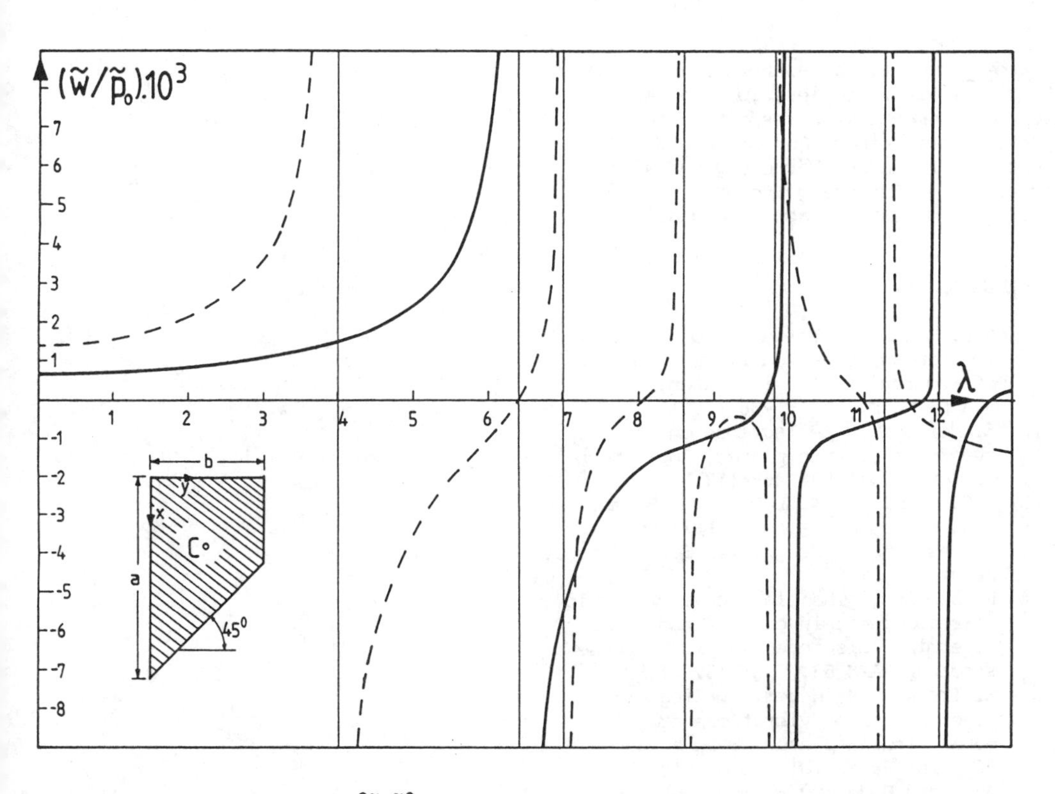

Fig. 1. $\tilde{w}/\tilde{p}_o = (w/p_o)[G(1+\tilde{e}\,\kappa^2\tilde{h})\tilde{h}^3/6(1-\nu)a]$ at a central point C(0.6b, 0.5b) as a function of $\lambda = \omega a\sqrt{\rho/G}$. $a/b = \sqrt{3}$, $\tilde{h} = 0.2$, $\kappa^2 = \pi^2/12$, $\nu = 0.3$.
—— $\tilde{d} = 0.2$, $\tilde{e} = 0.1$, - - - $\tilde{d} = \tilde{e} = 0$.

The nondimensional frequency response function is denoted by $\tilde{w}=w/a$, and the load factor $\tilde{p}=(p/G)6(1-\nu)/\tilde{h}(1+es)$, with $es=\tilde{e}/\kappa^2 \tilde{h}$).

Using this non-dimensional inputs, an advanced Boundary Element Method with Green`s functions of rectangular domains has been applied to the Dirichlet`s Helmholtz-problems of sec. 3, where right-angled trapezoidal plates are chosen as examples. (The corresponding numerical procedure, in which boundary conditions are satisfied exactly at the right-angled edges of the trapezoidal domain, is discussed in detail in /5/.)

Fig. 1 shows the ratio $\tilde{w}/\tilde{p}_0$ as a function of the dimensionless forcing frequency parameter $\lambda = \omega a \sqrt{\rho/G}$ at a central point of a right angled trapezoidal plate due to a spanwise constant time-harmonic loading with amplitude p_0. A plate without foundation, $\tilde{d}=0$, $\tilde{e}=0$, as well as the case of relatively high foundation moduli, $\tilde{d}=0.2$, $\tilde{e}=0.1$, are shown. Note the remarkable shift in the frequency content of the response.

Acknowledgement. The authors want to thank Professor Franz Ziegler, TU-Wien, Austria, for his comments with respect to the contents of this paper. Support of the Austrian "Fonds zur Förderung der wissenschaftlichen Forschung", central project S30/03 is gratefully acknowledged.

REFERENCES

/1/ R.D. Mindlin: Influence of rotatory inertia and shear on flexural motions of isotropic, elastic plates, J.Appl.Mech. 18, 31-38 (1951)

/2/ E. Reissner: Reflections on the theory of elastic plates, Appl. Mech. Rewievs 38, 1453-1464 (1985)

/3/ V.Z. Vlasov, U.N. Leonte`v: Beams, plates and shells on elastic foundations, Israel Progr. for Sci. Transl., Jerusalem, 1966

/4/ F. Ziegler: The elastic-viscoelastic correspondence in case of numerically determined discrete elastic response spectra, ZAMM 63, T135-137 (1983)

/5/ H. Irschik, R. Heuer, F. Ziegler: Free and forced vibrations of polygonal Mindlin-plates by an advanced BEM. In: Proc. IUTAM-Symposium Advanced Boundary Element Methods, San Antonio 1987, Springer-Verlag, Berlin-New York, in press

/6/ H. Irschik: Eine Analogie zwischen Lösungen für schubstarre und schubelastische Platten. ZAMM 62, T129-T131 (1982)

/7/ H. Irschik: Membrane-type eigenmotions of Mindlin plates, Acta Mechanica 55, 1-20 (1985)

/8/ F. Ziegler: Technische Mechanik der festen und flüssigen Körper, Springer-Verlag, Wien-New York, 1985

/9/ H. Irschik: Zur Berechnung thermisch beanspruchter dünner linear elastischer Platten, ZAMM 61, T97-99 (1981)

/10/ M. Marcus: Die Theorie elastischer Gewebe und ihre Anwendung auf die Berechnung biegsamer Platten, Springer-Verlag, Berlin, 1924.

Numerical Methods in Geomechanics (Innsbruck 1988), Swoboda (ed.)
© 1988 Balkema, Rotterdam. ISBN 90 6191 809 X

Practical considerations of blast loading in tunneling using numerical modeling

G.Swoboda & G.Zenz
University of Innsbruck, Austria

ABSTRACT: The loosening caused by blasting during driving, that is of such importance in tunneling, is studied with the help of a numerical model. An attempt is made to analyze the propagation of the loaded blast hole's detonation wave, which is then dealt with in a numerical mode. This analytical method is compared with measurements. The pressure wave's propagation is followed with the help of a dynamic finite element analysis, and the plastic zones thus occurring are determined in various steps. The influence exerted by rock loosening are examined in a subsequent static analysis.

1. INTRODUCTION

The static analysis of tunnels in rock generally neglects the load imposed by blasting during tunnel driving. Today, the influence of time on the redistribution of stresses is largely limited to the rock's rheology, whereby these are extremely slow load functions. It is precisely the extremely short effect of the blasting load that exerts an additional force on the rock and superposes on the loading from stress redistribution in the destroyed excavation zone. The result is a plastic change in the rock properties immediately behind the face, that has a major import on further static analysis. This loosening caused by blasting, that has been a known factor to design engineers for many years, was often used to advocate mechanical tunneling without it being possible to quantify its influence. Another question to be examined with the numerical model developed here is the blasting pattern. Nowadays, the blasting pattern is designed and the necessary explosives calculated by blasting experts under consideration of many factors [1], [2], [3], with the practicability of the explosive for construction operations, the rock's stress, the rock type and jointing and also the excavation area playing an important role. On the basis of experience, the blasting pattern is adapted to the conditions prevailing at the face in order to keep the blasting operations within economic bounds. The influence that blasting during driving has on the static support effect of the tunnel's primary lining is not considered in the analysis.

This work introduces a numerical model that permits first attempts at studying the change in the rock's support effect, which results from blasting operations during tunnel driving. In order to determine the plastic zones schematically shown in Fig. 1, the influence of the blasting load $p(t)$ and the excavation load $f(t)$ are needed. The excavation load results from removal of the supporting core by blasting. The blasting load is calculated with the help of an analytical model that gives the time-modified stresses at the excavation's assumed boundaries. The further analysis is performed with a finite element mesh, whose element size agrees with the load frequency, starting with an ideal, homogeneous, isotropic material behavior. This analysis shows regions where the maximum acceptable material strength is exceeded, thereby causing the rock's support behavior to deviate from the static analysis. Preliminary observations made in this connection will be described with a simple model and illustrated in an analysis.

2. BLASTING MECHANICS OF THE BOREHOLE'S NEAR AND FAR FIELDS

The explosive undergoes a chemical change in the blast hole, producing gases that are under high pressure and at a high temperature. These pressures are so high that the material strength of the surrounding rock is surpassed by far, and a crushed zone "A" and "B", as

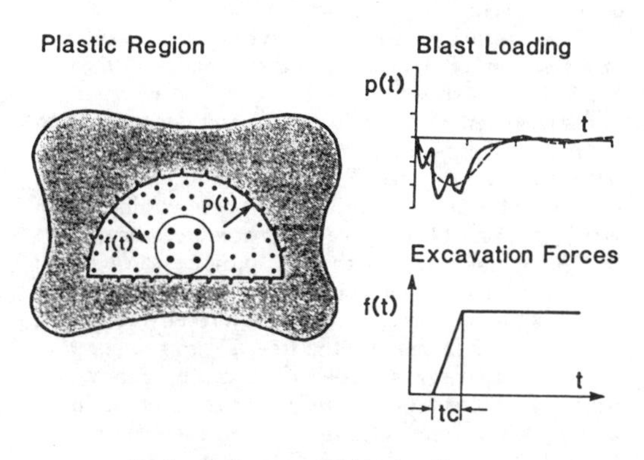

Fig.1 Influence of blast loading.

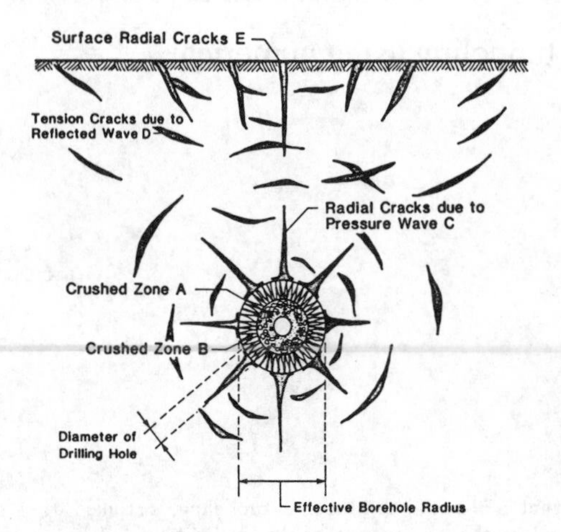

Fig.2 Failure zones around the blast hole.

shown in Fig. 2, is created [4], [5]. The larger zones A and B become, the more energy is used up in their creation and thereby lost for the actual blasting-free of rock. Too strongly crushed rock has an adverse effect on penetration into the cracks by the pressurized gases and therefore hinders the growth of cracks.

The pressure effects a stress wave that starts to propagate through the surrounding rock at a very high speed, about $4000 m/sec$ in granite. Depending on the strength of the rock, the shock energy makes up about $1/3$ of the total energy in hard rock and about $1/10$ of the total energy in soft rock. While the pressure wave propagates toward the boundary, the rock's compressive strength is no longer exceeded. In a tangential direction thereto, however, tensile stresses occur that exceed the maximum tension failure stresses, which can be between $1/50$ and $1/10$ of the compression failure strength, thus causing radial cracks C to occur. The cracks grow orthogonally to the maximum tensile strength.

The detonation front in the loaded blast hole continues to travel at detonation velocity. The detonation phase is followed by the so-called gas pressure phase that causes the radial cracks, that occurred as a result of the pressure wave, to grow.

The fracture process is influenced by the existence of a free surface, that also effects a change in the primary state of stress in the tunnel. The compressive stress wave reflects off this surface and goes back into the rock as a tensile stress wave, with fractures D from reflected waves and radial fractures E occurring on the surface. This process can also be observed during the first blast to start the excavation procedure, whereby the first firing time, namely the cut, produces an ejection cone. In all subsequent, properly arranged firing times, the blast can be directed against a further free surface and the debris forced into the free space, meaning that less blasting energy is needed. This can be clearly seen from the different maximum velocities recorded in vibration measurements for the cut and each subsequent firing time.

The numerical analysis does not take into consideration the near field, namely the immediate zone of destruction shown as the destruction caused by the propagating pressure wave. To describe this, various authors have attempted to analyze the course of pressure in the blast hole and the distribution of stress [6], [7], [8] in its surroundings, whereby the amount of stress exerted on the rock depends on the velocity at which the pressure builds up, namely the time it takes to reach maximum pressure. This is influenced by the type of explosive used and the conditions surrounding the borehole, under which the energy is converted. Another factor, one that plays an essential role in crack growth, is the spread of the pressurized gases. This can be determined in a numerical model, as for example in [9], using the appropriate flow rule.

3. BLASTING LOAD FOR NUMERICAL ANALYSIS

In order to determine the load function, the analytical model of Aimone [10] is referred to. This is based on an exponential assumption for the course of pressure after Duvall [6], that is used to find the pressure wave's course of stress. These exponential assumptions are derived from a sphere filled with a detonating explosive. The course of pressure is assumed after [10], as shown in equation (1).

$$P(t) = P_0(e^{-\alpha t} - e^{-\beta t}) \quad (1)$$

with

$$\alpha = f(n,c) \quad \beta = f(m,c) \quad (2)$$

In equation (1), P_o represents the theoretical detonation pressure that can be calculated from the explosive's characteristic values [11]. The parameters n and m are distance-dependent damping parameters that have a strong influence on the course of pressure. They are determined by measurement. The parameter c stands for the theoretical propagation velocity of the longitudinal wave, and t is the variable for time.

The zone of crush, A and B, is created in the immediate surroundings of the detonating explosive that fills a sphere, as shown in Fig. 2. The course of stress can be calculated as of a distance of about eight times the sphere's diameter. This distance corresponds to the size of the zone of crush B and is entered in the above equation (2) with its radius a. The damping of the shock wave can be obtained by introducing the distance-dependent damping parameters n and m. The sum of n spheres gives an almost cylindrical loaded blast hole. The summation of these n spheres to form a loaded blast hole and the superposition of the loaded blast holes is shown in [10]. This model can be used to obtain the course of velocity, distortion and of stress as a function of time. The assumptions made on the course of pressure and the application of distance-dependent damping parameters are confirmed by measurement and comparison calculations [12].

This model was developed to determine the rock particle size range as a function of the firing point and the blast hole pattern, in order to optimize blasting procedures with regard to the economy of rock extraction. The gas pressure phase, that is highly responsible for crack growth, was not taken into consideration. As a result, crack growth and therefore the nonlinear area around the blast hole were underestimated.

4. MEASUREMENTS AT THE MILSER TUNNEL

In the course of the construction work for the new Milser Tunnel on the A12 Inn Valley Freeway between Innsbruck and Landeck, measurements were taken during blasting operations for tunnel driving. The vibrations from two rounds were recorded at a time just before cut-through, in order to obtain readings that relate to the near field. The shortest distance between a measuring device and the blast site was $11\ m$. Until then, measurements [13] were performed at greater distances and therefore were not usable for this back analysis.

The two rounds recorded were carried out according to the blasting pattern given in Fig. 3. Twelve firing times were fired at $0.5 - sec$ intervals. The length of a round was $3.0\ m$ at this point in tunnel driving operations. A V-cut was used that consisted of six loaded blast holes fired at time zero. A total of $203\ kg$ of *Gelatine Donarit 1* was used for 100 blast holes. The blast holes on the boundary were loaded with *Gelatine Donarit 2E* for a good tunnel profile.

The measurement readings and back analysis were compared at time zero, because at this time it appeared to still be justifiable to determine the maximum velocities by means of an analytical model alone since the pressure wave can propagate in the undisturbed continuum. At any other time, wave propagation is disturbed by blasted rock. Since it was not possible to take these readings at various distances from the blast site, it was also not possible to adapt the damping parameters n and m to the prevailing rock. For this reason, the values proposed in [10] were used.

The measurement readings are shown in Fig. 4. The calculated course of the velocities is given in Fig. 5 for comparison. The relatively good correlation between the two curves was achieved by varying the firing point of the individual blast holes. This good correlation between the calculated and measured values was not able to be repeated again during the entire first blast because the damping parameters n and m were not adapted. Additional measurement values would have to be evaluated in order to arrive at a valid statement, whereby varying the damping parameters could improve correlation. This would bring the calculated values closer to the actual readings [12]. An attempt at back analysis was discontinued at this point because the range of the stresses occurring and an approximated course of the results were adequate for further numerical analysis.

The first blast to start the excavation procedure was used to evaluate and interpret the results because the correlation with all the other times was insufficient, as a curve in Fig. 6 shows. This is attributed to the fact that the $0.5 - sec$ interval between firing times was shown by the measurements to be unreliable. After comparing the measurement records and the blasting pattern, it must be assumed that various firing times overlapped. Nevertheless, the deviation from the actual firing time must have remained within practically justifiable limits, because the blasting results were good. This circumstance could be countered in future studies by recording the actual start of detonation. The prerequisites for back

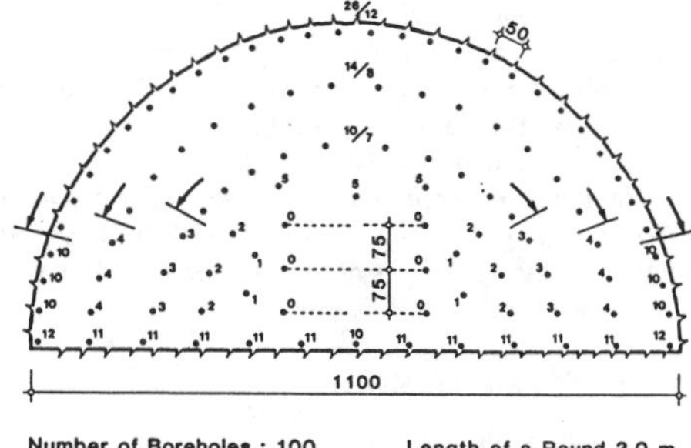

Fig.3 Blasting pattern for the Milser Tunnel.

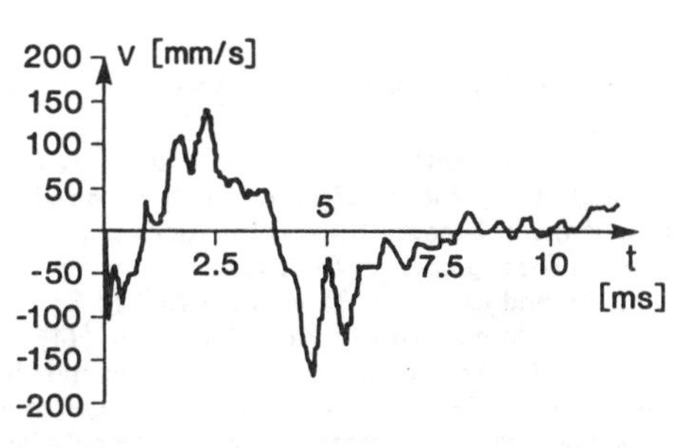

Fig.4 Measurement of course of velocity for time zero.

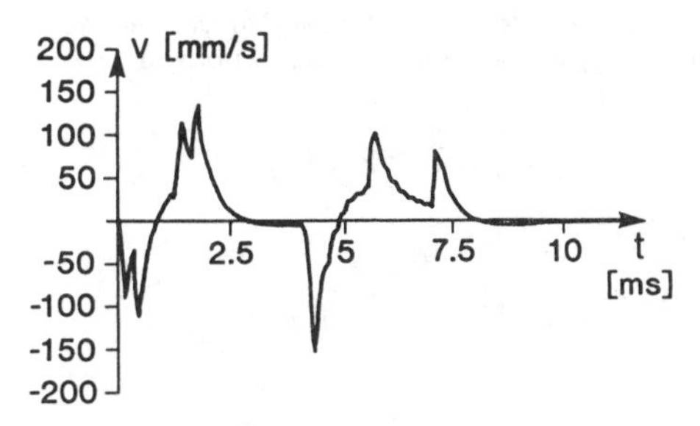

Fig.5 Analysis of course of velocity.

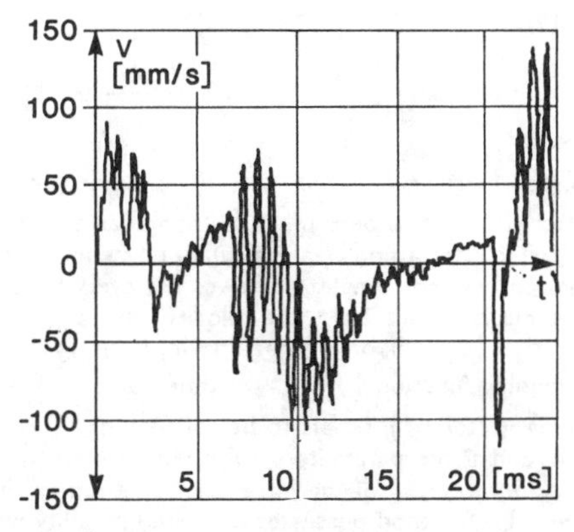

Fig.6 Measurement for an interim time.

analysis would thus be better met. When the first blast is detonated to start excavation, the ejection cone creates an additional free surface at the face. Once the first blast is made, the remaining material is blast into the free space created by it. As a result of this additional free surface, the reflections of the stress wave necessarily become stronger, which prevents back analysis with an analytical model [10], since reflections are not taken into consideration.

5. FINITE ELEMENT MODEL

5.1. Dynamic Finite Element Analysis

The relationship between the forces of a dynamically excited system is expressed by the following differential equation:

$$[M]\{\ddot{a}\} + [C]\{\dot{a}\} + [K]\{a\} = \{f\} \tag{3}$$

Here, the dynamic analysis of the structure was performed and, thus, the differential equation solved in the time domain, because these calculations permitted complex nonlinear properties to be taken into consideration. *Implicit* and *explicit* methods were studied for this purpose and incorporated into the *FE* program [14].

From the *implicit* methods, the *Newmark* and the *Wilson Theta* methods [15] were chosen. These have the advantage that results from (3) can still be obtained with even larger time increments, and that it is only for reasons of accuracy that small time steps have to be used as a function of loading and the structure in question. The *Newmark Method* [16] can be taken as an example, whereby the following set of equations has to be solved:

$$[\hat{K}]^t\{a\}^{t+\Delta t} = \{\hat{f}\}^{t+\Delta t} \tag{4}$$

with

$$[\hat{K}]^t = [K]^t + \frac{1}{\alpha\Delta t^2}[M] + \frac{\vartheta}{\alpha\Delta t}[C] \tag{5}$$

and

$$\begin{aligned}\{\hat{f}\}^{t+\Delta t} &= \{f\}^{t+\Delta t} + \\ &+ [M](\frac{1}{\alpha\Delta t^2}a^t + \frac{1}{\alpha\Delta t}\dot{a}^t + (\frac{1}{2\alpha} - 1)\ddot{a}^t) + \\ &+ [C](\frac{\vartheta}{\alpha\Delta t}a^t + (\frac{\vartheta}{\alpha} - 1)\dot{a}^t + \frac{\Delta t}{2}(\frac{\vartheta}{\alpha} - 2)\ddot{a}^t)\end{aligned} \tag{6}$$

In order to solve equation (4), the modified stiffness matrix $[\hat{K}]^t$ has to be subject to triangular decomposition, which, for a nonlinear calculation, means that the equations will be complex to solve. For every time step t, the displacement $\{a\}^t$, the velocity $\{\dot{a}\}^t$ and the acceleration $\{\ddot{a}\}^t$ will have to be known, if, using this and the loading function $\{f\}^{t+\Delta t}$ at time point $t + \Delta t$, the loading vector $\{\hat{f}\}^{t+\Delta t}$ is to be calculated.

α and ϑ are parameters, with whose help the stability and accuracy of the integration method are influenced. The method possesses unlimited stability when

$$\vartheta \geq 0.5 \qquad \alpha \geq 0.25 \cdot (0.5 + \vartheta)^2 \tag{7}$$

are chosen.

The *Central Difference Method* is used as an *explicit* method, whereby the following set of equations has to be solved:

$$[\hat{M}]\{a\}^{t+\Delta t} = \{\hat{f}\}^t \tag{8}$$

with

$$[\hat{M}] = \frac{1}{\Delta t^2}[M] + \frac{1}{2\Delta t}[C] \tag{9}$$

and

$$\begin{aligned}\{\hat{f}\}^t =& \{f\}^t - ([K] - \frac{2}{\Delta t^2}[M])\{a\}^t - \\ &- (\frac{1}{\Delta t^2}[M] - \frac{1}{2\Delta t}[C])\{a\}^{t-\Delta t}\end{aligned} \tag{10}$$

This method has the disadvantage of not always being numerically stable. This means that a time step size has to be chosen that is smaller than a certain value that results from the structure's properties and can be calculated with (11).

$$\Delta t_{krit} = \frac{T_n}{\pi} = \frac{1}{f_n\pi} \tag{11}$$

f_n is the discretized system's highest eigenfrequency. By choosing a *lumped* mass matrix [17], the method can become very advantageous, because given these prerequisites and neglected damping after (8) the n-coupled equations become n-independent and can be quickly solved.

6. DISCRETIZATION AS A RESULT OF HIGHER LOAD FREQUENCIES

The blasting load's high frequency content demands that the element's maximum dimension does not exceed a certain length [18], [19], [20]. This is dependent on the load's minimum wave length, the choice of mass matrix and the results' desired accuracy. From the publications by [21] and [22] a maximum element size for elements with a linear shape function can be taken as a function of the previously mentioned influence. This gives

$$\frac{\lambda_{min}}{l_{max}} \geq n \tag{12}$$

with $n = 6.0$, a solution with sufficient accuracy. λ_{min} is the smallest wave length of the wave propagating in the continuum and is obtained from the load frequency. l_{max} is the maximum element size of the discretized continuum. This means that for a given *FE* mesh a maximum load frequency can be used, and that higher frequencies of the load can no longer be put through the mesh. This in turn means that at higher load frequencies, insofar as these do not appear to be very significant and another mesh has to be chosen, these load frequencies have to be filtered out of the load function. In the

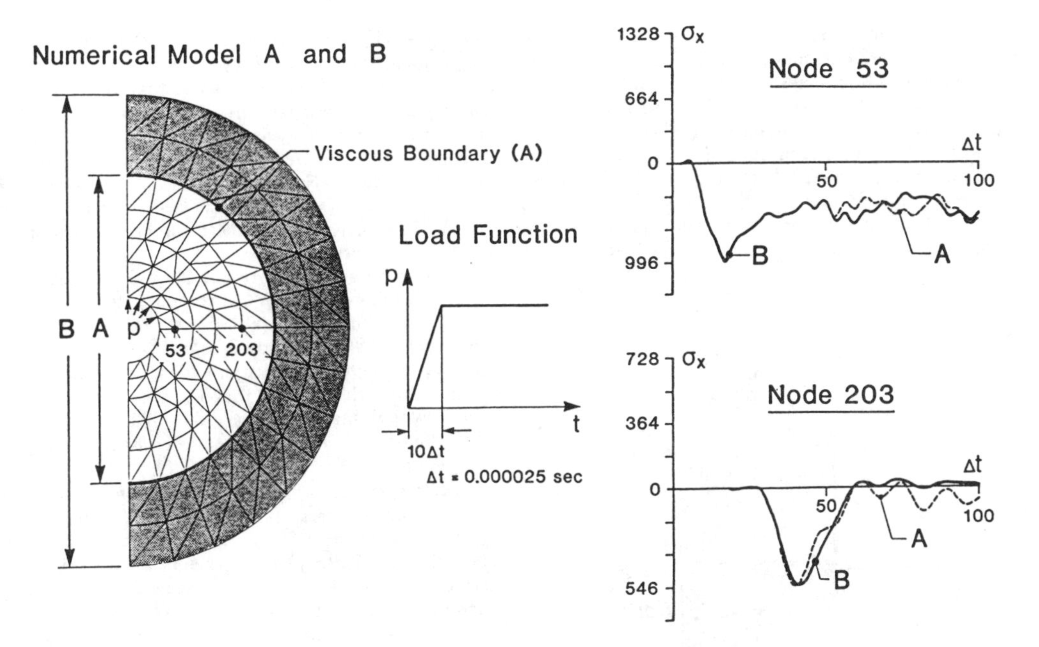

Fig.7 Wave propagation with and without system boundaries.

case of blasting, frequencies that are higher than $1\ kHz$ are filtered out of the calculated load function with the help of a Fourier transformation.

The difference in element sizes in the FE mesh produces reflections at the element boundaries, which is why the difference in the length of adjacent elements should not exceed a certain value [23]. Since this value also depends on the relationship between load frequency and element size, an FE mesh is used here with elements whose maximum difference in length is not greater than $1\ /\ 1.5$.

7. APPLICATION OF A VISCOUS BOUNDARY CONDITION TO THE BOUNDARY OF A FE MESH

In the FE analysis of a wave propagating in a continuum, wave reflections occur at the boundary of the continuum. In order to prevent the reflections, that occur as a result of the size of the necessary numerical model, from affecting the analysis results, a boundary condition has to be found that permits reflection of the wave's energy content. This is possible through the use of viscous dampers after [24]. If the wave comes in at an approximately orthogonal angle to the structure's boundary, the boundary can be successfully applied. This boundary condition is achieved by placing joint forces on the boundary elements.

$$\{f\} = [C]\{\dot{a}\} \tag{13}$$

The joint forces are calculated after [25] under consideration of the shape function and the stress components and applied opposite to the wave's direction of propagation. The two-dimensional wave propagation after Fig. 7 is given as an example of this, whereby the broken line A shows the result of analysis with a viscous boundary condition, and the solid line B the course of stress with no boundary influence.

If the direction of incidence falls in the boundary line (Rayleigh wave), this formula will not provide good results. For the two-dimensional model used to study blasting during tunnel driving, the mesh can be so modeled that the wave's direction of incidence is almost orthogonal to the mesh boundary.

The following analyses were performed using the blasting pattern for the first blast to start excavation of the Milser Tunnel. Fig. 8 shows a section of the blasting pattern's axis of symmetry. The numerical analysis was performed in a two-dimensional model in plane strain, because the present limited computer capacity does not

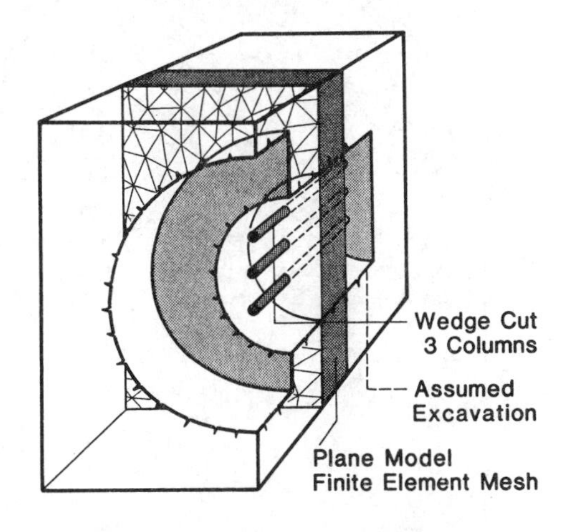

Fig.8 Numerical model of the ejection cone.

permit three-dimensional analysis. The stress is applied to the two-dimensional model that was discretized by finite elements. The stresses are calculated with the previously mentioned analytical model [10]. As illustrated in Fig. 9, only the stress components that act in the two-dimensional system are considered and not those that act orthogonally thereto. Fig. 10 shows the FE mesh used with the blast holes, the load function's course of stress and the stress wave propagating in the continuum. Throughout the entire area studied the element was given about the same size, because the influence of the distance on the frequency content of the stress wave was not studied. The analysis was first performed with linear, elastic, homogeneous, isotropic material behavior, with the load frequency being exactly adapted to the FE mesh and only frequencies up to 1 kHz permitted as load function. Higher frequencies are filtered out of the load function using FFT (Fast Fourier Transformation). The analysis shown with the broken line in Fig. 10 was used for this reason.

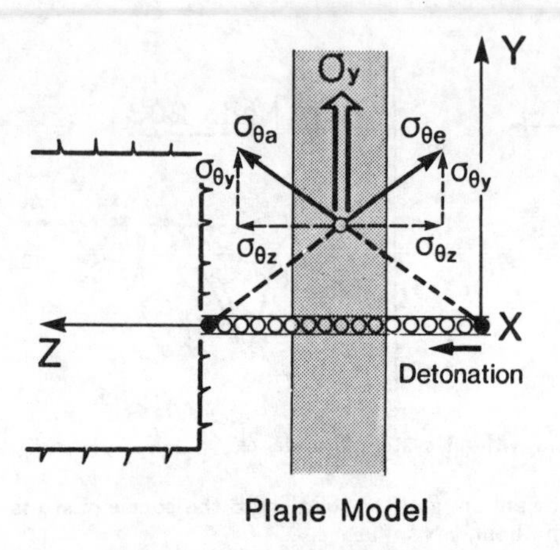

Fig.9 Load in a plane model.

It was possible to carry out the two solution methods with the same amount of computer time. Although the Central Difference Method with the help of lumped mass matrixes enabled the equation to be solved very quickly, stability considerations demanded the use of a time step for this method that was twice as small as for the Newmark Method, namely 0.00003 sec. A total of 125 time steps were calculated in order to analyze the entire load function.

Before excavation was started, the rock was in an initial state of stress. In order to transfer the load redistribution of the existing state of stress to the tunnel boundary elements, joint forces equivalent to the initial state of stress are assumed on the excavation boundary and activated in a bilinear function as a function of time (Fig. 7) in the course of analysis. The time taken for the increase is presently set as a function of the velocity of crack growth. After [11], this is approximately 0.001 sec for 30 cm depth, whereby the influence of the change in crack growth velocity was not studied , because it was not expected to be of any major significance for an overburden of 40 m.

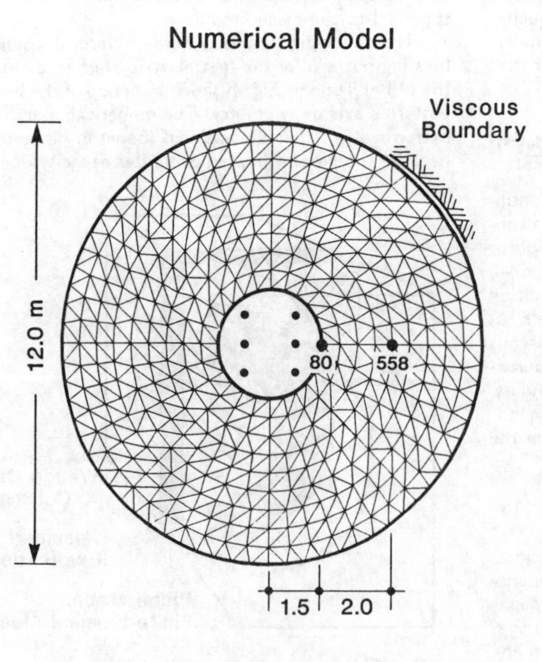

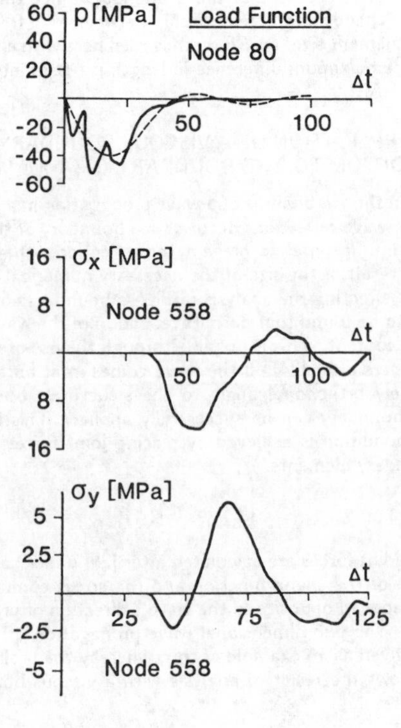

Fig.10 Course of stress with FE mesh.

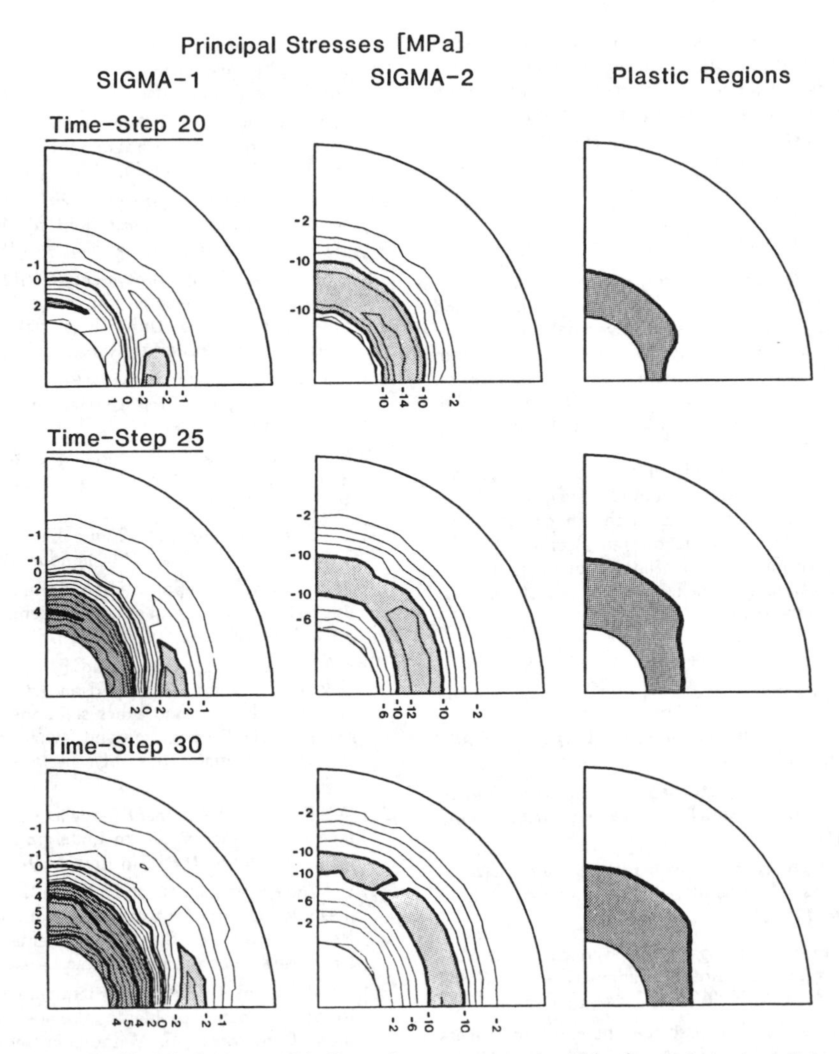

Fig.11 Principal normal isolines of stress as a result of blasting and plastic region.

For the nonlinear analysis whose results can be seen in Fig. 11, Mohr-Coulomb's Failure Condition was used as the flow rule. In order to find the blast wave's nonlinear affect on the rock, the plastic zones were defined as those zones where the failure body is exceeded. The characteristic values used here are:

$$E = 25GPa \qquad \nu = 0.2 \qquad \gamma = 28kN/m^3$$
$$c = 6MPa \qquad \varphi = 30^{\circ} \qquad h = 40.0m$$

Contrary to the scatter otherwise used for the firing point of the individual blasts, nonlinear analysis requires that the explosive in all six blast holes be fired simultaneously. The Central Difference Method was used here, whereby such a small integration step was used, namely half as large as for linear analysis, that the additional plastic forces that are applied after each time step as equivalent forces do not account for more than $3\ \%$ of the entire force. Fig. 11 shows the course of the plastic zones, whereby the shallow overburden of $40\ m$ causes these to be affected by static excavation loading to a very minor extent only. The region takes on a maximum size of approximately one and half times the excavation diameter. A comparative, purely static analysis of the excavation shows that elimination of the initial state of stress brought no plastic zones whatsoever. It must be noted, however, that the rounds, that come after the first blast to start excavation, will cause less strain because the blast can be directed against an additional free surface, meaning that considerably less explosives will be needed.

8. CONCLUSION

The aim of this study is to show the effect of blasting as compared to a static analysis. In this stage, this is performed by displaying the so-called plastic zones that have a strong influence on the formation of the primary support. Then the changes in the rock properties in these zones are studied in the subsequent static analysis. Further studies of the use of different blasting patterns, changed firing points and explosives of various strengths serve to compare the effect on changed lining forces in contrast to static analysis.

9. ACKNOWLEDGEMENT

Measurements of the blasting performed during driving at the Milser Tunnel were conducted with the kind consent of the owner's representative Dipl.-Ing. Temml and with the ready assistance of the construction firm Tiefbau Ltd. The measurements and their evaluation were performed at the Department of "Festigkeitslehre und Flächentragwerke" by Prof. Dr. J. Majer and Dr. G. Niederwanger.

This project was carried out as part of the research project FSP S 30/02c approved by the Austrian Fund for the Promotion of Scientific Research (Fonds zur Förderung der wissenschaftlichen Forschung) under the title "Lösungsalgorithmen für nichtlineare dynamische, numerische Berechnungsmodelle", whose assistance we gratefully acknowledge.

10. REFERENCES

[1] H.Jendersie, 'Sprengtechnik im Bergbau', VEB , Leipzig, 1983

[2] H. J. Wild, 'Sprengtechnik im Bergbau, Tunnel- und Stollenbau', Glückauf - Betriebsbücher, Band 10, 1977

[3] 'Sprengtechnische Ratschläge', Dynamit Nobel Wien Ges.m.b.H. und Schaffler und Co., Zehnte Auflage, (1986)

[4] W.Thum,'Über das physikalisch-mechanische Verhalten von Gestein unter Sprengeinwirkung', Nobel Hefte 37, ISSN 0029-0858, 1-24 (1970)

[5] H.K. Kutter, C. Fairhurst, 'On the Fracture Process in Blasting', Int. J. Rock Mech. Sci. 8, 181-202 (1971)

[6] W.I.Duvall, 'Strain-wave shapes in rock near explosions ', Geophysics 18, 310-326 (1953)

[7] H.L. Selberg, 'Transient compression waves from spherical and cylindrical cavities', Arkiv för Fysik 5 (hr. 7), 97-108 (1951)

[8] R.F.Favreau,'Generation of strain waves in rock by an explosion in a spherical cavity', Journal of Geophysical Research 74, 4267-4280, (1969)

[9] D.V.Swenson,L.M.Taylor, 'A Finite Element Model for the Analysis of Tailored Pulse Stimulation of Boreholes', Int. J. for Num. and Anal. Meth. in Geom. 7, 469-484 (1983)

[10] C.T. Aimone, 'Three - Dimensional Wave Propagation Model of Full-Scale-Rock Fragmentation', Ph.D.Dissertation, Northwestern University, 1982

[11] F.Scholz,'Über die Druckbeeinflussung von Sprengladungen durch die Schwaden früher detonierender Nachbarladungen beim Sprengen mit Millisekundenzündern im Karbongestein', Berichte der Versuchsgrubengesellschaft mbH, Dortmund Heft 16, ungekürzte Fassung der Dissertation, 1981

[12] C.H.Dowding and C.T.Aimone, 'Multiple blast-hole stresses and measured fragmentation', Rock Mech. and Rock Eng. 18, 17-36, (1985)

[13] P. Steinhauser, 'Sprengerschütterungen beim Tunnelvortrieb', Bundesministerium für Bauten und Technik, Straßenforschung, Heft 44 (1975)

[14] G.Swoboda, 'Programmsystem *FINAL*. Finite Elemente Analyse linearer und nichtlinearer Strukturen', Version 6.0, Univ. Innsbruck, Institut für Baustatik und verstärkte Kunststoffe, 1987

[15] K.J. Bathe, 'Finite Element Procedures in Engineering Analysis', Prentice Hall Inc., Englewood Cliffs, 1982

[16] N. M. Newmark, 'A method of computation for structural dynamics', J. of Eng. Mech. Div. ASCE 85, 67-94 (1959)

[17] O. C. Zienkiewicz, 'The Finite Element Method', Third Edition, McGraw-Hill, London 1985

[18] W.D.Smith, 'The application of finite element analysis to body wave propagation problems', Geophys., J. R. Astr. Soc. 42, 747-768 (1975).

[19] S.A.Shipley, H.G.Leistner and R.F.Jones, 'Elastic wave propagation - a comparison between finite element predictions and exact solutions', Proc. Int. Symp. Wave Propagation and Dynamic Properties of Earth Materials, Univ. of New Mexico, 509-519 (1967)

[20] D.P.Blair, 'Finite element modelling of ground surface displacements due to underground blasting', Int. J. for Num. Meth. in Eng. 5, 97-113 (1981)

[21] W.White,S.Valliappan and I.K.Lee,'Finite element mesh constraints for wave propagation problems', Proc. 3th Int. Conf. on Finite Element Methods, Univ. New South Wales, Sydney, Australia (1979)

[22] S.Valliappan, K.K. Ang, 'Dynamic analysis applied to rock mechanics problems', Proceeding of the 5th Inter. Conf. on Num. Methods in Geomechanics., Nagoya (Japan), 119-132 (1985)

[23] Z. Celep and Z. P. Bazant, 'Spurious reflection of elastic waves due to gradually changing finite element size', Int. J. for Num. Meth. in Eng. 19, 631-646, (1983)

[24] J.Lysmer,R.L.Kuhlemeyer, 'Finite Dynamic Model for Infinite Media', J. of the Eng. Mech. Div., ASCE, 95, 859-877 (1969)

[25] M.Cohen, P.C. Jennings, 'Silent Boundary Methods for Transient Analysis', Computation Methods for Transient Analysis, Eds. Belytschko and Hughes, Elsevier Science Pub., Ch. 7, 301-360, New York 1983

[26] G. Zenz, G. Swoboda, 'Numerische Analyse des Sprengvortriebes in der neuen Österreichischen Tunnelbauweise', Baudynamik, Forschung und Praxis, Bochum, SFB 151 Bericht Nr. 6, 54-59 (1987)

[27] G. Swoboda, G. Zenz, 'Tunnel Analysis of Rock Blasting', Int.Symp. on Geomech. Bridges and Struct., Lanzhou, 575-580 (1987)

Numerical Methods in Geomechanics (Innsbruck 1988), Swoboda (ed.)
© 1988 Balkema, Rotterdam. ISBN 90 6191 809 X

Response of underground openings to earthquake and blasting loading

H.-J.Alheid & K.-G.Hinzen
Bundesanstalt für Geowissenschaften und Rohstoffe, Hannover, FR Germany
A.Honecker
Control Data GmbH Hamburg, FR Germany
W.Sarfeld
Engineering Consult Berlin, FR Germany

ABSTRACT: A complete dynamic analysis of underground openings has to consider the structure as part of the whole system, consisting of the seismic source, the travel path and the structure itself. We suggest to separate the signal flow from the source to the structure into steps and to combine the solutions of the single steps to get the dynamic response of the structure. Examples demonstrate the efficiency of the suggested method: First, the response of an underground drift, deeply embedded in a layered medium, to an earthquake is calculated. Secondly the calculated response of a shallow underground drift to blasting loading is compared to results obtained from in-situ dynamic load experiments.

1 INTRODUCTION

Finite-Element (FE) analysis in rock mechanics is often applied to underground structures which are small compared to the surrounding geological characteristics and which respond to seismic motions induced by a remote source. The source and geological characteristics determine the nature of the excitation at the location of the structure. FE analysis of the dynamic response of underground structures gives an instructive insight into the effects of transient dynamic loading to the static prestressed vicinity of the structure. This knowledge is important in earthquake safety-analysis of underground terminal repositories for hazardous waste. If numerical dynamic analysis is combined with in-situ dynamic experimental results, the degree of damage of the rock mass surrounding the structure can be estimated. This is important even for static stability analysis.

To our experience it is essential that the dynamic excitation applied to a FE discretization does not contain any frequencies which result in a ratio of wavelength to node distance of less than 10. Otherwise unreal amplitudes, numerical dispersion and wrong wave velocities may result. These requirements on the spatial discretization and a complex geological characteristic as well as a complex geometry of the structure prohibit to include seismic source, travel path and structure in one single FE discretization.

To overcome the problems of limited storage capacities or computation time it is necessary to separate the signal flow from the source to the structure into steps and to combine the solutions of the single steps to get the dynamic response of the structure. In such a hybride concept different mathematical methods can be used to solve each partial problem and the most appropriate method can be applied (ALHEID et al. [2]). Three main steps in the signal flow are obvious:

- generation of the dynamic load in the source area
- transmission of the signal from the source region to the structure
- response of the structure to the dynamic excitation.

In the concept of submodels any source of dynamic excitation can be used as long as the farfield time-histories can be determined. This may be done analytically, numerically or even by measurements. In this way the farfield displacement functions from earthquakes with even complex source time functions and rupture geometries can be used.

The solution of wave propagation from the source to the structure must not only provide the complete wavefield at the depth of the embedded structure, but also allow to separate waves of different directions of propagation. A vertically inhomogeneous medium is adequate in most cases to model the geological characteristic. One dimensional FE analysis, reflectivity method or a ray theoretical approach can be applied to determine the development of the complete wavefield. The ray theoretical approach in time domain has the advantages of easy separation into up- and downgoing parts of the wavefield and small computational efforts.

The third submodel concentrates on that region where the seismic wavefield is assumed to be disturbed by the underground structure. Thus only a part of the fullspace including the structure is modeled. This reduces the numerical effort but requires suitable boundary conditions at all model boundaries. To analyse the response of the structure to the dynamic excitation the FE method in time domain is the appropriate tool, especially if nonlinear material behaviour has to be considered . Transmitting boundaries (ALHEID et al. [1]) and viscous boundaries are used.

In addition the correct excitation has to be applied to the four boundaries of the two dimensional FE model depending on the direction of wave propagation (ALHEID et al. [2]). This approach is essential in the case of earthquake loading because the wavefield is composed of waves with different directions of propagation. These waves act simultaneously on a structure.

2 THE MATHEMATICAL CONCEPT

Our mathematical model is based on the FE technique.It is implemented in the computer-system ANSALT developed co-operatively by Control Data GmbH and BGR (Bundesanstalt fuer Geowissenschaften und Rohstoffe). The time domaine non-linear analysis in ANSALT is based on the following features:

- solid element (4 - 8 nodes) for modeling rock and granular soil material. The nonlinear material behaviour is formulated for Mises and extended Drucker-Prager constitutive laws.
- viscous boundary element (2 - 3 nodes) to simulate energy absorbtion by propagating shear- and compressional waves. Vicous boundaries are perfectly absorbing in time domain, if the angle of incidence is normal to the boundary.
- transmitting boundaries developed by theory of semi-infinite element can be used to propagate arbitraryly incoming and outgoing waves at the FE boundary. These boundaries are defined as a stiffness and damping matrix and describe exactly any wave type in frequency domain. In order to use transmitting boundaries in time domain a practical approximation was introduced, by taking the arithmetic mean-value matrices for a chosen number of frequencies (ALHEID et el. [1]). Assembling the described element for a given dynamic problem leads to the well known equation of motion

$$m*a + c*v + k*u = p(t)$$

m: mass matrix
c: damping matrix
k: stiffness matrix
a: acceleration vector
v: velocity vector
u: displacement vector
p(t): load-time history

which will be solved numerically using implicit time integration (Newmark algorithm).

This procedure for solving dynamic problems in the area of rock mechanics has been verified extensively using simple models and is now used in standard calculations.

3 EARTHQUAKE LOADING

An earthquake focus is assumed to be situated 14 km below an underground drift embedded in a layered medium at a depth of about 1 km. The top of the bedrock is 3 km below the free surface. The drift has a maximum height of 6 m and a maximum width of 7 m. Following the concept of hybride modeling the complete model is separated into three submodels:

1. The source is situated in an elastic, homogeneous, non absorbing halfspace with constant p- and s-wave velocities, constant density and constant shear modulus. The rupture surface is assumed to be of rectangular shape with unilateral rupture process, i.e. the rupture growths only into one direction. The rupture velocity and the dislocation are constant and the source function is a simple step-function. In addition the rupture surface is normal to the rays traveling to the bottom of the layered medium. In this case the directivity effects of the source can be neglected.

The source time function and the farfield displacement function respectively are mainly determined by the rate of growth of the rupture surface. Following the prepositions, the displacement function is a simple boxcar. Between start- and stop phase there is no movement in the farfield (the velocity of the groundmovement is zero). The time duration of the displacement function, i.e. the width of the boxcar, is equal

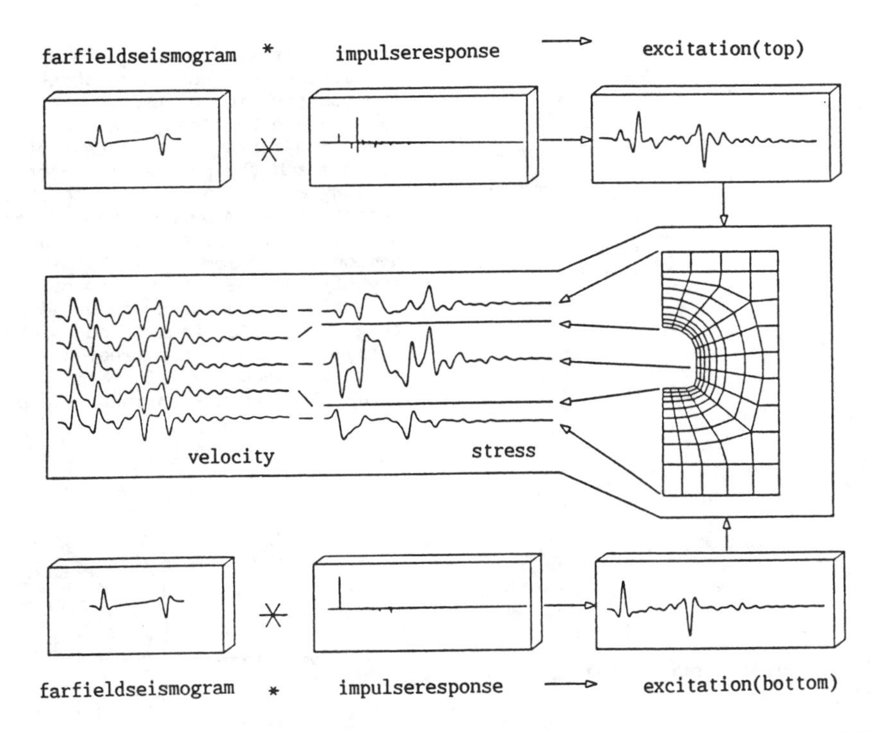

Figure 1. Earthquake stability calculation. Excitations at bottom and top of the Finite-Element discretization are obtained by a convolution of numerically calculated impulse responses and farfield seismogramms derived from analytical considerations.

to the time period which the rupture front needs to travel along the rupture area.

We used a duration of two seconds for the source time function. In order to avoid alising effects in the FE calculations the farfield displacement function was band limited to frequencies between 0.5 Hz and 5.0 Hz. The resultant farfield velocity time history is shown in figure 1.

2. The normal-incidence wavefield in the layered medium was calculated using a time-domain discrete state-space model (Mendel et al. [4]; Ferber [3]) with a sampling interval of 3 ms. The model consists of a horizontally layered medium of 10 geological layers with thicknesses from 10 m to 700 m. The vertical distribution of p-wave velocities in the layered medium is given in figure 2. The resultant pulse responses of the layered medium at the depth of the embedded structure for the up- and downgoing waves, respectively, are shown in figure 1. The pulse response of the upgoing wave is dominated by the peak of the direct wave at the beginning of the trace. The two wavelets with opposite signs in the corresponding velocity function describe the effect of the start- and stop phase of the source. The velocity function of the downgoing waves has a more complex structure. This is mainly caused by the reflexion from the interface at a depth of 700 m (see fig. 2), the strong reflexion from the surface and some multiple reflexions.

3. In the FE calculations we used the appropriate force-time histories derived from the velocity functions as input to the upper and lower boundaries of the FE model for down- and upgoing waves respectively. The time step used in the calculations is 12 ms. The response of the structure was calculated for a total time of 6.6 s, i.e. 550 time steps. The cross sectional two- dimensional discretization is shown in figure 1. Due to symmetric geometry and loading conditions (only compressional waves are used) only one half of the system is considered. The overall dimensions of the FE discretization are 12 m * 24 m.

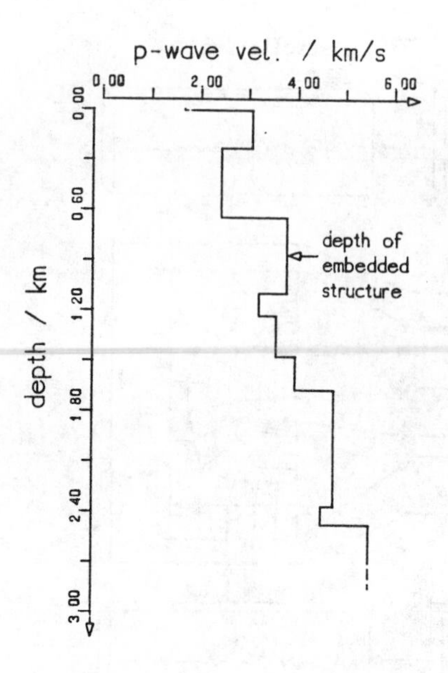

Figure 2. Vertical distribution of p-wave velocities.

The system consists of 107 8-node isoparametric elements with 361 nodes and 684 degrees of freedom. In order to simulate the fullspace surrounding the model transmitting boundaries are used at the right hand side. At top and bottom of the FE region and of the transmitting boundaries viscous boundary elements are placed.

In figure 1 the resultant vertical velocity and vertical stress time histories are presented. The calculated velocity time histories describe the simple superposition of the two velocity input functions. Obviously the model acts nearly as a rigid body due to the low frequency content of the seismograms. Comparing the stress time histories considerable differences are obviuos. The large stress concentration due to the drift at the wall-face is evident.

4 BLASTING LOADING

4.1 In-situ experiment

In-situ experiments were performed in order to validate the results of the FE-code. For this purpose a special underground opening was constructed. This opening is a drift of nearly 100 m length. It is 3.5 m high, 3.0 m wide and the maximum overburden is about 55 m. In the construction of the in-situ settings care had to be taken, that the spatial dimensions, frequency content of the signals and the parameters describing the geology were in a range which could reasonably be modeled in our FE numeric experiments. The material is a rather homogeneous dolomite with a p-wave velocity of 3800 m/s. A two dimensional measuring array was constructed in a range with minor faults and joints. This array consists of 6 boreholes. Figure 3 shows the geometry. In each borehole three 3D geophone-stations were planted.

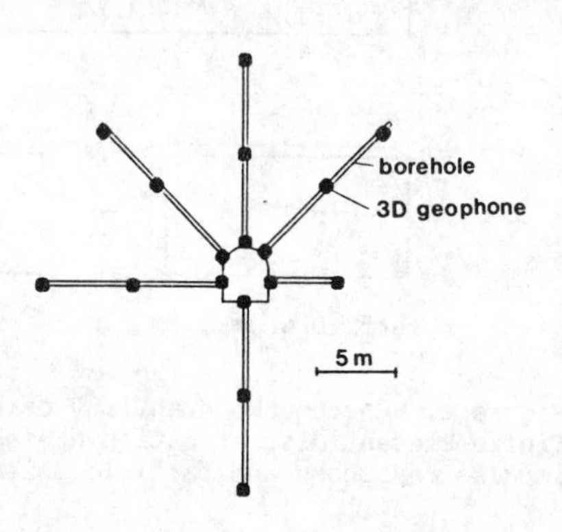

Figure 3. Geometry of in-situ measuring array

By shaking table testing the amplitude response of the geophones was found to be proportional to ground velocity within a frequency range from 30 Hz to 1500 Hz. The data were registrated with a computer controlled 64 channel digital recording system. The maximum sampling rate is 62.5 kHz per channel.

The dynamic load to the drift was generated by the detonation of explosive charges of a weight from 170 g to 1360 g. The charges were fired in a borehole which was drilled from the top of the overburden within the two dimensional measuring array. The overburden of the drift is about 40 m in this range. The charges were fired at depths between 15 m and 20 m.

The seismograms are 12bit data with a sampling frequency of 31 kHz. In addition to the signals from 20 geophone stations with 60 seismic channels, the exact moment of the detonation and the pressure time-history in the blasthole

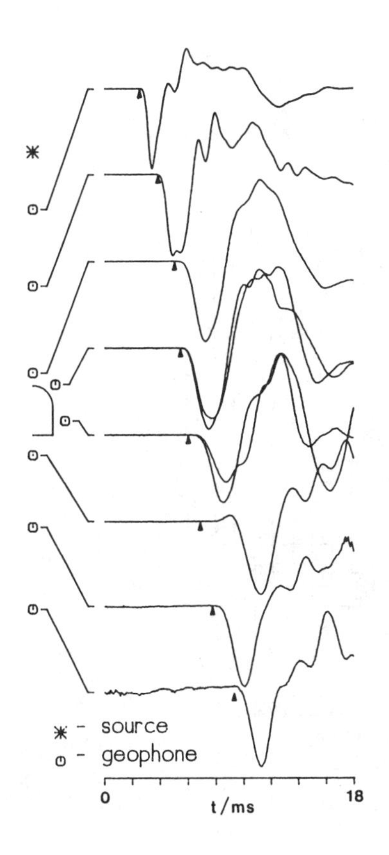

Figure 4. Velocity seismograms in the vicinity of a drift produced by the detonation of an explosive charge.

were recorded. The pressure history was used to model the excitation function in the numeric experiments.

Figure 4 shows a typical result of an in-situ dynamic load experiment (IDLE). The vertical component velocity seismograms in this figure are normalized to their maximum amplitudes. The * indicates the position of the 1360 g charge. The positions of the geophone-stations are given by small circles. Only the right half of the symmetric measuring array is shown. The lines indicate the location of the corresponding geophone station for each seismogram. In case of two seismograms at one place, the corresponding ground movements in the left and right part of the measuring array are given. In all IDELs the measured ground movements proved to be reproducible

The triangles at the seismic traces mark the arrival times, which would be expected, if the medium was a homogeneous halfspace with a p-wave velocity of 3800 m/s. While the first breaks for the two upper traces exactly meet the expected arrival times, a delay of p-arrivals is obvious at the measuring positions in the vicinity of the drift. The largest delay occures at the position close to the drift floor (6th trace in fig. 4). A zone of weakened material around the drift with a reduced p-wave velocity is a possible explanation for these observations. At an increased distance from the drift (lower trace) the influence of the weakened zone is no longer effective and the delay decreases.

The direction of the first motion is negativ (away from the source) for all but one of the seismograms. The exeption is the trace which was recorded close to the drift floor. The small upward movement at the beginning of the signal is an effect of the dynamic drift response.

4.2 Numerical Calculations

To calculate the response of the underground drift and its vicinity a two-dimensional FE discretization was chosen. The overall dimensions of the discretization are 7 m * 14 m. The system consists of three element groups with different elastic constants, a total of 457 4-node isoparametric elements with 505 nodes and 987 degrees of freedom. The fullspace surrounding was simulated by means of transmitting boundaries (vertical boundaries) and viscous boundary elements (horizontal boundaries). According to the results of the measurements two zones of weakened materials were assumed to surround the drift as indicated by the dotted areas in figure 5. This disturbed zones were simulated by linear elastic materials with Young's moduli being one quarter (inner zone) and one half (outer zone) of the Young's modulus of the intact rock.

As only downgoing waves had to be considered in this example, the excitation was only applied to the upper boundary of the discretization. The load-time history was found by a seperate FE calculation. In this analysis the stress-time history 5 m away from a borehole was calculated applying the measured pressure-time history as excitation to the borehole wall and using nonlinear material behaviour.

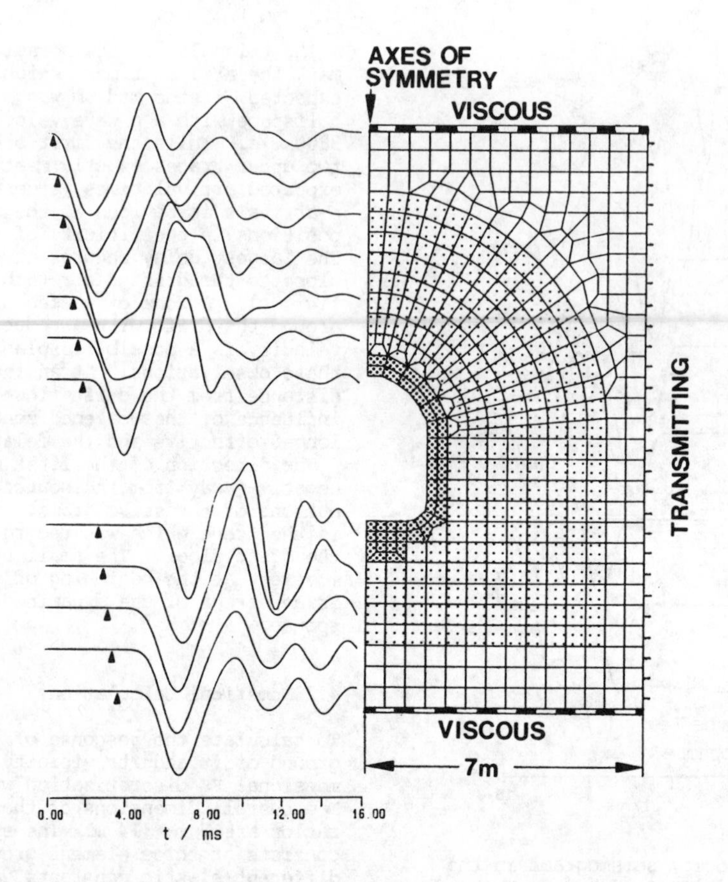

Figure 5. Velocity seismograms in the vicinity of a drift calculated by the Finite-Element method. Dotted areas indicate assumed weak zones around the drift.

As an example the vertical seismic profile along the axis of symmetry of the model is shown in figure 5. Each seismogram is normalized to the maximum amplitude. The triangles at the seismic traces mark the arrival times, which would be expected, if the medium was a homogenious halfspace with a p-wave velocity of 3800 m/s (compare fig. 4). The delay of p-arrivals in the vivinity of the drift is obvious. The largest delay occures at the position close to the drift floor (8th trace in fig. 5). At an increased distance from the bottom of the drift (lower trace) the delay of p-arrivals decreases. The seismogram calculated for the bottom of the drift exhibits a small upward movement. All these characteristics are the same as observed by the IDLEs.

This example demonstrates, that increased knowledge of the state and extent of the disturbed zone around an underground structure can be obtained in combining in-situ and numerical dynamic load experiments. However, in many cases it will not be possible to use a source situated in a separate borehole from the surface. In addition, if the source is situated only obove the underground structure the information about the bottom region is poor due to the shadowing effect of the drift. These considerations lead to the idea that small explosive charges fired at different positions in the borholes of the measuring array might provide further information. A numerical model was developed to demonstrate the influence of a weak zone arround an underground structure onto the wavefield induced by a local explosion. The source was assumed

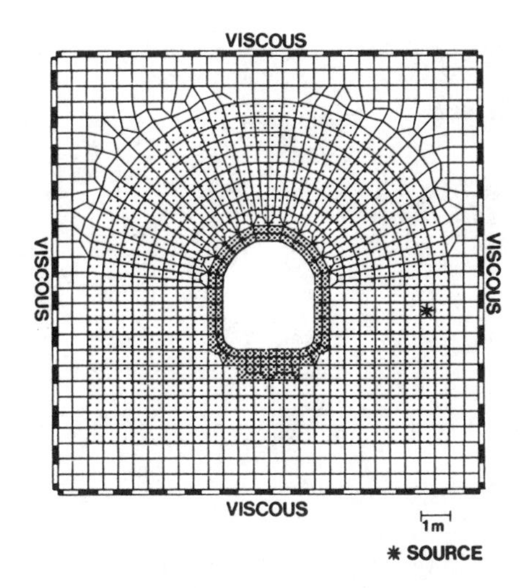

Figure 6. Cross-sectional discretization for non-symmetric loading conditions.

to be at a distance of 3.75 m aside the drift as indicated in figure 6. The overall dimensions of the discretization are 14 m * 14 m. The system consists of three element groups with different elastic constants, a total of 915 4-node isoparametric elements with 985 nodes and 1970 degrees of freedom. The fullspace surrounding was simulated by means of viscous boundary elements at all four boundaries of the two dimensional discretization. One calculation was performed using the same material parameters for all element groups. In a second calculation the weakened surrounding of the drift was modeled with three different materials as discussed above.

From the results of the FE calculations the velocity-time histories of only 45 nodes (from the total number of 985 nodes) were constructed. The limited number of seismograms was chosen in order to simulate a data base wich is close to that obtained from the planned IDLE. These synthetic seismograms were processed the same way as measured data. The first breaks were picked and isochron plans were assembled. They are shown in figures 7 and 8 for the homogeneous and heterogeneous material respectively. The delaying effect of the weak zones is obvious. The 3 ms isochron for example is close to the left wall-face in figure 8, while it is about 4m away from the left wall-face in figure 7. The detailed interpretation of the kinematics of the wave field, displayed by the shape of the isochrons close to the drift, promises an improved understanding of the state and extent of the disturbed zone and is subject of further research.

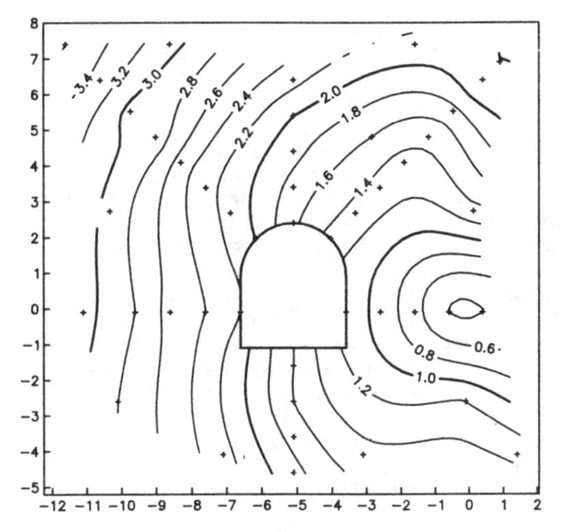

Figure 7. Isochron plan of the first breaks (intact material). Times are given in milliseconds, distances in meters.

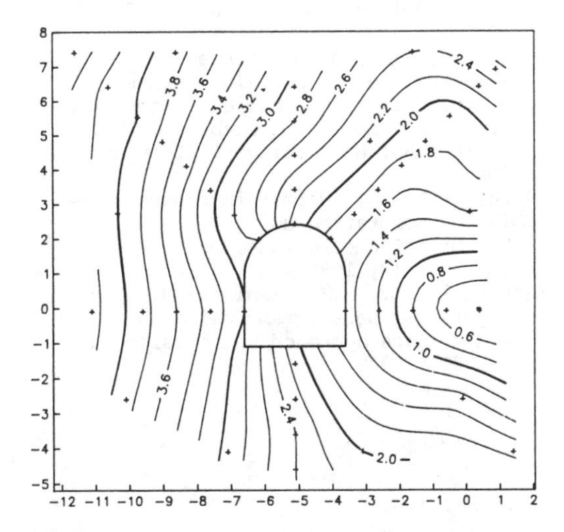

Figure 8. Isochron plan of the first breaks (weak zones). Times are given in milliseconds, distances in meters.

CONCLUSIONS

The first example demonstrates the possibility to analyse the dynamic response of an underground structure to earthquake loading. In order to obtain correct numerical results the state of the material which sourrounds the structure must be known. To evaluate this knowledge the interactive use of IDLEs and numerical dynamic load experiments (NUDLEs) is suggested. The increased knowledge of the state and extent of the disturbed zone around an underground structure is of general interest in many mining and civil engineering applications. The second example shows that numerical studies of physical effects help to understand measured data. If a confidential description of the material behaviour around an underground structure can be achieved the FE method is a suitable tool to calculate the effects of earthquake loading to underground structures, which of course cannot always be simulated experimentally to their whole extent.

REFERENCES

[1] Alheid,H.-J., A. Honecker, W. Sarfeld and H. Zimmer, 'Transmitting boundaries in time domain for 2-D nonlinear analysis of deep underground structures', Proc. NUMETA 85, Swansea, ed. J. Middelton and G. Pande, 117-127, 1985

[2] Alheid,H.-J., K.-G. Hinzen, A. Honecker and W. Sarfeld, 'transient analysis in rock dynamics - some problems and solution strategies',Proc. ECONMIG 86, University of Stuttgart, 1986

[3] Ferber,R.-G., 'Normal-incidence wavefield computation using vector-arithmetic', Geophys. Prosp. 33, 540-542, 1985

[4] Mendel,J.M., N.E. Nahi and M. Chan, 'Synthetic seismograms using the state-space approach, Geophysics 44, 880-891, 1979

Numerical Methods in Geomechanics (Innsbruck 1988), Swoboda (ed.)
© 1988 Balkema, Rotterdam. ISBN 90 6191 809 X

The effects of rock blasting with explosives on the stability of a rock face

Masantonio Cravero & Giorgio Iabichino
CNR, Centro di Studio per i Problemi Minerari, Torino, Italy
Renato Mancini
Politecnico di Torino e Centro di Studio per i problemi Minerari, Torino, Italy

ABSTRACT: This paper presents a method for the analysis of conditions of stability of rock blocks subjected to short lasting but intense seismic events such as impulses produced by rock blasting. The evolution of block motion caused by a seismic event on a rock block is analysed by the application, in its adequate succession, of the equilibrium equations corresponding to the single stages of the event. The law about the velocity of vibration forced on the block supporting base should be known; the form of the vibration may be either regular (a damped sinusoid) or fully arbitrary (the recording of a seismic event). The direction of vibration, however, is assumed to be contained within the plane on which the displacement of the block may occur. One also describes, for the damped sinusoid, the relationships that give the peak amplitude of the velocity of vibration that still does not induce relative motion between the block and the support. For the sake of example the procedure is then applied to the estimate of relative displacements. The said displacements were induced on rock blocks by nearby rock blasting operations carried out during the exploitation of a ballast quarry.

1 FOREWORD

The rock faces, where rock blasting works are carried out, feature conditions of stability wich are linked, up to a point, to both the intensity and the duration of seismic events. More precisely the critical disturbance factor in the conditions of stability of a jointed rock mass are the relative deformations (Dvorak 1965) which progress through the re-occurence of the vibratory action. They appear as successive dislocations of rock blocks on discontinuity planes and through the opening of fractures.

In its complexity the evolution of the static conditions of a whole rock wall following the re-occurence of impulsive events can only be foreseen through a broad outline ad in mainly qualitative terms even through the use of the most refined calculation techniques designed for the dynamic study of discontinuous systems (Cundall 1971, 1980) and granular materials (Walton 1983).

Apart from the analysis of the behaviour of the discontinuous system which in its entirety represents a rock wall, this paper presents a simplified scheme which allows to foresee the behaviour of a single block resting on a discontinuity plane shaken by a vibration featuring known characteristics. The procedure allows to describe the momentary state of instability induced by the blasts, locating the relative sliding movements or, if necessary, the relative separation movements of the block from its supporting base.

Supposing, as introduced in the beginning, a damped sinusoid law, one also defines peak parameters (amplitude, frequency) below which no relative movements occur, thus avoiding states of momentary instability.

2 BASIC HIPOTHESIS

The displacement through vibrations of a rock block may be considered as a special form of vibratory conveying where the conveying plane is replaced by the discontinuity surface upon which the block rests. Convenying problems through vibrations have been the object of careful studies. In our case, however, unlike the industrial application where importance is generally attached to the features of fully operational motion of conveyed material, one is interested in the succession of jerky dislocations with irregular amplitude which occur following an event which is essentially of the impulsive type. In short, if in vibratory conveying it is necessary to determine the average speed of conveying, here one is interested in the maximum amplitude of vibration which will correspond to the momentary instability and the extent - a small one hopefully - of the dislocation affecting the rock block at the end of the impulse.

Here are the suppositions our scheme is based on:

- The motion of a rigid body, without rotations, through sliding or separation of the block from the discontinuity which is its supporting base (movements take place in the representation plane which contains the vibrations of the supporting base);
- Resistance opposed to the motion of the block only through the base-block friction and dissipation of kinetic energy in case of impact;
- Size of the rock block smaller than the half wave length of the seismic impulse (all the points of the supporting base are affected by the same law of vibration);
- Law of vibration defined in terms of velocity according to a real vibrogram or a damped sinusoid.

The determination of the features of motion of a rock block is a problem for classic mechanics: for its solution it is opportune to consider the scheme shown in figure 1 (from Bykhovsky, 1972).

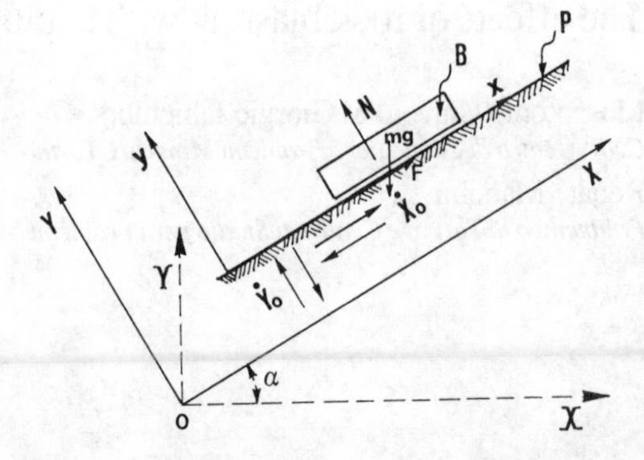

OXY : Fixed reference parallel to oscillating base P which is inclined with respect to the horizontal line X.

Oxy : Moving reference frame fixed on P.

B : Block featuring mass m; F, N support reactions provided by the base to the block.

$\dot{x}_o, \dot{y}_o$: components of the oscillation speeds forced onto the base.

Figure 1

x_o, y_o oscillations supply the supporting base with a translatory motion and the position of the block in the fixed reference is:

$$X = x + x_o, \quad Y = y + y_o \tag{1}$$

The equation governing the motion of the block follows conditions of equilibrium on X and Y:

$$\begin{aligned} m\ddot{x} &= -m\ddot{x}_o - m g \sin\alpha - F \\ m\ddot{y} &= -m\ddot{y}_o - m g \cos\alpha + N \end{aligned} \tag{2}$$

If the support reactions are known, the integration of (2) for a given vibrogram supplies the relative x,y position of the block which may:

a. Be fixed on the base ($x = y = 0$);
b. Slide ($x \neq 0$, $y = 0$);
c. Separate from the base ($x \neq 0$, $y \neq 0$).

a. The condition of a block in relative rest condition is expressed by:

$$F_o = m\ddot{x}_o + m g \sin\alpha \tag{3}$$

As long as

$$|F_o| < f\,N \qquad (4)$$

f = coefficient of base - block friction

with

$$N = m\,\ddot{y}_o + m\,g\,\cos\,\alpha > 0 \qquad (5)$$

b. The condition of sliding is expressed:

$$\ddot{x} = -\ddot{x}_o - g\,\sin\,\alpha - f_d(\ddot{y}_o + g\,\cos\,\alpha)\mathrm{sgn}\dot{x} \quad (6)$$

where $\mathrm{sgn}\dot{x}$ determines the sign of the F tangential reaction based on the direction of relative velocity.

c. The condition of separation is characterized by F = 0 and N = 0:

$$\ddot{x} = -\ddot{x}_o - g\,\sin\alpha \qquad (7a)$$

$$\ddot{y} = -\ddot{y}_o - g\,\cos\,\alpha \qquad (7b)$$

When the rock block falls again on the base it is assumed that the components of relative velocity are adjusted in accordance with Newton's theory of impact:

$$\dot{y}_+ = -R_y\dot{y}_- ; \quad \dot{x}_+ = R_x\dot{x}_- \qquad (8)$$

with R_x, R_y instant restituion coefficients ranging between 0 and 1.

The block motion is thus partly controlled by (6) or by condition (7) and it is the correct choice between the two possibilities which requires an adequate strategy: whereas integration can be carried out without difficulties.

The law of vibration of the base is given by a vibrogram of speed recorded during a seismic event or simulated in a simplified but easy-to-handle form such as a damped sinusoid.

3 STATE OF RELATIVE REST

In accordance with the supposition of vibration of the damped sinusoid type it is possible to locate the peak amplitude of vibration which separates the state of rest respectively from the block-base separation and from sliding. According to the scheme of figure 2 |A| is the maximum amplitude of vibration, with initial direction ß and the components of the velocity of translation are:

$$\dot{x}_o = |A|\cos(\beta-\alpha)\ \sin\omega\,t\ .\ e^{-k\omega t};$$

$$\dot{y}_o = |A|\sin(\beta-\alpha)\sin\omega\,t.\ e^{-k\omega t} \qquad (9)$$

with ω pulsation and k damping coefficient of the vibration.

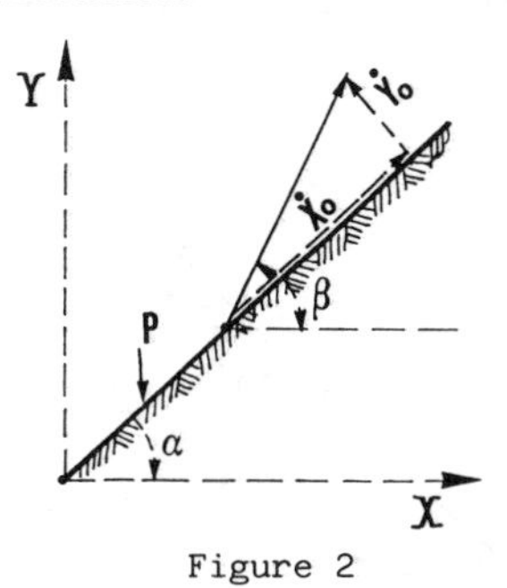

Figure 2

3.1 Condition of non separation

Thr rock block remains adherent to the base if y = 0 and N > 0.The equation (5) must result verified in particular when:

$$\sin(\beta-\alpha).(\cos\omega t - k\,\sin\omega t) < 0 \qquad (10)$$

The peak values of the amplitude |A| of velocity of vibration depend on its initial orientation:

$$|A| < g\,\cos\,\alpha\ /[\omega\ \ \sin(\beta-\alpha)\,D] \qquad (11)$$

for $\alpha < \beta < \alpha + \pi$

where $D = e^{-k[\tan^{-1}(2k/(k^2-1))+\pi]}$

This condition is obtained in accordance to the minimum of

$f(t) = (\cos\omega t - k\sin\omega t)\ e^{-k\omega t}$ and considering that the factor (k sin ω t-cosωt) for $t = t_{min}$ is identically 1 to the varying of k.

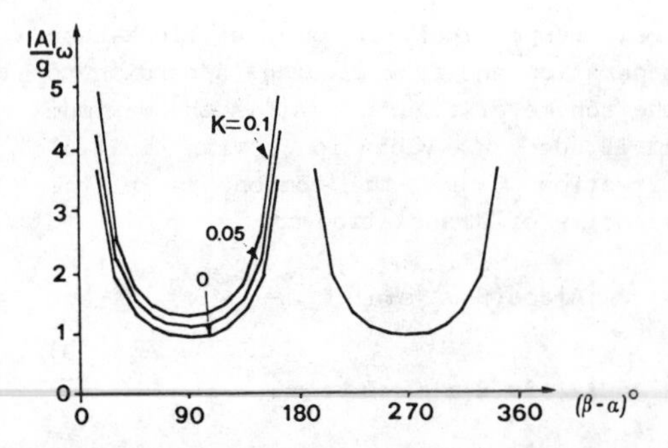

Fig.3 - Amplitude factor for non separation warying the vibration orientations (α = 20°).

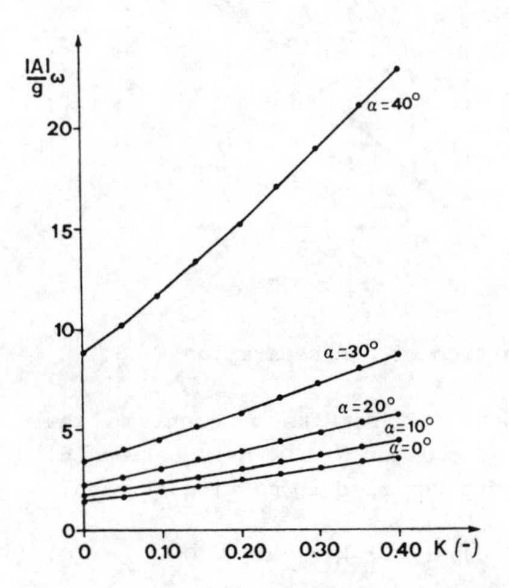

Fig.4 - Amplitude factor for non separation warying the damping values (ß = 45°).

$$|A| < -g \cos\alpha / [\omega \sin(\beta-\alpha)] \quad (12)$$

for $\alpha + \pi < \beta < \alpha + 2\pi$

this condition corresponds to the highest value of f(t) which one has for t = 0. In this way it is obvious that owing to a ß orientation interval of the initial vibration the non separation |A| peak amplitude is independent of the k damping coefficient (12).

In the (11) interval the peak amplitude on the contrary grows with k: this is represented in the diagram by figure 3. Diagrams of figures 4 and 5 represent the incidence on the peak amplitude of vibration respectively of the damping coefficient and of the inclination of the supporting base.

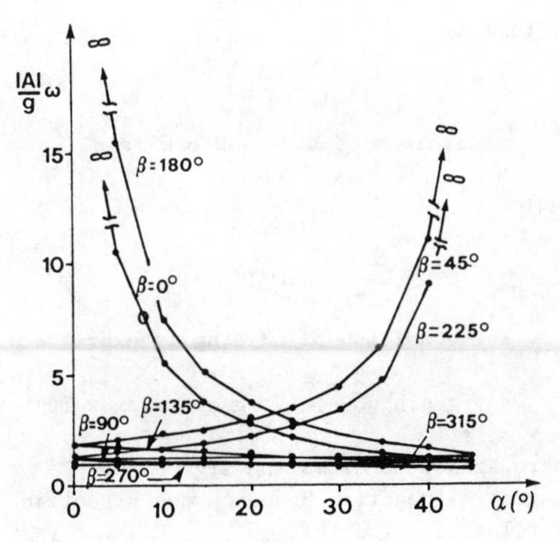

Fig.5 - Amplitude factor for non separation at various base inclinations (k = 0.1).

3.2 Condition of non sliding

In the absence of sliding (x=0) the value of the tangential reaction, which develops only when N > 0, is limited by the (4) relationship which splits into:

$$F_o < f N \quad \text{if} \quad F_o > 0 \quad (13a)$$

$$F_o > -fN \quad \text{if} \quad F_o < 0 \quad (13b)$$

The peak values of the amplitude of oscillation corresponding to the condition (13a) are:

$$|A| < \frac{g\,(f \cos\alpha - \sin\alpha)}{\omega[\cos(\beta-\alpha) - f\sin(\beta-\alpha)]} \quad (14)$$

for $[\cos(\beta-\alpha) - f\sin(\beta-\alpha)] > 0$

$$|A| < -\frac{g\,(f \cos\alpha - \sin\alpha)}{\omega[\cos(\beta-\alpha) - f\sin(\beta-\alpha)]D} \quad (15)$$

for $[\cos(\beta-\alpha) - f\sin(\beta-\alpha)] < 0$

Whereas the following values correspond to condition (13b)

$$|A| < \frac{g(f\cos\alpha + \sin\alpha)}{\omega[\cos(\beta-\alpha) + f\sin(\beta-\alpha)]\;D} \quad (16)$$

for $[\cos(\beta-\alpha) + f\sin(\beta-\alpha)] > 0$

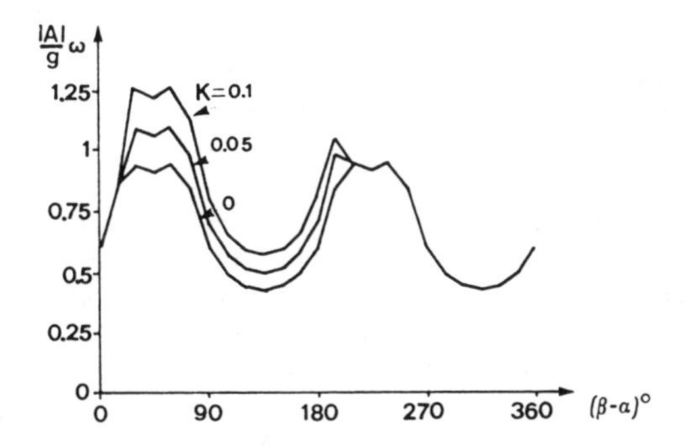

Fig.6 - Amplitude factor for non sliding warying the vibration orientation (α =20°, f = 1).

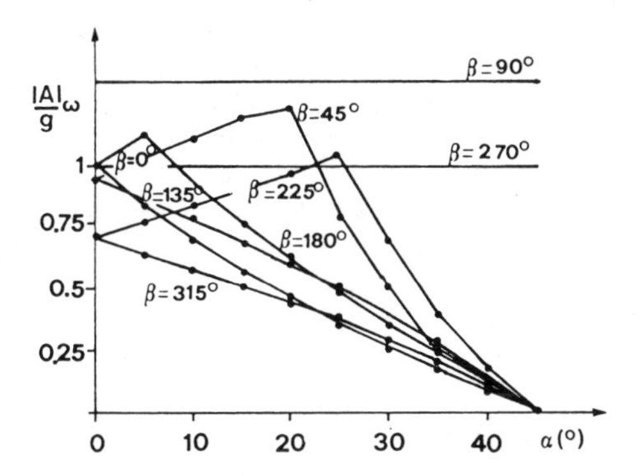

Fig.7 - Amplitude factor for non sliding at various base inclinations (k=0.1,f=1).

$$|A| < -\frac{g(f\cos\alpha + \sin\alpha)}{\omega[\cos(\beta-\alpha) + f\sin(\beta-\alpha)]} \qquad (17)$$

for $[\cos(\beta-\alpha) + f\sin(\beta-\alpha)] < 0$

Since from each of the two conditions (13a, 13b) a value of the amplitude of vibration results, it is obvious that one has to choose the lesser between the two as the peak value. The diagrams of the values of the peak amplitudes are indicated in step with the (β,k) parameters of vibration or with the characteristics (α,ϕ) of the supporting base in figures 6-9.

Now it is evident that for certain orientations of the initial vibration the peak amplitudes result independent of the damping coefficient; in addition to this the position of these intervals depends on the inclination of the supporting base.

In general the trend of the diagrams

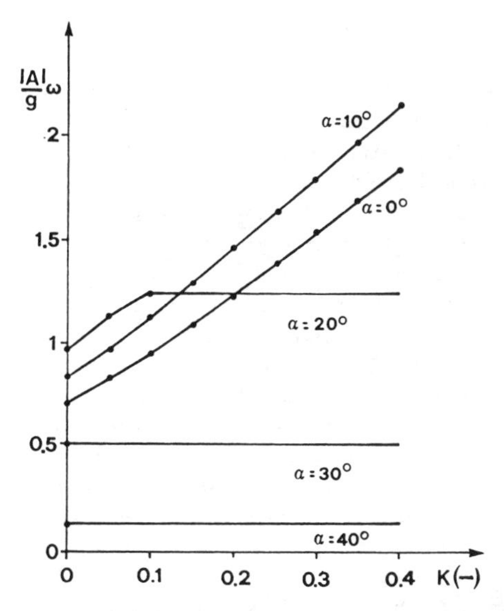

Fig.8 - Amplitude factor for non sliding warying the damping values (β=45°, f=1).

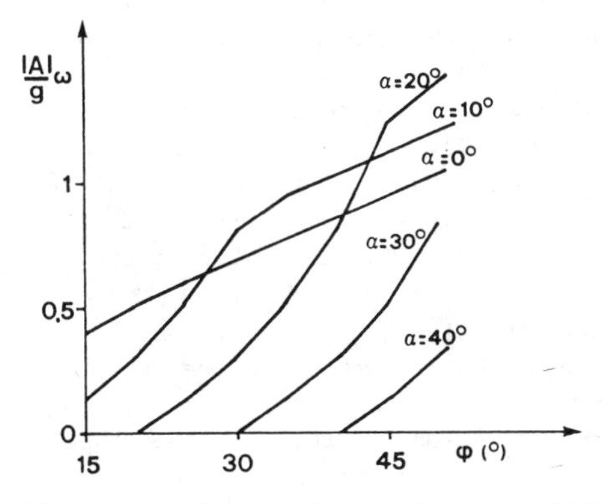

Fig.9 - Amplitude factor for non sliding warying the rock block-base friction values (β=45°, k=0.1).

may present variations of tendency in accordance with the different shape of the relationships which intervene in the determination of the peak amplitude.

4 DEFINITION OF MOTION OF THE ROCK BLOCK

Having imposed a law of velocity of vibration ($\dot{x}_o$, $\dot{y}_o$) on the supporting base, the determination of the relative movements of the block is obtained through direct integration of the (6) or (7) equations with the necessary initial conditions in terms of velocity and displacement. It is indeed the appropriate allocation of the initial conditions, together with the choice of the condition of motion to be analysed, sliding or separation, which is the basis of the method of analysis.

Schematically the procedure is structured in the following way:

1) Subdivision of the time interval, in which the analysis is performed, into intervals each of which is characterized by a specific condition: rest, incipient sliding, incipient separation.

The purpose of the choice is to draw up a list of successive conditions of motion to be applied to the block when, completed a given phase of movement, the current time t enters the successive interval of incipient motion present in the list. The limits of the individual intervals are represented by the zeroes of the functions (4 - incipient sliding) and (5 -incipient separation). The search for the zeroes is conducted by the "false position" method (Ralston 1965).

2) Choice of the condition of incipient motion 3) or 4). The initial time t_o corresponds with the base of the motion interval into which t has entered. As the block was at rest its initial velocity is zero and its initial sliding is the one already accumulated.

3) Condition of sliding: one makes t progress and one takes the integration of (6) for the velocity, then for the displacement. If between two consecutive times one ascertains a variation of sign of the $\dot{x}$ velocity one searches for the zero time (t_o) by the "false position" method and one goes to 5). At every increment of t one tests, if any, the entrance into a separation interval and if it is so one goes to 4), if not one continues with 3).

4) Condition of separation: one defines a progressive value of t and one takes the integration of (7) through the components of relative velocity and displacement. If between the two successive times one ascertains a variation of sign of the y displacement component one searches for the correspondent t_o. The zero condition, determined in this way, corresponds to impact and entails the definition of the initial rebound velocities to continue with 4) unless the kinetic term of the rebound is not deemed to be negligible and in that case one goes to 5).

5) One tests the time t_o against the list of the incipient conditions to locate the interval into which t falls. If one has entered or if one remains in an interval of incipient motion the time t_o, the corresponding displacement and, in case of rebound, the velocity from impact result new initial conditions for 3) or for 4).

The result of the analysis is the cumulated displacement of the rock block during the vibratory event, the time necessary to return to the condition of rest as well as the single "jerks" which correspond to the successive periods of instability originated by the impulse.

The general line of the described procedure is independent of the applied law of vibration which comes into play, however, in the integration phase as well as in the determination of the conditions of incipient motion.

4.1 Damped sinusoid

From the components of the velocity of vibration (9) one obtains the components $\ddot{x}_o$, $\ddot{y}_o$ of the acceleration of the supporting base then one proceeds to the integration of the equations (6) or (7) between the current time t and t_o in accordance to the above-mentioned scheme.

The equations of motion can be integrated in an elementary way and they lead to expressions such as:

$$v = A_1 \sin\omega t \, e^{-k\omega t} + A_2 t + A_3 \qquad (18)$$

for the relative velocities v_x, v_y;

$$d = -\frac{A_1}{\omega(k^2+1)}(k \sin\omega t + \cos\omega t)e^{-k\omega t} +$$

$$+ A_2 t^2/2 + A_3 t + A_4 \qquad (19)$$

for the relative displacements of the block d_x, d_y.

The form of the coefficients of the

expressions (18, 19) depends of the starting equation on which the integration is carried out, whereas the initial conditions (t_o, v_o, d_o) are contained in the integration constants A_3, A_4.

As an example figures 10, 11 indicate the relative displacements of the rock block in accordance with the varying of the parameters which define the oscillation of the supporting base.

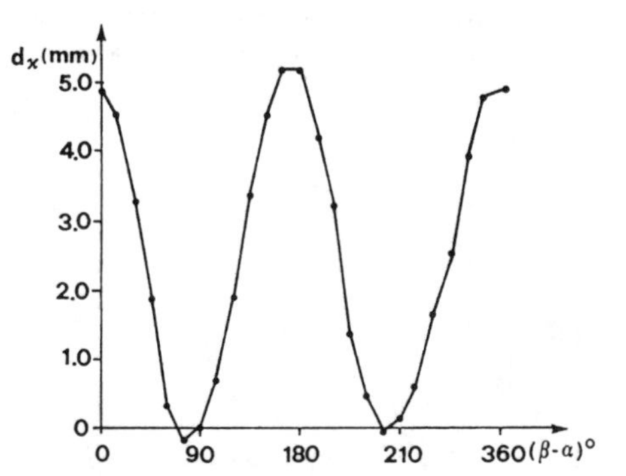

Fig.10 - Relative displacements varying the vibration orientations (α =20°, ϕ_s =35°, ϕ_d =25°, $|A|$ = 100 mm/s, f=20 Hz, k=0.1).

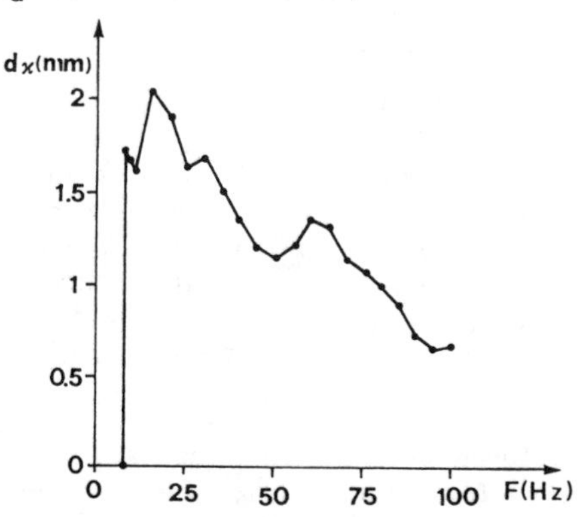

Fig.11 - Relative displacements varying vibration frequencies (α=20°, ϕ_s = 35°, ϕ_d= 25°, $|A|$ = 100 mm/s, ß=45°, k=0.1).

The use of a vibration of the damped sinusoid type does not present any particular difficulties, but the approximation linked with reducing a real event to an intrinsically regular periodic phenomenon may result rough and justifiable only in particular situations such as: single impulse, basic frequency corresponding to a decidedly prevailing amplitude of vibration, reduced phase shift between the longitudinal and vertical components of the vibrations. If these conditions occur one may find acceptable the reduction of the real vibrogram to a regular function such as the damped sinusoid.

In a more general way it is suitable to use the vibrograms as indicated in the following lines.

4.2 Ascertained vibrogram

The vibrograms of the vertical V_Y and longitudinal V_X component of the velocity ascertained during the seismic event must be known; for successive elaborations it is preferable to have a digital recording at one's disposal. Both components are dealt with by a Fourier's analysis in order to obtain their transformation in cosines and sines.

The analysis is conducted either by means of an FFT algorythm (Singleton 1967) or through Fourier's least-squares approximation (Ralston 1965). Both methods generate a sequence of the coefficients of the sine-cosine development which is quite similar from the practical point of view.

It suffices to observe that FFT generates (N/2)+1 coefficients giving back exactly the signal in the N sampling points whereas the approximation to the least-squares is a form of global interpolation, characterized by a generally lower number of coefficients, therefore in the sampling points one usually obtains an estimate of the observed value and not its replication.

Having obtained the coefficients A_{oX},A_{iX}, B_{iX} of the component V_X and A_{oY}, A_{iY}, B_{iY} of the component V_Y the components of the translation velocity in a direction both parallel and orthogonal to the supporting base are given by:

$$\dot{x}_o = \frac{a_o}{2} + \sum_{i=1}^{M} (a_i \cos i\,\Theta + b_i \sin i\,\Theta)$$

$$\dot{y}_o = \frac{a'_o}{2} + \sum_{i=1}^{M} (a'_i \cos i\,\Theta + b'_i \sin i\,\Theta) \qquad (20)$$

where the a_o, a_i's, b_i's coefficients are obtained in the reference frame parallel to oscillating base.

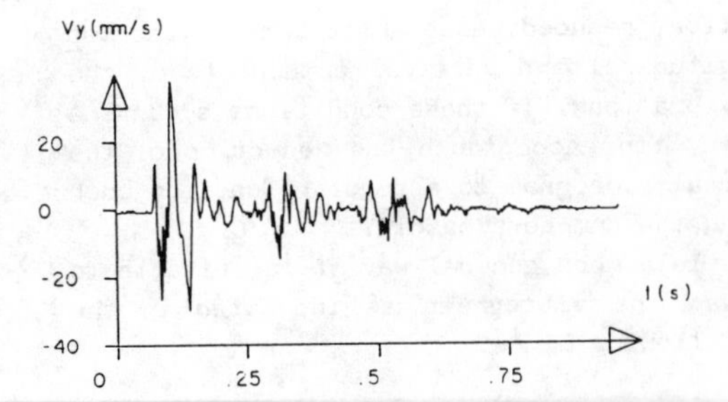

Fig.12a - Blast seismogram - vertical velocity component.

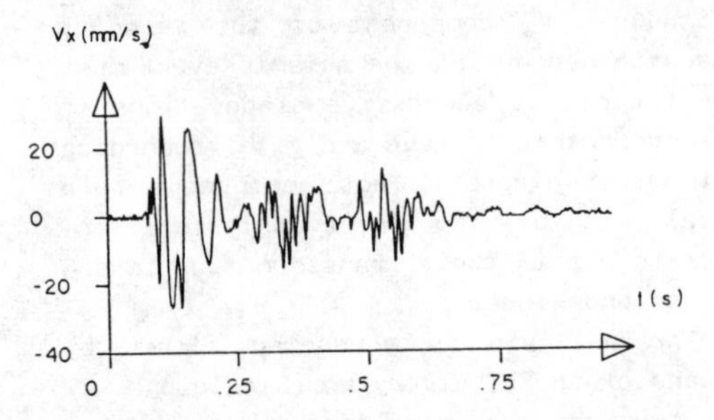

Fig.12b - Blast seismogram - longitudinal velocity component.

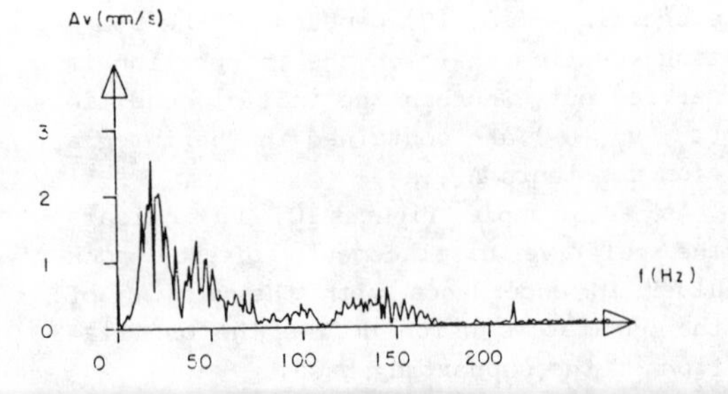

Fig.13a - Blast seismogram - amplitude spectrum for vertical velocity component.

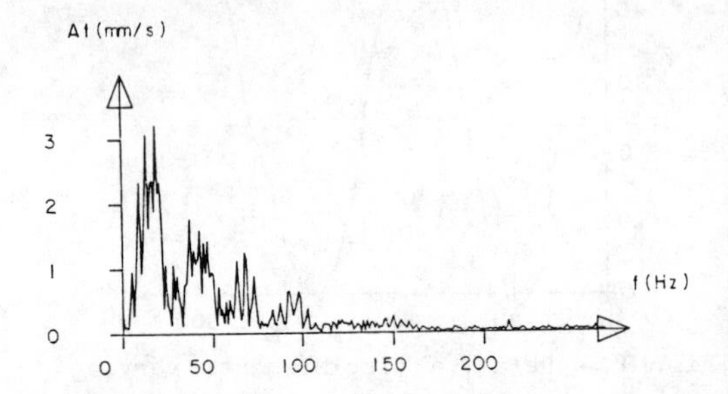

Fig.13b - Blast seismogram - amplitude spectrum for longituinal velocity component.

The relationship between the current time t and the angular interval Θ is: $\Theta = 2\pi t/t_c$, where t_c is the observation time of the signal.

The components of acceleration of the supporting base can be obtained by derivation from (20) whereas the expressions of the block motion always present, with the due transformations, similar form to (18, 19):

$$v = \sum_{i=1}^{M} (d_i \sin i\Theta - c_i \cos i\Theta) + A_2 t + A_3 \quad (21)$$

$$d = -\frac{t_c}{2\pi} \sum_{i=1}^{M} \frac{1}{i} (d_i \cos i\Theta + c_i \sin i\Theta) +$$

$$+ A_2 t^2/2 + A_3 t + A_4 \quad (22)$$

Now the coefficients of the relationships (21, 22) depend on the starting equation (sliding or separation) upon which one conducted the integration with the constants A_3 and A_4 defined by the appropriate initial conditions.

5 BLOCK MOTION UNDER A GIVEN VIBROGRAM

We refer to data obtained in a ballast quarry which exploits dolomitic limestone.

The recording of a typical seismic event is shown in figures 12a,12b which indicate respectively the progress of the vertical and longitudinal velocity induced by blasting.

The distance between the blast area and the recording point, fairly close to some rock blocks under control is ≃ 30 m.

In the signal one finds evident three distinct impulses corresponding to the microdelays which regulate the blast. The analysis of the frequencies (spectra) of figures 13a, 13b indicates a significant content of the frequencies up to 100 Hz with a 19 Hz peak amplitude frequency. The accelerations are shown in figures 14a, 14b and present some peak values greater than the gravity acceleration.

The rock blocks under control are placed, higher up than the blast site and present prevailing possibility of motion along stratification planar joints towards the face featuring average immersion for 20° - 25°.

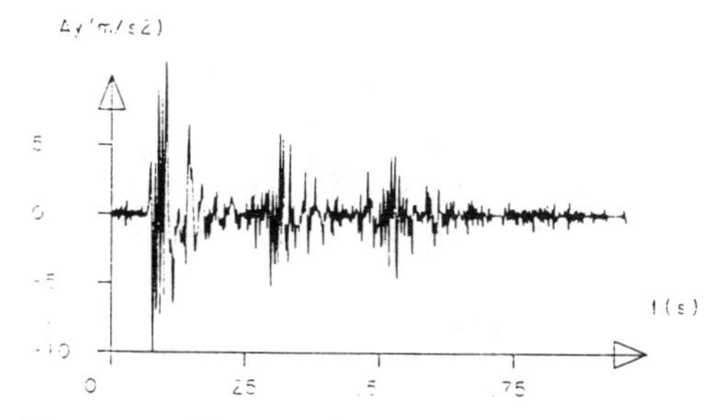

Fig.14a - Blast seismogram - vertical acceleration component.

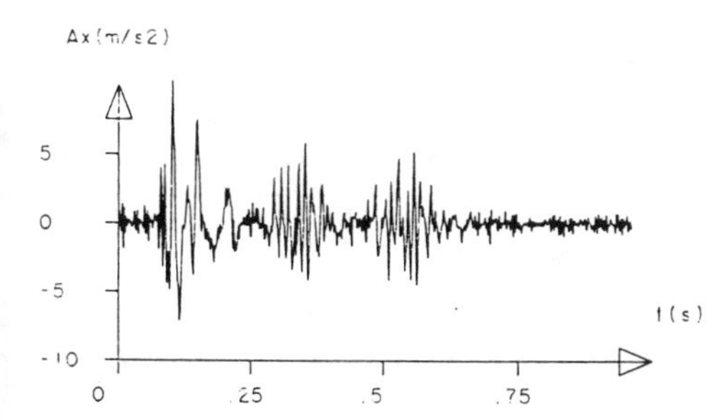

Fig.14b - Blast seismogram - longitudinal acceleration component.

The characteristics of contact of the discontinuity at the base of the blocks have allowed to estimate values of the friction angle mainly between 30° and 40°.

The observations carried out in the quarry did not allow to ascertain a clearly defined relationship between dislocations and distances of rock blocks from the blast area, nevertheless the values obtained at 10 ÷ 50 m are not over 3 mm.

The analysis was conducted using the event represented by the vibrogram and considering various situations, each of them corresponding to different inclinations of the supporting base and with varying values of the block-base friction angle. The obtained results are shown in a diagram in figure 15.

The analysis has shown that the only condition of movement present in practice is a jerky sliding, whereas the intensity of the cumulated displacement grows obviously with the inclination of the base and with the progressive reduction of the features of resistance of the block-to-base contact. For inclinations of the supporting base close to the ones of the layers, one obtains values of relative displacement comparable with the measured values.

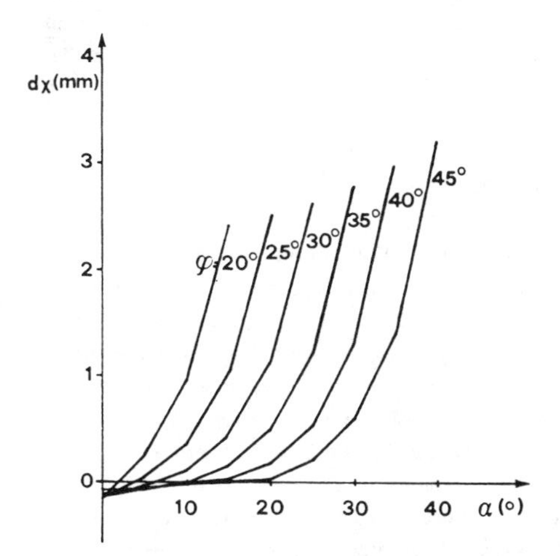

Fig.15 - Block relative displacement varying the base inclinations and at various friction values, for blast vibrogram.

6 CONCLUSIONS

The effects induced by blasts on the conditions of stability of the rock faces are enhanced by the relative dislocations which affected the rock blocks. The periodic succession of seismic impulses, as it happens during quarry work, entails an irreversible accumulation of deformations which may endanger the equilibrium of rock portions which, from a static point of view, can be considered safe.

The suggested analysis scheme allows to foresee the relative displacements acting on a single rock block which is stressed at its base by a vibration featuring either regular progression (damped sinusoid) or any other shape (blast vibrogram). In addition to this one can estimate the peak intensity of a seismic event which does not produce instability.

The conditions of the motion to be imposed on the block supporting base can be defined through Fourier's vibrogram analy sis or more simply by the damped sinusoid parameters which can be obtained at first from the vibrograms themselves.

The comparison between the block displacements ascertained in a quarry and the calculation estimates indicates that the procedure provides an acceptable forecast of the evolution of the conditions of stability of the same blocks subjected to vibrations induced by blasts.

REFERENCES

Bykhovsky, I. 1972. Fundamentals of vibration engineering, Moscow, Mir publishers.

Cundall, P. 1971. A computer model for simulating progressive large-scale movements in blocky rock systems, Proc.Symp. Int. Soc. Rock Mech., Nancy, V. 2.

Cundall, P. 1980, UDEC - A generalised distinct element program for modelling jointed rock, Final Technical Report, European research office, US Army, Report ADA087610.

Dvorak, A. 1965. Die stabilität von Felsböschungen bei sprengungen, Rock Mechanics and Engineering geology, Supplementum II, Springer Verlag: 136-142.

Ralston, A. 1965. A first course in numerical analysis, McGraw-Hill Book Company.

Singleton, R.C. 1967. On computing the fast Fourier transform, Communications of the ACM, 10(10)1967: 647-654.

Walton, O.R. 1983. Explicit particle dynamics model for granular materials, Proc. Fourth international Conference on numerical methods in Geomechanics, Edmonton, Canada: 1261-1268.

Numerical Methods in Geomechanics (Innsbruck 1988), Swoboda (ed.)
© 1988 Balkema, Rotterdam. ISBN 90 6191 809 X

Dynamic response analysis of ground using a coupled finite and boundary element method for time marching analysis

Terumi Touhei
Nuclear Engineering Division, Sato Kogyo Co., Ltd, Japan
Nozomu Yoshida
Engineering Research Institute, Sato Kogyo Co., Ltd, Japan

ABSTRACT: A coupled finite element and boundary element methods for the time marching analysis is presented. The equilibrium condition on the BE-FE interface boundary is relaxed using a method of weighted residuals so as to couple these two method, and a recurrence procedure similar to the Newmark's beta method is derived. This recurrence formula is a scheme to solve BE and FE region simultaneously using the same time increment in both region. Stability condition of this numerical integration scheme is examined and founded that time increment does not affect the stability. Numerical calculations are performed on a ground with topographic irregularity, which shows that the present method gives stable and adequately accurate solutions.

1 Introduction

Lately, boundary element method or coupled boundary and finite element method has come to be used in the dynamic response analysis of a ground. In almost of all these cases, the analysis for the boundary element region is formulated in the frequency domain. The reason seems to be the complicated formulation as well as the difficulty to obtain a stable solution exist in the time marching analysis. However, it is necessary to formulate them in the time domain to analyze transient response and/or nonlinear behavior because accurate result can not be obtained by the analysis in the frequency domain in these cases (Ohmi 1987).

One of the earliest study to formulate boundary element method for the time marching analysis was done by Cole et al (1978). They examined the applicability of the boundary element method on the 2-dimensional scalar wave motion problem and showed that, although only in a limited case, it is necessary to use a 0-th order interpolation function with respect to time for the traction. Mansur (1982 a,b) improved this study and discussed the relation between the length of the boundary elements and time increment for the numerical integration. Fukui et al (1986) showed the discretized method considering the characteristics of the wave propagation and obtained solutions for the 2-dimensional elastic wave motion and scalar wave motion problems, but it does not have sufficient accuracy.

Very recently, Fukui et al (1987) presented the method to combine finite and boundary element methods in the time marching analysis. They analyzed the 2-dimensional scalar wave propagation problem and obtained stable solution by using different analytical schemes in the finite element and boundary element regions, respectively, and employing iterative procedure and least square method to combine the solutions in both regions. The reason why they use different schemes is because of the two contracting requirements; the time increments in the finite element region should be small so as to obtain the stable solution whereas those in boundary element region can not be set so small because it is related to the length of boundary elements.

In this paper, a method which does not need iterative procedure is presented for the 2-dimensional scalar wave propagation problem for the time marching analysis. A method of weighted residuals is employed so as to satisfy the equilibrium and kinematic condition between the finite element and boundary element regions as well as in the finite element region. An recurrence formula in time similar to the Newmark's beta method is derived, which enable to solve finite element region and boundary element region simultaneously.

2 Basic Equations for Finite Element and Boundary Element Methods

The wave propagation equation for the scalar wave motion (SH wave) is represented as

$$\mu(\frac{\partial^2}{\partial x^2} + \frac{\partial^2}{\partial y^2})u - \rho\frac{\partial^2}{\partial t^2}u = 0 \qquad (2.1)$$

where μ denotes shear modulus, x and y are coordinate components, u denotes displacement, ρ denotes mass density, and t denotes time. Finite element expression of Eq. 2.1 is obtained by employing the Gelerkin's method in space as

$$[M]\{\ddot{u}\} + [K]\{u\} = \{P(t)\} \qquad (2.2)$$

Here, [M] and [K] indicate mass and stiffness matrices, respectively, $\{\ddot{u}\}$ and {u} are acceleration and displacement vectors, respectively, and {P(t)} indicate external force vector. If there is no body force except inertia force, only the component of {P} on the interface boundary between the boundary element region has non-zero value, which corresponds to the traction from the boundary element region, and other components are zero.

The problem considered here is shown in Fig. 2.1; the incident wave enters through boundary Γ and scattering waves are produced. By assuming zero displacement and velocity before time t less than zero, boundary integral equation in the time domain is expressed as

$$\varepsilon u(r,t) + \int_0^t\int_\Gamma T(r,t:r',t')\, u(r',t')\, dr'dt' = \int_0^t\int_\Gamma G(r,t:r',t')\, \sigma(r',t')\, dr'dt' + \bar{u}(r,t) \qquad (2.3)$$

where ε is a constant determined from the shape of the boundary, σ denotes traction, $\bar{u}$ indicates incident wave, and r is a position on the boundary. The terms G and T in Eq. 2.3 are the Green's function correspond to displacement and traction, respectively, and are given by

$$G(r,t:r',t') = \frac{C\, H[C(t-t') - |r-r'|]}{2\pi\mu\sqrt{C^2(t-t')^2 - (r-r')^2}} \qquad (2.4)$$

$$T(r,t:r',t') = -\frac{\partial(r-r')}{\partial n'}\frac{C\{C(t-t')-|r-r'|\}}{2\pi\sqrt{C^2(t-t')^2-(r-r')^2}^{\,3}} \cdot H[C(t-t')-|r-r'|] - \frac{\partial(r-r')}{\partial n'}\frac{G(r,t:r',t')}{C}\frac{\partial}{\partial t'}$$

Here, C denotes the phase velocity of the shear wave, H[ ] is an unit step function and n' indicates the coordinate component in the normal direction to the boundary.

So as to obtain the formulation for the time marching analysis, the boundary of the boundary element region is divided into M elements and the time axis between zero and current time is divided into N increments. Then the displacement and traction are approximated by the use of the interpolation functions ζ_k and η_k for time and ϕ_j and ψ_j for space, respectively, as

$$u(r,t) = \sum_{j=1}^{M}\sum_{k=1}^{N} \phi_j(r)\zeta_k(t)u_j^k$$
$$\sigma(r,t) = \sum_{j=1}^{M}\sum_{k=1}^{N} \psi_j(r)\eta_k(t)\sigma_j^k \qquad (2.5)$$

Here u_j^k and σ_j^k indicate the displacement and traction at the k-th time increment of the j-th node, respectively.

Discretized formulation of the wave propagation problem is obtained by substituting Eq. 2.5 into Eq. 2.3. Here it is noted that the integration in time can be done analytically but that in space should use approximate procedure such as Gauss's integral formula. The integral including singular points are performed analytically in both time and space domains. The discretized equation is expressed as

$$\sum_{j=1}^{M}\sum_{k=1}^{N} H_{ij}^{Nk}\, u_j^k = \sum_{j=1}^{M}\sum_{k=1}^{N} G_{ij}^{Nk}\, \sigma_j^k + \bar{u}_i^N \qquad (2.6)$$

where

$$H_{ij}^{Nk} = \varepsilon\delta_{ij}\delta_{Nk} + \int_0^{t_N}\int_\Gamma T(r_i,t_N:r',t') \cdot\phi_j(r')\zeta_k(t')dr'dt'$$
$$G_{ij}^{Nk} = \int_0^{t_N}\int_\Gamma G(r_i,t_N:r',t')\psi_j(r')\eta_k(t')\, dr'dt' \qquad (2.7)$$

Here δ_{ij} and δ_{Nk} are Kronecker's deltas.

From the nature of the Green's function, the following equations hold for the influence coefficients H_{ij}^{Nk} and G_{ij}^{Nk}.

$$H_{ij}^{Nk} = H_{ij}^{N+p\ k+p},\quad G_{ij}^{Nk} = G_{ij}^{N+p\ k+p} \qquad (2.8)$$

Here p can be an arbitrary integer greater than or equal to -k. Therefore, by putting p=-k, the influence coefficients H_{ij}^{N-k} and G_{ij}^{N-k} are used instead of H_{ij}^{Nk} and G_{ij}^{Nk}, respectively, hereafter.

The matrix form equation of the basic equation for the boundary element region is rewritten from Eq. 2.6 as

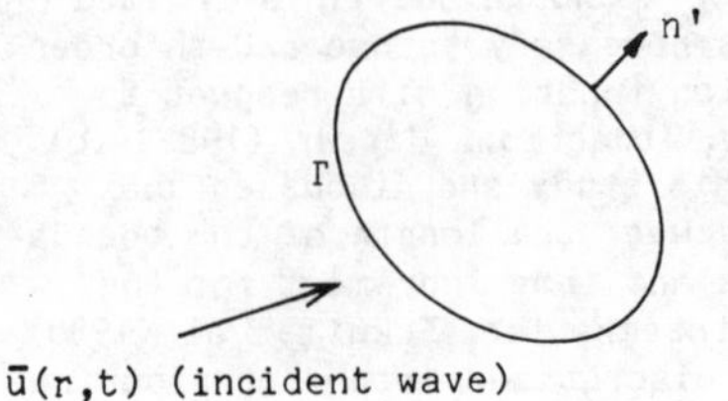

Fig. 2.1 Analyzed Model in the BEM

$$\sum_{k=1}^{N}[H^{N-k}]\{u^k\} = \sum_{k=1}^{N}[G^{N-k}]\{\sigma^k\} + \{\bar{u}^N\} \quad (2.9)$$

For the simplicity of the expression, $[H^{N-N}]$ and $[G^{N-N}]$ are written as [H] and [G], respectively, hereafter.

Equations 2.2 and 2.9 are the basic equations in the finite element and boundary element regions, respectively.

3 Formulation of Combined Method in Time Domain

The discretized formulation of the boundary integral equation, Eq. 2.9, gives the relation between the nodal displacements and the tractions. On the other hand, finite element formulation, Eq. 2.2, gives the relation between the nodal displacements and nodal loads. Therefore, they can not be combined directly; Eq. 2.9 should be changed to give the relation between the nodal displacement and nodal load.

The virtual work done at N-th time increment, $\delta\P$, is obtained by referring the Eq. 2.5 as

$$\delta\P = \delta\{u^N\}^T\int_\Gamma\{\phi\}^T\{\psi\}d\Gamma\ \{\sigma^N\} \quad (3.1)$$

Introducing nodal the load vector at the N-th time increment, $\{P^N\}$, the same virtual work is obtained as

$$\delta\P = \delta\{u^N\}^T\{P^N\} \quad (3.2)$$

By equating the right hand sides of Eqs. 3.1 and 3.2,

$$\{P^N\} = \int_\Gamma\{\phi\}^T\{\psi\}d\Gamma\ \{\sigma^N\} \quad (3.3)$$

The last term of Eq. 3.3, $\{\sigma^N\}$, is a traction vector at N-th time increment, and is obtained by solving Eq. 2.9 as

$$\{\sigma^N\} = [G]^{-1}[H]\{u^N\} - \{f^N\} \quad (3.4)$$

where

$$\{f^N\} = [G]^{-1}\Big(\sum_{k=1}^{N-1}[G^{N-k}]\{\sigma^k\} - \sum_{k=1}^{N-1}[H^{N-k}]\{u^k\} + \{\bar{u}^N\}\Big) \quad (3.5)$$

The relation between the nodal displacements and nodal loads is obtained from Eqs. 3.3 to 3.5 as

$$\{P^N\} = [\hat{K}]\{u^N\} - [D]\{f^N\} \quad (3.6)$$

where

$$[D] = \int_\Gamma\{\phi\}^T\{\psi\}d$$
$$[\hat{K}] = [D][G]^{-1}[H] \quad (3.7)$$

As can be seen in Eq. 3.6, $[\hat{K}]$ corresponds to the stiffness matrix* in the boundary element region and [D] corresponds to the distributed matrix which connects the traction and the nodal load.

The governing equation to solve finite element and boundary element region can be obtained by combining Eqs. 2.2 and 3.6 maintaining the kinematic and equilibrium conditions on the interface boundary,

$$\{u\}_B = \{u\}_F$$
$$\{P\}_B + \{P\}_F = \{0\} \quad (3.8)$$

where the subscripts B and F indicate the quantities which belong to the boundary element and finite element regions.

The governing equation satisfying the condition in Eq. 3.8 seems to be obtained by substituting Eq. 3.6 into Eq. 2.2, but it is still difficult because Eq. 2.2 is a differential equation whereas Eq. 3.6 is a recurrent formula; the condition may be satisfied in some instances but not always. To overcome this difficulty, the displacement or the traction is approximated by the use of interpolation functions and a method of weighted residuals is employed to relax the equilibrium conditions.

In the boundary element region, the nodal load is approximated. Employing the 0-th order interpolation function during each time increment, nodal load in the boundary element region between $t_N-2\Delta t\leq t\leq t_N$ is expressed as

$$\{P\}_B = \{P^N\}(H(\xi)-H(\xi-1)) + \{P^{N-1}\}(H(-\xi)-H(-1-\xi)) \quad (3.9)$$

Here, from Eq. 3.6,

$$\{P^N\} = [\hat{K}]\{u^N\} - [D]\{f^N\}$$
$$\{P^{N-1}\} = [\hat{K}]\{u^{N-1}\} - [D]\{f^{N-1}\} \quad (3.10)$$

and ξ is a dimensionless time parameter expressed as

$$\xi = t/\Delta t + 1 - N \quad (3.11)$$

In the finite element region, the displacement is approximated. Here it is noted that the order of the interpolation function should be greater than or equal to two so that the second derivative with respect to time does not disappear. By using the second order interpolation function, the displacement between the time $t_N-2\Delta t\leq t\leq t_N$ is expressed as

---

* The matrix $[\hat{K}]$ is, in general, unsymmetrical matrix. When combining to the finite element method, however, symmetric stiffness matrix is more convenient. Past study (Brebbia 1984) indicates that the error to replace the stiffness matrix $\hat{K}$ into the symmetric stiffness matrix K^* as follows is fairly small.

$$[K^*] = ([\hat{K}]^T + [\hat{K}])/2$$

Therefore, in the numerical examples shown later, K^* is used instead of $\hat{K}$.

$$\{u\}=\{u^{N-2}\}\frac{\xi(\xi-1)}{2}+\{u^{N-1}\}(1-\xi^2)+\{u^N\}\frac{\xi(\xi+1)}{2} \qquad (3.12)$$

where $\Delta t=t_N/N$ denotes time increment. From Eqs. 2.2 and 3.12, it is known that the force vector of the finite element region on the interface boundary is a second order function with time. On the other hand, that of the boundary element region is a piesewise constant function with time as shown in Eq. 3.9. Hence it is impossible to satisfy Eq. 3.8 at all instances. The same discussion can be made inside the finite element region; the equilibrium equation, Eq. 2.2, does not hold at all instances even inside the finite element region if Eq. 3.12 is used. Zienkiewicz (1977) showed a method to obtain a recurrence formula from Eqs. 2.2 and 3.12 in the finite element region. He relaxed the equilibrium condition by the use of the method of weighted residuals as

$$\int_{-1}^{1} w(\xi)([M]\{\ddot{u}\} + [K]\{u\})d\xi = \{0\} \qquad (3.13)$$

where $w(\xi)$ is a weight function. The same method is applied on the interface boundary. The relaxed equilibrium equation is written as

$$\int_{-1}^{1} w(\xi)(\{P\}_F + \{P\}_B)d\xi = \{0\} \qquad (3.14)$$

Taking an even function as a weight function, the governing equation is obtained from Eqs. 2.2 and 3.12 to 3.14 as

$$\begin{aligned}&[M+\beta\Delta t^2K+\Delta t^2\hat{K}/2]\{u^N\}\\ &=[2M-(1-2\beta)\Delta t^2K-\Delta t^2\hat{K}/2]\{u^{N-1}\}\\ &+[-M-\beta\Delta t^2K]\{u^{N-2}\}+\Delta t^2[D](\{f^N\}+\{f^{N-1}\})/2\end{aligned} \qquad (3.15)$$

Here β is a constant and is defined as

$$\beta = \int_{-1}^{1} w\frac{\xi(1+\xi)}{2}d\xi \Big/ \int_{-1}^{1} w d\xi \qquad (3.16)$$

Equation 3.15 is a recurrence formula to solve the finite element and boundary element regions simultaneously with the same time increment; the displacement $\{u^N\}$ can be calculated from the past displacements $\{u^{N-1}\}$ and $\{u^{N-2}\}$ and other known quantities. Once the displacement is obtained, the velocity and acceleration are obtained by differentiating Eq. 3.12 with respect to time.

4 Discussion

It is important to recognize the accuracy and the stability of the numerical integration procedure. The former will be discussed through numerical example in the next section, and the latter is discussed here.

Suppose that Eq. 3.15 can be separeted into the uncoupled equation through eigen value analysis, each equation will be written as

$$(m+\beta\Delta t^2k+\Delta t^2\hat{k}/2)u^N + \{-2m+(1-2\beta)\Delta t^2k +\Delta t^2\hat{k}/2\}u^{N-1} + (m+\beta\Delta t^2k)u^{N-2} = 0 \qquad (4.1)$$

Here, m, k and $\hat{k}$ are the effective quantities in this mode and u indicates participation factor. Putting

$$p = k/m\ \Delta t\ , \quad q = \hat{k}/m\ \Delta t \qquad\qquad u^N = \lambda u^{N-1}, \quad u^{N-1} = \lambda u^{N-2} \qquad (4.2)$$

Eq. 4.1 is rewritten as

$$\lambda^2(1+\beta p+q/2)+\lambda\{-2+(1-2\beta)p-q/2\}+(1+\beta p) = 0 \qquad (4.3)$$

The condition $|\lambda|\leq 1$ should be satisfied for the recurrence formula to be stable. When Eq. 4.3 has complex root or multiple root,

$$|\lambda|^2 = \frac{1 + \beta p}{1 + \beta p + q/2} \qquad (4.4)$$

in which case the stability is always guaranteed considering that both p and q are positive quantities. In other words, when the discriminant of Eq. 4.3 satisfies the condition

$$(1-4\beta)p^2 - 4p + q^2/4 - pq \leq 0 \qquad (4.5)$$

the recurrence formula is stable. Considering that p is positive, Eq. 4.5 holds if the following inequality holds.

$$(1-4\beta)p^2 + q^2/4 - pq \leq 0 \qquad (4.6)$$

Therefore the stability is discussed using Eq. 4.6 instead of Eq. 4.5. It is noted that time increment does not affect the stability of the formula; only the stiffnesses of the finite element and boundary element affect it. The condition is equivalent to the following condition.

$$\begin{aligned}&p \geq \frac{q}{2(1+2\sqrt{\beta})} && \text{for } \beta\geq\frac{1}{4}\\ &\frac{q}{2(1-2\sqrt{\beta})} \geq p \geq \frac{q}{2(1+2\sqrt{\beta})} && \text{for } \beta<\frac{1}{4}\end{aligned} \qquad (4.7)$$

In many actual situation, the upper condition is satisfied, hence $\beta\geq 1/4$ will usually give the stable numerical integration scheme.

5 Numerical Example

So as to know the accuracy of the proposed formula, numerical example is performed. In the examples, six points triangular element is used as the finite element model, and second order function is used as the interpolation function of the displacement in space in the boundary element model. The first order interpolation function is used for the traction in space.

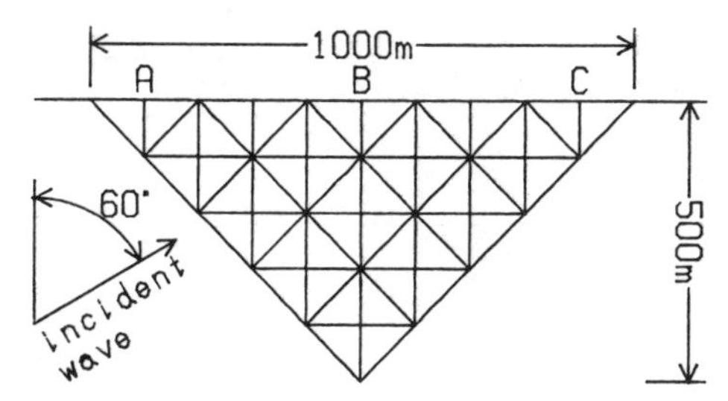

Fig. 5.1 Analyzed Model

In the first example, the result of the numerical calculation is compared with the exact solution. The analyzed model is a ground shown in Fig. 5.1. The shear modulus is set $10000kN/m^2$ and the mass density is set $1t/m^3$ for both regions. The incident waveform is shown in Fig. 5.2. The time increment is 0.05 second and the value of β is set 1/4 in the calculation. The result is shown in Fig. 5.3 comparing with the exact solution. Here, the exact solution is obtained by solving the scalar wave propagation equation for the homogeneous semi-infinite region. As shown in the figure, the agreement between both results are very good in spite of the frequently oscillating incident wave and large time increment.

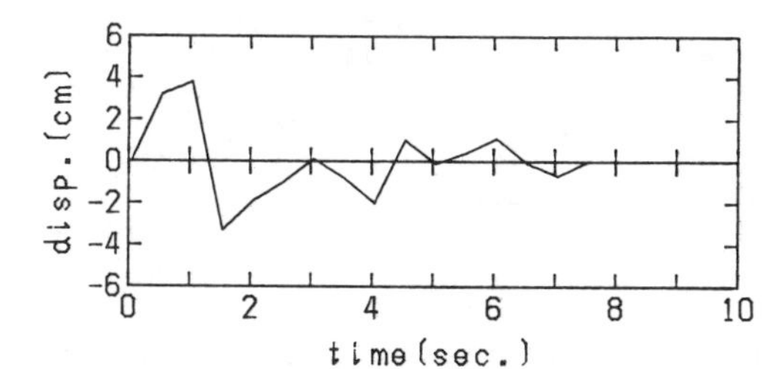

Fig. 5.2 Incident waveform

In the next example, the effect of the differences between the stiffnesses in the finite element and boundary element regions are examined. The same model shown in Fig. 5.1 is analyzed. A triangular pulse wave shown in Fig. 5.4 is used as an incident wave. The analysis in the frequency domain is usually not good at analyzing these pulse wave. The shear modulus of the boundary element region, μ_B is chosen $1000000kN/m^2$ and that in the finite element region, μ_F, is taken 100000, 10000 or $1000kN/m^2$. The mass

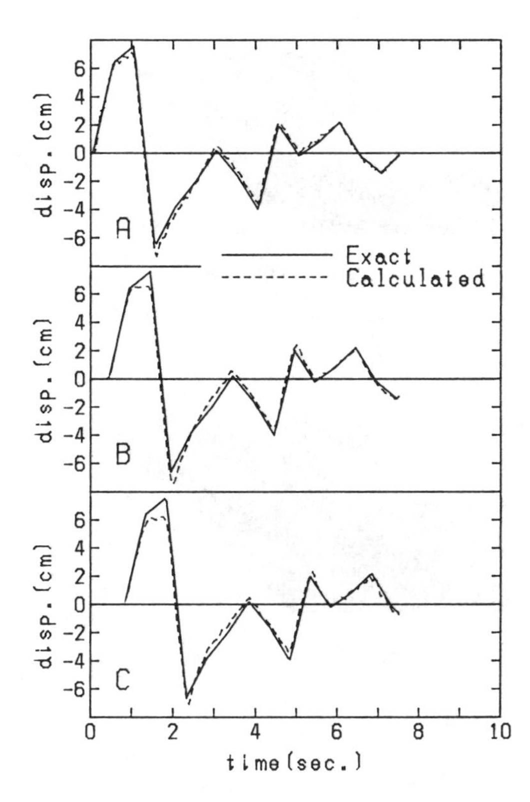

Fig. 5.3 Comparison of Displacement

density is selected $2t/m^3$ in both regions. The value of μ_F is only 1/1000 of μ_B in the last case. Therefore almost all the actual situation is covered in these example. The time increment is 0.06 second and the value of β is 1/4 in this case. The whole displacement time history on the surface of the finite element region up to 4 seconds is shown in Fig. 5.5, and the displacement at three points, A, B and C, are compared in Fig. 5.6. Here the time axis is zero when the wave front of the incident wave reaches the left upper corner of the finite element region.

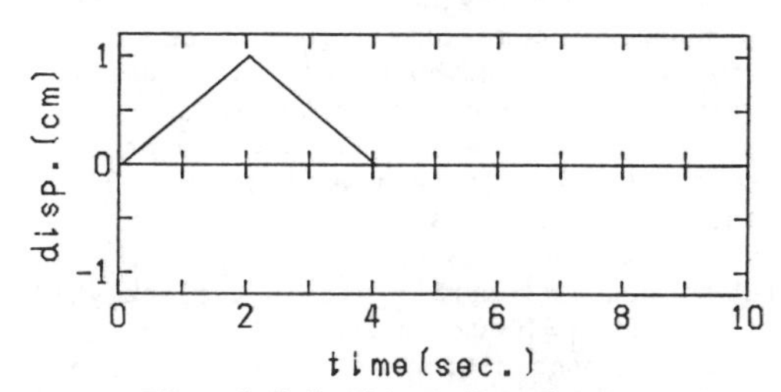

Fig. 5.4 Incident Waveform

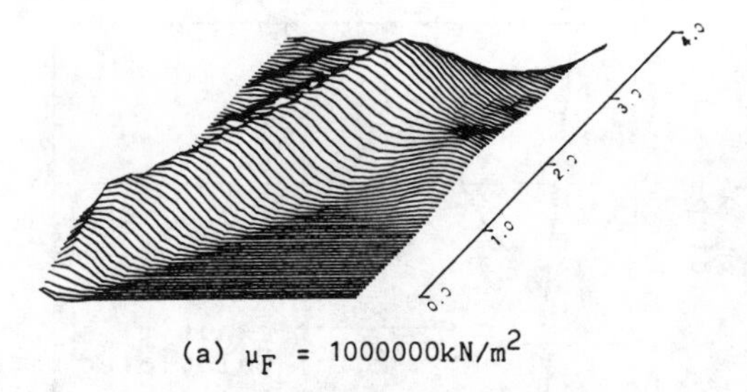

(a) μ_F = 1000000kN/m^2

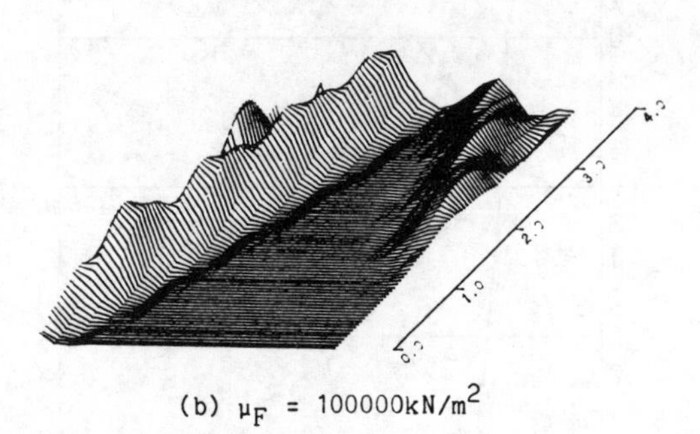

(b) μ_F = 100000kN/m^2

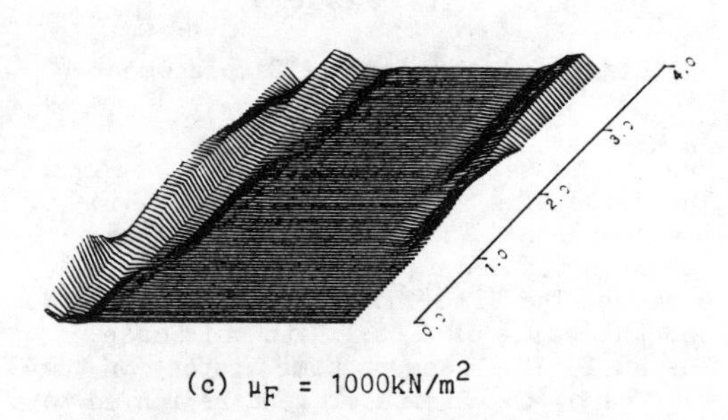

(c) μ_F = 1000kN/m^2

Fig. 5.5 Displacement on the Surface

At point A, the main shock arrives almost simultaneously after it enters the finite element region. The period of the main shock becomes smaller as μ_F becomes larger. It is also observed that a scattering wave appears after the main shock. It takes some time for the wave to arrive points B and C. It arrive at about 1.5 seconds at point B when μ_F=100000kN/m^2, and about 4 seconds when μ_F=10000kN/m^2, but it does not arrive within the analyzed time when μ_F=1000kN/m^2. This is also because of the difference of the stiffness of the finite element region. At point C, a diffracted wave arrives at about 2 seconds from the right side before the

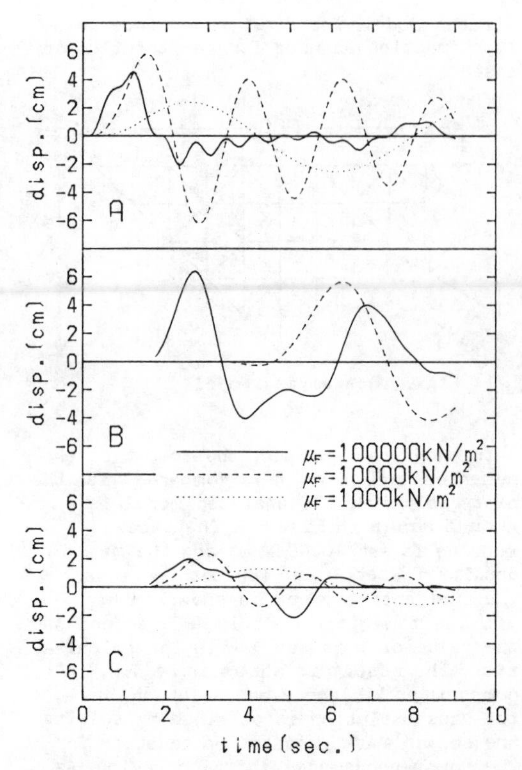

Fig. 5.6 Comparison of Displacement

wave through finite element region arrives. As shown in these figures, the behavior of the ground is well explained by the analysis, and it is also confirmed that, as described before, the present scheme is stable for almost all of the actual situation.

6 Concluding remarks

An analytical scheme is presented for the time marching analysis of the wave propagation problem by the combined finite element and boundary element methods. The combination is performed by relaxing the equilibrium condition on the interface between the finite element and boundary element region as well as inside the finite element region. As the result, it becomes possible to solve finite element region and boundary element region simultaneously under the same time increment. Through the examination on the stability condition of the numerical integration scheme, it is recognized that the magnitude of the time increment does not affect the stability.

REFERENCES

C.A. Brebbia, J.C.F. Telles and L.C. Wrobel, Boundary Element Techniques Theory and Application in Engineering, Springer-Verlag, 1984

D.M. Cole, D.D. Kosloff and J.B. Minister, A Numerical Boundary Integral Equation Method for Elastodynamics, I, Buli. Seism. Soc. Amer. 68, 1331-1357 (1978)

T. Fukui, Time marching Analysis of Boundary Integral Equation in Two Dimensional Elastodynamics, Proc. 4th Int. Symp. on Numerical Methods in Engineering, 1986

T. Fukui and Y. Ishida, Time Marching BE-FE Methods in Wave Problem, Proc. 1st Japan-China Symp. Boundary Element Methods, 95-106, 1987

W.J. Mansur and C.A. Brebbia, Formulation of the Boundary Element Method for transient Problems Governed by the Scalar Wave Equation, Appl. Math. Modelling 6, 307-312 (1982)

W.J. Mansur and C.A. Brebbia, Application of the Boundary Element Method to Solve the Transient Scalar Wave Equation, in Boundary Elements in Engineering (C.A. Brebbia, Ed.), Springer-Verlag, Berlin, 1982

M. Ohmi, S. Sasaki and Y. Tosaka, Dynamic Response of 3-D Embedded Foundations by Time Domain BEM, Journal of Structual Engineering 33B, 93-101 (1987)

Zienkiewicz, The Finite Element Method, third edition, MaGraw-Hill, 1977

Numerical Methods in Geomechanics (Innsbruck 1988), Swoboda (ed.)
© 1988 Balkema, Rotterdam. ISBN 90 6191 809 X

A numerical study of dynamic cavity expansion

M.F.Randolph
University of Western Australia

D.S.Pennington
De Beers Industrial Diamonds Ltd, Ireland
(Formerly University of Cambridge)

ABSTRACT: Recent advances in one-dimensional wave equation analysis of pile driving have emphasised the importance of radiation damping. At the base of the pile, radiation damping will occur during plastic penetration of the pile tip. This will entail that the bearing pressure will be greater than the ultimate static value, even in a non-viscous soil. In order to assess likely magnitudes of the dynamic bearing capacity, finite element analyses of cavity expansion have been undertaken at high expansion rates. Results are presented for spherical cavity expansion at different rates in cohesive soil. It is shown that the peak internal pressure is a function of the initial acceleration level and the rate of cavity expansion, while the limiting pressure at large strain is a function of cavity expansion rate. The consequences of the results for pile driving analysis are discussed.

1 INTRODUCTION

Numerical analysis of pile driving may be performed by means of three-dimensional (axisymmetric) finite element analysis, or, more simply, by a one-dimensional wave equation analysis of the pile. In the latter case, the soil is generally modelled by simple spring and dashpot components, with a plastic slider to limit the soil resistance at any particular node. Choice of appropriate spring, dashpot and slider parameters must allow for radiation damping (due to the inertia of the soil around the pile) and also viscous damping of the soil at high strain rates.

Randolph and Simons (1986) have discussed soil models for pile driving analysis, and propose that the response of soil along the pile shaft may be modelled as shown in Fig. 1(b). The radiation damping is confined to the response prior to slip between pile and soil (after which no further energy is radiated into the far field). However, at the base of the pile, the original model of Smith (1960) shown in Fig. 1(a) is appropriate, since radiation damping will continue even during plastic penetration of the pile.

The effect of radiation damping at the pile base will be to augment the bearing pressure over and above the static bearing capacity. This phenomenon has been explored by Smith and Chow (1982), using dynamic finite element analysis of the full axisymmetric problem. They show values of bearing capacity factor ranging from 9 for soil of high strength, up to 40 - 50 for low strength soil.

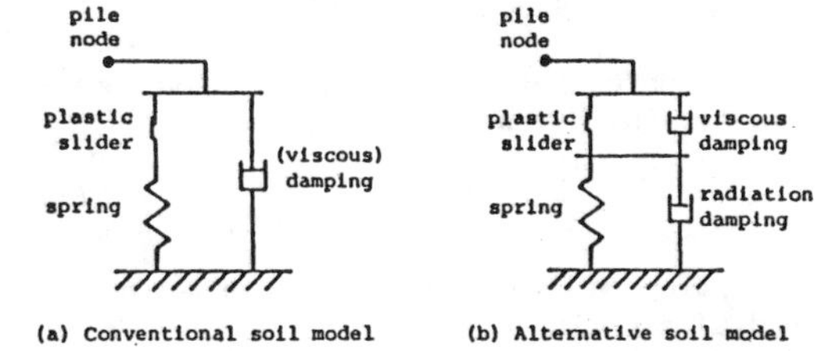

Figure 1. Soil models for one-dimensional analysis of pile driving

In order to explore the effects of radiation damping on the tip resistance of piles during driving, analyses have been undertaken of dynamic cavity expansion (cylindrical and spherical). The limitations of the cavity expansion analogue of bearing capacity have been emphasised by Baligh (1986). However, there is strong empirical evidence linking the bearing capacity of foundations to limit pressures measured in cylindrical or spherical cavity expansion (for example, Baguelin et al (1978), Yeung and Carter (1987)).

The simplicity of the cavity expansion approach is its main advantage. The approach may be used to provide general guidance on dynamic magnification of bearing capacity, while avoiding the complexity of a full axisymmetric finite element analysis.

2 CAVITY EXPANSION ANALYSIS

2.1 Elastic solution

Before describing the finite element analyses, it is helpful to consider the elastic response of a spherical cavity under dynamic loading conditions. Figure 2 shows the general arrangement, with a cavity of initial radius R, internal pressure p, and displacement δ. At a general radius r, the radial movement u, may be written in terms of a potential function ϕ, as

$$u = \frac{\partial \phi}{\partial r} \tag{1}$$

The strains are then

$$\epsilon_r = -\frac{\partial u}{\partial r} = -\frac{\partial^2 \phi}{\partial r^2} \tag{2}$$

$$\epsilon_\theta = -\frac{u}{r} = -\frac{1}{r}\frac{\partial \phi}{\partial r} \tag{3}$$

It is convenient to introduce the volumetric strain, given by

$$\epsilon_v = \epsilon_r + 2\epsilon_\theta = -\nabla^2 \phi \tag{4}$$

For soil of density ρ, the equation of radial equilibrium is

$$\frac{\partial \sigma_r}{\partial r} + 2\,\frac{\sigma_r - \sigma_\theta}{r} = -\rho \frac{\partial^2 u}{\partial t^2} \tag{5}$$

The elastic relationships may be expressed as

$$\sigma_r = 2G\epsilon_r + \lambda \epsilon_v \tag{6}$$

$$\sigma_\theta = 2G\epsilon_\theta + \lambda \epsilon_v \tag{7}$$

where G is the shear modulus of the soil and λ is Lame's constant.

Combining the above equations leads to a governing equation of the form

$$-\frac{\partial}{\partial r}\left[(\lambda+2G)\nabla^2\phi\right] = -\rho\frac{\partial^2 u}{\partial t^2} = -\rho\frac{\partial}{\partial r}\frac{\partial^2 \phi}{\partial t^2} \tag{8}$$

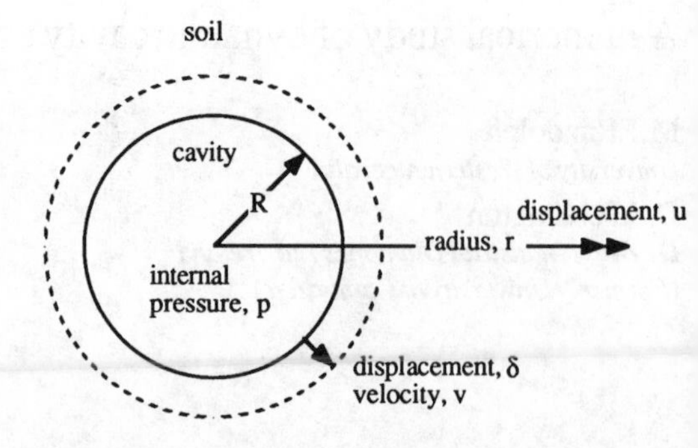

Figure 2. Notation for cavity expansion

This may be integrated to give

$$(\lambda+2G)\,\nabla^2\phi = \rho\,\frac{\partial^2 \phi}{\partial t^2} + C(t) \tag{9}$$

For an infinite domain, the function ϕ must tend to zero at large radius, and hence the function C(t) may be taken as zero.

The above equation leads to solutions of the form

$$\phi = -\frac{1}{r}\,F(t - \frac{r}{c}) \tag{10}$$

where c is the dilatational wave velocity, given by

$$c^2 = (\lambda+2G)/\rho \tag{11}$$

The form of the solution assumes outgoing waves only. The corresponding expressions for the radial displacement u and radial stress σ_r are

$$u = \frac{1}{r^2}\,F + \frac{1}{rc}\,F' \tag{12}$$

$$\sigma_r = (\lambda+2G)\frac{F''}{rc^2} + 4G(\frac{F}{r^3} + \frac{F'}{r^2 c}) \tag{13}$$

where F' and F" represent the first and second derivatives of the function F with respect to the argument (t-r/c).

For a spherical cavity of internal radius R, these expressions lead to a differential equation relating the internal pressure p and displacement δ of

$$p + \frac{R}{c}\dot{p} = \frac{4G}{R}\left[\delta + \frac{R}{c}\dot{\delta} + \left[\frac{R}{c}\right]^2\frac{(\lambda+2G)}{4G}\ddot{\delta}\right] \tag{14}$$

where the superscript dot represents the time derivative.

Consider now a harmonic internal displacement of the form

$$\delta = \delta_o \sin\omega t \qquad (15)$$

The magnitude of the resulting pressure variation is given by

$$|\Delta p| = 4G\frac{\delta_o}{R}\left[(1-Aa_o^2)^2 + (Ba_o^3)^2\right]^{1/2}/(1+a_o^2) \qquad (16)$$

where a_o is a non-dimensional frequency given by $a_o = \omega R/c$ and the two constants A and B are

$$A = \frac{\lambda-2G}{4G} \; ; \qquad B = \frac{\lambda+2G}{4G} \qquad (17)$$

It may also be shown that the magnitude of the deviator stress variation (at the cavity wall) to the peak internal pressure is

$$|\Delta q| = 2G\frac{\delta_o}{R}\left[(3+a_o+2a_o^2+a_o^3) + a_o^2(5+2a_o)^2\right]^{1/2}/(1+a_o^2) \qquad (18)$$

At low frequencies, equations (16) and (18) both revert to the static results - a stiffness of $\Delta p/\delta = 4G/R$ and a deviator stress of $\Delta q/\delta = 6G/R$. However, as the frequency of excitation increases, both these quantities show a marked variation, ultimately becoming proportional to the frequency. At high frequencies, the ratio of the deviator stress amplitude to pressure amplitude approaches 0.5/B.

As an illustration of the variation of these quantities, consider a soil with shear modulus G = 10 MPa, and Poisson's ratio ν = 0.49 (hence λ = 490 MPa). The constants A and B are equal to 11.75 and 12.75 respectively. Figure 3 shows the variation of the three quantites $|\Delta p|$, $|\Delta q|$ (both normalised by $4G\delta_o/R$) and $|\Delta p|/|\Delta q|$ with frequency of excitation. It may be seen that there is a minimum stiffness at a freqency close to $1/\sqrt{A}$, at which stage the pressure is $\pi/2$ out of phase with the internal displacement. At high frequencies, the stiffness increases sharply, while the deviator stress amplitude increases at a more moderate rate. One of the consequences of the low deviator stress ratio at high frequency is that it is possible for the soil to sustain large amplitude pressure variations - well in excess of the value to cause yielding under static conditions - while still remaining elastic.

2.2 Finite element analysis

Finite element analysis of the dynamic expansion of cylindrical and spherical cavities as been undertaken using the program DYCE. The structure of the program is based on the code FEACE, developed by Carter and Yeung (1985) for static cavity expansion.

The basic equation of virtual work approximated in the finite element analysis is

$$\int[\delta\underline{u}^T(\rho\ddot{\underline{u}} + c\dot{\underline{u}}) + \delta\underline{\epsilon}^T\underline{\sigma}]dV = 0 \qquad (19)$$

where $\delta\underline{u}$ is a vector of arbitrary displacements, $\delta\underline{\epsilon}$ are the corresponding strains and c is a damping parameter for the material. By expressing the strains and displacements in terms of the nodal displacements, $\underline{\delta}$, in the normal way, and writing the stresses as

$$\underline{\sigma} = \underline{\sigma}_o + \Delta t\, D\, \dot{\underline{\epsilon}} \qquad (20)$$

the equation may be re-written in an incremental form

$$\int[\rho N^T N\ddot{\underline{\delta}} + cN^T N\dot{\underline{\delta}} + B^T(\underline{\sigma}_o + \Delta t DB\dot{\underline{\delta}})]dV = 0 \qquad (21)$$

where N is a matrix of shape functions and B is the matrix relating strains to nodal displacements. In practice, a mass matrix M and stiffness matrix K are formed, where

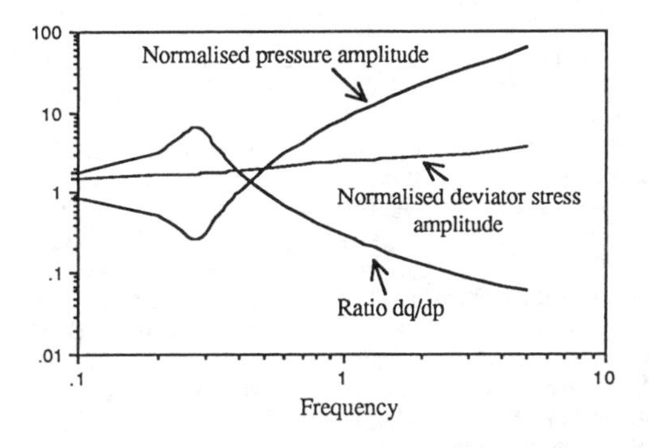

Figure 3. Cavity response to harmonic internal displacement (elastic conditions)

$$M = \int \rho N^T N \, dV \qquad (22)$$

$$K = \int B^T D B \, dV \qquad (23)$$

In the present work, Rayleigh damping is assumed, with a damping matrix C related to the mass and stiffness matrices by

$$C = \alpha M + \beta K \qquad (24)$$

The results presented later were all obtained for zero material damping (that is $\alpha = \beta = 0$).

At each time increment, the force vector for the right hand side is evaluated from the volume integral of $-B^T \underline{\sigma}_o$ from the previous time step.

Solution of the virtual work equation is effected using a time marching algorithm with the derivatives represented using Newmarks method. In that approach, the changes in a quantity x and its derivative in a time increment Δt are estimated from

$$\dot{x}_{t+\Delta t} = \dot{x}_t + \Delta t[\gamma \ddot{x}_{t+\Delta t} + (1-\gamma)\ddot{x}_t] \qquad (25)$$

$$\Delta x = \Delta t \dot{x}_t + \Delta t^2 [\beta \ddot{x}_{t+\Delta t} + (0.5-\beta)\ddot{x}_t] \qquad (26)$$

These equations may be re-arranged to give the derivatives at time $t+\Delta t$ explicitly in terms of the increment Δx and the known derivatives at the previous time t. Thus

$$\dot{x}_{t+\Delta t} = \frac{\gamma}{\beta}\frac{\Delta x}{\Delta t} - (\frac{\gamma}{\beta} - 1)\dot{x}_t - \Delta t \frac{\gamma}{\beta}(\frac{1}{2} - \frac{\beta}{\gamma})\ddot{x}_t \qquad (27)$$

$$\ddot{x}_{t+\Delta t} = \frac{\Delta x}{\beta \Delta t^2} - \frac{1}{\beta \Delta t}\dot{x}_t - (\frac{1}{2\beta} - 1)\ddot{x}_t \qquad (28)$$

The time marching process is stable for $\gamma \geq 0.5$, and $\beta \geq 0.25\sqrt{(\gamma + 0.5)}$. In the present work values of $\gamma = 0.6$ and $\beta = 0.3$ were adopted, thus providing some numerical damping.

In order to allow for large strains during cavity expansion, the nodal coordinates were continually updated after each increment. It was found worthwhile to use cubic strain elements (four nodes per element) in order to achieve sufficient accuracy without an excessive number of elements (Pennington (1986)). The size of each element was chosen so as to give a ratio of approximately 1.07 between the radii of inner and outer nodes. Thus 33 elements (100 nodes) would model a thick sphere (or cylinder) with an outer radius 10 times the inner radius, while 66 elements (199 nodes) would give a radius ratio of 100.

2.3 Transmitting boundary

It is important in a dynamic analysis to avoid any spurious reflection of elastic waves from the outer boundary of the finite element mesh. The solution described in section 2.1 may be used to form a 'transmitting boundary' which models an elastic region extending to infinity beyond the discretised region.

The required pressure displacement relationship for the outer node is given by equation (14). This equation may be rewritten using the Newmark finite difference relationships into a form

$$\Delta p = k \Delta x + f \qquad (29)$$

where the term R is a function of the pressure and displacement, and their derivatives, at the previous time step. The stiffness, k, may be added into the global stiffness matrix for the boundary node, while the term f is added to the force vector (allowing for the surface area of the outer node).

During the initial stages of a dynamic analysis, the perturbation is confined to the material immediately surrounding the cavity. It is therefore possible to use relatively few elements in order to capture the short term transient response. At any stage, the distance travelled by the perturbation may be calculated from the known dilatational velocity of the material. In the present work, the mesh was continually expanded (by adding additional elements up to a pre-set limit) such that the outer radius was always 40 % greater (that is 5 additional elements) than the theoretical limit reached by the perturbation.

3 SIMULATION OF PILE DRIVING

During pile driving, the impact of the hammer sends a velocity wave down the pile. A typical form of the velocity wave is shown in Fig. 4. The peak velocity is reached after a 'rise time' which will typically be of the order of 1 - 3 ms for a large diameter pile. After the peak, the velocity decreases, with the rate of decrease being a function of the mass of the hammer and the pile impedance.

In order to simulate the velocity wave,

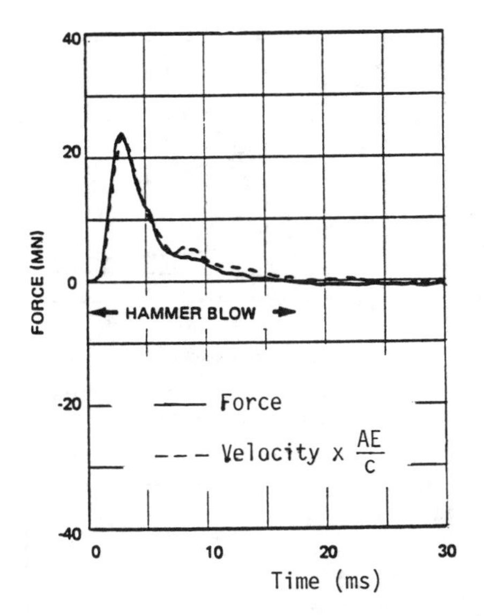

Figure 4. Typical form of force and velocity wave during pile driving

cavity expansion analyses have been undertaken where the cavity was expanded at a velocity given initially by

$$v = v_o \sin\omega t \quad (30)$$

Once the peak velocity has been reached, two alternative velocity variations have been adopted:

(a) continuing the sinusoidal variation down to zero velocity;

(b) maintaining the velocity constant.

These two alternatives were chosen to represent extremes on the form of the velocity wave. In practice, the peak cavity pressure is reached just before the peak velocity, so that the two alternatives lead to the same peak cavity pressure, but different subsequent responses.

4.1 Numerical results

In saturated soil, typical values of the dilatational wave velocity are in the range 800 - 1500 m/s (Abbis (1981)). An initial set of analyses was undertaken for a cavity of initial radius 1 m, with soil properties as given below:

Shear modulus G = 10 - 160 MPa
Poisson's ratio ν = 0.49
Density ρ = 20.4 kN/m³
Shear strength c_u = G/100

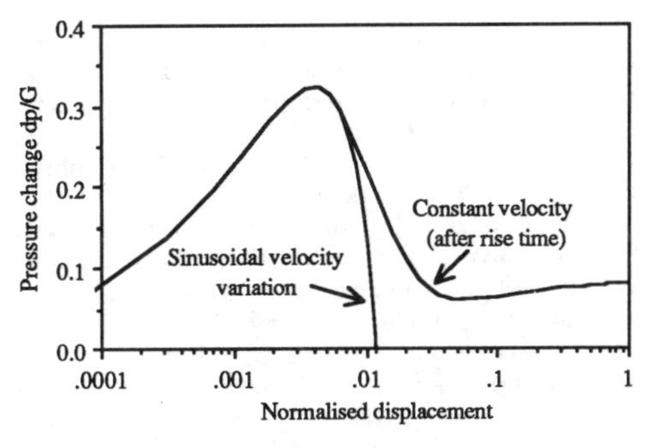

Figure 5. Response of cavity to velocity pulse

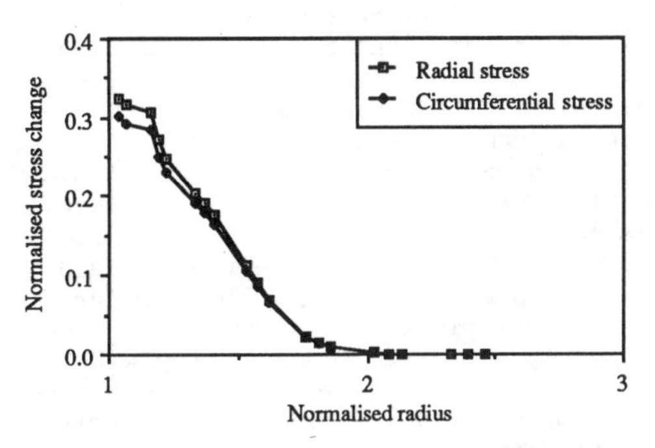

Figure 6. Stress variation in soil at peak cavity pressure

The range of shear modulus gives dilatational velocities in the range500 - 2000 m/s. The velocity of cavity expansion was varied from 1 to 50 m/s, with rise times varying between 1 and 10 ms (angular frequencies ω in the range 150 - 1500 rad/s).

Typical variations of the internal cavity pressure are shown in Fig. 5, for the case of G = 10 MPa (c = 500 m/s), a cavity expansion velocity of 5 m/s and a rise time of 2 ms (ω = 785 rad/s). For the sinusoidal velocity variation over the time period 0 - 4 ms, the computed pressure increase reaches a maximum of 0.324 G, before decreasing to zero and below. (It should be noted that negative values of the internal cavity pressure are a result of the extreme variation of velocity adopted; the pressure variation is not considered realistic beyond a time of about 3 ms (displacement of δ = 10 mm).) For the case where the cavity velocity is held constant at 5 m/s, the pressure decreases from its maximum value, and then starts to increase again due to the large cavity strains. Eventually, a limit pressure of about 0.081 G (8.1 times the shear strength) is reached, 14 % higher than the static value.

The radial variation of stress at peak cavity pressure is shown in Fig. 6 (the stresses are normalised by the shear modulus G). The stress difference close to the cavity wall is 0.02G (twice the shear strength). The peak pressure occurs at a time of 1.6 ms, by which time the dynamic effects have travelled 0.8 m from the cavity wall. Hence the stress changes decrease to zero at a normalised radius of approximately 1.8 (since R = 1 m).

The displacement at peak pressure is just over 4 mm (0.004 R), which is comparable to 'quake' values of 2 - 3 mm commonly adopted in pile driving analysis. For smaller diameter piles, the acceleration levels rise correspondingly. Dynamic expansion of a cavity of initial radius 0.1 m, with a rise time reduced proportionally to 0.2 ms, results in the same peak pressure change of 0.324 G. Thus, although peak dynamic pressures attained during pile driving should be largely independent of pile size, the displacement at which the peak pressure is mobilised is likely to scale with the pile diameter.

Although plastic straining of the soil takes place close to the cavity wall, it is interesting to note that the peak pressure of 0.324 G is relatively insensitive to the shear strength of the soil. Reducing the shear strength by a fator of 4 (to give a strength ratio of 400) only reduces the peak pressure change to 0.314 G. Similarly, increasing the shear strength until the response becomes purely elastic, gives a peak pressure change of 0.329 G. The insensitivity of the response to the strength of the soil is a function of the relatively small strains at the stage where the peak pressure change is observed, and thus the relatively smll deviator stress changes. One consequence of this is that the dynamic 'bearing capacity factor' (the ratio of peak pressure to undrained shear strength) may vary strongly depending on the ratio of stiffness to strength. Thus for G/c_u = 100, $\Delta p/c_u$ = 32.4, while for G/c_u = 400, $\Delta p/c_u$ = 125. For low ratios of G/c_u, the short term dynamic response becomes elastic, and it is then not meaningful to consider the ratio $\Delta p/c_u$.

Figure 7 shows the variation of peak pressure change $\Delta p/G$ with 'rise time' for different values of dilatational velocity c. All results are for a 1 m radius cavity expanded at a peak velocity of 5 m/s. The peak pressure change is a function of both the acceleration level and the rate of cavity expansion. These quantities may

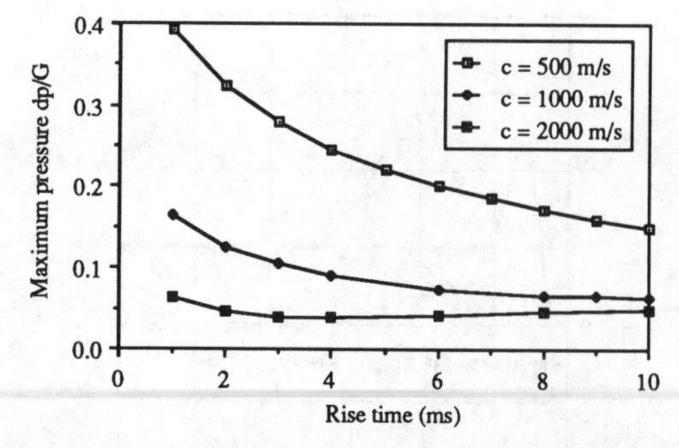

Figure 7. Variation of cavity pressure with rise time and cavity expansion rate

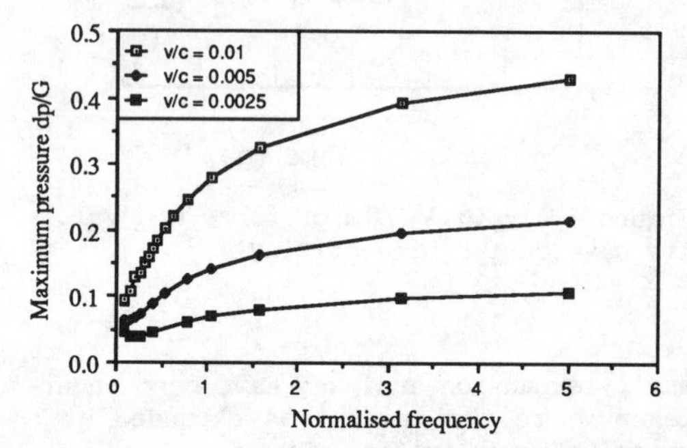

Figure 8. Variation of cavity pressure with normalised frequency and cavity expansion rate

be conveniently characterised by the ratio of cavity expansion rate to the dilational velocity (that is v/c) and the dimensionless frequency ωR/c. The dimensionless frequency may be obtained from the rise time Δt by

$$a_o = \frac{\omega R}{c} = \frac{\pi}{2\Delta t}\frac{R}{c} \tag{31}$$

The results from Figure 7 may be replotted as $\Delta p/G$ against dimensionless frequency, for different velocity ratios v/c - see Fig. 8. It may be seen that, for frequencies greater than about 0.5, the peak pressure change becomes proportional to the velocity ratio.

As was seen in Fig. 5, if the cavity is expanded continously at a steady rate, the internal pressure starts to rise again and eventually reaches a limit pressure somewhat greater than the static value. The value of the limit pressure is a function of the soil parameters (shear modulus and undrained shear strength) and also the velocity ratio v/c. Figure 9 shows the variation of limit pressure with rate of expansion. For typical values of velocity

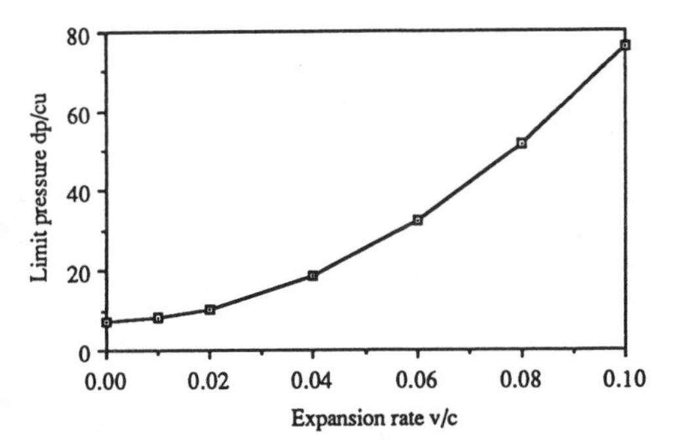

Figure 9. Variation of limit pressure with rate of cavity expansion

relevant for pile driving (generally 1 - 5 m/s) the velocity ratio will be such that the limit pressure will be close to the static value. However, as has been discussed above, a peak internal pressure is achieved at small strains due to the initially high acceleration levels.

5 DISCUSSION

This paper has considered the dynamic expansion of a spherical cavity in cohesive soil, using the cavity expansion as a analoueeaing capacity. The velocity wave that arrives at the tip of a pile during driving has been modelled as a sinusoidally increasing rate of cavity expansion up to some peak velocity, after which the velocity will decrease gradually.

The peak pressure change in the cavity occurs just prior to achieving peak velocity, and is a function of the rise time (or dimensionless frequency) and the ratio of cavity expansion rate to the dilatational velocity of the soil. At low rise times, the peak pressure becomes proportional to the cavity expansion rate.

The peak pressure is surprisingly insensitive to the shear strength of the soil. It has been shown that, for typical strength ratios, the soil response remains elastic apart from close to the cavity wall.

The high cavity pressures arise from the inertial resistance of the soil. For one-dimensional pile driving analysis, it is necessary to include dashpot (and/or lumped mass) elements in parallel with the spring and plastic slider elements that represent the static response. In principle, models of the form originally proposed by Smith (1960) - see Fig. 1(a) - are capable of capturing the effects discussed above. However, values of damping parameter for the dashpot element should be considerably larger than are currently used in practice (where the dashpot is assumed to represent primarily viscous effects).

It is considered that the results from dynamic cavity expansion may provide at least qualitative guidance on the dynamic end-bearing capacity of driven piles. However, further studies are needed in order to assess the direct relevance of the quantitative results. Such studies might comprise dynamic finite element studies of bearing capacity (in axial symmetry) or modifications of the strain path approach used for continuous penetration problems.

ACKNOWLEDGEMENTS

This work was initiated while the first author was a Visitor at the University of Sydney. The authors are indebted to J.P.Carter and S.K.Yeung for providing the source code of FEACE, and to J.R.Booker for assistance with some of the algebra.

REFERENCES

Abbis C.P. 1983. Shear wave meaasurements of the elasticity of the ground. Geotechnique 31(1):91-104.

Baguelin F., J.F.Jezequel & D.H.Shields 1978. The pressuremeter and foundation engineering. Transtech Publications.

Baligh M.M. 1986. Undrained deep penetration, I, II. Geotechnique 36(4):471-502.

Carter J.P. & S.K.Yeung 1985. Analysis of cylindrical cavity expansion in a strain weakening material. Computers and Geotechnics, 1:161-80.

Pennington D.S. 1986. Dynamic cavity expansion in soils. MPhil Thesis, University of Cambridge.

Randolph M.F. & H.A.Simons 1986. An improved soil model for one-dimensional pile driving analysis. Proc. 3rd Int. Conf. on Numerical Methods in Offshore Piling, Nantes, 1-17.

Smith E.A.L. 1960. Pile driving analysis by the wave equation. J. Soil Mech. & Found. Eng. Div., ASCE, 86:35-61.

Smith I.M. & Chow 1982. Three-dimensional analysis of pile drivability. Proc. 2nd Int. Conf. on Numerical Methods in Offshore Piling, Austin, 1-19.

Yeung S.K. & J.P.Carter 1987. An assessment of the bearing capacity of calcareous and silica sands. Submitted for publication in Int. J. Num. and Anal. Methods in Geomechanics.

Numerical Methods in Geomechanics (Innsbruck 1988), Swoboda (ed.)
© 1988 Balkema, Rotterdam. ISBN 90 6191 809 X

Vibration of thick plates on three-dimensional multilayered soil

Jerzy Kujawski & Czeslaw Miedzialowski
Bialystok Technical University, Poland
Nils-Erik Wiberg
Chalmers University of Technology, Göteborg, Sweden

ABSTRACT: The paper deals with vibration of thick plates interacting with three-dimensional multilayered soil. A simple semi-analytical finite element model for a plate-soil system is used. The total number of unknowns per plate-soil finite element node is equal to six. It allows to analyse large three-dimensional problems of dynamic plate-layered soil interaction. Numerical examples show the influence of varies factors which influence on the vibration of thick plates on layered soil.

1 INTRODUCTION

Problems of soil-structure interaction can be solved by means of a two or three dimensional finite element method. This can be very expensive, in case of large 3D problems [1-4] and hence seldom used in practise. The other approach is to use simple models for soil foundation such as: Winkler's, Vlasov's, Kerr's and others. However, the computational results are often not accurate enough for important structures.

In the References [5-7] a simple semi-analytical finite element model of layered soil foundation, resting on the rock, for plate-soil interaction has been proposed. This model leads to fairly good results and due to its simplicity can be used to the engineering analysis of large problems.

Some generalization of the numerical model of soil foundation were presented in papers [7-10]. This paper is a continuation of the work of the authors on subsoil numerical models. It considers vibration of thick plates interacting with multilayered soil foundation resting on rock. Varies types of soil non-homogeneity and its influence on vibration of thick plates are studied.

2 THEORY

2.1 Field equations

Consider a thick plate on multilayered soil (Fig. 1). The problem is three-dimensional and full interaction between plate and soil is accounted for. Small-deformation theory is assumed. The following equations hold:

Dynamic equations

$$-\tilde{\nabla}^T\sigma + c\dot{u} + U = \rho \ddot{u} \qquad (1)$$

and kinematic equations

$$\varepsilon = \tilde{\nabla}u = \begin{bmatrix} \varepsilon_x \\ \varepsilon_y \\ \varepsilon_z \\ \gamma_{xy} \\ \gamma_{xz} \\ \gamma_{yz} \end{bmatrix} = \begin{bmatrix} \partial x & 0 & 0 \\ 0 & \partial y & 0 \\ 0 & 0 & \partial z \\ \partial y & \partial x & 0 \\ \partial z & 0 & \partial x \\ 0 & \partial z & \partial y \end{bmatrix} \begin{bmatrix} u_x \\ u_y \\ u_z \end{bmatrix} \qquad (2)$$

$$(\partial x, \partial y, \partial z) = (\frac{\partial}{\partial x}, \frac{\partial}{\partial y}, \frac{\partial}{\partial z})$$

where the displacements u, the loads U and the stresses σ are defined by

$$u = [u_x, u_y, u_z]^T, \quad U = [U_x, U_y, U_z]^T$$

$$\sigma = [\sigma_x, \sigma_y, \sigma_z, \tau_{xy}, \tau_{xz}, \tau_{yz}]^T$$

ρ is the density of the material and $\ddot{u}$ is the acceleration vector.

Constitutive equations are

$$\sigma = S^d \varepsilon \qquad (3)$$

where S^d is a matrix of material parameters.

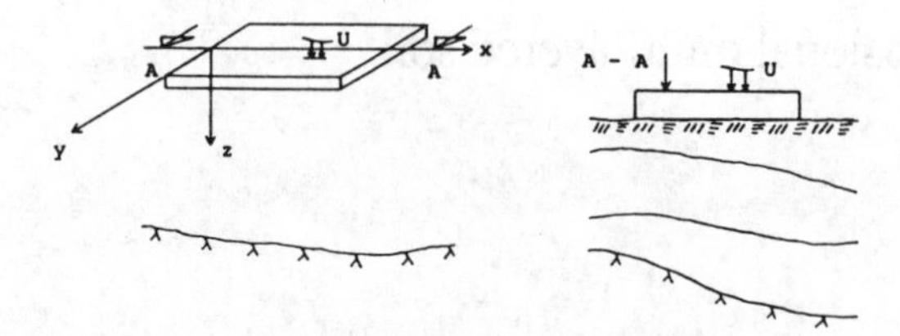

Fig. 1 Plate on multilayered soil

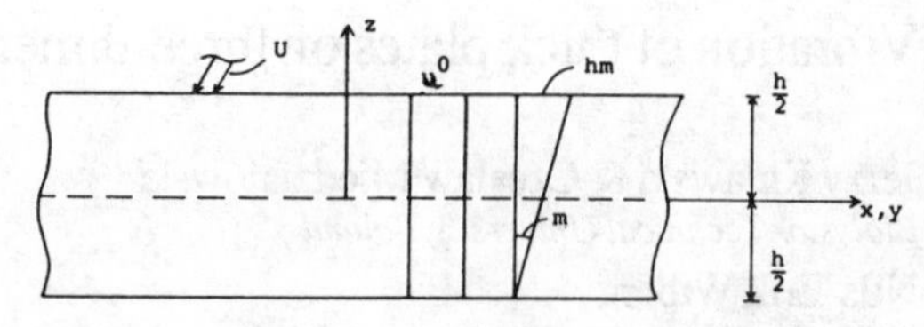

Fig. 2 Foundation plate. Deformation of cross-section

Finally there is a principle of virtual work

$$\int_t [\int_V (\delta\bar{\varepsilon})^T \sigma - \int_V (\delta\bar{u})^T (c\dot{u} - \rho\ddot{u}) dV - \quad (4)$$

$$- \int_A (\delta\bar{u})^T U dA] dt = 0$$

in which $\delta\bar{\varepsilon}$ and $\delta\bar{u}$ are the virtual strains and displacements respectively, and U is a load applied at the surface, c is the damping parameter and $\dot{u}$ is the velocity vector.

2.2 Plate equation

Let us consider a thick foundation plate (Fig. 2) loaded with a surface dynamic load U. For the displacements the basic assumption is that the displacement vector $u = u(x,y)$ can be expressed by the sum of zero-order components, which are constant along the z-direction, and the two angles of rotation $m = m(x,y)$ due to pure bending and shear deformations of the plate

$$u = u^0 + (\tfrac{1}{2}h + z)m \quad (5)$$

or

$$u = \begin{bmatrix} u_x \\ u_y \\ u_z \end{bmatrix} = \begin{bmatrix} u^0_x \\ u^0_y \\ u^0_z \end{bmatrix} + (\tfrac{1}{2}h + z) \begin{bmatrix} m_x \\ m_y \\ 0 \end{bmatrix} \quad (6)$$

By substitution of (5) into the stress-strain equations of the classical theory of elasticity (3) the following relationship is obtained:

$$\sigma = S^d \varepsilon = S^d [I_5 D] \begin{bmatrix} \tilde{\nabla}^0 & \tfrac{1}{2}h\tilde{\nabla}^1 \\ 0 & \tilde{\nabla}^m \end{bmatrix} \begin{bmatrix} u^0 \\ m \end{bmatrix} \quad (7)$$

where

$$\tilde{\nabla}^0 = \begin{bmatrix} \partial x & 0 & 0 \\ 0 & \partial y & 0 \\ \partial y & \partial x & 0 \\ 0 & 0 & \partial x \\ 0 & 0 & \partial y \end{bmatrix}, \quad \tilde{\nabla}^1 = \begin{bmatrix} \partial x & 0 \\ 0 & \partial y \\ \partial y & \partial x \\ 0 & 0 \\ 0 & 0 \end{bmatrix}$$

$$\tilde{\nabla}^m = \begin{bmatrix} \partial x & 0 \\ 0 & \partial y \\ \partial y & \partial x \\ 1 & 0 \\ 0 & 1 \end{bmatrix}, \quad D = \begin{bmatrix} zI_3 & 0 \\ 0 & I_2 \end{bmatrix}$$

$$\sigma = [\sigma_x, \sigma_y, \tau_{xy}, \tau_{xz}, \tau_{yz}]^T$$

$$S^d = E = \mathrm{diag}[E_f, E_s]$$

The matrices E_f and E_s for an orthotropic thick plate have the forms

$$E_f = \begin{bmatrix} B_{11} & B_{12} & 0 \\ B_{12} & B_{22} & 0 \\ 0 & 0 & B_{33} \end{bmatrix}, \quad E_s = \begin{bmatrix} B_{44} & 0 \\ 0 & B_{55} \end{bmatrix} \quad (8)$$

in which $B_{\alpha\beta} (\alpha, \beta = 1,2)$ and $B_l (l = 3,4,5)$ are material constants of an orthotropic body. The matrices I_5, I_3 and I_2 are the 5×5, 3×3 and 2×2 unit matrices, respectively.

The membrane forces, bending moments and shear forces are defined by expressions such as

$$N = \begin{bmatrix} N_x \\ N_y \\ N_{xy} \end{bmatrix} = \int_{-h/2}^{h/2} \begin{bmatrix} \sigma_x \\ \sigma_y \\ \tau_{xy} \end{bmatrix} dz \quad (9)$$

2.3 Layered soil

Let us consider an elastic layer of soil with thickness H resting on rigid rock (Fig. 3). The layer is composed of several sublayers of soil having different material properties and variable thickness. The contact surface of rock and layer can be variable as well. The layer of subsoil is loaded by a surface dynamic load U.

The basic assumption is that the horizontal displacements can be approximated by a linear distribution along the x-direction. However, the main vertical displacements can be expressed as the sum of a zero-order displacement having a linear distribution along the z-direction and a higher-order component given by a third degree polynomial (Fig. 4).

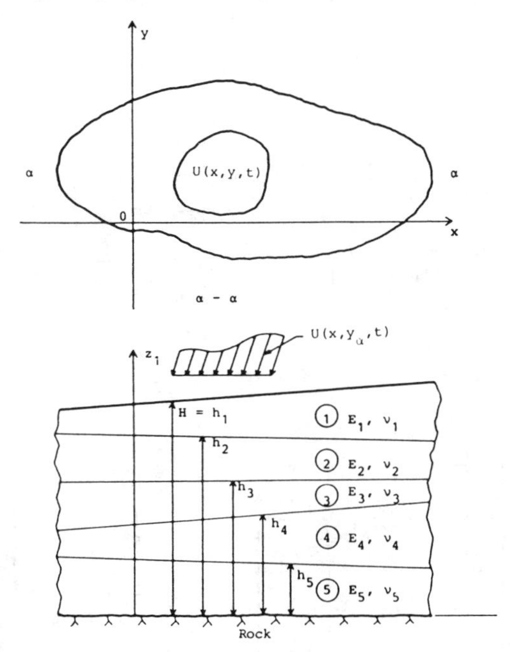

Fig. 3 Cross-section of soil resting on rock

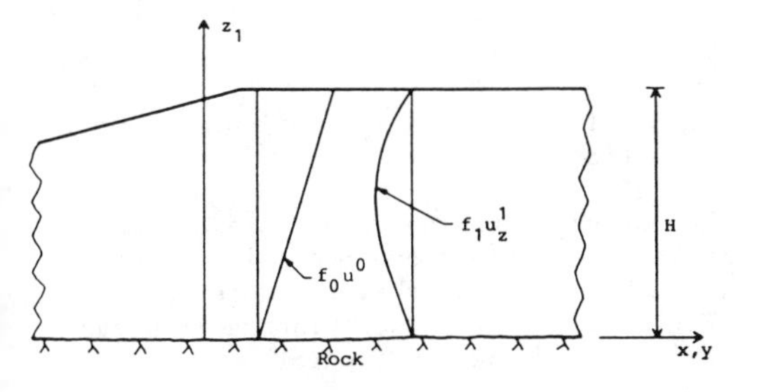

Fig. 4 Deformation of the cross-section of a soil

The displacement vector in the layer of soil can thus be expressed as

$$u = f_0 u^0 + f_1 u^1 = \begin{bmatrix} u_x \\ u_y \\ u_z \end{bmatrix} = f_0 \begin{bmatrix} u_x^0 \\ u_y^0 \\ u_z^0 \end{bmatrix} + f_1 \begin{bmatrix} 0 \\ 0 \\ u_z^1 \end{bmatrix} \tag{10}$$

where

$$f_0 = 1 - \zeta, \quad f_1 = (1-\zeta)(2-\zeta), \quad \zeta = z/H$$

in the global co-ordinate system (see Fig. 1) and

$$f_0 = \zeta_1, \quad f_1 = (1 - \zeta_1^2)\zeta_1, \quad \zeta_1 = z_1/H$$

in the local system of co-ordinates (Fig. 4).

The analysis presented in Reference 5 shows that the very simplified kinematic assumptions (10) appear to be sufficiently accurate for the analysis of plate-soil interaction problems in which the plate is subjected to a vertical load. This includes a homogeneous soil layer or a slightly layered soil foundation ($E_{max} \leq 3E_{min}$), where E is the Young's modulus of the appropriate sublayer.

A more accurate model of the soil foundation may be obtained by use of a higher-order approximation as suggested in Reference 6.

By substitution of (10) into the stress-strain equations (3) we obtain the stress vector

$$\sigma = S^d \varepsilon = S^d \left[\frac{z}{H}, I_6 \frac{1}{H}, I_6 D\right] \begin{bmatrix} \tilde{\nabla}^0 & 0 \\ \tilde{\nabla}^1 & 0 \\ 0 & \tilde{\nabla}^2 \end{bmatrix} \begin{bmatrix} u^0 \\ u_z^1 \end{bmatrix} \tag{11}$$

where

$$\sigma = [\sigma_x, \sigma_y, \sigma_z, \tau_{xy}, \tau_{xz}, \tau_{yz}]^T$$

$$S^d = E = \mathrm{diag}[E_f, E_s]$$

$$E_f = \frac{E(1-\nu)}{(1+\nu)(1-2\nu)} \begin{bmatrix} 1 & \frac{\nu}{1-\nu} & 0 \\ \frac{\nu}{1-\nu} & 1 & 0 \\ 0 & 0 & \frac{1-2\nu}{2(1-\nu)} \end{bmatrix}$$

$$E_s = \frac{E}{2(1+\nu)} \begin{bmatrix} 1 & 0 \\ 0 & 1 \end{bmatrix}$$

and where now

$$\tilde{\nabla}^0 = \begin{bmatrix} \partial x & 0 & 0 \\ 0 & \partial y & 0 \\ 0 & 0 & 0 \\ \partial y & \partial x & 0 \\ 0 & 0 & \partial x \\ 0 & 0 & \partial y \end{bmatrix}, \quad \tilde{\nabla}^1 = \begin{bmatrix} 0 & 0 & 0 \\ 0 & 0 & 0 \\ 0 & 0 & 1 \\ 0 & 0 & 0 \\ 1 & 0 & 0 \\ 0 & 1 & 0 \end{bmatrix}$$

$$\tilde{\nabla}^2 = \begin{bmatrix} 0 \\ 0 \\ 1 \\ 0 \\ \partial x \\ \partial y \end{bmatrix}, \quad D = \begin{bmatrix} f_1' I_4 & 0 \\ 0 & f_1 I_2 \end{bmatrix}$$

$$f_1' = H^{-1}(1 - 3\zeta_1^2)$$

The matrices I_6, I_4, and I_2 are 6×6, 4×4 and 2×2 unit matrices, respectively. Because of soil layering $E = E(z)$ and $\nu = \nu(z)$.

3 SOIL-THICK-PLATE INTERACTION

3.1 General

The theory of thick plates and of multilayered soil presented in the preceding section is formulated in such a way that it is very easy to analyse soil-plate interaction problems, such as the one shown in Fig. 5.

The surface of the subsoil is divided into finite elements. The finite element mesh is common to subsoil and plate (Fig.6). Using the finite element formulation, the displacement field may be expressed in terms of a set of discrete nodal displacements $\tilde{u} = [\tilde{u}^0, \tilde{m}, \tilde{u}_z^1]^T$ by means of a set of shape functions ϕ^T

$$u = [u_x^0, u_y^0, u_z^0]^T = \phi^T \tilde{u}^0 \qquad (12)$$

$$m = [m_x, m_y]^T = \phi^T \tilde{m}, \quad u_z^1 = \phi^T \tilde{u}_z^1$$

In the coupled analysis the chosen nodes are at the bottom surface of the plate and at the upper surface of the soil.

By means of the principle of virtual work (4) we obtain the stiffness matrices for the plate K_p and for the soil K_s.

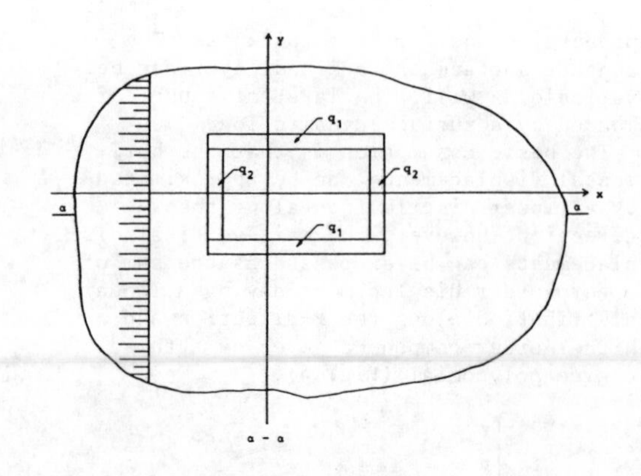

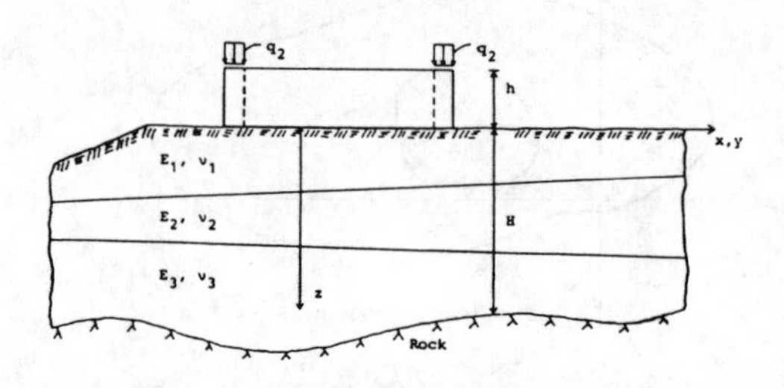

Fig. 5 Foundation plate and soil

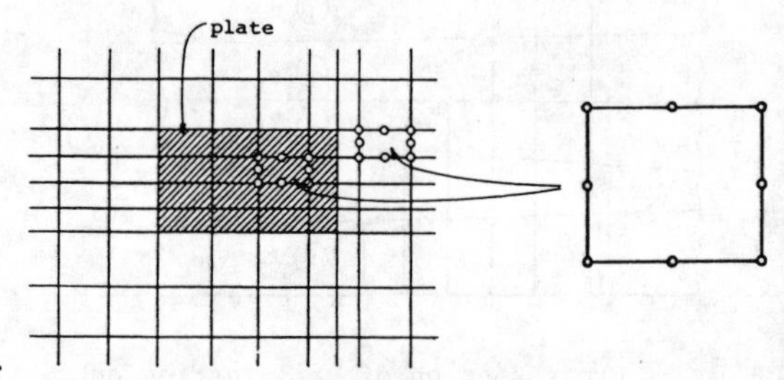

Fig. 6 Isoparametric elements in the horizontal plane

3.2 Stiffness matrix for soil-plate interaction

The choice of variables and element mesh for both the plate and the soil is made in such a way that they fit together (Fig. 7). The plate-soil interaction can thus be expressed in the following way:

$$Ku = U; \quad K = K_p + K_s \qquad (13)$$

$$\begin{bmatrix} K_{00} & K_{0m} & K_{01} \\ K_{m0} & K_{mm} & K_{m1} \\ K_{10} & K_{1m} & K_{11} \end{bmatrix} \begin{bmatrix} \tilde{u}^0 \\ \tilde{m} \\ \tilde{u}^1 \end{bmatrix} = \begin{bmatrix} \tilde{U}^0 \\ \tilde{U}^m \\ \tilde{U}^1 \end{bmatrix}$$

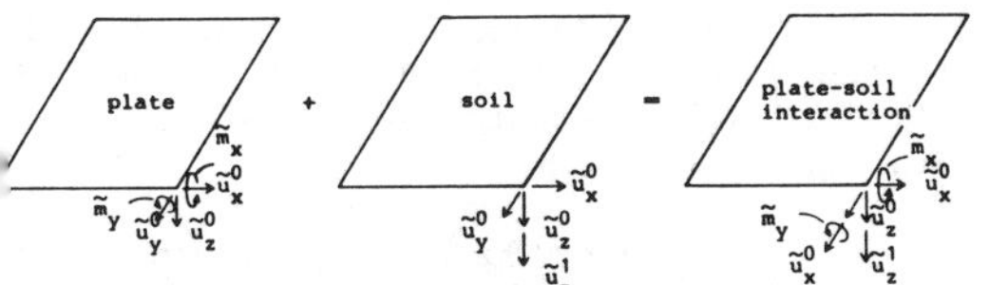

Fig. 7 Variables for soil-plate interaction analysis

where the nodal forces due to applied external loads $\tilde{U}$ and $\tilde{q}$ are given by

$$\begin{bmatrix} \tilde{U}^0 \\ \tilde{U}^m \end{bmatrix} = \int\int_A \phi \begin{bmatrix} \tilde{U} \\ \tilde{q} \end{bmatrix} dA \qquad (14)$$

3.3 Semidiscrete equation of motion

In a similar way the mass M and damping C matrices for plate-soil interaction can be established. The final semidiscrete equations of motion may be expressed as

$$M\ddot{u} + C\dot{u} + Ku = f \qquad (15)$$

where f is the vector of applied external dynamic load.

The dynamic equations of motion can be integrated by use of time stepping procedures. The well-known Newmark discrete time integration operators assume the form

$$Ma_{i+1} + Cv_{i+1} + Kd_{i+1} = f_{i+1} \qquad (16a)$$

$$d_{i+1} = d_i + \Delta t v_i + \frac{1}{2}\Delta t^2 [(1-2\beta)a_i + 2\beta a_{i+1}] \qquad (16b)$$

$$v_{i+1} = v_i + \Delta t[(1-\gamma)a_i + \gamma a_{i+1}] \qquad (16c)$$

Here d_{i+1}, v_{i+1} and a_{i+1} are displacement, velocity and acceleration vectors, respectively, at time t_{i+1}. Δt is the time step size, α and β are free scalar parameters which govern the accuracy and stability of the procedure.

For eigenvalue problems equation (15) may be transformed to

$$[K - \omega^2 M]u = 0 \qquad (17)$$

where ω is the natural frequency of the system under consideration.

4 NUMERICAL EXAMPLES

4.1 Problem description

The free vibration of an infinite plate resting on a layered soil is studied. The cross section of the vibrating system is shown in Fig. 8.

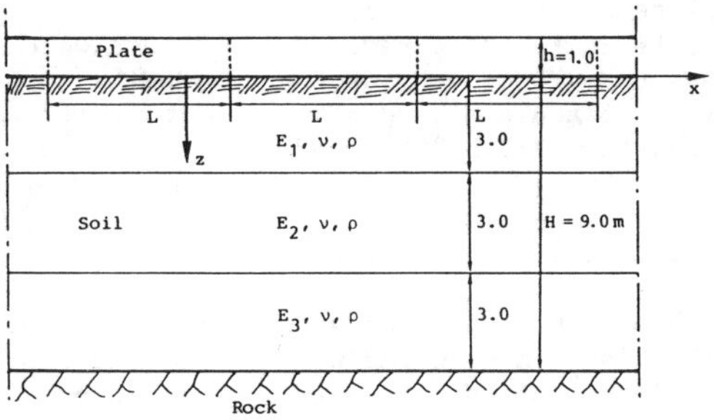

Fig. 8

It is assumed that the plate is vibrating as follow

$$u = \begin{Bmatrix} u_x \\ u_y \\ u_z \end{Bmatrix} = \begin{Bmatrix} \bar{u}_x \cos\alpha x \\ 0 \\ \bar{u}_z \sin\alpha x \end{Bmatrix} \qquad (18)$$

where $\alpha = \pi/L$.

In order to eliminate the secondary influence of the membrane vibration components for the fundamental frequency the following assumption was done $u_x^0 = -h/2m_x$. Thus the system has three degrees of dynamic freedom and the free vibration equation of the plate-soil interaction may be written as

$$\left(\begin{bmatrix} k_{11} & k_{12} & k_{13} \\ k_{21} & k_{22} & k_{23} \\ k_{31} & k_{32} & k_{33} \end{bmatrix} - \omega^2 \begin{bmatrix} m_{11} & 0 & m_{13} \\ 0 & m_{22} & 0 \\ m_{31} & 0 & m_{33} \end{bmatrix}\right) \cdot \begin{Bmatrix} \bar{u}_z^0 \\ \bar{m}_x \\ \bar{u}_z^1 \end{Bmatrix} = 0 \qquad (19)$$

Three cases of layered soil foundation are considered:

Soil 1 $(E_1, E_2, E_3) = (E_s, 2E_s, 6E_s)$

Soil 2 $(E_1, E_2, E_3) = (6E_s, 2E_s, E_s)$

Soil 3 $(E_1, E_2, E_3) = (6E_s, E_s, 2E_s)$

Poisson's ratio ν and soil density ρ were assumed the same for each sublayer. The average weighted Young's modulus of each soil is equal to $E_{av} = 3E_s$.

The computation of fundamental frequences was done with the influence of soil inertia forces (Plate + Soil Inertia) and without inertia forces in the soil (Plate + Soil). Full contact between the plate and soil foundation was assumed.

The following parameters for plate were assumed: Young modulus E = 200000 MPa, $\nu = 1/6$ and $\rho = 2400\ kG/m^3$; and for the soil $E_s = 333.333$ MPa and $\rho = 1800\ kG/m^3$.

4.2 Calculation results

Table 1 shows the fundamental frequency of the plate interacting with a homogeneous soil foundation having average Young's modulus $E = 3E_s = 1000$ MPa, different values of Poisson's ratio, and the distances l between lines of cylindrically vibrating infinite plates.

Table 1. Fundamental frequences $\omega_h(sec^{-1})$ of the plate on homogeneous soil foundation E = 1000 MPa.

	l = 6.0m		l = 4.0m	
	ν=0.25	ν=0.45	ν=0.25	ν=0.45
Plate + Soil Inertia	130.97	162.77	217.62	275.48
Plate + Soil	255.67	308.60	512.43	569.12

The influence of soil non-homogeneity for the plate frequency ω is shown in Tables 2 and 3 which stores the ratio of non-homogeneous soil frequency ω to the same frequency for average homogeneous soil ω_h.

Table 2. The ratio ω/ω_h for plate with l = 6.0 m.

		Soil 1	Soil 2
Plate + Soil Intertia	ν=0.25	0.954	1.059
	ν=0.45	0.933	1.060
Plate + Soil	ν=0.25	0.952	1.049
	ν=0.45	0.924	1.063

Table 3. The ratio ω/ω_h for plate with l = 4.0 m.

		Soil 1	Soil 2	Soil 3
Plate + Soil Inertia	ν=0.25	0.962	1.079	1.088
	ν=0.45	0.957	1.059	1.060
Plate + Soil	ν=0.25	0.970	1.033	1.029
	ν=0.45	0.940	1.054	1.046

Better approximation of the layered soil by the average homogeneous soil can be obtained using weighted soil properties. For instance if

$$E_W = \int_0^H z\,E(z)\,dz / \int_0^H z\,dz$$

the following values for soil 2 and 3 are obtained $(E_W^2, E_W^3) = (1370.37, 1296.30)$, they are greater than $E_{av} = 1000$ and should lead to better results. Table 4 shows the frequencies ω_1 for layered soil and corresponding frequencies ω_W for homogeneous soil with weighted Young's modulus.

Table 4. Fundamental frequencies of the plate resting on layered soil ω_1 and weighted homogeneous soil ω_W for l = 4.0 m.

		Soil 2		Soil 3	
		ω_1	ω_W	ω_1	ω_W
Plate + Soil Inertia	ν=0.25	234.90	235.36	236.74	232.28
	ν=0.45	291.80	295.53	291.92	291.79
Plate + Soil	ν=0.25	529.53	524.74	527.48	522.30
	ν=0.45	600.02	599.04	595.56	593.20

The differences between the frequencies of the plate interacting with a layered soil foundation and the same frequencies for the plate on a homogeneous soil with weighted physical properties are very small

5 CONCLUSION

The paper presents an analysis of vibration of thick plates interacting with a three-dimensional layered soil foundation. A simple semi-analytical model of a 3D layered soil foundation was used which possesses only four unknown functions. The simplicity of the model allows to compute large engineering problems.

The numerical examples of free cylindrical vibration of an infinite plate interacting with a layered soil foundation were solved. The analysis showed that it is a complex problem which depends on many factors.

The influence of the soil non-homogeneity for plate vibrations depends on the type of soil layers and their physical properties. In the considered cases of layered soil with the same average Young's modulus E the differences between thick plate fundamental frequencies are about 20% and depend on sublayer location. Larger differences may occur for stronger soil non-homogeneity

Replacement of the layered soil by an equivalent homogeneous layer with the average Young's modulus leads in the considered cases to frequency errors smaller than 10%.

Better approximation of the layered soil by the equivalent average homogeneous soil can be obtained with the following weighting of soil physical constants

$$(C,B,G) = \int_0^H z^n (C(z),B(z),G(z))dz / \int_0^H z^n dz$$

where

$$C = \frac{E(1-\nu)}{(1+\nu)(1-2\nu)}, \quad B = \frac{\nu E}{(1+\nu)(1-2\nu)},$$

$$G = \frac{E}{2(1+\nu)}$$

and n is a free power parameter.

The influence of soil inertia forces strongly reduces the fundamental frequency of the vibrating plate. For instance when $l = 4.0$ m and $\nu = 0.25$ the frequency with the influence of soil inertia is 512.43/217.62 = 2.35 times smaller than the frequency when soil inertia is neglected. That fact has strong influence for both free vibration and transient response of plates interacting with soil foundation.

Soil Poisson's ratio has also significant influence on the value of plate fundamental frequencies which increase for greater value of soil ratio ν.

6 REFERENCES

[1] D.R.J.Owen, E.M.Salonen, Three Dimensional Elasto-Plastic Finite Element Analysis, Int.J. for Num.Meth. in Eng. 9, 209-218 (1975)

[2] O.C.Zienkiewicz, The Finite Element Method, 3rd edition, McGraw-Hill, London, 1977

[3] D.R.J.Owen, E.Hinton, Finite Elements in Plasticity: Theory and Practise, Pineridge Press Ltd., Swansea, U.K., 1980

[4] H.J.Siriwardane, D.S.Desai, Computational Procedures for Non-Linear Three-Dimensional Analysis with Some Advanced Constitutive Laws, Int.J. for Num. and Anal.Meth. in Geomech. 7, 143-171 (1983)

[5] J.Kujawski, H.Tägnfors, N.-E.Wiberg, Situ-Plaso Computer Program for Three Dimensional Analysis of Multilayered Soil and Soil-Thick Plates Interaction by Use of a Semi-Analytical Finite Element Method, Publ. 82:1, Chalmers Univ. of Technology, Göteborg, Sweden, 1982

[6] J.Kujawski, N.-E.Wiberg, Thick Plates on Multi-Layered Soil, A Semi-Analytical 3D FEM Solution, Int. Symposium on Num. Models in Geomechanics, Zürich, 693-701 (1982)

[7] J.Kujawski, N.-E.Wiberg, Influence of a Wall System for Interaction Between Soil and Foundation Plates, Int.J. for Num. and Anal.Meth. in Geomech. 10, 283-309 (1986)

[8] J.Kujawski, M.Olejnik, Computation of Elastic Layer by a Semi-Analytical Finite Element Method, The Archieve of Civil Engineering (in Polish) 31, 413-422 (1985)

[9] J.Kujawski, N.-E.Wiberg, M.Olejnik, C.Miedzialowski, A Simple Numerical Model of 3D Elastic-Plastic Soil Foundation, XXXI Scientific Polish Conf. of Civil Eng., Krynica, 101-105

[10] J.Kujawski, N.-E.Wiberg, M.Olejnik, A Semi-Analytical FE-model for 3D Elastic-Viscoplastic Soil Foundation, Int. Symposium on Num. Models in Geomechanics, Ghent, Belgium, 1986

Numerical Methods in Geomechanics (Innsbruck 1988), Swoboda (ed.)
© 1988 Balkema, Rotterdam. ISBN 90 6191 809 X

Seismic response of ground under multi-directional shaking

L.Briseghella & A.Reginato
University of Padua, Italy

ABSTRACT: The effective stress method for the analysis of seismic response of layered sand deposits under multi-directional shaking, developed by J. Ghaboussi and S. U. Dikmen, is used to study the effect of various factors on the response and the liquefaction potential of saturated sands. Analysis are performed to study the effects of the earthquake type, peak base acceleration, pore pressure parameter λ , depth of water table and coefficient of permeability. The influence of these factors on the pore pressure build up, characteristics of ground surface response and the liquefaction potential are studied.

1 INTRODUCTION

The saturated sand was modeled as a two phase medium and the coupled equations of motion for the pore fluid and the granular solid phase are solved. This is a true effective stress analysis in the sense that the pore pressures and the stresses are computed directly and the behavior of sand is specified by a material model in terms of effective stresses. As the effective stresses change during the analysis, some of the material parameters are modified accordingly. In this way , the overall softening of the material due to reduction of the confining pressure is modeled. This overall softening of the material are significant influence on the seismic response, as will be demonstrated in the case studies presented in this paper.

The knowledge of the manner in which the response of the system depends on the input parameters can be an important aid in interpreting the results of analysis. The purpose of this paper is to present a study of the influence on the response and the liquefaction potential of the earthquake type, peak base acceleration, depth of water table and of two material parameters: coefficient of permeability and pore pressure parameter λ . These two material parameters are probably the most important factors in determining the potential for liquefaction in any site. The pore pressure parameter, λ , is directly related to the relative density; the smaller value of the parameter λ implies looser sand.

All the case studies are performed on a reference soil profile. This hypothetical soil profile is twenty meters deep with a uniform relative density of 60 percent and a water table at a depth of one meter. Initial shear moduli and other parameters are given in Figure 1.

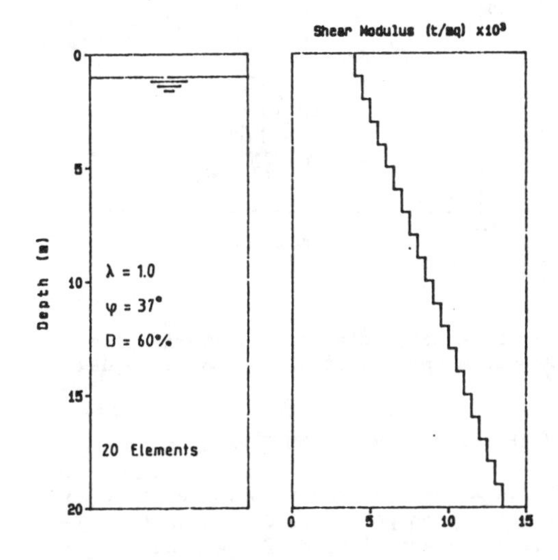

Fig.1 Hypothetical Soil Profile

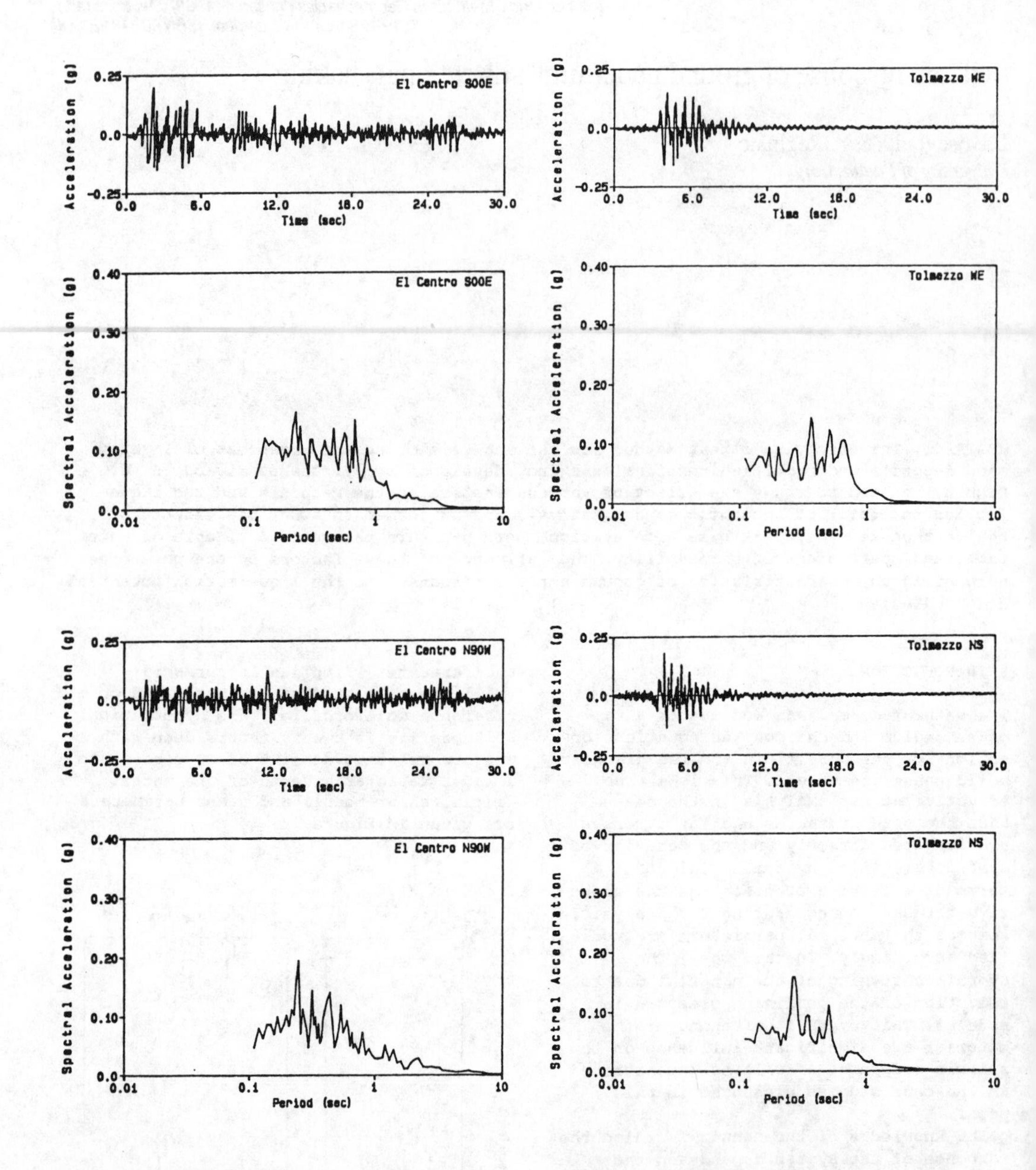

Fig.2 Acceleration Time Histories and Response Spectra of El Centro Earthquake

Fig.3 Acceleration Time Histories and Response Spectra of Tolmezzo Earthquake

2 EFFECT OF EARTHQUAKE TYPE

It is known that the seismic response and liquefaction potential depend on the specific characteristics of the base motion. The most important measures of earthquake type are: peak base acceleration, frequency content, and duration. The effect of peak base acceleration is studied in the next section. To study the effect of frequency content and duration, effective stress analyses of the reference soil profile are performed with four different earthquake records. These four records are: El Centro (May,1940), Tolmezzo (May, 1976), Sturno (November,1980), and Mexico (September,1985). All these records were scaled in way that the resultant peak base

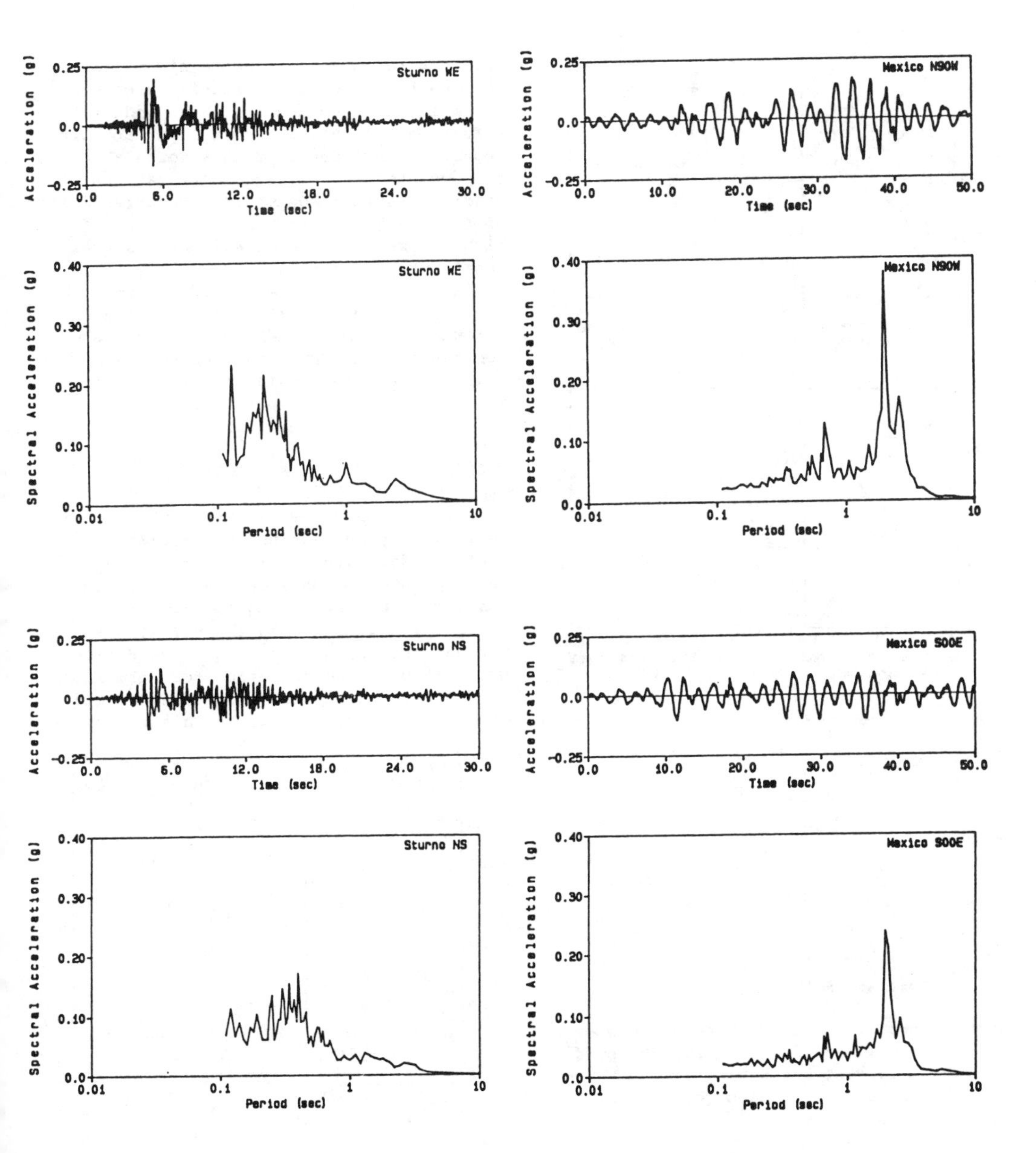

Fig.4 Acceleration Time Histories and Response Spectra of Sturno Earthquake

Fig.5 Acceleration Time Histories and Response Spectra of Mexico Earthquake

acceleration of two components is 0.2 g. The acceleration time histories and the response spectra for these four earthquake records are shown in Figures 2,3,4 and 5. From these figures it can be seen that that there are significant differences between the strong motion records selected. High amplitudes of acceleration are observed in the first 6.0 seconds of El Centro records. Whereas, in Tolmezzo and Sturno records, high amplitudes are after around 4.0 seconds. On the other hand, Sturno has moderate and Tolmezzo has relatively short duration of strong motion. Mexico records instead, high amplitutes are after 30.0 seconds and as lower frequency component. The earthquakes of El Centro, Tolmezzo and Sturno are predominant period between 0.1

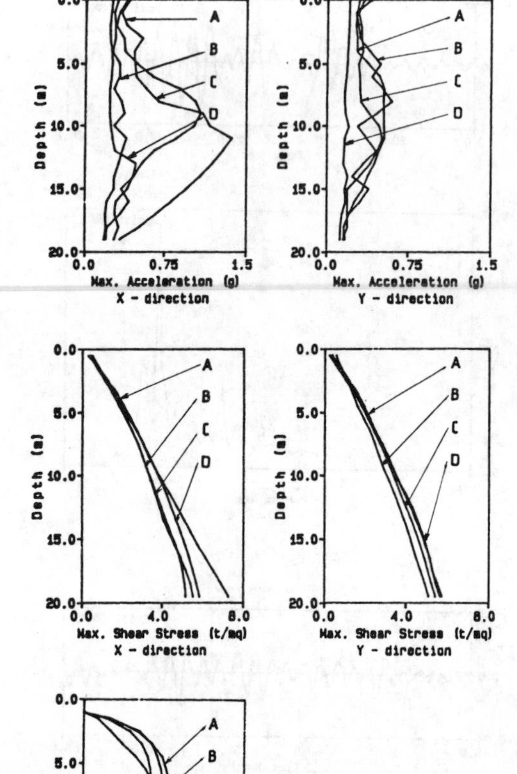

Fig.6 Comparison of Response Maximum of Reference Soil Profile (Fig.1) Subjected to Four Different Earthquake : (a) El Centro; (b) Tolmezzo; (c) Sturno; and (d) Mexico

and 1.0 second, whereas Mexico earthquake has predominant period about 3.0 seconds. Shown in Figure 6 are the variations through the depth of the maxima of pore pressure ratio, shear stress and acceleration for the four earthquake records. It is interesting to note that for all four earthquake records the pattern of excess pore pressure ratio ($\Delta\psi/\sigma'$) is very similar, with the maximum occurring between the depths of 6.0 and 12.0 meters. Only for Mexico earthquake the maximum of excess pore pressure occurring at a depth of around 15.0 meters. The high pore pressure ratio at this depth significantly reduces the effective confining pressure which in turn reduces the shear moduli, effectively creating a soft layer at this depth. The existence of this soft layer is responsible for the high accelerations gradients at this depth. The occurrence of of high peak accelerations in a zone of high excess pore pressure is always present in effective analysis.

Qualitatively, it can be staded that the liquefaction potential increases with the duration of strong shaking. The frequency content of the earthquake record also plays an important role in the response and the liquefaction potential.

3 EFFECT OF PEAK BASE ACCELERATION

A number of effective stress analyses were performed by subjecting the reference soil profile, described earlier, to Tolmezzo records, and the peak base acceleration was varied between 0.1 g and 0.3 g .

Shown in Figure 7 are the variations through the depth of maxima of pore pressure ratio and acceleration for Tolmezzo record with three peak base accelearations of 0.1 g, 0.2 g, and 0.3 g. This figures

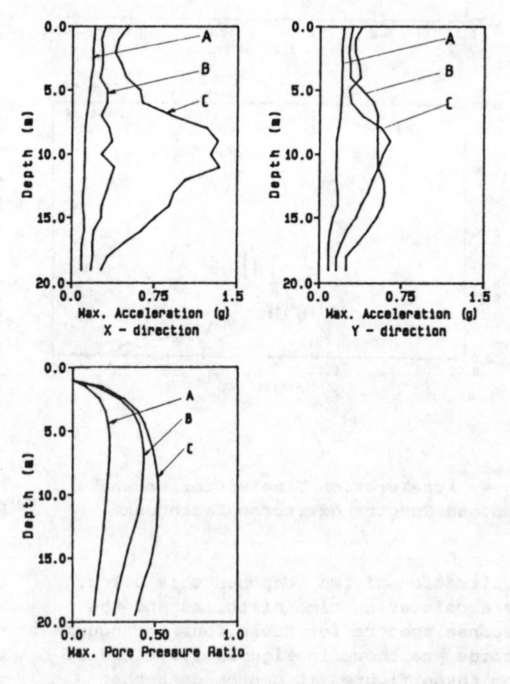

Fig.7 Comparison of Response Maximum of Reference Soil Profile (Fig.1) Subjected to Three Different Peak Base Acceleration (a) 0.1 g; (b) 0.2 g; and (c) 0.3 g

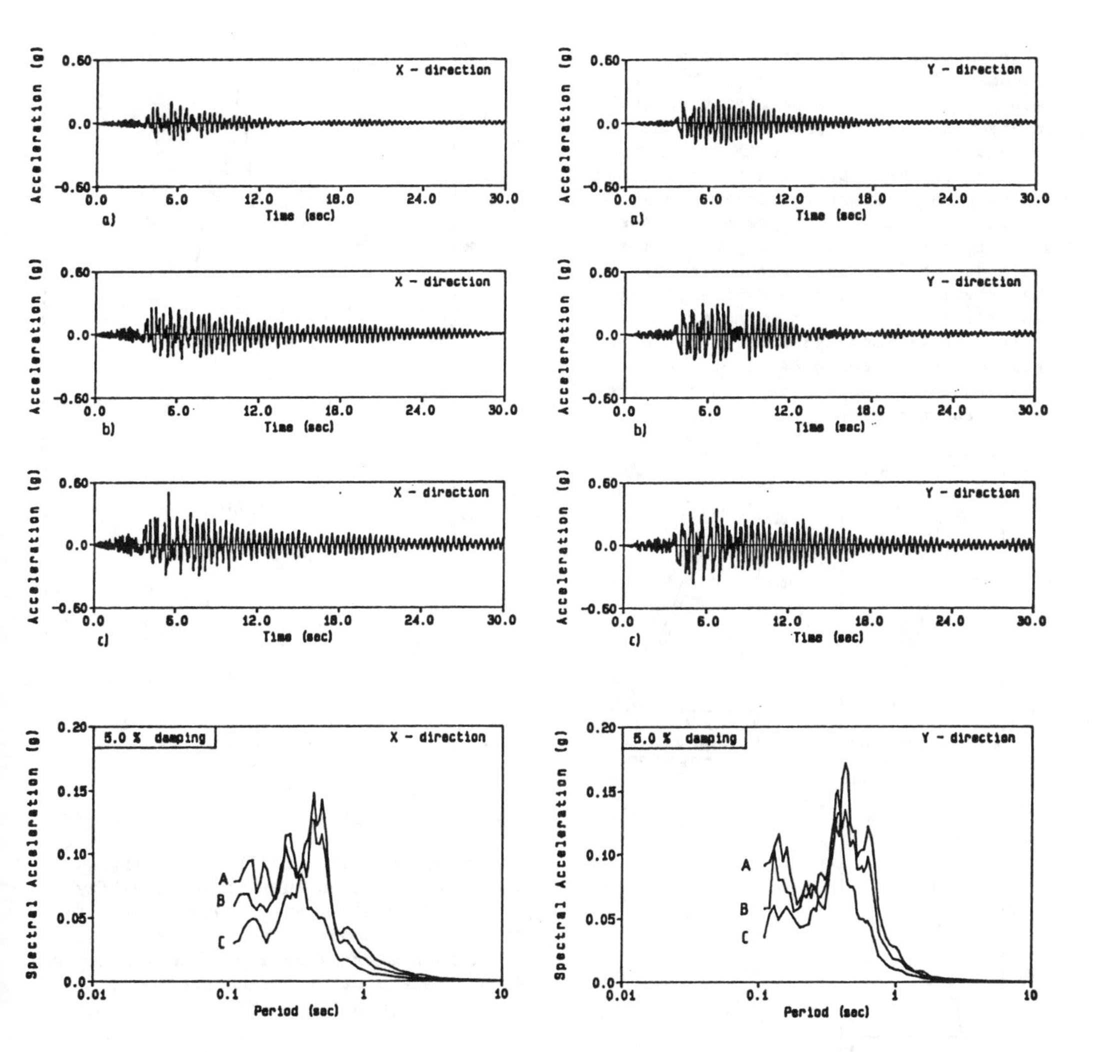

Fig.8 Influence of Peak Base Acceleration on Ground Surface Acceleration Time Histories and Response Spectra in X-Direction to Tolmezzo Record

Fig.9 Influence of Peak Base Acceleration on Ground Surface Acceleration Time Histories and Response Spectra in Y-Direction to Tolmezzo Record

quite clearly shows the differences in the response of the same profile when subjected to base motion with different peak base acceleration. The excess pore pressure increase increasing peak base acceleration. The surface acceleration time histories and response spectra in the two direction are shown in Figures 8 and 9. It is interesting to note that, increasing peak base acceleration, spectral accelerations increases, but also increases the predominant period. Moreover the ratio of the peak surface acceleration and the peak base acceleration is a decreasing function of base peak acceleration. This is due to the fact that the shear stresses increase with increasing peak base acceleration, thus, causing higher hysteretic damping.

4 EFFECT OF DEPTH OF WATER TABLE

The overall effect of the variation of the depth of water table is to vary the in-situ effective pressure and, thereby, reducing the susceptibility to liquefaction. Three cases are analyzed by effective stress method with the depth of water table at 1, 2, and 4 meters below ground surface. Base motion is Tolmezzo record with peak base acceleration scaled in way that peak acceleration of the resultant of two components is 0.2 g. The maximum shear stress and the maximum excess pore pressure ratios are shown in Figures 10. The shear stresses increase slightly with increasing depth of water table due to increase in shear moduli. The excess pore pressure ratios decrease significantly due to lowering of water table. The maximum excess pore pressure ratio decrease from 0.44 to 0.32 by lowering water table from 1 to 4 meters below ground surface.

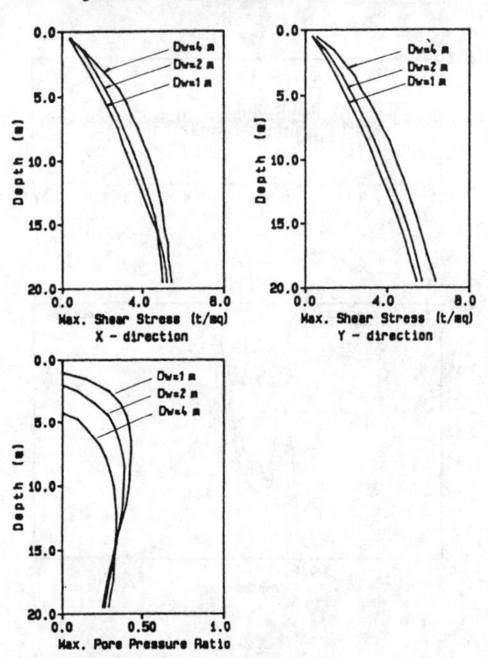

Fig.10 Influence of Depth of Water Table on Response Maxima

The ground surface response is not significantly influenced by lowering of the water table. The ground surface response spectra for three cases analyzed are shown in Fig. 11. Sligth differences can be observed in the high frequency range. Spectral values slightly increase due to lowering of water table. This is due, no doubt, to increase in in-situ effective pressure, producing higher shear moduli. As indicated by the results of the analysis, generally the lowering of the water table will reduce the potential for liquefation. But this effect is more drastic when the water table is originally close to ground surface.

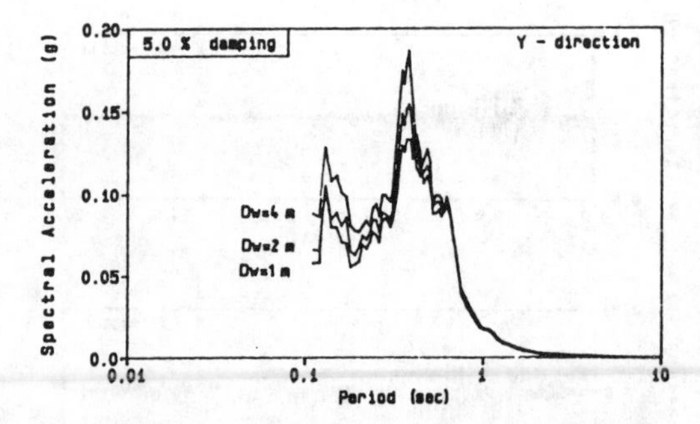

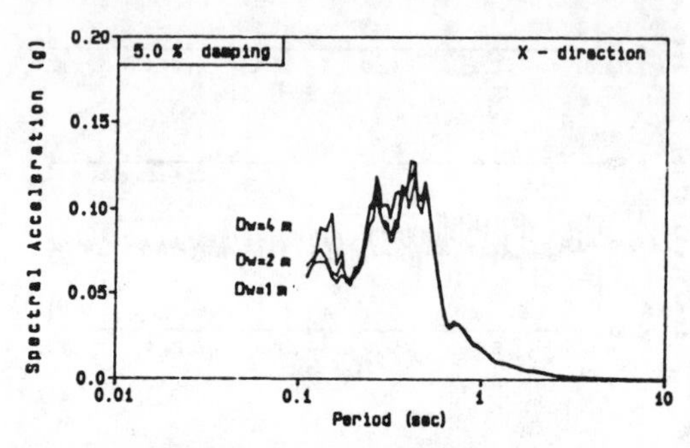

Fig.11 Influence of Depth of Water Table on Ground Surface Response Spectra

5 EFFECT OF COEFFICIENT OF PERMEABILITY

The reference soil profile was analyzed with three coefficients of permeability of 0.002, 0.0002, and 0.00002 m/sec. The maximum excess pore pressure ratios and the time histories of excess pore pressure at 5.5 meters below ground surface are shown in Figure 12 and 13.

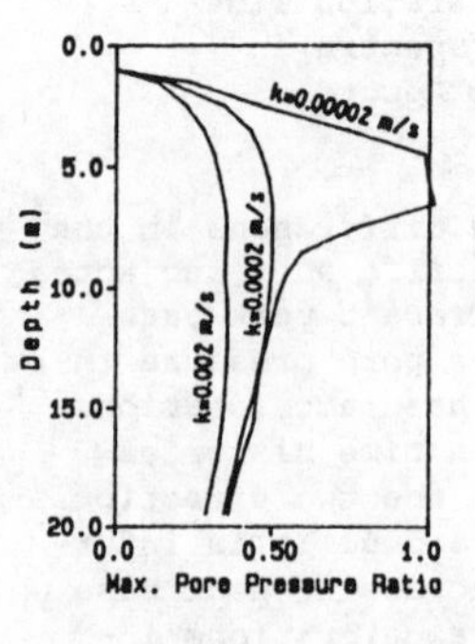

Fig.12 Influence of Coefficient of Permeability on Maxima Pore Pressure Ratio

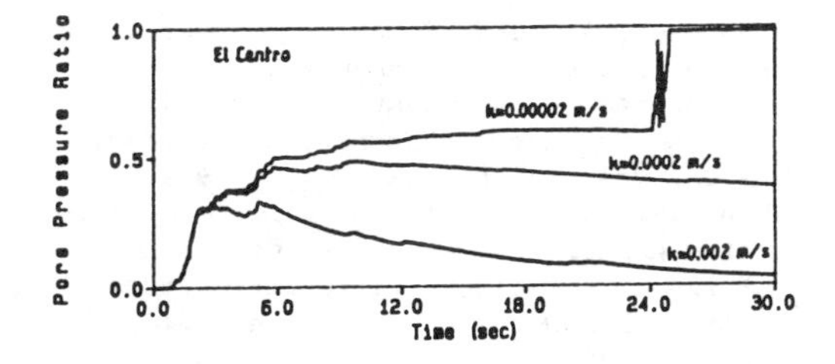

Fig.13 Comparison of Time Histories of Excess Pore Pressure Ratio Produced by Various Coefficient of Permeability at a Depth of 5.5 m Below Ground Surface

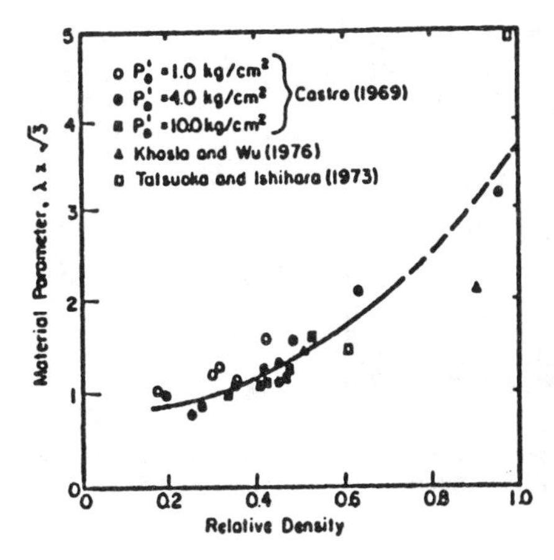

Fig.14 Relationship between Material Parameter λ and Relative Density

The coefficient of permeability, obviously, affects the dissipation of excess pore pressure. For higher coefficient of permeability, as the pore pressure are generated they can dissipate into adjacent zone, therefore reducing the likelihood of liquefaction. On the other hand, for the lowest coefficient of permeability, the excess pore pressure cannot dissipate quickly and thus they accumulate, leading to liquefaction.
The secondary effects, which develop later in the response, are due to influence of excess pore pressure on the shear moduli. Lower coefficient of permeability cause accumulation of excess pore pressure, leading to reduction of effective pressure and subsequently the shear modulus are reduced. In effect, after the initial portion of response, the system softens, and the degree of softening depends on the coefficient of permeability. These series of cases studies point to the importance of modeling of the pore pressure dissipation during the dynamic response analysis.

6 EFFECT OF PORE PRESSURE PARAMETER

In general, as the value of the pore pressure parameter λ decreases the potential for pore pressure build-up increases. The parameter λ is directly related to the relative density. A correlation between the value of pore pressure parameter λ and the relative density is given in Figure 14 for a number of experiments reported in the literature. To study the influence of pore pressure parameter three analyses are performed by subjecting the reference soil profile to El Centro record, with peak acceleration of resultant of two component scaled to 0.2 g and λ was varied between 0.75 and 1.25 . Shown in Figure 15 and 16 the variations through the depth of the maxima shear stress, acceleration, and pore pressure ratio for the cases studied.

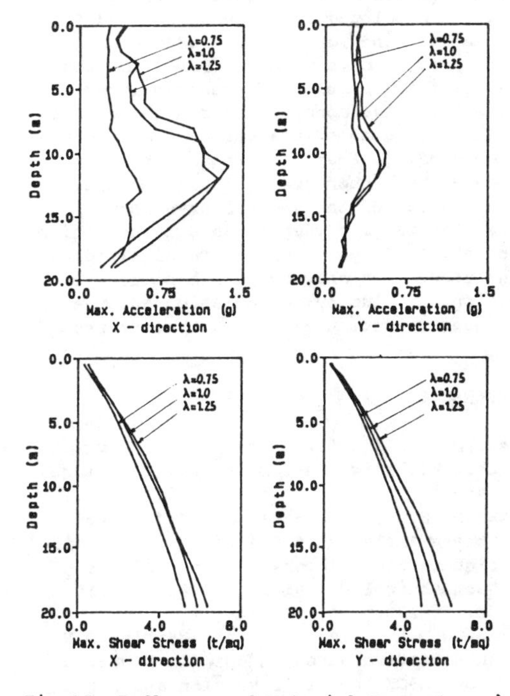

Fig.15 Influence of Material Parameter λ on Response Maxima

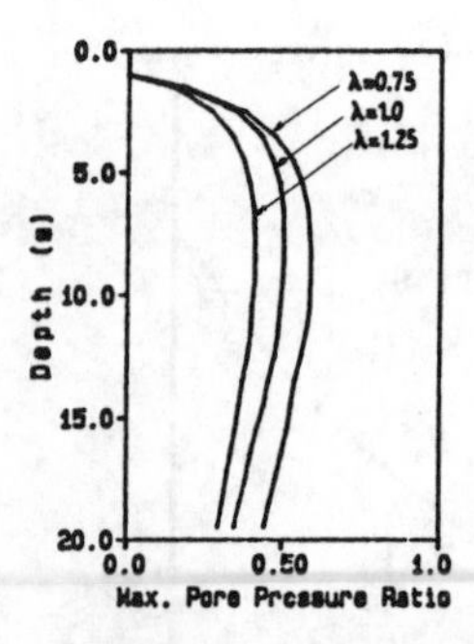

Fig.16 Influence of Material Parameter λ on Maximum Excess Pore Pressure Ratio

It can be seen that the lower values of λ, corresponding to lower relative densities cause generally higher excess pore pressure.

7 CONCLUSIONS

Using the effective stress method for the analysis of seismic response of layered sand deposits under multi-directional shaking depeloped by J. Ghaboussi and S.U. Dikmen a number of analyses of a standard soil profile have been performed. On the basis of the results of these analyses some tentative conclusions can be reached.

The influence of the earthquake type has been studied. Four different earthquake records have been used as base motions. The results of the anlysis have shown that the seismic response and liquefaction potential, in particular, are significantly influenced not only by peak base acceleration but also by the duration and the frequency content of the earthquake record.

REFERENCES

Castro,G., Liquefation of Sands. Harvard soil Mechanics Series No. 81, Cambridge, 1969

Dikmen, S.U., Ghaboussi, J., Effective Stress Analysis of Seismic Response and Liquefaction: Theory. Journal of the Geotechnical Division, ASCE, Vol. 110, No. 5, 1984, pp. 628-644

Finn, W.D.L., Lee, K.W., and Martin, G.R., An Effective Stress Model for Liquefaction, ASCE Annual Convention and Exposition, Philadelphia, Pa., 1976

Finn, W.D.L., Lee, K.W., and Martin, G.R., An Effective Stress Model for Liquefaction, Journal of the Geotechnical Division, ASCE, Vol. 103, No. GT6, June,1977

Ghaboussi, J., and Wilson, E.L., Variational Formulation of Dynamics of Fluid Saturated Porous Elastic Solids, Journal of the Engineering Mechanics Division, ASCE, Vol. 98, No. EM4, Aug., 1972

Ghaboussi, J., and Wilson, E.L., Liquefaction Analysis of Saturated Granular Soils, Proceedings, Fifth World Conference on Earthquake Engineering, Rome, Italy, 1973

Ghaboussi, J., and Dikmen, S.U., LASS-II, Computer Program for Analysis of Seismic Response and Liquefaction of Horizontally Layered Sands, Report No.UILU-ENG-77-2010 Dept. of Civ. Engrg. , Univ. of Illinois at Urbana-Champaign, Urbana, Ill., 1977

Ghaboussi, J., and Dikmen, S.U., Liquefaction Analysis of Horizontally Layered Sands , Journal of the Geotechnical Engineering Division,ASCE, Vol. 104, No. GT3, 1978

Ghaboussi, J., and Momen, H.,"Plasticity Model for Cyclic Behavior of Sands, Proceedings, Third International Conference on Numerical Methods in Geomechanics, Aachen, 1979

Ghaboussi, J., and Dikmen, S.U., LASS-III, Computer Program for Seismic Response and Liquefaction of Layered Ground under Multi-Directional Shaking",Report No. UILU-ENG-79-2012, Dept. of Civ. Engrg., Univ. of Illinois at Urbana-Champaign, Urbana, Ill., 1979

Ghaboussi, J., and Dikmen, S.U., Liquefaction Analysis for Multi-Directional Shaking, Journal of the Geotechnical Engineering Division, ASCE, Vol. 107, No. GT5, 1981

Ishihara, K., Tatsuoka, F., and Yasuda, S., Undrained Deformation and Liquefation of Sands under Cyclic Stresses", Soils and Foundations, Vol. 15, No. 1, March 1975

Kondner, R.L., and Zelask, J.S., A Hyperbolic Stress-Strain Formulation for Sands, Proceeding, 2nd Pan-American Conference on Soil Mechanics and Fountion Engineering, Brazil, Vol. 1, 1963

Tatsuoka, F.,Ishihara, K., Drained Deformation of Sand under Cyclic Stresses Reversing Direction", Soils and Foundation, Vol. 14, No. 3, December 1974

Numerical Methods in Geomechanics (Innsbruck 1988), Swoboda (ed.)
© 1988 Balkema, Rotterdam. ISBN 90 6191 809 X

Earthquake response of arch dams including the foundation-reservoir-dam interaction

Raimundo M.Delgado & Rui M.Faria
Oporto University, Portugal

ABSTRACT: This paper presents the application of the finite element method for analysing the earthquake response of foundation-reservoir-dam interaction. The foundation has been idealized using three-dimensional finite elements, and also for the reservoir has been used the same elements but with a special procedure that nullify the distorsional rigidity. The arch dam has been considered as a thick shell. Special elements has been included to connect the elements that describe the arch dam and the foundation. Joint elements has been used between the reservoir and the dam or foundation. The earthquake response of Lindoso dam was analysed computing the time-history of response to the horizontal component of El Centro, 1940, ground motion. The method is general, can be applied for systems with arbitrary geometry, and is appropriate for accurate representation of the hydrodynamic effects and interaction between the foundation and the dam.

1 INTRODUCTION

For the analysis of arch dams it is necessary include the deformability of the foundation. This aspect is usually taken into account using the Vogt coefficients, that relate the foundation area displacement with the corresponding actions as a function of the contact area dimensions and the rock mechanical characteristics.

When dynamic effects are included it is also necessary to analyse the interaction of the dam with the reservoir. The hydrodynamic effects can be included in an approximate way with the Westergaard added masses [1].

A more rigorous approach is adopted in this paper whereby the earthquake response of arch dams is studied using a finite element discretization for both the foundation and the reservoir.

2 STRUCTURAL ASPECTS

2.1 Arch dam

From the structural point of view arch dam are thick shells in which it is important to consider the shear deformation.

The shell element used for modelling the arch dam was developed by Ahmad et al. [2]. It is an 8-node element, with five degrees of freedom per node at the mid-surface, which can be used to model complex geometries in an economic manner, and also accounts for shear deformation.

2.2 Reservoir

Two alternative formulations can be employed for problems with small fluid displacements [3]: in the first one the fluid is characterized by one variable (the pressure or a potential) and the coupling between the two systems (structure and fluid) is done through the equilibrium of the interfaces forces between the two domains; in the second one the fluid is characterized by the displacements in the same manner as the solid, but making the shear modulus equal to zero. In this work the second approach is adopted because it is simpler and leads to a symmetric system of equations.

Twenty node 3D finite elements [4] are used for discretizing the reservoir.

In order to connect shell elements to 3D elements one has either to use special transition elements, or to provide a set of element transformations that allow for the assembly of the required degrees of freedom [5].

In this paper the latter technique is utilized which requires that the nodal displacements be measured in the shell upstream surface instead of in the middle surface.

Let $\underset{\sim}{\delta}'$, $\underset{\sim}{K}'$ and $\underset{\sim}{F}'$ be, respectively, the nodal displacements, stiffness matrix and load vector referred to the middle surface and $\tilde{\delta}$ and $\tilde{F}$ be the nodal displacements and load vector associated to the shell upstream surface. Then the two sets of nodal displacements can be easily related through the relationship

$$\tilde{\delta} = \tilde{T}' \delta' \quad (1)$$

and for the forces we have

$$\tilde{F}' = \tilde{T}'^T \tilde{F} \quad (2)$$

these two relations can be inverted to give

$$\tilde{\delta}' = \tilde{T} \tilde{\delta} \quad (3)$$

$$\tilde{F} = \tilde{T}^T \tilde{F}' \quad (4)$$

in which

$$\tilde{T} = \tilde{T}'^{-1} \quad (5)$$

Using standard transformations we obtain the new stiffness matriz referred to the upstream face as:

$$\tilde{K} = \tilde{T}^T \tilde{K}' \tilde{T} \quad (6)$$

Now the common degrees of freedom can be assembled as is shown in Fig. 1.

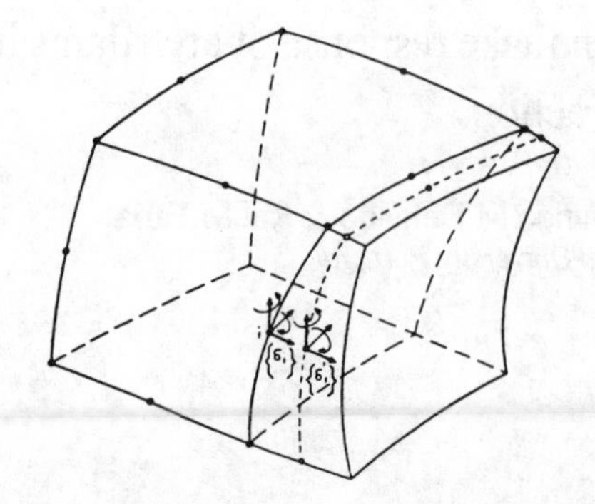

Fig.1 Coupling between 3D elements and shell elements (at the upstream surface)

2.3 Foundation

The foundation is essentially a three dimensional domain, so it can also be modelled with twenty noded bricks.

But in order to consider the dam-foundation interaction, we need to couple the shell element with the 3D brick in the manner showed in Fig. 2. This coupling imposes that the displacement of the j^{-th}, k^{-th}, etc. 3D element nodes be related those of the i^{-th}, etc. shell nodes. In other words the degrees of freedom of the j^{-th}, k^{-th} nodes, will be substituted by the degrees of freedom of the i^{-th} nodes.

A relationship can be easily established between the three displacement components δ' of the j^{-th} node and the five components $\tilde{\delta}_i$ of the i^{-th} node

$$\delta'_j = T_i \delta_i \quad (7)$$

If $\tilde{\delta}'$ and $\tilde{K}'$ are, respectively, the nodal displacement vector and the stiffness matrix of the 3D element (with 60 degrees of freedom) and $\tilde{\delta}$ the nodal displacement vector in which the degrees of freedom of the nodes j, k, etc. are substituted by those of nodes i, etc.. We can also define a $\tilde{T}$ matrix associating the T_i sub-matrices with identity sub-matrices (corresponding to those nodes where there are no transformation). And the transformation (6) can now be applied to obtain the stiffness matrix referred to the i^{-th}, etc. shell nodes.

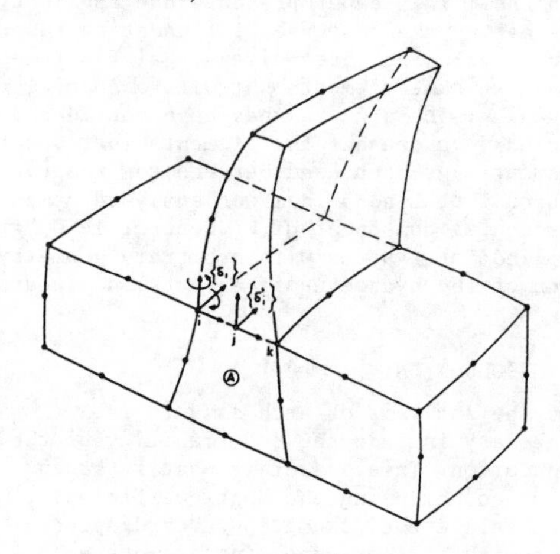

Fig.2 Coupling between 3D elements and shell elements (at the lateral surface)

The mass matrices are obtained in the same way

$$\tilde{M} = \tilde{T}^T \tilde{M}' \tilde{T} \quad (8)$$

2.4 Joint elements

In the dam-reservoir interface displacement compatibility is enforced only in the direction normal to the surface. To this end a special 3D joint element is used whose formulation is now described in 2D, for simplicity.

Fig. 3a) and b) shows an element with six nodes that are linked in pairs by

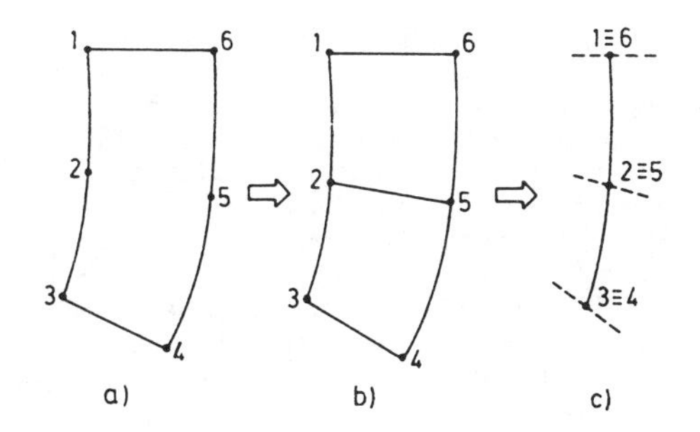

Fig.3 Joint element

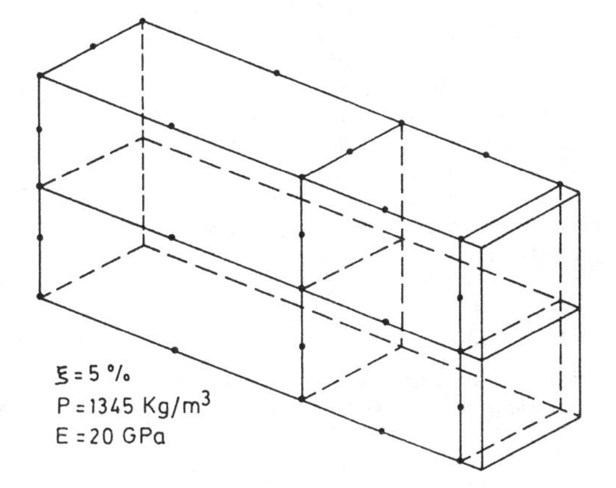

Fig.4 Cantilever retaining wall

three straight bars in the direction normal to the 1,2,3 element side. Large axial stiffness is assigned to the bars so that relative displacement normal to the joint is negligible but shear displacement is not restrained. In the limit as the length of the bar goes to zero, but with a finite axial stiffness, we have the element of Fig. 3c) where the dashed lines show the direction in which there is displacement compatibility.

3 SEISMIC ANALYSIS ASPECTS

To evaluate the earthquake response two different techniques are employed: using design response spectra or using a seismic time history, an accelerogram.

When response spectra are used the dynamic analysis is performed by means of modal analysis. To evaluate the corresponding eigenvalues the number of degrees of freedom can be reduced by recourse to static or dynamic condensation [6]. In order to compute the maximum total response a complete quadratic modal combination is used.

Table 1. Retaining wall frequencies (Hz)

	Added masses	Reservoir
1st	8.40	8.98
2nd	42.32	46.80
3nd	93.88	98.88

If the ground motion is characterized by an accelerogram the analysis is made via direct integration using the Newmark method.

4 APPLICATIONS

4.1 Cantilever retaining wall

Fig.4 shows the wall and the water of a tank submitted to seismic action. The earthquake effects were calculated by modal analysis, with a response spectra, and the water was modelled with both the element referred in 2.2 and the added masses techniques.

When the reservoir is discretized by finite elements the fluid nodes pick up a large number of frequencies with a small value associated to modal shapes involving negligible wall motion. In order to reduce this number of small frequencies, that have no interest at structural level, we can perform a dynamic condensation in some degrees of freedom of the reservoir.

As we can see in Table 1. the frequencies related to the modal shapes that have displacements in the wall are similar to those obtained with the added masses technique.

The maximum top displacement and base bending moment have been computed by the two techniques using the response spectra of the Portuguese code. With the 3D elements, the results are:

$$\delta_{3D} = 9.00 \times 10^{-2} \text{ cm} \; ; \; M_{3D} = 569 \text{ kN.m}$$

and with the added masses

$$\delta_{ad} = 9.46 \times 10^{-2} \text{ cm} \; ; \; M_{ad} = 595 \text{ kN.m}$$

As we can see both techniques give similar values although the 3D results are about 5% greater.

4.2 Alto Lindoso dam

Alto Lindoso dam is an arch dam with 110 m height, that is being built in the North of Portugal.

The foundation-reservoir-dam system is discretized by twelve, ten and five elements respectively, as we can see in Fig. 5. Joint

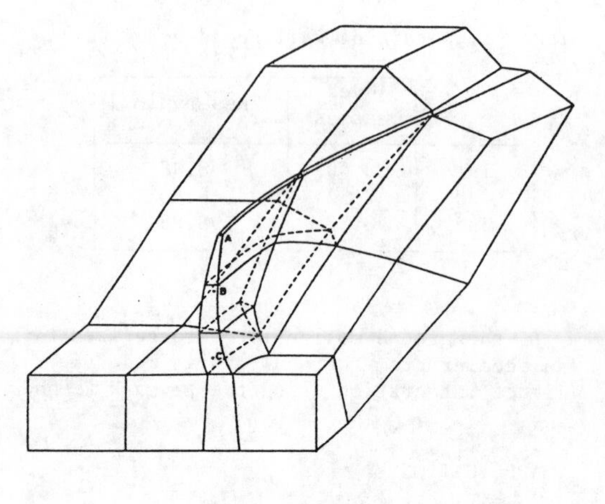

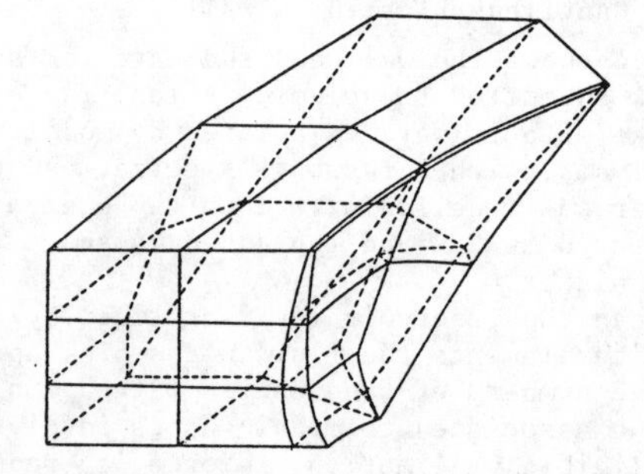

Fig.5 Lindoso arch dam

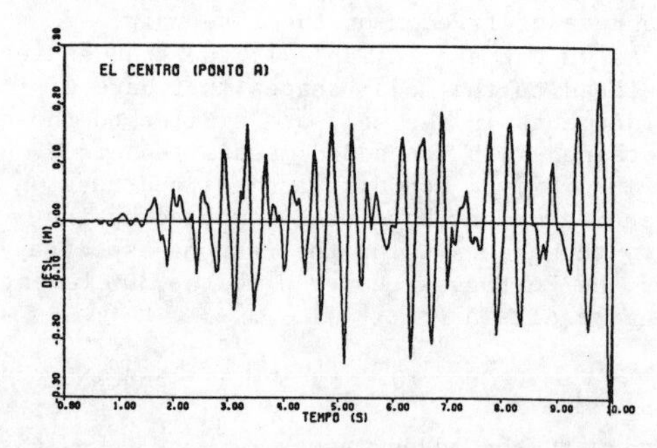

Fig.6 Displacement response at point A

elements are used at the reservoir dam interface. The El Centro earthquake (May 18, 1940), is used for a time domain analysis.

Fig. 6 shows the displacement response at point A, and we observe that they present technique can deal with seismic ground motion.

5 CONCLUSIONS

The method presented for three-dimensional linear analysis of earthquake response of arch dams is simple and effective for practical problems. The model includes the effects on dam response of dynamic interaction between dam, reservoir and foundation. The dam is discretized with thick shell finite elements and the foundation with twenty node 3D elements, which are also used in the reservoir, but with a special procedure that nulify the distorsional stiffness. The interaction between dam and reservoir is conveyed through joint elements.

A cantilever retaining wall is analysed and the results compared with those obtained via the added mass technique. The results are similar although the present technique leads to values about 5% greater.

The method is applicable to more complex systems with irregular geometry or nonhomogeneous material properties as we can see with the earthquake response of Alto Lindoso arch dam.

The substructure technique seems to be promising when we want to analyse large problems.

REFERENCES

[1] Westergaard, H.M. - Water pressures on dams during earthquakes; Transactions, ASCE, vol. 98, 1933.

[2] Ahmad, S. et al. - Analysis of thick on thin shell. Structures by curved finite elements. Int. J. Num. Meth. Eng., vol. 2, pag. 419-451, 1970.

[3] Zienkiewicz, O.C. and Bettess, P. - - Fluid-structure dynamic interaction and wave forces. An introduction to numerical treatment. Int. J. Num. Meth. Eng., vol. 13, pag. 1-16, 1973.

[4] Zienkiewcz, O.C. - The finite element method. Mc Graw-Hill, London, 1977.

[5] Delgado, R.M. - O Método dos elementos finitos na análise dinâmica de barragens incluindo a interacção sólido-líquido. Dissertação de doutoramento. Faculdade de Engenharia da Universidade do Porto, Outubro, 1984.

[6] Delgado, R.M., Faria, R.M. - The foundation influence in the earthquake response of arch dams. 6th International Conference on Numerical Methods in Geomechanics, 1988, Innsbruck, Áustria.

Numerical Methods in Geomechanics (Innsbruck 1988), Swoboda (ed.)
© 1988 Balkema, Rotterdam. ISBN 90 6191 809 X

Seismic stability of earth and rockfill dams

E.Baldovin
Geotecna Progetti, Milan, Italy
P.Paoliani
ENEA/DISP, Rome, Italy

ABSTRACT: In this paper the main steps of the dynamic response analysis of an existing earth dam are examined. It is pointed out the importance of numerical methods in dealing with the evaluation of the major factors related to the safety of embankment dams such as pore pressure build-up and permanent displacements due to earthquakes.

1 INTRODUCTION

During last decades the exploitation of the potential dam sites has been very large over the world. As a primary result of such trend it can be mentioned a general sharpening of the new structures difficulties, concerning either the foundations or the dam materials.
Furthermore the increased monitoring of ground seismic activities has shown the wide extension of the regions with such a critical peculiarity. Anyway the recent construction of a large amount of new earth and rockfill dams is a clue of the real adaptability of those structures. Obviously their spread has required and still requires a better understanding of the embankment dams behaviour in critical conditions, like the occurence of strong earthquakes.
Actually the re-appraisal of originally aseismic area often determines also the re-evaluation of existing structures. Some failures or near failures have been experienced in the past years during or immediately after important earthquakes. Well known examples can be mentioned, like the Sheffield dam failure and the slides in the Upper and Lower San Fernando dams.
Those and other accidents have clearly shown the limited reliability of the pseudo-static approach in the embankment dams seismic stability analysis.
Consequently new procedures, taking into account design accelerograms and spectra as well as the cyclic strength of the materials, have been developed and tested. They are quite complex and require the intervention of a multidisciplinary team experienced in several subjects: seismology, geology, earthquake soil mechanics, numerical and physical analysis. It is understood that the high costs of such procedures restrict their execution to important structures. At the present the International Commission on Large Dams (ICOLD) recommends the dynamic analysis approach for all high earth and rockfill dams whose failure could cause significant loss of lives or damage to properties.

2 NUMERICAL ANALYSIS PROCEDURES

Many factors are to be considered in the earthquake-resistant design of earth and rockfill dams, a lot of them requiring only the application of common-sense defensive measures to prevent the deleterious effects of shaking /6/.

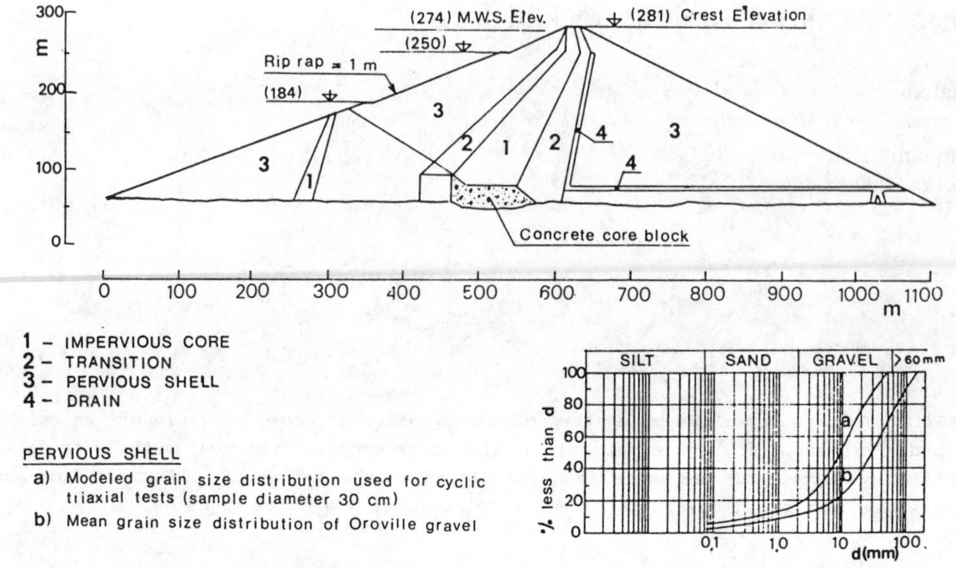

Fig. 1 - Oroville dam maximum section

However, the development of powerful computing facilities has made possible to face adequately some of those complex problems using numerical methods. In the following new procedures are proposed in order to evaluate through numerical methods two of the most important factors related to a safer design: slope failures due to pore pressure increase and loss of freeboard caused by permanent displacements.

For a sake of illustration, the procedures will be referred to an existing structure, Oroville dam in Northern California, because the amount of well-documented laboratory tests results and the available references /1/ make it a suitable benchmark for the proposed methods of analysis. The dam, whose cross-section is shown in fig.1 together with the mean grain size of the gravel material of the shells, is located in an active seismic region and could be shaken by an earthquake of magnitude M=6.5, occurring at the dam site. The design accelerogram has a peak acceleration of 0.6 g at a predominant frequency of 2.5 Hz and 0.3 g at a frequency of 0.9 Hz. In spite of its high density (Dr=84%), Oroville gravel exhibits a sharp increase in pore pressure when subjected to cyclic loading in laboratory undrained triaxial tests. In fig. 2 the stress ratio curves of the material, corresponding to the initial liquefaction condition (R_u=100%), are drawn for several values of the cell pressure σ_3' and the consolidation ratio Kc= σ_1'/σ_3'.

The undrained strenght of the soil in liquefaction condition is only about 15% less than the static strength, due to the dilatant behaviour of the material. Therefore a possible loss of strength can be related only to the drained behaviour of the material as high sustained excess pore-pressures are maintained for a certain time extent.

2.1 Slope failures due to pore pressure increase

The observed field performance of dams during past earthquakes leads to the conclusion that the pseudo-static approach is an acceptable method of analysis if the materials of the dam and the foundation soils do not develop high pore water pressure. When the increase in pore pressure becomes important, even if the undrained strength does not reduce dramatically, a more rational approach to study the stability of embankment dams during seismic loading is required.

In this case the Seed-Lee-Idriss analysis procedure /6/ overcomes the limits of the pseudo-static approach and can provide a considerable improvement in the design of earth and rockfill dams. A further enhancement of that procedure is the numerical modeling of the pore pressure build-up, that fills the gap existing in the original procedure between the numerical computation of the stresses induced in the embankment by the base excitation and the laboratory undrained cyclic tests results. In the proposed procedure the generation of pore water pressure is computed extending to the dynamic case the Skempton's equation, that relates the pore pressure increment Δu to the principal stress increment $\Delta\sigma_1$ and $\Delta\sigma_3$ according to the expression:

$$\Delta u = \Delta\sigma_3 + A(t)\ (\Delta\sigma_1 - \Delta\sigma_3) \qquad (1)$$

The parameter A(t) is defined as

$$A(t) = A_o + \psi A_1 / |\psi| \qquad (2)$$

and ψ, function of the shear stress history, is given by:

$$\psi = \tau\ \delta\tau / \delta t$$

While the parameter A_o has a relative small influence and the static value of A may be assumed for it, a minor change of A_1 can modify significantly the results of the analysis. Nevertheless the value of A_1 can be determined by assessing only the cyclic strength of the soil S by means of standard laboratory tests (cyclic triaxial or shear) or field tests (SPT or CPT). In fact A_1 can be expressed as /4/:

$$A_1 = (8\ N\ S)^{-1}$$

being N the number of cycles required to cause initial liquefaction for the stress ratio S. As an example, in fig. 3 the A_1 values for Oroville gravel are shown; the curves have been computed from the cyclic strength data previously provided.

The proposed procedure requires the static stress analysis of the dam, with non-linear laws governing the stress-strain behaviour of the soils. Hence the computed mean effective static stresses σ'_m are used for assessing the dynamic shear modulus at low levels of strain. Then the dynamic response of the dam can be performed: the equivalent linear method accounts for the non linear dynamic characteristics of the soil skeleton, while the model previously ilustrated makes allowance for the pore pressure increase.

The dynamic response of Oroville dam to the design accelerogram shows that the structure has a fundamental frequency of vibration of about 0.8 Hz. In fig. 4 the contour lines of the computed maximum acceleration values are represented. A large zone of the embankment is below the peak value of the design accelerogram (0.6 g) and only the upper part of the dam shows an amplification of the motion. The crest acceleration is close to 1.0 g. The computed excess pore water pressures are illustrated in fig. 5, normalized to

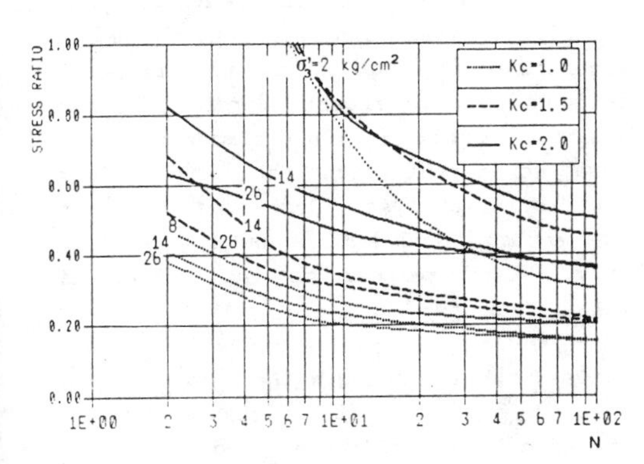

Fig. 2 - Stress ratio curves for Oroville gravel

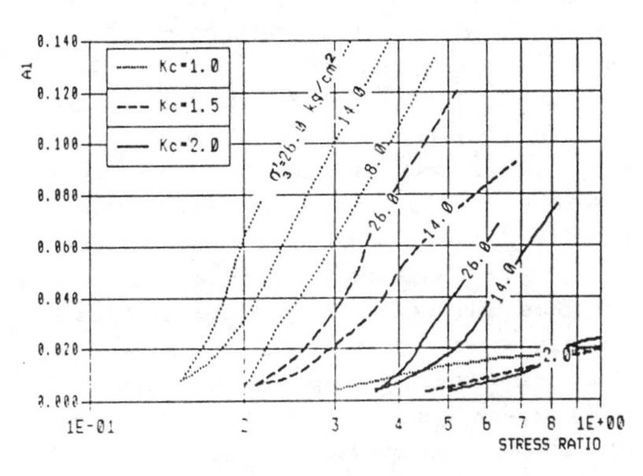

Fig. 3 - A_1 values for Oroville gravel

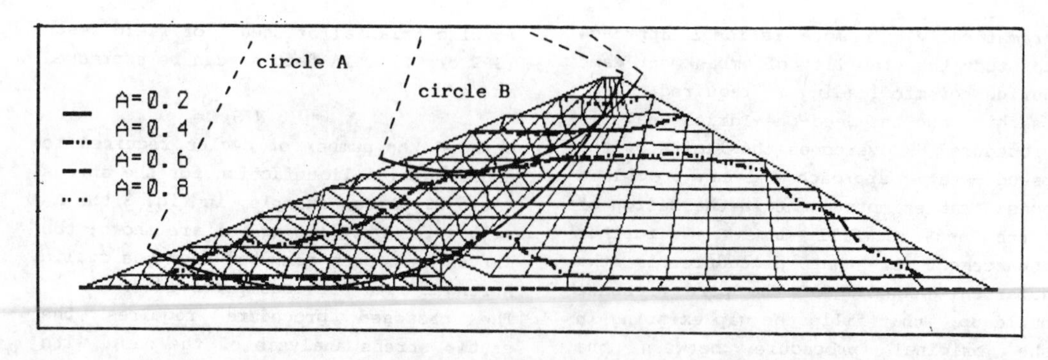

Fig. 4 - Computed maximum acceleration values and circles for permanent displacement analyses

the initial static effective horizontal stress. The R_u values reach 80% in two areas of limited extension, while larger zones of the submerged shell show values over 40%. In this condition a stability analysis with drained values of strength lets the safety factor drop from 3.1 (pre-earthquake condition) to about 1.4 (post-earthquake condition assuming no pore water redistribution). This substantial reduction of the safety factor, even if the examined dam remains in an amply safe state, gives an exact proportion of the importance of the pore pressure increase in the stability analysis of earth and rockfill dams.

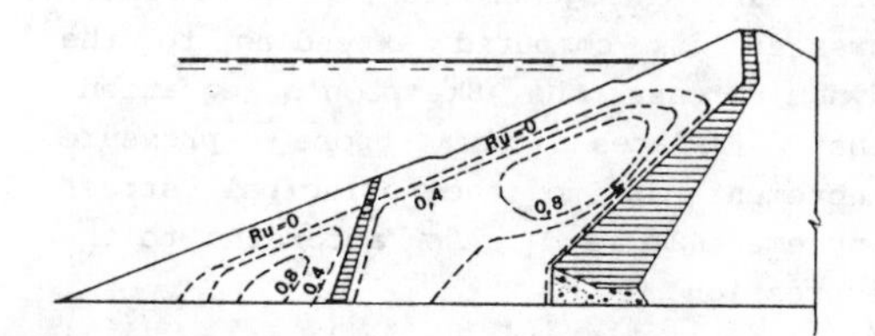

Fig. 5 - Pore pressure ratio values at the end of the earthquake

2.2 Development of permanent displacements

The integrity of the structure can be threatened also by the overtopping of the dam, owing to permanent displacements of the crest. They can be caused by the process of consolidation of excess pore water pressures, a certain while after the motion has stopped, and by the inertia forces, during the earthquake, acting on sliding portions of the embankment as well. In the former case a consolidation analysis can assess the final configuration of the dam when all the excess pore pressures have been dissipated. In many cases, i. e. a rigid foundation soil, these movements are small due to the stiffness of the materials used in the dam construction. But the permanent displacements caused by the motion of a portion of embankment along a slip surface can be large enough to cause serious damage to the entire structure. Newmark's analysis of sliding blocks is a first attempt to account for this problem /3/, but the sliding blocks technique gives displacements along a fictitious plan which is not properly defined. A more recent approach /5/ can deal with circular slip surfaces that are more consistent with the usual limit equilibrium analyses. Referring to fig. 6, the equation of motion of the sliding body, with respect to the center of rotation O, is:

$$\ddot{\vartheta} - \alpha\,(q - q_c)\ \sin\,(\vartheta_o + \vartheta) = 0 \qquad (3)$$

where $\alpha = Wd/I$, being W the weight of the sliding body, d the distance of the center of gravity G from O and I the mass moment of inertia. In eq. (3) q is the earthquake induced acceleration, function

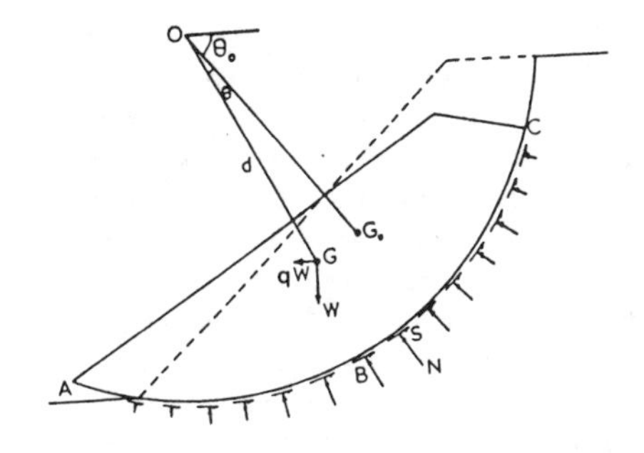

Fig. 6 - Rotational sliding surface

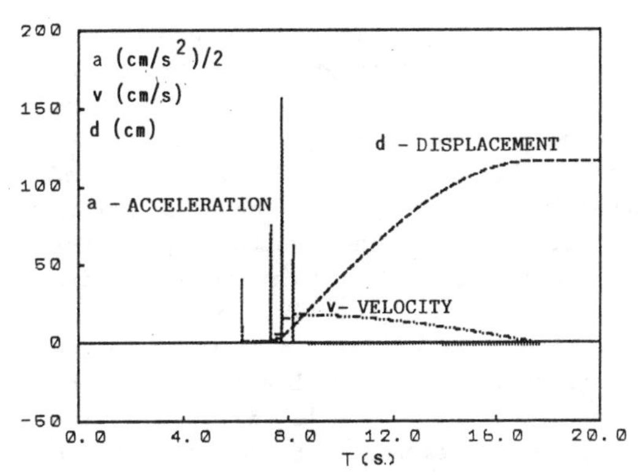

Fig. 7 - Computed values of a, v, d for the sliding surface A

of time, while q_c is the acceleration corresponding to a safety factor equal to one in a pseudo-static analysis (the critical acceleration). An analytical solution of eq. (3) has been obtained for rectangular pulses. The proposed procedure allows to evaluate the permanent displacements induced by acceleration time-histories, similar to those produced by earthquakes, solving eq.(3) by numerical integration. Taking into account the different resistance in the two directions, the equation (3) can be accordingly transformed in:

$$\ddot{\vartheta} + (1-\delta)\beta\dot{\vartheta} - \alpha(q-q_c)\delta\sin(\vartheta_o + \vartheta) = 0$$

where $\delta = 1$ for the mass sliding downhill

$\delta = 0$ " " " " uphill.

The parameter β has been set equal to the critical damping of the system to avoid the mass sliding in the uphill direction. A general purpose algorithm, particularly effective in dealing with non linear systems under impulsive loads /2/, has been used to solve the above equation of motion. So, referring again to Oroville dam, for the two circular slip surfaces indicated in fig. 4, it is possible to compute the permanent displacements of the crest. Displacement, velocity and acceleration along the slip surface A are drawn in fig. 7 for the base motion (q_{max} = 0.6 g) and for q_c = 0.3. The computed permanent displacement is 116 cm, that gives a vertical displacement of the embankment crest of about 1 meter. That value is amply acceptable, considering that the freeboard height is more than 6 meters. Performing again the analysis with the acceleration time history pertinent to the center of gravity of the sliding body A (q_{max} = 0.4 g), the computed permanent displacement is about 5 cm. On the contrary, for the sliding body B the maximum peak acceleration of the base is 0.4 g, while for the center of gravity it is higher (q_{max} = 0.54 g). In tab. 1 the permanent displacements

Table 1 Computed permanent displacements of the slides

SLIDING BODY	ACC. (g)	q_c	DISPL. (cm)
A	0.60 (BASE)	0.2	306
		0.3	116
		0.4	41
	0.40 (G_A)	0.2	195
		0.3	5
		0.4	0
B	0.40 (BASE)	0.2	281
		0.3	30
		0.4	0
	0.54 (G_B)	0.2	920
		0.3	313
		0.4	55

along the slip surfaces A and B are reported for different values of q_c and for the acceleration time-histories of the base of each circle and of the center of the sliding bodies. It can be observed that values of $q_c \geq 0.3$ mean a safety condition as far the overtopping of the dam is concerned. It is worth mentioning that 0.3 is the effective peak value of the input accelerogram in relation to the characteristic of vibration of the structure. So it is validated the Newmark's recommendation that the static resistance of the dam should be greater than the maximum earthquake acceleration.

3 CONCLUSION

On the whole the performance of embankment dams in highly seismic regions can be considered satisfactory. Earthquake induced failures or near failures are few and have affected basically hydraulic fill and tailing dams. A particular attention has then to be paid to those structures, that are not very diffused in Europe. As confirmed in 1987 ICOLD Executive Meeting, rolled earth and rockfill dams have shown an exceptional resistance to seismic events. Anyway a big effort has been done and is still in progress in order to understand and explain the embankment dams behaviour during and after earthquakes, as a necessary basis for correct design. The numerical analysis procedures that have been illustrated in the present paper can be considered part of this research. Coupled with serious laboratory and in situ geotechnical investigations on the foundations and construction soils they can be successfully used in the study of embankment dams seismic stability.

Table 2. CPU time consumed by the programs (min)

	COMPUTER	
	OH5430*	IBM3090VF
Dynamic Analysis	42.35	1.17
Displ. Analysis	0.67	0.064

(*)IBM4341 Performance

Actually it is well worth pointing out that thanks to the progressively increasing power of the computer facilities, both in speed performance and storage capacity, simulation programs can tackle many more aspects of the real behaviour of the dam, improving their contribution in producing a safer design. The CPU-time consumed by two different generation computers in carrying out the above described procedures, is presented in tab. 2. While the speed up is about 10 for the permanent displacement analyses, in the time-consuming assessment of the dynamic response of the dam by FEM it goes up over 35, owing to the high performance of the Vector Facility.

REFERENCES

/1/ N.G. Banerjee, H.B. Seed, C.K. Chan, Cyclic behaviour of dense coarse-grained materials in relation to the seismic stability of dams, Rep. No UCB/EERC-79/13 Un. of Cal., (1979).

/2/ R. Casciaro, Dyana: A general purpose algorithm for incremental analysis, Univ. of Rome, Ist. di Scienza delle Costruzioni, Pubbl. N. II-163, (1974).

/3/ N.M. Newmark, Effects of earthquakes on dams and embankments, Geotechnique 15, No 2, (1965).

/4/ P. Paoliani, A method for evaluating pore pressure rise during dynamic loadings of sands, 2nd Int. Conf. on Soil Dyn & Earthquake Eng., on board the QE2, (1985).

/5/ S.K. Sarma, Seismic displacement analysis of earth dams, Journ. Geot. Eng. Div., ASCE, 107, GT12 (1981).

/6/ H.B. Seed, Considerations in the earthquake resistant design of earth and rockfill dams, Geotechnique, 29 No 3, (1979).

Numerical Methods in Geomechanics (Innsbruck 1988), Swoboda (ed.)
© 1988 Balkema, Rotterdam. ISBN 90 6191 809 X

Two-dimensional nonlinear response analysis during earthquakes based on the effective stress method

Mamoru Kanatani & Koichi Nishi
Central Research Institute of Electric Power Industry, Japan
Makoto Kawakami & Masayuki Ohnami
Kozo Keikaku Engineering Inc., Japan

ABSTRACT: In order to examine the stability of foundation ground and structure during earthquakes, we have developed the nonlinear response analysis program based on the effective stress method (Nonlinear Analysis of Fluid-Solid System, PROGRAM NAFSS). In this analysis, equations of motion are introduced from Biot's porous elastic material theory and nonlinearity of soil properties expressed by constitutive relations based on the elasto-plastic theory. Application of this program to two dimensional ground-structure interaction problem indicate that the results of this analysis reasonably simulate the ground behavior during earthquakes.

1 INTROUDUCTION

In recent years various dynamic response analysis methods have been developed by many researchers to analyse the behavior of ground and structure during earthquakes. As is commonly known, the defromation and strength characteristics of soils are closely dependent on the effective stress, therefore to examine the stability of ground and structure in detail, dynamic effective stress analysis taking into account the changes of pore water pressure must be done. For the reasons above mentioned, nowadays several efforts have been made to explore the possibility of performing the dynamic effective stress analysis. And many codes for one dimensional problems have already been developed. On one hand, Zienkiewicz et al. (1982) and Ghaboussi et al.(1972) proposed to apply F.E.M. to Biot's wave propagation theory in the porous elastic materials (1962) and extended it to the earthquake response analysis for two dimensional fields.

In our procedure, equations of motion are constituted by treating the displacement of solid and fluid as valiables, what is called U-W analysis, and constitutive relations of soils are newly proposed ones based on the elasto-plastic theory by Nishi et al.(1988). And by using this code we try to analyse the two-dimensional earthquake response.

2 FIELD EQUATIONS FOR TWO PHASE SYSTEMS

In case of regarding the ground as two phase material which is constituted by solid particles and fluid with which its pore is filled, it is generally said that the kinematic mechanism of such materials is dependent on Biot's theory for porous elastis materials. Field equations of two phase materials based on the Biot's theory are represented as the followings

(1)Equation of motion for mixed soil

Equation of motion for soil as solid and fluid mixture is

$$\sigma_{ij,j} + \rho g_i = \rho \ddot{u}_i + \rho_f \ddot{w}_i \quad (1)$$

where σ_{ij} is the tatal stress in the conbined solid and fluid mixture, ρ and ρ_f are the mass density of mixed soil and pore fluid respectively. u_i is the displacement of solid skeleton and w_i is the relative displacement of fluid to solid skeleton. g_i is a body force.

(2)Equation of motion for pore fluid (general Darcy's law)

For the pore fluid we can give the equation of motion by adding the inertia terms to Darcy's law

$$-P_{,i} + \rho_f g_i = k_{ij}^{-1} \rho_f g \dot{w}_i + \rho_f \ddot{u}_i + \frac{\rho_f}{n} \ddot{w}_i \quad (2)$$

where P is fluid pressure, k_{ij} expresses the permeability coeffecient for isotropic and anisotropic conditions. And n denotes porosity of soil.

(3)Mass conservation equation for compressible fluid

Mass conservation equation, what is called constitutive equation of pore flulid, is applied to compressible fluid flow as following.

$$\dot{w}_{i,i} = -\dot{\varepsilon}_{ii} - \frac{(1-n)\dot{p}}{K_s} - \frac{n\dot{p}}{K_f} - \frac{\dot{\sigma}_{ii}'}{3K_s} \quad (3)$$

where ε_{ii} is volumetric strain of solid skeleton, K_s and K_f are the bulk modulus of solid particle and fluid respectively, and σ'_{ii} denotes the effective stress acting on the each particle. Further from eq.(3) we can obtain the following equation to calculate the pore fluid pressure at any instant.

$$\dot{p} = -M(\alpha \delta_{ij} \dot{\varepsilon}_{ij} + \dot{w}_{i,i}) \quad (4)$$

here

$$\begin{cases} \alpha = 1 - \dfrac{\delta_{ij} D_{ijk\ell} \delta_{k\ell}}{3K_s} \\ \dfrac{1}{M} = \dfrac{1-n}{K_s} + \dfrac{n}{K_f} \end{cases}$$

δ_{ij} : kronecker's delta

$D_{ijk\ell}$: stiffness matrix of solid skeleton

(4)Constitutive relation for solid skeleton and effective stress principle

We define the constitutive relation of solid skeleton in a incremental form for nonlinear properties depended on the effective stress.

$$\dot{\sigma}_{ij} = \dot{\sigma}'_{ij} - \alpha \delta_{ij} \dot{P} = D_{ijk\ell}(\dot{\varepsilon}_{k\ell} - \dot{\varepsilon}^{\circ}_{k\ell}) - \alpha \delta_{ij} \dot{P} \quad (5)$$

where ε_{ij} is the total strain and ε°_{ij} is the initial strain.

Above mentioned equations (1)-(5) are leaded by the assumption that the solid deformation and fluid flow is small, therefore convective term is neglected.

3 FINITE ELEMENT DISCRETIZATION

By appropriately transforming the equations which was introduced in the previous section, we can obtain the equations of motion denoted by solid and fluid displacement as variables. Then those equations are discretized by using finite element procedure (Galerkin method) and are described in the matrix forms as followings

$$\begin{bmatrix} M_s & M_c \\ M_c^T & M_f \end{bmatrix} \begin{Bmatrix} \ddot{u} \\ \ddot{w} \end{Bmatrix} + \begin{bmatrix} O & O \\ O & H \end{bmatrix} \begin{Bmatrix} \dot{u} \\ \dot{w} \end{Bmatrix} + \begin{bmatrix} K & C \\ C^T & E \end{bmatrix} \begin{Bmatrix} u \\ w \end{Bmatrix} = \begin{Bmatrix} F \\ G \end{Bmatrix} \quad (6)$$

$u = u_i^k$, $w = w_i^k$

$$M_s = \sum_{\nu=1}^{N} \int v_\nu \overset{(u)}{N}_m \rho \overset{(u)}{N}_k \, dV_\nu$$

$$M_c = \sum_{\nu=1}^{N} \int v_\nu \overset{(u)}{N}_m \rho_f \overset{(w)}{N}_k \, dV_\nu$$

$$M_f = \sum_{\nu=1}^{N} \int v_\nu \overset{(w)}{N}_m \frac{\rho_f}{n} \overset{(w)}{N}_k \, dV_\nu$$

$$H = \sum_{\nu=1}^{N} \int v_\nu \overset{(w)}{N}_m \frac{\rho_f \, g}{k_{ij}} \overset{(w)}{N}_k \, dV_\nu$$

$$K = \sum_{\nu=1}^{N} \int v_\nu \overset{(u)}{N}_{m,j} (D_{ijk\ell} + \alpha^2 M \delta_{ij} \delta_{k\ell}) \cdot \overset{(u)}{N}_{k,\ell} dV_\nu$$

$$C = \sum_{\nu=1}^{N} \int v_\nu \overset{(u)}{N}_{m,i} M \alpha \delta_{ij} \overset{(w)}{N}_{k,j} dV_\nu$$

$$E = \sum_{\nu=1}^{N} \int v_\nu \overset{(w)}{N}_{m,i} M \delta_{ij} \overset{(w)}{N}_{k,j} dV_\nu$$

$$F = \sum_{\nu=1}^{N} \{ \int s_\nu \overset{(u)}{N}_m \overline{T}_i dS_\nu + \int v_\nu \overset{(u)}{N}_m \rho g_i dV_\nu \}$$

$$G = \sum_{\nu=1}^{N} \{ \int s_\nu \overset{(w)}{N}_m \overline{P}_i dS_\nu + \int v_\nu \overset{(w)}{N}_m \rho_f g_i dV_\nu \}$$

Matrices F and G are the ones that express the volumetric body force and surface traction acting on the mixed material and fluid respectively.

On condition that the ground permeability is small (k<1cm/sec), the vibration motion of pore fluid is very small even in the field which the toltal stresses are hardly changed. What to neglect the interacting term of mass matrix does not give the serious effects to response was analytically checked up by comparing the resuls of linear analysis with exact solutions by Kanatani et al.(1987). Therefore to reduce the time for calculation of nonlinear analysis, we exchange the mass matrix into diagonal one by neglecting the mass matrix Mc and using the lumped mass concept.

On the other hand, boundary conditions are as follows,

deformation of solid skeleton

$u_i = \overline{u}_i$ ……for a known displacement of solid skeleton

$\sigma_{ij} \cdot n = \overline{T}_i$ …for a known stress of mixed soil

motion of pore fluid.

$$\begin{cases} w_i = \overline{w}_i \cdots\cdots \text{for a known relative displacement of pore fluid} \\ P \cdot n_i = \overline{P}_i \cdots \text{for a known pore fluid pressure} \end{cases}$$

4 CONSTITUTIVE RELATION FOR SOILS

As the constitutive relation for describing nonlinear properties of soils we adopt the newly proposed model which was developed by Nishi et al(1988). This model is based on the elasto-plastic theory and so as to make possible its application for two or three dimensional dynamic problems adopts the general Masing rule. Elasto-plastic cnstitutive relatin are proposed by paying attention to stress -dilatancy relations obtained from drained cyclic loading tests for saturated sand and the above mentioned general Masing rule as the hardening function.

By using this model, stress-srain behaviors under not only multi-directional cyclic loading condition but also the cyclic mobility which is inherent properties of dense sand are explained. For further particulars are described in session 2 in this proceeding by Nishi et al.

5 TIME INTEGRATION

Here we adopt the central difference method for step-by-step time integration to conduct the nonlinear analysis. Firstly rewrite eq.(6) more simply as is shown in eq.(7) and assume that it can be formed at time t.

$$[M]^t\{\ddot{u}\}+[C]^t\{\dot{u}\}+{}^t\{F\}={}^t\{R\} \quad (7)$$

and

$$\left.\begin{aligned} {}^t\{\ddot{u}\} &= ({}^{t+\Delta t}\{u\} - 2\cdot{}^t\{u\} + {}^{t-\Delta t}\{u\})/\Delta t^2 \\ {}^t\{\dot{u}\} &= ({}^{t+\Delta t}\{u\} - {}^{t-\Delta t}\{u\})/2\Delta t \end{aligned}\right\} \quad (8)$$

from eq.(7), (8) and the reason why [M], [C] are diagonal,

$${}^{t+\Delta t}\{u\}_k = {}^t\{\hat{R}\}_k/\{\hat{M}\}_k \quad (9)$$

(k : component of k-th freedom)

then increments of displacement and strain are denoted by

$${}^t\{du\}_k = {}^t\{u\}_k - {}^{t-\Delta t}\{u\}_k \quad (10)$$

$${}^t\{d\varepsilon\}_k = [B]_k {}^t\{du\}_k \quad (11)$$

therefore effective stress increments of solid are obtained by

$${}^t\{d\sigma'\}_k = {}^t[D]_k \cdot {}^t\{d\varepsilon\}_k \quad (12)$$

Pore pressure increments are calculated from strain of solid and fluid by using eq.(4).

6 APPLICATION FOR DYNAMIC GROUND-STRUCTURE INTERACTION PROBREM

In this section we will show the results of analysing the two-dimensional ground-structure system by our code.

6.1 Ground-structure model for analysis and finite element mesh

Ground-structure model and finite element mesh which is analysed this time is shown in Fig.1. The width and depth of ground model are 230m and 65m respectively. Structure model is a 30m squre and buried 10m of bottom into the ground. And ground water level is setted up the ground surface.

6.2 Boundary conditions

Boundary conditions for deformations of solid element are fixed one for bottom surface of ground and both side boundaries are treated in equal-displacement manner to express the behavior of free field. Side and bottom surface of buried part of structure are connected with soil. Boundary conditions for permeation are permeable one for only ground surface and rests are impermeable.

6.3 Input material constants

Input material constants of soil are shown in Table 1 and initial shear modulus Go is denoted in Fig.1 for each depth. Parameters α, K_s, K_f are always constant values of 1.0, 3.7×10^7(kg/cm²), 2.24 $\times10^4$(kg/cm²) for general soils. The Ko is 0.5 and permeability coefficient k is 0.01cm/sec. In this case it is assumed that the soil properties are the same for all layers. On the assumption that the structure is elastic, its young modulus, poison's ratio and mass density are given by E=1.464×10^4(kgf/cm²), ν=0.2 and ρ= 1.667(g/cm³) respectively.

6.4 Input earthquake motion

Input earthquake motion is a main shock of El-Centro NS component in 1940 duration 5

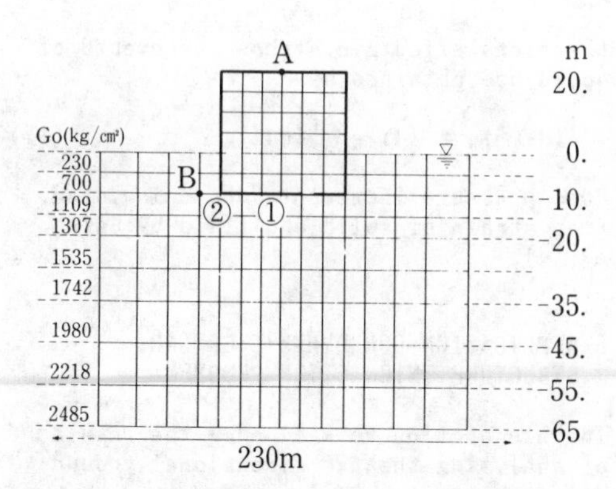

Fig.1 Model for analysis and finite element mesh

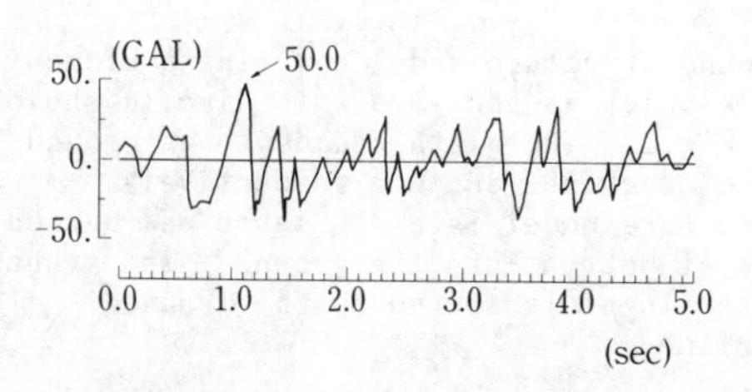

Fig.2 Input earthquake motion

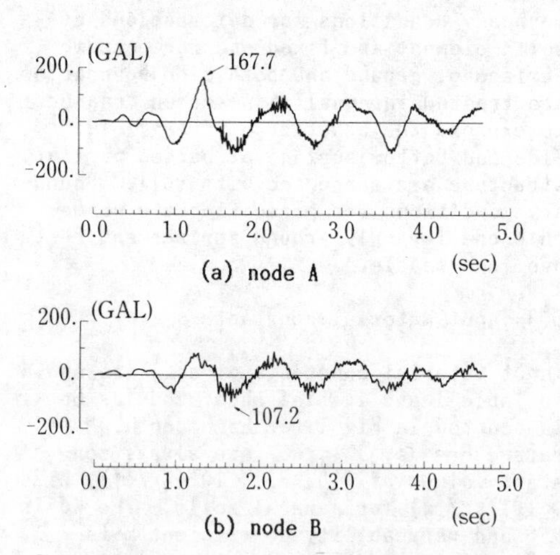

Fig.3 Time history of responce acceleration

Table 1 Material constants of soil

e_o	m*	κ	β	ϕ	ν	γ_s	γ_f
	(cm²/kg)			(°)		(g/cm³)	(g/cm³)
0.8	0.8	0.0015	200	36	0	2.4	1.0

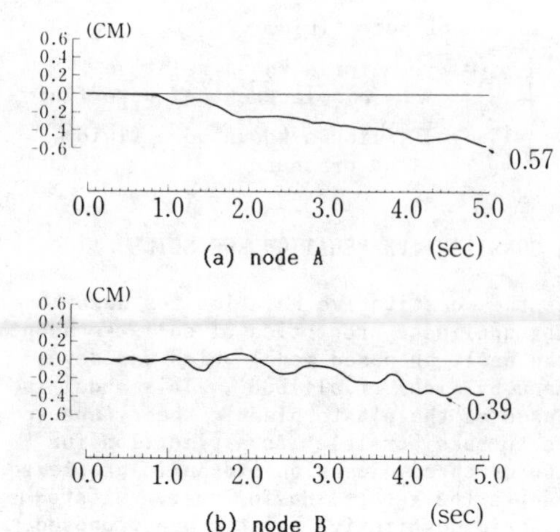

Fig.4 Time history of vertical displacement

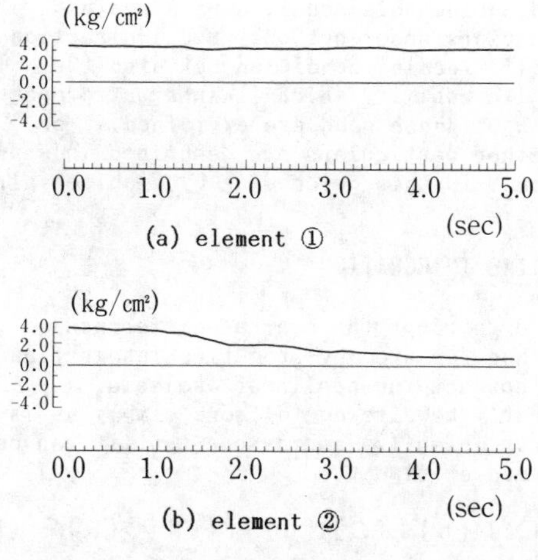

Fig.5 Time history of vertical effective stress

sec(from 1 to 6sec) as is shown in Fig.2. And maximum acceleration is 50 gal.

6.5 Meterial demping

It is generally said that soils and structures have a few percent damping, what is called material damping, at very low strain level. Therefore taking into account such damping we adopt the Rayleigh damping of 2% and 5% for ground and structure respectively which are proportional to the mass.

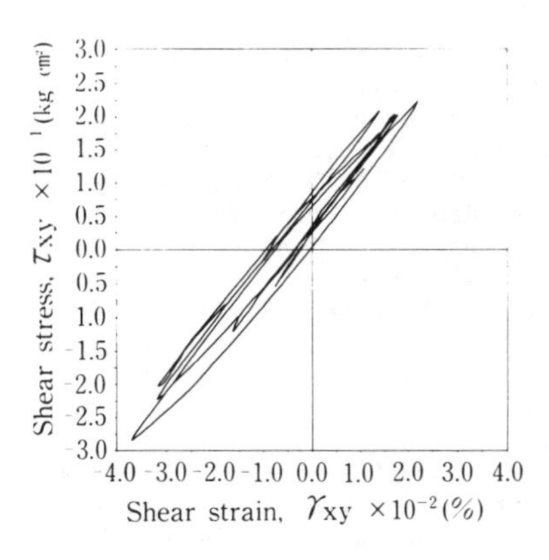

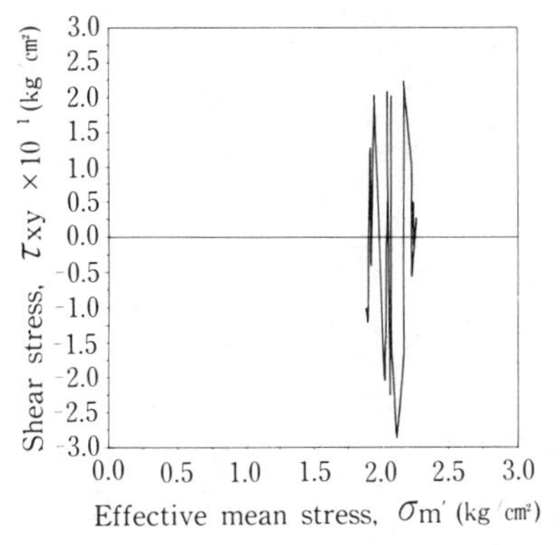

Fig.6(a) Stress-Strain curve and effective stress path of element ①

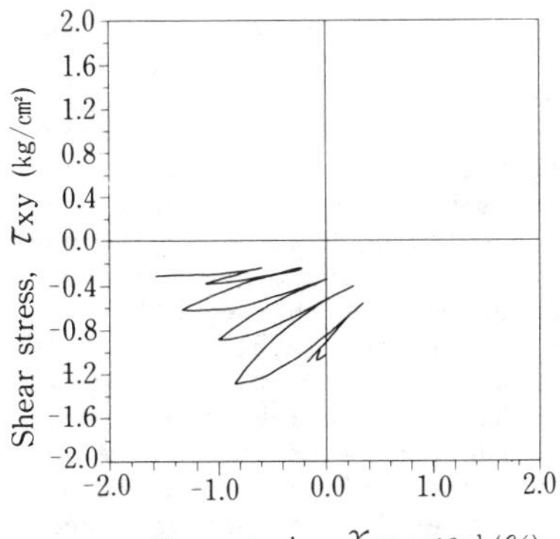

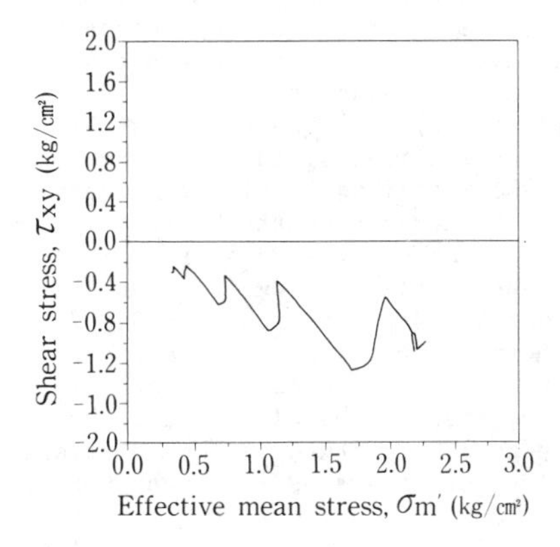

Fig.6(b) Stress-Strain curve and effective stress path of element ②

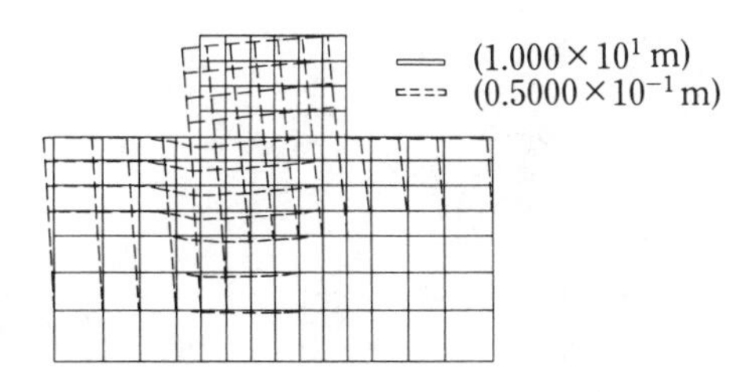

Fig.7 Deformation pattern at 5 sec

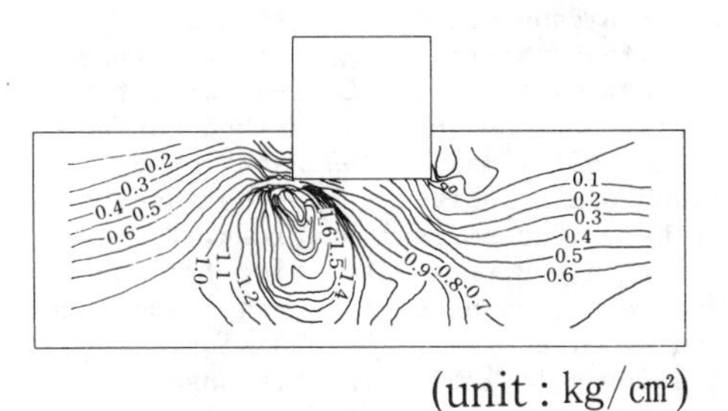

Fig.8 Contor of pore pressure at 5 sec

6.6 Results of analysis

Analysis of this time is carried out under the time step Δt=1/2000 sec. Although critical time step Δtcr is almost decided by the velocity of compressible wave of water, in this case above montioned time step is sufficiently covered Δtcr even from the width of mesh. Fig.3(a) and (b) show the time histories of response acceleration at the top of structure (node A in Fig.1) and the ground nearby the structure (node B) respectively. After 1.5 second high frequency components occur on both accelerations. Such phenomenon is frequently shown in case of analysing the nonlinear transient problems.
Although not sufficiently making clear its cause, we think it dependent on the analytical problem. But it is very impor-

tant to evaluate the maximum acceleration exactly. Time histories of vertical displacement at the node A and B are shown in Fig.4(a) and (b). Structure and ground are settled with time during earthquake by dissipation of pore water pressure. In particular upward displacement occurs at times at node B, this is by the connection of contact surface between structure and ground Fig.5(a) and (b) show the time histories of vertical effective stress at the right bellow(element ①) and near the structure (element ②). When it become zero means initial liquefaction. In element ① vertical effective stress almost unchange because of restriction effect of heavy structure. On the other hand, in element ② it fairly reduces until almost liquefaction.
Fig.6(a) and (b) show the stress-strain curve and effecttive stress path for each element. Nevertheless the initial shear stress is acted on element ②, they indicate the good expressons of soil behavior under cyclic loaing from the initial shear stress condition. In Fig.7 and 8 are shown deformation and pore pressure distributions at 5 second. Pore pressure is concentrated at the place under the bottom edge of structure by the rotational movement of structure as is shown in Fig.7.

7 CONCLUSIONS

We develope the nonlinear dynamic response analysis code which is based on the two-phase material theory and adopts the newly proposed constitutive relation and apply it to two dimensional ground-structure interaction problem. Consequently the results obtained by this method can generally well simulate the behavior of ground and structure during earthquakes. But hereafter several problems, for example boundary treatment, high frequency component existence of response acceleration etc., are left. In addition, expended time for calculation in this case study was 60 minutes by FACOM M-360.

8 REFERENCES

Zienkiewicz, O. C., Leung, K. H., Hinton, E. and Chang, C. T. (1982) : Liquefaction and permanent deformation under dynamic conditions-numerical solution and constitutive rerations, Soil Mechanics-Transient and Cyclic loads, Chapt.5, pp.71-103.

Ghaboussi, J. and Wilson, E. L. (1972) : Variational formulation of dynamic of fluid-saturated porous elastic solids, Proc. ASCE, Vol.98, EM4, pp.947-963.

Biot, M. A. (1962) : Mechanics of deformation and acoustic propagation in porous media, Jour. of Applid Physics, Vol.33, No.4, pp.1482-1498.

Nishi, K. and Kanatani, M. (1988) : Constitutive relations with general masing rule under multi-dimensional stress condition, Proc. 6th ICONMIG, (in preparation).

Kanatani, M. and Nishi, K. (1988) : Evaluation of stability of foundation ground during earthquakes (Part 4)-Development of response analysis with soil skeleton-pore water interaction, CRIEPI Report, in Japanese, (in preparation).

Numerical Methods in Geomechanics (Innsbruck 1988), Swoboda (ed.)
© 1988 Balkema, Rotterdam. ISBN 90 6191 809 X

Ultrasonic wave attenuation in dry and saturated rock under biaxial pressure

Yijun Zhao & Zhaodin Li
Mining Engineering Department, Northeast University of Technology Shenyang, Liaoning, People's Republic of China

ABSTRACT: An ultrasonic wave attenuation measurement system and a biaxial pressure loading system were designed and built. The quality factor (Q) values as well as velocities (V) of compressure (P) and Shear (S) wave in dry and saturated Hongyang standstone and diabase were studied in the laboratory at ultrasonic frequencies. A pulse trasmission technique and spectral ratios were used to determine Q value relative to a reference sample with very low attenuation. It was found that Q value is the function of differential pressure. The rate of the change of Q values versus diffential pressure varies deponding on rock type, crack porosity and its distribution and saturation of rock.

I. Introduction

One of the most important geophysical characteristics of rock is the existence of stress fields. Although many scientists and researchers have attempted to generate a theoretical model to calculate stress field, the results are equivocal. Lacking a theoretical model, the standard approach is to employ in-situ measurement. The most flexible and convenient method is to employ acoustic stimuli.

Studies of the parameter of acoustic velocity and its relationship to rock density and elasticity have yielded insights into stress in rock measurement (4,14,19 8).However, due to the limited reliabilty of acoustic velocity, scientists have turned to other parameters for study. In recent years studies of the correlation between the quality factor (Q) and physical and mechanical properties of rock have increased in rock acoustics and seismology.

Acoustic waves propagating through a medium are attenuated by the conversion of some fraction of the energy into heat because of the anelastic property of the medium. The amplitude of the wave decays exponentially with distance. The most commonly used measures of attenuation reported in the literature are the attenuation coefficient (α) and the quality factor (Q) or its inverse (Q^{-1}). The definition of (Q) is:

$$Q = \frac{\omega E}{-dE/dt} = \frac{2\pi W}{\Delta W} \qquad (1)$$

where (ω) is the angular frequency, (E) is the instantaneous energy in the system, dE/dt is the rate of energy loss, (W) is the elastic energy stored at maximum stress and strain, and (ΔW) is the energy loss (per cycle) of a harmonic excitation. The quality factor (Q) is dimensionless, and its relationship to (α) is

$$Q = \frac{\pi . f}{\alpha V} \quad (2)$$

where (V) is the velocity and (f) is the frequency. From the definition of (Q) we know that it only gives the rate of crest energy and the fraction of energy changed to heat, but reveals nothing of the specific mechanism whereby this occurs.

The methods generally used for measuring attenuation in the laboratory can be separated into four main categories (4): free vibration; forced vibration; wave propagation; and observation of stress-strain curves.

Attenuation measurements have been carried out on samples in the laboratory using a variety of techniques (1,12). In a most general way these expermients suggest that (Q) is independent of frequency, fluid saturation increases attenuation, and increasing pressure decreases attenuation (5,6,15).

Unfortunately, there has not been systematic study of the behavior of attenuation with pressure and saturation conditions. Additional laboratory data are needed under controlled conditions to ascertain the effects of fluides and pressures on the attenuation in rocks.

In this paper ultrasonic (Q) data are presented as functions of biaxial pressure for dry and saturated rock. The implication of the data to attenuation mechanisms are briefly discussed, and some conclusions put forth.

II. Experimental Method

The pulse transmission technique is most suitable for the pressured and saturated sample and the loss parameters involved closely parallel thos measured in the field experiments, provided correction can be made for geometric factor such as beam spreading and reflections.

The present study employed the pulse transmission technique and measured attenuation relative to a reference sample which has very low attenuation. The sample to be studied and the reference sample have exactly the same shape and geometry.

The amplitude of plane seismic waves for the reference and sample can be expressed as:

$$A_1(f) = G_1(x)\, e^{-\alpha_1(f)} e^{i(2\pi f - k_1 x)}$$

$$A_2(f) = G_2(x)\, e^{-\alpha_2(f)} e^{i(2\pi f - k_2 x)} \quad (3)$$

where (A) is amplitude, (f) is frequency, (X) is distance, ($k=2\pi f/V$) is the wave number, and (G(x)) is a geometrical factor which includes spreading, reflections, etc. Subscripts 1 and 2 refer to the reference and sample respectively. It is the opinion of Toksöz and knopoff that over the frequency range of the measurements, 0.1-1.0 M MHz, (α) is a linear function of frequency, although the method itself tests this assumption (Knopoff, 1964,Jackson and Anderson, 1970, Joston and Toksoz, 1980). Thus one can write:

$$(f) = \gamma f \quad (4)$$

where (γ) is constant and related to the quality factor (Q) by:

$$Q = \frac{\pi f}{\alpha V} = \frac{\pi}{\gamma V} \quad (5)$$

The ratio of the Fourier amplitudes of equation (3) is:

$$\frac{A_1(f)}{A_2(f)} = \frac{G_1(x)}{G_2(x)} e^{-(\gamma_1-\gamma_2)f\ x} \quad (6)$$

or

$$\ln \frac{A_1(f)}{A_2(f)} = (\gamma_2-\gamma_1)f\ x + \ln \left[\frac{G_1(x)}{G_2(x)}\right] \quad (7)$$

when both the sample and standard have some geometry (i.e. same sample dimensions, transducers,and arrangements).The $(\ln \frac{G_1(x)}{G_2(x)})$ is frequency-independent scale factor. Then ($\gamma_2 - \gamma_1$) can be found from the slope of the line fitted to $\ln (A_1/A_2)$ versus frequency. If the (Q) of the standard reference is known, (γ_2) can be determined.

In this study, aluminum was used as a standard reference. The value of (Q) for aluminum is approxumately 150,000 (Zamanek and Rudnick, 1961) as opposed to a $(Q) \leq 1000$ for rocks. Thus, taking ($\gamma_1 = 0$) (γ_2 can be easily determined) this never introduced more than 1% error. For typical rocks where (Q)=10-100 the error is less than 0.1% and is negligible. A more serious concern is the validity of the assumption that the geometric factors (G_1) and (G_2) have the same frequency dependence, and G_1/G_2 is independent of frequency. With polished rock surfaces and good coupling between the transducer holder and sample, frequency dependent reflection coefficients are not encountered at the interface. Repeated measurements at 10 MPa showed that pulse amplitudes, shapes, and spectra we were duplicated.

III. Experimental Design

1. Based on the work done by predecessors a system was developed to study ultrasonic wave attenuation in rocks under biaxial pressure. The components of the experimental system include: CCS-300 Microcomputer; BC-V Instantaneous Wave Memory Instrument; SYC-III Rock Acoustic Wave Parameters Measuring Instrument; X-Y Function Recorder; and necessary loading equipment. Figure 1 is a schematic outline of the system:

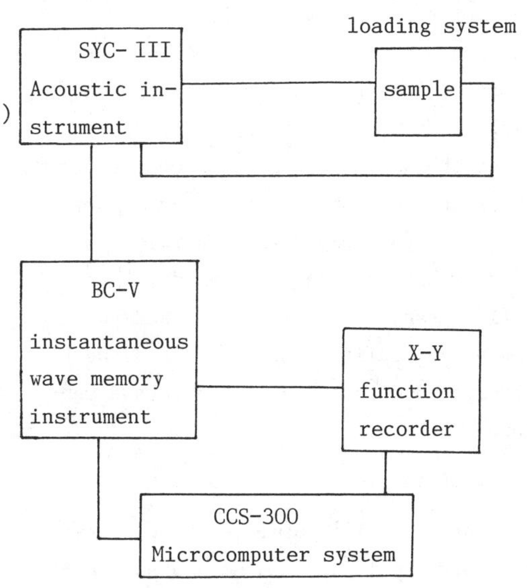

Fig.1 Experimental system of ultrosonic wave attenuation measurement.

The rock samples chosen for this study include sandstones and diabases. Relatively low (Q) values were expected in these rocks, making them amenable to the pulse transmission technique. Furthermore, these rocks are very common in mine sites and are quite commonly used in the laboratory. In fact, Berea sandstone is the a accepted standard rock of the U.S. Burau of Mines (Krech et al, 1974).

The samples used in this study were 10cm cubes. Transmitter and receiver transducers (each 5cm in diameter with natural frequencies of 1.3 MHz for P wave and 0.7

MHz for S wave respectively) were mounted at opposite ends of the sample.

Two sets of experiments were run, one with dry samples, the other with water-saturated samples(saturation used a vacuum techniques). During the experiment the principal pressure (at fixed lateral pressure) was increased to a maximum pressure of 50 MPa in discrete steps, at each pressure step. P or S wave velocities were measured and full P or S waveforms were recorded. Then the lateral pressure was changed and the procedure repeated.

When the waveform was recorded, it first went into the Instantaneous Wave Memory Instrument then was digitized into computer by A/D interface. A standard Fast Fourier transform routine was used to obtain the amplitude spectra. Then the (Q) values were determined by the method mentioned above.

IV. Experimental Results (Data)

The changes of (Q) value versus differential pressure in dry and saturated sandstone and diabase were studied in this paper and a large quantity of significant data were obtained.

1. Hunyang Sandstone:

Velocities for sandstone were plotted as a function of the principal pressure (at the fixed lateral pressure 10 MPa) and are shown in Figures 2 and 3. It is evident that the velocities increase with the increase of pressure regardless of condition of the sample (wet or saturated), (P) or (S) waves.

The (Q) values determined by the method mentioned above as a function of the principal pressure at different lateral pressure are shown in Figures 4 throught 7.

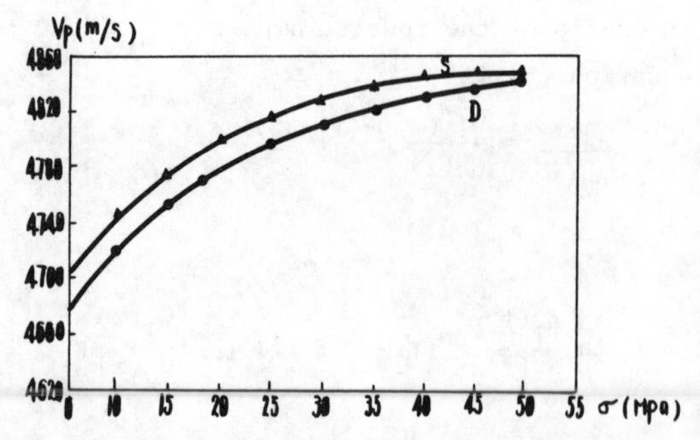

Fig.2 Velocities of P wave as a function of biaxial pressure (Lateral pressure 10 MPa)in the dry and saturated sandstone.

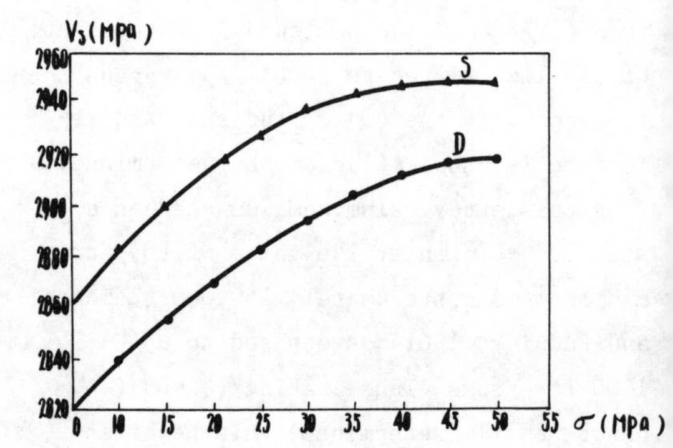

Fig. 3 Velocities of S wave as a function of biaxial pressure ( ) in the dry and saturated sandstone.

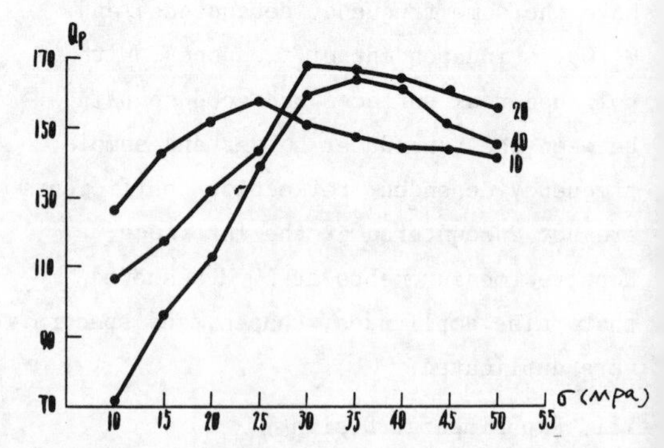

Fig. 4 Q values of P wave as a function of biaxial pressure (The number in the diagram are lateral pressure) in the dry sandstone.

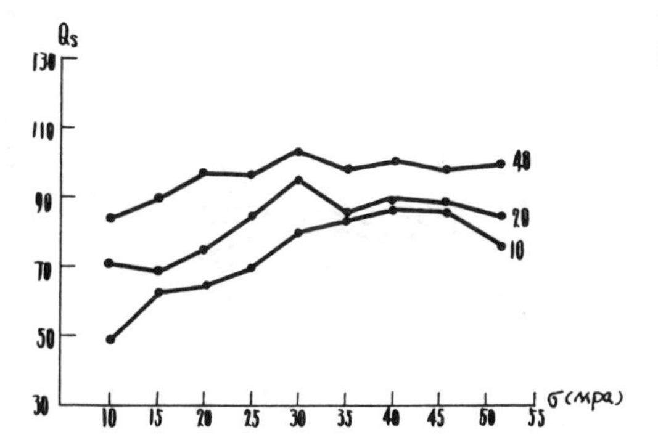

Fig.5 Q values of S wave as a function of biaxial pressure in the dry sandstone.

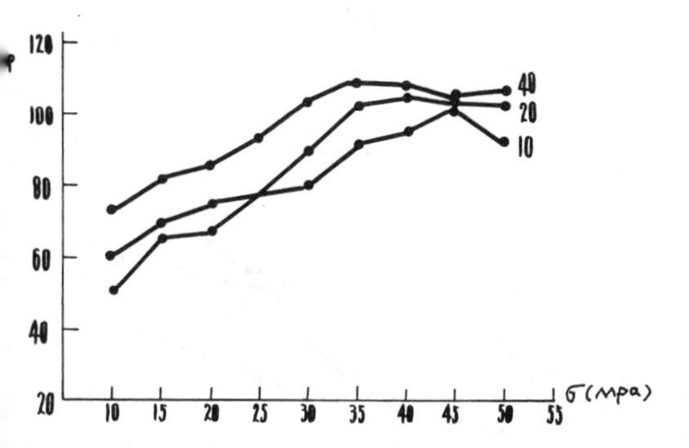

Fig. 6 Q values of P wave as a function of biaxial pressure in the saturated sandstone.

From these Figures we can find that at fixed lateral pressure the (Q) values for both dry and saturated samples increase rapidly with the increasing principal pressure. When the principal pressure reached a certain value the increasing rate of the increase of (Q) values were diminished and levelled off, with further increases of principal pressure producting slight decreases in (Q).

It is clear that both (Q_p)and (Q_s)for dry samples are substantially higher than their saturated counterparts and the tendency of increase with increasing principal pressure is much more obvious than with saturated samples. It appears that the effect of water saturation is to in-

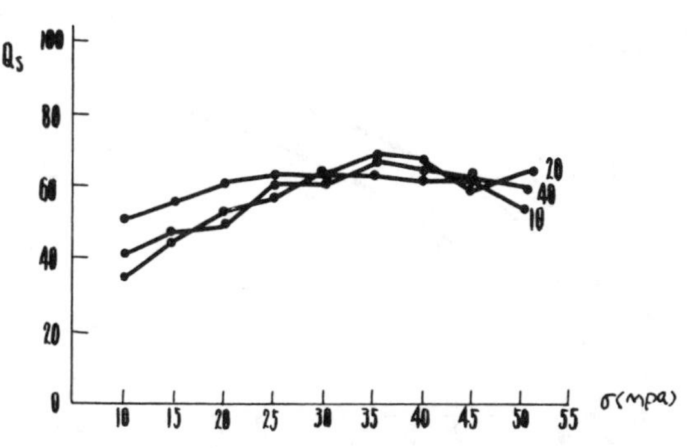

Fig. 7 Q values of S wave as a function of biaxial pressure in the saturated sandstone.

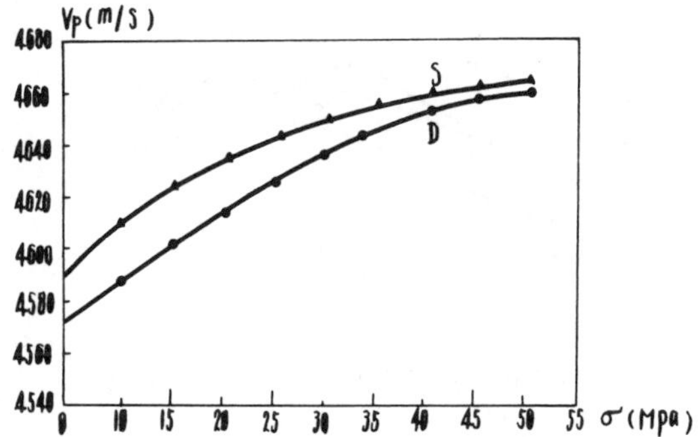

Fig.8 velocities of P wave as a function o of biaxial pressure (Lateral pressure 10 MPa)in the diabase.

crease the attenuation dramatically.

The data also suggest that the rate of (Q) values with increasing principal pressure becomes lower at high lateral pressure.

2. Diabase:

Velocities for dry and saturated diabase as a function of the principal pressure at fixed lateral pressure of 10 MPa are shown in Figures 8 and 9. for (P) and (S) waves.

In Figures 10 through 13 the (Q) values for the dry and water-saturated rock obtained from the spectral ratios are presented. Again the figures show some tendency but the absolute value of (Q_p)and (Q_s)for

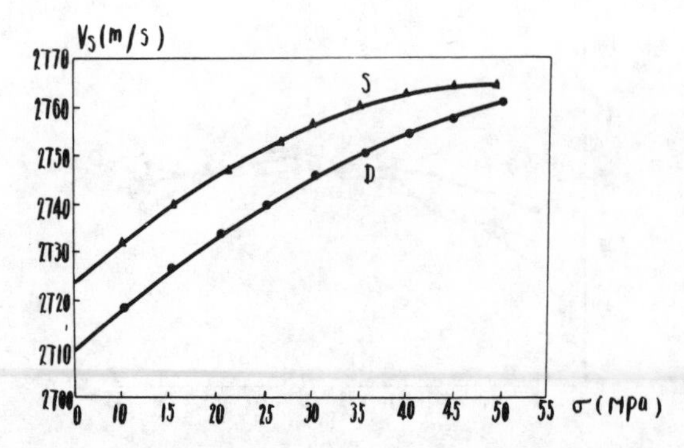

Fig.9 Velocities of S wave as a function of biaxial pressure in the diabase.

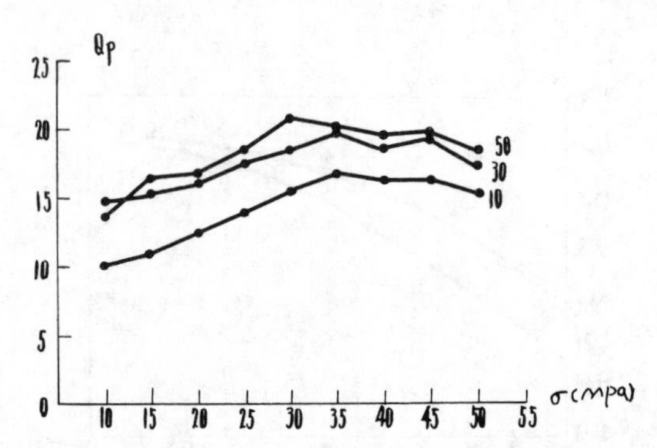

Fig. 10 Q values of P wave as a function of biaxial pressure in the saturated diabase.

diabase are substantially lower than for sandstone.

V. Implications for Attenuation Mechanism

In order to extrapolate these results to seismic frequencies it is necessary to specify the physical mechanisms for attenuation.

Numerous mechanisms have been proposed, including: Coulomb friction at crack and grain boundary contacts (Walsh, 1966), fluid flow (Biot,1956 a, b), squirting flow Marko and Nur, 1973, O'Connel and Budiansky, 1977), viscous shear relaxation in pores (Walsh, 1968), and scattering from

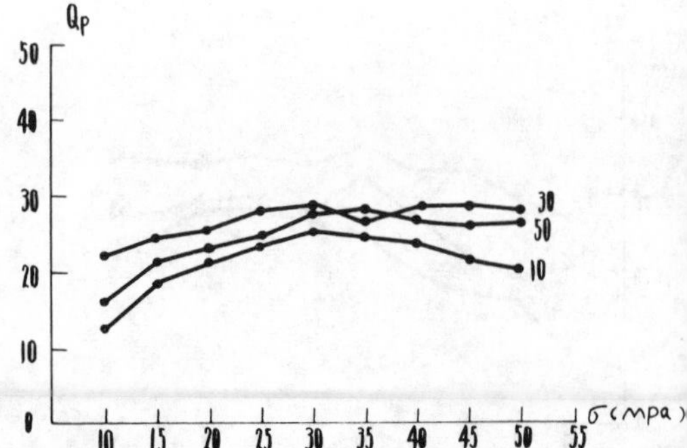

Fig. 11 Q values of P wave as a function of biaxial pressure in the dry diabase (The number in the diagram are lateral pressure) in the dry diabass.

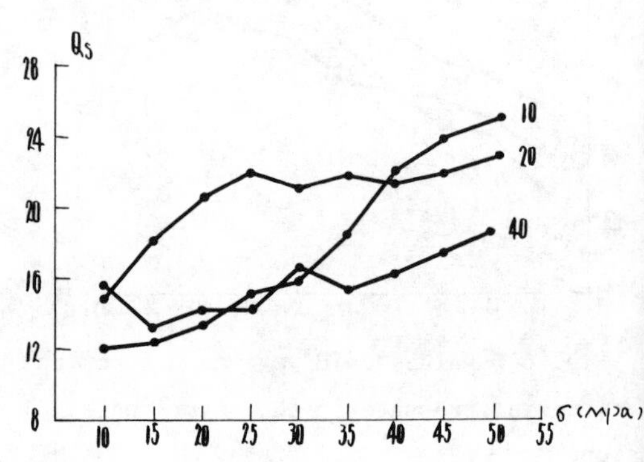

Fig. 12 Q values of S wave as a function of biaxial pressure in the dry diabase.

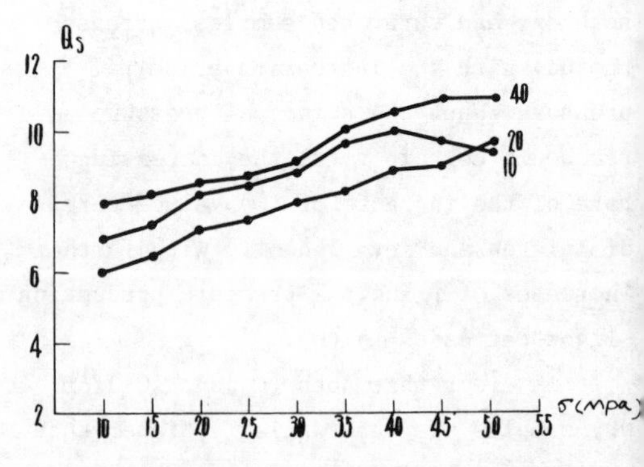

Fig. 13 Q values of S wave as a function of biaxial pressure in the saturated diabase.

pores (Yamakawa, 1962). It was found that the data presented in this paper could effectively be modeled by including a large contribution to attenuation due to grain surface friction and the rapid increases in (Q) observed at low principal pressures are due to the closing and locking of crack contacts. When the principal pressure reaches a certain value alomst all the cracks are closed and grain boundaries closely contacted,the attenuation due to grain boundary friction becomes less, so the rate of (Q) values reduces. Lateral pressure can also cause the close and the lock of the crack contacts, but only for those cracks whose long axials are vertical to the lateral pressuring axial. Therefore the increase of lateral pressure also causes the decrease of the attenuation, the (Q) values at the different lateral pressure are different than their counterparts of same principal pressure. The further increasing of principal pressure produces new cracks that will contribute to new attenuation, so the (Q) values are slightly decreased with the further compressing. In the saturated rocks the effect of friction is enhanced owing to wetting and lubricating of the sliding contacts by the pore fluid. The amount of increase in attenuation is probably determined by several complex factors including the amount of clay in the rock and wetting properties of the fluid relative to the solid matrix.

VI. Conclusions

In conclusion, P and S wave (Q) values presented for a suite of dry and saturated rocks including sandstones and diabases demonstrate several aspects of a attenuatio under the experimental conditions studied:

1. Under the conditions of the experiment in the frequency range 0.3-0.7 MHz α is a linear function of frequency, the quality factor (Q) is independent of frequency.
2. In ordinary pressure the (Q) value of dry rocks is greater than that of saturated rocks.
3. The (Q) value is a function of differential pressure. The rate of increase varies depending on the rock type, crack porosity and distribution and saturation of rock.
4. The changing law of (Q) value versus principal pressure under buaxial pressure is different from that of uniaxial and confining pressure. A possible explanation for these differences is that the opening and closing of cracks, the range and direction of crack expand vary under difference stress state.
5. Both velocity and attenuation of acoustic wave in rock are the function of pressure and saturation. However they have different relations with the properties of rock . The veloctiy represents the density and elasticity of rock while the attenuation reflects the unelastic characted of rock. The rate of the attenuatiuon of acoustic waves changes is much greater than that of the velicty.
6. Under the action of acoustic wave energy the relative sliding of crack surfaces casuing the change of surface area and the decrease of surface energy is the principal attenuation mechanism.
7. The adding of water makes the relative sliding much easier, so the (Q) value for a saturated rock is much lower than that for a dry one.

The experimental study of ultrasonic

wave under biaxial pressure is thefore the study of the attenuation of acoustic wave under differential pressure. It can be predicted that the attenuaiton technique will get more and more application in the measurement of mining rock stress state with the development of the measuring technique used in-situ and perfection of the condition of the laboratory.

Acknowledgments

The authors thank Professor Zhang Jia Lian for his advice. It was impossible to complete the experiments without the help of Jin Yin Dong and Lu Zhong Min.Especially we wish to thank Dr. Hutchson for his valuble assistance in preparing this manuscript.

REFERENCES

(1) Attewell, P.B., Y.V. Ramana, 1966, Geophysics, 31, 1049-1056.

(2) Biot, M.A., 1956a,J. Acoust. Soc. Amer., 28, 168-178.

(3) Biot, M.A., 1956b, J. Acoust. Soc. Amer., 28, 179-191.

(4) Du Pei Ying, 1982, (In Chinese)

(5) Gardner, G.H.F., M.R.J. Wyllie and D.M. Droschak, 1964, J.Petrol. Technol., 16, 189-198.

(6) Gordon, R.B. and L.A. Davis, 1968, J. Geophy. Res., 73, 3917-3935.

(7) Jackson, D.D. and Anderson, D.L., 1970, Rev. Geophy. Space. Phy., 8, 1, 1-63.

(8) Johnston, D.H. and Toksöz, M.N., 1980, J.G.R., 85, N. B2, 925-936.

(9) Knopoff, L. 1964, Rev. Geophy., 2, 625-660.

(10)Krech, W.W., F.A. Henderson and E.H Hjelmstad, 1974, U.S.Bur. Mines Rep. Invest., 7865.

(11) Marko, G.M. and A.Nur, 1979, Geophysics, 44, 161-178.

(12) Nur, A. and G. Simmons, 1969, Earth Planet Sci. Lett. 7, 99-108.

(13) Oconnel, R.J. and Budiansky, 1977, J.G.R., 82, 5719-5736.

(14) Tittona, B.R., R.M. Housely, G.A. Alers, and E.H. Cirlin, 1974, Geochim. et Cosmochin. Acta, Suppl., 5,3, 2913-2918.

(15) Toksöz, M.N., D.H. Johnston, and A Timor, 1977, Geophysics, 44, 681-690.

(16) Walsh, J.B., 1966, J.G.R., 71, 2591-2599.

(17) Walsh, J.B., 1968, J.G.R., 73, 2209-2216.

(18) Yamakawa, N, 1962, Geophy. Mag. 32, 63-103.

(19) Zamanek, J.Jr. and Rudnick, 1961, J. Acoust. Soc. Am.,33, 1283-1288.

Numerical Methods in Geomechanics (Innsbruck 1988), Swoboda (ed.)
© 1988 Balkema, Rotterdam. ISBN 90 6191 809 X

Bearing capacity of offshore piles by dynamic relaxation procedures

N.F.F.Ebecken, A.L.G.A.Coutinho & L.Landau
COPPE/UFRJ, Federal University of Rio de Janeiro, Brazil
A.Maia da Costa
PETROBRÁS/CENPES/DIPREX/SEDEM, Brazil

ABSTRACT

The purpose of this paper is to present the bearing capacity analysis of offshore steel pipe piles employing a three-dimensional finite element model. Soil and pile are discretized by a set of axisymmetric isoparametric finite elements, with variable number of nodal points. Infinite finite elements are used to simulate the boundary conditions. The soil is considered as an elasto-plastic material, whereas the soil-structue interaction is modelled by six-nodeded isoparametric interface elements. The solution is obtained by an optimum load time history dynamic relaxation method. This scheme is implemented on an explicit central difference program exploring the vectorizing Fortran compiler of the IBM 3090 computer.

1 - INTRODUCTION

The purpose of this paper is to discuss the possibility of performing the bearing capacity analysis of offshore steel piles by an optimum load time history dynamic relaxation method. This approach was implemented on a explicit central difference computer program developed to simulate the driving analysis, employing a three-dimensional finite element model.

The intention is to unify these explicit schemes, exploring the vectorizing Fortran compiler of the IBM 3090 computer /11,12/, to produce a code that can be used to simulate any static and dynamic analysis that are necessary for the design of offshore piles.

Soil and pile are discretized by a set of axisymmetric isoparametric finite elements. Proper boundary conditions in the soil mass are accomplished by infinite axysimmetric elements. The soil is considered as an elasto-plastic material, whereas the soil-structure interaction is modelled by 6-noded isoparametric interface elements.

2 - PILE-SOIL INTERACTION MODEL

In this work it is assumed that the soil satisfies the Modified Mohr-Coulomb failure criterium |9|. This criterium is represented by a yield surface without singularities, allowing the uniqueness in the definition of the plastic strain increments in associated plasticity /1, 2, 3/. The nonlinear pile behaviour is assumed to follow the Von Mises yielding criterium. The mechanism of wave transmission between pile and soil involves a kinematic boundary discontinuity, once pile-soil are not completely coupled. This can be properly simulated by using interface elements. The mathematical model of the interface elements is based upon the relative movements of the adjacents elements and a nondilatant behaviour is assumed /3, 4, 5/. Therefore, the strain field for the interface element can be written as,

$$\varepsilon = \begin{bmatrix} \varepsilon_t \\ \varepsilon_n \\ \varepsilon_\theta \end{bmatrix} = \begin{bmatrix} \dfrac{\delta u_t}{h} \\ \dfrac{\delta u_n}{h} \\ \dfrac{u_r}{R} \end{bmatrix} \qquad (1)$$

where the subscripts t, n and refer respectively to the tangential, normal and circunferencial directions, R is the radial coordinate, and h is the element thickness. The relative displacements u_t, u_n are obtained by a quadratic interpolation of the six nodal points displacement referred to the local coordinate system. The radial displacements u_r are evaluated at the midsurface of the interface element. Hence, the incremental constitutive relation for linear interface elements can be expressed as,

$$\Delta\sigma = \mathbf{D}_i \cdot \Delta\varepsilon \quad (2)$$

where $\mathbf{D}_i$ is the constitutive matrix of the interface, and it is given by,

$$\mathbf{D}_i = \begin{bmatrix} k_t & 0 & 0 \\ 0 & k_n & 0 \\ 0 & 0 & k_\theta \end{bmatrix} \quad (3)$$

The stiffness coeficients, k_t, k_n and k may be evaluated by experimental tests. However, it is often necessary to consider a nonlinear interface behaviour. Since the constitutive relation is uncoupled, it is possible to treat independently the nonlinearities. Therefore, it is assumed for the tangential direction a perfect-plastic constitutive relation, where the interface can be loaded until a maximum allowable shear stress (positive or negative), whereupon no increase of stress is carried. The nonlinear behaviour in the normal direction is associated to a hiperbolic constitutive relation.

The constitutive relation follows the equation;

$$\sigma_n = \sigma_0 \left(1 + \frac{\varepsilon_n}{\varepsilon_n^m - \varepsilon_n}\right) \quad (4)$$

This is an hiperbolic equation and allows the opening of the interface up to a minimum contact stress and a maximum closure when the normal strain reaches the maximum value ε_n^m. The tangent in σ_o is equal to the soil modulus.

For dynamic analysis, care must be taken with the interface parameters in the normal direction, once the travelling waves may be reflected in the interface or the development of gaps may yield to instabilities in the time integration procedure.

In any instant of the dynamic history of stresses inside the soil mass the total stress in a point is given by,

$$\sigma^t = \mathbf{D}\,({}^t\varepsilon - {}^t\varepsilon_p) + \sigma_0 \quad (5)$$

where $\mathbf{D}$ is the elastic constant matrix; ε and εp are respectively the total strains and the total plastic strains, both at time t and, σ_o are the gravitational initial stresses.

The gravitational initial stresses should be considered in an elasto-plastic analysis, where a Mohr-Coulomb yielding criterium is employed. This is true not only for the solid elements as so for the interface elements. As the initial stresses does not participate in the dynamic equilibrium equations, but is included in the total stress tensor, the nodal equivalent forces due to the initial stresses should be eliminated in the variational formulation of the problem.

The application of the Finite Element Method for infinite and semi-infinite media requires a special procedure for the continuum discretization. Traditionally, it is used a large number of elements and nodal points to keep the boundary far from the region of interest and to represent properly the infinite (or semi infinite) medium. In a dynamic analysis even restricting the total time analysis to minimize the effects of the impact waves reflection, the definition of the boundary is fundamental. If the time interval is not well defined, instead of permiting the dissipation of the incident waves through the medium it will promote its return, distorting the answer. Nowadays, special finite elements called "infinite" elements with all the basic characteristics of the conventional finite elements have been under intensive research. These elements provide an excellent modeling of the infinite and semi-finite continuous media. The parametric element formulated by Maia |6| used in this work, resembles that developed by Beer |7| without the incovenience of distortions related to the location of the integration points.

Even with this procedure there exists a problem in the determination of the minimum distance between the region of application of the external solicitation and the positioning of the boundary mesh for not having a partial wave reflection. The adopted solution follows White's /8/ conclusions and is based upon the application of a viscous damping surface at the boundary of the infinite element, in such way to absorb the incident wave that reaches it and reproduce the effect of wave propagation through the medium avoiding its reflection. The idea is as simple as efficient: the stresses induced by the incident wave over the element surface are calculated considering them proporcional to its velocity of propagation; by applying these forces in the opposite direction a reciprocal cancellation is obtained and then minimum reflection occurs. The expressions for these normal and shear stresses are given respectively by;

$$\sigma = a \rho V_p \dot{w} \tag{6}$$

$$\tau = b \rho V_s \dot{u} \tag{7}$$

where ρ is the specific mass, Vp is the velocity of the normal incident waves, Vs is the shear incident waves velocity, $\dot{w}$ is the normal velocity, $\dot{u}$ is the shear velocity and a, b are parameters for the degree of waves absorption at the boundary. The coefficients a and b may be adjusted to obtain the required absorption level. In general the results have shown to be very satisfactory.

3 - THE DYNAMIC SOLUTION METHOD

The discrete nonlinear differential equations of motion, can be written as,

$$\mathbf{R}({}^t\mathbf{U}) = {}^t\mathbf{F} - \mathbf{M} \cdot {}^t\ddot{\mathbf{U}} - \mathbf{C} \cdot {}^t\dot{\mathbf{U}} \tag{8}$$

where ${}^t\mathbf{U}$, ${}^t\dot{\mathbf{U}}$, ${}^t\ddot{\mathbf{U}}$ are the nodal displacements, velocities and accelerations vectors at time t, $\mathbf{R}({}^t\mathbf{U})$ is the vector of nodal forces equivalent to the current deformation state, ${}^t\mathbf{F}$ is the vector of nodal forces equivalent to the externally applied loads at time t, **M** is the mass matrix and **C** is the damping matrix. The dot denotes time differentiation. In this system of equation, **R** is a vector of elastic or elasto-plastic forces, that depends on the nodal unknowns **U**. The response of systems excited either by prescribed initial velocities or short transients time-force functions may be achieved by using the explicit time stepping procedure (central difference method) /8/, where velocities and accelerations are defined as,

$${}^t\ddot{\mathbf{U}} = \frac{1}{\Delta t^2} \left({}^{t+\Delta t}\mathbf{U} - 2\,{}^t\mathbf{U} + {}^{t-\Delta t}\mathbf{U}\right) \tag{9}$$

$${}^t\dot{\mathbf{U}} = \frac{1}{2\,\Delta t} \left({}^{t+\Delta t}\mathbf{U} - {}^{t-\Delta t}\mathbf{U}\right) \tag{10}$$

and Δt is the time step.

Substituting these equations on the dynamic equilibrium equations, the recursive procedure of the method can be expressed as,

$${}^{t+\Delta t}U_i = (\Delta t^2.(-{}^tR_i+{}^tF_i+{}^tR_{o_i})2.M_i.{}^tU_i- -(M_i-C_i.\Delta t/2).{}^{t-\Delta t}U_i)/(M_i+C_i.\Delta t/2) \tag{11}$$

where

${}^tR_{o_i}$ is associated to the gravitational initial stress /6/.

It must be stressed that equation (11) can be easily programmed taking full advantage of a fast array processor like IBM 3090 vector machine /11, 12/.

The viscous damping matrix **C** is assumed to be mass proportional, and it can be written as,

$$\mathbf{C} = \alpha\, \mathbf{M} \tag{12}$$

the proportionality constant, α, is determined by,

$$\alpha = 2\,\xi\omega \tag{13}$$

where ξ is the critical damping ratio related to the first natural mode shape of the pile-soil assembly and ω is the first natural frequency. The diagonal mass matrix is consistently lumped

4 - THE OPTIMUM LOAD TIME HISTORY DYNAMIC RELAXATION METHOD

Dynamic relaxation has been used in dynamic structural analysis programs based on explicit time integration schemes to make static possible. In this paper the modification of the dynamic relaxation method proposed in reference /10/ was adopted. In this way inertia and damping forces arising during the loading process are kept at a minimum using optimum load time history.

Supposing that

$$^{t}\mathbf{F} = \mathbf{F} \cdot f(t) \qquad (14)$$

where $\mathbf{F}$ determines the space distribution of external nodal forces and f(t) is a time-dependent factor of external load, and that the time interval of interest is $(0, \tau)$, the algorithm can be summarized as:

a) Evaluate the parameter which determines the rate of the loading process:

$$p = \mathbf{F}^{T} \cdot {}^{\tau}\mathbf{U} \cdot \tau\alpha^{3} / (\mathbf{F}^{T} \mathbf{M}^{-1} \mathbf{F}(-1.5 + \alpha\tau + 2\exp(-\alpha\tau) - 0.5\exp(-2\alpha\tau))) \qquad (15)$$

b) Calculate the time-dependent factor of external load:

$$f(t) = ({}^{t}\mathbf{R}\ {}^{t}\mathbf{M}^{-1}\ \mathbf{F})/(\mathbf{F}^{T}\mathbf{M}^{-1}\mathbf{F}) + p(1 - \exp(\alpha(t-\tau)))/(\alpha\tau) \qquad (16)$$

c) Compute the displacement incremental:

$$\Delta^{t}\mathbf{U} = \frac{f2\Delta}{(1+\quad t)}\ {}^{t-\Delta t}\mathbf{U} + \frac{f1 \quad \Delta t^{2}}{(1+\alpha\ \Delta t)} \cdot \mathbf{M}^{-1}\ \mathbf{F}f(t) - {}^{t}\mathbf{R} \qquad (17)$$

where f1 and f2 are real constants used for accelerate couvergence.

d) Update displacements:

$$^{t}\mathbf{U} = {}^{t-\Delta t}\mathbf{U} + \Delta^{t}\mathbf{U} \qquad (18)$$

5 - SAMPLE CASE

Just to evaluate the performance of the optimum load time history relaxation method when applied in a bearing capacity analysis the sample case from reference /13/ is presented. The finite element model is typified by Figure 1.

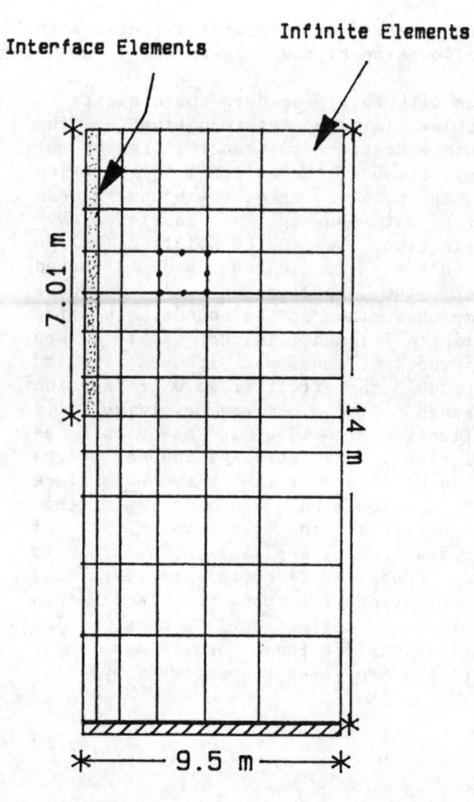

FIGURE 1

The pile data, the soil and interface parameters are the same from reference /13/. As shown in Figure 2 a continuos settlement-load curve is obtained, avoiding an incremental solution.

As indicated in this figure, the measured results are fitted quite well with the numerical solution and the reduced number of steps makes this approach very attractive.

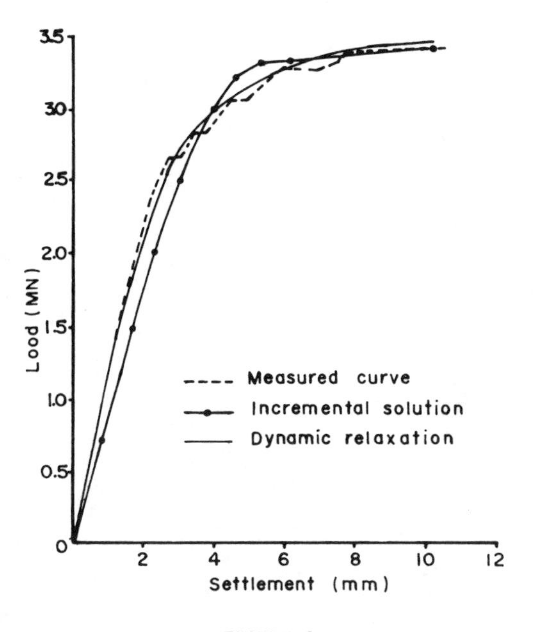

FIGURE 2

6 - CONCLUSIONS

The main objective of this paper was to show a unified numerical model for the dynamic and static analysis, of the pile-soil-interaction, that have resulted in an unique explicit computer program full oriented to computers with vector processors.

The FE solution developed herein allowed a realistic modelling of the several features involved, the geometric characteristics of the pile, the surrounding soil, including stratification, the pile-soil interface. This novel computational procedure was applied in the pile designing and the numerical results predicted fit with the results observed in a offshore test carried out by PETROBRÁS. Further, the FE analysis shows that no interference between piles during driving will be observed. Therefore, the confidence of such numerical predictions have proved that the pile with conic shape point is a reliable and economic engineering alternative for the foundation of the steel jacket platforms to be installed in the Northeast Pole of Campos Basin.

REFERENCES

01. Zienkiewicz, O.C., Pande, G.N., "Some Useful Forms of Isotropic Yield Surfaces and Rock Mechanics". Finite Elem. in Geom., Ed. by G. Gudehus, John Wiley & Sons, 1977, pp. 179-190.

02. Owen, D.R.J., Hinton, E., "Finite Elements in Plasticity: Theory and Practice" Pineridge Press Limited, Swansea, U.K., 1980.

03. Costa, A.M., "Dynamic Elasto-Plastic Analysis of Short Transients Including Soil-Fluid-Structure Interaction", M.Sc. Thesis, COPPE, Federal University of Rio de Janeiro, 1978.

04. Goodman, R.E., Taylor, R.L., Brekke, T.L., "A Model for the Mechanics of Jointed Rock", J. Soil Mech. Found. Div., ASCE, SM3, 1968, pp. 637-659.

05. Wilson, E.L., "Finite Elements for Foundations, Joints and Fluids", Finite Elem. in Geom., Ed. by G. Gudehus, John Wiley & Sons, 1977, pp. 319-350.

06. Costa, A.M., "An Application of Computational Methods and Rock Mechanics Principles to the Design and Analysis of Underground Mine Excavations", D.Sc. Thesis, COPPE/UFRJ, Federal University of Rio de Janeiro, 1984.

07. Beer, G. & Meek, J.L., "Infinite Domain Elements", Int. Journal for Num. Meth in Engineering, vol. 17, pp. 43-52, 1981.

08. White, W., Valliappan, S. & Lee, I.K. "Unified Boundary for Finite Dynamic Models", J. Eng. Mech. Div., ASCE, 103, pp. 949-964, 1977.

09. Bathe, K.J., "Finite Elements Procedures in Engineering Analysis", Prentice-Hall, NJ, USA, 1982.

10. Rericha, P., "Optimum Load Time History for Non-Linear Analysis using Dynamic Relaxation", Int. J. for Numerical Methods in Engineering, vol 23, 2313-2324, 1986.

11. Scarborough, R.G. , Kolsby, H.G.; "A Vectorizing Fortran Compiler", IBM J. Res. Develop., 3, 163-171, 1986.

12. Dubrulle, A.A., Scarborrugh, R.G., Kolsky, H.G., "How to Write Good Vectorizable Fortran", IBM Palo Alto Scientific Center Report No G320-3478; 1985.

13. Smith, I.M., "Analysis of Fixed Offshore Platforms 1972-1982", Proc. 4th Int. Conf. Num. Meth. Geomechanics, Canada, 1165-1179, 1982.

Numerical Methods in Geomechanics (Innsbruck 1988), Swoboda (ed.)
© 1988 Balkema, Rotterdam. ISBN 90 6191 809 X

Comparison of prediction methods for sand liquefaction

Y.Sunasaka, H.Yoshida, K.Suzuki & T.Matsumoto
Kajima Corporation, Tokyo, Japan

ABSTRACT :In recent years, several numerical methods for the liquefaction prediction have been developed. The purpose of this paper is the comparison and examination of these prediction methods for sand liquefaction. Seed's simplified method, Iwasaki · Tatsuoka's simplified method, two-dimensional effective stress analysis code DIANA-J and centrifuge model test are applied to liquefaction prediction of saturated ground and are compared each other.

1 INTRODUCTION

Since the Niigata earthquake (1964), the liquefaction of sand has been identified as a major cause of damage to ground and earth structures during earthquake. In recent years, the prediction technique for sand liquefaction has made a remarkable progress, and several numerical methods for the prediction of liquefaction phenomena have been developed.

Some of the prediction methods, such as Seed's simplified method (H.B. Seed and I.M. Idriss, 1971) and Iwasaki · Tatsuoka's simplified method (T. Iwasaki et al, 1978), have been introduced into several aseismic design methods. They are based on the total stress theory, and are used easily to predict the liquefaction of sand, although their applicability is not so wide. It is appropriate to use these methods, if the ground motion caused by seismic excitation at the base can be considered as the result only by shear deformation, and the theory of one-dimentional wave propagation through layered media can be used to compute the response of ground.

When these conditions are not satisfied, the effective stress analysis is needed. Recently, 2-dimensional effective stress analysis programs have been developed and applied to the liquefaction of sand ground whose profile is not composed of a series of horizontal layers or which is influenced by adjacent structures.

However, these methods have not been compared each other and it has not been clarified what kind of technique should be used in each situation. The purpose of this paper is to compare these prediction methods for the sand liquefaction and to make the feature of each technique distinct.

2 ANALYTICAL PROCEDURES

In order to prepare the prediction methods for sand liquefaction, Seed's simplified method, Iwasaki·Tatsuoka's simplified method and the effective stress analysis DIANA-J are chosen.

2.1 Seed's , and Iwasaki·Tatsuoka's simplified method

These methods are based on the comparison between the stress ratio in ground and the liquefaction resistance of sand during earthquake. The cyclic shear stress can be calculated by various methods, such as the multi-layer reflection theory and the finite element method. On the other hand, the liquefaction resistance is obtained in undrained cyclic shear tests or the empirical equations. In Seed's simplified method, empirical equations are functions of N-value, effective overburden pressure, and the magnitude of earthquake. In Iwasaki · Tatsuoka's simplified method, the empirical equations are functions of N-value, effective overburden pressure and mean grain size.

2.2 Effective stress analysis code DIANA-J

We developed the computer code DIANA-J (

Shiomi and Matsumoto et al, 1985) under the guidance of Drs. O.C. Zienkiewicz and T. Kawai, which is 2-Dimensional, static and dynamic FEM program for non-linear effective stress analysis.

In order to consider the interaction between soil skeleton and porewater, a generalized formulation of Biot's governing equation is used as the basic equation of motion for porous media.

For a constitutive relation of soil skeleton, Multi-Mechanism model was proposed by K. Kabilamany (1986) based on the model developed by C. Aubry et al (1982). This model can evaluate the dynamic liquefaction behaviour of the saturated sand, including the cyclic mobility.

There are some basic conditions in the Multi-Mechanism model as follows;

(a) Three difference shear mechanisms

Fig.1 shows the three-dimensional Mohr's circles of stress when three different principal stresses (σ_1, σ_2 and σ_3) act on soil element. The mobilized planes, on which shear-normal stress ratios (τ / σ_n) have the maximum values under the respective principal stresses, are located in three-dimensional space as shown in Fig.2. Based on the postulate proposed by H. Matsuoka and T. Nakai (1977), it is assumed that the linear summation of the plastic strain increments produced in every mechanism together with the elastic strain increments make up the total strain increments.

(b) Yield criteria

The yield surface for monotonic loading in the mobilized plane k is defined by the following equation ;

$$f_k^m = a_k - P_k \sin\phi \, (1 - \frac{\mu}{M} \ln P/Pc) \, h_k^m \qquad (1)$$

where a_k = radius of Mohr's circle in the mechanism k
P_k = mean principal stress in mechanism k
P = mean principal stress
P_c = the current critical state pressure
ϕ = angle of shear resistance
M = critical stress ratio
h_k^m = monotonic hardening parameter in mechanism k
μ = slope of the curve of shear work normalized by P versus shear strain

P_c and h_k^m are defined by the following equations;

$$P_c = exp\{(\Gamma - (1+e) - \kappa \ln P / (\lambda - \kappa)\} \qquad (2)$$

$$h_k^m = b\,\Omega_k / (a^m + \Omega_k) \qquad (3)$$

where Γ = specific volume of soil at critical state with $P=1$
κ = slope of over consolidation line
λ = slope of normal consolidation line
Ω_k = normalized shear work in mechanism k
b, a^m = experimental parameters

On the other hand, the yield surface for cyclic loading in the mobilized plane k is defined by the following equation ;

$$f_k^c = | a_k - C_k | - P_k \sin\phi \, (1 - \frac{\mu}{M} \ln P / P_c) \, h_k^c \qquad (4)$$

C_k = kinematic hardening parameter
h_k^c = hardening parameter similar to h_k^m

(c) Hardening rules

Fig.3 shows the stress space normalized with $F_k = \sin\phi \, (1 - \mu/M \ln P / P_c)$. In Fig. 3, the outer circle indicates monotonic yield surface. In case of reversal loading after point A in the mobilized plane k , it is postulated that the centers of the circles (the cyclic yield surfaces) during reversal loading are found along the inner normal at the reversed point until another reversal occurs or the current yield surface touches the monotonic surface.

3 1-DIMENSIONAL PROBLEM

The dynamic centrifuge tests on the model consisted of a horizontal bed of loose saturated sand were conducted by the Cambridge University Geotechnical Centrifuge (K. Venter, 1986). The liquefaction analysis technique and the dynamic centrifuge tests will be compared here.

3.1 Dynamic centrifuge tests

The plane strain model of a horizontal bed of loose saturated sand is shown in Fig.4. The model is 150mm high and 530mm wide. The centrifuge acceleration used in these tests series was 78g. The model, therefore, corresponds to a prototype 11.7m high. The model was instrumented by piezoelectric accelerometers and porewater pressure transducers.

The sand used here was the mixture of Leighton Buzzard sand. The relative density of sand was $D_r = 53\%$ and mean grain size was $D_{50} = 0.12$mm. A series of simple load-unload triaxial tests on this sand was also conducted (K. Venter, 1986).

The model was subjected to two earthquakes (Fig.5); EQ1 is small (5.75% of the vertical gravitation acceleration) and EQ2 large (24.4%). Liquefaction phenomena was observed by the second earthquake EQ2, but not observed by the first one EQ1.

3.2 Simplified methods

The stress ratio in the ground during excitation is computed by SHAKE (P.B. Schnabel et al ,1972). On the other hand, the liquefaction resistance is calculated by the empirical equations.

The parameters are obtained using the following equations ;

$$G = 330 \times \frac{(2.97-e)^2}{1+e} (\sigma_o')^{\frac{1}{2}} \quad (kgf/cm^2) \tag{5}$$

$$N = \left(\frac{D_r}{20}\right)^2 (\sigma_o' + 0.7) \tag{6}$$

where e = void ratio

σ_o' = effective overburden pressure

The magnitude of EQ1 and EQ2 in Seed's simplified method is assumed to be M=7.

3.3 Effective stress analysis DIANA-J

The parameters in this method are determined by the simple load-unload triaxial tests, and listed in Table 1.

The horizontal displacements of the side boundaries of FEM model (Fig.4) are fixed.

3.4 Comparisons

Fig.6 shows the safety factors of the liquefaction in the ground subjected to the excitation EQ1 and EQ2. The safety factor F_L by the dynamic centrifuge tests and DIANA-J, is obtained by the following equation.

$$F_L = \frac{P_o}{u} \tag{7}$$

where P_o = the initial mean effective stress

u = maximum excess pore pressure

According to the results from both the centrifuge test and DIANA-J, the whole bed of sand liquefied after about 60 ms. subjected to the excitation EQ2. The liquefaction resistance estimated by Iwasaki·Tatsuoka's method is greater than that by Seed's method, in particlar, in the shallow part of ground, because this sand is very loose, and N-value is determined from equation (6). Then, as shown in Fig.6, F_L from Iwasaki·Tatsuoka's method is greater than F_L from Seed's method. However, the simplified methods can well predict the liquefaction of sand.

Fig.7 shows the comparisons between the recorded porewater pressures from the dynamic centrifuge test (EQ2) and the computed results by DIANA-J at points A, B and C in Fig.4. It is seen that they agree very well. Then we can point out that DIANA-J can precisely simulate the dynamic behavior of this centrifuge test model effected by side boundaries.

4 2-DIMENSIONAL PROBLEM

Here we conduct the predictions of the sand liquefaction of 2-dimensional ground subjected to the El Centro earthquake (1940, M=7.1) with maximum acceleration 100 gal as shown in Fig.8. In these analyses we use the same sand as in the dynamic centrifuge tests. We assume that some parts of slope as shown in Fig.8 are not saturated and they can be treated as elastic material listed in Table 2.

4.1 Simplified methods

The stress ratio in the ground is computed by 2-dimensional elastic analysis model of DIANA-J. The liquefaction resistance is calculated by the empirical equations.

The parameters are obtained by from equations (5)and(6).

4.2 Effective stress analysis DIANA-J

The vertical displacements of the side boundaries of FEM model are fixed. The stiffness of the liquefied element is fixed at very small. We carried out the dynamic response analyses of this ground model subjected to the El Centro earthquake for 4.0 seconds.

4.3 Comparisons

Fig.9 shows the liquefied area by the simplified methods and DIANA-J subjected to the El Centro earthquake with maximum acceleration 100 gal.

The liquefaction area of Iwasaki·Tatsuoka's simplified method is smaller than that of Seed's simplified method, because the liquefaction resistance of this sand estimated by Iwasaki·Tatsuoka's method is greater than that by Seed's method, in particlar, in the shallow part of ground. The liquefaction occurred at the inner parts of the slope by both simplified methods.

On the other hand, the liquefaction didn't occur in the inner parts of the slope by DIANA-J , but the liquefied area especially extended over the toe of the slope.

The reason why the liquefaction occurs at the inner part of the slope in the results of the simplified methods may be that the maximum shear stress of the inner part of slope is relatively large because of the large initial shear stress. On the other hand, in the effective stress analysis DIANA-J, the accumulation of the porewater pressure is affected by the volume change of void, namely stress change subjected to earthquake.

5 CONCLUSIONS

Some liquefaction phenomena are analyzed by the dynamic effective stress analysis code DIANA-J and are compared with Seed's and Iwasaki·Tatsuoka's simplified methods.

(1) In the 1-dimensional problem, the simplified methods can well predict the liquefaction of sand as well as DIANA-J.

(2) In the 2-dimensional problem, the liquefied area computed by the effective stress analysis DIANA-J is different from that computed by the simplified methods.

REFERENCES

Seed H.B. and Idriss I.M. 1971.Simplified procedure for evaluating soil liquefaction potential. ASCE, Vol.97, N0.SM9:1249-1273.

Iwasaki T., Tatsuoka F., Tokita K., and Yamada S. 1978. A practical method for assessing soil liquefaction potential based on case studies at various sites in Japan.Proc. of The Fifth Japan Earthquake Engineering Symposium:641-648

Shiomi T., Yamamoto S.,and Matsumoto T. 1985. Application of numerical method on liquefaction problems. 5th ICONMIG: 1393-1400.

Kabilamany, K. 1986. Modeling of cyclic behaviour of sands and its application to seismic response analysis. dissertation, The University of Tokyo.

Matsuoka, H.and Nakai T. 1977. Stress strain relationship of soil based on the SMP. 9th ICSFME :153-162.

Aubry, C. Hujeux J.C.,Lassoudiere F and Meimon Y. 1982. A double memory model with multiple mechanisms for cyclic soil behaviour. Proc. Int. Symp. on Num. models in Geomech.: 3-13.

Schnabel P.B., Lysmer J and Seed H.B. 1972. SHAKE-a computer program for earthquake response analysis of horizontal layered sites. Report No.EERC 72-12, University of California,Berkeley.

Venter K. 1986. Report on two centrifuge tests KVV04 and KVV05.

Venter K. 1986. Report on seven triaxial tests.

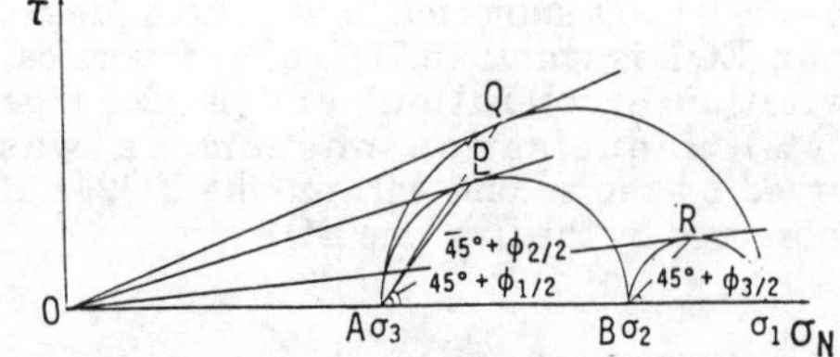

Fig.1 Stress state on three mobilized planes

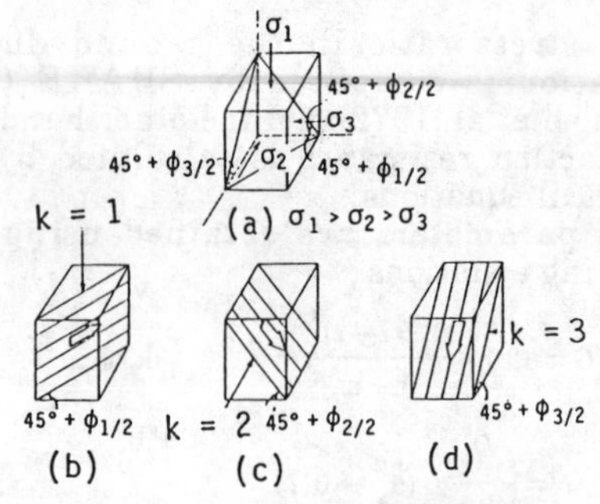

Fig.2 Three two-dimensional mobilized planes

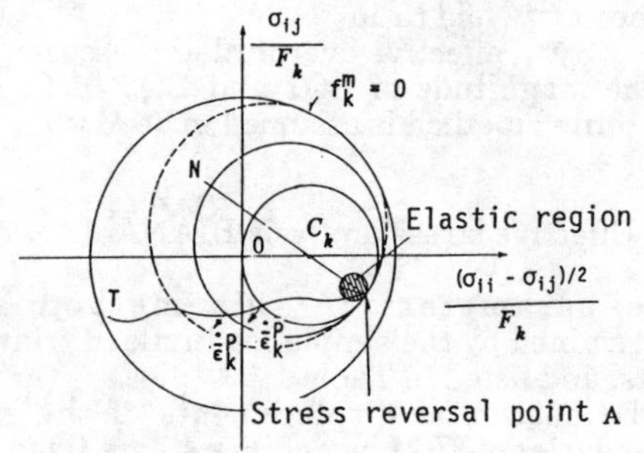

Fig.3 Yield surfaces in the stress space

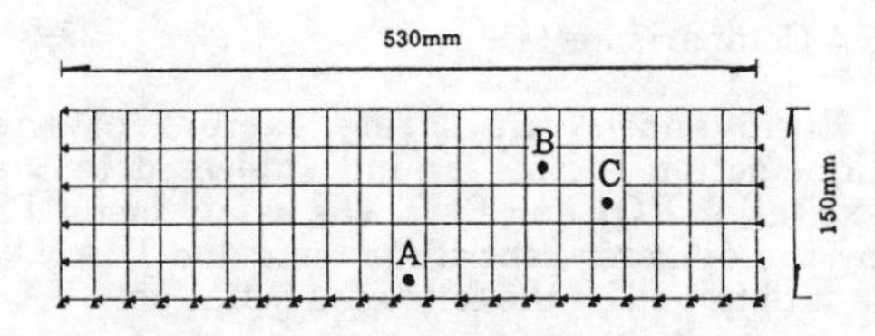

Fig.4 model of centrifuge test

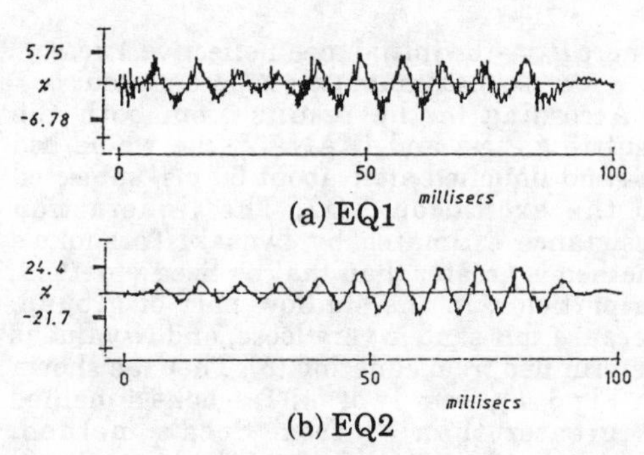

Fig.5 Input acceleration of centrifuge test

F_L
0 1 2 3 4 5
F_L
0 1 2 3
Depth(m)
5
10
(a) EQ1
(b) EQ2
Seed's simplified method
Iwasaki·Tatsuoka's simplified method
DIANA-J
centrifuge test

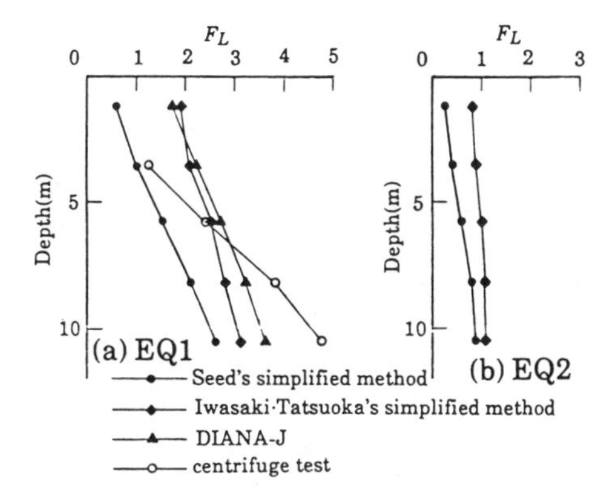

Fig.6 Comparisons of the safety factors F_L

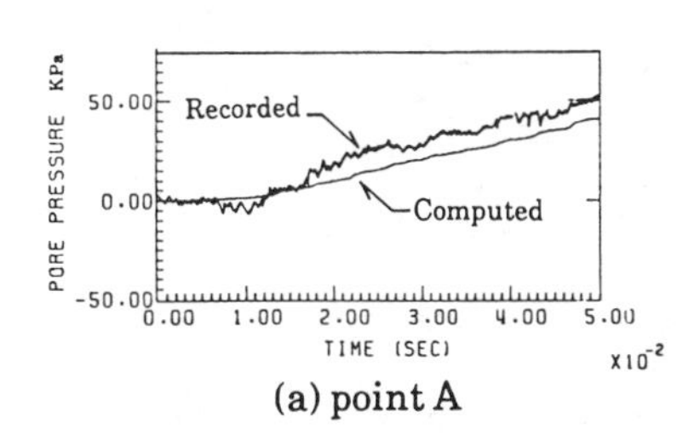

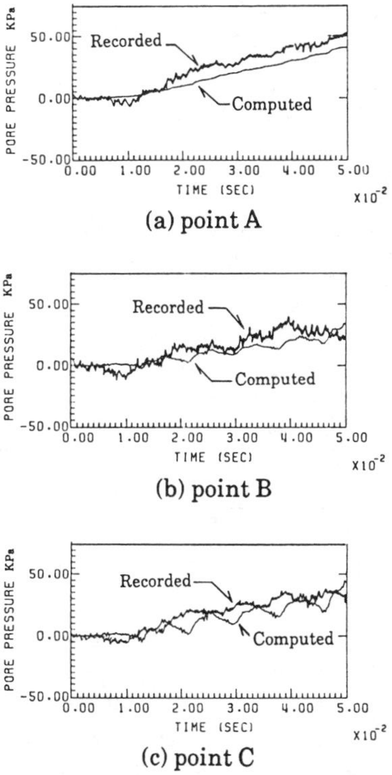

PORE PRESSURE KPa
50.00
0.00
-50.00
Recorded
Computed
0.00 1.00 2.00 3.00 4.00 5.00
TIME (SEC)
X10⁻²
(a) point A
(b) point B
(c) point C

Fig.7 Recorded and computed porewater pressure

60m
22.5m
60m
15m
25m
Elastic
Multi-Mechanism model

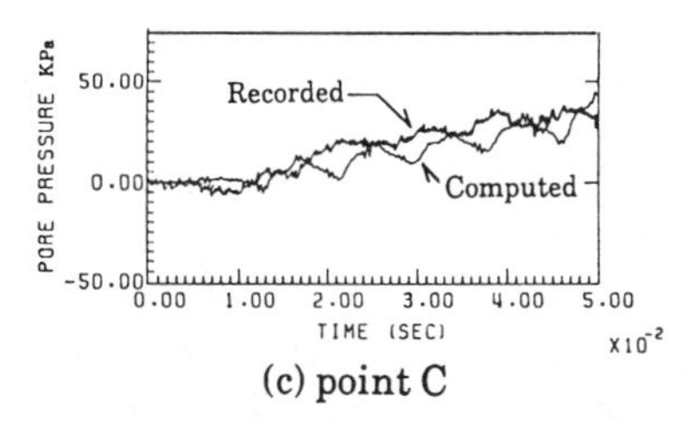

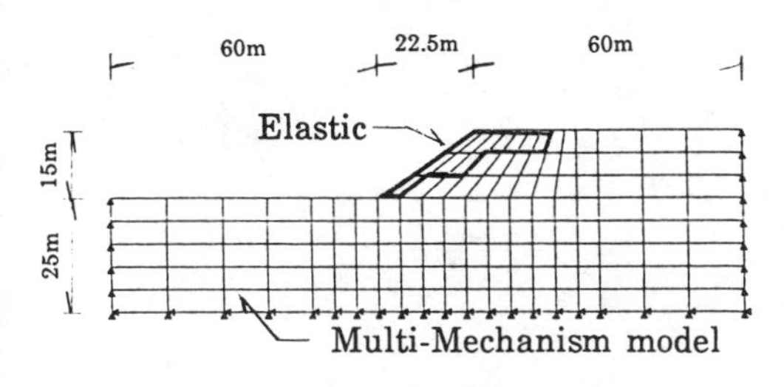

Fig.8 Finite element model for the slope

Table 1 Material properties

Young modulus		$2.94\times10^5 KP_a$
Poisson's ratio		0.234
Bulk moduli	soil	$1.0\times10^{40} KP_a$
	fluid	$1.092\times10^6 KP_a$
Densities	soil	$2.65 Mg/m^3$
	fluid	$0.98 Mg/m^3$
Permeability	k	2.1×10^{-6} m/sec
Slope of nomal consolidation	λ	0.006
Slope of over consolidation	κ	0.002
Critical stress ratio	M	1.35
Specific volume of soil at critical state with P=1	Γ	1.889
Experimental parameter	b	1.0
	a^m	0.00025
	a^c	0.00025
Mean stress at experiment	p_{ref}	$-191 KP_a$
Void ratio at experiment	e_{ref}	0.74

Table 2 Material properties

Young modulus	$1.715\times10^5 KP_a$
Poisson's ratio	0.243
Density	$1.96 Mg/m^3$

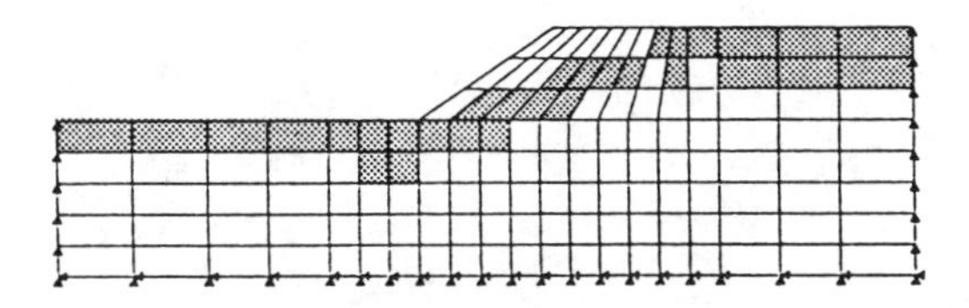

(a) Seed's simplified method
(b) Iwasaki·Tatsuoka's simplified method
(c) Effective stress analysis

Fig.9 Liquefied area

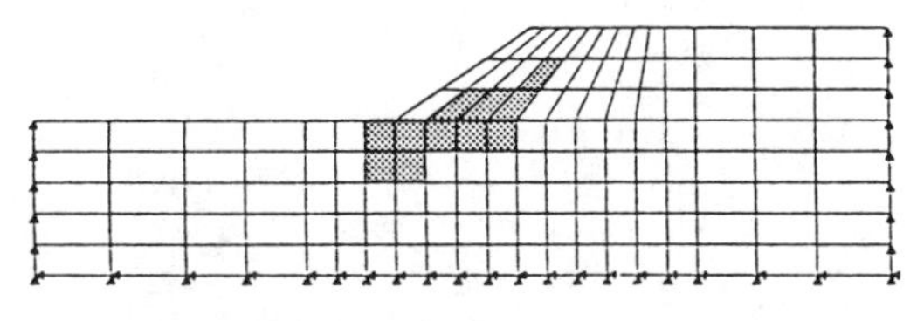

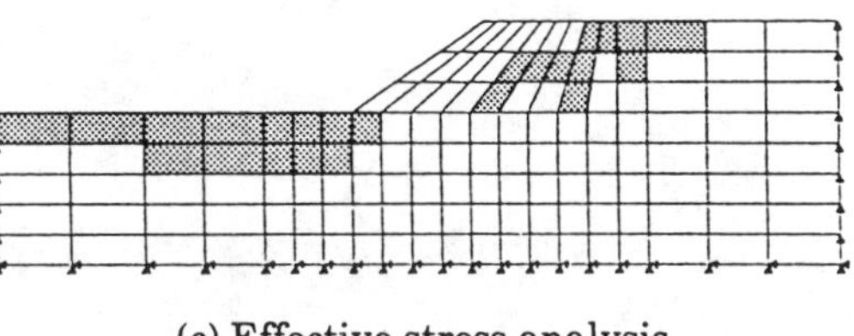

Numerical Methods in Geomechanics (Innsbruck 1988), Swoboda (ed.)
© 1988 Balkema, Rotterdam. ISBN 90 6191 809 X

Earthquake-induced translational and rotational motions of liquid-filled tanks based on a compliant soil

R.Seeber & F.D.Fischer
Institute of Mechanics, University of Mining and Metallurgy, Leoben, Austria

ABSTRACT: A three dimensional dynamic analysis of liquid-filled cylindrical tanks, excited by combined vertical and horizontal ground motions is presented. The interaction of the ideal fluid with the flexible tank and the compliant soil leads to a Neumann problem of potential theory which must be solved in conjunction with three equations of motion of shell and two equations of motion of ground. It is transformed analytically into a system of time dependent ordinary differential equations by a weighted residual approach and solved by the complex frequency reponse method. Natural frequencies and damping ratios of axisymmetric, antimetric and rocking vibrations are evaluated for different tanks and for different shear wave velocities of soil.

1. INTRODUCTION

Tanks as storage facilities for non-toxic, non-volatile, inflammable chemicals up to highly toxic, volatile and highly flammable chemicals, are matter of special importance. No assured water supply (San Francisco 1906), uncontrolled fires (Nigata and Alaska 1964) and spillage or clouds of toxic chemicals subsequent to earthqakes may cause much more damage than the earthquake itself.

Although the tanks are of modern design several have been damaged severely by recent earthquakes, both steel and concrete tanks. The more common failures are structural failures like bulging ("broad tanks") or buckling ("tall tanks") near the tank base ("elephant footing"), roof damage caused by sloshing, failure of the tank support system, failure of foundation and damage to connecting piping. They all are caused by dynamic loadings which mainly depend on the dynamic properties of the complete system tank-liquid-soil.

Dynamic investigations and experimental observations demonstrated the great influence of shell flexibility on natural frequencies for axisymmetric vibrations (Haroun 1983, Haroun 1984, Luft 1984, Veletsos 1984 and Hori 1987) and for antimetric vibrations (Veletsos 1974, Fischer 1979, Haroun 1980 and Sakai et al. 1984). A less frequently studied phenomenon is the influence of soil compliance on natural frequencies and damping ratios, caused by energy loss via radiation damping of soil. Moreover, rocking motions are activated in this case with natural frequencies mainly dependent of soil stiffness properties.

This paper presents a general formulation of the three dimensional tank-liquid-soil interaction problem including axisymmetric, antimetric, sloshing and rocking vibrations, gives an outline of the solution by a weighted residual approach in conjunction with a complex frequency response analysis and shows some interesting results with special regard to the influence of an elstic halfspace and an elastic layer resting on an elastic halfspace respectively.

2. TANK - LIQUID - SOIL SYSTEM

The system consists of three dynamically interacting domains: the tank the fluid and the soil. The tank and the fluid are bounded domains, the soil modeled as an elastic halfspace is unbounded. However, only the motion of the ground surface is of interest, so that problem can be transformed mathematically into a bounded domain problem using the corresponding complex-valued impedance functions (Tajimi 1984, J.E. Luco and A. Mita 1987, Th. Triantafyllidis et al. 1987) which represent the solutions of the halfspace.

The tank investigated is shown in fig 1. It is a thin walled, circular cylindrical, vertical shell of radius R and height H with a variable, axisymmetrically distributed wall-thickness h, a free edge at the top and a built-in or hinged edge at the bottom. Its material properties are assumed to be isotropic and ideal elastic. It is entirely filled with an incompressible ideal fluid and rests on a ground based bottom plate of neglegible mass and of zero or infinite stiffness.

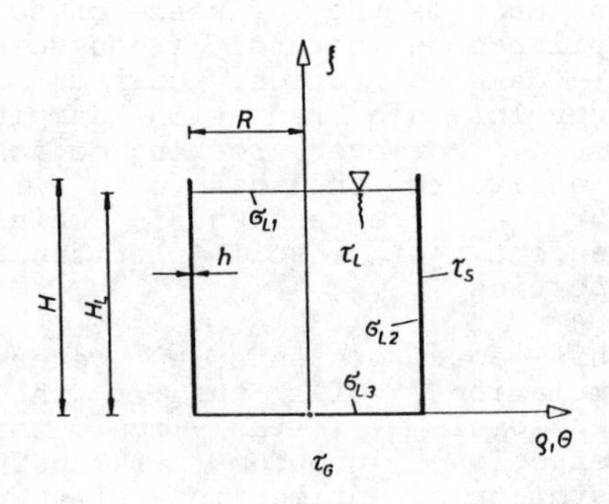

Fig. 1 Model of the tank

The soil is treated

1) as a uniform elastic halfspace and

2) as an elastic layer resting on an elastic halfspace.

The impedance functions of the uniform halfspace are fitted to the layered halfspace by modification of the static stiffness parameters as proposed by G. Gazetas (1983). Material damping is considered by Kausel's correspondence principle which should be reasonably accurate in case of a uniform halfspace. However, the sensitivity of shallow strata on rigid rock to material damping discredits this priciple to a large extent in the case of a layered soil (G. Gazetas 1983).

Two systems of coordinates are used (fig 2). Cylindrical coordinates with the unit vectors e_ρ, e_θ, e_ξ fixed with the tank, the center of the bottom being the origin describe quantities in the local system. It is moved translationally and rotationally with reference to a global cartesian system with the unit vectors e_i, $i=1,2,3$. Transformation of coordinates is done by:

$$\underline{x} = \underline{x}_A + \underline{Q}^{-1} \underline{x}_P \tag{01}$$

$\underline{x}_P$ being a vector measured in local cylindrical coordinates, $\underline{x}$ being a vector measured in global cartesian coordinates and $\underline{Q}$ being the time dependent matrix of rotation. Since only small rotations are of interest, $\underline{Q}$ may be linearized with respect to the angle and may be assumed to be independent of time.

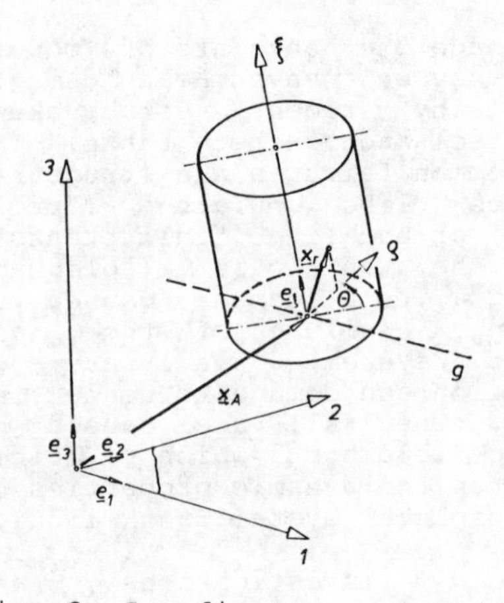

Fig. 2 Coordinate systems

3. FORMULATION OF THE PROBLEM

The problem can be specified as a boundary value problem of potential theory which must be solved in conjunction with three equations of motion of the shell and two equations of motion of the ground surface. The system is devided into three domains of different physical behaviour, but coupled by dynamic boundary conditions (fig 1):

1) the fluid τ_L with the boundary σ_L,
2) the shell τ_S with its boundary σ_S,
3) and the ground τ_G with its boundary σ_G.

3.1 The liquid

Due to the ideal properties of the liquid, a displacement potential Φ (ρ,Θ,ξ,t) is defind which must satisfy the "incompressible" potential equation,

$$\frac{\partial^2\Phi}{\partial\rho^2} + \frac{1}{\rho}\frac{\partial\Phi}{\partial\rho} + \frac{1}{\rho^2}\frac{\partial^2\Phi}{\partial\Theta^2} + \frac{\partial^2\Phi}{\alpha^2\partial\xi^2} = 0 \quad (02)$$

α being the ratio height/radius of the tank. The liquid displacement vector $\underline{u}_L$ and the liquid velocity vector $\underline{v}_L$ are defined by the gradient of the potential Φ,

$$\underline{u}_L = \underline{\underline{Q}}^{-1} \text{ grad } \Phi \quad (03)$$

$$\underline{v}_L = \underline{\underline{Q}}^{-1} \text{ grad } \frac{\partial\Phi}{\partial t} \quad (04)$$

Due to the second order of equation (02) one boundary condition must be satisfied. Since the liquid is assumed to be frictionless, this condition is of Neumann type, equalizing the normal components of the liquid and of the shell respectively the ground on their common interface. the movements of the ajacent domain at the boundary normally to their common spatial shape. Corresponding to the three different parts of the liquid boundary σ_L there exist three different types of conditions (fig 3):

$$1.\quad \underline{u}_L\, \underline{n}_{L1} = \underline{u}_F\, \underline{n}_{L1} \quad \text{at } \sigma_{L1}, \quad (05)$$

$$2.\quad \underline{u}_L\, \underline{n}_{L2} = \underline{u}_S\, \underline{n}_{L2} \quad \text{at } \sigma_{L2}, \quad (06)$$

$$3.\quad \underline{u}_L\, \underline{n}_{L3} = \underline{u}_G\, \underline{n}_{L3} \quad \text{at } \sigma_{L3}, \quad (07)$$

with u_F, u_S and u_G being the displacements of the free liquid surface, of the shell and of the ground surface and n_{Li} being the normal vectors of the deformed boundary shapes. The balances of momentum and of moment of momentum are satisfied automaticly in case of an existing velocity potential. The dynamic liquid pressure can be expressed by the equation of Euler in its linearized form without static terms,

$$p = -\rho_L \frac{\partial^2\Phi}{\partial t^2} \quad (08)$$

ρ_L being the density of liquid. With the pressure at the liquid surface equal to zero, the linearized sloshing condition reads

$$g u_{FA} - p_{\xi=1} = 0 \quad (09)$$

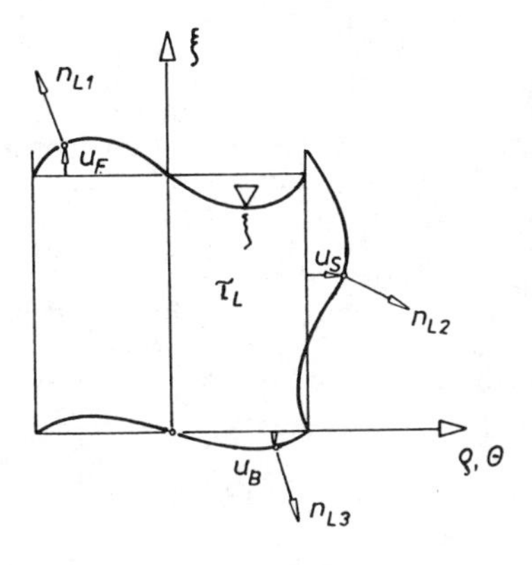

Fig. 3 Spatial shapes of the liquid boundary

3.2 The shell

The shell is governed by three equations of motion, two of second order governing the equilibrium for the axial and the circumferential direction and one of fourth order governing the equilibrium for the radial direction. Adding inertia and damping terms to Fluegge's equations (Fluegge 1973, eqs. 5.2a-d) and inserting there his approximate relations of elastic law (Fluegge 1973, eqs. 5.12a-f), these equations take the form

$$\underline{\underline{A}}\,\underline{u}_S = \underline{F}_I + \underline{F}_D + \underline{F}_P \qquad (10)$$

with $\underline{\underline{A}}$ being the matrix differential operator, $\underline{u}_S$ being the local displacements of the middle surface and $\underline{F}_I$, $\underline{F}_D$ and $\underline{F}_P$ being the inertia, damping and external forces of the shell. The elements of the differential operator A after elimination of the circumferential derivatives assuming a cos mθ distribution are shown in appendix A1.

3.3 The ground

The ground is modeled as an equivalent spring-dashpot system with frequency dependent, complex valued impedance functions which represent the solution of a massless circular foundation of infinite stiffness resting on a uniform halfspace when subjected to harmonic plane waves (J.E.Luco and A. Mita 1987). The balance of forces and of moments on the tank bottom leads to two equations of motion of ground surface. Left hand sides of the following equations (11) to (13) represent inertia terms determined by the liquid behaviour, right hand sides are complex stiffness terms determined by the soil behaviour.

$$\int_{\sigma_{L2}} p_S \, d\sigma_{L2} - (m_L + m_S)\frac{\partial^2 u_{GH}}{\partial t^2} = k_H u_{GH} \qquad (11)$$

$$\int_{\sigma_{L33}} p_G \, d\sigma_{L3} - m_L \frac{\partial^2 u_{GV}}{\partial t^2} = k_V u_{GV} \qquad (12)$$

$$\int_{\sigma_{L3}} R p_G \, \rho \cos\theta \, d\sigma_{L3} - J_L \frac{\partial^2 \delta_G}{\partial t^2} = k_{Rl} \delta_G \qquad (13)$$

k_H, k_V and k_{Rl} are the complex horizontal and vertical translational and the rotational stiffness, u_{GH}, u_{GV} and δ_G are the horizontal and vertical displacements and the rotational angle of the ground surface, m_L and m_S are the mass of liquid and shell, J_L is the inertia moment of liquid and p_S and p_G are the dynamic liquid pressure at the shell and at the bottom respectively.

4. METHOD OF SOLUTION

The time and space dependent equations of chapter 3 are transformed by a weighted residual approach in order to get a system independent of spatial functions which can be solved by a complex frequency response analysis. However, it is necessary that the transformed system is linear in time, i.e. that the time function appears only with power one. This requires some linearizations:

1. The equation of Euler is linearized neglecting velocity squared terms. This should cause a minor failure, since velocities are of small order of magnitude, see equation (08)
2. The liquid boundary condition, equations (05) up to (07), is linearized inserting the normal vectors of the undeformed boundary shapes instead of the deformed ones. This is no strong restriction to the solution, because both liqid sloshing and shell and ground displacements are small compared with the radius of the tank.

4.1 Transformation by a weighted residual approach

The shape functions are developed in a series of suitable trial functions. Separating time and space dependent quantities, these trial functions are:

Liquid displacement potential:

$$\Phi_S = \sum_m \sum_j \Phi_{Sjm}(t) \cos\lambda_j \xi \frac{I_m(\frac{\lambda_j \rho}{\alpha})}{I_m(\frac{\lambda_j}{\alpha})} \cos m\Theta \quad (14)$$

$$\Phi_G = \sum_m \sum_j \Phi_{Gjm}(t) \frac{\sinh(\varepsilon_j \alpha \xi)}{\sinh(\varepsilon_j \alpha)} J_m(\varepsilon_j \rho) \cos m\Theta \quad (15)$$

$$\Phi_F = \sum_m \sum_j \Phi_{Fjm}(t) \frac{\cosh(\varepsilon_j \alpha \xi)}{\cosh(\varepsilon_j \alpha)} J_m(\varepsilon_j \rho) \cos m\Theta \quad (16)$$

$$\Phi_E = (x_{EH}(t) R\rho\cos\Theta) + (x_{EV}(t) H(1-\xi) \quad (17)$$

$$\Phi(\rho,\Theta,\xi,t) = \Phi_s + \Phi_G + \Phi_F + \Phi_E \quad (18)$$

Liquid surface displacements:

$$u_{FA} = \sum_m \sum_j U_{FAjm}(t) \Pi_{FAjm}(\rho) \cos m\Theta \quad (19)$$

Axial, circumferential and radial shell displacements:

$$u_{SA} = \sum_m \sum_j U_{SAjm}(t) \Psi_{SAjm}(\xi) \cos m\Theta \quad (20)$$

$$u_{SC} = \sum_m \sum_j U_{SCjm}(t) \Psi_{SCjm}(\xi) \cos m\Theta \quad (21)$$

$$u_{SR} = \sum_m \sum_j U_{SRjm}(t) \Psi_{SRjm}(\xi) \cos m\Theta \quad (22)$$

Vertical ground displacements:

$$u_{GV} = \sum_m \sum_j U_{GVjm}(t) \Psi_{GVjm}(\xi) \cos m\Theta \quad (23)$$

Φ_{Sjm}, Φ_{Gjm}, Φ_{Fjm}, U_{FAjm}, U_{SAjm}, U_{SCjm}, U_{SRjm} and U_{GVjm} are complex-valued time dependent coefficients, I_m is the modified Bessel function of first kind of order m, J_m is the Bessel function of first kind of order m, λ_j and ε_j are the zeros of $\cos\lambda_j = 0$ and $J_m(\varepsilon_j) = 0$, $\Pi(\rho)$ and $\Psi(\xi)$ are radial and axial shape functions, m is the circumferential wave number and Φ_E is the potential due to ground excitation. It should be noticed that the trial functions are general solutions of the potential equation (02). All other shape functions only must satisfy the corresponding boundary conditions.

Inserting these trial functions into the equations (03) to (10) and multiplying them with suitable axial and radial test functions, $\Pi_W(\rho)$ and $\Psi_W(\xi)$, they can be transformed into a system of time dependent differential equations by a simple integration:

$$\int_{\sigma_{L1}} (\underline{u}_L \underline{n}_{L1} - \underline{u}_F \underline{n}_{L1}) \Pi_W(\rho)\, d\sigma_{L1} = 0 \quad (24)$$

$$\int_{\sigma_{L2}} (\underline{u}_L \underline{n}_{L2} - \underline{u}_S \underline{n}_{L2}) \Psi_W(\xi)\, d\sigma_{L2} = 0 \quad (25)$$

$$\int_{\sigma_{L3}} (\underline{u}_L \underline{n}_{L3} - \underline{u}_G \underline{n}_{L3}) \Pi_W(\rho)\, d\sigma_{L3} = 0 \quad (26)$$

$$\int_{\sigma_{L1}} (g u_{FA} - p_{\xi=1}) \Pi_W(\rho)\, d\sigma_{L1} = 0 \quad (27)$$

$$\int_{\tau_S} (\underline{\underline{A}} \underline{u}_S - \underline{F}_I - \underline{F}_D - \underline{F}_P) \Psi_W(\xi)\, d\tau_S = 0 \quad (28)$$

$$\int_{\sigma_{L2}} p_S\, d\sigma_{L2} - (m_L + m_S) \frac{\partial^2 u_{GH}}{\partial t^2} - k_H u_{GH} = 0 \quad (29)$$

$$\int_{\sigma_{L3}} p_G\, d\sigma_{L3} - m_L \frac{\partial^2 u_{GV}}{\partial t^2} - k_V u_{GV} = 0 \quad (30)$$

$$\int_{\sigma_{L3}} R p_G \rho \cos\Theta\, d\sigma_{L3} - J_L \frac{\partial^2 \delta_G}{\partial t^2} - k_{R1} \delta_G = 0 \quad (31)$$

4.2 Frequency response analysis

By a further separation of each time dependent coefficient into a time independent complex-valued amplitude and into a normalized time- function T(t), with T(t) being here the function of harmonic vibration, (Ω ... circular frequency)

$$T(t) = e^{i\Omega t} \qquad (32)$$

equations (24) to (31) can be transformed into the frequency domain. By means of a frequency sweep, the amplitude response curves are evaluated and investigated with a view to resonance phenomena. Damping ratios are found by the well-known method of the half-power band width (A.H. Hadjian et al. 1987) giving results of good accuracy in case of lightly damped systems. However, this means no great restriction, since accurate values of high damping ratios are nearly of no interest.

5. NUMERICAL EXAMPLE

The "broad" tank of Haroun (1980) has been investigated in the frequency domain between 0 and 8 Hz. It is a ground based steel tank with a constant wall thickness, a ratio height/radius of 0,667 and with a ratio radius/wall-thickness of 720. All other data of the tank and the liquid are shown in table 1. The soil is assumed:

1) as a halfspace with v_S = 150 m/s, see table 2
2) as a layer with v_S = 150 m/s on a halfspace with v_S = 450 m/s, see table 3.

Table 1. Data of tank and liquid.

Poisson's ratio	-	0.30
Young's modulus	N/m2	$2.07 \cdot 10^{11}$
Mass density	kg/m3	$7.84 \cdot 10^{3}$
Hyster. damping		2.00
Radius	m	18.30
Height	m	12.20
Wall thickness	m	$2.54 \cdot 10^{-2}$
Liquid density	kg/m3	$1.0 \cdot 10^{3}$

Table 2. Data of the halfspace (case 1)

Poisson's ratio	-	0.33
Shear wave veloc.	m/s	150.00
Mass density	kg/m3	$1.80 \cdot 10^{3}$
Hyster. damping	%	5.00

Table 3. Data of the layer resting on a halfspace (case 2)

LAYER:		
Poisson's ratio	-	0.33
Shear wave veloc.	m/s	150.00
Mass density	kg/m3	$1.80 \cdot 10^{3}$
Hyster. damping	%	5.00
HALFSPACE:		
Poisson's ratio	-	0.33
Shear wave veloc.	m/s	450.00
Mass density	kg/m3	$2.00 \cdot 10^{3}$
Hyster. damping	%	5.00

The natural frequencies and damping ratios are sumarized in table 4 (halfspace) and in table 5 (layer). In case of a rigid soil, Haroun (1980) gives a natural frequency for antimetric vibrations 6.18 Hz, Fischer et al. (1987) gives a natural frequency for axisymmetric vibrations of 6.14 Hz and the natural frequency of sloshing due to the exact formula (New Zealand Seismic Design Code 1986) is 0.132 Hz.

Table 4. Natural frequencies and damping ratios for a halfspace (case 1)

	f [Hz]	ζ [%]	f [Hz]	ζ [%]
axisymm.	0.25	-	5.80	5.40
antimetr.	0.12	66.0	5.38	2.03

Table 5. Natural frequencies and damping ratios for a layer resting on a halfspace (case 2)

	f [Hz]	ζ [%]	f [Hz]	ζ [%]
axisymm.	0.20	-	5.94	2.63
antimetr.	0.12	61.7	5.38	2.04

The frequency response curves for

an elastic halfspace (case 2) are shown in fig 5 (antimetric) and fig 4 (axisymmetric). Fig 6 shows the mode of radial shell diaplacements for the antimetric and axisymmetric vibrations.

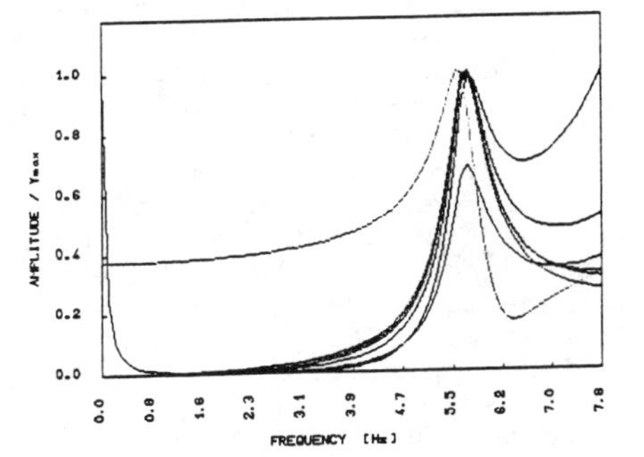

Fig. 4 Frequency response of the axisymmetric vibrations (case 2)

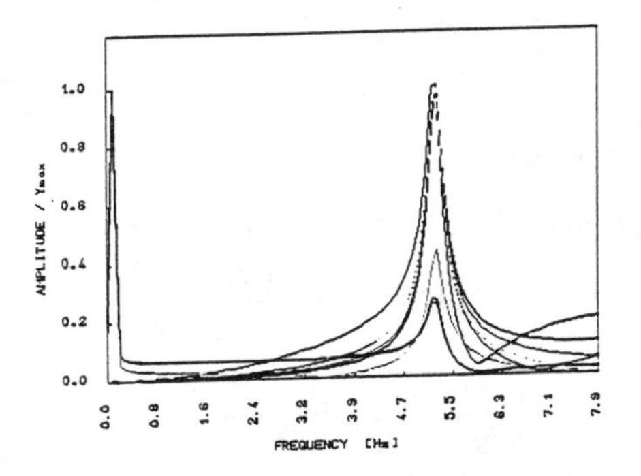

Fig. 5 Frequency response of the antimetric vibrations (case 2)

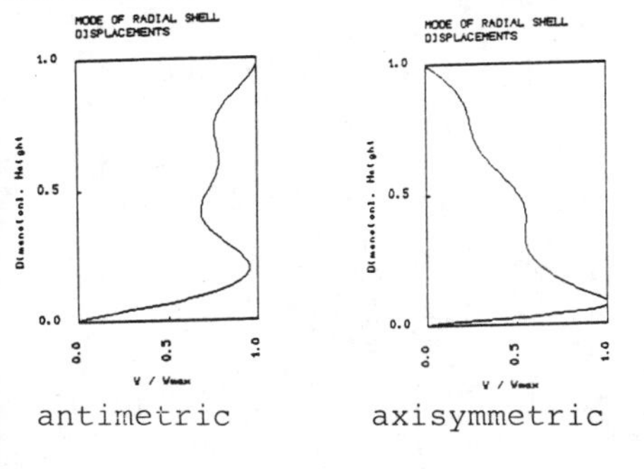

Fig. 6 Mode of radial shell displacements (case 2)

6. ACKNOWLEDGEMENTS

The financial support of the "Oesterreichischer Fonds zur Foerderung der wissenschaftlichen Forschung (FWF) under project number FWF S30-04 is gratefully acknowledged.

7. REFERENCES

1 Fischer, F.D., "Dynamic Fluid Effects in Liquid-Filled Flexible cylindrical Tanks", Journal of Earthquake Engineering and Structural Dynamics, Vol.7, 1979, pp. 587-601

2 Fischer, F.D., Rammerstorfer, F.G., "The Stability of Liquid-Filled Cylindrical Shells under Dynamic Loading", Buckling of Shells, Ramm E. (ed), Springer, Berlin-Heidelberg-New York, 1982, pp. 569-597

3 Fischer, F.D.; Seeber, R., "Dynamic Analysis of Vertically Excited Liquid Storage Tanks Considering Liquid-Soil Interaction", 1987, to appear in the Journal of Earthquake Engineering and Structural Dynamics

4 Fluegge, W., "Stresses in Shells",2nd ed., Springer, New York - Heidelberg - Berlin, 1973, pp. 206-213

5 Gazetas, G., "Analysis of Machine Foundation Vibrations: State of the art", Journal of Soil Dynamics and Earthquake Engineering, Vol. 2, No. 1, 1983, pp. 2-41

6 Hadjian, A.H., Masri, S.F., Saud, A.F., "A Review of Methods of Equivalent Damping Estimation From Experimental Data", Journal of Pressure Vessel Technology, Vol. 109, 1987, pp. 236-243

7 Haroun, M.A., "Dynamic Analysis of Liquid Storage Tanks", EERL 80-04, Feb. 1980, California Institute of Technology, Pasadena, California

8 Haroun, M.A., Tayel, M.A., "Numerical Investigations of Axisymmetrical Vibrations of Partly-Filled Cylindrical Tanks", Earthquake Behaviour and Safety of Oil and Gas Storage Facilities, Buried Pipelines and Equipment, Vol. 77, ASME, New

York, 1983, pp. 69-77
9 Haroun, M.A., Tayel, M.A., "Dynamic Behaviour of Cylindrical Liquid Storage Tanks under Vertical Earthquake Excitation", Proceedings of the 8th World Conference on Earthquake Engineering, San Francisco, Vol. 7, 1984, pp. 421-428
10 Haroun, M.A., Tayel, M.A., "Axisymmetrical Vibrations of Tanks - Analytical", Journal of Engineering Mechanics, ASCE, Vol. 111, 1985, pp. 346-358
11 Haroun, M.A., Ellaithy, A.M., "Model for Flexible Tanks Undergoing Rocking", Journal of Engineering Mechanics, ASCE, Vol. 111, 1985, pp. 143-156
12 Hori, N., " Dynamic Interaction of Liquid-Tank-Soil System under Vertical Seismic Excitation", Theoretical and Applied Mechanics, Univ. Tokyo Press, Vol. 35, Feb. 1987, pp. 75-85
13 Luco, J.E., Mita, A., "Response of a Circular Foundation on a Uniform Half-space to Elastic Waves", Journal of Earthquake Engineering and Structural Dynamics, Vol. 15, 1987, pp. 105-118
14 Luft, R.W., "Vertical Accelerations in Prestressed Concrete Tanks", Journal of Structural Engineering, ASCE, Vol. 110, 1984, pp. 706-713
15 Sakai, F., Ogawa, H., Isoe, A., "Horizontal, Vertical and Rocking Fluid-Elastic Response and Design of Cylindrical Liquid Storage Tanks", Proceedings of the 8th World Conference on Earthquake Engineering, San Francisco, Vol. 5, 1984, pp. 263-270
16 Tajimi, H., "Recent Tendency of the Practice of Soil-Structure Interaction Design Analysis in Japan and its Theoretical Background", Journal of Nuclear Engineering and Design, Vol. 80, 1984, pp. 217-231
17 Tani, S.; Hori, N., "Earthquake Response Analysis of Tanks Including Hydrodynamic and Foundation Interaction Effects", Proceedings of the 8th World Conference on Earthquake Engineering San Francisco, Vol. 4, 1984, pp. 817-824
18 Triantafyllidis, Th., Prange, B., Vrettos, Ch., "Circular and Rectangular Foundations on Halfspace: Numerical Values of Dynamic Stiffness Functions", Ground Motion and Engineering Seismology, A.S. Cakmak (ed.), Elsevier, Amsterdam-Oxford-New York-Tokyo, 1987, pp. 409-426
19 Veletsos, A.S., "Seismic Effects in Flexible Liquid Storage Tanks", Proceedings of the 5th World Conference on Earthquake Engineering, Rome, Vol. 1, 1974, pp. 630-639
20 Veletsos, A.S., Kumar, A., "Dynamic Response of Vertically Excited Liquid Storage Tanks", Proceedings of the 8th World Conference on Earthquake Engineering, San Francisco, Vol. 7, 1984, pp. 453-460
21 Veletsos, A.S., Tang, Y., "Interaction Effects in Vertically Excited Steel Tanks", Proceedings of the ASCE/IMD Speciality Conference on Dynamic Response of Structures, Los Angeles, 1986, pp. 636-643
22 Seismic Design of Storage Tanks, New Zealand National Society for Earthquake Engineering, M.J.N. Priestley (ed.), 1986

8. APPENDIX

Appendix A1 $\qquad (\ )' = \frac{\partial}{\partial \xi}$

$$A_{11} = -[\frac{D}{R}(\ )']' + \frac{D}{R}\frac{1-\nu}{2}(\ )$$

$$A_{12} = -[\nu\frac{D}{R}(\ )]' - \frac{D}{R}\frac{1-\nu}{2}(\ )'$$

$$A_{13} = -\nu[\frac{D}{R}(\ )]'$$

$$A_{21} = \nu D(\ )' + \frac{1-\nu}{2}[D(\ )]'$$

$$A_{22} = D(\ ) - \frac{1-\nu}{2}[D(\ )']'$$

$$A_{23} = D(\ ) + \frac{K}{R^2}[(\ ) - \nu(\ )'']$$

$$A_{31} = \nu D(\ )' \qquad A_{32} = D(\ )$$

$$A_{33} = D(\ ) + \frac{K}{R^2}[(\ ) - \nu(\ )''] + [\frac{K}{R^2}[(\ )'' - \nu(\ )]]''$$

Numerical Methods in Geomechanics (Innsbruck 1988), Swoboda (ed.)
© 1988 Balkema, Rotterdam. ISBN 90 6191 809 X

Study of numerical constitutive models for liquefaction problem

T.Shiomi, S.Tsukuni, Y.Tanaka, M.Hatanaka & Y.Suzuki
Takenaka Technical Research Laboratory, Tokyo, Japan
T.Hirose
Takenaka Komuten Co., Ltd, Tokyo, Japan

The four soil models which are capable to simulate liquefaction model are studied. The most interest is focused to 'how to determine the soil parameters and what parameters works for what'. Some comments are presented on this points. Comparative results are shown and commented in relation to the definition of the soil models. Discussion of models are focused to the specific characteristics of the each model. At the last section the simulation of the shaking table test by the one of the models is reported.

1 INTRODUCTION

Prediction of the ground liquefaction is of prime importance in the earthquake engineering. Numerical models for those problem are being extensively studied in the recent years and begin to be used in the geotechnical practice with dynamic finite element methods. There are many choices of the constitutive models which can be adapted to liquefaction phenomena. And there is only few comprehensive guidelines to choose a soil model and/or to determine the parameters which determine the model's behaviour. There are also very few comparison of the soil models to simulate an experimental test by the non-originators. So we conducted first the simulation of a triaxial test by the four soil models mentioned below in the same condition.

We performed two simulation tests. In the first case, a cyclic triaxial test is simulated by the above four models with the soil parameters which may be obtained through the ordinary soil constants. The procedure to determine the parameters is described for each models. And soil models are assessed by the comparative studies of the results of the numerical simulations and experimental data.

Secondly we simulate a shaking table test with these models using the finite element program DIANA-J. Characteristics of soil parameters are studied simulating the triaxial test before we proceed the simulation of the boundary problem.

2 SOIL MODELS

2.1 Basic formulation

we chose four soil models which were developed for the liquefaction phenomena for comparative studies. They may represent the following type of models.

In the group of endochronic type model

1. Densification model : pore pressure is depend on the accumulation of shear strain. Failure criterion is defined by Mohr-Coulomb or Drucker-Prager criterion [1].

In the group of models based on elasto-plastic theory

2. Single mechanism model with volumetric strain hardening (Reflecting surface model [2,3]).
3. Single mechanism model with double hardening rule (Pastor-Zienkiewicz model [4]).
4. Multi-mechanism model where each mechanism is defined on three mobilized planes and volumetric change (Hujeux-Kabilamany-Ishihara model [5,6]).

The following assumption is commonly adopted in the any soil models .

$$d\varepsilon_{ij} = d\varepsilon^{e}_{ij} + d\varepsilon^{p}_{ij} \quad (1)$$

where $d\varepsilon_{ij}$ and $d\varepsilon^{e}_{ij}$ are total and elastic strain respectively. And the effective stress is evaluated by the following equation.

$$d\sigma'_{ij} = D^{e}_{ijkl}\, d\varepsilon^{e}_{kl} \quad (2)$$

There is no well established approach to predict the plastic strain so that many assumption to evaluate it are proposed.

The endochronic type model introduce a plastic volumetric strain which corresponds with excess pore pressure (e.p.p. hereafter) as a empirical function. This plastic strain is named as 'Autogenous strain'. The empirical relationship is determined by a simple shear test.

The models based on elasto-plastic theory use a potential surface g to define the direction of the plastic strain vector. So it can be written for the single mechanism model as follows.

$$d\varepsilon^{p}_{ij} = \frac{\partial g}{\partial \sigma'_{ij}}\, d\lambda \quad (3)$$

For multi-mechanism model, the plastic strain vector is accumulated for each mechanism as

$$d\varepsilon^{p}_{ij} = \sum^{k} \frac{\partial g_k}{\partial \sigma'_{ij}}\, d\lambda_k \quad (4)$$

where $d\lambda$ is defined for each mechanism as follows.

$$d\lambda = \frac{\frac{\partial f}{\partial \sigma'_{ij}} D_{ijkl}\, d\varepsilon_{kl}}{\frac{\partial f}{\partial \sigma'_{ij}} D_{ijkl} \frac{\partial g}{\partial \sigma'_{kl}} + H_p} \quad (5)$$

where f is a yield function. The independent mechanisms are assumed and they consist of shear, volumetric and rotational mechanism.

In the most of recent models the plastic strain is also considered in unloading process so that the theory covers more general behaviour of material. So plastic strain can exist at any point of stress path.

The most of case elasto-plastic constitutive matrix (D matrix) is calculated to evaluate tangential stiffness matrix. That is determined by the elastic constitutive matrix D^e, the yield function f, the function of potential surface g and hardening parameter H_p by the similar equation in the case of the plastic strain.

$$D^{ep} = D^{e}_{ijkl} - \frac{\frac{\partial f}{\partial \sigma'_{mn}} D_{mnkl}\, D_{ijrs} \frac{\partial g}{\partial \sigma_{rs}}}{\frac{\partial f}{\partial \sigma'_{ij}} D_{ijkl} \frac{\partial g}{\partial \sigma'_{kl}} + H_p} \quad ----(6)$$

It must be noted that only three quantities such as the function f and g and a hardening parameter H_p are additional parameters to determine the elasto-plastic behaviour. And function g determine the direction of the plastic strain and this direction can be defined by an empirical equation.

2.2 Densification model (D-N model)

The densification model is proposed by Zienkiewicz et al [1]. This model is different from the other model on the definition of the pore pressure. The pore pressure or autogenous volumetric strain is independently determined from its yield criterion as follows.

$$d\varepsilon^{o} = \frac{A}{1 + Bk}\, dk \quad (7)$$

where k is a damage parameter defined as follows.

$$dk = e^{\gamma\theta}\, d\xi \quad (8)$$

And $d\xi$ is the incremental equivalent shear strain. Assuming the compressibility of fluid the incremental pore pressure can be calculated by the following equation.

$$dp = \frac{1}{n/K_f + 1/K_s}\, d\varepsilon^{o} \quad (9)$$

where K_f and K_s is the bulk modulus of the fluid and soil skeleton respectively.

2.3 Single mechanism model with volumetric hardening : Reflecting Surface model (R-S model)

The model was proposed by Pande and Pietruszczak and use a single mechanism. f and g are explicitly determined as

$$f = g = \sigma'^{2}_{m} + 2\sigma'_{m} a + \frac{\bar{\sigma}^{2}}{g(\theta)^{2}} = 0 \quad (10)$$

Hardening parameter is defined as:

$$H_p = \frac{\partial f}{\partial a} \cdot \frac{\partial a}{\partial e^p} \cdot \frac{\partial g}{\partial \sigma_{ij}} \delta_{ij} \qquad (11)$$

The functions f and g are chosen as the same function for both loading and unloading in this study. The variable a is a size parameter of yield and potential function and defined in the incremental form as a function of the volumetric plastic strain (eq. 12).

$$\frac{\partial \ln(a)}{\partial e^p} = (\frac{1}{W_1} - \frac{1}{W_2}) e^{z(e^p - e^p_o)} - \frac{1}{W_1} \qquad (12)$$

where W_1 and W_2 is the slope defined as the slope of plastic specific volume e^p against logarithmic mean stress σ_m.

2.4 Single mechanism model with shear and volumetric hardening : Pastor–Zienkiewicz model (P-Z model)

This model defines the potential surface as a function of dilatancy as follows.

$$\frac{\partial g}{\partial \sigma_{ij}} = \frac{\partial \sigma_l}{\partial \sigma_{ij}} \frac{\partial g}{\partial \sigma_l}$$

$$= \frac{1}{1+d_g^2} \frac{\partial \sigma_l}{\partial \sigma_{ij}} \{1, d_g\} \qquad (13)$$

where dg is the dilatancy defined as follows.

$$d_g = \frac{d\varepsilon_v^p}{d\varepsilon_s^p} = (1 + \alpha_g)(M_g - \eta) \qquad (14)$$

where $d\varepsilon_v^p$ and $d\varepsilon_s^p$ are the incremental plastic volumetric strain and the incremental plastic equivalent strain respectively and α_g and M_g are material constants. Eq. 14 can be determined from the experimental data.

The yield function is defined using the same type function as the potential function by introducing the material constants α_f and M_f. A empirical formula is proposed by the Pastor and Zienkiewicz as follows.

$$M_f = D_r M_g / 100 \qquad (15)$$

α_f is assumed equal to α_g in the most of case. Hardening parameter H_p is defined for loading and unloading process independently. For loading,

$$H_L = H_0 \sigma'_m \{1- \frac{\eta}{\eta_f}\}^4 \{H_v + H_s\} \qquad (16)$$

$$\text{where } H_v = 1 - \frac{\eta}{M_g}$$

$$H_s = \beta_0 \beta_1 \exp(-\beta_0 \xi)$$

ξ is the equivalent strain, β_0, β_1 and H_0 are additional soil constants, η is stress ratio and η_f is the maximum stress ratio. For unloading,

$$H_u = H_{uo} \{\frac{\eta_u}{M_g}\}^{-\gamma_u} \qquad (17)$$

where H_{uo} and γ_u are constants. As mentioned above, the hardening parameter is a function of the stress ratio and an accumulated equivalent strain. This shear hardening become less significant as shear strain increases.

2.5 Hujeux-Kabilamany-Ishihara model

This model was originally proposed by Hujeux and Aubry[5] and modified by Kabilamany and Ishihara [6]. In this model f is explicitly defined but g is defined in incremental form. And hardening parameter is expressed by an experimental formula. The plastic strain is divided into four parts. The one plastic strain is determined by the volumetric behaviour. The three parts are determined for three mechanism correspond with the three mobilized plane which are defined in the σ_1-σ_2, σ_2-σ_3 and σ_3-σ_1 planes.

$$f_k = q_k c_k - p_k \sin/_f h_k = 0 \qquad (18)$$

where c_k is positive for loading process and negative for unloading process, p_k and q_k is the mean stress and the deviatoric stress for each 2 dimensional principal stress space. h_k is the hardening parameter. The normal vector to the potential surface is directly determined by the following equation.

$$\frac{\partial g_k}{\partial \sigma_i} = {}^{+}_{-} c_k + \mu_k - \frac{q_k c_k}{p_k} \quad (19)$$

where μ_k is determined as a slope of normal work against plastic shear strain which can be obtained by a laboratory test. σ_i is a stress defined in the mechanism. The hardening rule is determined by

$$\frac{dh_k}{d\varepsilon_k^p} = \mu_k \frac{(1 - h_k)^2}{a_m} \quad (20)$$

where a is a$_m$property parameter which can be directly obtained from data of triaxial test (normalized work v.s. hardening parameter h_k.

3 A TRIAXIAL TEST

A series of triaxial tests have been performed with the Toyoura sand. A test result of Medium dense sand has been chosen for the simulation test. The conditions of the test is shown in Table **1** and the properties of Toyoura sand is shown in Table **2.**

Table 1

Diameter	5.0 cm
Height	12.4 cm
Weight(wet)	451.5 g
Weight(dry)	360.0 g
Confining pressure	0.5 kg/cm^2
Back pressure	3.0 kg/cm^2
B value	0.997
Dry unit weight	1.50 g/cm^3
Relative density ...	56.4 %
Void ratio	0.769
Axial dynamic stress	0.2 kg/cm^2

Table 2 Properties of Toyoura sand

Specific gravity	2.65 g/cm^3
Dry unit weight (max)	1.637 g/cm^3
Dry unit weight (min)	1.346 g/cm^3
Particle size (average)	0.135

Cyclic axial stress $\sigma_c + \sigma_d$ is applied to specimen through load cell where σ_c is a constant confining pressure and σ_d is a dynamic part of the stress. The stress of radial direction is kept to the constant confining pressure σ_c. The deviatoric stress is equal to σ_d and the total mean stress σ_m is $\sigma_c + \sigma_d/3$. The pore pressure p is measured through the bottom of the specimen. Then the effective mean stress is calculated as $\sigma'_m = \sigma_c + \sigma_d/3 - p$. The results of the triaxial test are shown in Figs. **1, 2** and **3.**

The stress path is shown Fig. **1.** The result shown up to the cyclic mobility started since the numerical simulation has been conducted to simulate the accumulation of the pore pressure. At the first cycle pore pressure is built up more than at the following cycles. This mechanism is explicitly modeled in the M-M model. The other models are also capable to simulate this but the mechanism is not implemented explicitly.

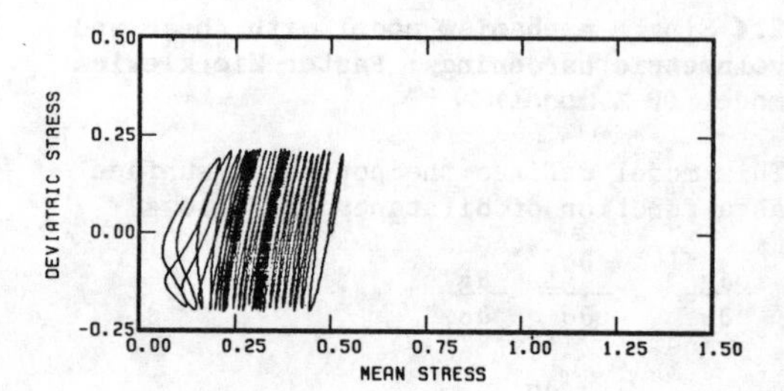

Figure 1 Stress path (mesured)

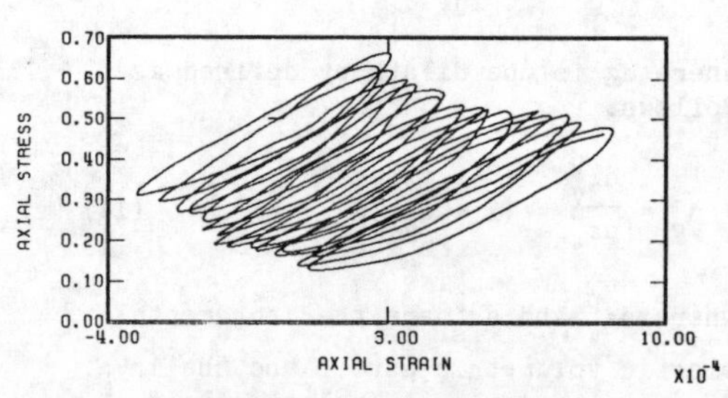

Figure 2 Axial stress v.s. Axial strain (measured)

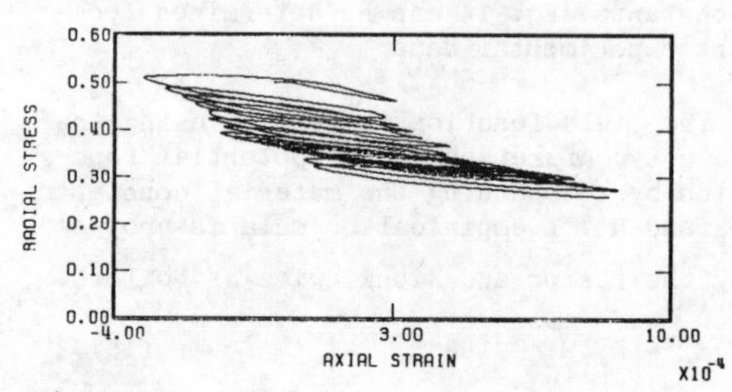

Figure 3 Radial stress v.s. Axial strain (measured)

Fig. 2 shows the relationship between axial stress and axial strain. The axial strain is shifted to the compression side as the axial stress decreases. The elastic modulus is determined from this figure.

Fig. 3 shows the relationship between radial stress and axial strain. The radial strain is half of the axial strain since the test is an undrained test.

4 TRIAXIAL TEST PROCEDURE

4.1 Algorithm

The simulation has been done by giving the incremental strain which was measured in the triaxial test and obtaining the effective stress. The procedure is shown below.

(1) read the material properties,
-- repeat from (2) to (4) --
(2) set the incremental strain
(3) calculate the effective stress
(4) plot/print the current stress

4.2 General procedure of soil model

In order to use the same model procedure for the local stress simulation and the finite element analysis a style of the procedure is proposed here. The four procedures are needed: i.e.

(1) Routine printing the material properties which are specific to the model: MDPROP
In-data : properties(number of data)
Out-data: none.

(2) Routine setting the initial value to the history memory of stress: MDHIST,
In-data : Initial stress(Nustr), Properties(Nupro),
Out-data: Stress memory(Numem).

(3) Routine calculating of the elasto-plastic constitutive matrix: MDSTIF,
In-data : Properties(Nupro), Current stress(Nustr), Stress memory(Numem), Preconditioned incremental strain(Nustr), Current strain(Nustr).
Out-data: D^{ep} matrix(Nustr,Nustr),

(4) Routine calculating of incremental stress corresponding with incremental strain MDSTRS,
In-data : Properties(Nupro), Current stress(Nustr), Stress memory(Numem), Predicted incremental strain(Nustr), Current strain(Nustr),
Out-data: Revised stress(Nustr),

where Nustr is the components of the stress (6 for 3D problem), Numem is the number of stress memory components (29 was the max. for the models studied) and Nupro is the number of the properties (Max. was 40).

It is convenient to pass the step number, iteration number and element number.

All the model routines have been written as the above style so that this form of the routines may be one of the general form.

5 SIMULATION OF TRIAXIAL TEST

5.1 Determination of general parameters

Elastic modula K and G were determined to fit the stress strain curve obtained by the triaxial test.

$E = 598\ kg/cm^2, \quad G = 230\ kg/cm^2.$

Failure angle of Toyoura sand is derived from the data of consolidation triaxial test and is 37.5°.

Phase transformation line is obtained by the diagram of dilatancy v.s. stress ratio.

5.2 Characteristic parameters for the models

For D-N model: the following parameters were determined by trial and error although initial value had been derived from the slope of e.p.p. against the damage parameter of an undrained test.

$\gamma = 2.0, \ A = 0.1, \ B = 550.$

For R-S model: the following parameters were determined by a triaxial consolidation test with rebound of stress. The parameter β is determined with parameter survey simulating the triaxial test.

$\lambda = 0.016575, \ k = 0.00453, \ \beta = 0.018.$

P-Z model: The phase transformation line M_g was determined as mentioned above. M_f was determined as $D_r M_g$ / 100. β_0 and β_1 were assumed as the average value of the other sands [5] and has not been studied.

$M_f = 0.68$, $\alpha_f = 0.35$, $\beta_0 = 4$,

$M_g = 1.2$, $\alpha_g = 0.35$, $\beta_1 = 0.2$,

$H_0 = 9000$, $H_{u0} = 6000$.

For M-M model: M is failure angle and μ is the phase transformation angle. The a_m and a_c were determined by parameter survey of a triaxial simulation. The same values are used for c and s to the Fuji river sand since this value does not differ with sands [6].

$M = 1.53$, $a_m = 0.00008$, $a_c = 0.000006$,

$c = 0.45$, $s = 0.0035$, $\mu = 1.2$.

5.3 Stress path

All the model simulated the number of cycles to reproduce the reduction of the effective stress from 0.5 kg/cm² to about 0.2 kg/cm² as shown in Fig. **4**. The D-N, the R-S and the M-M models showed similar tendency of stress path. The D-N and the M-M simulate well the first cycle in which the accumulation of e.p.p. is larger than one afterward. Specially the D-N model simulate well although this mechanism is not explicitly implemented. All the stress path drifted from the centre line. This is because the axial strain as input has the same tendency. The P-Z model showed little accumulation at the beginning and it become large when the effective mean stress reduced. This tendency is not observed in the triaxial test data.

5.4 Stress strain behaviour

Stress strain relationships are shown for axial stress and radial stress in Figs. **5** and **6** respectively. All simulation results are good at the average slope of the diagram but none of them reproduced good hysteresis loop. As the effective mean stress decrease during the cyclic loading the stiffness of the skeleton also decrease. These trend is seen the both measurement in Figs. 2, 3, 5 and 6. In Fig. 6 the M-M model show the very good agreement with the measured radial stress strain diagram at the first cycle.

5.5 Pressure accumulation

The cyclic history of the accumulation of e.p.p are shown for the four models in the Figs.**7**. The oscillation of the e.p.p. observed in both experiment and simulation may be induced by the compression of the specimen. The deviatoric component of the axial strain is kept almost same until cyclic mobility starts but the centre line shifted to the compression side. When cyclic mobility starts the significant plastic deformation is induced.

The envelop of the measured e.p.p shows slight convex upward. The D-N and the M-M model shows the same tendency but the e.p.p envelope of the P-Z is concave. Since the P-Z model use more parameters it was difficult to choose the all the parameters correctly and some parameters might have chosen inadequately.

As a whole the difference of the results between the models are much less than the diffrence between models and measured results. There must be further research to improve the models.

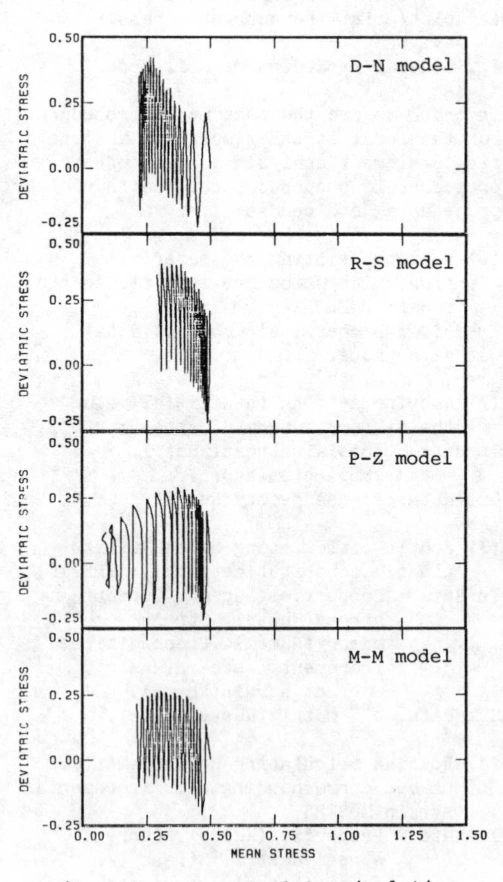

Figure 4 Stress path by simulation

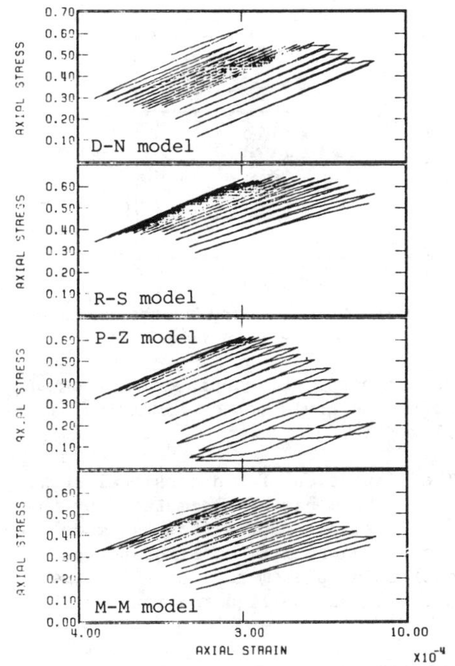

Figure 5 Axial stress v.s. Axial strain by simulation

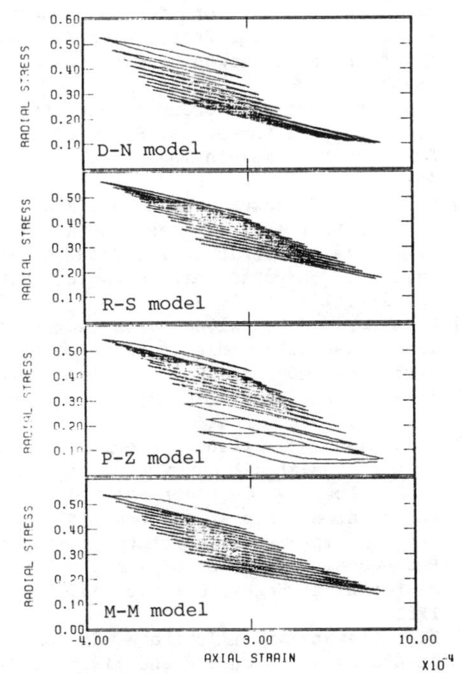

Figure 6 Radial stress v.s. Axial strain by simulation

6 SHAKING TABLE TEST SIMULATION

The shaking table test has been simulated by the R-S model as the first trial. The reason we chose the R-S model is that the model is the simpler and based on elasto-plastic theory. However the D-N model is interesting model because the accumulation of e.p.p is independently determined and easy to control.

The configuration of the shaking table test are shown in Fig. **8.** The tests are performed for the three problem, i.e. plane ground (model-1), soil-structure type problem (model-2) and soil-structure problem with a compacted area (model-3). Material properties are determined as same as the above approach which are very much different because of the low confining pressure. Fig. **9** shows the maximum e.p.p ratio against the initial vertical stress. The half upper layer show good agreement with the shaking table test but bottom layer show little difference from the experiment.

The final settlement is shown in Fig. **10** as the time history. The final settlement of the model-3 show good agreement with the experiment but little different for the model-2.

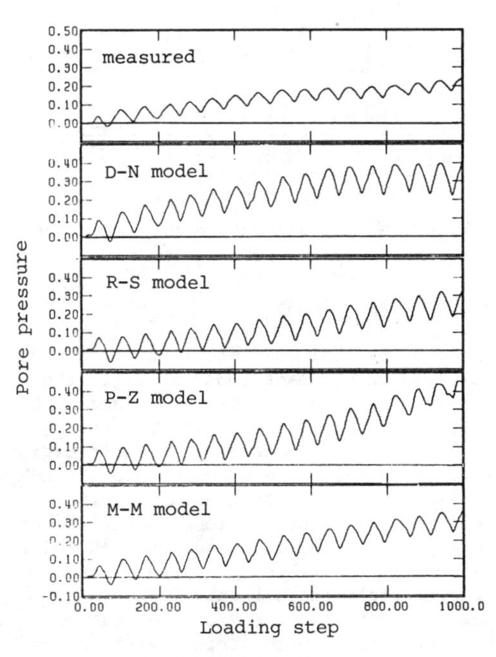

Figure 7 Excess pore pressure cyclic history

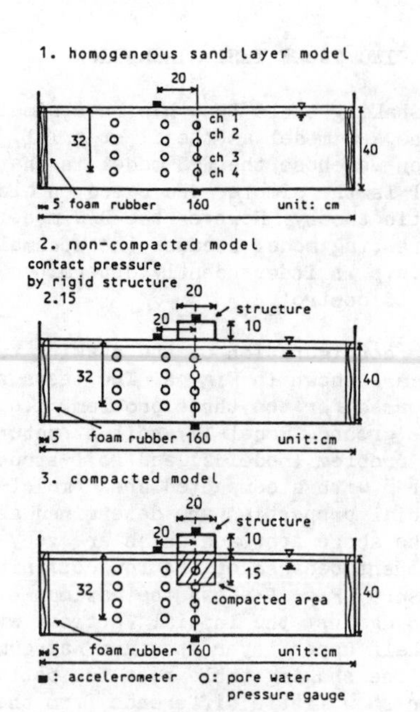

Figure 8 Section of experimental model

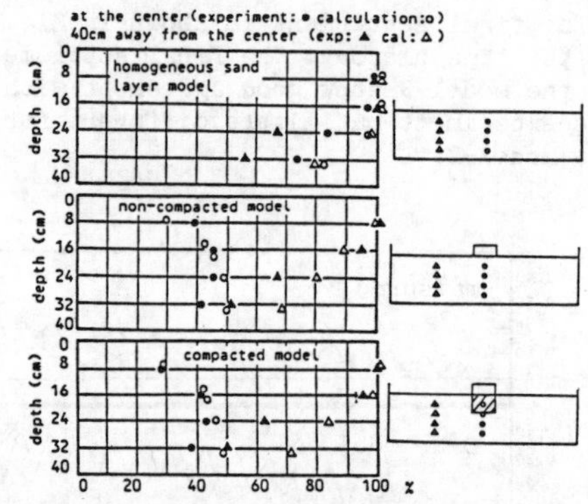

Figure 9 Maximum excess pore pressure ratio $(\Delta u/\sigma'_{vo})$

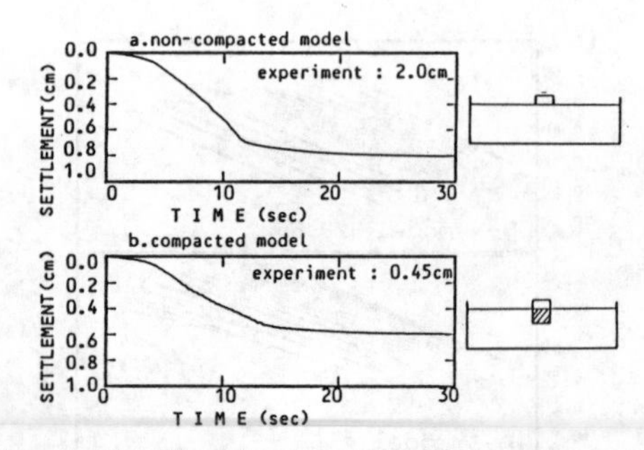

Figure 10 Time history of the settlement of the structure

7 CONCLUDING REMARKS

The four typical soil models have been compared on the same condition. All the models show good capability for accumulation of the excess pore pressure until stress reaches to the phase transformation line or failure line. The P-Z and the M-M models are capable to simulate cyclic mobility phenomena. As long as the excess pore pressure is concerned, the D-N model is most controllable since the mechanism is completely independent. In our study the M-M model and the R-S model showed good behaviour in comparison with the measurement. The stress strain relationship is well simulated for the tangential stiffness but the hysteresis loop are hardly agreed. The strain drift from the centre line could not be simulated by any soil models.

The simulation of 2 dimensional shaking table problem has been simulated by the one of the soil models (the R-S model) and shown its capability to simulate the liquefaction phenomenon including the settlement due to liquefaction.

REFERENCE

[1] O.C.Zienkiewicz, K.H.Leung, E.Hinton and C.T.Chang, Earth dam analysis for earthquakes: Numerical solution and constitutive relations for non-linear (damage) analysis, Proc, Design of Dams to Resist Earthquakes, I.C.E., London, 1980.

[2] G.N.Pande and S.Pietruszczak, Reflecting surface model for soils, Proc. of Int. Symp. on Num. Models in Geomech., Zurich 1982.

[3] T.Shiomi, S.Tsukuni and M.Hatanaka, Liquefaction analysis based on a dynamic effective stress analysis, Proc. of 1st symp. on computational mechanics, Tokyo 327-334, 1987.

[4] M.Pastor and O.C.Zienkiewicz, A generalized plasticity : Hierarchical model for sand under monotonic and cyclic loading, INME report C/R/534/86, Univ. College of Swansea, 1986.

[5] J.C.Hujeux and D.Aubry, A critical state type stress-strain law for monotonous and cyclic loading: Geotechnical and Numerical consideration, Proc. of the Sym. on Implementation of the Computer Procedures and Stress-Strain Laws in Geotechnical Engng, Chicago, 657-671, 1981.

[6] K.Kabilamany and K.Ishihara, Stress dilatancy relationship and yield surface for sand in the context of multiple mechanisms, (submitted to soil and foundation).

Numerical Methods in Geomechanics (Innsbruck 1988), Swoboda (ed.)
© 1988 Balkema, Rotterdam. ISBN 90 6191 809 X

Numerical analysis of the nonlinear earthquake response of an earth dam

S.J.Lacy
Rutgers University, Piscataway, N.J., USA
J.H.Prevost
Princeton University, Princeton, N.J., USA

ABSTRACT: The results of a nonlinear dynamic finite element analysis of the seismic response of an earth dam which is considered as a two phase system is presented. Coupled dynamic field equations model pore fluid-soil skeleton interaction. Multi-yield surface plasticity is used to accurately model the nonlinear, hysteretic behavior of the soil skeleton. Results of the numerical calculations are compared to calculations where the dam is modelled as a one phase system and to the recorded response of the dam to the San Fernando earthquake.

1 INTRODUCTION

One problem which has received considerable attention in geotechnical engineering is the dynamic response of earth dams to earthquakes. This exemplifies the kind of geotechnical problem in which there are many complexities; the material is saturated in part, although the extent of the saturated zone is a priori unknown. Further, the earthquake subjects the dam material to high amplitude cyclic loading causing inelastic, hysteretic behavior. The stress-strain behavior of the soil skeleton is complex and its modelling must be accurate and numerically efficient. Until recently, dynamic analyses of the response of soil systems to earthquakes have been solved by very simple numerical models. However, the disastrous consequences of a dam failure serve as a motive for making as accurate an analysis as possible of the problem and the current availability of computing facilties has made it possible to develop rigorous nonlinear dynamic analysis procedures.

Although it has been found that only dams susceptible to liquefaction have failed due to earthquakes (Ref. 16), no truly nonlinear dynamic analyses of earth dams which includes the effects of the pore fluid soil skeleton interaction and of the material strength deterioration during an earthquake have been yet reported. In saturated sands high pore water pressures build up during an earthquake making the soil vulnerable to liquefaction. These excess pore water pressures are subsequently redistributed, making the soil again vulnerable to liquefaction in the earthquake aftermath. The problem requires an integrated, coupled treatment of the dynamic interaction between the soil skeleton and the fluid phases.

This paper reports the results of a nonlinear dynamic finite element analysis of the response of the Santa Felicia Dam to the San Fernando earthquake. In the analysis performed, a two phase continuum formulation is used to model the saturated soil medium as a soil skeleton and a fluid phase(Refs. 4,10,12). The problem is solved using a general purpose large scale finite element program (Ref. 11) which is well suited to modelling a large non homogeneous structure with complicated goemetry such as a dam. The nonlinear hysteretic stress-strain behavior of the soil is modeled using nested conical yield surfaces, accounting for shear stress induced anisotropy and dilatancy effects. The earth dam and earthquake chosen for analysis are well documented: there are available detailed reports of the porperties of the dam materials (Ref. 1); recorded accelerations during the San Fernando earthquake are available at both the abutment and crest of the dam. Using the abutment record as an input motion at the base of the dam, the computed crest acceleration is compared with the recorded response.

2 PROBLEM GEOMETRY/FINITE ELEMENT MESH

The Santa Felicia dam is a modern rolled-fill earth embankment built in 1954-1955 and located in Ventura County, California, about 65 km. northwest of Los Angeles. The dam is 273 feet high above its lowest foundation. The crest is 30 feet wide and is 1,275 feet long. The upstream and downstream faces slope at 2.25:1 and 2:1 respectivley. The dam has an impervious core that rises from bedrock with slopes at 0.33:1. Figure 1 shows the cross section at midlength, a longitudinal cross section and a plan view of the dam.

Figure 2 shows the two-dimensional finite element model, representing the maximum cross section of the dam, used for the dynamic analysis. The mesh consists of 156 nodes. Each node is assigned four possible degrees of freedom: two perpendicular directions of soil displacement and two perpendicular directions of fluid velocity. For purposes of this analysis it was assumed that the reservoir was full to 240 feet. The free surface of fluid was located using the procedure described in Refs. (8,9). Then the dam was divided into 66 saturated elements and 67 dry elements with nodes having no fluid degrees of freedom. A plane strain assumption was used.

The canyon walls are made of rock. These walls were assumed rigid in all cases, and the nodes located along them were constrained to move in phase and with the same amplitude. Give the size of the dam, variation of the earthquake signal along the canyon boundary is likely, and the assumption of uniformity probably decreases the accuracy of the solution.

3. MATERIAL PROPERTIES

The Santa Felicia dam is made of an impervious clay core covered by a sand and gravel shell resting on a stiff foundation layer of gravel and sand down to bedrock. The material properties were extracted from tests reported in Ref. (1). Table 1 presents solid mass density (ρ_s), porosity (n^w), permeability (k), and friction angle (ϕ) for each material. The low strain elastic shear moduli for each material at each depth were back calculated from shear wave velocity measurements taken at the site (Ref. 1). These are listed in Table 2. The elastic bulk modulus is assumed to be $B_0 = 2G_0/3$. The shear and bulk moduli vary with pressure:

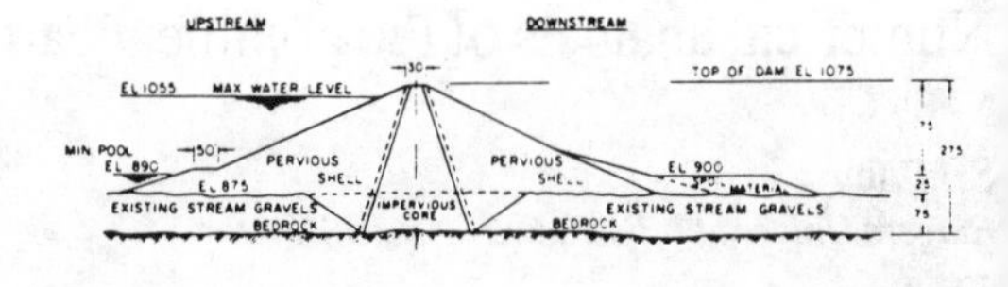

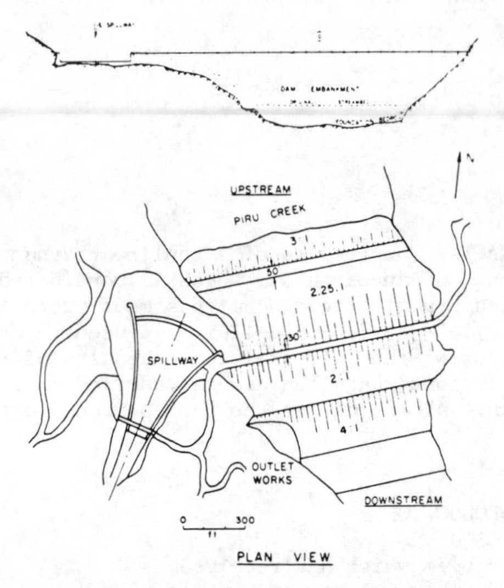

Figure 1. Santa Felicia earth dam Structural details

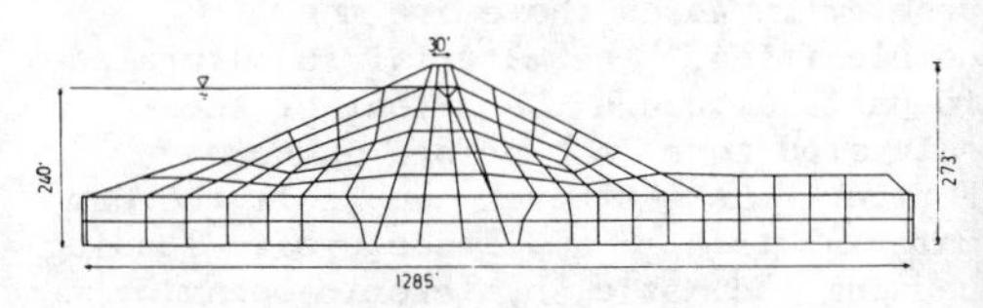

Figure 2. Santa Felicia earth dam Finite element mesh

$$G = G_0(p/p_0)^n \qquad B = B_0(p/p_0)^n$$

where n = 0.5 for sand, n= 1.0 for clay and p_0 is the computed overburden pressure for the element group. The pore fluid is assumed compressible, with a bulk modulus of $\lambda^w = 4.464 \times 10^7$ lbs./ft^2.

The constitutive theory presented in Refs (8) and (14) was used to model the nonlinear hysteretic stress-strain behavior of the dam materials. The material plasticity is pressure-sensitive and modeled by nested conical yield surfaces. A hyperbolic shear stress-strain curve was generated for each element group about the reference pressure for those elements. The reference strain,

γ_r, was assumed equal to 0.015 in tension and -0.015 in compression. Figure 3 shows the curve produced for saturated clay elements at 148.5 ft. below the dam crest. Ten conical yeild surfaces were used to approximate the shear stress strain behavior. A purely kinematic hardening rule was assumed, forcing the size of each yield surface to remain constant. One yield surface was used to model the hydrostatic compression vs. volumetric strain relation.

Table 1. Santa Felicia dam materials

Property	Core	Shell	Foundation
ρ_s (slugs/ft^3)	5.194	5.213	5.213
n^w	0.318	0.255	0.309
k (ft/day)	0.001	20.0	150.0
ϕ (degrees)	31.0	40.0	39.0

Table 2. Low strain shear moduli for Santa Felicia dam

	G_0 (lbs/ft^2)		
Depth below crest (ft)	Core	Shell	Foundation
254.25	–	4763929.0	39726600.0
216.75	–	4763929.0	39726600.0
181.5	3972660.0	4763929.0	–
148.5	3875524.0	4719510.0	–
115.5	3603646.0	4440157.0	–
82.5	3155159.0	3882862.0	–
49.5	2316342.0	3185766.0	–
16.5	1937243.0	2441492.0	–

4 INPUT GROUND MOTION

The model dam was subjected to the input ground motion recorded at the Santa Felicia dam site, near the outlet works, during the 1971 San Fernando earthquake (M_L = 6.3). The first 15 seconds of the recorded acceleration was used, with data at .02 second intervals and peak acceleration of 0.22g in the upstream-downstream direction. Figure 4 shows the input acceleration time histories in the upstream-downstream and vertical directions and their respective Fourier transforms. It should be noted that vibrations of a concrete standpipe underlying the floor of the valve house upon which the strong motion instrument was fixed introduced the peaks at about 10 Hz in the Fourier amplitude spectra. The two components of ground motion were

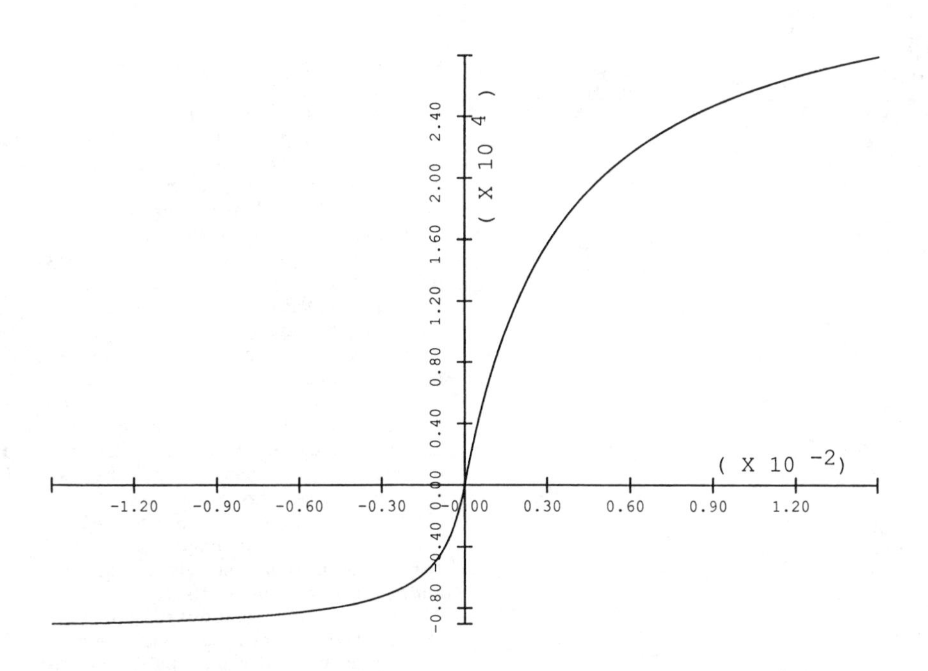

Figure 3. Stress-strain relation
Clay core at 0.33H

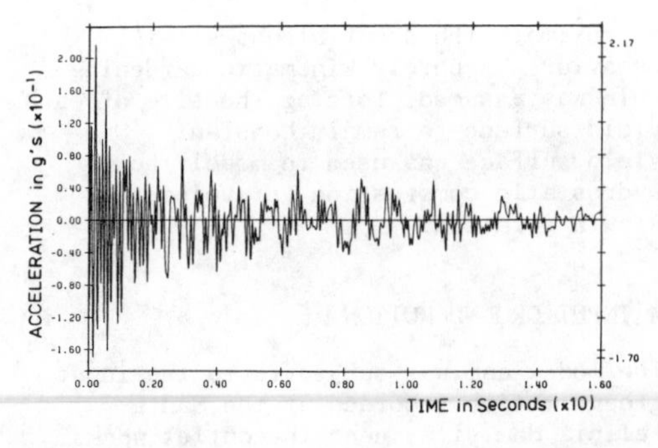

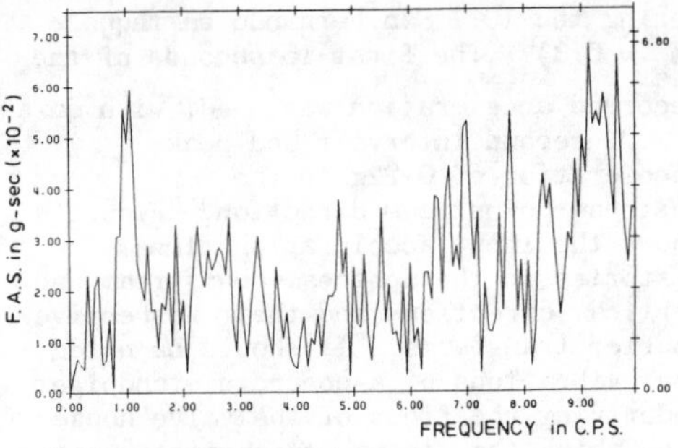

a. Upstream-downstream direction

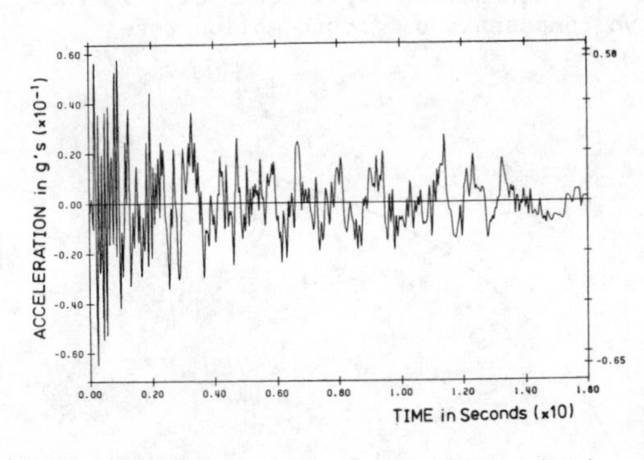

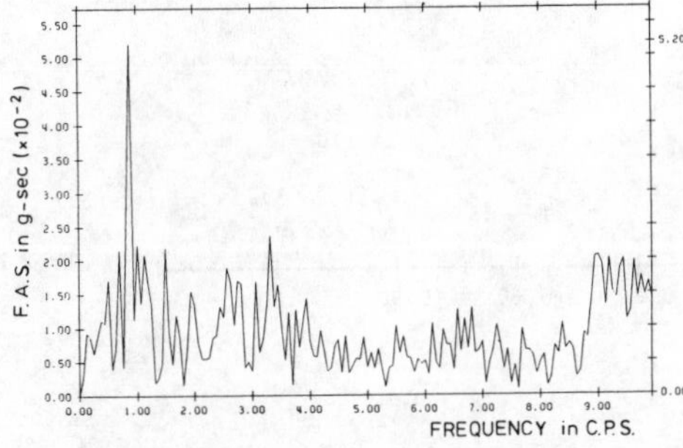

b. Vertical direction

Figure 4. Input ground motion

applied simultaneously as prescribed uniform input accelerations at the solid canyon boundary nodes.

5 IMPLEMENTATION

The solution was calculated in two steps. First, gravity loads and loads representing the fluid on the upstream face of the dam were applied and the dam consolidated during ten load steps at three day increments. When consolidation was completed, the nodal velocities and accelerations were zeroed, the time was reset to zero and the recorded earthquake signal was applied at the canyon boundary. Time integration of the semi-discrete finite element equations was accomplished using the implicit-explicit finite difference algorithm of Highes et al (Refs. 5,6,7). BFGS iterations were performed at each time step, with reformation of the coefficient matrix at the beginning of each time step. A maximum of 20 iterations was allowed and the convergence tolerance was set to 10^{-3}. The calculations were performed on a Cray computer.

6 RESULTS

Figures 5a and 5b compare the computed acceleration histories at the dam crest of the two phase model with acceleration histories computed by a one phase soil model (Ref. 15). Acceleration histories are shown in the upstream-downstream and vertical directions along with their respective Fourier transforms. In the upstream-downstream direction, maximum acceleration of the two phase model (.216g) is substantially lower than the maximum for the one phase model (.278g). As expected, introduction of pore-fluid soil skeleton interaction has damped the system response. The frequency content of the two signals is only slightly lower for the two phase system. In the vertical direction, the response of the two phase system is greater than that of the one phase system. It is interetsing to note that in a previous study (ref. 15) of a three dimensional nonlinear one phase model, the response in the vertical direction also unexpectedly exceeded that in the two dimensional one phase model.

Figures 6a and 6b compare the results of the two phase calculations to the recorded response at the dam crest during the San Fernando earthquake. The

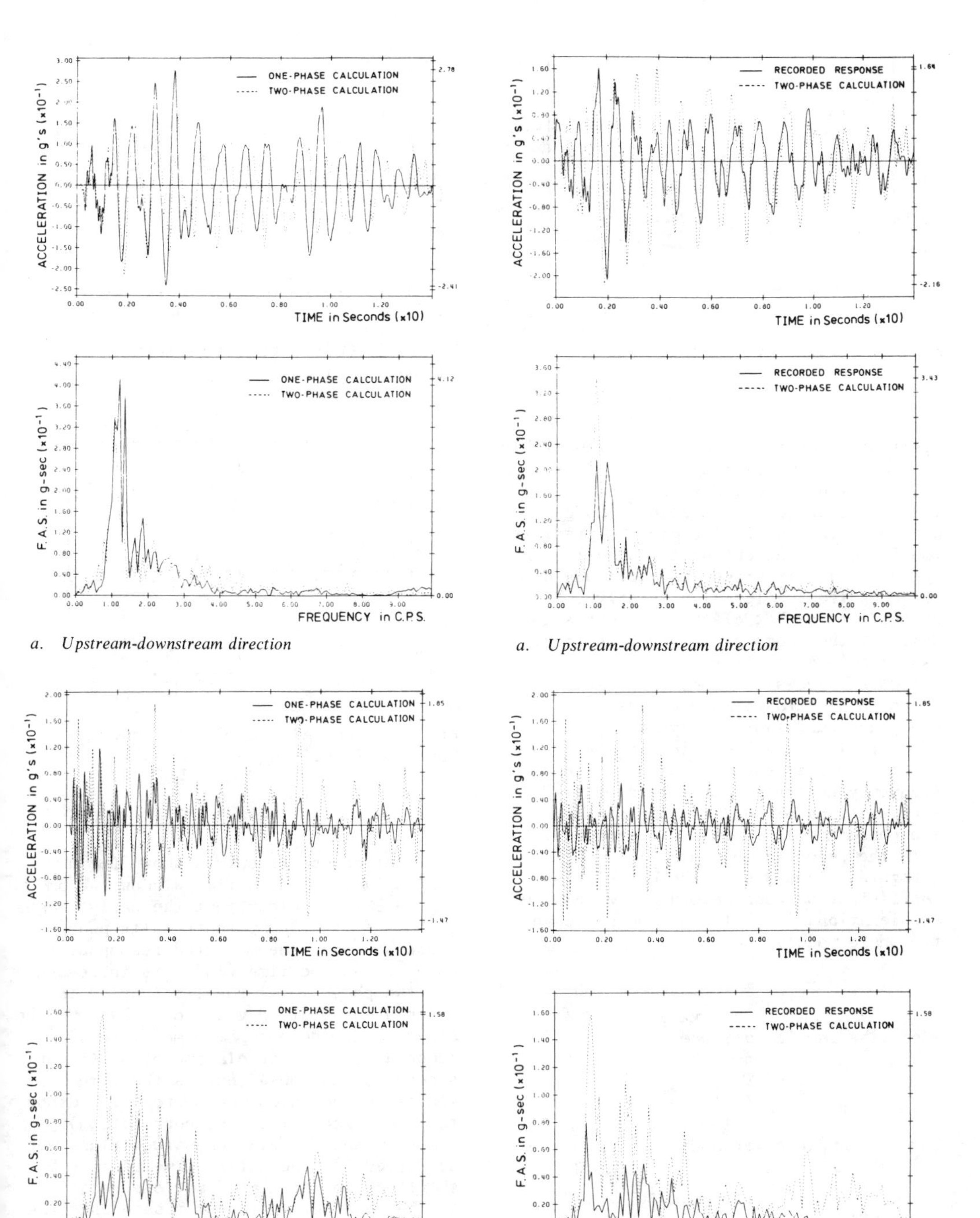

a. Upstream-downstream direction

b. Vertical direction

a. Upstream-downstream direction

b. Vertical direction

Figure 5. Acceleration of dam crest
1 Phase vs. 2 Phase calculations

Figure 6. Acceleration of dam crest
2 Phase calculations vs. recorded response

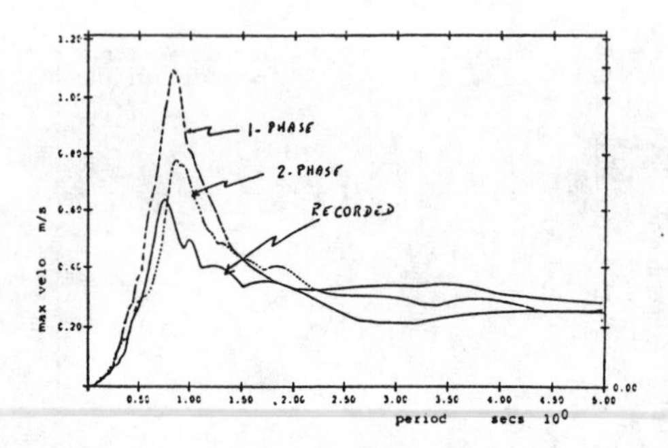

Figure 7. Velocity response spectra

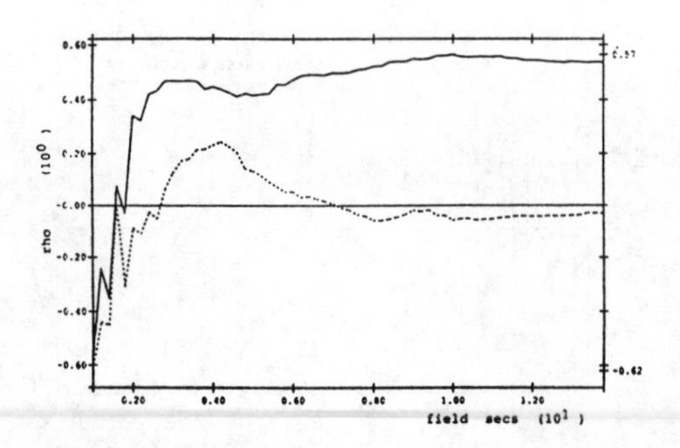

Figure 8. Correlation coefficient - Increasing time window

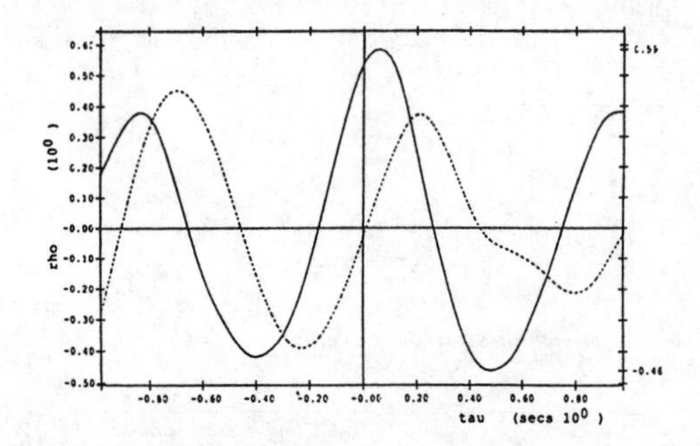

Figure 9. Correlation Coefficient Effects of Shifting

upstream-downstream signals are remarkably similar. The maximum computed acceleration (.216g) differs from the maximum recorded acceleration (.207g) by only 4%. The model also closely predicts the frequency content of the signal. In the vertical direction the magnitude of the computed acceleration greatly overestimates that of the recorded signal. However, the range of frequencies for the two signals is similar.

Figure 7 shows the computed velocity response spectra for the measured, one-phase, and two-phase model acceleration time histories in the upstream-downstream directions. The frequency content of the computed accelerations of both the one-phase and two-phase models is in close agreement with the recorded motion.

Figures 8 and 9 show the correlation coefficient between recorded and computed accelerations for both the one-phase and two-phase models,

$$r_{x,y} = \frac{s_{xy}}{s_{xx} s_{yy}}$$

where the sample variance is

$$s_{xx}^2 = \frac{1}{n} \sum_{i=1}^{n} (x_i - \bar{x})^2$$

and the sample covariance is

$$s_{xy}^2 = \frac{1}{n} \sum_{i=1}^{n} (x_i - \bar{x})(y_i - \bar{y})$$

with $\bar{x}$ = mean value ($\bar{x} = \frac{1}{n} \sum_{i=1}^{n} x_i$). The correlation coefficient is limited in value to $-1 \leq r_{xy} \leq 1$. In figure 8 the correlation coefficent is shown for a gradually increasing time window. From figure 8 it is clear that the early stages of both sets of computed results poorly correlate with the measured response. However, as the time window is increased, the two-phase model results improve to reach a correlation value of 0.54. On the other hand, the one-phase model results reach a maximum correlation of 0.24 (for 4 seconds of motions), but as the time window is increased the correlation drops to almost zero (i.e., no correlation). In figure 9 the correlation calculations were performed for the entire time history by shifting the calculated motions relative to the recorded motion by τ seconds. The results shown in figure 9 indicate that the two-phase model is virtually in phase with the measured motion and can hardly be improved upon by shifting. On the other hand, the one-phase model results could be improved by shifts of 0.25 seconds or -0.7 seconds.

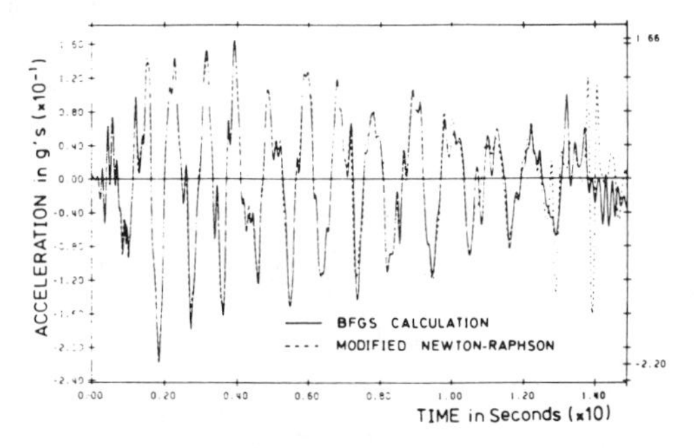

a. Upstream-downstream direction

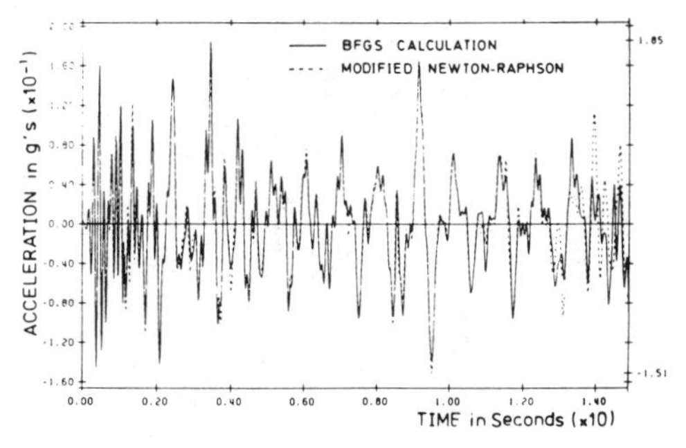

b. Vertical direction

Figure 10. Acceleration of dam crest BFGS vs. Modified Newton Raphson iterations

7 CONCLUSION

Elastoplastic response of a nonhomogeneous earth dam to an earthquake has been presented. The use of a rigorous nonlinear dynamic analysis produced a calculated response of the dam crest which compares well with the recorded response of the dam during an earthquake. Consideration of the dam as a two phase system with pore fluid soil skeleton interaction in the saturated zone improved the accuracy of the results.

Although the proposed methodology is sound, it is still far from having reached a mature stage. Nonlinear calculations are still difficult to complete successfully. This is illustrated in Figure 10 which shows a comparison of the computed acceleration time histories using BFGS and Modified Newton Raphson iterations with reforms at the beginning of each time step. The two computed responses are identical in both the upstream-downstream and vertical directions up to time 12 seconds after which they differ grossly. Although very stringent conservative convergence criteria had been used, the Modified Newton Rahpson iterations failed to provide the right answer after time 12 seconds (although the solution kept passing the convergence test.) Further research should therefore be devoted to devising better, sturdier and more reliable algorithmic strategies before such analyses can become "routine" and made widely available.

ACKNOWLEDGEMENTS

This work was supported in part by a grant from the National Science Foundation (#ECE-8512311 under the direction of Clifford J. Astill.) This support is gratefully acknowledged. S.J. Lacy also gratefully acknowledges the American Association of University Women for support provided by a Dissertation Fellowship.

8. REFERENCES

1. A.M. Abdel-Ghaffar and R.F. Scott, An Investigation of the Dynamic Characteristics of an Earth Dam, Report No. EERL 78-02, California Institute of Technology, Pasadena, California, August 1978.

2. A.M. Abdel-Ghaffar and R.F. Scott, Vibration Tests of a Full-Scale Earth Dam, Journal of the Geotechnical Engineering Division, ASCE, March 1981, 107(GT3), 241-269.

3. K.J. Bathe, and E.L. Wilson, Numerical Methods in Finite ELement Analysis, Prentice-Hall, Inc., NJ 1976

4. M.A. Biot, Theory of Elasticity and Consolidation for a Porous Anisotropic Solid," Journal of Applied Physics,1955, 26, 182-185.

5. H.M. Hilber, Analysis and Design of Numerical Integration Methods in Structural Dynamics, Report No. EERC 76-29, Earthquake Engineering Research Center, University of California, Berkeley, California, 1976.

6. T.J.R. Hughes, K.S. Pister and R.L. Taylor,Implicit-Explicit Finite Elements in Nonlinear Transient Analysis, Comp. Method App. Mech. Engr. 1979, 17/18, 159-182.

7. T.J.R. Hughes and W.K. Liu, Implicit-Explicit Finite Elements in Transient Analysis: Implementation and Numerical Examples, Journal of Applied Mechanics, June 1978, 45, 375-378.

8. S.J. Lacy Numerical Procedures for Nonlinear Transient Analysis of Two-Phase Soil Systems, Ph.D. Thesis, Department of Civil Engineering, Princeton University, Princeton, NJ, October 1986.

9. S.J. Lacy and J.H. Prevost, Flow Through Porous Media: A Procedure for Locating the Free Surface, Int. J. Num. Analyt. Meth. Geomechanics, 1987 (to appear).

10. S.J. Lacy and J.H. Prevost, Nonlinear Seismic Response Analysis of Earth Dams, Soil Dynamics and Earthquake Engineering, 6 (1), 1987, pp. 48-63.

11. J.H. Prevost, DYNAFLOW: A Nonlinear Transient Finite Element Analysis Program, Dept. of Civl Engineering, Princeton University, 1981, last update, 1987.

12. J.H. Prevost, Mechanics of Continuous Porous Media, International Journal of Engineering Science 1980, 18(5), 787-800.

13. J.H. Prevost Nonlinear Transient Phenomena in Saturated Porous Media, Comp. Meth. Appl. Mech. Eng. 1982, 30, 3-18.

14. J.H. Prevost A Simple Plasticity Theory for Frictional Cohesionless Soils, Soil Dynamics and Earthquake Engineering, 1985, 4(1), 9-17.

15. J.H. Prevost, A.M. Abdel-Ghaffar and S.J. Lacy, Nonlinear Dynamic Alanyses of an Earth Dam, Journal of Geotechnial Engineering, ASCE, July 1985, 111(7), 882-897.

16. H.B. Seed, Considerations in the Earthquake-Resistant Design of Earth and Rockfill Dams, Geotechnique 1979, 29(3), 215-263.

17. O.C. Zienkiewicz, The Finite Element Method, 3rd Edition, McGraw-Hill, London, 1977.

Numerical Methods in Geomechanics (Innsbruck 1988), Swoboda (ed.)
© 1988 Balkema, Rotterdam. ISBN 90 6191 809 X

Discrete element analysis of granular material under dynamic loading

J.R.Williams
Civil Engineering Department, MIT, Cambridge, USA

ABSTRACT: There has been recent interest in analyzing soil and other granular material on the microscopic level so that microscopic properties can be related to macroscopic phenomena, such as failure. The Discrete Element Method provides a numerical tool for conducting such analyses. Here the basic theory behind the method is reviewed and various formulations are derived from a finite element basis. Errors in some formulations are highlighted for problems involving large strains and rotations. Examples of applications of the method are described including projectile penetration into a soil, embankment collapse and flow of granular material from a hopper.

1.0 INTRODUCTION

The objective of the Discrete Element Method is to solve the dynamic equilibrium equations, for multiple interacting bodies, as illustrated in Figure 1. The dynamic equilibrium equations are given for individual elements or regions as

$$\rho \ddot{u}_i = R_i + \frac{\partial \sigma_{ij}}{\partial x_j} \quad \text{in region } \Omega \tag{1}$$

subject to boundary conditions $u_i = u_o$ on Γ_1 and $F_i = F_o$ on Γ_2 where ρ is the density, $\ddot{u}_i$ the acceleration in direction i, R_i the body force, σ_{ij} the stress tensor (compression positive), and x_j the cartesian coordinate of point p in the body at which all quantities are evaluated.

Let the displacement at any point in the element be given in terms of deformation modes, ϕ, such that

$$u_i = \sum_N \phi_i^N \alpha^N \tag{2}$$

where ϕ_i^N is the Nth modal vector component in direction i, and α^N is the corresponding time dependent participation factor. It is also assumed that the modes ϕ are chosen to be mass orthogonal so that $\int \rho \underset{\sim}{\phi}^M \cdot \underset{\sim}{\phi}^N dV = 0$ if $M \neq N$.

Substituting Equation (2) into Equation (1), weighting each with $\phi^L{}_i$ and integrating over the volume gives

$$\sum_M \int \rho\, \phi_i^L \phi_i^N \, dV\, \ddot{\alpha}^N = \int \phi_i^L R_i \, dV + \int \phi_i^L \frac{\partial \sigma_{ij}}{\partial x_j} dV \tag{3}$$

Using the orthogonality of modes, the equation for the participation of mode ϕ^L is derived

$$\int \rho\, \phi_i^L \phi_i^L \, dV\, \ddot{\alpha}^L = \int \phi_i^L R_i \, dV + \int \phi_i^L \frac{\partial \sigma_{ij}}{\partial x_j} dV \tag{4}$$

This can be combined with the force boundary condition $\int \phi^L{}_i(F_i - F_o) d\Gamma_2 = 0$ to give $m^L \ddot{\alpha}^L = S_A - S_I$ (5)

where
$$m^L = \int \rho\, \phi_i^L \phi_i^L \, dV$$
$$S_A^L = \int \phi_i^L R_i \, dV + \int \phi_i^L F_o \, d\Gamma_2$$
$$S_I^L = \int \phi_{i,j}^L \sigma_{ij} \, dV$$

The above derivation is quite general and the modes ϕ^N can be chosen from a wide selection of possibilities.

2.0 ALTERNATIVE FORMULATIONS FOR THE ELEMENT MODES

Equation (5) can be rewritten in matrix notation as

$$[\phi]^T[M][\phi]\{\ddot{\alpha}\} = [\phi]^T\{f\} - [\phi]^T[K][\phi]\{\alpha\} \tag{6}$$

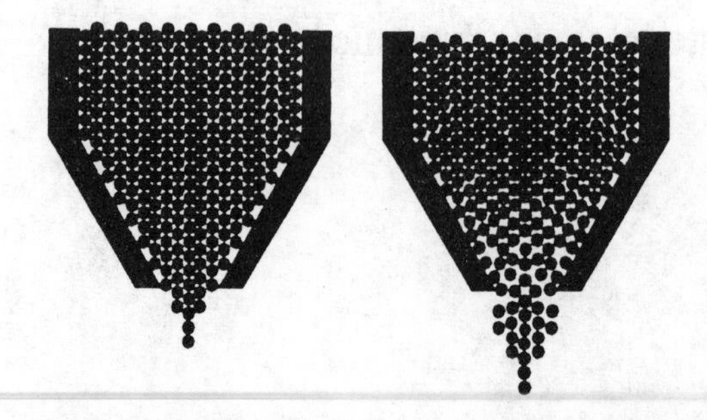

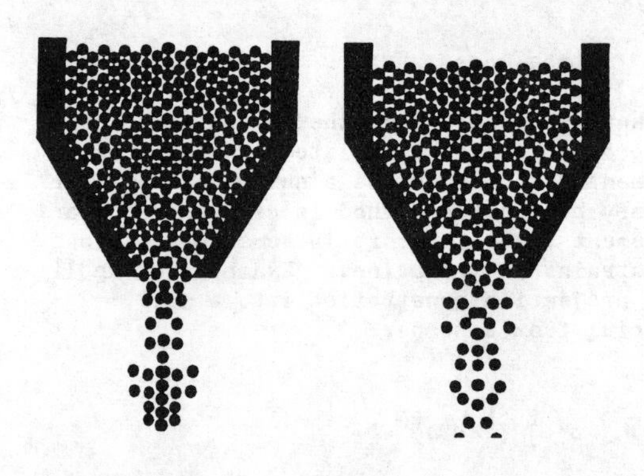

Figure 1 Discrete Element Analysis of Flow of Material from a Hopper Showing Formation of Dynamic Arch

where $\{u\} = [\phi]\{\alpha\}$. Now depending on the choice of $[\phi]$ various formulations are possible.

For clarity, let us consider a linear four noded quadrilateral where the nodal displacements are given by $\{u\}^T = \{u_1, u_2, u_3, u_4, u_5, u_6, u_7, u_8\}$ where $u_1 = u_x$ for node 1 and $u_2 = u_y$ for node 1, etc.

Choosing $[\phi] = [I]$, the identity matrix, returns the original explicit finite element form with $\{\alpha\} = \{u\}$. It should be noted that the nodal displacements still form an orthogonal set of modes, e.g.

$$\{\phi^1\}^T = \{1, 0, 0, 0, 0, 0, 0, 0\}$$

$$\{\phi^2\}^T = \{0, 1, 0, 0, 0, 0, 0, 0\}$$

such that $\{\phi^M\}^T\{\phi^N\} = 0$ if $M \neq N$.

The second possibility is to choose $[\phi]$ as the matrix of eigenmodes of the element, such that $[\phi]$ is the solution of the eigenproblem

$$[K]\,\alpha^N\{\phi^N\} = \lambda\,[M]\,\alpha^N\,\{\phi^N\}$$

As $[\phi]$ contains the eigenvectors of matrix $[K]$, $[\phi]^T[K][\phi]$ is in fact the stiffness matrix in its canonical form, and $[\phi]^T[M][\phi]$ is the diagonalized mass matrix, so that

$$[\phi]^T[K][\phi] = [\Lambda] \text{ and } [\phi]^T[M][\phi] = [I] \quad (7)$$

where $[\Lambda]$ is the diagonal matrix, with the eigenvalues λ_i on the diagonal, and $[I]$ is the identity matrix. The system of n coupled linear equations can now be decomposed into n independent equations

$$\ddot{\alpha}^N = \{\phi^N\}^T\{f\} - \lambda^N \alpha^N \qquad N = 1,n \quad (8)$$

from which we can solve for α^N directly.

The third possibility is to choose $[\phi]$ as a set of orthogonal modes, which are not eigenmodes, but which still diagonalize the mass matrix $[M]$. Originally it was proposed by Cundall et al., 1978, 1983, that element strains can be used as modes such that the displacement at any point p in the element is given by:

$$u_i = \bar{u}_i + \omega_{ij}\,\bar{x}_j + \epsilon_{ij}\,\bar{x}_j \quad (9)$$

where
- $\bar{x}_i$ is the coordinate of point p relative to the element centroid
- u_i is the displacement of any point p in the element
- $\bar{u}_i$ is the displacement of the centroid in direction i
- $\bar{x}_j$ is the coordinate of point p relative to the centroid
- ω_{ij} is the rotation tensor defined by $\omega_{ij} = e_{ijk}\omega_k$ where e_{ijk} is the permutation tensor and ω_k is the rotation vector
- ϵ_{ij} is the strain tensor

and summation over repeated indices is implied.

The modes for a linear quadrilaterial can be written as follows

$$\{\phi^1\}^T = \{1,0,1,0,1,0,1,0\} \quad \text{x translation}$$

$$\{\phi^2\}^T = \{0,1,0,1,0,1,0,1\} \quad \text{y translation}$$

$$\{\phi^3\}^T = \{-y_1, x_1, -y_2, x_2, -y_3, x_3, -y_4, x_4\} \quad \text{rotation}$$

$\{\phi^4\}^T = \{x_1,0,x_2,0,x_3,0,x_4,0\}$ strain x

$\{\phi^5\}^T = \{0,y_1,0,y_2,0,y_3,0,y_4\}$ strain y

$\{\phi^6\}^T = \{y_1,x_1,y_2,x_2,y_3,x_3,y_4,x_4\}$ shear strain

where x_i and y_i are the coordinates of the nodes with respect to the center of mass.

Two other modes are necessary to span the space, but they were not originally considered, and for the present they will be specified only as $\{\phi^7\}$ and $\{\phi^8\}$ and will be assumed orthogonal to the other modes. ($\{\phi^7\}$ and $\{\phi^8\}$ correspond to the so-called 'hourglass' modes, well known in the finite element literature.)

Now if the principle of modal superposition is to be valid, the strain deformation modes must be orthogonal to the rigid body translation and rotation modes (Bisplinghoff et al., 1962; DeVeubeke, 1976). These conditions can be written as

$$\int \rho\, \tilde{\phi}^N\, dV = 0 \quad \text{(for all modes N, excluding translation)} \qquad (10)$$

$$\int \rho(\tilde{r} \times \tilde{\phi}^N)dV = 0 \quad \text{(for all modes N, excluding rotation)} \qquad (11)$$

It has been shown by Williams et al., 1987 that for general shaped elements the shear strain deformation produces rotation of the element and cannot be used in a formulation dependent on superposition. Figure 2 illustrates graphically how shear strain and rotation are coupled for a rectangular element.

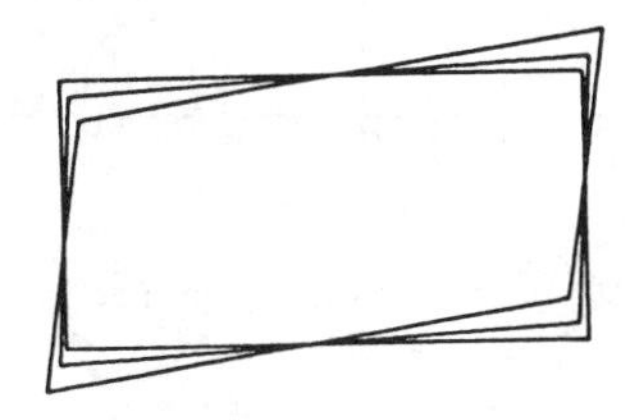

Figure 2 Illustration of Shear Giving Rise to Rotation of Rectangular Element

It is concluded that formulations which use strain as a basis for modal decomposition are in error.

Modes may be chosen to correspond closely to the strains of the element (Williams, et al., 1987), but they cannot be the actual strains.

If either these modes or the eigenmodes of the element are chosen, it should be noted that the modes must be specified at all times with respect to local axes which rotate with the element (Williams et al., 1985), since in global coordinates the mass matrix becomes non-diagonal when the element rotates. In many discrete element codes this is not done.

3.0 EXAMPLES

The DECICE discrete element computer code (Williams et al., 1985; Mustoe et al., 1987; Hocking et al., 1985, 1987) uses a modal formulation based on mass orthogonalized modes and was utilized to analyze the following problems.

Embankment Collapse

It has long been recognized that the wedge or embankment is a fundamental configuration for the study of self-weight or body force type problems which characterize major areas of interest in soil mechanics. It has direct relevance to such practical structures as earth and rock-fill dams, road and rail embankments, mine waste tips, ore stockpiles, etc., and can be extrapolated to the general case of slope stability.

The simplest arrangement is that of a symmetric wedge. Analytic solutions for the pressures under such a stable wedge are given by Lee and Herington (1971) and numerical solutions are given by Trollope and Burman (1980).

An embankment with a 30° angle of repose was idealized as 222 closely packed circular discrete elements (Figure 3).

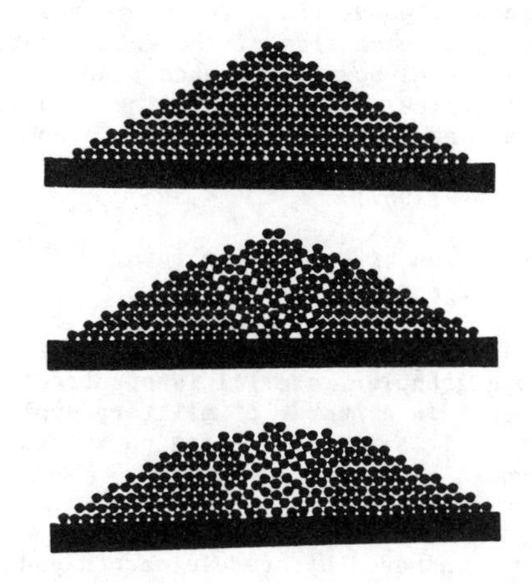

Figure 3 Dynamic Collapse of Embankment Showing Formation of Failure Planes

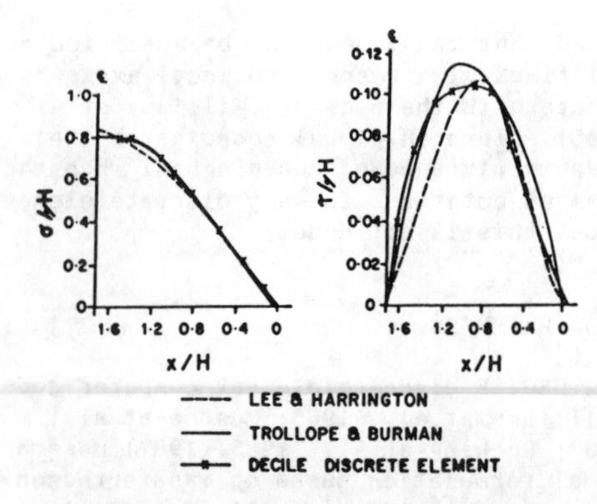

Figure 4 Comparison of Normal and Shear Stress Under Stable Embankment

The coefficient of friction was initially set to 0.7 and the system was brought to equilibrium under gravity loading.

The normal and shear stress along the base of the embankment at steady state is compared with the analytic results of Lee and Herington (1971) and numerical results of Trollope and Burman (1980) in Figure 4. The agreement is good for the normal stress distribution and lies between the analytic and numerical solutions for the shear stress distribution.

In the next phase of the analysis the inter-grain friction was reduced from 0.7 (35°) to 0.1 (6°) which is below the angle of repose of the embankment (30°).

Figure 3 shows the progressive failure of the embankment. The dynamics of stress waves passing through the soil are clearly visible. Two distinct failure planes develop on each side of the embankment. The material bounded by each plane moves initially as a rigid body. The principal stress trajectories and failure planes agree well with those predicted by Trollope (1957).

Penetration of Projectile into Granular Material

The launching of a projectile through or into a granular material is of direct interest in a number of military applications. It also has bearing on the cutting of rock.

The granular material was idealized as several hundred discrete elements and was contained in a box as shown in Figure 5. The material was initially consolidated to a given pressure by forcing a ram down

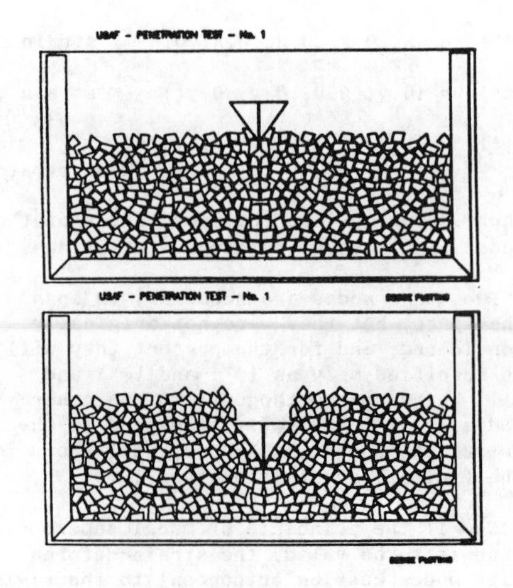

Figure 5 Penetration of Projectile into Granular Material

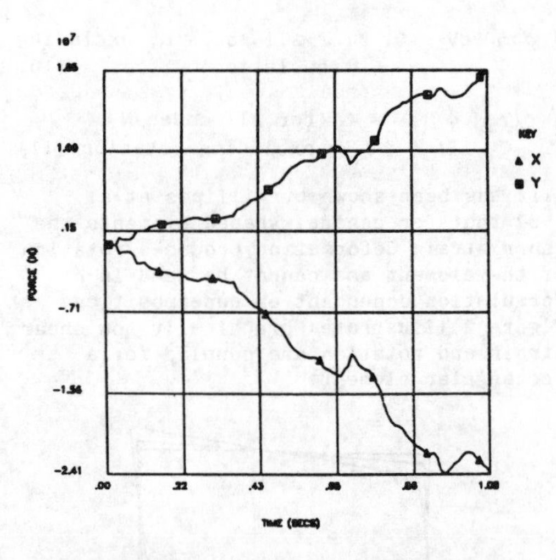

Figure 6 Force History on Projectile

onto the material. The ram was then removed and a wedge shaped projectile was driven vertically downwards at a constant velocity of 2 m/sec.

Snapshots of the system are shown at various times in Figure 5 and the force on the projectile in Figure 6.

Offshore Platform - Sea Ice Interaction

In many situations a full three dimensional analysis is necessary. The simulation of sea ice impacting a large

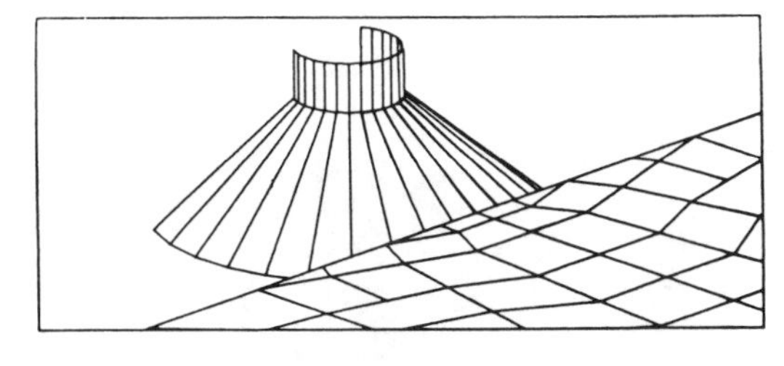

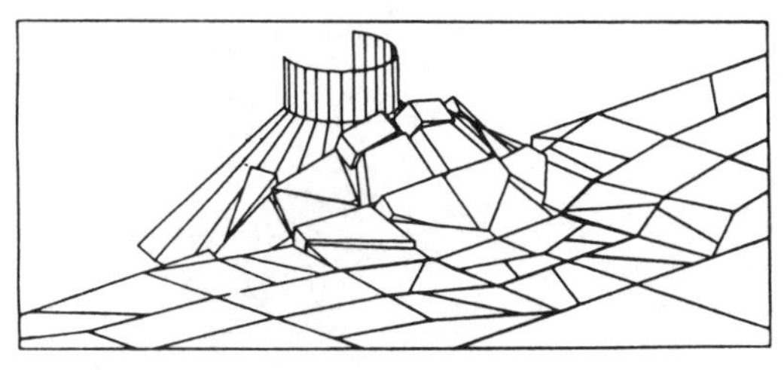

Figure 7 Ice Flow Fracturing Against Conical Drilling Structure

bottom founded offshore conical platform is illustrated in Figure 7, taken from Hocking et al. 1987. The ice sheet is initially flaw free and is driven against the platform by environmental driving forces. The sheet initially fractures radially followed by circumferential cracks. Upon further motion the fragmented ice sheet begins to clear around the structure. Note the override and submergence of many of the fragmented ice pieces.

REFERENCES

Bisplinghoff, R.L. and Ashley, H. 1962. Principles of aeroelasticity. Dover, NY.

Cundall, P.A. 1971. A computer model for simulating progressive large scale movements in blocky rock systems. Proceedings Int. Symp. on Rock Fracture, Nancy (ISRM), Paper II-8.

Cundall, P.A., Maini, T., Marti, J., Beresford, P.J., Last, N.C. and Asgian, M.I. 1978. Computer modeling of jointed rock masses. U.S. Army Engineers Waterways Experiment Station Technical Report N-78-4.

Cundall, P.A. and Hart, R.D. 1983. Development of generalized 2-D and 3-D distinct element programs for modeling jointed rock. U.S. Army Engineers Waterways Experiment Station, Final Technical Report.

DeVeubeke, B.F. 1976. The dynamics of flexible bodies. Journal of Engineeri Science, Vol. 14, pp. 895-913.

Hocking, G., Mustoe, G.G.W. and Williams J.R. 1985. Validation of the CICE discrete element code for ice ride up and ice ridge cone interaction. ARCTIC '85 Conf. ASCE, San Francisco, March.

Hocking, G., Mustoe, G.G.W and Williams, J.R. 1987. Dynamic analysis for three dimensional contact and fracturing of multiple bodies. NUMETA '87, 2nd Int. Conf. on Advanes in Numerical Methods in Engineering: Theory & Applications, Swansea, U.K. July 6-10.

Lee, I.K. and Herington, J.R. 1971. Stresses beneath granular embankments. Proc. 1st Australia-New Zealand Conf. Geomech., Vol. 1, pp. 291-297.

Mustoe, G.G.W., Williams, J.R., Hocking, G. and Worgan, K.J. 1987. Penetration and fracturing of brittle plates under dynamic impact. NUMETA '87, 2nd Int. Conf. on Advances in Numerical Methods in Engineering: Theory & Applications, Swansea, U.K. July 6-10.

Trollope, D.H. and Burman, B.C. 1980. Physical and numerical experiments wit granular wedges. Geotechnique, Vol. 3 No. 2, pp. 137-157.

Trollope, D.H. 1957. The systematic arching theory applied to the stabilit analysis of embankments. Proc. 4th In Conf. S.M. and F.E., London, Vol. 2, pp. 383-388.

Williams, J.R., Hocking, G., and Mustoe, G.G.W 1985. The theoretical basis of the discrete element method. NUMETA ' Numerical Methods in Engineering, Theo and Application, A.A. Balkema Swansea, January 7-11.

Williams, J.R. and Mustoe, G.G.W. 1987. Modal methods for the analysis of discrete systems. Computers & Geotechnics (in press).

Numerical Methods in Geomechanics (Innsbruck 1988), Swoboda (ed.)
© 1988 Balkema, Rotterdam. ISBN 90 6191 809 X

Probabilistic evaluation of soil-structure interaction in liquefaction

Achintya Haldar
Georgia Institute of Technology, Atlanta, USA
Shuh-Gi Chern
National University of Marine Science and Technology, Kellung, Taiwan

ABSTRACT: A method is proposed here to estimate structural damage considering the generation and dissipation characteristics of excess pore water pressure generated due to earthquake shaking. Anisotropic stress is used in the formulation. The residual and consolidation settlement are estimated accordingly. The information on differential settlement is used to develop a damage criterion. The influence of structural rigidity on the structural damage potential is also evaluated. A probabilistic model is developed to study the risk of structural damage since a considerable amount of uncertainty is associated with most of the parameters in the model. The methodology is illustrated with the help of an example.

1 INTRODUCTION

Historical evidence of the enormous damage associated with earthquake-induced liquefaction created a tremendous amount of interest among engineers for almost three decades. Unfortunately, most of the past work relates to the mechanism and parameters associated with liquefaction. However, merely estimating the liquefaction potential does not sufficiently address the problem. Liquefaction does not always cause damage, and a structure can suffer damage without liquefaction. Since damage is the main concern of engineers, controlling or limiting damage should be the design constraint. Very little work has been done in the area of quantification of damage associated with liquefaction. Recently, Haldar [1], Haldar and Luettich [2], and Haldar and Chern [3] tried to address this problem. Moreover, since loss of property is a pressing problem to engineers, liquefaction under anisotropic stress conditions (soil elements beneath a sloping surface or beneath an engineering facility) is more important than isotropic conditions (soil elements beneath level ground), which have attracted more attention in the past. In a very recent publication, Seed [4] suggested that "at the present time, the most prudent method of minimizing the hazards associated with liquefaction-induced sliding and deformations is to plan new construction or devise remedial measures in such a way that either high pore water pressures cannot build up in the potentially liquefiable soil, and thus liquefaction cannot be triggered, or, alternatively, to confine the liquefiable soils by means of stable zones so that no significant deformations can occur; by this means, the difficult problems associated with evaluating the consequences of liquefaction (sliding or deformations) are avoided." This statement clearly indicates the state-of-the-art in this area.

Although the problem is extremely complex, a method is proposed here to quantify damage associated with earthquake-induced liquefaction. Among many possible alternatives, differential settlement is used as the damage criterion. The method is based on the concept that any excess residual pore water pressure generated due to the earthquake shaking will eventually dissipate along some drainage route. The deformation of the soil deposit due to the generation and dissipation of pore water pressure is quantified. The interaction between the structure and the soil through the redistribution of vertical loads due to uneven settlement of the foundations is also considered. Since most of the parameters in the model are random in nature, the model is developed probabilistically. Due to page limitations, many feature of the model cannot be described here in detail; however, they will be discussed during the presentation.

2 PROPOSED METHODOLOGY

The differential settlement, which is the damage criterion used in the proposed model, can be estimated from the information on the total settlement, s_t, at various locations in the deposit. For earthquake-induced liquefaction, the total settlement has two components: the residual settlement, s_d, during an earthquake and the consolidation settlement, s_c, following the earthquake. Both s_d and s_c depend on the amount of excess pore water pressure developed during the earthquake shaking under anisotropic stress conditions. Thus, a pore pressure prediction model is a must at this stage.

2.1 Pore pressure generation model

Two phenomena take place simultaneously during an earthquake: (1) excess hydrostatic pressure, u_g, is generated in the sand by cyclic stresses induced by earthquake ground motions; and (2) the excess hydrostatic pressures tend to dissipate, with accompanying volume change, in accordance with the normal laws of diffusion. The general governing equation in three dimensions can be expressed as

$$\{\nabla^T\}\ [k]\ \{\nabla \frac{u}{\gamma_w}\} = m_v\ (\frac{\partial u}{\partial t} - \psi) \qquad (1)$$

where

$$\psi = \frac{\partial u_g}{\partial N}\frac{\partial N}{\partial t}\ ; \quad [k] = \begin{bmatrix} k_x & 0 & 0 \\ 0 & k_y & 0 \\ 0 & 0 & k_z \end{bmatrix} ;$$

$$\{\nabla^T\} = (\frac{\partial}{\partial x}, \frac{\partial}{\partial y}, \frac{\partial}{\partial z})^T ;$$

k = the coefficient of permeability; u = the excess pore-water pressure; γ_w = the unit weight of water; and m_v = the coefficient of volume compressibility. Equation (1) is a diffusion equation with a source term $\psi = \frac{\partial u_g}{\partial N}\frac{\partial N}{\partial t}$. $\frac{\partial u_g}{\partial N}$ can be estimated as [3].

$$\frac{\partial u_g}{\partial N} = \frac{u_f}{\pi}\frac{1}{\sqrt{1-((\frac{N}{N^*_{50}})^{1/\alpha^*}-1)^2}}\frac{1}{\alpha^* N^*_{50}}$$

$$\times (\frac{N}{N^*_{50}})^{\frac{1-\alpha^*}{\alpha^*}} = \frac{u_f}{\pi\alpha^* N^*_{50}}$$

$$\times \frac{1}{\sqrt{2(\frac{N}{N^*_{50}})^{1/\alpha^*} - (\frac{N}{N^*_{50}})^{2/\alpha^*}}}(\frac{N}{N^*_{50}})^{\frac{1-\alpha^*}{\alpha^*}} \qquad (2)$$

where u_f = the limiting value of residual pore pressure that can possibly occur for a given consolidation stress ratio; N^*_{50} = the number of cycles where the value of pore pressure build-up is 50% of the value of u_f; and α^* = a parameter whose value depends on both the consolidation stress ratio and the relative density. $\partial N/\partial t$ can be estimated as

$$\partial N/\partial t = N_{eq}/t_d\ , \qquad 0 < t \leq t_d \qquad (3)$$
$$= 0 \qquad \text{otherwise}$$

where N_{eq} = the equivalent cycles of the strong motion of an earthquake; and t_d = the duration of the strong motion.

For low values of the pore water pressure ratio, Seed et al [5] observed that m_v is fairly constant. However, for pore water pressure ratios larger than about 60%, m_v depends on the relative density and the pore pressure ratio. Seed et al [5] proposed a relationship to evaluate m_v as

$$\frac{m_v}{m_{v_o}} = \frac{e^{Ar_u^B}}{1 + Ar_u^B + \frac{1}{2}A^2 r_u^{2B}} \qquad (4)$$

where $A = 5(1.5 - D_r)$; $B = 3(2)^{-2Dr}$;

and m_{v_o} = the compressibility before excess pore water was generated in sands.

Equation (1) is formulated in terms of the minimum of the function:

$$\pi = \int\int\int \left\{ \frac{1}{2} \left[k_x \left(\frac{\partial u}{\partial x}\right)^2 + k_y \left(\frac{\partial u}{\partial y}\right)^2 + k_z \left(\frac{\partial u}{\partial z}\right)^2\right] + m_v \gamma_w \left[\left(\frac{\partial u}{\partial t} - \left(\frac{\partial u_g}{\partial t}\right)\right)u\right] \right\} \quad (5)$$

The solution for equation (5) can be sought if u is continuous and satisfies all the boundary conditions.

Consider a volume of soil V with surface area S. Suppose that a portion of the surface, S_D, is free to drain so that on S_D the excess pore pressure u will be zero. Suppose also that the remainder of the surface, S_I, is impermeable so that the component of the pore water velocity vector normal to S_I will vanish. In this study, the water table surface is assumed to be free to drain, i.e., the water table surface is the surface S_D. Thus, the pore pressure u will be zero on the water table surface. For a soil body under consideration, two sides and the bottom of the mass are considered impermeable, while the excess pore pressures in the media within the mass are considered to be continuous. Moreover, before the earthquake loadings are applied, the pore pressures in the media are assumed to be zero, i.e., u = 0 at t = 0. A solution to find a pore pressure field u, which satisfies equation (5) and these boundary conditions, can be sought if equation (5) is formulated in a variational form as:

$$\int \left(\frac{1}{\gamma_w} \{\nabla \delta u\}^T [k] \{\nabla u\} + m_{v3} \delta u\left(\frac{\partial u}{\partial u} - \psi\right)\right) dv = 0 \quad (6)$$

Equation (6) can be solved by the finite element method. The quadrilateral elements are used for this purpose. The details of this formulation cannot be shown here due to lack of space. A computer program was developed for this purpose.

Thus far, the generation of pore pressures during the earthquake loading has been discussed. After the earthquake has stopped, pore water pressure generation ceases and the saturated sand deposit begins to deform due to the dissipation of the pore water pressure. This phase can be modeled adequately by considering equation (1) without the porepressure generation term. It was also suggested [6] that during the pore pressure dissipation, , the value of m_v would remain constant and equal to the maximum value reached during pore pressure build-up.

This situation also arises during the excitation for the layers located above any layer which develops a condition of initial liquefaction. In such a case, shear stresses would no longer be transmitted above the liquefied zone, and thus the pore pressure distribution would be identical to the dissipation case.

2.2 Residual and consolidation settlements

Once the pore pressure generation and dissipation models are available, the residual settlement during an earthquake and the consolidation settlement can be estimated in the following way.

The change in vertical strain, $\Delta\varepsilon_1$, caused by the pore pressure increment, Δu, for anisotropic samples can be shown to be [7]:

$$\frac{\Delta\varepsilon_1}{\Delta u} = \frac{R_f \left(\dfrac{\sigma'_d}{\sigma'_{ult} - \sigma'_d} \right)^2 \left(\dfrac{2 \sin \phi'}{1 - \sin \phi'} \right)}{K\, P_a \, (\sigma'_3/P_a)^n} + \frac{n \left(\dfrac{\sigma'_d}{\sigma'_{ult} - \sigma'_d} \right) \left(\dfrac{\sigma'_{ult}}{\sigma'_3} \right)}{K\, P_a \, (\sigma'_3 / P_a)^n} \quad (7)$$

in which $\sigma'_3 = \sigma'_{3c} - u$; P = atmospheric pressure; σ'_d = deviatic stress = $(K_c - 1)\,\sigma'_{3c}$, which is assumed to be constant; ϕ' = friction angle; K, n, R_f = soil parameters that can be estimated from a set of static consolidated drained triaxial tests; and

$$\sigma'_{ult} = \frac{1}{R_f} \left[\sigma'_3 \frac{2 \sin \phi'}{1 - \sin \phi'}\right] \quad (8)$$

As discussed earlier, the pore water pressure build-up in anisotropically consolidated sand deposits can be estimated. Consequently, the undrained residual strain between two consecutive loading cycles can also be obtained by using equation (7). After the vertical strain is accumulated to the Nth cycle using equation (7), the residual settlement, s_d, can be determined provided the thickness of the soil layer, h, is known, i.e.,

$$s_d = h \cdot \varepsilon_1 \quad (9)$$

Assuming that the sand layer is compressible, that no later deformation is possible during the dissipation of excess pore water pressure, that there is enough time for the pore pressure to dissipate, and that during the pore pressure dissipation the volume compressibility m_v remains constant and equal to the maxmimum value reached during the pore water pressure build-up, then the consolidation settlement of the layer can be obtained as

$$s_c = m_v h\, u \quad (10)$$

All the parameters were described earlier.

2.3 Soil-structure interaction

In the previous section, the loading system was idealized as a set of independent loads applied at ground level, and structural continuity was ignored. If a particular foundation of a column of the structure is very heavily loaded, then the settlement underneath it is expected to be large. This will cause a redistribution of forces and part of the load will be transferred to less stressed support points, thus changing the settlement profile. In some cases, the rigidity of a structure will influence its settlement characteristics.

A methodology is proposed here for structures supported on shallow foundations. The method considers only the consolidation settlement after the earthquake has ceased. Elastic structural behavior is implied in the analysis.

The redistribution of vertical loads is modeled in terms of load transfer coefficients [8]. These are structure-dependent parameters, and are elastic constants of the entire structure which can be easily calculated using structural theories for statically indeterminate structures. They can be represented as a square matrix T of order n for a structure supported on n points.

For a two-bay, three-column framed structure underlain by a single compressible sublayer, the soil-structure equations can be written as

$$\begin{bmatrix} f_{11} & 0 & 0 \\ 0 & f_{22} & 0 \\ 0 & 0 & f_{33} \end{bmatrix} \begin{bmatrix} \alpha_{11} & \alpha_{12} & \alpha_{13} \\ \alpha_{21} & \alpha_{22} & \alpha_{23} \\ \alpha_{31} & \alpha_{32} & \alpha_{33} \end{bmatrix} \begin{bmatrix} Q_1 \\ Q_2 \\ Q_3 \end{bmatrix} + \begin{bmatrix} T_{11} & T_{12} & T_{13} \\ T_{21} & T_{22} & T_{23} \\ T_{31} & T_{32} & T_{33} \end{bmatrix} \begin{bmatrix} s_1 \\ s_2 \\ s_3 \end{bmatrix} = \begin{bmatrix} s_1 \\ s_2 \\ s_3 \end{bmatrix} \quad (11)$$

in which f_{jj} = soil flexibility at mid-depth in the layer under support j; α_{ij} = stress coefficient under support i due to column load j; and s_j = settlement under support j.

Assuming that the soil layer has reached equilibrium after the earthquake has ceased, the only vertical stress induced in the soil layer is the increase in the effective stress, p', due to the dissipation of pore water pressure u. The increased p'_i will be equal to u_i beneath a column i. If s_{ci} is the calculated consolidation settlement of column i, then the redistribution stress due to the influence of structural rigidity beneath column i must be corrected as:

$$q_i = Q_i/A_i = u_i + \frac{1}{A_i} [T_{ii} s_{ci} + \Sigma\, T_{ij} s_{cj}], \quad i \neq j \quad (12)$$

in which A_i = ith footing area, and $p'_i = u_i$.

The compressive stress at a point underneath column i can be shown to be:

$$\sigma_i = \alpha_{ii} q_i + \sum_{j=1}^{n} \alpha_{ij} q_j, \quad i \neq j \quad (13)$$

The consolidated settlement beneath column i will be:

$$s_{ci} = m_{vi}\, h_i [\alpha_{ii} q_i + \sum_{j=1}^{n} \alpha_{ij} q_j] \tag{14}$$

If the assumed compressible sand deposit is divided into k sublayers, the general form of equation (14) may be formulated in matrix form as:

$$(\sum_{i=1}^{k} m_{vi} h_i \alpha_i)\ (\underset{\sim}{p}' + t\ \underset{\sim}{s}) = \underset{\sim}{s} \tag{15}$$

in which $\underset{\sim}{x}$ = the vector with n elements if n columns are considered; and $t = T_{ij}/A_i$.

$$\text{Let } (\sum_{i=1}^{k} m_{vi} h_i \alpha_i) = \omega \tag{16}$$

Then equation (15) becomes

$$\omega\ (\underset{\sim}{p}' + t\ \underset{\sim}{s}) = \underset{\sim}{s} \tag{17}$$

and s can be solved by rewriting equation (17), i.e.,

$$\underset{\sim}{s} = (I - \omega\ t)^{-1} \omega\ \underset{\sim}{p}' \tag{18}$$

in which I is an n x n unit matrix. Evaluating equation (18) at the mean values of all the parameters involved, the mean consolidation settlement considering the soil-structure interaction can be estimated.

2.4 Structural damage

The method proposed here can be applied to estimate structural damage. As mentioned earlier, the structural damage can be estimated in terms of the induced maximum differential settlement, δ_{max}, measured from the deformed shape of the foundation after the uniform settlement and the tiltcomponents have been removed. For the three-footing structure shown in Figure 1, and assuming that the middle support settles more than the exterior supports, δ_{max} can be approximated as

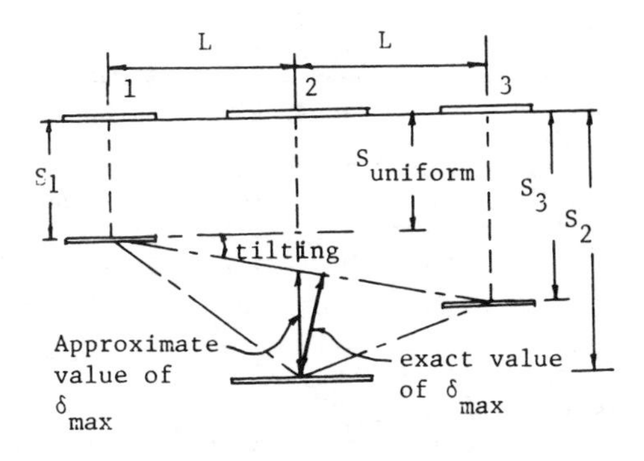

Figure 1. Maximum differential settlement

$$\delta_{max} = S_2 - \frac{1}{2}(S_1 + S_3) \tag{19}$$

Within a probabilistic framework, δ_{max} is a random variable and can be denoted as Δ_{max}. The mean and the variance of Δ_{max} can be obtained in terms of the means and variances of the total settlements assuming that the settlements are independent of each other. Thus,

$$E(\Delta_{max}) = E\ (S_2) - \frac{1}{2}[E\ (S_1) + E\ (S_3)] \tag{20}$$

and

$$Var(\Delta_{max}) = Var(S_2) + \frac{1}{4}[Var\ (S_1) + Var(S_3)] \tag{21}$$

Detailed estimation procedures for the means and variances of S_i's cannot be shown here due to lack of space. Once the statistics of Δ_{max} are known, the damage potential of a structure can be evaluated if the allowable differential settlement, δ_{all}, of the structure is given.

3. EXAMPLE

A structure with three symmetrically loaded footings as shown in Figure 1 is considered here. The allowable bearing pressure is 2000 ksf. The structure is assumed to be sitting on a level ground surface of a hypothetical site having soil properties similar to Oosterschelde sand. The site consists of seven sublayers. All the sublayers are assumed to be homogeneous. The site is subjected to an earthquake of magnitude 7.5 for a duration of 30 seconds and an estimated acceleration of 0.20g at the ground surface.

A finite element mesh consisting of forty-two quadrilateral elements are used to obtain a numerical solution. Initially, each element is considered to have compressibility of 1.0×10^{-6} ft^2/lb and vertical permeability of 3.28×10^{-4} ft/sec. The footing dimensions are 8 ft x 8 ft for the exterior columns and B ft x B ft for the interior column. The example will be described in more detail during the presentation.

Considering that the serviceability constraint requires limiting the maximum net slope $(\delta/L)_{max}$ to a value less than or equal to 1/300, the probability of structural damage can be expressed as

$$P(\Delta_{max} \geq \delta_{all}) =)\ (\Delta_{max} \geq 0.8\text{"}\ \Big|\ L = 20') \qquad (22)$$

In Table 1, the probability of structural damage as a function of interior footing width B is given for two cases: when the structural rigidity is considered and when it is neglected. The probability of structural damage increases as B increases and decreases as the soil-structure interaction effect is considered. Other major observation will be discussed during the presentation.

Table 1. Probabilistic Parameters of Δ_{max} as Functions of Interior Footing Width B

Interior Footing width B (ft)	8	10	12	14
$E(\Delta_{max})$ (in)	0.725	1.015	1.248	1.626
	0.651*	0.902*	1.101*	1.500*
$\sigma_{\Delta_{max}}$ (in)	0.450	0.448	0.447	0.455
$P(\Delta_{max} \geq 0.8")$	0.436	0.751	0.841	0.968
	0.380*	0.587*	0.750*	0.942*

*means structure rigidity is considered.

ACKNOWLEDGEMENT

This material is based upon work partly supported by the National Science Foundation under Grants No. CEE-8312181, MSM-8352396, MSM-8544166, MSM-8644348 and MSM-8746111. Any opinions, findings, and conclusions or recommendations expressed in this publication are those of the writers and do not necessarily reflect the views of the National Science Foundation.

REFERENCES

[1] A. Haldar, Probabilistic evaluation of damage potential in earthquake-induced liquefacation in a 3-D soil deposit, Report No. SCEGIT-83-117, Georgia Institute of Technology, 1983.

[2] A. Haldar and S. M. Luettich, Subsidence approach to damage in earthquake-induced liquefaction, Report No. SCEGIT-85-106, Georgia Institute of Technology, 1985.

[3] A. Haldar and S. Chern, Probabilistic analysis of pore pressure-induced damage potential for structures subjected to earthquake motions, Report No. SCEGIT-86-113, Georgia Institute of Technology, 1986.

[4] H. B. Seed, Design problems in soil liquefaction, J. of Geotechnical Engineering, ASCE, Vol. 113, No. 8, 827-845, 1987.

[5] H. B. Seed, P. P. Martin, and J. Lysmer, Pore-water pressure changes during soil liquefaction, J. of Geotechnical Engineering, ASCE, Vol. 102, No. GT4, 1976.

[6] H. B. Seed, Evaluation of soil liquefaction effects of level ground during earthquakes, ASCE Annual Convention and Exposition, Philadelphia, Pa., 1976.

[7] C. S. Chang, Residual deformation of undrained samples during cyclic loading, J. of Geotechnical Engineering, ASCE, Vol. 108, No. GT4, 637-646, 1982.

[8] S. Chamecki, Structural rigidity in calculating settlements, J. of Soil Mechanics and Foundation Division, ASCE, Vol. 82, No. SM1, 1-19, 1956.

Numerical Methods in Geomechanics (Innsbruck 1988), Swoboda (ed.)
© 1988 Balkema, Rotterdam. ISBN 90 6191 809 X

Two-dimensional seismic response analysis based on multiple yield surface model

K.Kabilamany
University of California, Davis, USA
K.Ishihara
University of Tokyo, Japan

ABSTRACT: The constitutive equations based on the multiple yield surfaces are described briefly and incorporated into the numerical algorithm to conduct 2-D seismic response analysis. The numerical scheme was used to delineate the behavior of submerged sand embankment model tested in the centrifuge.

INTRODUCTION

The constitutive laws based on the multiple shear mechanism which were developed by Aubry et al. (1982) appear to have an advantage in that it considerably simplifies the numerical implementation particularly for 2-D plane strain condition. Some modifications are introduced in the present study into this model for further refinement. This renewed model was integrated into the 2-D numerical computer program, Diana, for conducting seismic response analysis.

STRESS DILATANCY RELATIONSHIP

The actual plastic strains are assumed to consist of summation of three strain components which are produced in three independent shear mechanisms designated by k = 1,2,3. Then, the plastic strain increment in i-th direction, $\dot{\varepsilon}^p_i$, is decomposed as,

$$\dot{\varepsilon}^p_i = (\dot{\varepsilon}^p_i)_1 + (\dot{\varepsilon}^p_i)_2 + (\dot{\varepsilon}^p_i)_3 + (\dot{\varepsilon}^p_i)_4 \quad \cdots\cdots (1)$$

where $(\dot{\varepsilon}^p_i)_4$ indicates the plastic strain component resulting from the pure compression. The volumetric plastic strain due to dilatancy in k-th mechanism, $(\dot{\varepsilon}^p_{vd})_k$, and the plastic shear strain in k-th mechanism, $(\dot{\bar{\varepsilon}}^p)_k$, are defined as,

$$(\dot{\varepsilon}^p_{vd})_k = (\dot{\varepsilon}^p_i)_k + (\dot{\varepsilon}^p_j)_k,$$

$$(\dot{\bar{\varepsilon}}^p)_k = (\dot{\varepsilon}^p_i)_k - (\dot{\varepsilon}^p_j) \quad \cdots\cdots (2)$$

where i ≠ j ≠ k. It is assumed that the plastic strain in the direction of k-th mechanism is always equal to zero, i.e.,

$$(\dot{\varepsilon}^p_i)_k = (\dot{\varepsilon}^p_j)_k = 0 \quad \text{for} \quad i = k$$

$$\text{or} \quad j = k \quad (3)$$

The above condition indicates that the plastic strain can take place in each mechanism under the condition of plane strain.

The variables specifying the state of stress in each mechanism, k, are chosen as,

$$p_k = \frac{1}{2}(\sigma_i + \sigma_j),$$

$$q_k = \frac{1}{2}(\sigma_i - \sigma_j) \quad \cdots\cdots (3)$$

With the definitions of stress and strain as above, the increment of plastic shear work, dD_k, normalized to p_k is obtained as

$$d\Omega_k = \frac{dD_k}{p_k} = (\dot{\varepsilon}^p_{vd})_k + \frac{q_k}{p_k}(\dot{\bar{\varepsilon}}^p)_k \quad \cdots\cdots (4)$$

where $d\Omega_k$ is the normalized shear work suggested by Moroto(1976). For the sand consisting of rigid particles, it may be assumed that

$$d\Omega_k = \mu_k(\dot{\bar{\varepsilon}}^p)_k \quad \cdots\cdots (5)$$

where μ_k is a material parameter which is shown to be independent of stress history, anisotropy and density of the sand. The value of μ_k tends to increase slightly with increasing shear strain, approaching a value at the phase transformation when the shear strain grows to a few percent. Introducing Eq.(5) into Eq.(4), one obtains a flow rule as follows which holds true within the realm of each mechanism,

$$\frac{(\dot{\varepsilon}_i^p)_k + (\dot{\varepsilon}_j^p)_k}{\mu_k - \zeta_k \cdot q_k/p_k} = \frac{(\dot{\varepsilon}_i^p)_k - (\dot{\varepsilon}_j^p)_k}{\zeta_k} = \lambda_k \quad \cdots\cdots (6)$$

$$\zeta_k = q_k/|q_k|$$

where ζ_k takes a value of +1 when q_k is positive and $\zeta_k = -1$ when q_k is negative. Eq.(6) leads to the expressions as follows for individual strain component in k-th mechanism.

$$(\dot{\varepsilon}_i^p)_k = \frac{\lambda_k}{2}\left(\zeta_k + \mu_k - \frac{q_k}{p_k}\zeta_k\right)$$

$$(\dot{\varepsilon}_j^p)_k = \frac{\lambda_k}{2}\left(-\zeta_k + \mu_k - \frac{q_k}{p_k}\zeta_k\right) \quad \cdots\cdots (7)$$

In the triaxial stress condition, two mechanisms, k = 2 and 3, are considered to be activated equally while the mechanism k = 1 in the axial direction is not activated. Therefore, the value of $\mu_2 = \mu_3$ is a key parameter for specifying the flow rule in the mechanism, k = 2 and 3.

On the other hand, through the same line of reasoning as indicated in Eqs.(4) and (5), the relationship between the increment of normalized shear work $d\Omega$ and shear strain $\dot{\bar{\varepsilon}}^p$ in the global variables can be postulated as,

$$d\dot{\Omega} = \mu \dot{\bar{\varepsilon}}^p \quad \cdots\cdots (8)$$

where μ is a material parameter which can be determined by plotting the triaxial test results in terms of Ω versus $\bar{\varepsilon}^p$ (Ghaboussi and Momen, 1982). For the undrained triaxial condition, it is shown that there exist a unique relationship between the values of μ_3 and μ as follows,

$$\mu_3 = \frac{3\mu}{\mu + 6} \quad \cdots\cdots (9)$$

The analyses of many triaxial test results have indicated that the value of μ can be numerically given by the following empirical formula,

$$\mu = \mu_0 + \frac{2}{\pi}(M - \mu_0)\tan^{-1}\left(\frac{3\bar{\varepsilon}^p}{2s_c}\right) \quad \cdots\cdots (10)$$

where μ_0 and M are the values of μ at small and large strains, respectively. S_c is a constant which takes a value of 0.0035 for Fuji river sand.

YIELD FUNCTION

The yield function in k-th mechanism is assumed to take a form,

$$f_k(\sigma_{ij}, h_k) = q_k\zeta_k - p_k \sin\phi_f h_k = 0 \quad \cdots\cdots (11)$$

where ϕ_f is the angle of internal friction at failure and the function, h_k, allows for the effects of hardening due to the evolvement of plastic shear strain. Since the stress ratio, q_k/p_k, is assumed to become equal to $\sin\phi_f$ at failure, the function, h_k, varies in the range $0 \leqq h_k \leqq 1.0$. The yield function of the type of Eq.(11) was used by Pietruszczak and Poorooshasb (1983) in their constitutive model for the sand.

HARDENING RULE

It is assumed with good reasons that the increment in the hardening parameter, Δh_k, is related to the increment of the normalized shear work, $\Delta\Omega_k$, as follows.

$$\frac{dh_k}{d\Omega_k} = \frac{(1 - h_k)^{1/2}}{a} \quad \cdots\cdots (12)$$

Thus, with reference to Eq.(5), the rule for evolution of h_k can be obtained as,

$$\frac{dh_k}{d(\bar{\varepsilon}^p)_k} = \mu_k \frac{(1 - h_k)^{1/2}}{a} \quad \cdots\cdots (13)$$

where a is a hardening parameter.

ELASTIC DEFORMATION

The total strain increment is composed of elastic part and plastic part. The elastic bulk modulus, K, and the elastic shear modulus, G, are assumed to be functions of the current value of the mean principal stress, p, and given by

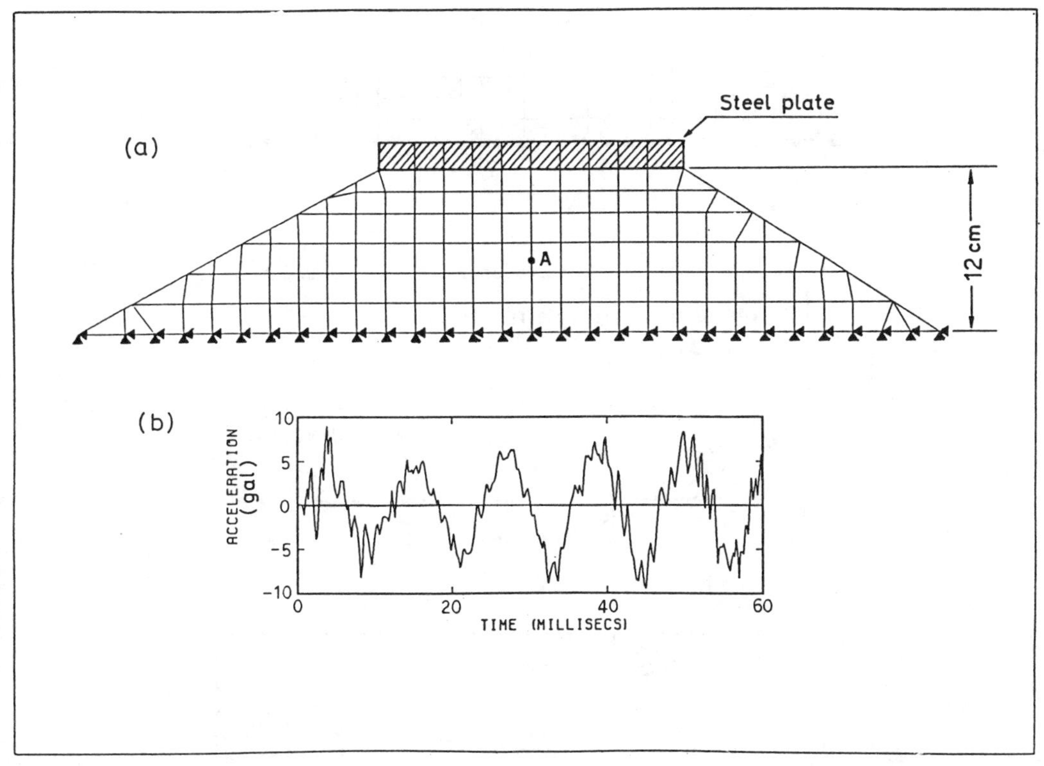

Fig. 1 Finite element grids and input motions

$$K = K_0(\frac{p}{p_0})^{1/2} \ , \ G = G_0(\frac{p}{p_0})^{1/2} \cdots (14)$$

where p is the mean principal stress at which K and G are measured.

CONSISTENCY CONDITION

This is the condition requiring any stress point to remain on the yield surface as the stresses are changed in the course of load application. The consistency equation is obtained by differentiating the yield function in Eq.(11),

$$f_k = \frac{\partial f_k}{\partial \sigma_{ij}} \dot{\sigma}_{ij} + \frac{\partial f_k}{\partial h_k} \frac{\partial h_k}{\partial (\bar{\varepsilon}^p)_k} (\dot{\bar{\varepsilon}}^p) = 0 \quad \cdots\cdots(15)$$

Introducing Eqs.(6) and (13) into Eq.(15), one obtains a set of equation for unknown variable, λ_k, for each mechanism. Once the value of λ_k is determined in each step of load increment, the corresponding plastic strain component is obtained for each mechanism. The plastic strains in each mechanism thus obtained are added to yield the actual plastic strains. The details of numerical scheme for determining the value of λ_k are different depending upon whether the numerical calculation is stress-controlled or strain-controlled, but the procedure is essentially the same as that suggested by Lade and Nelson (1984).

UNLOAD-RELOAD CYCLES

The set of constitutive equations as developed above for the monotonic loading condition is extended to cover the deformation behavior in unloading and reloading branches in a time history of cyclic load application. The basic principle for doing this is what is called Masing rule in which a skeleton curve of stress-strain relation in monotonic virgin loading is enlarged to an appropriate scale and attached to points of stress reversal to form a stress-strain curve for

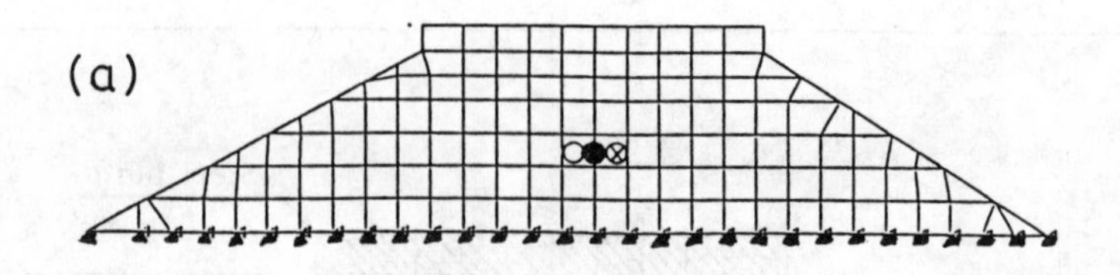

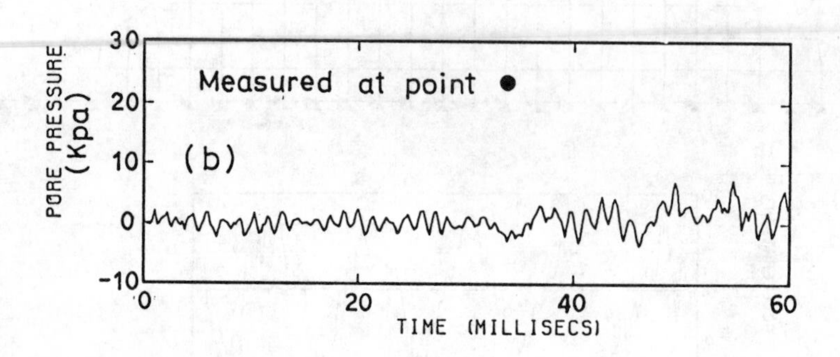

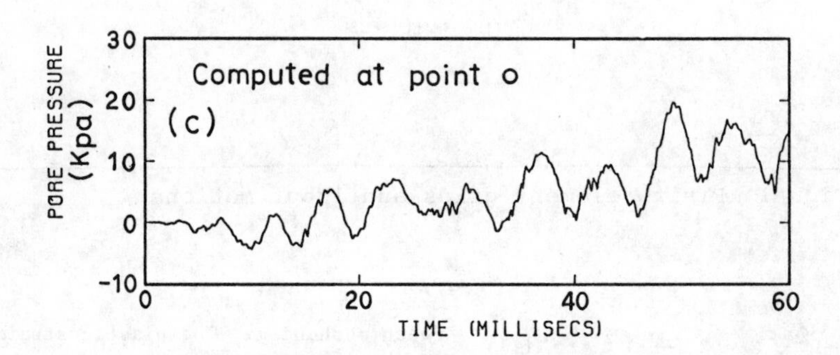

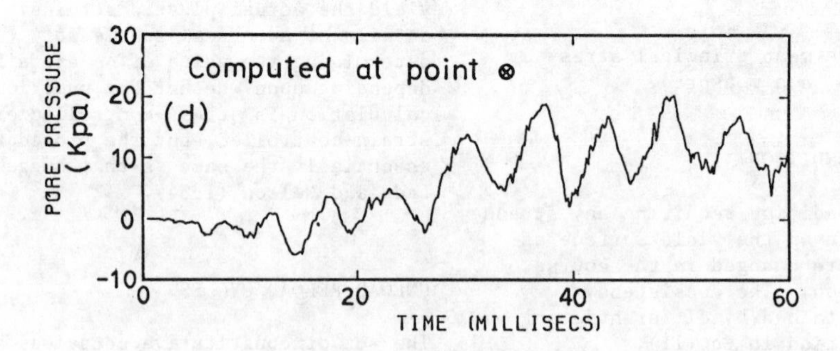

Fig. 2 Comparison between computed and measured pore water pressures.

Table 1 Material constants in the model

Elastic constant:
$K_0 = G_0 = 37500$ kpa
at $P_0 = 100$ kpa
Constants in the model:
$\mu_f = 30^\circ$
$\mu_0 = 0.45$
$M = 1.42$
$a = 4 \times 10^{-4}$
$a_r = 8 \times 10^{-6}$

the forthcoming stress changes in unload or reload phase. In the unload-reload cycles, a hardening parameter, a_r, which is different from the parameter in the virgin loading was adopted. The details of numerical implementation scheme are not described herein.

NUMERICAL ANALYSES AND COMPARISON WITH TEST RESULTS

The constitutive model outlined above was incorporated into the two-dimensional numerical analysis computer program, Diana, developed at the University of Swansea (Zienkiewicz and Shiomi, 1984). This analysis procedure is used to calculate the non-linear response of a model embankment of sand tested in a centrifuge at the Cambridge University. A model bank 12 cm in height capped with a rigid steel plate as shown in Fig.1(a) was subjected to a shaking during its flight in the centrifuge. The model bank composed of fine uniform sand was prepared with a relative density of about 60%, submerged under water and made to undergo a shaking having on acceleration time history as shown in Fig.1(b). The pore water pressure recorded at a point A beneath the center of the cap plate is shown in Fig.2(b). Its maximum corresponds to about 50% of the effective overburden pressure at that point. The numerical analysis was made for the finite element grid shown in Fig.1(a). First of all, the static stress analysis was made to determine the state of stress prior to the cyclic shaking. Then the dynamic analysis was conducted by applying the motion shown in Fig.1(b) at the base and the pore water pressure was calculated by the step-by-step integration technique using the Diana program. The material constants assumed for the constitutive modeling are shown in Table 1. The pore pressures calculated at the finite element grids in the vicinity of point A are shown in Figs.2(c) and 2(d). The comparison of these values with the measured pore water pressures shown in Fig.2(b) indicates that the agreement is qualitatively satisfactory although there is some degree of discrepancy. At present, without knowing exact relative density of sand at which the centrifuge test was conducted, any efforts for comparison in further detail will not be rewarding.

REFERENCES

Aubry, D., Hujeux, J.C., Lassoudiere, F., and Meimon, Y. (1982)," A Double Memory Model with Multiple Mechanisms for Cyclic Soil Behavior," Proc. of the International Symposium on Numerical Models in Geomechanics, Zurich, pp.3-13

Ghaboussi, J. and Momen, H. (1982),"Stress Dilatancy and Normalized Work for Sands," IUTAM Conference on Deformation and Failure of Granular Materials, Delft, pp.265-274

Lade P.V. and Nelson R. B. (1984), "Incrementalization Procedure for Elasto-Plastic Constitutive Model with Multiple, Intersecting Yield Surfaces," International Journal for Numerical and Analytical Methods in Geomechanics, Vol.8, PP.311-323.

Moroto, N. (1976), "A New Parameter to Measure Degree of Shear Deformation of Granular Material in Triaxial Compression Tests," Soils and Foundations, Vol.16, No.4, PP.1-9.

Zienkiewicz, O.C. and Shiomi, T. (1984), "International Journal for Numerical and Analytical Methods in Geomechanics, Vol.8, PP.71-96.

ACKNOWLEDGEMENTS

Preparation of this paper was greatly assisted by Dr. T. Shiomi of Tadenaka Co. to whom acknowledgement is extended.

Numerical Methods in Geomechanics (Innsbruck 1988), Swoboda (ed.)
© 1988 Balkema, Rotterdam. ISBN 90 6191 809 X

Random response of structures to earthquake

A.Y.T.Leung
Department of Civil and Structural Engineering, University of Hong Kong

ABSTRACT: Earthquake are naturally nonstationary stochastic. To a good approximation, the description can be simplified a great deal by means of time evolution random processes. A recently developed method for single-degree-of-freedom system response to exponentially modulated random excitation is extended to multi-degree-of-freedom systems. The evolutionary function is also extended to a more common type, $(a+bt)e^{\beta t}$. Exact integration formulae for the evolutionary mean square responses are given explicitly. It is shown that the usual approximation by decomposition to background and resonance parts can be unacceptably erroneous.

1. INTRODUCTION

For many buildings, dams, bridges and elevated highways, a major design loading is that due to an anticipated earthquake. An important characteristic of the earthquake motion is its non-stationary nature. Investigations of response statistics for non-stationary random excitation have centered on frequency domain analysis with use of the power spectral density function [R.L.Barnowski and J.R.Maurer 1969, T.K.Caughey and H.J.Stumpf 1961, R.B.Corotis and E.H.Vanmarcke 1975, R.B.Corotis, E.H.Vanmarcke and C.A. Cornell 1972, M.Shinozuka and J.H.Yang 1969, J.S.Bendat and A.G.Piersol 1966]. Mean-square response and reliability of a system relative to a specified response level are of major interest.

Due to the complexity of the analysis, time modulated stationary random excition is usually assumed. It is the purpose of this paper to point out that if the modulating function is exponential, most response formulae for stationary excitation remain valid, with slight modification. This is possible because formulae for harmonic excitation of the type $e^{i\omega t}$ can be extended to exponentially varying harmonic excitation of the type $e^{\alpha t}$, $\alpha=i\omega+\beta$, simply by replacing $i\omega$ by α [A.Y.T.Leung 1985].

2. DYNAMIC FLEXIBILITY AND STEADY STATE FLEXIBILITY

The steady state response of a system is that due to an excitation which has been continuously applied for all time up to the present. The factor which is to be multiplied by the excitation amplitude to obtain the steady state response amplitude is called the steady state flexibility. If an arbitrary point on the time axis is taken as the initial time then non-zero initial conditions will be obtained, in general. If zero initial conditions are enforced, the dynamic response will be a superposition of transients and the steady state response. The dynamic flexibility is the excitation and response relation when zero initial conditions are enforced.

The response of the system

$$\ddot{u} + 2\zeta\omega_1\dot{u} + \omega_1^2 u = f(t),$$
$$u(0)=u_0 \text{ and } \dot{u}(0)=\dot{u}_0, \tag{1}$$

is conventionally given by

$$u(t) = u_d(t) + g(t)u_0 + h(t)\dot{u}_0, \tag{2}$$

where $u_d(t) = \int_0^t h(t-\tau)f(\tau)d\tau$, $h(t)=\lambda^{-1}e^{-\zeta\omega_1 t}\sin\lambda t$, $g(t) = e^{-\zeta\omega_1 t}(\cos\lambda t + \zeta\omega_1\lambda^{-1}\sin\lambda t)$, and $\lambda^2=(1-\zeta^2)\omega_1^2$ are the Duhamel integral, the impulse response, the indicial response and the damped natural frequency, respectively. When f(t) is harmonic, the steady state $u_s(t)$ can be found by assuming

$$f(t) = Fe^{i\omega t}, \; u_s(t) = Ue^{i\omega t} \quad (3)$$

to obtain

$$U = F/(\omega_1^2-\omega^2+2\zeta\omega\omega_1), \quad (4)$$

whose polar form gives the magnitude and phase angle. With the initial conditions accounted for, the complete solution is given by

$$u(t)=u_s(t)+g(t)[u_0-u_s(0)]+h(t)[\dot{u}_0-\dot{u}_s(0)]. \quad (5)$$

Since the complete solution is unique, one obtains a relationship between the Duhamel integral $u_d(t)$ and the steady state response $u_s(t)$ by comparing equations (2) and (5):

$$u_d(t)=u_s(t)-g(t)u_s(0)-h(t)\dot{u}_s(0). \quad (6)$$

Usually, the steady state response is simpler than the Duhamel integral. If required, the Duhamel integral can be obtained by equation (6).

The steady state method can be extended to exponentially varying harmonic excitation simply by letting

$$f(t) = Fe^{\alpha t} = F(\cos\omega t+i\sin\omega t)e^{\beta t} \quad (7)$$

and $u_s(t) = Ue^{\alpha t}$, $\alpha = i\omega+\beta$, to obtain

$$U = H(\alpha)F, \; H(\alpha) = 1/(\alpha^2+2\zeta\omega_1\alpha+\omega_1^2), \quad (8)$$

and the complete solution has the same form as equation (5).

If the system is initially at rest, then

$$u(t) = u_d(t) = u_s(t)-g(t)u_s(0)-h(t)\dot{u}_s(0)$$
$$= z(\alpha,t)f(t). \quad (9)$$

The relationship between the dynamic flexibility $z(\alpha,t)$ and the steady state flexibility $H(\alpha)$ is obtained by comparing equations (8) and (9);

$$z(\alpha,t) = H(\alpha)[1 - e^{-\alpha t}\{g(t)+\alpha h(t)\}]. \quad (10)$$

This can further be extended to compute the steady state response subject to the excitation

$$f(t) = (\bar{F} + Ft)(\cos\omega t+i\sin\omega t)e^{\beta t}$$
$$= (\bar{F} + Ft)e^{\alpha t} \quad (11)$$

where $\alpha=i\omega+\beta$ again. Assuming the steady state response be

$$u_s(t) = (\bar{U} + Ut)e^{\alpha t} \quad (12)$$

and substituting into equation (1), one has

$$\begin{bmatrix} \alpha^2+2\zeta\omega_1\alpha+\omega_1^2 & 2\zeta(\omega_1+\alpha) \\ 0 & \alpha^2+2\zeta\omega\alpha+\omega_1^2 \end{bmatrix} \begin{Bmatrix} \bar{U} \\ U \end{Bmatrix} = \begin{Bmatrix} \bar{F} \\ F \end{Bmatrix} \quad (13)$$

or $U = H(\alpha)F$ and $\bar{U} = H(\alpha)[\bar{F}-2\zeta(\omega_1+\alpha)H(\alpha)F]$ (14)

The steady state response is completely determined by equation (12). If the system is initially at rest, then

$$u(t) = u_d(t) = u_s(t)-g(t)u_s(0)-h(t)\dot{u}_s(0)$$
$$= u_s(t)-g(t)\bar{U}-h(t)(\alpha\bar{U}+U) \quad (15)$$

Unfortunately, because there are two independent excitation coefficients $\bar{F}$ and F, an explicit expression for the dynamic flexibility similar to equation (10) is not possible.

In the following, an exponentially modulated stationary random process is formulated. Multi-degrees-of-freedom systems are discussed. And the formulation is extended to modulating function of the type $(\bar{F} + Ft)e^{\beta t}$.

3. EXPONENTIALLY EVOLUTIONARY STATIONARY RANDOM PROCESS

Consider the case when the forcing function is an evolutionary stationary random process,

$$f(t) = f_m(t)f_s(t), \quad (16)$$

where $f_m(t)$ is a given modulating function and $f_s(t)$ is a stationary random process with power spectral density $S_s(\omega)$. The autocorrelation of the response is given by

$$R_u(t_1,t_2) = \int_0^{t_1}\int_0^{t_2} h(t_1-\tau_1)h(t_2-\tau_2)f_m(\tau_1) f_m(\tau_2)R_s(\tau_1-\tau_2)d\tau_1 d\tau_2$$
$$= \int_{-\infty}^{\infty} S_s(\omega)I(\omega,t_1)I^*(\omega,t_2)d\omega, \quad (17)$$

where * denotes the complex conjugate and

$$I(\omega,t) = \int_0^t h(t-\tau)f_m(\tau)e^{i\omega t}d\tau \quad (18)$$

is the response to $f_m(\tau)e^{i\omega t}$ with $u(0)=\dot{u}(0) = 0$: i.e., the Duhamel integral. In particular, if $f_m(\tau)=e^{\beta\tau}$ then

$$I(\omega,t) = z(\beta+i\omega,t)e^{\beta t} = z(\alpha,t)e^{\beta t}. \quad (19)$$

The mean square response $\sigma_u^2(t)$ is usually of importance and is obtained from

$$\sigma_u^2(t) = R_u(t,t) = e^{2\beta t}\int_{-\infty}^{\infty} S_s(\omega)\,|z(\alpha,t)|^2 d\omega, \quad (20)$$

where

$$|z(\alpha,t)|^2 = |H(\alpha)|^2[1-\{g(t)+\alpha h(t)\}e^{-\alpha t}] [1-\{g(t)+\alpha h(t)\}e^{-\alpha^* t}]$$

$$= |H(\alpha)|^2(1-2e^{-\beta t}\{[g(t)+\beta h(t)]\cos\omega t+\omega h(t)\sin\omega t\}+e^{-2\beta t}[g^2(t)+(\beta^2+\omega^2)h^2(t)+2\beta g(t)h(t)]), \quad (21)$$

$$|H(\alpha)|^2 = 1/(\alpha^2+2\zeta\omega_1\alpha+\omega_1^2)(\alpha^{*2}+2\zeta\omega_1\alpha^*+\omega_1^2)$$
$$= 1/[(\nu_1^2-\omega_1^2)^2+4\zeta_1^2\nu_1^2\omega^2],$$
$$\nu_1^2= \beta^2+2\zeta\omega_1\beta+\omega_1^2, \quad \zeta_1^2=(\beta+\zeta\omega_1)^2/\nu_1^2. \quad (22)$$

The steady state transfer function $|H(\alpha)|^2$ is written here in a form similar to that of $|H(i\omega)|^2$ in harmonic analysis. This is to illustrate that most formulae in stationary analysis are also valid for exponentially modulated stationary random excitation when the modified natural frequency ν_1 and damping ζ_1 are used.

4. INTEGRATION FORMULAE

The integrand of equation (20) and its moments can be integrated in closed form. It can be proved [I.S. Gradshteyn and I.W. Ryzhik 1965] that

$$\int_{-\infty}^{\infty}|H(\beta+i\omega)|^2d\omega = \pi/2\zeta_1\nu_1^3,$$
$$\int_{-\infty}^{\infty}\omega^2|\beta+i\omega)|^2d\omega = \pi/2\zeta_1\nu_1,$$
$$\int_{-\infty}^{\infty}\cos\omega t|H(\beta+i\omega)|^2d\omega = \pi g_1(t)/2\zeta_1\nu_1^3,$$
$$\int_{-\infty}^{\infty}\omega\sin\omega t|H(\beta+i\omega)|^2d\omega = \pi h_1(t)/2\zeta_1\nu_1, \quad (23)$$

where $g_1(t)=e^{-\zeta_0\nu_1 t}(\cos\lambda_1 t+\zeta_1\nu_1\lambda_1^{-1}\sin\lambda_1 t)$,

$h_1(t)=e^{-\zeta_0\nu_1 t}\lambda_1^{-1}\sin\lambda_1 t$ and $\lambda_1^2=(1-\zeta_1^2)\nu_1^2$.

In particular, applying the integration formulae to modulated white noise, $S_s(\omega)=S_0$, one has, from equations (20) and (23),

$$(2\zeta_1\nu_1^3/\pi S_0)\sigma_u^2(t)$$
$$= e^{2\beta t}+[g(t)+\beta h(t)]^2+\nu_1^2h^2(t)$$
$$-2e^{\beta t}\{g_1(t)[g(t)+\beta h(t)]+\nu_1^2h_1(t)h(t)\}. \quad (24)$$

When $f_m(\tau)=1$, i.e., $\beta=0$, $\zeta_1=\zeta$, $\nu_1=\omega_1$, $g_1(t)=g(t)$, $h_1(t)=h(t)$, then

$$\sigma_u^2=(\pi S_0/2\zeta\omega_1^3)[1-g^2(t)-\omega^2h^2(t)], \quad (25)$$

the same result as given by [Caughey and Stumpf 1961].

5. PARTIAL FRACTION

In many engineering applications, the excitation power spectral density can conveniently be expressed as

$$S_s(\omega)=(a+b\omega^2)/[(\omega_0^2-\omega^2)^2+4\zeta_0^2\omega_0^2\omega^2], \quad (26)$$

where a, b, ω_0 and ζ_0 are given constants and the subscript 0 denotes excitation parameters. Therefore, it is expedient to be able to integrate the expressions

$$P(\omega^2)/Q(\omega^2),\ \cos\omega t P(\omega^2)/Q(\omega^2),\ \omega\sin\omega t P(\omega^2)/Q(\omega^2) \quad (27)$$

over $-\infty \le \omega \le \infty$, where

$$P(\omega^2)=a+b\omega^2+c\omega^4+d\omega^6,$$
$$Q(\omega^2)= \prod_{j=0}^{m}[(\omega_j^2-\omega)^2+4\zeta_j\omega_j^2\omega^2]= \prod_{j=0}^{m}Q_j(\omega^2). \quad (28)$$

Terms up to ω^6 in $P(\omega^2)$ are included for the evaluation of spectral moments and the products up to m factors in $Q(\omega^2)$ deal with multi-degrees-of-freedom systems. No distinction notationally between ν_j and ω_j is made here. If exponential modulating functions are considered, the natural frequencies and dampings are modified prior to integration.

All terms in expressions (27) can be integrated by means of formulae (23) provided P/Q can be put in partial fraction form,

$$\frac{P(\omega^2)}{Q(\omega^2)} = \sum_{j=0}^{m}\frac{a_j+b_j\omega^2}{(\omega_j^2-\omega^2)^2+4\zeta_j\omega_j^2\omega^2} = \sum_{j=0}^{m}\frac{P_j(\omega^2)}{Q_j(\omega^2)} \quad (29)$$

Now, one of the roots of $Q_j(\omega^2)$ is

$$x_j=\omega_j^2[(1-2\zeta_j^2)+i2\zeta_j\sqrt{1-\zeta_j^2}\,] \quad (30)$$

and the numerators can be evaluated at $\omega^2=x_j$:

$$P_j(x_j)=P(x_j)/\prod_{j\neq k}Q_k(x_j)=a_j+b_jx_j. \quad (31)$$

Since both a_j and b_j are real, they can be determined from

$$b_j=\mathbf{Im}P_j(x_j)/\mathbf{Im}x_j,\quad a_j=\mathbf{Re}P_j(x_j)-b_j\mathbf{Re}x_j. \quad (32)$$

Therefore the following integration formulae are sufficient to integrate all the terms in expression (27)

$$\text{(i)}\quad \int_{-\infty}^{\infty}\frac{P_j(\omega^2)}{Q_j(\omega^2)}d\omega = \frac{\pi}{2\zeta_j\omega_j^3}(a_j+b_j\omega_j^2);$$

$$\text{(ii)}\quad \int_{-\infty}^{\infty}\frac{P_j(\omega^2)}{Q_j(\omega^2)}\cos\omega t\,d\omega$$

$$= \frac{\pi}{2\zeta_j\omega_j^3}\,[(a_j+b_j\omega_j^2)g_j(t) - 2\zeta_j\omega_j^3 b_j h_j(t)] \quad ;$$

(iii) $$\int_{-\infty}^{\infty}\frac{P_j(\omega^2)}{Q_j(\omega^2)}\sin\omega t\,d\omega$$

$$= \frac{\pi}{2\zeta_j\omega_j^3}\,[2\zeta_j\omega_j b_j g_j(t) + \{a_j+\omega_j^2 b_j(1-4\zeta_j^2)\}h_j(t)] \tag{33}$$

Note that subscript j=0 denotes excitation and j>0 denotes natural modes. Excitations having more than one peak in the spectral density curve can be included by taking j<0. If the excitation spectral curve is modulated by an exponential function, ζ_j and ω_j, j>0, are taken as the modified damping and natural frequency as given in equations (22).

6. MULTI-DEGREES OF FREEDOM SYSTEMS

Consider the governing equation of a system which has been derived from approximating methods such as finite element,

$$[\mathbf{M}]\{\ddot{u}\} + [\mathbf{C}]\{\dot{u}\} + [\mathbf{K}]\{u\} = \{\mathbf{f}\} \tag{34}$$

with initial conditions $\{\mathbf{u}(0)\}=\{\dot{\mathbf{u}}(0)\}=\{\mathbf{0}\}$, where $[\mathbf{M}]$, $[\mathbf{C}]$ and $[\mathbf{K}]$ are respectively the nxn matrices of mass, damping and stiffness and $\{\mathbf{u}(t)\}$ and $\{\mathbf{f}(t)\}$ are the response and excitation vectors. For a lightly damped system, equation (34) can be solved by means of the natural modes ω_j and $\{\boldsymbol{\phi}_j\}$ of the corresponding undamped system

$$[\mathbf{M}]\{\boldsymbol{\phi}_j\}\omega_j^2 = [\mathbf{K}]\{\boldsymbol{\phi}_j\}$$

so that $[\boldsymbol{\Phi}]^T[\mathbf{M}][\boldsymbol{\Phi}]=[\mathbf{I}]$ and $[\boldsymbol{\Phi}]^T[\mathbf{K}][\boldsymbol{\Phi}]=[\boldsymbol{\Omega}^2]$ (35)

where $[\boldsymbol{\Phi}] = [\boldsymbol{\phi}_1\ \boldsymbol{\phi}_2\ \dots\ \boldsymbol{\phi}_n]$ and $[\boldsymbol{\Omega}^2] = \mathrm{diag}[\omega_1^2\ \ \omega_2^2\ \dots\ \omega_n^2]$. Assuming modal damping so that $[\boldsymbol{\Phi}]^T[\mathbf{C}][\boldsymbol{\Phi}]=\mathrm{diag}[2\omega_j\zeta_j]$ one has the solution

$$\{\mathbf{u}(t)\} = [\boldsymbol{\Phi}]^T\int_0^t[\mathbf{h}(t-\tau)][\boldsymbol{\Phi}]^T\{\mathbf{f}(\tau)\}d\tau \tag{36}$$

where $[\mathbf{h}(t)] = \mathrm{diag}[h_j(t)]$ and $h_j(t) = \lambda^{-1}e^{-\zeta_j\omega_j t}\sin\lambda_j t$ (37)

If the excitation $\{\mathbf{f}(t)\}$ is a random process of the form

$$\{\mathbf{f}(t)\} = [\mathbf{f}_m(t)]\{\mathbf{f}_s(t)\} \tag{38}$$

where $[\mathbf{f}_m(t)]$ is a matrix of evolutionary functions and $\{\mathbf{f}_s(t)\}$ is stationary with power spectral density $[\mathbf{S}(\omega)]$, then the correlation matrix of the response is given by

$$[\mathbf{R}_u(t_1,t_2)] = \langle\{\mathbf{u}(t_1)\}\ \{\mathbf{u}(t_2)\}^T\rangle$$

$$=\langle[\boldsymbol{\Phi}]^T\int_0^{t_1}[\mathbf{h}(t_1-\tau_1)][\boldsymbol{\Phi}]^T[\mathbf{f}_m(\tau_1)]\{\mathbf{f}_s(\tau_1)\}d\tau_1$$

$$\int_0^{t_2}\{\mathbf{f}_s(\tau_2)\}^T[\mathbf{f}_m(\tau_2)][\boldsymbol{\Phi}][\mathbf{h}(t_2-\tau_2)]d\tau_2[\boldsymbol{\Phi}]^T\rangle$$

$$=\int_{-\infty}^{\infty}[\mathbf{I}(\omega,t_1)][\mathbf{S}(\omega)][\mathbf{I}^*(\omega,t_2)]d\omega \tag{39}$$

where * denotes the complex conjugate transpose and

$$[\mathbf{I}(\omega,t)]=[\boldsymbol{\Phi}]\int_0^t[\mathbf{h}(t-\tau)]\{\mathbf{f}_m(\tau)\}e^{i\omega\tau}d\tau[\boldsymbol{\Phi}]^T \tag{40}$$

Since $[\mathbf{h}]$ is a diagonal matrix, the integrals in equation (40) can be evaluated one by one similar to the integral in equation (18).

7. MORE COMPLICATED MODULATING FUNCTION

The method can be generalised to modulating functions of the form $f_m(t) = (\bar{F}+Ft)e^{\beta t}$. The integral (18) is in fact given by equation (15),

$$I(\omega,t) = \int_0^t h(t-\tau)f_m(\tau)e^{i\omega\tau}d\tau$$
$$= (\bar{U}+Ut)e^{\alpha t}-g(t)\bar{U}-h(t)(\alpha\bar{U}+U)$$

where $\bar{U}$ and U are obtained by equation (14) when $\bar{F}$ and F are given. In this case, equation (17) can not be integrated in closed form and numerical integration is inevitable.

8. EXAMPLE

Consider a single degree of freedom system subject to an excitation with spectral density

$$S_s(\omega) = (\omega_0^2+4\zeta_0^2\omega^2)/[(\omega_0^2-\omega^2)^2+4\zeta_0^2\omega_0^2\omega^2]$$
$$= P(\omega^2)/Q(\omega^2) \tag{41}$$

with time modulating function $f_m(t)=e^{\beta t}$. The natural frequency and damping ratio are thus modified to be, from equations (22), $\bar{\omega}^2=\beta^2+2\zeta_1\omega_1\beta+\omega_1^2$ and $\bar{\zeta}_1^2=(\beta+\zeta_1\omega_1)^2/\bar{\omega}_1^2$. The modified quantities are used in the rest of the example, and the bars are omitted for simplicity. It is obvious that if β=0 the natural frequency and damping are unchanged. The transfer function is given by

$$|H_1(\omega^2)|^2=1/[(\omega_1^2-\omega^2)+4\zeta_1^2\omega_1^2\omega^2]=1/Q_1(\omega^2).$$

Using partial fractions, one has

$$S_s(\omega)|H_1(\omega^2)|^2 = (a_{10}+b_{10}\omega^2)/Q_0(\omega^2) + (a_{11}+b_{11}\omega^2)/Q_1(\omega^2)$$

$$\omega^2 S_s(\omega)|H_1(\omega^2)|^2 = (a_{20}+b_{20}\omega^2)/Q_0(\omega^2) + (a_{21}+b_{21}\omega^2)/Q_1(\omega^2)$$

$$\omega^4 S_s(\omega)|H_1(\omega^2)|^2 = (a_{30}+b_{30}\omega^2)/Q_0(\omega^2) + (a_{31}+b_{31}\omega^2)/Q_1(\omega^2)$$

$$\omega^6 S_s(\omega)|H_1(\omega^2)|^2 = (a_{40}+b_{40}\omega^2)/Q_0(\omega^2) + (a_{41}+b_{41}\omega^2)/Q_1(\omega^2)$$

where

$$b_{k0} = \mathbf{Im}[x_0^{k-1}P(x_0)/Q_1(x_0)]/\mathbf{Im}x_0,$$

$$b_{k1} = \mathbf{Im}[x_1^{k-1}P(x_1)/Q_0(x_1)]/\mathbf{Im}x_1,$$

$$a_{k0} = \mathbf{Re}[x_0^{k-1}P(x_0)/Q_1(x_0)] - b_{k0}\mathbf{Re}x_0,$$

$$a_{k1} = \mathbf{Re}[x_1^{k-1}P(x_1)/Q_0(x_1)] - b_{k1}\mathbf{Re}x_1,$$

$$x_j = \omega_j^2[(1-2\zeta_j^2)+i2\zeta_j\sqrt{1-\zeta_j^2}\ ],\quad k=1,2,3,4,\ j=0,1$$

The generalized impulsive and indicial responses are defined as

$$h_j(t)=\lambda_j^{-1}e^{-\zeta_j\omega_j t}\sin\lambda_j t,\quad \lambda_j^2=(1-\zeta_j^2)\omega_j^2,$$

$$g_j(t)=e^{-\zeta_j\omega_j t}(\cos\lambda_j t+\zeta_j\omega_j\lambda_j^{-1}\sin\lambda_j t).$$

The mean square displacement is obtained from equation (20):

$$\begin{aligned}\sigma_u^2(t)&=e^{2\beta t}\int_{-\infty}^{\infty}S_s|Z_1|^2d\omega\\ &=(\pi/2)[e^{2\beta t}+g^2(t)+\beta^2h^2(t)+2\beta g(t)h(t)]\\ &\quad\sum\{(a_{1j}+b_{1j}\omega_j^2)/\zeta_j\omega_j^3\} - \pi e^{\beta t}[g(t)+\\ &\quad\beta h(t)]\sum\{[a_{1j}+b_{1j}\omega_j^2)g_j(t)-2\zeta_j\omega_j^3b_{1j}\\ &\quad h_j(t)]/\zeta_j\omega_j^3\} - \pi e^{\beta t}h(t)\sum\{[2\zeta_j\omega_jb_{1j}\\ &\quad g_j(t)+(a_{1j}+\omega_j^2b_{1j}-4\omega_j^2b_{1j}\zeta_j^2)h_j(t)]/\\ &\quad\zeta_j\omega_j\} + (\pi/2)h^2(t)\sum\{(a_{2j}+b_{2j}\omega_j^2)/\\ &\quad\zeta_j\omega_j^3\},\qquad j=0,1\end{aligned}\tag{42}$$

From equation (17), the autocorrelation for the velocity response is

$$R_{\dot u}(t_1,t_2)=\int_{-\infty}^{\infty}S_s(\omega)\dot I(\omega,t_1)\dot I^*(\omega,t_2)d\omega$$

and when $f_m(t)=e^{\beta t}$, then $\dot I(\omega,t)=\alpha I(\alpha,t)=\alpha Z(\alpha,t)e^{\beta t}$, $\alpha=\beta+i\omega$. Therefore, the mean square velocity response is

$$\begin{aligned}\sigma_{\dot u}^2(t) &= R_{\dot u}(t,t)\\ &= e^{2\beta t}\int_{-\infty}^{\infty}(\beta^2+\omega^2)S_s|H|^2d\omega\\ &= \beta^2\sigma_u^2(t)+e^{2\beta t}\int_{-\infty}^{\infty}\omega^2S_s|H|^2d\omega,\end{aligned}\tag{43}$$

and, similarly, the mean square acceleration response is

$$\begin{aligned}\sigma_{\ddot u}^2(t) &= R_{\ddot u}(t,t)\\ &= e^{2\beta t}\int_{-\infty}^{\infty}(\beta^2+\omega^2)S_s|H|^2d\omega\\ &= \beta^4\sigma_u^2(t)+2\beta^2\sigma_{\dot u}^2(t)+e^{2\beta t}\int_{-\infty}^{\infty}\omega^4S_s|H|^2d\omega\end{aligned}\tag{44}$$

The integral in equation (43) is readily obtained from that of equation (42) by changing $(a_{1j},b_{1j},a_{2j},b_{2j})$ to $(a_{2j},b_{2j},a_{3j},b_{3j})$ and the integral in equation (44) by changing it to $(a_{3j},b_{3j},a_{4j},b_{4j})$, respectively.

To provide a simple case for comparison purposes, one can assume that $\beta=0$, so that $f_m(t)=1$, and then

$$\begin{aligned}\sigma_u^2(t)&=(\pi/2)[1+g^2(t)]\sum\{(a_{1j}+b_{1j}\omega_j^2)/\zeta_j\omega_j^3\}\\ &\quad-\pi g(t)\sum\{[a_{1j}+b_{1j}\omega_j^2)g_j(t)-2\zeta_j\omega_j^3b_{1j}\\ &\quad h_j(t)]/\zeta_j\omega_j^3\} - \pi h(t)\sum\{[2\zeta_j\omega_jb_{1j}\\ &\quad g_j(t)+(a_{1j}+\omega_j^2b_{1j}-4\omega_j^2b_{1j}\zeta_j^2)h_j(t)]/\\ &\quad\zeta_j\omega_j\} + (\pi/2)h^2(t)\sum\{(a_{2j}+b_{2j}\omega_j^2)/\\ &\quad\zeta_j\omega_j^3\},\qquad j=0,1,\end{aligned}$$

etc. For the stationary response, $t \longrightarrow \infty$, $g(t)=h(t)=0$ and hence

$$\begin{aligned}\sigma_u^2(t) &= (\pi/2)\sum\{(a_{1j}+b_{1j}\omega_j^2)/\zeta_j\omega_j^3\},\\ \sigma_{\dot u}^2(t) &= (\pi/2)\sum\{(a_{2j}+b_{2j}\omega_j^2)/\zeta_j\omega_j^3\}\\ \sigma_{\ddot u}^2(t) &= (\pi/2)\sum\{(a_{3j}+b_{3j}\omega_j^2)/\zeta_j\omega_j^3\}\qquad j=0,1\end{aligned}\tag{45}$$

If the damping is small, it is usual practice to assume

$$\begin{aligned}\sigma_u^2(t) &= \pi S_s(\omega_1)/2\zeta_1\omega_1^3,\\ \sigma_{\dot u}^2(t) &= \zeta S_s(\omega_1)/2\zeta_1\omega_1,\\ \sigma_{\ddot u}^2(t) &= [\pi S_s(\omega_1)\omega_1/2\zeta_1](1+4\zeta_1^2).\end{aligned}\tag{46}$$

If $\omega_0=1$ for the excitation, the percentage errors evaluated by equation (46) as compared to the exact formulae (45), for this simple case, are tabulated in Table 1. It is interesting to note that the root mean square values of displacement are more accurate when the natural frequency is less than ω_0 and those of the acceleration are more accurate when the natural frequency is greater than ω_0. The system damping ratio is not as important as $\omega_0-\omega_1$ in the approximation, as is usually the case.

9. CONCLUSION

The spectral method in stationary random process is extended to exponentially modulated stationary and non-stationary random processes in a straighforward manner by replacing the frequency parameter $i\omega$ by $\alpha=\beta+i\omega$ where $\exp(\beta t)$ is the modulating function. Therefore, the transfer function is now $H(\alpha)$ instead of $H(i\omega)$, and the formulation follows the conventional method. Integration formulae for the response spectra for the root mean square (r.m.s.) response, velocity and acceleration are presented. The results are compared to approximate formulae, for resonance conditions. For the case of a constant modulating function ($\beta=0$), the usual approximate formula for the r.m.s. displacement is more accurate when the natural frequency ω_1 is less than that, ω_0, at the peak of the excitation spectral density curve. for the approximation, the behaviour of the acceleration is just the reverse of that of the displacement, and the influence of the damping is not as significant as that of $\omega_0-\omega_1$, as is usually the case. The method is suitable for applications in earthquake engineering and blasting analysis.

10. REFERENCES

R.L.Barnowski and J.R.Maurer 1969. Mean square response of simply mechanical systems to nonstationary random excitation. Journal of Applied Mechanics **36**, 221-227.

T.K.Caughey and H.J.Stumpf 1961. Transient response of a dynamic system under random excitation. Journal of Applied Mechanics **28**, 563-566.

R.B.Corotis and E.H.Vanmarcke 1975. Time dependent spectral content of system response. American Society of Civil Engineers, Journal of the Engineering Mechanics Division **101**, 623-637.

R.B.Corotis, E.H.Vanmarcke and C.A.Cornell 1972. First passage of nonstationary random processes. American Society of Civil Engineers, Journal of the Engineering Mechanics Division **98**, 401-414.

M.Shinozuka and J.N.Yang 1969. Random vibration of linear structures. International Journal of Solids and Structures **5**, 1005-1036.

J.S.Bendat and A.G.Piersol 1966. Random Data: Analysis and Measurement Procedures. New York: Wiley.

A.Y.T.Leung 1985. Structural response to exponentially varying harmonic excitations, Earthquake Engineering and Structural Dynamics, vol 13, 677-681.

I.S.Gradshteyn and I.W.Ryzhik 1965. Tables of Integrals Series and Produces. New York: Academic Press.

TABLE 1

Comparison of results

ω_1	ζ_1	Sig (Displ.) Exact	Sig (Displ.) Approx.	Sig (Displ.) % Error	Sig (Veloc.) Exact	Sig (Veloc.) Approx.	Sig (Veloc.) % Error	Sig (Accel.) Exact	Sig (Accel.) Approx.	Sig (Accel.) % E[partial]
$\zeta_0=0\cdot2$										
0·40	0·02	125·43	125·23	0·2	50·94	50·09	1·7	22·59	20·04	1
0·40	0·04	88·82	88·55	0·3	36·59	35·42	3·2	17·55	14·17	19
0·40	0·06	72·62	72·30	0·4	30·31	28·92	4·6	15·49	11·57	2
0·40	0·08	62·96	62·21	0·6	26·61	25·05	5·9	14·32	10·02	3
0·40	0·10	56·38	56·00	0·7	24·10	22·40	7·0	13·56	8·96	33
0·80	0·02	80·42	81·49	−1·3	64·87	65·19	−0·5	53·07	52·15	1
0·80	0·04	56·07	57·62	−2·8	45·51	46·10	−1·3	37·93	36·88	2
0·80	0·06	45·10	47·05	−4·3	36·79	37·64	−2·3	31·14	30·11	3
0·80	0·08	38·47	40·75	−5·9	31·49	32·60	−3·5	27·01	26·08	3
0·80	0·10	33·88	36·44	−7·6	27·80	29·16	−4·9	24·13	23·32	3
1·20	0·02	35·65	34·66	2·8	41·55	41·60	−0·1	49·14	49·92	−
1·20	0·04	25·68	24·51	4·6	29·22	29·41	−0·7	34·15	35·30	−3
1·20	0·06	21·22	20·01	5·7	23·66	24·02	−1·5	27·37	28·82	−5
1·20	0·08	18·51	17·33	6·4	20·27	20·80	−2·6	23·26	24·96	−7
1·20	0·10	16·62	15·50	6·7	17·91	18·60	−3·8	20·40	22·32	−9
1·60	0·02	10·67	9·31	12·8	15·58	14·89	4·4	24·07	23·82	1
1·60	0·04	8·35	6·58	21·2	11·44	10·53	8·0	17·15	16·85	1
1·60	0·06	7·38	5·37	27·2	9·64	8·60	10·8	14·10	13·76	2
1·60	0·08	6·81	4·65	31·7	8·57	7·45	13·2	12·26	11·91	2
1·60	0·10	6·42	4·16	35·2	7·84	6·66	15·1	11·00	10·65	3
2·00	0·02	4·88	3·90	20·0	8·26	7·80	5·6	15·77	15·60	1
2·00	0·04	4·01	2·76	31·2	6·14	5·52	10·1	11·25	11·03	2
2·00	0·06	3·66	2·25	38·5	5·23	4·50	13·9	9·27	9·01	2
2·00	0·08	3·47	1·95	43·8	4·70	3·90	17·1	8·08	7·80	3
2·00	0·10	3·34	1·74	47·7	4·35	3·49	19·7	7·28	6·98	4
$\zeta_0=0\cdot4$										
0·40	0·02	103·73	103·84	−0·1	41·76	41·54	0·5	17·99	16·61	7
0·40	0·04	73·27	73·43	−0·2	29·67	29·37	1·0	13·60	11·75	13
0·40	0·06	59·76	59·95	−0·3	24·33	23·98	1·4	11·75	9·59	18
0·40	0·08	51·69	51·92	−0·4	21·15	20·77	1·8	10·70	8·31	22
0·40	0·10	46·18	46·44	−0·6	18·98	18·58	2·2	10·00	7·43	25
0·80	0·02	50·08	50·82	−1·5	40·03	40·66	−1·6	32·60	32·52	0
0·80	0·04	34·92	35·94	−2·9	27·87	28·75	−3·2	23·08	23·00	0
0·80	0·06	28·14	29·34	−4·3	22·41	23·47	−4·8	18·84	18·78	0
0·80	0·08	24·06	25·41	−5·6	19·11	20·33	−6·4	16·31	16·26	0
0·80	0·10	21·27	22·73	−6·9	16·84	18·18	−8·0	14·56	14·55	0
1·20	0·02	22·55	22·46	0·4	26·51	26·95	−1·6	31·78	32·34	−1
1·20	0·04	15·99	15·88	0·7	18·45	19·06	−3·3	22·10	22·87	−3
1·20	0·06	13·07	12·97	0·8	14·83	15·56	−4·9	17·76	18·67	−5
1·20	0 08	11·33	11·23	0·9	12·64	13·48	−6·6	15·15	16·17	−6
1·20	0·10	10·13	10·04	0·9	11·13	12·05	−8·2	13·36	14·46	−8
1·60	0·02	9·37	8·95	4·6	14·34	14·31	0·2	22·75	22·90	−0
1·60	0·04	6·90	6·33	8·3	10·15	10·12	0·3	15·98	16·19	−1
1·60	0·06	5·82	5·16	11·3	8·28	8·26	0·3	12·96	13·22	−2
1·60	0·08	5·19	4·47	13·9	7·17	7·16	0·2	11·15	11·45	−2
1·60	0·10	4·77	4·00	16·1	6·40	6·40	−0·0	9·91	10·24	−3
2·00	0·02	4·77	4·41	7·6	8·90	8·83	0·9	17·61	17·65	−0
2·00	0·04	3·60	3·12	13·4	6·34	6·24	1·6	12·42	12·48	−0
2·00	0·06	3·11	2·55	18·1	5·21	5·10	2·2	10·11	10·19	−0
2·00	0·08	2·83	3·21	21·9	4·54	4·41	2·8	8·73	8·83	−1
2·00	0·10	2·64	1·97	25·1	4·08	3·95	3·2	7·79	7·89	−1
$\zeta_0=0\cdot6$										
0·40	0·02	101·31	101·51	−0·2	40·65	40·60	0·1	17·51	16·24	7
0·40	0·04	71·50	71·78	−0·4	28·77	28·71	0·2	13·19	11·48	12
0·40	0·06	58·27	58·61	−0·6	23·50	23·44	0·3	11·37	9·38	17
0·40	0·08	50·37	50·76	−0·8	20·36	20·30	0·3	10·34	8·12	21
0·40	0·10	44·96	45·40	−1·0	18·21	18·16	0·3	9·65	7·26	24

TABLE 1 (*cont.*)

ω_1	ζ_1	Sig (Displ.)			Sig (Veloc.)			Sig (Accel.)		
		Exact	Approx.	% Error	Exact	Approx.	% Error	Exact	Approx.	% Error
0·80	0·02	41·95	42·32	−0·9	33·42	33·86	−1·3	27·37	27·09	1·0
0·80	0·04	29·41	29·93	−1·7	23·34	23·94	−2·6	19·53	19·15	1·9
0·80	0·06	23·82	24·44	−2·6	18·82	19·55	−3·9	16·08	15·64	2·7
0·80	0·08	20·48	21·16	−3·4	16·10	16·93	−5·1	14·02	13·54	3·4
0·80	0·10	18·18	18·93	−4·1	14·23	15·14	−6·4	12·63	12·11	4·1
1·20	0·02	19·86	19·84	0·1	23·45	23·81	−1·5	28·39	28·57	−0·6
1·20	0·04	14·05	14·03	0·2	16·35	16·83	−3·0	19·96	20·20	−1·2
1·20	0·06	11·48	11·45	0·2	13·16	13·75	−4·5	16·21	16·49	−1·8
1·20	0·08	9·94	9·92	0·2	11·24	11·90	−5·9	13·96	14·28	−2·3
1·20	0·10	8·89	8·87	0·2	9·92	10·65	−7·3	12·43	12·78	−2·8
1·60	0·02	9·88	9·68	2·0	15·36	15·50	−0·9	24·67	24·79	−0·5
1·60	0·04	7·11	6·85	3·7	10·77	10·96	−1·7	17·36	17·53	−1·0
1·60	0·06	5·90	5·59	5·3	8·72	8·95	−2·6	14·11	14·31	−1·4
1·60	0·08	5·19	4·84	6·6	7·49	7·75	−3·5	12·17	12·40	−1·9
1·60	0·10	4·70	4·33	7·8	6·64	6·93	−4·4	10·84	11·09	−2·3
2·00	0·02	5·56	5·36	3·6	10·67	10·72	−0·4	21·37	21·44	−0·3
2·00	0·04	4·06	3·79	6·7	7·51	7·58	−0·9	15·07	15·16	−0·6
2·00	0·06	3·41	3·09	9·3	6·11	6·19	−1·4	12·27	12·38	−0·9
2·00	0·08	3·03	2·68	11·7	5·26	5·36	−1·8	10·60	10·72	−1·1
2·00	0·10	2·78	2·40	13·7	4·68	4·79	−2·3	9·45	9·59	−1·4
$\zeta_0 = 0·8$										
0·40	0·02	103·90	104·13	−0·2	41·62	41·65	−0·1	18·06	16·66	7·8
0·40	0·04	73·31	73·63	−0·4	29·40	29·45	−0·2	13·67	11·78	13·8
0·40	0·06	59·74	60·12	−0·6	23·98	24·05	−0·3	11·84	9·62	18·8
0·40	0·08	51·63	52·06	−0·8	20·74	20·83	−0·4	10·79	8·33	22·8
0·40	0·10	46·09	46·57	−1·0	18·52	18·63	−0·6	10·10	7·45	26·3
0·80	0·02	39·79	40·00	−0·5	31·68	32·00	−1·0	26·15	25·60	2·1
0·80	0·04	28·00	28·29	−1·0	22·18	22·63	−2·0	18·85	18·10	4·0
0·80	0·06	22·75	23·10	−1·5	17·93	18·48	−3·0	15·67	14·78	5·7
0·80	0·08	19·61	20·00	−2·0	15·38	16·00	−4·0	13·79	12·80	7·2
0·80	0·10	17·46	17·89	−2·4	13·63	14·31	−5·0	12·52	11·45	8·6
1·20	0·02	19·60	19·59	0·0	23·21	23·51	−1·3	28·34	28·21	0·5
1·20	0·04	13·86	13·85	0·1	16·21	16·62	−2·5	20·13	19·95	0·9
1·20	0·06	11·32	11·31	0·1	13·08	13·57	−3·8	16·50	16·29	1·3
1·20	0·08	9·80	9·79	0·1	11·20	11·75	−5·0	14·35	14·10	1·7
1·20	0·10	8·77	8·76	0·1	9·90	10·51	−6·2	12·89	12·62	2·1
1·60	0·02	10·72	10·61	1·0	16·80	16·98	−1·1	27·19	27·17	0·1
1·60	0·04	7·65	7·51	1·9	11·75	12·01	−2·2	19·24	19·21	0·2
1·60	0·06	6·30	6·13	2·7	9·50	9·80	−3·2	15·72	15·69	0·2
1·60	0·08	5·50	5·31	3·5	8·14	8·49	−4·3	13·63	13·59	0·3
1·60	0·10	4·95	4·75	4·2	7·21	7·59	−5·3	12·20	12·15	0·4
2·00	0·02	6·46	6·33	2·0	12·55	12·66	−0·9	25·33	25·33	−0·0
2·00	0·04	4·65	4·48	3·7	8·80	8·95	−1·7	17·91	17·91	−0·0
2·00	0·06	3·86	3·66	5·3	7·13	7·31	−2·6	14·62	14·62	−0·0
2·00	0·08	3·39	3·17	6·7	6·12	6·33	−3·5	12·66	12·66	−0·0
2·00	0·10	3·08	2·83	8·0	5·43	5·66	−4·3	11·33	11·33	0·0
$\zeta_0 = 1·0$										
0·40	0·02	108·26	108·36	−0·1	43·67	43·35	0·7	16·51	17·34	−5·0
0·40	0·04	76·47	76·64	−0·2	31·10	30·66	1·4	11·03	12·26	−11·2
0·40	0·06	62·38	62·58	−0·3	25·56	25·03	2·1	8·43	10·01	−18·8
0·40	0·08	53·97	54·19	−0·4	22·29	21·68	2·7	6·73	8·67	−28·8
0·40	0·10	48·23	48·47	−0·5	20·06	19·39	3·4	5·45	7·76	−42·4
0·80	0·02	39·76	39·93	0·4	32·17	31·95	0·7	25·27	25·56	−1·1
0·80	0·04	28·00	28·24	−0·8	22·90	22·59	1·4	17·64	18·07	−2·4
0·80	0·06	22·77	23·06	−1·2	18·84	18·44	2·1	14·19	14·76	−4·0
0·80	0·80	19·64	19·97	−1·6	16·43	15·97	2·8	12·09	12·78	−5·7
0·80	0·10	17·50	17·86	−2·0	14·81	14·29	3·5	10·61	11·43	−7·7
1·20	0·02	19·98	20·13	−0·8	24·29	24·16	0·6	28·87	28·99	−0·4
1·20	0·04	14·01	14·24	−1·6	17·28	17·08	1·1	20·32	20·50	−0·9
1·20	0·06	11·35	11·62	−2·4	14·20	13·95	1·8	16·50	16·74	−1·4
1·20	0·08	9·75	10·07	−3·3	12·38	12·08	2·4	14·21	14·50	−2·0
1·20	0·10	8·64	9·00	−4·2	11·15	10·80	3·1	12·62	12·97	−2·8
1·60	0·02	11·43	11·56	−1·1	18·57	18·49	0·4	29·53	29·59	−0·2
1·60	0·04	7·98	8·17	−2·3	13·20	13·08	0·9	20·84	20·92	−0·4
1·60	0·06	6·44	6·67	−3·7	10·83	10·68	1·5	16·97	17·08	−0·6
1·60	0·08	5·50	5·78	−5·1	9·44	9·25	2·0	14·66	14·79	−0·9
1·60	0·10	4·85	5·17	−6·6	8·49	8·27	2·6	13·07	13·23	−1·2
2·00	0·02	7·14	7·24	−1·5	14·54	14·48	0·4	28·94	28·96	−0·1
2·00	0·04	4·97	5·12	−3·1	10·32	10·24	0·8	20·44	20·48	−0·2
2·00	0·06	3·99	4·18	−4·9	8·46	8·36	1·2	16·67	16·72	−0·3
2·00	0·08	3·39	3·62	−6·8	7·37	7·24	1·7	14·41	14·48	−0·5
2·00	0·10	2·97	3·24	−9·0	6·62	6·48	2·2	12·87	12·95	−0·6

Numerical Methods in Geomechanics (Innsbruck 1988), Swoboda (ed.)
© 1988 Balkema, Rotterdam. ISBN 90 6191 809 X

Dynamic stiffness of rigid strip footing on nonhorizontally layered soil stratum

H.T.Chen
National Central University, Chungli, Taiwan
J.L.Kow
Yue Loong Motor Engineering Center, Taoyuan, Taiwan

ABSTRACT: The effect of nonhorizontal soil layers on the dynamic stiffness of a rigid surface strip footing supported by a soil stratum of finite depth is investigated using the finite element method in the frequency domain. For the case where the rigid bedrock is inclined, the results will shift, causing significant differences at certain range of frequencies for stratum of shallow depth, and this effect will decrease when the depth increases, as compared with those of the case with flat bedrock. When the soil layers are inclined, over the range of frequencies considered, the increase in the degree of inclination will cause reduction in the real stiffnesses of all modes of motion and the imaginary part (damping) of vertical stiffness, while the opposite trend is observed for the damping of swaying and rocking stiffnesses.

1 INTRODUCTION

An important step in the dynamic soil-structure interaction analysis using the substructure method is the determination of foundation stiffnesses. Over the past several decades, different approaches have been proposed to compute the dynamic stiffnesses of foundation of arbitrary shape resting on, or embedded in, a layered halfspace or a layered soil stratum of finite depth. In general, these approaches can be divided into two categories: continuous solution and discrete solution. In a paper by Gazetas (1983), the characteristics of these approaches and the results obtained were discussed.

The behavior of strip footing subjected to vibration has attracted many research interests. Luco and Westmann (1972) obtained pairs of Cauchy-type integral equations, the solution of which is exact for a strip footing resting on a halfspace for Poisson's ratio equal to 0.5 and is only approximate in the low range of frequencies for other values of Poisson's ratio. Gazetas and Roesset (1976) used the Fourier transform technique to compute the dynamic compliances of a strip footing on the surface of a layered soil medium and this approach allows one to study the effect of rigidity of bedrock on the results. Using the boundary element method, Von Estorff and Schmid (1984) and Abascal and Dominguez (1984) investigated the behavior of a strip footing resting on a one-layer soil stratum underlain by rigid bedrock and on a soil deposit underlain by a viscoelastic half-plane, respectively. On the other hand, Chang-Liang used finite element method and consistent boundary to study the behavior of surface and embedded strip footings supported by a soil stratum of finite depth. Tassoulas (1983) developed a hyper element which avoids the need to discretize the soil medium into typical finite elements, an improvement over Chang-Liang's approach (1974).

In the studies mentioned above, the soil layers and the rigid bedrock, if any, were assumed to extend horizontally. However, in reality, there may be cases where the soil layers and/or the rigid bedrock are not horizontal. Thus, the purpose of this study is to investigate the effect of nonhorizontal soil layers or rigid bedrock on the dynamic stiffnesses of a rigid strip footing resting on the surface of a soil stratum of finite depth in the frequency domain.

2 METHOD OF ANALYSIS

In this study the finite element method

is employed to compute the dynamic stiffnesses. In the followings, the description of model, finite element discretization scheme and computational procedure will be presented.

2.1 Description of model

Figure 1 depicts the model used in the present study. The strip footing is resting on the surface of a soil stratum underlain by a rigid bedrock which consists of three zones: a left zone, a right zone and an irregular zone. The soil layers and the bedrock of the left and the right zones are assumed to be horizontal and the surfaces of these two zones are free of stresses. In the irregular zone, the soil layers and/or the rigid bedrock can be inclined and the surface of which is also free of stresses, except the part occupied by the footing. The strip footing which is perfectly bonded to the soil medium is assumed to be rigid and massless, and it is subjected to harmonic motions.

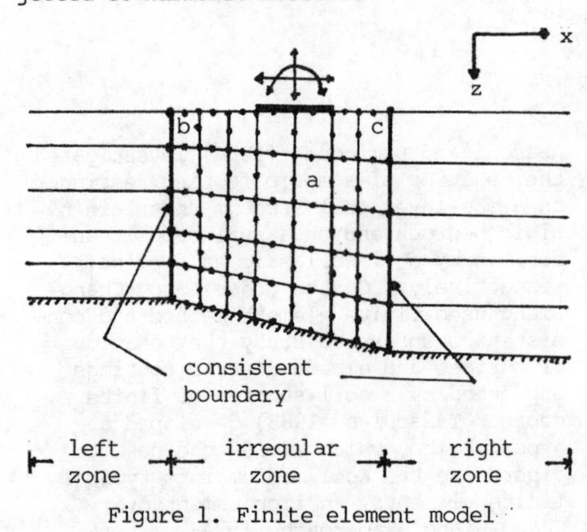

Figure 1. Finite element model

2.2 Finite element discretization

Since the finite element method is used in this study, fititious lateral boundaries must be established; however, improper use of such boundaries can trap the energy in the system and contaminate the results. different boundaries have been proposed: viscous boundary (Lysmer et al. 1969), unified boundary (White et al. 1977) and consistent boundary (Waas 1972). Roesset and Ettouney (1977) compared the viscous boundary and the consistent boundary and found that the consistent boundary is more efficient; in fact, it has been shown that the consistent boundary can give very good results even when placed at the edges of the footing. The consistent boundary was derived for the horizontally layered system underlain by a rigid boundary with stress-free surface. From the assumption adopted for the left and the right zones, it was then decided to use the consistent boundary to model these two zones. Detailed formulation about this boundary can be found in the work by Waas (1972), by Christian et al.(1977) and by Tassoulas (1981) and will not be repeated here.

For the irregular zone, the hyper element developed by Tassoulas (1981) can not be applied due to the nonhorizontal layers and /or bedrock. As a result, it was discretized into typical plane strain finite elements for the entire zone due to asymmetry of the model. In this study, quadratic quadrilateral elements were used; however, due to the assumption of linear variation in the vertical direction for the consistent boundary, there are three types of element employed, as denoted in figure 1 by a, b and c, and the shape functions of which can be derived easily (Bathe 1982).

2.3 Computational procedure

Once the boundary matrices of the left and the right zones have been obtained, they can be added to the stiffness matrix of the irregular zone through direct stiffness method. Then, the frequency domain analysis requires the solution of the following equations.

$$([\overline{K}] - \omega^2[M])\ \underline{U} = \underline{F}$$

where ω is the frequency of excitation, $[\overline{K}]$ the combined stiffness matrix for the irregular zone and the left and the

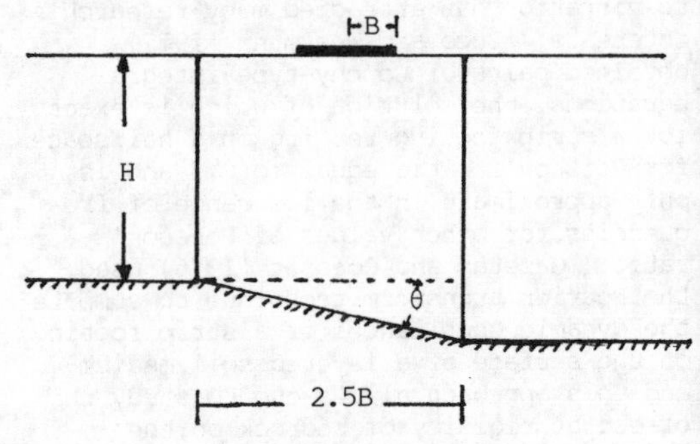

Figure 2. Soil stratum with inclined rigid bedrock

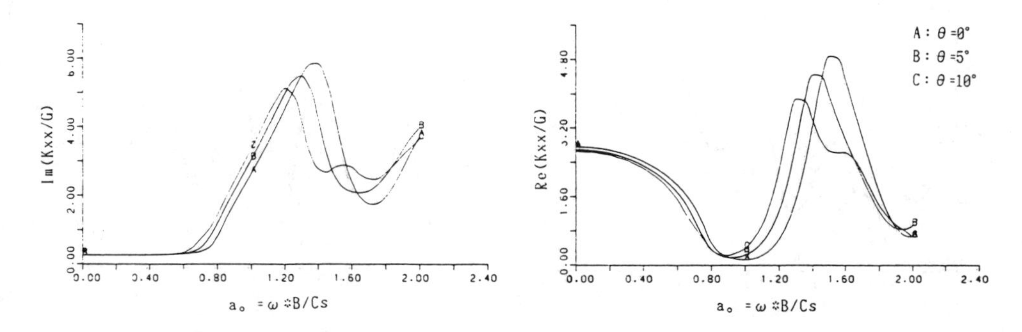

Figure 3. Effect of inclined rigid bedrock on dynamic swaying stiffness(H/B=2)

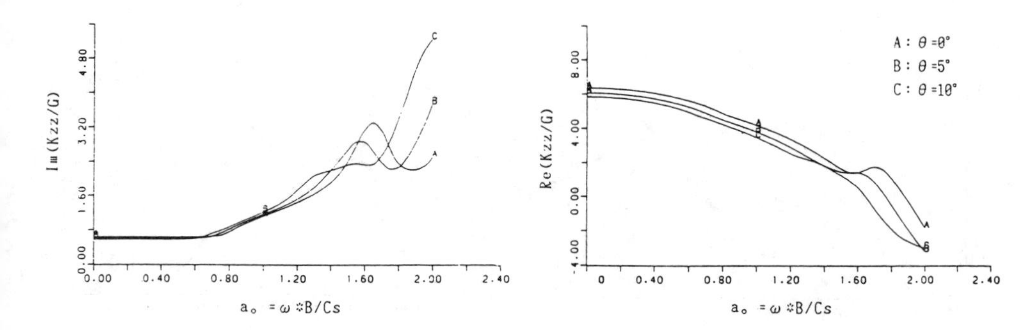

Figure 4. Effect of inclined rigid bedrock on dynamic vertical stiffness(H/B=2)

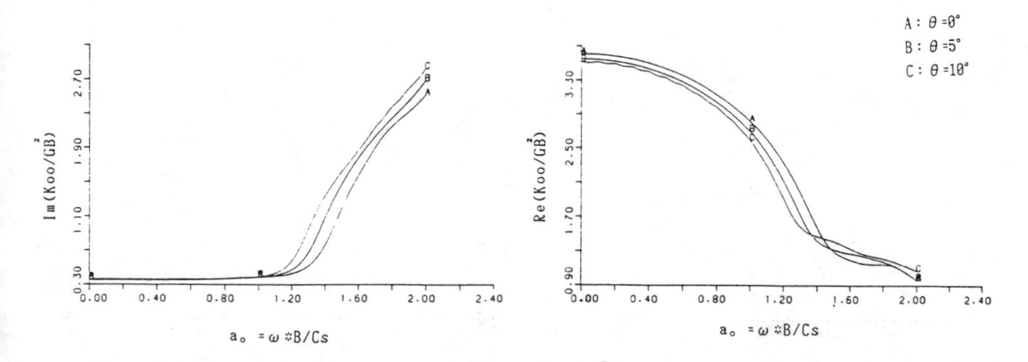

Figure 5. Effect of inclined rigid bedrock on dynamic rocking stiffness(H/B=2)

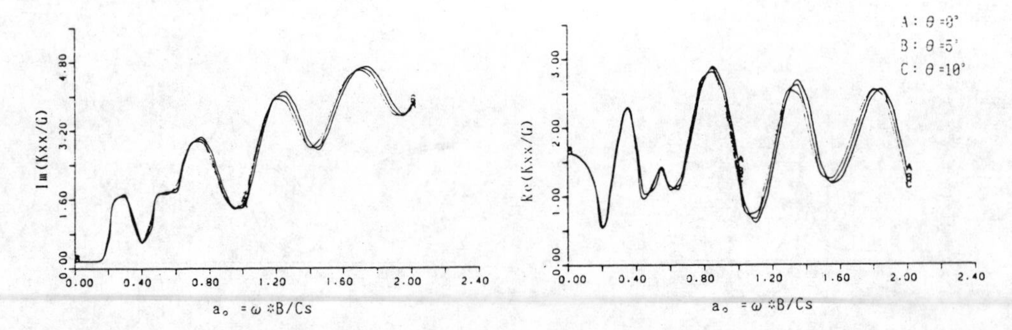

Figure 6. Effect of inclined rigid bedrock on dynamic swaying stiffness(H/B=8)

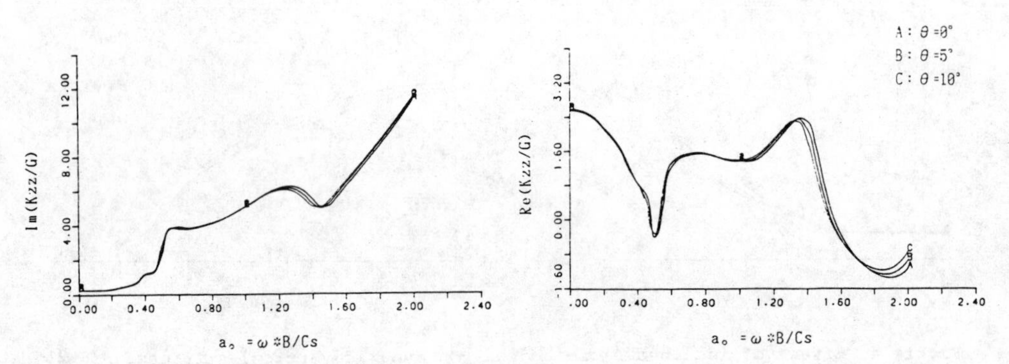

Figure 7. Effect of inclined rigid bedrock on dynamic vertical stiffness(H/B=8)

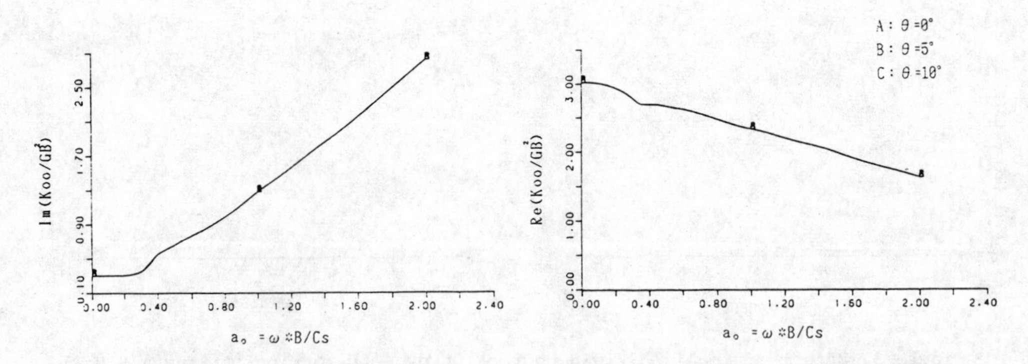

Figure 8. Effect of inclined rigid bedrock on dynamic rocking stiffness(H/B=8)

right zones, [M] mass matrix for the irregular zone, $\underline{U}$ displacement vector, and $\underline{F}$ the force vector. Then, by applying unit harmonic displacements along the soil-footing interface, the sum of reaction forces along this interface is the dynamic stiffness of the footing.

3 RESULTS AND DISCUSSIONS

Based on the procedures described in the previous section, the effect of nonhorizontally layered soil stratum of finite depth can be investigated. In this study, the soil medium is assumed to be homogeneous, isotropic and linearly viscoelastic with material damping of hysteretic type (independent of frequency). The dynamic stiffnesses which are complex functions of excitation frequency are presented in terms of real and imaginary parts and are plotted with respect to dimensionless frequency, the product of excitation frequency and half width of strip footing divided by the shear wave velocity of the soil. It has been shown that for horizontally layered or homogeneous soil medium, only the cross stiffness between swaying and rocking exists; however, due to the asymmetry of the model used in this study, the cross stiffnesses between vertical motion and swaying, and between vertical motion and rocking also appear (Chen 1987). Since their magnitudes are relatively small, as compared with the stiffnesses associated with each mode of motion, they are not presented here; instead, only the swaying stiffness (K_{xx}), the vertical stiffness (K_{zz}) and the rocking stiffness (K_{oo}) are displayed, which are also made dimensionless as follows.

$$K_{xx}/G;\ K_{zz}/G;\ K_{oo}/GB^2$$

where G is the shear modulus of soil and B the half width of strip footing.

3.1 Effect of inclined rigid bedrock

Figure 2 depicts the model used to study the effect of inclined bedrock on the dynamic stiffnesses of a surface strip footing. In this study, the depth of left zone is always kept constant, while that of the right zone varies, depending on the values of angle of inclination, θ. The soil stratum consists of only one layer, the Poisson's ratio and material damping ratio of which are 0.4 and 0.05, respectively. Two different depths are considered for the left zone: H/B=2 and H/B=8, and two values of Q which is positive when measured clockwise are used to investigate the effect: 5° and 10°.

Figures 3-5 are the results for the case H/B=2. It can be observed that as Q increases, the curves will shift toward left, i.e., lower frequency with decrease in amplitude, indicating that the natural frequencies of soil stratum decrease; this is due to the fact that more soil masses are included. In addition, such shifts cause large differences at certain ranges of frequency; in general, differences are small for the frequency below the fundamental frequency.

Displayed in figures 6-8 are the results for the case H/B=8. For the swaying and the vertical stiffnesses, the curves fluctuate with decreasing amplitude, as compared with those of H/B=2, while smooth curves are observed for the rocking stiffness. Although it is still notable that the curves shift toward left for the swaying and the vertical stiffnesses, their differences are comparatively smaller than those of H/B=2. For the rocking stiffness, the change in the angle of inclination do not affect the solution; this is due to the fact that for such a depth of stratum, the half-space solution is obtained.

In general, the trend described above can be observed for the embedded strip footing as well (Chen 1987).

3.2 Effect of nonhorizontal soil layers

For most of the results published previously, the soil layers were assumed to be horizontal. In this study, the effect of nonhorizontal soil layers is examined.

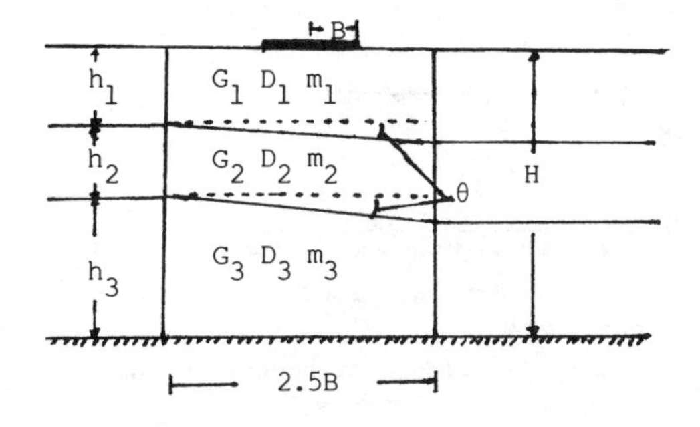

Figure 9. Soil stratum with nonhorizontal layers underlain by flat rigid bedrock

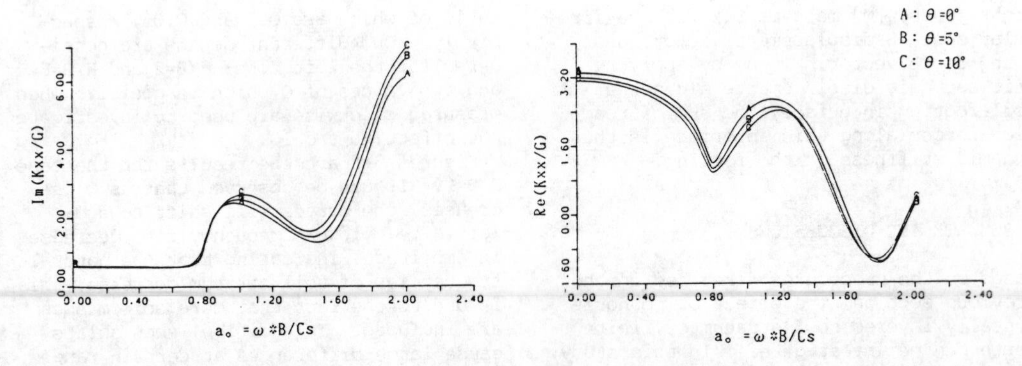

Figure 10. Effect of nonhorizontal soil layers on dynamic swaying stiffness

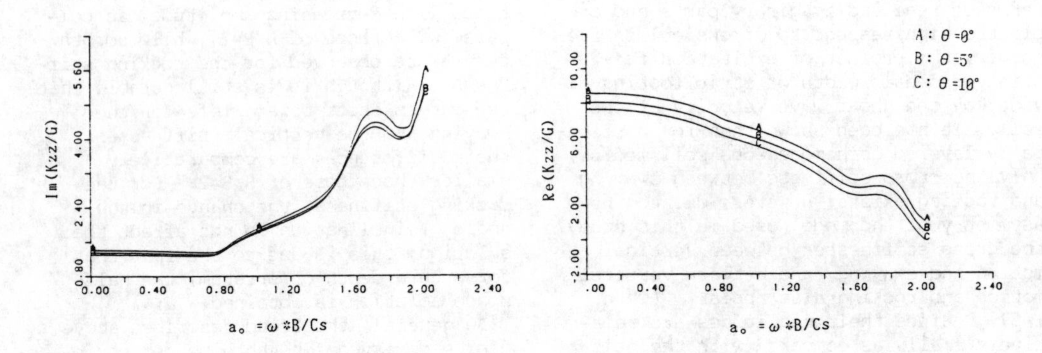

Figure 11. Effect of nonhorizontal soil layers on dynamic vertical stiffness

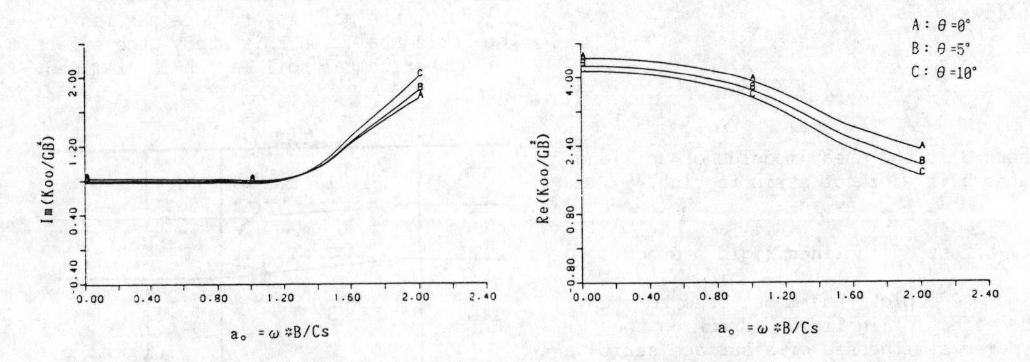

Figure 12. Effect of nonhorizontal soil layers on dynamic rocking stiffness

Shown in figure 9 is the model adopted for this study, which consists of three layers underlain by a rigid bedrock extending to infinity horizontally. To simplify the situation, the angles of inclination for the soil layers are taken to be equal. The soil properties are $G_2/G_1=4$, $G_3/G_1=8$, $D_2/D_1=0.5$, $D_3/D_1=0.3$, $m_2/m_1=2$, $m_3/m_1=3$, where G, D, m are modulus, material damping ratio, and mass density, respectively, and the subscript denotes the layer number. The Poisson's ratio is 0.4 for all three layers and D_1 is 0.1. The depth of the stratum is H/B=4. When the angle of inclination varies, the thicknesses of soil layers of the left zone are kept constant, i.e., $h_2/h_1=1$ and $h_3/h_1=2$. The dimensionless quantities are defined using the properties of the first layer.

The results for this study are shown in figures 10-12. For all three stiffnesses, it can be observed that as the angle of inclination varies, there is no shift in the curves, except at the higher frequencies for the real swaying stiffness. For the real stiffness, the values decrease with increasing angle of inclination, and the same trend is observed for the imaginary part (damping) of the vertical stiffness, while the opposite trend is found for the imaginary parts of the swaying and the rocking stiffnesses.

4 CONCLUSION

Under the plane strain condition, the effect of nonhorizontally layered soil stratum of finite depth is investigated in this study using the finite element method and the concept of consistent boundary in the frequency domain. The conclusions drawn from this study are as follows.

(1). For a one-layer soil stratum underlain by an inclined rigid bedrock, the change in angle of inclination will lead to the change in the natural frequencies of the system, resulting in a shift of the responses. In general, such a shift does not cause significant differences in the frequency below fundamental frequency; however, it can induce significant differences at certain ranges of frequencies, as compared with the results of the soil stratum with a flat rigid bedrock.

(2). The curves for the results of a nonhorizontally layered soil stratum underlain by a flat rigid bedrock do not shift except at the higher frequencies for the real swaying stiffness over the range of frequencies considered in this study, as the angle of inclination increases. In general, the increase in angle of ininclination will reduce the real stiffnesses and the imaginary part (damping) of the vertical stiffness and increase the values of the imaginary parts of the swaying and the rocking stiffnesses.

ACKNOWLEDGEMENTS

The results presented in this paper are part of a research sponsored by the National Science Council of the Republic of China under the grant number NSC-76-0410-E008-03. The authors would like to express their appreciation for the support.

REFERENCES

R.Abascal and J.Dominguez: Dynamic behavior of strip footings on non-homogeneous viscoelastic soils, Proceedings of the International Symposium on Dynamic Soil-Structure Interaction, Minneapolis, 4-5 September, 25-35 (1984)

K.J.Bathe, Finite element procedures in engineering analysis, Prentice-Hall, Inc., Englewoods Cliffs, New Jersey, U.S.A., 1982

V.Chang-Liang: Dynamic response of structures in layered soils, Research Report R74-10, Department of Civil Engineering, M.I.T., January (1974)

H.T.Chen: Dynamic stiffness of a strip footing on a sloping soil stratum, Research Report to National Science Council, September (1987)

J.T.Christian, J.M.Roesset and C.S.Desai: Two- and three-dimensional dynamic analysis, Numerical Methods in Geotechnical engineering (C.S.Desai and J.T.Christian, eds), McGraw-Hill, New York, U.S.A. 1977

O.Von Estorff and G.Schmid: Application of the boundary element method to the analysis of the vibration behavior of strip foundation on a soil layer, Proceedings of the International Symposium on Dynamic Soil-Structure Interaction, Minneapolis, 4-5 September, 11-17 (1984)

G.Gazetas and J.M.Roesset: Forced vibrations of strip footings on layered soils Proceedings, National Structural Engineering Conference, ASCE, Madison, Wisconsin, Vol. 1, 115-131 (1976)

G.Gazetas: Analysis of machine foundation vibrations: state of the art, Soil Dynamics and Earthquake Engineering, 2, 2-42 (1983)

J.E.Luco and R.A.Westmann: Dynamic response of a rigid footing bonded to an elastic half space, Journal of Applied Mechanics, 527-534 (1974)

J.Lysmer and R.L.Kuhlemeyer: Finite dynamic model for infinite media, Journal of the Engineering Mechanics Division, 95, 859-877 (1969)

J.M.Roesset and M.M.Ettouney: Transmitting Boundaries: a comparison, International Journal for Numerical and Analytical Methods in Geomechanics, 1, 151-176 (1977)

J.L.Tassoulas: Elements for the numerical analysis of wave motion in layered media, Research Report R81-2, Department of Civil Engineering, M.I.T., January (1981)

G.Waas: Linear two-dimensional analysis of soil dynamics problems in semi-infinite layered media, Ph. D. Thesis, University of California at Berkeley (1972)

W.White, S.Valliappan and I.K.Lee: Unified boundary for finite dynamic models, Journal of the Engineering Mechanics Division, 103, 949-964 (1977)

Numerical Methods in Geomechanics (Innsbruck 1988), Swoboda (ed.)
© 1988 Balkema, Rotterdam. ISBN 90 6191 809 X

Seismic modelling and design of underground structures

J.E.Monsees
Parsons Brinckerhoff Quade & Douglas, Los Angeles, Calif., USA
J.L.Merritt
Redlands Operations, BDM Corporation, Calif., USA

1 INTRODUCTION

This paper is the outgrowth of work done by the authors to prepare the seismic design criteria for underground structures on the Southern California Rapid Transit District (SCRTD) Metro Rail Project. It is a follow-on to similar papers presented at the October, 1984, ASCE National Convention (20) and the June, 1985, RETC (21). The first five sections follow closely the material in the RETC paper (21). The remaining sections provide additional analytical results. For design, the earthquake-induced distortions recommended herein are to be added to the normal static loading conditions for the underground structure. This design approach has been used by 18 section designers on the SCRTD Project. Results from two- and three-dimensional finite element solutions have verified the approach. Areas where additional research is needed for special cases are identified.

2 EFFECTS OF EARTHQUAKE

The effects of earthquakes on underground structures may be broadly grouped into two general classes -- shaking and faulting. In response to earthquake motion of bed-rock (shaking), the soil transmits energy by waves. Seismologists identify various types of earthquake waves, but underground engineers are generally interested in the effects of transverse shear waves which produce a displacement of the ground transverse to the axis of wave propagation. The orientation of propagation is generally random with respect to any specific structure. Waves propagated parallel to the long axis of a linear structure, such as a tunnel, will tend to enforce a corresponding transverse distortion on the structure. Waves traveling at right angles to the structure will tend to move it back and forth longitudinally, and may tend to pull it loose at zones of abrupt transitions in soil conditions, where wave properties may vary. Diagonally impinging waves subject different parts of a linear structure to out-of-phase displacements.

Faulting includes direct primary shearing displacements of bedrock, which may carry through the overburden to the ground surface. Such physical shearing of the rock or soil is generally limited to relatively narrow, seismically active fault zones, which may be identified by geological and seismological surveys. In general, it is not feasible to design underground structures to restrain major ground faulting. Useful design measures are limited to identifying and avoiding fault crossings, or, if this is not possible, to accepting the displacement, localizing and minimizing damage, and providing means to facilitate repairs.

The major contribution to deformations and corresponding stresses in long linear structures, such as tunnels, is traveling wave effects. These can be accounted for by assuming that the tunnel and surrounding soil move together as the wave passes.

It should be recognized that although the absolute amplitude of earthquake displacement may be large, this displacement is spread over a long length. The gradient of earthquake distortion is generally small, and is often within the elastic deformation capacity of the structure. If it can be established that the maximum deformation imposed by the earthquake will not strain the structure beyond the elastic range, no further provisions to resist the deformation are required. If certain parts of the structure are strained into the plastic range, the ductility of such parts must be investigated. If continuity of the structure has been assumed in the design for static loads, the residual effects of plastic distortions induced by earthquake motions may require special consideration.

3 DESIGN GROUND MOTIONS

Design ground motions and geotechnical properties will be project- and site-specific. The values used in this paper (Table 1) are adopted from the SCRTD Project (8), and are given for illustrative purposes only. Two levels of earthquake are considered:

a. The operating design earthquake (ODE) has an estimated return period of several hundred years. It may have a probability of being exceeded over a design life of 100 years of approximately 50 percent and a Richter magnitude of approximately 6.5.

b. The maximum design earthquake (MDE) has an estimated return period of several thousand years. It may have a probability of being exceeded over a design life of 100 years of approximately 5 percent and a Richter magnitude of approximately 7.0.

Shearing distortions in the ground are given by Figures 1 and 2, which are reproduced here from Converse (8) for materials with seismic velocities as indicated. The authors and Dr. A.J. Hendron, Jr. modified the Converse curves by adding the dashed lines for "New Alluvium". Data points for these additions were generated by means of the SHAKE code (31) and specific results are given in Table 2.

4 STIFFNESS OF UNDERGROUND STRUCTURES

The general approach taken in assessing the static or dynamic behavior of an

Table 1. Design Earthquake Parameters

Design Earthquake	Foundation Condition	Design Ground Motion Parameters: Acceleration (a_{max}) (g) Hor.	Vert.	Velocity (v_{max}) (ft/sec) Hor.	Vert.	Displacement (ft) Hor.	Vert.
ODE	Soil	0.30	0.20	1.4	1.0	1.6	1.0
	Rock	0.30	0.20	0.8	0.6	0.5	0.3
MDE	Soil	0.60	0.40	3.2	2.1	3.3	2.2
	Rock	0.60	0.40	1.9	1.3	1.0	0.7

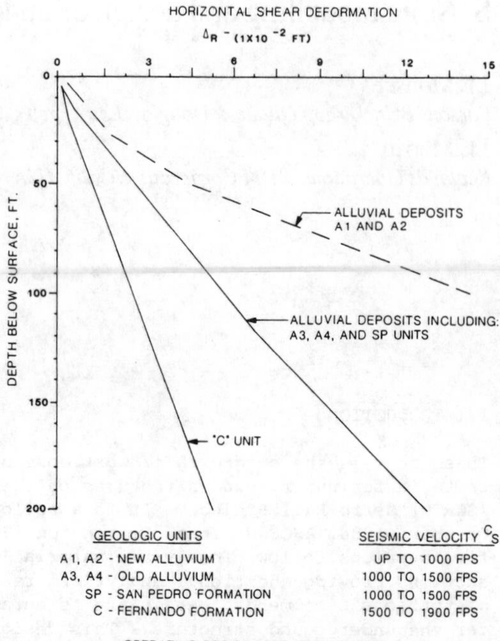

Figure 1. Horizontal Shear Deformation in Various Geologic Units

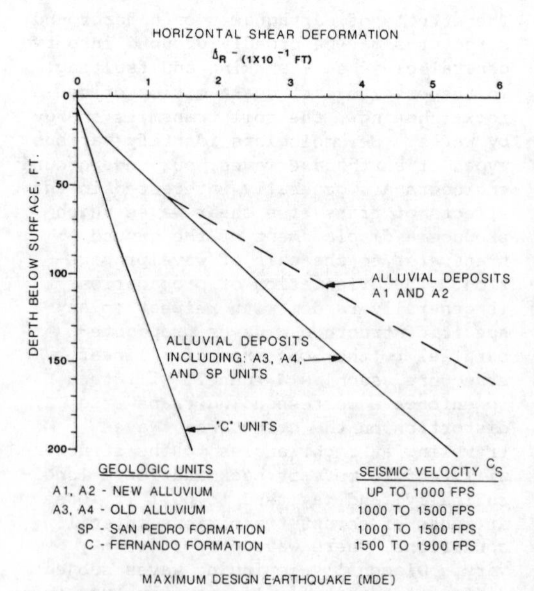

Figure 2. Horizontal Shear Deformation in Various Geologic Units

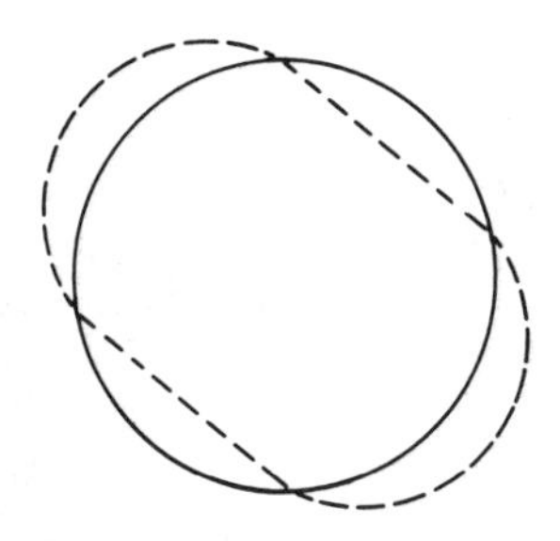

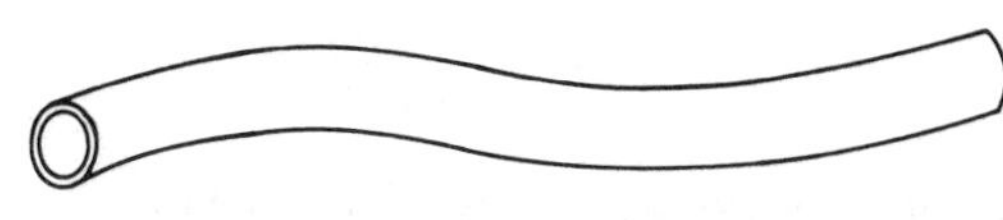

Figure 3. Primary Types of Deformation

Table 2. Results Of "Shake" Calculations Lateral Displacement Relative To Ground Surface, 10^{-2} FT

Depth, Ft	0.3g Input	0.6g Input
5	0.07	0.10
10	0.23	0.36
15	0.47	0.79
20	0.79	1.41
24	1.18	2.19
37.5	2.58	4.98
50	4.68	9.35
67.5	8.10	17.01
85	11.66	25.56
110	16.54	37.16
140	22.13	52.05
175	28.5	67.45
210	35.1	86.64

underground structure depends on the relative stiffness of the structure and the surrounding soil. For tunnel linings this relative stiffness has been described in terms of two separate and distinct ratios. The compressibility ratio is a measure of the extentional stiffness of the medium relative to that of the lining and the flexibility ratio is a measure of the flexural stiffness (resistance to ovaling) of the media relative to that of the lining (26, 27).

Calculations, observations, and experience have shown that the compressibility has very little effect on the behavior of the tunnel and that the liner will behave essentially as a perfectly flexible structure if the flexibility ratio is larger than 20. For each application, it is necessary to check the flexibility, but experience has shown that for most cases the flexibility ratio will exceed 20 (14, 15, 16, 20, 26, 28) except for reinforced concrete linings in soft soils. Thus, most properly designed linings are flexible and will conform to the distortions imposed by the medium, i.e., liner distortions can be estimated by calculating the free field distortions. For very soft soils, interaction between the soil and the structure may be considered, but for any reasonably competent ground this interaction may be ignored. It is always a conservative assumption to assume that the structure experiences the distortions of the free field as described more fully below. Consideration of interaction of a stiffer structure will always give structural distortions less than the free-field distortions.

4.1 Snaking Mode for Tunnel Linings and Stations

Flexibility of the subway structure in plan view, where the structure appears as a long narrow tube or "snake" subjected to the earthquake ground wave motion (Figure 3), has not been defined mathematically in the literature. However, based on the following considerations, it is recommended that the structure be investigated by imposing on the structure, as a design bound, the strains produced in the ground by the earthquake (15,16):

a. If the structure exhibits the same flexibility as the ground replaced, assuming the structure strains are equal to the strains in the ground, is the exact solution.

b. If the structure is more flexible than the ground, the structure will certainly follow exactly the ground motions so that once again assuming the structure strains are equal to the strains in the ground is the exact solution.

c. If the structure is less flexible than the ground, it will attempt to move less than the ground moves in the free field. By moving less than the ground, the structure necessarily is strained less than the ground. Therefore imposing the strains in the ground on the structure is conservative.

Thus, for all cases of structural flexibility, the strain in the ground is either equal to or a bound of the strain in the structure, and imposing the strain in the ground on the structure is a conservative assumption.

5 REQUIRED CONDITIONS FOR PSEUDOSTATIC TREATMENT OF WAVE RESPONSE

Analytical studies by Paul (24), Yoshihara (38), and Akbarian and Johnson (5) show that the dynamic amplification of stresses associated with a stress wave impinging on a tunnel are negligible if the rise time of the pulse is more than about 2 times the transit time of the pulse across the opening; or in other words if the wave length of peak velocities, λ in Figure 4, is at least 8 times larger than the width of the opening. In these cases, the free field stress gradient across the opening is relatively small and the seismic loading can be considered as a pseudostatic load. The local stress and strain increase near the boundary of the opening can be estimated by using the peak stresses associated with the stress wave and the static stress concentration factors for the shape of the opening.

The transit time of a shear wave across the opening can be estimated as

$$t_T = \frac{B}{C_s}$$

where B is the span of the opening and C_s is the shear wave velocity of the soil or rock surrounding the underground opening. The rise time may be taken as about 1/4 of the period associated with the frequency, f, at which peak velocities would occur in an earthquake motion in the same medium which is considered to be surrounding the tunnel.

The rise time and duration of the shaking motion characteristic of an earthquake can be approximated by counting the zero crossings of an earthquake accelerogram during the total time of significant motion. This number divided by the total time is nearly equal to twice the frequency of motion in the velocity pulses. Since the amplitudes of velocity and strain are proportional to one another, this frequency applies to velocity and strain. Earthquake shaking is random motion both in amplitude and frequency. Thus, the frequency defined in this fashion is an average.

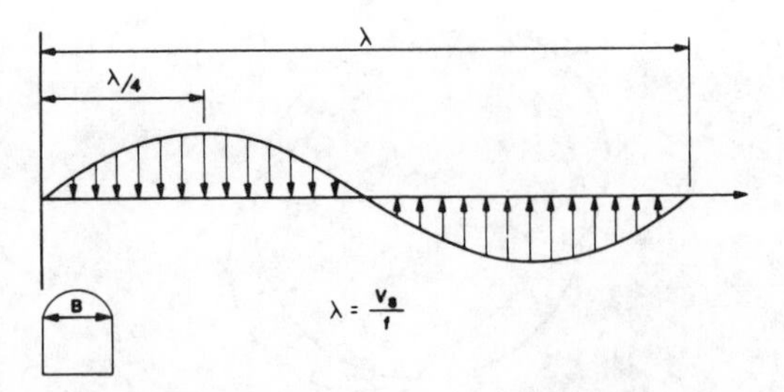

Figure 4. Comparison of Wave Length of Particle Velocities and Size of Underground Opening

Use of this average is generally satisfactory since the random nature of shaking is such that significant resonances cannot develop.

Zero crossings were counted in records from El Centro, 1940 (22) and Brawley Airport, 1979 (Imperial Valley) (31) to represent earthquake records for soil conditions similar to the stiff alluvial deposits in Los Angeles. The peak accelerations for these records ranged from 0.19 to 0.33 g and the number of zero crossings of the acceleration-time history ranged from 7.9 to 8.8 crossings per second. Thus the average frequency of the acceleration pulses ranged from 4.0 to 4.4 cycles per second, and the rise time of the partial velocity pulses would be greater than 1/4 the periods associated with these frequencies.

Thus the shortest rise time to be expected would be

$$\frac{1}{4.4 \text{ cycles/sec.}} \times 1/4 = .057 \text{ sec.}$$

The station structures have a span of about 50 ft. in stiff alluvium. The transit time is about

$$\frac{50 \text{ ft.}}{1,200 \text{ ft./sec.}} = .042 \text{ sec.}$$

Thus the ratio of rise time to transit time should be as low as .057/.042 = 1.36, which is less than the ratio of 2.0 for which there is no dynamic amplification, but the ratio is high enough that the dynamic amplification is not significant for design. The ratio of rise time to transit time is even greater for stations in rock or stiffer soils.

6 DESIGN OF STATION FOR RACKING

If it can be established that the maximum deformation imposed by the earthquake will not strain the structural frame beyond yield at any point no further provisions to resist the deformation are required. If certain joints are strained into the plastic range, the structure must be checked and redesigned as necessary to assure that no plastic hinge combinations capable of leading to a collapse mechanism can be formed. The general procedure is as follows:

1. Base initial size of members on static design and appropriate strength requirements.

2. Impose earthquake deformation (racking) on the structure using data from Figures 1 and 2 and following the concept shown in Figure 5. These racking deformations induce moments and other internal forces in the structure. These forces are added to those from the static analysis. Each racking component (horizontal and vertical) is considered separately, not summed.

3. Evaluate conditions in the structure considering the following situations with appropriate load factors.

a. Where the permanent subway structure is designed for "at-rest" earth pressures, no increase in pressures during or subsequent to an earthquake need be considered. Where structures are designed for static active earth pressures, both active and at rest pressures will be used in the above equations for dynamic loading (8, 28).

b. Vertical loads for cut-and-cover structures from soil backfill, water in soil, structural components, or other materials, over cut-and-cover structures will be increased by 20 percent and 40 percent for ODE and MDE respectively to account for vertical acceleration, thus, the total vertical load is 1.2 or 1.4 times the static value for the two specified design earthquakes.

c. Evaluate possible mechanisms, see Figure 6. Conditions such as those in Figure 6-a., are acceptable because a failure (collapse) mechanism has not formed. Conditions such as those in Figure 6-b, are acceptable because collapse is prevented by the surrounding soil. Mechanisms such as 1, 2, 3, 4, or 5 in Figure 6-c, however, would lead to collapse and are, therefore, not acceptable.

d. Redistribution of moments in accordance with ACI 318 is acceptable (1).

e. Consideration of plastic hinges is acceptable (6, 18, 25).

f. Under static or pseudo-static loads, the maximum usable compressive concrete strain for flexure is 0.004 and for axial loading is 0.002^c (10, 11, 12, 17, 30, 33).

4. Check the structure for strain in the longitudinal direction resulting from frictional soil drag.

5. Design completed if ultimate conditions in the context of plastic design are not exceeded at any point for the reinforcement selected in initial static design; however, note item 6b. below.

6a. Modify the sizes of structural elements as necessary so that an acceptable design results.

6b. Select the reinforcing steel percentages to avoid brittle behavior. Follow ACI (1, 2) or the Uniform Building Code (34).

7 VERIFICATION OF APPROACH

To demonstrate the validity of the approach suggested herein, a series of computer analyses of representative problems has been initiated by the SCRTD Project. These analyses consist of two- and three-dimensional finite element runs of selected geometries using an input earthquake record consisting of the 1971 San Fernando earthquake scaled up to the SCRTD design earthquakes. Results are available for four sets of models (4):

1. A 3-D model (Figure 7).

2. A 2-D model that corresponds to a vertical slice through the cross-section of the station.

3. A 2-D model that corresponds to a vertical slice through the length of the station and soil medium.

4. A 2-D model that corresponds to a horizontal slice through the length of the station and soil medium.

Procedures for modeling material properties had to be developed in three areas, as follows:

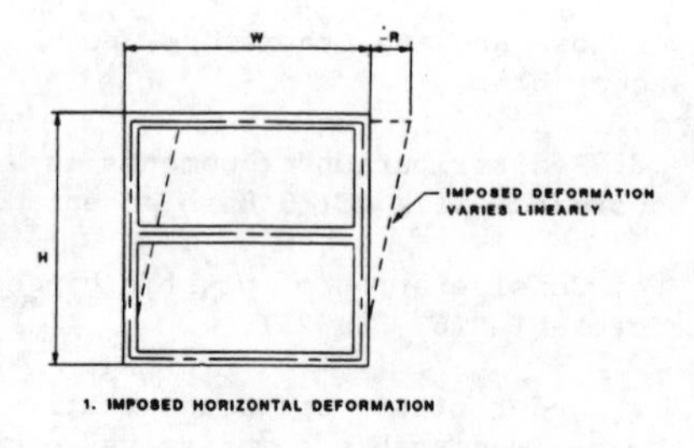

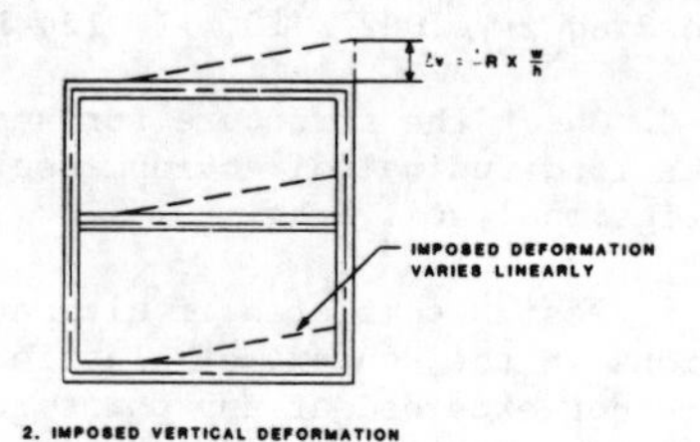

Figure 5. Earthquake Racking of Structure Application of Imposed Earthquake Deformation

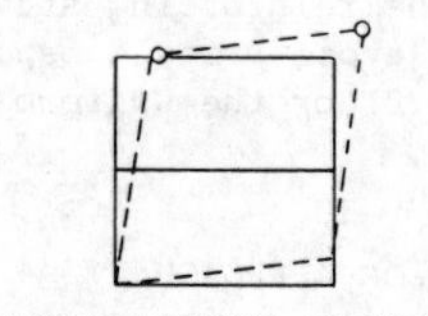

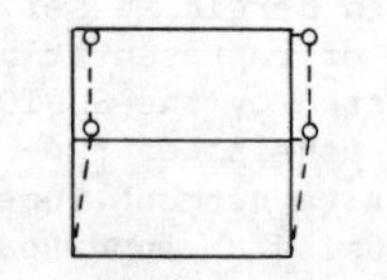

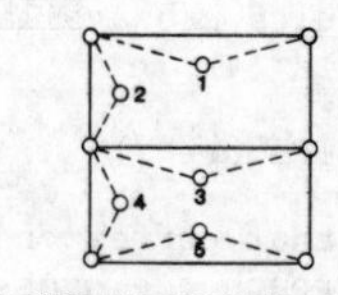

Figure 6. Structure Mechanisms

1. Cracked Section Properties - Design requirements provided that the station structure's resistance to racking deformation be based on cracked section properties for the reinforced concrete structure. To meet this requirement, the elastic modulus of the concrete (E_s) was

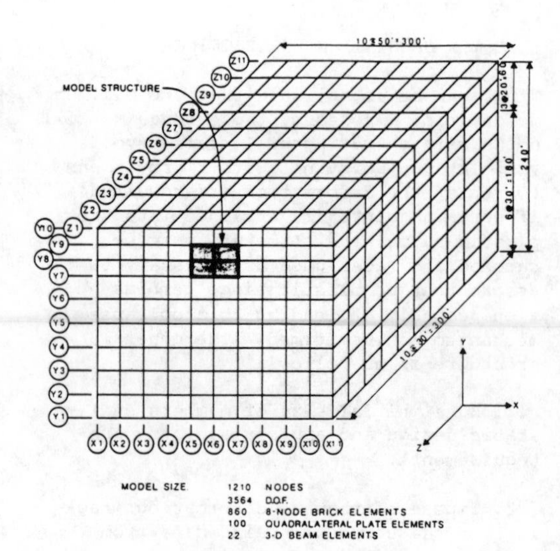

Figure 7. Three Dimensional Finite Element Model

calculated following the American Concrete Institute (ACI) Code (1). This modulus was increased by 10% for dynamic effects ($E_d = E_s$ x 1.10) and then the cracked section modulus of $0.5E_dI$ was calculated following the ACI Commentary (2).

2. Internal Damping Ratios - A relationship was developed that resulted in an internal damping ratio of 5 to 7% of critical over the frequency range of interest for this soil/structure system.

3. Energy-Absorbing Boundary Conditions - The TRI/SAC Code (3) used for this study uses energy-absorbing boundary dashpots of the type developed by Lysmer et al., (19). It is recognized that these dashpots introduce degrees of approximation into the model that may, for example, result in unwanted absorption of certain stress waves. One remedy is to extend the model boundaries beyond the range of interest. This strategy was considered impractical. The approach selected was to apply the dashpots selectively as pioneered by Lysmer et al., (19), to exercise care and judgement in developing the model, and to interpret results so as to identify and isolate any boundary effects that may occur (4).

By using these techniques, a series of runs of the four models were executed to allow comparison of static vs dynamic and 2-D vs 3-D results. Two of those comparisons will be discussed since they illustrate the major points.

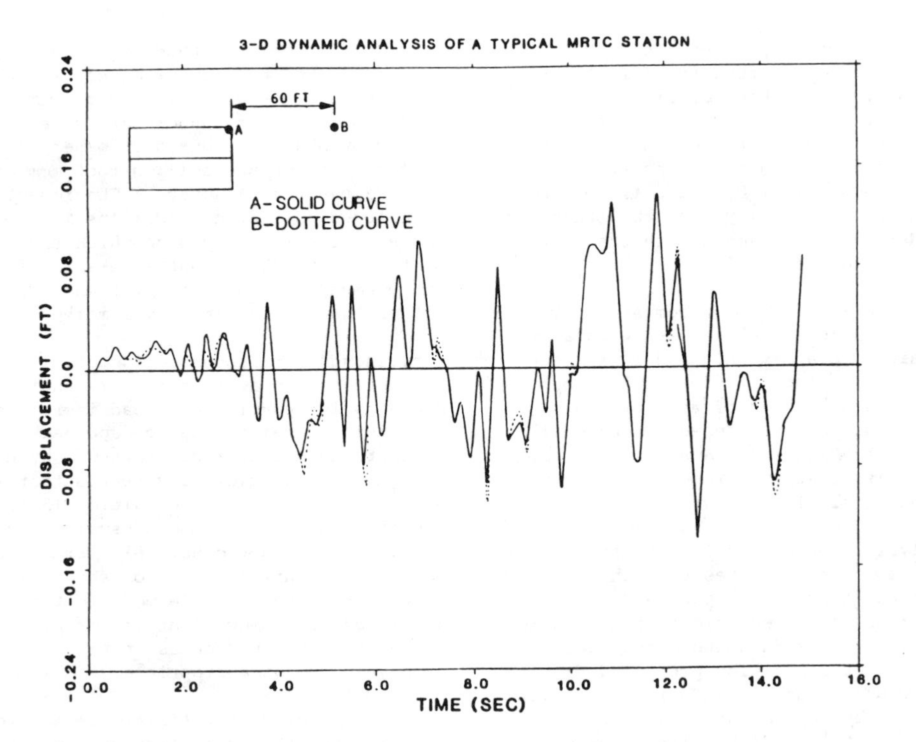

Figure 8. Comparison of Movement-Structure and Grounds

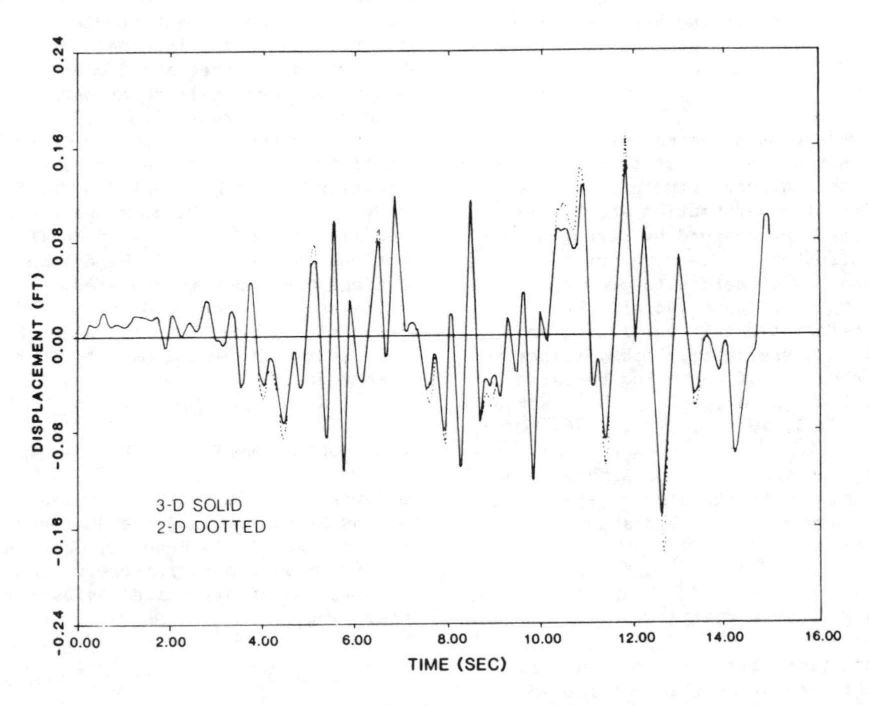

Figure 9. Comparison of of 2-D and 3-D Analyses

In the first comparison, the model is that of a buried structure of rectangular cross section as shown in Figure 7. Plotting the relative deformations for the upper right-hand corner of the structure and those 60 feet away in the free field, the time history of Figure 8 is obtained. Inspection will verify that the structure does deform in agreement with the free-field ground deformations.

In a similar manner, a comparison run of the three-dimensional results with a two-dimensional plane-strain model was made. The plot of relative deformations for the corresponding corner of the structure (two-dimensional vs. three-dimensional) is given in Figure 9. Again, inspection will verify that the two plots are essentially identical.

Analysis of the data from these preliminary runs thus indicates that the structure does follow the ground and that a two-dimensional simplification is justified. That same data indicates that the magnitude of distortion predicted from SHAKE runs (Figures 1 and 2) may be overestimated by approximately 25 percent in comparison to the distortion given by the finite element solution (4). This difference, however, may be due to the degradation of stiffness with shear strain level used in the SHAKE program.

8 CLOSURE

The design approach recommended herein recognizes that the effect of the earthquake on underground structures is the imposition of a deformation which generally cannot be changed by strengthening the structure. The structural design goal is therefore to provide sufficient ductility to absorb the imposed deformation without losing the capacity to carry static loads, rather than a criterion of resisting resisting inertial loads at a specified unit stress (1, 2, 6, 7, 9, 18, 23, 29, 34, 36).

It has been shown by approximate methods and preliminary finite element analyses that this approach is valid and conservative.

9 SUGGESTED FUTURE WORK

In conducting this work, the authors have identified areas that require additional development for the general case:

a. Static design methods for tunnels, especially to incorporate concepts such as loosening loads, require more work. Loosening loads on tunnels are those caused by shaking induced or other partial failure of the soil on rock immediately above the structure. The impact of such loads on static conditions may not be known to the accuracy with which the distortion approach estimates seismic effects for elements constructed by tunneling rather than by cut-and-cover methods.

b. Though not discussed in detail herein, the authors are aware that some engineers would apply dynamic soil load increments as an alternate seismic design approach for underground structures. Further, even though it is specifically stated that it does not apply to this condition (32), the Mononobe-Okabe solution is usually suggested for want of any other (8). Either the validity of this application should be proven and an applicable modification developed, or a new solution for underground structures such as stations should be developed, as appropriate.

c. This preliminary finite element work demonstrates that the structural distortion follows the soil, but it also indicates that results from SHAKE yield soil distortion about 25 percent larger than those from the finite element studies. This is probably due to the internal sub program in SHAKE which utilizes a different shear modulus-versus-strain relationship, thereby yielding a different effective shear modulus for the soil than was input to the elastic finite element analyses. This situation should be checked by further study. Our observation, supported by preliminary study of designs on 18 SCRTD sections, shows that if the design distortions given herein are used, the effect of the seismic racking on design of underground structures is to increase the required reinforcement by 5 to 10 percent at some joints.

10 ACKNOWLEDGEMENT

The authors thank Mr. T. R. Kuesel, Parsons Brinckerhoff Quade and Douglas, and Professor A. J. Hendron, Jr., University of Illinois, for their contributions to and constructive reviews of the underlying.

11 REFERENCES

(1) ACI Committee 318 (1983), Building Code Requirements for Reinforced Concrete (ACI 318-83), American Concrete Institute, Detroit.

(2) ACI Committee 318 (1983), Commentary on Building Code Requirements for Reinforced Concrete (ACI 318-83), American Concrete Institute, Detroit.

(3) Agbabian Associates, (1980), "User's Guide for TRI/SAC Code," Rev. R-7128-4-4102, El Segundo, California.

(4) Agababian Associates, (1985), "Baseline Dynamic Analyses in Support of Supplemental Seismic Design Criteria - Metro Rail Project," Report to Metro Rail Transit Consultants, Los Angeles.

(5) Ali-Akbarian, M. and Johnson, J. (1969), "Oblique Incidence of Plane Stress Waves on a Thick Cylindrical Shell", Technical Report No. AFWL-TR-69-56, Air Force Weapons Laboratory, Kirtland Air Force Base, New Mexico.

(6) ATC (1978), "Tentative Provisions for the Development of Seismic Regulations for Buildings. A Cooperative Effort with the Design Professions, Building Code Interests and the Research Community," Applied Technology Council Publication ATC 3-06, NBS Special Publication 510, NSF Publication 78-8.

(7) Blume, J.A. et al. (1961), Design of Multistory Reinforced Concrete Buildings for Earthquake Motions, Portland Cement Association, Chicago.

(8) Converse Consultants (1983), "Seismological Investigation & Design Criteria," Southern California Rapid Transit District, Metro Rail Project.

(9) Ellingwood, B. et al. (1980), "Development of a Probability Based Load Criterion for American National Standard A58," NBS Special Publication 577, Washington.

(10) Ford, J.S. et al. (1981a), "Behavior of Concrete Columns Under Controlled Lateral Deformation", Journal of the American Concrete Institute, January - February 1981, No. 1, Proceedings V.78, pp. 3-20.

(11) Ford, J.S. et al. (1981b), "Experimental and Analytical Modeling of Unbraced Multipanel Concrete Frames," Journal of the American Concrete Institute January - February 1981, No. 1, Proceedings V.78, pp. 21-35.

(12) Ford, J.S. et al. (1981c), "Behavior of Unbraced Multipanel Concrete Frames," Journal of the American Concrete Institute, March - April 1981, No. 2, Proceedings V.78 pp. 99-115.

(13) Not used.

(14) Hendron, A.J., Jr. and Fernandez, G. (1983), "Static and Dynamic Considerations for Underground Chambers," in Seismic Design of Embankments and Caverns, Proceedings of a Symposium, T. R. Howard, Ed., Philadelphia, 1983, published by ASCE, N.Y.

(15) Hendron, A.J., Jr., (1978), "Considerations for Tunnel Lining Design - Sanegran Interceptor Tunnels," draft report.

(16) Hendron, A.J., Jr., (1973), "Evaluation of Cooling Water Tunnels - Seabrook Nuclear Station," draft report to United Engineers and Constructor.

(17) Hognestad, E. (1951), "A Study of Combined Bending and Axial Load in Reinforced Concrete Members," University of Illinois Bulletin, Volume 49, Number 22, November 1951, Bulletin Series No. 399.

(18) Housner, G.W. and Jennings, P.C. (1982), "Earthquake Design Criteria," Earthquake Engineering Research Institute, Volume Four of a series titled: Engineering Monographs on Earthquake Criteria, Structural Design, and Strong Motion Records.

(19) Lysmer et al, (1975), "Flush--A Computer Program for Approximate 3-D analysis of Soil-Structure Interaction Problems," EERC 75-30, University of California, Berkley, California.

(20) Monsees, J.E., and Merritt, J.L. (1984), "Seismic Design of Underground Structures - Southern California Metro Rail Project," in Lifeline Earthquake Engineering: Performance, Design, and Construction, Proceedings of a Symposium, J.D. Cooper, Ed., San Francisco, 1984, Published by ASCE, N.Y.

(21) Merritt, S.J.E. et al, (1985), "Seismic Design of Underground Structures," Rapid Excavation and Tunneling Conference (RETC) Proceedings, N.Y., N.Y., 1985, Published by ASCE-AIME, Littleton, Colorado.

(22) Newmark, N.M. and Rosenblueth, E. (1971), Fundamentals of Earthquake Engineering, Prentice-Hall, Inc., Englewood Cliffs, N. J.

(23) Nyman, D.J. (1983), Principal Investigator, "Guidelines for the Seismic Design of Oil and Gas Pipeline Systems," Committee on Gas and Liquid Lifelines, ASCE.

(24) Paul, S.L. (1963), "Interaction of Plane Elastic Waves with a Cylindrical Cavity", Ph.D. Thesis, University of Illinois.

(25) Park, R. and Paulay, T. (1975), Reinforced Concrete Structures, John Wiley & Sons, New York.
(26) Peck, R.B. et al. (1972), "State of the Art of Soft Ground Tunneling," First North American Rapid Excavation and Tunneling Conference, ASCE-AIME, Chicago.
(27) Peck, R.B. (1969), "Deep Excavation and Tunneling in Soft Ground," State of the Art Volume, Seventh International Conference on Soil Mechanics & Foundations, Mexico City.
(28) Ranken, R.E. et al. (1978), "Analysis of Ground-Liner Interaction for Tunnels," U.S. Department of Transportation Report No. UMTA-IL-06-0043-78-3, Department of Civil Engineering, University of Illinois at Urbana-Champaign Report No. UILU-ENG-78-2021.
(29) Rosenblueth, E. (1980), Design of Earthquake-Resistant Structures, John Wiley & Sons, New York.
(30) Roy, H.E.H. and Sozen, M.A. (1964), "Ductility of Concrete," Flexural Mechanics of Reinforced Concrete, Proceedings of the International Symposium, Miami, Florida, 10-12 November 1964, sponsored by ASCE/ACI, pp. 213-235.
(31) SCRTD (1984), Supplemental Criteria for Seismic Design of Underground Structures, Los Angeles.
(32) Seed, H.B. and Whitman, R.V. (1970), "Design of Earth-Retaining Structures for Dynamic Loads," presented at the 1970 Specialty Conference Lateral Stresses in the Ground and Design of Earth-Retaining Structures, sponsored by ASCE, 22-24 June 1970 at Cornell University, Ithaca, N. Y., pp. 103-147.
(33) Sozen, M. A., (1985), Private Communication.
(34) <u>Uniform Building Code</u> (1982), International Conference of Building Officials, Whittier, California.
(35) Not used.
(36) Wiegel, R.L. et al. (1970), Earthquake Engineering, Prentice-Hall, Inc., Englewood Cliffs, N.J.
(37) Not used.
(38) Yoshihara, T. (1963), "Interaction of Plane Elastic Waves with an Elastic Cylindrical Shell", Ph.D. Thesis, University of Illinois.

Numerical Methods in Geomechanics (Innsbruck 1988), Swoboda (ed.)
© 1988 Balkema, Rotterdam. ISBN 90 6191 809 X

The foundation influence in the earthquake response of arch dams

Raimundo M.Delgado & Rui M.Faria
Oporto University, Portugal

ABSTRACT: In the context of the earthquake response of arch dams, it will be presented two different ways for modelling the soil foundation: the Vogt's coefficients (with an addicional mass) and the three-dimensional finite elements. It will be shown how to reduce the dimension of the stiffness and the mass matrices, using a static or dynamic condensation. A simplified Rayleigh's method to obtain the first modal shapes of vibration will be proposed. Finally, some applications of these subjects will be done to Caldeirão Dam.

1 INTRODUCTION

It is of great interest to analyse the modifications that can be verified in the dynamic behaviour of arch dams, when the foundation is considered as infinitely rigid or with deformation. In fact, some differences can be reported, either in the frequencies and modal shapes of vibration, or in the stresses, which represent the final response of the arch dam to the earthquake.

The Vogt's coefficients technique is widely used to reproduce the effects of an elastic foundation, not only when static solicitations are present (such as the hydrostatic impulse), but also when a dynamic behaviour is to be analysed. In the latter situation, an additional mass of soil must be added to the nodal points (of the mesh of finite elements in which the dam is divided) close to the foundation.

An alternative model for the foundation can be performed, using the three-dimensional finite elements of soil. In this case, it is possible to take into account singularities which can be present at the foundation. Meanwhile, the dimensions of the problem will increase rapidly (taking by reference the Vogt's technique).

One possible way of reducing the dimension of the stiffness and mass matrices is to make a static or dynamic condensation. The former will allow a great reduction of time of computer calculation. Although, a dynamic condensation will be preferable, everytime it might be necessary a greater (or even complete) accuracy; time of calculation will grow with the increment of the desired accuracy. With any kind of condensation, the dimensions of the matrices involved in the calculations will be significantly lower, and so it will be possible a great economy of internal computer memory.

Sometimes it will be enough to obtain only the order of magnitude of the stresses due to the first modal shapes of vibration. This can happen, for instance, when it is still being performed an optimization of the shape of the dam and, at that stage, a complete dynamic calculation would not be worth while. A Rayleigh's technique can then be improved, so that it might be obtained an approximation of the first modal shapes of vibration, using a classic program of finite elements, prepared only to consider static solicitations.

2 MODELS OF STRUCTURAL ANALYSIS

2.1 Dam modelling

The arch dam is represented through the use of a mesh of thick shell finite elements. In particular, it was used the Ahmad's degenerated thick shell element, in which are considered eight nodal points in the central surface. Each nodal point has got five degrees of freedom: three displacements along the axis of reference, and two rotations of the normal to the central surface at that point. Though, each finite element will have forty degrees of freedom. The stiffness and the mass matrices are calculated accordingly. The mass matrix is of consistent kind.

2.2 Model of the impounded water

The water that is present in the reservoir is taken into account with an extension of the Westergaard's added mass. As the Westergaard's concept is only valid when the seismic motion is normal to the dam's surface, some modifications shall be necessary to take into account the fact that in an arch dam, for a given input direction of an earthquake, the direction of vibration at each nodal point is not necessarily normal to the surface at that point (Priscu 1985: Earthquake engineering for large dams). So, after having evaluated the resultants of the hydrodynamic pressures at the upstream face of a given finite element, the correspondent vector of nodal forces will permit (after the appropriate transformations just referred) to calculate the additional mass of water at each nodal point. This mass will then be added to the mass matrix that was previously determined for the concrete mass, to constitute the mass matrix of the ensemble concrete + impounded water.

2.3 Soil modelling

Utilizing known formulas for deformations of an infinite foundation with a plane surface, Dr. Frederic Vogt has determined the relations between the deformations and the corresponding efforts applied to a rectangular area on the foundation. This technique is equivalent to introduce elastic supports at each nodal point of the arch dam, just connected to the foundation. Its corresponding deformability matrix relates the three displacements and the three rotations (associated to a set of axis in the plane of the foundation) to the corresponding forces and moments.

Using a similar procedure it is possible to estimate the mass and the dumping factor associated to a rectangle with unit length, whose deformability is represented by the Vogt's coefficients, and that is subjected to a dynamic vibration. That mass and dumping factor depend on the elastic characteristics of the soil, its unit weight, on the geometry of the foundation, and on the frequency of the vibration (O. Pedro 1977: Dimensionamento das barragens abóbada pelo método dos elementos finitos).

A set of three-dimensional finite elements of twenty nodes constitutes a powerful way to model the soil - with a significative increase in terms of computer memory and time of CPU. Beasides, some additional procedures shall be necessary to assemble the two different finite elements that are to be used at the same time: with eight nodes for the dam and with twenty nodes for the soil (R. Delgado, 1984: O método dos elementos finitos na análise dinâmica de barragens incluíndo a interacção sólido-líquido). In this paper it will be made a comparison between these two models.

3 DYNAMIC ANALYSIS

To evaluate the dynamic response of the dam, a modal analysis has been used. A modification of the Moler-Stewart subroutines was used to obtain the frequencies and modal shapes of vibration of the dam. The influence of the earthquake is taken into account by the use of the design response spectra established at the portuguese regulation (1985, Regulamento de Segurança e Acções para Estruturas de Edifícios e Pontes). The maximum response corresponding to each mode of vibration is obtained effectuating the square root of the sum of the squares of the responses due to each direction of the earthquake. Finally, a complete quadratic combination (CQC) is made, to compute the maximum total response of the dam.

4 CONDENSATION OF DYNAMIC PROBLEMS

4.1 Dynamic condensation

As has been referred at 2.3, the inclusion of three-dimensional finite elements on the foundation will increase significantly the size of the mass and stiffness matrices. There are frequently other reasons, similar to that, which can cause a large number of degrees of freedom: modelling the impounded water with finite elements, changes in geometry, and so on.

A reduction of the dynamic problem will appear, sometimes, very convenient.

In the dynamic condensation method, having assigned an approximate value to the i-th eigenvalue ω_i^2, it will be determined the ith and (i+1)th eigenvalues of a reduced problem, in which are involved two reduced matrices: $[\bar{K}]$ and $[\bar{M}]$ (respectively the reduced stiffness and mass matrices), of a lower size than the total [K] and [M] matrices of the structure. The (i+1)th eigenvalue just calculated will permit to repeat the iteration process, and hence to determine the (i+2)th eigenvalue. The dynamic condensation, if wanted, will produce exact results, even when a great reduction is made to the dynamic problem. In fact, it is possible to improve each eigenvalue ω_i^2: it is only necessary to produce so many iterations for each eigenvalue, so that the error at its value, between two consecutive iterations, would be lower than an established limit.

According to Mario Paz (M. Paz, 1984: Structural Dynamics - Theory and computation) if the simplification $\{y\} = \{Y\} \sin \omega_i t$ is introduced, the classic dynamic equations of equilibrium $[K].\{y\} + [M].\{\ddot{y}\} = \{0\}$ will be transformed into

$$\left[\begin{array}{c|c} [K_{ss}]-\omega_i^2[M_{ss}] & [K_{sp}]-\omega_i^2[M_{sp}] \\ \hline [K_{ps}]-\omega_i^2[M_{ps}] & [K_{pp}]-\omega_i^2[M_{pp}] \end{array}\right] \left\{\begin{array}{c} \{Y_s\} \\ \hline \{Y_p\} \end{array}\right\} = \left\{\begin{array}{c} \{0\} \\ \hline \{0\} \end{array}\right\}$$

(letter p associated to the remaining or primary degrees of freedom, and letter s associated to the secondary degrees of freedom which are to be reduced). Giving to ω_i^2 the value determined at the precedent iteration (or the zero value when at the beginning of the process), and making a Gauss-Jordan elimination of $\{Y_s\}$, it will be produced the following:

$$\begin{bmatrix} [I] & -[\bar{T}_i] \\ [0] & [\bar{D}_i] \end{bmatrix} . \begin{Bmatrix} \{Y_s\} \\ \{Y_p\} \end{Bmatrix} = \begin{Bmatrix} \{0\} \\ \{0\} \end{Bmatrix}$$

At this moment, two important matrices are determined: $[\bar{T}_i]$ and $[\bar{D}_i]$. [I] is the identity matrix.

The reduced mass matrix $[\bar{M}_i]$ is calculated with

$$[\bar{M}_i] = [T_i]^T [M] [T_i]$$

in which

$$[T_i] = \left[\begin{array}{c} [\bar{T}_i] \\ \hline [I] \end{array}\right].$$

The reduced stiffness matrix is determined accoding to

$$[\bar{K}_i] = [\bar{D}_i] + \omega_i^2 [\bar{M}_i] .$$

With $[\bar{K}_i]$ and $[\bar{M}_i]$ it will be solved the eigenproblem

$$[[\bar{K}_i] - \omega^2 [\bar{M}_i]] \{Y_p\} = \{0\}$$

and so the eigenvalues ω^2 and corresponding eigenvectors $\{Y_p\}$ are calculated. The secondary degrees of freedom shall be evaluated with $\{Y_s\} = [\bar{T}_i] \{Y_p\}$.

4.2 Static condensation

Notice that if it is made $\omega_1^2 = 0$ and no iterative process is stated, that is, if the desired n first eigenvalues and eigenvectors are taken as the n first eigenvalues and eigenvectors of the eigen-problem $[[\bar{K}_1] - \omega^2[\bar{M}_1]].\{Y_p\} = \{0\}$, the process will be considerably less time consuming. In this situation $[\bar{K}_1] = [T_1]^T [K] [T_1]$ and $[\bar{M}_1] = [T_1]^T [M] [T_1]$. This is the theory of the static condensation. Experience reveals that attention must be given to the kind and the number of degrees of freedom which are reduced. Otherwise, some signifivative errors can be introduced.

5 THE RAYLEIGH'S METHOD

If a given arch dam is geometrically and mechanically symetrical, its first two modes of vibration have a quite simple shape. This suggests that an attempt to reproduce them and obtain their frequencies with a simplified method would be practicable and very convenient (at least for the preliminary stages of the project). It is well known that during a vibration process, the displacements $\{\tilde{d}\}$ associated to a given modal shape can be expressed as $\{\tilde{d}\} = \{\tilde{y}\}\, Z \sin \omega t$, $\{\tilde{y}\}$ representing the shape of the mode, Z being a modal co-ordinate and ω the angular frequency. It is possible to express the maximum speed $\{\dot{\tilde{d}}\}_{max}$ and the maximum acceleration $\{\ddot{\tilde{d}}\}_{max}$ at the degrees of freedom as a function of $\{\tilde{y}\}$, Z and ω:

$$\{\dot{\tilde{d}}\}_{\bar{max}} = \omega.\{\tilde{y}\} . Z$$

$$\{\ddot{\tilde{d}}\}_{\bar{max}} = \omega^2.\{\tilde{y}\} . Z$$

If an acceleration equal to the gravity acceleration g is applied to the dam in the direction of each axis of reference, the corresponding deformation will be an approximation to the modal shape $\{\tilde{y}\}$. At the same time, the nodal force vector at each finite element will contain a set of forces which are proportional to a simplified diagonal mass matrix for that element. Assembling those nodal forces will permit to obtain a diagonal mass matrix to the total dam, whose non-nule terms will be named as $\{\tilde{f}_1\}$.

The maximum kinetic energy of the dam is then given by

$$T = \frac{1}{2\times g}\{\underset{\sim}{f_1}\}^T\{\underset{\sim}{\dot{d}^2}\}_{max} = \frac{\omega^2\times Z^2}{2\times g}\{\underset{\sim}{f_1}\}^T\{\underset{\sim}{y^2}\},$$

and the maximum potential energy is equal to

$$V = \frac{Z}{2}\{\underset{\sim}{f_1}\}^T\{\underset{\sim}{d}\}_{max} = \frac{Z^2}{2}\{\underset{\sim}{f_1}\}^T\{\underset{\sim}{y}\}$$

Now it is possible to apply the Rayleigh's method, which is founded on the principle of conservation of energy, in a freely vibrating system (Clough and Penzien 1975: "Dynamics of Structures"):

$$T = V$$

This will permit (after some manipulation of the expressions) to calculate the angular frequency:

$$\omega = \sqrt{\frac{g\,\{\underset{\sim}{f_1}\}^T\{\underset{\sim}{y}\}}{\{\underset{\sim}{f_1}\}^T\{\underset{\sim}{y^2}\}}}$$

It is important to note that for application of this method to an arch dam, only a common programme of static calculations with finite elements will be needed. If wanted, the calculation of ω can be automatically introduced in the program, with little modifications. Of course, the accuracy on ω will depend on the shape that is proposed for the mode of vibration $\{y\}$. Significative improvement will be obtained if an iterative process is established: after having determined ω_r in the rth iteration, the force of inertia (associated to the ith degree of freedom) to be used in the (r+1)th iteration is calculated according to

$$(f_{r+1})_i = \frac{(\omega_r)^2\times Z_r}{g}\times(f_1)_i\times(y_r)_i$$

Introducing the $\{\underset{\sim}{f_{r+1}}\}$ vector as a new solicitation applied to the arch dam, a new approximation $\{y_{r+1}\}$ to the modal shape will be obtained. The principle of conservation of energy once more being applied, will impose to ω_{r+1} the new expression:

$$\omega_{r+1} = \sqrt{\frac{g\,\{\underset{\sim}{f_{r+1}}\}^T\{\underset{\sim}{y_{r+1}}\}}{Z_{r+1}\,\{\underset{\sim}{f_1}\}^T\{\underset{\sim}{y^2_{r+1}}\}}}$$

(Note the simultaneous presence of $\{\underset{\sim}{f_1}\}$ and $\{\underset{\sim}{f_{r+1}}\}$). The maxima modal amplitudes Z_r and Z_{r+1} can be easily determined at each step.

6 APLLICATIONS TO CALDEIRÃO DAM

Caldeirão Dam is about 34m high (above the foundation, at the central cantilever), and 120m long (at the upper arch). It is 10m thick at the basis of the central cantileve and 4m thick at the crest.

6.1 The Vogt's technique

The calculations were made with a mesh of 12 elements, applied to half of the dam (due to its geometrical symmetry). Full reservoir was considered. To analyse the foundation's effect, the results obtained with a rigid foundation were compared to those obtained with an elastic foundation (Vogt's technique was used, with Young's modulus of the soil equal to the Young's modulus of the concrete). It was observed that the shapes of the first five modes of vibration were not significantly different, but the higher modes (in the case of elasti foundation) exhibited greater vertical displacements. The consideration of the soil's deformation has reduced the frequencies in about 20% (at the lower modes) in comparison to the rigid foundatio

The soil's deformability has increased in about 25% the maxima horizontal stresses. At the bottom of the central cantilever, the vertical stresses have decreased 15%.

6.2 The three-dimensional finite elements on the foundation

To compare with the Vogt's technique, the foundation has been considered through the use of a mesh of 9 three-dimensional finite elements, each one 20m thick in upstream-downstream direction. The dam was modelled with a mesh of 5 thick shell finit elements (Figure 1).

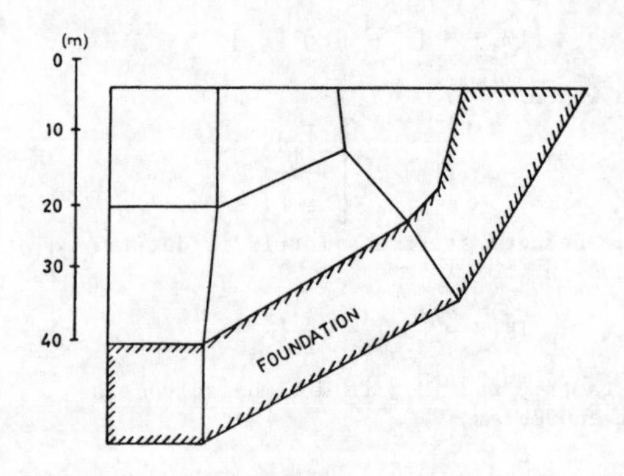

Figure 1. Mesh of finite elements.

Three calculations were made (with full reservoir), each one having a different Young's modulus (EF) in the soil's mesh: EF=EC, EF=EC/2, EF=EC/4 (EC is the Young's modulus for the concrete). Another calculation was made, with a mesh of 5 elements at the dam's concrete, and with the Vogt's coefficients to model the foundation. Table 1 shows the values of the frequencies that have been obtained in the four calculations.

Table 1. Values of the frequencies (Hz) of the symmetrical modes.

	VOGT	EF=EC	EF=EC/2	EF=EC/4
1st	8.2	8.6	8.1	7.4
2nd	11.5	12.9	12.2	11.1
3rd	15.1	17.5	16.4	14.7

It can be seen that, for the same Young's modulus, Vogt's technique gave lower frequencies, in comparison with the 3D finite elements (EF=EC). Figure 2 allows to compare the stresses which were obtained with the four hypothesis just described. The first conclusion is that along the

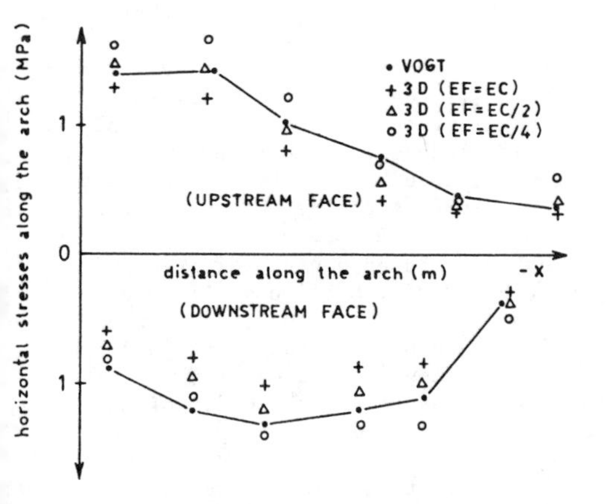

Figure 2. Stresses along the arch at the crest.

central cantilever only sligt differences can be observed in the vertical stresses, when EF is changed. However, along the arch the horizontal stresses increased more significantly, as the EF decreased (a 20% variation to the maxim horizontal stresses calculated with Vogt's coefficients (EF=EC) was noticed, when considering a 3D mesh at the soil, with EF varying between EC and EC/4). This result is a consequence of the fact that rock's deformability, causing tilting and displacements upstream-downstream at the bases of the cantilevers will increment the impulse on the arches.

6.3 Static and dynamic condensation

To analyse the static condensation, some calculations of the first five frequencies of vibration of Caldeirão Dam were done. Full reservoir and rigid foundation were considered, and in this situation a total of 174 degrees of freedom were associated to the mesh of 12 elements described at 6.1. In figure 3, several static condensations are compared (NR means the number of remaining or primary degrees of freedom).

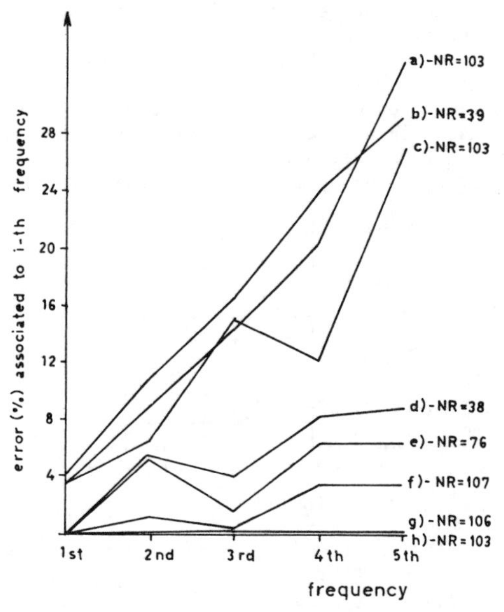

Figure 3. Static condensation.

Noting that (u,v,w) are respectively displacements across the valley, along the valley and vertical, and (α,β) are the rotations of Ahmad's finite element, some important conclusions can be stated:
-curve g), associated to the condensation of all rotations, shows that no visible error is introduced;
-the error will become greater, as the condensation of a "certain kind" increases. See sequence of curves g)-e)-d), in which were condensed respectively: (α,β)-(α,β,u)-(α,β,u,w);
-comparison between curves b) and d) shows that approximately equal NR can produce large differences in the errors. The reason is that in d) were condensed (u,w,α,β), which

have little importance to characterize the modal shape, while in b) many "v" displacements were reduced. An analogous reason explains the curious fact that in d), with NR=38, the results are much better than with NR=103, in curve a): in d) no "v" displacement is eliminated, while in a) some "v" are condensed. The same conclusion applies to curves c) and f): in the former (v,α,β) were reduced and in the latter (u, α,β) were condensed.

Beyond the advantage of reducing computational memory, condensation can also minimize time calculations. Figure 4 shows the time needed to calculate the first 3 modes with EDC (exact dynamic condensation), ADC (approximate dynamic condensation) or SC (static condensation). Line L represents a standard calculation of the eigenvalues. Imposing an appropriate limit to the error associated to each frequency, it will be produced curve EDC, with absolutely exact results. Curve ADC corresponds to a dynamic condensation with an higher error permited to each eigenvalue. The results are not virtually exact, but they are still very good, with less time consuming than EDC.

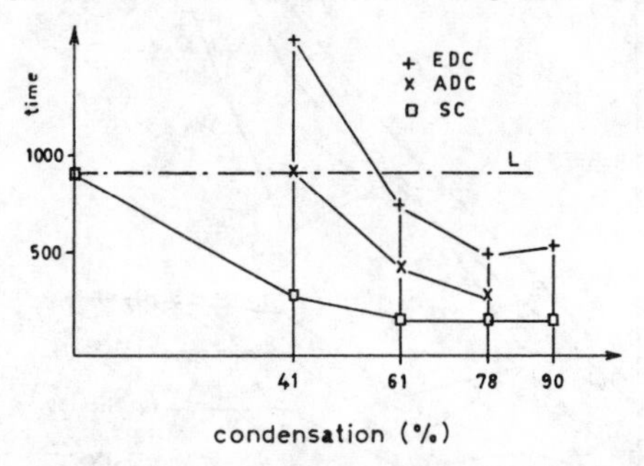

Figure 4. CPU time/degree of condensation.

6.4 Rayleigh's method

The improved Rayleigh's method provided the frequencies of the two first modes of Caldeirão Dam with the errors 0.8% and 5%. Figure 5 represents the stresses of Caldeirão Dam when considering: 10 exact modes, 2 exact modes or 2 approximate Rayleigh's modes. The results prove the accuracy of the proposed method.

7. CONCLUSIONS

1)- The analysis of soil's deformability on the dynamic behavior of Caldeirão Dam has shown that when varying the EF modulus from EF=EC to EF=EC/4, the modifications at the extreme values of stresses on the arch were less than 20%, when compared to Vogt's

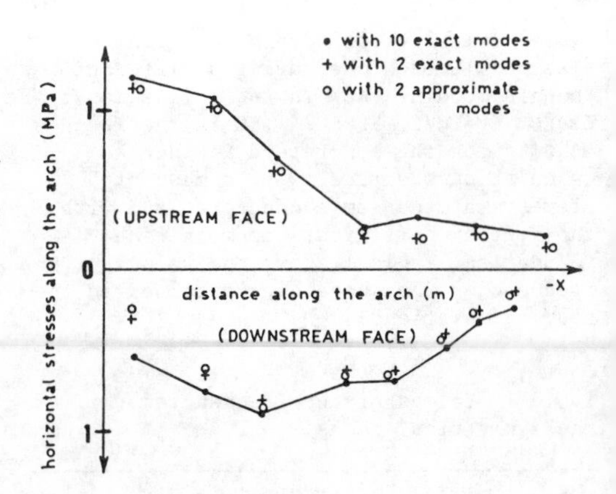

Figure 5. Stresses in the upper arch.

technique (with EF=EC). No significative variation was noticed on the stresses at the central cantilever. The Vogt's coefficients model, with an additional mass on the foundation, constitutes an economical and considerably safe tool for representing soil's deformability at Caldeirão Dam.

2)- Computer time consuming can be significantly reduced when using a static or dynamic condensation. These powerful methods shall be specially advisible in dynamic problems involving the foundation and the water in the reservoir (where many wasteful degrees of freedom require large computer's memory). The less significative degrees of freedom in the modal shape must be the first to be reduced. Preferably, a dynamic condensation must be adopted: as intended, enough accurate or even completely exact solutions can be obtained. With dynamic condensation, whatever may be the number of reduced degrees of freedom, the first modes of vibration shall be exact; yet, a degree of condensation between 60-80% would be preferable, at Caldeirão Dam, to reduce time calculation.

3)- As a way to get the order of magnitude of frequencies and maxima stresses on arch dams (under a seismic action), the Rayleigh's method, as proposed at this paper will be useful. Almost coincident results shall be obtained, when using the two first Rayleigh's modes, or the exact ones.

Numerical Methods in Geomechanics (Innsbruck 1988), Swoboda (ed.)
© 1988 Balkema, Rotterdam. ISBN 90 6191 809 X

A numerical analysis on the shape of the response spectra in cohesive and granular soils with non-linear behaviour

S.Lagomarsino
Istituto di Scienza delle Costruzioni, University of Genoa, Italy

ABSTRACT: The problem of the local amplification of surface deposits in earthquake engineering is full dealt with and much importance is given to the non-linear behaviour of the ground. A numerical analysis, carried out on homogeneous sites of sand or clay, which are different in thickness and geotechnique characteristics, has evaluated the influence of these parameter on the free field response of the ground. The magnitudes taken into account are the peak acceleration of the time history on the surface of soil and the response spectrum in the same point, which are the ones most used at the design level. In particular this contribution sets out to give useful indications for the definition of response spectra in studies of seismic microzonation.

1 INTRODUCTION

Dynamic structural analysis of the effects of earthquake ground motion can be carried out by following two different paths:

1. time history analysis;
2. response spectrum analysis.

The first technique consists in the direct integration of the equations of the dynamic system, subjected to one or more time histories representing the earthquake expected in that site. It makes it possible to determine the structural response as a temporal function and therefore the absolute maximum desired. The second technique starts off from the response spectra given by the normative or processed in some cases for the site concerned (seismic microzonation); by means of modal analysis one arrives at the maximum value as a combination of the maxima on every mode, following more or less approximate rules. This method is without doubt the most used for its simplicity, but it has noteworthy uncertainties both for the modal combination and for the characterization of the response spectrum.

This paper sets out to deal with the latter aspect by analysing the influence of local conditions on the problem. It is in fact widely recognized by now that damage to constructions depends on given considerations, both morphological (depth of the layers, crest effects) and geomechanical (speed of shear waves V_s, damping factor β, density ρ); there are many cases differentiated of damage to buildings of the same kind, quite close to each other and subjected to the same event [7].

2 THE SITE DEPENDENT SPECTRA

The two seismic input data for a check with the modal analysis technique are:

1. maximum ground acceleration a_0;
2. shape of the response spectrum.

The Italian standards, at present in force, divide the territory into zones with three different degrees of seismicity S. The shape of the response spectra is otherwise the only one for the three above categories and for every local soil conditions of the site (except for a foundation coefficient equal to 1.3 for soft deposits).

The National Group for Defense against Earthquake (GNDT) has produced a document [1] proposing a new code for constructions in seismic zones. This still maintains the differentiation of the territory according to three different earthquake motion intensities, corresponding to the following values for a_0: 0.15, 0.25, 0.35 [g's]. Also the shape of the response spectrum is moreover differentiated (Figure 1, dashed lines) to keep in mind the geotechnique and stratigraphic characteristics of the foundation soil. Two different kind of sites are considered:

1. SOIL S_1 - Rock with a covering of less than 5 m; dense sands and gravels or stiff cohesive soils of depth greater than 90 m.

2. SOIL S_2 - Deep alluvial deposits (V_s < 250 m/s and 5 < H < 30 m; V_s < 350 m/s and H > 50 m); deep soils of sand or clay lying on bedrock with shear waves speed smaller than 500 m/s.

This kind of approach is in line with a tendency wich leads to the definition of site-dependent spectra obtained by a statistical processing over accelerograms recorded on deposits with diverse characteristics [4]. The spectra of Figure 1 are obtained by processing Italian recordings because, apart from the local conditions of sites, the frequency content of an earthquake is influenced by the geologic and geotectonic features of the region.

3 ARTIFICIAL GENERATION OF TIME-HISTORIES

The influence of local characteristics is investigated by means of a monodimensional analysis of amplification, carried out with the SHAKE code [6] wich operates in the frequency domain. The input accelerogram to assign to the rock outcropping was generated artificially so as to have the nearest possible response spectrum to the one proposed by GNDT for ground of type S_1. This is done so that it is effectively meaningfull, in terms of the frequency content, of the accelerograms occuring on rocks or on stiff ground. In order to obtain these time histories it was used the computer code TIMHIS which, starting from a real accelerogram having a spectrum sufficiently similar to the one required, repetitively modifies that time history, adjusting the frequency content with the aid of the FFT algorhythm.

From the many Italian earthquake recordings available, I have chosen for the generation those of Tolmezzo and Auletta relative, respectively, to the events in Friuli (1976) and in Irpinia (1980).

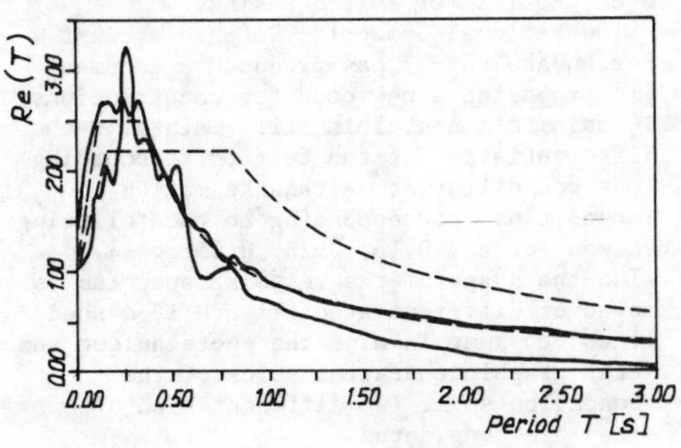

Figure 1. Response spectra of the Tolmezzo recording and of the artificially generated time history, compared with those of GNDT.

The choice of using two time histories is due to considerations of the author and others in two previous papers on this subject ([2],[3]). These in fact showed considerable differences in the response of the same deposit subjected to accelerograms from Friuli or Irpinia, even though scaled to the same maximum acceleration. The lenght of the time histories generated is 20 s, with a time step of 0.01 s; the baseline correction is applied to these.

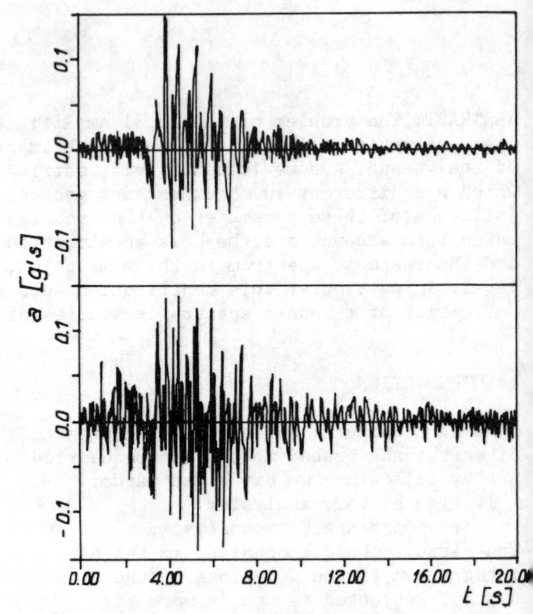

Figure 2. Comparison of the accelerogram recorded in Tolmezzo with the one generated artificially by the computer code TIMHIS.

4 LAY-OUT OF THE PARAMETRIC ANALYSIS

Deposits of sand or clay with thickness variable between 7 and 70 m are analysed.

For the sand relative density values Dr of 30, 45, 60, 75, 90 [%] are considered; this is due to the fact that curves giving shear modulus G and damping ratio β in terms of shear strain γ are available for such values in the literature [5]. The density ρ for the sands in question are respectively 1700, 1775, 1850, 1925, 2000 [Kg/m^3]. For the clay the cohesion value Cu is made to vary from 24 to 200 [kN/m^2], the density is assumed equal to 1800 Kg/m^3.

For the rock base a shear wave velocity of 2400 m/s and a density of 2600 Kg/m^3 are chosen. One of the purpose of this contribution is that of verifying whether the fundamental period of the deposit, wich for the linear behaviour of the ground is given by the well-known formula:

$$T_0 = \frac{4\ H}{V_s} \quad (1)$$

can be assumed like single parameter characterizing the response spectrum expected in a given deposit. In particular reference is made to the fundamental equivalent period T*, evaluated in non linear field and defined as the period for which the amplification function of the deposit shows its first peak. In [2] and [3] empirical formulae are proposed for the calculation of T* in terms of the soil characteristics and of the intensity a_0 of the earthquake. For cohesive sites:

$$T^* = \alpha_C(T^*, a_0, Cu)\ T_0 \quad (2)$$

where the coefficient of non-linearity α_C is obtained iteratively from Figure 3.

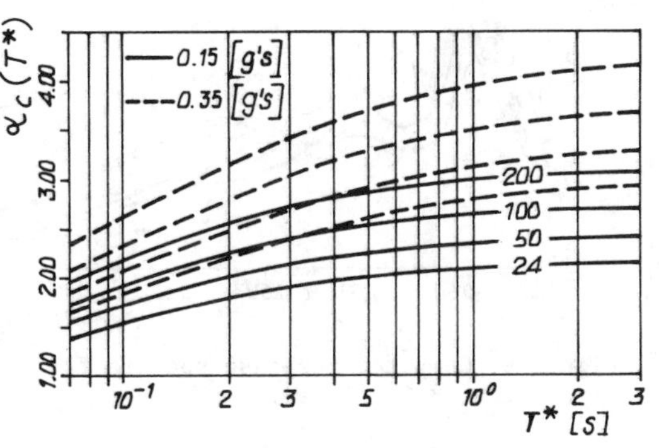

Figure 3. Coefficient of non linearity α_C.

For layers of sand:

$$T^* = \alpha_S(a_0, Dr)\ T_0 \quad (3)$$

In this case α_S can be obtained directly from the graph of Figure 4.

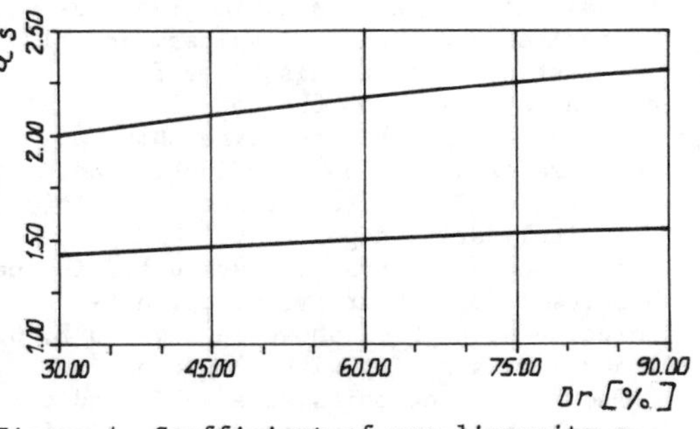

Figure 4. Coefficient of non linearity α_S.

The choice of cases to analyze is made so as to have clusters of different deposits, endowed with the same value of T*; 17 sand layers and 14 clay sites are examined.

The accelerograms considered are scaled to two different levels of maximum acceleration a_0: 0.15 and 0.35 [g's].

5 THE RESULTS OF THE ANALYSIS

Tables 1 and 2 give the list of the layers examined with the relative values of the linear period T_0 and of the fundamental equivalent period T*, compared with T_T and T_A, periods resulting from the numerical analysis with the two artificial histories (Tolmezzo and Auletta). The errors ε in comparison with the T* are also showed.

Table 1. Periods resulting from the analysis over sand sites (a_0 = 0.15 g's).

H	Dr	T_0	T^*	T_T	ε[%]	T_A	ε[%]
10	90	0.172	0.268	0.263	1.7	0.257	3.8
7	30	0.181	0.262	0.240	8.2	0.238	8.8
15	90	0.233	0.363	0.354	2.4	0.359	1.2
10	30	0.237	0.342	0.320	6.3	0.332	2.8
20	90	0.289	0.450	0.434	3.5	0.436	3.0
15	30	0.321	0.463	0.433	6.5	0.435	6.0
26	75	0.373	0.573	0.552	3.7	0.560	2.3
20	30	0.398	0.575	0.533	7.2	0.534	7.0
35	75	0.467	0.716	0.688	3.9	0.683	4.7
26	30	0.485	0.700	0.650	7.0	0.648	7.3
35	45	0.545	0.806	0.748	7.1	0.745	7.5
50	90	0.575	0.895	0.820	8.4	0.816	8.9
35	30	0.606	0.874	0.783	10.	0.777	11.
50	60	0.654	0.987	0.928	5.9	0.909	7.8
70	90	0.739	1.152	1.120	2.7	1.034	10.
50	30	0.792	1.143	1.074	6.0	1.008	11.
70	45	0.916	1.355	1.303	3.8	1.223	9.7

Table 2. Results of analysis over clay deposits (a_0 = 0.15 g's).

H	Cu	T_0	T^*	T_T	ε[%]	T_A	ε[%]
7	49.	0.112	0.264	0.269	1.8	0.264	0.1
7	24.	0.161	0.457	0.449	1.6	0.457	0.0
15	100.	0.169	0.371	0.398	7.3	0.397	7.0
20	84.	0.245	0.572	0.580	1.3	0.584	2.0
26	93.	0.301	0.700	0.694	0.8	0.693	0.9
35	118.	0.361	0.813	0.802	1.3	0.809	0.4
35	88.	0.417	0.994	1.007	1.3	0.970	2.4
50	177.	0.421	0.988	0.890	9.9	0.890	9.9
50	123.	0.505	1.147	1.228	7.0	1.117	2.6
70	200.	0.551	1.151	1.230	6.8	1.117	2.9
50	100.	0.565	1.338	1.405	5.0	1.296	3.1
35	42.	0.603	1.654	1.702	2.9	1.731	4.6
70	118.	0.722	1.670	1.716	2.7	1.764	5.6
70	49.	1.118	3.028	3.025	0.1	2.966	2.0

A first consideration concerns the substantial correspondence of the periods T_T and T_A, obtained with the two time histories generated. Similar agreement has also been verified on the response spectra; this has made it possible to go ahead for the rest of the analysis with the Tolmezzo artificial accelerogram alone.

Another mention should be made about the validity of the formulae (2) and (3) for the evaluation of T*; average errors of 6 % for sand and 2.5 % for clay are committed.

Figures 5, 6, 7 and 8 show the response spectra, on the surface of the deposits, grouped together in bands of the T* period.

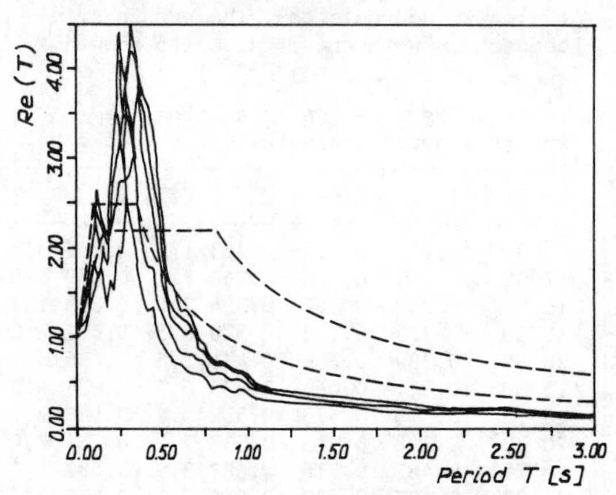

Figure 5. Computed response spectra for sites with T* between 0.24 and 0.4 [s].

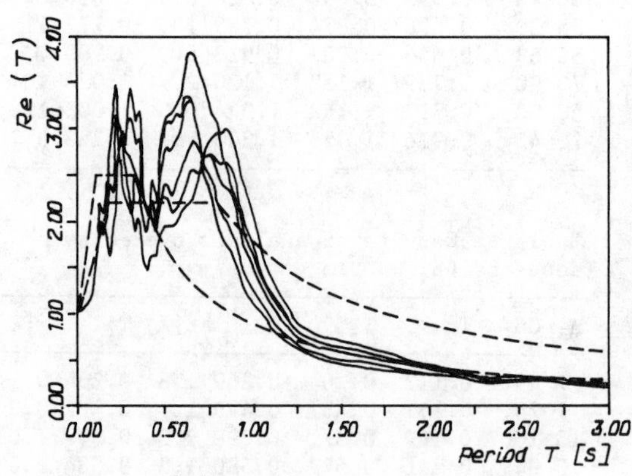

Figure 6. Computed response spectra for sites with T* between 0.65 and 0.8 [s].

From these graphs one can see how T* is effectively an optimum parameter to characterize the shape of the expected spectrum. The other result to observe attentively is the maximum acceleration a_1 on the surface of the deposit; the values of A, defined as ratio between a_1 and a_0,

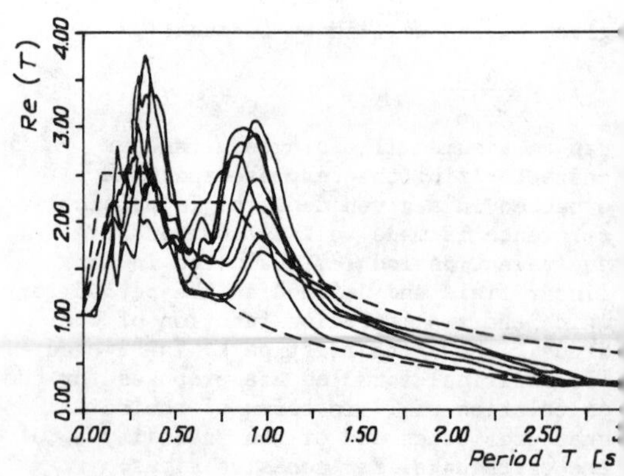

Figure 7. Computed response spectra for sites with T* between 0.9 and 1.1 [s].

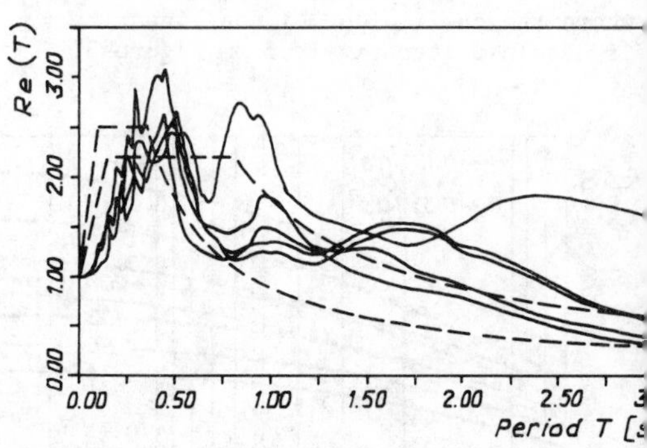

Figure 8. Computed response spectra for sites with T* > 1.4 [s].

are shown in a diagram (Figure 9, symbols) in terms of T*; one can see a very regular trend of this relationship which becomes less than 1 for layers gifted with T* greater than 1.5 s.

The continuous curve with symbols, on the other hand, is relative to results obtained by considering the shear modulus G constant. The same deposits in this case reveal, as one can easily imagine, lesser T* periods and greater A values. What is interesting is that this curve follows the course obtained from the non-linear calculation. Further the line shows a decrease of A for low T* values, tending to $a_1 = a_0$ for T* = 0 (as it should be for infinitely stiff deposits).
The response spectra too, evaluated in the hypothesis of linearity, practically coincide with those shown in Figures 5, 6, 7 and 8 at a parity with T* period.

The same 31 deposits were subjected to the accelerogram generated from that of Tolmezzo, scaled to the intensity of 0.35 g's, the maximum value envisaged by the GNDT document. Analogous considerations to those relative to the case a_0 = 0.15 g's, can be made both for the validity of formulae (2) and (3) and for the shape of the response spectra and the course of the ratio A. The same deposit, subjected to different seismic intensities, obviously reveals greater T^* periods as the parameter a_0 increases; it has however been noted how the shape of the response spectra is still influenced by the single T^* value. On the contrary, in spite of a similar course, the A values are a little less in the case of high intensity (Figure 9, dashed lines); this fact is due to the relevant non-linear behaviour of the ground under strong earthquake. Figure 9 also shows the course of A values obtained from an analysis carried out with constant modulus G (continuous line); the latter follows the trend of the one obtained from the linear analysis with a_0 = 0.15 g's, with less A values, due to the increase of damping β.

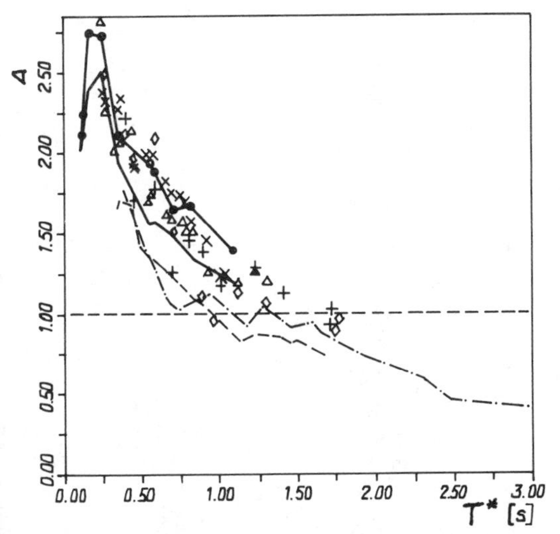

Figure 9. Amplification factor A (a_1/a_0).

6 CONCLUSIONS

The results shown up to now, even though derived from a relatively limited number of cases, make it possible to draw certain interesting conclusions.

The first important result concerns the confirmation of the fundamental period of the deposit as a meaningful parameter for the definition of the response spectrum. Considering therefore a design earthquake, defined by the maximum acceleration a_0 and by its response spectrum on rock or on very stiff ground, a seismic analysis of a structure built on a flexible deposit must take two factors into account:

1. a different maximum acceleration a_1;
2. the modification of the shape of the response spectrum.

Both corrections can be evaluated by means of the value of the period of the stratum, better if evaluated taking account of non-linear behaviour (T^*).

As for first point, the graphs of Figure 9 can give an idea of this phenomenon, in terms of the reference intensity a_0.

As for the form of the response spectrum, from the curves presented (Figures 5, 6, 7, 8) one can obtain, with good approximation, the one relative to a specified site.

Finally, about the spectrum proposed by GNDT for ground S_2, it works quite well for flexible deposits, for which above all the maximum acceleration a_1 is not amplified; some caution however should be observed in presence of a structure with a period T comparable with that one of the deposit.

There exist furthermore a set of layers (with period between 0.2 and 0.5 [s]) for which there are consistent values of the amplification factor A, accompained by very high peaks in the spectrum. As regard to be in favour of or against safety, considering the non-linear behaviour of the ground, this depends on the fundamental period of the structure to be built; usually in the case of flexible structure it is better to take non-linearity into account.

REFERENCES

[1] Gruppo Nazionale per la Difesa dai Terremoti, Norme Tecniche per le Costruzioni in Zone Sismiche, Ingegneria Sismica II/1, 23-36 (1985)

[2] R.Carpaneto, A.Del Grosso, S.Lagomarsino and G.Solari, Un Procedimento per la valutazione dell' Amplificazione di Terreni in Regime Non-Lineare, II Convegno Nazionale: L' Ingegneria Sismica in Italia, 4/1-9, Rapallo 1984

[3] S.Lagomarsino, L'amplificazione nei Depositi di Terreno Incoerente in Regime Non-Lineare, Atti dell'Istituto di Scienza delle Costruzioni 5/87, Genova 1987

[4] H.B.Seed, C.Ugas and J.Lysmer, Site-Dependent Spectra for Earthquake Resistant Design, EERC 74-12, Berkeley 1974

[5] H.B.Seed and I.M.Idriss, Soil Moduli and Damping Factors for Dynamic Response Analysis, EERC 70-10, Berkeley 1970

[6] P.B.Schnabel and J.Lysmer, SHAKE: A Computer Program for Earthquake Response Analysis of Horizontally Layered Sites, EERC 72-12, Berkeley 1972

[7] E.Vitiello, Impressioni sul Terremoto di Città del Messico del 19 Settembre 1985, L'Industria Italiana del Cemento 595, 794-808 (1985)

Numerical Methods in Geomechanics (Innsbruck 1988), Swoboda (ed.)
© 1988 Balkema, Rotterdam. ISBN 90 6191 809 X

Dynamic failure of rockfill models simulated by the distinct element method

Tatsuo Omachi
Graduate School, Nagatsuta, Tokyo Institute of Technology, Yokohama, Japan
(Currently Visiting Scholar: John A.Blume Earthquake Engineering Center, Stanford University, USA)
Yasuhiro Arai
Nihon Doro Kodan, Japan
(Formerly Tokyo Institute of Technology, Japan)

ABSTRACT: Idealized models of rockfill dams use two types of regular arrangement of numerous circular particles; one a uniform type and the other a zoned type. The total failure of these models simulated numerically by the distinct element method is discussed in comparison with experimental results. Special emphasis is placed on the validity of the simulation method, and on the dynamic failure characteristics of the particle arrangement. As for the latter, an opening which takes place in the arrangement is found to be a key process in the failure mechanism, whether the failure is induced by gradual base inclination or by impulsive ground acceleration. The fact that the opening requires finite duration of time to grow into a total failure suggests that a rockfill dam might exhibit different features of failure as a result of the change in frequency component of earthquakes.

1 INTRODUCTION

In the 1984 Morgan Hill Earthquake (M6.2), ground motion acceleration with a maximum of 1.3g was observed at the left abutment of the Coyote Lake Dam (Shakal et al 1984). Despite such a large acceleration, the rockfill dam built in 1935 was free of any damage. This is a surprising fact which forces us to reconsider earthquake resistance of rockfill dams. In fact, there have been many examples of rockfill dams which demonstrated their great earthquake resistance much to our surprise. To account for these facts, our present knowledge mainly based on the mechanics of continua is likely to meet a limit. It requires us to formulate a theory based on the mechanics of discontinua to explain the behavior of rockfill dams which are essentially discontinuous structures.

According to previous works on earthquake resistance of rockfill dams, the resistance is largely associated with the sliding characteristics of rockfill particles. This is because excessive sliding of the rockfill particles exposes impervious core, which can lead to the development of fatal cracks in the core. From this context, the present study is focused mainly on the rockfill sliding induced by ground motion excitation, to which the distinct element method is applied.

2 FAILURE DUE TO BASE INCLINATION

2.1 Outline of Simulation Procedure

The distinct element method used here is a conventional one first proposed by P. A. Cundall (Cundall 1974). For transmission of forces between two elements that are in contact, it utilizes a set of an elastic spring and a viscous dashpot in the direction normal to the contact face, and in the tangential (shear) direction, a Coulomb slider is added to another set of a spring and a dashpot. In order to check the validity of the method, the simplest failure pattern, that of catastrophic failure induced by a gradual increase in inclination of the basement of a rockfill dam is modeled.

Parameters listed in Table 1 are those used for the numerical simulation of an actual experiment in which a rockfill dam was modeled by regular arrangement of a number of brass rods of the same size. In Table 1, the static and dynamic coefficients of friction are measured values, but spring constants and damping constants are not measured ones. The spring constants listed in Table 1 are smaller than 1/100,1000 of the actual brass rods and the rate of inclination actually used in the experiment is smaller than 1/10 of the value listed in the table. If

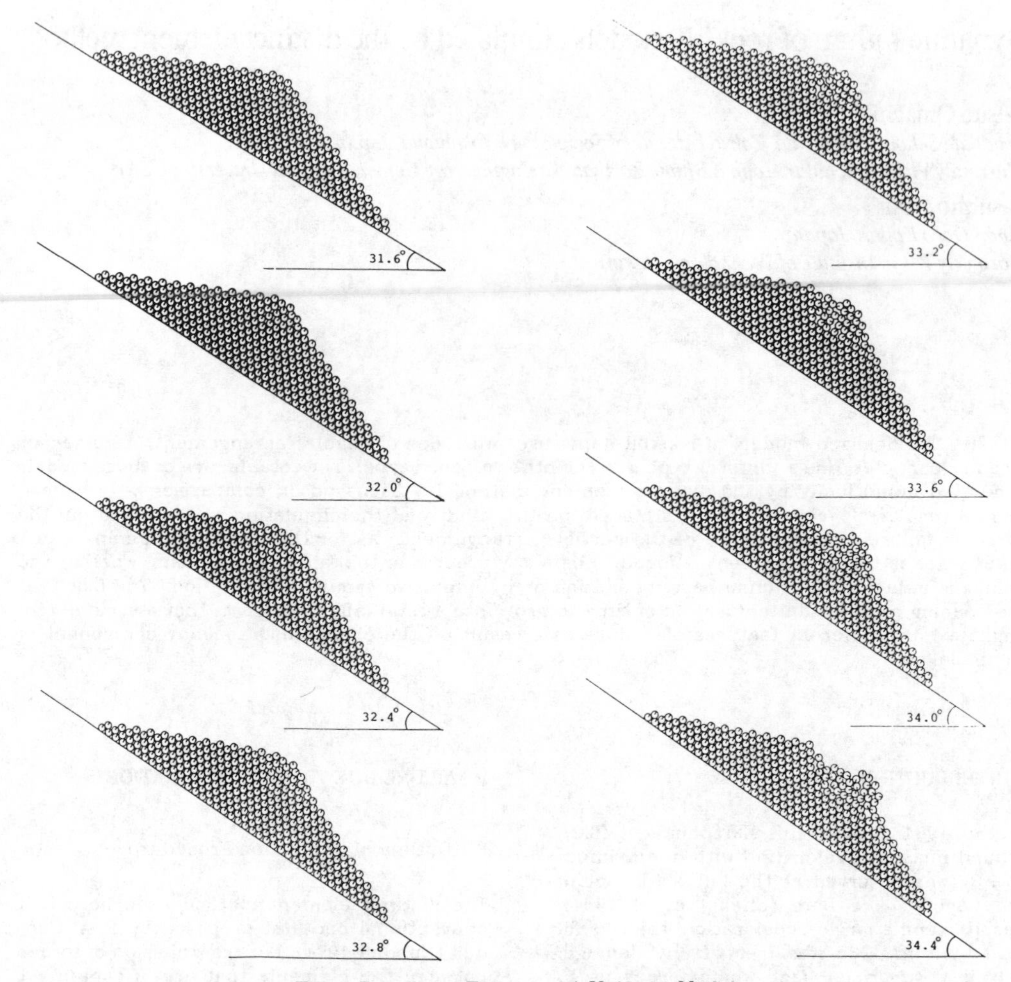

Fig. 1 Simulated Failure of A Uniform Model

Table 1 Parameters for Numerical Simulation

Mass of an Element	M = 66g
Diameter of an Element	D = 10mm
Length of an Element	L = 100mm
Normal Spring Constant	$K_n = 1.0\text{x}10^5$ N/m
Shear Spring Constant	$K_s = 2.5\text{x}10^4$ N/m
Normal Damping Ratio	$h_n = 0.10$
Shear Damping Ratio	$h_s = 0.05$
Static Coef. Friction	$\mu_s = 0.32$
Dynamic Coef. Friction	$\mu_d = 0.24$
Cal. Time Step	$\Delta t = 1.0\times10^{-5}$ sec
Rate of Inclination	$d\theta/dt$ = 10 deg/sec

we apply realistic values, we have to use a time step much smaller than that shown in Table 1 for the sake of stability and reliability of the simulation (Omachi and Arai 1986). To avoid the resulting enormous computational job, the parameters in Table 1 were employed.

2.2 Results

In Fig. 1 are shown some of the results of the simulation in which all the particles at the bottom are fixed to the base. One can see both translational and rotational movements of each particle in Fig. 1. Some differences in the failure characteristics between the simulation and the experiment are;

1. the angle of inclination at failure is 32 degrees in the simulation while it is 37 degrees in the experiment,

2. a wide opening which triggers total failure of the model takes place at the upper portion of the model in the simulation while it takes place at the lower portion in the experiment,

3. a bridge between the particles along the right slope surface can be seen in the last two figures of Fig. 1. This kind of bridge is not observed in the experiment.

The above mentioned differences in the failure characteristics may be partly attributed to the the difference in conditions between the simulation and the experiment such as spring constants and the rate of inclination. Some other trial simulations suggested that formation of the unrealistic bridge could be caused by inappropriate values of coefficients of friction, as might be suspected from the following simulation. The wide opening of the particle arrangement is found to be a key process pertaining to the failure of the model. This is seen in both the inclination and vibratory experiments.

Photo 1 shows some features of the openings observed in a vibratory experiment (the time step between each frame is 1/64 sec.). It can be noticed in Photo 1 that the widest opening appears around the bottom, and that it disappears as the direction of the table motion is reversed. At this stage, an amplitude of acceleration of the sinusoidal table motion is about 0.7g. Mechanism of occurrence of the wide opening such as its location and timing is discussed in detail elsewhere (Omachi & Arai 1987).

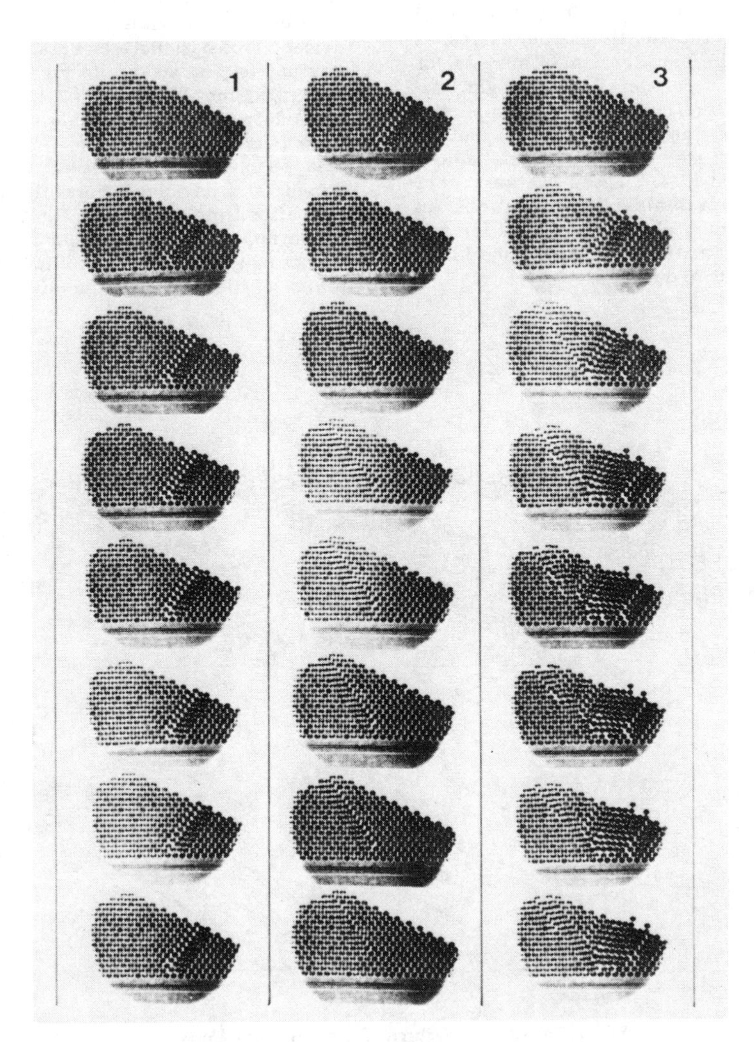

Photo 1 Observed Vibratory Failure

3 FAILURE DUE TO GROUND ACCELERATION

3.1 Model and Parameters

In the simulation, use is made of a zoned type rockfill dam model made up of particles 8 mm in diameter in a central portion and 10 mm in diameter 10 mm in the upstream- and downstream sides. The number of the particles amounts to 439 in total. All the particles are assigned a static coefficient of friction of 0.64 so as to give an angle of repose of 45 degrees. The dynamic coefficient of friction is set to half of the static value. The remaining parameters such as spring and damping coefficients are set to the same values as those used in the earlier simulation. As for horizontal ground acceleration, a single box-car shaped acceleration is used for simplicity. The duration of the acceleration is set to 0.004sec, 0.04sec and 0.2sec, with the amplitude fixed at 1.0g. These values were selected from preliminary simulations in which the fundamental period of vibration and the angle of inclination at failure were found to be 0.03sec and 26 degrees, respectively.

Fig. 2 shows a simulated failure of this model caused by base inclination. Note that in Fig. 2 the opening takes place at the lower portion of the model, and there is no formation of a bridge on the sloping surface.

3.2 Results

In Figs. 3 - 5 are shown results of the numerical simulation. The model apparently shows quite different features of failure in accordance with a change in duration of the ground acceleration, even when the amplitude of the acceleration remains unchanged. In Fig. 3, neither translational nor rotational motion of any particle is visible. In Fig. 4, several wide openings can be seen in the arrangement around the bottom left side of the model at time between 0.04sec and 0.08sec, but they disappear at 0.10sec. Instead, larger openings appear and develop around the bottom right side after time 0.08sec. In Fig. 5, the openings which appear at the bottom left side show a monotonous growth, resulting in complete failure of the model at time after 0.12sec.

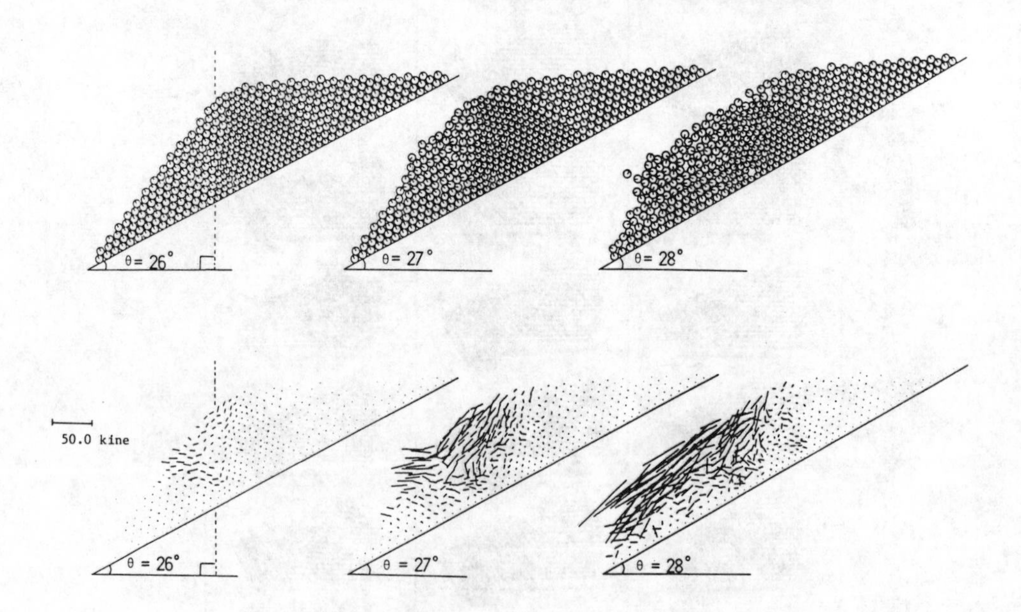

Fig. 2 Simulated Failure (A Preliminary Trial)

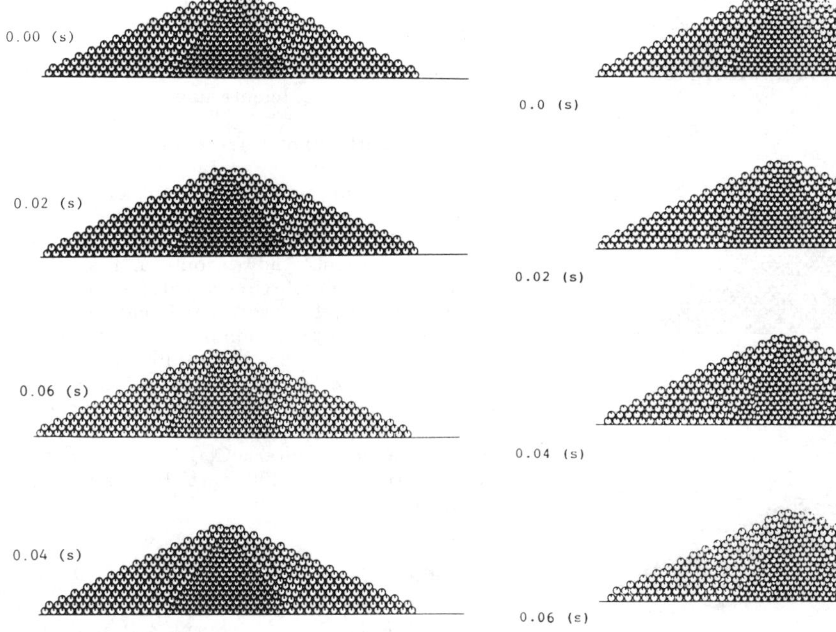

Fig. 3 Simulated Result No.1 (T=0.004sec)

Evidently, these different features of failure are the result of the nature of motion of the particles. As was pointed out by Newmark (Newmark 1965), each particle requires sufficient amount of momentum (in other words, acceleration of sufficient level as well as of sufficient duration) to initiate and continue motion, whether it is translational or rotational. Consequently, if the direction of ground motion acceleration is reversed before the opening develops to a harmful extent, the opening is closed without triggering a total failure of the particle arrangement. This situation is demonstrated in Photo 1 and is well simulated in Figs. 3 - 5. Moreover, this seems to give a reasonable explanation for the case history of the Coyote Lake Dam.

4 CONCLUSIVE REMARKS

Two types of failure of rockfill models were simulated by the distince element method; one caused by base inclination and the other caused by ground motion acceleration. Some findings obtained from the present study are:

1. the method is apparently capable of simulating complicated motion of particles as long as a set of appropriate parameters are used.

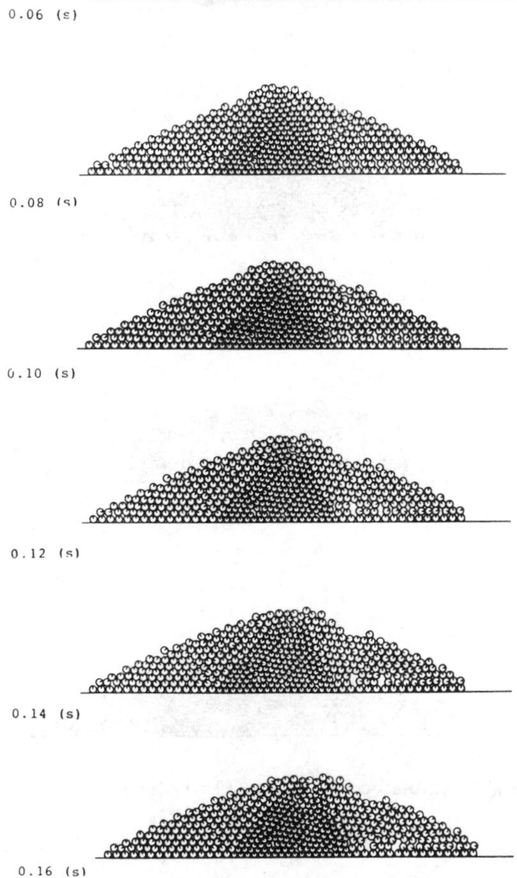

Fig. 4 Simulated Result No.2 (T=0.04sec)

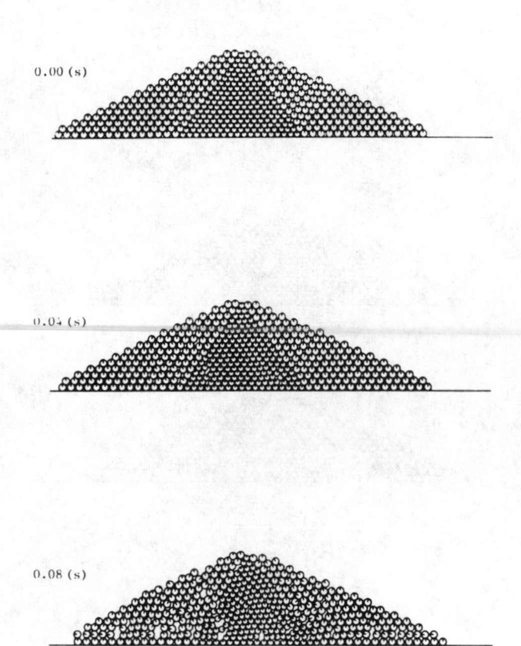

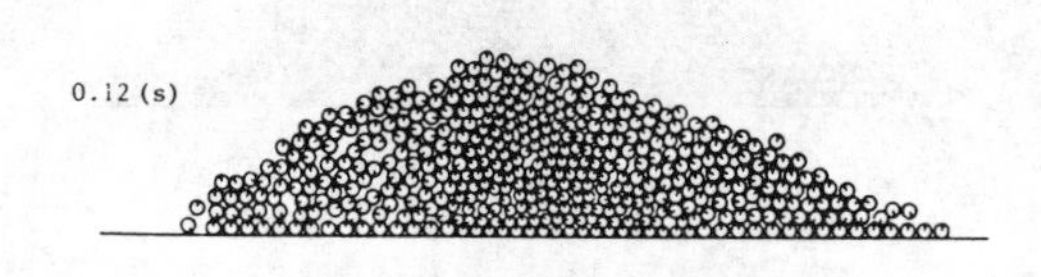

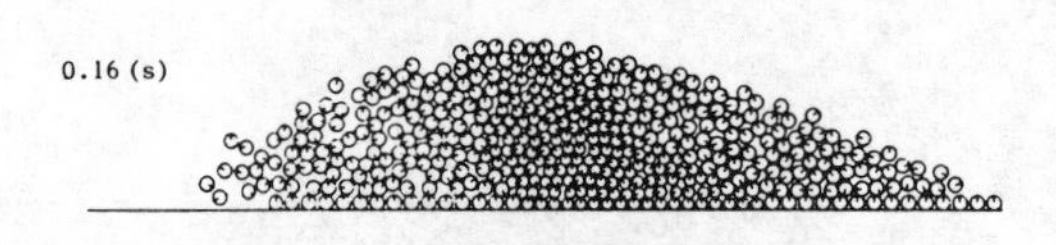

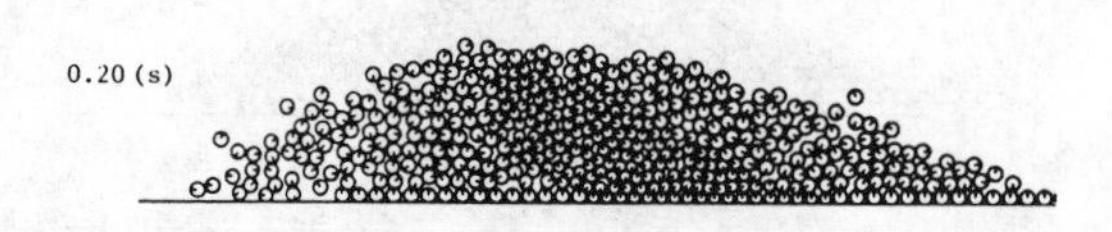

Fig. 5 Simulated Result No.3 (T=0.20sec)

2. of the various parameters, the coefficients of friction have major effects on the failure characteristics of rockfill mass. The method for selection of appropriate values for the coefficients needs to be established,

3. the pattern of total failure of rockfill mass varies with the characteristics of the ground motion acceleration. Although this looks quite natural, a detailed relationship between them is left for future study.

4. the present study is only a first step towards a more realistic simulation of rockfill dams subjected to severe earthquake excitation. In addition to randomness of earthquake acceleration, there are several important factors which need to be included in future studies. These are, for example, random arrangement of irregular particles, degradation of particles, interaction between reservoir water and rockfill particles, and fatal cracks in impervious core.

REFERENCES

* Shakal, A. F., R. W. Sherburne & D. L. Parke 1984. Principal Features of the Strong-Motion Data from the 1984 Morgan Hill Earthquake, The 1984 Morgan Hill, California Earthquake, Special Publication 68.
* Cundall P. A. 1974. A Computer Model for Rock Mass Behavior Using Interactive Graphics for the Input and Output of Geometrical Data, Dept. Army Contr., No.DACW 45-74-C-006.
* Omachi, T. & Y. Arai 1986. On the Distinct Element Parameters Used for Earthquake Ground Motion Simulation, Computers and Geotechnics, Vol. 2, No. 6.
* Omachi, T. & Y. Arai 1987. Dynamic Failure Mechanism of Rockfill Dams Composed of Regularly Arranged Circular Particles (to be published).
* Newmark, N. M. 1965. Effects of Earthquakes on Dams and Embankments, Geotechnique 15, No. 2.

Numerical Methods in Geomechanics (Innsbruck 1988), Swoboda (ed.)
© 1988 Balkema, Rotterdam. ISBN 90 6191 809 X

Deformation behaviour of artificial saturated soil bases under seismic action

V.A.Ilyichev, V.M.Likhovtsev & A.Y.Kurdyuk
Research Institute of Bases and Underground Structures, Moscow, USSR

ABSTRACT: The method for quantitative evaluation of intensity and spectrum of seismic vibrations of an improved soil base has been proposed in the present report. This method allows to predict the seismicity level of a building spot including the account of elevation of ground water level.

The study of deformation behaviour of saturated soil bases under seismic action is caused by the necessity in building in seismic regions including saturated soils. In an effort to improve a soil base there are employed technological techniques which allow to decrease the seismicity level of the site under construction. For this reason a special analysis is called for problems of evaluation of the influence of soil base artificial preparation on the intensity and the spectrum of seismic vibrations. The abovesaid evaluation may be carried out on the basis of the realization by finite elements of problems in deformation behaviour of multi-phase heterogeneous soil masses under dynamic (seismic) loading. In those conditions the soil-base is considered to be a multi-phase medium being consisted of three phases (liquid, gas, mineral particles) with the behaviour of each of these phases under dynamic loading to be adhered to definite relationships.
The formulation of a mathematical model for describing the motion of saturated mediums have been performed by many researchers. In conformity with the results of these studies the motion of a quasi-two phase medium is described by the following system of the equations (Biot 1941):

$$
\begin{aligned}
& m_s \rho^s \ddot{u}_i + (1-m_s)\rho^w \ddot{u}_i^w = \delta_{i2}\rho g + \sigma_{ij}^{ef} + \delta_{ij} P_{,j}^w \\
& \dot{u}_i^w - \dot{u}_i = \frac{K_F}{1-m_s}\left[\delta_{ij}\frac{P_{,j}^w}{g\rho^w} + \delta_{i2} - \frac{\ddot{u}_i^w}{g}\right] \qquad (1) \\
& \dot{u}_{i,i}^w + \frac{m_s}{1-m_s}\dot{u}_{i,i} = \frac{1}{M}\dot{P}^w
\end{aligned}
$$

where
$m_s = \frac{V_s}{V}$ - ration of mineral particles volume to soil volume;
ρ - soil density; $\ddot{u}_i$, $\dot{u}_i$, $\ddot{u}_i^w$ and $\dot{u}_i^w$ - acceleration and velocity of motion of soil skeleton and water in i-direction;
K_f - coefficient of filtration;
M - coefficient of compressibility of water-air mixture;
g - free fall acceleration;
σ_{ij}^{ef} - effective stresses in soil skeleton;
P^w - pressure in porous fluid;
δ_{ij} - Kroneker's symbol.
In the system of equations (I) the compressibility of mineral particles is not taken into account and it is assumed that the compaction occurs at the expense of variations in porosity. In the system of equations (I) the first set of equations is obtained from the law of the change of motion quantity for a quasi-two phase soil. The second set of equations describes the motion of a water-air mixture in a porous medium. The third set of equations defines the law of mass conservation for a quasi-two phase soil.
For two-dimentional formulation there are known numerical realizations base on the finite element approach (Gaboussi 1972; Illarionov 1979; Hardar 1979). To present the system of equations (I) in the finite element form let's use Galerkin's principle assuming that functions of the form are chosen as a weight function. Then the system of equations (I) with respect to new unknowns

(u^{sk} displacement of soil skeleton; W - displacement of water in relation to the soil skeleton) will take the following form:

$$\begin{bmatrix} m_{sk} & m_s \\ m_s^T & m_w \end{bmatrix} \begin{Bmatrix} \ddot{U}^{sk} \\ \ddot{W} \end{Bmatrix} + \begin{bmatrix} d & 0 \\ 0 & c \end{bmatrix} \begin{Bmatrix} \dot{U}^{sk} \\ \dot{W} \end{Bmatrix} + \begin{bmatrix} k & h \\ h^T & h \end{bmatrix} \begin{Bmatrix} U^{sk} \\ W \end{Bmatrix} = \begin{Bmatrix} f \\ f' \end{Bmatrix}, \quad (2)$$

where

$[D]$ -matrix of elastic characteristics;

$[I]$ -2x2 unit matrix;

$[N]$ -functions of the form;

$[B]$ -matrix of gradients;

$\{U^{sk}\}$-vector of nodal displacements,

$$[m_{sk}] = \int_S \rho [IN]^T [IN] dS ;$$

$$[m_s] = \int_S \rho^w [IN]^T [IN] dS ;$$

$$[m_w] = \int_S \frac{\rho^w}{M} [IN]^T [IN] dS ;$$

$$[c] = \int_S \frac{\rho^w g}{K_F} [IN]^T [IN] dS ;$$

$$[h] = \int_S \frac{M}{m} [A]^T [A] dS ;$$

$$[k] = \int_S ([B]^T [D][B] + \frac{M}{m} [A]^T [A]) dS ;$$

$$\{f\} = -\int_S \rho g \delta_{i2} [N] dS ; \{f'\} = -\int_S \rho^w g \delta_{i2} [N]^T dS ;$$

$$[A] = [\frac{\partial N_1}{\partial x_1}, \dots, \frac{\partial N_2}{\partial x_2}].$$

Integration of the system of equations (2) is accomplished through the use *and* step-by-step procedure--Newmark's method -according to which the following relationships (Baute 1982) are used

$$\dot{u}_{t+\Delta t} = \dot{u}_t + [(1-\delta)\ddot{u}_t + \delta \ddot{u}_{t+\Delta t}]\Delta t , \quad (3)$$
$$u_{t+\Delta t} = u_t + \Delta t \cdot \dot{u}_t + [(0.5-\alpha)\ddot{u}_t + \alpha \ddot{u}_{t+\Delta t}]\Delta t^2 ,$$

where

α, δ-parameters that define the accuracy of the stability *and* integration.

The absolute difference scheme is realized when δ =0,5 and α =0,25. With numerical realization of problems for semi-infinite domains by the finite element method it is segregated a bounded computational domain. When modelling a semi-infinite domain on the boundary of restricted computational domain it is set boundary conditions that allow the waves to pass through the conventional boundary without reflection. With this aim in view the boundary conditions are used- the standard viscous boundary presented in (Idriss/Seed 1970). For a quasi-two phase soil it is assumed that formulation of boundary conditions is accomplished through substituting outer domain reaction(with respect to the computational domain) by a distributed load having intensity:

$$\sigma = a \rho c_p u ; \quad \tau = b \rho c_s v ,$$

where

u,v - velocities of motion of boundary points along x and y coordinates respectfully;

C_p, C_s-velocities of propagation of longitudinal and transverse waves;

a,b - dimensionless parameters.

Such boundary conditions may be treated as setting up a viscous damper on the boundary. The approach proposed for taking into account the boundary conditions allows to transform the system of equations (2) to the following form:

$$[M]\{\ddot{U}\} + [C]\{\dot{U}\} + [K]\{U\} = \{R\} - [P]\{\dot{U}\}, \quad (4)$$

where $[M]$, $[C]$ *and*

$[K]$ - matrices of masses, damping, rigidity; $\{\ddot{U}\}, \{\dot{U}\}, \{U\}$ *and*

$\{R\}$ - vectors of accelerations, velocities, displacements and outer load;

$[P]$ - matrix of outer damping dimensionality of which is coincident with dimensionality of the matrices $[M], [C], [K]$.

Therewith its diagonal zero terms are related to the standard viscous boundary.

With regard to the proposed method of solution (Newmark's method -the relationships (3) the system of equations (4) is transformed to the following form:

$$[\tilde{M}]\{\ddot{U}\}_{t+\Delta t} = \{\tilde{R}\}_{t+\Delta t} , \quad (5)$$

where

$[\tilde{M}]$ - effective matrix of masses;

$\{\tilde{R}\}$ - effective vector of forces;

$\{\ddot{U}\}_{t+\Delta t}$- vector of unknown accelerations at time moment t + Δ t.

The algorithm described was approbated on model problems(Kurdyuk / Likhovsev 1987). A satisfactory correlation of the results of numerical studies with the solutions of model problems that has been achieved allows to use this algorithm for the prediction of behaviour of a heterogeneous saturated soil base under dynamic (seismic) loading. On a basis of analysis of the results of numerical studies (Kurdyuk/Ilyichev/Likhovtsev 1986) the rectangular domain with dimensions 24x324 m (Fig.I) being digitized by triangle elements of the first order (homogeneous grid of I8 columns with 8 elements in each) was assumed to be a computational domain. A layer of surface deposits ABKL is underlaid by a layer of more dense soils ADEL which is modelled by the standard viscous boundary. In numerical studies there were modified the dimensions of a compacted area (width and depth), the physical and mechanical characteristics of an improved soil base by which different methods of soil engineering reclamation are being modelled. There were carried out studies incorporating inclusions of I8,36,54,72 m wide. An improved soil base is modelled by physical and mechanical characteristics that specify soil base preparation by ground piles, chemical grouting. The record of the San Fernando earthquake accelergram was assumed to be an initial seismic action. For evaluating the seismicity level of a building spot it is proposed to use a method which is based on comparison between the recorded with time seismic vibrations characteristics of an improved base soil and the bedrock. For similar analysis being run with the object of micro-zoning the horizontal vibrations with period from 0,I to I,0 sec. are of the prime importance. Since the long-period side of the range to be of our interest exhibits the values of 2-3 sec., the duration of design accelergram is assumed to be IO sec. The result of finite element studies is records of vibrations which are processed in compliance with the known processing techniques used in seismic micro-zoning.

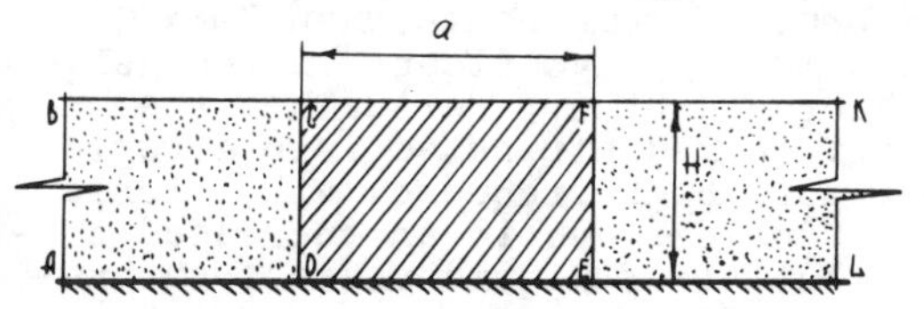

Fig.1 Design scheme of a layer with an artificial base (the width of inclusion - a) resting on gravel deposits.

In Fig.2 is shown the ratio of maximal average accelerations at the inclusion surface to maximal accelerations at the layer without the inclusion depending on the value of a/h (a - the size of inclusion; h - the thickness of layer). From the analysis of the relationships given in Fig.2 it follows that the value of maximal average accelerations at the inclusion surface depends on the size of inclusion. And the optional width of inclusion at which accelerations have the least values is equal to one and a half of the layer width. Increasing of the inclusion width for more than the three thicknesses of the layer does not change the amplitude of accelerations.

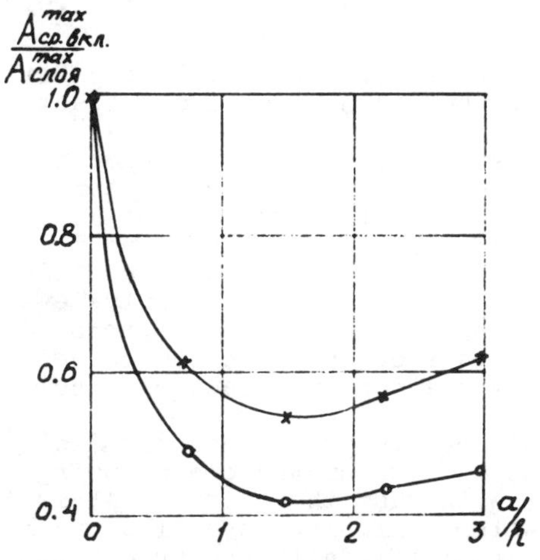

Fig.2 Ratios of maximal amplitudes of recording of vibrations obtained at the surface of an artificial base(x - the inclusion of type I; o - the inclusion of type 2) to maximal amplitude of vibrations obtained at the surface of a layer depending on the ratio a/h.

From the given relationships it follows that the minimal value of the ration of maximal amplitudes for the inclusion of type I is equal to 0,55 and for the inclusion of type 2 it is equal to 0,4I. In conformity with the seismic rigidity method the incremental increase in the seismicity level of a natural layer with respect to a layer being consisted of the soil of type I is equal to 0,6 and in the case of the soil of type 2 it is equal to I,0. Thus base preparation carried out in the form of ground piles (soil bedding and chemical grouting allows to decre-

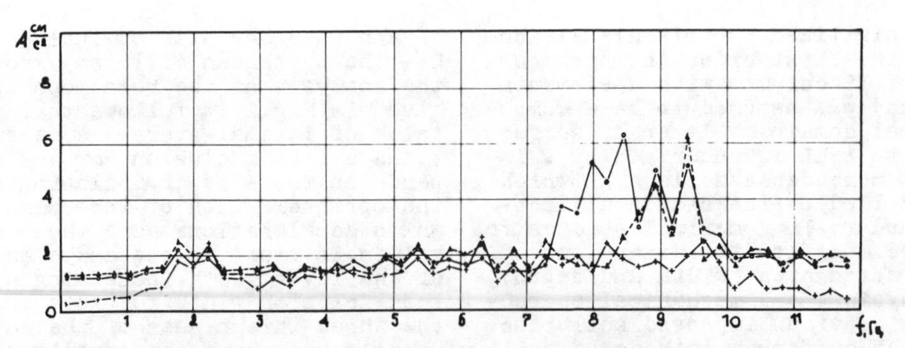

Fig.3 Fourier smoothed spectrum of recording of vibrations at the surface of an artificial base in case of inclusion of type I (a/h= I,5) with different ground water levels: o -UGV = 6 m. ▲ - fully saturated; + - non-saturated.

ase the seismicity level of a building spot by 0,5 and I,0 respectfully.

The spectral analysis of numerically obtained vibrations is carried out through the use of fast Fourier transform. The analysis presented in (kurdyuk/Ilyichev/Likhovtsev I986) allows to make a quantitative evaluation of the effect of improved soil base on the spectrum of seismic vibrations for each particular case.

Let's consider the case when soil base is a saturated one assuming herewith that the boundary AL is a water confining layer. The width of an artificially prepared inclusion is chosen to be equal I,5 h. As a design seismic effect it was used the model of seismic vibrations, which characterizes the spectrum of their effect, in terms of

$$U = \sum_{n=1}^{N} V_n \cos(\omega_n t + \varphi_n),$$

where V_n - crest value of velocity; ω_n - cyclic frequency; N -whole number. It was assumed that the design values were:

for non-saturated natural soil- $C_s = 200\ m/s$, $C_p = 489.9\ m/s$, $\rho = 1400\ kg/m^3$, $\nu = 0.4$

for saturated natural soil - $C_s^{sk} = 149.67\ m/s$, $C_p^{sk} = 366.61\ m/s$, $\rho^{sk} = 2500\ kg/m^3$, $C_w = 1000\ m/s$, $K_F = 0.27 \cdot 10^{-3}\ m/s$, $\rho^w = 1000\ kg/m^3$, $m = 0.44$, $\nu = 0.4$

for underlayer soils - $C_s = 871.78\ m/s$, $C_p = 1600\ m/s$, $\nu = 0.28$, $\rho = 2000\ kg/m^3$

for non-saturated soils of artificially prepared base- $C_s = 398.53\ m/s$, $C_p = 972.2\ m/s$, $\rho = 1700\ kg/m^3$, $\nu = 0.4$

for saturated soils of artificially prepared base- $C_s = 328.63\ m/s$, $C_p = 807.42\ m/s$, $C_w = 1000\ m/s$, $\rho^{sk} = 2500\ kg/m^3$, $\rho^w = 1000\ kg/m^3$, $\nu = 0.4$, $K_F = 0.27 \cdot 10^{-4}\ m/s$, $m = 0.33$

Fourier smoothed spectrum of vibrations in points at the surface of an artificially prepared soil base are given in Fig.3. From analysis of the given relationships it follows that in case of non-saturated as well as fully saturated natural soil layer at approximately 2 Hertz it is observed a certain increase in the amplitudes of vibration components which are typical of the least natural frequency of a natural soil layer. For a non-saturated soil the increase in vibration components was observed at approximately 8-IO Hertz. Saturation of a soil base results in a sharp increase of vibration components to approximately three times at these frequencies. Therewith the maximal values of vibration components when $h_{UGV} = 0$ are mostly significant in the neighbourhood of 9,5 Hertz. Thus the energy is redistributed into a region of more high-frequency vibrations. This occurs due to an increased contribution of liquid phase to vibrations of the system as a whole.

Artificial preparation of a soil base causes redistribution of the energy of seismic action and it is necessary to carry out the analysis of its spectral characteristic with consideration for seismic stability of a building and the equipment installed in it.

The proposed method for quantitative evaluation of intensity and spectrum of seismic vibrations of an improved soil base allows to predict the seismicity level of a building spot with the method of soil base preparation chosen including with regard to elevation of the ground water level.

REFERENCES

Biot M.A. I94I.General theory of three-dimensional consolidation. Journal of Applied Physics I2: I55-I64.

Gaboussi J.,Wilson E.L.I972.Avariational formulation of dynamics of fluid saturated porous elastic solids.Proc. ASCE,v.98 EM4:546-554.

Illarionov E.D. I979.Calculation of dynamic reaction of saturated zones in dams under seismic actions. Energetical constructions 2:62-66.

Hardar A.K.,Reddy D.V. I979.Dynamic finite element formulation for a fluid-saturated porous media. Proc. CANCAM-79.Sherbrook.Canada, 2 : 90I-902.

Baute K.,Wilson E. I982.Numerical methods of analysis and the finite element method.Stroiizdat.Moscow: 447.

Idriss I.M.,Seed H.B. I970.Seismic responce of soil deposits.Proc. ASCE,v.96,SM2:63I-638.

Kurdyuk A,Yu.,Likhovtsev V.M. I987. Algorithm of numerical realization fo evaluation of the intensity of vibrations of saturated soil bases.The applied problems of strength and plasticity.Statics and dynamics of deformable systems.Proc. of the Gorkiy University.

Kurdyuk A.Yu.,Ilyichev V.A.,Likhovtsev V.M. I986. Evaluation of the influence of soil base artificial preparation on the intensity and spectrum of seismic vibrations. Proc. of Research Institute of Bases and Underground Structures, No.86.

12 Mining applications

Numerical Methods in Geomechanics (Innsbruck 1988), Swoboda (ed.)
© 1988 Balkema, Rotterdam. ISBN 90 6191 809 X

Simulation of hydraulic filling of large underground mining excavations

R.Cowling & A.G.Grice
Mount Isa Mines Limited, Queensland, Australia
L.T.Isaacs
University of Queensland, Australia

ABSTRACT: The application of a finite difference seepage model to the prediction of pore water pressure and water levels during filling of underground mining excavations is demonstrated. Results show that maximum pore water pressure is dependent only on excavation size and the location of blocked drains. The coefficient of permeability is markedly different to laboratory derived values. Effective porosity is shown to be an important parameter.

1. INTRODUCTION

Mount Isa Mines Limited mines about 5 million tonnes per year each of copper and lead-zinc-silver ores by underground methods, to a depth of 1000 metres. The majority of the ten million tonnes is extracted by a method known as open stoping. In this method large excavations are created in the orebodies, and then backfilled to enable recovery of adjacent ore. Open stopes vary in plan area from 15 m by 15 m to in excess of 40 m by 40 m. Heights range from 50 m to 300 m.

Once the broken ore is extracted, the voids are filled by a variety of backfill materials, depending on location, sequence in the mining schedule and economics. One of the more important backfill materials is a product known locally as hydraulic fill which, as its name suggests, is produced and reticulated in slurry form. Hydraulic fill is usually used with cementing agents and in conjunction with some form of rockfill.

To contain the backfill within the stoped voids until the cement has reacted, and the water drained, brick bulkheads are constructed at each of the entrances to the stopes. Several hundred bulkheads are constructed each year and there has been no difficulties arising from their use until recently, when several failed and one in particular resulted in a spillage of sufficient quantities to potentially endanger men and facilities. This resulted in wide-ranging investigations into the stability of bulkheads and the factors contributing to their in-stability.

This paper records the investigations into the factors influencing the generation of pore pressure within, and the drainage of water from, stopes.

2. FILLING PRACTICES

Accounts of mining and filling practices at Mount Isa Mines have been reported previously, (1,2,3). This section will only summarise those practices which are necessary for an understanding of the factors affecting water distribution and pore pressure generation.

2.1 Fill materials and reticulation

Hydraulic fill is the product resulting from de-sliming, and partial de-watering, of crushed and ground ore, after the minerals have been removed. Occasionally this material is used on its own to fill stopes. More typical practice is to add cementing agents to produce cemented hydraulic fill. Both materials are reticulated, by gravity, through pipelines and boreholes, usually 150 mm diameter, to the top of the stopes.

Cemented hydraulic fill is usually used in conjunction with one of two other materials which are locally named aggregate and rockfill.

Aggregate is a gravel-size, waste material produced by pre-concentration of lead ore in a heavy medium process. Rockfill is the product of crushing and screening quarried rock. The size distribution curves of aggregate, rockfill and hydraulic fill are shown in Figure 1. Aggregate is added to cemented hydraulic fill at about 20 to 25% by weight, to form cemented aggregate fill, and reticulated through the same boreholes and pipelines as cemented hydraulic fill.

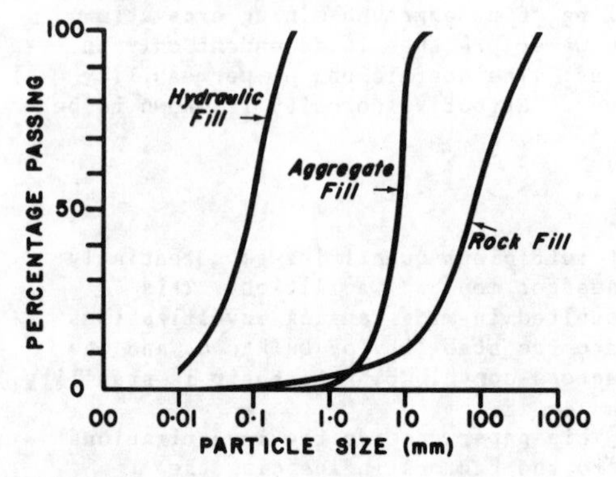

Figure 1. Size distribution curves for fill materials

Rockfill is conveyed from the quarry/crusher/screen area to one of two vertical passes which reach from surface to 600 m below surface. These passes are kept full at all times, more rockfill being added at the top as it is drawn-off at the bottom. Once underground, the rockfill is conveyed to the top of the stopes and introduced simultaneously with cemented hydraulic fill to form cemented rockfill. The ratio of rockfill to cemented hydraulic fill varies from stope to stope, but 2:1 is considered typical.

A summary of placement rates is presented in Table 1.

Volume filling rates and in-situ densities are complicated by the fact that there is large scale segregation of aggregate and rockfill from the cemented hydraulic fill in their respective fill types. Cemented rockfill and cemented aggregate fill are used in copper and lead open stopes respectively.

Table 1. Fill components and placement rates

Fill Type	Placement Rate Solid t/hr	Components	% Weight
Hydraulic Fill	200	Hydraulic Fill	100
Cemented Hydraulic Fill	220	Hydraulic Fill Cementing Agents	91 9*
Cemented Aggregate Fill	275	Cemented Hydraulic Fill Aggregate	80 20
Cemented Rockfill	600	Cemented Hydraulic Fill Rockfill	33 67

(* The cementing agents are:
Portland Cement 3%
Copper Reverberatory Furnace Slag 6%)

Idealised views of the two types of stopes are presented in Figure 2, along with an indication of the segregation. For the compositions presented in Table 1, typical properties are given in Table 2.

Table 2. Fill properties

	Hydraulic Fill	Cemented Hydraulic Fill	Cemented Aggregate Fill	Cemented Rockfill
In-situ Density (t/m^3)	1.37	1.37	2.16	2.33
Moisture Content %	25	35	6	4
Bulk Density (t/m^3)	1.71	1.85	2.29	2.42

Hydraulic and cemented hydraulic fill are reticulated at pulp densities of about 69 weight percent solids which, for their specific gravity (2.9 to 3.0), means that the slurry is more than 50% water by volume. The majority of this

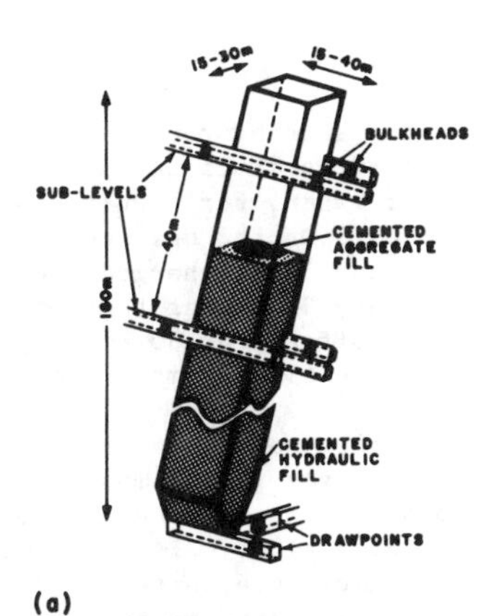

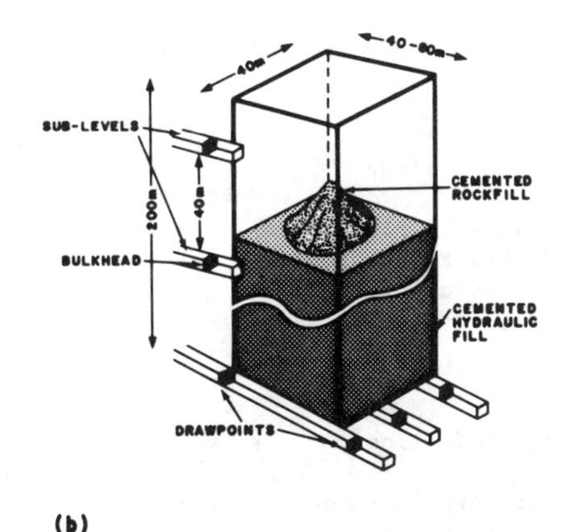

Figure 2. Idealised views of (a) Lead and (b) Copper open stopes, showing location of bulkheads and segregation of fill materials

water remains in the fill as moisture content and the remainder either decants as the fill surface passes the sublevels (Figure 2) or drains through the bulkheads.

After each pour of slurried fill the lines and boreholes are flushed clean by running only water for up to 20 minutes, depending on location. As this can amount to 40 tonnes of water, there are distinct advantages in reducing this quantity and ensuring that the scheduled quantity of fill material is delivered in the minimum number of pours possible.

2.2 Bulkheads and Drainage

The location of bulkheads is shown in Figure 2 and details of their construction is given in Figure 3. The bricks are specified to have a compressive strength of 10 MPa and to remain porous. Immediately behind each bulkhead are drainage pipes which facilitate the rapid removal of water which has decanted on top of the fill, and assist in drainage after the fill surface has passed that level. In the recent past, various types of flexible, porous drainage pipes suspended between sublevels have been tried. Their contribution to drainage was erratic and further trials of geotextiles are in progress.

3. INVESTIGATIONS

Following the bulkhead failure and spillage a number of investigations were

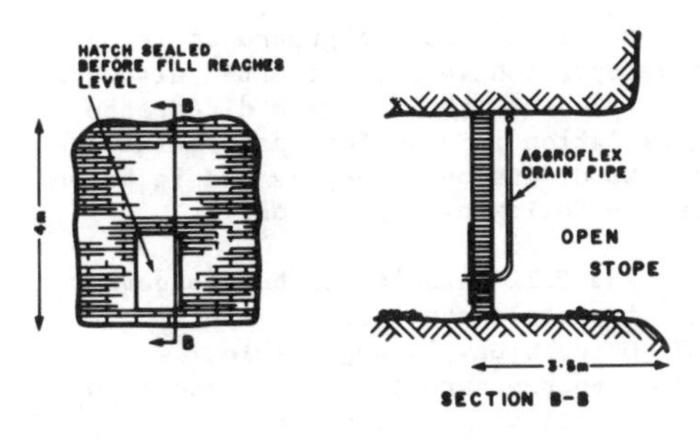

Figure 3. Details of bulkhead

initiated. These concentrated on a number of topics as summarised below.

3.1 In-stope Measurements

Comprehensive measurements during stope filling, including

1. pore pressure
2. earth pressure
3. fill and water quantities entering the stope
4. water and fill levels within the stope
5. water quantity leaving the stope
6. bulkhead loading and deformation.

Several stopes have been monitored in this fashion and one exercise has been reported by Patterson (4).

3.2 Bulkhead Testing

Several bulkheads have been constructed

in remote locations and hydraulically loaded to failure.

3.3 Bulkhead Modelling

Three dimensional finite element modelling of brick bulkheads has been carried out, including parameter studies of properties, location, shape, and use of strengthening ribs, (5).

3.4 Drainage Modelling

This topic is covered in detail in the remainder of the paper.

4. DRAINAGE MODEL

The model and the background to its development have been reported elsewhere (6). In summary, a finite difference formulation is used for the simulation of two dimensional seepage and is based on the following assumptions:

1. the fill material is homogeneous, isotropic and incompressible,
2. Darcy's Law is applicable,
3. pore pressure is zero adjacent to functioning drains, and
4. fill and water levels are horizontal.

Information which has to be supplied for each analysis includes:

1. stope geometry, represented as a two-dimensional finite difference grid
2. location of drains (bulkheads)
3. fill properties - permeability
   - porosity
4. filling rates - solids per unit time
   - water per unit time
5. time parameters - total simulation time
   - time increment
   - duration of pour
   - duration of rest.

Earlier work with the model (6) was based on fill property values considered to be typical for sand of similar size grading as fill - field values were not available at that time. Subsequent work has provided field values for the coeffecient of permeability and estalished guidelines for handling water content with respect to free water and porosity.

4.1 Porosity

When 200 tonnes of hydraulic fill is placed in a stope it is accompanied by 90 tonnes of water, for a pulp density of 69% solids. During drainage only 40 tonnes of water is discharged from the stope for every 90 tonnes that are input. The difference, 50 tonnes, is retained in the fill as moisture content.

In calculating porosity for the various fill types it is important to take account of the water which is tied up in the interstices. If no account is taken of the moisture content, the porosity is calculated to be:

porosity = volume of voids/total volume
= 1 - (bulk density/specific gravity)
= 1 - 1.4/2.9
= 0.517

However, if allowance is made for the contained moisture content the, effective, porosity is 0.167. Correspondingly, the amount of water available for drainage, the free water, has to be reduced by the moisture content, (7). When these adjustments are made there is excellent agreement between measured and computed values.

5. APPLICATION OF THE MODEL

A previous paper (6) has shown the potential uses of the model, but concluded that the results were limited by the non-availability of property values and field measurements. The following sections report the results of several field experiments and parameter studies.

5.1 Back Analyses of Field Measurements

Three stopes have been monitored, to varying degrees, during filling and a fourth is currently being monitored.

Stope 1	Cemented Rockfill	Fill and water in, water out
Stope 2	Cemented Aggregate Fill	Fill and water in, water out, fill and water height, total and pore pressure, bulkhead deformation

Stope 3	Cemented Aggregate Fill	Fill and water in, water out, fill and water height, pore pressure
Stope 4	Hydraulic Fill	Fill and water in, water out, total and pore pressure

Figure 4 shows the measured and computed heights of the fill and water surfaces in Stope 1. That the real stope took longer to drain than the model can probably be explained by the use of a constant value of permeability, whereas it is likely that permeability reduces with time, (8). Analyses of Stopes 2 and 3 produced similar results.

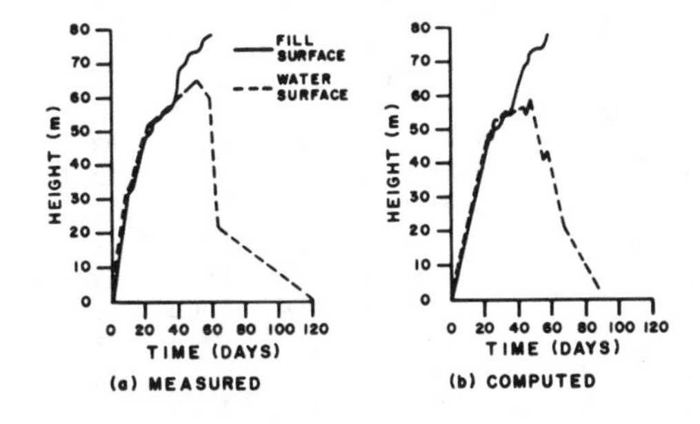

Figure 4. Comparison of measured and computed fill and water heights in stope 1

Accurate modelling of total water balance is a precursor to deriving realistic values for the coefficient of permeability and confirming the role of moisture content in effective porosity. Table 3 summarises the drainage from the three stopes completed to date. The values of porosity and permeability used to obtain this agreement are presented in the same table.

Table 3. Comparison of computed and actual quantities of water (tonnes) to drain from the stopes

Stope	Drained Water		Perm.	Porosity
	Computed	Actual		
1	12693	12650	0.0054	0.038
2	6540	6554	0.0050	0.035
3	13801	13737	0.0140	0.130

(Perm. = Permeability)

Comparisons of pore pressure were not so good although the general pictures were similar. Importantly, the maximum pore pressure measured was 140 KPa, which was only marginally higher than the computed value of 120 KPa. (Work by Beer (5) and field measurements indicate that initial failure of a bulkhead (Figure 3) is initiated at less than 200 KPa and that ultimate failure occurs at 800 KPa.)

5.2 Parameter Studies

Based on the values presented in Table 3 a series of parameter studies were carried out to assess the relative influences of:

1. fill type
2. pulp density
3. pour and rest time
4. stope dimensions
5. blocked bulkheads
6. flushing time.

Details of the parameter studies are summarised in Table 4.

A summary of the results is presented in Table 5, and Figures 5 to 7.

5.2.1 Pore Pressure

For the range of parameters considered it is apparent that pore pressure is only affected by stope geometry and the occurrence of a blocked bulkhead at the bottom of the stope. Figure 5 shows pore pressure distribution for runs 1 and 18 – the extremes of parameters for a 15 m stope, Figure 6 the effect of a bulkhead blocked at different sublevels, and Figure 7 shows the effect of stope size. Interestingly, a blocked bulkhead on the first sublevel shows only a marginal increase in pore pressure compared with the free-draining situation. A number of the bulkheads which have indicated the onset of failure have been located at levels above the base of the stope. These analyses would suggest that, for free draining bulkheads, pore pressure is not a major contributor to failure, and that the cause of the problem lies elsewhere. (Investigation by Grice as reported in (4) indicates that piping (9) may be the cause of bulkhead failure.)

Table 4. Detail of parameter studies

Run	Fill	Density	Pour	Rest	Size	Bulkhead	Flush
1	HF	68	8	16	15	nil	20
2	HF	68	12	12	15	nil	20
3	HF	68	16	8	15	nil	20
4	HF	70	8	16	15	nil	20
5	HF	70	12	12	15	nil	20
6	HF	70	16	8	15	nil	20
7	HF	72	8	16	15	nil	20
8	HF	72	12	12	15	nil	20
9	HF	72	16	8	15	nil	20
10	CAF	68	8	16	15	nil	20
11	CAF	68	12	12	15	nil	20
12	CAF	68	16	8	15	nil	20
13	CAF	70	8	16	15	nil	20
14	CAF	70	12	12	15	nil	20
15	CAF	70	16	8	15	nil	20
16	CAF	72	8	16	15	nil	20
17	CAF	72	12	12	15	nil	20
18	CAF	72	16	8	15	nil	20
19	CAF	70	12	12	20	nil	20
20	CAF	70	12	12	30	nil	20
21	CAF	70	12	12	40	nil	20
22	CAF	70	12	12	15	#a	20
23	CAF	70	12	12	15	#b	20
24	CAF	70	12	12	15	#c	20
25	CAF	70	12	12	15	#d	20
26	CAF	70	12	12	15	nil	0
27	CAF	70	12	12	15	nil	10

Notes: HF = hydraulic fill
CAF = cemented aggregate fill
Density = % solids by weight
Pour = pour time, hours per day
Rest = rest time, hours per day
Size = stope size, metres
Bulkhead = blocked bulkhead, location number according to Figure 6
Flush = flushing time, minutes per pour
Properties: HF Permeability 0.03
Porosity 0.18
CAF Permeability 0.005
Porosity 0.03

5.2.2 Water Levels

Generally, the results show the water surface to be above the top of the fill at the completion of filling. The only exceptions are at the highest pulp density for hydraulic fill in a 15 m stope, and for the 40 m stope. This closely simulates practice where, under normal circumstances, there is usually water on top of the fill.

Table 5. Summary of results

Run	Water In	Drained	Remaining	Fill/ Height	Water Height	Maximum Pore Pressure
1	9327	4366	4961	119.3	120.0	106
2	9024	4082	4942	118.0	120.0	106
3	8866	3940	4926	117.4	120.0	106
4	7691	3359	4332	120.0	106.0	106
5	7392	2466	4926	119.9	120.0	106
6	7236	2324	4912	119.3	120.0	106
7	6171	3094	3077	120.0	75.2	106
8	5873	2116	3757	120.0	92.3	105
9	5719	1611	4108	120.0	101.0	105
10	5014	4108	906	117.9	120.0	106
11	4792	3850	942	117.9	120.0	106
12	4594	3721	873	115.8	120.0	106
13	3543	2663	880	117.9	120.0	106
14	3309	2405	904	117.9	120.0	106
15	3132	2276	856	117.9	120.0	106
16	2126	820	1306	119.2	120.0	106
17	1880	1012	868	119.2	120.0	106
18	1723	883	840	119.2	120.0	106
19	5818	4205	1613	117.5	120.0	143
20	12880	9518	3362	119.4	120.0	213
21	23011	17602	5409	120.0	115.2	272
22	3309	2406	903	117.9	120.0	204
23	3309	2406	903	117.9	120.0	109
24	3309	2406	903	117.9	120.0	106
25	3309	2406	903	117.9	120.0	106
26	2779	2148	631	118.5	120.0	106
27	3044	1889	1155	119.2	120.0	106

6. DISCUSSION OF RESULTS

Values of permeability derived from back analyses of field measurements are significantly different from laboratory values, (8). That this is so supports the comments by Thomas (10) about laboratory derivation of the coefficient of permeability. The importance of moisture content to the value of effective porosity is critical.

The results derived from the parameter studies are in very good agreement with observations generally, and with the limited number of field measurements.

From an applications point of view, the necessity for free draining bulkheads is obvious. The more likely cause of bulkhead failure is the generation of pipes within the stope. This can be minimised or eliminated by the use of higher density slurries which will reduce the height of the phreatic surface within the stope. This paper only covers one aspect of a multi-faceted investigation.

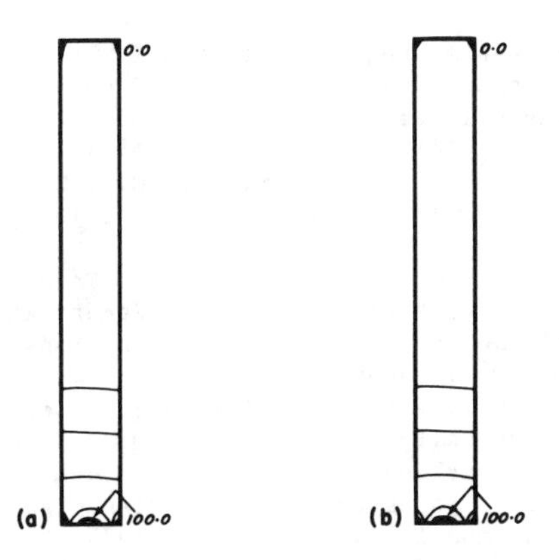

Figure 5. Pore pressure distribution for runs 1, (a) and 18, (b)

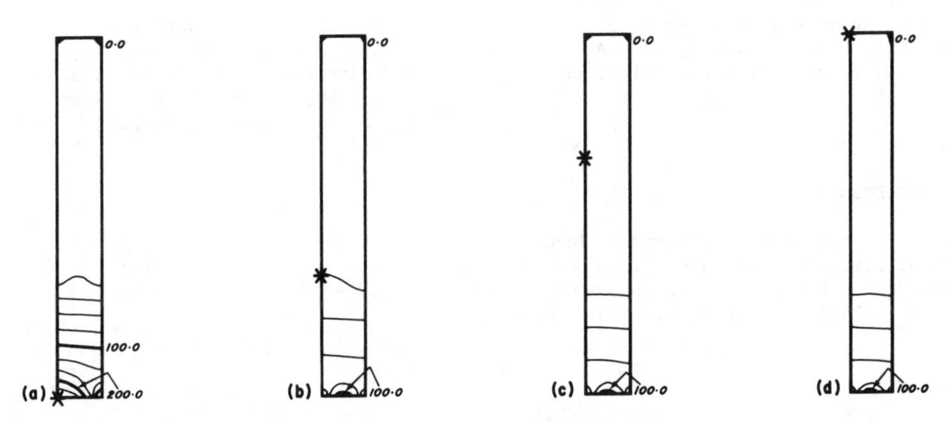

Figure 6. Pore pressure distribution for a bulkhead blocked at different levels (* = location of blocked bulkhead)

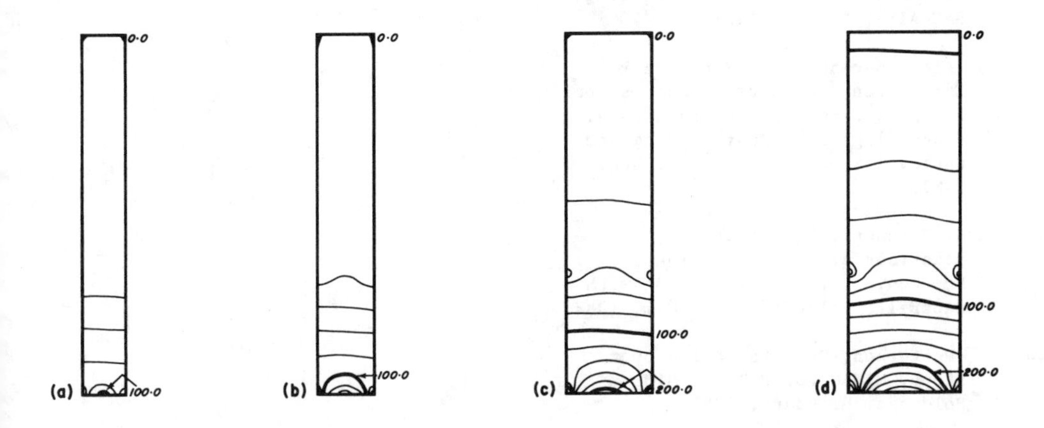

Figure 7. Pore pressure distribution for (a) 15 m, (b) 20 m, (c) 30 m and (d) 40 m stopes

When the results of the components are combined more definitive statements on pour times and pulp densities will be possible.

7. CONCLUSIONS

Back analyses of field measurement have confirmed the application of the seepage model reported by Isaacs. It would appear that the coefficient of permeability for fill materials can only realistically be derived by back analyses. The influence of moisture content on effective porosity is paramount to the use of the model and, when taken account of, produces good agreement both in terms of pore pressure and water balance. When combined with the results of the other investigations, the model will be an essential tool in establishing filling strategies. A three dimensional version of the model is under development.

ACKNOWLEDGEMENTS

The permission of the management of Mount Isa Mines Limited to prepare this paper is acknowledged. Numerous colleagues were involved in this and other investigations.

REFERENCES

(1) I. Goddard, The development of open stoping in lead orebodies at Mount Isa Mines Limited, Proceedings Int. Conf. Caving and Sublevel Stoping, SME-AIME, Denver, 1981.

(2) E. Alexander and M.W. Fabjanczyk, Extraction design using open stopes for pillar recovery at Mount Isa, Proceedings Int. Conf. Caving and Sublevel Stoping, SME-AIME, Denver, 1981.

(3) L.B. Neindorf, Fill operating practices at Mount Isa Mines, Proceedings Int. Sym. Mining with Backfill, Balkema, Rotterdam, 1983.

(4) J. Patterson, Secrets of the open stope, MIMAG Vol 2 1987, MIM Holdings, Brisbane, 1987.

(5) G. Beer and R. Cowling, Computer modelling of brick bulkheads at Mount Isa, in preparation.

(6) L.T. Isaacs and J.P. Carter, Theoretical study of pore water pressure developed in hydraulic fill in mine stopes, Trans. Instn Min. Metall. (Sect.A:Min.industry), 92, 1983.

(7) D.K. Todd, Groundwater Hydrology - Second Edition, p26-37, John Wiley and Sons.

(8) R.J. Mitchell, J.D. Smith and D.J. Libby, Bulkhead pressures due to cemented hydraulic backfills, Canadian Geotechnical Journal, 12, 1975.

(9) M.E. Harr, Mechanics of Particulate Media, McGraw-Hill, New York, 1977.

(10) E.G. Thomas, L.H. Nantel and K.R. Notley, Fill Technology in Underground Metalliferous Mines p55-60, International Academic Services Limited, Kingston, 1979.

Numerical Methods in Geomechanics (Innsbruck 1988), Swoboda (ed.)
© 1988 Balkema, Rotterdam. ISBN 90 6191 809 X

A theoretical analysis of the stability of underground openings due to storage of LPG

Y.Inada
Ehime University, Matsuyama, Japan
K.Takaichi
Taisei Corporation, Tokyo, Japan

ABSTRACT: As the temperature of L.P.G. is -43°C, the rock mass around the openings will be affected by low temperatures. Namely, steep temperature gradient and thermal stress will occur around the openings and the plastic zone will be increased with time.
This paper presents the results of theoretical analysis for the plastic zone occurring around underground openings, and the results of an investigation of stability.

1 INTRODUCTION

The demand for energy resources has increased in recent years. Especially the demand for L.P.G., which has an important role as an energy substitute for petroleum, has rapidly increased in Europe, America and Japan.
Double structural tanks made of special steel or semi-underground concrete tanks have been used for the storage of L.P.G., and many have been built on land reclaimed from the sea. But these storage methods require a vast area for prevention of disasters.
They not only destroy the beauty of the shoreline, but also, from a military standpoint are not always desirable. So, storage in underground openings excavated in rock mass is another method to be considered.
As the temperature of L.P.G. is -43°C, the rock mass around the openings will be affected by low temperatures. Namely, steep temperature gradient and thermal stress will occur around the openings, and the plastic zone will increase with time.
This paper presents the results of theoretical analysis for the plastic zone occurring around underground openings, and the results of an investigation of stability.
Temperature distribution around the openings was calculated by using the Finite Divided Element Method (F.D.E.M.).
Then, changes over a period of time in the plastic zone and the range of cracks around the openings were estimated by using "Crack Analysis Method","No Tension Analysis Method" and "Anisotropic Analysis Method" which are applied method of the Finite Element Method (F.E.M.). From the results of the analysis, ways to use the openings safely after storage of L.P.G. are discussed.

2 TEMPERATURE DISTRIBUTION AROUND OPENINGS

2.1 Temperature distribution calculation method

In this paper, the temperature distribution around underground openings used for storage of L.P.G. was calculated by using a hybrid numerical analysis method termed the Finite Divided Element Method (F.D.E.M.) (Inade & shigenobu, 1983).

2.2 Results and considerations

It is assumed that the openings are excavated in fresh rock mass at a depth of 100 meters from the ground surface.
The temperature distribution which was obtained by using F.D.E.M. is shown in Figure 1.
From this figure, it is found that the temperature gradient is extremely sharp in early time, but becomes gentler with time. Also the change of temperature distribution becomes slower with time. It is seen that the temperature distribution reaches a semi-steady state after 1 year.
On the other hand, theoretical solutions of heat conduction for the area around the openings (confined to early time) are shown by the dotted line in Figure 1. Comparing the results obtained by F.D.E.M.

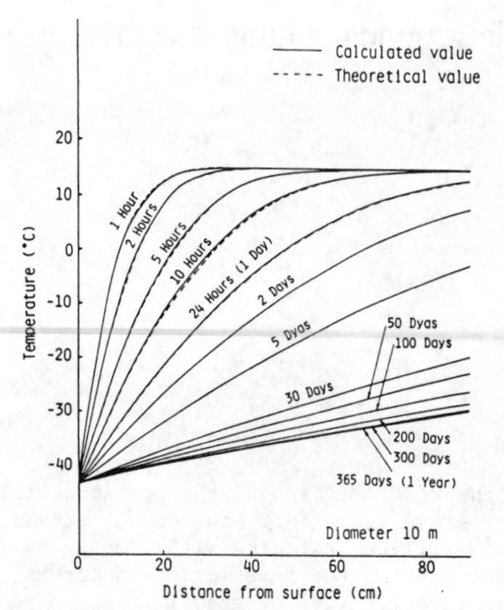

Fig.1 Temperature distribution around circular openings (Diameter:10 m).

and strict solution, it is found that they are in agreement.

3 PLASTIC ZONE AROUND OPENINGS

3.1 Physical constants of rocks

For analysis, the strength, elastic constants and thermal properties of rocks at low temperature must be known. As we had obtained these values by other experiments (Inada et al., 1979, 1980), they were used for the analysis. These values are shown in Table 1.

Table 1 Physical constants of rocks used in this analysis (in the wet state).

Temperature (°C)	Expansion coefficient $(1/°C)\times10^{-6}$	Young's modulus $(kgf/cm^2)\times10^6$	Poisson's ratio	Compressive strength (kgf/cm^2)	Tensile strength (kgf/cm^2)
20~ 10		0.494	0.25	-1670	79
10~ 0	15.1	0.494	0.25	-1670	79
0~ -10	14.9	0.495	0.25	-1678	95
-10~ -20	14.5	0.495	0.25	-1689	104
-20~ -30	14.0	0.496	0.25	-1711	111
-30~ -40	13.6	0.496	0.25	-1741	116
-40~ -50	13.2	0.497	0.24	-1778	121

3.2 Stress analysis method

In the case of circular openings, it is supposed that the openings are excavated to a depth of 100 meters from the ground, surface, with a diameter of 10 meters, and homogeneous and elastic rock mass around the openings.
A part of the analysis model, hereinafter referred to as Model I, is shown in Figure 2 (a) and part of the analysis model used for the Anisotropic Analysis Method, hereinafter referred to as Model II is shown in Figure 2 (b).
Stress changes around the underground openings were analyzed using the F.E.M. assuming an arbitrary coefficient of lateral pressure:

$$\lambda_o = \sigma_h/\sigma_v$$

where : σ_h = horizontal stress
σ_v = vertical stress
λ_o = coefficient of lateral pressure

In this paper, the results for λ_o= 0.5 (for model I) and for λ_o= 0.0 (for model II) are shown as a typical example. To estimate the change over a period of time in the plastic zone around the openings, the theory of Mohr's envelope was adopted as a failure criteria (Inada & Taniguchi, 1987). On the three matually perpendicular principal stresses at the center of gravity in the devided element, when only one direction's principal stress exceeds tensile strength, it is estimated that the element is fractured as a thin cracker would be and the other directions in which the element is not fractured, are considered to behave as an elastic body. So we call this state of the element a plastic zone. When the three directions' principal stresses exceed fracture strength, we call this state of the element the fracture zone. Plane strain conditions were assumed for the analysis.
Time-step analysis using the "No Tension Analysis Method", "Crack Analysis Method" and "Anisotropic Analysis Method" were adopted. The methods differ only if the tensile strength of the rock is exceeded. The analysis procedures can also be used to analyze the effect of low temperature storage on the fracturing and stress patterns around underground openings.

1) No Tension Analysis Method

This method was proposed by Zienkiewicz et al. (1968) for analyzing "no tension" materials.
This simulates a material which is able to resist and transmit a compressive force but is not able to withstand a tensile force. In this study, we adopted the divided element whose tensile stress culculated by F.E.M. had exceeded its tensile strength.
The procedure in this study is shown in Figure 3.

2) Crack Analysis Method

The method of analysis which traces the

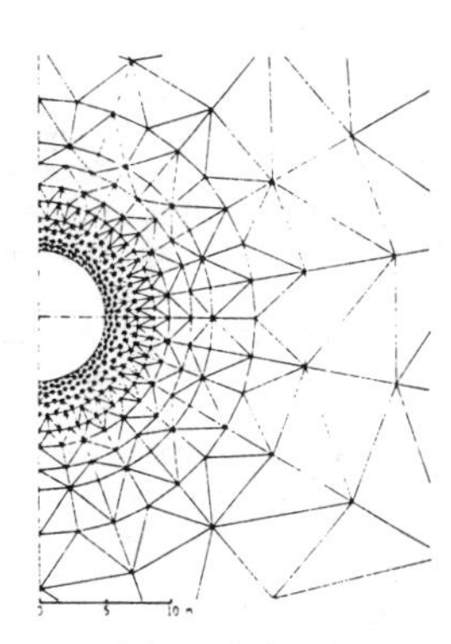

(a) Model I

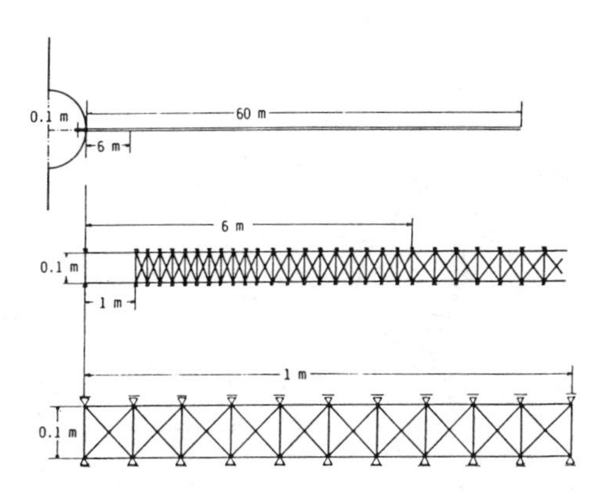

(b) Model II

Fig.2 Part of the analysis model.

advance of cracks with time, is termed the "Crack Analysis Method". The principle of the crack Analysis Method is shown in Figure 4. That is, when the stress in the element exceeds the tensile strength, it is assumed that the element itself is not fractured, but cracks occur between adjacent elements. The procedure in this study is shown in Figure 5.

3) Anisotropic Analysis Method

This method is one of the methods in addition to the No Tension Analysis Method for analyzing "no tension" materials. It is used when one direction's principal stress exceeds tensile strength, setting up that direction's Young's modulus to zero, and other mutually perpendicular principal stresses' directions are considered to behave as an elastic body.
That is, it is supposed that material changes to, what we call, perpendicular anisotropy. The procedure in this study is shown in Figure 6.

3.3 Results and considerations

Plastic zone and cracks around openings calculated by each method are compared in Figure 7.
After 1 year, as shown in Figure 7, it was found that the plastic zone calculated by No Tension Analysis Method and Anisotropic Analysis Method are approximately in agreement but the range of cracks is smaller than by other methods.
To compare the results by theoretical analysis and actual phenomena, an experiment using plate glass with a circular hole made at the center, and using Liquefied Nitrogen (-196°C) was performed.
The size of the plate glass used was 300× 300×5 mm and the diameter of the hole was

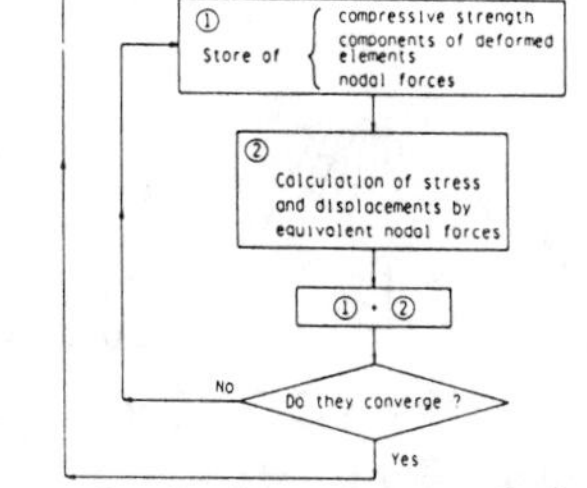

Fig.3 Flow chart of No Tension Method.

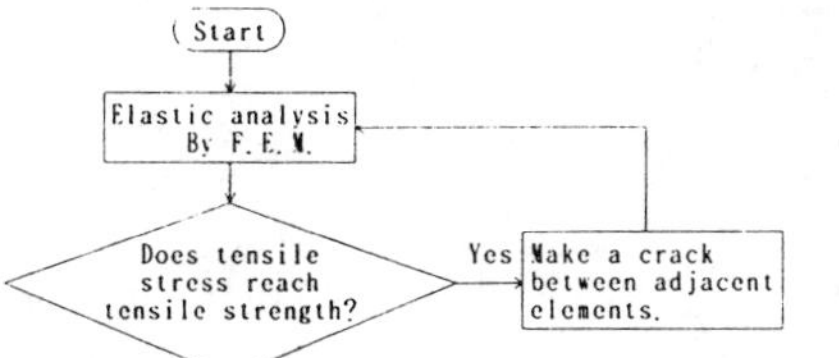

Fig.4 Principle of Crack Analysis Method.

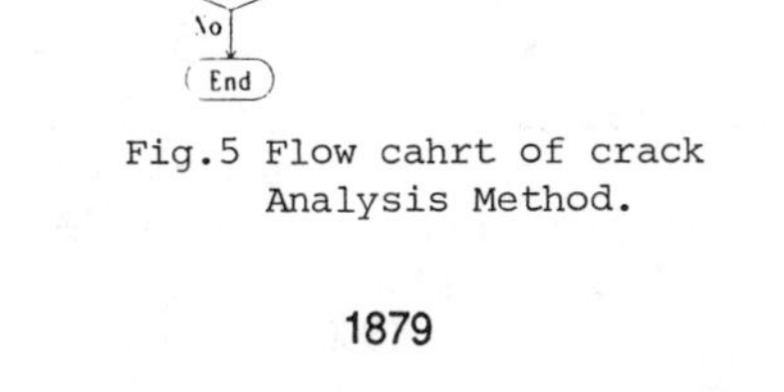

Fig.5 Flow cahrt of crack Analysis Method.

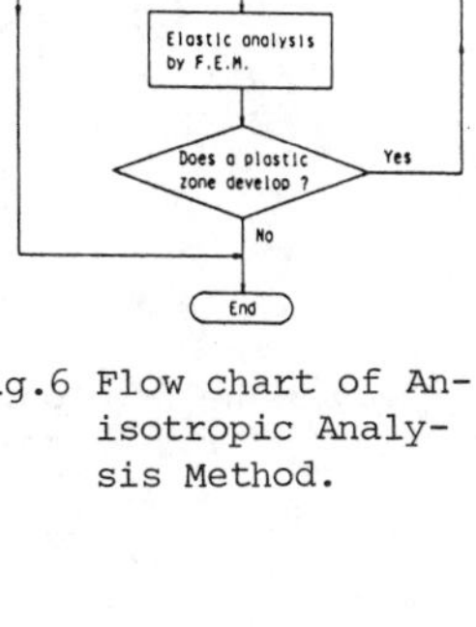

Fig.6 Flow chart of Anisotropic Analysis Method.

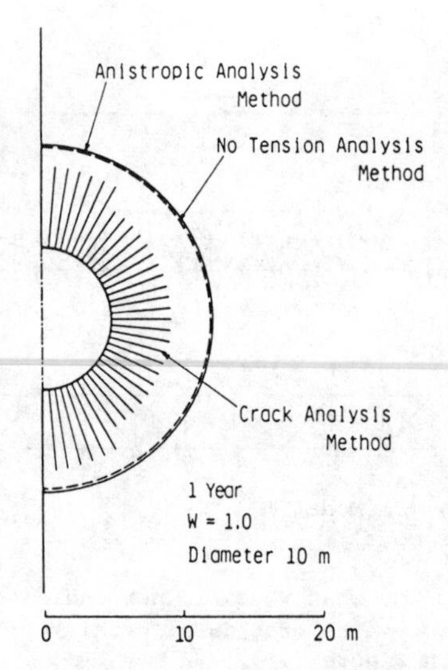

Fig.7 Comparison of plastic zone and cracks around openings (after 1 year).

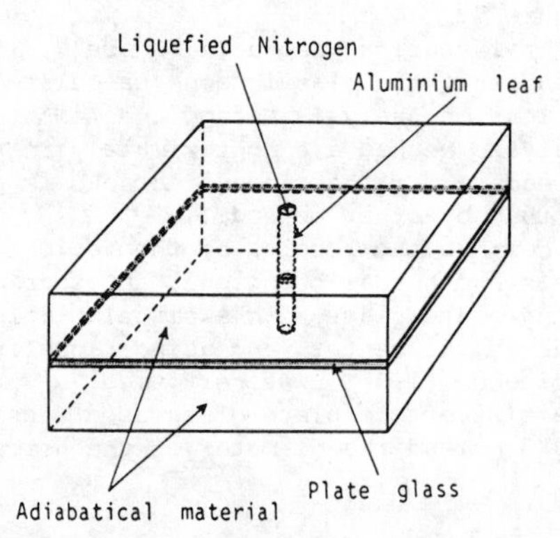

Fig.8 Schematic diagram of the test.

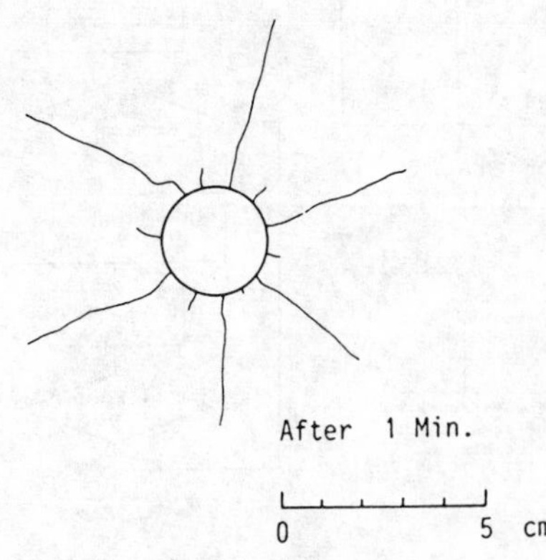

Fig.9 The developing cracks around circular hole.

26 mm.

In this test, in order to avoid the complicated affects of thermal conduction and the occurrence of a plastic zone due to Liquefied Nitrogen entering into micro-cracks of the hole's surface, the hole was protected by aluminium leaf (thickness : 15 $\times 10^{-6}$ meters). A schematic diagram of the test is shown in Figure 8. And the results of the test is shown in Figure 9.

It can be seen that the cracks occurred roughly at reqular intervals around the hole, and that the long cracks and the short cracks occurred roughly in a line alternatively.

From these facts, it is reasoned that the first crack would occur at the weakest point around the hole and stress relaxation occurred around the cracks.

Similarly the stress relaxation occurred around the long cracks. Therefore, the short cracks are seen near the long cracks. It seems that these phenomena occurred in an extremely short time.

To investigate the above, the range of the crack where the latent cracks had not existed, and that where they had existed, were compered.

In this study, it was supporsed that the latent cracks were located 40 cm from the openings' surface and it's length was 60 cm towards mountain.

The results after 1 year are shown in Figur 10(a),(b). From these figures, it is found that in the case where latent cracks did not exist, the cracks advance to 5.7 meters at the top and bottom of the openings and 4.2 meters at the side walls. In the case where latent cracks existed, the cracks advance to maximum 3.0 meters, minimum 2.2 meters at the top and bottom of the openings and 1.4 meters at the side walls. Moreover, the cracks advance to 5.7

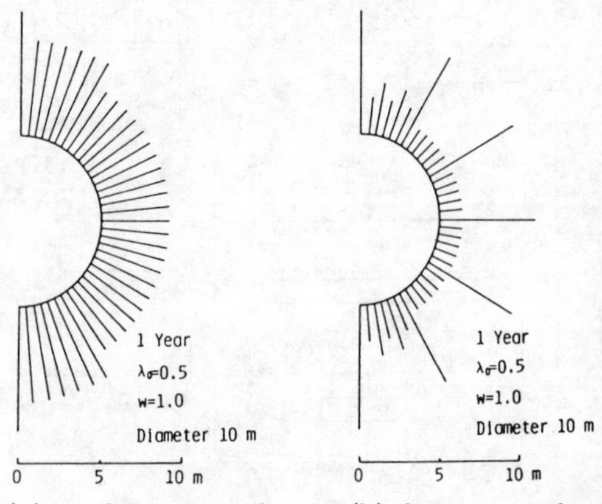

(a) no latent cracks (b) latent cracks

Fig.10 Comparison of cracks around openings (after 1 year).

meters by connecting with latent cracks. On the whole in the case where latent cracks had existed, the range of the cracks is smaller except as part of the latent cracks, compared with the case where latent cracks had not existed.
Thus, analysis which considers the latent cracks is closer to actual phenomena seen from the results of the test which used plate glass, as mentioned before.
That is, rock mass should contain many weak planes such as actual cracks and joints, and cracks caused by thermal stress connect with these weak planes.
Then, it is assumed that the stress concentration should occur at the top of the crack and stress relaxation should occur around cracks. Considering these phenomena, it seems to be possible to control the range of the plastic zone by artificially making cracks around the openings.
From the results of three kinds of analysis mentioned before and by test using plate glass, the Crack Analysis Method is able to simulate closely to actual phenomena. However, as mentioned before, rock mass should contain many weak planes and it is difficult to know their location and direction. So in general, we cannot disregard the No Tension Analysis Method except in cases where information about the weak plane in the rock mass is well known.

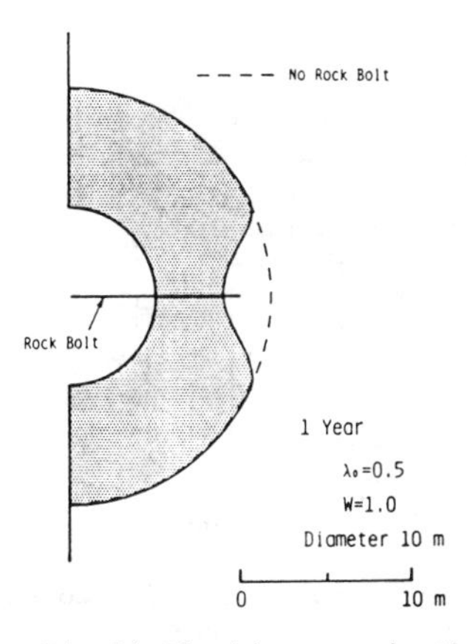

Fig.11 Plastic zone in the case using rock bolt (after 1 year).

4 A FEW REMARKS ON APPLICATION TO THE FIELD

4.1 Rock bolt

From some results as mentioned before, it is found that the plastic zone and the cracks occurring by thermal tensil stress, advance with time.
In this study, control of the plastic zone and cracks and how to secure the safety of openings are discussed.
As the openings shrink to the mountain due to storage of L.P.G., it is necessary to resist that. So, in analysis was performed for the case in which the rock bolts are connected to each other at the center of the openings. Thermal conduction from the rock bolt to the mountain is neglected.
The result is shown in Figure 11. From this figure, it was found that the plastic zone decreased 2 meters using the rock bolt.
It is necessary to use a special steel rock bolt (otherwise brittle fracture will occur) and ways how to cut off thermal conduction from rock bolt to mountain should be considered.

4.2 Storage by pressure

As vapor pressure of L.P.G. is 8 kgf/cm^2, the case in which this pressure was applied from the inside wall of the openings was calculated. From the results, it was found that the stress distribution did not change remarkably compared with the openings where L.P.G. was not stored. Therefore it was found that openings was stable, but one of the problems is airtightness. Actually, as rock mass has many joints and cracks, it is difficult to neglect this problem.

4.3 Enclosure by water

In the case in which L.P.G. is stored at low temperature, one problem is to prevent gas and liquid leaks.
One of the techniques is enclosure by water which surrounds the openings artificially.
A test using cement mortar and Liquefied Nitrogen was performed.
The schematic diagram is shown in Figure 12.
The size of the specimen was as follows : cement mortar : 250×250×200 mm, the diameter and depth of the hole : 30 mm, 150 mm, width and depth of the slit : 5×130 mm.
The slit was filled with water.
The results are shown in Figure 13. From the observation of the test, as the area around the hloe is frozen, leakage of liquid from hole was not seen. That is, the effect of enclosure by water was confirmed, however another important problem, namely the behavior of ice in the slit has been left unsolved.

5. CONCLUSIONS

In this paper, the theoretical analysis of the plastic zone and cracks occurring

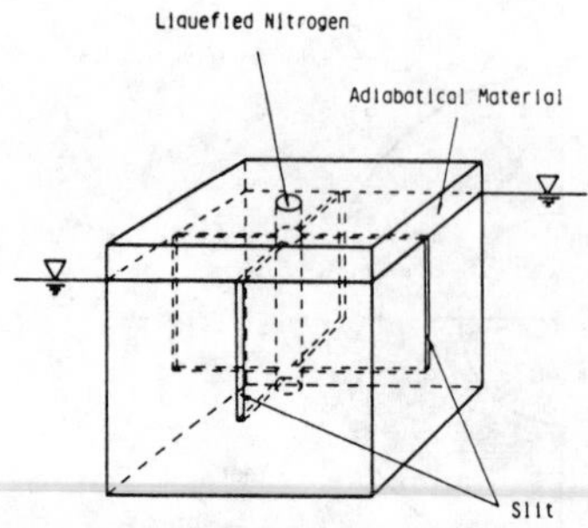

Fig.12 Schematic diagram of the test.

around underground openings due to storage of L.P.G. has been discussed.
The principal results obtained are as follows :

1. It has been shown that the temperature distribution around openings can be simply and accurately modeled using the Finite Divided Element Method. The temperature distribution around openings reached a semi-steady state after 1 year.

2. In the case of storage of L.P.G. at -43°C in 10 m diameter underground openings at 100 m below the ground surface, the analysis indicated that the plastic zone and cracks after 1 year advance to 6.8 m by the No Tension Analysis Method, 6.5 m by Anisotropic Analysis Method, and 5.7 m at the top and bottom, 4.2 m at the side wall of the openings by the Crack Analysis Method.

3. From the results of the test using plate glass and Liquefied Nitrogen, cracks around hole are similar to the cracks obtained by Crack Analysis Method.

4. It seems to be possible to control the range of the plastic zone by artificially making cracks around the openings.

5. A few remarks on application to the field, for instance the cases using a rock bolt, storage by pressure, and enclosure by water are discussed.

REFERENCES

Inada, Y. and Yagi, N., 1979, "Mechanical Characteristics of Rocks Related to Cooling," J.Soc. Mate. Sci. Japan, Vol. 28, No.313, Apr., pp.979-985.
Inada, Y. and Yagi, N., 1980, "Mechanical Characteristics of Rocks at Low Temperatures," J.Soc. Mate. Sci. Japan, Vol.

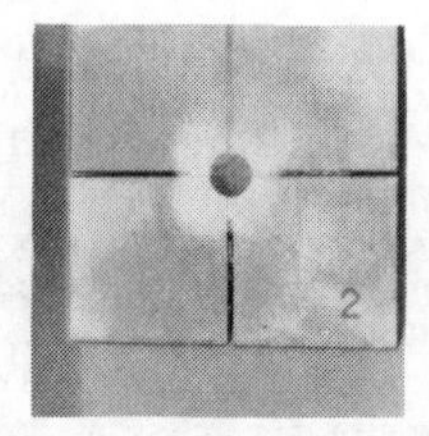
Fig.13 Results of the test.

29, No.327, Jun., pp.1221-1227.
Inada, Y. and Yagi, N., 1980, "Thermal Properties of Rocks at Low Temperatures," J.Soc. Mate. Sci. Japan, Vol.29, No.327, Mar., pp.1228-1233.
Inada, Y. and Shigenobu, J., 1983, "Temperature distribution around Underground Openings excavated in Rock Mass due to Storage of Liquefied Natural Gas," J. Min. Mate. Inst. Japan, Vol.99, No.1141, Mar., pp.179-185.
Inada, Y. and Taniguchi, K., 1986, "A Theoretical Analysis of the Range of the Plastic Zone around Openings due to Storage of L.N.G.", U.S. Symp. Rock Mech. Proc. 27th, pp.782-788.
Zienkiewicz. O. C., Valliappan, S. and King, I. P., 1986, "Stress Analysis of Rock as a 'No Tension' Material," Geotechnique, 18, pp.56-66.

Numerical Methods in Geomechanics (Innsbruck 1988), Swoboda (ed.)
© 1988 Balkema, Rotterdam. ISBN 90 6191 809 X

Constitutive behaviour of rock salt: Power law or hyperbolic sine creep

G.Borm & M.Haupt
Institute of Soil Mechanics and Rock Mechanics, Department of Rock Mechanics, University of Karlsruhe, FR Germany

ABSTRACT: For long term prediction of creep convergence of underground excavations in rock salt, stationary power law creep approaches have widely been applied in the past and present.

On the other hand, stationary creep of rock salt at moderate temperatures has nowhere been observed in situ nor in laboratory tests. Moreover, in numerical calculations according to the finite element method based on initial strain procedures, even the adoption of a purely stationary creep law results into stresses and strains which are generally of transient nature.

The measured creep convergence of a borehole in a German saltdome could equally well be described by power law, hyperbolic sine, or strain hardening power law creep approaches. The best agreement to the results of laboratory creep and relaxation investigations, however, has been obtained through application of the strain hardening power law.

The presented analytical solution of the time dependent closure of a borehole in rock salt, which has been verified by both laboratory testing and long-term convergence measurements in situ, offer an excellent means for calibrating numerical models prior to treating more complex rheological boundary value problems.

INTRODUCTION

Stress relaxation plays a dominant role in underground excavation works in rock salt. Representative examples are the release of supplementary stresses during the course of time, when former extraction chambers are filled up with debris, the time dependent compression of cavity fillings with fossile raw material or special waste in salt-domes, or the relaxation of differential stresses in side walls, pillars and roofs leading to an overall stress redistribution and to wide span stress arches around the entire mine outlay.

The rheological stresses, stress rates and strain rates are generally of transient nature. Instationary creep convergences of underground cavities in rock salt may accordingly be described by transient creep laws and stationary deviatoric stresses on one side, or by stationary creep laws induced by relaxing stresses on the other side of a wide spectrum of constitutive behaviour and of initial and boundary conditions.

In the following, three selected constitutive laws of rock salt are compared with respect to results of laboratory creep and relaxation tests as well as to convergence measurements in situ at a deep research borehole in a saltdome.

Notations

σ_{ij}	Cauchy stress tensor
s_{ij}	deviatoric Cauchy stress tensor
ε_{ij}	Euler strain tensor
e_{ij}	deviatoric Euler strain tensor
r	radial coordinate
r_0	initial radius of borehole
r_i	radius of borehole
r_a	external radius of cylinder
ρ	relative radial coordinate
φ	tangential coordinate
t	time coordinate
t_0	initial time
$\hat{t}$	generalized time coordinate
τ	relaxation time
Ψ, Φ	relaxation functions
σ	uniaxial stress
σ_{eff}	effective stress
ε	uniaxial strain
ε_{eff}	effective strain
e_0	reference creep strain
E	Young's modulus of elasticity
G	Kirchhoff's shear modulus
ν	Poisson's ratio
Q	activation energy
R	universal gas constant
T	Kelvin temperature

POWER LAW CREEP

Uniaxial Creep and Relaxation

Stationary creep laws are normally combined with a linear elastic part inducing stress relaxation. In this case, the constitutive equation for power law creep according to Heard (1972) is given by

$$\dot{\varepsilon} = \frac{\dot{\sigma}}{E} + A\left(\frac{\sigma}{\sigma_0}\right)^n \quad , \quad n > 1 , \tag{1a}$$

where A is conveniently set as

$$A = A' \exp\left(-\frac{Q}{RT}\right) .$$

σ_0 is a standardization parameter chosen here as $\sigma_0 = 1$ MPa, and E is Young's modulus of elasticity [1]. For stationary creep, the elastic strain rate is ignored, and the creep rate becomes

$$\dot{\varepsilon} = A\left(\frac{\sigma}{\sigma_0}\right)^n \quad . \tag{1b}$$

For pure relaxation (i. e. $\varepsilon = 0$), integration of eq. 1a with respect to time and to the initial condition $\sigma(t = t_0) = \sigma_a$ yields

$$\frac{\sigma}{\sigma_a} = (1 + \hat{t})^{\frac{1}{1-n}} \tag{2a}$$

where

$$\hat{t} = \frac{t - t_0}{\tau} \tag{2b}$$

and

$$\tau = \frac{\sigma_a}{(n-1)EA}\left(\frac{\sigma_0}{\sigma_a}\right)^{n-1} \tag{2c}$$

Eqs. 2 characterize the uniaxial response of a totally relaxing material, since $\lim_{t\to\infty} \sigma = 0$.

Borehole Creep Closure

An analytical solution of the rheological equation of borehole closure in rock salt according to eq. 1a was introduced by Borm (1987):

Consider the stresses at the borehole wall (fig. 1, $r = r_i$) under plane strain conditions, where

$$\sigma_r = 0, \; \sigma_\varphi = \sigma_\varphi(t) \text{ and } \sigma_z = \sigma_\varphi/2.$$

Let $\sigma := \sigma_\varphi(r = r_i)$. The relaxation function of this tangential stress is obtained as

$$\frac{\sigma}{\sigma_a} = \frac{1}{n}\left(1 + (n-1)\Psi(t)\right) = \Phi(t) \tag{3a}$$

---

[1] *In the following, the symbols used in the equations will not be redefined unless otherwise stated. The reader is referred to chapter* ***Notations*** *of the introduction.*

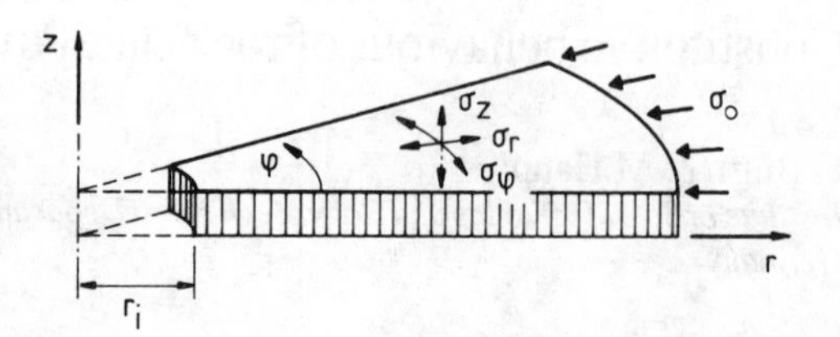

Fig. 1: Geometry of borehole model and notations

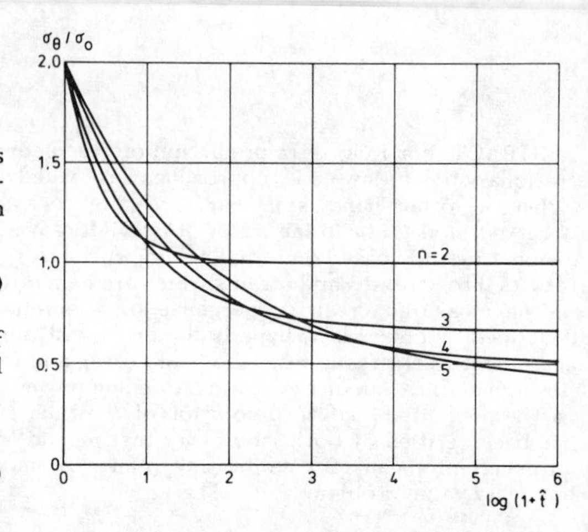

Fig. 2: Relaxation of tangential stresses at the borehole wall as a function of time and different values of creep exponent n

where

$$\sigma_a = 2\sigma_0 \tag{3b}$$

$$\Psi(t) = (1 + \hat{t})^{\frac{1}{1-n}} \tag{3c}$$

and

$$\hat{t} = \frac{t - t_0}{\tau} \tag{3d}$$

$$\tau = \frac{1}{(n-1)}\,\frac{e_0}{A} \tag{3e}$$

$$e_0 = \frac{\sigma_0}{2G} \tag{3f}$$

Fig. 2 shows the relaxation of the tangential stresses at the borehole wall as a function of time for different values of the creep exponent n in dimensionless representation. In the special case of linear viscosity (n = 1), no stress relaxation would occur.

The strain rates of the borehole wall in the rock salt under plane strain conditions and incompressible flow are given by

$$\dot{e}_r = -\dot{e}_\varphi \, , \quad \dot{e}_z = 0 .$$

Let $\dot{e} := \dot{e}_r(r = r_i)$. The time dependent elastic strain rate $\dot{e}^{ela}$ is obtained from eqs. 1 and 3 as

$$\dot{e}^{ela} = -\left(\frac{e_\infty}{\tau}\right)\Psi^n(t) \qquad (4a)$$

where $$e_\infty = \frac{e_0}{n} \quad . \qquad (4b)$$

The viscous strain rate follows from the power law and eq. 3a as

$$\dot{e}^{vis} = \dot{e}_\infty \Phi^n(t) \qquad (5a)$$

where

$$\dot{e}_\infty = A\left(\frac{3}{4}\right)^{\frac{n+1}{2}}\left(\frac{2}{n}\right)^n \sigma_0^n \qquad (5b)$$

The initial values for $t = t_0$ are determined by

$$\dot{e}_0^{tot} = \dot{e}_0^{ela} + \dot{e}_0^{vis}$$

$$\dot{e}_0^{ela} = \frac{1-n}{n}\dot{e}_0$$

$$\dot{e}_0^{vis} = n^n \dot{e}_\infty$$

The total strain e^{tot} is assumed to be composed of an elastic strain e^{ela} and a viscous strain e^{vis} as

$$e^{tot} = e^{ela} + e^{vis} .$$

The elastic and viscous strains are obtained from integration of the corresponding strain rates (eq. 4 and eq. 5) with respect to time. These integrations can be performed analytically for integer values of the creep exponent n, and numerically else:

$$e^{ela} = \int_{t_0}^{t} \dot{e}^{ela}\,dt = e_\infty \Phi(t)$$

$$e^{vis} = \int_{t_0}^{t} \dot{e}^{vis}\,dt = \dot{e}_\infty \int_{t_0}^{t} \Phi^n(t)\,dt \quad .$$

The borehole convergence, i. e. the radial displacement $u(r = r_i)$ of the borehole wall, results from the total strain by integration with respect to the radial coordinate, which in case of incompressible flow is equivalent to multiplying the tangential strain by the radial coordinate

$$u(r = r_i) = r_i \varepsilon_\varphi = -r_i e \quad . \qquad (6a)$$

In large strain analysis (Jaeger, 1969), one would obtain instead

$$u(r = r_i) := r_0(1 - \exp(-e)) \qquad (6b)$$

where the sign of the borehole closure is taken here as positive for positive radial strain e.

Since the elastic convergence at time $t = t_0$ is given by $u_0 = r_0(1 - \exp(-e_0))$, the ratio of creep convergence and initial radius is given by

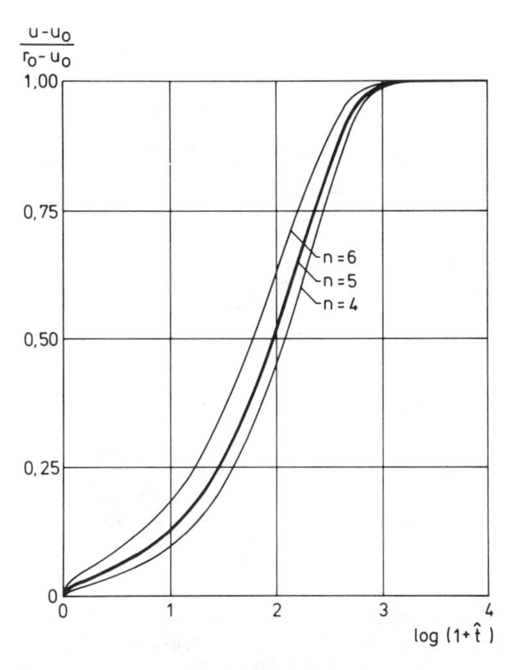

Fig. 3: Normalized borehole convergences as a function of generalized time $\hat{t}$ and selected values of creep exponent n

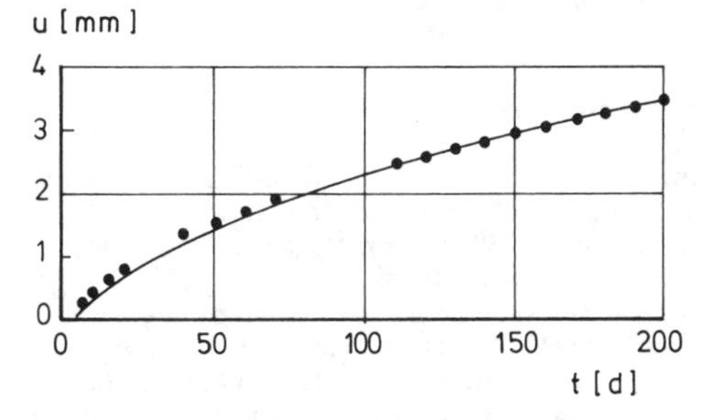

Fig. 4: Borehole closure measurements and analytical approach

$$\frac{u - u_0}{r_0 - u_0} = 1 - \exp(e_0 - e) \quad . \qquad (6c)$$

The relative convergence according to eq. 6c, is shown in fig. 3 as a function of time and selected values of the creep exponent n. In order to condense the spectrum of convergence characteristics, the relaxation time as given by eq. 3e is multiplied here by the factor of n^n.

In a bi-logarithmic diagram, the convergence curves for small strains would appear as straight lines, whereas in the common linear representation they show the familiar parabolic shape (e. g. fig. 4)

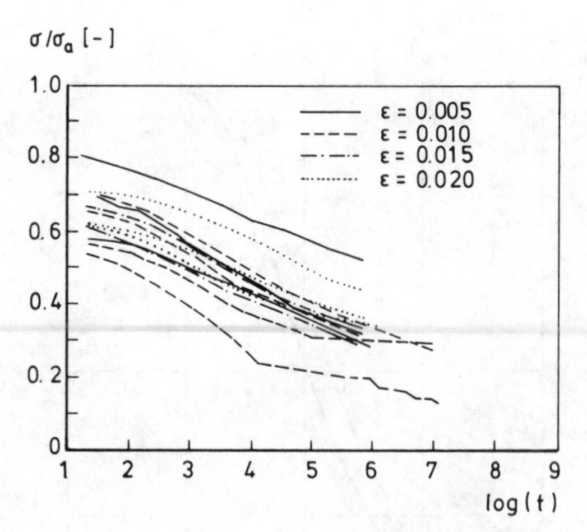

Fig. 5: Stress relaxation curves of rock salt samples subjected to different initial strains

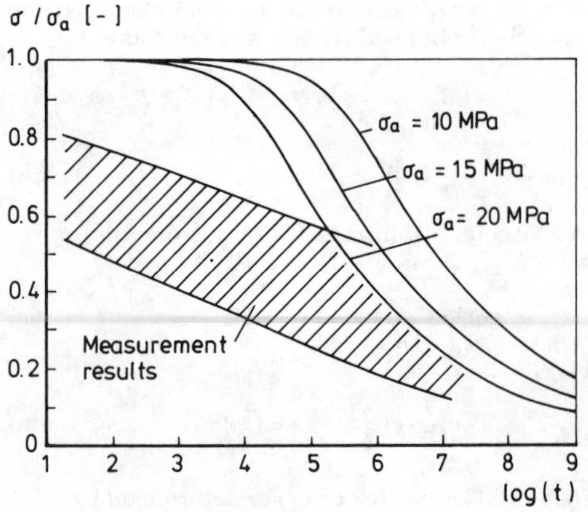

Fig. 6: Results of stress relaxation tests compared to theoretical predictions according to the stationary power law creep.

The borehole convergence measured by Doeven et al. (1983), can adequately be approached (fig. 4) by eqs. 3 to 6, adapting the parametric values as follows

$n = 5.5$,
$A = 1*10^{-15}/s$,
$G = 400$ MPa,
$\sigma_0 = -22$ MPa.

Experimental Results

Uniaxial relaxation tests on rock salt samples ($d = 54$ mm, $h \approx 13o$ mm) were carried out at a temperature of 35°C using a specially developed relaxation device (Balthasar et al., 1987). During relaxation measurement, the length of the sample was kept constant within a tolerance of ± 0.5 μm. Tests were performed at 4 different strain levels as shown in fig. 5, where the time t is taken in seconds. It can be recognized from that figure that there is a strong instantaneous stress release of 20 to 40 percent. The stress rate decreases monotonously. An unambiguous strain dependency of the relaxation curves can not be perceived.

The determination of representative material parameters from the data obtained in laboratory creep tests for use in eqs. 1 and 2 appeared to be rather difficult, since stationary creep has not been observed even during very long-term test periods. Least square approaches to some 30 creep tests yielded the values $A = 1.4*10^{-15}/s$ and $n = 5.0$, for $T = 308°K$. The elastic modulus was taken as $E = 20000$ MPa and the reference stress as $\sigma_0 = 1$ MPa.

Comparison of theoretically predicted relaxation curves according to eq. 2a to those responses measured in laboratory tests show relatively poor agreement (fig. 6). Obviously, the stationary power law does not hold for transient stresses.

For the sake of simplicity and convenience, the majority of finite element calculations is still performed on the basis of stationary power law creep for long term stress and strain predictions for underground openings in rock salt.

Hence, it cannot be surprising when numerical precalculations based on stationary creep tests in laboratory do not match the field measurements later on sufficiently.

For instance, Preece (1987) reports very frankly the disappointing comparison of creep closure measurements in a 1433 m deep borehole in rock salt to numerical calculations through the finite element method:

Though the creep model parameters were derived from extensive triaxial laboratory testing, the calculated radial closure was a factor of 4 below the average measured with borehole caliper logs.

The author refers also to Morgan et al. (1986) who likewise experienced underprediction of creep displacements in salt when calculating the creep closure of drifts in bedded rock salt: They observed a factor of approximately 3 between measured and calculated data.

It is often argued that though the stationary power law does not necessarily hold for the excavation phase and during the relatively short transition time of stress rearrangement, it is nevertheless well apted for prediction of the very long-term mechanical behaviour of the cavities. Even if this might be true for uncased openings, it is still misleading for all geotechnical applications associated with mechanical support or compaction of backfill. As will be shown below, comparable conclusions can also be drawn for stress-strain-analyses based on the hyperbolic sine creep law.

HYPERBOLIC SINE CREEP

Uniaxial Creep and Relaxation

Heard (1972) used the steady-state flow equation originally proposed by Nadai (1938) in order to extrapolate his experimental data into the field of geologically likely strain rates:

$$\dot{\varepsilon} = \dot{\varepsilon}_0 \exp\left(-\frac{Q}{RT}\right)\sinh\left(B\frac{\sigma}{\sigma_0}\right) \quad .$$

At intermediate to high stress levels or at higher temperatures, the experimental data fitted closely the theoretical curves of stationary hyperbolic sine creep.

Writing the creep equation analogously to eq. 1

$$\dot{\varepsilon} = \frac{\dot{\sigma}}{E} + A\sinh\left(B\frac{\sigma}{\sigma_0}\right) \quad , \tag{7}$$

the relaxation function is obtained by integrating the homogeneous differential eq. 7 with regard to the initial condition $\sigma(t = t_0) = \sigma_a$

$$\frac{\sigma}{\sigma_a} = -\frac{\sigma_0}{B\sigma_a}\ln\left[\frac{1 - C\exp(-\hat{t})}{1 + C\exp(-\hat{t})}\right] \tag{8}$$

where

$$C = \tanh\left(\frac{B\sigma_a}{2\sigma_0}\right)$$

$$\hat{t} = \frac{AE}{\sigma_0}(t - t_0) \quad .$$

If $(B\sigma/\sigma_0) > 5.0$, eq. 7 can be approached by

$$\dot{\varepsilon} = \frac{\dot{\sigma}}{E} + \frac{A}{2}\exp\left(B\frac{\sigma}{\sigma_0}\right) \tag{9}$$

which is equivalent to the exponential creep law proposed originally by Ludwik (1909) for metals.

Integration of the homogeneous differential eq. 9 yields the relaxation function as

$$\frac{\sigma}{\sigma_a} = 1 - D\ln(1 + \hat{t}) \tag{10}$$

where

$$D = \frac{\sigma_0}{B\sigma_a}$$

$$\hat{t} = \frac{EAB}{2\sigma_0}\exp\left(B\frac{\sigma_a}{\sigma_0}\right)(t - t_0) \quad .$$

Eq. 10 is valid only for intermediate time ranges. For long time $\hat{t}$, the second expression on the right hand side of eq. 10 would exceed the value of 1.0, i. e. the approximation leading to the creep eq. 9 is no longer valid.

Stress Relaxation Monitoring In Situ

Stress measurements in creepable rock mass are at best carried out through the hard inclusion method (Natau et al., 1986). High pressure injection of synthetic resin achieves the prestressing of the hard inclusion pressure cells. The overstress releases with the lapse of time

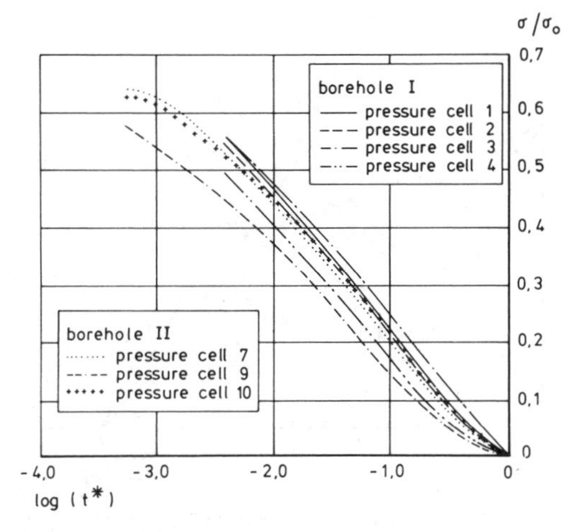

Fig. 7: Stress relaxation curves derived from prestressed hard inclusion measurements in rock salt in situ (Natau et al., 1986; $t^* = 1/300\ t\ [d]$)

approaching monotonously a stationary stress level.

The radial and tangential stresses were registered in different depths of the boreholes over a long period of time. The observed relaxation function could closely be approached by a hyperbolic sine function.

The various measurements of stress relaxation in two different boreholes resulted in a nearly uniform type of curves, as shown in fig. 7. The pressure cells were located in the differently orientated boreholes with different prestressings. However, the slopes of all relaxation curves in fig. 7 appear to be rather similar. The slight shift of the curves reflects only small differences in relaxation times between the two boreholes.

Experimental Results

Graphical representation of eq. 9 in a semi-logarithmic diagram leads to a straight line of the (σ/σ_0) vs. $\log(\dot{\varepsilon})$-relationship, from which the material parameters A and B can be derived. By statistical analysis of about 30 creep tests, the following values have been obtained: $A = 1.4 * 10^{-12}/s$, $B = 0.46$, for $T = 308°K$. The elastic modulus was taken as $E = 20\,000$ MPa, and the reference stress as $\sigma_0 = 1$ MPa.

In fig. 8, the calculated stress relaxation curves according to eq. 8 and to the above material parameters are much steeper than those based on the power law shown in fig. 6. The approximately linear descent of the relaxation curves in the semi-logarithmic diagram appears similar to that of the stress relaxation recordings in situ through the hard inclusion technique (fig. 7).

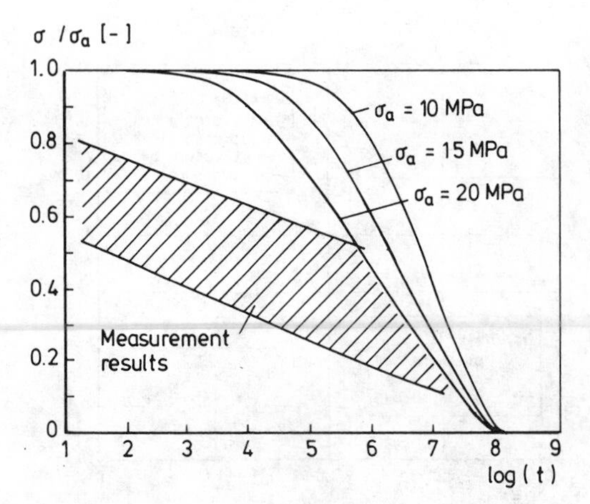

Fig. 8: Results of stress relaxation tests compared to theoretical predictions according to the stationary hyperbolic sine creep law as a function of time t [sec].

On the other hand, especially in the initial phase of the stress relaxation measurement on rock salt samples in laboratory, the calculated curves do not sufficiently agree to the experimental data indicated by the shaded region in fig. 8. A better fit could be obtained only by the application of a transient creep law, as will be shown in the following.

COMBINED STRAIN HARDENING POWER CREEP AND RELAXATION LAW

Constitutive Law of Rock Salt

Lomenick and Bradshaw (1969) conducted laboratory tests on rock salt samples at low and elevated temperatures. They found the creep rates to continue to decrease even after more than three years of testing. Also Heard (1972) reported pronounced long term strain hardening when testing polycrystalline halite at moderate temperatures. Comprehensive creep and relaxation tests of Haupt (1987) have lead to a combined strain hardening power creep and relaxation law written as

$$\frac{\dot{\varepsilon}_{ij}}{k(\sigma_{eff}, \varepsilon_{eff})} = \frac{s_{ij}}{\sigma_0} - \frac{\dot{s}_{ij}}{r(\sigma_{eff}, \varepsilon_{eff})} \tag{11a}$$

where k( ) is the creep function

$$k(\,) = \frac{3}{2}\,\alpha\,\varepsilon_{eff}^{-\beta}\left(\frac{\sigma_{eff}}{\sigma_0}\right)^{\delta-1} \tag{11b}$$

and r( ) is the relaxation function

$$r(\,) = -A\,\varepsilon_{eff}^{-B}\left(\frac{\sigma_{eff}}{\sigma_0}\right)^{D-1} \quad . \tag{11c}$$

The unified creep and relaxation law as given by eqs. 11 is linear with respect to the rates of stress and strain but highly nonlinear with regard to the cumulated stresses and strains.

The primary advantage against the traditional approaches is seen in the separated differential equations for both creep and relaxation behaviour.

If $\beta = B$, $\delta = D$ and $A/\alpha = E$, the proposed constitutive law reduces to the customary strain hardening approach with an additive elastic strain rate (e. g. Menzel/Schreiner, 1977). Moreover, if $\beta = B = 0$, the purely stationary power law corresponding to eq. 1b is obtained as one special case.

Inserting eqs. 11b and 11c into eq. 11a leads to

$$\frac{2}{3\alpha}\,\varepsilon_{eff}^{\beta}\left(\frac{\sigma_{eff}}{\sigma_0}\right)^{1-\delta}\dot{\varepsilon}_{ij} = \frac{s_{ij}}{\sigma_0} + \frac{1}{A}\,\varepsilon_{eff}^{B}\left(\frac{\sigma_{eff}}{\sigma_0}\right)^{1-D}\dot{s}_{ij} .$$

That combined creep and relaxation law requires six independent material parameters. A, B and D can be derived from relaxation tests whereas α, β and δ are obtained from creep tests. The reference stress σ_0 was adopted as σ_0 = 1 MPa.

In uniaxial creep tests, eqs. 11 reduce to

$$\dot{\varepsilon}_1 = \alpha\,\varepsilon_1^{-\beta}\left(\frac{\sigma_1}{\sigma_0}\right)^{\delta} . \tag{12}$$

Integration of eq. 12 with respect to time and initial conditions $\varepsilon_1 (t = t_0) = \varepsilon_a$ yields

$$\varepsilon_1 = \varepsilon_a (1 + \hat{t})^{\frac{1}{\beta+1}} \tag{13}$$

where

$$\hat{t} := \left[\frac{\alpha(\beta+1)}{\varepsilon_a^{\beta+1}}\left(\frac{\sigma_1}{\sigma_0}\right)^{\delta}\right](t - t_0)$$

If $\hat{t} \gg 1$, a time hardening creep law is obtained which agrees excellently to the experimental results of e. g. Lomenick/Bradshaw (1969) or Mellegard et al. (1983).

The uniaxial pure relaxation behaviour of the rock salt can correspondingly be derived from eqs. 11: Let $\dot{\varepsilon}_1 = 0$, then

$$\dot{\sigma}_1 = -A\,\varepsilon_1^{-B}\left(\frac{\sigma_1}{\sigma_0}\right)^{D} \tag{14}$$

and

$$\frac{\sigma_1}{\sigma_a} = (1 + \hat{t})^{\frac{1}{1-D}} \tag{15a}$$

where $D < 1$ and

$$\hat{t} := \left[(D-1)\frac{A}{\sigma_a}\,\varepsilon_1^{-B}\left(\frac{\sigma_a}{\sigma_0}\right)^{D}\right](t - t_0)$$

In case of $D = 1$, integration of eq. 14 would lead to

$$\frac{\sigma_1}{\sigma_a} = \exp(-\hat{t}) \tag{15b}$$

where $\hat{t} := \frac{A\,\varepsilon_1^{-B}}{\sigma_0}\,(t - t_0)$.

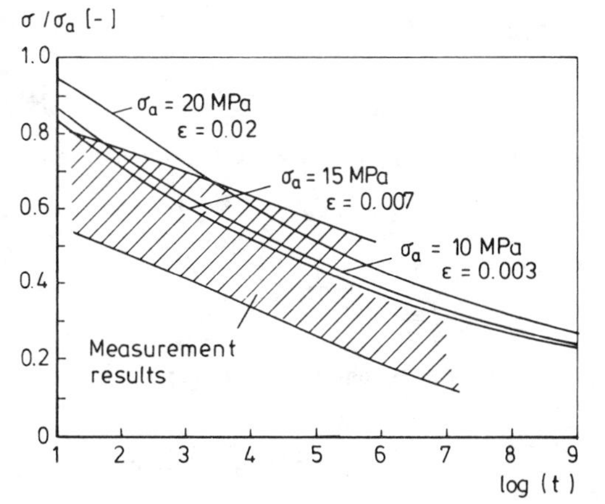

Fig. 9: Results of stress relaxation tests compared to theoretical predictions based on the strain hardening power law

Experimental Results

The determination of the relaxation parameters A, B and D according to eqs. 11 is based on the analysis of pure relaxation tests (eq. 14), where D appears as the slope of the $(-\dot{\sigma})$ vs. (σ/σ_0)-relationship in a bi-logarithmic diagram, and A and B can be derived from the intersection of that line with the coordinate axes. Corresponding investigation of creep curves based on eq. 12 provides the second set of parameters α, β and δ. The following rounded-off values have been obtained from least square fits to the laboratory creep and relaxation measurements: $A = 3*10^{-31}$ MPA/s, $\alpha = 2*10^{-36}$/s, $B = \beta = 6.0$, $D = \delta = 15.0$, $\sigma_0 = 1$ MPa.

As can be seen from fig. 9, the combined strain hardening creep and relaxation power law matches the experimental data much more closely than the pure power or hyperbolic sine creep law considered before. Further, the creep strain responses to stationary stresses agree well to those measured by Menzel/Schreiner (1977).

In the following, an approach to in-situ borehole convergence measurements will be given by use of the transient power creep law.

Borehole Creep Closure

In cylindrical plane strain analysis, the strain rate is given by

$$(\dot{\varepsilon}_r, \dot{\varepsilon}_\varphi, \dot{\varepsilon}_z) = \left(\frac{\partial \dot{u}}{\partial r}, \frac{\dot{u}}{r}, 0\right) \tag{16}$$

For incompressible flow

$$\dot{\varepsilon}_r + \dot{\varepsilon}_\vartheta = 0 , \quad \varepsilon_r + \varepsilon_\vartheta = 0 , \tag{17a}$$

and, hence

$$\frac{\partial \dot{u}}{\partial r} + \frac{\dot{u}}{r} = 0 , \quad \frac{\partial u}{\partial r} + \frac{u}{r} = 0 . \tag{17b}$$

Integration of eq. 17b with respect to r yields

$$\dot{u} = \frac{a}{r} + f , \qquad u = \frac{b}{r} + g$$

where a, b, f and g are integration constants to be determined by boundary conditions. Since f and g are relevant only to translatoric motions, they are zero in the problem considered here.

From eqs. 16 and 17

$$\dot{\varepsilon}_r = -a r^{-2} = -\dot{\varepsilon}_\varphi , \tag{18a}$$

$$\varepsilon_r = -b r^{-2} = -\varepsilon_\varphi . \tag{18b}$$

The deviatoric stress components are given by

$$s_r = \frac{1}{2}(\sigma_r - \sigma_\varphi) = -s_\varphi , \tag{19}$$

$$s_z = 0 .$$

Referring to Odqvist/Hult (1962), the equivalent stress σ_{eff}, equivalent strain ε_{eff} and equivalent strain rate $\dot{e}_{eff}$ are respectively defined as

$$\sigma_{eff} = \sqrt{\frac{3}{4}}\, |\sigma_r - \sigma_\varphi| ,$$

$$e_{eff} = \sqrt{\frac{4}{3}}\, |e_r| ,$$

$$\dot{e}_{eff} = \sqrt{\frac{4}{3}}\, |\dot{e}_r| ,$$

Assuming stationary stresses in the rock, the generalized strain hardening power creep law reads

$$\dot{e}_r = \frac{3}{2} k \sigma_{eff}^{(n-1)} e_{eff}^{-\mu} s_r = -\dot{e}_\varphi \tag{20a}$$

Squaring eq. 20a yields

$$\dot{e}_{eff} = k \sigma_{eff}^{n} e_{eff}^{-\mu} \tag{20b}$$

which is analogous to the uniaxial strain hardening power creep law as given by eq. 12 $(n = \delta, \mu = \beta)$.

Combining eqs. 18, 19 and 20, one obtains

$$-\frac{a}{r^2} = \left(\frac{3}{4}\right)^{\frac{1+n+\mu}{2}} k |\sigma_r - \sigma_\varphi|^n |b r^{-2}|^{-\mu} \operatorname{sign}(\sigma_r - \sigma_\varphi) . \tag{21a}$$

Let

$$c^n = -a |b|^\mu \left(\frac{3}{4}\right)^{\frac{1+n+\mu}{2}} k^{-1} \operatorname{sign}(\sigma_r - \sigma_\varphi)$$

so that eq. 21a is rewritten as

$$|c| r^{-\frac{2}{n}(1+\mu)} = |\sigma_r - \sigma_\varphi| . \tag{21b}$$

Since $(\sigma_r - \sigma_\varphi) > 0$ in the case of an external radial compressive stress intensity σ_0 larger than the internal radial compressive stress intensity σ_i, eq. 21b is equivalent to

$$c r^{-\frac{2}{n}(1+\mu)} = \sigma_r - \sigma_\varphi \tag{21c}$$

Integration of the equilibrium equation

$$r \sigma_r' = \sigma_r - \sigma_\varphi = -c r^{-\frac{2}{n}(1+\mu)} \tag{22a}$$

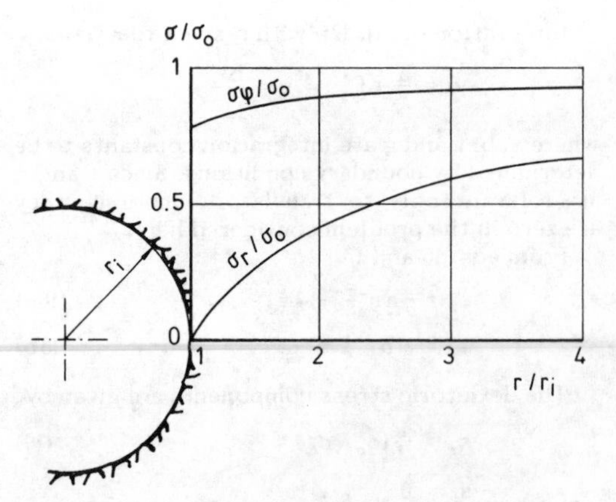

Fig. 10: Stationary tangential and radial stresses around the hole in rock salt ($n = 5.0$, $\mu = 1.0$)

yields the radial stress σ_r as

$$\sigma_r = A\, r^{-\frac{2}{n}(1+\mu)} + B \tag{22b}$$

where

$$A = \frac{c\,n}{2(1+\mu)} \;,$$

and B is an arbitrary integration constant.

The tangential stress σ_φ is obtained by back-substitution of eq. 22b into eq. 22a

$$\sigma_\varphi = A\left(1 - \frac{2}{n}(1+\mu)\right) r^{-\frac{2}{n}(1+\mu)} + B \;.$$

The coefficients A and B are determined by the boundary conditions

$$\sigma_r = \sigma_\varphi = \sigma_0 \,, \quad \text{for } r \to \infty \,,$$

$$\sigma_r = \sigma_i \,, \quad \text{for } r = r_i \;.$$

Hence

$$A = (\sigma_i - \sigma_0)\, r_i^{-\frac{2}{n}(1+\mu)} \,, \quad B = \sigma_0$$

and finally

$$\sigma_r = (\sigma_i - \sigma_0)\, \rho^{-\frac{2}{n}(1+\mu)} + \sigma_0 \;, \tag{23a}$$

$$\sigma_\varphi = (\sigma_i - \sigma_0)\left(1 - \frac{2}{n}(1+\mu)\right)\rho^{-\frac{2}{n}(1+\mu)} + \sigma_0 \tag{23b}$$

where

$$\rho = \frac{r}{r_i} \;. \tag{23c}$$

Fig. 10 shows the radial and tangential stress distribution, respectively, as a function of radial distance for selected values $n = 5$ and $\mu = 1$.

The radial strain rate $\dot{e}_r$ is evaluated according to eqs. 20

$$\dot{e} \equiv \dot{e}_r = \left(\frac{3}{4}\right)^{\frac{1+n+\mu}{2}} k\,|\sigma_r - \sigma_\varphi|^n\, e_r^{-\mu} \tag{24a}$$

where $e_r > 0$ and $\sigma_r - \sigma_\varphi > 0$.

The stress difference $(\sigma_r - \sigma_\varphi)$ is obtained from eqs. 23 as

$$\sigma_r - \sigma_\varphi = \frac{2}{n}(1+\mu)\,\rho^{-\frac{2}{n}(1+\mu)} \;. \tag{24b}$$

Consequently, the radial strain rate is given as

$$\dot{e} = C\,k\,|\sigma_i - \sigma_0|^n\, \rho^{-2(1+\mu)}\, e^{-\mu} \tag{24c}$$

where

$$C = \left(\frac{3}{4}\right)^{\frac{1+n+\mu}{2}} \left(\frac{2}{n}\right)^n (1+\mu)^n \;.$$

Integration of eq. 24c with respect to time t yields

$$\int_{e_0}^{e} e^{\mu}\, de = C\,k\,\rho^{-2(1+\mu)} \int_{t_0}^{t} |\sigma_i - \sigma_0|^n\, dt \tag{25a}$$

where e_0 is the radial strain at $t = t_0$, and e is synonymous to e_r.

For an open borehole $\sigma_i = 0$, assuming the primary stress σ_0 as constant, eq. 25 is integrated to

$$e = e_0\left(1 + \rho^{-2(1+\mu)}\hat{t}\right)^{\frac{1}{1+\mu}} \tag{25b}$$

where

$$\hat{t} = (1+\mu)\,C\,k\,|\sigma_0|^n\, e_0^{-(1+\mu)}\,(t - t_0)$$

Since $e = e_r = -e_\varphi$, and $e_\varphi = u/r$, the radial convergence u of the borehole wall ($r = r_i$, $\rho = 1$) is finally given by

$$u(r = r_i) = -e_0\, r_i\,(1 + \hat{t})^{\frac{1}{1+\mu}} \;. \tag{26a}$$

In large strain analysis, the borehole closure is obtained as

$$u = r_0\left[\exp\left(-e_0(1+\hat{t})^{\frac{1}{1+\mu}}\right) - 1\right] \tag{26b}$$

where r_0 is the initial radius of the borehole at time $t = t_0$.

Adaption to Borehole Creep Measurements

Fernandez/Hendron (1984) investigated long-term in situ measurements of volume change which were carried out in a borehole test drilled 1850 m deep into a bedded salt formation. The monitoring of borehole closure indicated rather conclusively that steady-state creep did not develop around the well. The authors carried out a finite element analysis based on a time hardening transient creep law. The time hardening exponent was chosen in the range of 0.38 to 0.46 where the larger values were associated with higher deviatoric stress levels.

In the following, the strain hardening approach, which does not include the time coordinate t explicitely in the constitutive law, is used to fit the convergence curve of a borehole at

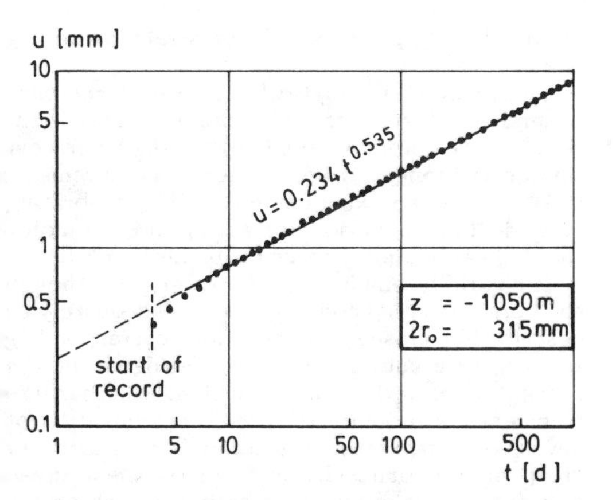

Fig. 11: Borehole closure measurements and analytical approach by strain hardening power law creep

a depth of about 1050 m below surface, measured by Doeven et al. (1983) through caliper logs over a period of about 850 days. Prij/Mengelers (1981) fitted the measured creep data on the first 200 days by a stationary power law, whereby the external boundaries of the applied model were subjected to transient loading.

By drawing the radial displacements as a function of time in a bi-logarithmic diagram, the convergence characteristics is approximately a straight line (fig. 11). From the slope of this line, the time hardening parameter according to eq. 26a can be determined, which in the present case is obtained as $\mu = (0.234^{-1} - 1) \approx 3.275$.

Relaxation time τ and initial strain e_0 are evaluated from the intercept of the convergence curve with the coordinate axes:

Choosing $\tau = 1$ d, the initial strain e_0 is obtained from eq. 3f and from the radial displacement $u(t = 1) = 0.234$ mm $= \varepsilon_0 r_0$. Since $2r_0 = 315$ mm, the initial strain e_0 is about 0.15 %. By assuming a primary stress of $\sigma_0 = 24$ MPa, the shear modulus is calculated as $G = 8\,000$ MPa, which is much more realistic than the corresponding value obtained earlier by back-analysis on the basis of stationary power law creep.

More rigorous evaluation of material parameters from the transient creep record, however, must take into consideration also the additive unit number in eq. 26a, and furthermore, the primarily transient nature of the rock stress according to the right hand side of eq. 11a.

Error due to Finite Lateral Boundary of Numerical Model

Finite element or finite difference models are generally restricted to finite external boundaries.

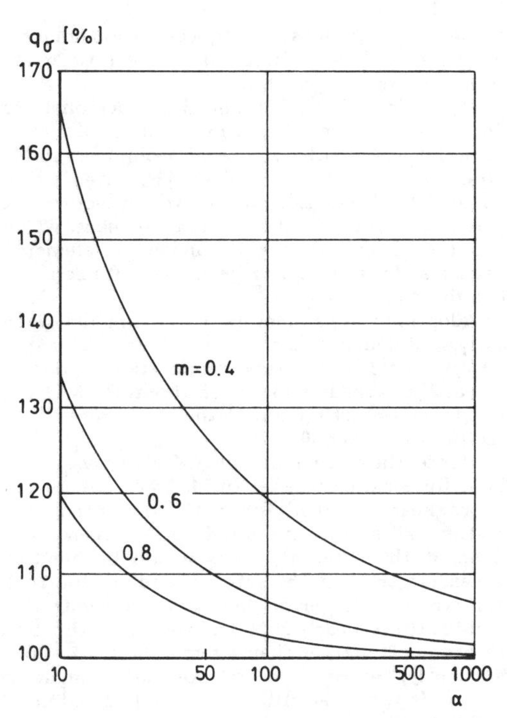

Fig. 12: Stress concentration enlargement factor vs. radial aspect ratio of thick-walled cylinder

The error due to finiteness of the numerical model can be assessed by analyzing the stresses and strain rates at the wall of a borehole in a thick-walled cylinder with internal radius r_i and external radius r_a.

Adaption of the radial and tangential stresses in eq. 23a and eq. 23b, respectively, to the boundary condition $\sigma_r = \sigma_0$ at $r = r_a$ yields

$$\sigma := \sigma_r - \sigma_\varphi = \frac{m(\sigma_i - \sigma_a)r^{-m}}{r_i^{-m} - r_a^{-m}} \qquad (27a)$$

where

$$m := \frac{2}{n}(1 + \mu) \; . \qquad (27b)$$

Comparison of the stress difference according to eq. 27a to the corresponding stress difference of the infinitely thick cylinder as given by eq. 24b, yields an enlargement factor of

$$q_\sigma = (1 - \alpha^{-m})^{-1} \qquad (28a)$$

where

$$\alpha := \frac{r_a}{r_i} \qquad (28b)$$

for the stress difference σ at the wall of the borehole in the cylinder of finite thickness.

Fig. 12 shows the stress concentration enlargement factor q_σ as a function of the ratio α for some selected values of m.

Nipp/McNulty (1987) aimed to demonstrate how boundary conditions can affect the accuracy of solutions involving creep around a single borehole in an infinite medium. The authors give an analytical solution to derive an equivalent elastic stiffness for an edge finite element, which ought to help reduce the number of elements needed and yet still produce an accurate solution in a finite region model.

Adopting the parameters of their model given as $r_i = 0.5$ mm, $r_a = 15$ mm and $n = 5.0$, $\mu = 0.0$, the overestimation of stress concentration according to eq. 28a exceeds a factor of about 135 % giving rise to an overestimation of the strain rate ε by a factor of almost 450 %.

Hence, the method of Nipp/McNulty does only hold for elastic material ($n = 1$, $\mu = 0$), where the approximation error for both stress concentration and strain rate is indeed less than 0.1 %. Even, if the external radius would be hundred times larger than the internal radius, for rock salt ($n = 5$) the error in stress concentration would still be nearly 20 %, and the overestimation of strain rate larger than a factor of 235 %. Consequently, the finite lateral boundaries must be selected very carefully to avoid substantial overestimations of stress concentration and borehole creep velocities.

CONCLUSIONS

For long term prediction of creep convergence of underground excavations in salt rock, stationary power law creep approaches have widely been applied in the past and present.

On the other hand, stationary creep of rock salt at moderate temperatures has nowhere been observed in situ nor in laboratory tests. Moreover, in numerical calculations according to the finite element method based on initial strain procedures, even the adoption of a purely stationary creep law results into stresses and strains which are generally of transient nature.

Comprehensive creep and relaxation tests on natural rock salt have lead to the proposal of a combined strain hardening power creep and relaxation law. This very general constitutive law gives rise to a substantial sensitivity of numerical models of underground openings in rock salt with respect to the inherent finiteness of their external boundaries. These boundary effects are even more pronounced for pure power law or hyperbolic sine creep analyses.

The measured creep convergence of a borehole in a German saltdome could equally well be described by power law, hyperbolic sine, or strain hardening power law creep approaches. The best agreement to the results of laboratory creep and relaxation investigations, however, has been obtained by application of the strain hardening power law.

The presented analytical solution of the time dependent closure of a borehole in rock salt, which have been verified by both laboratory testing and long-term convergence measurements in situ, offer an excellent means for calibrating numerical models prior to treating more complex rheological boundary value problems.

It should be emphasized, however, that though the borehole creep convergence can adequately be described by closed form solutions corresponding to the different creep laws considered in the present paper, the final decision, whether the transient convergence results from relaxation of deviatoric stresses in connection with stationary creep on the one side or from transient creep induced by stationary stresses in the rock on the other side, cannot be drawn, unless long-term stress recordings have been evaluated as a function of time and radial distance. Only the monitoring and comparison of both, the timely and spatial distribution of deviatoric stresses in the surrounding of the borehole as well as the transient deformation of the hole, can provide the necessary information for linking the results of in situ measurements to those obtained in laboratory creep and relaxation tests. Through this, the most realistic approach to the long-term stress and strain behaviour of underground openings in rock salt during and after excavation as well as during and after compaction of the filling, may be obtained by computer models which are calibrated with respect to both the transient constitutive behaviour of the rock salt and the effect of finite lateral boundaries.

ACKNOWLEDGEMENT

The authors gratefully acknowledge the support of Prof. Dr.-Ing. O. Natau, Head of the Department of Rock Mechanics at the Institute of Soil Mechanis and Rock Mechanics, University of Karlsruhe, to the research work presented in this paper.

REFERENCES

Balthasar, K., Haupt, M., Lempp, Ch., Natau, O.: Stress relaxation behaviour of rock salt: Comparison of in situ measurements and laboratory test results. Proc. 6th Int. Congr ISRM Montreal, 11-13, Balkema, Rotterdam

Borm, G. (1985): Wechselwirkung von Gebirgskriechen und Gebirgsdruckzunahme am Schachtausbau. Felsbau 3, No. 5, 153-158

Borm, G. (1987): Bohrlochkonvergenz und Spannungsrelaxation im Steinsalzgebirge. Proc. 6th Int. Congr. ISRM Montreal, 819-823, Balkema Rotterdam

Doeven, I., Soullie, P. P., Vons, L. H. (1983): Convergence measurements in the dry-drilled 3oo m borehole in the Asse II - saltmine. Europ. Appl. Res. Rept. - Nucl. Sci. Technol., Vol. 5, No. 2, 267-324

Fernandez, G., Hendron, A. J. (1984): Interpretation of a long-term in situ borehole test in a deep salt formation. Bull. Ass. Eng. Geol., Vol. 21, No. 1, 23-38

Haupt, M. (1987): Untersuchungen zum Relaxationsverhalten von Steinsalz. Unpublished Report, Inst. of Soil Mech. and Rock Mech., Univ. of Karlsruhe

Heard, H. C. (1972): Steady state flow in polycrystalline halite at pressures of 2 kilobars. Flow and Fracture of Rocks, 191-209, A. G. U., Washington D. C.

Jaeger, J. C. (1969): Elasticity, fracture and flow. With engineering applications. Methuen & Co., Science Paperbacks, London.

Lomenick, T. F., Bradshaw, R. L. (1969): Deformation of rock salt in openings mined for the disposal of radioactive wastes. Rock Mechanics 1, 5-30

Ludwik, P. (1909): Elemente der technologischen Mechanik. Springer Verlag, Berlin.

Mellegard, K. D., Senseny, P. E., Hansen, F. D. (1983): Quasi-static strength and creep characteristics of 100-mm-diameter specimens of salt. RE/SPEC Report ONWI-250, Batelle Memorial Inst., Columbus, Ohio

Menzel, W., Schreiner, W. (1977): Zum geomechanischen Verhalten von Steinsalz verschiedener Lagerstätten der DDR, Teil II: Das Verformungsverhalten. In: Neue Bergbautechnik 7, No. 8, 565-574

Morgan, H. S., Stone, C. M., Krieg, R. D. (1986): An evaluation of WIPP structural modeling capabilities based on comparisons with south drift data, SAND85-0323, Sandia National Laboratories, Albuquerque, NM

Nadai, A. (1938): The influence of time upon creep, the hyperbolic sine creep law. Timoshenko Anniversary Volume, 155-170, Macmillan, New York

Natau, O., Lempp, Ch., Borm, G.(1986): Stress relaxation monitoring by prestressed hard inclusions. Proc. Int. Symp. ISRM on Rock Stress and Stress Measurements, Stockholm, 509-514, Centek Publ., Lulea

Nipp, H.-K., McNulty, E. G. (1987): Effects of mesh size and boundary conditions on creep analysis of rock salt. In: Farmer, I. W. et al. (eds.): Proc. 28th US Symp. on Rock Mech., Tucson, 653-661

Odqvist, F. K. G., Hult, J. (1962): Kriechfestigkeit metallischer Werkstoffe. Springer Verlag, Berlin

Pfeifle, T. W., Mellegard, K. D., Senseny, P. E. (1981): Constitutive properties of salt from four sites. RE/SPEC Report ONWI-314, Batelle Memorial Inst., Columbus, Ohio

Preece, D. S.: Borehole creep closure measurements and numerical calculations at the Big Hill, Texas SPR storage site. Proc. 6th Int. Congr. Rock Mech. ISRM, Montreal 219-224, Balkema, Rotterdam

Prij, J., Mengelers, J. H. J. (1981): On the derivation of a creep law from isothermal bore hole convergence. Netherl. Energy Res. Found., Rep. ECN-89

Numerical Methods in Geomechanics (Innsbruck 1988), Swoboda (ed.)
© 1988 Balkema, Rotterdam. ISBN 90 6191 809 X

3-D FEM analysis of a room and pillar mine housing industrial waste

G.P.Giani
Dipartimento di Georisorse e territorio, Politecnico, Torino, Italy

ABSTRACT: Two FEM three-dimensional models in elasto-plastic field for the study of the static behavior of room and pillar mining exploitations is presented. This paper refers to an almost exausted exploitation, the openings of which will be filled with industrial sludge, and to a design for a second quarry. The reported numerical results o f the FEM analyses are the stresses in the pillars which are compared with experimental and numerical results of a previous study.

1 INTRODUCTION

Gypsum orebodies of the Messianist formation in the Asti Province in North West Italy are underground exploited at low depth, using room and pillar mining methods.

The exploitation have a depth to about 80-90 m. The sections of the pillars vary between 5 and 6 m in diameter and from 5 to 6 m in height. The geometrical disposition of the pillars is fairly regular and the dip of the orebody is almost vertical.

From the engineering point of view the problem of the two quarries are different. The first exploitation is almost exhausted and the examined problem is a feasibility study of the static behavior of the stopes filled with industrial sludge. The second exploitation problem is the design od two new room and pillar levels using a more ordered geometry and giving a larger section size to the pillars.

The numerical model has been carried out using a 3-D Finite Element Method (FEM) model and describing the stress - strain behavior of the gypsum mass by means of an elastic - ideally plastic law.

Two FEM simulations have been carried out, examing the excavation and filling phases.

The numerical results here reported are the principal stresses in the centered pillars at two different exploitation levels. The numerical results are compared with the numerical results of a previous 2-D FEM analysis and the experimental measured stresses cited in a previous study (Barla and Innaurato 1974).

General description of the Geology and mining methods at the Montiglio (Asti) orebody have been published previously (Occella 1958).

Moreover, previous study of possible creep problems of the gypsum mass, dissolution problems due to leakage produced by the sludge and water circulation in the upper part of the orebody connected with limited fessuration or karst phenomena are here not examined (Bortolami 1984; Bortolami and Di Molfetta 1987).

The results of the overall stability of the gypsum quarries are not reported here.

2 GEOMECHANICAL FEATURES AND MODEL INPUT DATA

The exploited gypsum mass is subdivided from the point of view of industrial production, in two classes: one has crystal of few millimeters; the other has crystal up to a few centimeters. The fine grain gypsum is present in a larger percentage in the almost exhausted quarry (Codana quarry), while in the second quarry (Lavone quarry) the two classes are found in about the same percentage.

Previous study on the Codana quarry, characterizing the Gypsum mass, from a geomechanical point of view, concentraded on the fine grain material only, which was found in the quarry's upper part in quantitaties 3 times greater than the coarse grain material.

This study considered both fine and coarse material; results of laboratory tests are reported in table 1.

A set of drills having a lenght varying from 20 to 25 m allowed one to determine new stratigraphies and RQD values.

As to the differenting qualities of the Gypsum mass, an average RQD value has been assigned to those parts where the fine grain material is prevalent and a second average RQD value to those parts where the coarse grain material is prevalent.

A geostructural survey has involved the main discontinuities of the exploited ore-body which are clay-shale intercalations with a waved surface and a contact generally smooth between the surface.

The fillings are almost always a few centimeter thick.

The mechanical features given to the mass and the clay-shale intercalations are given in table 1.

The gypsum mass is considered as a continuum and the mass deformability modulus is determined utilizing a reduction factor of the elastic intact material modulus relating from the RQD value.

The intercalations having a larger thickness are dealt with for their uniqueness, their mechanical properties are assigned to them.

The shear strength parameters used are the more conservative parameters obtained for natural unfilled discontinuity of the gypsum mass.

The behavior model assumed for the mass is elastic - ideally plastic (Naylor et al 1981). The limit between the elastic and plastic behavior is given by the Drucker - Prager plasticity condition. The yield function utilized is

$$F = \sigma_q - 2\sigma_m \sin\Phi - 2c \cos\Phi$$

where c and Φ are the cohesion and the friction angle of the material, σ_m is the first stress invariant and σ_q is the second invariant of the deviatoric stress.

Table 1

	Γ (t/m^3)	C_o (MPa)	E_s (MPa)	E_t (MPa)	T_o (MPa)	RQD %
clay - size gypsum	2.09	13.0	11500	13200	0.14	40
grain size gypsum	2.19	6.3	4500	2570	0.06	50

	E_m (MPa)	m (-)	C_m (MPa)	Φ_m (-)	T_o
coarse mass	1500	0.33	0.15	35	0
clay-shale	750	0.33	0.07	30	0
industrial sludge	100	0.33	-	-	-

3 NUMERICAL MODEL

a) Codana quarry

The Codana quarry has 7 room and pillar levels. The rock slab separating the openings have an average height of 5 m. The pillars have a 5 m side square section for the first 6 levels and are 5 m high.
The pillars of level 7 have rectangular section (5 x 15 m).

The regularity of the pillars - openings system is often interrupted by rock diaphram and room and pillar areas never include more than 15 pillars.

The FEM model discretized a structure formed by 7 superimposed levels (figure 1). The model is 120 m high and for reasons of symmetry represented only 1/4 of the real structure; this part has a rectangular section of 40 x 30 m.

The whole structure represented an exploiting area of 15 pillars and surrounded by rock mass.

The FEM model is composed of 1420 isoparametric or subparametric hexaedric solid elements (Bathe 1982) and has 1970 nodal points.

The filling of the openings with industrial sludge is represented in the model by new elements which have weight and mechanical properties of the industrial sludge.

The static analyses carried out involved the gravitational analysis for the initial state of stress computation and the numerical simulation of the excavation and the filling phases. 7 excavation and 3 filling phases were carried out for the Codana quarry.

Three computation steps with 40 iterations each were carried out for excavation or filling phase. The stiffness matrix was reformed the first step of each new phase.

b) Lavone quarry

The Lavone quarry has two already exploited levels and two in the design phase. The model reproduces (figure 2) the geometry of the existing and of the future levels. Different from the Codana quarry, a larger horizontal extension of the exploitation is foreseen for the Lavarone quarry. The model, which represents only 1/4 of the real structure discretizes the level 3 and level 4 as a indefinite extension of rooms and pillars.

The pillars have a section reducible to a 7/m side square section and are about 5.5 m high. The rock slabs separating the levels have a thickness of 5.5 m

The model is 104 m high and has a rectan-

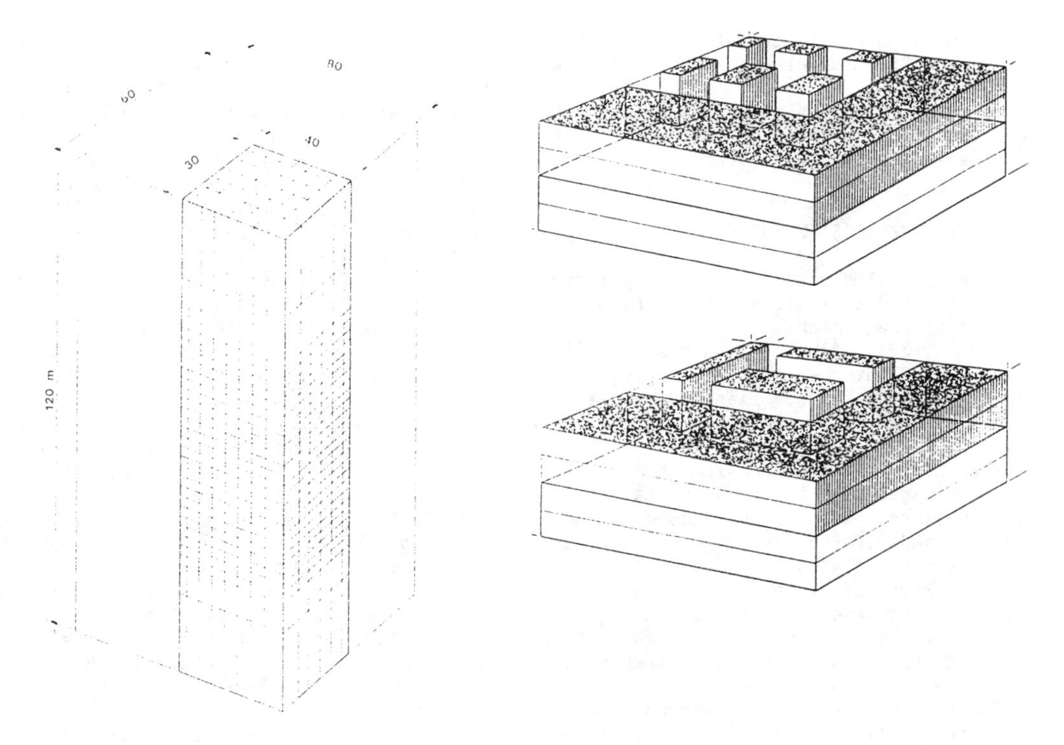

Figure 1. Codana quarry - FEM model and view on detail of exploitation levels (1-6,7)

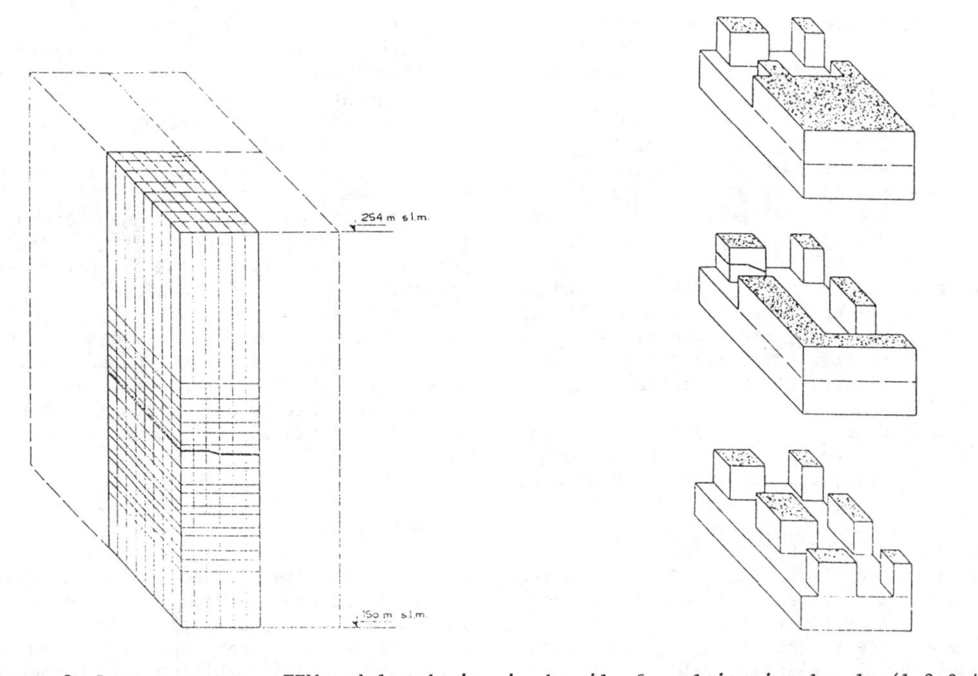

Figure 2. Lavone quarry - FEM model and view in detail of exploitation levels (1,2,3-4).

gular horizontal section of 28 x 21 m. The model has 969 3-D elements and 1360 nodal points.

The numerical simulation was carried out exploiting 4 levels after the application of the initial state of stress.

4 STRESS ANALYSIS IN THE PILLARS

The analyzed FEM results were the stresses and the yield function value in the Gauss' points of the elements.

The yield function values showed, especially for the deepest exploitation levels of the two quarries, some plastic zones limited to the external part of the pillars.

The computed stresses in the pillars were compared with the experimental stresses measured with the CSIR door stopper method and with the computed FEM 2-D and 3-D elastic analyses.

The examined exploitation levels are level 5 and level 6 of the Codana quarry which are at the depth of 70 and 80 m respectively and level 3 of the Lavone quarry which is at a depth of 70 m.

The figures 3, 4, 5 e 6 showed, for the centered pillar, the principal stress diagram at the middle of the height of the pillar, along the median and diagonal line.

The reported diagram are as follows:

figure 3: Codana quarry, end of the excavations, level 5: 2-D and 3-D elastic and 3-D elasto-plastic analysis;

figure 4: Codana quarry, end of the excavations, level 5: 3-D elasto-plastic analysis; Lavone quarry, end of the excavations, level 3: 3-D elasto-plastic analysis;

figure 5: Codana quarry, level 5, end of the excavations: elasto-plastic analysis, experimental CSIR door stopper stresses; end of filling phases: elasto-plastic analysis;

figure 6: Codana quarry, end of the excavations, level 6: elasto-plastic analysis, experimental CSIR door stopper stresses.

The results of 2-D FEM elastic analysis are referred to 2 plane models crossing through vertical and perpendicular sections of the exploitations.

The average values of the maximum stresses in the 2-D elastic field are approximately two times greater than the natural stresses.

In the 2-D field the values of the stresses are almost null near the excavation face. As the section of the pillars was assumed to be square, stresses in the corners and in the median line are examined; the stress concentrations are larger in the corners. The increase of stress concentration on the corners is about 1 MPa for the two principal stresses.

The comparison between elastic and elasto-plastic stresses puts in evidence the limited plastic areas and the ridistribution of the plastic stresses.

The average values of the maximum stress measured in the Codana quarry are about 5 and 6 MPa at the depth of 70 and 80 m, respectively. These values are comparable with the computed FEM 3-D elasto-plastic values at the end of the excavation simulation.

Is difficult to interpret the measured values of the minimum stress at the contour of the pillar which reach values greater than 2.5 MPa for the two situation. Unlike the 2-D model and from the value in the middle side of the pillar in 3-D model, the minimum stress in the corner of the pillar reaches rather high values (1.6 MPa at 70 m and 2 MPa at 80 m).

The sides of the pillars are greater than the pillars height, thus they are rather squat and the stress gradient is higher starting from the corner.

The effect of the weight and the stiffness of the filling produces a variation of the stresses, increasing both the principal stresses. These rather limited (0.2 ÷ 0.5 MPa) augment the mean stress, maintaining rather constant the deviatoric stress and producing an improuvement of the pillar static condition, at least as it concerns the assumed yield criterium.

The geometrical scheme of new Lavone exploitation levels foresees larger section for pillars and rooms and doesn't foresee any rock diaphram left in place.

A comparison between Lavone and Codana quarry analysis puts in evidence a lower difference in the squatest Lavone pillars, between the stress in the corner and in the median line. The plastic zones are still limited to some external part of the pillars.

5 CONCLUSIONS

A three-dimensional model for the static analysis of room and pillar exploitations in gypsum rock has been presented.

The model was set up calibrating the strenght and deformability mass parameters by means of a comparison with tress measurements obtained in a previous work which

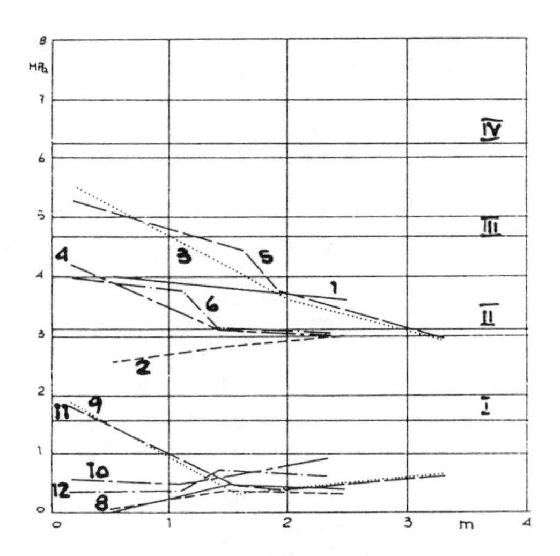

Figure 3. Stresses in the pillar, Codana quarry, end of the excavations, level 5, 2-D (1,2,7,8), 3-D elastic (3,9 diagonal line - 4,10 median line) and 3-D elasto-plastic analysis (5,11 diagonal line - 6,12 median line); σ_1 : 1-6; σ_3: 7-12.

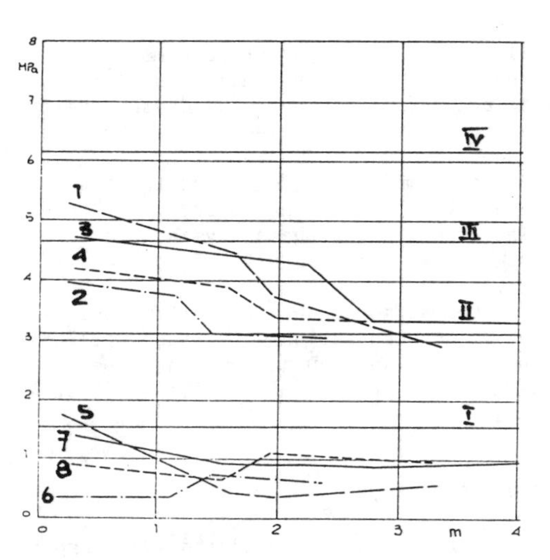

Figure 4. Stresses in the pillar, Codana quarry, end of excavations, level 5, 3-D elasto-plastic analysis (1,5 diagonal line 2,6 median line); Lavone quarry, end of the excavations, level 3, 3-D elasto-pla_ stic analysis (3,7 diagonal line - 4,8 me_ dian line); σ_1 : 1-4; σ_3: 5-8.

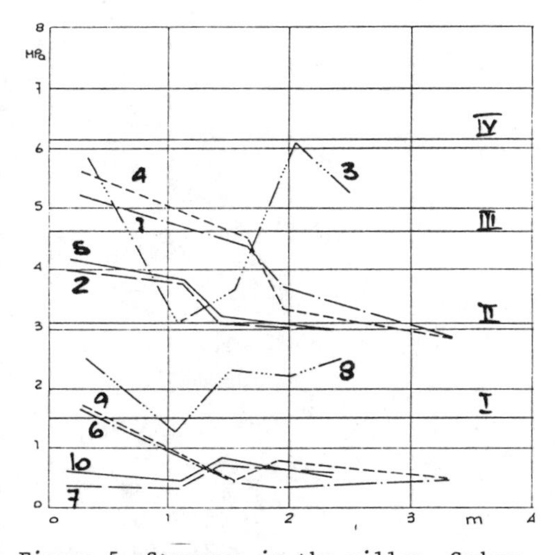

Figure 5. Stresses in the pillar, Codana quarry, level 5, 3-D elasto-plastic analysis (1-6 diagonal line - 2,7 median line) expe_ rimental measurements (3,8); end of the filling phases, 3-D elasto-plastic analysis (4,9 diagonal line - 5,10 median line); σ_1: 1-5; σ_3: 6-10.

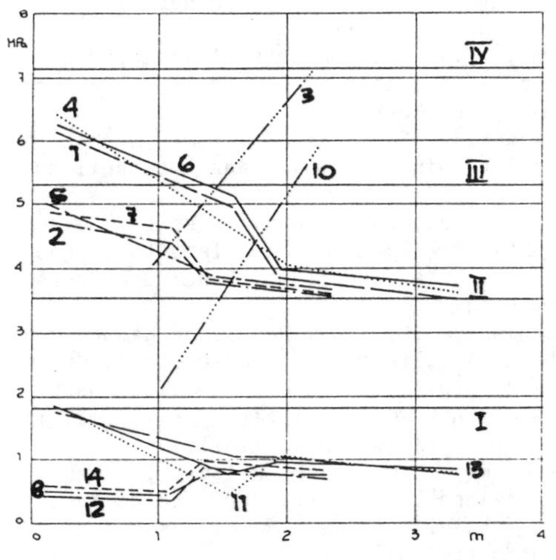

Figure 6. Stresses in the pillar, Codana quarry, end of the excavations, level 6, elasto-plastic analysis (1,8 diagonal line 2,9 median line), experimental measurements (3,10), 3-D elastic analysis (4,11 diago_ nal line - 5,12 median line); end of the filling phases (6,13 diagonal line - 7,14 median line); σ_1: 1-7; σ_3: 8-14.

utilized the CSIR door stopper method.

The model was then utilized for the static behavior study of the Codana quarry; the openings were filled with industrial sludge.

A second model was developed for the enlargement of the Lavone quarry.

The results of the analysis are also compared with the results of the elastic FEM 2-D and 3-D analyses.

The comparison showed the difference in stress concentration in the pillars between 2-D and 3-D analysis and between the corner and the median line of the pillar.

ACKNOWLEDGEMENTS

The author is indebted to Professors Bortolami, Del Greco, Di Molfetta, Innaurato and Stragiotti who allowed him to use experimental and bibliographic data.

REFERENCES

Barla, G. and Innaurato, N. 1974. Stato tensionale nei pilastri di una cava di pietra da gesso, considerazioni sulla stabilità. Convegno Internazionale sulla coltivazione di pietre e minerali litoidi. Torino vol III: 1 - 25

Bathe, K.J. 1982. Finite element procedures in engineering analysis, Prentice Hall, New York.

Beniawski, Z.T. 1984. Rock Mechanics design in mining and tunnelling, Balkema, Rotterdam.

Naylor, D.J., Pande, G.N., Simpson, B. and Tabb, R. 1981. Finite elements in geotecnical engineering. Pineridge Press, Swansea.

Bortolami, G.C. 1984. Relazione geologico-tecnica sulla cava di gesso Codana della IECME S.p.A., situata in località Montiglio d'Asti da utilizzare per lo smaltimento di fanghi industriali.

Bortolami, G.C. and Di Mplfetta, A. 1987. Relazione tecnica sul deposito di fanghi industriali nella cava di gesso Codana in Montiglio d'Asti.

Occella, E. 1958. Caratteristiche di stabilità dei cantieri sotterranei nelle coltivazioni di pietra da gesso nell'Astigiano. Industria mineraria. 9 - 10: 1 - 23

Stragiotti, L., Innaurato, N. and Del Greco, O. Relazione tecnica sulle condizioni di stabilità di scavi sotterranei adibiti al deposito di fanghi industriali nella cava di gesso Codana della IECME S.p.A. in comune di Montiglio (AT), Politecnico di Torino, Italy.

Numerical Methods in Geomechanics (Innsbruck 1988), Swoboda (ed.)
© 1988 Balkema, Rotterdam. ISBN 90 6191 809 X

Modelling of subsidence caused by longwall mining using finite element and displacement discontinuity methods

H.J.Siriwardane
Department of Civil Engineering, West Virginia University, Morgantown, USA
J.Amanat
Geotechnology, Inc., St. Louis, Mo., USA
(Formerly West Virginia University)

ABSTRACT: Numerical techniques based on finite element method and displacement discontinuity method have been developed for predicting subsidence over longwall mine panels. A new approach based on history matching (back analysis) concept, called as "displacement approach" is presented. Material bulking due to roof collapse and subsequent failure of overlying strata is also introduced. A modified version of the displacement discontinuity method is used to simulate discrete cracks caused by mining in the overburden. The developed procedure has been verified by comparing predictions with published field observations at several mine panels. Here, the results for a mine with shallow overburden depth and for a deep mine located in northern West Virginia are presented. The study indicates that the shape and the magnitude of subsidence can be predicted reasonably well by simulating major mechanisms of roof collapse and crack propogation.

1 INTRODUCTION

Due to fast growing needs of energy supplies, a concerted effort is being made to increase coal production in the United States. Underground mining is an inevitable method for extracting coal resources that are located several hundred feet below the ground surface. One major consequence of underground mining is the undesirable subsidence (settlements) in the overlying strata and at the ground surface. In fact, it has been estimated that about 2 million acres of U.S. land have been affected by subsidence due to mining; a majority (95 percent) of this is associated with bituminous coal. The subject of subsidence has received increased attention in the U.S. during the last two decades. The influence of subsidence can cause damages to houses, sewerage and drainage systems, highways, water liners, gas lines, and other engineering structures. Increasing awareness of the magnitude of the subsidence problem has resulted in state and federal law that requires the mining industry to take remedial measures to control subsidence. This in turn requires reliable predictive models that can be used by the mine operators and regulatory agencies associated with the problem.

There have been two broad categories of methods for predicting surface subsidence, namely empirical and phenomenological methods. The empirical methods are based mainly on available field data and experience, while phenomenological methods are based on mathematical and numerical modelling of the behavior of materials that are assumed to follow the principles of continuum mechanics. Most of the empirical methods used in U.S. coal fields are mere extensions of the empirical methods developed for European coal fields. In general, empirical methods do not account for variations in geologic profile, topography, material properties and methods of mining. In fact, it has been reported [7] that the popular empirical procedure developed by the National Coal Board of Britain, known as NCB [8], do not provide predictions that are in good agreement with observations in U.S. coal fields.

In general, solutions based on phenomenological methods are obtained by idealizing the rock mass by a continuous medium that deforms according to a chosen constitutive law, and that satisfies compatibility and equilibrium conditions. Depending on the complexity of the geometry and chosen constitutive models, either closed form or numerical solutions can be obtained for predicitng subsidence. Finite element method has been proved to be a powerful tool for solving complex problems in several disciplines in science and engineering and appears to be a viable tool in subsidence prediction. There have been a number of applications of the finite element analysis for studying problems associated with underground mining. These have been reviewed elsewhere [13] and details are not included to avoid undue length of this paper.

Displacement discontinuity method originally proposed by Crouch [4] is based on the analytical solution to the problem of a constant displacement discontinuity over a finite line segment in an infinite or semi-infinite elastic body. This procedure is applicable to general cavity-type problems, such as thin slit, or crack which is of particular interest in this study. This paper presents the general details of an improved procedure for predicting subsidence based on fintie element method and displacement discontinuity method.

2 METHODOLOGY

The analysis of subsidence over longwall mine panels was performed by using the finite element method which is well documented in the literature [14]. The element equilibrium equations can be written as:

$$[k]\ \{q\} = \{Q\} \tag{1}$$

where [k] is the element stiffness matrix, {q} is the element displacement vector, and {Q} is the element load vector. These can be expressed as:

$$[k] = \iiint_v [B]^T[C][B]dv \tag{2}$$

and

$$\{Q\} = \iiint_v [N]^T\{\bar{X}\}dv + \iint_{S_1} [N]^T\{\bar{T}\}ds - \iiint_v [B]^T\{\sigma_o\}dv \tag{3}$$

where [B] is the strain-displacement relationship, [C] is the constitutive relationship, $\{\sigma_o\}$ is the vector of initial stresses such as insitu or tectonic, $\{\bar{X}\}$ is the body force vector, $\{\bar{T}\}$ is the surface traction vector, [N] is the matrix of interpolation functions, S_1 is the surface on which the surface traction is applied, and v is the volume of the element. In this study, four-noded isoparametric elements were used in the finite element analysis.

The finite element program [9] used in this study has the capability of using a number of user-selected constitutive models for the geological materials present in the overburden. These include linear and nonlinear elastic, and elastoplastic models based on ideal plasticity as well as hardening concepts. Details of derivations of incremental forms of constitutive equations associated with different models and aspects of computational algorithms are presented elsewhere [11]. The constitutive model adopted for modelling the behavior of overburden rock is based on the Drucker-Prager [6] yield criterion which can be considered as a generalization of the well known Mohr-Coulomb failure criterion.

The tensile failure conditions in overburden elements are checked by comparing the computed stresses in each element with the corresponding tensile strength of that element. The failure occurs when the tensile state of stress exceeds the tensile strength of the material; this can be expressed as:

$$J_1 = T_{cut} \tag{4}$$

where T_{cut} is the value of the tension cutoff, which is related to the tensile strength, σ_{yt}, of the material, and J_1 is the first invariant of stress tensor.

2.1 Simulation of Underground Mining

In a paper presented [10] in the previous symposium held in Nagoya, Japan, the conventional stress equilibrium approach was used to simulate the stress-free boundaries at the mine. The above method produced reasonable predictions for the shape of the subsidence profile, but no general guidelines could be established for the prediction of maximum subsidence based on the above approach. Therefore, a new model based on the ideas of history matching using the finite element method for the prediction of maximum subsidence as well as the shape of the subsidence profile was used in this study. The new approach is called the "displacement appraoch" in this paper. The general concept of this method is based on the assumption that total roof collapse occurs behind mine face as it advances. This assumption appears to be true for almost all longwall mining conditions. In this approach, the displacements at the mine roof are prescribed to be equal to a certain percentage of the seam thickness. The displacements and stresses in the overburden resulting from the roof movements are then computed using the finite element method. The induced stresses are then checked against yield criterion and their corresponding tensile strength, to obtain the tensile/plastic zones caused by mining.

2.2 Displacement Discontinuity Formulation

In contrast to continuous media, the equations of compatibility can not be applied to discrete media, while equilibrium equations are the same for both discrete and continuous media. The primary objective in this study is to simulate discrete cracks caused by mining activities in the overburden and to investigate their influence on subsidence predictions. It is well known that high stress areas are generated over the mine edge due to mine excavation. This could lead to cracks initiating from rib side and propagating toward the surface. Such cracks have been observed in scaled laboratory models used to simulate subsidence mechanisms [12]. The displacement discontinuity method proposed by Crouch [4] appears to be suitable for the analysis of ground movements addressed in this study. A detailed derivation of governing equations is given by Crouch [4], and Amanat and Siriwardane [3] and will not be included here to avoid undue length of this paper. A brief description of this method is given below.

In the displacement discontinuity method a number of, displacement discontinuity segments with unknown displacement magnitudes, D_1^j and D_2^j for values of j ranging from 1 to N are placed along the boundaries of the region to be analyzed. Then, a system of algebraic equations subject to prescribed displacements or stresses at the boundaries can be constructed. Solving these equations, the discontinuity values which produce the prescribed boundary values can be found. In general the shear and normal stresses (σ_1 and σ_2) at the discontinuity segments could be written in terms of displacement discontinuities (D_1 and $D2$) as:

$$\sigma_1^i = \sum_{j=1}^{N} F_{11}^{ij} D_1^j + \sum_{j=1}^{N} F_{12}^{ij} D_2^j$$

$$\sigma_2^i = \sum_{j=1}^{N} F_{21}^{ij} D_1^j + \sum_{j=1}^{N} F_{22}^{ij} D_2^j$$

$$i = 1 \text{ to } N \qquad (5)$$

where F_{11}^{ij}, etc in these equations are called influence coefficients which are known values. The coefficient F_{21}^{ij}, for example, give the normal stress at the i-th segment (σ_2^i) in terms of the shear component of displacement discontinuity at the j-th segment (D_i^j). For stress free boundaries such as a cavity or a crack the boundary conditions at the i-th element can be taken as,

$$\sigma_1^i = 0$$

$$\sigma_2^i = 0 \qquad (6)$$

for i = 1 to N

Then Equation (5) represents a system of 2N simultaneous linear equation with 2N unknowns, i.e., D_1^i, and D_2^i for i = 1 to N. These equations can be solved by standard methods of numerical analysis such as Gaussian elimination. A similar set of algebraic equations could be constructed for displacement boundary conditions, and displacements values could be computed for a set of prescribed displacements at the boundary. Therefore, both the conventional stress equilibrium approach and displacement approach could be employed with the displacement discontinuity technique. Numerical solution for cracks in semi-infinite bodies obtained by using this method is given by Crouch [4]. These were limited to regions with horizontal ground surfaces only. An improved computer code was developed in this study based on the displacement discontinuity formulation to include factors such as material non-homogenity. This code also has the capability of handling hilly ground terrains which is essential for analysis of subsidence in coalfields located in the Appalachian coal basin.

3 RESULTS

The displacement approach introduced earlier was verified with twelve cross-sections of nine longwall mines. Two of these case histories are presented here.

Finite Element Analysis of Shoemaker II Mine (West Virginia)

This mine is located in Northern West Virginia. The information pertinent to this mine was obtained from Reference [1]. Shoemaker II mine is a longwall mine with an average panel width (w) of 600 feet (183m), and a seam thickness (t) of 5.5 feet (1.68m). The overburden height (h) in the geologic profile was shown as 745 feet (227m). The measured maximum subsidence, S_{max} was reported [1] as 3.28 feet (1.0m) in an instrumented panel. The idealized geologic profile used in the analysis is shown in Figure 1. Although the geologic profile was available, engineering properties of geologic strata were not available for Shoemaker II mine. Therefore, the first step in this study has been to compile reported engineering properties of different rock types present in a typical geologic profile in the state of West Virginia. The material properties for Shoemaker II mine selected on this basis are shown in Table 1. The value of cohesion for different rock types were then calculated on the basis of known values of unconfined compressive strength, σ_{yc}, and the angle of internal friction as described in the literature [2].

In order to perform a finite element analysis of subsidence deformations associated with an underground mine, the geometric extent of the boundaries has to be first established. In order to have consistency of results from stress and displacement approaches, a parametric study was performed with respect to the distances to the boundaries. For this purpose, a hypothetical mine with mine geometry analogous to that of Shoemaker II mine was selected. The overburden was assumed to be homogenous having material properties of shale as shown in Table 1. Based on the results of the parametric study, the extent of vertical boundary can be selected as eight times the half-width of the mine away from the centerline of the mine panel. The fixity conditions at this distance did not have a significant influence on predicted deformations. Therefore, stress-free boundary conditions were assumed for that boundary. The lower boundary was placed at the mine floor level and movements in both x- and y-directions were fixed. A typical finite element mesh and boundary conditions used for Shoemaker II mine are shown in Figure 2.

The normalized subsidence profiles obtained from linear elastic finite element analysis based on stress equilibrium approach and displacement equilibrium approach are compared with field observations in Figure 3. The magnitidues of computed maximum subsidence are also shown in this figure. The comparison of normalized profiles can be considered as poor. However, the magnitude of predicted maximum subsidence based on the displacement approach can be considered as very good in comparison to field value. The predicted maximum subsidence based on the stress approach is poor compared to field values. In the displacement approach a bulking factor, α, equal to 0.10 was assumed in the finite element analysis. This means that the displacements specified at the mine roof would be restricted to 90 percent, $(1 - \alpha)$, of the seam thickness. This value for bulking factor was selected on the basis of available data at a number of Longwall panels located in the Appalachian coal basis.

The induced stresses were then checked against Drucker-Prager yield criterion and tenile strengths of the overburden materials, to obtain the tensile/yeild zones in the medium. Tensile and yield zones obtained from the finite element analysis of Shoemaker II mine based on displacement approach are shown in Figure 4. Analysis based on stress equilibrium approach did not generate any tensile or yield zones. As could be seen, two separate tensile zones could be observed. One immediately above the mine roof which shows the extent of the caved (failed) zone surrounded by yielded zone. Another tensile zone is observed near the ground surface where the horizontal tensile strains exist.

Finally, a sensitivity analysis of material properties using the displacement approach was performed. This study indicated that for a change of as much as 70 percent in material properties, less than 10 percent change in predicted deformations at the surface was observed. The major factor contributing to the predicted deformation values is the prescribed displacements at the mine roof (i.e. simulation of roof collapse). Therefore, problem of scaling laboratory determined material properties to field values is eliminated by using the displacement approach.

Finite Element Analysis of Mine A (West Virginia)

Mine A (panel 2) is a longwall mine located in Northern West Virginia with an average panel width of 550 feet (168m), panel length of 3700 feet (1128m), and an average seam thickness of 5.5 feet (1.68m). The ground surface is hilly at the site. The overburden height ranges from 100 feet (30.5m) to 300 feet (91.4m) over the mined area. Thus, Mine A could be categorized as a mine with a shallow cover. Field measurements at two transverse cross-sections and a longitudinal section through middle of panel 2, and material properties of different rock types at the mine, were made available to the writers. Details of the analysis of a transverse cross-section (F-F′), are given here. The geologic profile [5] for this section is shown in Figure 5. The finite element mesh used for section F-F′ is shown in Figure 6.

The subsidence profile obtained from the finite element analysis based on the displacement approach is compared with field observations in Figure 7. A bulking factor, α, of 0.25 was used in this analysis. As could be seen excellent agreement between predicted and field values is observed over the gob area. However, near the rib sides and over the solid coal areas predictions tend to deviate from field measurements and comparison cannot be considered as good.

The tensile and plastic zones detected in the finite element analysis for transverse cross-section F-F′ is shown in Figure 8. As could be seen different shapes of tensile and yield zones are observed here in comparison to those corresponding to Shoemaker II mine (deep mine). Two vertical tensile zones are observed immediately above the edges of the mine, which are surrounded by plastic zones propogated to the surface. However, the middle portion over the mine remained elastic. This might be explained as a result of block movement of the middle portion as a whole due to cracks initiated at the edges of the mine which propogate toward the surface. This trend was not observed for the deep mine (Shoemaker II mine) analyzed earlier. The ideas of cracks initiating at the mine edge and propogating toward the surface was studied using the displacment discontinuity method as described below.

Displacement Discontinuity Analysis of Mine A

In order to analyze Mine A which is located at a site with hilly ground terrain, the displacement discontinuity method proposed by Crouch [4] had to be modified to account for hilly nature of the surface. This was done by placing displacement discontinuity segments with prescribed stress free boundary conditions, to shape the hilly form of the ground terrain. Two cases were analyzed with the modified displacement discontinuity method. First, the mine was simulated by displacement discontinuity segments with displacement boundary conditions prescribed at the roof and bulking factor, α, of 0.25. The hilly ground surface was also simulated with discontinuity segments as stated earlier. For the second case, cracks initiating at each panel edge and propogating toward the surface at an angle of 15 degrees from vertical was simulated. This angle was selected on the basis of availabe data from model studies. A comparison of predicted and measured subsidence values are shown in Figure 9. Predictions were very close to those obtained from the finite element analysis for the first case where no cracks were simulated in the overburden. For the case with cracks propogating from rib sides to the surface, excellent agreement between preditions and field measurements for both the magnitude of maximum subsidence and the shape of subsidence profile was observed. This suggests that cracks induced in the overburden could be a major factor contributing to the shape of the subsidence profile. This hypothesis was studied with respect to the Shoemaker II mine as described below.

Displacement Discontinuity Analysis of Shoemaker II Mine

The influence of cracks in the overburden on the subsidence was also studied for Shoemaker II mine using the modified Displacement Discontinuity method. First, similar to Mine A, inclined cracks initiating from the rib sides and propogating toward the surface were analyzed. Results show a "bath-tub" shape predicted subsidence profile which is obviously not in good comparison with the "bowl" shaped field profile. These results suggests that different subsidence mechanisms contribute to the actual subsidence profile in deep mines. This was also apparent from the tensile and yield zones detected for Shoemaker II mine compared to those of Mine A (Figures 4 and 8). Therefore, a parametric study was performed to investigate the influence of different locations, numbers, orientation, and propogation height of the cracks induced in the overburden due to mining. The details of this study are given in Reference [3]. In this study, three cracks initiating at the rib side were simulated. The cracks were chosen to propogate only up to a certain distance below the surface. In this case it was taken as 85 feet (25.9m). The crack pattern was selected such that, they propogated over equal spans between panel centerline and a line over the mine edge. The comparison of field and predicted profiles is shown in Figure 10. The comparison can be considered as very good. The above case was also re-analyzed by finite element method using special interface elements to simulate cracks. Similar results were obtained.

4 CONCLUSIONS

The analyses of the Shoemaker II mine and Mine A indicate that the shape and magnitude of subsidence can be predicted reasonably well by simulating major mechanisms of roof collapse and crack propogation. Results indicate that roof collapse and material bulking are major factors contributing to the magnitude of maximum subsidence observed at the surface, while induced cracks in the overburden due to mining activities contribute mainly to the shape of subsidence profile. Subsidence mechanisms for mines with shallow cover appear to be different from those for deep mines.

5 ACKNOWLEDGEMENTS

This research was performed as Task No. 5 of an integrated project entitled "An Improved Longwall Mining Technique" directed by the Longwall Mining and Ground Control Research Center and sponsored by West Virginia University's Energy and Water Research Center. The data on subsidence at Mine A were obtained by Dr. Khair under Task No. 4. His assistance in making this data available for this research is appreciated.

6 REFERENCES

1. Adamek, V., and P.W. Jeran, "Evaluation of Existing Predictive Methods for Mine Subsidence in the United States," Proceedings, Workshop on Surface Subsidence Due to Underground Mining, Morgantown, WV, 1981.
2. Amanat, J. and H.J. Siriwardane, "Finite Element Analysis and Prediction of Subsidence Caused by Underground Mining", Report CE/GEO-83-2, Department of Civil Engineering, West Virginia University, Morgantown, 1983.
3. Amanat, J. and H.J. Siriwardane, "Analysis and Prediction of Subsidence Caused by Longwall Mining Using Numerical Techniques," Report, Department of Civil Engineering, West Virginia University, Morgantown (Under Preparation)
4. Crouch, S.L., "Solution of Plane Elasticity Problems by the Displacement Discontinuity Method," Int. J. for Num. Meth. in Eng., Vol. 10, 1976, pp 301-343.
5. Donaldson, A.C., Personal Communication, West Virginia University, Morgantown.
6. Drucker, D.C., and W. Prager, "Soil Mechanics and Plastic Analysis of Limit Design," Quarterly of Applied Mathematics, Vol. 10, No. 2, July, 1952.
7. Munson, D.E., and W.F. Eichfeld, "European Empirical Methods Applied to Subsidence in U.S. Coal Fields," Report, SAND80-1920, Sandia National Laboratories, 1980.
8. National Coal Board, Subsidence Engineers Handbook, London, 1975.
9. Siriwardane, H.J., "User's Manual and Background for a Computer Code for Nonlinear Two-Dimensional Finite Element Analysis of Some Geotechnical Problems," Report, Department of Civil Engineering, West Virginia University, Morgantown, West Virginia, 1983.
10. Siriwardane, H.J., "A Numerical Procedure for Prediction of Subsidence Caused by Longwall Mining," Proceedings, The Fifth International Conference on Numerical Methods in Geomechanics, Nagoya, Japan, April 1985, pp. 1595-1602.

11. Siriwardane, H.J., and C.S. Desai, "Computational Procedures for Nonlinear Three-Dimensional Analysis with Some Advanced Constitutive Laws," Int. J. of Num. and Analy. Methods in Geomechanics, Vol. 7, No. 2, 1983.

12. Sutherland, H.J., and K.W. Schuler, "A Review of Subsidence Predicting Research Conducted at Sandia National Laboratories," Proceedings, Workshop on Surface Subsidence due to Underground Mining, Morgantown, West Virginia, 1981.

13. Voight, B., and W. Pariseau, "State of the Predictive Art in Subsidence Engineering," Journal of the Soil Mechanics and Foundations Division, Proc., ASCE, Vol. 96, No. SM2, 1970, pp. 721-750.

14. Zienkiewicz, O.C., The Finite Element Method in Engineering Science, 3rd Edition, McGraw-Hill, London, 1977.

Table 1 Material Properties used for the Analysis of Shoemaker II Mine

Rock Type	Elastic Modulus $EX10^6$ psi	Poisson's Ratio ν	Density ρ pcf	Cohesion c psi	Angle of Internal Friction ϕ (deg)	Unconfined Compression Strength σ_{yc} (psi)	Indirect Tensile Strength σ_{yt} (psi)	Coefficient of Earth Pressure at Rest (k_o)
Shale	2.473	0.232	159	4315	29.27	14729	650	0.51
Limestone	3.872	0.19	166.5	6739	30	23346	921	0.50
Coal	0.340	0.346	90	672	31	2377	207	0.48
Soil	0.007	0.30	120	2	30	--	2	0.5

1 psi = 6.89 KN/m^2

1 pcf = 16.02 kg/m^3

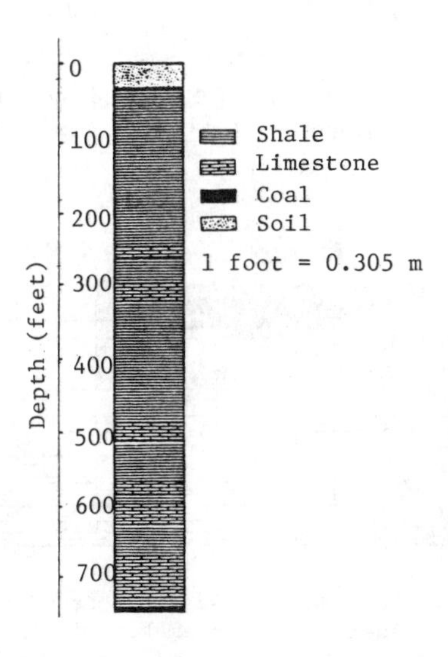

Fig. 1. Idealized Geologic Profile at Shoemaker II Mine.

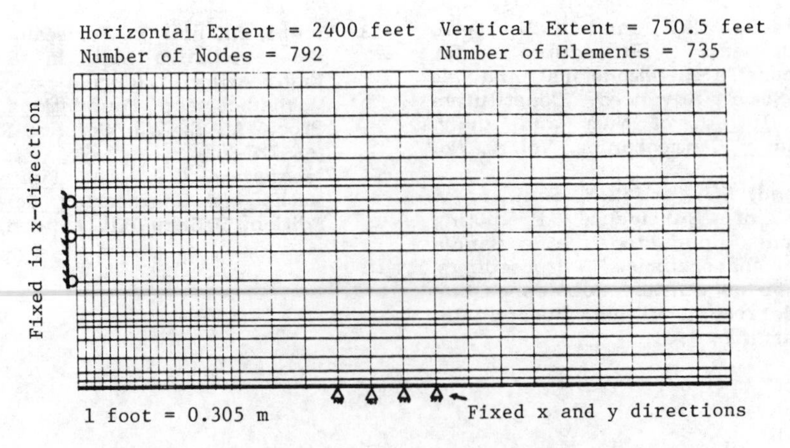

Figure 2. Finite Element Mesh for the Shoemaker II Mine

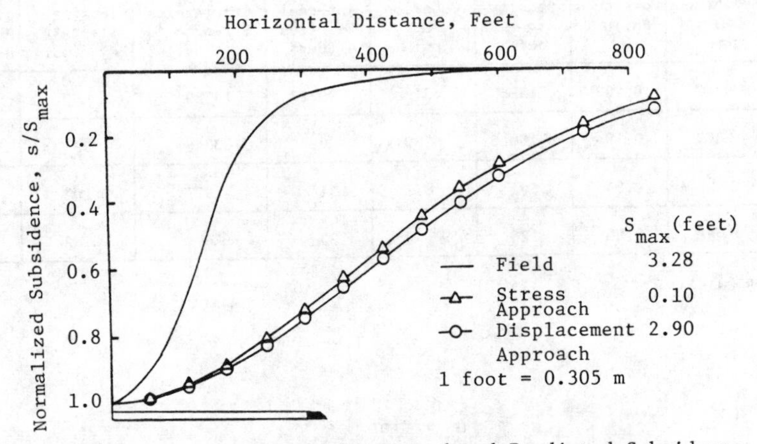

Figure 3. Comparison of Measured and Predicted Subsidence at Shoemaker II Mine

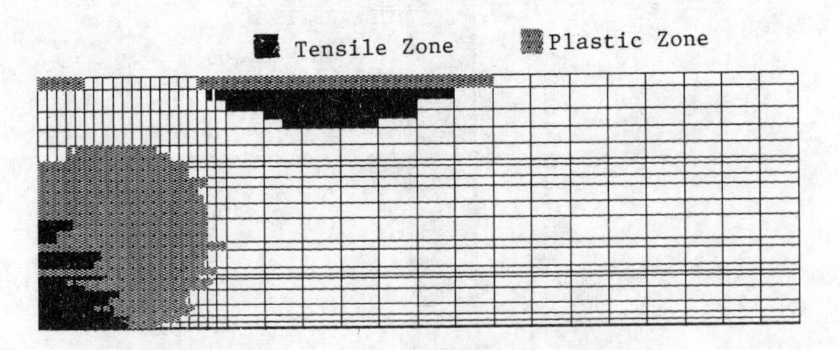

Figure 4. Tensile and Plastic Zones Detected in the Analysis of Shoemaker II Mine.

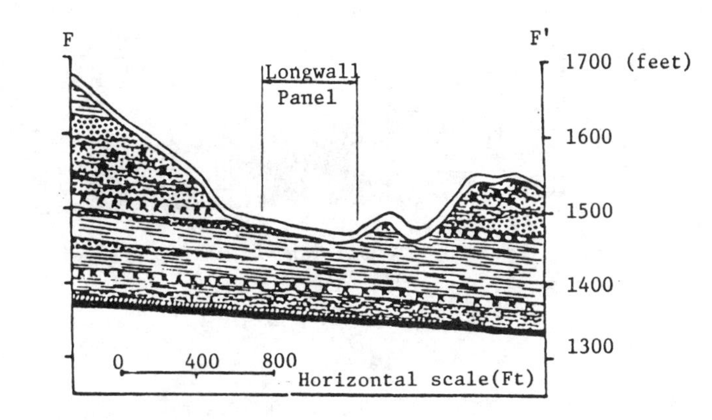

Figure 5. Geologic Profile at Mine A

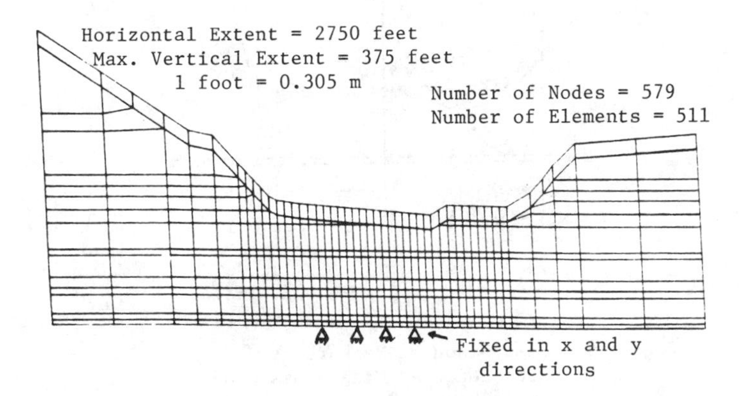

Figure 6. Finite Element Mesh for Mine A.

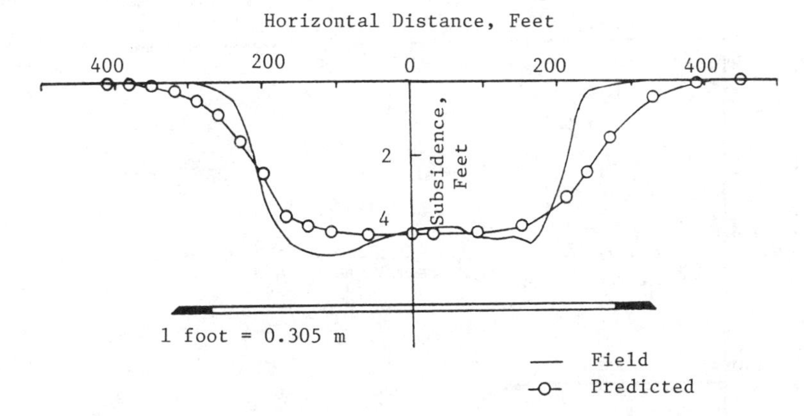

Figure 7. Comparison of Measured and Predicted Subsidence at Mine A.

Figure 8. Tensile and Plastic Zones Detected in the Analysis of Mine A.

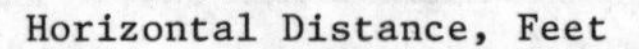

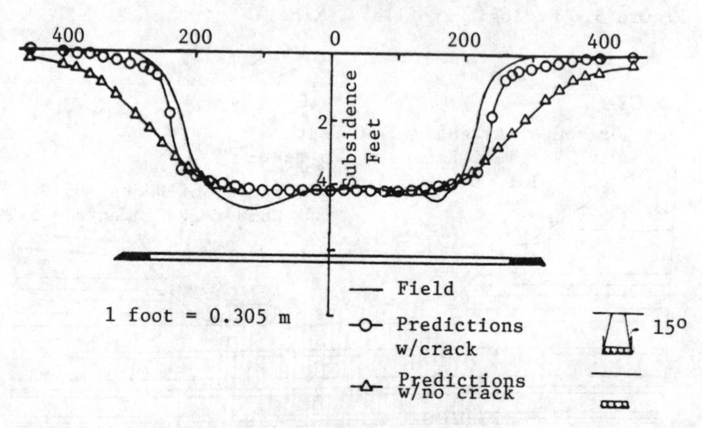

Figure 9. Comparison of Measured and Predicted Subsidence at Mine A Based on Displacement Discontinuity Method.

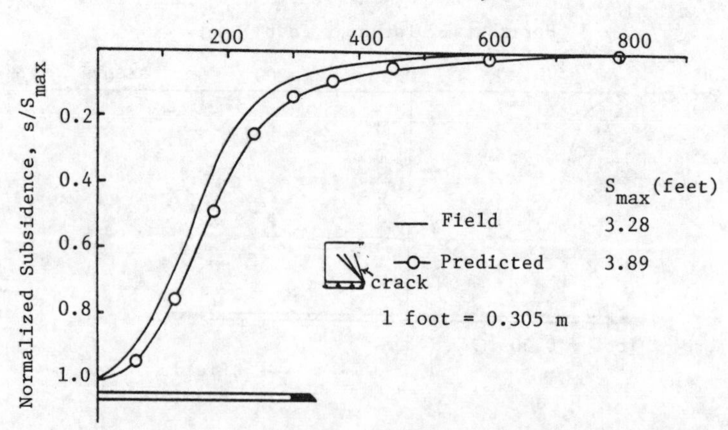

Figure 10. Comparison of Measured and Predicted Subsidence at Shoemaker II Mine, Based on Displacement Discontinuity Method.

Numerical Methods in Geomechanics (Innsbruck 1988), Swoboda (ed.)
© 1988 Balkema, Rotterdam. ISBN 90 6191 809 X

On the solution of large 3-D geomechanical problems

R.Blaheta & Z.Dostál
Mining Institute of the Czechoslovak Academy of Sciences, Ostrava

ABSTRACT: The development of an efficient software for solutions of three-dimensional geomechanical problems is discussed. The attention is focused on finite-element analysis based on linear tetrahedron elements. The use of mapping functions, the regular grid concept and a nodal description of the considered region as well as fast iterative solution procedures of large linear systems of equations are described in some detail.

1 INTRODUCTION

An analysis of many geomechanical problems requires efficient software for the computation of stress fields in a three-dimensional region. The elastic solution is either of independent interest or is required in the process of the solution of nonlinear problems. However, to obtain the solution, one has to overcome many difficulties connected with three dimensions, namely discretization of a large 3D heterogeneous region, description of the problem, post-processing of the results and the solution of the systems of many linear equations.

In our contribution we discuss the solution of the above problems by means of the finite element method based on a linear tetrahedron element. This element seems to be convenient for the following reasons:

- the shape of the region, its heterogenity and resulting regularity of the solution [1],
- the possibility to improve the accuracy of computed stresses by averaging or other post-processing procedure [2],
- the sparsity and other properties of resulting stiffness matrix.

On the other hand, exploiting linear tetrahedron elements we encounter difficulties connected, firstly, with a simple and illustrative division of the region and description of the problem, and, secondly, with an effective solution of large systems of linear equations.

We shall show that the first difficulty can be overcome by the use of mapping functions and the regular grid concept. Especially, we shall show that all input information can be connected with the nodes of the grid and not with hardly imaginable elements as the tetrahedrons. Secondly, we shall discuss some fast iterative methods for the solution of large systems of linear equations arising from the finite element analysis of the geomechanical problems.

The ideas presented here were involved in the finite element system GEM3 which was exploited to solutions of a number of practical problems up to 32 000 nodal unknowns. We provide some examples of such practical computations.

2 MAPPING FUNCTION AND REGULAR GRID

Let n_x, n_y, n_z be natural numbers. Then we can start with the following easy imaginable objects - the index parallelepiped

$$Q^* = \langle 1, n_x \rangle \times \langle 1, n_y \rangle \times \langle 1, n_z \rangle$$

and the index grid

$$G^* = \{(i,j,k) \in Q^*,\ i,j,k \text{ are integer}\}.$$

The index grid can be equipped by the topology : two points (i_1, j_1, k_1) and (i_2, j_2, k_2) are neighbours iff

$$\max\{|i_1-i_2|, |j_1-j_2|, |k_1-k_2|\} = 1.$$

Moreover, the index cells B^* and the index tetrahedrons T^* can be defined as cubes and tetrahedrons, the vertices of which are neighbouring points from G^*.

Now, we can introduce an one-to-one mapping function ϕ which represents a deformation of the index parallelepiped Q^* to the region Q. We shall suppose that the deformation of any index tetrahedron T^* is a tetrahedron again.

By means of the mapping function ϕ we can define the regular grid G, $G = \phi(G^*)$, with the topology induced from the uniform pattern G^*. Moreover, the G-cells $B = \phi(B^*)$ and the G-tetrahedrons $T = \phi(T^*)$ can be defined as a deformation of index cells and index tetrahedrons.

The mapping function must be defined in a suitable manner to enable the approximation of the boundary of the discretized region as well as the interfaces of different materials. The definition of the mapping function consists in assigning coordinates to all nodes of G. To this end, we can exploit a number of techniques, e.g. the interpolation, the isoparametric or conformal mapping, cf. [3].

3 NODAL DESCRIPTION OF THE REGIONS

At the finite element analysis of a geomechanical problem, we are interested in the region Ω which consists of homogeneous parts Ω_i, i.e.

$$\bar{\Omega} = \bigcup_i \bar{\Omega}_i \;,$$

where $\bar{\Omega}$ denotes the closure of Ω. We shall suppose that $\Omega \subset \phi(Q^*)$ and, moreover, that the region Ω and all the subregions Ω_i can be expressed as unions of nonoverlapping G-tetrahedrons.

For easy and illustrative description of the regions Ω, Ω_i we introduce the following notion. We say that the region Ω_j is cell convex iff

$$\Omega_j = \bigcup_B \text{conv}\,(N_j \cap B),\; N_j = \bar{\Omega}_j \cap G.$$

Let us note that conv (M) denotes the convex hull of the set M and B denotes the G-cells.

It is important that the regions of our interest Ω, Ω_i can be expressed as unions of the nonoverlapping cell convex regions. For illustration of this see Fig. 1, where all the regions Ω_0, $\Omega_1, \ldots, \Omega_4$ are cell convex but the region $\Omega = \Omega_1 \cup \ldots \cup \Omega_4$ is not.

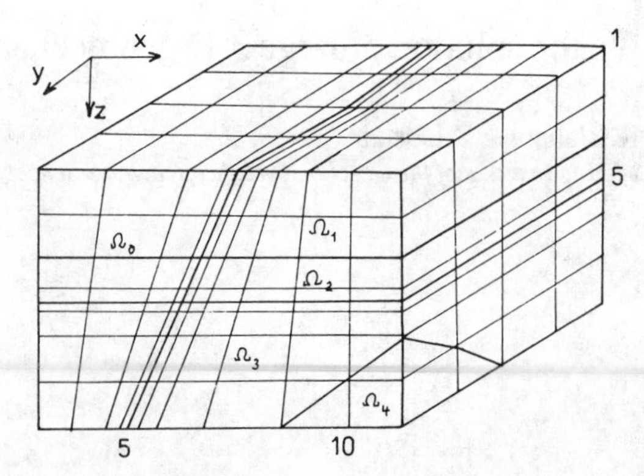

Figure 1 Example of a regular grid and the (slope stability) region $\Omega = \Omega_1 \cup \ldots \cup \Omega_4$.

The cell convex region Ω_j can be easily described by the set of nodes N_j. For this description, we can use input data of the following form:

identifier of the material
+ list of nodes.

The list of nodes may be prepared by means of the mapping function and three-dimensional indices.

A significant saving of input data can be obtained by an automatic generation of the difference $\Omega_j \setminus \Omega_k$ of two cell convex regions. In other words, we have the possibility to overlap a part of the previously described region. Let us note that these differences are not necessarily cell convex. As a result, the algorithm for their description is fairly complicated [4].

The overall information obtained directly from input data or generated from the differences of cell convex regions can be stored in the records of the structure: the cell, the material, the missing vertices of the described part of the cell.

Finally, let us suppose that the following materials correspond to the regions Ω_i shown in Fig. 1:

Ω_1, Ω_3 ... material No. 1 (M = 1),
Ω_2 ... material No. 2 (M = 2),
Ω_4 ... material No. 3 (M = 3),
Ω_0 ... material No. 0 (void space).

Then, the presented ideas can be illustrated by the following input data for the description of the above regions:

M = 1;   FOR = 010101, UP = 090511;
M = 0;   FOR = 010101, UP = 050505;
M = 2;   FOR = 030105, UP = 050511;
M = 3;   FOR = 070511;
         FOR = 080510, UP = 080511;
         FOR = 080411;
         FOR = 090509, UP = 090511;
         FOR = 090410, UP = 090411;
         FOR = 090311;

Here, for example, the item FOR = 030105, UP = 050511 represents the list of nodes $\phi(i,j,k): 5 \leq i \leq 11,\ 1 \leq j \leq 5, 3 \leq k \leq 5$. Let us note that the regions Ω_1, Ω_3 are defined by the difference $Q \setminus (\Omega_0 \cup \Omega_2 \cup \Omega_4)$.

4 DECOMPOSITION INTO TETRAHEDRONS

Having defined the mapping function and all homogeneous subregions Ω_i, the decomposition of the whole region Ω into G-tetrahedrons can be carried out automatically. The algorithm for this decomposition proceeds cell by cell and decomposes all homogeneous parts of G-cells into G-tetrahedrons.

The homogeneous G-cells are divided into 5 tetrahedrons in two ways. The obtained local stiffness matrices as well as the computed stresses are averaged. The nonhomogeneous cells are divided into 5 or 6 tetrahedrons in several ways to obtain the "symmetric" discretization again. The discretization algorithm is based on the actualization of the incidental matrix of homogeneous parts of the cells [4].

Let us note that the conformal decomposition of the region Ω may be violated on the interface of two materials. But this is a relatively rare case which concerns few elements only. Some error analysis of this nonconformity can be found in [5].

5 SOLUTION OF THE FINITE ELEMENT SYSTEM OF EQUATIONS

The discretization of the truly three-dimensional geomechanical problems (Figures 2 and 3 show examples of such problems) with the use of linear tetrahedron elements leads to the solution of large systems of linear algebraic equations. For the solution of these systems, we must take into account the following characteristics: 10 000 - 30 000 or even more unknowns, half-bandwidth frequently greater than 1 000 and up to 81 nonzero elements per equation of the stiffness matrix.

The above characteristics exclude the possibility of an effective solution of such systems using the direct methods [6]. For that reason, iterative methods were pursued.

Our first experience was connected with the conjugate gradient (CG) iterative method [6] combined with the initial approximation computed by aggregation of the unknowns. This initial approximation is given by the formula

$$u_0 = P\,(P^T AP)^{-1}P^Tf$$

where $Au = f$ is the solved system and P is a prolongation matrix given by aggregation of the unknowns. P is $n \times m$ full rank matrix, $m < n$, the columns of P are orthogonal vectors of 0's and 1's. If $P = (p_{ij})$, $p_{ik} = p_{jk} = 1$ for some k, then the unknowns u_i and u_j will assume the same value in the initial approximation. Moreover, we can assign the zero values to some unknowns (e.g. to the unknowns which represent horizontal displacements).

The formula for u_0 can be easily deduced by minimizing the energy functional

$$J(u) = \frac{1}{2}\ u^T Au - u^Tf$$

on the range of P. Let us note that the use of regular grids enables an easy description of the aggregation [7].

The examples of the behaviour of the iterative process combining the initial approximation by aggregation and the CG method are shown in Figs. 2 and 3. The solved systems arise from the discretization of practical geomechanical problems in very heterogeneous regions. Two graphs in the Fig. 2 correspond to the solution of the problem without an underground opening (Fig. 2A) and with an underground opening (Fig. 2B). Here, the solution of the problem in 2A serves as initial approximation for the solution of the problem in 2B.

Some characteristics of the solution of the above mentioned problems are summarized in Table 1. Here, the accuracy of the iteration u^k is defined by the ratio

$$\varepsilon = \|Au^k - f\|_{1_2} / \|f\|_{1_2}\ .$$

The number of iterations to obtain the reported accuracy can be seen in the third column of the Table 1. Let us

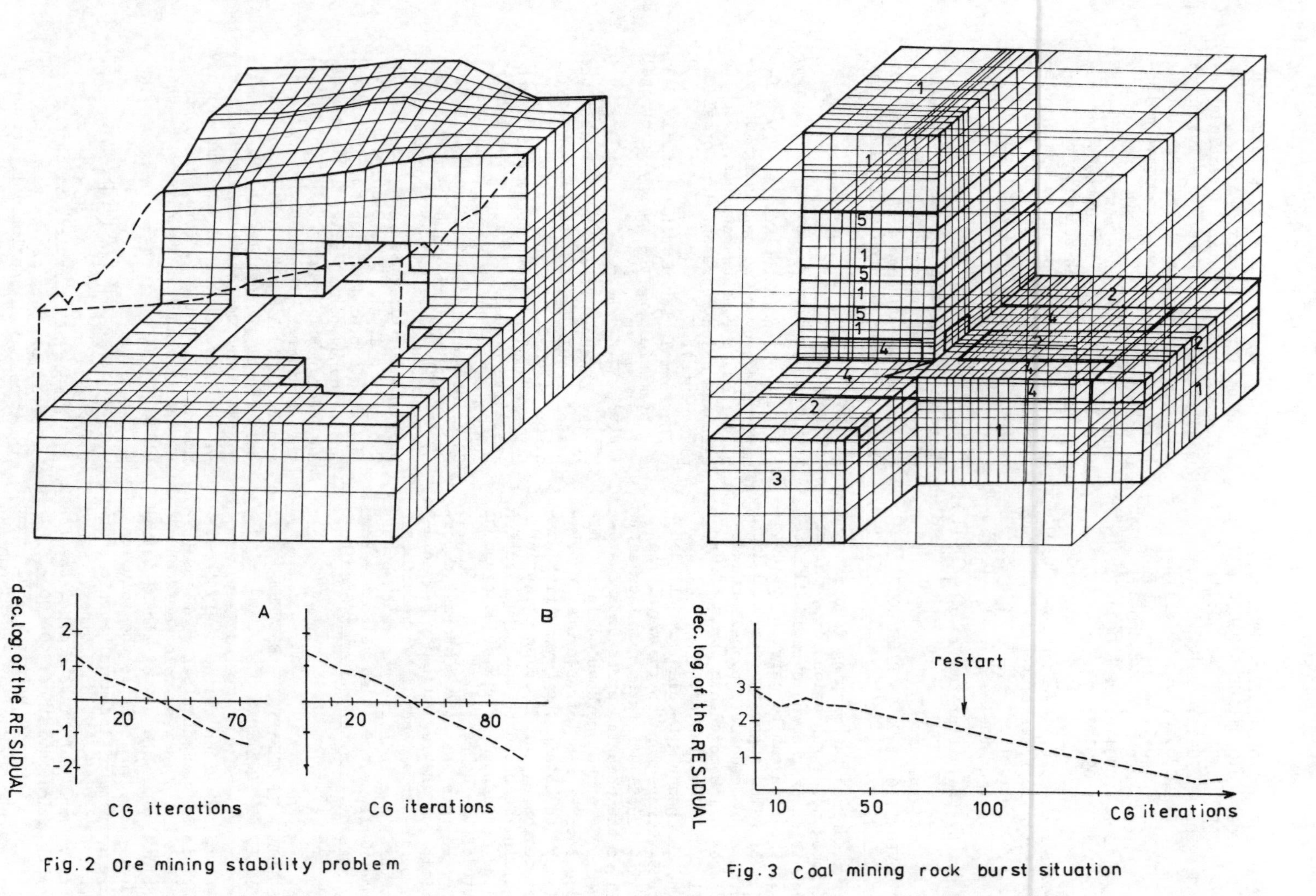

Fig. 2 Ore mining stability problem

Fig. 3 Coal mining rock burst situation

Table 1 Characteristics of iterative processes

Problems in	Number of equations	Number of iterations	Accuracy	Time	Convergence factor
Fig. 2A	11 806	73	.0018	30'	.919
Fig. 2B	10 732	89	.0019	32'	.930
Fig. 3	31 104	200	.0046	250'	.973

note that the accuracy $\varepsilon \in \langle .001, .01 \rangle$ was sufficient in all problems that we solved. The convergence factors are connected with the l_2-norm of the residuals. The computer time represents the CPU time on EC 1040 computer (about 10^5 flops/sec, 640 KB in core) in multiprogramming environment.

From our experience we can say that the combination of the conjugate gradient method and the initial approximation by aggregation of the unknowns represents a good equation solver. Nevertheless, the solution of the finite element system is the most time consuming part of the computation. As a consequence, we sought a more effective method. Some examples of such methods are described in the following sections.

6 PRECONDITIONED CG METHOD

To accelerate the convergence of the conjugate gradient method, we can use the preconditioning of the system being solved by a positive definite preconditioning matrix C, cf. [6]. To this end, we suggested the use of the matrix C obtained from the modified incomplete factorization of the separated displacement componentpart A_s^s of the stiffness matrix A. The matrix A^s is created from A by canceling the couples between the unknowns which represent the displacements in different directions.

In the most simple situation of a 2D elasticity problem and using a rectangular grid, we can write

$$C = (D - L)\, D^{-1}\, (D - L^T)$$

where $A^s = \Lambda - L - L^T$, Λ is a diagonal and L is the lower triangular matrix and D is the diagonal matrix defined by the condition of equal row sums of the matrices A^s and C.

An example of the application of such preconditioning is shown in Fig. 4. Here, CG-DS denotes the CG method with diagonal scaling and CG-SDC-IF denotes the preconditioned CG method described above. More details as well as applications to 3D problems can be found in [8, 9].

7 MULTILEVEL METHODS WITH CORRECTION BY AGGREGATION

The graphs in Figs. 2A and 3 show that the greatest convergence rate is reached immediately after the initial approximation by aggregation of the unknowns. This fact is exploited in the two-level method, one iteration of which consists of the following steps:

1. - perform ν iteration of some standard iterative method as Gauss-Seidel or conjugate gradient method,
2. - perform the correction

   $$u \longleftarrow u + P\,(P^T\, AP)^{-1}\, P^T\, (Au - f)$$

   where P is the prolongation by aggregation defined in the section 5, u on the rhs stands for the approximate solution computed in step 1 and the u on the left hand side of the arrow is the new iteration.

For details and further variants of the algorithm see [8, 7, 10, 11]. An example of the application of two-level methods to the solution of a 2D elasticity problem is shown in Fig. 5. Here, GS denotes Gauss-Seidel iterative method, AC 2,2 and AOC 2,1,0 are two-level methods with the correction and the overcorrection by means of the aggregation of the unknowns [8].

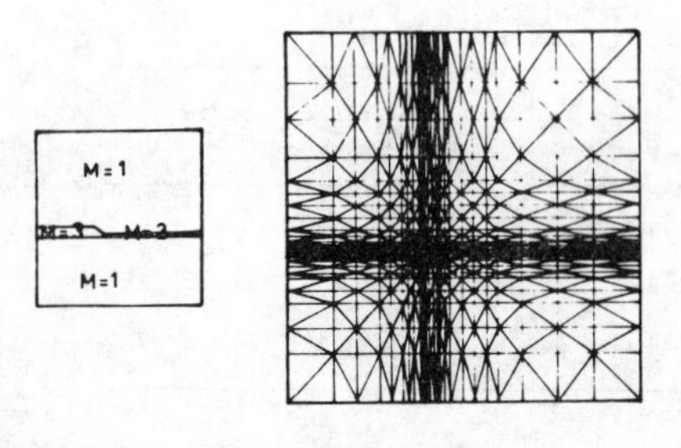

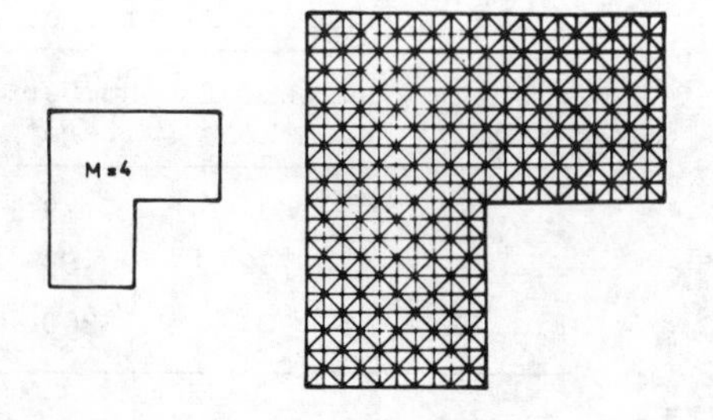

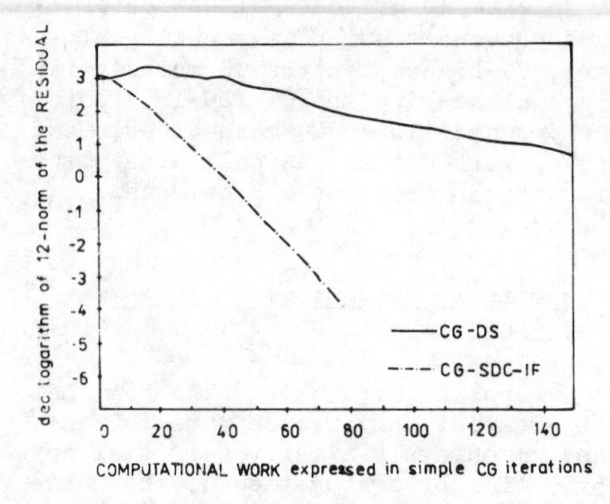

Fig. 4 Longwall mining situation

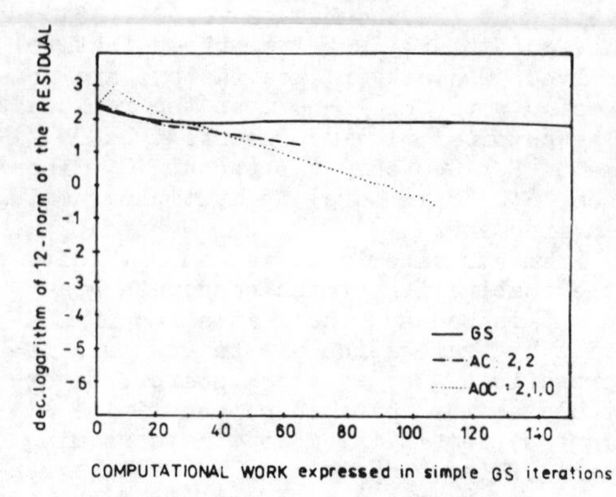

Fig. 5 Pillar design problem

REFERENCES

1 V.G.Korneev, U. Langer, Approximate solution of plastic flow theory problems, Teubner, Leipzig, 1984

2 M.Křížek, P.Neittaanmäki, On superconvergence techniques, Preprint No. 34, Univ. Jyväsylä, 1984

3 J.E.Akin, Application and implementation of finite element methods, Academic Press, London, 1982

4 R.Blaheta, Z.Dostál, Discretization of 3D problems-theory and algorithms (in Czech), Sb.prací HOÚ ČSAV, Ostrava, 1986

5 J.Stříbná, Nonconformal finite elements and their use (in Czech), Master Thesis, Charles University, Prague, 1987

6 O.Axelsson, V.A.Barker, Finite element solution of boundary value problems, Academic Press, Orlando, 1984

7 R.Blaheta, Multilevel methods for elasticity problems, Proc. I. Conf. on Mechanics, Prague 1987

8 R.Blaheta, Iterative methods for numerical solution of elasticity problems, Ph.D. Thesis, Charles University, Prague, 1987

9 R.Blaheta, Solution of systems of linear algebraic equations arising from the modelling of geomechanical situations (in Czech), Proc. Conf. Numer. Meth. in Geomechanics, V. Tatry, 1987

10 W.Hackbusch, Multi-grid methods and applications, Springer, Berlin, 1985

11 R.Blaheta, Multilevel method with overcorrection by aggregation, 2nd Int.Symp.Numer.Anal.,Prague, 1987.

Numerical Methods in Geomechanics (Innsbruck 1988), Swoboda (ed.)
© 1988 Balkema, Rotterdam. ISBN 90 6191 809 X

Finite elements investigation of the peculiarities of three-dimensional state of stress of rock mass in the process of mining

A.A.Kozyrev, S.N.Savchenko & V.A.Maltsev
Mining Institute of the Kola Branch of the USSR Academy of Sciences

ABSTRACT: The results are presented of investigations of rock stress state in the vicinity of the pillar located between worked-out areas in the process of mining thick deposits when gravitational - tectonic state of stress acts in the rock. The investigations were carried out with the finite elements method in "three-dimensional" solution. A comparison was made with the results of an analogous problem in "plane strain" solution.

Many geomechanical problems are solved by analitical and numerical methods in a plane formulation as a rule because a volumetric formulation is derived in a complicated manner. In this case mutually orthogonal sections of the region under the investigation are used, assuming that this region, being in normal, according to the section, direction, strikes at an unlimited distance. Normal stresses obtained in a plane formulation are considered to be quasi-principle stresses of volumetric state of stress. Bat such simplifications of a problem are not always obvious and sunstantiated sufficiently.

When thick deposits are mined by the opposite face lines method, the most complicated conditions, connected with control over rock pressure, take place in mining a pillar situated between mined-out areas under the hanging wall, especially in case when complex gravitational-tectonic stress field acts in a rock-mass, i.e. when the horizontal stresses exceed the vertical ones. Such conditions are characteristic of deposits of apatite-nepheline ores of the Khibini rock massif in the Kola Peninsula. It has been established by the measurements that at the depth of 100-300 m from the surface in the virgin ground the horizontal stresses exceed the stresses caused by the weight of rocks (H) by factors between 5 and 10. Owing to it, even at small depth (100-300 m from the surface) sudden rock destructions occur in workings and pillars in a kind of rock burstings, microshocks and at great depths rock bursts are detected. Especially dangerous phenomena occur in the vicinity of the boundaries of the mined-out areas, in the zone of their mutual influence when mining takes place under the hanging wall.

Use of solutions in a plane formulation may lead to rather substantial errors when overlying rocks have formed overhangings and the borders of mined-out spaces of complex configuration are very close to each other.

Developed modern numerical methods, the boundary integral equations method /1/ and the finite elements method /2/ make it possible to solve geomechanical problems in a vilumetric formulation.

The paper discusses the investigation results of state of stress of rocks obtained with the help of a mathematical model in the neighbourhood of the pillar situated between worked-out areas approached to each other, for conditions resemble the conditions of mining of the Khibini apatite-nepheline deposits. These investigations are carried out by the finite elements method in a volumetric formulation. In the solution of the problem the parallelepipeds are taken as the finite elements of division of the region under the investigations; the results of the in situ stress measurements carried out in a tight rock-mass are used as the boundary conditions.

Five stages of the level mining are discussed:

- initial state, before the formation of the mined-out areas, taking into account the effect of mining of the overlying levels;
- the formation of a single mined-out area in the central part of the level;
- the formation of the second mined-out area and of a pillar between them;
- the beginning of the sectional sill cut mining by the opposing face lines method under the border of the caved rocks;
- the sill cut is completely mined out.

Fig.1 shows a scheme of the model and the conditions of its loading on the border. In accordance with in situ measurement data, the T_x- and T_y-forces increasing with the depth from the surface in a step-like manner, have been applied to the sides normal to the OX- and OY-axes (table 1).

Table 1. Change of the border conditions of loading of the model along the OX- and OY-axes with the depth from the surface

Depth from the surface (m)	T_x (MPa)	T_y (MPa)
0–160	2	10
160–320	5	15
320–400	15	20
400–440	20	21.5
440–480	22	23
480–520	26	24
680–840	40	30

The influence of the rocks' own weigh (=3 t/m^3) was modulated in the direction of the vertical axis. Zero displa-

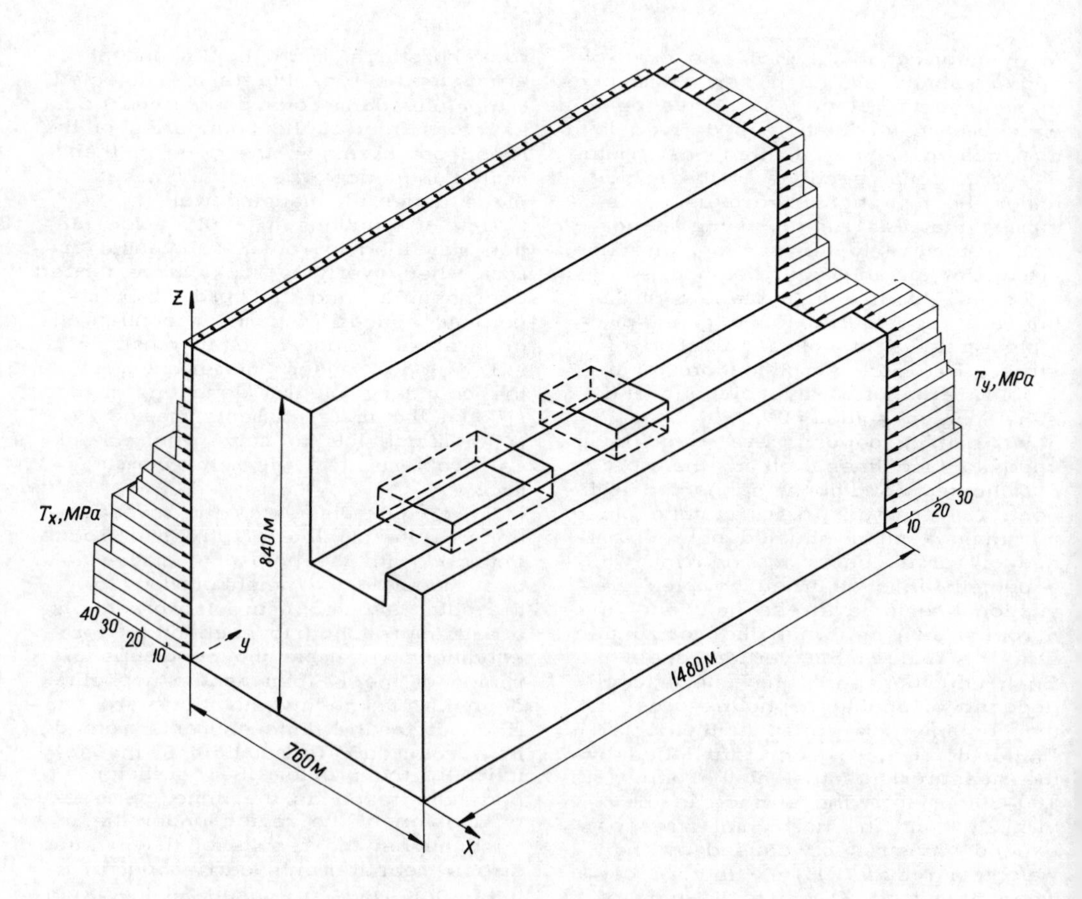

Figure 1. A scheme of the model and conditions of its loading in the border.

Figure 2. Distribution of normal stresses σ_x, σ_y, σ_z in a cross-section at the depth of 10 m lower than the borders of the mined-out area at different stages of the level mining.
a – initial state up to the formation of the mined-out area;
b – formation of a single mined-out area in the central part of the level;
c – formation of a second mined-out area and of the pillar between them;
d – the beginning of a sill cut mining by the opposing face lines method near the border of the caved rocks;
e – the sill cut is completely mined out.

cements U=0, V=0, W=0 were designed in the planes YOZ, XOZ and XOY, respectively.

The distribution of normal stresses O_x, O_y, O_z in a haulage horizon in the vicinity of a block-pillar at different stages of the deposit mining is shown in Fig.2. The stresses O_x, O_y, O_z in the initial state (Fig.2a) depend substantially on the surface relief and have the concentration with respect to forces acting at the given depth: $O_x/T_x=(1.35-1.73)$; $O_y/T_y=(1.25-1.35)$; $O_z/\ H=(1.33-1.67)$, respectively.

At the first stage of mining (Fig.2b), when a single worked-out area is formed, the highest concentration of stresses $O_x=2.3T_x$, is brought about underneath the worked-out area, under the cantilever of the overhanging rocks. In the vicinity of the future block-pillar stresses $O_x=(1.8-1.92)T_x$. Under the mined-out area stresses O_y increase up to $1.45T_y$ and stresses are halved in comparison with those of the initial state. In the vicinity of the block-pillar stresses O_y fall off slightly as against the stresses in the initial state and stresses O_z increase up to 2 H.

The formation of the second mined-out area (Fig.2c) results in further increase of stresses O_x concentration under the mined-out areas and in the region adjacent to the block-pillar stresses O_x $1.92T_x$ everywhere. At this stage of mining under the mined-out areas stresses O_y also increase up to $(1.45-1.67)T_y$ and in the region adjacent to the block-pillar they are at the same level. Stresses O_z increase up to 2.33 H in the block-pillar under the caving border.

The beginning of the sill cut mining (Fig.2d) in the block-pillar results in further increase of stresses O_x under the mined-out areas, in expansion of the zone of increased concentration of stresses O_x, and in increase of stresses O_x in a rock-mass under the block-pillar. Stresses O_y increase up to $(1.45-1.67)T_y$ below the sill cut faces and under the border of the caved rocks stresses O_y increase up to $1.45T_y$. The vertical stresses O_z increase up to 2.67 H under the caving border in the block-pillar. In the rocks of the foot wall of the block-pillar, stresses $O_z=(1.33-2)$ H.

At the last stage of mining (Fig.2e), when the connection of the sill cut is finished, stresses O_x, acting in the hanging wall rocks under the cantilever of the overhanging rocks below the sill cut, have the concentration of $2.7T_x$; in the rocks of the hanging wall, when the sill cut is mined out completely in the block-pillar, stresses $O_x=2.3T_x$, and in the footwall rocks stresses $O_x=1.92T_x$. Under the mined-out areas and the sill cut, the concentration of stresses O_y is $1.67T_y$, in the hanging wall stresses $O_y=1.45T_y$, and in the footwall, stresses $O_y=1.25T_y$. The level of vertical stresses O_y does not change here and is the same one like at the previous stage of mining, slightly decreasing in the footwall rocks only.

On the basis of the previous investigations of state of stress of rocks in the vicinity of the mined-out areas, the formulae have been obtained by the finite elements method in a plane formulation which allow to evaluate the stresses in the pillar formed between the two mined-out areas. Averaged values of stresses O_x and O_z obtained in the pillar are expressed by the formulae:

$$O_x=2.2T_x+0.3\ H$$

$$O_z=\frac{(1.35+2.4\ l/H)\ H(0.55+0.95\ L/H)}{1-e^{-0.55\ d/h}}+0.7T_x \qquad (1)$$

where T_x – intensity of the tectonic forces acting across the strike
– volumetric weight of the overlying rocks
H – depth of mined out spaces
L – span of the mined-out spaces
l – length of the cantilever of the overhanging rocks
d – width of the pillar
h – height of the pillar.

Fig.3 shows the plots of averaged values of stresses O_x and O_z obtained by the formulae (1) for the boundary conditions, marked in a Table 1 (curves 1). Averaged values of stresses O_x and O_z, obtained from volumetric modelling data for the same boundary conditions, presented by the curves 2.

Performing analysis on data (Fig.3) we may make a conclusion that stresses O_x and O_z obtained by the solution of the problem in plane and volumetric formulations are similar in case for wide pillars (d/h 2). For all this, vertical stresses in a plane formulation are smaller than those in a volu-

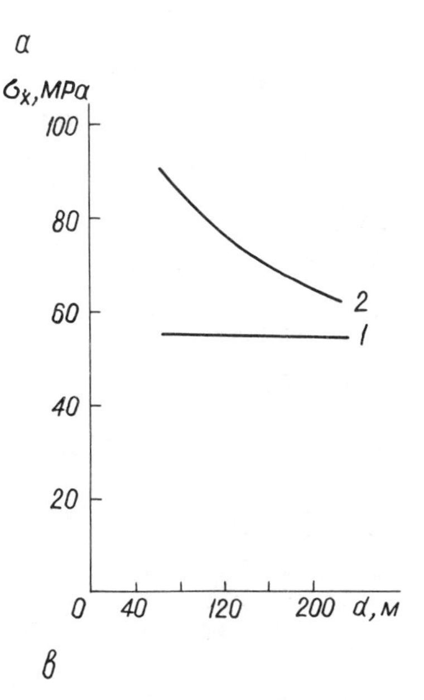

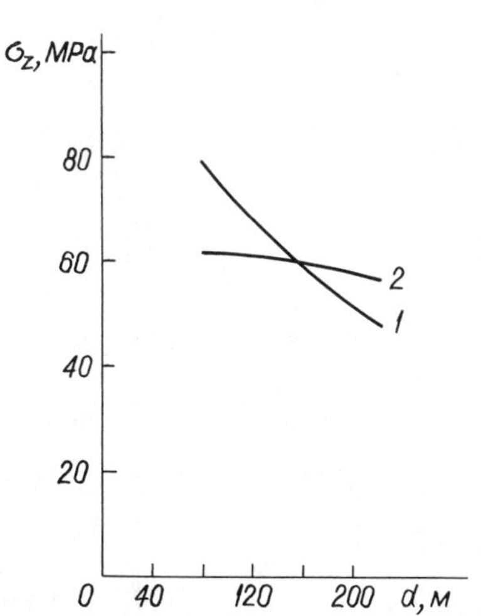

Figure 3. The plots of the averaged values of normal stresses O_x and O_z obtained in accordance with the data of calculation in plane (1) and in volumetric (2) formulations.

metric one, and horizontal stresses O_x in a volumetric formulation are greater tha than those in a plane one. As the pillar's width decreases (d/h 2), stresses O_z in a volumetric formulation are smaller than those of a plane one, and stresses O_x in a volumetric formulation are greater than those of a plane one. Besides, the coefficient of the concentration of the vertical stresses near the mined-out area in a plane formulation of the problem is smaller than that one obtained by a plane formulation by a factor of 1.5-2.0. At same time, the coefficient of horisontal stresses O_x acting across the strike of the ore body is greater than that one obtained by a plane formulation by a factor of 1.2-1.5.

Thus, it may be concluded from the calculation results that:

When the width of the pillar between worked-out areas is smaller than 200 m (d/h 3), the application of the concentration zones from each of the mined-out areas takes place.

When the dimensions of the pillar are d/h 2, vertical and horizontal stresses, acting across the strike of the ore body and obtained by volumetric and plane formulations, are close to each other, and when d/h 2, vertical stresses in a volumetric formulation are smaller and horizontal stresses are greater if compared with those obtained by a plane formulation. Moreover, the difference in values of horizontal stresses, acting near the borders of the mined-out areas, may be 20-50% and in values of vertical stresses - 50% and even more.

When d/h 2 and horizontal stresses, compared with/or exceeding the vertical ones, are acting in a tight rock-mass, then the calculation of strength of the edge parts of the pillar should be carried out taking into account the action of these horizontal stresses in the pillar, because , according to the absolute value, they may be greater than the vertical ones. In this case the notions about the stability of workings situated in the foot of the pillar are changed.

The laws obtained have been used in determination of the dimensions of the block-pillar and of the stages of its mining in one of the Khibini mines in the Kola Peninsula.

REFERENCES

/1/ Boundary integral equations method. M., Mir, 1978

/2/ R.H.Gallagher, Finite element analysis: Fundamentals, Prentice - Hall, INC., Englewood Cliffs, New Jersey, 1975

Numerical Methods in Geomechanics (Innsbruck 1988), Swoboda (ed.)
© 1988 Balkema, Rotterdam. ISBN 90 6191 809 X

Stress analysing of the rock surrounding a shaft and its application

Hua Anzeng & Zhuo Zhaowei
China Institute of Mining and Technology, Xuzhou, Jiangsu

ABSTRACT: In the present paper, stress and displacement of the rock surrounding a shaft approaching a coal seam are studied on the basis of 3 - D nonlinear elasto-plastic analyses by means of ADINA finite element program. Based on such analyses, the authors state the causes for coal and gas burst, work out the measures for preventing and controlling coal and gas burst, and select the relevent parameter. Three seams prone to coal and gas burst have been passed safelly one after another with the aid of the shaft's own releasing effect and the method of energy release through borehole. The practice proves that the theory and selected parameter coincide with the actual conditions.

1 PREFACE

At LuLing Colliery, the new auxiliary shaft encountered three seams prone to coal and gas burst in 464-500m. One of the coal seams is 11m thick, and gas pressure in the coal seam is 3.0 MPa, which is unique at present in china. The China Institute of Mining and Technology, cooperated with the Jiangsu Mining Capital Construction Company, has studied the shaft's safety and technical measures, the three coal seams have been safely passed through by means of the method of energy release through boreholes.

2 NUMERICAL RESULTS

For the clarification of the causes for coal and gas burst, mechanism of measures to prevent and control it, and selected relevent parameter, stress and displacement of the rock surrounding a shaft approching a coal seam are studied on the basis of 3-D nonlinear elasto-plastic analyses by means of ADINA finite element program. The model is totally divided into 721 3-D elements, and 1059 nodes. Virgin stresses of rock mass are its gravity. The condition of Drucker- Prager-Cap is regarded as the yield condition of rock mass and coal. The calculation parameter set by experimental results in labs and statistics is seen in Table 1. The parameter in Table 1 is the ratio to virgin stresses of rock mass, because, under the present conditions, the absolute value is not accurate enough, and it can only represent isolated cases, but the relative value can be obtained according to statistics and analyses based on a lot of facts, and it is therefore dependable. Besides, the relative value represents a series of cases, it is favourable for reguar researches. The research method of the relative relation in the numerical analysis is known as a dimensionless numerical simulation method. The following analyses have been carried out: The different thickness of coal seams, different distances between the shaft and coal seams, different parameter of rock, different inclinations of rock and their influence on stresses and displacement of the rock surrounding a shaft. All cited examples are based on virgin stresses and displacement for the driving of a shaft, and the relative variations are analysed afterwards. Fig 1 - Fig 5 show the plastic area and maxium principal stress distribution of longitudinal section passing through the shaft center and along the inclination of coal seams. In the figures, the two coordinates, direction of radius and that of plumb represent the different distances of any point from the centre of the shaft. The dotted line represents the plastic area border near the shaft. The thin lines represent the equal value line of the maxium principal stress, the numbers

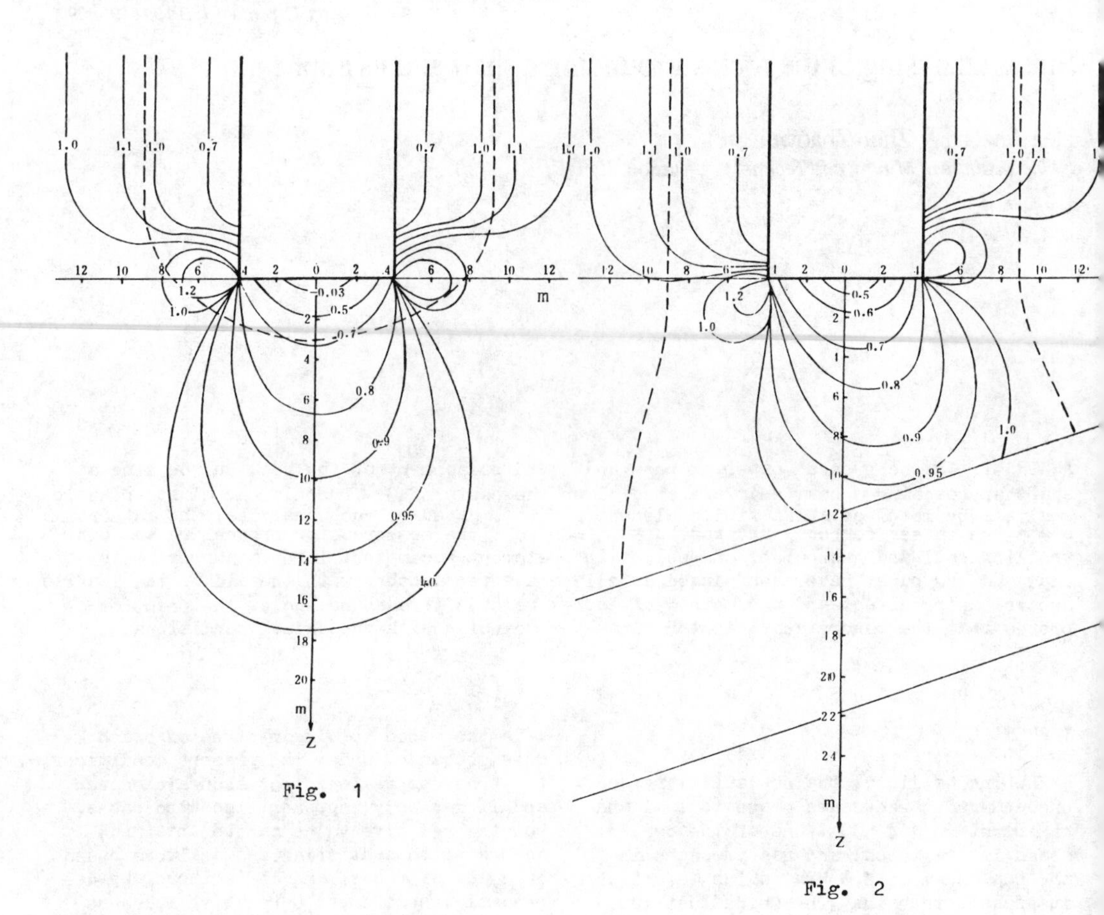

Fig. 1

Fig. 2

Table 1. Calculation parameter

Item	Rock mass	Coal
Ratio of elastic modulus to stress	800	80
Ratio of initial cap location to virgin stress	3.2	0.32
Ratio of K($\frac{6c \cos \varphi}{\sqrt{3}(3-\sin \varphi)}$) to virgin stress	0.2	0.08
$\alpha = 2\sin\varphi/\sqrt{3}(3-\sin\varphi)$	0.08	0.03

on the thin line represent the ratio of the maxium principal stress of the rock after being driven to the maxium principal stress of virgin stress. In Fig 1, the heading of the shaft does not approache the coal seam. The different distances between the heading of a shaft and the coal seam(10m thick) are shown in Fig2–Fig 4. In Fig 5, the coal seam is 2 m thick. Other cases are omitted.

3 CALCULATION AND RESULTS OF FIELD MEASUREMENTS

In order to observe gas pressure in coal seams, and the variations of gas pressure in coal seams as well as the relative deformation between the roof and the floor of coal seams in process of driving, the driving of a shaft is stopped at a place where the distance between the heading of the shaft and coal seam is 11 m, (measured by the intersection between the centeral-axial line of the shaft and the roof of coal seam). The boreholes are drilled towards coal seams around the shaft, pressure gauges are installed for observing gas pressure, deep reference points are fixed in the roof and floor of coal seams for observing the relative deformation. The actual measured gas pressure distribution is show in Fig 6 when the heading of a shaft is 11 m from coal seams (11 m thick). The ordinate represents gas pressure, and the abscissa the distance between the observing point and the shaft centre. The figure shows that gas pressure

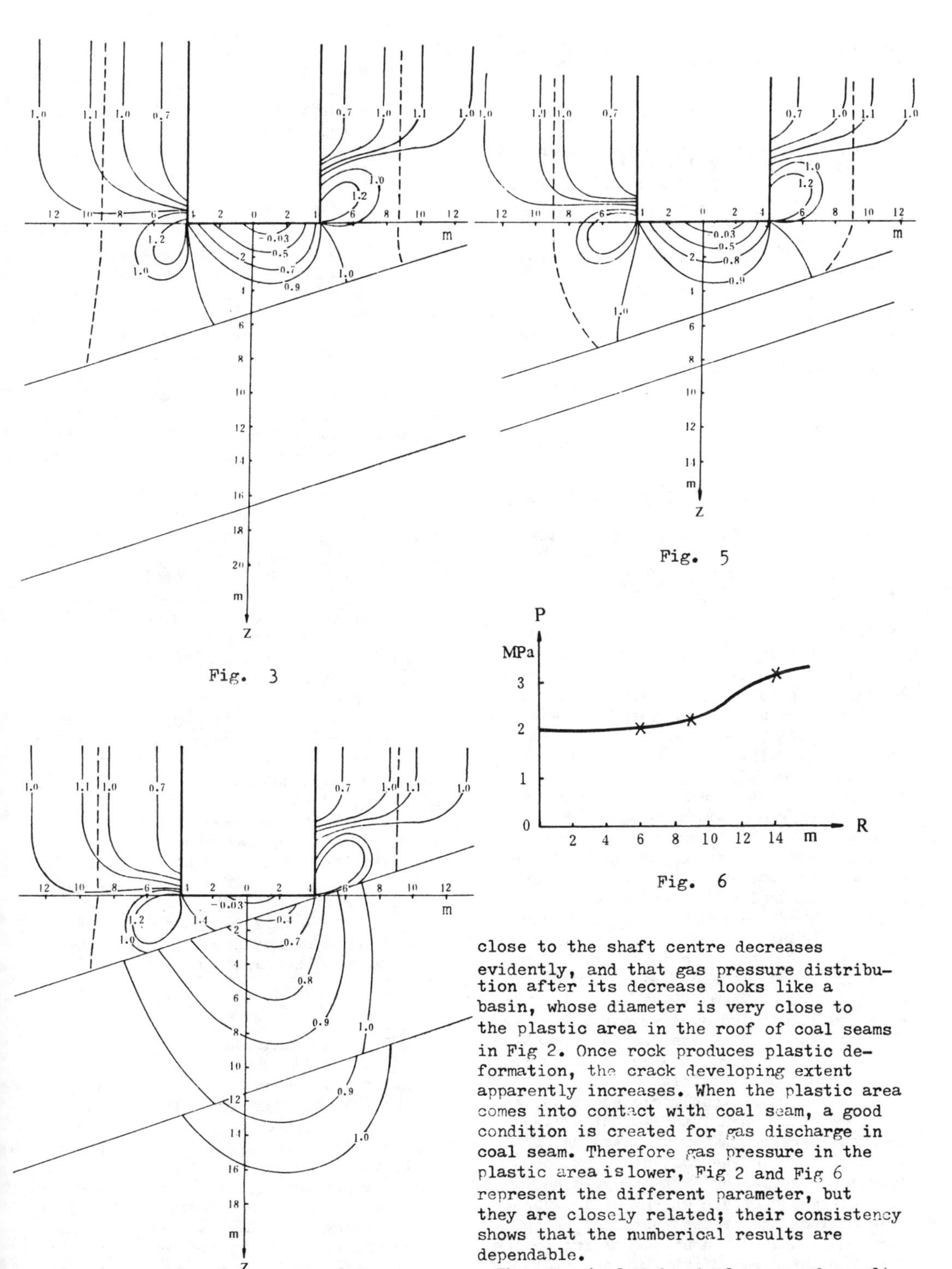

Fig. 3

Fig. 5

Fig. 6

Fig. 4

close to the shaft centre decreases evidently, and that gas pressure distribution after its decrease looks like a basin, whose diameter is very close to the plastic area in the roof of coal seams in Fig 2. Once rock produces plastic deformation, the crack developing extent apparently increases. When the plastic area comes into contact with coal seam, a good condition is created for gas discharge in coal seam. Therefore gas pressure in the plastic area is lower, Fig 2 and Fig 6 represent the different parameter, but they are closely related; their consistency shows that the numberical results are dependable.

The numerical and actual measured results of the relative displacement in the roof and floor of coal seam 8 are shown in

Table 2, and its data prove that the numerical results coincide with the measured results.

Table 2 The relative displacement in the roof and floor of coal seam 8

Item	Calculation	Observation
Relative roof heave in driving	1×10^{-4}	$0.4\times2\times10^{-4}$
Relative deformation of coal seam in driving	$<1\times10^{-4}$	$0.5-1.3\times10^{-4}$
After borehole being drilled and gas discharge		$0.5-3.1\times10^{-4}$

4 RELATIONSHIP BETWEEN BURST AND STRESS

When coal seam 7 (1.5 m thick) of LuLing Coliery was first disclosed in December 1963, gas suddenly burst from seam and kept burning for 16 hours. When coal seam 8 (10 m thick) was disclosed in February 1964, the coal seam suddenly heaved up for 3 m, 119 ton in weight. In order to prevent such accident, scientific analyses have to be made. According to the above numerical results, when a shaft is very far away from coal seam, the stress in the area below the working face 18 m begins to decrease gradually (Fig. 1), i.e, energy has already been released. Therefore the strain energy of rock mass near the heading remains little and there will occur no breaks. When the heading of the shaft is approaching coal seam, the grade of the maxium principal stress on the central-axial line of the shaft increases gradually (Fig. 2 and 3) because of the obstruction of coal seam. This means energy grade will increase. It shold be, especially, noted that coal seam always keep virgin stresses in rock mass, i.e, strain energy in rock mass. Though gas pressure somehow decreases, it occurs only in part close to the roof. Before stress release, microholes are in a tightly closed state, gas in microholes can not sufficiently discharge. Therefore, when the heading of a shaft is approaching coal seam, it does approach a dangerous high energy zone, where gas energy and strain energy will suddenly release when they reach the limit, bringing about coal and gas burst. According to the variation of stress distribution, it is possible to explain the cause for accidents. Thus appropriate preventive measures can be worked out bassed on the numerical results of stresses. The detailed analyses for mechanism of coal and gas burst is not the subject of the present paper, and you may refer to other papers by Hua Anzeng, coauthor of the present paper.

5 ROCK CAP AND RELEASE AREA

According to the above analyses and our experience of burst prevention and control in crosscut, accidents can be prevented if strain energy and gas energy in coal seams are released evenly and gradually before coal seams are disclosed. In Fig 4. Stress in coal seam has already decreased, i.e, strain energy has suddenly released on itself. In Fig 3 stress has not decreased. Hence, the heading of a shaft is decided to be retained at the location of Fig 3. So that we can adopt preventive measures. The rock mass which serves as the protection between the heading and coal seam is known as a rock cap. According to the calculations, Fig 3 gives the thinest rock cap under LuLing conditions; the radius of roluntary stress release areas is about 7 m when the thickness of coal seam is less than 2 m; while the radius is twice the shaft radius when the thickness of coal seam is more than 2 m, as shown in Fig 4, at the location where the value of equal stress curve is 1. If the artifical measures are adopted to release stress in advance, stress in coal seams will not suddenly release either, and the accidents are avoided. For safety sake, it's better to have larger artificial releasing areas when the thickness of coal seam is over 2 m. As shown in Fig 2, 3 and 4, the stress affecting radius surrouding a shaft is about 35 times the shaft's radius. Thus, it is not necessary that the maxium artificial releasing stress areas exceed the shaft's radius 3.5– 4 times.

6 SHAFT'S OWN RELEASING EFFECT

The first seam prone to coal and gas burst that LuLing new auxiliary shaft encountered is coal seam 7 which is 1.6 m thick. In accordance with the preceeding set principle, it is requested that the release area be 7 m. According to Fig 5, the projecting radius of intersection between plastic area boundary and coal seam roof is 7 m. Though stress in coal seam covered by the plastic rock area does not reduce at first, it will become loose gradually so as to release strain energy slowly with the creep of plastic rock, while gas energy may be released as well. When coal seam is not thick, we should make full use of the shaft's own releasing effect and regard the shaft as the extracting protective seam of coal seams prone to coal and gas burst. The shaft is stoped at the location as in Fig 5,as the cracks in sandstone roof of coal seam 7 are quite developed, in 7 days the safe state has been reached after the all-round inspection and identification. Then coal seam 7 has been passed safely by means of normal driving method(The measures such as concussion blasting are not used) without abnormal phenomena. All this proves that our expection is correct.

7 ENERGY RELEASE THROUGH BOREHOLE

The second coal and gas burst prone seam that LuLing new auxiliary shaft encountered is coal seam 8 which is 11 m thick. According to the above principle, it is requested that the release area be 14 m.

The shaft's own releasing area can not meet the demand. Besides the coal seam is too thick and the energy near the floor is difficult to release slowly. So the borehole releasing method is adopted to release strain energy and gas energy. First the working face of a shaft is stopped at the location as in Fig 3, then boreholes are drilled in the demanded releasing area around the shaft, the diameter of a borehole is 100 mm, the distance between boreholes(measured by the distance between intersections of a borehole and coal seam's interface) is 1.3 m, totalling 177 boreholes(including observation boreholes). The amount of coal extracted is 2/1000 of the release area. With the high creep rate and big creep deformation of burst prone coal seams, a borehole may be crowded tightly due to the volume deformation in a short time of 1-2 days. Magnitude of strain in volume is basicly equal to that of coal extracted from boreholes under the conditions that the roof and floor of coal seams produce certain deformation, volume strain is over 2 times the relative approximation of the roof and floor, strain energy may be reduced to the safe extent. As the relative approximation of the roof and floor is linear strain, linear strain $\mathcal{E}_a$ and volume strain $\mathcal{E}_v$ have the following relationship: $\mathcal{E}_v=(1+2\nu)\mathcal{E}_a$, possion's ratio ν is considered to be equal to 0.5 when the coal seam is in a plastic state. If $\mathcal{E}_v=2\mathcal{E}_a$, the approximation of the roof and floor is in keeping with volume strain of the coal seam, and strain energy in the coal seam does not change. If $\mathcal{E}_v>2\mathcal{E}_a$ the coal seam will become loose due to the decrease in volume, causing strain energy to release in certain extent. At the same time, the cracks in coal seams are extended, coefficient of ventilation increase and the flow of gas raised, resulting in gas discharge through boreholes and rock cracks, and reduction of gas energy, so that the danger of coal and gas burst are avoid.

The relative deformation of the roof and floor in Table 2 is an order less than volume strain (2×10^{-3}) of coal seams, which shows that the extracted coal magnitude is enough for strain energy to be released The examination of gas parameter also proves that the safety state has already been reached, coal seam 8 has also been passed safely by means of normal chriving method. The third seam prone to coal and gas burst is coal seam 9 which is 3 m thick, and between coal seam 8 and 9 there is only a rock stratum of 3 m, thus, the both seams are drilled through at the same time, and their energy released simultaneously, resulting in safe pass of coal seam 9. The facts prove that, whether in crosscut or in shaft, the method of energy release through boreholes can be adopted for thick seams prone to coal and gas burst to be passed through safely. It is unique at present in china that in a shaft downword boreholes are drilled passing through the thick seam of 11 m.

8 CONCLUSION

1. When a shaft approaches coal seams stain energy in coal seams does not decrease. Onee coal seams are disclosed by a shaft, a quantity of strain energy and gas energy suddenly release, resulting in coal and gas burst.
2. For the sake of safe driving, strain energy and gas energy should be released evenly and slowly, and the radius of releasing area is 2-4 times that of a shaft.
3. When the thickness of coal seams is less than 2 m, the shaft's own releasing effect can be used and the safe state can be reached in a certain period of time.
4. When the thickness of coal seams is more 2 m, it is better to extract(1-3)/1000 amount of coal in release area through evenly scattered boreholes and the safe state can be reached after a short period of gas discharge.

The correct numerical method provides a good means for the above conculations. Both the field measurements and the practice in production prove that the theory is correct, the parameter suitable and the measure efficient.

Numerical Methods in Geomechanics (Innsbruck 1988), Swoboda (ed.)
© 1988 Balkema, Rotterdam. ISBN 90 6191 809 X

A new approach to borehole stability based on bifurcation theory

J.Sulem
Ecole Nationale des Ponts et Chaussées, Paris, France
I.Vardoulakis
University of Minnesota, Minneapolis, USA

ABSTRACT: In this paper a simple rigid-plastic, pressure sensitive constitutive model for dilatant rock is presented and applied to predict the various failure modes of dee boreholes. This is done by using a simplified bifurcation analysis for both surface instabilities and shear-band formation at the borehole wall. The theory is illustrated by a computational example that is based on existing experimental results.

INTRODUCTION

Borehole breakouts and exfoliations are important phenomena that influence the engineering design of drilling hardware and can become critical for the progress of the drilling process. Breakouts lead in general to progressive deterioration of the borehole. Wellbore breakouts are attributed to the existence of significant deviatoric stresses that act in the horizontal plane at great depth and to the stress concentration around the borehole (Bell and Gough 1979). It should be noticed, however, that not only the stress deviator but also temperature and pore-fluid pressure gradient influence borehole stability.

Most of the existing work on borehole stability is based on elastic-perfectly plastic models that are calibrated on test data taken from conventional triaxial compression experiments. Rupture is assumed to occur when the stresses reach the elastic limit that is usually set at the peak of the experimental stress-strain curve; see Cheatham (1984). Despite the fact that this procedure is widely used in the design of underground structures, it is widely viewed as being insufficient to quantitatively predict failure and to describe various failure modes observed in underground excavations as well as in instrumented hollow cylinder tests. The assumption of linear elastic behavior up to failure leads to excessive stress concentrations at the borehole wall. Cases of abnormal stability as compared to the theoretical stress concentration given by linear elasticity are frequently observed in situ or in hollow cylinder experiments. This point of view is sheared in Guenot's (1987) recent review on experimental results from hollow-cylinder test and for a great variety of materials (sandstone, limestone, coal, marble, granite, etc.). It was found that with the assumption of linear elastic behavior the critical load for which failure must occur at the borehole wall is usually underestimated by a factor 2 to 8. Maury (1987) on the other hand has given recently several examples of in situ abnormal stability cases as compared to predictions arising from linear elasticity.

A major drawback of the classical procedure is its inadequacy to describe some surface rupture modes usually referred to as 'axial cleavage fracture' (Gramberg and Roest 1984) or 'extension rupture' (Maury 1987). This deficiency is related to the ad hoc assumption that failure is an intrinsical property of the material that should naturally be associated to the elastic-plastic limit. An alternative way to describe rupture phenomena in rocks is presented here by means of equilibrium bifurcation and a more realistic constitutive modeling. This approach allows to differentiate between the rheological behavior of the material and the rupture phenomenon. Furthermore bifurcation theory can by used to describe and predict the occurrence of the various observed failure modes.

MATERIAL BEHAVIOR

As already mentioned, linear elastic theory is not sufficient to describe realistically the behavior of rocks. Various phenomena like pressure sensitivity, strain hardening, non-linear volumetric strains even in the so-called 'linear' part of the stress-strain curve, must be taken into account. In this paper we will refer to a constitutive model that was recently proposed by Vardoulakis et al. (1988) for a rigid-plastic, strain hardening, pressure sensitive and dilatant material. One should keep in mind that the quantitative results from any bifurcation analysis depend very strongly on the constitutive law. As it is shown however by Vardoulakis et al. (1988), the application of the present model to real cases was in good agreement with experimental and field observations. It should be mentioned also that rigid-plastic behavior is not an essential restriction and that similar analysis can be done for elastoplastic material.

Finite Constitutive Equations

Let σ_{ij} and ε_{ij} be the stress and strain tensors, which are decomposed into deviatoric and into spherical part as follows:

$$\sigma_{ij} = s_{ij} + \sigma_{kk}\,\delta_{ij}/3$$
$$\varepsilon_{ij} = e_{ij} + \varepsilon_{kk}\,\delta_{ij}/3\ . \tag{1}$$

Let also

$$p = \sigma_{kk}/3 \quad , \ \tau = (s_{ij}s_{ij}/2)^{1/2}$$
$$\varepsilon = \varepsilon_{kk} \quad , \ \gamma = (2e_{ij}e_{ij})^{1/2} \tag{2}$$

denote the mean pressure, the shearing stress intensity, the volumetric strain and the shearing strain intensity, respectively.

For a cohesive-frictional material with a Mohr-Coulomb type yield surface F the state of (plastic) strain hardening is described by the following condition:

$$F = \tau/(q - p) - \mu(\gamma) = 0\ , \tag{3}$$

where q is a constant related to the cohesion of the material and μ is the so-called mobilized friction function.

According to Vardoulakis and Mühlhaus (1986) the finite constitutive equations of a deformation theory of plasticity for rigid-plastic, cohesive-frictional, strain-hardening and dilatant materials are defined as follows:

$$e_{ij} = \frac{1}{2h_s}\ \frac{s_{ij}}{q - p} \tag{4}$$

$$\varepsilon = D(\gamma)\ . \tag{5}$$

Equation (4) describes the deviatoric behavior of the rock whereas equation (5) provides a kinematic constraint for the volumetric strains. The material function $h_s(\gamma)$ is identified as a secant modulus to an appropriate stress obliquity-shear strain curve.

Incremental Constitutive Equations

Constitutive equations for the stress increments can be derived from equations (4) and (5) through differentiation. For plane-strain deformations and in the coordinate system of principal axes of initial stress, the components of the incremental Jaumann stress tensor $\delta\sigma_{ij}$ are related to the infinitesimal strain tensor $\Delta\varepsilon_{ij}$ through the following equations (Vardoulakis et al. 1988):

$$\delta\sigma_{11} = (1-f_1)\delta p + C_{11}\Delta\varepsilon_{11} + C_{12}\Delta\varepsilon_{22}$$
$$\delta\sigma_{22} = (1+f_2)\delta p + C_{21}\Delta\varepsilon_{11} + C_{22}\Delta\varepsilon_{22}$$
$$\delta\sigma_{33} = (1+f_3)\delta p + C_{31}\Delta\varepsilon_{11} + C_{32}\Delta\varepsilon_{22} \tag{6}$$
$$\delta\sigma_{12} = 2G_s\Delta\varepsilon_{12}$$

$$\Delta\varepsilon_{11} + \delta^2\Delta\varepsilon_{22} = 0\ , \tag{7}$$

where

$$f_1 = 2\mu\cos\alpha/\surd 3$$
$$f_2 = 2\mu\cos(\pi/3-\alpha)/\surd 3 \tag{8}$$
$$f_3 = 2\mu\cos(\pi/3+\alpha)/\surd 3$$

and

$$C_{11} = 2\{2G_s - 2(G_s - G_t)\cos^2\alpha\}/3$$
$$C_{12} = 2\{-G_s + 2(G_s - G_t)\cos\alpha\cos(\pi/3-\alpha)\}/3$$
$$C_{21} = C_{12} \tag{9}$$
$$C_{22} = 2\{2G_s - 2(G_s - G_t)\cos^2(\pi/3-\alpha)\}/3$$
$$C_{31} = 2\{-G_s + 2(G_s - G_t)\cos(\pi/3+\alpha)\cos\alpha\}/3$$
$$C_{32} = 2\{-G_s - 2(G_s - G_t)\cos(\pi/3+\alpha)\cos(\pi/3-\alpha)\}/3$$

α is the orientation angle of the deviatoric stress vector in stress space:

$$\cos 3\alpha = -\sqrt{6}\,\frac{s_{ij}s_{jk}s_{ki}}{(s_{mn}s_{nm})^{3/2}} . \qquad (10)$$

G_s and G_t are secant and tangent shear moduli,

$$G_s = (q-p)h_s \; ; \; h_s = \mu/\gamma$$
$$G_t = (q-p)h_t \; ; \; h_t = d\mu/d\gamma \; , \qquad (11)$$

and δ^2 is a dilantancy parameter,

$$\delta^2 = \frac{\sqrt{3} + 2\,\beta\,\cos(\pi/3-\alpha)}{\sqrt{3} - 2\,\beta\,\cos\alpha} \qquad (12)$$

with

$$\beta = dD/d\gamma \qquad (13)$$

being the mobilized dilatancy function.

BOREHOLE STABILITY ANALYSIS

Problem Statement

We consider a borehole with radius r_o in a rock formation and we are interested in a deep section of it. Under these conditions we may assume that any deformation of the rock is taking place in a plane normal to the borehole axis. We are dealing thus with a plane-strain problem and all mechanical quantities are described with respect to a single, fixed-in-space cylindrical coordinate system $(x_i)=(r,\theta,z)$ with the z-axis coinciding with the borehole axis. In addition, we assume that the far field is isotropic, characterized by the stress σ_∞ at infinity. The drilling process is simulated by a gradual reduction of the support pressure σ_o at the borehole wall. For a given value of the support pressure ($\sigma_o \leq \sigma_\infty$) the corresponding configuration of the borehole is denoted by C. As soon as (σ_∞-σ_o) is small as compared to the uniaxial compression strength σ_c of the rock, a small reduction of σ_o at configuration C will result to a uniquely determined uniform closure of the borehole. In this case the deformation of the borehole is stable and is described by a (trivial) axisymmetric displacement vector field:

$$\mathring{u}_r = -u_{ro} \; , \; \mathring{u}_\theta = 0 \; . \qquad (14)$$

The borehole is assumed to be unstable at C as soon as in addition to the above trivial solution $\mathring{u}_i$ another non-trivial solution $\hat{u}_i$ exists that fulfills homogeneous boundary conditions. In this case an equilibrium bifurcation is said to

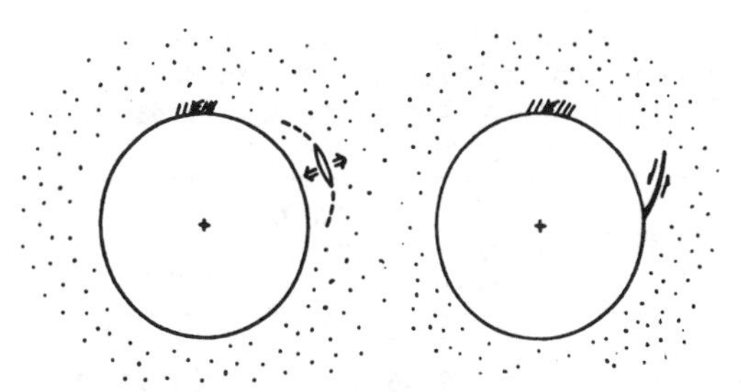

Figure 1. Borehole failure modes: (a) exfoliation; (b) shear banding

be taking place at C. The corresponding bifurcation mode is then a linear combination of the trivial and of the non-trivial mode:

$$u_i = c_1\mathring{u}_i + c_2\hat{u}_i \; . \qquad (15)$$

One possibility for non-axisymmetric deformation is warping of the borehole that will open and activate surface parallel microcracks (Figure 1a) and eventually lead to spalling failure. The physical components of the warping mode are given in terms of two unknown amplitude functions of the dimensionless radius $\rho=r/r_o$:

$$\hat{u}_r=\hat{u}(\rho)\cos(m\theta) \; , \; \hat{u}_\theta=\hat{v}(\rho)\sin(m\theta). \qquad (16)$$

With increasing modal number (m=1,2..) the wavelength of the corresponding warping mode decreases. If warping instabilities are not possible then shear-banding is assumed to be the dominant failure mode (Figure 1b).

Simplified Bifurcation Analysis

Vardoulakis and Papanastasiou (1988) discussed recently warping bifurcations in deep boreholes by using two different numerical techniques: (a) a Finite Element Method, similar to the one presented by Bassani et al. (1980) who studied bifurcations of spherical cavities in an infinite elastoplastic medium, and (b) a semi-analytical Transfer Matrix Method based on the eigenfunctions solution that is holding for a finite ring under constant state of friction hardening (Graf 1984). The solution of the problem consists in determining the lowest stress at infinity σ_∞ that for a given modal number m will cause warping of the borehole. Table 1 summarizes the comparison between the two numerical techniques for the case of a friction hardening and incompressible material with a linear stress-strain curve (ϕ_c=33°).

Table 1. Comparison between the Finite Element and the Transfer Matrix Method (Vardoulakis and Papanastasiou 1988)

m	σ_∞/σ_c	
	F.E.M.	T.M.M.
10	9.9867	9.9907
15	1.2266	1.2272
20	1.0625	1.0629
50	0.8498	0.8502
100	0.7927	0.7932
1000	0.7533	0.7457
∞(*)	0.7406	

(*) solution corresponding to surface instability.

With increasing modal number m the critical bifurcation stress decreases and the corresponding eigendisplacement field gradually localizes close to the borehole wall. This finding suggests that the critical bifurcation stress corresponds to the short wavelength limit ($m \rightarrow \infty$) that is affecting a vanishingly narrow ring of the material in the vicinity of the borehole wall. Consequently, for sufficiently large m one can restrict the analysis to a small domain (d) in the neighborhood of the borehole wall (Figure 2) and neglect the stress-gradient. For the same reason domain (d) can be replaced by a half-space of material loaded by the boundary stresses σ_{ro} and $\sigma_{\theta o}$ that may cause plane-strain surface instabilities. If surface instabilities are not possible than the dominant failure mode is shear-band formation at the borehole wall.

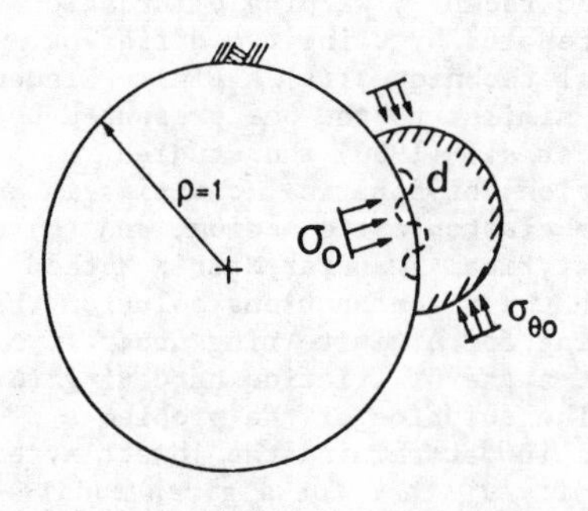

Figure 2. Surface instability at the borehole wall

Shear-Band Formation

Shear-band bifurcation analysis has been explained in many recent papers; e.g. Darve (1983) and Vardoulakis (1984). By using the incremental constitutive equations (6) and (7), the condition for localization of the deformation into a shear-band can be derived from the so-called geometrical and statical compatibility conditions; cf. Vardoulakis 1988). This procedure yields finally into the well known form of the characteristic equation for the orientation of the shear-band (Figure 3):

$$a \tan^4\theta + b \tan^2\theta + c = 0 \qquad (17)$$

where

$$a = G_s+t \; ; \; c = \lambda^2\delta^2(G_s-t)$$

$$b = C_{22}+\lambda^2\delta^2 C_{11} - \lambda^2(C_{12}+G_s+t)+\delta^2(C_{21}+G_s-t) \qquad (18)$$

$$\lambda^2 = \frac{1+f_2}{1-f_1} = \tan^2(45°+\phi_m/2) \qquad (19)$$

$$t = (\sigma_{ro}-\sigma_{\theta o})/2 \; . \qquad (20)$$

In equation (19) ϕ_m is the so-called mobilized friction angle which is related to the friction coefficient μ through the following equation:

$$\sin\phi_m = \frac{\sqrt{3}\,\mu\cos(\pi/6-\alpha)}{\sqrt{3} - \mu\sin(\pi/6-\alpha)} \; . \qquad (21)$$

The shear-band bifurcation condition is derived from the requirement that the characteristic equation (17) has real solutions. This condition is firstly met at a state C_B for which

$$b/a < 0 \text{ and } D = b^2 - 4ac = 0 \; . \qquad (22)$$

Above condition can be solved numerically in order to determine the critical bifurcation strain γ_B, for which shear banding is possible. For this state the computed shear band inclination angle is:

$$\theta_B = \pm \arctan\left[\lambda^2\delta^2\,\frac{1-t/G_s}{1+t/G_s}\right]^{1/4} . \qquad (23)$$

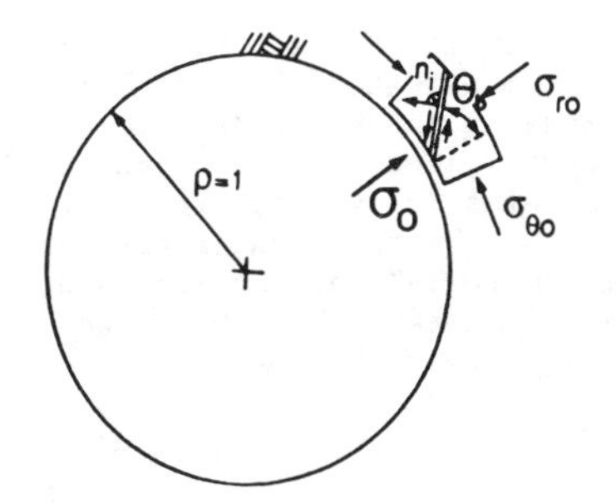

Figure 3. Shear-band formation at the borehole wall

Surface Instabilities

Surface instabilities have been first discussed by Biot (1965) and have been related to rock slabing and exfoliation phenomena by Bazant (1967), Vardoulakis (1984) and Vardoulakis and Mühlhaus (1986). For the considered constitutive equations the bifurcation condition for plane-strain surface instabilities reads according to Vardoulakis et al. (1988) as follows:

$$P[\surd(c/a)-\delta^2] + \delta^2[\surd(c/a)+b/a]+c/a = 0 \ , \qquad (24)$$

where

$$P=C_{22}+\delta^2(C_{11}-C_{12}-C_{21}-G_s+t) \ . \qquad (25)$$

Solutions of the above bifurcation condition in terms of the critical strain γ_S for which surface instability is possible are sought in the elliptic regime of the governing differential equations ($D<0$ or $D>0$ with $b/a>0$ and $c/a>0$; see Vardoulakis 1984).

APPLICATION OF THE THEORY

An experimental program with uniaxial and triaxial compression tests and thick wall cylinder tests has been performed recently by Santarelli (1987) in order to study the mechanical behavior of a carboniferous sandstone. Data from the uniaxial compression tests have been used here in order to calibrate the present constitutive model. In Figure 4 the experimental results are plotted in terms of shearing stress intensity vs shearing strain intensity, $\tau=T(\gamma)$, and have been fitted by the following function:

$$T = c_1\gamma^{c_2}\exp(-c_3\gamma) \qquad (26)$$

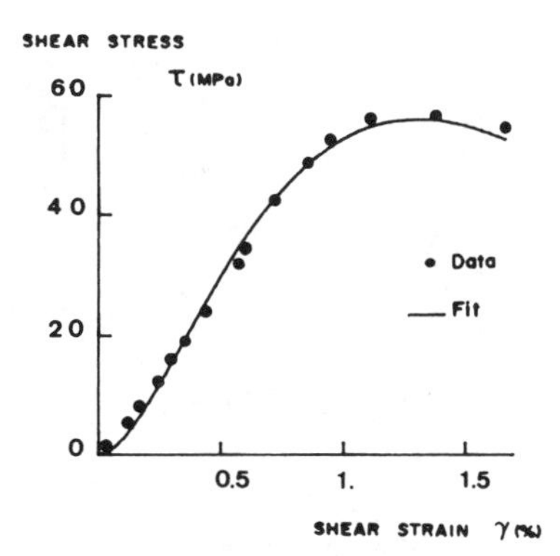

Figure 4. Stress-strain curve, $\tau=T(\gamma)$; Santarelli (1987)

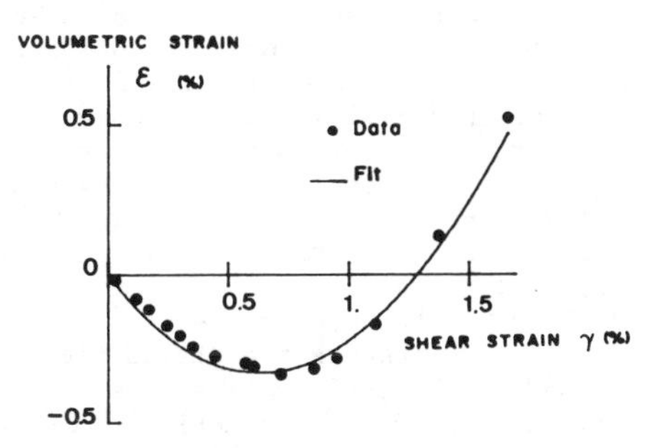

Figure 5. Volumetric strain-shear strain curve, $\varepsilon=D(\gamma)$; Santarelli (1987)

where c_1=1.098603E6 MPa, c_2=1.85, c_3=1.4231E2

The experimental results for volumetric strain $\varepsilon=D(\gamma)$ were approximated by a function:

$$D = c_4\gamma - \ln(1 + c_5\gamma) \qquad (27)$$

where c_4=12.69, c_5=13.76 (Figure 5).

A shear-band bifurcation analysis of the uniaxial compression test has been performed in order to determine the uniaxial compression strength σ_c and the mobilized friction angle ϕ_c at shear failure. It was found that σ_c=88.7 MPa and ϕ_c=33.1°. With these values the parameter q that is appearing in the constitutive equations could be computed from the following expression:

$$q = \sigma_c \cot\phi_c \cot(45°+\phi_c/2)/2 \ , \qquad (28)$$

resulting to q=36.8 MPa.

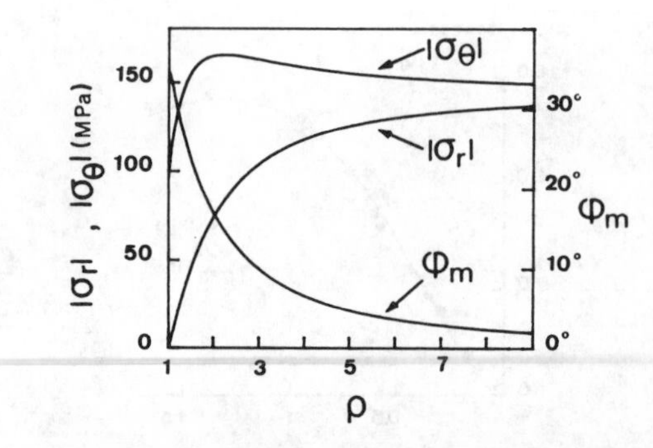

Figure 6. Stresses and mobilized friction angle distribution in the vicinity of a borehole in carboniferous sandstone at the state of surface instability

A bifurcation analysis of the borehole problem showed that for the considered material the dominant failure mode is surface instability. Surface buckling is predicted for γ_S=1.2% i.e. in hardening regime of the assumed stress-strain curve $\tau=T(\gamma)$. The value of the mobilized friction angle at the borehole wall for plane-strain surface instabilities is ϕ_S=34.5°. With these values it is then possible to compute the stress field around the borehole wall at the bifurcation state (a numerical integration of the equilibrium equation is given in Vardoulakis et al. 1988). In particular for an unsupported borehole (σ_{ro}=0) the computed critical hoop stress is $\sigma^S_{\theta o}$=96.4 MPa, and the corresponding critical stress at infinity $\sigma^B_{\infty,cr}$=143 MPa (Figure 6).

In thick-walled cylinder tests with the same rock Santarelli (1987) found that the failure mode is indeed spalling of the borehole wall and that the critical external stress that corresponds to internal failure is $\sigma^E_{\infty,cr}$=135 MPa. Consequently, the critical value of the stress at infinity and the failure mode obtained within the frame of the present theory is in good agreement with experimental observations. Similar results, corroborating the present theory, have been presented by Vardoulakis et al. (1988), who found that for a different rock shear-banding was the dominant failure mode.

It should be emphasized that linear elastic theory would give $\sigma^{el}_{\theta o}=2\sigma^E_{\infty,cr}$=270 MPa. Considering the fact that the uniaxial compressive strength of this rock is about 90 MPa, the result from a linear elastic analysis is paradoxical. Present analysis demonstrated that in reality the hoop stress at the borehole wall is significantly lower than the one predicted by linear elasticity. This result is justified by the fact that a pressure sensitive materials are stiffer and consequently more stable than linear elastic materials.

SCALE EFFECT

We considered here a borehole in a deep rock layer. Geometrically the considered problem has only one characteristic length, the radius r_o of the borehole. From this point of view all boreholes are geometrically similar. The dependency of the borehole stability on the radius of the borehole is usually termed as scale effect. Physical evidence and the empirical observations suggest that a scale effect does exist and that large holes in highly stressed or weak rocks are less stable than small holes.

If gravity effects are insignificant, as it is the case in deep boreholes, then the needed internal length must come from an appropriate constitutive model. By assuming a classical constitutive theory, like elasticity, elastoplasticity etc., we are introducing into the problem no material property with the dimension of length. As a result we find that the critical bifurcation mode is the infinitely small wavelength mode and that this mode affects actually an infinitesimal ring of material close to the borehole wall. Since the short wavelength limit is an accumulation point of bifurcations, almost all wavelengths of the warping mode are possible, and consequently there is again no influence of the radius. If we want to accommodate the influence of the radius of the borehole and we want to determine a finite depth for the the region in which the skin effect is taking place, then we must introduce into the material model properties that have the dimension of length. This can be done as demonstrated recently by Mühlhaus and Vardoulakis (1986 and 1987) by means of Cosserat-type constitutive models. These models contain bending stiffness moduli M, which when normalized by an elastic modulus G introduce an internal length in $(M/G)^{½}$ into the problem. The critical stress that causes surface instabilities is then found to depend on the wave length of the warping mode and the scale effect can be appropriately modelled.

ACKNOWLEDGEMENT

The authors wish to thank Société Nationale Elf Aquitaine (Production) for supporting this research.

REFERENCES

J.L.Bassani, D. Durban and J.W. Hutchinson, Bifurcations at a spherical hole in an infinite elastoplastic medium, Math Proc. Camb. Phil. Soc. 87, 339-356 (1980)
Z.P.Bazant, L' instabilité d' un milieu continu et la resistance en compression, Bulletin Rilem, 35, 99-112 (1967)
J.S.Bell and D.I.Gough, D.I., Northeast-southwest compressive stress in Alberta: Evidence from oil wells, Earth Planet Sci. Lett. 45, 475-482 (1979)
M.A.Biot, Mechanics of Incremental Deformations, Wiley, New York, 1965.
J.B.Cheatham, Wellbore Stability, J. of Petr., 889-896 (1984)
F.Darve, Rupture d'aigiles naturelles par surfaces de cisaillement à l'eau triaxial. Revue Française de Géotechnique, 23, 27-38 (1983)
B.Graf, Theoretische und experimentelle Ermittlung des Vertikaldrucks auf eingebettete Bauwerke, Dissertation, Univ. Karlsruhe, 1984
J.Gramberg and J.P.A.Roest, Cataclastic effects in rock salt laboratory and in-situ measurements, Comm. of Eur. Com., Nuclear Science and Technology, Rapport EUR 9258 EN (1984)
A.Guenot, Contraintes et ruptures autour des forages pétroliers, 6th Cong. Int. Soc. Rock Mech., Montreal, Vol. I, 109-118 (1987)
V.Maury, Observations, recherches et résultats recents sur les mécanisme de rupture autour de galenes isoleés. 6th Cong. Int. Soc. Rock Mech., Montreal, Vol. II, 1119-1128 (1987)
H.B.Mühlhaus and I.Vardoulakis, Axially symmetric buckling of the surface of a laminated half-space with bending stiffness, Mech. of Mat. 5, 109-120 (1986)
H.B.Mühlhaus and I.Vardoulakis, The thickness of shear bands in granular materials, Gèotechnique, 37, 271-283 (1987)
F.Santarelli, Theoretical and experimental investigation of the stability of the axisymmetric well bore, Ph.D. Thesis, Univ. of London, 1987
I.Vardoulakis, Rock bursting as a surface instability phenomenon, Int. J. Rock Mech. Min. Sci. and Geomech. Abstr. 21, 137-144 (1984)
I.Vardoulakis, Stability and bifurcation in geomechanics, 6th Int. Conf. on Num. Meth. in Geomechanics, Balkema, Rotterdam, 1988.
I.Vardoulakis and H.B.Mühlhaus, Local rock surface instabilities, Int. J. Rock Mech. and Min. Sci. and Geomech. Abstr. 23, 379-383 (1986)
I.Vardoulakis and P.Papanastasiou, Bifurcation analysis of deep boreholes. Int. J. Num. Anal. Meth. Geomech. in print (1988)
I.Vardoulakis, J.Sulem and A. Guenot, Borehole instabilities as bifurcation phenomena, Int. J. Rock Mech. Min. Sci. and Geomech. Abstr., submitted (1988)

Numerical Methods in Geomechanics (Innsbruck 1988), Swoboda (ed.)
© 1988 Balkema, Rotterdam. ISBN 90 6191 809 X

Parameter identification of the unified model of methane gas flow in coal

V.U.Nguyen
Department of Mining and Civil Engineering, University of Wollongong, Australia

ABSTRACT: The paper describes a new and unified theory of methane gas transport in coal. Methane gas flow in coal is characterized by strong sorptive behaviour of gas molecules to the internal surfaces of coal. The "permeability" is thus largely influenced not only by the mechanics of laminar flow (Darcy's law) and diffusion (Fick's law) but also by the rate of interchange between the adsorbed and the free gas phases. The paper also illustrates that the new model is most appropriate for solution using the finite element technique involving parallel and coupled codes, one for gas and moisture flow and the other stress-deformation analysis, like the consolidation problem. A new optimisation scheme has been devised after the simplex reflection technique to identify flow parameters and to validate the proposed model.

INTRODUCTION

Methane gas is a by-product of the coalification process. In many collieries , methane gas (CH_4) has continually posed serious hazards causing disasters such as explosions, and instantaneous outbursts of coal and gas, or fires. Drainage of methane gas has thus become a practicable measure to increase safety in many "gassy" coal mines throughout the world. Furthermore, with the recognition that proven reserves of petroleum fuels could be exhausted within the foreseeable future, increasing attention has been focused on methane gas drainage from coal seams as an economic source of natural gas production.

A complete understanding of the mechanisms underlying the flow process of methane gas in coal, essential for safe coal mining and design of ventilation and gas drainage systems, however still appears to be outstanding. The literature is not in total consensus on the mechanics of methane flow in coal (e.g. [17]), as exemplified by some models being based on Darcy's law involving flow down a pressure gradient, some on Fick's law of gas diffusion along a concentration gradient, and some involving diffusion in the micropores followed by Darcy's flow in the macropores (e.g. coal cleats). Validation of the models with laboratory test results and field data is often compounded by the fact that the flow permeability or diffusion coefficient is highly dependent on the magnitude of the local confining stress, and the functional dependence of flow parameters on stress changes induced by mining excavations is generally not suitable for adaptation to the uncoupled (or "flow-only") numerical models.

THE UNIFIED THEORY

Methane gas in coal

Methane gas present in coal exists in two states:

the free gas state and the adsorbed state. Free gas is present in the fractures and macropores of the coal, and its behaviour can be described by Boyle's law and the kinetic theory of gases. In the adsorbed state, gas molecules adhere to the large internal surface area of coal (20 to 200 m^2/g). At normal coal bed pressures, adsorbed gas occupies roughly 90 to 95% of the total gas content in coal, and there is a continual interchange of molecules between the free gas and the adsorbed gas. The content of the adsorbed gas is related to the free gas pressure in equilibrium by experimentally determined isotherms. There are many mathematical models used to describe the process of methane adsorption, or desorption in coal, and the Langmuir model, as described below, appears to be favoured by many workers (e.g. [2], [8]).

The quantity of free gas present in the macropore space of coal can be described by Boyle's law :

$$PV = ZnRT \tag{1}$$

where P is the pressure (kPa), V is the macropore volume (m^3), n is the number of moles, R is the universal gas constant (8.30438 kPa/kmol/$^\circ K$), T is the absolute temperature ($^\circ K$), and Z is the gas deviation factor defined as the ratio of volume actually occupied by a gas at a given pressure and temperature to the volume it would occupy if it behaved ideally. At temperatures existing in most mines, the variation of Z is negligible at low pressures. For methane the ideal gas equation is acceptable at pressures up to 4000 kPa, but for carbon dioxide, the ideal gas equation is inapplicable at pressures beyond 1100 kPa [10].

In terms of coal porosity Φ, and standard conditions of pressure and temperature, the quantity of free gas per unit volume of coal, $C_v (cm^3/cm^3)$ is computed by (e.g. [2]):

$$C_v = \Phi \frac{P}{P_\circ} \frac{273}{T} \tag{2}$$

in which Φ is the porosity, i.e. fraction of pore space to total coal volume, $P_\circ$ is the standard atmospheric pressure (101.3 kPa), T is the absolute Kelvin temperature.

It is widely known (e.g. [2], [8]) that the free gas makes up only a small fraction (5 to 10%) of the total gas content. The volumetric free gas content, C_V, is proportional to the porosity, Φ, which is affected by changes in in-situ stresses and the moisture content of coal or rock. Φ can vary between 0.01 to 0.15 (cm^3/cm^3) for coal, as commonly reported in the literature.

At equilibrium, a unique relationship exists between the quantity of adsorbed gas and the free gas pressure, commonly described by the Langmuir equation :

$$C_a = \frac{ABP}{1 + BP} \tag{3}$$

where C_a is the quantity of adsorbed gas at pressure P, cm^3/cm^3 of coal, or more commonly m^3/ton of coal. A and B are Langmuir's constants, in units of cm^3/cm^3 (or m^3/t) and kPa^{-1}, respectively. Note that in the case of m^3/t-unit being used, the conversion to cm^3/cm^3 is obtained by multiplication with the coal dry density (t/m^3). Figure 1 shows a typical experimentally-determined Langmuir curve for methane gas in coal.

Darcy's law for compressible systems

The relative bulk permeability of methane gas in rock is defined as:

$$V_i = \Phi\left(\frac{\delta u_{if}}{\delta t} - \frac{\delta u_i}{\delta t}\right) \tag{4}$$

in which Φ is the rock porosity, u_{if} and u_i are displacements of the fluid and rock, respectively. If V_{if} and V_{is} are used to denote velocities of fluid and solid, equation (1) can be rearranged as:

$$\Phi V_{if} = V_i + \Phi V_{is} \tag{5}$$

where V_i can be described by Darcy's law as:

$$V_i = -\frac{K}{\mu}\left[\frac{\delta P}{\delta x_i} + \rho g_i\right] \tag{6}$$

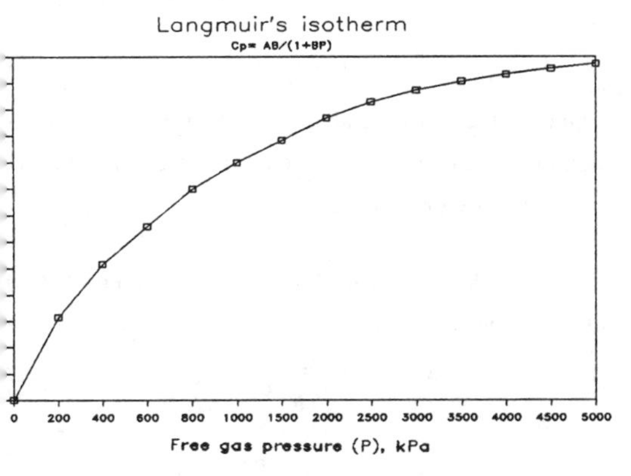

Figure 1. Typical Langmuir isotherm of methane in coal

in which K is the intrinsic permeability of the porous medium (assumed to be isotropic), μ is the fluid viscosity, P is the fluid pressure, ρ is the fluid density and g_i is the gravitational acceleration in the i-th direction.

The concentration of adsorbed gas in coal is related to the free gas pressure, at equilibrium, by an isotherm such as the Langmuir relation (equation (3)). It is thus expected that under normal flow or dynamical equilibrium, there is a continual phase interchange between adsorbed and free gas. The quantity or rate of phasal gas exchange is complex but expected to be a function of the difference between the existing free gas pressure and the gas pressure in equilibrium with the adsorbed phase.

The equation of continuity for the free gas phase can be written as:

$$-\frac{\delta}{\delta x_i}(\rho \Phi V_{if}) = \frac{\delta(\Phi\rho)}{\delta t} + \Gamma \qquad (7)$$

where ρ is the fluid (gas) density and Γ is the rate of gas transfer from the adsorbed state to the free phase.

Combining equations (5) and (7), and the continuity condition for rock solid:

$$-\frac{\delta}{\delta x_i}[(1-\Phi)\rho_s V_{is}] = \frac{\delta}{\delta t}[(1-\Phi)\rho_s] \qquad (8)$$

where ρ_s is the rock solid density,
we have, after rearranging:

$$\frac{\delta}{\delta x_i}[\rho \frac{K}{\mu}\frac{\delta P}{\delta x_i}] = \rho(\Phi\beta_f + \alpha)\frac{dP}{dt} + \rho\frac{\delta V_{is}}{\delta x_i} + \Gamma \qquad (9)$$

in which, the rock matrix compressibility is defined as:

$$\beta_s = \frac{1}{\rho_s}\frac{d\rho_s}{dP} \qquad (10)$$

the effective rock compressibility:

$$\alpha = (1-\Phi)\beta_s \qquad (11)$$

and the fluid compressibility:

$$\beta_f = \frac{1}{\rho}\frac{d\rho}{dP} \qquad (12)$$

Detailed derivation of equation (9)-(12) is given elsewhere [11].

Movement in the adsorbed phase

Concurrent with gas movement in the free phase, is surface molecular diffusion of adsorbed gas described by Fick's law:

$$q_{Fi} = -D_F\frac{\delta C_a}{\delta x_i} \qquad (13)$$

where D_F is the diffusivity of methane gas in coal having dimensions of length squared over time, C_a is the volumetric concentration of adsorbed gas (corresponding to free gas pressure P), and q_{Fi} is the surface diffusion rate per unit cross-sectional area.

The continuity equation for the adsorbed phase can be written as:

$$-\frac{\delta}{\delta x_i}[\rho q_{Fi}] = \frac{\delta}{\delta t}(\rho C_a) - \Gamma \qquad (14)$$

where ρ is the gas density.

Combining (13), (14) and the Langmuir relation (3), we obtain:

$$\frac{\delta}{\delta x_i}[\rho D_F \lambda_a \frac{\delta P}{\delta x_i}] = C_a \rho \beta_f \frac{\delta P}{\delta t} + \rho \lambda_a \frac{\delta P}{\delta t} - \Gamma \quad (15)$$

where

$$\lambda_a = \frac{\delta C_a}{\delta P} = \frac{AB}{(1 + BP)^2} \quad (16)$$

The combined gas flow equation can now be obtained by adding equations (9) and (15):

$$\frac{\delta}{\delta x_i}[\rho(D_F \lambda_a + \frac{K}{\mu})\frac{\delta P}{\delta x_i}] = [\rho(\beta_f(\Phi + C_a) + \alpha + \lambda_a)]\frac{\delta P}{\delta t} + \rho \frac{\delta V_{is}}{\delta x_i} \quad (17)$$

We can further define the effective composite permeability of methane gas in coal:

$$K_g = \rho(D_F \lambda_a \mu + K) \quad (18)$$

and total compressibility comprising gas, rock matrix compressibilities and equivalent phase interchange:

$$c_T = \rho(\beta_f(\Phi + C_a) + \alpha + \lambda_a) \quad (19)$$

The differential equation describing flow of methane gas in coal becomes:

$$\frac{\delta}{\delta x_i}(\frac{K_g}{\mu}\frac{\delta P}{\delta x_i}) = c_T \frac{\delta P}{\delta t} + \rho \frac{d\epsilon_V}{dt} \quad (20)$$

where ϵ_V is the bulk volumetric strain.

Equation (20) requires the coupling of another stress-deformation equation (see [11]) for the complete solution of the methane gas flow problem in coal mining. Its formulation as such is appropriate for numerical simulation by the finite element technique. Work is continuing to modify the coupled finite element code used in soil consolidation modelling for application to the methane problem.

PARAMETER IDENTIFICATION

The flow equation

By approximation, equation (20) can be re-written as:

$$\frac{\delta}{\delta x_i}(D \frac{\delta P}{\delta x_i}) = \frac{\delta P}{\delta t} \quad (21)$$

which is an uncoupled flow equation having a diffusion form, and

$$D = \frac{K_g}{\mu c_T} \quad (22)$$

Equation (22) is the fundamental form of this formulation and it can be used to validate the model in conjunction with a desorption test [15].

In cylindrical coordinate system, equation (22) assumes the form:

$$\frac{\delta P}{\delta t} = \frac{1}{R} \cdot \frac{\delta}{\delta R}(D.R \frac{\delta P}{\delta R}) \quad (23)$$

The mathematical model given by (23) was expanded by the Crank-Nicholson scheme for the history matching of desorption test data on coal samples [16].

Figure 2, reproduced from the published results in [16], indicates that the analytical solution to equation (23), using constant values of equivalent diffusivity D [23], gives poor history matching to experimental data. A sharp kink is observed at around 85-90%of the total gas content desorbed, in contrast with a more gentle gas desorption pattern, normally reported in the literature (e.g. [4], [17], [24]). The slope of desorption curve is a measure of the gas diffusivity or equivalent permeability K_g, and a gently changing slope with increasing time or decreasing gas pressure can thus be modelled by the permeability being a direct function of gas pressure, i.e.

$$D = D_\circ(\frac{P}{P_\circ})^q \quad (24)$$

or

$$k = k_\circ(\frac{P}{P_\circ})^q \quad (25)$$

where $k = \frac{K_g}{\mu}$, $P_\circ$ is the original seam gas pressure or sorption pressure of a laboratory coal specimen, and q is an exponent. Validation of the unified model of gas flow therefore is primarily centered on parameter identification of equation (23) (or its permeability form (20)), i.e. determination of $D_\circ$ (or $k_\circ$) and exponent q.

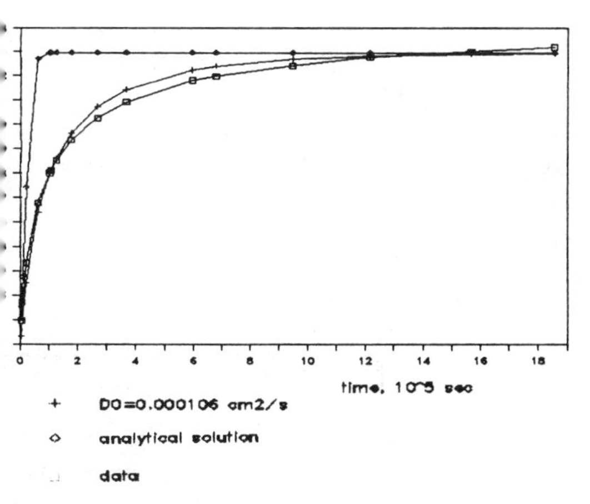

Figure 2. History matching by analytical solution [23] and by the unified model

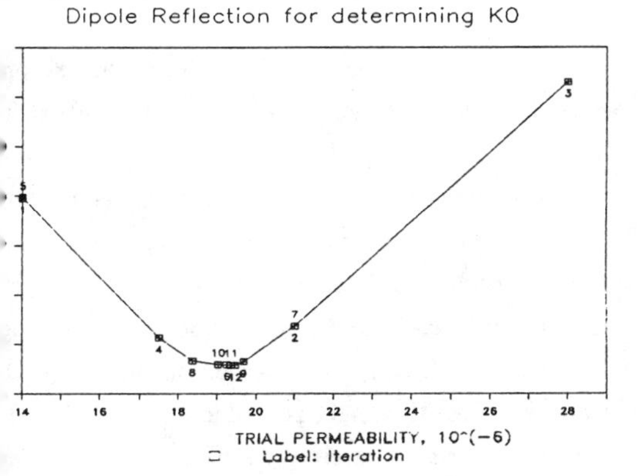

Figure 3. Illustration of the Dipole Reflection method

The Dipole Reflection method

Parameter identification , also called back analysis, currently receives much attention by the geotechnical community. It is however not a well-posed problem in the mathematical sense, since it involves not a unique but a multitude of solutions. One can differentiate two types of parameter identification approaches, direct and indirect.

The direct approach involves reformulation of the matrix system of equations, such as the potential flow equation or elastic stress-deformation equation, and the required parameters form the variable vector, e.g. vector of flow transmissivities in hydrology, of elasticily moduli in stress analysis, whilst the state parameters such as hydraulic heads and deformation or strain measurements are inverted and built into the coefficient matrix (see [26] for example).

Indirect parameter identification problems involve unknown parameters not explicitly used in the forward solution system; for example, back-figuring the shear strength parameters or like in the present problem, determining flow parameters. Solution these problems normally requires the use of optimization techniques, especially those by-passing evaluation of partial derivatives such as: Powell's method, Rosenbrock's method, the simplex reflection method [25]. Most of these optimization routines can be readily found from software libraries. It is considered that the indirect problem can be further sub-divided into two categories. One category involves the minimization of a standard error function:

$$f = \sum_{i=1}^{n} [t_i - m_i]^2 \qquad (26)$$

in which n is the number of data points, t_i represents the ith computational (or predicted) value and m_i its measured value. The optimum of the error function f is not normally zero. The other category is the back analysis of a failure which is equivalent to locating the minimum of function f (normally a limit state function), which is equal to zero. Optimizing such limit state function can be effectively carried out by the Newton-Raphson method in conjunction with a Taylor's series expansion (e.g. [28]).

It is interesting to note that parameter identification, as normally found in electrical engineering literature, involves first data or curve smoothing involving recursive relations. This is most appropriate with state variable or parameter being represented as a time series.

Curve smoothing or filtering is also used to account for the uncertainty nature of the variables and parameters. As such, its formulation tends to suppress the mechanistic nature of parameter relations. Under this principle of "filtering", and in view of the poor history matching of analytical solution to desorption data shown in Figure 1, it is considered that parameter identification of equation (20) or (23) is best carried out in two stages, the first to determine the shape of the desorption curve as represented by exponent q, and the second to determine $k_\circ$ or $D_\circ$. This approach has been found to give more meaningful results to the parameter values required in the formulation above.

A new optimization scheme has been devised to determine $D_\circ$ (or $k_\circ$) from desorption data. The scheme, hereafter called the Dipole Reflection technique, is a simplfied and single-parameter version of the simplex reflection technique, commonly used in geomechanics applications in recent years [25]. Basically the dipole reflection algorithm begins with evaluation of the objective function at two trial values of the unknown parameter. The values of the two "poles" are then compared with each other, and the one having a higher value is reflected to the other side of the other pole by the same separation distance. The function is again evaluated at the reflected pole and the process of comparing and reflecting is repeated until convergence. To avoid vacillating one pole across and back against the other, a second reflection on the same pole uses only half the separation distance in reflection. Figures 3 and 4 illustrate the dipole reflection technique used in determining $k_\circ$. The objective function used in optimization and reflection is:

$$f = \sum_{i=1}^{n} [d_{Ci} - d_{Ei}]^2 \qquad (27)$$

where d_{Ci} is the predicted desorption quantity obtained by the Crank-Nicholson scheme, and d_{Ei} is the ith experimentally determined desorption quantity (see [15], [16]).

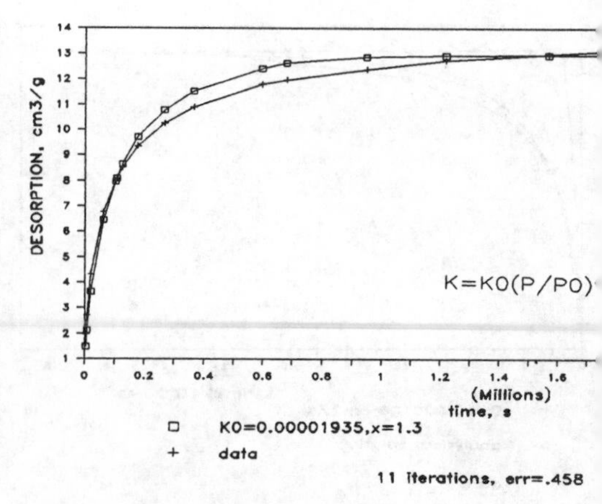

Figure 4. Equivalent permeability determination by the Dipole Reflection method

Reconciliation of the unified theory with Darcy's model

The composite permeability by definition (18) comprises both the Darcy term K and the surface diffusion term D_F. Darcy's K is determined by the permeability test involving application of a pressure differential at the two ends of a cylindrical core. In a permeability test there is virtually no interaction between adsorbed gas and free gas, whereas in a desorption test, according to the present theory, flow mechanism involves both Darcy's laminar flow in the pores and surface diffusion. Such mechanism is conceivably almost identical to that underlying methane gas flow in coal seams, in-situ. Using the results given in [16], and reproduced in Figure 4 under a different form, the Darcy permeability can be computed from (18), assuming that D_F is negligible:

$$K = \frac{K_g}{\rho} = \frac{\mu k}{\rho}$$

$$K = \frac{\mu}{\rho} k_\circ \left(\frac{P}{P_\circ}\right)^q \qquad (28)$$

in which $\mu = 1.087 \times 10^{-8}$ kPa-s, $k_\circ = 19.36 \times 10^{-6}$, $q = 1.3$, and $\rho = 0.023$ g/cm^3. Using $P = 3500$ kPa and initial sorption pressure is 4000 kPa, we have:

$$K = 0.915 mD$$

(mD is abbreviated from milliDarcy, $1D = 10^{-8}cm^2$) which is of the same order of magnitude as reported in [16], [17], and [21].

CONCLUDING REMARKS

The unified theory presented here has been focused only on the basic mechanism of methane gas flow in coal, where the gas is the only fluid present. In real-life, water and some other gases (e.g. CO_2) are also present in coal and surrounding rock strata. The coupling interaction between gas flow and stress though demonstrated in the formulation for completeness was not dealt with due to space limitation.

The proposed theory was shown to provide an excellent history matching tool for desorption test data. It is strictly based on fundamental theories of fluid flow in porous media, and has a form analogous to the soil consolidation problem rendering it most suitable for application of modern numerical techniques, such as the finite and boundary element techniques. Application of the theory in simulation analysis of field situations requires, as shown in the paper, the proper identification of parameters from basic desorption tests. The dipole reflection technique has been devised for this purpose.

The same model can also be used to determine the Fick coefficient of surface diffusion using concurrent diffusion and permeability tests. Further work is continuing.

ACKNOWLEDGEMENTS

The work was supported by the CSIRO-University Grant Scheme. It was originally an outgrowth of a project initiated by Dr. R.D. Lama, Manager (Technology and Development) of Kembla Coal and Coke ltd, to whom the writer is indebted for many fruitful discussions. Any opinions expressed in the paper however do not necessarily reflect those of others.

REFERENCES

1. Airey, E.M. (1969) Diffusion of firedamp in mine airways. The Mining Engineer, No. 100, Vol. 128, Jan. 1969.

2. Boxho, J., Stassen, P., Mucke, G., Noack, K., Jeger, C., Lescher, L., Browning, E.J., Dunmore, R., Morris, I.H. (1980) Firedamp Drainage Handbook for the Coal Mining Industry in the European Community. Verlag Gluckauf GmbH, Essen (Germany) , 415pp

3. Cupps, C., Brow, G.T. and Fry, J. (1971) Mathematical simulation of historic gas distribution by diffusion in an oil reservoir. U.S.B.M. Report of Investigations RI 7511.

4. McColl Stewart, I. (1971) Diffusional analysis of seam gas emission in coal mines. The Canadian Mining and Metallurgical Bulletin. April 1971, pp 62-70.

5. Mordecai, M. and Morris, L.H. (1974) The effects of stress on the flow of gas through coal measure strata. Mining Engineering, Inst. Min. Eng. , Vol. 133, pp 43.5-443

6. Ostensen, R.W. (1982) The effect of stress dependent permeability of gas production and well testing. Proc., 57th Tech. Conf., Soc.

7. McKee, C.R. and Bumb, A. (1983) Pressure response of a stress sensitive aquifer. Topical report, Contract No. 5081-214-0729, Gas Research Institute, 8600 West Bryn Mawr Ave., Chicago, IL 60631.

8. Curl, S.J. (1978) Methane prediction in coal mines. IEA Coal Research, London, December 1978, 80 pp.

9. Hargraves, A.J. (1984) Particular gas problems of Australian deep coal mining. Proc. 3rd Int. Mine Ventilation Congress, Harrogate (UK), pp 127-133.

10. Lama, R.D. and Bartosiewiccz, H. (1982) Determination of gas content of coal seams, IN : Seam gas drainage with particular reference to the working seam (editor : A.J. Hargraves). Aus. I.M.M. Symposium (Wollongong) , pp 36-52

11. Nguyen, V.U. (1988) On the unified theory of methane gas transport in coal. (in preparation)

12. De Wiest, R.J. (1969) Flow through porous media. New York, Academic Press.

13. Giron, A., Pavone, A.M., and Schwerer, F.C. (1984) Mathematical models for production of methane and water from coal seams. Quarterly Review of Methane from Coal Seams Technology (published by The Gas Research Institute U.S.A.), Vol. 1, No. 4, March 1984 , pp 19-34.

14. Klinkenberg, L.J. (1941) The permeability of porous media to liquids and gases. API Drill Prod. Pract., pp 200-213.

15. Vutukuri, V.S. and Lama, R.D. (1986) Environmental engineering in Mines. Cambridge University Press.

16. Lama, R.D. and Nguyen, V.U. (1987) A model for determination of methane flow parameters in coal from desorption tests. APCOM 87 Conference, Johannesburg (South Africa).

17. Gray, I. (1987) Reservoir engineering in Coal seams: Part 1 - The physical process of gas storage and movement in coal seams. SPE Reservoir Engineering, February 1987, pp 28-34 . Part 2 - Observations of gas movement in coal seams. SPE Reservoir Engineering, Feb. 1987, pp 35-87.

18. Airey, E.M. (1968) Gas emission from broken coal. An experimen- tal and theoretical investigation. Int. Jour. Rock Mech. & Mining Eng. Sci., Vol. 5, pp 475-494

19. Kissell, F.N. and Bielecki, R.J. (1972) An in-situ diffusion parameter for the Pittsburg and Pocahontas No.3 coal-beds, U.S.B.M. Report of Investigations RI 7668.

20. Bielicki, R.J., Perkins, J.H. and Kissell, F.N. (1972) Methane diffusion parameters for sized coal particles. A measuring apparatus and some preliminary results. U.S.B.M. Report of Investigations RI 7697.

21. Hemala, M.L., Chapman, G. and Reichman, J. (1982) Mathematical model for simulation of seam gas drainage. IN : Seam gas drainage with particular reference to the working seam (editor : A.J. Hargraves). Aus. I.M.M. Symposium (Wollongong), pp , 53-61.

22. Jones, A.H., Ahmed, U., Abou-Sayed, A.S., Mahyera, A., and Sakashita , B. (1982) Fractured vertical wells versus horizontal boreholes for methane drainage in

advance of mining U.S. coals. IN : Seam gas drainage with particular reference to the working seam (editor : A.J. Hargraves). Aus. I.M.M. Symposium (Wollongong), pp , 172-201.

23. Crank, J. (1956) The Mathematics of Diffusion. London, Oxford University Press.

24. Higuchi, K., Ohga, K. and Isobe, T. (1982) Considerations to increase gas drainage ratio basing upon methane drainage practice in Japan. IN: Seam Gas Drainage with Particular Reference to the Working Seam (editor: A.J. Hargraves), Aust. I.M.M. Symp. (Wollongong), pp 232-241

25. Nguyen, V.U. (1985) New Applications of Classical Numerical Techniques in Slope Analysis. Proc. 5th Int. Conf. on Numerical Methods in Geomechanics, Vol. 2, pp 949-958, (Nagoya, Japan).

26. Gioda, G. and Jurina, L. (1981) Numerical identification of soil structure interaction pressures. Int. Jour. Numer. Anal. Methods in Geomechanics. Vol. 5, pp 33-56.

27. Hargraves, A.J. (1983) Instantaneous outbursts of coal and gas - A review. Proc. Australasian Inst. Min. Metall. No. 285 , pp 1-37.

28. Nguyen, V.U. (1985) Reliability index in geotechnics. Computers and Geotechnics, Vol. 1, No. 2, pp 117-138

29. Smith, D.M. and Williams, F.L. (1982) Diffusional effects in the recovery of methane from coal beds. SPE-DOE paper 10821. Symp. Unconventional gas recovery. Pittsburg, PA, May 16-18.

30. Kissell, F.N. and Deul, M. (1974) Effect of coal breakage on methane emission. Trans. Soc. of Min. Engineers AIME, Vol. 256, , June, 182-185.

Numerical Methods in Geomechanics (Innsbruck 1988), Swoboda (ed.)
© 1988 Balkema, Rotterdam. ISBN 90 6191 809 X

A technique and program for computing stress-strain state of mining system elements and mining openings in solid and fractured rock mass

N.P.Vlokh & O.V.Zoteyev
Institute of Mining, Sverdlovsk, USSR

ABSTRACT: In this paper application of finite element method for computing the stress-strain state of rock masses around mining openings is discussed. Methods of specifying the boundary conditions are considered. A program allowing to use up to 8000 nodes at required internal memory space of not more than 400 kilobytes is presented.

In order to determine the stability of mining openings and open-pit slopes and to determine zones of sliding and caving in underground mining and distribution of strains in these zones one must know the stress-strain state (SSS) of a rock mass and physical and mechanical properties of rocks forming this rock mass.

When calculating SSS one must take into account the rock mass structure defined by the shape, size and spatial relationship of structural elements (unit blocks, volumes, rocks) and of their couplings. The unit blocks are blocks identified due to existence of background (general) fractureness and other discontinuities.

The following models may be used for computing the stress-strain state of a real rock mass:

- solid or bulk homogeneous or heterogeneous medium;
- heterogeneous or homogeneous medium with one system of through joints;
- heterogeneous or homogeneous medium with two systems of through joints;
- heterogeneous or homogeneous medium with a "masonry" fractureness, i.e. a medium with one system of through joints and one system of discontinuous joints.

When there are more than two systems of regular joints or when there is intensive random fractureness as well as bulk medium, calculation of SSS may be based on the model of solid medium /1/.

Finite-element method (FEM) is an universal numerical method allowing application of all above mentioned mathematical models of media. Digitization of a medium in the form of finite elements of limited size coupled by a limited number of nodal couplings provides ability to retain the properties of the medium because each structural block can have its own physical and mechanical characteristics and the existence of weakness planes is accounted for by introducing special contact-elements.

For solving two-dimensional problems triangular finite elements approximating structural blocks and Goodman's contact element approximating weakness planes have been selected /2/. Usage of triangular finite elements (FE) lead to a loss of accuracy in calculations but allow to remove numerical integrations thereby significantly reducing computing time. The size of contact elements (CE) is assumed according to intensity of fractureness. In order to provide higher accuracy it is reasonable to divide structural blocks into more small FE and to equalize computational results for example by weighted summing of nodal values.

When using FEM two problems arise:

- too complex and time consuming

preparation of input data for computations (coordinates and numbers for FE and CE etc.);
- necessity of large working storage.

Bearing these problems in mind the following goals have been pursuited when preparing algorithm:
- development of automatic dividing of computational area into finite elements;
- input and storing of global stiffness matrix in external storage.

The program package consist of three modules:

GRID program provides automatic divisioning of computational area into finite elements, adjustment of computational area (adding, removing and changing of nodal coordinates and coordinates of FE and CE) as well as minimization of the width of tape for global stiffness matrix.

Divisioning of computational area into finite elements is based on the idea suggested by Segerlind whereby computational area is specified by separate zones each with its own characteristics (strain modulus, Poisson's ratio, density, inclination of joints, joint spacing etc.). For subareas of a solid medium an automatic joining is provided while subareas of fractured media are joined "manually" this being done by a subprogram of computational area adjustment.

The principle of minimization of tape width for global stiffness matrix is presented in Fig.1. After divisioning of computational area into finite elements a direction in which the number of nodes is minimal has to be selected. Coordinate axes are given in this direction and new coordinates of nodes (x' - y') are computed in this direction too. Then nodes are numbered as following: for minimal y' the numbers of nodes are rising from x'_{min} to x'_{max} and then this operation is repeated for the next level of y'.

Correctness of area divisioning is controlled by an AREA program that provide testing for identifying the following errors: duplication of FE and CE, existence of "free" nodes i.e. nodes that are not assigned to any FE, and improper joining of FE and CE (each FE must have two CE joined to it - one CE to each side of FE).

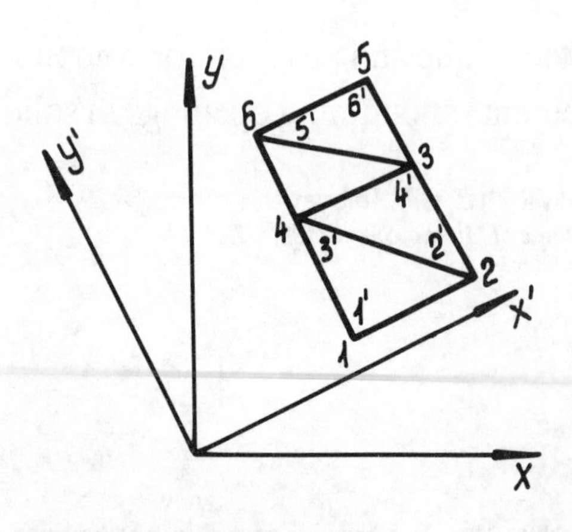

Figure 1. Minimization of tape width for global stiffness matrix. In the new coordinate system x'-y' tape width has been reduced from 8 to 6. 1-6 are "old" numbers of nodes and 1'-6' are "new" numbers.

On completion of running these programs data for all FE and CE along with nodal coordinates and their "old" and "new" numeration become stored in external memory.

Program STRESS provide computation of rock mass SSS and stability of mining openings. A method of direct stiffness is used for compiling global matrix. This matrix is stored in a direct-access file of an external memory. In order to reduce input, output and search operations this matrix is stored in the form of doubled columns (only the lower triangular of the matrix is stored).

Since under certain circumstances the stiffness matrix of a contact element is degenerated and since under conditions of intensive fractureness the global stiffness matrix loose its positive definitions application of Holessky method or iterative methods for solution of linear equation sets has been found impossible for a general case. Therefore Gauss method has been used with due account of symmetry and profile of the matrix.

Computation of SSS for fractured rock masses represent a set of iterations for determination of normal and shear stiffness of joints. In the zero iteration joints are as-

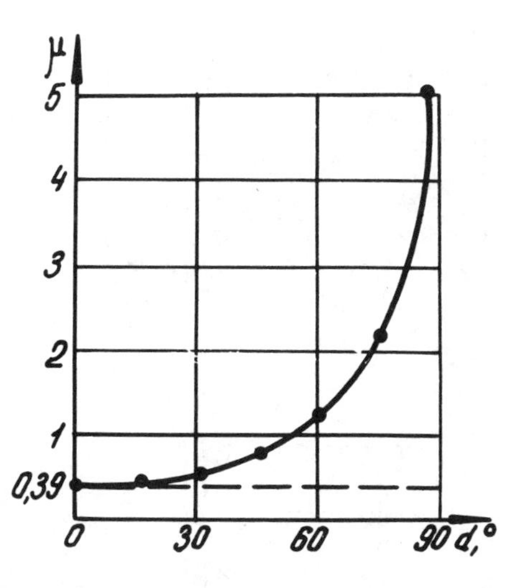

Figure 2. Lateral outward thrust μ vs. dip angle α of joints.

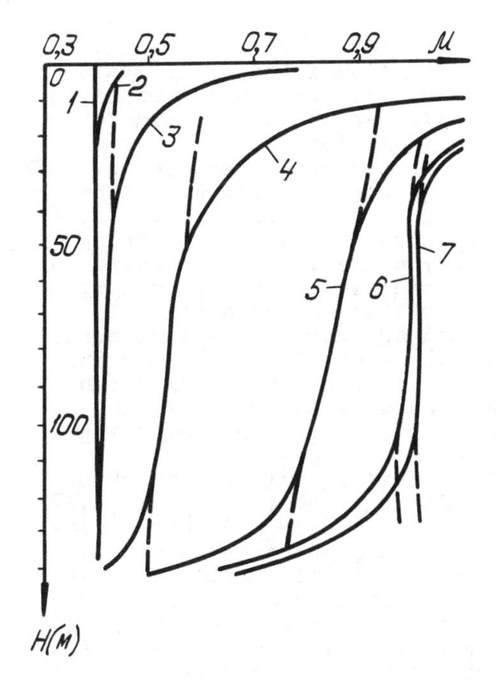

Figure 3. Lateral outward thrust μ vs. depth H and elastic displacement limit $\mathcal{E}_{пр}$. Curves 1-7 are functions μ (H) at $\mathcal{E}_{пр} = 10^{-4} - 10^{2}$ with steps of 10.

sumed to be noncompressible and nonshearable (normal stiffness and shear stiffness are at least an order of magnitude higher than the strain moduli of structural blocks) i.e. a solid medium is assumed. During successive iterations "new stiffness" is determined each time from the results of previous iterations while assuming that compressive stresses are negative

$$K_H = \begin{cases} E \cdot 10^{n}, & \sigma_n \leq 0 \\ 0, & \sigma_n > 0 \end{cases}$$

$$K_C = \begin{cases} \dfrac{\sigma_n \, tg \, \varphi + C}{\mathcal{E}_{пр}}, & \mathcal{E} \leq \mathcal{E}_{пр} \\ \dfrac{\sigma_n \, tg \, \varphi}{\mathcal{E}}, & \mathcal{E} > \mathcal{E}_{пр} \end{cases} \qquad (1)$$

where K_H and K_C are normal stiffness and shear stiffness, respectively; n = 1÷4; φ and C are an angle of friction along the contact and cohesion on contact, respectively; $\mathcal{E}_{пр}$ is a limit for elastic displacement along the joint; $\mathcal{E}$ is displacement along the joint according to previous iteration; and E is strain modulus of structural blocks.

After computing new stiffness values for CE's the global stiffness matrix is updated and new iteration step is accomplished. This iterative procedure continues until relative displacement along joints become lower than a specified value (this difference represent the accuracy of calculations) or until this displacement become higher than limiting displacement defined by a stability criterion. Output data are provided (strains and stresses) both for finite element and (after weighted summing) for nodes.

This program package allow to use up to 8000 nodes while the width of global stiffness matrix and the number of finite elements are practically unbounded. The required capacity of internal memory do not exceed 400 kilobytes. The programs are written in FORTRAN and are run under control of operational systems OC-6.1 and OC-7.1.

So this program package is a reasonably powerful tool for computing SSS practically in any medium but it require careful selection of computing area size and careful specifying of boundary conditions. So while one may select the size of computing area for a solid medium according to St.Venant's principle this size has to be enlarged for a fractured medium.

Furthermore when thereare sets of joints with dip angles other than 0° and 90° one is faced with serious difficulties when specifying boundary conditions for a computing area. Some violation in symmetry of the stress state in relation to vertical and horizontal axes result in disturbance of a rock mass SSS practically over entire computing area caused by stringent fixing of side boundaries in the computing area.

The practical solution in this situation will be loading the sides. In order to update and correct values of measured stresses one may use unbounded (horizontally) areas (Zoteyev and Nozhin, 1977). Due to joining the area boundaries by means of identical numeration of appropriate boundary nodes one can get undistorted stress state of a blocky mass. Moreover usage of such areas allows adjustment of friction characteristics on contacts: under conditions of two steeply dipping sets of through joints and under conditions of underestimation of friction angle along contacts a portion of blocks having wedge form will be pressed out in the upward direction. When the sides of the computing area are loaded lower than it is actually loaded the computing area will "ravel out".

Estimates of SSS for blocky masses are presented in Figs.2 and 3. As can be seen in these figures the coefficient of lateral outward thrust μ for a blocky mass is a function of dip angles of joints (Fig.2), depth H of the rock mass point being considered (Fig.3) and coefficient of shear stiffness K_c. According to (1) shear stiffness is linearly related to σ_n, $tg\,\varphi$, $\mathcal{E}_{np}$ and $\mathcal{E}$.

Normal stress σ_n on contacts of layers increase with depth but the amount of mutual sliding of blocks along the contacts of layers, $\mathcal{E}$, decrease with depth thereby causing increase of K_c with depth of a joint in a rock mass.

Upper and lower boundaries of the computing area have significant effect on SSS of a rock mass. In

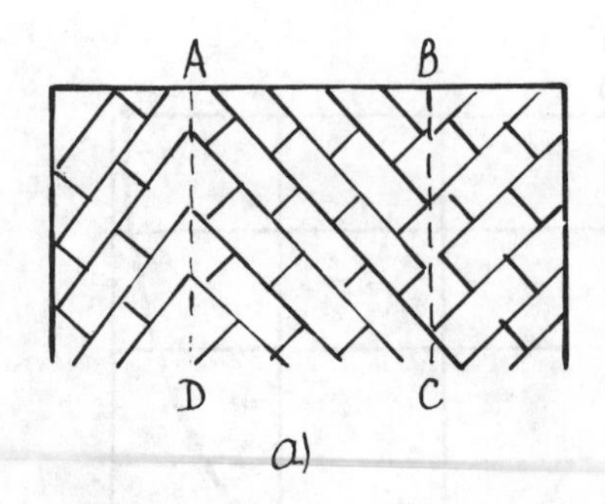

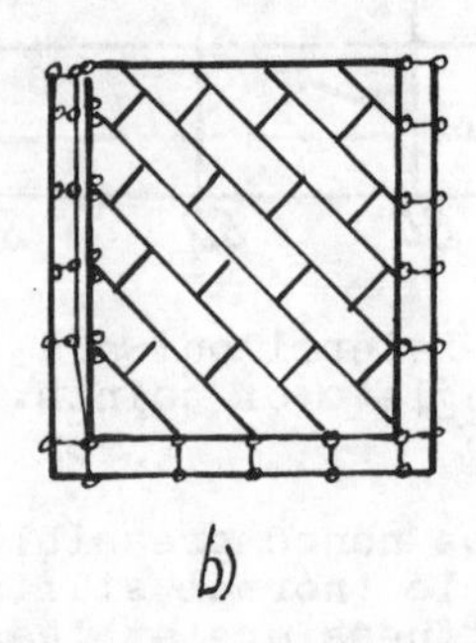

Figure 4. Specification of computing area for a blocky mass. a - blocky mass with folded structure; b - specification of boundary conditions.

the medium-depth part of the area coefficient μ decrease linearly, this decrease being related with similar increase of K_c with depth. Near lower boundary decrease of become more intensive and it approaches to the value that is characteristic for a solid medium (μ = = 0.39). And on the contrary the free face causes sharp increase of μ. At surface μ is infinite becaus the vertical stress component has a zero value and the horizontal stress component has a value other than zero one.

Fig.3 shows that disturbing effect of the surface increase with dip angle of joints. This fact can be explained simply when considering the selfwedging effect of layers near surface.

When the limit of elastic displacement along the joint is large ($\mathcal{E}_{np}$ = 100 m) we have hydrostatic state (μ = 1) in the middle part of the computing area. In this case the disturbing effect of the boundaries reach its maximum.

When calculating SSS for a blocky rock mass with folded structure as shown in Fig.4a it is sufficient to select a rock mass section between points A and B and fix the computing area as shown in Fig.4b. Since SSS is symmetrical in respect to straight lines AD and BC the horizontal fixing of boundary nodes does not introduce distortions into the pattern of stresses and strains. Vertical joint along the left boundary AD provide free displacement of blocks along bedding planes by means of gravitational forces. Calculations show that strict fixing of boundary AD in the computing area generate horizontal tensile stresses that is not true in reality.

After determination of boundary conditions one may begin calculations of SSS for mining system elements. In order to reach this goal a new computing area is defined where mining openings are specified. The selection of the computing area size is based on ST.Venant's condition for a solid medium. In case of fractured media this size has to be increased based on conditions presented in Figs.2 and 3 until the effect of side and lower boundaries on the area considered become zero.

Another run of calculations is done for selected boundary conditions that give us SSS of the rock mass around mining openings and in the mining system elements.

Thus the main difficulty in calculations of SSS for mining system elements in fractured rock masses is determination of the computing area size and boundary conditions.

REFERENCES

Zoteyev O.V. Mathematical modelling of heterogeneous masses of fractured rocks. Moscow, VINITI, No.8732, VDEP, 1986: 14p.

Goodman R.E., Taylor R.L., Brekke T.L. A model for the mechanic of jointed rock. J.of the Soil Mech. and Found.Div., 1968, Vol.94, No.3:637-659.

Segerlind L. Application of finite element method. Moscow, MIR publishing office, 1979: 392p.

Zoteyev V.G., Nozhin A.F. Calculation of the stress-strain state for a fractured rock mass around mining openings: a program. Inst. of Mining, Ministry of Ferrous Metallurgy, USSR, No.GR 1003900, Sverdlovsk, 1977: 94p.

Numerical Methods in Geomechanics (Innsbruck 1988), Swoboda (ed.)
© 1988 Balkema, Rotterdam. ISBN 90 6191 809 X

Numerical stability investigations of drifts in stratified salt rock

W.Hüls, A.Lindert & W.Schreiner
Institut für Bergbausicherheit, Leipzig, German Democratic Republic

ABSTRACT: Issuing from the MOHR's fracture hypothesis and the strain-hardening theory reproducing laboratory tests, material laws are developed. Thus, a basis is established for the stability assessment of mining cavities in salt rock under short-term and long-term behaviour. On the example of drifts (deep tunnels or underground roadways) in stratified salt rock, a stability evaluation is carried through.

1 INTRODUCTION

Numerical methods are a substantial tool for parameter studies aimed at determining the contour behaviour of mining cavities as well as the geomechanical behaviour of the whole rock mass. For plastic and rheological constitutive equations, incremental procedures of the finite-element-method stood well /7/. At the Institute for Safety-in-mines (Institut für Bergbausicherheit), the programme family GEOFEM was developed for boundary value problems of salt rock mechanics /4/. A confirmation of the calculated stress and deformation fields, however, can only be obtained by in-situ measuring procedures. In addition to the traditional methods of deformation measurement, use was recently made of direct investigations of the stress components by means of the hydraulic frac technology /6/. Vital importance in the solution of geomechanical boundary value problems falls to the choice of the constitutive law both allowing for the dominating material parameters and yielding sufficient accuracy of the result.

These requirements can only be met if modelling is done with knowledge of the external conditions (rock pressure), the internal conditions (geological structure) and the influence factors due to procedure and exploitation, respectively.

2 GOVERNING EQUATIONS FOR SALT ROCK

The caracterizing qualities such as strength, deformation behaviour and fracture behaviour of salt rocks are different due to the geological conditions varying in each deposit, even in working areas of the same deposit.

Salt rocks, for instance those of the Upper Permian series, have geomechanical properties distinguishing them from other natural rocks /3/:

- low tensile strength of about 2 to 5 % of the uniaxial compressive strength;
- high increase in compressive strength within the triaxial loading range;
- considerable creep deformation as a function of lifespan with simultaneous evident decrease in strength after long lifespans.

As far as fracture behaviour is concerned, salt rocks present determinant qualitative differences decisively affecting the geomechanical behaviour. It has to be distinguished

- salt rock of the type rock salt (sylvinite, hard salt) presenting pronounced plastic behaviour as well as very great deformations up to fracture;
- salt rock of the type carnallitite presenting pronounced elastic behaviour up to fracture under spontaneous fracturing without substantial increase in deformation rate. This behaviour especially occurs with rapid loading.

Consequently, a visco-plastic approach dominates in the determination of rock properties for rock salt, and a visco-elastic one with carnallitite.

The description of the strength behaviour of a solid implies that a failure hypothesis is assumed. In rock mechanics, the theory of the boundary state according to MOHR is widely held.

For the determination of the boundary stress state, uniaxial compressive tests,

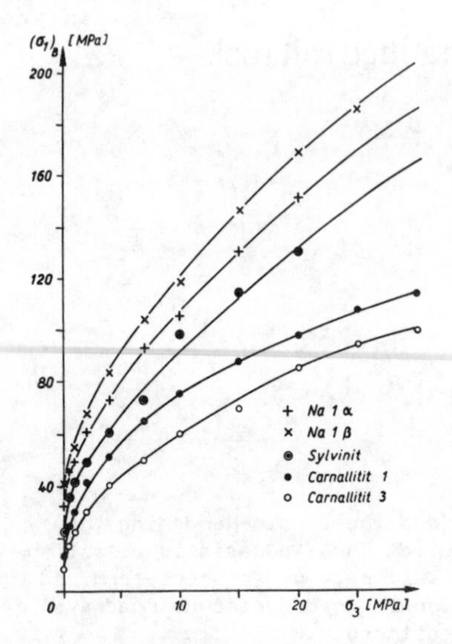

Fig.1 Triaxial compressive strength of salt rock

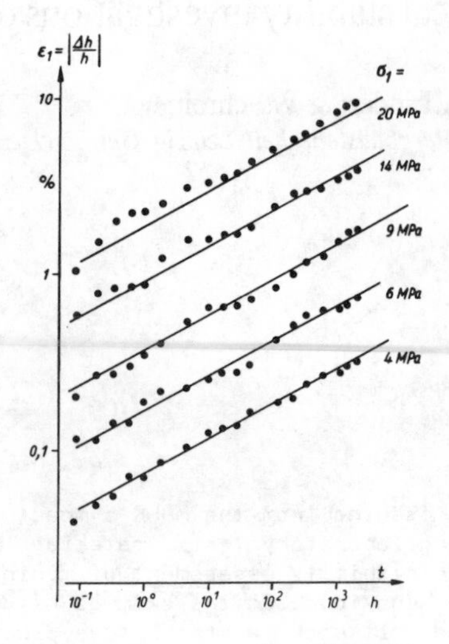

Fig.2 Creep curves of rock salt

uniaxial tensile tests and triaxial compressive tests are carried out. Bending tests and Brazilian tests are used for the description of special loading conditions. In these tests, the value of the stress maximum and/or the ultimate load bearing capacity is defined as fracture point. These force-controlled short-term tests are carried out at a loading rate of $\dot{\sigma}$ = 500 MPa/h.

The vertical fracture stresses σ_{1B} determined during the triaxial test are a function of the confining pressure σ_3. For salt rocks of different strength, a functional connection between maximum and minimum principal stress is revealed by the fracture of the specimen in the following form (fig. 1)

$$\left(\frac{\sigma_{1B}}{\sigma_D}\right)^{\varkappa} = 1 + \frac{\varkappa \cdot \sigma_3}{\sigma_Z} \qquad (1)$$

with
σ_D - uniaxial compressive strength
σ_Z - uniaxial tensile strength
$\varkappa$ - parameters of the triaxial compressive strength.

This equation describes with sufficient accuracy the short-term behaviour of all studied types of salt rock.
Table 1 represents the parameters for triaxial strength of different types of salt rock of the Upper Permian series.

Various theoretical governing equations are used in scientific writing for the description of the deformation behaviour of rock salt. Individual dominating properties can well be represented both with the elastic after-effect theory and the rheological structure models. By means of the strain-hardening theory used in this case, it is however possible to describe short-term tests at constant loading rate just as well as long-term tests at constant loading /2/ (fig. 2).

Application is made of the following assumptions:
- incompressibility at creep;
- independence of the creep rate on a superimposed hydrostatic pressure;
- isotropy and collinearity of the principal directions of strain and stress tensor, resulting therefrom.

In a axi-symmetric system, the state equation takes the form

$$\dot{\varepsilon}_{1CR} = A \cdot (\sigma_1 - \sigma_3)^{\beta} \cdot \varepsilon_{1CR}^{-\mu} \qquad (2)$$

and in the uniaxial stress state at σ_3 = 0, it leads to the form

$$\dot{\varepsilon}_{CR} = A \cdot \sigma^{\beta} \cdot \varepsilon_{CR}^{-\mu} \qquad (3)$$

The overall deformation shall be composed of an elastically reversible part ε_E and of an irreversible creep deformation ε_{CR}.

$$\varepsilon(\sigma,t) = \varepsilon_E + \varepsilon_{CR} \qquad (4)$$

The methodology of the strain-hardening theory gets extended to the whole deformation behaviour by the following assumptions:

- The mechanical state is completely described by the acting effective stresses

Table 1. Strength parameters of types of salt rock

Types of salt rock	E [$\cdot 10^4$ MPa]	ν	σ_D [MPa]	σ_Z [MPa]	$\varkappa$	$\varrho \cdot g$ [Mpm^{-3}]
Werra-type rock salt						
Na 1β(Middle)	2.5	0.48	43	1.8	2.4	2.15
Na 1α(Lower)	1.8	0.48	30.5	1.3	2.15	2.15
Sylvinite	1.8	0.48	35.5	4.2	1.67	2.05
Carnallitite 1 (65 %)	1.4	0.33	12	0.61	2.0	1.7
Carnallitite 2 (80 %)	0.8	0.33	9.0	0.47	2.0	1.65
Carnallitite 3 (100 %)	0.8	0.33	6.5	0.31	2.0	1.65

and strains.

- The relaxation behaviour at static loading is represented with sufficient precision by the conversion of the effective stresses and strains into elastic stress portions and/or reversible strain portions, using a material parameter E.

By means of the integration of (2) under the conditions of static loading

$(\sigma_1 - \sigma_3)_0$ = constant for $t > 0$

it follows that

$$\varepsilon_{1CR} = C \cdot (\sigma_1 - \sigma_3)^n \cdot t^m \qquad (5)$$

with

$$m = \frac{1}{\mu+1} \quad n = \frac{\beta}{\mu+1} \quad C = [A \cdot (\mu+1)]^{\frac{1}{\mu+1}}$$

Paying attention to the assumptions made above

$$(\sigma_1 - \sigma_3) \longrightarrow \varepsilon_{1E} \cdot E$$

$$\varepsilon_{1CR} \longrightarrow \frac{\sigma_1 - \sigma_3}{E}$$

it follows for the elastically reversible part of deformation

$$\varepsilon_{1E} = \left(\frac{1}{C}\right)^{\frac{1}{n}} \cdot \left(\frac{1}{E}\right)^{\frac{n+1}{n}} \cdot (\sigma_1 - \sigma_3)^{\frac{1}{n}} \left(\frac{1}{t}\right)^{\frac{m}{n}} \qquad (6)$$

At constant loading rate

$$\dot{\sigma}_0 = \frac{\sigma_1 - \sigma_3}{t} = \text{const.},$$

there is set

$$t \longrightarrow t' = \frac{m}{n+m} \cdot t$$

Hence, there are obtained overall deformations

- under the condition of constant loading

$$\varepsilon_1 = \left(\frac{1}{E}\right)^{\frac{n+1}{n}} \cdot \left(\frac{1}{C}\right)^{\frac{1}{n}} \cdot (\sigma_1 - \sigma_3)^{\frac{1}{n}} \cdot \left(\frac{1}{t}\right)^{\frac{m}{n}} + \qquad (7)$$

$$+ C \cdot (\sigma_1 - \sigma_3)^n \cdot t^m$$

- under the condition of constant loading rate

$$\varepsilon_1 = \left(\frac{1}{E}\right)^{\frac{n+1}{n}} \cdot \left(\frac{1}{C}\right)^{\frac{1}{n}} \left(\frac{n+m}{m} \cdot \dot{\sigma}_0\right)^{\frac{m}{n}} \times \qquad (8)$$

$$\times (\sigma_1 - \sigma_3)^{\frac{1-m}{n}} +$$

$$+ C \left(\frac{m}{n+m} \cdot \frac{1}{\dot{\sigma}_0}\right)^m \cdot (\sigma_1 - \sigma_3)^{n+m}$$

with giving t in hours, σ and E in MPa.

In order to find out the elastically reversible part of deformation, the parameter of elasticity E has to be determined. On evaluating the hysteresis loops of intermediate stress relievings in the short-term test, a dependence on the acting confining pressure experimentally comes to light in the form (fig.3)

$$E = E_0 \left(1 + \frac{\sigma_3}{\sigma_0}\right)^2 \qquad (9)$$

Long-term parameters are determined by means of uniaxial and triaxial creep tests carried out up to 1500 hours at constant loading. As deformations occur well over 10 %, the LUDVIC's "natural"

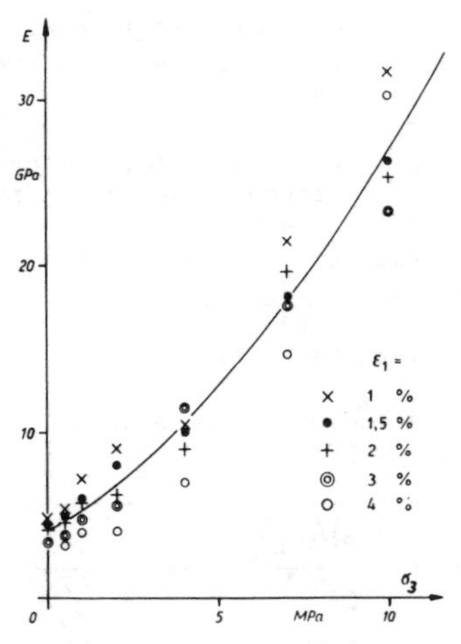

Fig. 3 Variable modulus of elasticity of rock salt

deformation $\varepsilon = |\ln(h/h_0)|$ is introduced for the relative deformation instead of the "technical" relation $\varepsilon = |\Delta h/h_0|$.

When loadings of $\frac{\sigma_1 - \sigma_3}{(\sigma_1 - \sigma_3)_B} > \frac{1}{3}$ occur with carnallitite, the creep deformation mechanics changes. The stable creep process passes over to an instable creep.

On comparing the studied types of rock salt with each other, no significant differences can be made out in creep deformation at low confining pressure (σ_3 = 0 up to 12 MPa).

Table 2 indicates the deformation parameters of the studied types of salt rock.

Table 2. Creep parameters

Type of salt rock	$\frac{\sigma_1-\sigma_3}{(\sigma_1-\sigma_3)_B}$	C	n	m
Carnallitite				
(unstable)	$> \frac{1}{3}$	$7 \cdot 10^{-8}$	3.0	0.5
(stable)	$< \frac{1}{3}$	$1.3 \cdot 10^{-5}$	1.5	0.25
Rock salt		$2.0 \cdot 10^{-5}$	2.14	0.14

Starting from the principle that a non-elastic work of deformation performed in the solid up to the point of fracture is equal to the fracture energy, it follows for the point of fracture

$$t_B^m = \xi \cdot \left(\frac{(\sigma_e)_B}{(\sigma_e)_{crit}} \right)^{n+1} \quad \text{in h}$$

with being

ξ = 1.5 proportionality coefficient derived from the tests and

$\sigma_e = \sqrt{\frac{3}{2} I_2}$ effective stress from a second invariant of the stress deviator.

From this relation, a time strength coefficient $f_t = \frac{(\sigma_e)_{crit}}{(\sigma_e)_B}$ can be derived (table 3).

Table 3. Fracture parameters of carnallitite

Lifespan	$S_t = (\sigma_e)/(\sigma_e)_{crit}$	f_t [%]
1 hour	1.50	67
1 day	2.23	45
1 week	2.85	35
1 month	3.42	29
3 months	3.92	25
1 year	4.67	21

The calculated parameters correlate with the creep tests on specimens subjected to uniaxial and triaxial loading after a test period of half a year where the temporal fracture occured in the loading range

$$20\,\% < \left[\frac{\sigma_1 - \sigma_3}{(\sigma_1 - \sigma_3)_B} \right]_{\sigma_3 = \text{const.}} < 50\,\%$$

Analogous parameters of rock salt were given in /4/.

3 PARAMETER STUDY FOR DEEP DRIFTS

The example given in fig. 4 is called carnallitite dome-and-room method (in this case, dome is a synonym of uparched strata of great thickness), a specific modification of the back stoping with pillars of temporally unlimited stability, accomplished in a twophase system of mining /1, 5/. This procedure particularly has regard to the irregular strata conditions in uparched carnallitite domes of great thickness in the potash area "Werra" at a depth of about 900 m. During the second mining phase, the drifts are partially upwards enlarged into a funnel-shaped trough by steep overhead caving in the room-like retreating system. Then, mining heights up to 50 m can be originated.

By means of the geological preliminary exploration of a bloc to be mined, it must be decided upon the position of the drifts along the seam basis. This decision depends on the carnallitite variety concretely found on the planned drift level. As roof bolting is normally only used in salt mining, special attention has to be given to the problem of the needed contour safety. This means to apply a constitutive model for salt rocks allowing

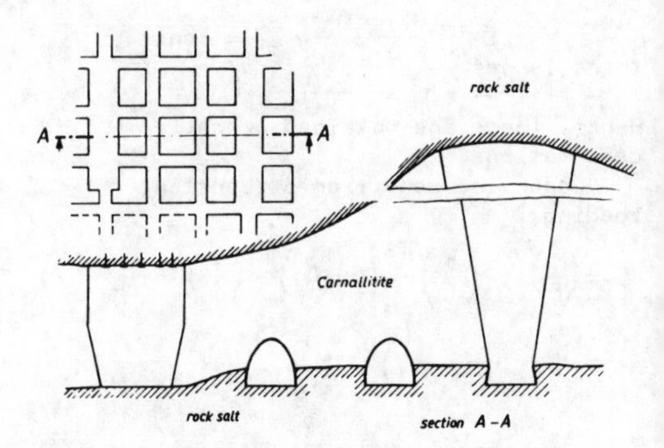

Fig. 4 Representation of dome mining method (dome = uparched strata of great thickness)

for the simulation of the contour fracture both as boundary state and also in the temporal development.

3.1 Cross section of drift

The model is based on parallel drifts to be advanced at a spacing of at least 40m. Thus, the determination of the stress state can be considered as a single cavity problem. The shape parameters intended to exist are as follows:

Shape of cross section	Bottom width (m)	Vertex height (m)
(a)	6	5
(b)	7	7
(c)	8	10

In fig. 5, the stress redistribution is shown in two profiles for carnallitite variety 3. The profiles placed 1.5 m high in the wall and also those leading from the vertex into the roof exhibit qualitatively known stress states. It can be observed that a spacing of 8 m, the initial stress state practically restores irrespectively of the vertex height of the drift, revealing that the cavity influence died away. The immediate contour is pearly equally stressed with all cross sectional shapes. The influence zone up to the stress peak of the vertical stress extends along the wall up to about 1 m with shape (a), and up to about 1.5 m with shape (c). The twofold value of the overburden pressure is not exceeded. Above the vertex, there is a stress maximum at about 1 m while dying-away of the stress into the roof takes place slowlier than in the wall.

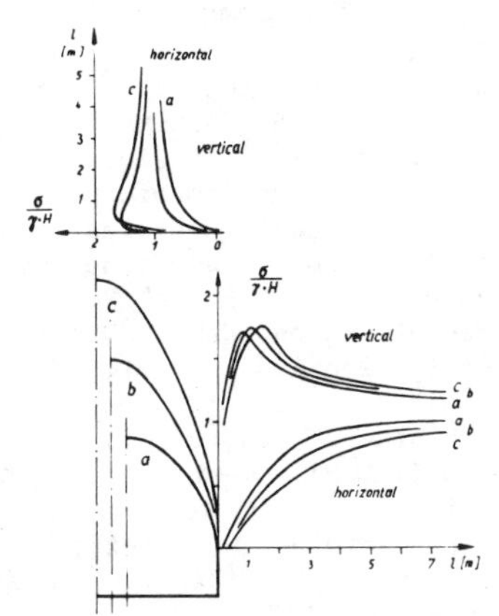

Fig. 5 Stress profiles around drifts in the carnallitite

An analogous stress distribution becomes evident for the other carnallitite varieties.

However, an evaluation of the stress state by means of the time strength quotient shown in chapter 2 results in a more differenciated assessment.

Considering the measures taken for safety-in-mines, such as scaling, and the necessary lifespan of drifts, the statements made for depths of 900 m, are as follows:
Drifts in Ca1 do not loose their contour stability before some month, with fracture zones being situated in the range of decimeters. By scaling in time intervals, the needed contour stability can be maintained. The fracture zones in Ca3 can reach the order of magnitude of 0.5 m along the contour already after some days. The fracture process gets accelerated with the cross sections (b) and (c). Therefrom, the conclusion has to be drawn that drifts in Ca3 should be avoided.

3.2 Stratified salt rock

In order to be able of assuring the stability of drifts necessary for their utilization phase, stabilizing measures typical of the mining branch must be taken. An "improvement" of the strength properties of contour is obtained by advancing drifts to a certain extent into the horizontal rock salt.

At the same time, it must be investigated to which thickness an interbedded stratum Ca3 impairs the contour stability. Model variants with a rock salt toe of 2 m shall be considered. For the drift cross section, a bottom width of 5 m and a height of 4.5 m are assumed.

Using a two-strata-model (fig. 6), it can be observed that fracture zones in Ca3 reach an extent that does not assure the required contour stability beyond the period of weeks. Consequently, in analogy with the statement made for § 3.1 where the whole contour is situated in Ca3, such a drivage of drifts has to be avoided.

With drifts intended to be advanced in stratified types of salt rock, the contour behaviour is substantially influenced by strata limits and salt varieties. Under the conditions considered in the present case, rock salt responds to the exceeding of the yield strength by plastic deformations controllable by miners' methods. With carnallitite, however, fracture phenomenons getting intensified by the increase in carnallitite content occur.

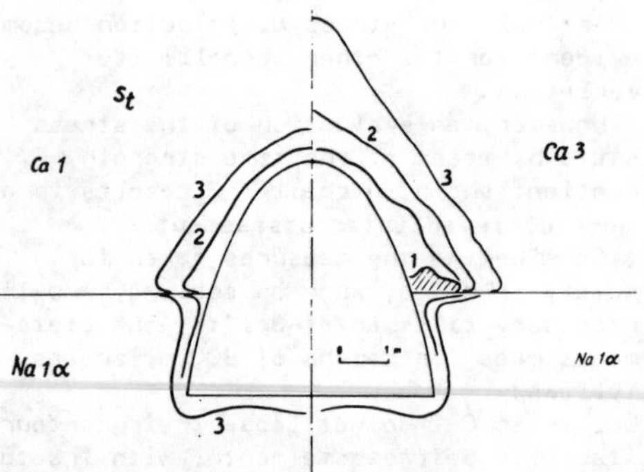

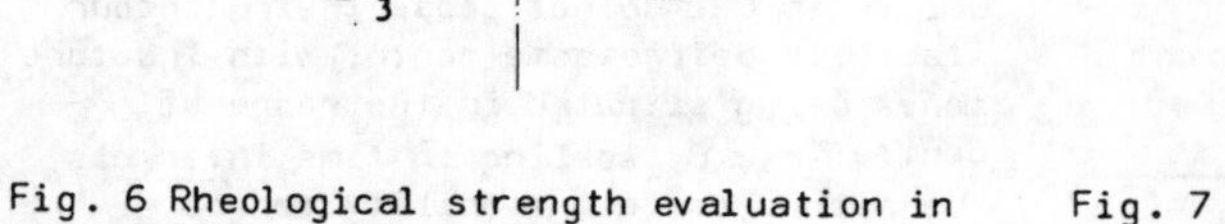

Fig. 6 Rheological strength evaluation in the two-strata-model

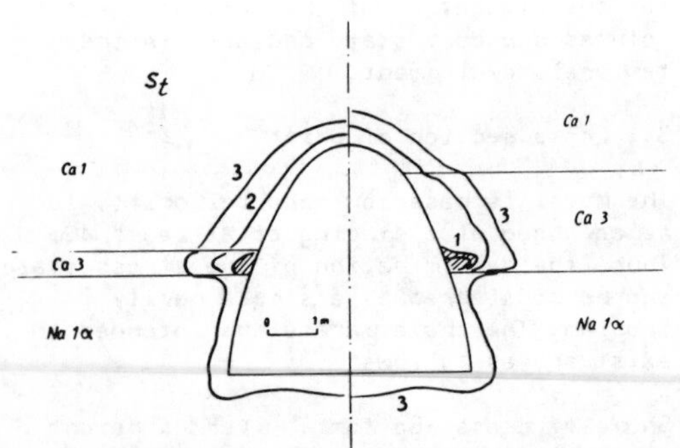

Fig. 7 Rheological strength evaluation in the three-strata-model

Fig. 7 compared with the previous considerations reveals that the variations of the fracture behaviour along the contour with interbedded Ca3 are essentially restricted to this interbedded stratum. Hence, the influence of the Ca3 stratum on the contour stability of the whole drift depends on its thickness.

By means of the temporal course of the fracture model, it can be stated that the fracture of the contour starts in carnallite along the stratum limit towards the rock salt toe while loosening continues along the stratum limit into the wall.

An advancing contour fracture can be avoided only if sufficiently strong material is in the solid along the contour and in the rock surrounding the drift.

Obviously, the height of an interbedded stratum Ca3 must no exceed 0.5 m. At minor thickness, its influence on the needed contour stability seems then to be neglectable.

The rock salt toe hasgot the required strength.

4 CONCLUSIONS

The present paper demonstrates that the behaviour of salt rock can with sufficient accuracy be described in the pre-failure zone, using the strain-hardening theory. Thus, the formulation of geomechanical problems in which theses properties have a dominating influence can also numerically be treated by means of constitutive laws. The non-linearity of the constitutive equations is simulated by an incremental-iterative process, using the finite-element-method.

The evaluation of the calculated stress redistributions can be accomplished by the connection of a post-processor. On the example of deep tunnels (drifts) in potash mining, the conditions for driving such drifts in the stratified salt rock are determined in this way by means of a time strength coefficient.

REFERENCES

/1/ Duchrow, G.; Schilder, Ch.: Reduction of mining losses brought about by the room-and-pillar method. Neue Bergbautechnik 15(1985)12, 441-447

/2/ Erzanov, Z.S. a.o.: Rheology and seismomechanics of the rock mass. Nauka, Alma Ata, 1984.

/3/ Erzanov, Z.S. a.o.: Computing methods for salt pillar systems and seismotectonic stress states in the rock mass. VEB Deutscher Verlag für Grundstoffindustrie, Leipzig, Freiberger Forschungshefte A 744, 1986.

/4/ Hüls, W.: Finite-element-modelling of stress transfer around large openings in salt rock, ICONMIC 85, BALKEMA, Rotterdam, 1986, Vol. IV, 1859-1864.

/5/ Hüls, W.; Menzel, W. and Schreiner, W.: Results of geomechanical investigations in carnallitite mining at great depths. Neue Bergbautechnik 10 (1980)4, 203-209

/6/ Hüls, W.; Menzel, W. and Weber, D.: Stress investigations in salt pillars and their numerical modelling. Int. Symp. on Rock Stress Measurements, Stockholm, 1986, CENTEK-Publ., Luteä 1986, 537-541.

/7/ Owen, D.R.J. and Hinton, E,: Finite Elements in Plasticity. Pineridge Press, Swansea U.K., 1980.

Numerical Methods in Geomechanics (Innsbruck 1988), Swoboda (ed.)
© 1988 Balkema, Rotterdam. ISBN 90 6191 809 X

Modelling of cut-and-fill mining systems – Nasliden revisited

Ye Qian Yuan
Jiangxi Institute of Metallurgy, People's Republic of China
William Hustrulid
Colorado School of Mines, USA

ABSTRACT, This paper presents one-demensional load-deformation curves for the fill and revised Liberman's formula to calculate the pressure and convergence in cut-and-fill mining system. In Naslidn mine actual and predicted fill pressures, wall convergence and pillar stresses as a fuction of mining sequence are in good agreement by modified Liberman's method

1. INTRODUCTION

The Nasliden Mine located in the municipality of Norsjo in the county of Vasterbotten in northen Sweden, was brought into production by the Boliden Company in 1970. The ore body being mined by modern cut-and-fill techniques is a complex sulfide ore. It dips at about 70° to the west, has a length which varies between 110m and 220m, and a maximum width of 25m (the average being 18m). Although the vertical extent is known to a depth of 770m. Each stope will be about 100m high. Filling is carried out with deslimed tailing sand which is transported hydralically to the mining stopes.

An extensive rock mechanics research and development program was conducted at the Nasliden Mine, Sweden by a team of investigators from the Boliden Company, the university of Lulea and the Royal institute of technology. A series of finite element and other models were used to describe the response of the ore body, the hanging and foot walls and the fill during the cut-and-fill mining process. Although the values predected by final model were in good general agreement with the measurements, some variation in both trend and amount in the fill pressures, wall convergence and pillar stresses were noted. One possible explanation for this is the inherent difficulty of adequtely simulating the non-linear behavior of the fill.

The approach to the problem described in this paper is to employ a model that allows detailed modelling of the non linear fill behavior and the stope pillars. One dimension load deformation curves for the fill have been incorporated into a filled slot analysis of the Nasliden cut-and-fill systems. Actual amd predicted fill pressures, wall convergence and pillar stresses as a function of mining sequence are found to be good agreement. It is noted this type of model does not provide information outisde of the vein. It can however be used to provide vein boundary pressure displacements data that can be used as input for the finite element models.

2. THE FILLED INCLUSION MODEL

An alternative way of modelling the cau-and-fill sequence at the Nasliden Mine is to approximate the vein by a filled slot in a gravity loaded elastic medium. The general procedure for determination of the boundary pressure and closure distribution in a slot filled with a non-linear material in a gravity loaded elastic medium has been outlined by Liberman and Khaimova. The extension of their approach to vein mining systems in which pillars fill a portion of the slot has been discussed by Hustrulid and Moreno.

2.1 Libermen Approach

The approach taken by Liberman is to approximate the slot by an ellipse. The slot itself is then divided into a number of intervals over which different boundary pressure can be applied. The relationship between the closure of the slot and the applied pressure depends upon the load-deformation relationship of the material in the interval. It has been found that the equation

$$\varepsilon = \varepsilon_o \left[1 - e^{-\left(\frac{P}{P_o}\right)^n}\right] \qquad (1)$$

where

ε =strain

ε_o=the maximum strain required to

reduce the void volum to zero.

P =pressure acting on the fill material

P_o=a parameter characterizing the shape of the curve

N =exponent

The basic formula for determing the boundary pressure $p(x)$ and closure $\delta_v(X)$distribution for a slot filled with a nonlinear material in a gravity loaded elastic medium as presented by Liberman and Kraimova is given below

$$\sum_{k=1}^{n} P_k D_k(X_l)=V_{l,k}(X_L)-\varepsilon_o h\left[1-e^{-\left(\frac{P_l}{P_o}\right)^N}\right] \quad (2)$$

where

P_k =pressure acting on the k-th element

X_L =the corrdinate of middle point of the L-th element

$D_k(X_L)$ =the influence coefficient such that the product of $P_k D_k(x_L)$ is the displacement induced by the pressure P_k at the midpoint of L-th element

$V_{Lk}(x_L)$=the displacement induced by the gravity at the L-th element

h =the seam thickness

ε_o =strain at which the void volume becomes zero

This system of equations can be solved for the individual pressures and convergences in each interval of the slot.

When using this equation to calculated fill pressures in steeply dipping veins, one finds that the calculated pressure are smaller then the in-situ measurements. The reason for this is that the pressure in the hydraulic backfill consists of two parts, one caused by the dead weight of the backfill and the second induced by the convergence of the hanging wall and the footwall. Application of the Liberman formula yields only the pressure induced by convergence, in the following section the modifications to Liberman's formula required for calculating the total pressure and convergence in the fill are pressnted.

2.2 Modification for gravity loading of fill

For the calculation of the normal pressure in the slot due to the weight of the fill, the classical silo theory will be used. The basic formula is shown below

$$P_s=\frac{\psi\rho}{tg\delta}A\left[1-e^{\frac{\nu k_o}{A}tg\delta}\right]$$

$A=\frac{a}{4}$ (pressure against short side of silo)

$A=\frac{2ab-a^2}{4b}$ (pressure against long side of silo

where

ψ =angle of friction between the rock wall and the hydraulic backfill

ν =depth below ground surface or backfill (height of backfill above the bottom)

ρ =density

k_o =$1-\sin\phi'$

ϕ'=internal friction angle

The total stress (p_t) in the hydraulic backfill is the sum of that caused by the dead weight of the backfill (p_s) and that induced by the convergence (p_c)of that hanging wall and the footwall.

$$p_t=p_s+p_c$$

where

p_s= dead weight pressure.

p_c = convergence pressure.

Equation (1) can be rewritten as

$$\varepsilon_{total}=\varepsilon_o\left[1-e^{-\left(\frac{P_s+P_c}{p_o}\right)^N}\right]$$

From the Liberman equation one obtains the strain due to convergence (ε_c) and not the total strain (ε_t).

however

$$\text{Since}\quad \varepsilon_{dead}=\varepsilon_o\left[1-e^{-\left(\frac{P_s}{P_o}\right)^N}\right]$$

$$\text{Then}\quad \varepsilon_c=\varepsilon_t-\varepsilon_d$$

$$\text{or}\quad \varepsilon_c=\varepsilon_o\left[e^{-\left(\frac{P_s}{P_o}\right)^N}-e^{-\left(\frac{P_s+P_c}{P_o}\right)^N}\right]$$

Replacing the term $\varepsilon_o(1-e^{-\left(\frac{P_l}{P_o}\right)^N})$in Libermans equation (2) by

$$\varepsilon_o\left[e^{-\left(\frac{P_s}{P_o}\right)^N}-e^{-\left(\frac{P_s+P_c}{P_o}\right)^N}\right]$$

One obtains the more genaral equation

$$\sum_{k=1}^{n} P_k D_k(X_l)=V_{l,k}(X_l)-\varepsilon_o h\left[e^{-\left(\frac{P_s}{P_o}\right)^N}-e^{-\left(\frac{P_s+P_c}{P_o}\right)^N}\right] \quad (3)$$

which now includes both the effect of dead weight fill loading plus stope wall convergence.

2.3 Modification for pilla interval

Although in the origenal Russian articles, the ellipse was used to found only the fill zone, it is of couse possible to include the solid areas at the ends of the fill and pillars, the term

$$\varepsilon_o h\,[e^{-(\frac{P_s}{P_o})^N} - e^{-(\frac{P_s+P_c}{P_o})^N}]$$

in equation (3) is replaced by $\frac{P_i h}{E_i}$

where

E_i = elastic modulus of the material in interval

2.4 Numerical solution of equation

The vein is divided into M elements and the modified Liberman's formula is written for each element. The resulting M simulataneous nonlinear equations can be solved by following iteration method,

step 1. Provide an initial guess for each P and substitute these initial values into right hand side of the modified Liberman's equation, and get M simultaneous linear equation.

step 2. The Gussian elimination technique is used to obtain a first approximation to the pressures in each interval.

step 3. Because of the pressure of the term

$$\varepsilon_o h\,[e^{-(\frac{P_s}{P_o})^N} - e^{-(\frac{P_s+P_c}{P_o})^N}]$$

The equation is ill-conditioned and in genaral one cannot get a convergent solution through the repeated use of Gussian elimenation for the fill elements. One uses the results from the Gussian elimenation as input for Newton-Rephson evaluation of each equation. For the solid elements a new guess of the pressure is given because of normolly large difference to the fill value.

The Newton-Raphson equation is shown below

$$P_{i(n+1)} = P_{i(n)} - F/G$$

where

$$F=\sum_{k=1}^{n} P_k D_k(X_i) - V_{im}(X_i) + \varepsilon_o h\,[e^{-(\frac{P_s}{P_o})^N} - e^{-(\frac{P_s+P_i}{P_o})^N}]$$

$G = \delta F/\delta P_i$

$P_{i(n)}$ = nth approximation

$P_{i(n+1)}$ = (n+1)th approximation

3. NASLIDEN MODELLED AS A FILLED SLOT

3.1 Properties of the fill

One-dimensional load-deformation curves were obtained for samples of the fill from the Nasliden Mine using a special split-platen consolidation device. The results of three tests conducted on dry fill samples of different initial void ratios are shown in figure 1. and described in table 1.

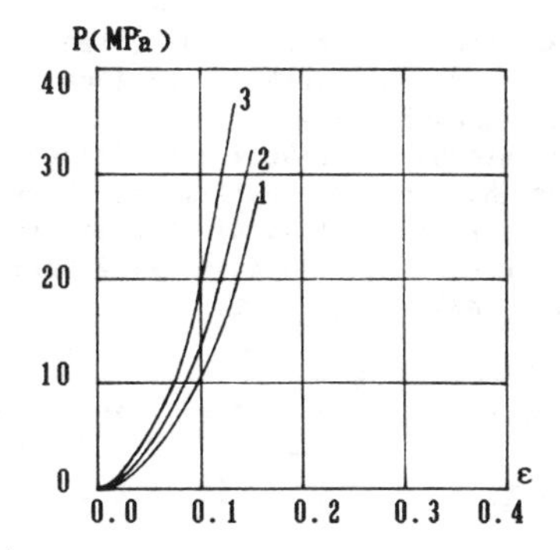

Fig.1

Table 1

test	initial void ratio	volume ratio ε_o
1	0.932	0.485
2	0.835	0.455
3	0.754	0.43

The curves can be will described by an equation of the form

$$\varepsilon = \varepsilon_o(1 - e^{-(\frac{P}{P_o})^N}) \quad (4)$$

The values for the constants in equation 4 are given in table 2.

Table 2

test	P_o(MPA)	N
1	126	0.54
2	345	0.43
3	256	0.49

In comparing the initial void ratios with those suggested to be appropriate for Nasliden, the test 3 will be used for the simulations. In addition it will be assumed that

Bulk density ρ= 2.2 tons/m³

Angle of internal friction ϕ = 30°

Friction angle between fill and wall ϕ'=36°

3.2 Elastic properties of the Nasliden rock and in situ-stress field

The elastic properties used in the final model are summarized below .

Yonges Modulii, footwall 27.6 GPA, Hanging wall 27.6 GPA, Orebody 110 GPA.

Leigon found that the measured vertical field stress (σ_v) could be approximated by that due to gravity

σ_v = 0.026H

where H = depth below the surface

The horizontal stress (σ_H) oriented perpendicular to the orebody was found to be
$\sigma_H = 6.1+0.045H$

The nature of the Liberman's formulation requires that use of a horizontal to vertical stress ratio instead of absolute values. In the mining zone (100-400m below surface) one finds that

H-depth(m)	100	200	300	400
σ_H/σ_v	4.1	.29	2.5	2.5

For the purpose of analysis the ratio $\beta = \sigma_H/\sigma_v = 3.05$

3.3 Geometry used for simulations

The model used to represent the mining of the orebody is shown diagrammatically in figure 2

The ellipse is made up of 98 elements each having a length along the vein of 3.72m and a vertival height of 3.5m. The dip of the vein has been assumed to be constant at 70° and the vein thickness is 18.3m. This geometry is the same as used by Groth. As can be seen, stopes 1, 2, and, 3 are represented by elements 68-85, 40-56 and 11-33 respectively.

Although one can 'mine' the computer model in exactly the same cut and fill sequence as was actually done, for present purposes, the mining has been divided into 10 sequences shown in table 3. The nomenclature indicates those elements which are solid and filled at any time. It is also possible to use unfilled elements but this was not done.

3.4 Comparison of predicted and field results

A great deal of fill pressure and stope wall convergence data was collected at Nasliden, and the number of comparison that could be made with model prediction is large. In this paper comparisons are presented for the same approximate location as discussed by Krauland.

Figures 3 through 4 for stope 3 and figure 5 for stope 2 reveals a good agreement between measured and predicted fill pressure value as a function of mining geometry. A comparison of predicted and measured wall convergences for several positions in stope 2 and 3 are shown in figure 6 through 7. In genaral the agreement is good. Another comparison which can be made is of the predicted and measured stresses in roof of stope 3. As can be seen the agreement is also quite good (Fig. 8)

Table 3

Sequence	solid element	fill element
1	1-10 13-19 42-67 70-90	11-12 40-41 68-69
2	1-10 15-39 43-67 71-98	11-14 40-42 68-70
3	1-10 18-39 43-67 73-98	11-17 40-42 68-72
4	1-10 20-39 44-67 75-98	11-19 40-43 68-74
5	1-10 23-39 46-67 77-98	11-22 40-45 68-76
6	1-10 25-39 50-67 80-98	11-24 40-49 68-79
7	1-10 27-39 50-67 80-98	11-26 40-49 68-79
8	1-10 29-39 53-67 82-98	11-28 40-52 68-81
9	1-10 31-39 55-67 84-98	11-28 40-54 68-83
10	1-10 33-39 57-67 86-98	11-32 40-56 68-85

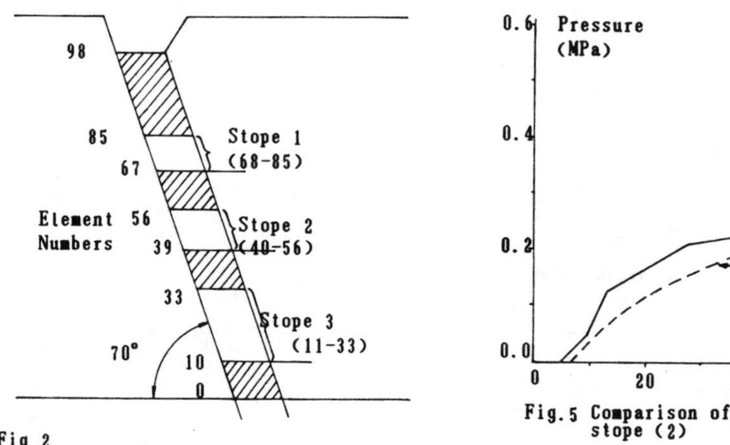

Fig.2

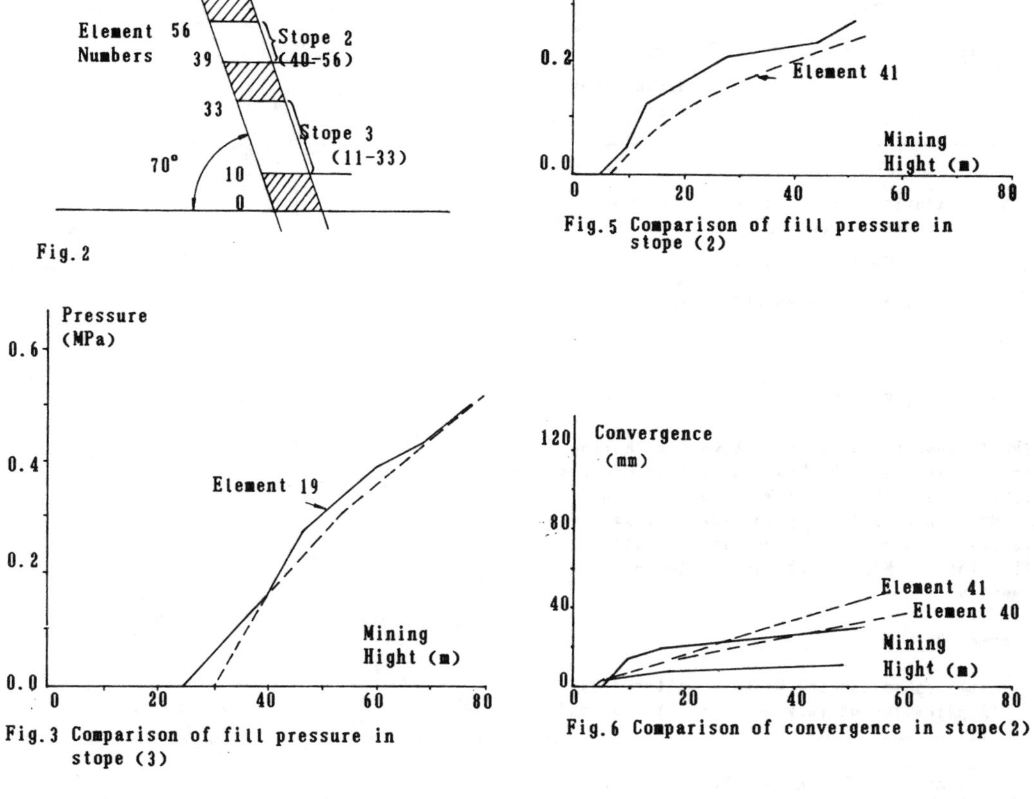

Fig.5 Comparison of fill pressure in stope (2)

Fig.3 Comparison of fill pressure in stope (3)

Fig.6 Comparison of convergence in stope(2)

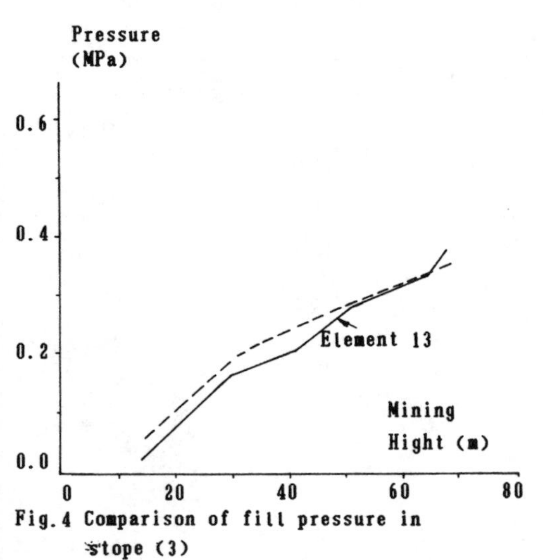

Fig.4 Comparison of fill pressure in stope (3)

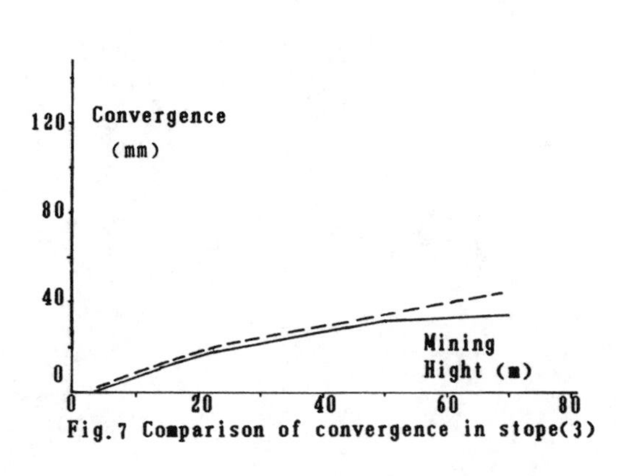

Fig.7 Comparison of convergence in stope(3)

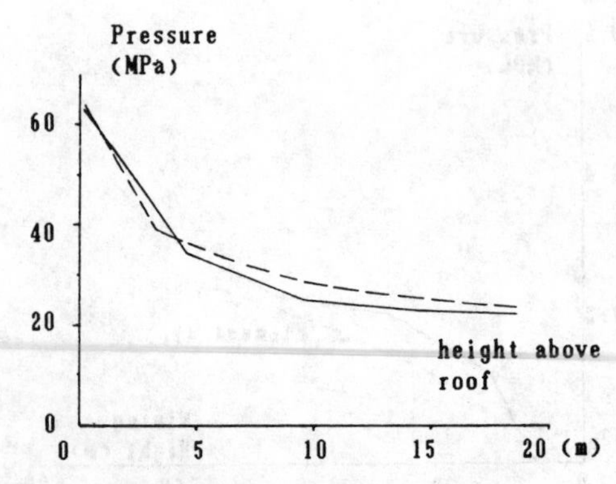

Fig.8 Comparison of stress distribution in roof of stope (3)

-------Predicted values

———Measured values

4. ACKNOWLEDGEMENTS

The fundamental works have been done during Mr. Ye visiting CSM. Now new fortran program is fixed in PDP-11 in JIM. Nobert Kreuland of Boliden Company supplied the fill material for use in one dimensional cosolidation tests. Mr. Falah helped to test the samples.

REFERENCES

1. Ove Stephansson and Michael, editors, 1981, 'Application of rock mechanics to cut and fill mining'

2. Liberman, and Khaimova,' Rock pressure on backfill with a non - linear shrinkage characteric', Soviet Min. Sci, 9, No, 2 1973 PP109-112

Numerical Methods in Geomechanics (Innsbruck 1988), Swoboda (ed.)
© 1988 Balkema, Rotterdam. *ISBN 90 6191 809 X*

The numerical simulation of floor heave in deep roadway on Long-feng mine

Yishan Pan, Cengdan Liu & Mengtao Zhang
Fuxin Mining Institute, People's Republic of China

ABSTRACT: In this paper, the finite element method for viscous fluid has been used to analyse the floor heave in deep roadway on Longfeng Mine. The back analysis method has been used. By adjusting the boundary conditions and viscosity coefficient of wall rock, we make the calculated values approach to the values measured by insitu experiment and the equivalent material scale-down test. The viscosity coefficient of mudstone thus obtained is 10^{15}-10^{16} poise.

1 INTRODUCTION

The floor heave in deep roadway is one of the most difficult problems when coal mining gets deeper and deeper. With the increasement of depth of coal mining and ground pressure, rock presents rheological characteristic. In deep roadway with soft rock, a great amounts of surrounding rockmass extrude to the space of the roadway, especially the floor lift. The depth of road 625 of Longfeng Mine in Fushun Bureau, is 700 meters and the wallrock is mudstone. This road has heaving nearly 20 years since 1968 when it is formed. Every year the floor of this road must be cut down 4-6 times and 0.3-0.5 meters every time. The speed of floor heave is nearly uniform, approximately 3-5 mm every day. The floor heave is a course of rock rheology. The viscoelastical and viscoplastic analysis on road floor heave by numerical method had been done by the authors before[5]. But the amount of floor heave calculated is attending towards stability as time goes on. The nearly uniform velocity of floor heave can not be simulated. In this paper, the wallrock of roadway 625 of Longfeng Mine is looked as Newton Viscous Fluid, the finite element method has been used to calculate the stress and velocity. By adjusting the boundary conditions and viscosity coefficient of wallrock, this paper uses back analysis method to make the calculated values fit the insitu experiment values and the values measured in the equivalent material creep scale-down test.

2 THE INSITU EXPERIMENT AND SIMULATION TEST

2.1 The insitu experiment

In november of 1984, Fushun Mine Bureau and Fuxin Mining Institute have begun to study the floor lift on road 625 of Lonfeng Mine. The velocity of floor heave is 3-5 mm every day[6].

2.2 The equivalent material creep scale-down test

Some equivalent material scale-down tests of roadway have been carried out, but the focus of study is on the deformation and failure of wallrock with the increasement of road depth and ground pressure. When the roadway is loaded with constant stress, the course of wallrock deformation and failure as time goes on has not been studied. In fact, the floor heave is a time-effected. There is no point in studying floor heave if the time effect of mudstone has been neglected.
The eqiupment for the equivalent material creep scale-down test has been made by the authors. Fig. 1 is its sketch map.
We have found a kind of equivalent material which can simulate the creep property of mudstone properly. We arrange measure points around the roadway. The flow law of wallrock has been gotten. The velocity of floor heave is 3.5 mm every day.

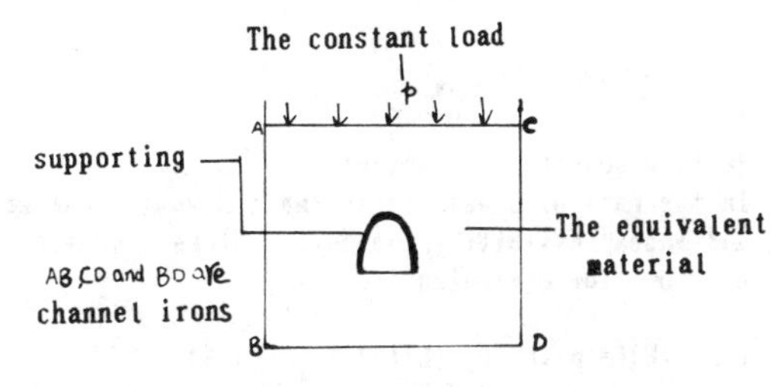

Fig. 1 The sketch map of simulation test.

3 THE NUMERICAL SIMULATION

3.1 The basic equations of Newton viscous fluid [1]

constitutive relations,

$$\{\sigma\}=\{m\}p+[D']\{\dot{\varepsilon}\} \qquad (3\text{-}1)$$

equilibrium equations,

$$\{L\}^T\{\sigma\}+\{b\}=0 \qquad (3\text{-}2)$$

continuity equation,

$$\partial\rho/\partial t + \nabla^T(\rho\{u\})=0 \qquad (3\text{-}3)$$

relationships between strains and velocities,

$$\{\dot{\varepsilon}\}=\{L\}\{u\} \qquad (3\text{-}4)$$

Where (in plane condition),

$$\{b\}=\{b_o\}-\rho\{c\}$$

$$\{c\}=\partial\{u\}/\partial t+(\nabla\{u\}^T)^T\{u\}$$

$\{\sigma\}^T=[\sigma_x,\ \sigma_y,\ \tau_{xy}]$ ----stresses

$\{\dot{\varepsilon}\}^T=\{\dot{\varepsilon}_x,\ \dot{\varepsilon}_y,\ \dot{\varepsilon}_{xy}\}$ rate of strains

p is fluid pressure.

ρ is fluid density.

$\{u\}^T=\{u_x, u_y\}$ fluid velocities.

$\{m\}^T=[1,1,0]$

$\{c\}^T=\{\dot{u}_x, \dot{u}_y\}$ acceleration vector.

$\{b_o\}^T=\rho\,[F_x, F_y]$ volume forces

$[D']=\mu\begin{bmatrix}2&0&0\\0&2&0\\0&0&1\end{bmatrix}$ viscosity coefficient matrix

$$\{L\}^T=\begin{bmatrix}\partial/\partial x, & 0, & \partial/\partial y\\ 0, & \partial/\partial y, & \partial/\partial x\end{bmatrix}$$

$$\nabla^T=[\partial/\partial x,\ \partial/\partial y]$$

μ is viscosity coefficient.

In the case of steady-state, small Reynolds number and incompressibility, the Navier-Stokes equation of slow flow Newtonian fluid is,

$$\{b_o\}+\{L\}^T\{m\}p+\{L\}^T[D']\{L\}\{u\}=0 \qquad (3\text{-}5)$$

3.2 The discretization of viscous flow equations

In this we shall discretize the velocity and pressure in terms of independent parameters,

$$\{u\}=\{N_u\}\{\alpha_u\},\quad p=\{N_p\}\{\alpha_p\} \qquad (3\text{-}6)$$

where, N_u and N_p are shape functions of velocity and pressure matrix respectively. $\{\alpha_u\}$, $\{\alpha_p\}$ are nodal velocity and nodal pressure vector respectively.

$$\{\dot{\varepsilon}\}=\{L\}\{N_u\}\{\alpha_u\}=[B]\{\alpha_u\} \qquad (3\text{-}7)$$

$$\{\sigma\}=\{m\}p+[D']\{\dot{\varepsilon}\}=\{m\}\{N_p\}\{\alpha_p\}+[D'][B]\{\alpha_u\} \qquad (3\text{-}8)$$

We use variational priciples and get the following equations,

$$([K]+[\bar{K}])\{\alpha_u\}+[K_p]\{\alpha_p\}+\{M\}\partial\{\alpha_u\}/\partial t+\{f\}=0 \qquad (3\text{-}9)$$

where,

$$[K]=\iint_\Omega [B]^T[D'][B]\,d\Omega$$

$$[\bar{K}]=\iint_\Omega \rho\{N_u\}^T(\nabla(\{N_u\}\{\alpha_u\})^T)^T\{N_u\}d\Omega$$

$$[K_p]=\iint_\Omega [B]^T\{m\}\{N_p\}d\Omega$$

$$[M]=\iint_\Omega \{N_u\}^T\rho\{N_u\}d\Omega$$

$$\{f\}=-\iint_\Omega \{N_u\}^T\{b_o\}d\Omega-\int_s\{N_u\}^T\{T\}ds$$

Ω is calculated area.
s is boundary line.
For incompressible fluid, the rate of volume strain equals 0

$$\{\dot{\varepsilon}_v\}=\{m\}^T[B]\{\alpha_u\}=0 \qquad (3\text{--}10)$$

we pre-multiply equation (3-10) by $\{N_p\}^T$ and then integrate in area Ω,

$$\iint_\Omega\{N_p\}^T\{m\}^T[B]\{\alpha_u\}d\Omega=0 \qquad (3\text{--}11)$$

$$\{K_p\}^T=\iint_\Omega\{N_p\}^T\{m\}^T[B]d\Omega=0$$

$$[K_p]^T\{\alpha_u\}=0 \qquad (3\text{-}12)$$

(3-12) and (3-9) can be written together as following,

$$\begin{bmatrix}[K]+[\bar{K}] & [K_p]\\ [K_p]^T & 0\end{bmatrix}\begin{bmatrix}\{\alpha_u\}\\ \{\alpha_p\}\end{bmatrix}+\begin{bmatrix}[M] & 0\\ 0 & 0\end{bmatrix}\frac{\partial}{\partial t}\begin{bmatrix}\{\alpha_u\}\\ \{\alpha_p\}\end{bmatrix}+\begin{bmatrix}\{f\}\\ 0\end{bmatrix}=0 \qquad (3\text{-}13)$$

We introduce penalty function [3] to simplify equation system (3-13).
Let us write p as following, $p=\lambda\,\dot{\varepsilon}_v$.
where λ is a large number. As $\dot{\varepsilon}_v\rightarrow 0$, p will be a

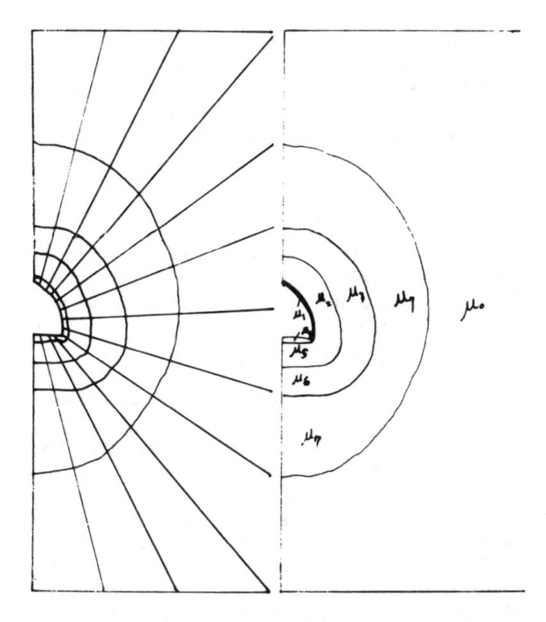

Fig. 2, The mesh of road   Fig. 3, The distribution of viscosity coefficient

finite quantity.

$$p=\lambda\{m\}^T[B]\{\alpha_u\} \quad (3\text{-}14)$$

$$[K]_p\{\alpha_p\}=\int_\Omega\int [B]^T[m]\lambda[m]^T[B]\{\alpha_u\}d\Omega$$

$$[\bar{\bar{K}}]=\int_\Omega\int [B]^T\{m\}\lambda[m]^T[B]d\Omega$$

$$[K]_p\{\alpha_p\}=[\bar{\bar{K}}]\{\alpha_u\} \quad (3\text{-}15)$$

In the case of steady-state and small Reynolds number, we get the discretization equations;

$$\{[K]+[\bar{\bar{K}}]\}\{\alpha_u\}+\{f\}=0 \quad (3\text{-}16)$$

3.3 The finite element division

The 40m×40m region around the roadway has been selected to calculate. Because of symmetry, the half of region is enough to calculate. we use 8-node elements with 60 elements and 215 nodes (as Fig. 2 shown)

3.4 The choice of parameters

As Fig. 3 shown, we choose different viscosity coefficient in different reigions.

(1) The viscosity coefficient of original rock is μ_0.

(2) The support of road 625 is arch-stone. Its viscosity coefficient μ_1 is higher than the viscosity of orginal rock μ_0. From reference [2], [3], we choose $\mu_1=10^{16}-10^{17}$ poise.

(3) The viscosity coefficient in the two failure regions of roof and sidewall is μ_2, μ_3

(4) The failure region in floor is 4m measureed by Super-Wave method. It can be divided into three subareas. The viscosity coefficient in three areas are μ_4, μ_5, μ_6 respectively.

(5) In region outside the floor failure region and roof and sidewall failure region, the viscosity coefficient is μ_7.

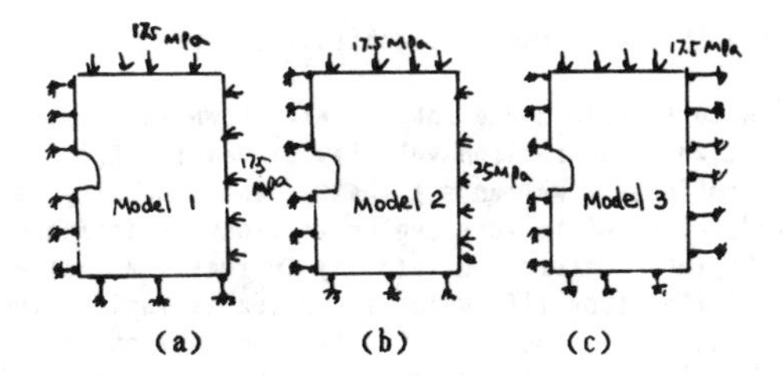

Fig. 4 The finite element calculation models.

3.5 The boundary conditions

In the upper boundary, the model bears 17.5 Mpa stress caused by the weight of 700 meters depth rock. The lower boundary is fixed in vertical direction. The left boundary is fixed in horizontal direction. we have studied three models;

First, in the right boundary, the model bears 17.5 Mpa. As Fig. 4 (a) shown.

Second, in the right boundary, the model bears 25. Mpa. As Fig. 4 (b) shown.

Third, The right boundary is fixed in horizontal direction. This model is correspond to the equivalent material creep scale-down test. As shown in Fig. 4 (c).

4 THE ANALYSIS OF CALCULATED RESULTS

4.1 The approach of calculated values to measured values

As mentioned above, measured by the equivalent material creep scale-down test, the velocity of floor lift is 3.5mm every day. By adjusting the viscosity coefficient of model 3, the calculated floor heave velocity is 3.52 mm every day (as table 3 shown). The floor heave velocity meassured by insitu experiment is 3-5mm every day. By adjusting the viscosity coefficient of wallrock of model 1, the calculated velocity of floor heave is 2.4mm, 3.7mm and 4.3mm every day. From the results mentioned above, we can conclude that it is feasible we use Newtonian viscous fluid model to simulate the floor heave problem.

4.2 The floor and roof contraction velocity

Table 1, Table 2 and Table 3 are shown the floor and roof contraction velocity of model 1, 2, 3. From Table 1, we can see that if the viscosity coefficient of failure region in floor is decreased, the roof contraction velocity increases a little. But the floor lift velocity increases rapidly. The velocity of floor heave is far exceed than the velocity of roof contraction. Which are correspond to the field experiment. The field observation shows that, in the serious floor heave roadway, the failure scope in floor is very deep. The rock in failure scope has no ability to bear load. The viscosity coefficient in this scope must be smaller. So it is feasible to prevent floor heave by reinforcing the floor rock[6].

From Table 2, it can be seen that if there is horizontal structural stress, the velocity of floor lift is far exceed roof contraction.

4.3 The flow law of wallrock

Fig.5 shows the flow law of wallrock of model 3. The calculated flow velocity of wallrock is correspond to the values measured in equivalent material creep scale-down test.

The floor of road 625 in Longfeng Mine has lifted nearly twenty years. The accumulative height of floor-heave rock is approximately 30 meters. Fig.4 can illstrate the source of so-enormous rock. There are three parts, In area A, the rock flows vertically into the road space. In which the floor heave rate is bigger; In area B, the rock flows horizontally into area A. In which the flow rate is smaller. But the scope of this region is larger, a little flow will give great amount of rock to region A; The third part is the replenishment flow of C region in sidewall.

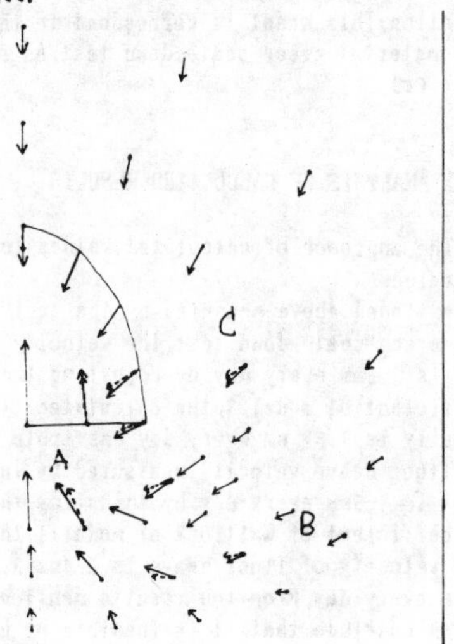

Fig.5 The flow law of wallrock of model 3

Notes, The real lines stand for calculated velocity vectors and the doted lines stand for velocity vectors measured in equivalent material creep scale down test. Only real lines have been drawn if the calculated vector and measured vector are same.

5 CONCLUSION

(1) It is feasible to adopt Newtonian viscous fluid model to analyse the roadway floor heave problem. This model can properly reflect the flow law of wallrock in floor heave roadway.

(2) If we can accurately measure the viscosity coefficient of mudstone by experiment, it will be more better to use viscous fluid model to calculate floor heave problem.

(3) Beause this paper only approach calculated velocity to the velocity of floor and roof contraction, the accuracy of back analysis results is effected. It will be more better if we make the calculated values approach to the much more values of road surface measured by insitu experiment. Because the measured results in simulation test is very many, the approach results is better in model 3.

(4) Because there is no measured stress result, we have not analysed the stress result calculated by Newtonian viscous fluid model.

REFERENCES

[1] zienkiewicz, O.C., The Finite Element Method, 3rd (1977).

[2] Griggs, D. and Handin, J. et al, Rock Deformation (1960)

[3] Eirich, F.R., et al, Rheology, Vol.3.(1960).

[4] Wang Ren, Ling beiyuan, Sun Xunyin, An analysis of tunnel deformations by the finite element method. ACTA MECHANICAL SINICA, Vol.17, No.2(1985).

[5] Pan Yishan, Li Guozhen and zhang Mengtao, Experimental study and numerical simulation on floor lift in deep roadway, Proceeding of international symposium on tunneling for water resources and power, India (1988).

[6] Lin Daogong, Cao Ronghuan, Xu Zhenhe, wei Hongxuan and Pan Yishan (1986), An experiment on the control of floor heave in a deep roadway with soft surrounding rock at Longfeng Mine, Proceedings (II) of International symposium on Modern Coal Mining Techanology, Fuxin, China.

Table 1, The velocity of floor and roof contraction (model 1)

Viscosity coefficient 10^{18} (poise)								contraction velocity (mm/day)	
μ_0	μ_1	μ_2	μ_3	μ_4	μ_5	μ_6	μ_7	roof	floor
1.363	2.745	1.190	1.276	0.015	0.075	0.147	1.270	2.519	6.146
1.363	2.745	1.190	1.276	0.075	0.092	0.412	1.270	2.030	4.320
1.363	2.745	1.190	1.276	0.084	0.233	0.438	1.270	1.876	3.705
1.363	2.745	1.190	1.276	0.084	0.265	0.671	1.270	1.737	2.848
1.363	2.745	1.190	1.276	0.147	0.758	0.844	1.363	1.571	1.881
1.363	1.363	1.363	1.363	0.931	1.017	1.103	1.363	1.572	1.702
1.363	2.745	1.363	1.363	0.752	0.931	1.017	1.363	1.300	1.650

Table 2, The velocity of floor and roof contraction (model 2)

Viscosity coefficient 10^{18} (poise)								contraction velocity (mm/day)	
μ_0	μ_1	μ_2	μ_3	μ_4	μ_5	μ_6	μ_7	roof	floor
1.363	2.745	1.190	1.276	0.084	0.179	0.438	1.276	3.580	7.499
1.622	2.745	1.190	1.276	0.084	0.179	0.438	1.363	1.543	6.233
1.708	2.745	1.190	1.276	0.084	0.179	0.524	1.449	1.498	5.912
1.795	2.745	1.190	1.276	0.084	0.179	0.524	1.535	1.286	5.614
1.788	2.745	1.190	1.276	0.084	0.222	0.611	1.535	1.302	5.219
1.881	2.745	1.190	1.276	0.084	0.308	0.611	1.622	1.207	4.937

Table 3, The velocity of floor and roof contraction (model 3)

Viscosity coefficient 10^{18} (poise)								contraction velocity (mm/day)	
μ_0	μ_1	μ_2	μ_3	μ_4	μ_5	μ_6	μ_7	roof	floor
1.535	2.745	1.190	1.276	0.023	0.086	0.179	1.449	2.057	5.437
1.987	2.745	1.363	1.449	0.084	0.173	0.265	1.728	1.662	3.744
2.246	2.745	1.535	1.622	0.084	0.173	0.265	1.987	1.557	3.532
1.795	2.745	1.190	1.276	0.084	0.222	0.524	1.535	1.619	2.931
1.881	2.745	1.190	1.276	0.084	0.308	0.611	1.616	1.547	2.568

Numerical Methods in Geomechanics (Innsbruck 1988), Swoboda (ed.)
© 1988 Balkema, Rotterdam. ISBN 90 6191 809 X

A study of computer engineering classification method

Y.M.Lin
Northeast University of Technology, Shenyang, People's Republic of China

Abstract: In the world there are many different rock mass engineering classification method, which can be used to give an estimate of classes for the total sampling probabilities. Most traditional classification methods belong to one type, which aims to build various kinds of tables based on engineering judgement without critical mathematical analysis. There has been virtually no change in rock mass classification procedure which still rely on manual interpretation of classes despite the huge advances being made in microcomputer data base systems. This paper introduces the research results of an alternative classification approach. This approach is to collect a significant number of measurements for a wide rang of parameters from surrounding rock mass of openings, and then develop several computer engineering classification systems using statistical and cluster analysis methods according the requirement for each mine. The important advantage of this method is that, compilers of systems don't previously or artifically indicate the standard of engineering classification, merely run special computer procedures to select the optimum critirions and estimate optimum number of classes, finally a reasonable standard of engineering classification will be given by computer based on selected critirions. The application of this method in mine reports significant benefits due to the satisfactory performance of the method.

1. INTRODUCTION

With the development in mining engineering, there has been a need to divide rock mass into limited classes and give some simple numerical indexes reliable enough for mining or geological engineer to determine the stability of rock engineering and design an appropriate support system.

In the recent years a large number of classification systems have been put forward and improved upon. However, there is a large range of qualitative and quantitative parameters in both field and laboratory measurements which has been used in the formation of existing rock classification systems. Often the choice of parameters is made subjectively and even may be based on the experience of only few persons. Thus, the efficacy of a particular rock classification system will by influenced by the subjective judgement of the compiler of the system.

An alternative approach is to collect a significant number of measurements for a wide range of parameters from rock constructions and rock samples, selected and obtained in the same place of construction, then develop a classification system using statistical and cluster analysis methods. Such a system named 'DYNAMIC CLUSTER CLASSIFICATION METHOD' has been developed by author (1984) and reported in the International Symposium on mining Technology and Science, which was held in P.R.C. . This paper is better to read in conjunction with previous published papers (References 1 and 2), in which details of various aspects of the work are described.

2. A SHORT DESCRIPTION OF DYNAMIC CLUSTER CLASSIFICATION METHOD

The essence of this method is to make an approximate initial classification, then repeated revisions will be made based on some principles until an optimal classification system is obtained. The procedure of the dynamic cluster analysis for rock mass performance is shown in Figure 1.

The initial classification of each sample can be estimated arbitrarily. However, the following formula has been developed to provide a convenient initial estaimate:

$$NC(I)=IFIX(-\frac{(K-1)*(AMAX-sum(I))}{(AMAX-AMIN)}+0.5)+1 \quad (1)$$

where IFIX denoted the integer of the expression in the brackets, sum (I)= $\sum_{j=1}^{m} X_{ij}$, AMAX and AMIN are the maximum and minimum values of the sum(I), K is the suggested number of classes, i represents each sample, i=1,2......n, j represents the j-th variable of each sample.

The concept of nearest distance is very important in dynamic cluster analysis. Suppose we have n samples with two variables, then it is easy to plot them using X-Y coordinate system as shown in figure 2. Suppose we have n samples with m variables. They have to be regarded as n points in the m dimension space from the geometrical point of view. The nearest distance is estimated using the mathematical formula as follows:

$$D_{ij} = \left(\sum_{k=1}^{m} (X_{ik} - X_{jk})^2 \right)^{1/2} \quad (2)$$

where i and j are two different samples, k is the k-th variable of each sample, D is the Euclidean distance.

The magnitude of the Euclidean distance is dependent on the dimensional units for each sample variable. In order to make the Euclidean distance independent of units. All sample data must be standardised. A standard value is obtained by calculating the factor which represents the variable's value with respect to the mean value in terms of a number of standard deviations. Values are converted back to appropriate units for the presentation of results.

After the initial classification is obtained for each sample, the centre of gravity for eachgroup (class) is calculated and used as initial clustering point. The calculation of the Euclidean distance from each sample to various gravity centres is made. A special value, called ' the classification function DS ' is obtained using:

$$DS= \sum_{i=1}^{n} (D_{i}, B(k))^2, \; k=1,2.....kk; \quad (3)$$

where B(k) is the gravity centre of each group, i represents each sample, i=1,2... n, D is the Euclidean distance from each sample to appropriate gravity centre.

The classification function value DS is used to assess the suitability of the classification.Whenever the classification is amended. The function DS should be calculated also.The principle for the amendment of classification is the minimization of the classification function value DS. All the above concepts and processes have been incorporated in a computer program* specially developed for a design classification method suitable for geotechnical application.

3. TEORITICAL STUDY OF NEW CLASSIFICATION METHOD.

3.1 Comparasion with existing rock mass classification methods

In the world there are many different rock mass classification systems, which can be used to give an estimate of classes for the total sampling probabilities. Most classification methods belong to one type, which aims to build various kinds of tables based on engineering judgement without critical mathematical analysis. In the general case, a classification table consist of three elements, namely:

1) the number of classes;
2) the parameters for which the data is accumulated.
3) the class boundary values.

Till now no classification system has been put forward to analyse the components of these tables or to build theoritical basis for the selection of optimal three elements.

The most significant advantage of our classification system based on dynamic cluster analysis, which has over traditional systems is objectivity in the determination of three elements based on modern mathematical analysis.

Using existing microcomputer leased data handling facilities, the relevance of individual parameters, an optimal number of classes and appropriate class boundary values of system can be easily obtained.

3.2 Choice of geotechnical parameters

There are numerous factors which effect the host rock stability, such as magnitude and distribution of natural stress, rock structure, mechanical property of rock mass and its structure plane, shape and dimension of drift section and construction method, etc. All these factors are intricate and condition each other. Among these factors, the rock structure and mechanical properties are intrinsic in the rock mass, which reflect the quality of rock mass itself; while the size of drift section and construction method are artifical and changeable factors. Therefore, different factors will cause different effect on the host rock stability.

The rock structure and mechanical properties of the rock mass play an important role in the host rock stability, so a basic index should be selected from them.

*Programme is only to be sent to somebody,, who is interested by this new classification method and connected with the author at address 'Northeast University of Technology, Shenyang, Liaoning, P.R.C.'

The natural stress field (dimension and distribution) of the rock mass is an intrinsic factor and probably varies with orientation of drifts. This factor will be shown in the displacement of surrounding rock of the drift, so it is unnecessary to be considered separately.

The shape and dimension of drift section are important, as well as the construction method, but they are sensible mainly in case of soft rock. Therefore, viewing from a whole mine, it will not change dramatically and can be fixed in the case of field measurements of parameters. During the case that section sizes are quite different, an amendable index containing this factor is recommended to use in our system.

Analyse some dozens of parameters which have been used for different systems of rock mass stability classification both at home and abroad (see Table 1), it is discovered that these parameters could be basically divided into four categories physically.

Meanwhile, a large number of measurements for a lot of parameters selected from the four different categories in table 1 have been accomplished. The original data list of a representative mine are prepared with following 14 parameters:

1) I_s -Point load strength;
2) n_b/N -Proportion of fissure sample of the total samples;
3) V_{mass} -Acoustic velocity of rock masses;
4) V_{rock} -Sonic wave velocity of rock samples;
5) I -Completness coefficient;

$$I = \left(\frac{V_{rock}}{V_{mass}}\right)^2$$

6) Ls -Fissure coefficient;

$$Ls = \frac{(V_{rock})^2 - (V_{mass})^2}{(V_{rock})^2}$$

7) $\overline{AB}$ -Acoustic attenuation coefficient;
8) T -Displacement-stabilizing time;
9) $\frac{du}{dt}$ -Initial displacement rate;
10) U_{max} -Maximum value of displacement;
11) d_p -Average of joint spacings;
12) J_v -Block size;
13) N_m -Rock mass structure modulus;

$$N_m = \frac{A.1}{d_1.d_2.d_3}$$

A=cross section area of drift;
d_1,d_2,d_3= rock block size;

14) R.B. -Aspect ratio of rock block.

Using statistical correlation of various parameters as a guide for the selection of a reasonable set of parameters for the dynamic cluster analysis, the correlation coefficients for linear regression between each two parameters for all combinations are calculated and listed in table 2. For pairs of parameters in table 2, which show high correlation according the number of samples (here we take 0.7 as a critical numerical value), one of the pair can be rejected from the cluster analysis to improve the efficiency of the procedure. For instance: one of the most readily avaliable pairs of table data is completness coefficient I and fissure coefficient Ls (their correlation coefficient is -1.0). Finally, through probability analysis, four indexes have been selected as a optimal set of classification parameters based on existing data base, which are:

1) Rock mass structure index;
2) Point load strength;

Table 1. Parameters for various systems.

Category of rock mass property	Category of completeness of rock mass	Category of deformation of rock and rock mass	Category of structure plane and rock structure
Index name, sym.	Index name, sym.	Index name, sym.	Index name, sym.
1. Uniaxial compressive strength Rc	Acoustic velocity in rock mass Vmass	Elastic modulus E	Rock structure type
2. Uniaxial tensile strength Rt	Rock quality designation RQD	Poisson's ratio μ	Number of joint set Jn
3. Cohesion C	Completness Coefficient I	Resistance coefficient Ko	Joint Roughness Jr
4. Angle of internal friction ϕ	Fissure coefficient Ls	Displacement coefficient E_o/E (in plate bearing test / modulus of elasticity indoors)	Joint Alteration Ja
5. Protodjakonov's coefficient f	Modulus of dynamic elasticity Ed	displacement Stabilizing time T	Fracture frequency Jo
6. Point-load strength Is	Sound attenuation coefficient $\overline{AB}$	Initial displacement rate $\frac{du}{dt}$	The average of joint spacing dp (dimension index of rock block)
7. Density γ	Vadose rate Vw	Maximum displacement Umax	Rock mass structural modulas Nm
8. Weathering coefficient Ky	Nortar absorption during grout Vc	Stress effect coefficient SRF	Aspect ratio of rock block R.B.
9. Proportion of fissure samples among total n_b/N	Effect of fissure water Jw	Modulus ratio E/R	
10. Poisson's ratio μ	Acoustic velocity of rock block Vrock	Self-bearing time of rock wall Tc	

Table 2. Table of Coefficient Correlation.

	Is	n_b/N	V_{mass}	V_{rock}	I	Ls	$\overline{AB}$	T	$\frac{du}{dt}$	U_{max}	d_p	J_v	N_m	R.B.
Is	1.000	-0.848	0.819	-0.605	-0.812	-0.812	-0.900	-0.520	0.477	-0.292	0.213	-0.401	-0.570	-0.461
n_b/N	-0.848	1.000	-0.776	0.455	-0.707	0.706	0.734	0.676	-0.314	0.401	-0.274	0.365	0.423	0.046
V_{mass}	0.819	-0.776	1.000	0.766	0.967	-0.967	-0.937	-0.690	0.159	-0.522	0.121	-0.352	-0.388	-0.326
V_{rock}	-0.605	0.455	0.766	1.000	-0.896	0.896	0.760	0.257	-0.325	0.276	-0.040	0.301	0.309	0.368
I	-0.812	-0.707	0.967	-0.896	1.000	-1.000	-0.850	-0.557	0.260	-0.443	0.112	-0.376	-0.397	-0.396
Ls	-0.812	0.706	-0.967	0.896	-1.000	1.000	0.949	0.559	-0.259	0.445	-0.113	0.376	0.398	0.393
$\overline{AB}$	-0.900	0.734	-0.937	0.760	-0.850	0.949	1.000	0.502	-0.317	0.399	-0.172	0.441	0.424	0.531
T	-0.520	0.676	-0.690	0.257	-0.557	0.559	0.502	1.000	0.400	0.771	-0.162	0.281	0.391	-0.249
$\frac{du}{dt}$	0.477	-0.314	0.159	-0.325	0.260	-0.259	-0.317	0.400	1.000	0.546	0.225	-0.218	-0.046	-0.583
U_{max}	-0.292	0.401	-0.522	0.276	-0.443	0.445	0.399	0.771	0.546	1.000	0.461	-0.282	-0.043	-0.300
d_p	0.213	-0.274	0.121	-0.040	0.112	-0.113	-0.172	-0.162	0.225	0.461	1.000	-0.934	-0.007	-0.015
J_v	-0.401	0.365	-0.352	0.301	-0.376	0.376	0.441	0.281	-0.218	-0.282	-0.934	1.000	0.913	0.194
N_m	-0.570	0.423	-0.388	0.309	-0.397	0.398	0.424	0.391	-0.046	-0.043	-0.007	0.913	1.000	-0.009
R.B.	-0.461	0.046	-0.326	0.368	-0.396	0.393	0.531	-0.249	-0.583	-0.300	-0.015	0.194	-0.009	1.000

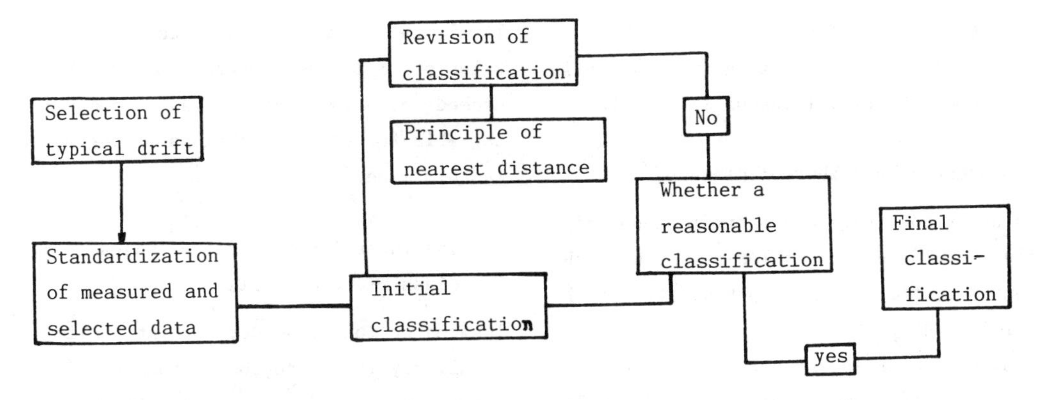

Figure 1. Flowchart for the dynamic cluster analysis.

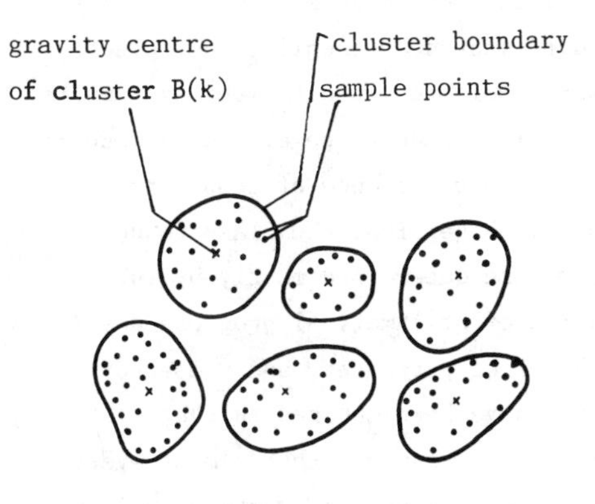

Figure 2. Concept of nearest distance

Table 3. Summary classification table

Divided To	1	2	3	4	5	6	7	8	9	10	11	12	13	14
2	1	1	1	1	1	1	1	2	2	2	2	2	2	1
3	1	1	1	1	1	1	1	2	3	3	3	2	3	1
4	2	2	2	2	2	2	2	3	4	4	4	3	4	1
5	2	2	2	2	2	2	2	4	5	5	5	3	5	1
6	3	2	2	2	2	2	2	4	6	6	5	4	5	1

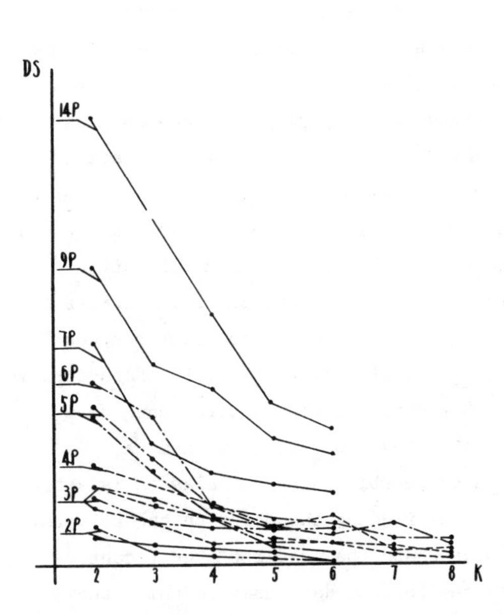

Figure 3. Classification function value DS versus caterory number K for Lai Wu Iron Mine

K-number of classes

P-parameters

2P-a set of 2 parameters

3) Acoustic velocity;

4) Displacement-stabilizing time, which is a comprehensive index of multifactors.

3.3 Acceptable number of classes

In order to obtain the optimal number of cluster. The upper and lower limits of the number of classes are input as part of the original data. Results appropriate to each number of classes are output automatically in a form of summary classification table (Table 3).

Just as the statistical correlation of various parameters is used to give an objective selection of reasonable set of parameters, the classification function value DS can be used as a indicator for the selection of acceptable number of classes. Because after an initial estimate, the arrangement of samples into classes was adjusted to minimize the classification function value DS for each number of classes between the specified upper and lower bounds. Meanwhile, the value of classification function DS is a function of the number of classes, therefore the relation between DS and number of classes D may be characterized by DS-K curve on figure 3 for each data set obtained from Lai Wu iron mine.

An acceptable number of classes is determined by finding the point on the DS vers. K plot where the gradient significantly reduces.For the data used in this study, the plots in Fig.3 showed that no matter which combination of available parameters is inputed to the computer under the sampling conditions of Ma Zhuang Mine Area in Lai Wu iron mine, its decline gradient is remarkablly changed when K=5. As pointed out previously, the decreases of DS means the further reasonable adjustment of the classification system. Regarding to the little change which take place in the value of DS if the number of classes K exceeds 5, a value of K=5 was judged acceptable for all combinations of parameters in this mine.

3.4 The inner law of three elements of a classification table

The contention of academic point of view was lively in the region of rock stability classification. Some compilers insist to use single index because of its simplicity and another held just opposite point of view, considering that the more parameters in use, the more accurate rock stability could be judged. Meanwhile, the selected number of classes in different systems is altered in a wide range of choice according to the experience of compilers (from 3 to 20). Otherwise, the class boundary values were determined mostly by interpolation of arithmetical progress despite which set of parameters and number of classes have been choiced in that system. Therefore, we can say that almost every compiler of traditional systems outline his three elements individually and consider this is a correct way to determine and compile a classification system.

After a study of the inner law of three elements based on a significant number of measurement and computer calculation, it is discovered that these three elements are interdepedent, if any two elements of them had been estimated, the third can not be developed subjectively, otherwise a classification table are very likely to contain inconsistances or redundancies. Dynamic cluster classification method takes only two of these three elements, usually the optimal set of parameters and optimal number of classes as described

Table 4. The four-categories classified with four parameters

Categories / Classification index	Gravity centre of first category	Gravity centre of second category	Gravity centre of third category	Gravity centre of fourth category
Point-load-strength Is, in MPa	6.79	10.32	5.64	4.36
Sonic wave velocity V_{mass}, in M/sec.	5202.000	4787.857	3070.000	2946.000
Stable displacement time T, in day	12.0	15.40	17.110	22.840
Module of rock mass structure N_m, in blocks	75.0	144.00	119.340	763.130

above, and then uses cluster analysis method to obtain the third element as a function of the other two. A representative table of classification obtained from computer and satisfied with the inner law of elements is given in table 4. The boundary values in this table are determined objectivity, being rid of oneself the subjective judgement of the compiler of the system.

REFERENCES

Lin Yun-mei, Li zhao-guan, zhang jing-yang and Wang wei-gang. 1985. Dynamic cluster analysis applied to engineering stability classification of rock mass. International Symposium on Mining Technology and Science. Proceeding of Mine Construction Session, cd 8: 1-6.

Lin Yun-mei. 1986. Cluster analysis through computer and its application. Proceedings of the Internaional Symposium on Engineering in Complex Rock Formations. Science Press, Beijing, China: 82-90.

Lin yun-mei and Wang wei-gang, 1983. A study on engineering stability classification of rock masses. Journal of Northeast University of Technology (Quarterly), No.1 (Sum 34): 23-38.

Fang kai-tai and Pan enpei. 1982. Cluster analysis. Geological Press, Beijing, China.

Numerical Methods in Geomechanics (Innsbruck 1988), Swoboda (ed.)
© 1988 Balkema, Rotterdam. ISBN 90 6191 809 X

The application of digital modelling to geomechanical problems of Polish mining

H.Filcek & J.Walaszczyk
Institute of Mining Geomechanics, Krakau, Poland

ABSTRACT: On the basis of the Institute own investigations as well as of other scientific institutions in Poland the authors have presented the examples of applications of numerical methods,especially the method of finite elements,to modelling of: rheological phenomena in the rockmass, phenomena of rockmass relief defined as a violent liberation of energy in elastic rocks and of considerable energy, phenomena of explosions , stability of the pillar-chamber field while taking into account the post-damage characteristics of pillars,elastic vibrations in heterogenous media, phenomena similar to crumps.

This paper presents examples of the application of digital simulation for the geomechanical problems of the Polish mining.It has been based on the results of investigations which have been carried out in the Institute of Mining Geomechanics of the St.Staszic Academy of Mining and Metallurgy in Cracow.As this paper is of informative character,the details of numerical solutions will not be discussed while their general ideas as well as the way of application in geomechanics will be indicated.

The first example concerns the digital forecasting of the rock - mass deformations in the course of the bore-hole exploitation of sulphur deposits /1/.This type of exploitation is accompanied by the erruptions,i.e.uncontrolled breakouts and outflows of deposit waters onto the surface.Theoretical investigations and field experiments proved that the damages of exploitation bore-holes,caused by the rock-mass deformations were the main cause of the erruption formation.Assuming that the surface deformations are the reprezentations of the phenomena occurring inside the rock-mass,a thesis was put forward that the moderation of these deformations by means of an appropriate time-space control of the exploitation process would contribute to the decrase of the number of erruptions.The optimalization of the bore-hole exploitation system required a new methodology of rock-mass deformation forecasting to be worked out.On the basis of the experiences of underground mining,an attempt of the application of the Budryk-Knothe's theory for this purpose and the solutions based on the discrete media and the theory of elasticity was made.
A file of computing programs was worked out which enable the deformation indeces to be determined at any point of the surface area and, in case of the programs which apply the solutions of the theory of elasticity and theory of discrete media,also inside the rock-mass. These programs assumed that the exploitation zone around a single bore-hole is of the shape of a cubicoid,cylinder,cone ar an other rotational solid whose dimensions depend on the resources and output. The principle of superposition was used to determine the values of deformation indeces which result from the exploitation with many bore - holes.The programs were written in Fortran for the ODRA 1305 computers,in Basic for the Spectrum microcomputers and in Pascal for the IBM PC microcomputers.

These computers carried out the computations of the formations of the rock-mass deformation indeces at verious displacements of exploitation bore-holes and various ways of introducing the bore-holes into exploitation. The values of inclinations curvatures and horizontal deformations in the region of exploitation bore-holes were considered to be the optimization criterion. On this basis it was observed that the highest moderation of the subsiding trough can be obtained by means of the multirow exploitation and the appropriate control of introducing the successive rows. The application of the proposed system in the Jeziórko mine contributed to a significant decrease of the erruption and, in some regions, to their total elimination. The discussed programs can be also applied for the forecasting of rock-mass deformation resulting from the bore hole exploitation of other raw materials /e.g. salt/and in underground mining. The enclosed computation example presents the forecast of rock-mass deformation according to the Budryk-Knothe´s theory in the initial exploitation period of one of the fields. The data input consists in giving the parametrs of the theory being applied, the depth of strata deposition, the bore-hole coordinates in the accepted of coordinates, resources and the output. The dimensions of the exploited zone around the bore-hole are computed automatically due to appropriate empirical formula. The results of the computations of deformations indeces in chosen points or lines are given in the form of tables. The prepared programs also enable the graphs of chosen indeces to be obtained by means of a printer or plotter. The enclosed example gives only results /Figures 2, 3/ in the form of settlement graphs W and total inclination graphs T in two lines which overlap the bore-hole rows /Fig. 1/. In Fig. 1 the system x-y denotes the location of the bore-holes /circles/and the exploited zones /rectangulars/ around respective bore-holes. Figure 2 concerns the bore-holes placed along the line y=100/m/ while. Fig. 3 concerns the holes along the line ȳ=155.6/m/ .

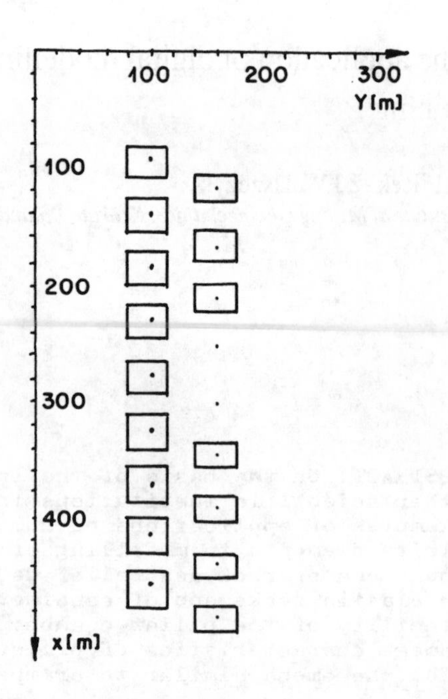

Figure 1

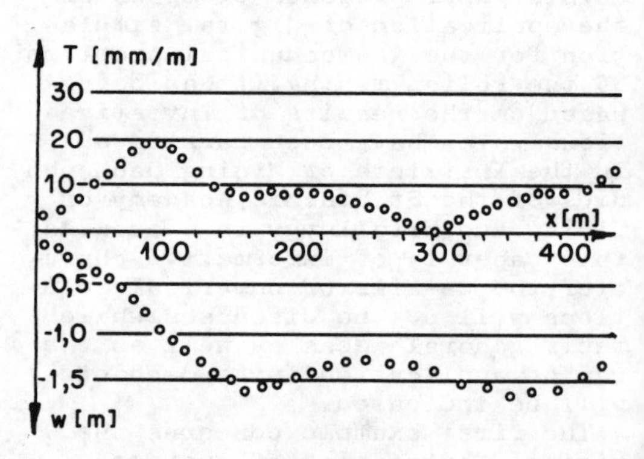

Figure 2

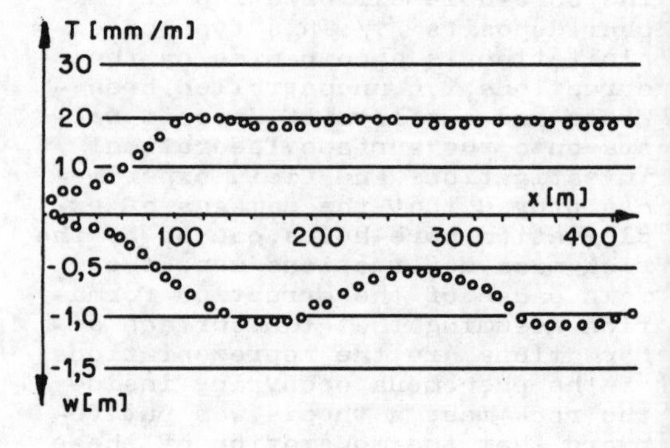

Figure 3

An instance of the narrowing of a dog heading in time in one of the mines of the Lublin Cooal Region can prove the significance of the rock mass rheological properties /fig.4/.

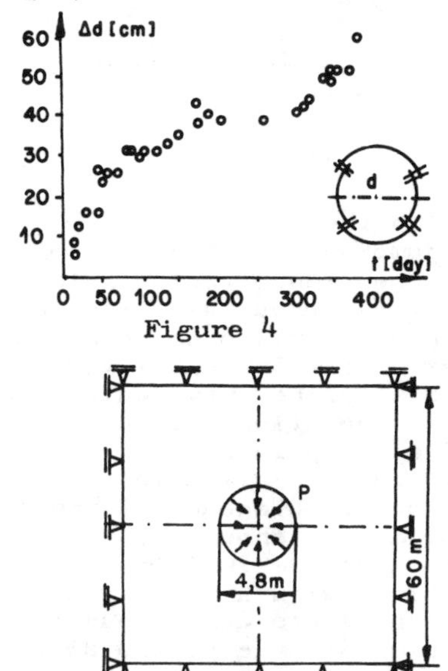

Figure 4

Figure 5

A relatively simple rheological model was used as a mathematical description of the phenomenon ilustrated by Fig.4. Its behaviour in the uniaxical case can be described, as follows:

if $\sigma < \sigma_{gr}$, then $\varepsilon = \frac{\sigma}{E}$

while if $\sigma \geqslant \sigma_{gr}$, then $\varepsilon = \frac{\sigma}{E} + \sigma \frac{t}{\eta}$

where: σ - stress, σ_{gr} - yield point, E - Young's modulus, t-time, η - Trouton's viscosity coefficient, ε - axial strain.

The boundary conditions for a digital rock-mass model in the vicinity the dog heading of Fig.4 are presented at Fig.5. This model is used for determining the effect of the dog heading execution on the state of stress and rock-mass displacement therefore imaginary forces P act on the dog heading contour which are equal to the primary state of stress in the rock-mass. The rock-mass model was verified on the basis of accessible information concerning the behaviour of rocks in the vicinity of dog headings in the Lublin Coal Region as well as on the basis of laboratory tests of the physicomechanical properties of rockes from this Region which have been carried out so far. The results of pilot rheological investigations carried out in the Institute of Mining Geomechanics of the Academy of Mining and Metallurgy as well as the results of the displacement of the dog heading contour in the Bogdanka mine in the Lublin Coal Region were used in this paper. A digital rock-mass model of the Bogdanka mine was constructed in the direct vicinity of the dog heading in the V29/8 lining. The interaction between the lining and the rock-mass was accepted according to the V29/8 lining characteristics, given by the GIG laboratory. It was also assumed that the rock-mass behaved according to Maxwell's rheological model. Detailed computation results are given in the work /2/. They announce, among other things, that: a/The numerical model of the rock-mass creep gives the results which are qualitatively similar to the results of mine observations and this respect the verification was postive. b/A quantitative compatibility of the model with a real phenomenon is still not fully satisfactory. The reason of this is the impossibility of answering univocally the question: which material constans should be introduced into a numerical model? It is expected that the future laboratory investigations, and the investigations in situ mainly, will improve this situation.

The next example illustrates the application of special simplified /plate-bar/ elements for the rock-mass modelling /3/. The fragment of the pillar-chamber field /Fig.6/

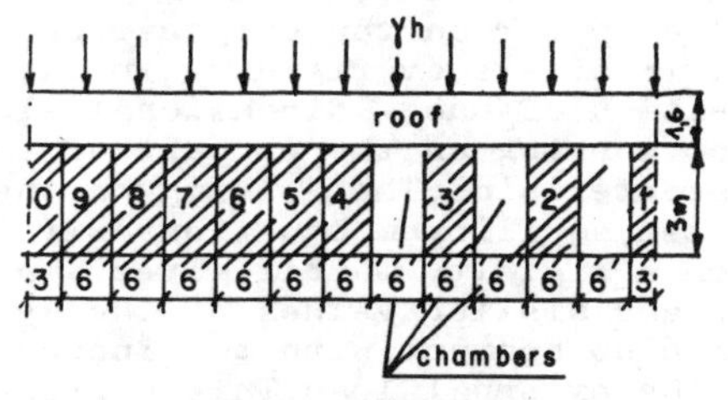

Figure 6

was modelled with bar elements /pilars and bed/ and with plate elements /roof/. It has been showed /3/ that the simplified plate-bar model can be used for the evalu-

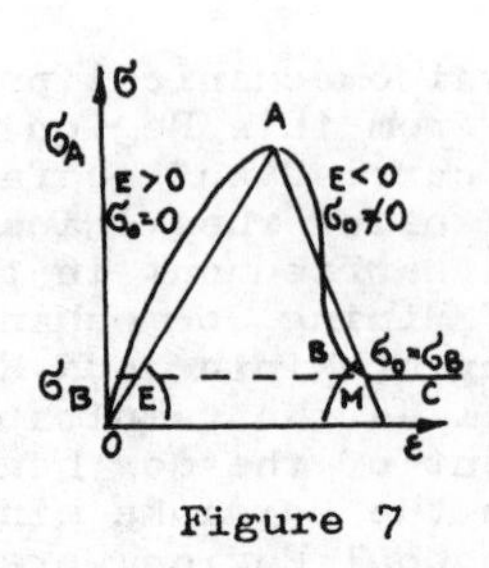

Figure 7

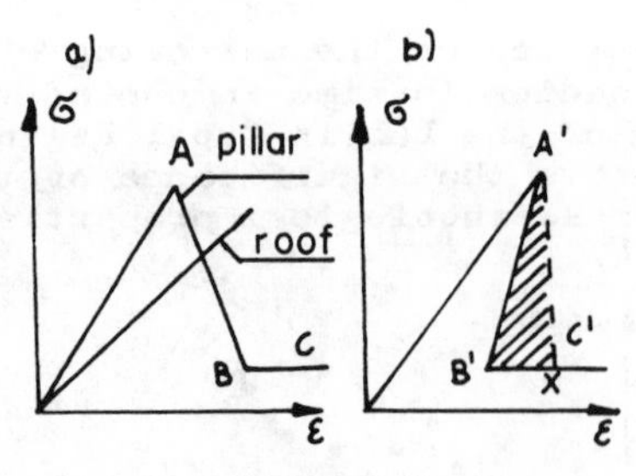

Figure 8

ation of stress in pillars.This model was then applied /3/ to evaluate the stability of the pillar - chamber field,taking into consideration the results of experimental investigations according to which the σ-ε characteristics for the rocks of the Legnica-Głogów Cooper Region is the same as in Fig.7 /curve OAC/.This characteristics was linearized in sections /3 sections in Fig.7 OA, AB, B C/. Generally,each section of the strain stress characterisics can be described by the equation: $\sigma = E\varepsilon + \sigma_o$ in which on the AB and BC section a non-zero value σ_o occurs which is treated as the initial stress and which is introduced into the computations in the form of nodal forces influencing the system. These forces are added at nodes to the forces resulting from the internal load.The solution consists in solving the equilibrium equation of the method oh finite elements F=k.u where: F-matrix of node forces , k-matrix of system rigidity, u-matrix of displacements of node points.Let us consider a pillar-chamber field,modelled by means of an elastic plate and one-dimensional elements.If we assume a sectionally linear characteristics of a high weakening modulus,then a negative value appear on the diagonal of the matrix of system rigidity in the position of one - dimensional element,working on the AB part of the characterisics.The solving of the system equilibrum equations with such a rigidity matrix gives smaller and smaller values of the system displacement with the increase of the external load.This corresponds to the work on the A B section of the global characterstics /Fig.8/of the system which is physically impossible because the system has a surplus of energy /the A B K area/ which causes the system to jump to the point X.

Such a jump has the features of a dynamic phenomenon /potential energy is transformed into kinetic and at a new equilibrum point the node has a certain velocity which causes its further movement/which, in mining conditions,can be of a frrm of a crump.This a numerical model in which a negative value appears on the diagonal of the main rigidity matrix should be considered as a model simulating the mining field with a local instability.If there are conditions in the rock-mass for a gentle discharge of the energy surplus then the system may transmit to the point X and continue working on the XC section of the characteristics. This is connected with a considerable drop of stresess in the pillar subjected to jumping.This stress must be received by the surrounding rock-mass /neighbouring pillars,undisturbed soil/.This may be the reason of local instabilities in neighbouring pillars and this process may grow very rapidly, leading to strong deformations of the rock-mass.The work /3/contains a part of the results obtained according to the above presented model;these results have been to a large extent confirmed by the mine observations.

The application of numerical methods for the analysis of the so-called dynamic rock-mass relief/4/ is another example.It occurs,for example,in the rock-mass with the mining exploitation as a result of local destruction of rock on the surfaces of weakend coherence of the local loss of stability.This phenomenon is modelled in the following way:

a/ the state of stress and strain of the body loaded with the forces constant in time is determined according to the theory of elasticity /or elastoplasticity/in a static approach; b/ it is evaluated whether the computed static state of stress can cause a local loss of continuity /by means of testing the stability criterion/; c/ after it is observed that the static state of stress causes the loss of continuity, the phenomenon of relief being formed is described as a non - stationary dynamic problem of the theory of elasticity in which the boundary and initial conditions result from the taking into consideration of the formed discontiunity and from the critical static state of stress.

To describe the phenomenon of the propagation of the wave of stresses a dynamic differential equation of a solid body movement was used which is presented in the matrix notation

$$B\ddot{g} + C\dot{g} + Kg = F \quad /1/$$

where:

B-matrix of inertia, C=Cz+Cw-matrix of damping which is the sum of a matrix of external damping Cz and internal Cw, K-matrix of regidity, g, $\dot{g}$, $\ddot{g}$ -vectors of generalized displacements, their first and second derivatives to the time t, F-vector of forces. The numerical integration of the equation /1/in a discrete set of points was carried out by means of the method of spacetime finite elements /7/but the method of finite elements /8/was applied to the space discretization. The mining situation shown at Fig.9 can be an example.

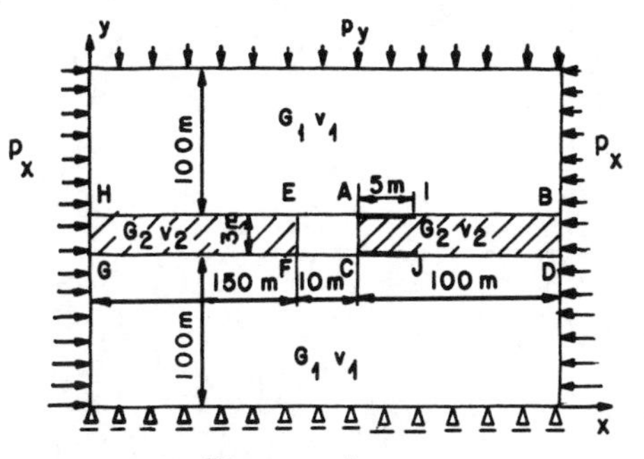

Figure 9

This figure shows a model of rock-mass composed of a partially exploited bed /FEAC/with the waste rock over and under it. The model is a shield in a plane strain state in which the waste rock properties are described by Kirchhoff's modulus G_1 and Poisson's ratio V_1, the bed properties by the G_2 and V_2 modulus. There is a load at the shield edges which results the primary state of stress of the rock-mass, modelled in two stages:

a/ the first stage determined the static state of displacement and stress of the model, resulting from the action of gravity forces and the geometry of the dog heading,

b/ at the second stage it was assumed that there was an abrupt destruction of rock continuity in the roof /AI/and floor /CJ/parts of the bed on the 5m length from the AC edge which caused a discharge of elastic energy and the rock-mass movement towards a chosen part. The equation movement was solved by means of the time-space finite element method, disregarding the damping of the medium.

The time step accepted for the computations corresponds to the time of passing of the elastic wave thorough the smallest finite elements of the model and is $1.1.10^{-4}$ s. The rock-mass movement was observed in 50 first time steps. Figure 10 presents a part of obtained results These are the diagrams of the SX horizontal stresses in the ACIJ area for the 1st, 10th, 25th and 50th time step.

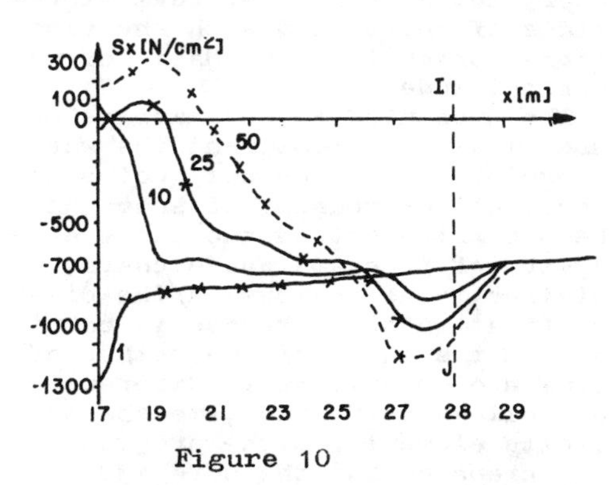

Figure 10

The digital modelling of elastic vibrations in heterogeneous media is an example of the application of the method of finite elements. The details of this problem can be found in literature /5/while here only one aspect will be presented wlich is shown at Fig.11. The dog heading, placed not very deep, can be located by means of special equipment whose receiver /O/observes the course of vibrations caused by the shock /p.U/.These vibrations were simulated numerically applying the equation /1/and taking into consideration the dumping properties of the medium according to the dependence:

$$v^i = \frac{\partial}{\partial t} g^i e^{-\beta n}$$

where v^i is the velocity for the i-time step, t-is time and n is a number of the time step.

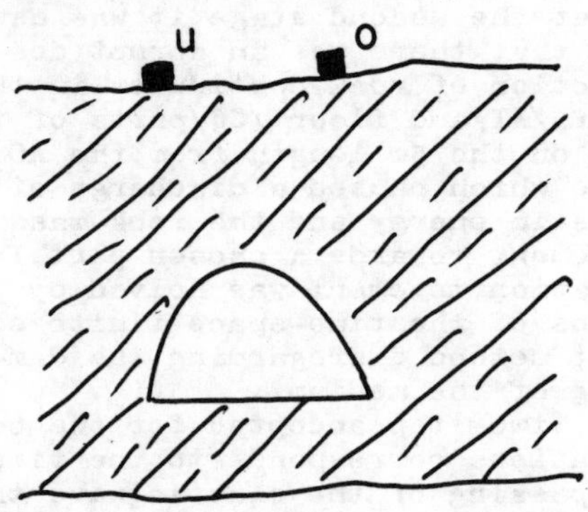

Figure 11

The β value was determined by comparing the record of real vibrations with the vibrations obtained by means of the numerical simulation.The obtained results allowed the authors to observe a satisfactory compatibility of real vibrations of the medium with the vibrations foracast on the basis of numerical model.

The presented above examples of numerical solutions /and the whole range of other ones,not quoted because of the reasons of space/have bean carried out on the basis of a system of programs whose characteristics is as follows: a/the discretization of rock-mass is carried out on the basis of the method of finite elements,finite diferences as well as stiff and time-space finite elements; b/the programs are prepered for the Odra 1300 computer and written in the algorithmic language FORTRAN 4; c/ the programs of data preparation assume a halfautomatic generation of the element network,checking of the data and the generation of the model boundary conditions; d/ the programs forming the systems of equations and solving these systems are provided for the linear and non -linear elastic,elastoplastic and rheological medium as well as for dynamic and static problems together with the investigations of inital stability; e/the programs of result derivation tabulate and draw graphs or contour line plans of results; f/ it is possible to solve plane,axial-symetrical and special problems.

BIBLIOGRAPHY

/1/ H.Filcek and....,The mechanism of formation eruptions,Proc. Univ.Min.Met.Cracov,1980

/2/ J.Walaszczyk and....,The verification of rheological model of rock mass.Proc.Univ.Min.Met. Cracov 1984

/3/ J.Walaszczyk,L.Gawlik, The digital method of stability testing of room and pillar workings with nonlinear characteristic of pillars,Proc.Inst.Civ.Eng. Univ.of Wrocław,No 45/18 1985

/4/ J.Walaszczyk, Certain application possibilities of numerical methods for problems of realief of the orogeny,Proc.Univ.Min. Met.Cracov,No.107,1980

/5/ J.Jarzyna, The modelling of waves diagram by acoustic grading,Nafta 1986

/6/ Z.Kączkowski, The space-time finite element method,Arch.Civ. Eng.No 22,1976

/7/ O.C.Zienkiewicz, The finite element method,Mc Graw.Hill. London 1977

Numerical Methods in Geomechanics (Innsbruck 1988), Swoboda (ed.)
© 1988 Balkema, Rotterdam. ISBN 90 6191 809 X

Three-dimensional finite element and discrete element coupling method applied to shaft deformation and earth's surface subsidence caused by coal mining

Liang Bing & Wang Laigui
Fuxin Mining Institute, People's Republic of China

Abstract: In this paper, three dimensional finite element and discrete element coupling method is applied to calculation and analysis for shaft deformation and earth's surface subsidence caused by coal mining. Discrete elements are used to simulate the area near the coal face and finite elements are used to simulate the area far from the coal face where the medium is continuous. By calculation, we obtain the regularity of shaft deformation and earth's surface subsidence when the coal face is in a state of stability, which are very important to design the coal column of shaft and choose the best mining plan in future, and it is possible for computer to analogue the shaft deformation and the earth's surface subsidence.

1 INTRODUCTION

When the seams surrounding shaft are mined, shaft deformation and the earth's surface subsidence will occur. If mining is done, the shaft deformation and the earth's surface subsidence will be more serious and lead to the damage of normal hoist, ventilation and buildings on the earth's surface. It is very important for prevention and cure of shaft deformation and earth's surface subsidence to gain the regularity of the shaft deformation and the earth's surface subsidence. A lot of researching work has been done on the shaft deformation and the earth's surface subsidence. Many scholars have made theoretical analysis for this problem which the model of shaft deformation and the earth's surface subsidence is plane strain. In fact, the model of the shaft deformation and the earth's surface subsidence should be three mechanical model. In addition, because caving method and multiple-horizon mining method are applied, cap rock is mined repeatly and the roof of seams will cave. The medium of falling areas is uncontinuous, which is very difficult to simulate by finite elements. This paper puts forward the coupling method of finite element and discrete element that uncontinuous medium of the caving areas is treated by discrete elements, continuous medium far from the caving areas is treated by finite elements. The result computed by coupling method approaches the practical engineering and which provides a theoretical basis for the design of coal column of the shaft and the choice of the best mining plan hereafter.

2 BASIC PRINCIPLE OF DISCRETE ELEMENT

Discrete element method has been developed in recent years. The method which fully considers the uncontinuous rock body and coal body caused —by the mining of seams surronding breakage face finds the solution of the problems by dynamic flabbiness method of time-step iteration. The fundamental equation of discrete element method consists of two sets. One is physical equation and the other is moving equation.

2.1 Physical equation

1. The relationship between normal force and normal displacement.

Suppose the normal force F among block bodies is direct ratio with their normal displacement, which can be seen in Fig. 1, that is,

$$F_n = K_n \delta_n \qquad (1)$$

where K_n is the stiffness coefficient, δ_n is much smaller than the size of block boby.

2. The relationship between shearing force and shearing displacement. Because the shearing force of block body have something to do with its movement and the history of loading, tangential increment F is direct ratio with tangential displacement of two block bodies in a time-step which can been seen in Fig. 2, i.e.

$$\Delta F_s = K_s \delta_s \qquad (2)$$

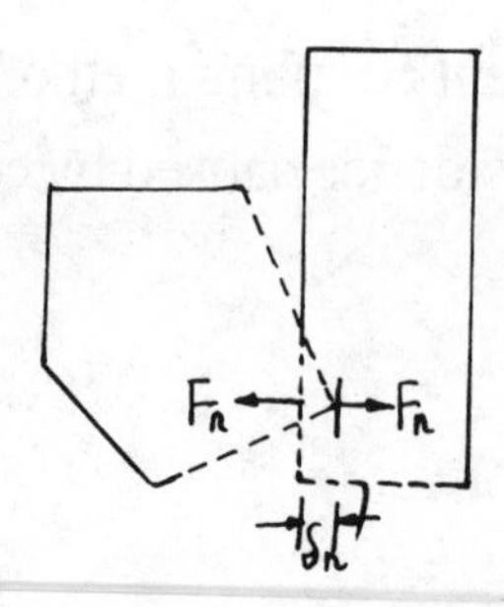

Fig.1 Relationship between normal force and normal displacement

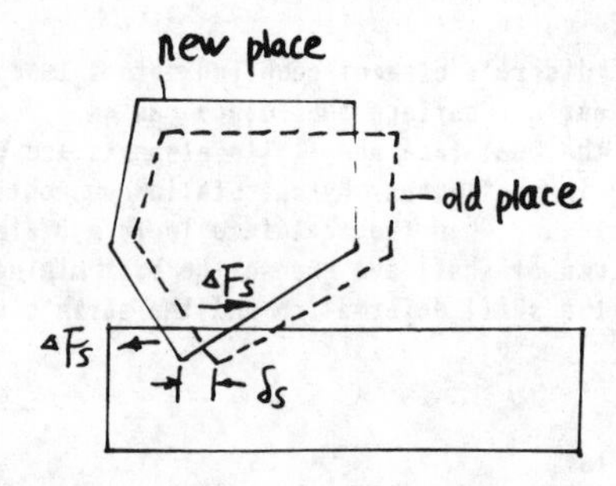

Fig.2 Relation between shearing force and shearing displacement

where K is the tangential stiffness coefficient of contact points.

Tangential force can be expressed in the following formulation.

$$F_S = F_S^{\circ} + \Delta F_S \qquad (3)$$

According to the Coulomb-Mohr's criterion, the maximum tangential force depends on the following formulation,

$$|F_S| \leq C + F_n \operatorname{tg}\phi \qquad (4)$$

in which F_S is the added shearing force, C is the cohesive force, and ϕ is the angle of friction.

2.2 Moving equation

They are a group of forces which act on one block body. Their resultant of forces and moment of resultant of forces that can be seenin Fig. 3 depends on the following formulation.

$$F_x = \sum_{i=1}^{n} F_{xi}$$
$$F_y = \sum_{i=1}^{n} F_{yi} \qquad (5)$$
$$F_z = \sum_{i=1}^{n} F_{zi}$$
$$M_{xy} = \sum_{i=1}^{n} [F_{yi}(X_i - X_c) - F_{xi}(Y_i - Y_c)]$$
$$M_{yz} = \sum_{i=1}^{n} [F_{zi}(Y_i - Y_c) - F_{zi}(Z_i - Z_c)]$$

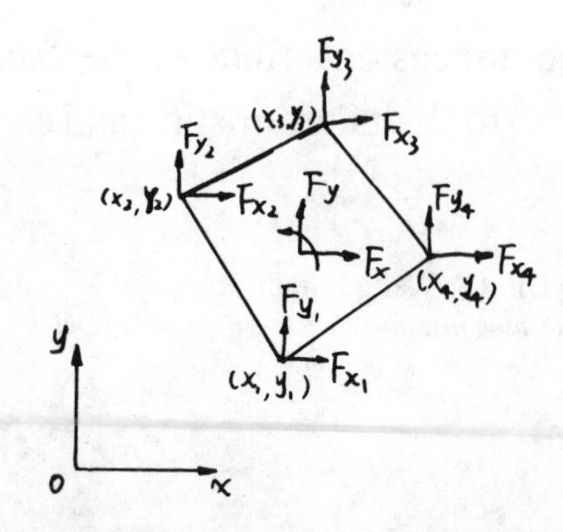

Fig.3 The force acting on the block body and its composition

$$M_{zx} = \sum_{i=1}^{n} [F_{xi}(Z_i - Z_c) - F_{zi}(X_i - X_c)]$$

where X_c, Y_c, Z_c denote the co-ordinates of the block centroid respectively.

The resultant of forces and its moment acting on block body make the block body produce linear accelerator $\ddot{x}$, $\ddot{y}$, $\ddot{z}$ and angle accelerator $\ddot{\theta}_x$, $\ddot{\theta}_y$, $\ddot{\theta}_z$ at the time t_N. From the time $t_{N-\frac{1}{2}}$ to the time $t_{N+\frac{1}{2}}$, we employ the Newton's second law, we can gain,

$$m\ddot{x} = F_x$$
$$m\ddot{y} = F_y$$
$$m\ddot{z} = F_z \qquad (6)$$
$$I_x \ddot{\theta}_x = M_{xy}$$
$$I_y \ddot{\theta}_y = M_{yz}$$
$$I_z \ddot{\theta}_z = M_{zx}$$

where m and I denote the mass of block body and rotation inertia respectively.

Suppose $\ddot{x}$, $\ddot{y}$, $\ddot{z}$ and $\ddot{\theta}_x$, $\ddot{\theta}_y$, $\ddot{\theta}_z$ are fixed, value in time Δt step. Making use of the central difference method, we can obtain the velocity and angle velocity,

$$(\dot{x})_{N+\frac{1}{2}} = (\dot{x})_{N-\frac{1}{2}} + [F_x/m]_N \Delta t \qquad (x, y, z) \qquad (7)$$
$$(\dot{\theta}_x)_{N+\frac{1}{2}} = (\dot{\theta}_x)_{N-\frac{1}{2}} + [M_{xy}/I_x]_N \Delta t$$

where footnote N denotes the time t_N, 1 and 1/2 denotes one time-step and half time-step.

Above expression is integrated which gives new state of block body element, i.e. x, y, z and θ_x, θ_y, θ_z

$$(x)_{N+1} = (x)_N + (\dot{x})_{N+\frac{1}{2}} \Delta t \qquad (x, y, z) \qquad (8)$$

$$(\theta_x)_{N+1} = (\theta)_N + (\dot{\theta}_x)_{N+\frac{1}{2}} \Delta t$$

2.3 Damping

To find the static stable solution, damping must be added to absorb kinetic energy of block body, otherwise the system will be unstable. Two kinds of cohesive dampings can be employed in discrete element method. One is the damping of mass which makes the absolute movement of body to have been hindered. The other is the damping of stiffness which limits the relative movement of body. This paper applies damping of mass to treat with the unstable system.

The damping of mass can be regarded as a series of equpment of cohesive damping D which links the centre of mass of every block body. The equpment of cohesive damping produce the force whose direction is opposite from the velocity of block body and is direct ratio with the mass of block body (see Fig . 4). The moving eqution can be expressed by the following formulation.

$$m\ddot{x} = F_x - m\alpha\dot{x}$$

$$m\ddot{y} = F_y - m\alpha\dot{y}$$

$$m\ddot{z} = F_z - m\alpha\dot{z} \qquad (9)$$

$$I_x \ddot{\theta}_x = M_{xy} - I_x \alpha \dot{\theta}_x$$

$$I_y \ddot{\theta}_y = M_{yz} - I_y \alpha \dot{\theta}_y$$

$$I_z \theta_z = M_{zx} - I_z \alpha \dot{\theta}_z$$

where α is coefficient of damping.

i.e.

$$(\dot{x})_{N+\frac{1}{2}} = \{(\dot{x})_{N-\frac{1}{2}}[1-\alpha\Delta t/2] + F_x \Delta t/m\}/(1+\alpha\Delta t/2)$$

$$(\dot{\theta}_x)_{N+\frac{1}{2}} = \{(\dot{\theta}_x)_{N-\frac{1}{2}}[1-\alpha\Delta t/2) + M_{xy}\Delta t/I_x\}/(1+\alpha\Delta t/2$$

$$(x, y, z) \qquad (10)$$

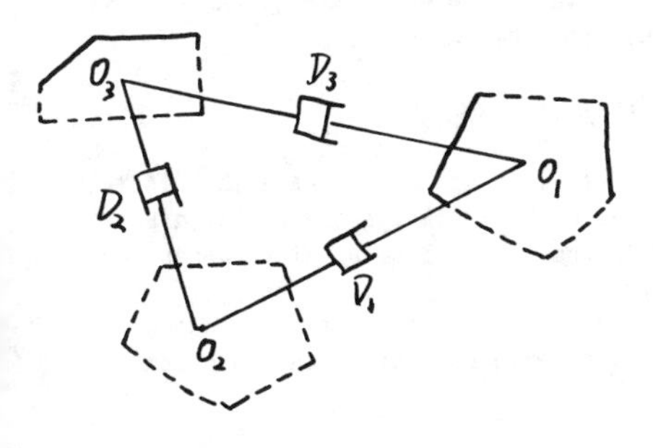

Fig.4 The damping of mass

2.4 The choice of time step

When the time step is taken as , $\Delta t < 2\sqrt{m/k}$, the displacement solution of the system is stable which has a concentrate mass and a coefficient of stiffness.

2.5 Average stress

The average stress $\overline{\sigma_{ij}}$ in the volume V can be definite as ;

$$\overline{\sigma_{ij}} = 1/V \int_V \sigma_{ij}\, dv \qquad (11)$$

According to the equilibrium condition and Gauss' theorem , above formulation can be replaced by equation,

$$\overline{\sigma_{ij}} = 1/V \int_S x_i t_j dS \qquad (12)$$

where S is the surface of block body , x_i is the co-ordinates of the point i of the surface S , and t_j is the j components of surface force.

Because the surface force of a typtical rock body is discrete force acting on the surface of rock body , above expression can be replaced by the following formulation.

$$\overline{\sigma_{ij}} = 1/V \sum_{i=1}^{n} x_i F_j \qquad (13)$$

2.6 The process of computing circulation of the discrete element method

Fig. 5 indicates the computing circulation of discrete element method in every time-step .The arrowhead in Fig. 5 designates the order of the computing.

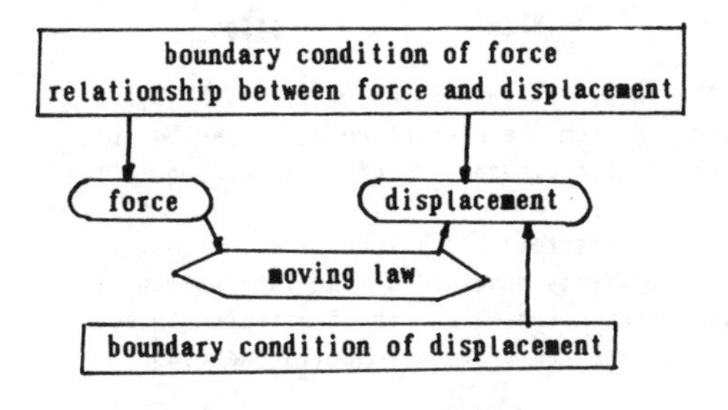

Fig . 5 computing circulation of time step

3 PRINCIPLE OF DISCRETE ELEMENT AND FINITE ELEMENT COUPLING METHOD

Governing equations,

$$[M]\{\ddot{u}\} + [C]\{\dot{u}\} = \{f\}$$

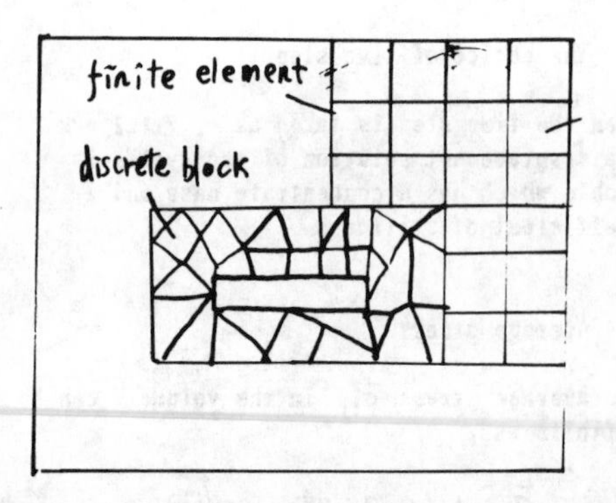

Fig.6 Coupling of discrete element to finite elements for from working face

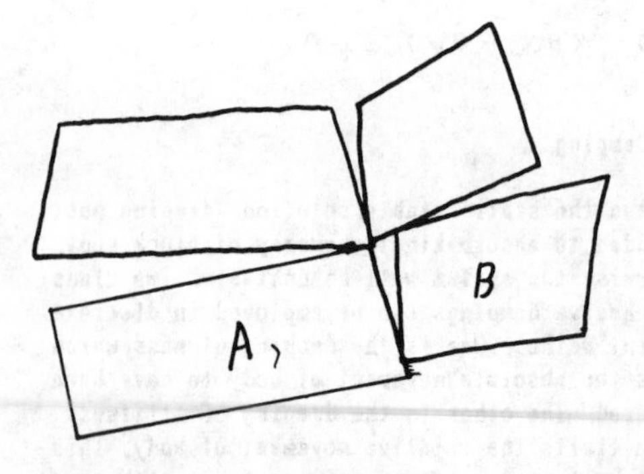

Fig.7 Discrete elements

$$\{f\}=\{f^{ext}\}-\{f^{int}\}$$

where

$\{u\}$=nodal displacement column matrix
$\{f^{ext}\}$=external force column matrix
$\{f^{int}\}$=internal force column matrix
$[M]$=mass matrix, $[C]$=damping maxtrix

and superposed dots denote time derivatives, thus $\{\ddot{u}\}$ is the column matrix of nodal accelerations.

In the finite element, the internal forces arise from the resistance of the element to deformation, and are given by,

$$\{f^{int}\}_E=\int_{V_E}[B]^T\{\sigma\}dV$$

where a subscript E indicates that the variable pertains to element E, i.e. continuous element, $\{\sigma\}$ is the stress tensor, V_E is the volume of element E, and [B] is the matrix which relates the velocity strain (rate of deformation) $\{\dot{\varepsilon}\}$ to the nodal velocities by ;

$$\{\dot{\varepsilon}\}=[B]\{\dot{u}\}_E \qquad (16)$$

The total internal force matrix $\{f^{int}\}$ is obtained from the element nodal forces by the usual finite element procedure according to the node numbers.

In the discrete block model, the internal forces are completely governed by the interaction of block boundaries which are illustrated in Fig. 6 and represented by equation (17) and (18).

$$f_n=k_n u_n \quad \text{if } u_n<0 \qquad (17a)$$

$$f_n=0 \quad \text{if } u_n>0 \qquad (17b)$$

$$f_s=k_s u_s \quad \text{if } |f_s|<f_n\tan\phi \qquad (18a)$$

$$f_s=\text{sign}(\dot{u}_s)f_n\tan\phi \quad \text{if} |f_s|>F_n\tan\phi \qquad (18b)$$

where f_n and f_s are the components of the contact force, u_n and u_s are the normal and tangential components of the relative displacement between discrete blocks and ϕ is the angle of friction.

Equations (17a) and (17b) limit the penetration of the coner of block A into the edge of block B to be compressive. (Fig . 7). Equations (18a) and (18b) limit the maximum allowable shear force f_s. The shear law can take many forms ;however in the current formulation, the Coulomb-Mohr friction law is applied . For computational convenience, equations (17a) and (18a) are replaced by the icremental relation in equations (19a) and (19b);

$$\Delta f_n=k_n\Delta u_n+\alpha k_n\Delta\dot{u}_n \qquad (19a)$$

$$\Delta f_s=k_s\Delta u_s+\alpha k_s\Delta u_s \qquad (19b)$$

where α is the stiffness proportional damping parameter necessary to dampen block vibration.

Finite element nodes which are connected to discrete block are called slave nodes ;the time histories of these nodes are not governed by equation (14), but are obtained by assuming that they are rigidly connected to the block. Because of the assumption of a rigid interface, the velocity of the slave nodes are given in terms of the discrete block velocity by ;

$$\{\dot{u}\}_s=[T]\{\dot{u}\}_R \qquad (20)$$

where subscripts R and S designate the discrete block nodes (take as the center of the discrete elements) and finite element slave nodes respectively.

The internal forces at the slave nodes are considered as external forces on the blocks, so that

$$\{f^{ext}\}_R=\{f^{ext}\}_R-\sum_{S=1}^{N}[T]_S^T\{f^{int}\}_S \qquad (21)$$

where N is the number of slave nodes connected to block R.

Both the finite element mesh and discrete blocks are integrated explicitly by the central difference method ,which gives ;

$$\{\ddot{u}\}^n=[M]^{-1}(\{f\}^n-[C]\{\dot{u}\}^{n-\frac{1}{2}}) \qquad (22)$$

$$\{\dot{u}\}^{n+\frac{1}{2}}=\{\dot{u}\}^{n-\frac{1}{2}}+\Delta t\{\ddot{u}\}^n \qquad (23)$$

$$\{u\}^{n+1}=\{u\}^n+\Delta t\{\dot{u}\}^{n+\frac{1}{2}} \qquad (24)$$

where Δt is the tme increment and superscripts denote the time step number.

The stability limit for the time increment is ;

$$\Delta t<2\sqrt{m/k}$$

4 PROGRAM STEPS OF FINITE ELEMENT AND DISCRETE COUPLING METHOD

This paper puts forward the coupling method that links the discrete block model to the finite element model .Coupling method arises from explicitive integration .Space 8-node isoparameter finite element is employed to denote continuous medium far from mining face and discrete element to designate the cracke areas and roof falling areas near the mining face .The program steps are ;

1. n=0 , time=0 .
2. Compute internal force for all finite element at both master and slave nodes .
3. Transform all internal at slave nodes to master of discrete block by equation (21) .
4. For all the discrete blocks ;
(1) Compute forces $\{f\}^n$ and added to internal forces obtained in step 3 ;
(2) Compute accelerations $\{\ddot{u}\}_R^n$ by equation (22);
(3) Get new velocities $\{\dot{u}\}_R^{n+\frac{1}{2}}$ and displacements $\{u\}_R^{n+1}$ by equations (23) and (24) .
5. Obtain velocities for continuous slave nodes from $\{\dot{u}\}_R^{n+\frac{1}{2}}$ by equation (20) .
6. For all continuous master nodes , find accelerations $\{u\}$ by equation (22) .
7. For all continuum nodes , get new velocities $\{\dot{u}\}_c^{n+1/2}$ and displacements $\{u\}_c^{n+1}$ by equations (23) and (24) .
8. $n \leftarrow n+1$, go to step 2 .

5 THE COMPUTATIONAL EXAMPLE AND ANALYSIS

This paper takes for DongFeng shaft of Wu Long coal mine in Fuxin as example and the calculation completed on WANG VS-100 computer in our college. The regularity of shaft deformation and earth's

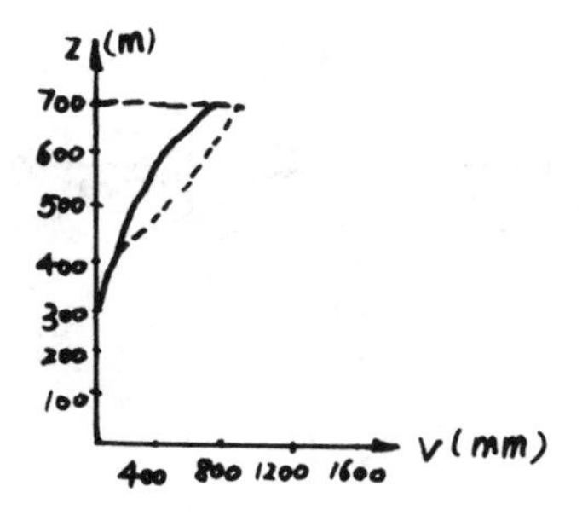

Fig.8

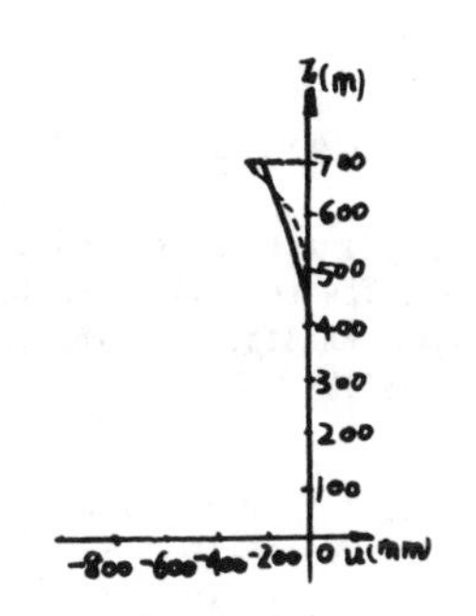

Fig.9

surface subsidence at the state of stability is obtained .

Fig .8 and Fig .9 are the curve of the shaft deformation . In the figures , the real line is the result of compution by finite element and discrete element coupling method and the dot line stands for tested data in work-situ . We can find out from the curve that the deformation value on top of shaft is lager . At the depth of some 270 meters , the aslant value of shaft along inclination is about zero . At the depth of some 300 meters , the aslant value of the shaft along the run is about zero . The maximum displacement is ;

$$u_{es}=\sqrt{u_e^2+u_s^2}=\sqrt{(980)^2+(210)^2}=1002.2 \text{ (mm)}$$

The direction is ;

$$\theta_{es}=tg^{-1}(u_s/u_e)=12^{\circ}5'$$

The direction of shaft aslant is southeast , which is caused by large mining amount along the southeast and by nonsymmetry mining .

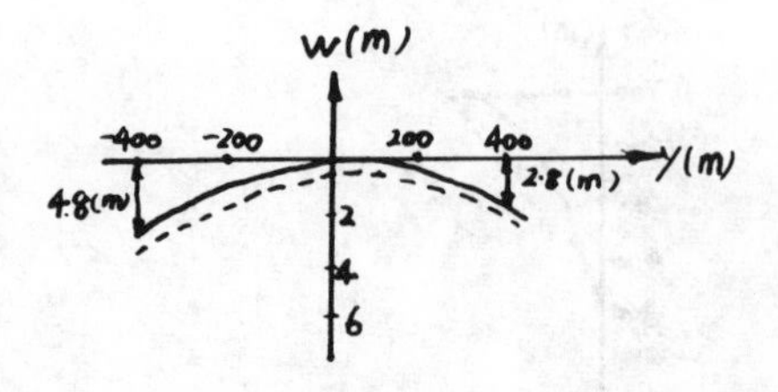

Fig.10

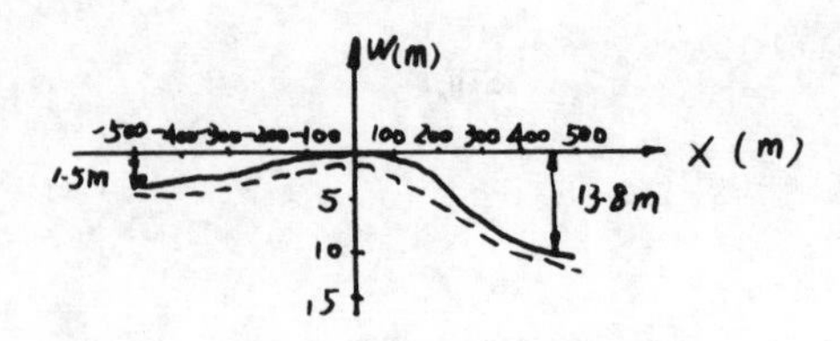

Fig.11

Fig . 10 and Fig . 11 are the curves of earth's surface subsidence . The real line denotes the result of calculation and the dot line is the tested data in work-situ .

6 CONCLUSION

This paper puts forward the coupling method of finite element and discrete element that the uncontinuous medium of the caving areas is treated by discrete elements , continuous medium far from the caving areas is treated by finite elements . The results computed by coupling method approach the practical engineering and which provide a theoretical basis for the design of coal column of the shaft and the choice of the best mining plan in future .

REFERENCES

Dowing, C.H. Damage to rock tunnels from earthquake shaking.Journal of the Geotechnical Division ,ASCE G02,104,171-191(1978).

Zienkiewicz O.C .Time-dependent multilaminate model of rocks--a numerical study of deformation and failure of rock masses .Int.Journal for Numerical and Analytical Methods in Geomechanics ,1,219-247 (1977).

Belytschko, T.Mixed method for time integration. Computer Methods in Applied Mechanics and Engineering , North-Holland Publishing Company, 1979,pp.259-275.

Liang, B.Elasto-plastic Analysis of Deformation of DongFeng shaft in Fuxin WuLong coal mine. Coal Technology of Northeast ,6,25-28 (1986).

Numerical Methods in Geomechanics (Innsbruck 1988), Swoboda (ed.)
© 1988 Balkema, Rotterdam. ISBN 90 6191 809 X

Rock movement and surface subsidence of South Xikuangshan mine

Zhang Zhihua & Zhou Jifu
Xikuangshan Bureau of Mines, People's Republic of China
Hu Qinghuai
Changsha Institute of Mining Research, People's Republic of China

ABSTRACT: This paper deals with the investigations of rock movement and surface subsidence in the mine for two aspects: the historical disasters of strata failure and the new subject of mining the safety pillar of the river bed. It covers a theorectical analysis of elastical successive foundation beam with lateral force and that of finite element.

1 INTRODUCTION

The mineral deposit of South Xikuangshan Mine, China, is a flat antimony sulphide one with a tabular shape and a flat dip, formed by hydrotheamal process at a low temperature accompanied with metasomatism. The orebodies came into existence in Chilijian silicified, Chizichao system, above which there is a formation of Changlenjia shale with thickness of about 80 m, as well as Touzidan limestone, Devonian red hematite and Margolou limestone with various thicknesses and dip angles of 10-20 degrees. The payable seams are of thicknesses from 2 to more than 25 m. The orefield is mountainous and has a rugged topography. On the surface, a small seasonal river named Feihui flows down south with water fall in upstream and flow volume varied from 0.55 to 47 M^3/se. Origionally room and pillar mining method is used, but with bolting in mining the thiner seams. However from 1973 to date the filling method is preponderant.

Historically, the disasters happened three times in east, centre and west of the mine successively during the years from 1965 to 1971 when, the large mined-out areas suddenly and intensively fell down and it subsequently caused the surface rapidly subsided and fractured. Afterward, the new problem of rock movement and surface subsidence in mining the safety pillar of the river bed was urgently put forward for the sake of production and personal safety. Thus two aspects are covered in this report.

2 SURFACE SUBSIDENCE OF THE CENTRAL AREA IN EXISTENCE OF LATERAL PUSH

The caving of large-scale mined-out area is caused by the damage of pillars. Thus it is possible to give rise to a series of responses and cause the pillars over a large area broken, so that the strata failure is developed and the ground surface in turn effected by the rock movement, forming a complete or incomplete basin of subsidence on the surface untill that the caved area is filled up and packed tightly by the broken rocks. Typically, the central region is the case.

This region in neighbour of the shaft pillar was mined out in 1964 by room and pillar method. The orebody there has an average thickness of 8 m and a dip angle of 18°. The capping over it is about 150m thick. Here is just where the breaking-down of the strata heavily took place in December, 1965. The caved area covered an area of more than 30,000 m^2. The surface was rapidly sunk and fractured two hours later after the breaking had occured. There were 43 tensile cracks, most of them were orientated in the north-east direction. The longest one was 260m long and the widest one had a 2.1m width. The surface uplift above the pillar was obviously observed. As a result, the shaft headframe was out of vertical and inclined towards north-west. From analysing all of these appearences in-situ, we realized that there existed a great lateral force towards the north-west, pushing the safety pillar during the intense development of breakingdown and surface movement. The push may relate to the topograph in which the east is much higher than the west. The surface situation and the observation lines in this area is shown in Fig. 1. The in-situ observed curves of subsidence, declination and curvature in the observation line III set up in dip direction of the strata is shown in Fig. 2. The subsidence curve at the third day after breaking is comparatively drawn in the same figure, which has a maximum subsidence 1075 mm and takes 65% of the final result (1682mm). By conjoining the centre of the fully underground caved area and the point of maximum subsidence on surface, the so-called maximum subsidence angle $\theta=85^{o}$ was obtained.

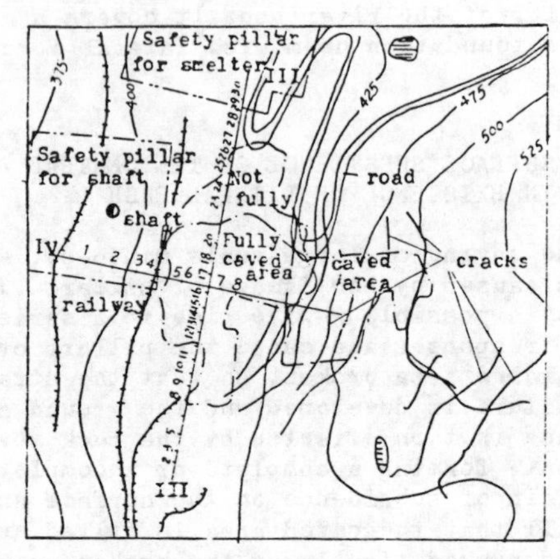

Fig. 1. Plan of surface observation lines in the central area

Theoretically, a pertinent analysis of the surface subsidence by use of the theory of elastic foundation beam described in appendix were made, which results in the equations of subsidence(W), declination (T) and curvature (K) on the surface as follows:

1. for the surface of upper respect $(x>0)$

$$W=1464\delta(58.6-|x|)+845e^{-0.0206|(58.6-|x|)|}\sin[2.05(58.6-|x|)-\gamma(58.6-|x|)\cdot60]+210\delta(40-|z|)+120e^{-0.03|(40-|z|)|}\sin[3(40-|z|)-\gamma(40-|z|)\cdot60] \quad (1)$$

$$T=-34.9e^{-0.0206|(58.6-|x|)|}\cos[2.05(58.6-|x|)-\gamma(58.6-|x|)\cdot30]-7.8\gamma(z)e^{-0.03|(40-|z|)|}\cos[3(40-|z|)-\gamma(40-|z|)\cdot30] \quad (2)$$

$$K=-1.44e^{-0.0206|(58.6-|x|)|}\sin[2.05(58.6-|x|)]-0.48e^{-0.03|(40-|z|)|}\sin[3(40-|z|)] \quad (3)$$

$$z=x-120 \quad (4)$$

2. for the surface of lower respect $(x\le0)$

$$W=1464\delta(58.6-|x|)+845e^{-0.0206|(58.6-|x|)|}\sin[2.05(58.6-|x|)-\gamma(58.6-|x|)\cdot60]+450\delta(54-|z|)+260e^{-0.022|(54-|z|)|}\sin[2.2(54-|z|)-\gamma(54-|z|)\cdot60] \quad (5)$$

$$T=34.9e^{-0.0206|(58.6-|x|)|}\cos[2.05(58.6-|x|)-\gamma(58.6-|x|)\cdot30]+11.6\gamma(z)e^{-0.022|(54-|z|)|}\cos[2.2(54-|z|)-\gamma(54-|z|)\cdot30] \quad (6)$$

$$K=-1.44e^{-0.0206|(58.6-|x|)|}\sin[2.05(58.6-|x|)]-0.52e^{-0.022|(54-|z|)|}\sin[2.2(54-|z|)] \quad (7)$$

$$z=x+100 \quad (8)$$

where

$$\delta(y)=\begin{cases}1 & \text{when } y>0\\ 0 & \text{when } y\le0\end{cases} \quad (9)$$

$$\gamma(y)=\begin{cases}1 & \text{when } y>0\\ -1 & \text{when } y\le0\end{cases} \quad (10)$$

Note that, the point of maximum subsidence (No.16) in caved area is taken as origion of x axis, and the direction from lower respect to upper respect as its positive in these equations. They consist of the main terms with parameters $\varphi=60^{o}$ and $\frac{\pi}{L}=0.0357$ comming from the average of the surface measurements on both upper and lower respects of the caved area, and the modifying terms in consideration of the effects of the not fully caved area in upper respect and the irregular shape in lower respect,etc. The calculated curves are also plotted in Fig. 2 for a comparison with those in situ observed. Obviously, the coincidence is quite satisfactory.

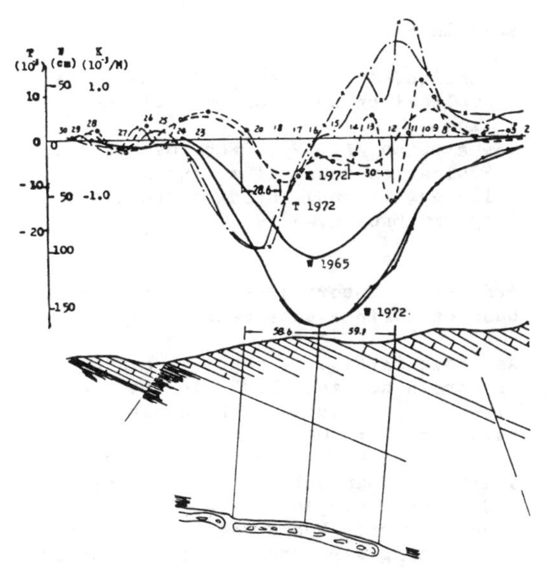

Fig. 2 The observed and calculated curves of surface subsidence, declination and curvature in observation line III

From the analysis above, we can easily explain all the occurrence such as the ground uplight and the headframe devision. The subsidence angles of 60° in upper respect and 65° in lower respect are obtained.

3 SURFACE SUBSIDENCE IN SAFETY PILLAR AREA OF RIVER BED

In order to utilize the resource left for pillars and maintain a regular production in the mine, it was decided to mine the ore blocks of 4 – 25 m thicknesses within the safety pillar of the river bed from 1973. The depth of the deposit from the surface varies from 110 to 150 m. The regional geological structure is some what complex with well developed joints and fissures which exert a negative effect to the stability of the river bed. Hopefully, the Changlenjia shale as the main overburden of 90 m provides a better water insulation because of its low seepage.

The coverage area of extraction is 250 m along the strike and 350 m along the dip, envolving 6 levels (from No.6 to No.11). Up to the end of 1986, 94 ore blocks have been extracted out including 39 pillar blocks and 55 rooms with a total extraction volume of 591,000 m^3. The mining method used is a cemented filling method in which the pillar blocks are filled up with cemented rocks to create artificially the protective walls for stoping the rooms, and then the rooms are filled up with tailings. Commonly, the pillar is 8 m wide and the room is of 10 m width. Both are 60 m long along the strata dip.

With the purposes of ground control and safety guarantee during mining the blocks in the pillar area, a large research program has been carried out since 1973 including the observations of rock movement both on surface and inside the rock. Three observation lines were set out on surface among which one is parallel to the strike of the strata, the second is in the dip direction and the third, along the river bed. Several surface bareholes and underground holes drilled into the roof constitute the observation net for measuring the inside rock movement and seam cracking. The plan with the observation layout and the underground extracted areas in the pillar region is shown in Fig. 3

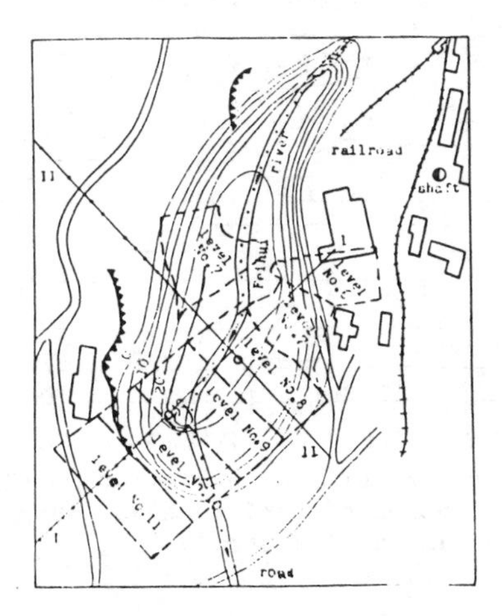

Fig.3 Plan of the safety pillar area of river bed and the surface subsidence contours

Usually, it is measured that the deepest crack is occured at a distance of about 18 m above the roof, but the major seam cracking is 4 – 10 m deep from the roof. The reason of the cracking is mainly due to the long time open before filling up the space of the extracted pillar blocks. As for the surface subsidence, it goes very slow. It was firstly surveyed in July, 1979 with a basin shape on surface and a maximun value of 12 mm at the center of the basin. Now a maximum subsidence of 31 mm has been reached through more than ten years. The contours of surface subsidence surveyed in situ are also illustrated in Fig.3.

To investigate the distribution of the subsidence theorectically, a two-dimensional analysis of finite element on section of the observation line along the strike is made. Fig. 4 is the model mesh plotted by computer. Considering the roof subsidence due to the seam cracking and the inperfect contact of the cemented walls with the roof before that the walls entirely bear the ground gravity,some initial displacements known based on the in situ observation are preapplied to those roof points at the central position of the extracted ore body in the model. The curves of surface subsidence distribution both computed and measured in situ are also given in Fig. 4. Similarily, they are of basin shapes but with maximum values of 21 and 25 mm respectively.

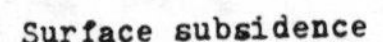

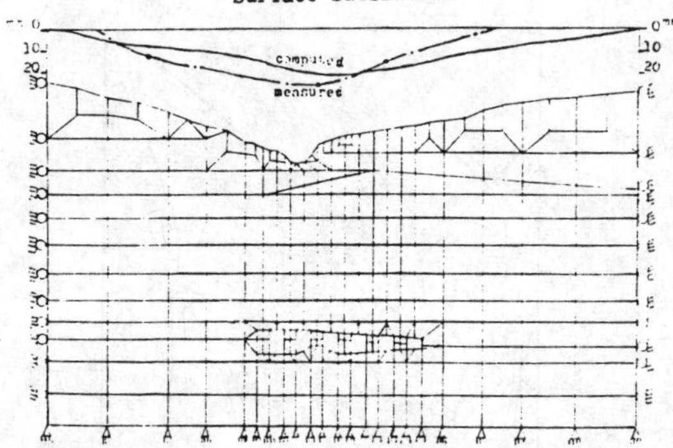

Fig. 4 Mesh of finite element model and distribution of surface subsidence along the strike of the strata

4 CONCLUSIONS

The ground contral is very important to productive mines. In condition of South Xikuangshan Mine, while using the caving method for the control, the continuous caved area should be more than 120x120 m^2 in forming a complete subsidence basin on surface and making the caved rocks to bear the ground gravity. However, the filling method with cemented rocks and tailing is of great safety in mining the ore within the pillar range for the river bed, etc. The effectiveness of the mining method used in this case has been well proven by both the mining practice and the investigation.

REFERENCE

Xu Zhilon, 1963. Theory of elasticity. Publication house of people's education

Sawustowich, A. 1955. Rock mechanics. Pubication house of mining and metallurgy, Poland, Abridged traslation by Lin Guoxia, 1959, China.

APPENDIX Theory of elastic foundation beam on surface subsidence

As shown in Fig. 1, consider a section of strata horizontally laid with one side being infinite rock masses in uncaved area ($x<0$) and another, loose chops of rock in caved area ($x\geq 0$), as a elastic foundation beam of a unit width under conjunctive action of the ground gravity P and the lateral force P_z. Letting W represent the surface subsidence, W_o, the subsidence at the boundary point ($x=0$), and J the moment of inertia of the cross section of the beam, according to Winkeler's hypothesis, the differential equations of the subsidence of the bended strata are as follows:

$$1.\ \frac{d^4W}{dx^4}+\frac{P}{EJ}\frac{d^2W}{dx^2}-\frac{1}{EJ}\left[P_z-k(W-W_o)\right]=0\quad (x\geq 0)\quad (1)$$

$$2.\ \frac{d^4W}{dx^4}+\frac{P}{EJ}\frac{d^2W}{dx^2}-\frac{1}{EJ}\left[P_z-cW\right]=0\quad (x<0)\quad (2)$$

Solving these equations with consideration of the continuity at point x=0 and W=0 at t=-∞ on uncaved area, we obtain

$$1.\ W=W_\infty-\frac{\beta^2}{\alpha^2+\beta^2}W_\infty e^{-\alpha x\sin\frac{\varphi}{2}}\left[\frac{-\alpha^2\sin\frac{3}{2}\varphi-2\alpha\beta}{-\alpha^2\cos\frac{3}{2}\varphi+2\alpha\beta}\cdot\frac{\cos\varphi\sin\frac{\psi}{2}+\beta^2\sin\frac{\varphi}{2}}{\sin\varphi\sin\frac{\psi}{2}+\beta^2\cos\frac{\varphi}{2}}\sin\left(\alpha x\cos\frac{\varphi}{2}\right)+\cos\left(\alpha x\cos\frac{\varphi}{2}\right)\right]\quad (x\geq 0)\quad (3)$$

$$2.\ W=\frac{\alpha^2}{\alpha^2+\beta^2}W_\infty e^{\beta x\cos\frac{\psi}{2}}\left[-\frac{-\beta^2\sin\frac{3}{2}\psi-2\alpha\beta\cos\psi}{-\beta^2\cos\frac{3}{2}\psi+2\alpha\beta\sin\psi}\cdot\frac{\sin\frac{\varphi}{2}+\alpha^2\sin\frac{\psi}{2}}{\sin\frac{\varphi}{2}+\alpha^2\cos\frac{\psi}{2}}\sin\left(\beta x\cos\frac{\psi}{2}\right)+\cos\left(\beta x\cos\frac{\psi}{2}\right)\right]\quad (x<0)\quad (4)$$

Where W_∞ is the surface subsidence at infinite on the caved area. Parameters α and β characterizing the mechanical property of the rock and angles φ and ψ, the effect of the lateral force are as follows:

$$\alpha = \sqrt[4]{\frac{k}{EJ}} \quad (5)$$

$$\beta = \sqrt[4]{\frac{c}{EJ}} \quad (6)$$

$$tg\varphi = \sqrt{\frac{k}{EJ} - (-\frac{P}{2EJ})^2} \Big/ \frac{P}{2EJ} \quad (7)$$

$$tg\psi = \sqrt{\frac{c}{EJ} - (-\frac{P}{2EJ})^2} \Big/ \frac{P}{2EJ} \quad (8)$$

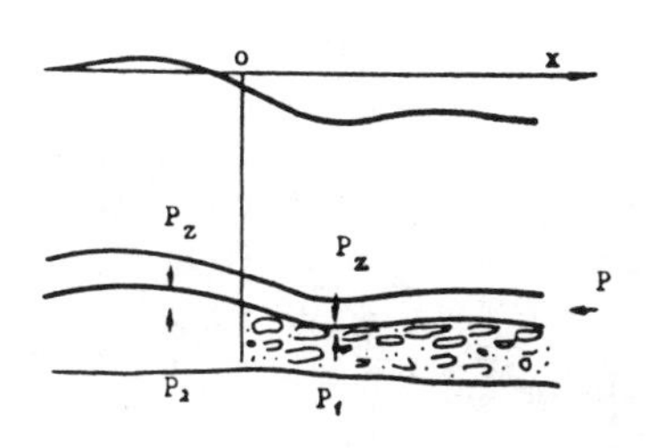

Fig. 1 The beam of elastic foundation

For simplicity, let $\alpha \approx \beta$ approximately, thus the equations of surface subsidence (W), declination (T) and curvature (K) in caved area ($x \geq 0$), become as follows:

$$W = W_\infty + \frac{1}{2}W_\infty e^{-\frac{\pi}{L}x}\, tg\frac{\varphi}{2}\, \sin(\frac{\pi}{L}x - \varphi)/\sin\varphi \quad (9)$$

$$T = \frac{dW}{dx} = \frac{1}{2}W_\infty \frac{\pi}{L} e^{-\frac{\pi}{L}x}\, tg\frac{\varphi}{2} \cos(\frac{\pi}{L}x - \frac{\varphi}{2})/\sin\varphi\cos\frac{\varphi}{2} \quad (10)$$

$$K = \frac{d^2W}{dx^2} = -\frac{1}{2}W_\infty (\frac{\pi}{L})^2 e^{-\frac{\pi}{L}x}\, tg\frac{\varphi}{2}\, \sin(\frac{\pi}{L}x)/ \sin\varphi \cos^2\frac{\varphi}{2} \quad (11)$$

$$2L = \frac{2\pi}{\cos\frac{\varphi}{2}} \quad (12)$$

The similar equations in uncaved are ($x < 0$) are as follows.

$$W = \frac{1}{2}W\, e^{\frac{\pi}{l}x}\, tg\frac{\psi}{2}\, \sin(\frac{\pi}{l}x + \psi)/ \sin\psi \quad (13)$$

$$T = \frac{1}{2}W_\infty \frac{\pi}{l} e^{\frac{\pi}{l}x}\, tg\frac{\psi}{2}\, \cos(\frac{\pi}{l}x + \frac{\psi}{2})/\sin\psi\cos\frac{\psi}{2} \quad (14)$$

$$K = -\frac{1}{2}W_\infty (\frac{\pi}{l})^2 e^{\frac{\pi}{l}x}\, tg\frac{\psi}{2}\, \sin(\frac{\pi}{l}x)/ \sin\psi\cos^2\frac{\psi}{2} \quad (15)$$

$$2l = \frac{2\pi}{\cos\frac{\psi}{2}} \quad (16)$$

Fig.2 shows the curves of these equations. The parameters φ, $\frac{\pi}{L}$ and W_∞ etc can be obtained from the in-situ surveyed curves by means of the coordinates of maximum subsidence and curvature and W_{max} as follows:

$$\varphi = \frac{x_1 - x_2}{x_1 + x_2}\pi = \frac{x_1 - x_2}{x_1 + x_2}180 \quad (17)$$

$$\frac{\pi}{L} = \frac{\pi}{x_1 + x_2} \quad (18)$$

$$W = W\, max/1 + \frac{e^{-\frac{\pi}{L}x_1} tg\frac{\varphi}{2}}{4\sin\frac{\varphi}{2}} \quad (19)$$

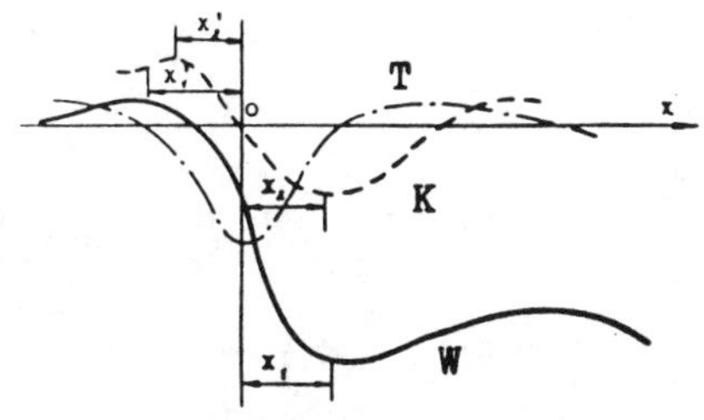

Fig. 2 Theoretical curves of subsidence, declination and curvature

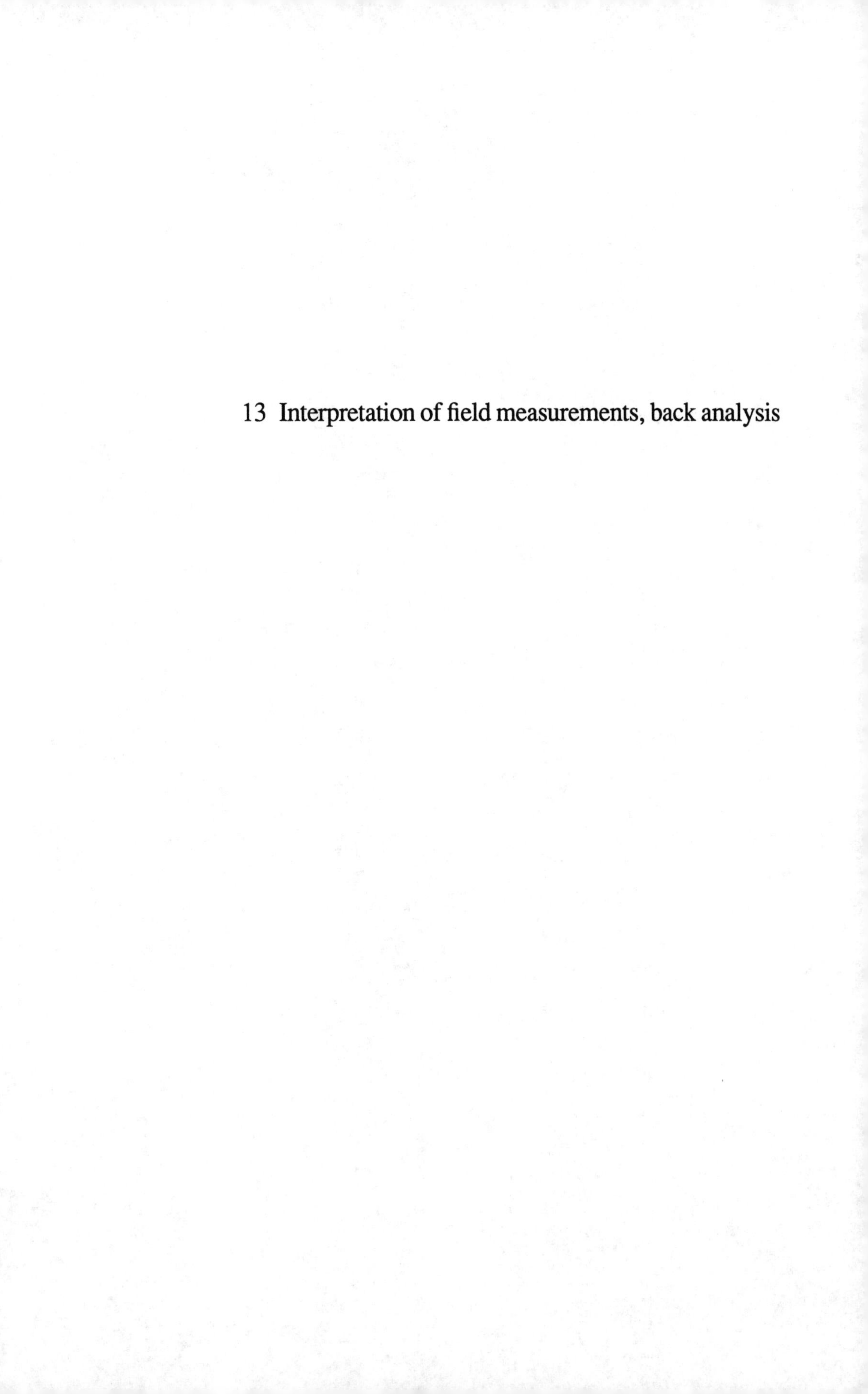

13 Interpretation of field measurements, back analysis

Numerical Methods in Geomechanics (Innsbruck 1988), Swoboda (ed.)
© 1988 Balkema, Rotterdam. ISBN 90 6191 809 X

Reliability of constitutive parameters for a soil obtained from laboratory test data

M.M.Zaman, A.Honarmandebrahimi & J.G.Laguros
School of Civil Engineering and Environmental Science, Norman, Okla., USA

ABSTRACT: A series of laboratory tests are conducted using a fluid-cushion cubical device on clay soil sampled from the U.S. 77 highway project near Ponca City, Oklahoma. Three sets of test data are selected to evaluate and study sensitivity of material constants for two consitutive models (δ_o and δ_1) based on the hierarchical approach developed by Desai and co-workers [1-5]. The first data set includes only two tests along two stress paths, while the other sets include additional tests performed along different stress paths. A relatively low variance is observed for the elastic, failure, and hardening parameters, compared with the non-associative parameter for the δ_1 model. Parameters determined from the data set I showed a close comparison between the back-predicted and observed stress-strain response. It is concluded that the evaluation of material constants for a constitutive model can be highly sensitive to a number of factors such as consistency, processing and discretization of test data.

1 INTRODUCTION

Determination of material constants of a constitutive model is as important as the development of the model, if not more. Usefulness of a model can be extremely limited if it fails to represent the basic characteristics of a material behavior, even though the model is mathematically sophisticated. This is particularly true for soil because of its complex response under three-dimensional loading and because it is extremely difficult to obtain consistent experimental data with low variance. Although substantial efforts have been made in recent years for the development of constitutive models [2,9] for soil, the reliability of material constants determined from typical experimental data has not been addressed adequately.

The main objective of this paper is to discuss the reliability of constitutive parameters for a clay soil which has been extensively tested in the laboratory using cylindrical triaxial and cubical devices. Questions pertaining to reliability of test data and material constants for selected constitutive models are addressed from the view point of (i) equipment/devices used in laboratory testing, (ii) processing of raw test data, and (iii) procedures used for evaluating the desired constitutive parameters for a model. The equipment sensitivity issue is addressed by using two different types of equipment: (1) a cylindrical triaxial device, and (2) a newly fabricated fluid-cushion multiaxial cubical device. Four constitutive models, namely Mohr-Coulomb and Drucker-Prager models, and two generalized plasticity models developed by Desai and his co-workers [1-5] are selected as benchmark models for parameter evaluation. Sensitivity of material constants is studied by varying the number of tests included in parameter evaluation. Some results pertaining to sensitivity of material constants for Mohr-Coulomb and Drucker-Prager models have been presented in an earlier paper by Honarmandebrahimi and Zaman [6]; the present paper will focus on the evaluation and sensitivity of material constants for the generalized plasticity models.

2 CHARACTERISTICS OF SOIL USED IN THE STUDY

The soil used in this experimental study was sampled from the U.S. 77 highway project near Ponca City, Oklahoma. The site constitutes an experimental pavement project sponsored by the Oklahoma

Department of Transportation (ODOT) and the University of Oklahoma [8]. The soil used in the present experimental study is non-stabilized and was sampled from an area near the control section. It consists of plastic weathered shale which belongs to the lower Wellington Formation of Permian age. Pedological information on these shales is found in the United States Department of Agriculture's Soil Survey for Kay County, Oklahoma [12]. This shale varies in color from yellowish gray to gray and grayish brown, the AASHTO classification ranging between A-7-6 (25) and A-7-6 (39), and the plasticity index varying from 26 to 37. For all the tests performed, the dry density was maintained approximately constant at 0.0012 kg/cm^3 and the moisture content was kept constant at 15%. Only undrained tests were performed. No pore pressure measurements were taken; however, from previous studies [7,8] it is estimated that at the moisture content considered here such effects would be negligible.

3 EXPERIMENTAL STUDY

A photographic view of the multiaxial fluid (air)-cushion cubical device used in this study is shown in Fig. 1. Tests along any desired stress path in the three-dimensional principal stress space, with any number of loading, unloading and reloading sequence can be performed by this device. In the present experimental study, a total of 24 tests were conducted along eight different stress paths shown in Fig. 2. Some additional tests were conducted to ensure reproducibility of test data along selected stress paths. At least two unloading and reloading cycles were included in each test. A test was terminated when the normal strain along one or more axes became large (about 15%). The failure stresses were estimated by extrapolation of plotted experimental stress-strain data. After each load increment, the specimen was allowed to stabilize for about 10 minutes before applying the following increment. The confining pressures were kept in the range of 10 psi (68.9 KPa) to 25 psi (172.25 KPa).

4 BENCHMARK CONSTITUTIVE MODELS

The benchmark constitutive models used here to study sensitivity/reliability of associated material constants were developed by Desai and his co-workers [1-5]. The models are based on a hierarchical approach that allows for progressive development of models of higher grade corresponding to different levels of complexities [3]. The model for initially isotropic material, hardening isotropically with associative plasticity, is treated as the basic model, δ_o. The model of immediate higher grade involving isotropic hardening with nonassociative response due to friction is regarded as δ_1 model.

Both δ_o and δ_1 models are used here for evaluation of appropriate material constants and back-prediction of stress-strain response. Detail description of these constitutive models can be found in various publications [1-5,11], only a brief description is included here to facilitate the discussion of pertinent material constants.

A large number of plasticity models [1-5,10,11] are presently available to represent stress-strain behavior of soil. Of these, critical state and cap models [10] are widely used. One of the major drawbacks of all these models is that the yielding is controlled by two separate yield functions which intersect each other with a slope discontinuity, resulting in nonuniqueness of the normal at the point of intersection.

To eliminate this problem, Desai and his co-workers [1-5] proposed a plasticity formulation based on single yield surface concept. The yield function was expressed in the form [3]

$$F \equiv J_{2D}/p_a^2 - F_b F_s = 0 \tag{1}$$

where, J_{2D} = second invariant of the deviatoric stress tensor, F_b = basic function, describing the shape of the yield function in the $J_1 - \sqrt{J_{2D}}$ space, and F_s = shape function describing the shape in octahedral planes, and p_a = atmospheric pressure. For the δ_o and δ_1 models used in this study, F_b and F_s were of the following form

$$F_b = -\alpha\,(J_1/p_a)^n + \gamma\,(J_1/p_a)^2 \tag{2}$$

$$F_s = (1 - \beta\,S_r)^m \tag{3}$$

where, J_1 = first invariant of the stress tensor, α = hardening parameter, γ, β, m = material response functions associated with the ultimate behavior, n = phase change parameter, and S_r = a stress ratio defined by

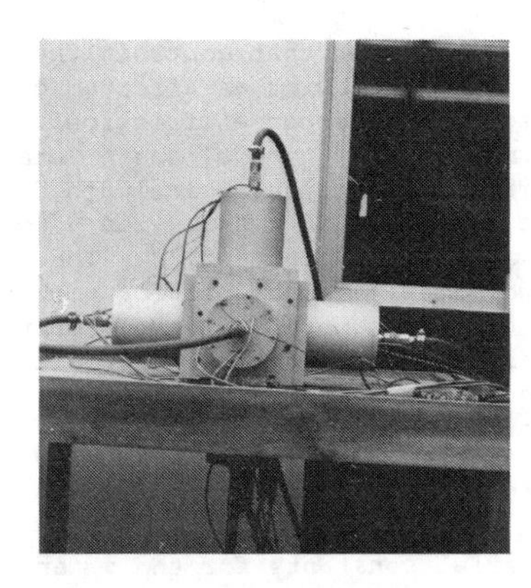

(a) Cubical device

(b) Data acquisition system

Figure 1 Photographic view of the cubical device and data acquisition system

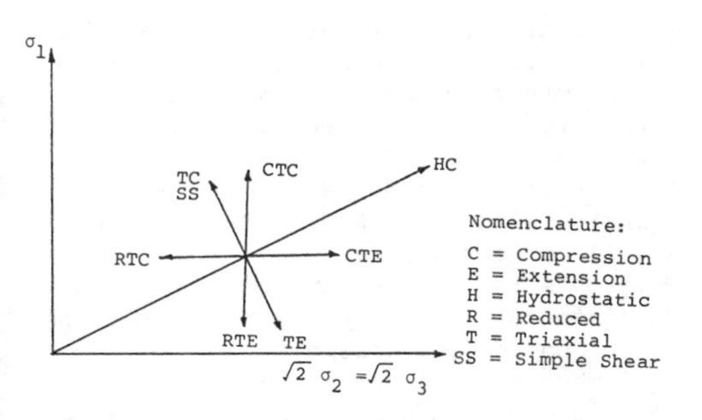

Figure 2 Stress paths on triaxial plane

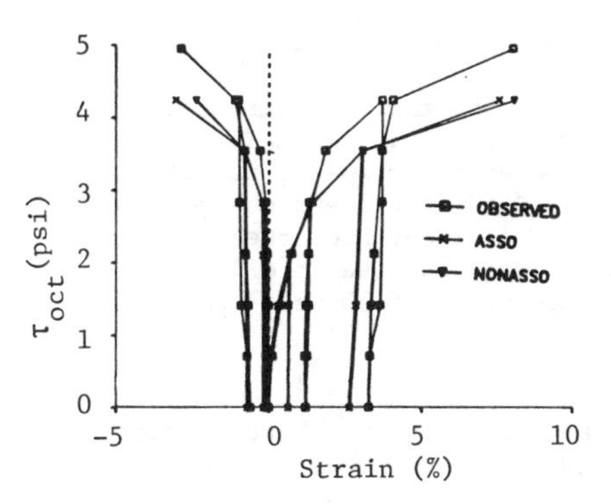

Figure 3 Comparison of observed and back-predicted response for set I (TC 22)

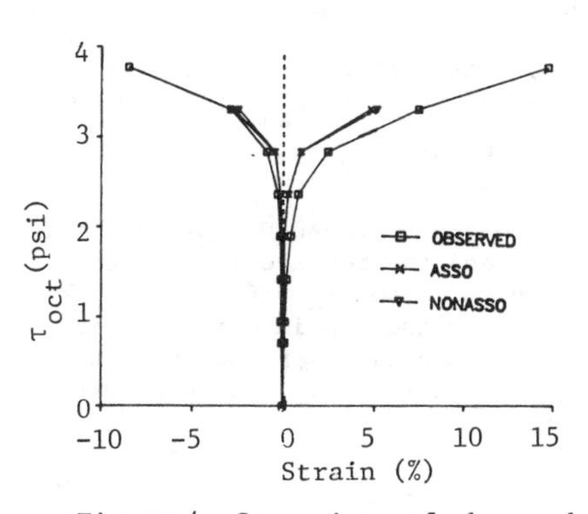

Figure 4 Comparison of observed and back-predicted response for set I (RTC 20)

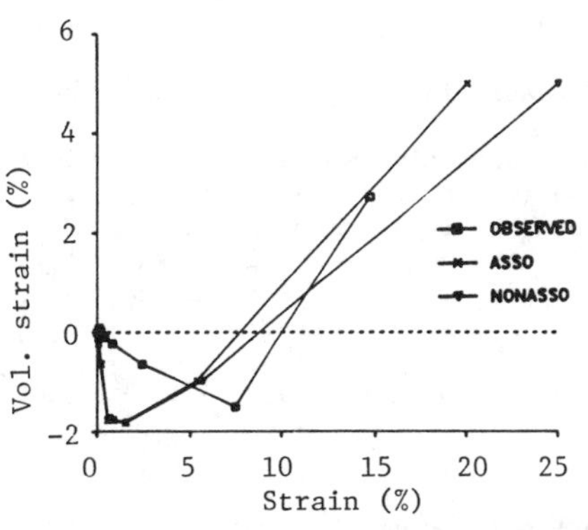

Figure 5 Comparison of measured and predicted volumetric strain for set I (RTC 20)

1.0 psi=6.89 kpa

$$S_r = \sqrt{27}\,[J_{3D}\, J_{2D}^{-3/2}]/2 \,, \qquad (4)$$

J_{3D} being the third invariant of the deviatoric stress tensor. In Eq. (3), the parameter m is usually assigned a value of -0.5. The phase change parameter, n, identifies the stress at which the material starts dilating, and is defined by the following expression:

$$n = 2\gamma\, F_s/[\gamma\, F_s - (J_{2D}/J_1^2)] \qquad (5)$$

n has an effect on the shape of yield surfaces in the $J_1 - \sqrt{J_{2D}}$ space.

4.1 Hardening function, α

The hardening function α is expressed in a general form as

$$\alpha = \alpha\,(\xi,\, \xi_v,\, \xi_D,\, r_v,\, r_D) \qquad (6)$$

in which, ξ is the trajectory of the plastic strain tensor, ε^p_{ij}, defined by

$$\xi = (\varepsilon^p_{ij}\;\; \varepsilon^p_{ij})^{\frac{1}{2}} \,, \qquad (7)$$

and the subscripts v and D represent the volumetric and the deviatoric part of ξ, respectively. Also, $r_v = \xi_v/\xi$ and $r_D = \xi_D/\xi$. For a situation in which the influence of hydrostatic and proportional loading is significant, α was expressed in the form [5]

$$\alpha = b_1 \exp\,[-b_2\, \xi\, (1-A)] \qquad (8)$$

where, b_1 and b_2 are hardening parameters. For initial hydrostatic loading of isotropic materials, A = 0, then [3]

$$\alpha = b_1 \exp\,[\,-\, b_2\, \xi] \qquad (9)$$

4.2 Non-associative parameter, κ

For the associative model (δ_o), the yield function (F) and the plastic potential function (Q) are assumed identical. However, for the non-associative model (δ_1), Q is defined as in Eq. (1), but the basic function, F_b, is expressed as

$$F_b = -\,\alpha_Q\,(J_1/p_a)^n + \gamma\,(J_1/p_a)^2 \qquad (10)$$

where,

$$\alpha_Q = \alpha + \kappa\,(\alpha_1 - \alpha)\,(1 - r_v)\,, \qquad (11)$$

α_1 being the value of α at which non-associative response starts, and κ is the material parameter that controls the volume change behavior as affected by non-associative response (friction).

In summary, the δ_o model has a total of eight material constants, including two elastic constants, E and ν. The δ_1 model has an additional constant, κ. The required material constants for both the δ_o and δ_1 models were evaluated for the Ponca City clay and their sensitivity was studied as discussed in the following section.

5 EVALUATION OF MATERIAL CONSTANTS

The material constants for the δ_o and δ_1 models were evaluated using a computer code developed by Desai, et al., [3]. The code was procured from the University of Arizona, Tucson and was modified to run on the University of Oklahoma VAX 11/780 system. The measured stress-strain values were plotted to identify any inconsistency in the test data. Any test which appeared to have inconsistent data was not included in the parameter evaluation. The plotted stress-strain curves were discretized to obtain the desired input for the code [3]. The sensitivity of material constants is studied by varying the number of tests included in parameter evaluation. Three sets of data are considered; the first set consisting of only two tests representing two stress paths, while the second and the third sets consist of five and ten tests, respectively, representing additional stress paths. Details of the tests included for each set are given in Table 1 and the corresponding material constants are listed in Table 2.

5.1 Elastic constants (E, ν)

The elastic constants are determined from the unloading/reloading slopes of shear tests. Average values from all tests considered in a given set are reported in Table 2. For set I, the average value for the Young's modulus was 22.4 MPa, compared with 42.1 MPa and 34.8 MPa for set II and set III, respectively. Poisson's ratio determined from the test data ranged between approximately 0.3 and 0.4. Considering possible human and equipment errors, the subjective judgement involved in discretizing the unloading/reloading curves, and the number of points used in discretization, the aforementioned variations in E and ν are considered acceptable.

Table 1 Tests included in evaluation of material constants

Set No.	No. of Tests	Stress Paths*	Confining Pressure KPa
Set I	2	RTC	137.8
		TC	151.6
Set II	5	CTC	75.8
		†CTC	103.4
		†CTE	137.8
		RTC	137.8
		RTC	151.6
Set III	10	CTC	75.8
		CTC	89.6
		†CTC	105.4
		†CTE	68.9
		CTE	103.4
		†CTE	137.6
		RTC	103.4
		RTC	117.1
		RTC	137.8
		TC	151.6

* see Fig. 2 for details
† dilation did not occur

Table 2 Material Constants for δ_o and δ_1 models

Parameter	Values		
	Set I	Set II	Set III
E (MPa)	22.4	42.1	34.8
ν	0.406	0.317	0.368
m	-0.5	-0.5	-0.5
γ	0.008	0.009	0.010
β	0.333	0.565	0.760
n	3.937	7.437	5.018
b_1	0.68×10^{-4}	0.34×10^{-4}	0.48×10^{-4}
b_2	0.433	0.717	0.378
κ	1.101	0.406	0.229

5.2 Ultimate state parameters, m, γ and β

The computer code [3] used for the material constant evaluation, assigns m = -0.5, based on the observations for many geologic materials (soils and rocks). The other two parameters (γ and β) are related to conventional friction angles, for both compression (ϕ_c) and extension (ϕ_E) tests. From the plot of ultimate/failure stresses (for different stress paths) in the conventional p - q stress space (see Fig. 6 of ref. [6]), The average values of ϕ_c and ϕ_E were found to be approximately 16.7° and 11.3°, respectively. The corresponding β can be evaluated from the following equations:

$$\beta = (1 - TR)/(1 + TR) \tag{12}$$

where $TR = (\tan \theta_c/\tan \theta_E)^{2/m}$. The angles θ_c and θ_E are related to ϕ_c and ϕ_E by

$$\tan \theta_c = (2/\sqrt{3}) \sin \phi_c/(3 - \sin \phi_c) \tag{13a}$$

and

$$\tan \theta_E = (2/\sqrt{3}) \sin \phi_E/(3 + \sin \phi_E) \tag{13b}$$

The β is found to be 0.7. The computer code [3], predicted β varying between 0.33 and 0.76. These parameters show acceptable variance and are within the range of values reported by Desai and Wathugala [3] for various geologic materials.

For $\beta = 0.7$, the parameter γ can be predicted from

$$\sqrt{\gamma} = \tan \theta_c/(1 - \beta)^m = \tan \theta_E(1 + \beta)^m \tag{14}$$

and is found to be in the range of 0.027 and 0.068. The code predicted γ varies between 0.008 and 0.01, data set I yielding the lowest value and data set III yielding the highest.

5.3 Phase change parameter, n

The value of n is determined from the state of stress at which the plastic volume change is zero, i.e., from the state when the material starts dilating. For a clay soil, dilation can be insignificant even at ultimate state of stress. Analysis of computer output showed that for several tests dilation did not occur. These tests are indicated in Table 1 by (†). For the other tests, the code predicted dilation but the magnitude was insignificant. For the tests with dilation, the value of n was generally in the range of 2.0 to 3.4, however, for the tests with no dilation, the values of n were substantially higher, similar to those for rocks and plain concrete [3].

Since the phase change parameter occurs as exponent in the yield and plastic potential functions, it can significantly influence the quality of predicted stress-strain response. For data set I, the value of n (3.9) appears to be more realistic than those for set II (7.4) and set III (5.0). Comparison of the observed and back-predicted stress-strain response in Figs. 3 through 10 shows that the parameters determined from set I were more consistent than the others.

5.4 Hardening parameters, b_1 and b_2

The hardening parameters (b_1 and b_2) are determined from the hydrostatic compression (HC) tests. The same hydrostatic test data are used with all three sets. Taking the natural log of both sides in Eq. (9) yields

$$\ln \alpha = \ln b_1 - b_2 \ \xi_v \qquad (15)$$

A set of simultaneous equations can be obtained by considering several points on the HC stress-strain curves, which are then solved using the least square procedure to obtain $\ln b_1$ and b_2. The values of b_1 and b_2 in Table 2, are substantially smaller than those reported in Ref. [3] for various sands; however, the variance for the three sets are appreciably low from the statistical view point.

It may be noted that the number of data points used for discretization of the HC test varied from 2 to 4, the majority of the points being in the high volumetric strain-range. Different values of b_1 and b_2 could be obtained if more data points are included in the discretization.

5.5 Non-associative parameter, κ

This parameter is calculated from the following equation:

$$\kappa = [(\alpha_1 - \alpha)\ (1 - r_v)]^{-1}\ (Y/Z - \alpha) \qquad (16)$$

where the expressions for Y and Z are given in Ref. [3]. The values of κ determined from the three data sets varied between 0.23 and 1.1. The lowest values were obtained from set III, which contained maximum number of tests (10) and stress paths (4).

It may be noted that the variance for this case is substantially larger, compared with those for the other parameters. This could be partly attributed by the selection of data points from the stress-strain curve. Even though κ could be calculated for any stress point, the portion near the ultimate condition can be emphasized since $(\alpha_1 - \alpha)$ is larger in the ultimate zones [3]. Preferential selection of data points was not followed here. A study currently underway would investigate the effects of discretization of stress-strain response on material constants.

6 COMPARISON OF OBSERVED AND BACK-PREDICTED RESPONSE

A viable constitutive model must be able to back-predict the laboratory tests used in evaluating its material constants, and other stress paths which are not used in determination of material constants [1-5]. For both constitutive models, selected back-predicted and observed responses are compared in Figs. 3 through 10. It may be noted that the parameters determined from set I produced the most reasonable back-prediction, even for a test that was not used in parameter evaluation (see Fig. 6). For all stress paths considered here, δ_o and δ_1 models predicted very close response, indicating that the effects of non-associative plasticity were negligible, except for the volumetric strain (Fig. 5). For some stress paths, δ_o and δ_1 models overpredicted the response (see e.g. Fig. 5), while for other stress paths predicted response underestimated the laboratory observation.

7 CONCLUDING REMARKS

In this study, a total of 24 laboratory tests along various stress paths were performed on clay soil using a fluid-cushion cubical device. The test data were used to determine the associated material constants for two selected constitutive models, based on the hierarchical approach and developed in the framework of plasticity theory [3]. Three sets of data are considered in the material constant evaluation, each with different number of tests and stress paths. It is observed that the material constants for a constitutive model can be sensitive to test data included in parameter evaluation, as well as on the nature of discretization of (input) stress-strain curve. For the cases studied herein, variance for the elastic and ultimate parameters were low, compared with the non-associative parameter. Back-predicted response showed varied correlation with the observed response depending the set of material constants used in back-prediction.

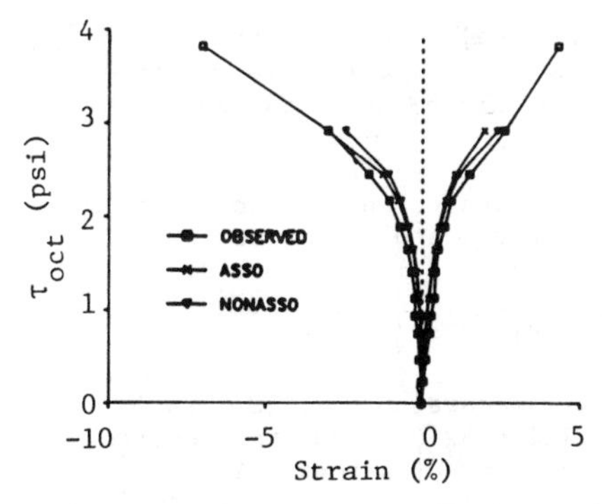

Figure 6 Comparison of observed and predicted response for set I (CTE 15) *

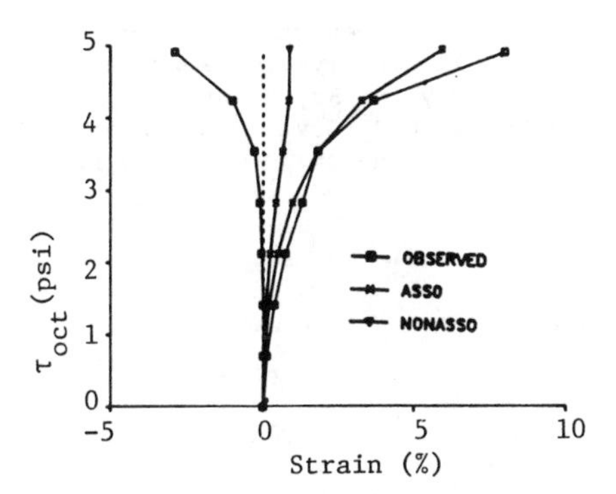

Figure 7 Comparison of observed and predicted response for set II (TC 22)

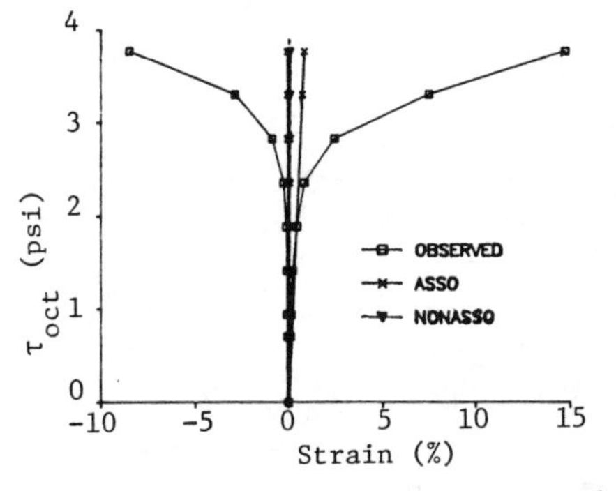

Figure 8 Comparison of observed and predicted response for set II (RTC 20)

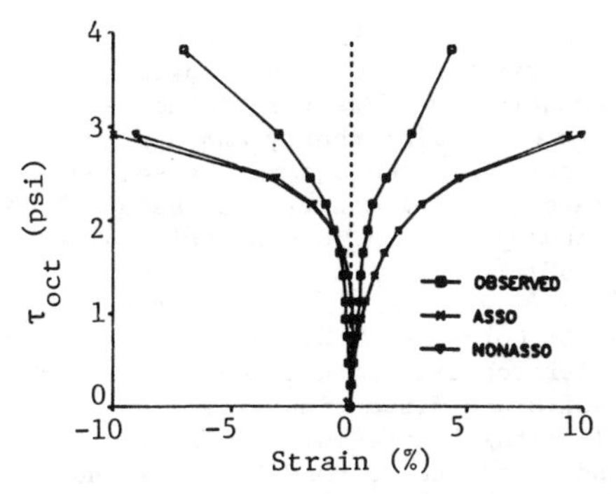

Figure 9 Comparison of observed and predicted response for set II (CTE 15)

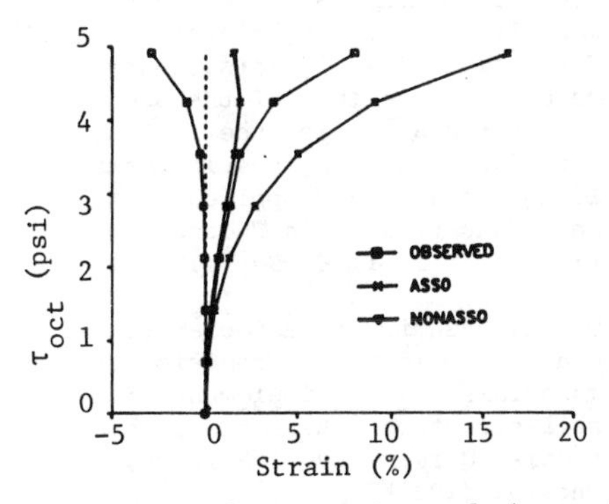

Figure 10 Comparison of observed and predicted response for set III (TC 22)

1.0 psi=6.89 kpa

* Indicates stress path and confining pressure in psi.

ACKNOWLEDGEMENTS

The computer code used for evaluation of material constants was developed by Prof. C.S. Desai and his co-workers at the University of Arizona, Tucson.

REFERENCES

[1] C.S. Desai and M.O. Faruque, Constitutive model for (geological) materials. Journal of Engineering Mechanics Division, ASCE, Vol. 110, No. 9: 1391-1408, (1984).

[2] C.S. Desai, E. Kremple, P.D. Kiousis, and T. Kundu, (Editors) Constitutive laws for engineering materials: theory and applications. Proceedings of the Second International Conference, held January 5-8, in Tucson, Arizona, (1987).

[3] C.S. Desai and G.W. Wathugala, Hierarchical and unified models for soils and discontinuities (joints/interfaces). Notes for short course, Second International Conference on Constitutive Laws for Engineering Materials: Theory and Application, held January 5-10, in Tucson, Arizona, (1987).

[4] C.S. Desai, H.M. Galagoda, and G.W. Wathugala, Hierarchical modelling for geologic materials and discontinuities - joints, interfaces. Proceedings of the Second International Conference on Constitutive Laws for Engineering Materials: Theory and Applications, held January 5-8, in Tucson, Arizona, Vol. 1: 81-94, (1987).

[5] Q.S.E. Hashmi, Nonassociative plasticity model for cohesionless materials and its implementation in soil-structure interaction, Ph.D. thesis, University of Arizona, Tucson, (1987).

[6] A. Honarmandebrahimi and M.M. Zaman, Significance of testing procedure and equipment on determination of constitutive parameters of soil. Proceedings of the International Conference on Numerical Methods in Engineering: Theory and Applications, held July 6-10 at Swansea, Vol. 2: C18/1-8, (1987).

[7] M.S. Keshawarz, Field stabilization of Ponca City shale, Ph.D. Dissertation, University of Oklahoma, (1985).

[8] J.G. Laguros and M.S. Keshawarz, Fly ash in shale stabilization for highway construction. Proceedings, Materials Research Society, Vol. 65: 37-47, (1986).

[9] G.N. Pande and J. Middleton, (Editors). Transient/dynamic analysis and constitutive laws for engineering materials. Proceedings of the International Conference on Numerical Methods in Engineering: Theory and Applications, NUMETA '87, held July 6-10, at Swansea, U.K., (1987).

[10] I.S. Sandler and D. Rubin, An algorithm and a modular subroutine for the cap model, Int. J. Num. Anal. Met. in Geomech., Vol. 3, (1979).

[11] A. Varadarajan and C.S. Desai, Multiaxial testing and constitutive modelling of a rock salt. Proceedings of the Second International Conference on Constitutive Laws for Engineering Materials: Theory and Applications, held January 5-8, in Tucson, Arizona, Vol. 1: 465-473, (1987).

[12] United States Department of Agriculture, Soil Conservation Service, Soil survey for Kay County, Oklahoma. U.S. Government Printing Office, Washington, D.C., (1967).

Numerical Methods in Geomechanics (Innsbruck 1988), Swoboda (ed.)
© 1988 Balkema, Rotterdam. ISBN 90 6191 809 X

Back analysis using prior information – Application to the staged excavation of a cavern in rock

A.Gens, A.Ledesma & E.E.Alonso
Technical University of Catalunya, Barcelona, Spain

ABSTRACT: The paper presents a maximum likelihood formulation which provides a framework for the performance of back analysis to estimate geotechnical parameters. The formulation allows the error structure of the measurements and prior information on the parameters to be introduced in a consistent and straightforward manner. Some details related to the numerical implementation of the procedure are also described. The formulation is applied to a case involving the excavation of a large underground cavern. Analysis with and without prior information are performed and the results are compared. The prior information analysis yields more consistent results with previously determined parameters values without increasing noticeably the differences between computed and observed measurements.

1 INTRODUCTION

In geotechnical engineering, back analysis from field measurements is often one of the most reliable methods to obtain soil and rock parameters. It is not surprising, therefore, that an increased effort is being made to devise numerical methods that can be used to perform back analysis in a systematic manner (e.g. [1],[2],[3],[4]). More recently the same type of methods have been applied to problems using nonlinear constitutive laws, namely the hyperbolic model ([5],[6]). In most cases, the estimation process is performed using the least squares criterion. The optimum parameters are obtained by minimizing the sum of the squared differences between computed and observed measurements.

There are advantages, however, in approaching the problem from a probabilistic viewpoint. In this way, it is possible to introduce easily in the analysis the error structure of the measurements and prior information on the parameters. Adopting a probabilistic framework, Cividini et al [7] used a Bayes approach to perform the estimation analysis. In this approach the unknown parameters, $\underline{p}$, are considered as random quantities and the optimum parameters are obtained by maximizing

$$f(\underline{p}/\underline{x}^*) \qquad (1)$$

the probability density of $\underline{p}$, given a certain set of measured quantities $\underline{x}^*$.

An alternative approach exists, though, in wich the model is assumed deterministic and the best estimation is found by maximizing the likelihood, L, of an hypothesis, $\underline{p}$, given some set of measured values, $\underline{x}^*$.

$$L(\underline{p}) = k\ f(\underline{x}^*/\underline{p}) \qquad (2)$$

The likelihood of an hypothesis [8] is proportional to the conditional probability of $\underline{x}^*$ given a set of parameters $\underline{p}$.

There are some conceptual advantages in this approach [9]:

a) The model parameters are considered fixed but uncertain due to lack of information. This allows to introduce the prior information on the values of the parameters based on the probability density function of prior information in a simple manner [10].

b) It does not require the model to be capable of reproducing the true system exactly [11]. This has advantages if model identification is to be attempted.

It must be stressed, however, that the objective function to be minimized which results from this approach is the same as that obtained from a Bayesian approach in which the probability density function is maximized. Only the conceptual basis is changed.

In this paper, the maximum likelihood

approach is formulated and applied to the estimation of rock parameters using field measurements obtained during the excavation of an underground cavern. The rock is assumed linear elastic and the Young's modulus, E, and horizontal stress coefficient K_0 are identified taking into account parameter information available before construction. Excavation was carried out in phases and this has also been considered in the analysis. Some relevant details of the numerical implementation are also described.

2 BASIC FORMULATION

An explicit model $\underline{M}$ relates parameters, $\underline{p}$, and observations, $\underline{x}$, as

$$\underline{x} = \underline{M}(\underline{p}) \qquad (3)$$

Although the parameters are unknown there may be some prior information on them that can be introduced in the formulation.

Assuming that the probability distributions of the prior information of the parameters and measurements are multivariate Gaussian, it is possible to write:

$$P(\underline{p}) = |\underline{C}_p^0|^{-1/2}(2\pi)^{-m/2}\exp[-\frac{1}{2}(\underline{p}-\langle\underline{p}\rangle)^t(\underline{C}_p^0)^{-1}(\underline{p}-\langle\underline{p}\rangle)] \qquad (4a)$$

$$P(\underline{x}) = |\underline{C}_x|^{-1/2}(2\pi)^{-n/2}\exp[-\frac{1}{2}(\underline{x}^*-\underline{x})^t(\underline{C}_x)^{-1}(\underline{x}^*-\underline{x})] \qquad (4b)$$

where

- $\underline{C}_p^0$ is the "a priori" parameter covariance matrix based on the available prior information
- $\underline{C}_x$ measurements covariance matrix
- $\langle\underline{p}\rangle$ "a priori" estimated parameters
- $\underline{x}^*$ measured variable values
- m number of parameters
- n number of measurements
- $()^t$ is used to indicate a transposed matrix.

If measurements and the "a priori" estimations of the parameters are independent, the basic likelihood postulate states that the likelihood of an hypothesis, $\underline{p}$, is

$$L(\underline{p}) = k\,P(\underline{p})\,P(\underline{x}) \qquad (5)$$

where k is an arbitrary constant.

The problem of parameter estimation is now equivalent to finding the set of parameters, $\underline{p}^*$, that maximizes (5). This is the same as minimizing the function S:

$$S = -2\ln L(\underline{p}) \qquad (6)$$

$$S = (\underline{x}^*-\underline{M}(\underline{p}))^t\underline{C}_x^{-1}(\underline{x}^*-\underline{M}(\underline{p})) + (\underline{p}-\langle\underline{p}\rangle)^t(\underline{C}_p^0)^{-1}(\underline{p}-\langle\underline{p}\rangle) + \ln|\underline{C}_x| + \ln|\underline{C}_p^0| + n\ln(2\pi) + m\ln(2\pi) - 2\ln k \qquad (7)$$

If the error structure of measurements and parameters are considered fixed, only the first two terms must be used in the minimization process, the rest being constant. The case in which the error structure is not fixed has been treated elsewhere [12].

If measurements and "a priori" parameters estimates are independent the covariance matrices $\underline{C}_x$, $\underline{C}_p$ will be diagonal. If the n observed values are obtained from r independent instruments with individual covariance matrices $(\underline{C}_x)_i$ and the m parameters can be divided in s groups with individual "a priori" covariance matrices $(\underline{C}_p^0)_j$, equation (7) becomes:

$$S = \sum_{i=1}^{r}(\underline{x}_i^*-\underline{x}_i)^t(\underline{C}_x)_i^{-1}(\underline{x}_i^*-\underline{x}_i) + \sum_{j=1}^{s}(\underline{p}_j-\langle\underline{p}_j\rangle)^t(\underline{C}_p^0)_j(\underline{p}_j-\langle\underline{p}_j\rangle) \qquad (8)$$

There is a wide range of algorithms available to find the minimum of S. In this paper, Marquardt's algorithm [13] has been used. The parameter correction is computed from

$$\Delta\underline{p} = \Delta\underline{p}^0 + [\underline{A}^t\underline{C}_x^{-1}\underline{A} + (\underline{C}_p^0)^{-1} + \mu\underline{I}]^{-1}\underline{A}^t\underline{C}_x^{-1}(\Delta\underline{x} - \underline{A}\Delta\underline{p}^0) \qquad (9)$$

where

- $\Delta\underline{p}^0 = \langle\underline{p}\rangle - \underline{p}$
- $\Delta\underline{x} = \underline{x}^* - \underline{x}$
- $\underline{A} = \partial\underline{x}/\partial\underline{p}$ (sensitivity matrix)
- μ is a scalar multiplier which must be reduced to zero on approaching the minimum.

Note that if μ equals zero, Marquardt's procedure becomes the standard Gauss - Newton algorithm.

3 NUMERICAL IMPLEMENTATION

In order to analyze realistic problems, it is necessary to define the model numerically. Usually a finite element model is adopted. Then,

$$\underline{K}\,\underline{u} = \underline{f} \qquad (10)$$

where $\underline{K}$ is the global stiffness matrix, $\underline{u}$ the nodal displacement vector and $\underline{f}$ the nodal force vector.

$$\underline{K} = \int_V \underline{B}^t\underline{D}\,\underline{B}\,dV \qquad (11)$$

$$\text{and} \quad \underline{f} = \int_S \underline{N}^t \underline{\sigma} \, dS \qquad (12)$$

where $\underline{B}$ is the matrix that relates strains and nodal displacements, $\underline{D}$ contains the incremental relationship between stresses and strains given by the constitutive law, $\underline{N}$ the shape function matrix and $\underline{\sigma}$, for an excavation case, is the vector of stresses acting on the excavation boundary.

The evaluation of the parameter correction $\Delta\underline{p}$ in (9) requires the computation of the sensitivity matrix $\underline{A} = \partial\underline{x}/\partial\underline{p}$. It should be noted that the measurements $\underline{x}$ do not necessarily correspond to nodal displacements $\underline{u}$. In this paper, the case in which $\underline{x}$ is a relative displacement between two arbitrary points along a direction which does not coincide with any of the global coordinate axes will be considered. This is the type of measurements to be used in the example of application presented later. In this case, an individual measurement x_i will be obtained

$$x_i = \underline{L}_i \underline{T}_i \underline{N}_i \underline{u}_i \qquad (13)$$

where

$\underline{u}_i$ is the vector of nodal displacements of the elements which contain measurement points.

$\underline{N}_i$ is the shape function vector for the same elements.

$\underline{T}_i$ is the rotation matrix required to transform the components of displacements in the global coordinate system to the measurement direction.

$\underline{L}_i$ is the matrix containig the linear combination of displacements implied by the type of measurement.

In case of relative measurement between two points

$$\underline{L}_i = [\ 1\ ,\ -1\ ]$$

The vector of all measurements will be given as

$$\underline{x} = \underline{L}\,\underline{T}\,\underline{N}\underline{u} = \underline{R}\underline{u} \qquad (14)$$

$$\text{and} \quad \underline{A} = \frac{\partial\underline{x}}{\partial\underline{p}} = \underline{R}\frac{\partial\underline{u}}{\partial\underline{p}} \qquad (15)$$

For a linear elastic material $\partial\underline{u}/\partial\underline{p}$ can be obtained by derivating (10) with respect to the parameters $\underline{p}$ and, rearranging

$$\frac{\partial\underline{u}}{\partial\underline{p}} = K^{-1}\left[\frac{\partial\underline{f}}{\partial\underline{p}} - \frac{\partial\underline{K}}{\partial\underline{p}}\underline{u}\right] \qquad (16)$$

In the case that p is the Young's modulus, E

$$\frac{\partial\underline{K}}{\partial E} = \int_V \underline{B}^t \frac{\partial\underline{D}}{\partial E}\underline{B}\, dV \quad \text{and} \quad \frac{\partial\underline{f}}{\partial E} = 0 \qquad (17)$$

If p is the value of K_0

$$\frac{\partial\underline{K}}{\partial K_0} = 0 \qquad \frac{\partial\underline{f}}{\partial K_0} = \int_S \underline{N}^t \frac{\partial\underline{\sigma}}{\partial K_0} dS \qquad (18)$$

In the first phase of excavation $\underline{\sigma} = \underline{\sigma}^0 = (K_0\sigma_y, \sigma_y, 0)$ (initial "in situ" stresses), and

$$\frac{\partial\underline{\sigma}^0}{\partial K_0} = (\sigma_y, 0, 0)$$

In subsequent phases $\underline{\sigma} = \underline{\sigma}^0 + \Delta\underline{\sigma}$ where $\Delta\underline{\sigma}$ is the increment of stresses during the previous excavation stages. In this cases, an additive term appears:

$$\frac{\partial\underline{f}}{\partial K_0} = \int_S \underline{N}^t \left[\frac{\partial\underline{\sigma}^0}{\partial K_0} + \frac{\partial(\Delta\underline{\sigma})}{\partial K_0}\right] dS \qquad (19)$$

The evaluation of $\partial(\Delta\underline{\sigma})/\partial K_0$ is cumbersome. Fortunately, in practice, it can often be neglected without impairing significantly the rate of convergence to the minimum of the objective function.

There is an alternative way to obtain an exact expression for $\partial\underline{u}/\partial K_0$ at phase j which does not involve computing $\partial(\Delta\underline{\sigma})/\partial K_0$. It is based on considering the displacements of phase j, $\underline{u}^j$, as $(\underline{u}^J - \underline{u}^{J-1})$, where $\underline{u}^J$ indicates the nodal displacements caused by all the excavation phases up to stage j. Then,

$$\frac{\partial\underline{u}^j}{\partial K_0} = \underline{K}_J^{-1}\frac{\partial\underline{f}_J}{\partial K_0} - \underline{K}_{J-1}^{-1}\frac{\partial\underline{f}_{J-1}}{\partial K_0} \qquad (20)$$

Now $\partial\underline{f}_J/\partial K_0$ can be evaluated exactly from

$$\frac{\partial\underline{f}_J}{\partial K_0} = \int_S \underline{N}^t \frac{\partial\underline{\sigma}^0}{\partial K_0} dS \qquad (21)$$

Finally it should be noted that, in general, the measurement vector $\underline{x}$ will refer to different excavation stages: $\underline{x}^1, \underline{x}^2, \ldots\ \underline{x}^n$. Therefore the basic finite element equation representing the model is in fact

$$\begin{bmatrix} \underline{f}^1 \\ \underline{f}^2 \\ \vdots \\ \underline{f}^n \end{bmatrix} = \begin{bmatrix} \underline{K}^1 & \underline{0} & \ldots & \underline{0} \\ \underline{0} & \underline{K}^2 & \ldots & \underline{0} \\ & & \ldots & \\ \underline{0} & \underline{0} & \ldots & \underline{K}^n \end{bmatrix} \begin{bmatrix} \underline{u}^1 \\ \underline{u}^2 \\ \vdots \\ \underline{u}^n \end{bmatrix} \qquad (22)$$

where $\underline{K}^1, \underline{K}^2, \ldots\ \underline{K}^n$ are the stiffness matrices corresponding to the different stages and include the effect of geometry changes due to excavation, $\underline{f}^1, \underline{f}^2, \ldots \underline{f}^n$ and $\underline{u}^1, \underline{u}^2, \ldots \underline{u}^n$ are the nodal forces and displacements of each excavation stage.

Once calculated the sensitivity matrix, $\underline{A}$, the parameter correction, $\Delta\underline{p}$, can be computed from (9). Because of nonlinearity, the correction will no be exact and the procedure has to be repeated until convergence to the minimum of (8) is achieved. The iterative process has been described in some detail in [14].

4 EXAMPLE OF APPLICATION

4.1 Description of the problem

As an example of application of the formulation described in this paper, the analysis of the excavation of an underground cavern will be presented (Fig. 1) The problem considered is based on a real field case concerning the construction of an underground powerhouse in the Spanish Pyrenees. The dimensions of the cavern are 37.5 m high, 20 m wide and 89 m long. By analyzing a section near the central part, plane strain conditions can be assumed.

In that area the rock is an schist which shows no oriented texture due to the high degree of metamorphism to which it has been subjected.

For the back-analysis, the rock is assumed isotropic linear elastic with Young's modulus E unknown and Poisson's ratio $\nu = 0.28$ derived from laboratory tests. One of the major uncertainties of the project concerned the value of the horizontal stresses. Consequently, the value of the ratio between horizontal and vertical stress, K_0, will be the other parameter to be estimated. The direction of the initial principal stresses is assumed vertical and horizontal, which is consistent with measurements of "in situ" stress carried out prior to the excavation of the cavern. The initial vertical stress is considered equal to the overburden pressure.

The location of the measurements used in the parameter estimation analysis are indicated in Fig. 1. Some of them (I,J,K,L) are convergence measurements. The rest are displacement measurements obtained from bar extensometers. Therefore, in all cases, the observations used are relative measurements between two points.

The excavation was carried out in several stages. For the purposes of the

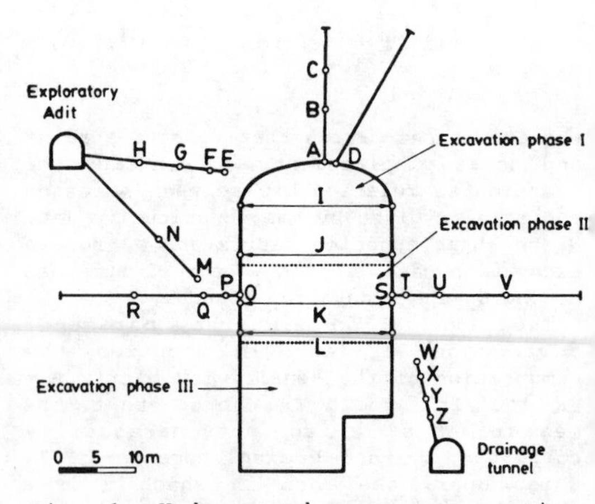

Fig. 1. Underground cavern excavation. Geometry and measurement locations.

analysis, three phases have been considered which are indicated in Fig. 1. In consequence, some measurements (for instance convergences K and L) will only be available for the third phase of excavation. In contrast, extensometers H - E and M - N provide measurements for all the excavation phases, as they were installed before excavation. In total, 36 measurements have been used, 8 correspond to the first phase, 7 to the second phase and 21 to the third phase. It should be stressed that the computations presented in this paper are the results of an analysis in which significant simplifications have been made involving, among others, the effect of discontinuities, stress - strain non linearity, time and 3-D effects which will be considered in the future. Nevertheless, the results presented provide an interesting insight into the nature of the problem.

4.2 Estimation of parameters with no prior information

The first estimation analysis to calculate E and K_0 is performed without considering any prior information. Given the nature of the measurement operations, the observations are assumed independent and with constant variance. Therefore, the measurements covariance matrix will be

$$\underline{C}_x = \sigma_x^2 \, \underline{I} \tag{23}$$

where σ_x^2 is the measurement variance and $\underline{I}$ the identity matrix. In consequence, the objective function to be minimized,

$$J = (\underline{x}-\underline{x}^*)^t(\underline{x}-\underline{x}^*) \tag{24}$$

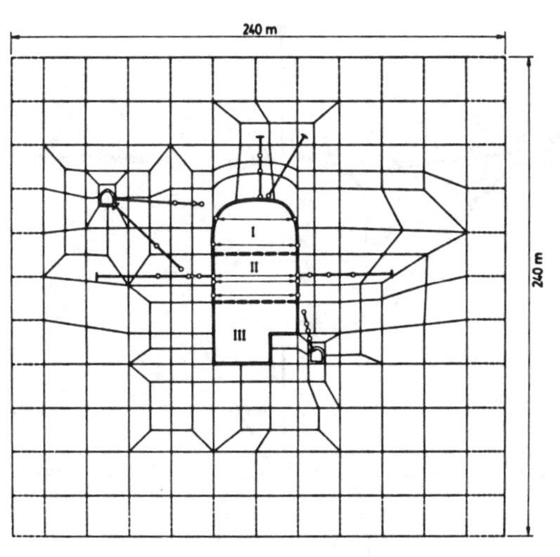

Fig. 2. Finite element mesh.

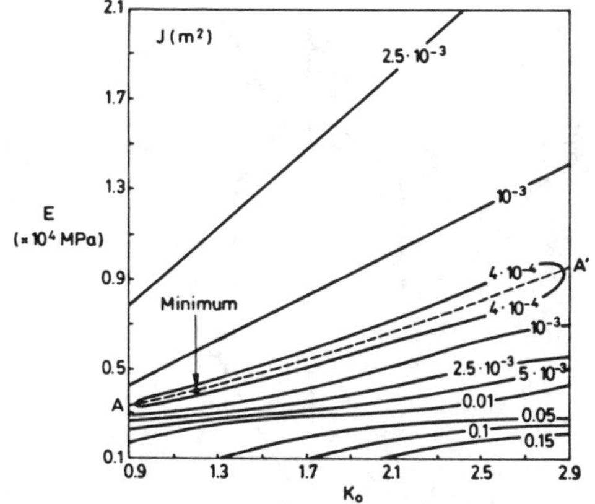

Fig. 3. Contours of values of the objective function, J.

corresponds to the least squares criterion.

The finite element mesh used in the analysis is shown in Fig. 2. It has been checked that varying the position of the boundaries has only a marginal effect on the results.

For illustration purposes, contours of the objective function values have been plotted in Fig. 3 which show the relative effect of variations of E and K_0. The estimation analysis has yielded the values of $E = 0.39 \times 10^{-4}$ MPa and $K_0 = 1.24$ giving the minimum value of $J = 2.89 \times 10^{-4}$ m^2. The position of this point is also indicated in Fig. 3. Computing the estimated standard deviation

$$\sigma_x = \sqrt{\frac{J}{n-m}} \tag{25}$$

a value of $\sigma_x = 2.9$ mm is obtained. The parameters found, therefore, achieve a good general approximation to the observed desplacements. A comparison between computed and observed results is presented in Fig. 5.

However, the parameters obtained do not seem to be very consistent with the geological and experimental information existing on the project. The values of both E and K_0 appear to be too low. Inspecting Fig. 3, it is apparent that, in fact, more than a clearly defined minimum, there exists a number of sets of parameters (indicated by line A - A' in Fig. 3) giving very similar values of the objective function, J. This can be more clearly seen in Fig. 4 where the values of J along line A - A' have been plotted.

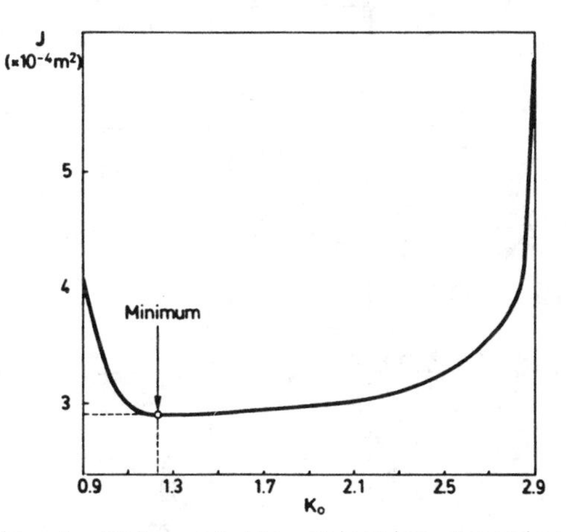

Fig. 4. Value of the objective function along line A - A'.

Therefore, any parameter set lying on this line will imply very similar differences between computed and measured displacements and it is no possible to discriminate satisfactorily between them. To increase the reliability of the estimated parameters, it is necessary to take into account the available prior information on the parameters.

4.3 Estimation of parameters considering prior information

If prior information is introduced in the analysis, the objective function to be minimized is now (8). Assuming that the prior information on E and K_0 is uncorrelated, the objective function will become

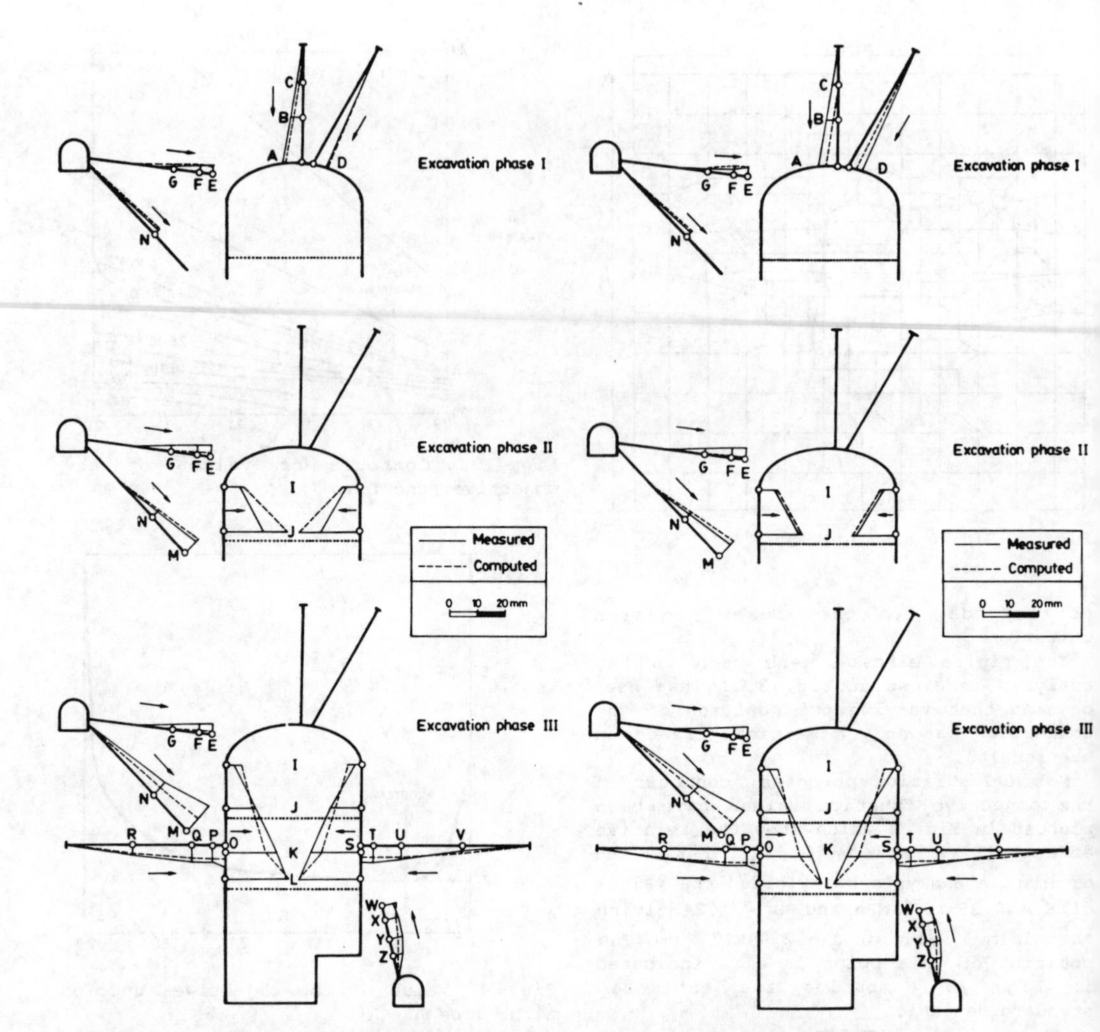

Fig. 5. Measured vs. computed displacements. Parameters estimated with no prior information.

Fig. 6. Measured vs. computed displacements. Parameters estimated considering prior information.

$$S = \sigma_x^{-2}(\underline{x}-\underline{x}^*)^t\,(\underline{x}-\underline{x}^*) + \sigma_E^{-2}(E-\langle E\rangle)^2 + \sigma_K^{-2}(K_0-\langle K_0\rangle)^2 \qquad (26)$$

$\langle E\rangle$ and $\langle K_0\rangle$ are the parameter estimations based on prior information only and σ_E^2 and σ_K^2 are the variances of the prior information on E and K_0. In fact, only the variance ratios are relevant and the objective function to be minimized can be written as,

$$J = (\underline{x}-\underline{x}^*)^t\,(\underline{x}-\underline{x}^*) + (\sigma_x^2/\sigma_E^2)\,(E-\langle E\rangle)^2 + (\sigma_x^2/\sigma_K^2)\,(K_0-\langle K_0\rangle)^2 \qquad (27)$$

The Young's modulus of the rock was determined using "in situ" tests. Both flat jack tests and dilatometer tests were performed. The mean value of all these determinations was 1.5×10^4 MPa. Flat jack tests carried out at different orientations indicated that the rock stiffness was isotropic. The value of (σ_E^2/σ_x^2) to be used in the analysis depends not only on the scatter of the results of the "in situ" tests used to determine E but, also, on the weight that is to be assigned to the prior information. In this particular case and given the usual uncertainties involved in determining rock deformability, it was decided not to assign an excessive weight to the prior information. A value of $(\sigma_E^2/\sigma_x^2) = 10^4$ was adopted. If σ_x is approximately 3 mm (as estimated in the analysis without prior information), the range of likely E values will be

$$E = \langle E \rangle \pm 2\sigma_E = 1.5\times10^4 \pm 0.6\times10^4 \text{ MPa}$$

The "in situ" stresses were measured using both flat jack tests and solid inclusion triaxial cells. In both cases the vertical stress was approximately equal to the overburden stress. However the results were different regarding the value of the horizontal stresses. Flat jack tests gave values of K_0 in the region of 1 whereas the solid inclusion cells gave values of K_0 in the range of 2 - 3. Due to this scatter, an estimated value of $K_0 = 2$ and a value of $(\sigma_K^2/\sigma_x^2) = 4\times10^4$ was used. Assuming again the value of $\sigma_x = 3$ mm, this gives a range of K_0:

$$K_0 = \langle K_0 \rangle \pm 2\sigma_K = 2 \pm 1.2$$

which spans the whole range of reasonable K_0 values.

In this conditions, the estimation procedure achieves a minimum of the objective function, J for values of $E = 0.77\times10^4$ MPa and $K_0 = 2.36$. Now, the sum of squared errors is 3.24×10^{-4} and the estimated standard deviation 3.1 mm, practically the same as in the former analysis with no prior information.

The comparison between the computed and the observed measurements using this new set of parameters is presented in Fig. 6. The pattern of movements is in fact quite similar to that obtained with the other set of parameters (Fig. 5). The same set of parameters, therefore, gives practically the same prediction error but they incorporate, in some measure, the available information on deformability and "in situ" stresses, resulting in a more reliable estimation.

The value of K_0 obtained in the identification analysis lies well within the "a priori" estimated range but the value of E lies outside. An important reason is that field measurements involve larger times and include, hence, viscous effects, whereas "in situ" tests are comparatively fast. Also possible stress-strain nonlinearity (including localized plastification) and effects of discontinuities have not been included in this analysis.

5 CONCLUDING REMARKS

The problem of parameter estimation has been formulated in a maximum likelihood framework. This allows the introduction of the error structure of the measurements and of prior information in a consistent and straightforward manner. Some details of the numerical algorithm required in the numerical analysis are described.

The formulation is applied to a field case based on the excavation of an underground cavern. It is shown that a back analysis using a least squares criterion fails to discriminate between different sets of parameters giving similar values of the objective function. By introducing the information available on deformability and "in situ" stresses, new values for the parameter estimates are found more consistent with the results of the measurements prior to the excavation. The magnitude of the differences between computed and observed displacements is similar to that found using the set of parameters obtained in the least squares analysis.

The identification procedure outlined can be generalized by including the variance ratios in the set of parameters to be estimated. In this way, there is no need to make any assumptions on their likely values and a more systematic analysis ensues. This topic is outside the scope of this paper, but a description of the resulting formulation and the method of solution can be seen in [12].

6 ACKNOWLEDGEMENTS

The support given by FECSA to this research project is gratefully acknowledged.

7 REFERENCES

[1] K.Arai, J.Ohta and K.Kojima, Estimation of soil parameters based on monitored movement of subsoil under consolidation, Soils and Foundations, 24,95-108 (1984)

[2] G. Gioda, Some remarks on back analysis and characterization problems in geomechanics, Proc. 5th Int. Conf. Numerical Meth. in Geomech., Nagoya, 1,47-61 (1985)

[3] G.Maier and G.Gioda, Optimization methods for parametric identification of geotechnical systems, Numerical Methods in Geomechanics, J.B. Martins (ed.), Reidel, Dordrecht, 1982

[4] S.Sakurai and K.Takeuchi, Back analysis of measured displacements of tunnels, Rock Mechanics and Rock Engineering, 16,173-180 (1983)

[5] K.Arai,H.Ohta and K.Kojima, Estimation of nonlinear constitutive parameters on monitored movement of subsoil under consolidation, Soils and Foundations, 27,35-49 (1987)

[6] A.Ledesma, Identificación de parámetros en Geotecnia. Aplicación a la excavación de túneles, Ph.D. Thesis, University of Catalunya (1987)

[7] A.Cividini, G.Maier and A.Nappi, Parameter estimation of a static geotechnical model using a Bayes' approach, Int. J. Rock Mech. Min. Sci. and Geomech. Abstr., 20,215-226 (1983)

[8] A.W.F. Edwards, Likelihood, Cambridge Univ. Press, Cambridge, 1972

[9] J.Carrera, State of the art of the inverse problem applied to the flow and solute transport equations, NATO Adv. Res. Workshop on Advances in Analytical and Numerical Groundwater Flow and Quality Modelling, Reidel (in press)

[10] J.V.Beck and K.J.Arnold, Parameter estimation in engineering science, Wiley, New York, 1977

[11] Y.Baram and N.R.Sandell, An information theoretic approach to dynamical systems modelling and identification, IEEE Transactions on Automatic Control, AC-23,61-66 (1978)

[12] A.Gens, A.Ledesma and E.E.Alonso, Maximum likelihood parameter and variance estimation in geotechnical back analysis, Proc. 5th Int. Conf. Applications of Statistics and Prob. in Soil and Struct. Eng., 2, 613-621 (1987)

[13] D.W.Marquardt, An algorithm for least square estimation of nonlinear parameters, J. Soc. Indust. Appl. Math., 11, 431-441 (1963)

[14] A.Ledesma, A.Gens and E.E. Alonso, Identification of parameters in a tunnel excavation problem, Proc. 2nd Int. Conf. Numer. Models in Geomech., Ghent, 333-344 (1986)

Numerical Methods in Geomechanics (Innsbruck 1988), Swoboda (ed.)
© 1988 Balkema, Rotterdam. ISBN 90 6191 809 X

Identification of the damage tensor for jointed rock mass by an inverse analysis method

T.Kyoya, Y.Ichikawa, O.Aydan & T.Kawamoto
Nagoya University, Japan

ABSTRACT: The rock mass involves several sets of joints, which essentially characterize mechanical behaviour of the rock mass. Such distributed joints can be characterized by a second-order symmetric tensor, called tha damage tensor, and their mechanical effects can be treated in a damage mechanics theory proposed by authors. However, it is often there are no sufficient informations of the jointing, so that the damage tensor cannot be determined exactly. In order to determine the damage tensor exactly under such circumstances, we here propose a method for identification of the dmage tensor from in-situ loading tests by the inverse formulation. An application of this method to a numerical plate loading test is shown.

1 INTRODUCTION

The effects of joints distributed in rock mass are too complicated to be estimated by a simplified manner. We have proposed a damage mechanics theory and its numerical analysis method by finite elements for treating such distributed discontinuities [4,7]. The theory describes mechanical behaviour of jointed rock mass based on material properties of the intact rock and geometry of the joints. In the damage theory, distributed joints are characterized by a second order symmetric tensor, called the damage tensor which is determined by in-situ observation, or more conveniently, by surveying photographs of outcrops [6,7,8].

However, in many cases we do not obtain sufficient informations of the joints from surface observation, so that the damage tensor cannot be determined exactly. Under such circumstances, informations obtained by field tests for rock mass such as a plate loading test or a borehole jack test may be helpful. We here present an inverse analysis method to identify the damage tensor from data of in-situ loading tests. The mechanical characteristics of a rock mass are thus determine easily and exactly from measured data, while the intact rock properties are specified by laboratory test.

2 DAMAGE THEORY FOR DISCONTINUOUS ROCK MASS

We here outline the damage mechanics theory and its numerical scheme by finite element method.

2.1 Damage tensor for rock mass

In the creep damage theory for metals, Murakami and Ohno[3] characterized a set of planar defects distributed in a material body with the density Ω and the unit normal vector $\boldsymbol{n}$ by

$$\boldsymbol{\Omega} = \Omega(\boldsymbol{n} \otimes \boldsymbol{n}) \qquad (1)$$

where $\otimes$ denotes the tensor product.

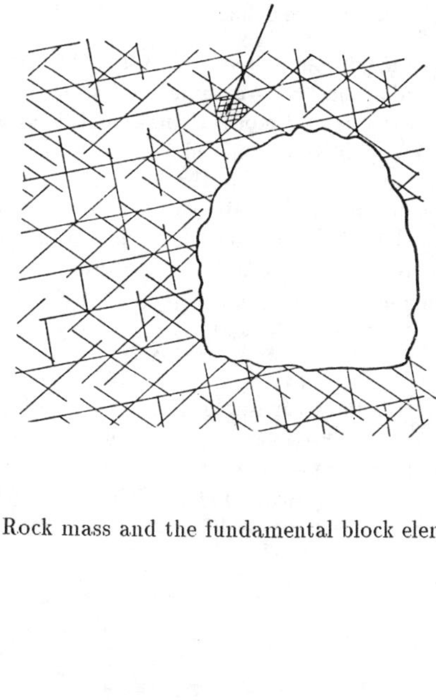

fundamental block element

Fig.1 Rock mass and the fundamental block element

For the rock mass, the jointing is assumed as follows :

(1) Each joint is of plane shape.

(2) The rock mass potentially consists of intrinsic elements of the intact rock. Discontinuities exist on interfaces of these fundamental elements, and will propagate along the interfaces if the applied load exceeds some limit.

Then, the damage tensor for rock mass is defined as

$$\boldsymbol{\Omega} = \frac{l}{V}\sum_{k=1}^{N} a^k(\boldsymbol{n}^k \otimes \boldsymbol{n}^k) \quad (2)$$

where V is a volume of the rock mass, N the number of joint surfaces contained in V, a^k and $\boldsymbol{n}^k$ the surface area and its unit normal vector of the k-th joint, respectively, and l the representative length of the fundamental rock block [4,7].

However, it is impossible to measure a^k and $\boldsymbol{n}^k$ for all jonits in rock mass. For an actual joint set, from in-situ observation, we can evaluate the average area of joint surfaces $\bar{a}$, the dominant direction of the set $\boldsymbol{n}$, and the density N/V. The representative length of the fundamental element l can be also evaluated from the average spacing of the joints. Then, the damage tensor for the i-th set of joints is given by

$$\boldsymbol{\Omega} = \frac{l}{V}\bar{N}^i\bar{a}^i(\boldsymbol{n}^i \otimes \boldsymbol{n}^i) \quad (3)$$

If the rock mass involves J sets of joints, the damage tensor is given by summing up the tensors for each set :

$$\boldsymbol{\Omega} = \sum_{i=1}^{J} \boldsymbol{\Omega}^i \quad (4)$$

2.2 Net stress in rock mass and constitutive equation

In a damaged material body, the effective surface area supporting the stress vector is reduced by the defects, so that the stress vector is changed and that the stress tensor acting in a damaged body becomes different from the one in a perfectly continuous body. The stress in a damaged body is called the net stress and denoted by $\boldsymbol{\sigma}^*$. If the defects do not support any force, the net stress tensor is given through a tranformation of the Cauchy stress tensor $\boldsymbol{\sigma}$ as follow:

$$\boldsymbol{\sigma}^* = \boldsymbol{\sigma}\,(\,\boldsymbol{I} - \boldsymbol{\Omega}\,)^{-1} \quad (5)$$

where $\boldsymbol{I}$ is the second-order identity tensor [3].

In rock mass, however, a part of shearing stresses and of compressive normal stresses are transmitted along/through discontinuities. Then, the net stress $\boldsymbol{\sigma}^*$ in rock mass is modified as

$$\begin{aligned}\boldsymbol{\sigma}^* = \; & \boldsymbol{T}^t\,[\,\boldsymbol{\sigma}'_t\,(\,\boldsymbol{I} - C_t\,\boldsymbol{\Omega}'\,)^{-1} \\ & + \boldsymbol{\sigma}'_n\,\{\,\boldsymbol{H}\,(\,\boldsymbol{\sigma}'_n\,)(\,\boldsymbol{I} - \boldsymbol{\Omega}'\,)^{-1} \\ & + \boldsymbol{H}\,(-\,\boldsymbol{\sigma}'_n\,)(\,\boldsymbol{I} - C_n\,\boldsymbol{\Omega}'\,)^{-1}\}]\,\boldsymbol{T}\end{aligned} \quad (6)$$

where $\boldsymbol{T}$ is the coodinate transformation tensor which transforms the damage tensor $\boldsymbol{\Omega}$ to its diagonal tensor $\boldsymbol{\Omega}'$ by

$$\boldsymbol{\Omega}' = \boldsymbol{T}\boldsymbol{\Omega}\boldsymbol{T}^t$$

The Cauchy stress tensor is similarly transformed into the principal direction of the damage :

$$\boldsymbol{\sigma}' = \boldsymbol{T}\boldsymbol{\sigma}\boldsymbol{T}^t$$

The $\boldsymbol{\sigma}'$ is decomposed into the diagonal components $\boldsymbol{\sigma}'_n$ and the shearing components $\boldsymbol{\sigma}'_t$ as

$$\boldsymbol{\sigma}' = \boldsymbol{\sigma}'_n + \boldsymbol{\sigma}'_t$$

The components of $\boldsymbol{\sigma}'_n$ are normal forces acting on joint surfaces and those of $\boldsymbol{\sigma}'_t$ are shearing force along the surfaces.

The characteristic function $\boldsymbol{H}$ is defined as

$$H_{ij}(x_{ij}) = \begin{cases} 0 & \text{if} \quad x_{ij} \leq 0 \\ 1 & \text{if} \quad x_{ij} > 0 \quad (i,j;\text{not summed}\,) \end{cases}$$

The coefficients C_n and C_t have values between 0 to 1, and controll the rate of transformation of $\boldsymbol{\sigma}'_n$ and $\boldsymbol{\sigma}'_t$, respectively.

The constitutive law is given between the net stress $\boldsymbol{\sigma}^*$ and $\boldsymbol{\epsilon}$:

$$\boldsymbol{\epsilon} = \boldsymbol{\Phi}(\,\boldsymbol{\sigma}^*\,) \quad (7)$$

Since the intact rock specimens are smaller than the fundamental element of the rock mass, no damage exist in the specimens in the sense of intactness of the fundamental element. This implies that $\boldsymbol{\Omega} = \boldsymbol{o}$ for these specimens, and $\boldsymbol{\sigma}^* = \boldsymbol{\sigma}$. Thus, the constitutive law obtained from laboratory tests is directly used in the damage theory, and no constitutive relation of the rock mass is needed.

If the intact rock is linear elastic, it is given in the form :

$$\boldsymbol{\epsilon} = \boldsymbol{C}\boldsymbol{\sigma}^* \;,\boldsymbol{\sigma}^* = \boldsymbol{D}\boldsymbol{\epsilon} \quad (8)$$

2.3 Finite element analysis by the damage theory

The virtual work equation holds in terms of the Cauchy stress $\boldsymbol{\sigma}$ as

$$\int_V \boldsymbol{\sigma}\cdot\delta\boldsymbol{\epsilon}dV = \int_{St} \boldsymbol{t}^o\cdot\delta\boldsymbol{u}dS + \int_V \boldsymbol{f}\cdot\delta\boldsymbol{u}dV \quad (9)$$

where $\boldsymbol{t}^o$ is the stress vector given on the boundary S_t, $\boldsymbol{f}$ the body force vector, and $\delta\boldsymbol{\epsilon}$ and $\delta\boldsymbol{u}$ are the variations of strain and displacement, respectively.

Both Eqns(5) and Eqn(6) are rewritten in the form

$$\sigma^* = \sigma + \psi \tag{10}$$

where

$$\psi = \sigma (I - \Omega)^{-1} \Omega$$

for Eqn(5), and

$$\begin{aligned} \psi = \; & T^t [\sigma'_t (\phi_t - I) \\ & + \sigma'_n \{ H (\sigma'_n) \phi \\ & + H (- \sigma'_n) \phi_n - I \}] T \end{aligned}$$

$\phi = (I - \Omega')^{-1}$, $\phi_n = (I - C_n \Omega')^{-1}$, $\phi_t = (I - C_t \Omega')^{-1}$

from Eqn(6), respectively.

Substituting Eqn(10) into Eqn(9), we obtain the virtual work equation for a damaged body as

$$\int_V \sigma^* \cdot \delta\epsilon dV = \int_{St} t^o \cdot \delta u dS + \int_V f \cdot \delta u dV + \int_V \psi \cdot \delta\epsilon dV \tag{11}$$

By discretizing the Eqn(11) in the standard manner of finite element method such as

$$\{u\} = [N]\{U\}, \quad \{\epsilon\} = [B]\{U\} \tag{12}$$

where $[N]$ is a shape function matrix, $[B]$ the strain-displacement matrix, $\{U\}$ the nodal displacement vector, and by using the linear constitutive law of Eqn(8)

$$\{\sigma^*\} = [D]\{\epsilon\} \tag{13}$$

we have the following simultaneous equation system :

$$[K]\{U\} = \{F\} + \{F^*\} \tag{14}$$

where

$$[K] = \int_V [B]^t [D][B] dV$$

$$\{F\} = \int_{S_t} [N]^t \{t^0\} dS + \int_V [N]^t \{f\} dV$$

$$\{F^*\} = \int_V [B]^t \{\psi\} dV$$

The stiffness matrix $[K]$ depends only on the material properties of the intact rock, and the mechanical effects of discontinuities are represented by an additional force vector $\{F^*\}$. This is an intrinsic characteristic of the damage mechanics analysis for rock mass.

3 INVERSE ANALYSIS FOR IDENTIFICATION OF THE DAMAGE TENSOR OF ROCK MASS

As outlined the previous sections, the damage analysis needs precisely data of joints, such as density, spacing, trace length, direction and their mechanical properties. However, it is often that we cannot obtain sufficient informations of the above items. In many cases, we can only obtain informations about the dominant directions of joints, and the results of some in-situ tests, such as plate loading tests. Then, we here develop a numerical analysis method to identify the damage tensor from such availabl informations by the inverse formulation method [1,2].

3.1 Identification of the damage tensor

We assume the followings :

(1) Results of some in-situ tests exist.

(2) A linear constitutive law of the intact rock :

$$\sigma = D \epsilon \tag{15}$$

is obtained from laboratory tests.

(3) Joints are uniformly distributed in the rock mass around the testing site.

(4) At the first stage of any in-situ loading test, joints are almost open, so that the net stress can be approximated by

$$\sigma^* = \sigma (I - \Omega)^{-1}$$

Then, our problem is formulated as follows:

(equation of equilibrium)

$$\nabla \cdot \sigma = o \quad \text{in V} \tag{16}$$

(constitutive law)

$$D \epsilon = \sigma^* \tag{17}$$

(net stress-Cauchy stress relation)

$$\sigma^* = \sigma + \sigma (I - \Omega)^{-1} \Omega \tag{18}$$

(boundary conditions)

$$t = \sigma \nu = t^o \quad \text{on } S_t \tag{19}$$

$$u = u^o \quad \text{on } S_u \tag{20}$$

(measured boundary)

$$u = \bar{u} \quad \text{on } S_t \tag{21}$$

Discretization procedure as mentioned in Sec.2.3 leads to the following numerical model

$$[K]\{U\} = \{F\} + \{F^*\} \tag{22}$$

where

$$\{F\} = \int_{S_t} [N]^t \{t^0\} dS$$

Eqn(22) can be solved by separating as follows:

$$[K]\{U_F\} = \{F\} \tag{23}$$

$$[K]\{U_F^*\} = \{F^*\} \tag{24}$$

$$\{U\} = \{U_F\} + \{U_F^*\} \tag{25}$$

In the above equations, the stiffness matrix $[K]$, the applied force vector $\{F\}$ are the known terms, and a part of the displacement vector $\{U\}$ is measured.

We can solve Eqn(23), and from the displacement vector $\{U_F\}$ the Cauchy stress $\{\sigma\}$ is obtained through the constitutive relation:

$$\{\sigma\} = [D][B]\{U_F\} \tag{26}$$

since the $\{U_F\}$ is the displacements corresponding to the non-damage state of the rock mass.

In Eqn(18), if we set

$$\boldsymbol{\Phi} = (\boldsymbol{I} - \boldsymbol{\Omega})^{-1} \boldsymbol{\Omega} \tag{27}$$

then, the additinal force vector :

$$\{F^*\} = \int_V [B]^t \{\psi\} dV$$

can be represented in the form:

$$\{F^*\} = [G]\{\Phi\} \tag{28}$$

where $\{\Phi\}$ is a column vector collecting m independent components of the tensor $\boldsymbol{\Phi}$, and, concequently, $[G]$ becomes a matrix having m columns and l rows if the numerical model have l freedoms. The matrix $[G]$ is a known term since the Cauchy stress $\{\sigma\}$ is known.

Substituting Eqn(28) into Eqn(24), we have

$$[K]\{V\} = [G]\{\Phi\} \tag{29}$$

in which, we set as

$$\{V\} = \{U\} - \{U_F\}$$

And, Eqn(29) can be patitioned as follows:

$$\begin{bmatrix} K_{11} & K_{12} \\ K_{21} & K_{22} \end{bmatrix} \begin{Bmatrix} V_1 \\ V_2 \end{Bmatrix} = \begin{bmatrix} G_1 \\ G_2 \end{bmatrix} \begin{Bmatrix} \Phi_1 \\ \Phi_2 \end{Bmatrix} \tag{30}$$

where $\{V_1\} = \{U_1\} - \{U_{F1}\}$ and $\{U_1\}$ collects measured displacements.

A condensation to vanish the unknown vector $\{V_2\}$ leads to

$$[R]\{\Phi\} = \{H\} \tag{31}$$

where

$$[R] = [G_1] - [K_{12}][K_{22}]^{-1}[G_2]$$

$$\{H\} = ([K_{11}] - [K_{12}][K_{22}]^{-1}[K_{21}])\{V_1\}$$

Applying the least squre method to Eqn(31) gives

$$[R]^t[R]\{\Phi\} = [R]^t\{H\} \tag{32}$$

By solving Eqn(32), we can easily obtain $\{\Phi\}$.

While, from Eqn(27) we have

$$\boldsymbol{\Omega} = \boldsymbol{I} - (\boldsymbol{\Phi} + \boldsymbol{I})^{-1} \tag{33}$$

Using the above realtion, we can obtain the damage tensor from $\boldsymbol{\Phi}$.

If there are J sets of joints in the rock mass, and if all the unit normal vectors of each joint set are known, for example, from stereo net projection, the damage tensor can be represented as

$$\boldsymbol{\Omega} = \sum_{i=1}^{J} \Omega^i (\boldsymbol{n}^i \otimes \boldsymbol{n}^i) \tag{34}$$

In Eqn(34), parameters to be determined are the areal densities of each joint set, Ω^i ($i = 1, 2, \cdots J$).

By representing the tensors in Eqn(34), $\boldsymbol{\Omega}$ and $(\boldsymbol{n}^i \otimes \boldsymbol{n}^i)$, in the form of column vectors as

$$\{\Omega\} = \{ \; \Omega_{11} \quad \Omega_{22} \quad \cdots \; \}^t$$

$$\{m^i\} = \{ \; n_1^i n_1^i \quad n_2^i n_2^i \quad \cdots \; \}^t$$

Eqn(34) can be written in the following form:

$$\{\Omega\} = [M]\{\omega\} \tag{35}$$

where $[M]$ is the matrix composed of the column vector $\{m^i\}$:

$$[M] = \{m^1 m^2 \cdots m^J\}$$

and the $\{\omega\}$ collects the unknown areal densities of each set Ω^i as

$$\{\omega\} = \{\Omega^1 \Omega^2 \cdots \Omega^J\}^t$$

Applying the least squre method to Eqn(35), we have

$$[M]^t[M]\{\bar{\Omega}\} = [M]^t\{\Omega\} \tag{36}$$

Thus, we can decompose the damage tensor into the ones of each joint set.

3.2 Numerical example

A two-dimensional damage analysis for plate loading test are carried out. The finite element model calculated is shown in Fig.2. Elastic constants given to the intact rock are

Young's modulus $E = 100 Mpa$
Poisson's ratio $\nu = 0.25$

Two joint sets are assumed to exist, of which areal densities and unit normal vectors are as follows:

Set 1;

$$\Omega^1 = 0.3, \; \boldsymbol{n}^1 = (\cos 45°, \; \sin 45°, \; 0)$$

Set 2;

$$\Omega^2 = 0.4, \; \boldsymbol{n}^2 = (\cos 60°, \; \sin 60°, \; 0)$$

therefore, the damage tensor of the rock mass becomes

Table 1 Relations between measured points and identified areal densities of joint sets

Measured Points							
Identified value of areal density	Ω^1	0.997	0.301	0.299	0.300	0.301	0.301
	Ω^2	−2.301	0.399	0.401	0.400	0.399	0.399

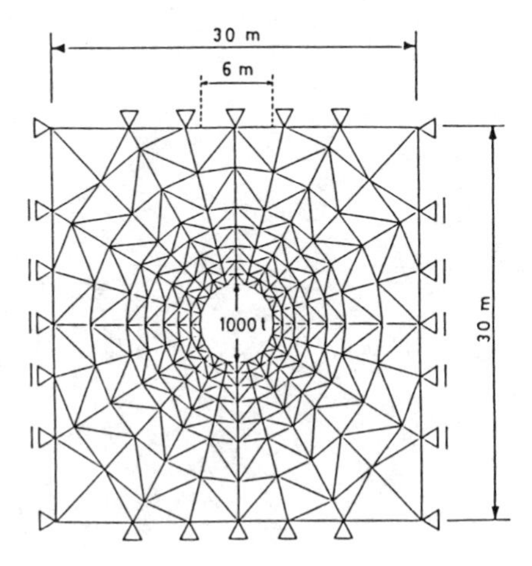

Fig.2 Finite element model of the plate loading test

$$\boldsymbol{\Omega} = \begin{bmatrix} 0.283 & 0.339 & 0.0 \\ 0.339 & 0.417 & 0.0 \\ 0.0 & 0.0 & 0.0 \end{bmatrix}$$

Regarding the displacements calculated in the foreward analysis as measured ones, the above damage tensor and the areal densities are identified by the proposed method. Table 1 shows relations between input measured displacements and the areal densities identified. The letter x and y mean that a horizontal or a vertical displacement is assumed to be measured at that point.

It is observed in Table 1 that the areal densities, consequantly the damage tensor, are precisely identified when more than three measured displacements are input. This is because in this two-dimensional case the number of components to be determined of the vector $\{\Phi\}$ in Eqn(31) is three. In the next step of this study, the stability of this identification procedure have to be examined.

4 CONCLUSION

Distributed joints in rock mass have essential effects on mechanical behaviour of rock mass. The damage mechanics theory treats the mechanical effects of the joints by characterizing them by the damage tensor, while by specifying the constitutive relation from the laboratory tests of the intact rock specimens.

We present an inverse analysis method to identify the damage tensor of rock mass from obtainable informations such as results of in-situ tests, material properties of intact rock, and dominant directions of joints observed in-situ. The propcedure is formulated by an inversion of the damage mechanics analysis, and is applied to a numerical model of plate loading test.

By using this method, the damage mechanics analysis may be able to provide usuful informations in design of rock mass structures.

REFERENCES

[1] G.Gioda : Some remarks on back analysis and characterization problems in geomechanics, Numerical Methods in Geomechanics Nagoya, vol.1, Balkema, pp.47-61 (1985)

[2] G.Maier and G.Gioda : Optimization methods for parametric identification of geotechnical systems, Numerical Methods in Geomechanics, Proc. of the NATO Advanced Study Institute, Reidel, pp.273-304 (1982)

[3] S.Murakami and N.Ohno : A continuum theory of creep and creep damage, Proc. 3rd IUTAM Symp. on Creep in Structures, Springer-Verlag, pp.422-443 (1980)

[4] T.Kyoya, Y.Ichikawa and T.Kawamoto : A damage mechanics theory for discontinuous rock mass, Numerical Methods in Geomechanics Nagoya, vol.1, Balkema, pp.469-480 (1985)

[5] T.Kawamoto, Y.Ichikawa and T.Kyoya : Rock mass discontinuities and damage mechanics, Computational Mechanics '86, vol.2, Springer-Verlag, p.IX-27 (1986)

[6] T.Kyoya, Y.Ichikawa, M.Kusabuka and T.Kawamoto : A damage mechanics analysis for underground excavation in jointed rock mass, Proc. Int. Symp. Eng. Complex Rock Formations, Beijing, China, pp.506-513 (1986)

[7] T.Kawamoto, Y.Ichikawa and T.Kyoya : Deformation and fracturing process of discontinuous rock mass and damage mechanics theory, Int. J. Numerical and Analytical Methods in Geomechanics, vol.11 (1987) to be appeared.

[8] Y.Ichikawa, Y.Nakamura, T.Kyoya and T.Kawamoto :Identification of damage field of rock mass, Proc. 2nd Int. Conf. Education, Practice and Promotion of Comp. Methods in Eng. Using Small Computers, Guangzho, (1987) to be appeared.

Numerical Methods in Geomechanics (Innsbruck 1988), Swoboda (ed.)
© 1988 Balkema, Rotterdam. ISBN 90 6191 809 X

Calculation and measurement of stress changes induced in a single-heater test in rock salt

S.Heusermann & N.Jacob
Federal Institute for Geosciences and Natural Resources, Hannover, FR Germany

ABSTRACT: A large-scale heater test was carried out by BGR in the Asse salt mine to study the thermomechanical response of rock salt. Extensive theoretical and experimental studies were made to determine whether thermal loading causes critical stresses and rock failure. Finite-element calculations of the heater test taking into account non-steady-state temperature distribution and nonlinear creep of rock salt show that significant stress changes occur during heating but that no tensile stress for rock failure appear. Thermally induced stress was measured in situ using hydraulic pressure cells. For all measurement horizons the same results were obtained. Maximum stress changes occur with radial orientation to the heater. Calculated and measured changes in stress and temperature during heating show good agreement. Thus, the numerical models can be validated and are suitable for prediction of rock stability.

1 INTRODUCTION

The design of high-level radioactive waste disposal repositories in rock requires extensive theoretical and experimental studies of the thermomechanical response of the host rock under realistic conditions. For this reason, the Federal Institute for Geosciences and Natural Resources (BGR) has performed several heater tests in rock salt over the last several years. A single-heater test was made in 1985 at the 800-m level of the Asse salt mine. In conjunction with this test, numerical calculations were made using the finite-element method.

The purpose of the calculations was:

- precalculation of thermomechanical response of the rock salt to establish design criteria for the in-situ test;
- simulation of real test conditions to study critical stresses and zones of failure;
- development of numerical models to predict critical stresses;
- evaluation and interpretation of measurements.

The purpose of the field measurements was:

- development, testing, and demonstration of the applicability of several in-situ test methods;
- experimental study of the thermomechanical response of rock under thermal loading;
- validation of numerical models.

2 DESCRIPTION OF TEST SITE

Numerous geomechanical and geophysical measurements e.g. extensometer (Ex), inclinometer (In), convergence, temperature, fluid level (P), ultrasonic (M), electrical logging, permeability, televiewer, and stress measurements (Sp) were made (see Kopietz & Meister [1]). The borehole configuration consisted of about 20 observation boreholes around a central borehole containing the electrical heater (Figure 1). The boreholes were drilled vertically, 20 to 35 m deep in the floor of a gallery 5.5 m high and 6.0 m wide.

The heater test lasted 64 days. After 57 days the initial heating power of 1800 W/m was increased to 2350 W/m. After the heater was turned off, measurements were continued for 336 days.

3 NUMERICAL CALCULATIONS

The FE program packages ADINA/ADINAT (Bathe [2]) and ANSALT/ANTEMP (Wallner & Nipp [3]) were used to calculate the far-field thermal and thermomechanical response of the rock around the test site and to calculate the near-field thermo-elastic response of the pressure cells used for stress measurements.

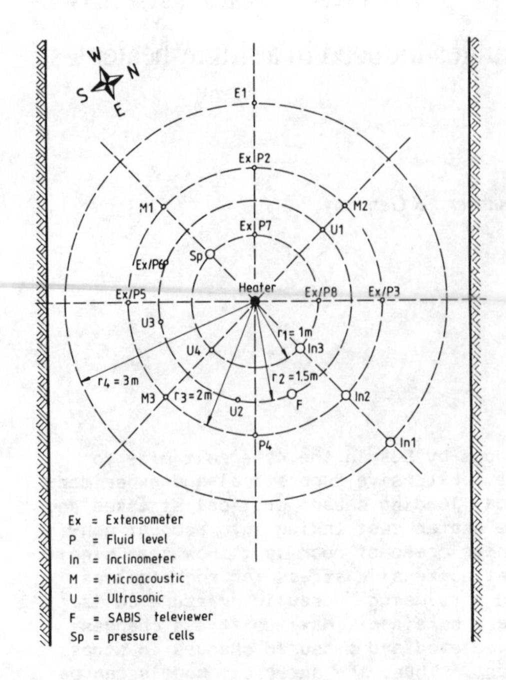

Figure 1. Configuration of heater and test boreholes

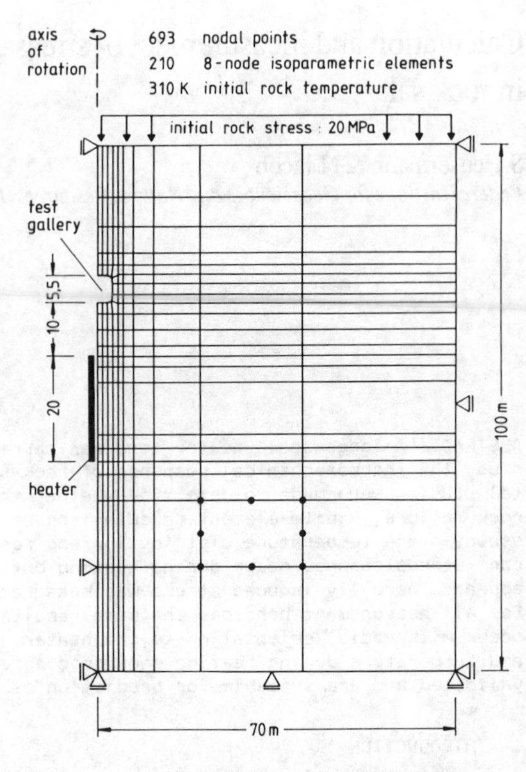

Figure 2. FE modelling of test site

3.1 FE-modelling of test site

The test site, including gallery, heater, and surrounding rock, was modelled by a rotation symmetric finite-element mesh of 210 8-node isoparametric elements (Figure 2). It was assumed that the rock is homogeneous and isotropic with an initial rock pressure of 20 MPa and an initial temperature of 310 K, corresponding to a total depth of about 800 m.

The change in temperature with time was calculated assuming constant heat flow into the surrounding rock over the whole length of the heater. The following values were used for the thermal parameters for rock salt: temperature-dependent heat conductivity $\lambda = 6.1/(1 + 0.0045\ T)$ (in $W \cdot m^{-1} \cdot K^{-1}$) and specific heat $c_p \cdot \rho = 1843\ kJ \cdot m^{-3} \cdot K^{-1}$. Air ventilation was roughly modelled by keeping the temperature of the gallery walls, floor, and ceiling at a constant 310 K. Isotherms calculated for the 57th day after heating was started are shown as an example in Figure 3. The maximum temperature of 410 K occurs around the heater.

The thermomechanical response of the salt rock mass was calculated as a function of time using the Arrhenius formula for nonlinear steady-state creep of rock salt (Langer [4]):

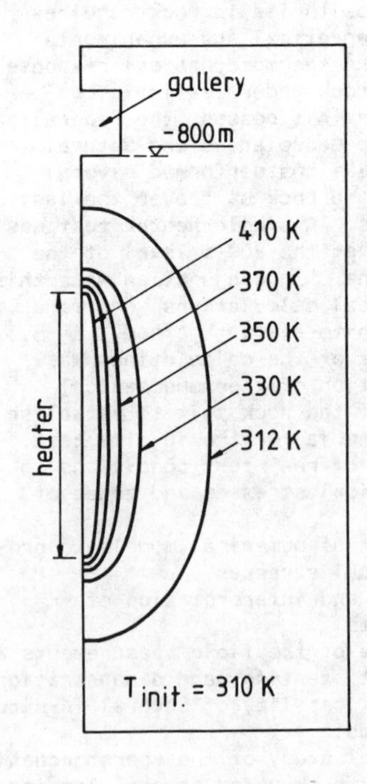

Figure 3. Isotherms after 57 days

$$\epsilon^{c}_{eff} = A \cdot e\ (-Q/RT) \cdot (\sigma_{eff}/\sigma^{*})^{n} \quad (1)$$

where

ϵ^{c}_{eff} is the effective creep rate (in d^{-1}),
σ_{eff} is the effective stress (in MPa),
σ^{*} is the reference stress (in MPa),
A is a scale factor (= 0.18 1/d),
n is a stress exponent (= 5.0),
Q is the activation energy (= 54 kJ/mol),
R is the universal gas constant (= $8.3143 \cdot 10^{-3}$ kJ/K·mol)
and T is the temperature (in K).

To take real test conditions into account, a period of 100 days without heating, followed by a period of 57 days with a heating power of 1800 W/m, a period of 7 days at 2350 W/m, and a period of 336 days without heating was considered. The change in effective stress is plotted, as an example, in Figure 4 for the times t = 0 (beginning of heating), t = 1 d, t = 5 d and t = 10 d at a distance of 1.25 m from the heater. The maximum effective stresses of about 17 MPa appear 15 to 25 m from the gallery floor a short time after beginning of heating (t ≅ 1 d).

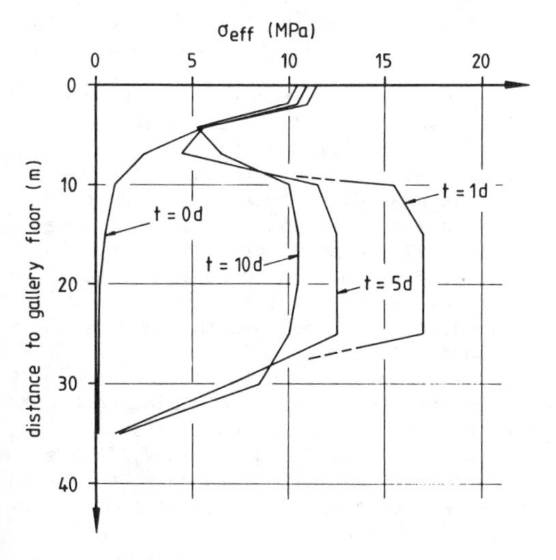

Figure 4. Effective stresses at a distance of 1.25 m from the heater

To determine whether critical stresses and rock failure occur, two criteria were taken into account: First, tensile stresses should not appear with respect to the low tensile strength of rock salt. Second, an empirical model for the limit of failure strength was used based on laboratory tests (Liedtke & Meister [5]):

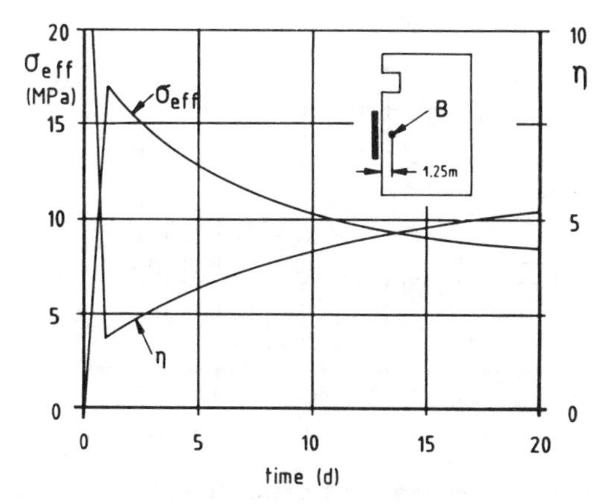

Figure 5. Effective stresses and local safety factors

$$\sigma_{eff} = \frac{\sigma_m + c}{a + b\ (\sigma_m + c)} \quad (2)$$

where a = 0.19, b = 0.012 (1/MPa), and c = 0.64 (MPa). This hyperbolic formula gives the limit of effective stress σ_{eff} that may occur in the rock as a function of isotropic stress σ_m. Possible failure of rock must be taken into account if the effective stress values calculated using the FE method are larger than those obtained by Equation 2.

A local safety factor η can be defined by the following equation (see [5]:

$$\frac{\sigma_{eff}\ \text{(according to Equ. 2)}}{\sigma_{eff}\ \text{(according to FE calc.)}} \quad (3)$$

If values of $\eta < 1$ are calculated the occurence of failure is indicated. As an example, the change of effective stress with time obtained by FE calculations and the change of η with time according to Equation 3 is plotted in Figure 5 for a selected location (middle of the heater) with a distance of 1.25 m. The smallest value of η = 3.8 appears short time after beginning of heating. Then, the safety factor increases continuously with time.

Numerical calculations show neither tensile stress nor critical effective stresses (as defined by Eq. 2 and Eq. 3) during the heating stage of the test. Low tensile stresses appear around the heater during cooling of the rock after heating is stopped. With respect to real temperature conditions in a radioactive waste disposal - no rapid drop in temperature - these tensile stresses are not relevant.

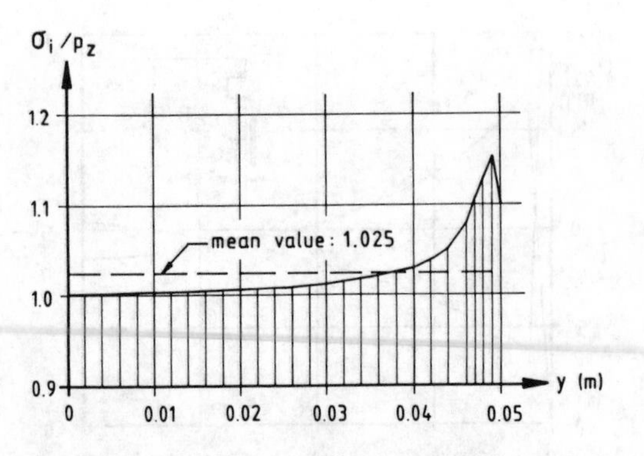

Figure 6. Normalized pressure cell response induced by ΔT = 100 °C

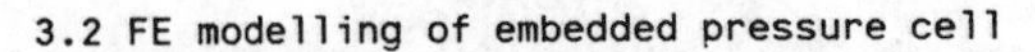

3.2 FE modelling of embedded pressure cell

Since the thermomechanical response of embedded hydraulic pressure cells is influenced by different thermal expansion of cell and surrounding rock, finiteelement calculation was made assuming plane strain conditions to study the thermoelastic behavior of pressure cell, borehole grout and surrounding rock salt. The cell was considered assuming uniform material properties E = 150 GPa, ν = 0.3, and $\alpha_T = 2.0 \cdot 10^{-5}$ 1/K for steel plates and liquid (Magiera [6]). Borehole grout and rock salt were modelled assuming E = 25 GPa, ν = 0.27, and α_T = $4.0 \cdot 10^{-5}$ 1/K. Considering a temperature change ΔT = 100 °C the normalized pressure cell response σ_i/p_z (σ_i = cell response, p_z = stress change in rock) is shown in Figure 6. It can be seen that values of 1.004 to 1.15 and an average value of 1.025 are calculated. Thus, a reduction of the cell response is necessary to obtain the real thermally induced stress in the surrounding rock (see chapter 4).

4 RESULTS AND EVALUATION OF STRESS MEASUREMENTS

To monitor stress changes during heating of the rock mass, hydraulic pressure cells (Glötzl type) were installed in borehole SP (see Figure 1) at different depths (10, 15, and 20 m) at various orientations. After installation, the hole was backfilled with a mixture of crushed rock salt, magnesite, and brine. Each stress monitoring probe included three pressure cells to measure three horizontal stress components so that maximum horizontal stress changes and their orientation can be determined (see Heusermann [7]).

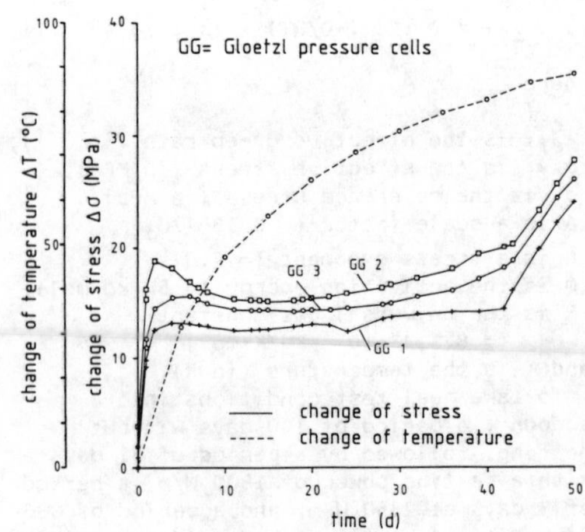

Figure 7. Measured stress changes induced by temperature

A detailed description of Glötzl pressure cells is given by Grainger [8].

Thermally induced stress changes measured in three horizontal directions at a depth of 20 m are shown, as an example, by the solid lines in Figure 7. The change in temperature measured in borehole P7 at the same depth is plotted as a dashed line. Immediately after the beginning of heating an increase in rock stress up to values of about 19 MPa occurs without a significant increase in rock temperature. After several days, rock stress decreases and then increases again. Maximum values of about 27 MPa occur after 50 days, corresponding to a measured temperature change of 35 °C. All directions of measurement show the same response by the pressure cells.

Since the measured data includes not only stress changes in the surrounding rock but also stress changes caused by thermal expansion of the liquid (mercury) inside the pressure cell, the data must be corrected for temperature changes as follows (Rehbinder [9]):

$$\Delta\sigma_T = \Delta\sigma_M - c \cdot \Delta T \qquad (4)$$

where

$\Delta\sigma_T$ is the true change in stress (in MPa),
$\Delta\sigma_M$ is the measured change in stress (MPa),
ΔT is the measured change in temperature (in K), and
c is a correction factor (in MPa/K).

The results of calculations described above show that c can be assumed to be equal to 0.025.

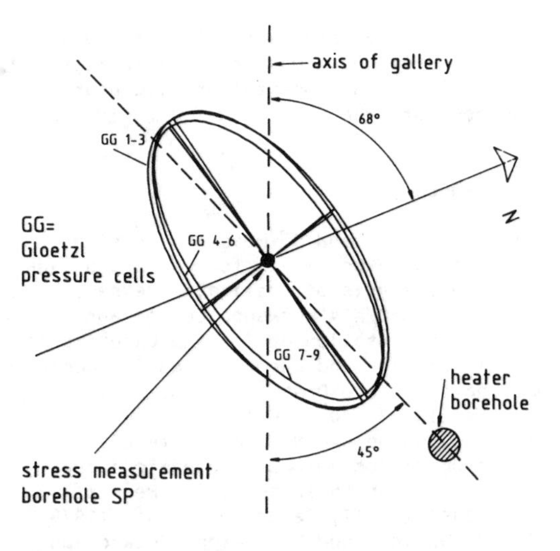

Figure 8. Ellipses of stress changes obtained from measured data

Comparison of the ellipses of stress changes (see Figure 8) calculated from data measured after 1 d at three horizons shows the following: All of the depths of measurement yield the same size and orientation of the ellipses; the maximum horizontal component (18 MPa) of the change in stress occurs radially to the heater, and the mimimum horizontal component is half of the maximum.

5 COMPARISON OF THEORETICAL AND EXPERIMENTAL RESULTS

5.1 Temperature

A comparison of measured and calculated temperatures shows that the calculated values are generally 10 - 15 % higher than the measured ones. For example, the change in temperature with time is plotted in Figure 9 for a distance of 1 m to the heater and 25 m to the gallery floor (middle of heater). It can be seen that the shape of the measured temperature curve agrees well with the calculated curve. The highest temperature values (calculated T = 158 °C and measured T = 126 °C) occur at the end of the first test period (heating power 1800 W/m). The difference between the theoretical and empirical results is probably caused by the difference between theoretical and real heat conductivity of rock salt.

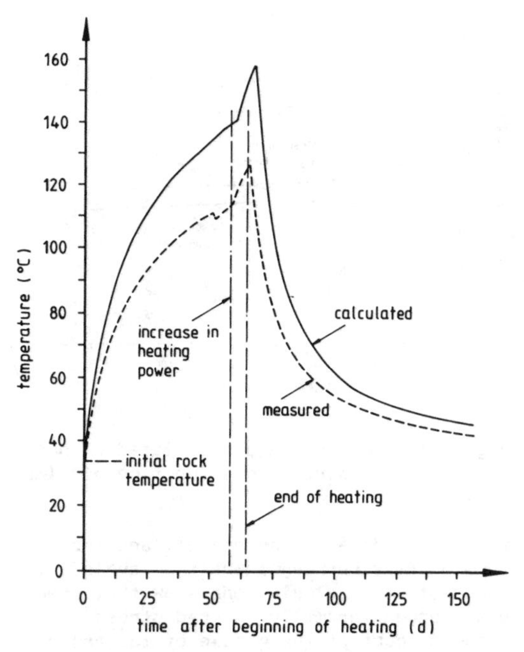

Figure 9. Measured and calculated temperature as a function of time

5.2 Thermally induced stresses

A plot of the thermally induced change of radial stress as a function of time is shown in Figure 10. Solid lines show the values calculated for points at the middle and the top of the heater, each 1.25 m from the heater. Dashed lines show the values measured at the middle and the top of the heater at a distance of 1.00 m. It can be seen that at the middle of the heater, the measured values are lower than the calculated ones. The shapes of the curves are in good agreement at the beginning of heating. At the end of the heating period, the measured values increase and become larger than the calculated values. The measured and calculated curves for the top of the heater are not in agreement at the beginning of heating. Measured values are generally higher than calculated. After 20 days, the measured and calculated curves agree well.

In terms of the direction of maximum horizontal stress changes, both calculated and measured results show that the maximum component of the change in stress generally arises radially to the heater, as expected. The measurements show no significant difference between the results obtained at the top of the heater, those

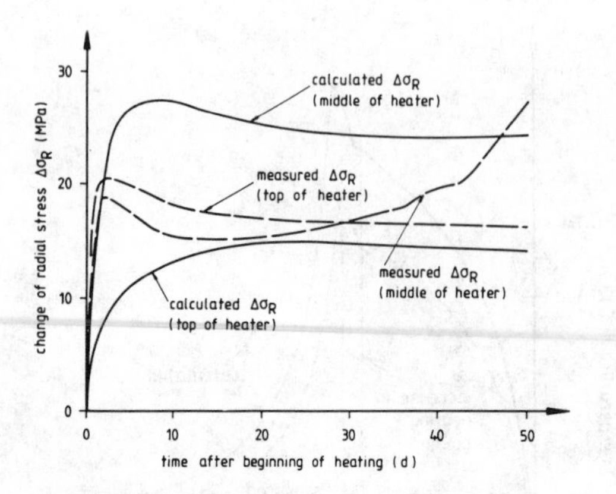

Figure 10. Measured and calculated changes in stress as a function of time

from the middle of the heater, and those from halfway between these two levels. In contrast, the calculations show that somewhat larger thermally induced stress changes occur at the middle of the heater.

6 CONCLUSIONS

To study the thermomechanical response of rock salt a large-scale single-heater test was performed. In conjunction with this test, finite-element calculations were made. Results of those calculations show that significant changes of effective stresses occur around the heater during the heating stage, but no critical stresses or zones of failure appear.

Measurement of thermally induced stress changes were made using hydraulic cells. The data demonstrate the applicability of these cells for short-term and long-term monitoring of stress changes caused by temperature. All of the measurement horizons yield the same results.

To validate the numerical models used for calculation of the test area experimental and theoretical results are compared. Generally, good agreement between measured and calculated stresses and temperatures was obtained. The numerical models developed for calculating stress and temperature fields in the test area are suitable to simulate the thermomechanical response of rock salt and to study the stability of the salt rock mass.

Because the actual conditions in a radioactive waste disposal repository in salt rock will not have the rotational symmetry, the results obtained from the single-heater test have to be applied to more complex arrangements of heat sources. For this reason, a multiple-heater test with a linear arrangement of five heaters has been started.

REFERENCES

[1] J. Kopietz and D. Meister, In-situ heating experiments for the production and analysis of critical stresses, Proc. CEC/NEA Workshop Design and Instrumentation of In-Situ Experiments in Underground Laboratories for Radioactive Waste Disposal, Brussels, May 15-17, 405-416 (1984)

[2] K.J. Bathe, ADINA - A finite element program for automatic dynamic incremental nonlinear analysis, Report 82448-1, MIT, Cambridge, Mass. (1978)

[3] M. Wallner and H.K. Nipp, Entwicklung eines optimalen Finite-Element-Programms (ANSALT) zur Berechnung thermomechanischer Vorgänge bei der Endlagerung hochradioaktiver Abfälle, Endbericht zum BMFT-Forschungsvorhaben KWA 20708, BGR, Hannover (1984)

[4] M. Langer, Main activities of engineering geologists in the field of radioactive waste disposal, Bulletin IAEG 34, Paris, 25-38 (1986)

[5] L. Liedtke and D. Meister, Stability analysis of underground structures in rock salt utilizing laboratory and in-situ testing and numerical calculations, Solution Mining Research Institute, Woodstock, Ill. (1982)

[6] G. Magiera, Weiterentwicklung des hydraulischen Kompensationsverfahrens zur Druckspannungsmessung in Beton, BAM, Forschungsbericht 102, Berlin (1984)

[7] S. Heusermann, Evaluation of stress measurements in the Asse salt mine, Joint U.S. DOE/FRG Workshop on Geotechnical Instrumentation, Asse Mine, May 4-8 (1987)

[8] B.N. Grainger, Evaluation of the Glötzl stress gauge for use in concrete structures, Central Electricity Research Lab., Letherhead, Surrey (1978)

[9] G. Rehbinder, Strains and stresses in the rock around an unlined hot water cavern, Rock Mechanics 17, 129-145 (1984)

Numerical Methods in Geomechanics (Innsbruck 1988), Swoboda (ed.)
© 1988 Balkema, Rotterdam. ISBN 90 6191 809 X

Finite element analysis of discontinuous geological materials in association with field observations

S.Sakurai
Kobe University, Japan

T.Ine
New Japan Engineering Consultants, Inc., Osaka

M.Shinji
OYO Corporation, Tokyo, Japan

ABSTRACT: In this paper, a new constitutive equation for finite analysis is proposed for analyzing the mechanical behaviour of discontinuous geological materials. The use of this equation makes it possible to clarify the difference between the loosened and plastic zones occurring in the ground due to its excavation. The proposed equation is simple to apply to practice. As an application of this equation to practical problems, two examples are shown, particularly for the occurrence of a loosening zone. One is for a tunnel excavated in shallow depth, and the other is a cut slope problem.

1 INTRODUCTION

The ground consisting of soils and/or jointed rock masses is classified into three groups, i.e., (a) continuous, (b) discontinuous, and (c) pseudo continuous types, as shown in Fig. 1 (Sakurai 1987). Type (a) may be for ground consisting of intact rocks or soils, Type (b) represents jointed rock masses, and Type (c) is for highly fractured and/or weathered rock masses, thus, for this Type (c), the global behaviour of the ground seems to be a continuous body. We call this the pseudo continuous type of ground.

The mechanical behaviour of the Type (a) ground can be analyzed by means of a mechanical model based on continuum mechanics, while a discontinuous model such as those proposed by Cundall (1971) and Kawai (1980) is used for analyzing the Type (b) ground, where joint elements in the finite element analysis are also useful. Concerning the Type (c) ground, one can of course adopt a discontinuous model similar to that for Type (b). In engineering practices, however, it is almost impossible to explore all of the joint systems or to investigate all their mechanical characteristics. Moreover, it seems that this type of ground behaves, in a global sense, just like a continuous body. Therefore, a continuum mechanics model can be used for this type. It should be noted, however, that in this model the effect of discontinuities must be taken into account.

In this paper, a method for analyzing Types (a) and (c), i.e., the continuous and the pseudo continuous types of ground, is discussed. The method is based on continuum mechanics, and an anisotropic constitutive equation is proposed in the finite element analysis. As an application of the proposed constitutive equation to practical problems, shallow tunnel and cut slope problems are described.

2 LOOSENED AND PLASTIC ZONES

When excavating the ground, such as tunnelling and cutting slopes, a loosened zone and/or a plastic zone may occur in the vicinity of the excavated free surface, depending on geological conditions, the joint system, the mechanical characteristics of the materials, initial stress, and even on the excavation method. The loosened zone may be defined as the zone in which all the discontinuities tend to open or slide along certain slip surfaces due to stress relief caused by excavation. On the other hand, the plastic zone occurs in ground consisting of weak materials under large initial stress at a great depth in the ground. In this plastic zone, the state of stress satisfies a yielding criterion such as Mohr-Coulomb, Drucker-Prager, etc., which are generally given in terms of stress. It should be emphasized, however, that the loosened and plastic zones are hardly distinguished in an ordinary finite element analysis. This is mainly due to the fact that the constitu-

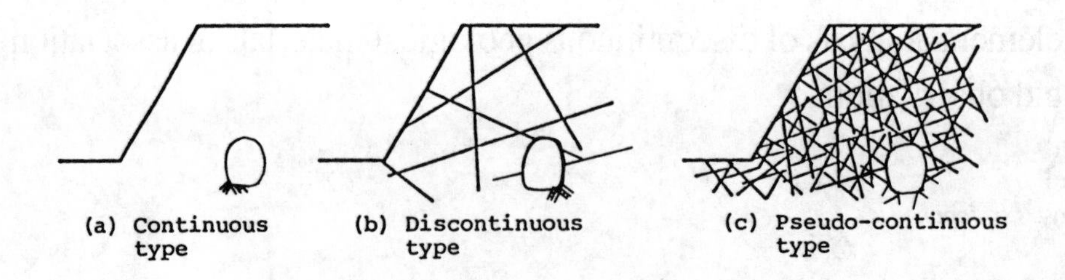

Fig. 1 Classification of the ground

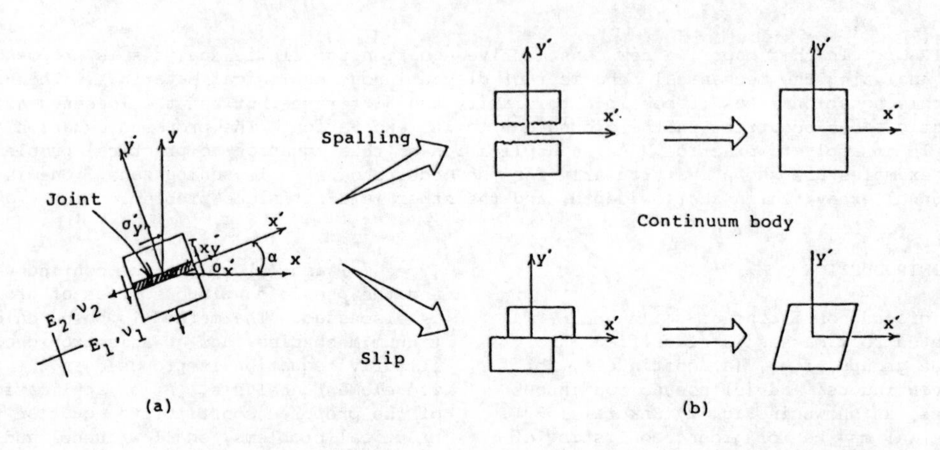

Fig. 2 Modeling for discontinuous deformation in continuum mechanics

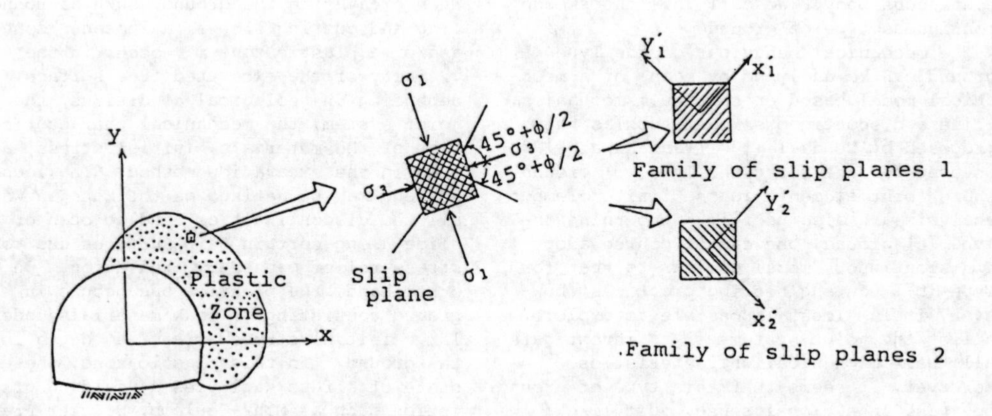

Fig. 3 Families of slip planes in plastic zone

tive law of representing the mechanical behaviour of loosened materials is ambiguous, and the same type of yielding criterion is generally used for analyzing both the loosened and plastic zones. Since the mechanism of occurrence for the two different zones is not the same, even in analyses a special consideration must be made for distinguishing one from the other.

Following in this paper a constitutive equation used in the finite element analysis, which can distinguish between the loosened and plastic zones, is introduced. It is based on continuum mechanics, and may belong to a family of constitutive equations for anisotropic materials. All the material properties contained in this proposed equation can be determined by a back analysis of the results of field measurements performed during mining, excavation and construction of structures.

3 CONSTITUTIVE EQUATION

As described earlier, the scope of this study is to develop a method for analyzing the mechanical behaviour of the continuous type of ground. Therefore, the effect of discontinuities must be taken into account in a constitutive equation which is for a continuous body equivalent to the discontinuous materials.

Let us consider a small element that contains a discontinuous plane as shown in Fig. 2. Assuming the element as an equivalent continuous body, the incremental stress and strain relationship can be expressed in the x'-y' local coordinate system as follows;

$$\{ \Delta\sigma' \} = [D'] \{ \Delta\varepsilon' \} \tag{1}$$

where

$$[D'] = \frac{E_2}{(1+\nu_1)(1-\nu_1-2n\nu_2^2)} \cdot \begin{bmatrix} n(1-n\nu_2^2) & n\nu_2(1+\nu_1) & 0 \\ n\nu_2(1+\nu_1) & 1-\nu_1^2 & 0 \\ 0 & 0 & m(1+\nu_1)(1-\nu_1-2n\nu_2^2) \end{bmatrix}$$

$$(n = E_1/E_2 , m = G_2/E_2) \tag{2}$$

Hence, it is transformed into the x-y global coordinates as follows;

$$\{ \Delta\sigma \} = [D] \{ \Delta\varepsilon \} \tag{3}$$

where

$$[D] = [T] [D'] [T]^T \tag{4}$$

$[T]$ is a transformation matrix expressed as;

$$[T] = \begin{bmatrix} \cos^2\alpha & \sin^2\alpha & -2\sin\alpha\cos\alpha \\ \sin^2\alpha & \cos^2\alpha & 2\sin\alpha\cos\alpha \\ \sin\alpha\cos\alpha & -\sin\alpha\cos\alpha & \cos^2\alpha-\sin^2\alpha \end{bmatrix} \tag{5}$$

α is the angle between the x'-and x-axes. It should be noted that Eq. (3) can represent both the loosened and plastic behaviours of materials by changing the material constants, particularly n and m, which are called anisotropic parameters.

3.1 Constitutive equation representing the loosened zone

As already described, the loosened zone is defined as the zone in which the discontinuous planes tend to open and/or to slide along slip surfaces. Therefore, the anisotropic parameters are determined such that the constitutive equation can represent the spalling of discontinuous planes and/or the sliding along slip surfaces.

(a) Spalling of the discontinuous plane
Spalling of the discontinuous plane shown in Fig. 2(b) can be represented by increasing the anisotropic parameter n, i.e., by reducing the value of E_2 against E_1. Poisson's ratio ν_2 is taken to be zero, because spalling in the direction of the y'-axis makes no movement in the x'-axis. In this case, the other anisotropic parameter m must be taken as $m = 1/2(1+\nu)$.

(b) Sliding along the slip surface
When sliding occurs along the slip surface parallel to the x'-axis, the anisotropic parameter m can be reduced to a small value i.e., $m<1/2(1+\nu)$, while $n = 1.0$ and $\nu_1 = \nu_2$ are assumed.

3.2 Constitutive equation representing the plastic zone

It is obvious that ordinary constitutive laws such as the associated or non-associated flow rule, can be effected by the

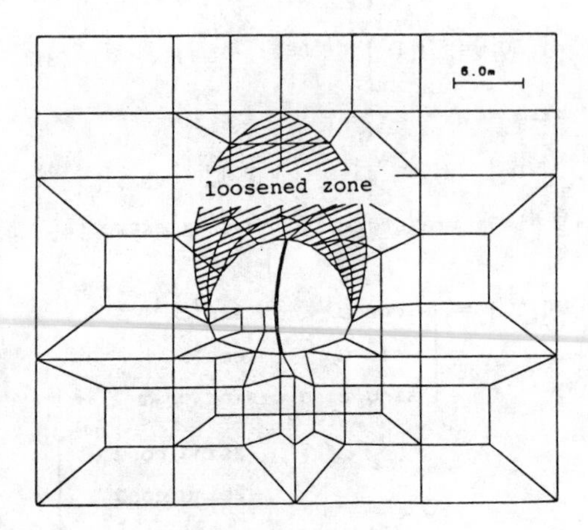

Fig. 4 Finite element mesh and the extent of the loosened zone

Table 1. Input data used in finite element analysis

Initial stress	σ_{x0}	-1.0 kg/cm^2
	σ_{y0}	-3.1 kg/cm^2
	τ_{xy0}	-0.8 kg/cm^2
Young's modulus	E	6900 kg/cm^2
Poisson's ratio	ν	0.3
Anisotropic parameter	m	0.1

plastic zone. In this paper, however, an alternative constitutive equation is proposed. It is simple and easy to apply to engineering practices.

The theory of plasticity demonstrates that a material under a plastic state tends to slide along the two families of potential sliding planes with an angle of $\pm(45°+\phi/2)$ from the direction of the maximum principal stress, as shown in Fig. 3, where ϕ denotes the internal friction angle.

Let us take two different local coordinate systems for consideration of the mechanical behaviour of the two conjugate slip planes, as shown in this figure 3. The stress-strain relationship for each family of slip planes is given in the same form as Eqs. (1) and (3) for the local and global coordinate systems, respectively. The total strain, thus, is assumed to be given as,

$$\{\Delta\varepsilon\} = \frac{1}{2} [\{\Delta\varepsilon_1\} + \{\Delta\varepsilon_2\}] \qquad (6)$$

where $\{\Delta\varepsilon_1\}$ and $\{\Delta\varepsilon_2\}$ are incremental strains due to the families of two slip planes, respectively.

Considering Eq. (3), Eq. (6) becomes;

$$\{\Delta\varepsilon\} = \frac{1}{2} [[D_1]^{-1} + [D_2]^{-1}] \{\Delta\sigma\} \qquad (7)$$

Eq. (7) is rewritten in the following form.

$$\{ \Delta\sigma \} = [D] \{ \Delta\varepsilon \} \qquad (8)$$

where

$$[D] = [\frac{1}{2} \{ [D_1]^{-1} + [D_2]^{-1} \}]^{-1} \qquad (9)$$

Eq. (8) is a constitutive equation that represents the plastic behaviour of materials. It has already been demonstrated that, for a deep tunnel problem, Eq. (8) can provide a good agreement with the results obtained through use of the ordinary elasto-plastic constitutive equation (Sakurai and Ine 1986)

4 EXAMPLE PROBLEMS

4.1 Shallow tunnel

A shallow tunnel excavated in ground consisting of sandy deposits is used as an example to demonstrate the validity of the proposed constitutive equation. The tunnel diameter is approximately 12 m, and the height of overburden is 17 m. The input data for material properties and initial stress are given in Table 1. It is assumed that a loosened zone appears around the tunnel crown, and the slip surfaces are mobilized in this zone. The loosened zone may be evaluated by analyzing the results of field observations and/or measurements. In Fig. 4 the loosened zone is given as a shaded area, where the anisotropic parameter m is taken into consideration. The finite element mesh used for analyzing one of the steps of excavation is also indicated in Fig. 4. The results of numerical analysis are given in Fig. 5, which shows the vertical displacements along three different reference lines. In this figure, the displacements measured by the Sliding Micrometer (ISETH) are also shown. It is seen from this figure that the computed displacements agree well with the measured values. It should be noted that the results obtained by the ordinary finite element analysis assuming isotropic elastic materials, are largely different from the measured values. The difference is particularly dominant for the displacements above the tunnel crown. Fig. 6 gives the maximum shear strain distribution around the tunnel.

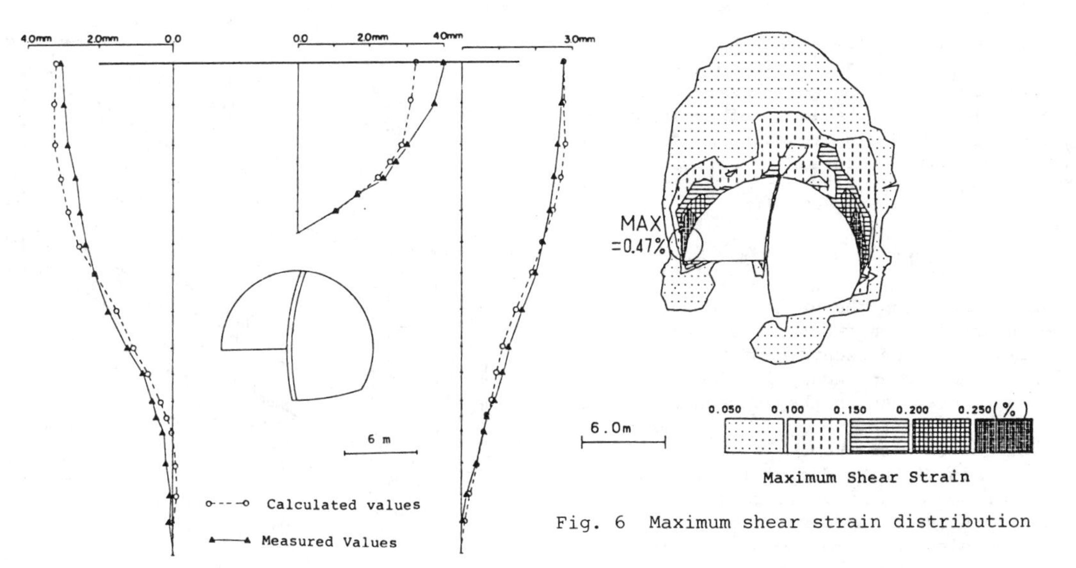

Fig. 6 Maximum shear strain distribution

Fig. 5 Comparison between calculated and measured displacements

4.2 Cut slope

The case study for the cut slope shown here is for a railway tunnel constructed by the cut-and-cover method in ground consisting of layers of sedimentary soft rock and detritus deposits as shown in Fig. 7. In finite element analysis, four different layers almost parallel to the slope surface are assumed, as shown in Fig. 8. Each layer has different values for the anisotropic parameter. All the initial stress and material properties including the anisotropic parameter can be evaluated from the field measurement results. In this project, displacement measurements were carried out using an inclinometer. The first readings were taken before the excavation, so that the total displacements due to excavation were measured. One of the results is shown in Fig. 8, which is for the final phase of excavation. Evaluating these measurement results, the following initial stress and material constants were obtained. The initial stress shown here is that for the floor of the excavated area.

$$\sigma_{x0}/E = -0.871\times10^{-2}$$
$$\sigma_{y0}/E = -0.148\times10^{-2}$$
$$\tau_{xy0}/E = 0.473\times10^{-3}$$
$$\nu = 0.3 \text{ (assumed)}$$

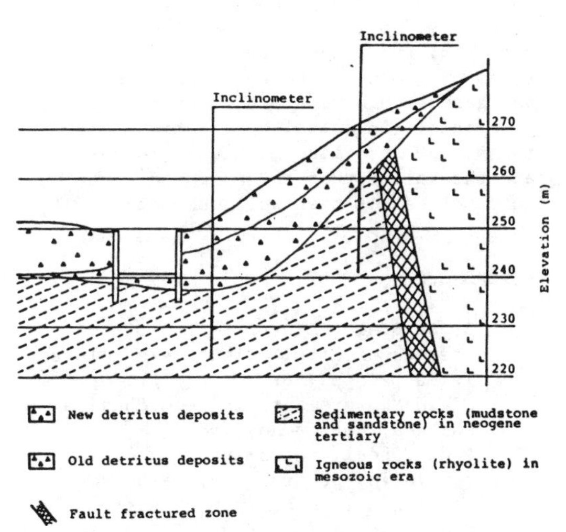

Fig. 7 Geological condition

$$m_1 = 0.01 \quad , \quad m_5 = 0.006$$
$$m_2 = 0.05 \quad , \quad E_L/E = 400$$
$$m_3 = 0.1 \quad , \quad E_B/E = 0.01$$
$$m_4 = 0.004$$

These values are used as input data for the finite element method to calculate the stress, strain and displacements. The calculated displacements are compared with the measured values. This comparison is given in Fig. 8, which indicates a good agreement between the two.

The maximum shear strain distribution is given in Fig. 9, and it is obvious that there is a clear potential sliding surface occurring along the layer of sedimentary rocks. Fig. 9 also reveals the fact that a large strain occurs at the floor of excavation.

5 CONCLUSIONS

When excavating the ground, loosened and/or plastic zones generally appear in the vicinity of the excavated free surface. In finite element analysis, however, it is hard to distinguish them because the same form of constitutive equation is usually used for both of them.

In this paper, a definition is given to the loosened and plastic zones, and the constitutive equations are proposed to distinguish one from the other. The proposed equations are based on continuum mechanics, so that they are effective in both continuous and pseudo continuous types of ground. The equations are simple and easy enough to apply to engineering practices. In actual applications, shallow tunnel and cut slope problems are analyzed, and the validity of the proposed equations to practice has been well demonstrated.

REFERENCES

1 S. Sakurai, Interpretation of the results of displacement measurements in cut slopes, Proc. 2nd Int. Sympo. Field Measurements in Geomechanics, Balkema, Rotterdam, 1987.

2 P.A. Cundall, A computer model for simulating progressive, large-scale movements in blockly rock systems, Symposium of Int. Soc. of Rock Mech., Nancy, 1971.

3 T. Kawai, Some considerations on the finite element method, Int. J. Num. Meth. Engng., 16, pp. 81-120, 1980.

4 S. Sakurai and T. Ine, Strain analysis of jointed rock masses for monitoring the stability of underground openings, Int. Sympo. on Computer Aided Design and Monitoring in Geotechnical Engineering, Bangkok, 1986.

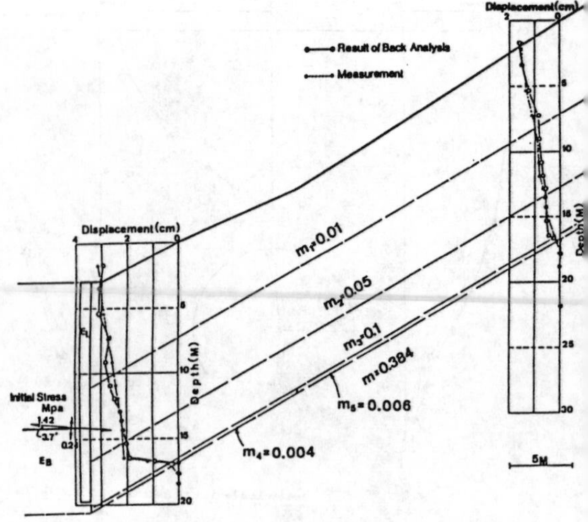

Fig. 8 Comparison between calculated and measured displacements

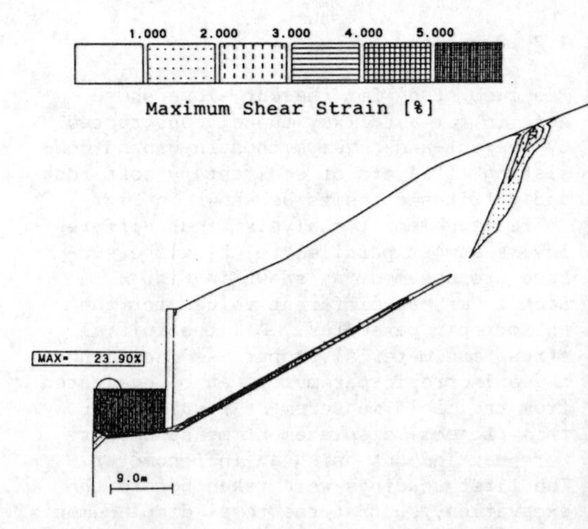

Fig. 9 Maximum shear strain distribution

Numerical Methods in Geomechanics (Innsbruck 1988), Swoboda (ed.)
© 1988 Balkema, Rotterdam. ISBN 90 6191 809 X

The importance of soil yield in the performance of an embankment on soft ground

A.B.Fourie
University of Queensland, St. Lucia, Australia
P.Klibbe
Cameron McNamara Consultants, Brisbane, Australia

ABSTRACT: The measured performance of a large embankment founded on very soft sediments is compared with the results of three numerical techniques of analysis. Measurements of surface settlement, pore pressure and settlement with depth were made. The finite element method was found to provide the best estimate of these parameters, while the conventional Terzaghi approach was significantly in error. Problems in using the finite strain approach were also encountered; settlement rates were markedly overpredicted initially and pore pressure variations were not accurately modelled. Possible reasons for deficiencies in some of the analytical techniques are offered.

1 INTRODUCTION

Development of a waterfront recreation project in Indonesia required the reclamation of an area of approximately 90 hectares extending some 1100m from the previous shoreline and spanning a width of 650m. The site is located immediately north of metropolitan Jakarta and lies on marine alluvial deposits of Quarternary and Recent ages. The coastal geomorphology indicates that the area is subject to deposition of the fine grained sediment transported into the bay during wet seasons.

The existing site had been reclaimed for approximately 400 metres from the original shoreline during the period 1979 to 1981. The source of the fill material was from the adjacent reservoir to the south-east of the site, and was similar to the underlying natural strata. In order to obtain a better understanding of likely consolidation settlements a 68m x 68m trial embankment was constructed to failure at the site in question. This paper describes the analysis of the embankment performance using three different approaches.

2 SITE CHARACTERISATION

An extensive programme of laboratory and field testing was undertaken, including:

2.1 Field Testing:

The field testing consisted of logging, sampling, Standard Penetration testing, shear vane and cone penetrometer testing. Approximately 12m of sediments were found to underlie the site, varying in thickness from 9 to 14m in thickness. Underlying these sediments to a depth of 18m were alternating layers of silts and sandy clays varying from firm to stiff in consistency. Below this strata were predominantly silts and sands of dense to very dense consistency. The SPT tests gave zero readings in the top 12m, and penetrometer soundings gave a cone resistance of less than 100 kPa over the top 5m approximately, and did not exceed 1 MPa until deeper than 12m. Thereafter a very rapid increase occurred. Vane shear test results vary from 2 kPa at 1m depth to 14 kPa at 10m, with a stiffer crust of about 10 - 15 kPa strength. Below 12m depth strength values increased rapidly from 10 to 60 kPa.

2.2 Laboratory Testing:

Aside from standard index tests, an extensive series of conventional laboratory tests were conducted. These included oedometer tests, and unconsolidated undrained (U.U.) and consolidated undrained (C.U.), triaxial tests. The

U.U. tests resulted in an undrained shear strength profile in very good agreement with the vane test results. The C.U. tests consistently gave an angle of shearing resistance of 35°. A shear modulus of 350 kPa was obtained from the consolidated undrained tests.

The oedometer tests (about 30 in number) gave an average C_c value of 0.64 and a C_v value of between 10^{-2} and 2×10^{-2} cm²/sec. A wide scatter of permeability values ranging between 10^{-8} and 10^{-7} cm/sec was obtained from the oedometer tests. The void ratio profile was estimated from laboratory testing on undisturbed samples. This profile, together with the Atterberg Limit and natural moisture content variations are shown in Figure 1.

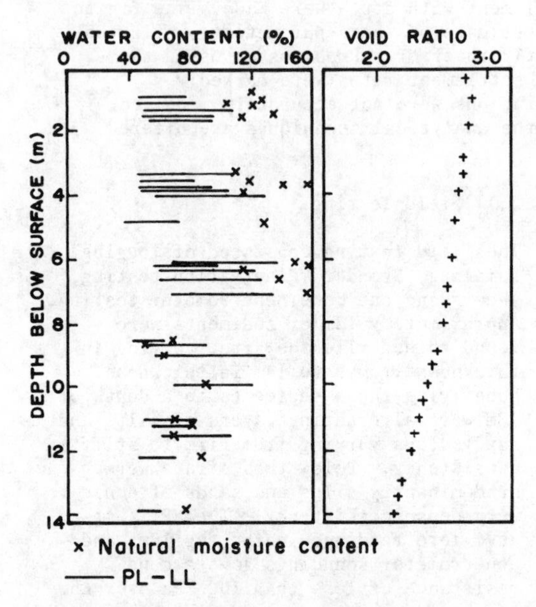

Figure 1 In-situ conditions prior to construction; (a) Natural moisture contents and Atterberg Limits, and (b) Assumed void ration profile.

2.3 In-situ stress conditions:

The oedometer tests gave a normally consolidated coefficient of compression, C_c, value of 0.64 with a void ratio of 3.11 at unit mean effective stress. The above void ratio values were then plotted on this figure, and from the assumed swelling line slope C_r of 0.08 the overconsolidation ratio was determined. The K_o value, and thus the horizontal effective stress, was estimated from the equation postulated by Mayne and Kulhawy [1]

$$K_o = (1 - \sin\phi') \, OCR^{\sin\phi'}$$

where OCR is the overconsolidation ratio.

3 FIELD MEASUREMENTS

In addition to surface settlement measurements, sub-surface movement was monitored by a series of magnet extensometers installed in a borehole on the centreline of the embankment. Five piezometers, at depths of 2m, 4m, 6m, 8m and 11m below original ground surface, were located adjacent to this borehole. The trial embankment, which was 68m x 68m in plan, was built up in uniform layers until failure by instability of the side slopes. The embankment height at this time was approximately 3.0 metres.

4 NUMERICAL ANALYSIS

This section describes the procedures that were used to analyse the embankment performance. All analyses were restricted to one dimension only to simplfy comparison, and because predictions were to be compared with measured performance on the centreline of a large loaded area.

4.1 Analysis using Terzaghi consolidation theory.

The consolidation theory proposed by Terzaghi [2] has been incorporated into a computer program CSETT, developed by the Waterways Experiment Station, Vicksburg [3].

The program CSETT calculates the stress distribution within the consolidating layer using either Boussinesq or Westergaard elastic theory, and then solves the Terzaghi consolidation differential equation for an interpolated system of soil layers. The consolidation settlement is then calculated from the change in void ratio.

The parameters required for the CSETT program were calculated from the oedometer tests and field tests. They were:

C_c (the coefficient of consolidation) = 0.64
C_r (coefficient of recompression) = 0.08
C_v (coefficient of consolidation) = 31.5m²/yr
γ_b (bulk unit weight) = 14.7kN/m³

4.2 Finite strain consolidation theory.

As described earlier, the top 10-12m of in-situ material consists of very soft sediments of low density. Previous work by Cargill [4] indicated that the one-dimensional finite strain consolidation theory proposed by Schiffman et al [5] would likely be appropriate for such a material. This theory was used as the basis for a computer program CONFIN [6]. The program is comprehensively described in the above manual and only brief details are given here. The governing differential equation for one-dimensional consolidation of a fully saturated clay layer as derived by Gibson et al [7] is solved by approximating it in a finite difference equation. This equation, given by,

$$\pm \left(\frac{\rho_s}{\rho_w} - 1 \right) \frac{d}{de} \left[\frac{k(e)}{1+e} \right] \frac{\partial e}{\partial z} + \frac{\partial}{\partial z} \left[\frac{k(e)}{\rho_w (1+e)} \cdot \frac{d\sigma'}{de} \cdot \frac{\partial e}{\partial z} \right] + \frac{\partial e}{\partial t} = 0$$

may be solved in a number of ways. The approach used in the program CONFIN is based on the work of Mesri and Rokhsar [8], where the void ratio is assumed to be a function of both effective stress and time. The input parameters used in the analysis of the trial embankment were calculated from the oedometer and field tests as well as the single field pumping test. These parameters were:

C_c = 0.64 and C_r = 0.08 (as before)

$C\alpha$ (secondary compression coefficient) = 0.03

C_k (permeability coefficient) = 0.5

e_o (void ratio at unit effective stress) = 3.0

SG (Specific gravity) = 2.7,

k_o (permeability at e_o) = 1.0m/year

4.3 Finite element analysis using Critical State constitutive law.

The program used in this analysis was originally written by Litwinowicz and Carter [9], and analyses consolidation of a one-dimensional strata using the Modified Cam-Clay soil model (Roscoe and Burland [10]). The analytical procedure is coupled, i.e. a numerical solution for the displacement and pore pressure fields is obtained simultaneously. As described in the above report the equations governing these distributions are derived from the expression of Virtual Work, and an expression based on Darcy's law and continuity of volume respectively.

The constitutive law is elasto-plastic, volumetric strain-hardening and based on the critical state concepts of soil mechanics. The yield surface describes an ellipse in normal stress-shear stress space, and an associated flow rule was used, resulting in a symmetric finite element stiffness matrix. The finite elements consist of single nodes since the analysis is one-dimensional, and two degrees of freedom exist at each node-vertical displacement and excess pore pressure. Solution of the finite element equations is by the "tangent stiffness" method, and for the time increment calculations a forward marching solution is adopted.

The parameters used in this analysis were computed from the laboratory oedometer and triaxial test results. These parameters were:

λ (slope of normal consolidation curve in e - ℓn p' space) = 0.28

K (slope of recompression curve in e-ℓnp' space) = 0.03

M (the critical state strength parameter) = 1.5

G (the shear modulus) = 350 kPa

e_{cs} (the value of the void ratio at unit p' on the critical state line) = 2.96

k (permeability) = 0.5 m/year.

The soil sub-strata was discretised into 16 finite elements, each one metre in length. The embankment was simulated by applying a surface traction to the upper boundary equal to the vertical load. The sequence of loading was specified to simulate the measured surcharge loadings with appropriate time intervals to allow for delays in fill placement. After completion of the loading phase of the analysis additional time increments, applied in equal logarithmic intervals, were specified for the consolidation phase of the analysis.

5 COMPARISON OF MEASURED AND PREDICTED EMBANKMENT PERFORMANCE

This section describes the results obtained using the three numerical procedures described previously, and compares them with the measured values. Comparisons are presented in terms of surface settlement and pore pressure variation versus time, and variation of settlement with depth.

5.1 Variation of surface settlement with surcharge.

The variation of applied surcharge

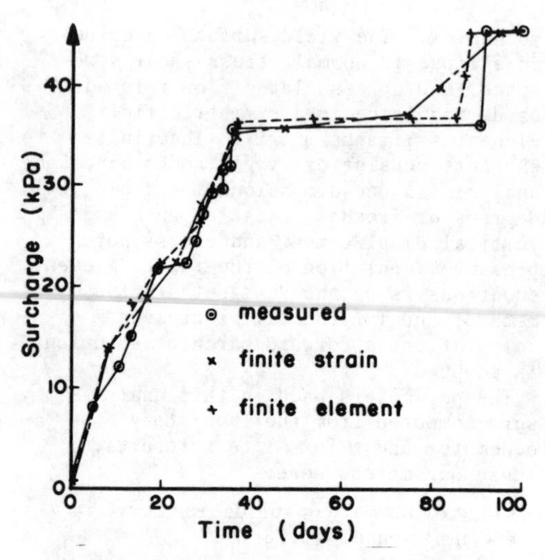

Figure 2 Variation of surface surcharge with time.

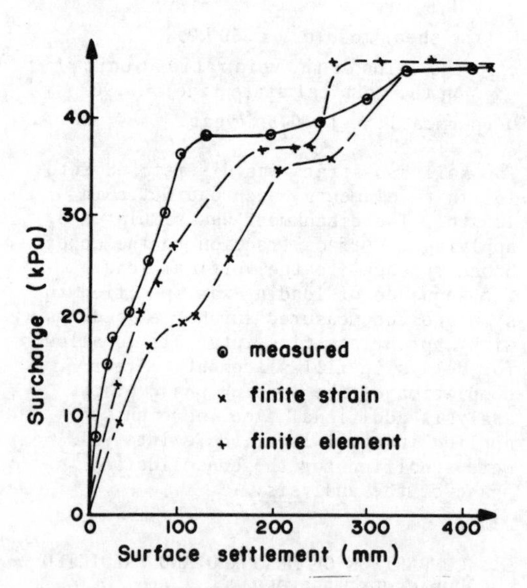

Figure 3 Surface settlement variation with applied surcharge

with time is given in Figure 2. When the fill height reached approximately 2m, further construction was halted for about 50 days. Construction to the final fill height of 2.8m (± 50 kPa) then proceeded. The assumed variation in two of the numerical analyses are included on the figure. The corresonding variation of measured surface settlement with applied surcharge is given in Figure 3.

It can be seen that once the applied load reached approximately 20 kPa the rate of settlement continued at a fairly constant rate, until construction re-started, whereupon the settlement rate again increased. Included in Figure 3 are the predicted variations of surface settlement with applied surcharge using the programs CONFIN and CONCAM. At any particular value of surcharge the finite strain theory can be seen to substantially overestimate the field settlement. It is also interesting to note that this method predicts less settlement during the period the fill level is unchanged than either the finite element method or the measured settlement. The final values of settlement are all within 12% of one another, and it is therefore in the rate of settlement that disagreement occurs. This becomes more obvious when the surface settlement with respect to time is considered.

5.2 Variation of surface settlement with time.

Figure 4 shows a plot of surface settlement against time for the measured values as well as all three predictions. Once again the finite element method provides distinctly the best correlation with the measured values, with the two curves diverging only at about 90 days after commencement of contruction. The finite strain method overpredicts settlements (by up to 95%) up until about 3 months whereafter it predicts slightly less settlement than was measured. Although the final predicted settlement is of the same order as the measured value, there is an important difference between the two curves. The rate of settlement (ie the tangent to the curve) predicted by the finite strain method decreases continuously. The fastest rate of settlement can be seen to occur in the first ten days. This method does not therefore predict the accelerating rate of settlement which was observed to occur between 20 and 30 days after start of loading.

The conventional method based on the solution to the one-dimensional Terzaghi consolidation equation provides a very poor estimate of both the final consolidation settlement and the rate of settlement. As was found with the finite strain method, this technique does not predict a change in the settlement rate as was observed in the field. The large underestimate of settlement which occurs with this method cannot be ascribed to

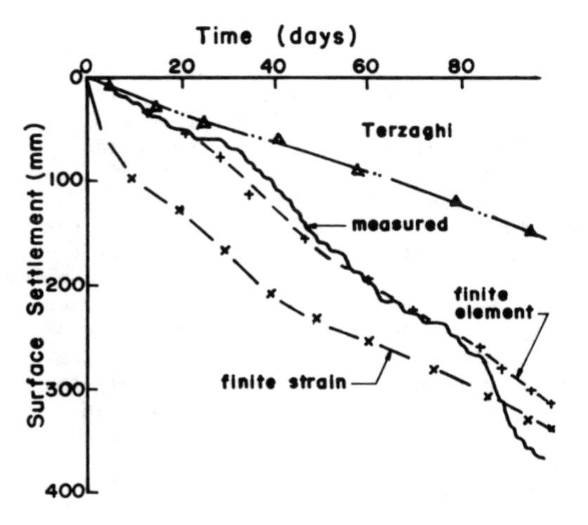

Figure 4 Variation of surface settlement with time.

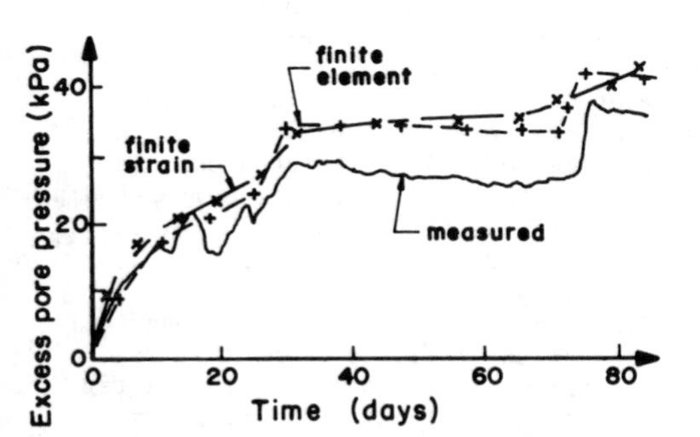

Figure 5 Excess pore pressure at a depth of 6m below original ground surface.

unrealistic soil parameters since the same laboratory and field tests were used to calculate parameters for the finite strain and finite element methods as was used in the Terzaghi approach. Possible reasons for these differences will be discussed in a subsequent section.

5.3 Excess pore pressure change versus time.

The variation of excess pore pressures with time are given in Figure 5 for a piezometer located at a depth of 6m below ground level on the centreline of the embankment. Results from the Terzaghi analysis are not included. It can be seen that the finite element method accurately estimates the trend of the field values, although it over-predicts the magnitudes. It reflects accurately the rather sudden increases in

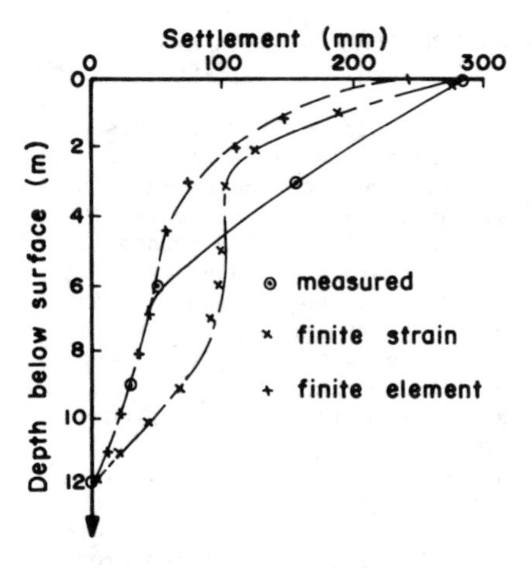

Figure 6 Settlement profiles with depth.

pore pressure at about 25 and 85 days after start of construction. The finite strain method provides reasonable estimates of the pore pressure magnitude, but the variation of pore pressure with time is completely different from the measured performance.

The results shown in Figure 5 do not offer any definitive conclusions about the embankment performance, or which method of analysis is most appropriate. The most useful conclusion that can be drawn is that the assumed permeability values must be representative of the field value as both the finite element and finite strain methods provide reasonably accurate predictions of the pore pressure response.

5.4 Settlement variation with depth.

Figure 6 compares the measured variation of settlement with depth with the predictions of the finite strain and finite element methods. Above 6m depth the finite element method underpredicts settlement although the trend of the results is similar to the measured values. It should also be noted that there was only one measuring station between the surface and 6m depth and inaccuracies in this single reading would significantly alter the measured profile. The finite strain method produces a profile which is distinctly different from the other two. Between the depths of 3m and 8m virtually no settlement occurs, with most movement being concentrated at the two boundaries. These differences, as well as others previously mentioned, are discussed in the following section.

6 EVALUATION OF APPROPRIATE NUMERICAL TECHNIQUE

The results presented earlier in this paper indicated the large differences between the measured settlement magnitudes and rates, and those predicted using the Terzaghi method. This latter method is of course a standard technique for analysing consolidation settlements of structures of the type described in this paper. The validity of using the method in the present context is, however, questionable. It seriously underpredicts the magnitude of settlement, and erroneously predicts rate of settlement.

The finite strain method has previously been successfully used to model the rate of consolidation of loosely placed dredge spoil (Cargill [4]). In the present application, however, it has not provided a representative prediction of sub-strata performance. In particular, during the early stages of embankment loading (first 25 days) it substantially overpredicts the magnitude and rate of surface settlement. Thereafter it continuously underpredicts settlement rate. This has been ascribed to the relationship between effective stress and permeability inherent in the method. Upon initial loading, settlements occur rapidly as the void ratio decreases rapidly. This decrease in void ratio substantially decreases the soil permeability and thus the rate of settlement slows. Another distinctive disagreement between this method and the measured values was found to occur in the variation of excess pore water pressure with time. The finite strain method predicts a regular increase of excess pore pressure with time, whereas the measured values show a definite, sudden increase at 25 and 85 days after start of construction which corresond to increases in the rate of fill placement.

The finite element method provides a more accurate and realistic estimate of embankment performance than either of the other two methods. The rate of surface settlement is accurately modelled up until about 85 days after the start of loading. In addition the pore pressure variation, including the two sudden increases mentioned above, is realistically estimated, and the profile of settlement with depth are similar in form to the measured variation. An improved correlation between the finite element method and the measured variation could of course be produced by judicious choice of the input parameters, but this was not the objective of the present study. All parameters used in the three numerical analyses were obtained directly from conventional laboratory and field tests. One parameter of uncertainity in the finite element study was the choice of the shear modulus G. An additional analysis was carried out with a shear modulus of 2000 kPa (as opposed to 350 kPa in the original analysis) and a difference of only 6% in predicted surface settlement at 80 days after start of loading occurred. The elastic shear modulus was therefore of only minor importance in the present study, and settlement was due almost entirely to yield and thus plastic straining within the soil sub-strata. This factor provides an indication as to the differences in predicted results from the Terzaghi and finite element methods.

In the finite element method the majority of soil elements were either at, or close to, the yield surface. Increments of compressive load thus often resulted in yield and plastic flow. Figure 7 shows the stress path for the centroid of the element of soil at 1.5m below original ground level predicted by the finite element method. Also sketched on this figure are the yield surfaces corresponding to the stress states shown. The yield surface expands continuously and thus plastic strain occurs throughout the loading process. Point A corresponds to the cessation of fill placement at 25 days after start of loading. Thereafter the stress path moves to the right as consolidation occurs. This is accompanied by further expansion of the yield surface and thus additional plastic strain. Points B and C on this curve correspond to the re-commencement and final completion of fill placement.

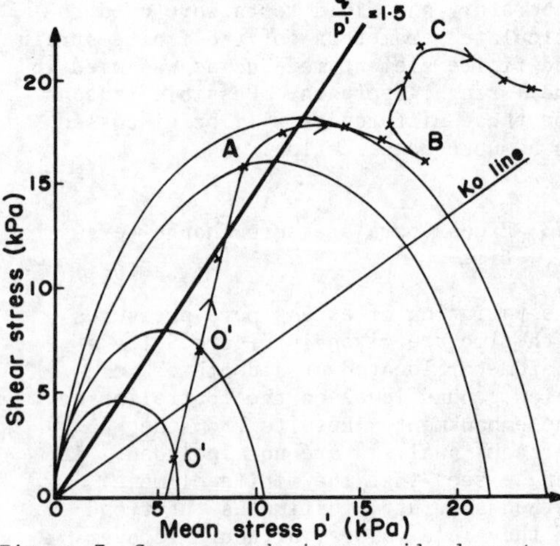

Figure 7 Stress path for a soil element at 1.5m below ground surface.

Referring to Figure 4 again, the Terzaghi analysis predicts substantially less settlement than the finite element method. The Terzaghi technique, however, assumes changes in void ratio, and thus settlement, are due entirely to changes in the mean effective stress p'. In the finite element method, however, assuming the vertical and horizontal directions remain principal directions, the vertical strain is given by

$$\varepsilon_z = \tfrac{1}{2}\,\upsilon^P + \tfrac{1}{2}\,\varepsilon^P$$

where υ^P and ε^P are the plastic volumetric and shear strains respectively.

Therefore shear strains also contribute towards settlement, and as can be seen from Figure 7 during loading from point 0 to A the increase in shear stress exceeds the increase in mean effective stress. A similar conclusion may be drawn about increments in shear strain and volumetric strain if it is remembered that in the present analysis fully associated plasticity has been assumed. A simple hand calculation shows that the ratio of plastic strain increments for yielding at point 0' for example is equal to 1.6, i.e. the shear strain increment exceeds the volumetric strain increment. Furthermore, using a shear modulus of G = 350 kPa, the plastic shear strain increment resulting from loading from 0 to 0' in Figure 7 was found to be more than fifty times the elastic shear strain increment.

CONCLUSIONS

The results of the present study appear to indicate conclusively that the finite element method provides the best representation of the consolidation of a thick, soft sediment. However, it must be cautioned that the present study is limited to comparison of numerical prediction with a single set of field data. The temptation always exists to believe field results implicitly, but a strong possibility of error should be recognised. The general conclusion mentioned above should therefore be tested by comparison with additional laboratory and field data. Specific conclusions which may be drawn from the present study are:

(i) The conventional Terzaghi approach to the one-dimensional consolidation of a thick layer may severely underestimate surface settlements if imposed shear stresses cause the loaded clay to yield.

(ii) This yield may substantially increase surface settlements even if it is localised. In the example presented in this paper the top ±1m of soil did not yield as it had a higher undrained shear strength than the soil immediately below it.

(iii) The finite strain method should be used with caution. The assumption that permeability decreases with void ratio may result in overestimates of settlement rates of soil adjacent to free-draining boundaries.

(iv) The finite element method provides an attractive alternative to the conventional Terzaghi approach. Settlements induced by changes in both mean effective and deviatoric stresses are accounted for, and excess pore pressures may be realistically predicted.

ACKNOWLEDGEMENTS

The computational work described in this paper was made possible by a grant from the Prentice Computer Centre, University of Queensland. The authors gratefully acknowledge this assistance.

REFERENCES

[1] P.W. Mayne and F.H. Kulhawy. K_o - OCR relationships in soil. Jnl of the Geot. Eng. Div., ASCE, 108, GT6, 851-872, (1982).

[2] K. Terzaghi. Theoretical Soil Mechanics. John Wiley and Sons. New York.

[3] A.E. Templeton. Computer program for determining induced stresses and consolidation settlements (CSETT). U.S. Army Corps of Engineers, Report K-84-7, Washington, DC (1984).

[4] K.W. Cargill. Prediction of Consolidation of Very Soft Soil. Jnl. of the Geotech. Eng. Div., ASCE, 110, GT6, 775-795. (1984).

[5] R.L. Schiffman. Finite and Infinitesimal Strain Consolidation. Jnl. of the Geotech. Eng. Div., ASCE, 106, GT2, 115-119 (1980).

[6] Cameron McNamara. CONFIN - One Dimensional Finite Strain Consolidation Analysis. User Manual. Cameron McNamara Consultants. Brisbane. Australia (1985).

[7] R.E. Gibson, G.L. England and M.J.L. Hussey. The Theory of One-Dimensional Consolidation of Saturated Clays. I. Finite Non-Linear Consolidation of Thin Homogeneous Layers. Geotechnique 17, No.3, 261-273 (1967).

[8] G. Mesri and A. Rokhsar. Theory of Consolidation for Clays. Jnl. of the Geotech. Eng. Div., ASCE, 100, GT8,

889-904 (1974).
[9] A. Kitwinowicz and J.P. Carter. CONCAM, a computer program for the finite element analysis of one-dimensional consolidation utilising the modified Cam-clay soil model. School of Civil and Mining Engineering. University of Sydney (1980).
[10] K.H. Roscoe and J.B. Burland. On the Generalised Stress-Strain Behaviour of 'Wet' Clay. Engineering Plasticity (eds) J. Heyman and F.A. Leckie, Cambridge University Press, 535-609, (1968).

Numerical Methods in Geomechanics (Innsbruck 1988), Swoboda (ed.)
© 1988 Balkema, Rotterdam. ISBN 90 6191 809 X

Soil-structure interaction related to actual construction sequences

H.Ohta
Kanazawa University, Japan

A.Iizuka
Kyoto University, Japan

Abstract: Since soils are much softer than other construction materials such as steel and concrete, the stress state in the subsoils often goes beyond the elastic limit during construction works resulting in irrecoverable deformation of subsoils. The mechanical qualities of soil-structure overall system thus depend on the mechanical histories subjected to the soils during construction as well as on the past histories that subsoils experienced prior to the construction works. However, simply because the grasp of change in the mechanical states of subsoils is not easy, the current design methods of foundations have been developed in such a way that the estimate of the effect of construction procedures is not needed. Aiming at tackling this untouched but essential influence of construction sequence on the final quality of the structures, the authors have tried to analyse the behaviour of soil-structure systems under construction. The tool which is employed in the analysis is a finite element programme capable to deal the soil-water coupling boundary value problems based on the constitutive model consisting of Cam-clay type of elasto-plastic model being furnished by a rate dependent viscous model. After a series of trial of post-mortem analysis of case histories, a total system of analysis starting from the parameter determination to the specification of boundary conditions imposed by the progress of construction works was developed.

1 INTRODUCTION

In a construction work dealing with some elastic material such as a steel bridge, the quality of constructed structure does not depend on its construction procedure and sequence, because the mechanical behaviour of the elastic material is recoverable and mechanical state does depend only on the final stage of construction unless the material experiences the stress state beyond the elastic limit.

On the contrary this is not the case of soils. The final quality of soil structures is highly dependent upon the construction sequence, which could produce a wide variety of mechanical history of soils, because soil materials have such low elastic limit that soils yield at very low stress level resulting in the irreversible behaviour under a wide range of external mechanical conditions. The movement of pore water constituting the soil material often affects the mechanical behaviour of soils as a whole making the phenomena even more complicated to analyse.

Stress Path Method[10],[12], ADP Method [2],[3],[8], and SHANSEP Method[9] are typically listed as the attempts to take such an influence of stress history into account.

Since the design parameters needed in these methods are determined by using the laboratory shear tests under certain stress and boundary conditions, the design parameters determined are not always applicable to the complicated stress state which is often encountered in the actual construction work.

On the other hand the finite element method has wider applicability to various stress and boundary conditions. If the elasto-plastic/elasto-viscoplastic constitutive model is used in the finite element analysis to describe the history-dependent behaviour of soils, the actual construction work can be simulated considering the construction procedure and sequence.

In this paper the authors adopt the soil/water coupling finite element technique using the elasto-viscoplastic constitutive model[15]. This model can be distinguished from the other critical state models in its describability of the induced anisotropy and the creep and relaxation characteristics of soils but shares the used parameters in common with the Cam-clay type of constitutive model.

After brief introduction of the constitutive model and the finite element technique, the authors propose the practical determination procedure of input parameters. Post-mortem analyses of a trial embankment on soft foundation and an excavation of soft subsoil are demonstrated in comparison with the monitored in-situ performance. The influence of construction sequence on the mechanical state of subsoil being connected with the behaviour of structure under construction is finally demonstrated for the case ofmaginary construction works.

2 MATHEMATICAL STRUCTURE OF MODELLING

2.1 Modelling of soil element

The basic assumptions needed in the incremental elasto-viscoplastic constitutive model[15] are different from those of the Cam-clay model[13], but the final mathematical form is found to be the extension of the Cam-clay model. The principal feature of the constitutive model[15] is briefly summarized in Fig.1.

volumetric strain of clays			continuum mechanics
consolidation	dilatancy		non-linear elasticity
$\dot{\varepsilon}_v^e = \frac{\kappa}{1+e_o}\cdot\frac{\dot{p}'}{p'}$	$\dot{\varepsilon}_v^e = 0$	elastic (recoverable)	elastic - limit (yield condition)
$\dot{\varepsilon}_v^p = \frac{\lambda-\kappa}{1+e_o}\cdot\frac{\dot{p}'}{p'}$	$\dot{\varepsilon}_v^p = D\dot{\eta}^*$	plastic (irreversible)	$MD\ln\frac{p'}{p_o'} + D\eta^* - \varepsilon_v^p = 0$
$\varepsilon_v^{vp} = \alpha\ln[1+\dot{v}_o t/\alpha\cdot\exp(\frac{f}{\alpha})] = F$ $f = \frac{\lambda-\kappa}{1+e_o}\cdot\ln\frac{p'}{p_o'} + D\eta^*$		viscous (time - dependent)	flow rule $\dot{\varepsilon}_{ij}^p = E\cdot\frac{\partial f}{\partial\sigma_{ij}'}$ or $\dot{\varepsilon}_{ij}^{vp} = H\cdot\frac{\partial F}{\partial\sigma_{ij}}$

Fig.1 Summary of constitutive model[15]

Table 1 Summary of input parameters and laboratory tests to determine them

		analysis parameter	main laboratory test	remarks
material properties	Λ	irreversibility ratio	triaxial consolidation test	$\Lambda = 1 - \kappa/\lambda$
	M	critical state parameter	triaxial CU test	$M = \frac{6\sin\phi'}{3-\sin\phi'}$
	D	coefficient of dilatancy	triaxial CD (p'=const.) test	$D = \frac{\lambda-\kappa}{M(1+e_o)}$
	ν'	effective poisson ratio	triaxial CU test	G
	α	coefficient of secondary compression	triaxial consolidation test	$\alpha = dv/d(\ln t)$
	$\dot{v}_o$	initial volumetric strain rate	triaxial consolidation test	$\dot{v}_o = \alpha/t_c$
preconsol stress	σ_{vo}'	preconsolidation vertical pressure	oedometer test	
	K_o	coefficient of earth pressure at rest	triaxial K_o-consolidation test	
initial stress	σ_{vi}'	effective overburden pressure	unit weight test	$\sigma_{vi}' = \gamma' z$
	K_i	coefficient of in-situ earth pressure at rest	triaxial K_o-swelling test	
	k	coefficient of permeability	permeability test	$k = \gamma_w m_v c_v$

stress parameter $\eta^* = \sqrt{\frac{3}{2}(\eta_{ij}-\eta_{ijo})(\eta_{ij}-\eta_{ijo})}$, $\eta_{ij} = s_{ij}/p'$, $s_{ij} = \sigma_{ij}' - p'\delta_{ij}$, $p' = \sigma_{ii}'/3$

This model describes the inviscid characteristics of elasto-plastic soils as well as the time dependent characteristics of elasto-viscoplastic soils. The inviscid part of this constitutive model is reduced to the model[11] when it is applied to the axisymmetric stress condition such as seen in conventional triaxial tests. It is further reduced to the original Cam-clay model when it is applied to the isotropic initial stress condition. The viscid part of this constitutive model is developed on the basis of investigation[16]. The input soil parameters needed in this constitutive model are summarized in Table 1. In this investigation the irreversibility ratio Λ, the critical state parameter M and the coefficient of dilatancy D are chosen as input parameters instead of λ, κ, e_o and D, where λ and κ are the compression and swelling indices, eo is void ratio and the coefficient of dilatancy D was originally introduced by Shibata[17]. The coefficient of secondary compression α and the initial volumetric strain rate $\dot{v}_o$ are needed in modelling the viscid characteristics[14]. The subscript o denotes the values at the reference state just after Ko-consolidation.

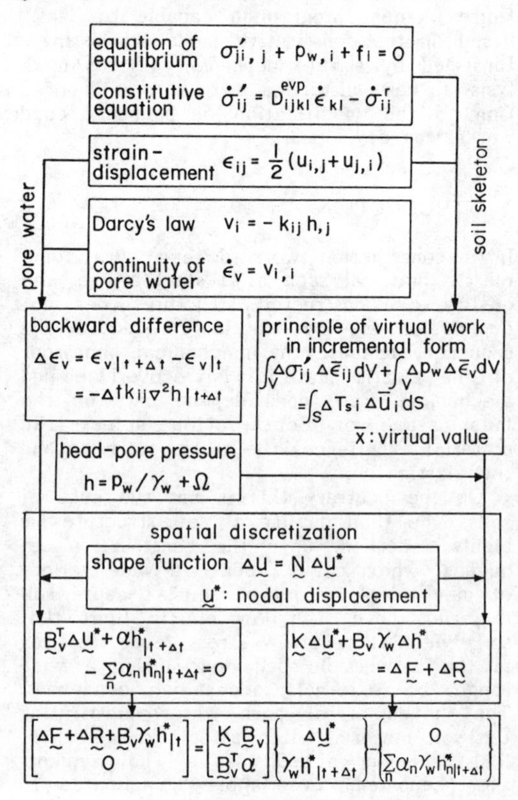

Fig.2 Finite element formulation of DACSAR

2.2 Modelling of soil mass

Soil mass is in general regarded as the two phase mixture of soil skeleton and pore water. The governing equations describing the behaviour of the soil skeleton coupled with the pore water flow are formulated by Biot[4].

The finite element programme used in this investigation was developed by the authors and is named DACSAR[7]. The mathematical structure of DACSAR is summarized in Fig.2. The technique employed in discretizing the pore water flow was originally proposed by Akai and Tamura[1], which is a sort of the so-called Christian's techniques[5] Since DACSAR solves the boundary value problems in an incremental manner, it makes possible to simulate the actual construction work considering the time effect and the change in some boundary conditions.

3 INPUT PARAMETER DETERMINATION

The determination procedure of input parameters is designed as shown in Fig.3. This procedure requires at least three kinds of information:

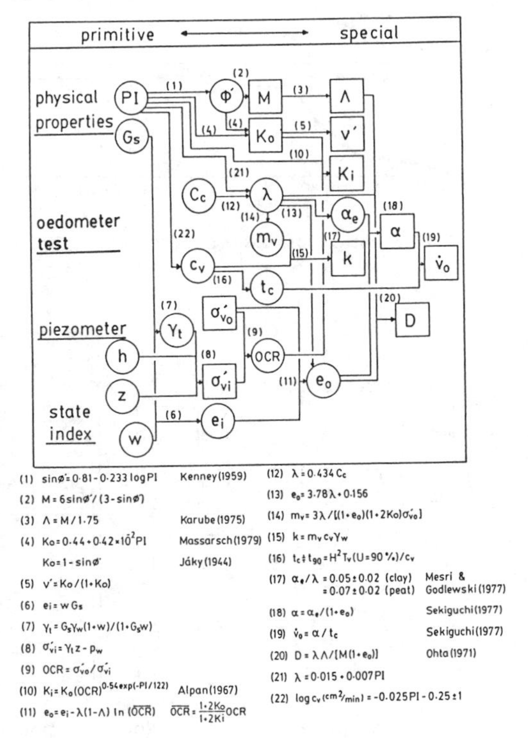

Fig.3 Determination procedure of input parameters

(1) physical properties of soils (the plasticity index and the unit weights of soil particles); (2) the results of oedometer tests; (3) the in-situ initial stress (effective stress and pore pressure) distribution in the subsoil. Fig.3 is recommendable for the primary determination of the input parameters needed in the elasto-viscoplastic soil/water coupling finite element analysis.

The empirical relationship between parameters are specified in Fig.3. These relationships are selected through a number of trial case studies carried out not only from the viewpoint of the accuracy of the relationship between any of two parameters but also from the viewpoint of the overall appropriateness of analytical results for a wide variety of subsoil profile and boundary condition, in detail, see Ref.[7].

Input parameters needed in analysis should be directly determined, in their nature, from laboratory tests, but it is not always feasible to carry out the sufficient number of such soil tests on the undisturbed samples. Even in case that input parameters are estimated from the primary soil properties, typically plasticity index, based on this determination procedure, at the present stage these input parameters give the most agreeable correspondence between computed results and monitored field performance.

4 CASE HISTORIES

4.1 Highway embankment

An example of application to a trial embankments (Fig.4) on the soft ground at Kanda in Ibaragi prefecture in Japan is introduced. The trial embankment is placed on a fine sand layer (about 8 m thick) underlain by soft clay layers (7 m thick). The subsoil profile and its soil properties are summarized in Fig.5.

All of input parameters, except elastic constants of the sandy layer, are determined from the plasticity index by using the proposed parameter determination proce-

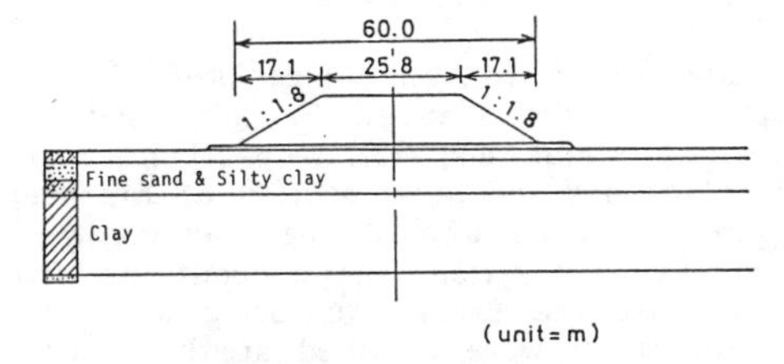

Fig.4 Profile of the test embankment

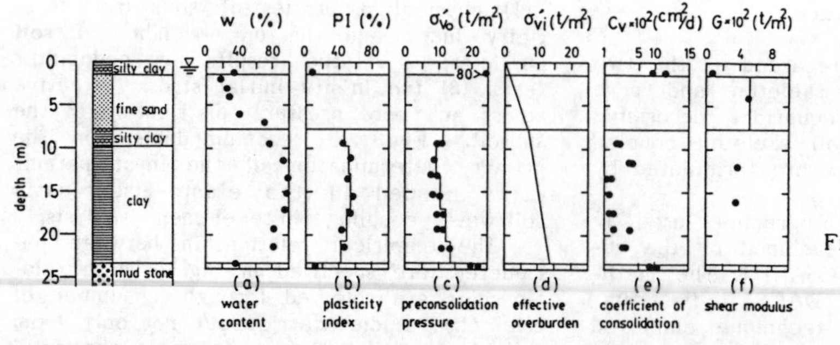

Fig.5 Soil properties of the embankment foundation

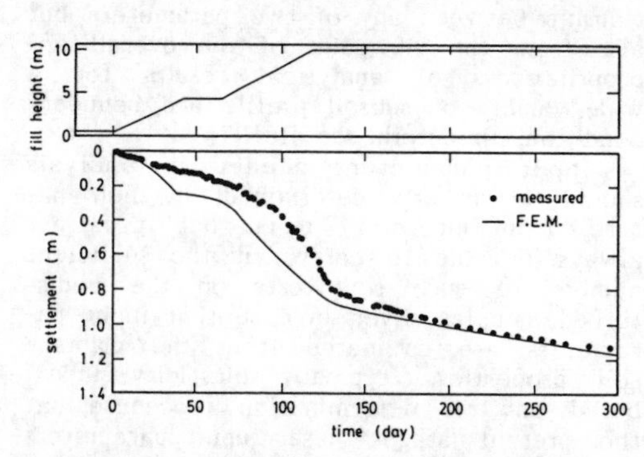

Fig.6 Settlement of ground surface under the centre of embankment

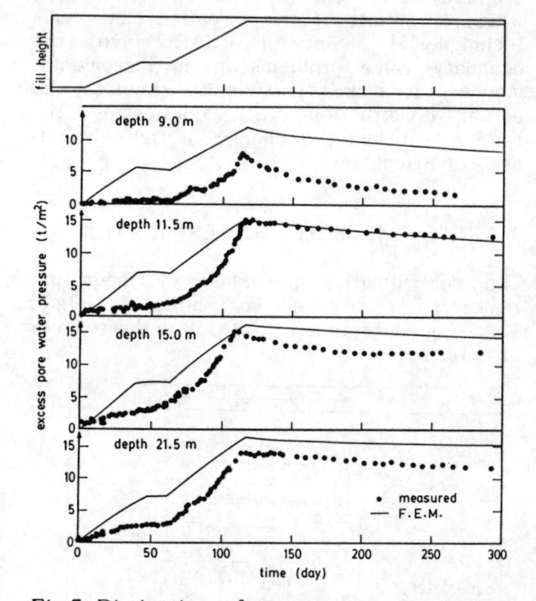

Fig.7 Dissipation of excess pore water pressur

dure(Fig.3). The elastic constants of the sandy layer are determined based upon the pressuremeter tests with assumption that Poisson's ratio in term of effective stress equals to 0.4. The fill body (unit weight 1.7 t/m^3) is represented by finite elements placed on the ground surface in such a way that the actual loading rate is simulated. The computed settlement of the ground surface under the centre of the embankment is shown in Fig.6 and the dissipation of pore water pressure in typical elements in clay layers under the centre of the embankment are shown in Fig.7. These computed values are compared with the values monitored in the field. The computed curves fairly agree with the measured plots.

4.2 Deep cut in soft layers

An excavation work in Akita prefecture in Japan is summarized in Figs.8(a) and (b), where Figs.8(a) and (b) respectively show the plane and the cross section of the deep square pit excavated during the construction. The excavation was divided into five stages. At the every excavating stage the bracing struts were installed at the position indicated by the arrow signs in Fig.8(b). Excavation is represented by eliminating the

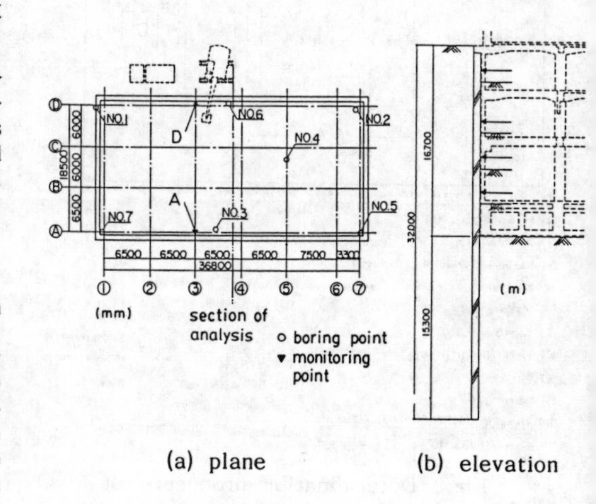

Fig.8 Plane and elevation of the pit[6]

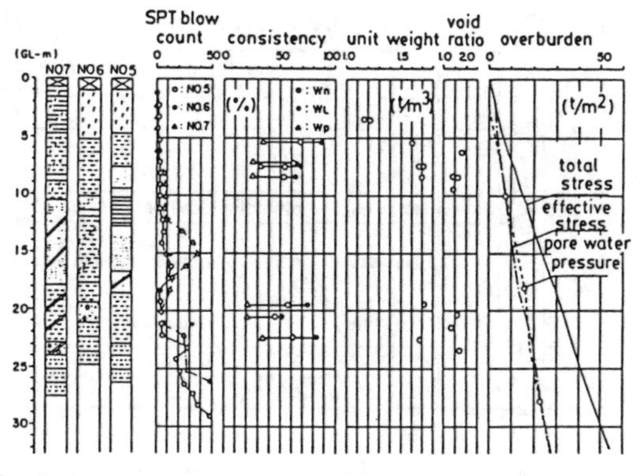

Fig.9 Profile of subsoil properties of the pit[6]

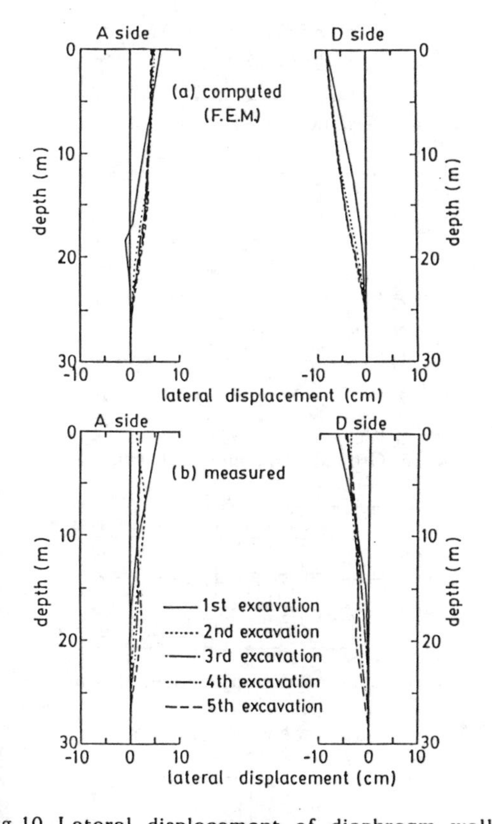

Fig.10 Lateral displacement of diaphragm wall

finite elements in such a way that simulate the actual construction sequence. The bracing struts and diaphragm wall are modelled by the truss and beam elements.

The profile of soil properties is summarized in Fig.9. All input parameters of clay layers except the irreversibility ratio are determined from the plasticity index by using the chart (Fig.3). The irreversibility ratios of clay layers are determined by assuming Cs/Cc=1/12 as Honda[6] suggested. The coefficients of permeability of the finite elements located close to the diaphragm wall are chosen small enough to simulate the water cut-off effect of the diaphragm wall.

The computed result of deformation of diaphragm wall is shown in Fig.10(a) and the measured one in Fig.10(b). The computed results of deformation well agree with the measured values except at the top edge of the diaphragm wall.

5 EFFECT OF CONSTRUCTION SEQUENCE

5.1 Analysis of shield construction

An imaginary series of construction works, the construction of building, tunnel excavation and excavation of a pit using diaphragm wall, is analysed, see Fig.11. Three hypothetical cases shown in Fig.12. are analysed.

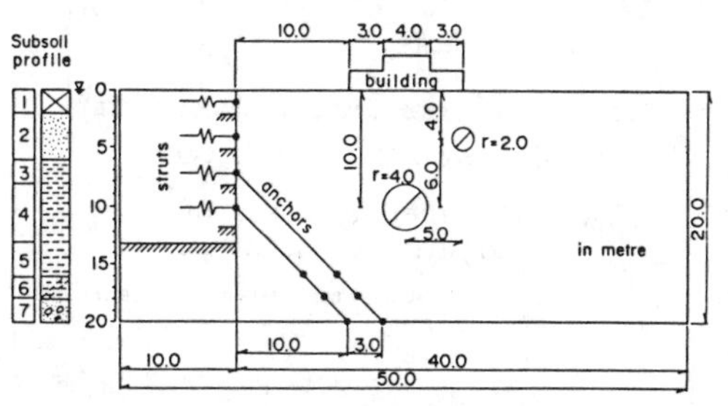

Fig.11 Profile of an imaginary shield construction

The building is treated as nodal forces in the finite element computation and beam elements are used in the bottom of building to take the stiffness of the building into account. The tunnel excavation is performed 3,500 days after the construction of the building. The upper tunnel is assumed to be driven by an ideal jacking method causing no disturbance in surrounding subsoil. The lower tunnel is assumed to be constructed by using the shield machine with the tail void ratio of 6%, back-filling ratio of 50%. Tunnels are perfectly water-proof. The diaphragm wall is assumed to be constructed 1,000 days after construction of tunnels. The diaphragm wall is 1 m thick and the tensile crack due to bending is assumed not to appear. The bracing struts are installed in 6.5 m pitch. No pre-loading of the struts is assumed. The earth anchors, JIS PC stand 12.7 mm diameter, are installed in 1.6 m pitch pretensioned by a force of

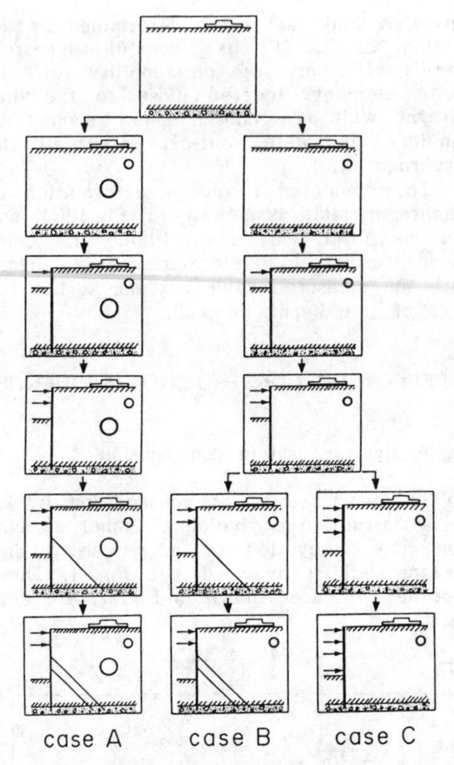

Fig.12 Hypothetical cases of construction sequence

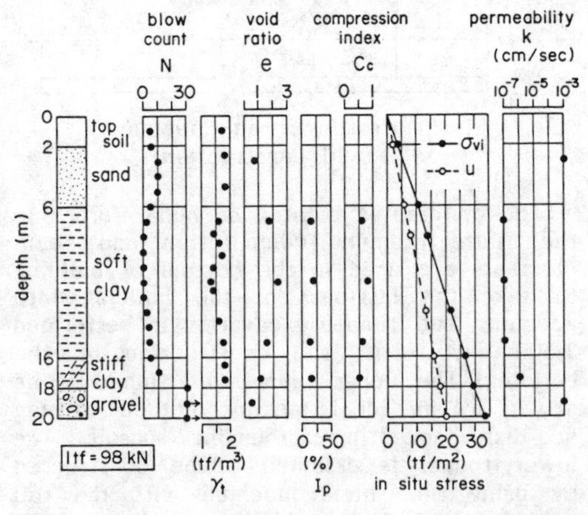

Fig.13 Subsoil properties of the shield construction

100 tonne for each anchor. Fig.13 indicates the soil properties of subsoil. The top layer (0-2 m deep), the upper sand layer (2-6 m deep) and the lower gravel layer (18-20 m deep) are treated as elastic materials. Soil parameters are summarized in Table 2.

Fig.14 indicates the construction sequence of Case A, see Fig.12. As seen in Fig.15 the immediate deformation occurs just after

Table 2 Soil input parameters used in analys

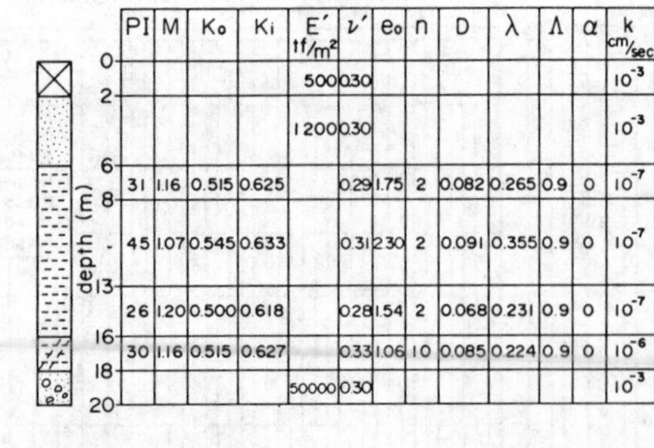

depth (m)	PI	M	K_o	K_i	E' tf/m²	ν'	e_o	n	D	λ	Λ	α	k cm/sec
0–2					500	0.30							10^{-3}
2–6					1200	0.30							10^{-3}
6–8	31	1.16	0.515	0.625		0.29	1.75	2	0.082	0.265	0.9	0	10^{-7}
8–13	45	1.07	0.545	0.633		0.31	2.30	2	0.091	0.355	0.9	0	10^{-7}
13–16	26	1.20	0.500	0.618		0.28	1.54	2	0.068	0.231	0.9	0	10^{-7}
16–18	30	1.16	0.515	0.627		0.33	1.06	10	0.085	0.224	0.9	0	10^{-6}
18–20					50000	0.30							10^{-3}

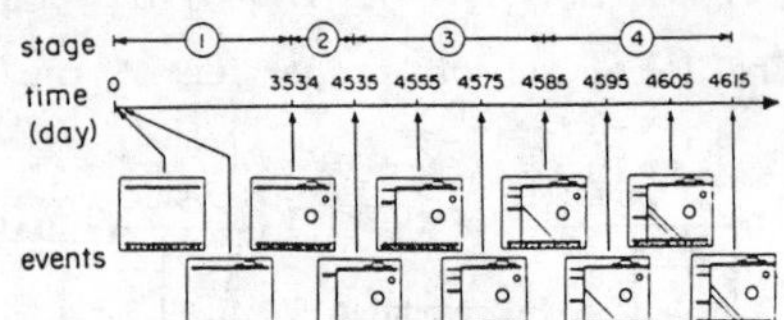

Fig.14 Construction sequence of Case A

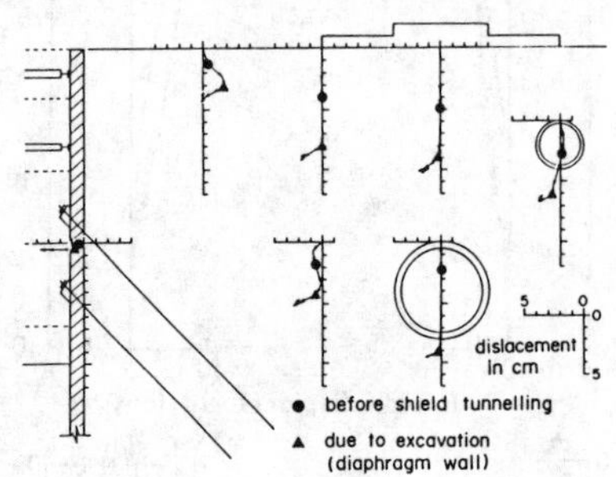

Fig.15 Ground movement of Case A

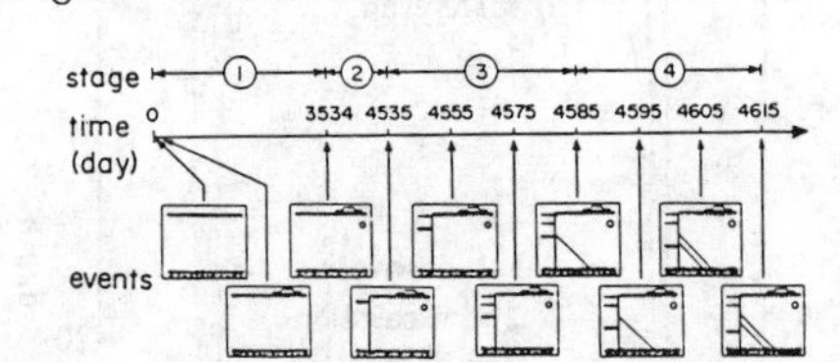

Fig.16 Construction sequence of Case B

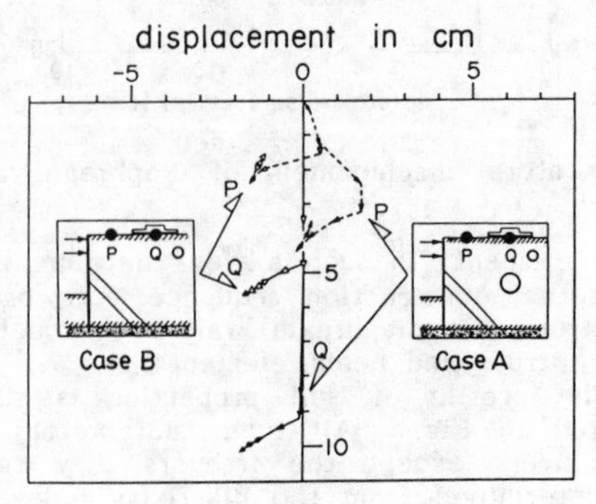

Fig.17 Effect of excavating the lower shield tunnel

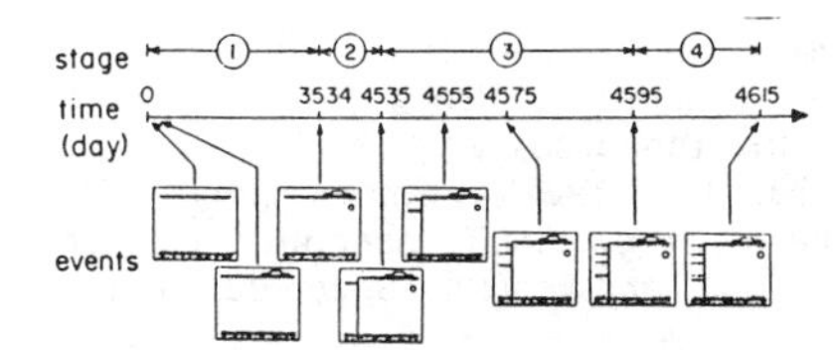

Fig.18 Construction sequence of Case C

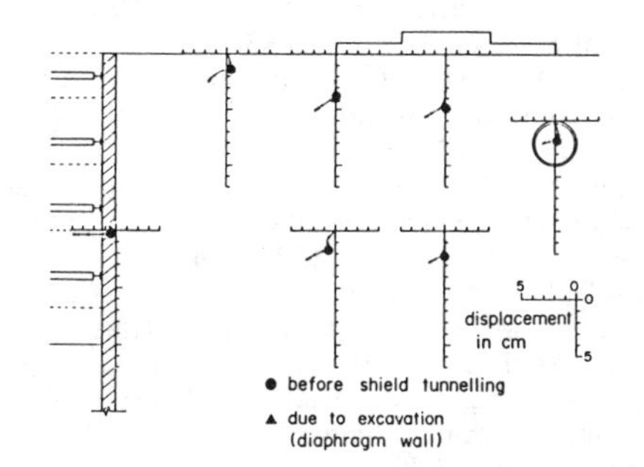

Fig.19 Ground movement of Case C

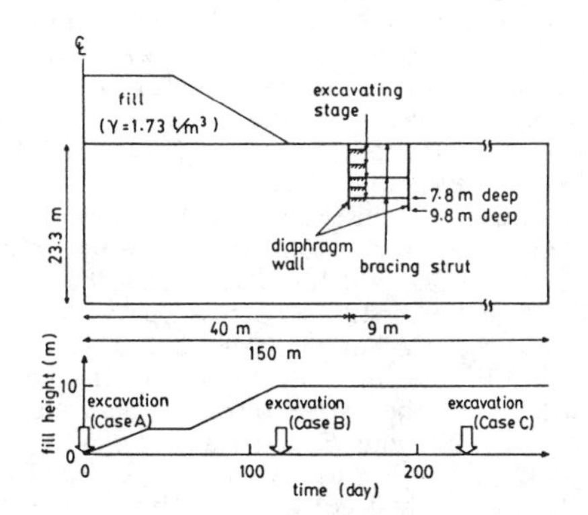

Fig.20 Profile of excavating trench and hypothetical construction sequence

construction of two tunnels followed by the deformation due to consolidation that continues. Fig.16 shows the construction sequence of Case B. If Case A and Case B are compared, the movement of points P and Q are different in Case A and Case B, as shown in Fig.17. The excavation of the lower tunnel (Case A) induces larger movements at points P and Q, compared with Case B in which the lower tunnel is not driven. Fig.18 indicates the construction sequence of Case C in which the lower tunnel is not excavated and the bracing struts are installed in stead of the earth anchors. The ground movements in Case C shown in Fig.19 are much less compared with Case A.

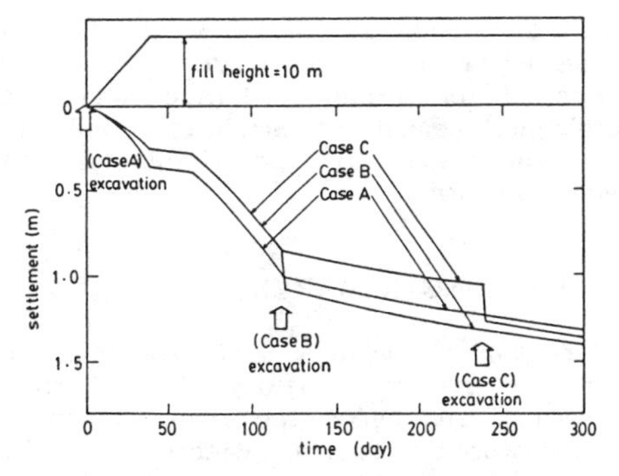

Fig.21 Effect of construction sequence on settlement of ground surface

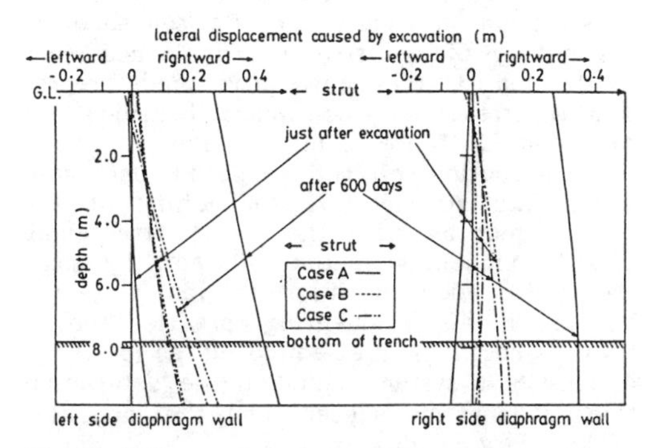

Fig.22 Effect of construction sequence on lateral displacement of diaphragm wall

5.2 Analysis of embankment construction

An imaginary construction work is analysed. In addition to the actual embankment work introduced in Section 4.1, it is hypothetically assumed to excavate a deep trench as shown in Fig.20. The lower part of Fig.20 shows the loading rate of the fill together with timing of trench excavation and three different construction are considered.

The subsoil properties are the same as shown in Fig.5. The diaphragm walls and bracing struts are treated as line elements (beam and truss elements).

Fig.21 indicates the settlement of ground surface under the centre of the embankment. The timings of excavation are also indicated in Fig.21 by arrow signs. Fig.22 indicates the lateral displacements of diaphragm walls at the time just after excavating work and 600 days after excavating work.

Since three case studies are distinguished from each other only by the construction timing relative to embankment work, the final ground movement should be the same to each other if the ground is elastic.

However, because of the irreversibility of soil behaviour during construction, the final settlement of ground surface and the lateral displacement of diaphragm wall are largely influenced by the timing of the trench excavation.

6 CONCLUDING REMARKS

The mechanical behaviour of soil material depends on the stress history and the stress distribution, and hence the final quality of soil structure is much affected by the choice of construction procedure and its sequence.

In order to investigate the effect of construction procedure and its sequence on the quality of soil structure, it is necessary to develop (1) the constitutive model representing irreversible non-linear behaviour of soils and (2) finite element technique taking the interaction of soil skeleton and pore water into account. The modelling of the soil element by using the constitutive model and the modelling of the soil mass by using the finite element discretization may not be useful in engineering practice unless these modellings are employed as parts of a consistent system including the determination of input parameters and the modelling of boundary conditions.

From such a point of view, a theoretical framework of modelling the soil element and the soil mass is introduced together with the practical determination procedure of input parameters.

After showing the acceptability of thus proposed system by comparing the computed results with the monitored field performance of subsoil foundations, the effect of construction sequence on the quality of soil structures is demonstrated by analysing imaginary construction works.

ACKNOWLEDGMENT

The authors wish to acknowledge the continuing encouragement of Professor Shojiro Hata, Kyoto University, and to express their thanks to Mr. Hirokazu Yuno, formerly postgraduate student of Kanazawa University, for his calculation. This study was supported in part by a Grant-in-Aid for Science Research, No. 61550350 from the Ministry of Education, Science and Culture.

REFERENCES

[1]Akai,K. & T.Tamura 1978. Numerical analysis of multi-dimensional consolidation accompanied with elasto-plastic constitutive equation, Proc. JSCE, No.269: 95-104 (in Japanese)

[2]Bjerrum,L. 1972. Embankment of soft ground, Proc. ASCE Specialty Conference on earth and earth supported structures, Vol.2: 1-54

[3]Bjerrum,L. 1973. Problems of soil mechanics and construction on soft clays and structurally unstable soils, Proc. 8th. ICSMFE, Vol.2: 549-566

[4]Biot,M.A. 1941. General theory of three-dimensional consolidation, J. Appl. Phys. Vol.12: 155-164

[5]Christian,J.T. 1968. Undrained stress distribution by numerical method, Proc. ASCE, Vol.94, SM6: 1333-1345

[6]Honda,T. 1986. The fundamental study on application of elasto-plastic finite element analysis to the construction controll of retaining wall, Dr.Eng. Thesis, Kyoto University (in Japanese)

[7]Iizuka,A. & H.Ohta 1987. A determination procedure of input parameters in elasto-viscoplastic finite element analysis, Soils and Foundations, Vol.27, No.3

[8]Janbu,N. 1977. Slope and excavation in normally overconsolidated clays, State-of-the-Art Report, Proc. 9th. ICSMFE, Vol.2: 549-566

[9]Ladd,C.C. & R.Foot 1974. New design procedure for stability of soft clays, Proc. ASCE, Vol.100, GT7: 763-786

[10]Lambe,T.W. 1967. Stress path method, Proc. ASCE, Vol.93, SM6: 309-331

[11]Ohta,H. & S.Hata 1971. A theoretical study of the stress-strain relations for clays, Soils and Foundations, Vol.11, No.3

[12]Poulos,H.G.,L.P.de Ambrosis & E.H.Davis 1976. Method of calculating long-term creep settlements, Proc. ASCE, Vol.102, GT7: 787-804

[13]Roscoe,K.H., A.N.Schofield & A.Thurairajah 1963. Yielding of clays in states wetter than critical, Geotechnique, Vol.13, No.3: 12-38

[14]Sekiguchi,H. 1977. Rheological characteristics of clays, Proc. 9th. ICSMFE, Tokyo, Vol.1: 289-292

[15]Sekiguchi,H. & H.Ohta 1977. Induced anisotropy and time dependency in clays, Proc. 9th. ICSMFE, Tokyo, Specialty Session 9: 229-238

[16]Sekiguchi,H. & M.Toriihara 1976. Theory of one-dimensional consolidation of clays with consideration of their rheological properties, Soils and Foundations, Vol.16, No.1: 27-44

[17]Shibata,T. 1963. On the volume changes of normally consolidated clays, Annuals, Disaster Prevention Research Institute, Kyoto University, No.6: 128-134 (in Japanese)

Numerical Methods in Geomechanics (Innsbruck 1988), Swoboda (ed.)
© 1988 Balkema, Rotterdam. ISBN 90 6191 809 X

Back analysis using Kalman filter-finite elements and optimal location of observed points

Akira Murakami & Takashi Hasegawa
Kyoto University, Japan

ABSTRACT: This paper focuses on the following two topics: one is a new back analysis method using Kalman filter-finite elements; the other is a method for determining the optimal location of observed points for back analysis problems. A formulation is shown on FEM in conjunction with the Kalman filter for two-dimensional and plane strain problems, where Lamé's constants of non-homogeneous regions are solved. In the numerical procedure, the location of observed nodal points is determined taking into account the 'sensitivity' coefficients. These coefficients are defined as the partial derivatives of observed displacements with respect to unknown parameters. The role and the meaning of 'sensitivity' coefficients in the Kalman filter-finite elements are also discussed. The numerical performances for various problems in the field of geomechanics are described, and the applicability of the above procedure to some back analysis problems is examined in comparison to the results derived from other algorithms.

1 INTRODUCTION

An increasing interest in the techniques for solving back analysis problems has been shown in the field of geomechanics. Numerical identification has been successfully performed on practical applications. According to Gioda[1], such techniques can be subdivided into three categories: the "Inverse" method[2-3] giving a least squares solution for condensed system equations; the "Direct" method[4-5] based on a linear programming technique for error function minimization; and the Bayes' approach[6-7] taking into account a priori error information involved in the observation.

The "Inverse" method has been applied to the identification of elastic moduli and initial stresses in tunnel problems, and to that of the load acting on the retaining structure. In this approach, stiffness equations are equivalently transformed into the condensed inversion and, in general, the resultant equations are solved by the least squares method. The "Direct" method uses an iteration technique until the error function is minimized. The error function is selected as the square of the discrepancy between the measured quantities (displacement, pore pressure) and the correspondingly calculated ones. Neither approach can take account of the measurement error which may exist in the observation. It should be pointed out that the quality of the observed data has an influence on the identified results. Bayes' approach is one of the methods which permits a priori consideration of measurement error during the solution procedure. In all the above mentioned approaches, observational location has not been considered, whereas the results of finite element identification are thought to be highly sensitive to the location of the observed points. This is the remaining problem of the existing back analysis methods.

Kalman filter[8] is another powerful method which allows the error information in the observation for numerical implementation. Kalman filter incorporated with finite element discretization is presented herein as the special purpose method. Also, the sensitivity coefficients of observed displacements at the nodal points are introduced in order to expedite determination of an optimal location for the observation.

This paper begins with a discussion of the Kalman filter-finite element formulation for two-dimensional plane strain problems. Numerical results obtained from the proposed procedure are then compared to those obtained from other algorithms for the sample problem. An approach to the determination of observed points collocation follows. The results of both sample and practical problems are presented and discussed.

2 FORMULATION

A formulation of the Kalman filter with finite element discretization is given in this section. Kalman filtering requires two system equations: the observation equation (eq. (1)) and the state equation (eq. (2)).

$$\underset{\sim}{y}_t = H_t \underset{\sim}{x}_t + \underset{\sim}{v}_t \tag{1}$$

$$\underset{\sim}{x}_{t+1} = F_t \underset{\sim}{x}_t + G_t \underset{\sim}{w}_t \tag{2}$$

where, $\underset{\sim}{y}_t$ is the observation vector and $\underset{\sim}{x}_t$ is the state vector. Step by step measurement $\underset{\sim}{y}_t$ gives the estimation of $\underset{\sim}{x}_t$ through the filtering algorithm. The key to the formulation is the correspondence between the above filter equations and the finite element discretized governing equations.

2.1 Observation equation

A finite element equation, which is written for the two-dimensional linear elastic problem under the plane strain condition, is derived.

$$K \underset{\sim}{u} = \underset{\sim}{f} \tag{3}$$

We seek here the discrete finite element system denoted by observed displacement. The direct strategy is to use the condensation technique to eliminate the unknown and the boundary displacements.

We block-partition eq. (3), separating the observed displacement $\underset{\sim}{u}^*$ from that to be eliminated $\underset{\sim}{u}$, thus,

$$\begin{bmatrix} K_{11} & K_{12} \\ K_{21} & K_{22} \end{bmatrix} \begin{Bmatrix} \underset{\sim}{u}^* \\ \underset{\sim}{u} \end{Bmatrix} = \begin{Bmatrix} \underset{\sim}{f}^* \\ \underset{\sim}{f} \end{Bmatrix}, \quad K_{mn} = \begin{bmatrix} K_{mn}^{11} & K_{mn}^{12} \\ K_{mn}^{21} & K_{mn}^{22} \end{bmatrix} \tag{4}$$

where K_{mn}^{ij} is the partitioned stiffness submatrix; λ_i and μ_i are Lamé's constants in region i, and are the unknown parameters which correspond to the vector $\underset{\sim}{x}_t$ in the filtering equation. Eq. (4) constructs the condensed system:

$$\underset{\sim}{f}^* - Q(\lambda_i,\mu_i)\underset{\sim}{f} = [(\lambda_j K_{11}^{j1} + \mu_j K_{11}^{j2}) - Q(\lambda_i,\mu_i)(\lambda_j K_{21}^{j1} + \mu_j K_{21}^{j2})]\underset{\sim}{u}^* \tag{5}$$

where, $Q(\lambda_i,\mu_i) = (\lambda_j K_{12}^{j1} + \mu_j K_{12}^{j2})[\lambda_j K_{22}^{j1} + \mu_j K_{22}^{j2}]^{-1}$ (j:sum)

and is equivalently rewritten as[9]

$$\underset{\sim}{f}^* - Q(\lambda_i,\mu_i)\underset{\sim}{f} = [(K_{11}^{11} - QK_{21}^{11})\underset{\sim}{u}^*, (K_{11}^{12} - QK_{21}^{12})\underset{\sim}{u}^*, \cdots, (K_{11}^{n1} - QK_{21}^{n1})\underset{\sim}{u}^*, (K_{11}^{n2} - QK_{21}^{n2})\underset{\sim}{u}^*] \begin{Bmatrix} \lambda_1 \\ \mu_1 \\ \vdots \\ \lambda_n \\ \mu_n \end{Bmatrix} \tag{6}$$

This equation corresponds to the observation equation (1) by adding the noise term.

2.2 State equation

The following stationary condition of parameters describes the state equation (2) for the Kalman filtering, because λ_i and μ_i are constant.

$$\begin{Bmatrix} \lambda_1 \\ \mu_1 \\ \vdots \\ \lambda_n \\ \mu_n \end{Bmatrix}_{K+1} = \mathbf{I} \begin{Bmatrix} \lambda_1 \\ \mu_1 \\ \vdots \\ \lambda_n \\ \mu_n \end{Bmatrix}_K \tag{7}$$

This equation denotes the stationary condition of parameters.

2.3 Solution procedure

The solution strategy consists of the following filtering scheme. Because of the nonlinearity in the observation equation, an iterative procedure should be adopted under the constant observed displacement, until the convergence of the estimated parameters can be achieved. In this case, suffix k in eqs. (6) and (7) is not the time axis but the 'iterative axis'. We will briefly review the Kalman filtering algorithm to complete a numerical procedure.

Filtering equation

$$\underset{\sim}{x}_{t+1/t} = F_t \underset{\sim}{x}_{t/t} \tag{8}$$

$$\underset{\sim}{x}_{t/t} = \underset{\sim}{x}_{t/t-1} + K_t[\underset{\sim}{y}_t - H_t \underset{\sim}{x}_{t/t-1}] \tag{9}$$

Kalman gain

$$K_t = P_{t/t-1} H_t^T [H_t P_{t/t-1} H_t^T + R_t]^{-1} \tag{10}$$

Estimate error covariance matrix

$$P_{t+1/t} = F_t P_{t/t} F_t^T + G_t Q_t G_t^T \tag{11}$$

$$P_{t/t} = P_{t/t-1} - K_t H_t P_{t/t-1} \tag{12}$$

Initial condition

$$\underset{\sim}{x}_{0/-1} = \bar{\underset{\sim}{x}}_0, \quad P_{0/-1} = \Sigma_0 \tag{13}$$

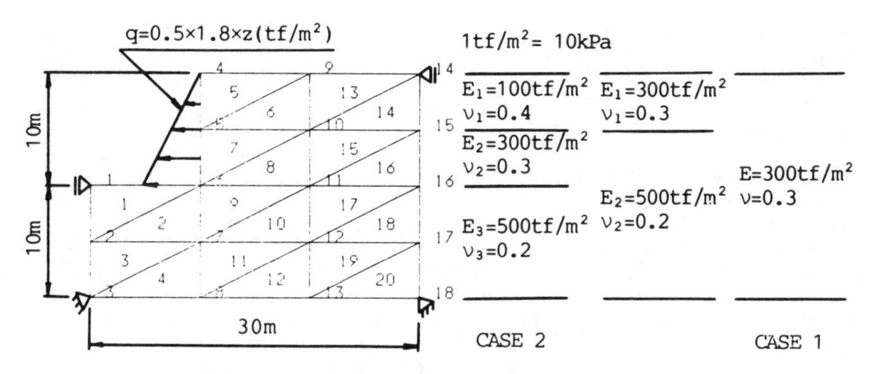

Fig. 1 Analytical model (after Arai et al.[4])

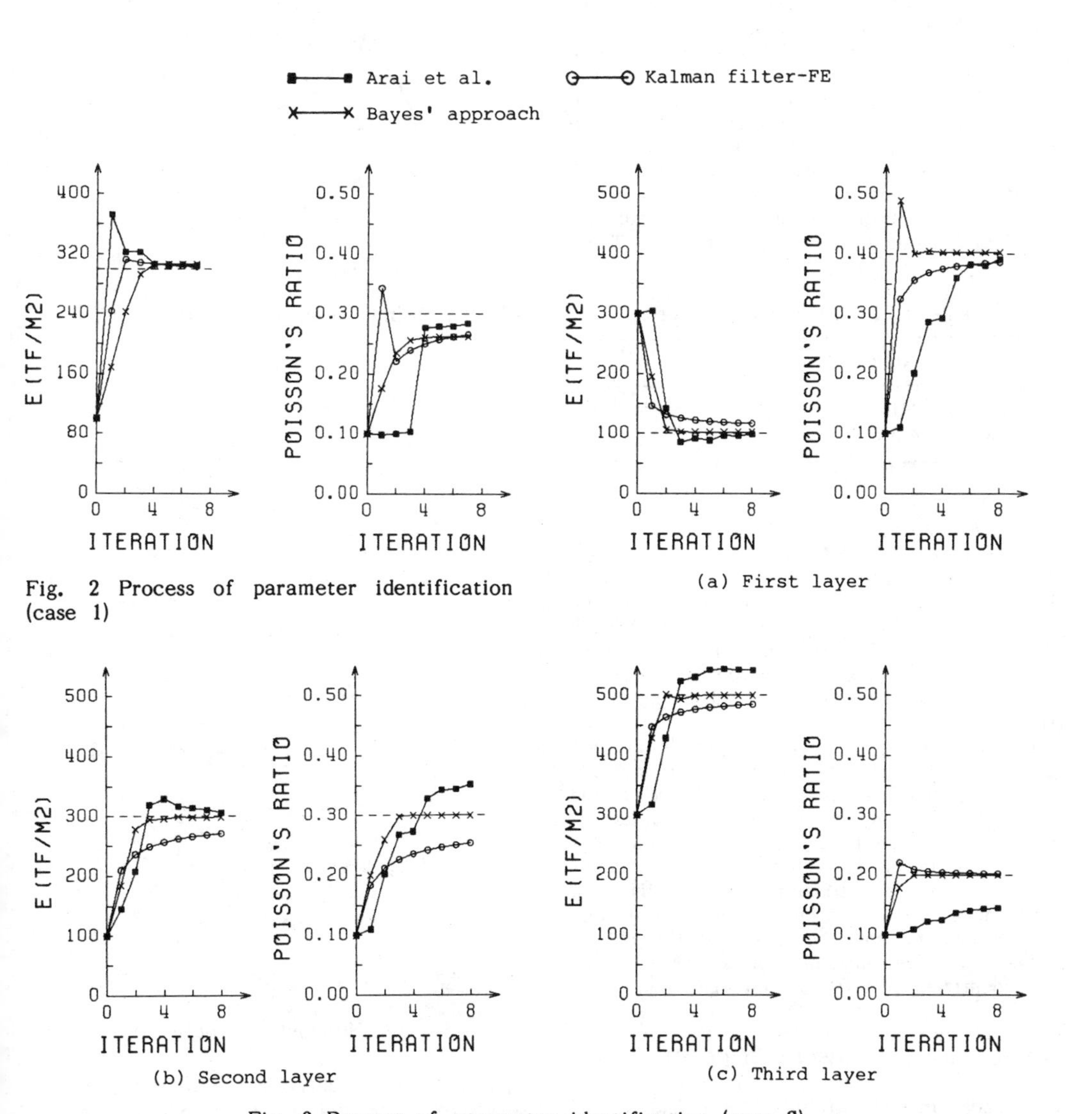

Fig. 2 Process of parameter identification (case 1)

Fig. 3 Process of parameter identification (case 2)

Further details of the filtering algorithm are available in the references[10-11].

3 NUMERICAL EXAMPLES

To examine the validity of the foregoing procedure, the accuracy of the present method is assessed by a comparative analysis using some algorithms for the problem appearing in reference[4]. Figure 1 shows the problem setup and provides the finite element discretization of the system. All geometric conditions and material properties used are shown in this diagram. For this problem, Arai et al.[12] have compared their own identification with the results obtained through Bayes' approach. In this paper, a comparison is also made between these results and a solution obtained by the formulation outlined in the previous section. It is carried out for two cases presented in Figure 1: a homogeneous soil deposit (case 1); and a heterogeneous deposit made up of three layers (case 2). Observational nodal points whose displacements are the numerical input in such cases, are listed in Table 1.

Table 1. Observed nodal points.

	Nodal Point No.
Case 1:	1,14 (Ver.);4,5,6,9 (Hor. & Ver.)
Case 2:	4,5,6 (Hor. & Ver.)

Figures 2 and 3 illustrate the processes of parameter identification in two cases, case 1 and case 2, respectively. From Figure 2, in the homogeneous case, it can be seen that the identified results through different procedures lie close to each other and rapidly converge to the correct value.

In case 2, however, the discrepancy of the identified results for the second and the third layers is found in Arai's method and the proposed procedure, but not in Bayes' approach. As seen in Figure 1 and Table 1, the number of observed points involved in these layers is 2, whereas the number of observed points in the first layer is 4. It can be recognized that even such poor numerical performances are influenced by the quality and the quantity of the observation. The influence of observation on the identified results is then discussed.

4 LOCATION OF OBSERVED POINTS

For back analysis problems, the matter of an optimal location for buried measurement devices is of great interest to design engineers. In fact, the quality and the quantity of observations exert a major influence on the accuracy of the identified parameters. Here we propose a method to determine the measurements collocation when specifying a certain region in the analytical domain[13].

As shown in Figure 4, we choose a certain region on which special emphasis is to be placed: for instance, the material property of this region should be sought with good accuracy; the variation in material property of this region should be in-

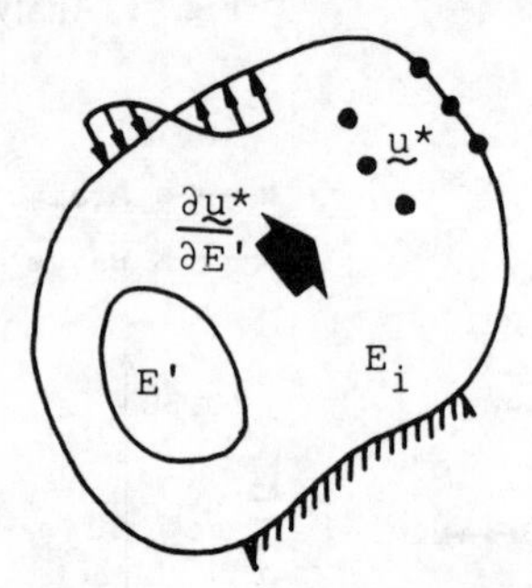

Fig. 4 Definition of the sensitivity vector

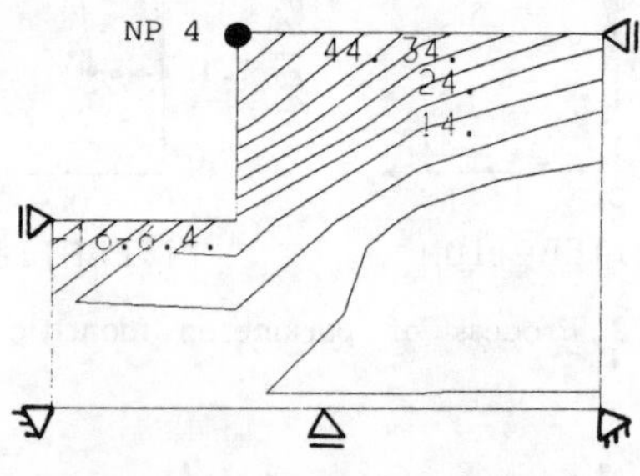

(a) Vertical direction

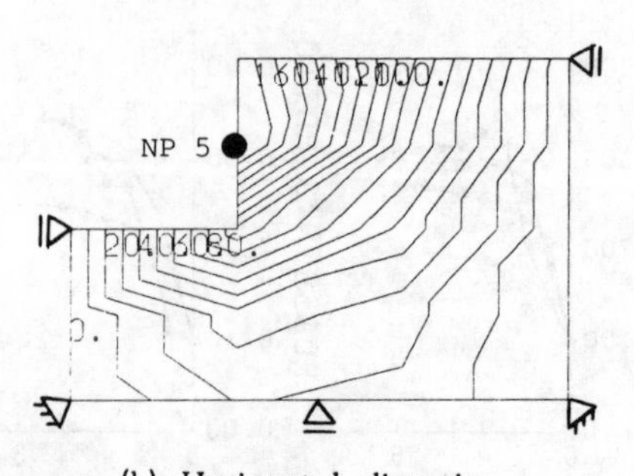

(b) Horizontal direction

Fig. 5 Entire sensitivity distribution on observed displacement (case 1) $\times 10^{-6}$ m/tf/m^2

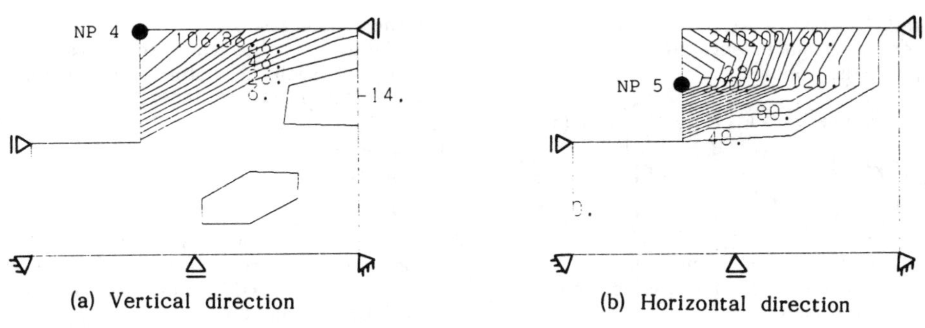

Fig. 6 Specified sensitivity distribution on observed displacement (case 2)

$\times 10^{-6}$ m/tf/m^2

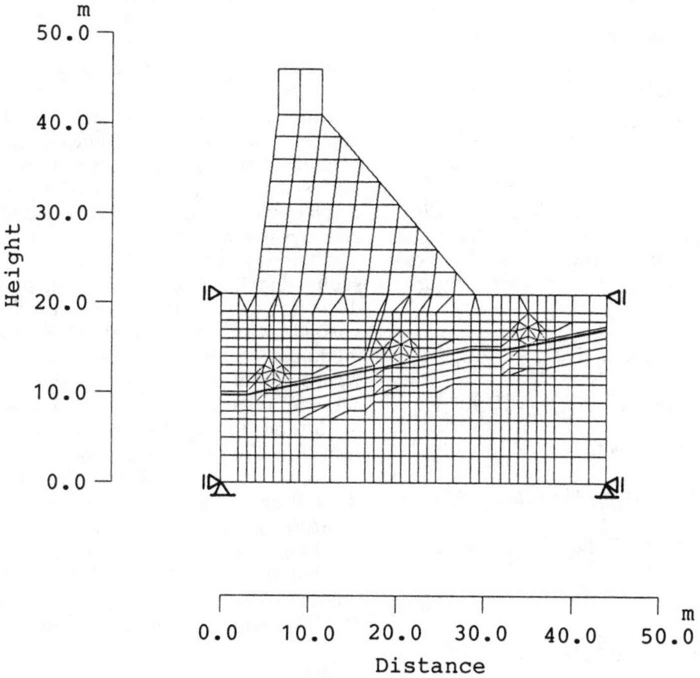

m
50.0
40.0
30.0
Height
20.0
10.0
0.0

0.0 10.0 20.0 30.0 40.0 50.0 m
Distance

Fig. 7 Finite element mesh of the concrete gravity dam site

$\times 10^{-6}$ m/tf/m^2

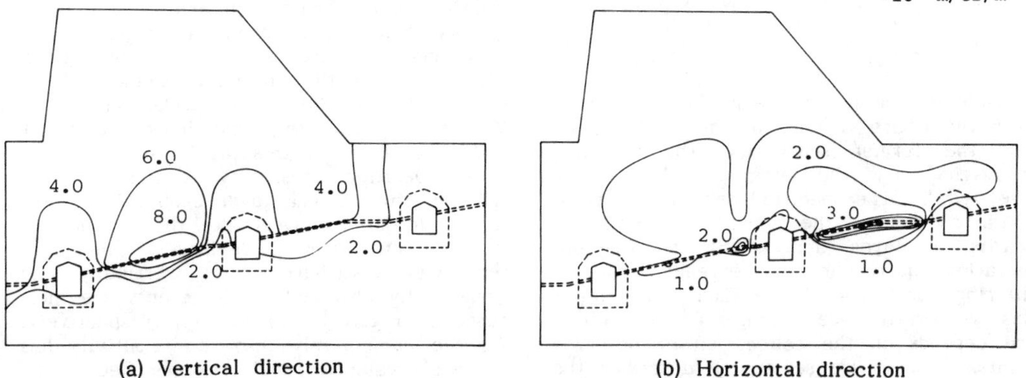

Fig. 8 Specified sensitivity distribution on observed displacement

vestigated. After that, the sensitivity of the observed displacements to such a region is computed as follows:

$$[E_1K^1+\cdots+E_nK^n+E'K']\begin{Bmatrix} \underset{\sim}{u}^* \\ \underset{\sim}{u} \end{Bmatrix} = \begin{Bmatrix} \underset{\sim}{f}^* \\ \underset{\sim}{f} \end{Bmatrix} \quad (14)$$

$$K^i=\begin{bmatrix} K^i_{11} & K^i_{12} \\ K^i_{21} & K^i_{22} \end{bmatrix}, \quad K'=\begin{bmatrix} K'_{11} & K'_{12} \\ K'_{21} & K'_{22} \end{bmatrix}$$

where K_i is the stiffness matrix for the i-th region, K' is the stiffness matrix for the specified region.

From the above equation, the unknown and the boundary displacements are eliminated.

$$\underset{\sim}{u}^*=[(E_jK^j_{11}+E'K'_{11})-(E_jK^j_{12}+E'K'_{12})(E_jK^j_{22}+E'K'_{22})^{-1}(E_jK^j_{21}+E'K'_{21})]^{-1}\{\underset{\sim}{f}^*-(E_jK^j_{12}+E'K'_{12})(E_jK^j_{22}+E'K'_{22})^{-1}\underset{\sim}{f}\}=\underset{\sim}{h}(E_i,E')$$
$$(j\text{:sum}) \quad (15)$$

The partial derivative of observed displacements with respect to E', Young's modulus in the specified region, introduces the following sensitivity vector (see Figure 4):

$$\frac{\partial \underset{\sim}{u}^*}{\partial E'} = -A(E_i,E')^{-1}[K'_{11}-K'_{12}(E_jK^j_{22}+E'K'_{22})^{-1}(E_jK^j_{21}+E'K'_{21})+B(E_i,E')K'_{22}(E_jK^j_{22}+E'K'_{22})^{-1}(E_jK^j_{21}+E'K'_{21})-B(E_i,E')K'_{21}]$$
$$A(E_i,E')^{-1}\{\underset{\sim}{f}^*-B(E_i,E')\underset{\sim}{f}\}+A(E_i,E')^{-1}\{-K'_{12}(E_jK^j_{22}+E'K'_{22})^{-1}\underset{\sim}{f}+B(E_i,E')K'_{22}(E_jK^j_{22}+E'K'_{22})^{-1}f\} \quad (16)$$

$$\text{where, } A(E_i,E')=-[(E_jK^j_{11}+E'K'_{11})-(E_jK^j_{12}+E'K'_{12})(E_jK^j_{22}+E'K'_{22})^{-1}$$
$$B(E_i,E')=(E_jK^j_{12}+E'K'_{12})(E_jK^j_{22}+E'K'_{22})^{-1} \quad (j\text{:sum})$$

Such a vector can also be superposed over the entire region: In what follows, we call the former sensitivity 'the specified sensitivity', and the latter one 'the entire sensitivity'. The specified sensitivity vector is the element of the sensitivity matrix appearing in Bayes' approach[6] and the observation matrix in the extended Kalman filtering formulation[13]. Each element of this vector is independent of the others, and depends on the collocation of observed points. The entire sensitivity describes the contribution of the observed displacements over the whole region.

For a given collocation, the norm of the elements of such a vector is obtained, and we have 'the influence index':

$$\frac{\left(\frac{\partial \underset{\sim}{u}^*}{\partial E'}\cdot\frac{\partial \underset{\sim}{u}^*}{\partial E'}\right)^{\frac{1}{2}}}{N} \quad (17)$$

After computing such an influence index among the trial collocations of measuring devices, we seek an optimal location which has a larger value index.

4.1 Calibration problem

In order to demonstrate the foregoing procedure, a sensitivity analysis is carried out for the excavation problem appearing in section 3. Figure 5 (a) and (b) show the entire sensitivity distribution in the homogeneous soil deposit (case 1 in the preceding section), for the horizontally observed displacement and the vertically observed one, respectively. Geometrical parameters and the structure of the model determine such distribution. For homogeneous soil deposits especially, such distribution is invariant with the changing values of Young's modulus. It can be seen from these figures that the horizontal and excavated surface measurements (e.g. Nodal Point Nos. 4, 5 and 6) should provide useful information for the back analysis.

Figure 6 (a) and (b) describe the specified sensitivity distribution in the three layered ground (case 2 in the preceding section) for the horizontally observed displacement and the vertically observed one, respectively. A similar conclusion can be pointed out in this case also.

4.2 Practical problem

Further application concerns a concrete gravity dam site as shown in Figure 7[14]. Its foundation includes a weak, thin 5-30cm thick layer in the horizontal plane. It is constructed in the tunnel style, known as 'the dowel concrete', and links the upper and lower parts of this layer.

A 'specified sensitivity' coefficient distribution for such a layer should be effective in order to investigate the location of installed displacement devices. Calculated sensitivity distributions for vertically and horizontally observed displacements are indicated in Figure 8 (a) and (b), respectively. We see immediately that the sensitivity has its peak value around the specified layer. It can also be recognized that Young's

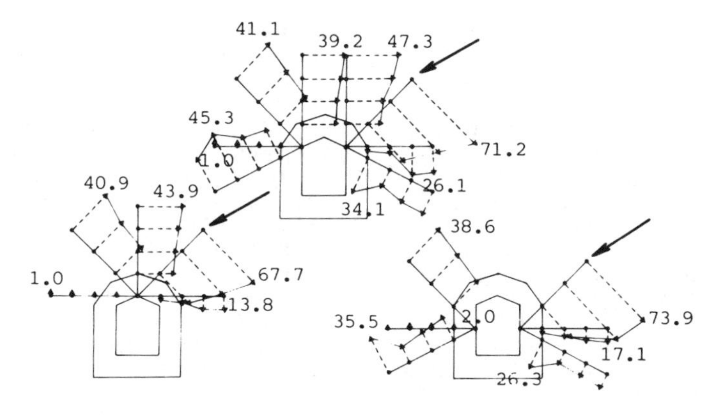

(a) dowel concrete structure in the upper stream

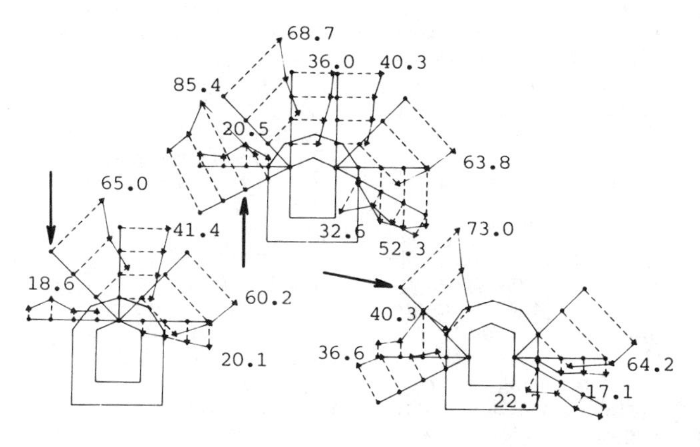

(b) dowel concrete structure in the middle stream

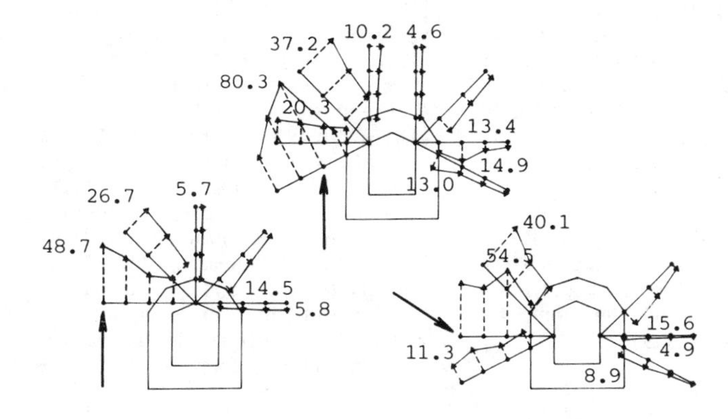

(c) dowel concrete structure in the downstream

Fig. 9 Hypothetical location of measuring devices $\times 10^{-6}$ m/tf/m^{2}

modulus of the thin layer has little influence on the observed displacements in the dam concrete and the lower part of the layer.

The plots in Figure 9 show hypothetical cases of displacement measurement and the sensitivity values at the location of devices. It is assumed that the relative displacement can be measured with two devices. Figure 9 (a), (b) and (c) correspond to the case of the dowel concrete in the upper stream, middle stream and downstream, respectively. The setup pointed out with an arrow is thought to be an optimal one for each hypothetical case, because it has higher sensitivity than the others.

5 CONCLUSION

A method has been presented which may be used to back-analyze the piecewise linear elastic moduli of the ground. A comparative analysis for excavation problems reveals that the proposed programme also provides the identified parameters with good accuracy.

The advantages of the method shown herein are:

- proposed procedure is capable of considering the observational error as error covariance while renewing the estimate error covariance in the solution algorithm;
- a sensitivity distribution, using a priori geometric and material knowledge based on in-situ investigations, provides us with information on the locations to be observed.

It should be pointed out that the sensitivity analysis presented herein will no longer apply to cases in which there is absolutely no knowledge of the material and geometric properties a priori. It will be the subject of later investigations.

ACKNOWLEDGEMENT

The authors wish to acknowledge Messrs. Kazuhito Tomioka and Etsuo Hasegawa for their assistance in a part of the computation involved in this investigation.

REFERENCES

[1]G.Gioda, Some remarks on back analysis and characterization problems in geomechanics, Proc. 5th Int. Conf. Numer. Methods Geomech., Nagoya, 1, 47-61 (1985)

[2]S.Sakurai and K.Takeuchi, Back analysis of measured displacements of tunnels, Rock Mechanics and Rock Engineering, 16, 173-180 (1983)

[3]G.Gioda and L.Jurina, Numerical identification of soil structure interaction pressures, Int. j. numer. anal. methods Geomech., 5, 33-56 (1981)

[4]K.Arai, H.Ohta and T.Yasui, Simple optimization techniques for evaluating deformation moduli from field observations, Soils and Foundations, 23, 1, 107-113 (1983)

[5]K.Arai, H.Ohta and K.Kojima, Estimation of soil parameters based on monitored movement of subsoil under consolidation, Soils and Foundations, 24, 4, 95-108 (1984)

[6]A.Cividini, G.Maier and A.Nappi, Parameter estimation of a static geotechnical model using a Bayes' approach, Int. J. Rock Mechanics and Mining Sciences, 18, 487-503 (1983)

[7]A.Cancelli and A.Cividini, An embankment on soft clays with sand drains: numerical characterization of the parameters from in situ measurements, Proc. Int. Conf. Case Histories Geotech. Eng., St. Louis (1984)

[8]R.E.Kalman, A new approach to linear filtering and prediction problems, Trans. ASME, J. of Basic Eng., 82, 35-45 (1960)

[9]G.Gioda, Indirect identification of the average elastic characteristics of rock masses, Proc. Int. Conf. Structural Foundations on Rock, Sydney (1980)

[10]A.H.Jazwinski, Stochastic processes and filtering theory, Academic Press, New York, 1970

[11]T.Katayama, Applied Kalman filter, Asakura-shoten, Tokyo, 1983 (in Japanese)

[12]K.Arai and M.Miyata, On applicability of statistical back analysis method, Proc. 21st Japan National Conf. Soil Mechanics & Foundation Eng., 2, 1089-1090 (in Japanese) (1986)

[13]A.Murakami and T.Hasegawa, Back analysis by Kalman filter-finite elements and optimal location of observed points, Proc. 22nd Japan National Conf. Soil Mechanics & Foundation Eng., 2, 1033-1036 (in Japanese) (1987)

[14]T.Hasegawa and A.Murakami, Technique of monitoring for dam safety and its application, Proc. 16th Int. Congress on Large Dams, San Francisco (to appear in 1988)

Numerical Methods in Geomechanics (Innsbruck 1988), Swoboda (ed.)
© 1988 Balkema, Rotterdam. ISBN 90 6191 809 X

Determination of optimum tunnel shape by three-dimensional back analysis

M.Hisatake
Kinki University, Osaka, Japan

ABSTRACT: A method of finding out the optimum tunnel shape as well as the optimum excavation and support system is shown after evaluating the equivalent mechanical properties by a three dimensional back analysis. The method has been applied to a practical tunnel in an extremely squeezing ground and the optimum tunnel shape has been determined. Appropriateness of the method has been shown by comparing with the field results. Also a simple figure to back analyze the maximum and the minimum principal stresses of the ground and the direction of these stresses in a tunnel cross section has been proposed, by which engineers can directly determine these values in situ.

1. INTRODUCTION

Intensity of tunnel damage depends on parameters such as geological conditions, virgin stresses, shape of a tunnel as well as excavation and support methods. In the geological conditions, the strength and deformation characteristics of a rock are affected not only by stress-strain relationships of the rock material itself but also by structural conditions such as cracks and joints which exist in the rocks before the tunnel excavation. Also, tunnel excavations mean stress relief. Therefore, in the estimation of tunnel behavior, it becomes very important to find out mechanical characteristics of a rock under stress relief condition by taking the structural characteristics into accounts.

In a stage of tunnel planning, however, it is impossible to find out the distribution and the direction of the cracks and the joints in an entire tunnel line, because of long distant tunnel execution. If displacements measured after excavations of the tunnel face are able to be back analyzed three dimensionally, the mechanical properties back analyzed may be called as equivalent mechanical properties, because these values reflect the structural characteristics and the stress relief condition.

In this study, a method of finding out the optimum tunnel shape as well as the optimum excavation and support system is shown after evaluating the equivalent mechanical properties by the three dimensional back analysis [Hisatake & Ito, 1985].

The proposed method has been applied to a practical tunnel in an extremely squeezing ground to determine the optimum tunnel shape. Appropriateness of the method has been shown by comparing with the field results.

In order to decrease costs and time in three dimensional back analysis, a simple method is proposed, by which the maximum and the minimum principal stresses of the ground and the direction of these stresses in a tunnel cross section are directly determined by a figure.

2. OUTLINE OF BACK ANALYSIS BY OPTIMIZATION

In the following, the ground is treated to be elastic. This assumption, however, is set in order to evaluate the actual tunnnel movements with equivalent linear materials [Hisatake, 1987], which are reflected by joints, cracks, the stress relief and the non-linear mechanical characteristics.

The vertical(σ_v) and the horizontal(σ_h) virgin stresses are defined as

$$\sigma_v = wh, \qquad \sigma_h = K\sigma_v \tag{1}$$

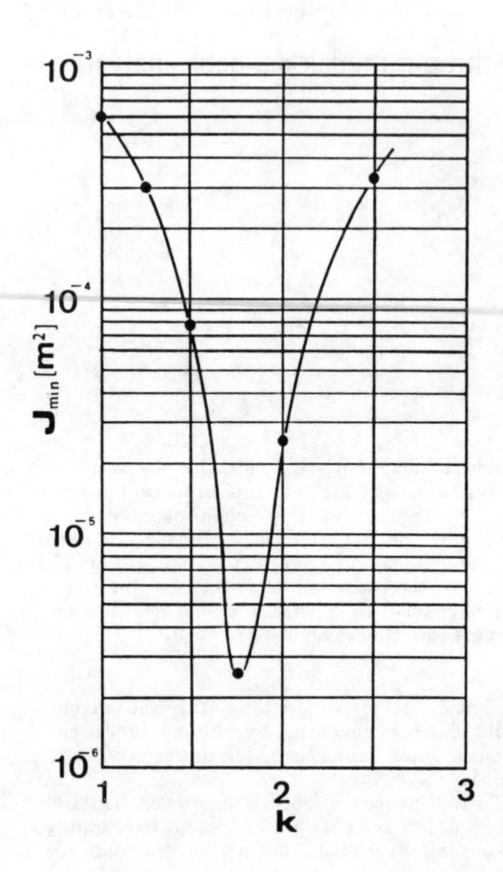

Fig.1 Relationship between Jmin and K (Eca=Ec/5)

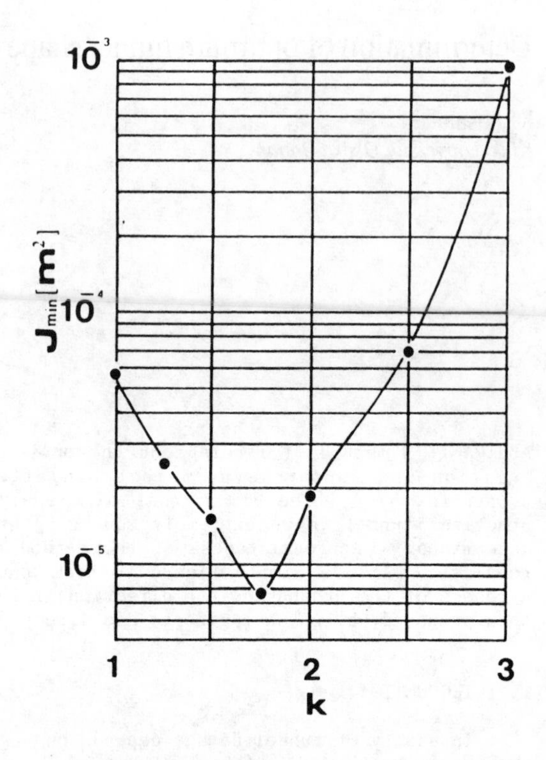

Fig.2 Relationship between Jmin and K (Eca=Ec/10)

where, w is unit weight of the ground, h is tunnel depth and K is the ratio of the initial principal stresses in a tunnel cross section.

If the deformation is not symmetrical to the vertical section in a tunnel axis, shearing component of virgin stresses should be taken into consideration.

The general concept of an optimization method is that analytical displacements u_i may fit with measured displacements u_i^*, if the equivalent input data can be given to the analysis. Even if a linear analysis is employed, this fitting can be accomplished [Hisatake & Ito, 1985] by three dimensional analysis. The equivalent input data can be back analyzed by the optimization method in which the following objective function is made minimum.

Objective function:

$$J=\sum_{i=1}^{n}(u_i-u_i^*)^2 \longrightarrow \text{Minimum} \quad (2)$$

Constrains:

$$E > 0 , \quad 0.0 < \nu < 0.5 , \quad K > 0$$

where, n is a number of measured displacements and u_i is a function on unknowns of E, ν and K.

3. DETERMINATION OF OPTIMUM TUNNEL SHAPE

In this section, the optimum shape of the upper part of an extremely squeezing tunnel is determined by an analysis and that is compared with the actual shape determined by executive records of tunnel movements.

By appling the three dimensional back analysis method to an extremely squeezing tunnel site and obtained is Fig.1 which shows a relationship between K and the minimum value of $J(=J_{min})$ calculated by Eq.(2). Fig.1 indicates that the optimum K value$(=K_{opt})$ which gives the minimum value of J_{min} is 1.75. The result that K-value is greater than 1.0 is reasonable because the horizontal displacements at the spling line is greater than the settlements at the crown, even though the tunnel width is greater than the tunnel height. In the analysis, the Poisson's ratio of the ground is assumed as 0.3, because this value does not give much influence on estimation of the equivalent modulus of elasticity. As the convergence values measured exceed 1m and destruction of the shotcrete is recognized, one fifth of the modulus of elasticity of the shotcrete(Ec) measured in a labolatory is used in the analysis(Eca). But the same value of K_{opt}(=1.75) can be obtained when one tenth of Ec is used in the analysis(Fig.2), so K_{opt} can be recognized to be not much influenced by the Eca value. This distinctive feature will be utilized in order to make back analysis simple in the following chapter.

The back analyzed value of E_{opt} of the ground becomes 60 tf/m^2 under K_{opt}=1.75 and Eca=Ec/5, which value is very small. But qualitative check can be done to this value by calculating a plastic radius around the tunnel by the following elasto-plastic theory presented by Kastner [Kastner, 1971].

$$R=r_i[(1-\sin\varphi)(P\tan\varphi/C+1)]^a \quad (3)$$

where,

R :radius of plastic region
r_i :tunnel radius
C :cohesion
φ :angle of internal friction
P :overburden pressure
a =$(1-\sin\varphi)/(2\sin\varphi)$

The value of R/r_i calculated by Eq.(3) is about 28, in which the values of C and φ are determined by rock specimens in a laboratory, so the value of E_{opt}(=60tf/m^2) obtained by back analysis may not be unreasonable.

Fig.3 shows the several types of tunnel shape used by sequence analysis in which back analyzed values of K_{opt} and E_{opt} are used. The optimum tunnel shape defined here is the tunnel shape in which the maximum shearing stress produced in the shotcrete becomes the minimum in all tunnel shapes. As the shotcrete stresses increase with increasing distance between the tunnel face and measuring tunnel cross sections, progress of the tunnel face is taken into account in calculation of shotcrete stresses.

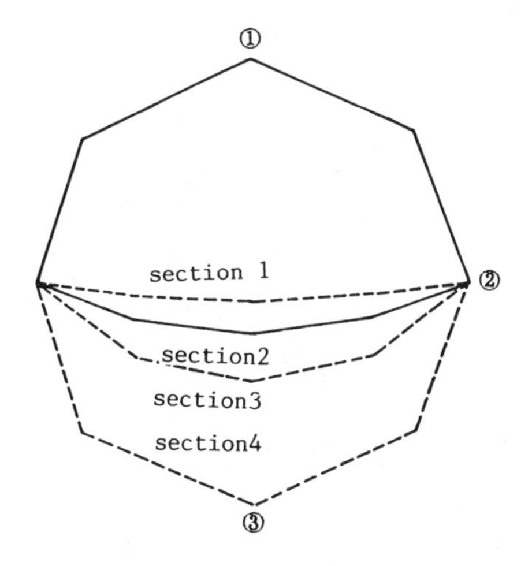

Fig.3 Tunnel shapes in analysis

As the tunnel shape changes from section 1 to section 4, tangential stress at points ① and ③ increases but decreases at point ②. In the four case calculations, the maximum shearing stress in case 3 becomes the minimum. Therefore, the section 3 is determined as the optimum tunnel shape in this site.

Before doing the back analysis, the tunnel shape was changed eight times to seek out the optimum tunnel shape from actual records of tunnel movements. The optimum tunnel shape determined by the analysis coinsides almost with that (section 2) determined by the actual records of tunnel movements.

Therefore, the optimum tunnel shape as well as the optimum excavation and support system may be determined with the data based on the three dimensional back analysis.

4. A SIMPLE METHOD TO FIND OUT THE MAXIMUM AND THE MINIMUM PRINCIPAL STRESSES AND THEIR DIRECTIONS

The above example shows that the lining stresses are affected by tunnel shape in tunnel cross section. In seeking out the optimum tunnel shape, it is, of course, necessary to determine the ratio of the minimum and the maximum principal stresses(K= $\sigma 3/\sigma 1$, where $\sigma 1$: the maximum principal stress, $\sigma 3$: the minimum principal stress) and their directions θ (Fig.4) beforehand.

In the method of the back analysis coupled with the optimization technique ,

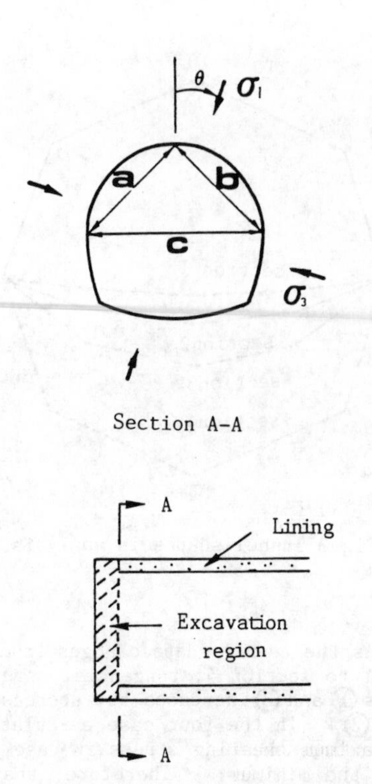

Fig.4 The maximum and the minimum principal stresses and their directions

however, it costs a lot of time and money to determine virgin stresses and E_{opt}. Because the problem is three dimensional and this method bases on an iteration procedure. But it was made clear in the above analysis that K-value and the direction of the two principal stresses in a tunnel cross section were not much affected by the modulus of elasticity of the shotcrete(Ec). By using this characteristics, the values of K and θ are easily estimated in the followings.

Recently displacement measurements are commonly performed as shown in Fig.4, where a, b and c are initial distances before excavations of the tunnel face, and da, db and dc are variations of these distance due to the excavations. If the initial stress in vertical direction is assumed to be overburden pressure, the principal stresses are shown as

$$\sigma 1 = \frac{2wh}{(1+K)+(1-K)\cos 2\theta} \quad (4)$$

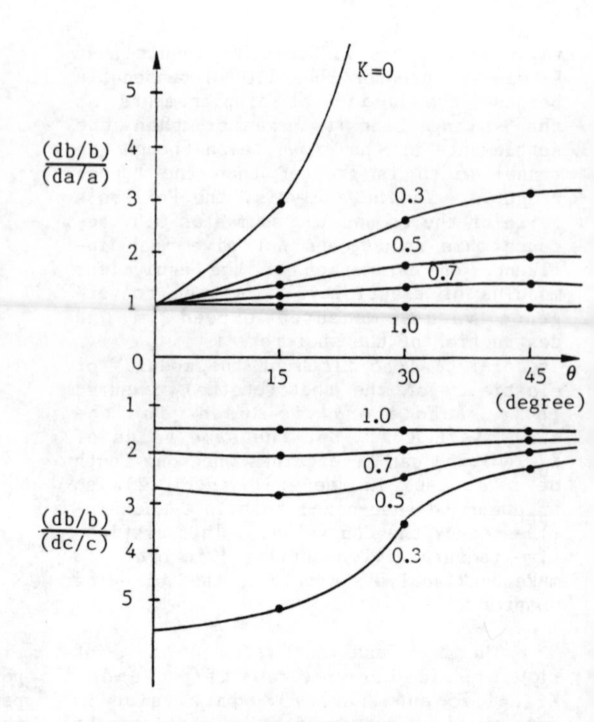

Fig.5 θ and K are back analyzed by measured strain ratios

$$\sigma 3 = \frac{2whK}{(1+K)+(1-K)\cos 2\theta} \quad (5)$$

Provided that the tunnel axial stress is assumed as Kσ1.

One excavation nodal force released at the face are calculated in a finite element method(FEM) under the arbitrary initial stress state. Therefore, by performing three dimensional ordinary sequence analysis of FEM, the relationship among K, θ and the strain ratio such as (db/b)/(da/a), (db/b)/(dc/c) can be obtained as shown in Fig.5. The values of parameters used are indicated in Table 1.

Table 1

E of the ground	5,000 kgf/cm^2
Ec of shotcrete	100,000 kgf/cm^2
Poisson's ratio of the ground	0.3
Poisson's ratio of shotcrete	0.15
Overburden pressure	17.5 kgf/m^2

By applying the two sets of measured strain ratios (db/b)/(da/a) and (db/b)/(dc/c) to Fig.5, the values of K and θ ,which satisfy the respective value of these strain ratios simultaniously at both the upper and the lower parts in Fig.5, are directly determined.

In short, engineers can back-analyze the values of K and θ at tunnel sites directly, provided that Fig.5 has been made beforehand.

The curves in Fig.5 are not much affected by the ratio of modulus of elasticities of the ground and the shotcrete.

In fact, after calculation of da, db and dc by FEM in which K=0.5, θ =30 and 5 times of modulus of elasticity(=250,000 tf/m^2) are used, application of strain ratios to Fig.5 leads K=0.47 and θ =33 which are excellent estimation from engineering view point.

After knowing the values of K and θ , engineers can estimate qualitative distribution of stresses and strains of the ground and the shotcrete. Knowledge of the qualitative tendency for stresses and strains around the tunnel may lead engineers to reasonable control of supports and to reasonable planning in the following excavation and support system.

In case of necessity to investigate stresses and strains of the supports and the ground quantitatively, one time of back analysis is only necessary to find out E_{opt}. After obtaining the values of K, θ and E_{opt}, the optimum tunnel construction may be accomplished by FE analysis with parameters determined.

This procedure is explained in the followings.

Step 1: Make a figure, beforehand, like Fig.5 by 3-D FE analysis for the relation among K, θ and strain ratios.

Step 2: Apply measured strain ratios to the above figure in sites and get values of K and θ.

Step 3: Calculate the maximum and the minimum principal stresses by substituting K and θ to Eq.(4) and (5). Then qualitative distribution of stresses and strains is determined.

Step 4: Perform back analysis only one time to get value of E_{opt}.

Step 5: Perform sequence analysis with parameters determined previously and check support stability with stresses and strains calculated. Also optimum planning such as optimum tunnel shape and support system is determined for the position where the tunnel face newly advances.

CONCLUSIONS

1) It is possible to find out the optimum tunnel shape for respective tunnel site by using parameters determined by three dimensional back analysis.

2) Appropriateness of the back analysis method proposed has been recognized by comparing field results and analytical ones.

3) A simple method to find out the two set of the principal stresses in a tunnel cross section and the directions of them has been proposed. These values are directly determined with a figure made by sequence analysis previously. By following this procedure, one time of back analysis with one unknown is only performed, so it becomes very economical.

REFERENCES

1) Hisatake,M. & Ito,T.: Back analysis for tunnels by optimization method, Proc. 5th Int. Conf. Numeri. Method Geomech., pp.1301-1307, 1985.

2) Hisatake, M.: Assessment of tunnel face stability by back analysis, Proc. 2nd Int. Symp. Field Measure. Geomech. pp.947-954, 1987, Kobe, Japan.

3) Kastner, H.: Statik des tunnel-und stollenbaues, Springer-verlag, 1971.

Numerical Methods in Geomechanics (Innsbruck 1988), Swoboda (ed.)
© 1988 Balkema, Rotterdam. ISBN 90 6191 809 X

FEM analysis for rockburst and its back analysis for determining in-situ stress

Lu Jiayou, Wang Changming & Huai Jun
Institute of Water Conservancy and Hydroelectric Power Research, Beijing, People's Republic of China

ABSTRACT: This paper describes the rockburst of a circular headrace tunnel of a hydroelectric power project in limestone. The mechanism of rockburst is analysed on the basis of the rockburst phenomena observed in this tunnel. The paper points out that rockburst will occur in brittle rock when the stress reaches the value specified by the Griffith criterion and the rockburst will become a strong one if the stress reaches its ultimate state. The possibility for predicting the initiation and the intensity of rockburst and back analysis for determing the in-situ stress by FEM is also discussed.

1 INTRODUCTION

The headrace tunnel at the Tianshengqiao hydroelectric power project has a diameter of 10 m and a length of 10 km, of which 7 km pass through the limestone formation. Its average buried depth is about 400 m and the maximum buried depth is about 800 m. Rockburst occurred in tunnel No. 2, which is a construction branch tunnel also in the limestone, nearly perpendicular to the main tunnel. This branch tunnel is 1.3 km in length, 10 m in diameter and also circular in cross section. Both the main tunnel and this branch tunnel were excavated by TBM. The rockburst occurred in sections where the buried depth ranges from 200 m to 250 m. Five major rockbursts occurred, as shown in Fig. 1., at places where the limestone is rather hard and brittle, with no fissures detectable by visual observation. The rock in this region is very dry since the groundwater level is below the tunnel.

The rockburst took place mostly in the top part of the tunnel. The positions of rockburst were somewhat towards the left in the first three cross sections and were at the centre of the top in the other two cross sections (Fig. 1). The distance from the excavated face to the location where rockburst first occurred is the same as the diameter of this tunnel. The rockburst was very active in 24 hours and it weakened and ceased after two months. The mechanism of rockburst is cleavage fracture and the fracturing sound is audible. The thickness of the rockburst zone is found to be dependent on the overbunden of the tunnel. In four of the cross sections (Fig. 1), the fractured zones were in slab form, but in cross section 0+938, the failure zone was in arch shape.

No measurement of tectonic stress has been made, but the direction of the maximum tectonic stress σ_1 can be known from the position of the rockburst. For example, the direction of the maximum tectonic stress in cross section 0+938 is horizontal. The vertical tectonic stress is assumed to be $\sigma_v = \gamma H$ (Herget, G., 1974). It is known that in this zone the horizontal tectonic stress σ_H is about 1.2 - 2 times the σ_v. The ultimate compressive strength of the limestone in Tianshengqiao is about 60 - 80 MPa. The maximum tangential stress σ_θ during rockburst is estimated to be less than one half of the ultimate compressive strength. Hence, the rockburst in the limestone of Tianshengqiao is a phenomenon of low stress fracture in high strength material.

2 BEHAVIOR OF BRITTLE ROCK

Generally, the failure mechanism of rock specimen in laboratory test may be classified into three kinds, namely, brittle cleavage failure, shearing failure accompanied by tensile failure, and shearing failure (Fig. 2). Especially in low confining σ_3 or in uniaxial stress ($\sigma_3 = 0$), the failure of the specimen is in a

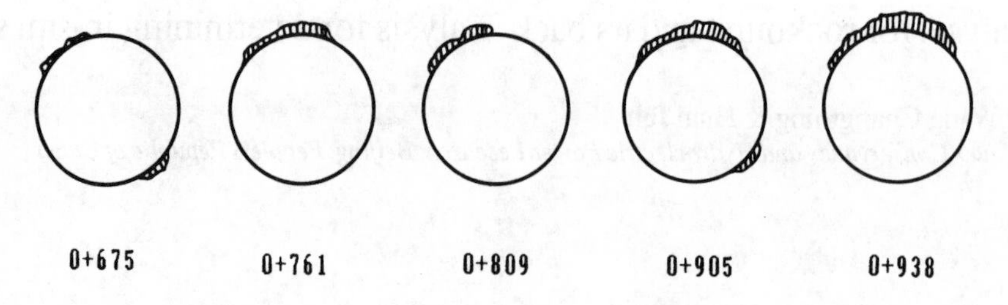

Fig. 1 The positions of rockburst in five cross section

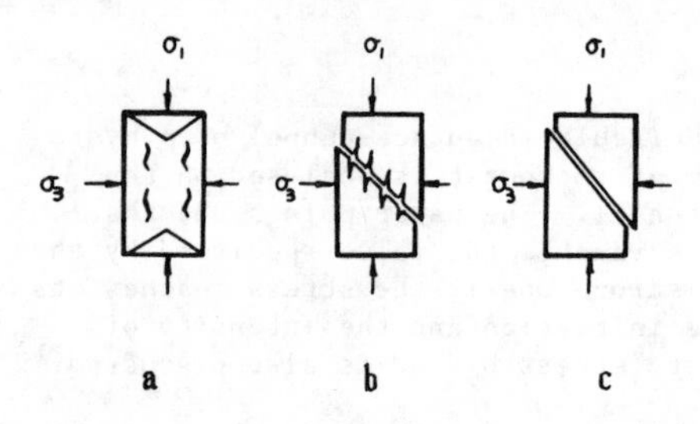

Fig.2 Failure mechanism of rock specimens

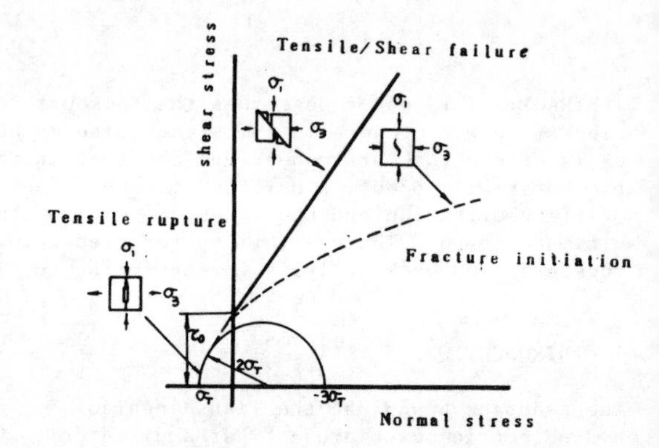

Fig.3 Failure criterion of brittle rock

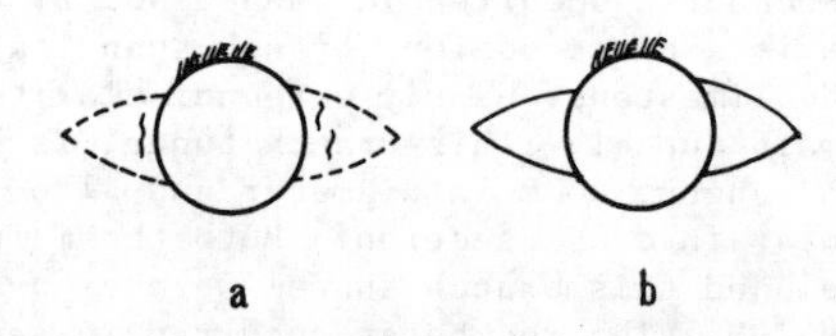

Fig. 4 Possible failure mechanism of adjoining rock of circular tunnel

disorderly pattern (Fig. 2-a). If the confining stress σ_3 is rather high, the failure is of shearing type (Fig. 2-c).

In the fracturing process of brittle rock, a series of acoustic emission signals can be recorded. Every acoustic emission signal is a response of the growth of a microfracture point. But during the process of loading the specimens do not exhibit the cleavage fracture as the rockburst and the failure surface consisting of a lot of micro fractures appears only after the macro failure has been formed. It is reported in Scholz's work (Scholz, C.H., 1968) that the fractured points concentrate near the macro failure surface of the specimens.

Under the condition of compression, the Griffith criterion can be used to reflect the initiation of fracture. Under tension condition, the failure of specimen happens at once when microfracture is initiated. A lot of rock specimen test results show that the Coulomb-Navier criterion can fairly well describe the macro failure of rock. In certain cases, the Modified Griffith criterion may be better because it takes into consideration the close up of the cracks in rock. Thus, in brittle rock the Grifith criterion is a criterion for initial fracture and the Coulomb-Navier or Modified Griffith criterion is the criterion for ultimate failure, as illustrated in Fig. 3.

3 FAILURE MECHANISM AND CRITERION OF ROCKBURST

According to the limit equilibrium theory, the boundar surface after failure of circular tunnel is a logarithmic spiral line as illustrated in Fig. 4-b. Some model tests also give the same results, because the models are made of elastoplastic material, which obeys the Coulomb-Navier criterion in ultimate state and the brittle fracture is not obvious. Because the radial stress σ_r at the surrounding of the tunnel is rather small and σ_r equals zero at the boundary of the tunnel, so the major principal stress σ_1 will be σ_θ. The stress state near the boundary is like the uniaxial stress state. If the behavior of the rock is brittle, fracture will be initiated near the boundary under low stress condition, leading to cleavage spall. Failure might be formed by one crack or by the connection of a series of cracks and the fractured surface must be parallel to the boundary (Fig. 4-a) (lu, etc., 1984). The mechanism of rockburst in Tianshengqiao tunnel is typical of brittle fracture.

By comparing the failure mechanism of rockburst with that of specimen of brittle rock, it can be seen that:

1. The cracks at the adjoining rock of tunnel are much more in quantity and in size than the microcracks in the specimen, so the adjoining rock has smaller fracture toughnesses, K_{IC} and K_{IIC}, and larger stress-intensity factors K_I and K_{II} than the rock specimen.

2. The boundary conditions for the specimen and the tunnel are different. The speciment is laterally constrained at the top end while the tunnel is free to move in the radial direction.

3. The stress σ_1 in the specimen has a uniform distribution but the σ_θ in the rock surrounding the tunnel shows variation with great gradient.

Hence, the microfracture and fracture at the boundary of the tunnel are liable to spall. Conclusion may be drawn that the initiation of rockburst and the fracture in the specimen are similar in characteristics, and they both obey Griffith criterion. The differences are that the magnitudes of cracks are not of the same order and the strength parameters in the Griffith criterion are different. The macro failure of rock surrounding the tunnel and the ultimate failure of the specimen are the same and they obey the Coulomb-Navier criterion or Modified Griffith criterion. In the prototype the rock mass surrounding the tunnel is cut by joints and fissures and the boundary surface after failure is very irregular, no longer being logarithmic spiral line.

4 THE INTENSITY OF ROCKBURST

The intensity of rockburst can be divided into four classes as shown in Table 1. The first class is low rockburst activity. which means the initiation of microfracture and fracture. The release of strain energy due to fracture initiation is limited. The stress state reaches Griffith criterion and causes eleavage. Because spalling is the result of extension and connection of some cracks, the acoustic emission must produce earlier. The rockburst of Tianshengqiao tunnel as mentioned above belongs to this class.

According to the failure analysis of rock specimen and the adjoining rock of the tunnel, it can be seen that when the stress state reaches Coulomb-Navier or Modified Griffith criterion, the failure zone will be large. The rockburst occurs like gun shot when large strain energy is released in the ultimate state. Before the rock collapses, the acoustic emission must be very strong. This is the third class rockburst. The rockburst in between these two cases mentioned above is moderate rockburst. At some sections of Tianshengqiao headrace tunnel, the rockburst continued for seveal months. This is categorized as the second class.

It appears to be possible to predict the initiation of rockburst and evaluate the intensity of rockburst by using FEM. The classification of rockburst as described by Russenes (Broch, E., 1984) will then be extended as illustrated in Table 1.

5 FEM ANALYSIS FOR ROCKBURST

In the fracturing process of brittle rock, the relationship between stress and strain is nearly linear elastic. The stress analysis of rockburst can be simplified, and a linear FEM program is used. The FEM program is developed with Griffith criterion and Coulomb-Navier criterion. The rockburst will occur in brittle rock when the stress reaches the value specified by the Griffith criterion and the rockburst will become a strong one if the stress reaches its ultimate state.

In this program, four nodals isoparametric element is set up.

Here is an example of FEM analysis, at the Tianshengqiao, where the buired depth is about 453 m. We assum that the vertical initial stress is the weight of the overlying rock mass and the horizontal initial stress σ_H is 1.3 times σ_V. The mechanical parameters of rock are:

$$c = 6.6 \text{ MPa}, \quad f = 2.2$$

and the tensile strength

$$\sigma_T = 3 \text{ MPa}.$$

Table 1 Rockburst classes as described by Russenes (1974) and extended in this paper

Rockburst class	Description	Stress state
0	NO ROCKBURSTING No stability problems caused by rock stresses. No noises from the rock	
1	LOW ROCKBURSTING ACTIVITY Some tendencies to cracking and loosening of rock. Light noises from the rock.	Griffith criterion
2	MODERATE ROCKBURSTING ACTIVITY Considerable slabbing and loosening of rock. Tendencies to develop a deformed periphery with time. Strong cracking noises from the rock.	
3	HIGH ROCKBURSTING ACTIVITY Severe rockfalls from roof and walls immediately after blasting. Slabs pop from the floor, or the floor may heave. Considerable over-breaks and deforming of the periphery. Rock noises of gun shot strength may be heard.	Coulomb-Navier criterion or Modified Griffith criterion

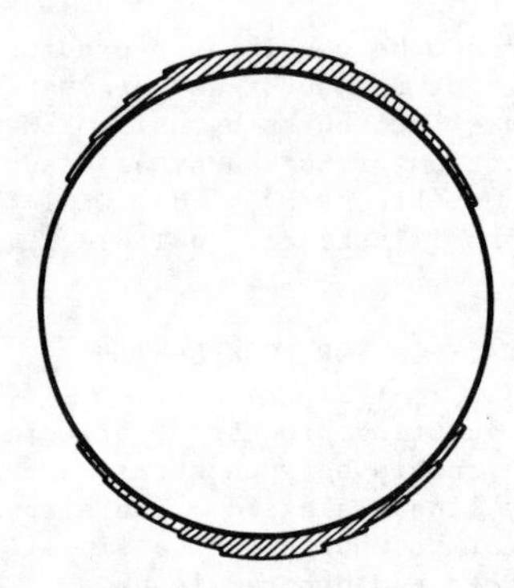

Fig. 5 The failure region of FEM analysis result

The maximum thickness of failure region is about 43 cm and the minimum thickness is about 5 cm. The stress reaches the value specified by the Griffith criterion, no strong rockburst occurs. It is similar in characteristic with the tunnel of Tianshengqiao.

6 BACK ANALYSIS FOR DETERMINING IN-SITU STRESS

The FEM computational program has been developed for predicting the rockburst. It can be used to determine the in-situ stress according to the critical brittle failure, i.e., the first class rockburst. As an example, the computation of the in-situ stress at cross section 0+905 is made by using the mechanical and geometrical parameters of that section. As mentioned above, the maximum stress is in the horizontal direction and the vertical stress σ_V is the weight of the overlying rock mass. Because of the wide range of tensile strength $[\sigma_T]$ of rock in Griffith criterion, $[\sigma_T]$ = 1.5 —— 3.7 MPa is chosen. The result gives that σ_H is 1.1 —— 1.7 times σ_V.

REFERENCE

[1] G. Herget, Ground stress determinations in Canada, Rock Mechanics, 6, 1974

[2] C.H. Scholz, Experimental of fracturing process in brittle rock, J. Geophy, Res. Vol. 73, No. 4, 1968

[3] E. Hoek, Brittle failure of rock, Rock Engineering Practice, Charpter 4. 1968

[4] Lu Jiayou etc., Review and prospect of rock mechanics in underground engineering, Chinese Journal of Rock Mechanics and Engieering, Vol. 3, No. 1, 1984

[5] E. Broch., S. SØrheim, Experiences from the planning, construction and supporting of a road tunnel subjected to heavy rockbursting, Rock Mechanics and Engineering, 17, 1984

Numerical Methods in Geomechanics (Innsbruck 1988), Swoboda (ed.)
© 1988 Balkema, Rotterdam. ISBN 90 6191 809 X

Back analysis for determining nonlinear mechanical parameters in soft clay excavation

G.X.Zeng, X.N.Gong, J.B.Nian & Y.F.Hu
Zhejiang University, Hangzhou, People's Republic of China

ABSTRACT: A calculation programme of back analysis has been developed in order to determine the deformation parameters of soft clay in excavation with sheet piling. The soil constitutive relationship is nonlinear-elastic, in which two different basic stress paths are taken into account. The effect of location of measurement is studied by using the programme. The programme can also be used in layered ground. Finally, the programme is applied to analyze a case of excavation project and the results are satisfactory.

1 INTRODUCTION

The determination of deformation parameters is one of the most important problems in geotechnique. Usually, the parameters are determined by laboratory and insitu tests and field measurement. In recent years, the observational method has been widely used to supplement the labortory and field investigations. Soil parameters can be determined by back analyzing the field measurements during construction. Though some problems have been studied (e.g. Gioda,1979,1985; Cividini et al., 1981; Arai et al.,1983,1986 and so on), many problems, especially the problem of excavation in soft clay, remain unsolved.

In this paper, a calculation programme of nonlinear finite element method has been developed in order to back analyze the soil deformation parameters of excavation in soft clay. Two stress-path dependent modulus equations are derived. The effect of locations of measurement in layered ground is studied by the programme. Finally, a case of excavation project is analysed by the programme. The results of back analysis and laboratory test are in good agreement.

2 THE FINITE ELEMENT ANALYSIS OF EXCAVATION

2.1 Nonlinear stress-path dependent modulus equations

It is known that stress-strain relationship of clay is not only nonlinear but also stress-path dependent. Stress paths during excavation are complicated and can not be fully considered. Two different basic stress paths have been taken into account herein, as shown in Fig.1.

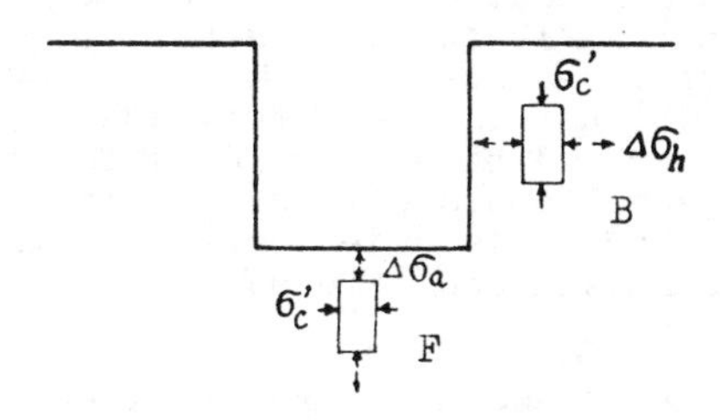

Fig.1 Typical stress paths during excavation

Corresponding modulus equations are developed. Deformation modulus is defined as follows:

$$E_d = \frac{\Delta\sigma_a - \Delta\sigma_h}{\Delta\varepsilon_a} \quad (1)$$

in which ε_a is the vertical strain

in triaxial test; σ_a and σ_h are the vertical and horizontal stresses, respectively.

From the theory of elasticity, for the stress condition in the triaxial test, the value of tangent Young's modulus, E_t, may be related to stress and strain increments as

$$E_t = \frac{(\Delta\sigma_a+2\Delta\sigma_h)(\Delta\sigma_a-\Delta\sigma_h)}{\Delta\sigma_h(\Delta\varepsilon_a-2\Delta\varepsilon_h)+\Delta\sigma_a\cdot\Delta\varepsilon_a} \qquad (2)$$

in which ε_h is the horizontal strain in triaxial test.

For condition F (Fig.1)

$$E_t = E_d \qquad (3a)$$

and for condition B(Fig.1)

$$E_t = 2\nu_t E_d \qquad (3b)$$

in which ν_t is tangent poisson's ratio.

For normally consolidated clay the stress-strain relationship usually can be fitted by a hyperbolic function (Kondner,1963) and normalized (Zeng and Gong,1985). Therefore, Eqs.(3a) and (3b) can be expressed as

$$E_t = \overline{E}_i\sigma_c'(1-R_f S)^2 \qquad (4a)$$

$$E_t = 2\nu_t\overline{E}_i\sigma_c'(1-R_f S)^2 \qquad (4b)$$

in which $\overline{E}_i$ is the initial tangent modulus of normalized stress-strain relationship; S is shear stress lever; R_f is failure ratio.

For undrained test, ν_t =0.5, thus Eq.(4b) has the same form as Eq.(4a). However,the parameters used for the two equations should be determined from the tests under corresponding stress path conditions.

2.2 Basic assumption of finite element analysis

1. Plane strain condition is assumed in the analysis. Since excavation is usually symmetrical, only a half section is adopted in the analysis.
2. Initial stress field in the ground is calculated under the at rest condition with a lateral stress coefficient K_o. The changes of initial stress condition and soil properties due to the driving of sheet piling and pile groups within the excavation site are neglected.
3. Since excavation is usually temporary, undrained condition is generally appropriate. Thus total stress method is used.
4. Excavation process is simulated step by step in the analysis. Young's modulus of each element is first calculated according to the initial stresses and then altered after each step to the new stress condition.
5. Quadratic isoparametric elements are used for soil mass and elastic beam elements for sheet piling. Possible cracking and slipping between soil and sheet piling are not considered in the analysis.

3 NUMERICAL PROCEDURE OF BACK ANALYSIS

It can be seen from Eqs.(4a) and (4b that the parameter $\overline{E}_i$ is of primary importance. Thus, back analysis was applied to determine the normalized initial tangent moduli $\overline{E}_i$'s of different stress path and layered ground. Back analysis is generally divided into two different types. Direct method needs no complicated mathematical formulations.
In this method, optimized parameters directly minimize the discrepancy between the field measurements and the corresponding quantities obtained by means of finite element analysis.

The following error function ε is adopted as a practical definition of the mentioned discrepancy:

$$\varepsilon = \sqrt{\frac{1}{n}\sum_{i=1}^{n}(S_i - S_i^*)^2} \qquad (5)$$

in which S_i^* is the measured displacement of i-th point, and S_i the corresponding calculated quantity.

The error defined by Eq.(5) is a highly nonlinear function of the material parameters. Therefore, the optimization algorithm adopted for the solution of the problem must be capable to handle nonlinear function Since the analytical expression of ε can not be defined, the algorithm must not require the analytical evalution of the function gradient. In this paper, SIMPLEX method has been used, and considering calculation cost, only two parameters are back analyzed.

4 NUMERICAL RESULTS

For simplicity, in cases 1-6 the same parameters are used in the two different stress path zones in Fig.1, and the excavation of a soil deposit consists of two layers. The effect of initial calculation point and location of measurement is studied. In case 7 two different stress path zones are considered. Details of all cases are shown in Table I and Fig.2, in which u is the horizontal displacement and v is the vertical one. Cases 1-3, have the same measuring points, but their initial back calculation points are not the same. The optimization sequence of cases 1-3 in the space of the lower layer normalized initial modulus $\bar{E}_{i1}$ and the upper layer one $\bar{E}_{i2}$ are shown in Figs.3-5. Their results are shown in Table I. It is obvious that they all converge to accurate values. Therefore, we can see that the convergence is not dependent on the value of initial back calculating point.

In case 4, measuring points, as shown in Table I. are the horizontal displacements of sheet piling. It can be seen from Fig.6 that for all layer parameters $\bar{E}_i$'s converge to accurate values.

Arai(1983) used conjugate gradient method to back calculate E and ν of soil mass in excavation, and he

Table I Data of back analysis

case	measuring points	initial point $\bar{E}_{i1},\bar{E}_{i2}$ (×50)	back analysis $\bar{E}_{i1},\bar{E}_{i2}$ (×50)	accurate values $\bar{E}_{i1},\bar{E}_{i2}$ (×50)
1	20,24,32,37,41,	4.00, 3.00	3.00, 1.51	3.00, 1.50
2	45,51,54,58,62	5.00, 3.00	3.04, 1.53	3.00, 1.50
3	(u and v)	0.40, 0.40	3.03, 1.50	3.00, 1.50
4	37,38,39,40,41,42,43,44, 45 (u)	4.00, 3.00	2.98, 1.51	3.00, 1.50
5	43,45,51,60,62 (u and v)	4.00, 3.00	3.04, 1.48	3.00, 1.50
6	19,24,30,41,45,47,53,36, 58,62 (u and v)	4.00, 3.00	4.24, 1.07	3.50, 1.00
7	20,24,32,37,41,45,51,54, 58,62 (u and v)	4.00, 3.00	1.61, 2.01	1.60, 2.00

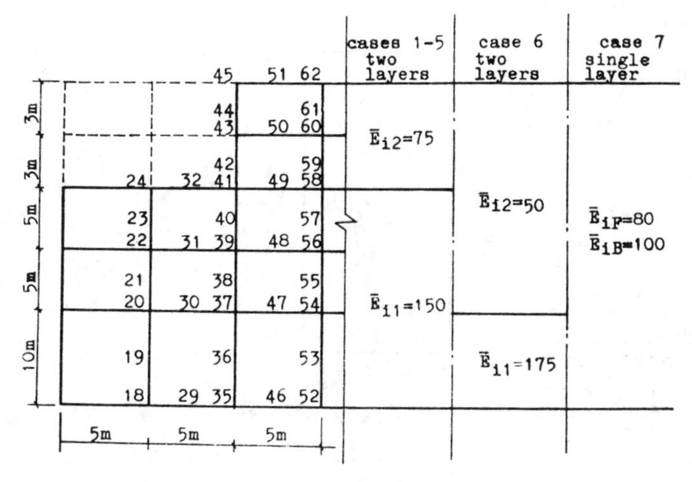

Fig.2 Data of back analysis

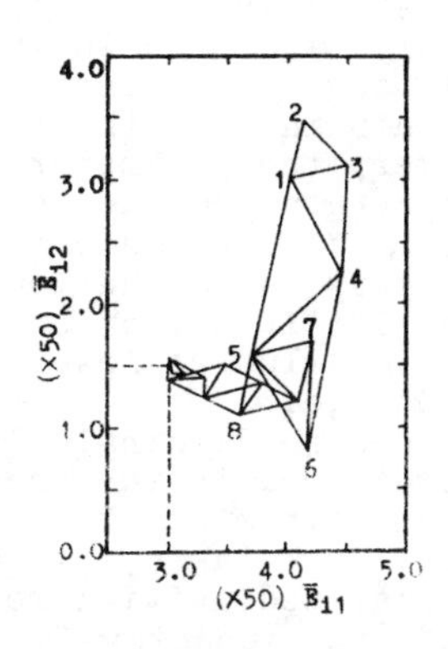

Fig.3 Optimization sequence (case 1)

considered that the given data of displacements (at the nodal points located at where they are restained by much stiffer structure) may contribute little to the estimation of the elastic constants of the soil deposit, and suggested to select the measuring points apart from the sheet piling. However, the authors consider that the excavation problem is a soil-piling interaction problem. The displacement of sheet piling depends on soil property. Therefore, piling displacement can be used to back calculate parameter E_i. Furthermore, back calculation can also be performed when only horizontal displacements are ready.

In case 5, measuring points are within the upper layer. It can be seen from Fig.7 that all layer parameters E_i converge to accurate values. Arai (1983) considered that the displacements of upper layer can not be used to back calculate parameters of lower layer. However, this investigation shows that though the upper layer has main effect, the behavior of lower layer soil has an effect on displacements of upper layer to some degree. From Fig.7, it can be seen that parameter E_i's firstly approach the upper layer accurate value, and then approach the lower layer one. When the lower layer is shallow, it has greater effect and its displacements can back calculate lower layer E_i.

In case 6, lower layer is deep, as shown in Fig.2, the results of analysis are shown in Fig.8. Although the selected measuring points include the ones of lower layer, the parameter of upper layer converges and the one of lower layer not. It is because that lower layer is very deep, the load of excavation has little effect on it. Therefore, the modulus of lower layer can not converge.

In case 7, single layer soil is considered in the analysis. The two basic types of stress path shown in Fig.1 are taken into account. Their normalized initial moduli are $\bar{E}_{iF}$ and $\bar{E}_{iB}$ respectively. The result converges to accurate values as shown in Fig.9.

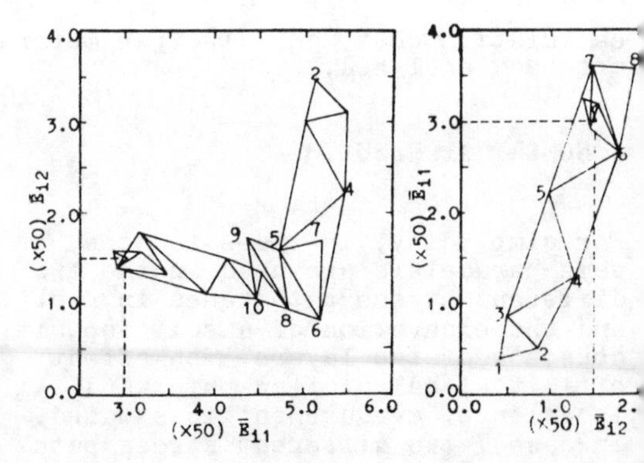

Fig.4 (case 2) Fig.5 (case 3)

Optimization sequence

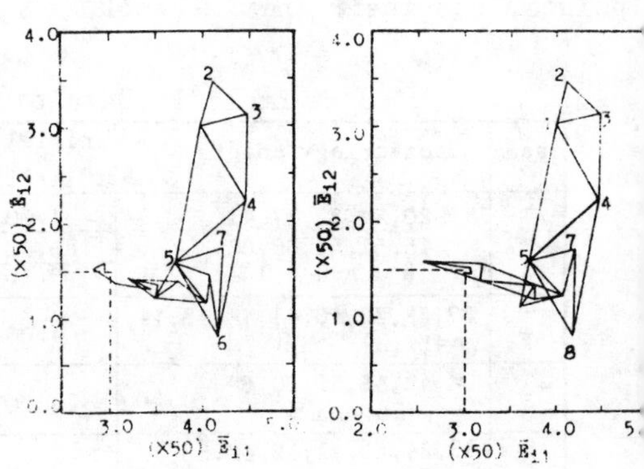

Fig.6 (case 4) Fig.7 (case 5)

Optimization sequence

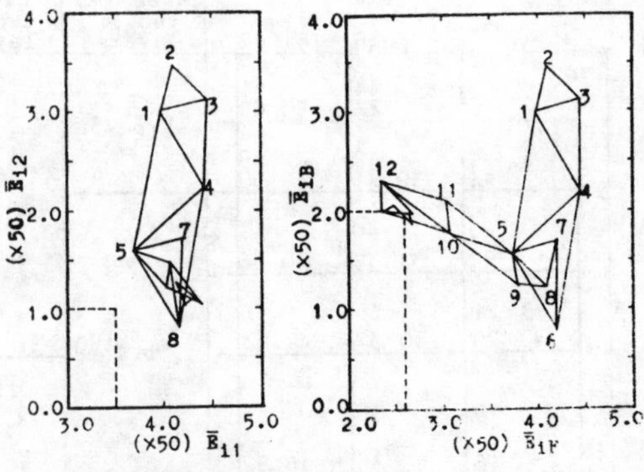

Fig.8 (case 6) Fig.9 (case 7)

Optimization sequence

5 A CASE STUDY

In 1985, the Friendship Hotel, twenty three story high, was built in Hangzhou. According to the design, an excavation, 7.0m in depth, for construction of the foundation was required. The plane size of the excavation is 30×60m. Supports by cantilever sheet pilings were used. Some of the results of measurements and analysis are presented below.

5.1 Ground condition

Some of the soil properties of the four layers are listed in Table II. Ground water table was about 1.5m below the ground surface. Layer No.4 is weathered rock where no deformation is assumed. The field instrumentation was described elsewhere (Zeng et al, 1986).

Table II Physical properties of soils

No.	Depth (m)	w (%)	e	γ (kN/m^3)	w_L	w_P	I_P
1	4.5	-	-	18.6	-	-	-
2	15.1	41.5	1.13	17.6	37.5	18.5	19.0
3	27.3	28.3	0.76	19.1	41.0	20.6	20.4
4	34.6	51.2	1.48	15.9	48.6	37.0	11.6

5.2 Back analysis

The first layer is nonhomogeneous miscellaneous fill, its normalized initial tangent modulus is assumed to be the same as layer No.2. According to the results of laboratory test, modulus ratio of region F to region B (Fig.1) is chosen to be 0.8. The region affected by stress path is assumed to be up to the bottom of sheet piling, as shown in Fig.10.

The values of horizontal displacements I-5 and I-6 are used in back analysis. Since the displacements of upper layer are significantly affected by the soil moduli, the displacements of the upper layer are weighted to accelerate convergence. Then, error function are taken as

$$\varepsilon = \sqrt{\frac{1}{n+w\cdot m}\left(\sum_{i=1}^{n}(S_i-S_i^*)^2+\sum_{i=1}^{m}w(S_i-S_i^*)^2\right)} \quad (6)$$

in which n is the unweighted points number, m is the weighted points number, w is weight, w = 3.

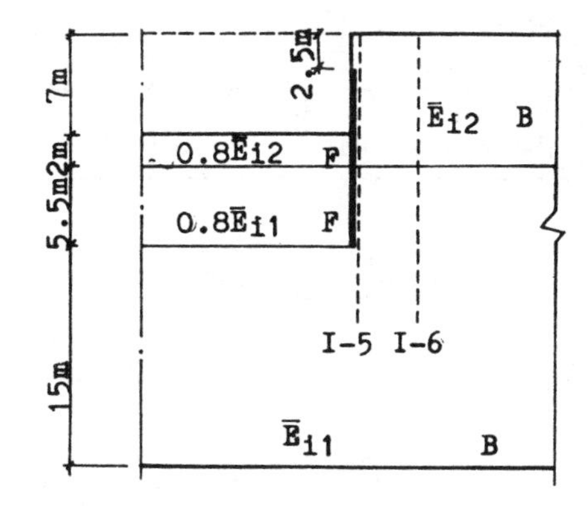

Fig.10 Geometry of analysis

Optimization sequence is shown in Fig.11. The moduli calculated by using back analysis are in good agreement with those of triaxial test results, as shown in Table III. In Fig.12 a comparison is made between real deflections of inclinometers I-5 and I-6 and those obtained by the numerical analysis where the normalized initial tangent moduli are back calculated, and they are in good agreement. Therefore, it can be concluded that back analysis can be used to predict the deformation parameters in situ effectively.

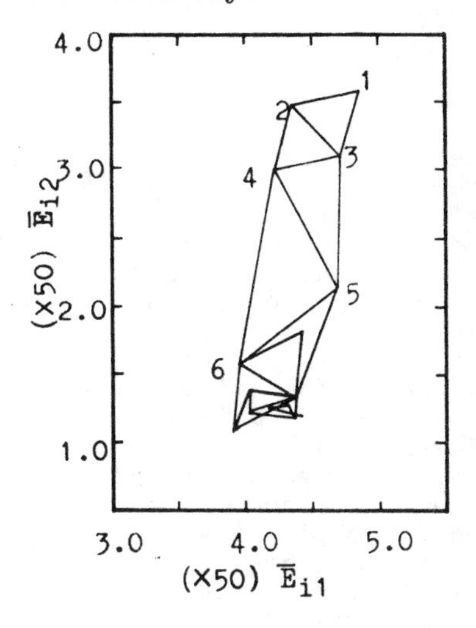

Fig.11 Optimization sequence

Table III Final results of $\bar{E}_i$

		Region F	Region B
Upper layer	Back analysis	49.1	61.4
	Lab. test	48.0	62.0
Lower layer	Back analysis	177.0	221.3

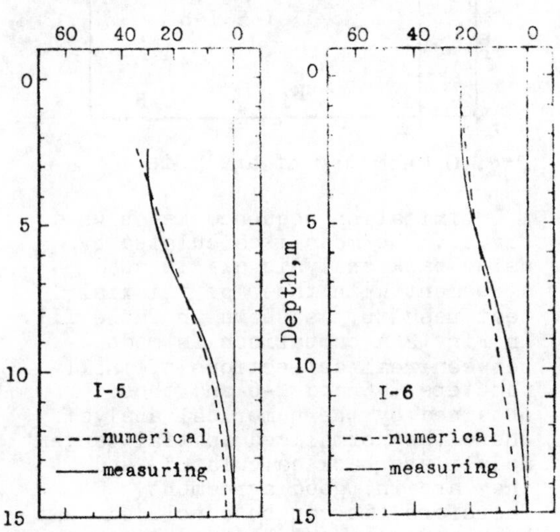

Fig.12 Comparison of numerical and measuring

6 CONCLUSION

1. The values of soil deformation parameter in excavation can be successfully direct back calculated from the field measuring displacements or only horizontal ones.

2. The convergence of back analysis in nonlinear elastic model has little relation to the value of initial back calculating point.

3. In excavation problem, when lower layer is shallow, the parameters of layered ground can be calculated from upper layer displacement. For very deep lower layer, although the selected measuring points include the displacements of lower layer, only the parameters of upper layer converges and that of lower layer not.

4. Finally, a case history is analysed by the method. Good agreements are obtained between the deformation parameters by using back analysis and that of laboratory test results. The displacements obtained by the numerical calculation and the ones of measuring are in good agreement.

REFERENCES

Arai,K.,Ohta,H., and Yasui,T. 1983. Simple optimization techniques for evaluating deformation moduli from field observations. Soils and Foundations. 23:107-113.

Arai,K.,Ohta,H.,Kojima,K., and Wakasugi,M. 1986. Application of back-analysis to several test embankments on soft clay deposits. Soils and Foundations. 26:60-72.

Cividini,A.,Jurina,L., and Gioda,G. 1981. Some aspects of 'characterization' problems in geomechanics. Int.J.Rock Mech. Min. Sci. & Geomech. 18:487-503.

Gioda,G.,1979. A numerical procedure for defining the values of soil parameters affecting consolidation. Proc. of the 7th European Conf. on Soil Mech.&Found. Engg. 1:169-172.

Gioda,G. 1985. Some remarks on back analysis and characterization problems in geomechanics. 5th Int. Conf. on Num. Mech. in Geomech. 47-61.

Zeng,G.X., and Gong,X.N. 1985. Geotechnical aspects of soft clay ground under tanks. Proc. 11th Int Conf. Soil Mech. Found. Engg. (4), 2291-2294.

Zeng,G.X.,Pan,Q.Y.,Hu,Y.F. 1986. Behavior of excavation with sheet piling in soft clay. Proc. of Int. Conf. on Deep Found., Beijing. 3.1-3.6.

Numerical Methods in Geomechanics (Innsbruck 1988), Swoboda (ed.)
© 1988 Balkema, Rotterdam. ISBN 90 6191 809 X

Use of the calculated soil seeping consolidation in the analysis of the measured pore pressures

S.V.Bortkevich & N.A.Krasilnikov
Scientific Research Centre of the 'Hydroproject' Institute, Moscow, USSR

ABSTRACT: In a comparison of the results of calculation of the soil seeping consolidation by the finite difference method and the observed data it was possible to verify the method of calculation, to specify the estimated soil performances, and to find defects in pressure cells.

It is feasible to apply Florin's solution (Florin 1961) to the soil seeping consolidation theory modified by introducing some empirical positions substantiated by full-scale observations of dams (Krasilnikov 1973) to analyse the full-scale measurements of pore water pressure in earth dams and foundations, formed of saturated plastic clay soils.

The following conditions can be considered in the pore pressure calculations:

a) the arbitrary configuration of the calculated area;

b) the actual schedule of the dam construction;

c) the actual schedule of the impoundment and, if necessary, the drawdown;

d) the dam body in height can be divided into several zones with different seeping, and physical and mechanical soil properties. The soil in each zone can be anisotropic due to seepage. In particular, it makes it possible to differentiate in height the density and humidity requirements for dam cores;

e) the dam design scheme can incorporate a drainage of arbitrary configuration, which works constantly or till the given moment of time;

f) the limit pressures can be time-dependent. The curtain grouting in the foundation can be taken into account.

The redistribution of pore pressure between the core and the shells can be also considered in the calculation of pore pressures in dam cores.

Pore pressures are calculated for the construction and initial operation periods until the steady-state seepage in the dam body is established. The typical calculation is presented in Fig.1. The pore pressure is found by solving a plane problem of the three-phase soil seeping consolidation theory by the finite difference method. For this purpose the area of calculation is divided into equal rectangular sections and the pore water piezometric heads are defined in the nodes of the net obtained, Fig.2a. There may be some 2000 nodes. The horizontal step of the net ΔX is constant for the whole calculated area, the vertical step ΔZ can be variable. The interpretation of the calculated net indices is shown in Fig.2b.

The piezometric heads in all the nodes are found in the equation:

$$H_{t+\Delta t,i,K} = H_{t,i,K}\left[1 - 2\alpha\left(\frac{z}{m^2} + 1\right)\right] + \frac{1}{\omega}\Delta\sigma^{*}_{t+\Delta t,i,K} + \alpha\left[\frac{z}{m^2}\left(H_{t,i-1,K} + H_{t,i+1,K}\right) + H_{t,i,K-1} + H_{t,i,K+1}\right]$$

where

$$z = \frac{K_X}{K_Z};\ m = \frac{\Delta X}{\Delta Z};\ \alpha = \frac{c}{\omega}\cdot\frac{\Delta t}{\Delta Z^2};\ c = \frac{K_Z(1+\varepsilon)}{a\cdot\rho_b}$$

$H_{t+\Delta t,i,K}$ is the piezometric head of water in the node i,K at the moment of time $t+\Delta t$;

$H_{t,i,K}$ is the piezometric head of water in the same node i,K at the moment of time t;

$H_{t,i-1,K}$ are the piezometric heads of water in the points nearest to the node $i-1,K$ at the moment of time t;

ε, a are the average values of void ratio and the coefficient of clay soil consolidation;

ρ_b is the water density;

Δt is the estimated time span;

K_X, K_Z are the average values of the clay soil seepage coefficient in the direction of axes X, Z;

ω is the coefficient which incorporates three-phase soils with $\omega = 1$ for

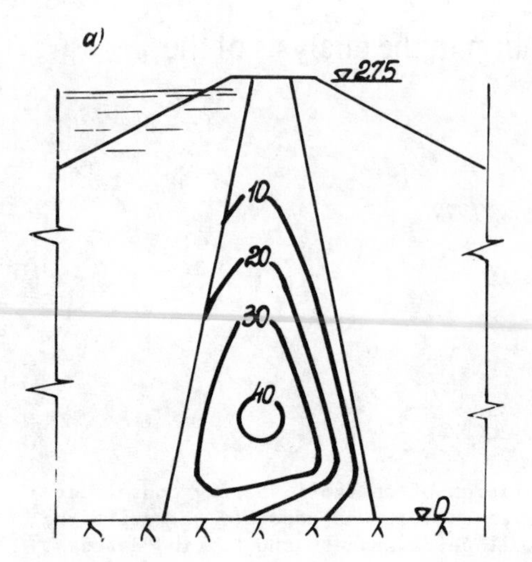

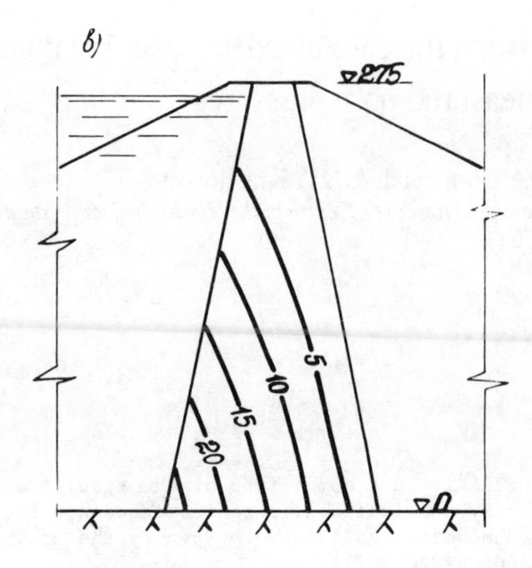

Fig.1 The calculated pressures in the core of the earth-and-rockfill dam
a) piezometric heads (10^5Pa) when the construction is over; b) when the seepage regime is stabilized.

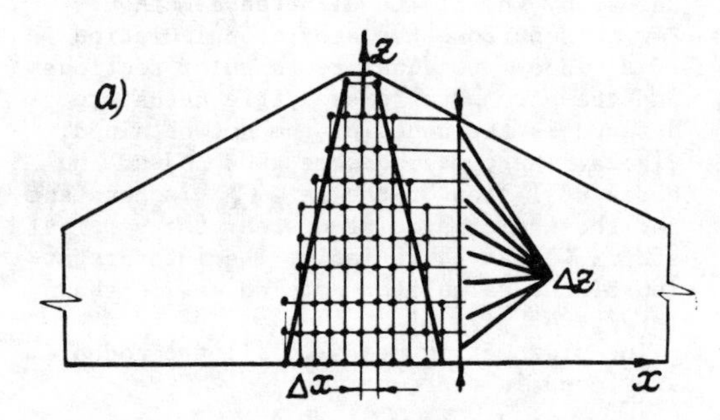

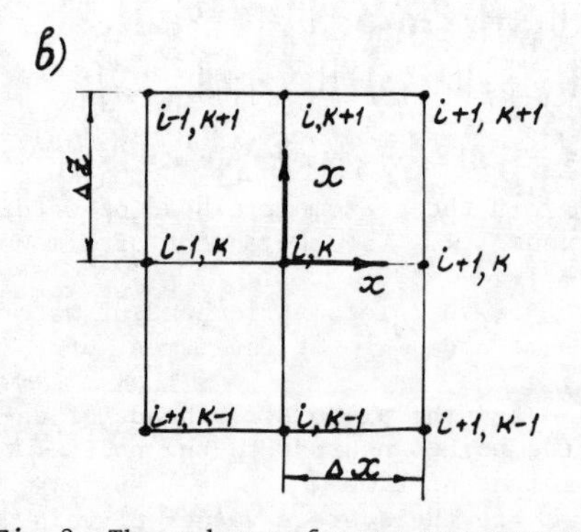

Fig.2 The scheme of pore pressure calculations in the dam core (a) and the designation of the rectangular net nodes when the pore pressure is calculated (b)

a two-phase saturated soil;
$\Delta\sigma^*$ is the variation of the medium stress σ_{med} in the node i,k at the moment of time from t till $t+\Delta t$ (an increment in load).

In the majority of cases the value $\Delta\sigma^*$ can be defined as the increment in load owing to the mass of the overlaid soil and water. When calculating the pore pressure of consolidation in the cores of earth-and-rockfill dams with regard to stresses redistributed in their bodies, the value $\Delta\sigma^*$ must be found by solving the continuum theory.

The calculation starts under the following conditions:

Initial conditions. The soil is saturated across the whole width, but the water pressure along the lower contact of the dam impervious element is zero. The increment ΔH in pore pressure (piezometric head) for the dam impervious element at time Δt is assumed to be equal to the increment of the medium stress value $\Delta\sigma^*$ divided by ω. In the specific case $\Delta\sigma^* = \rho\Delta h$ may be adopted, where Δh is the increment in height of the dam body at time Δt, and ρ is the soil density.

Boundary conditions.

1. The piezometric head of the pore water is taken as zero at dam crest, downstream face and part of the upstream face of the impervious element being not reached by the water level in the reservoir.

2. For the part of the upstream face of the impervious element which is currently below the upstream water level, the piezometric head is equal to the vertical distance of the point in question to the water level (at each given moment of time).

3. The basement of the impervious element can be pervious and water-resisting, or pervious at one part of the surface and

water-resisting at the other one.

4. The pore pressure in the nodes located on the horizontal interface of the areas with different seepage, and physical and mechanical properties is calculated with the formula:

$$H_{t,i,k} = H_{t,i,k+1} - \left(H_{t,i,k+1} - H_{t,i,k-1}\right) : \left(1 + \frac{1}{p}\right)$$

where $p = K_{z\text{в}} \cdot \Delta Z_H / K_{zH} \cdot \Delta Z_\text{в}$ where H is the lower area, and в is the upper area.

The whole calculation can be represented schematically in the following way: the selected step ΔZ determines the height of a lift laid with a definite time interval Δt_n according to the schedule of the dam construction. t_n and Δt_n are interpolated for each ΔZ ($n = 0,1,2\ldots$), then the value $\Delta\sigma^*$ is estimated for each net node, giving the value $H_{t+\Delta t,i,k}$ in each net node with $t+\Delta t$.

Using the impoundment schedule the corresponding water levels h_n and Δh_n are interpolated with t_n (an increment in water level as compared to the previous moment of time).

The value $\Delta\sigma^*$ for each net node during the construction period depends upon the dam body configuration and the impervious element, soil densities, and the water level in the reservoir.

For the initial operation period the pore pressure is estimated as the water level in the reservoir is fluctuating in an up-and-down manner.

The calculations are based on:

1. The dam diagram. Height, foundation length, spacing and configurations of inland drainage systems, their operation time.
2. The schedule of the dam construction and the data on the stress value in the impervious element of the dam for the whole period under review until the steady-state seepage is maintained.
3. The schedule of the impoundment in the construction period and the water-level fluctuations in the initial operation period.
4. The data on the boundary conditions in the basement of the impervious element.
5. The physical and mechanical properties of the soil in the dam body, namely

a) density, porosity and dry soil density for non-cohesive soils;

b) saturated soil density, average values of void ratio and the coefficient of consolidation, seepage coefficient in the horizontal and vertical directions, compression curve equation and the coefficient ω for clay soils in the impervious element.

There are some possibilities in the design procedure and program as to affect the pore pressure value in the core, and consequently structural stresses and deformations which are dependent upon pressure. Drainage systems, change in dam construction rate, etc., may help the cause.

In this context the calculated and measured results can be compared to define more precisely the calculation basis and estimated soil performances for the initial construction period.

The said design procedure of pore pressure studies is applied to interpret the field observations of clay core consolidation on two earth-and-rockfill dams.

One dam has a clay soil central symmetric core with a crushed aggregate and gravelly pebble side slopes. The yield strength of soil in the core is 0.19, the plasticity limit is 0.15, and the soil particles density is 2750 kg/m^3. The soil was placed in layers 0.3 - 0.5 m wide and roller-compacted. The typical results of the geotechnical control in the core is shown in Fig.3. The average density of dry fine earth is 2160 kg/m^3, its average humidity (for particles less than 5 mm in size) is 0.095 - 0.10, and the level of saturation is some 0.70. The optimum humidity is 0.10 - 0.14.

The dam core rests upon a concrete pad. Its surface under the core is covered by a special mastic, which makes it water-resisting. The pore water can only be forced out along the pipes of two strain

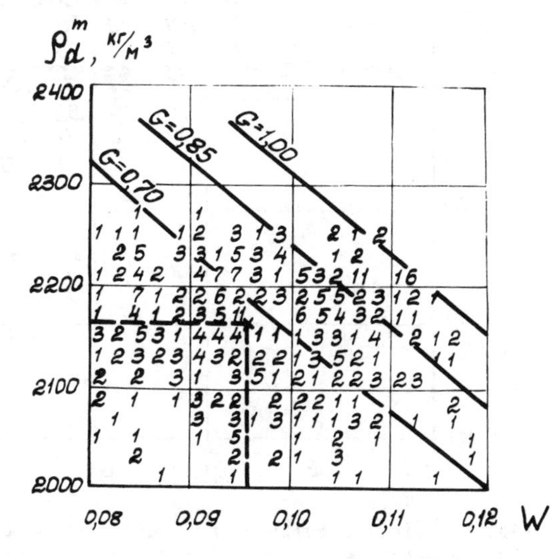

Fig.3 Typical results of the geotechnical control in the dam core: density ρ_d^m versus humidity W ($\times$ represents the average density and humidity for fine earth, G is the level of humidity)

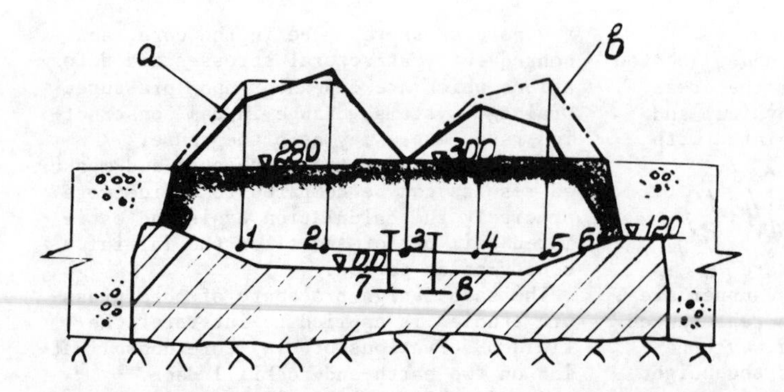

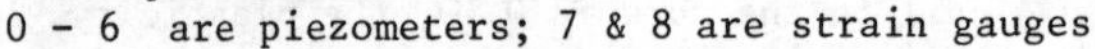

Fig.4 Comparison of the measured (a) and calculated (b) values of pore pressure in the lower part of the dam
0 - 6 are piezometers; 7 & 8 are strain gauges

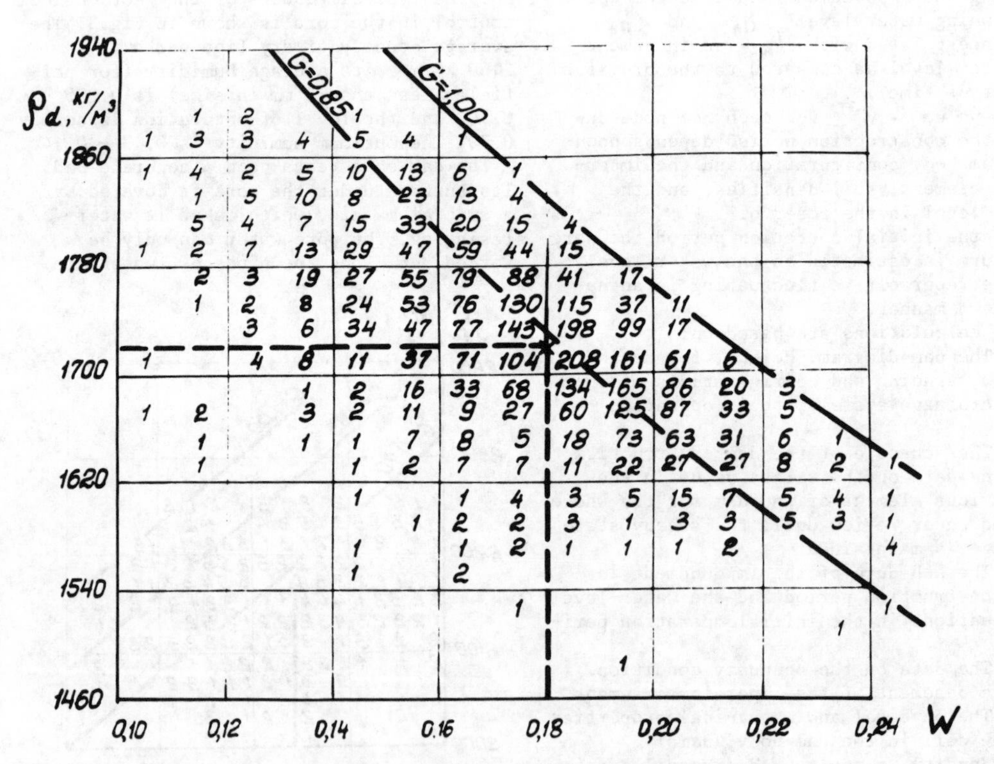

Fig.5 Typical results of the geotechnical control in the dam core: density ρ_d versus humidity W (x represents the average density and humidity for soil, G is the level of humidity)

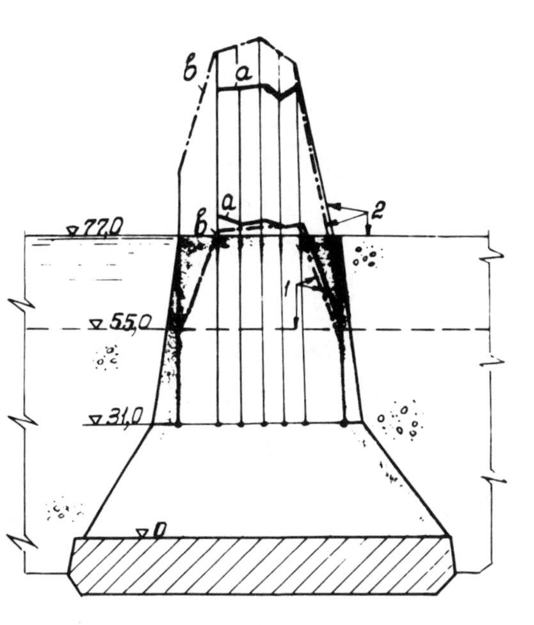

Fig.6 Comparison of the measured (a) and calculated (b) values of pore pressure in the core of the dam
1 & 2 are the stages of construction

gauges, Fig.4. Seven wire piezometric strain gauges were embedded into the lower part of the core to register pore pressure. Some results of the measurements and the arrangement of the gauges in the core are presented in the same figure. It was found that the pore pressure was 23-48 m of the water column or 0.45-0.95 of the overlaid soil mass above the concrete pad in the lower part of the core that was 28 m high at that moment. The upstream and downstream water levels were below the top of the pad through all the measurements. The actual coefficient of soil consolidation was $3.10^{-6} m^2/s$. This corresponds to the seepage coefficient, that is a little more than $1.4.10^{-9}$ m/s for the average density of dry fine earth, 2160 kg/m^3, and the initial void ratio
= 0.278.

The other dam has a loam central core with sandy gravel transition zones and stone lateral shells. The core also rests upon a concrete pad. The physical properties of the soil are the yield strength, 0.30, the plasticity limit, 0.19, and the soil particles density, 2710 kg/m^3. The loam was placed into the dam core in layers, 0.30 m thick, and then rolled by dump-trucks MAZ-525. The average density of the compacted soil skeleton is 1730 kg/m^3, and the average humidity is 0.15-0.16. The characteristic results of the geotechnical control in one of the dam sections are shown in Fig.5. The comparison of the calculated and measured pore pressure results for the 77 m high core is presented in Fig.6.

The observed results revealed that the vertical stresses σ_z, in the core mass including the horizontal stresses σ_x and those on site σ_α, inclined at 45° to the horizontal, are roughly equal and close to $\rho \cdot h$, where h is the height of the overlaid soil column. In this case the medium stress is

$$\sigma_{med} = \frac{\sigma_z + \sigma_x}{2} \approx \sigma_z$$

It was found that in the middle of the core at 20.0-40.0 levels the pore pressure along with stresses is getting closer to the overlaid soil mass. The highest value of the pore pressure coefficient is 0.85-0.90. After the impoundment the pore pressure was slightly increased at the upstream face of the core and continued to dissipate at the downstream. It was observed that the rock foundation of the dam is pervious and should be considered as a drainage with a variable head. The coefficient of the core soil consolidation is practically close to 1.10^{-6} m^2/s, that is the value obtained in laboratory.

The comparison of the calculated and measured values obtained during the construction and operation of the dams showed that pore pressure in the core was changing in a manner that conforms to the theory of the seeping consolidation of a saturated soil.

REFERENCES

Florin, V.A. 1961. Soil mechanics fundamentals. Vol.2. Gosstroiizdat.

Krasilnikov, N.A. 1973. Experience in calculating pore pressures in the core of earth-and-rockfill dams. No.32. Moscow, Trudy Gidroproekta.

Numerical Methods in Geomechanics (Innsbruck 1988), Swoboda (ed.)
© 1988 Balkema, Rotterdam. ISBN 90 6191 809 X

Prognosis determination of rock state of stress deformations and stability development on models in connection with the construction of tunnels

F.Nazari
Chair of Geotechnics TU, Brno, Czechoslovakia

ABSTRACT: The article presented summary of the results obtained on models, solving some problems of the Strahov-tunnel construction, in Prague. The methodology of simulating and prediction of rock state of stress, deformation and stability in the course of the technological driving were elaborated complexly. These models indicate the problem solution procedure from the illustration of the inputs, in situ measurements, to physical and mathematical modelling.

1 INTRODUCTION

The Strahov-tunnel the largest road tunnel in Czechoslovakia, is an important part of the main traffic circuit of the Prague basic road network. The main technological problem of this working consists in the relatively complicated rock state of stress and resulting the requirements on the tunnel driving method. The tunnel structure proper is 2005 m long; it consists of three two-lane tunnels; of the entire length 1536 m are driven by the core driving method and 469 m are being excavated. Through the tunnel tube a two-lane road - 8 m wide - of the medium city traffic circuit is led. The minimum vertical headroom clearance is specified to 4.80 m. The driven tunnels have a constant upgrade of 3.2 percent from south to north; the roadway in the tunnels has a cross-slope of 2 percent. The three tunnel tubes are parallel; the distance between them was established with regard to the rock pillar bearing value to 35 m.

2 GENERAL CONSIDERATIONS

At the problem formulation it was necessary to start from data known. For a tunnel in general the basic data are the tunnel dimensions, the geological conditions in its alignment and the tunnel construction technology. But the natural conditions have a dominant importance, as they may in principle influence the tunnel dimensions and the choice of the construction technology. This turned finally out even in the case of the Strahov-tunnel, where is became necessary to change the project from the initially designed two three-lanes tunnels to three two-lane profiles. The (engineering) geological tunnel alignment evaluation was based on a detailed geological survey. Its results have shown, that the geological terrain structure is very complex and that the tunnel construction will have to be realized in difficult conditions. The bedrock of the special-interest territory consists of Ordovicien rocks and Upper Cretaceons rocks. The tunnel will traverse along its line the following layers of the Ordovicien: Skalec-, Šárka-, Bohdalec-, Záhořany- and Vinice strata. Between the Šárka and Bohdalec - layers the tunnel will traverse the so called Prague fault. The rock cover of the Strahov tunnel consists of three significantly different layers. The tunnels proper will be driven in the ordovicien rock layers. The ordovicien rocks are superimposed by the cretaceons strata group and on it - the covering formations are bedded. By the tunnel construetion the most influenced will be of course its immediate surroundings, i. e., the ordovicien strata groups, in this case concretely - the Bohdalec strata group. Their relative and real small strength will be the cause of a considerable area of loosened rock, which will result in pressure on the tunnel lining. The precondition for rock vault formation, i. e. sufficient rock cover height is fulfilled but by the action of stress concentration - the rock will be considerably strained even before the strength limit is attained. The more strong-cretaceons strata group will act statically as a slab, loaded by the weight of the rock cover and own weight, the subgrada of which in a certain area will subside by the effect of tunnel driving. The cretaceous formation has aquifers, and at its deterioration by fissures, the arises the danger, that the water will penetrate into the Bohdalec argillaceous slates; their transfer into the overburden cretaceons strata group requires a special attention for its relevancy. When two or more parallel tunnel tubes are constructed at the same time, they interact. When constructing the second tube the loading of the first tube lining riges substantially as well as the rock deformation. The second tube is driven in more difficult conditions, because the initial undisturbed rock stressing increase by the stresses concentrated in the surroundings of the first tube. It is necessary to find out the pillar width, which would still fulfill the condition that the limit rock strength in it is not exceeded. The condition of rock bearing value in the pillar is decisive for the distribution of stresses and deformations around the tunnel breakings. When the limit rock strength is exceeded, the pillar loses its bearing function and this manifests itself very unfavourably on the stress and strain magnitude in the tunnel overburden. The problem of the dimensions, i. e. of the rock pillar bearing value is thus closely bound to the foregoing problem of deformation distribution and their transfer into the overburden layers. It is however also important for the driving of the second tunnel, in which the increased stresses may impair the stability conditions. The problem of rock stability is significant for the tunnel construction from the operation safety point of view on the one hand, but above all from the point of view of work mechanization and its maximum effectiveness. The transition from full face heading to driving by divited working faces manifests itself by greater labouriousness and lowering of the output. The endeavour to open the entire section at once, or to derive the section as little as possible and to utilize at maximum the rock bearing value is therefore comprenensible. A realistic prediction of rock stability is however very difficult, because it is necessary to take into account quite a series of interacting factors. These are first of all the dimensions of the not wainscotted tunnel face, the rock strength, its original state of stress and the size of stress concentrations called forth in the rock by excavation, and the stiffness of the temporary tunnel lining /8, 9, 10/.

3 MEASUREMENT OF ROCK PRESSURES IN SITU

The measurement of rock pressure on the pilot adit outfit should give a first picture of its magnitude and time behaviour. Two measurement profiles have been installed in the area of south section and two profiles in the area of the pilot adit berth section. The length of south part of the pilot adit is 539 m. The temporary support consists of steel-arc support of the type K-24-00-0-06, built-into a steel band of rolled channel. Concrete casints serve for lagging. Contact stress sensing units with remote read-off, Geo-Brno-83-D-S, were used for the measurements. The sensing units are placed between the steel arc and the reinforced concrete chasing, always five pickups is one measuring section. Their detailed location is evident from fig. 1. The arc footings, carrying the manometric boxes - are bedded on mechanical dynamometers Geo-Brno-D-79, registering the magnitude of the support reaction. The rock pressure is the force action of the rock on the tunnel lining /3/. It begins to act in the moment of outfit activation and its magnitude changes

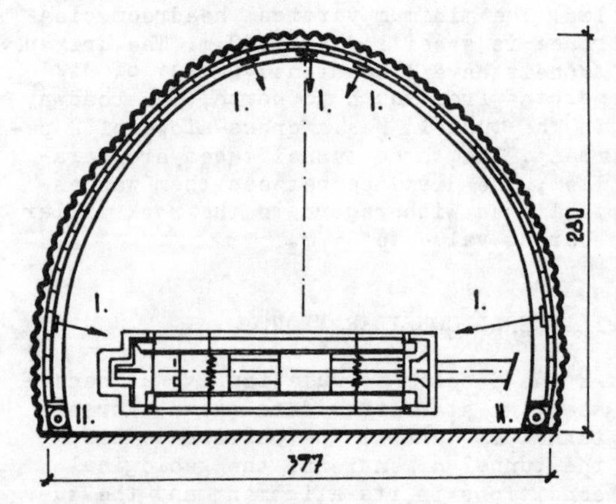

Figure 1. Pilot adit cross section, in situ measurement: I - manometric box, II - dynamometer tubes

with the progress of further tunnel excavation and with time. In order to take its real magnitude the measuring sections were established (I-manometric box II-dynamometer tubes - B) on the adit face immediately after its excavation. On the basic of measurement results and their evaluation the following conclusions can be accepted: In the first stages after installation of the gauges, the rise of stress is influenced in the first place with the distance of the tunnel face consequently by the tunnelling progress. At the daily progress speed of two to three meters the drifting effects cease to act and the pressure changes are influenced by the pure time factor only. In the north section of the Strahov-tunnel a maximum contact stress component of 0.09 MPa has been measured. The maximum vertical force in the arch apex has the magnitude of 0.045 MN. The support reactions are: Right 0.075 MN, left 0.080 MN. In the profile Strahov-South the contact stress component was measured 0.187 MPa total arch load 0.227 MN and the right side registers reactions of 0.096 MN.

4 THE PHYSICAL MODELS

At the overburden of max. height 85 m, five physical models with different equivalent material were used in various scales from 1:50 up to 1:100 of the standard section. The experiments were executed in the laboratory of physical modelling of the Institute of Geotechnics of the TU Brno. The entire complex of problems, connected with road tunnel drifting in rocks of minor strength and stability was consecutively solved on the described models. The results of the modelling may be formulated as follows: From the shape point of view - the convenient section is of circular or of horseshoe shape with invert. With rapid drifting - the section with the diameter of 3 m had longterm stability and the length without outfit may be relatively large. When the excavation diameter is enlarged to 5 m, the face distance without lining must not surpass one meter. The optimum rock pillar thickness between a pair of tunnel tubes was determined by model tests. Such a pillar thickness was looked for, which would warrant its bearing value, but at the same time - would be acceptable from the point of view of tunnel alignement. As minimum width - the width of one tunnel diameter was studied. The pillar of such tunnel tubes is not sufficiently dimensioned; at drifting of the second tube the rock in the pillar is crushed and the overburden deformations would increase considerably. When the pillar width is enlarged to 1.5 tunnel width, the stress in the pillar drops to 57.5 percent. Sliding surfaces appeared in the model material, which partly affected the pillar mass integrity. From the pillar bearing value point of view its limit bearing value is evidently given by the thickness equal to 1.5 excavation width. Within the framework of physical models, the problem of the tunnel surrounding rock deformation influence to the straining of the overburden cretaceons strata group was studied. It follows explicity from the results of tests on models, that at normal tunnel drifting procedure - the cretaceons strata group is deformed without failure. With two models, was therefore the failure of the cretaceons formation called forth artificially by model overloading and deliberate sinking first of one and then of the second tunnel tube. At two fold model load increase and complete overburden sinking over one tunnel - the cretaceous strata group fails by a tension fissure up to two third of its thickness. At complete sinking of the second tube - additional vertical fissures are forming and the Trompeter zone begins to form itself, i. e. the horizontal shear failure. On sinking of both tunnel tubes - the rock pillar gets also disturbed. As valuable we consider the establishment of the stress components course on the pillar bettween the tunnels, on the basis of which the optimum pillar thickness was established and was decided to construct three two-lane tubes instead of the initially intended two three-lane tubes. The height of overburden at the north portal is 18 m. With regard to the location of underground engineering networks, traffic routes and in order to limit the deformations and the total risk, it is necessary to realize special measures, i. e. to construct a protective umbrella. In the broader surroudings of the area of interest, 50 meter in length, there are the ordovicien slates, constituted of Vinice-layers. On the first 38 meters - there are weathered slates, consisting of a rock, disintegrating into chips, with soft fragments and loany filling. On the next 22 metres there are partly weathered slates, disintegrating into fragments and pieces. In the northern transitory section of the Strahov-tunnel

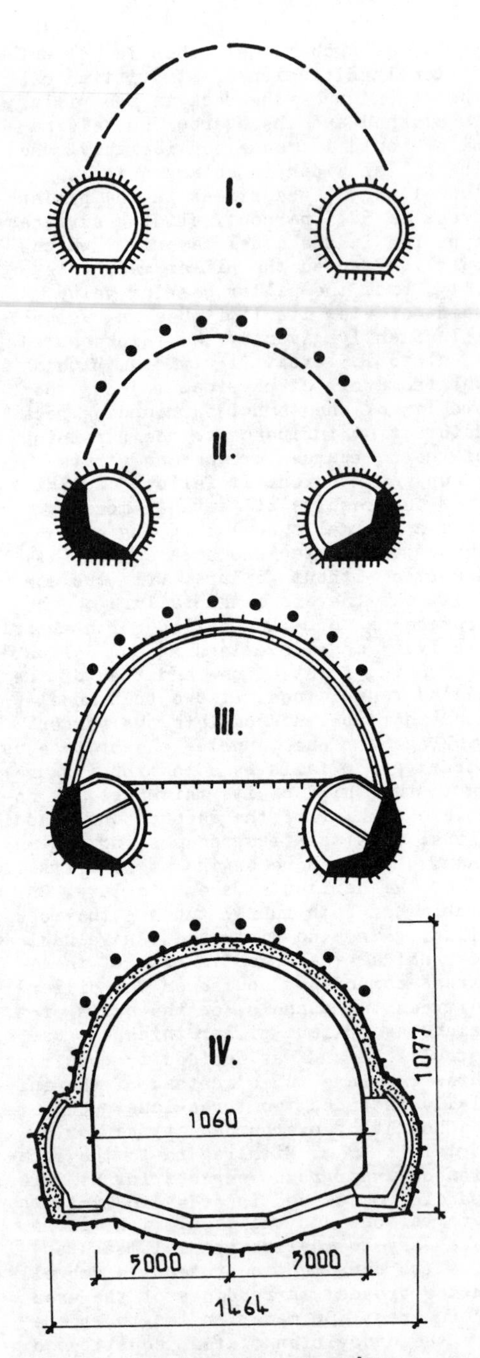

Figure 2. Construction procedure:
I - excavation of the abutment tunnels, lining. II - abutment concreting and pile jacking. III - callote breaking, prefabricated lining and tunnel excavation. IV - excavation of tunnel invert and prefabrication of lining.

it is necessary to reinforce the rock medium by means of an umbrella of horizontal jacked piles. It is a matter of 8 piles with 1 m diameter, jacked through in a length of 50 m behind the tunnel portal. For the solution of this problem - experimental physical modelling is most suitable. It was therefore realized on the Institute of Geotechnics of TU Brno for the section mentioned, named Geo-Brno-86. The models were built in the scale without the protective umbrella - and with this umbrella. The model was built in a stand with the dimensions of 200x200x100 cm. The drifting technology was executed according to fig. 2. After the execution of the protective umbrella structure the drifting was started by means of top shield and arch area material excavation. The tunnel drifting was executed in four phases. The modelling task was to determine objectively the lining system state of stress and strain with the surrounding rock mass and to utilize it for a safe and economic tunnel execution. The autor considers as valuable findings by the comparison with the model without protective umbrella: the course of settlement in dependence of the drift progress, when the value of settlement decreases by 41 percent; at limit state of loading by 182 percent, after rotation of abutments, displacements on the face wall and solution of the structure system (mechanical) behaviour i. e. of the tunnel lining and umbrella and rock /4, 5, 6, 7/.

5 MATHEMATICAL MODELLING

The essence of the finite elements method consists in the distribution of the continuum solution into discrete elements for which the conditions of behaviour under the influence of loading are determined. By arraingoing the mutual linkages of these elements for the entire continuum - a system in a set of equations is obtained. Their primary unknowns are - according to the model type chosen - (compatible, balanced, hybrid and mixed) either the fields of connection stress model displacement or a mixed field of displacements and stresses. At the formulation of mathematical problems it was necessary to go out from the mentioned inputs from in situ measurements and physical modelling. The mathematical model for the evaluation of the Strahov-tunnel cross section was executed by means of the Automatic

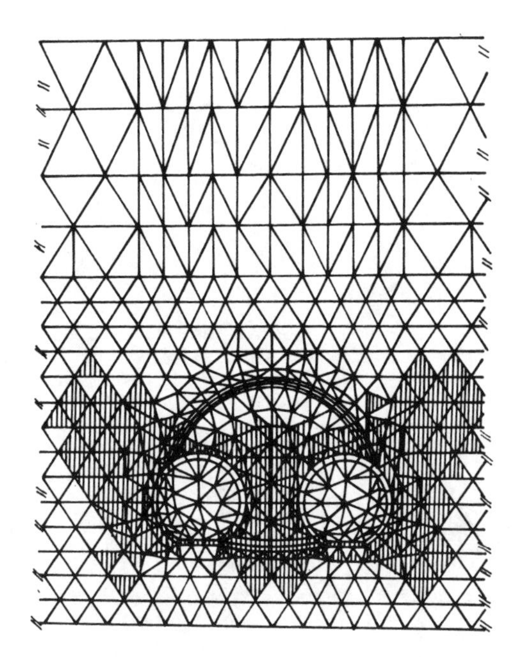

Figure 3. Mathematical model by means of finite elements method

design system, i. e. SAPRO-PODIS, fig. 3. The mathematical model has the shape of a rectangle 104 m wide and 59 m high. By triangulation of the area, thus established, a network resulted, in the extent of 1912 elements and 992 nodes /1, 3/. The contemporary version of the PODIS subsystem uses for the solution of obligatory and special tasks a hygrometric model with elements in the form of linear triangles and generally curvilined quadrangles with linear, qudratic and cubic representation. The form functions for these element types take into consideration two unknown quantities in the nodal points - the displacements in the directions of the global axes x and y and secure the continuity of the first derivations according to the displacements. Evaluation of the gained results of input values are the stress and vector components of nodal point network displacement in the plane of the coordinate system. In an elastic isotropic medium - the deformations are in linear relation with the stress, but if the medium behaves also plastically, it is necessary take into consideration also nonlinear relations. Kondner formulated for cohesive and cohesionless soils a hyperbolic relation of stress - deformation, Kondner's soil model was adjusted for the method of finite elements - as first - by Duncan and Chang, who derived from the dependence an expression for the tangential modulus of elasticity and used this dependence together with Mohr's criterion of limit plastic equilibrium for the formation of a more general mathematical soil model. The writer verified the convenience of this model and used it for the solution of the concrete model Geo-Brno.

6 CONCLUSION

A complex methodology for the problem solution of the structure interaction with the rock medium was elaborated, consisting of the mathematical and experimental solution and of measurements in natural conditions and on the structures in execution for the years 1986-1990 /2/.

REFERENCES

/1/ I.Kameníček: Mathematical modelling at Strahov-tunnel constructions, ČSVTS Prague, 106-115 (1985)

/2/ I.Kameníček, F.Nazari: Project of checking measurement at Strahov-tunnel construction, PÚDIS, 1-85 (1985)

/3/ I.Kameníček, F.Nazari, V.Zadinová, J.Hudek, O.Tesař, J.Vorel: The Strahov-tunnel, Annual reports, PÚDIS, 1-180 (1982-1987), 1-75 (1985-1987).

/4/ F.Nazari: The umbrella vault method at tunnel drifting under the protection of semicircular shield, ČSVTS, 200-205 (1987)

/5/ F.Nazari: Modelling of geomechanical problems, Czechoslovak Academy of Sciences, Prague, 126-129 (1987)

/6/ F.Nazari: Stress and strain investigation in the umbrella arch method on a tunnel, IMEKO, 429-435 (1987)

/7/ F.Nazari: Possibilities analysis of mathematical modelling application at the solution of underground walls, NUMEG 87, 50-52 (1987)

/8/ F.Nazari: Conception problems solution comented with the Strahov-tunnel construction on models, ČSVTS Brno, 145-148 (1984)

/9/ F. Nazari: Geotechnics I., textbook, Brno, VUT, 1-112 (1980)

/10/F.Nazari: Tests on site, application of and comparison with theoretical consideration, ITA - Prague, 98-105 (1985)

Numerical Methods in Geomechanics (Innsbruck 1988), Swoboda (ed.)
© 1988 Balkema, Rotterdam. ISBN 90 6191 809 X

Numerical calculation for synthetic analysis of the crustal stress field at Laxiwa dam site

Liu Shi-Huang
Northwest China Hydropower Investigation and Design Institute, MWREP, Xi'an

Abstract, Using numerical solution to redo the regional tectonic and crustal stress fields, the anthor has studied the regional geological stability, dominant direction of tectonic stress and distribution of crustal stress in the Laxiwa project area. These findings are quite agreeable to the results of other analysis.

1. Introduction

The proposed Laxiwa Hydropower Project is located in Qinghai Provice of Northwest China, 3 2 . 8 Km downstream from the Longyangxia Hydropower Station which is near its completion and 4 0 Km upstream from the Lijiaxia Hydropower Project to be constructed in the near future. It will be the largest hydropower station on the Yellow River, with a 250 m high double curvature arch dam and an installed capacity of 3720 MW (a final installed capacity of 7000MW).

The investigation of the Shimen damsite indicates that the measured crustal stress amounts to 200kg/cm², over two-four times the dead weight of the superstratum. In 40 per cent of boreholes, 5 6 0 4 core cakes have been discovered. During excavation rock debris spurted. The above mentioned facts indicate the Laxiwa Hydropower Project will be founded in rockmass with high tectonic stress.

The following questions might be presented to construction a large hydropower project in high crustal stress field,

1). The amount prevailing direction and distribution law of crustal stress in this area;

2). How to evaluate the differences in amount and direction of the measured crustal stress at Laxiwa and nearby Longyangxia and Lijiaxia;

3). The effect of high strain energy at Laxiwa on regional geological stability;

4). In given crustal stress field, how to assess the effect of crustal stress on layout, design, construction and operation of hydraulic structures.

The answers of the foregoing questions lie in study of the relationship between the project and the crustal stress field at Laxiwa. In order to solve this problem, we have collected a lot of tectonic stress at Laxiwa; conducted analysis of core-cakes and back analysis of tectonic stress field and crustal stress field at the project site by means of surface tectonic traces, mechanism of seismic focus and measured crustal stress. Based on the results of above analysis, a deepgoing study has been made on the layout and design of the project. The paper only deals with the numerical calculation for synthetic analysis of the crustal stress field.

2. Geological Background

The dam and reservoir of the Laxiwa Hydropower Station situate in the southern Qinghai Mountain India-China geosynclinal synclinorium of the Songpan-Ganzi geosynclinal system. It can been seen clearly in ~~satellite photo that the~~ NWW Buqingshan and Animaqingshan Mountains, the NW southern Qinghai Mountain and laji Mountain, the NNW Elashan and Gangchasishan Mountains form into a rhombohedral net geomorphologically and cut the Qinghai-Tibeten massif into a rhombohedral block along with their corresponding faults at bottom. Laxiwa is located in the northeast of the block.

From tectonic traces and evolution of earth crust, it has been found that the dominant direction of tectonic stress in this region has changed for five times from the Proterozoic era to recent, SN→ NNE→NE→NEE→NE→ or NNE. from investigation of seismic deformation zone, topographical deformation survey and mechanism of seismic focus, it can be seen that the seismic origin in this mainly the NWW faults. The prevailing direction of tectonic stress is in NE direction, P-axis elevation angle 20°.

Table 1. Measured crustal stresses at Laxiwa Longyangxia and Lijiaxia

Location	NO of data grout	Thickness of overlying strata (m)		σ_1	σ_2	σ_3
LaXiWa	1	278	quantitykg/cm	228.7	132.9	94.7
			direction	N10W	N69E	N33X
			dipdirection	NW	SW	SE
			dip angle	41°	11°	46°
	2	236	quantitykg/cm	227.0	186.4	131.4
			direction	N22W	N88E	N28E
			dipdirection	NW	NE	SW
			dip angle	33°	27°	45°
	3	200	quantitykg/cm	205.0	140.3	57
			direction	N12E	SW262	151°
			dipdirection	NE	SW	SE
			dip angle	39°	22°	42°
LongYangXia		68	quantitykg/cm	41.1	18.6	4.1
			direction	NE83	SW202	NW310
			dipdirection	NE	SW	NW
			dip angle	47°	23°	32°
			quantitykg/cm	18.1	10.2	6.9
			direction	245°	57°	155°
			dipdirection	SW	NE	SE
			dip angle	14°	75°	2°
LiJiaXia		130	quantitykg/cm	55	45	28
			direction	58°	275°	168°
			dipdirection	NE	NW	SE
			dip angle	43°	41°	20°

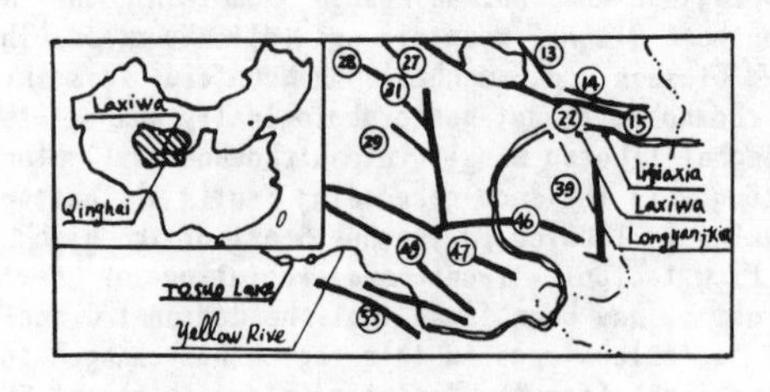

Fig 1. The tectonic systems at Laxiwa

(46) Zhong tie WE fault.
(47) Tosuo Lake-Maxin Curve fault.
(49) Bugingshan-Chama fault.
(51) Maduo-Gande fault.
(31) Elashan Hot spring fault.
(13) Riyueshan fault zone.
(34) Gangchasi-Duofuton fault.
(22) Daotanghe-Xunhuananshan fault.
(14)(15) Lajjishan fault zone.

Aeromagnetic survey also indicates that for nearly two million years the gravity faults have shifted north 30-40 km (maximum 90 km) compared with the surface deep faults. Meanwhile the southern of block has shifted east or southeast as compared to the northern of block. his movement is differential. The movement of the west is greater than that of the east.

3. Numerical Analysis of Regional Tectonic Stress Field

It is a major subject of geomechanics to redo the tectonic stress field by using surface tectonic traces and data of crustal stress and to probe into the internal forces that caused deformation. Since the western part of China lacks data of topographical deformation rate and the data of measured crustal stress at Laxiwa etc, are subsurface crustal stress effected by the topograpyh of river vally, it can not be based on these data to redo the regional tectonic stress field. Therefore, the calculation is only a qualitative analysis. In order to make the deformation of this region as close to its geological background as possible and the direction of theoretical principal stress coincide with that of P-axis of the solution of mechanism of seismic focus, a calculation domain of over 300 km in WE length and over 200 km in NS width is considered and a numerical model is suggested. The assumptions are;

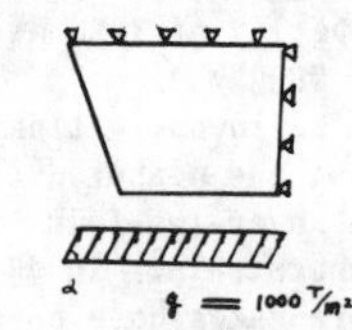

1). This rhombohedral block is only a small part of the Qinghai-Tibet Massif. It can be simplified that load is unifomly distributed on the force transfering boundary.

2). In view that the block moves northward along with the massif and the east part of the block is squeezed by the Sichuan and Gansu blocks, its east and north boundaries are confined.

3). This block is assumed to be composed of isotropic elastical materials and data of in-situ point load tests are adopted for their mechanical parameters.

4). The effect of topographical deformation on tectonic stress is neglected and a deep horizontal section is selected as calculating section.

By repeatedly adjusting the boundary confinement and load direction, it has been found that if $\alpha=60°$, the theoretical direction of tectonic stress at Hualong is NE 39.42° which is close to the principal stress direction NE 44° of the earthquake occurred on December 22, 1968. The rhombohedral block moves northward, but its west part moves larger than its east part and the east part relative to the west part tends to shift towards southeast. The NWW, NW Toso Lake Maxin fault and Lajishan fault are compressed and their relative displacement is larger than that of the

NNW Elashan and Gangchasi faults. Thus the result from the foregoing computations are considered practicable.

Through computation the following conclusions can be drawn.

1). The tectonic stress of this region is compressive stress. The dominant direction of regional tectonic stress changes gradually NE 30° →NE 40° from west to east. The value of principal compressive stress decreases slightly from west to east, but the decreasing magnetude is small.

2). the NNW Longyang fault group in this region exerts more or less adjustment on regional tectonic stress either in direction or in amount, but the effect of adjustment is small.

Laxiwa is not a high stress concentrated area independently of Longyangxia and Lijiaxia, its strain energy slightly lower than Longyangxia. Only because the rock mass at Longyangxia and Lijiaxia is relatively fractured with poor conditions for energy storage and the measuring points of crustal stress are distributed shallowly, the crustal stress at Laxiwa appears higher.

3). Located in the southeast of the rhombohedral block, the NWW Toso Lake-Maxin fault zone under the action of the NE tectonic stress becomes "a line of defence" to be firstly affected by transferring force. Strain energy concentrates at Toso Lake, resulting in large earthquakes. Just because energy concentrates and releases at Toso Lake, and the Toso Lake earthquake occurred in concert with the 歹-system Longmenshan earthquake in Sichuan and independent of the NNW tectonic movement, the internal stress is reduced and the most part of the region is under moderate energy condition. Therefore it is considered that the so called high crustal stress at Laxiwa will not exert remarkable effect on the regional stability of this region.

The above mentioned conclusions on tectonic stress direction are in agreement with the conclusions on mechanism solution of seismic focus and on the tectonic stress direction derived from the river flow direction. Our knowledge about the regional stability is also in concert with the conclusions on regional geological survey, observation of fault activity, determination of the age of recent fault activity and law of earthquake distribution.

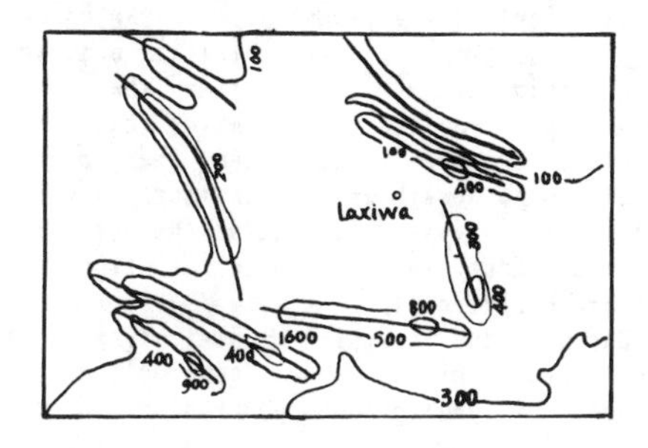

Fig 2. Isogram of strain energy in the eastern Qinghai 10^{11} erg

4. Multielement Regression Calculation of the Crustal Stress at the Project Site.

In addition to the calculation of regional tectonic stress field, calculation of the crustal stress field perpendicular to the river vally plane has also been carried out. In calculation the Shimen No. 2 exploring section is chosen as calculation section. The domain of calculation ranges 1725 m deep in river bed, 1400 m in left bank and 2510 m in right bank. The assumptions are as follows. The block is composed of different isotropic elastic materials and is calculated as a plane strain problem. The measured crustal stress data is considered as the basis of regression. The effects of temperature, seepage pressure etc are neglected. And the measured crustal stress is simplified as the sum of tectonic stress and dead weight stress.

$\sigma_k = b_1 \sigma_{k1} + b_2 \sigma_{k2}$

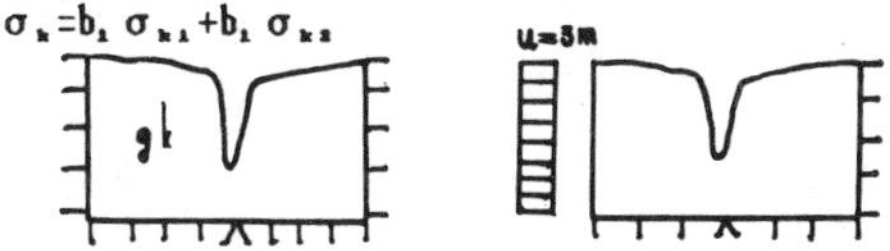

dead wight stress field   tectonic stress field

Where σ_{k1} theoretical stress in dead weight stress field;

σ_{k2} theoretical stress in tectonic stress field;

$b_1 b_2$ regression coefficient;

σ_k measured crustal stress.

There are n equations for n measured crustal stresses, and n dualistic equations with b_1, b_2 as unknown numbers. When sum of residue squared

$$Q = \sum_{K=1}^{n} e_k^2 = \sum_{K=1}^{n} [\sigma_k - (b_1 \sigma_{k1} + b_2 \sigma_{k2})]^2$$

is the least, $b_1 \sigma_{k1} + b_2 \sigma_{k2}$ is considered close to the practical crustal stress.

Through calculation, if $b_1 = 1.6$, $b_2 = 0.3733248$, the result of complex correlation coefficient R=0.9837 is good.

The results of calculation indicate,

1). The measured crustal stress at Laxiwa σ_1 / σ_3 is 2.58, while that at Ertan hydropower project and Lubuge Hydropower project is 4.819 and 3.148 respectively. The spherical stress tensor of the crustal stress at Laxiwa is relatively large while the stress deviator is small. The stress ellipse is a near spherical one.

2). The water surface of the Yellow River at Laxiwa is only 50 m wide and the river vally is 60°-65° in slope. Due to the effect of topography of the river valley, tectonic action decreases gradually. The direction of principal stress changes gradually from NE to SN, perpendicular to the river valley, and the dip angles approaches to the slope angle of the river vally.

3). Due to the effect of horizontal tectonic stress, there is an elliptical stress concentrated area in the bottom of the river valley. It ranges 300 m in length horizontally and 150 m in depth, σ_{1max} is 600*.kg/cm^2 and τ_{max} is 284.9kg/cm^2.

Table 2. results of multielement regression Calculation

poin No	Stress	Measured stress $[\sigma_k]$	theoretical stress $b_1 \sigma_{k1}$	$b_2 \sigma_{k2}$	regressed value $[\sigma_k^*]$
N01	σ_x	-152.82	-155.22	- 9.48	-166.70
	σ_y	-174.93	- 57.65	- 90.84	-148.49
	τ_{xy}	- 66.91	- 45.31	- 33.56	- 78.81
N02	σ_x	-171.40	-117.70	- 21.50	-139.20
	σ_y	-188.62	- 51.76	-106.54	-158.30
	τ_{xy}	- 40.80	- 43.12	- 44.86	- 87.98
N03	σ_x	-126.4	-105.12	- 33.10	-138.22
	σ_y	-144.88	- 50.30	-121.63	-171.83
	τ_{xy}	- 66.50	- 42.54	- 52.18	- 94.72
1#	σ_y	-174.7	- 59.68	- 94.03	-153.71
2#	σ_y	-165.0	- 54.90	-114.34	-169.24

Since stress is highly concentrated in this region, core cakes are richly deposited.

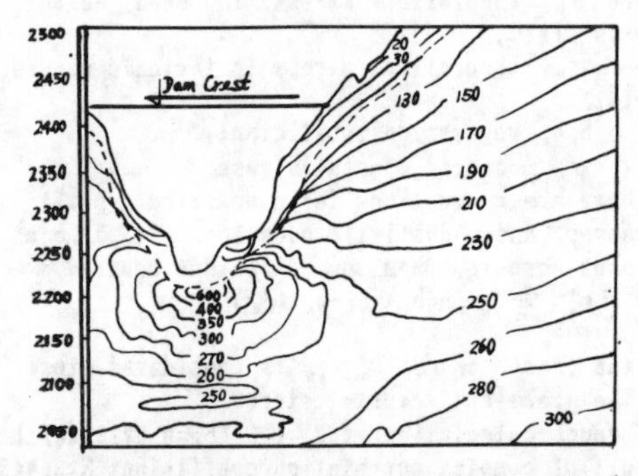

Fig.3. σ_1 isogran at Laxiwa shiman dam site kg/cm²

4). The data of measured stress and result of computation indicate that under the action of horizontal tectonic stress and the effect of the stress concentrated area in the river valley, the stress increases at the foot of bank slopes and the stress increase disappears gradually with the distance from the foot of bank slope. The concentration of crustal stress varies with the elevation of bankslops, stress concentration seems to disappear above 70 m from the river bottom, that is to say, there is no stress concentrated zone along bankslopes.

5). The confined pressure factor $k = \sigma_{total\ horizontal}/\sigma_{v\ total}$ is not a constant but increases with depth, and for a given point, that of different direction is also different.

6). The pluripoint mean value of $[\sigma_{(x)}+\sigma_{(y)}]_{dead\ weight}/[\sigma_{(x)}, \sigma_{(y)}]_{total}$ in the crustal stress field at Laxiwa is 0.68. Neverther less according to global contribuion analysis, V (dead weight)= 178429, V (tectonic) =175264, that is to say, the tectonic action and the dead weight action are comparable or the dead weight action is slightly larger than tectonic action.

5. The Effect of Crust Stress on Hydraulic Structures

The synthetic analysis of crustal stress is aimed at studying the effect of crustal stress on hydraulic structures, including layout of project, orientation of cavern axis, stability of rock around openings, stability of high rock bankslopes and potentiality of induced earthquakes in high crustal stress area etc.

1). Effect of foundation deformation on riverbed powerhouse.
The excavated plane of foundation inevitably lead to stress relief and stress readjustment, and this will result in foundation deformation.

The Gezhouba Hydropower Station is founded on clayey siltstone. Although the horizontal crustal stress is only about 30kg/cm², the bedrock rebounded 13 mm after excavation and some weak intercalations deformed for 7-8 cm along normal direction. The Grand Coolee Hydropower Station is founded on granite. The granite fractured in horizontal layers due to its stress relief. The measured horizontal crustal stress at Laxiwa is 170 kg/cm², σ_{1max}=600 kg/cm² in the riverbed. In addition, there is a fissure with low dip angle H_{f4} near the river valley. It is very difficult to avoid foundtion deformation similarly occurred at Grand Coolee and Gezhouba. In view that foundation deformation does not occur immediatly after stress relief but usually lags and sustains for a period, this will increase the construction quantity of foundation treatment for riverbed powerhouse.

2). Stress concentration in the foot of bank slope and semi-underground powerhouse.

Since the river valley is rather narrow, if the scheme of semi-underground powerhouse is adopted, two generating units each with a capacity of 620 MW have to be place at the foot of the bankslope and the 50 m long semi-underground cavern will happen to be located in high crustal stress field in the foot of the bankslope. σ_1=350 kg/cm², τ_{max}=150 kg/cm², 1.4 times that in bankslope, and the strain energy is 2.5 times that in the bankslope. From the view point of load conditions, it is much more complicated than overall underground powerhouse. Furthermore, the increase of crustal stress will highten the potentiality of rock burst, this will make construction all the more difficult.

Based on the foregoing crustal stress analysis including project layout, contruction etc, the scheme of underground powerhouse is finally selected.

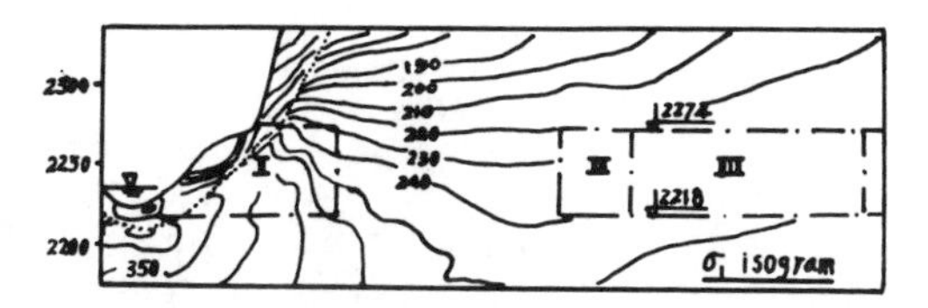

Fig 4. Effect of crustal stress on powerhouse.

I. Sime-undergroand powerhouse

II. underground powerhouse 200 m from weathered Line

III. underground powerhouse 250 m from weathered Line

3). The proper distance from uderground to powerhouse river bank.

From the concept that there are stress relief, stress concentrated and stress stablized zones in river bankslope of high crustal stress region, the underground poworhouse must be located in the stress stablized zone of a bankslope. Therefore, referring to the engineering experience of other projects, during the feasibility study of the Laxiwa project, the underground powerhouse was located 250 m from the river bank. However, the measured data of crustal stress at Laxiwa and computation indicate that there is no stress concentrated zone in the bankslope as the stress increase in the foot of the bank slope disappears gradually with elevation. Therefore, the underground powerhouse can be shifted to 200 m from the river bank. If the tailwater and load conditions can be satisfied, this scheme is more economic than the original one.

4). Prediction of rock burst.

From the measured data and computation results, it has been found that the crustal stress near the underground powerhouse and diversion tunnel amonts to 200 kg/cm². As the axis of diversion tunnel is normal to σ_1, and due to the effect of stress concentration, the crustal stress at arch foot and sidewall rims of the powerhouse and diversion tunnel might amount to 400-600 kg/cm², and the crustal stress at the intersection of the main powerhouse and the tailwater tunnel might be up to 1000 kg/cm². It is predicted that rock bursts of class II-III might occur during excavation.

Referenes

1. Supplemented preliminary design report on the Longyangxia Hydropower Station on the Yellow River-engineering geology, Northwest China Hydropower Investigation and Design Institute, MWREP.
2. Feasibility study repot on the Laxiwa Hydropower Project on the Yellow River intermediate report on engineering geology, Northwest China Hydropower Investigation and Design Institute, MWREP.
3. Map and interpretation of the tectonic systems and earthquke distribution in Qinghai Province, Geological Research Institute, Qinghai, China.
4. Summary on the regional stability and reservoirinduced earthquake for the Longyangxia Project, Chengdu Geology college and Northwest China Hydropower Investigation and Design Institute.
5. Effects of modern massive and faults in Qinghai Provonce, Li Changhui.
6. Discussion on the crustal stress field at Longyangxia and its surrounding area in Qinghai Province, Lanzhou Seismic Institute.
7. Report on the initial plane stress field for the Laxiwa Hydropower Project on the Yellow River, Tianjin University.

Numerical Methods in Geomechanics (Innsbruck 1988), Swoboda (ed.)
© 1988 Balkema, Rotterdam. ISBN 90 6191 809 X

Development of a computer system on monitoring geotechnical measurements

E.Lindner
Büro für Planung und Ingenieurtechnik GmbH, Grenzach-Wyhlen, FR Germany

ABSTRACT: GEOCONTROL is a newly developed software package in order to record and process geotechnical data of tunnels under construction. At the tunnel-construction-site a computer is installed that is connected to central-unit to provide communication between a specific construction-site, the building constructor and geotechnical surveyors or engineers. So the actual situation of a specific construction-site can be checked any time on terminals via remote data transfer. The geotechnical data obtained by GEOCONTROL are incorporated into a central data base in order to achieve long-term-analysis of geotechnical behaviour of the tunnel construction.

1 INTRODUCTION

Generally speaking, geotechnical measurements have the following tasks to fulfil in foundation and subsurface engineering

- checking the effectivness of the means of support used
- indicating major deformations or those which do not minimise during the course of time
- monitoring the influence of existing buildings
- subsequent assessment of the stability.

Accordingly, it is essential that data evaluation, processing, and assessment is carried out quickly and reliably.

In order to fulfil these tasks, the GEOCONTROL EDP system was developed on behalf of the Hanover/Würzburg Project Group - Centre of the Federal Railway Construction Office by the Büro für Planung und Ingenieurtechnik GmbH - b p i in order to evaluate and monitor geotechnical measurements. It was essential for practical application that the development of the programme took place through a working group of programmers and geotechnicans in constant touch with the construction site over a period of more than one and a half years.

The development of the system for application for traffic tunnels was completed and the system is in use in the case of 13 tunnel sites of the Federal Railway new route between Hanover and Würzburg and two tunnel sites of the new route between Mannheim and Stuttgart as well as the Dortmund Underground railway and Heslach road tunnel Stuttgart.

2 OBJECTIVES

The information and decision-making sequence during the evaluation and interpretation of geotechnical data should be supported through the development and application of the EDP system GEOCONTROL to realize the following objectives:

- securing a complete collection of the substantial data material which is always up to the minute and corresponds to the requirements
- objective selection as well as the rapid and standard evaluation and processing of all important data
- short-term procurement of standard and lucid principles for the interpretation of the data
- establishing a warning system independent of the interpretation
- determining the necessary measurements and monitoring their execution
- direct data transfer facilities to external decision makers.

3 THE EDP SYSTEM

The measured value evaluation is carried out either via direct data transfer of the electronic surveying instrument via cassette interface and data converter to the EDP unit or via manual input and mask-controlled operator guidance on the display.

The following data can be recorded daily:

- date of the measurements day
- drivage level of the crown, bench, floor
- for each read-off measured cross-section: measured cross-section number (station), time of the reading, data (settlements and convergences or horizontal displacements)
- in the case of newly set-up cross-sections: measured cross-section number, type of cross-section, excavation category, registration page number, roof overlay as well as alarm group and measurement interval control.

The logical programme sequence with card-controlled operator guidance assures fast and safe handling of the programme without special programming knowledge. In the process, numerous evaluation and output possibilities are available, which can if necessary be summarised as automatically running data packages.

Generally speaking, the following graphic charts can be produced:

- current tunnel longitudinal presentation of the latest measured roof settlements of all measured cross-sections with details of roof settling speed, measured value age class, roof overlay, position of drivage and marking of critical cross-sections for a selective tunnel section of 250 m length (Fig.1).
- path-time diagram of the roof and floor settlements, of the horizontal movements including information on the drivage level of crown, bench and floor, roof overlay, excavation class and number of plan sheet (Fig.2).
- Presentation of the cross-sectional deformations for required (possible) cross-sections and periods (Fig.3).

During evaluation, all obtained data are examined on the basis of prescribed multi-stage limit values and marked in the graphic charts and lists as critical should they be excessive. From the current measured value alterations, the necessary measurement intervals are determined and their observation monitored with the aid of an algorithm relating to face state, settlement speed, limit value and age of a measurement station.

In addition, the support registration pages of the selected means of supporting can also be evaluated. These data are then available if required at any time.

All measured data can be called up for any required measurement stations and periods thank to lucid lists.

GEOCONTROL provides free selection of single or multicolour output via display screen, printer or plotter.

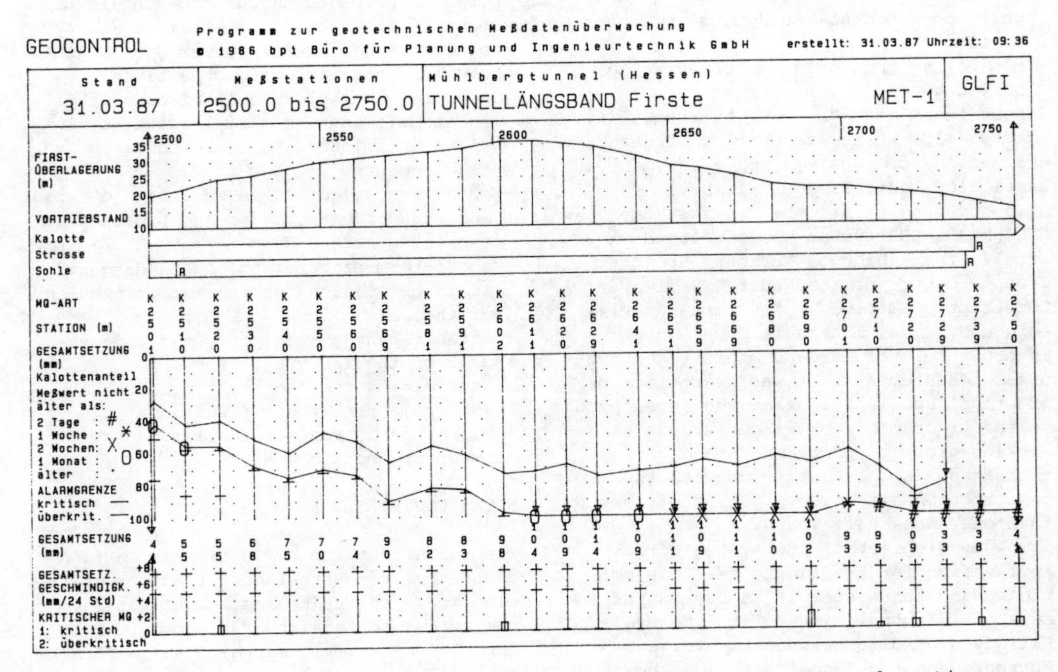

Fig.1 Longitudinal presentation of the roof settlements for a given tunnel section

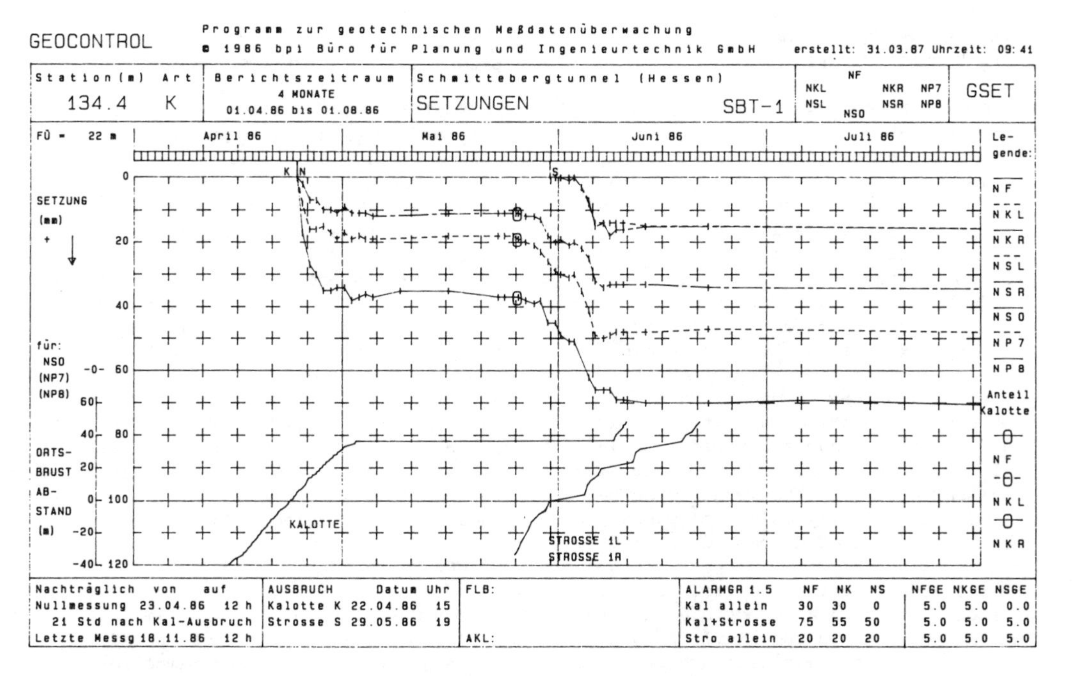

Fig.2 Path-time-diagram of the roof and floor settlements

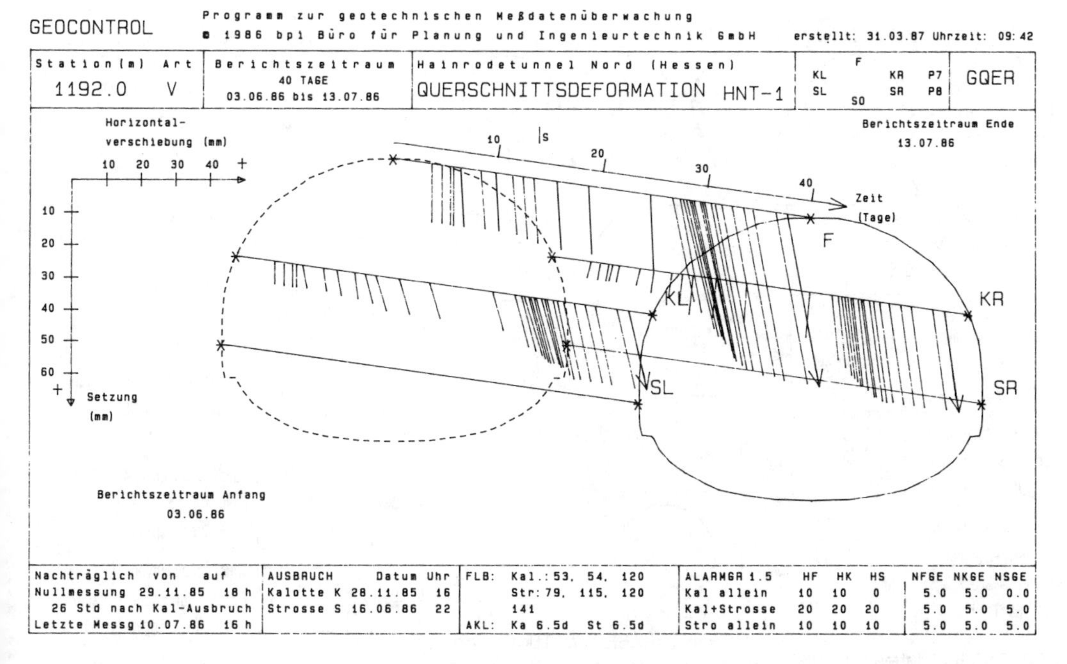

Fig.3 Presentation of the cross-sectional deformations

4 USING THE SYSTEM FOR NEW FEDERAL RAILWAY ROUTE TUNNELS

The GEOCONTROL EDP system has been in use for monitoring tunnel structures on the Hanover - Würzburg Federal Railway new line since 1985.

In this connection, the modularly built-up programme system was adapted to conform with the special requirements of the principal and for tunnelling.

In addition, a data remote transfer network was set up, which enables the principal and the experts involved to obtain information on the latest level of the measured data of the relevant tunnelling projects via external data stations (Fig.4).

The data processing is carried out via a small computer in the office of the site supervising engineer. The results can be provided automatically as a data package without any complicated commands having to be fed in.

But individual evaluations according to individual selection are possible directly on the site thanks to the simple programme handling. Once the daily data are fed in, they are transferred to the central data bank via a modem and telephone line by means of specially developed communication software. The data are automatically fed into the central computer and stored there.

External display stations have been set up at the H/W Project Group - Centre - in Frankfort as well as with the control engineers and experts - corresponding to the data station on site.

These external display stations can call off the measured results of the connected tunnelling site at any time. In this connection, the values available at the display station are brought up-to-date by the central computer. This is carried out via telephone line. The call is passed on automatically to the central unit via the postmodem and the data transfer controlled via the special communication software.

In this way, all the latest measured data from the tunnels involved are available at the visual station and can again be called up either as a data package or individually in accordance with individual preference.

5 FUTURE PROSPECTS

The modular set-up of the programme and the flexible form of the input permit rapid adjustment of the programme for varying tasks, conditions and profile types. Newly, the programme was adapted for use in urban tunnelling, e.g. for the S-Bahn (rapid transit railway) as well as

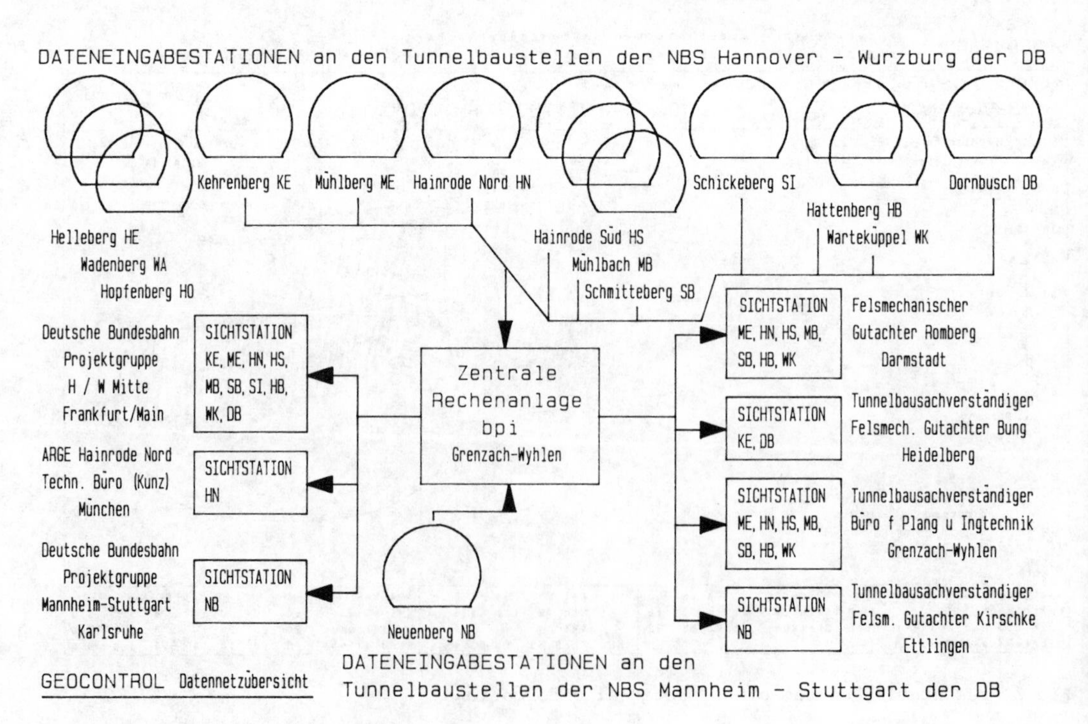

Fig.4 Data remote transfer network

for monitoring earth and rock embankments.

Thanks for the EDP storage of the data obtained during the period of construction, a far-ranging and authoritative data memory is available after construction is completed as the basis for systematic longterm monitoring. This can be optimized both technically and economically by setting up interfaces for data transfer e.g. on the monitoring units of pressure cell cross-sections and by using mobile data reading units in conjunction with central evaluating computers.

6 SUMMARY

For the evaluation, processing and monitoring of geotechnical data an EDP system was developed - for use at ten tunnel sites along the German Federal Railway new line between Hanover and Würzburg in the Fulda-Kassel area.

The EDP system distinguishes itself as follows:

- reliability of the obtained values thanks to a lucid and standard means of presentation
- increasing the validity of the evaluation and interpretation of the results for assessing the safety.

With the EDP system, a data remote transfer network was established which makes it possible for the principal and the experts involved to inform themselves of the latest level of measured data relating to the connected tunnelling projects at any time via external data display stations.

The data collected during the construction period are available as the basis for systematic long-term observation of the structure.

14 Use of microcomputers

Numerical Methods in Geomechanics (Innsbruck 1988), Swoboda (ed.)
© 1988 Balkema, Rotterdam. ISBN 90 6191 809 X

A microcomputer program for modelling large-strain plasticity problems

Peter Cundall
University of Minnesota, Minneapolis, USA
Mark Board
Itasca Consulting Group, Inc., Minneapolis, Minn., USA

ABSTRACT: This paper describes the FLAC (Fast Lagrangian Analysis of Continua) microcomputer program. This code uses the explicit finite difference method to model a variety of geotechnical problems, including large-strain plasticity, strain-softening, interfaces, and support-structure interaction. The theoretical development of the method is described, and a number of simple example problems are given which illustrate the capabilities of the code. FLAC operates in two dimensions.

1 INTRODUCTION

The paper shows that it is possible to use a standard micro computer to model large, nonlinear problems in continuum mechanics. It is also possible to obtain accurate collapse loads with low order (constant strain) elements, using the technique of mixed discretization (see Section 2). Program FLAC uses a form of dynamic relaxation (Otter et al., 1966) adapted for arbitrary grid shapes and large strains. Since the solution is by relaxation, matrices are not formed, which means that very large grids (2000 elements) can be accommodated on a standard 640K personal computer. Furthermore, large strains are modelled for negligible overhead, because coordinates can be updated at each step. The dynamic solution method is well-suited for collapse problems since continuous plastic flow is represented by a field of steady velocities (see Cundall, 1986).

2 CALCULATION METHOD

FLAC embodies an explicit, time-marching solution scheme: the right-hand-sides of all equations consist of known values, because the time step is chosen to be small enough that information cannot physically propagate from one element to the next within the time step. The calculation cycle of FLAC consists of the following operations:

1. **For all zones, derive:**
   - strain increments from known nodal velocities
   - new stresses from strain increments using constitutive model
2. **For all nodes, derive:**
   - nodal forces from known stresses in surrounding zones
   - updated velocities from forces, using law of motion
   - updated coordinates

The equations corresponding to these operations are presented below.

Each FLAC quadrilateral is modeled as two overlayed pairs of constant-stress triangles:

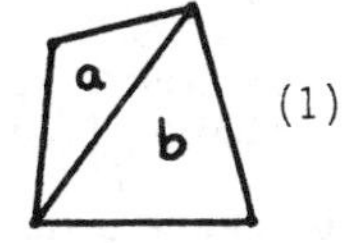

(1)

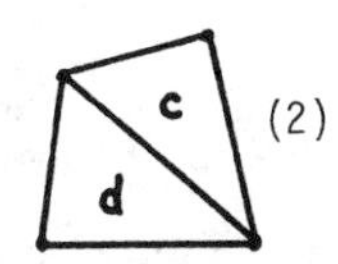

(2)

Each triangle maintains its stress components independently of the other triangles —i.e., twelve stress components are stored for each quadrilateral (in plane-strain mode). The forces exerted on each node are taken to be the mean of those exerted by the quadrilaterals (1) and (2). The overlay scheme thereby ensures symmetry where symmetry should exist.

The difference equations for a generic triangle are derived from the generalized form of Gauss' divergence theorem (e.g., see Malvern, 1969):

$$\int_s n_i\, f\, ds = \int_A \frac{\partial f}{\partial x_i}\, dA \qquad (1)$$

where $\int_s$ is the integral around the boundary of a closed area,

$\int_A$ is the integral within the area,

n_i is the unit vector normal to the surface,

x_i is a position vector, and

f is a scalar, vector or tensor.

Defining the average value of the gradient over the area as

$$< \frac{\partial f}{\partial x_i} > = \frac{1}{A} \int_A \frac{\partial f}{\partial x_i}\, dA \qquad (2)$$

we obtain

$$< \frac{\partial f}{\partial x_i} > = \frac{1}{A} \int_s n_i\, f\, ds \qquad (3)$$

If the closed area is a triangle and f varies linearly along each side, we obtain the finite difference formula:

$$< \frac{\partial f}{\partial x_i} > = \frac{1}{A} \sum_s <f>\, n_i\, \Delta s \qquad (4)$$

where the summation is over the 3 sides, and <f> is the average value of f over the side.

This formula, suggested by Wilkins (1969), enables strain increments, Δe_{ij}, to be written in terms of nodal velocities for a zone by substituting the velocity vector for f:

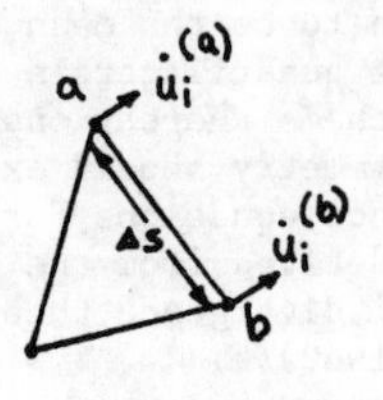

$$\frac{\partial \dot{u}_i}{\partial x_j} \simeq \frac{1}{2A} \sum_s (\dot{u}_i^{(a)} + \dot{u}_i^{(b)})\, n_j\, \Delta s \qquad (5)$$

$$\Delta e_{ij} = \frac{1}{2} \left[\frac{\partial \dot{u}_i}{\partial x_j} + \frac{\partial \dot{u}_j}{\partial x_i} \right] \Delta t \qquad (6)$$

where Δt is the timestep.

The volumetric strain is then averaged over each pair of triangles, according to the mixed discretization scheme of Marti and Cundall (1982):

$$\Delta e_m = \frac{\Delta e_{11}^{(a)} + \Delta e_{22}^{(a)} + \Delta e_{11}^{(b)} + \Delta e_{22}^{(b)}}{2}$$

$$\Delta e_d^{(a)} = \Delta e_{11}^{(a)} - \Delta e_{22}^{(a)}$$

$$\Delta e_{11}^{(a)} := \frac{\Delta e_m + \Delta e_d^{(a)}}{2} \qquad (7)$$

$$\Delta e_{22}^{(a)} := \frac{\Delta e_m - \Delta e_d^{(a)}}{2}$$

and similarly for the strains in (b) and for the triangle-pair (c)-(d).

At this stage, new zone stresses are computed by means of a specified constitutive model:

$$\sigma_{ij} := M(\sigma_{ij}, \Delta e_{ij}, S_1, S_2 \ldots) \qquad (8)$$

where M() is the constitutive model, and

S_i are state variables.

Mixed discretization is again invoked to equalize the isotropic stress between the two triangles in a pair:

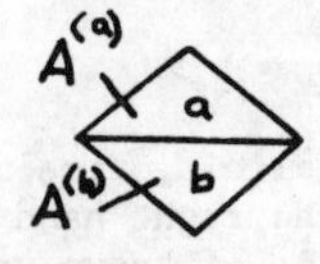

$$\sigma_o^{(a)} = \sigma_o^{(b)} := \left[\frac{\sigma_o^{(a)} A^{(a)} + \sigma_o^{(b)} A^{(b)}}{A^{(a)} + A^{(b)}} \right] \qquad (9)$$

This equalization only has an effect for constitutive laws that involve shear-induced dilatation.

In the explicit method, Eq. (8) is evaluated once per zone per timestep. No iterations are necessary because information cannot physically propagate from one zone to the next within one timestep. The various constitutive models available in FLAC are mentioned in Section 3.

The net force on a node is calculated from the summation of forces imposed by the zones surrounding the node. The force that one triangle contributes to the node is found from the stress vector acting on the edges of the triangle:

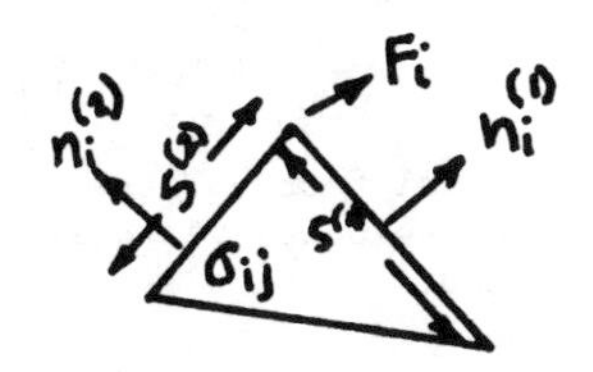

$$F_i = \frac{\sigma_{ij}^{(1)} \left[n_j^{(1)} S^{(1)} + n_j^{(2)} S^{(2)} \right]}{2} \tag{10}$$

At equilibrium, or in steady-state flow, the net force ΣF_i on a node is zero; otherwise, the node is accelerated. The acceleration is integrated numerically to obtain velocities:

$$\dot{u}_i := \dot{u}_i + \left[\Sigma F_i - \alpha \, |\Sigma F_i| \, \mathrm{sgn}(\dot{u}_i) \right] \frac{\Delta t}{m} + g_i \, \Delta t \tag{11}$$

where m = inertial mass of the node (see below),

g_i = gravitational acceleration, and

α = damping factor.

Because FLAC is intended to model quasi-static problems, the inertial mass can be treated as a relaxation factor and set for optimum convergence. For a single mass-spring system, the critical timestep is

$$\Delta t_c = 2 \, (m/k)^{1/2} \tag{12}$$

where k is the spring stiffness.

Assuming that optimum convergence is obtained when Δt_c is the same for all nodes, the inertial mass is set equal to the sum of the stiffnesses connected to the node (includes zone stiffness, structural connections, and interface connections).

The term containing α in Eq. (11) is a form of damping that vanishes when the net force on a node is zero—i.e., it imposes no body forces on material in steady-state flow. This is advantageous when modeling plastic flow, but the solutions tend to be under-damped for realistic values of α (e.g., 0.8).

Equation (11) may be integrated again to update coordinates in large-strain mode:

$$x_i := x_i + \dot{u}_i \, \Delta t \tag{13}$$

3 SUMMARY OF PROGRAM ATTRIBUTES

The formulation for continuum zones has been presented in detail. Other elements, such as interfaces and structures, operate in a similar fashion, but there is not enough space to discuss them in detail. These, and other features of FLAC, are summarized below.

- Constitutive models for continuum zones:
  1. Null (void)
  2. Elastic - isotropic
  3. Elastic - orthotropic
  4. Elastic/plastic: associated or non-associated flow rule
  5. Ubiquitous joint + elastic/plastic for intact material
  6. Strain-softening or -hardening
  7. Two types of creep model (in special version)
  8. Double-yield (cap) model (in special version)
- Large-strain formulation for grid, structures and interfaces
- Construction/excavation easy to perform
- Structural elements:
  1. Straight beams (bending & axial deformation)
  2. Cables (bonded, with yield in bond and cable)
- Interfaces: slip; opening; bond breakage
- Plotting: 16 color screen plots & pen plots of
  1. contours, vectors, tensors, grid, state variables, structures.
  2. histories of all variables
- Mesh generation:
  1. graded grid
  2. circles, arcs, lines
- User interface:
  1. Free-format, free-sequence input
  2. Remote input
  3. Output logging

4. Save/restore facility
Heat transfer: conduction with convective and radiative boundaries; may be coupled to any material model for thermal stress analysis (special version)

The program executes at a speed of 50 zone-steps per second on an 8 MHz IBM AT with 80287 floating-point chip. The number of steps (cycles) necessary depends very much on the nature of problem to be modeled, but between 200 and 2000 steps are commonly needed. With the problem-oriented input of FLAC, it is easy to do a small-grid run to get an idea of the mechanical response, and then increase the number of zones for greater accuracy.

For self-similar problems (i.e., problems with identical geometry, properties, etc.) the solution time increases as $N^{3/2}$, where N is the number of zones. This is in contrast to many implicit schemes, where the solution time may increase as N^2, or even N^3, for some boundary element plasticity codes.

Relaxation schemes are inefficient, however, for solid bodies that have long natural periods compared to the natural period of the smallest, stiffest element (e.g., beams).

4 VALIDATIONS AND EXAMPLES

A large number of comparisons have been made between FLAC results and exact solutions for plasticity problems [see Itasca (1987) and Lin (1987)]. The results show uniformly good agreement with theoretical results. In particular, an explicit method has no trouble modeling physical instability or strains well past the collapse limit.

4.1 Strain-softening example

The localization effect of strain softening is illustrated by this example, in which a biaxial test is modeled with rigid, rough platens and constant confining pressure. The simplest possible assumption is made about the softening: that the friction angle of the material drops linearly from 30° to 20° over a 10% range of accumulated plastic strain.

Figures 1(a) and 1(b) shows the deformed state and velocity components at a sample strain of 11%. A localized shear band has formed. Figures 2(a) and 2(b), in contrast, show the results of an identical test in which the material retains its 30° friction angle independently of strain.

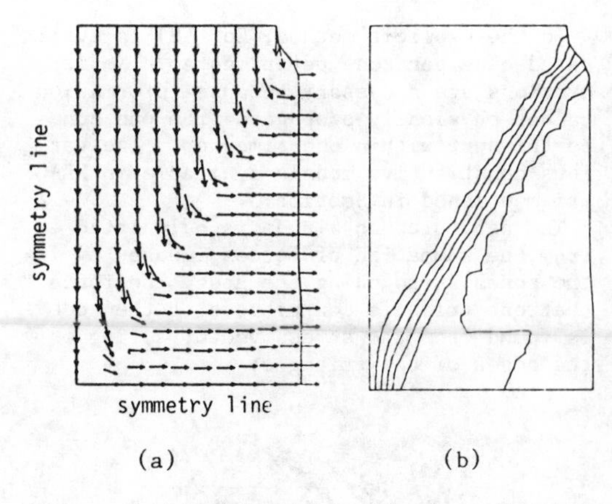

Figure 1. Biaxial test on strain-softening material: (a) velocity vectors; (b) x-velocity contours

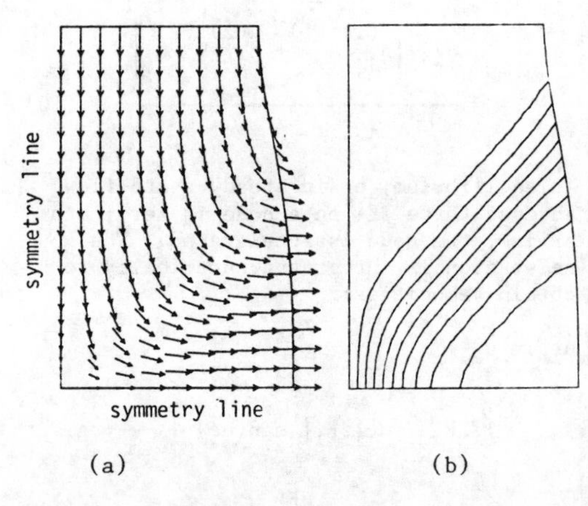

Figure 2. As Fig. 1, but with constant-strength material

The displacement field is much more diffuse. Note that the strain-softened sample was not "seeded" with soft-spots or other devices.

4.2 Stabilization of an embankment in cohesionless soil

This example demonstrates the use of structural elements in FLAC. Initially, a sheet wall [consisting of beam elements, as shown in Fig. 3(a)] is used to stabilize an embankment under gravity loads only. A loading of 1 MPa is then applied to the surface near the face, to simulate

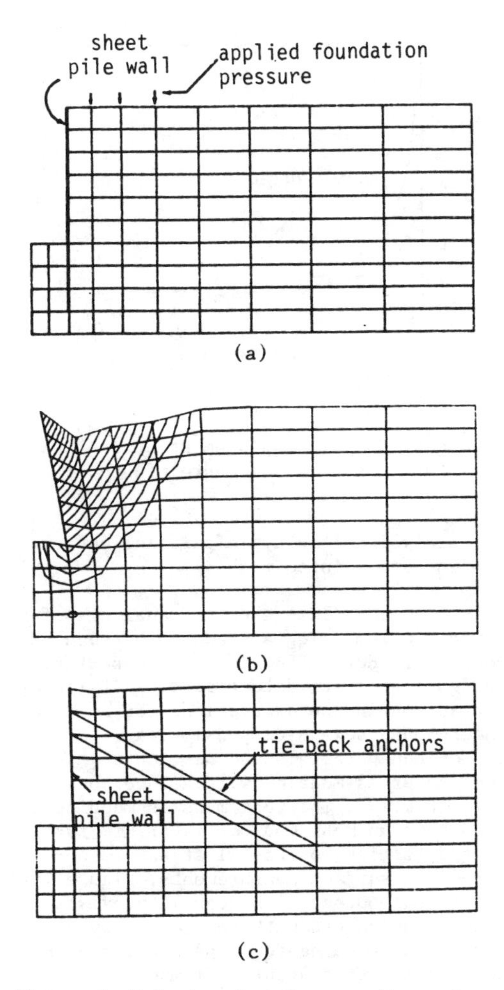

Figure 3. Embankment under gravity and foundation load: (a) grid plot showing sheet pile wall; (b) collapse under footing pressure; (c) addition of tie-back anchors

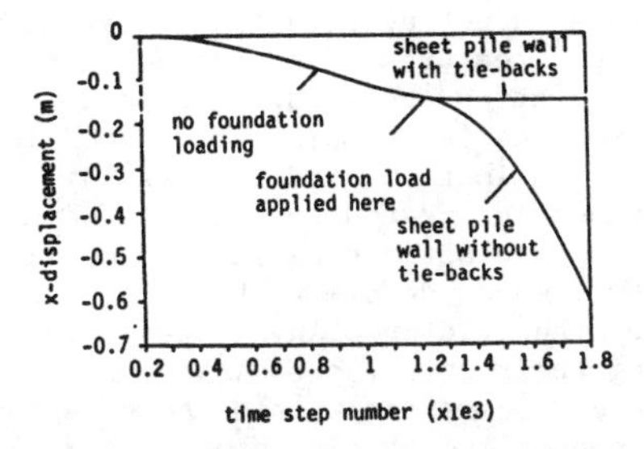

Figure 4. History of x-displacement at embankment face showing stability after tie-backs installed

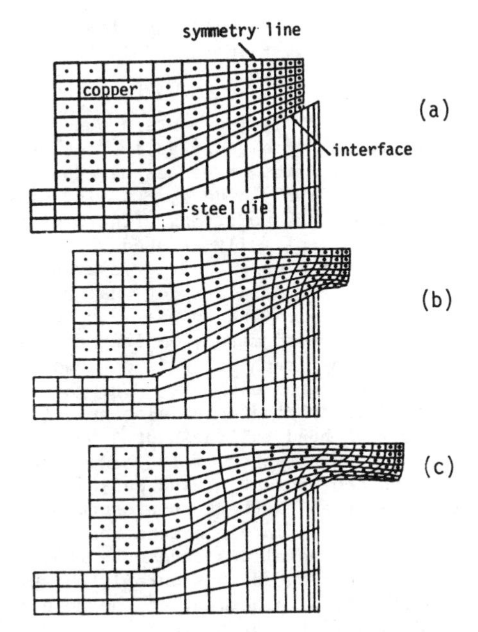

Figure 5. Extrusion of a copper billet through a steel die along a frictional interface. A constant velocity is applied to the left face of the billet.

pressure from a foundation. As seen in Fig. 3(b), the slope collapses under this load: the x-velocity contours indicate the failure area. Two tie-back anchors (32mm diameter) are then installed, modeled here as point-anchor cables [Fig. 3(c)]. The displacement histories at the embankment face clearly indicate the stabilizing effect of the tie-back anchors (Fig. 4).

4.3 Solid extrusion through a die

The problem of extrusion of a copper billet through a die involves very large strains in the billet as well as frictional sliding along the billet/die interface. Figure 5(a) shows the initial disconnected finite difference grid, which represents the lower half of a symmetrical problem. An interface is defined at the boundary between the disconnected meshes. The copper billet is given a fixed velocity at its left-hand face. The billet is modeled as a von Mises material, and the die as elastic. Figures 5(b) and 5(c) illustrate the successive mesh distortions calculated by FLAC as the billet is extruded.

4.4 Fully-grouted rock-bolting of a circular excavation in an elasto-plastic rock mass

The stresses and displacements around a circular tunnel in elastoplastic rock under hydrostatic field stresses have been given by Bray (1967). Here, FLAC is used to investigate the modifying effects of a regular pattern of fully grouted rock-bolts. Figure 6 shows a quarter symmetry tunnel of 1m radius in a frictional and cohesive material. The tunnel is reinforced with 15mm-diameter, 3m long cables on 0.3m spacing; each cable is divided into six structural elements. The grout-cable bond is treated as an elastoplastic material with bond stiffness of $4.5*10^7$ N/m/m and bond strength of 10^5 N/m. Figure 7 compares the tangential and radial stresses with and without reinforcement. A small reduction in yield zone radius is seen, but the radial displacements (see Fig 8) are influenced considerably by the reinforcement: displacement at the hole is reduced by a factor of two.

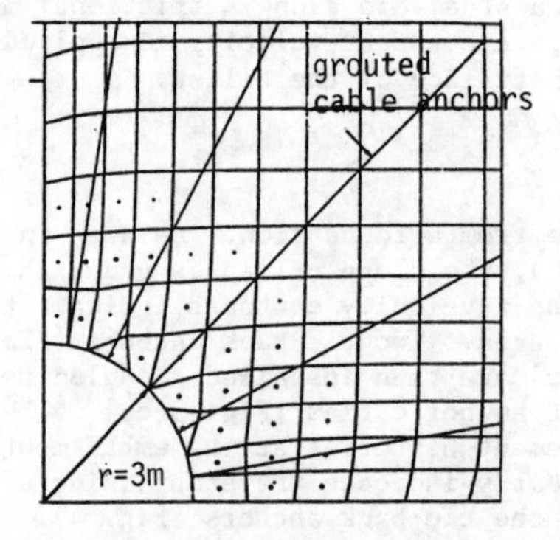

Figure 6. Circular hole reinforced by 3m long grouted-cable anchors

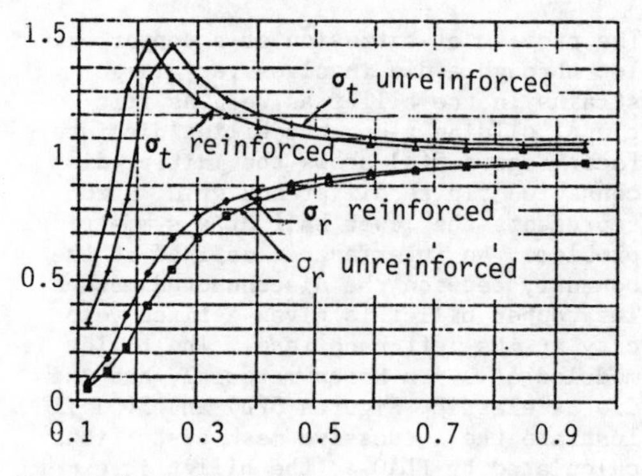

Figure 7. Radial and tangential stress with radial distance for reinforced and unreinforced cases

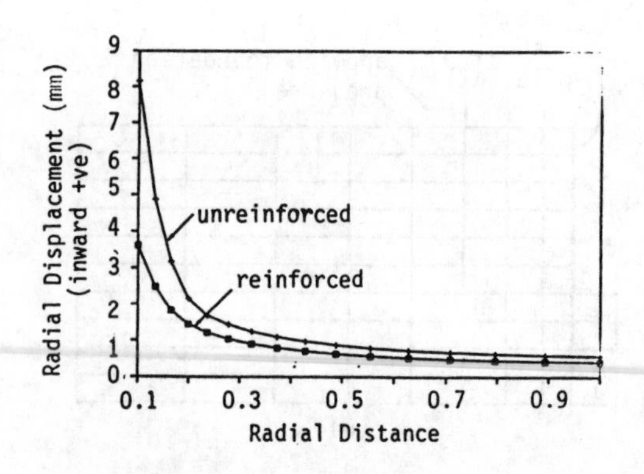

Figure 8. Radial displacement at the hole wall for reinforced and unreinforced cases

4.5 Analysis of multiple parallel caverns in hard jointed rock

This example illustrates the analysis of a large problem using a standard personal computer. Here, FLAC is used to perform design studies to detect possible failure mechanisms in the excavations for a large underground sewage disposal plant. A system of ten parallel caverns with a thin rock cover is modeled with a 2000 zone grid (Fig. 9). The rock mass is modeled as a no-tension Mohr Coulomb material, subjected to high horizontal stresses, observed in surface measurements. Figure 10 shows that potential instability exists around the cavern with the least cover. Work is continuing to compare the continuum model with discontinuum models and field measurements (Johansson et al., 1987).

4.6 Example grid generation

Grids are generated in FLAC from an initial rectangular mesh which is then deformed to the desired shape. Internal boundaries, denoting potential excavations or geologic features, are created with shape generation commands. The data file below produces the complex grid shown in Fig. 11. An initial grid of 31(x) by 25(y) zones is created by the **GRID** command and given an elastic constitutive model. The **GEN** commands deform rectangular sections of the grid into any shape by redefining their corner coordinates. The grading of zone sizes within these regions may be controlled by giving an x- and y-ratio of expansion or contraction (RATIO keyword). Circular, arc or line shapes may be created with a series of "shape

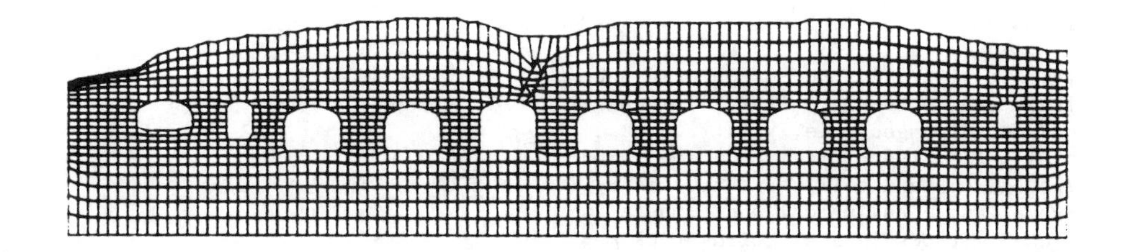

Figure 9. Typical problem geometry for design analysis of multiple caverns (Analyses employ 2,000 zones with Mohr-Coulomb no-tension constitutive behavior.) [Johansson et al., 1987]

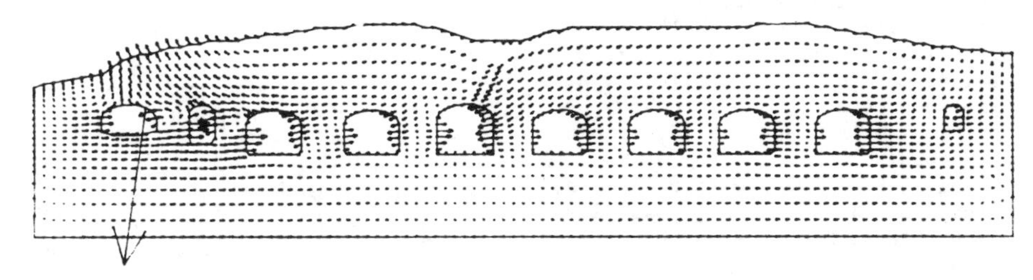

Figure 10. Gridpoint displacements show potential instability in cavern at far left with elastic rock cover [Johansson et al., 1987]

generators", which require information such as the centroid, radius or end points of a line. Any internal or external boundary may define a REGION, which can be addressed simply by referring to one zone within it. The **ADJUST** command further deforms the grid to produce a more uniform mesh.

5 CLOSING REMARKS

FLAC has been in routine use for over two years. It is possible to perform large, nonlinear simulations that previously required a large minicomputer. FLAC runs may be done at almost no machine cost, since most engineering offices already have several microcomputers, which are usually idle at night. Although a large FLAC run takes several hours, the apparent speed is increased when a number of machines are employed simultaneously to do the parameter studies that must be part of any engineering project.

It is often overlooked that even though a minicomputer may be ten times faster than a micro, the mini's apparent speed drops below the micro's when the mini is shared by ten similar users.

```
grid  31 25
model elas  i=1,20 j=1,20
gen (0,-25)  (0,0)  (20,0)  (20,-25)  i=1,14 j=1,21
gen (20,-25) (20,0) (40,0)  (30,-25)  i=14,21 j=1,21
gen arc  (0,-10)  (0,-20)  90
gen arc  (20,-10) (20,0)  90
gen circ (20,-15) 6
model null, region=(1,20)
model null, region=(14,8)
gen adjust
model elas, i=22,31
gen (30,-25) (40,0) (70,0) (70,-55) ratio=(1.2,1) i=22,32 j=13,26
gen (0,-25) (30,-25) (70,-55) (0,-55) ratio=(1.2,1) i=22,32 j=1,13
model elas, i=1,20 j=22,25
gen (30,0) (30,5) (45,5) (45,0) i=1,21 j=22,26
```

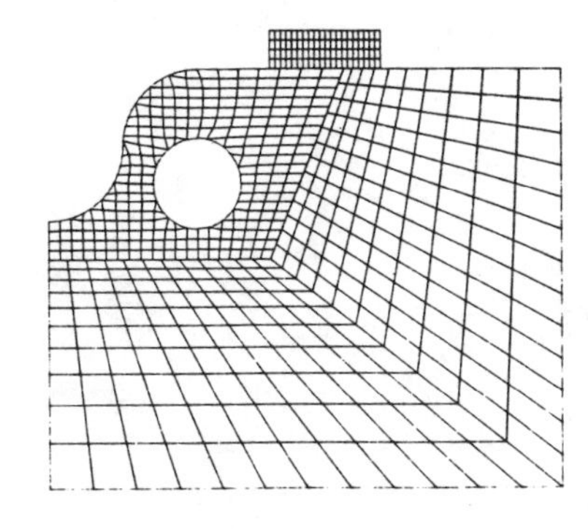

Figure 11. Complex grid created from above input commands

REFERENCES

Bray, J.W. 1967. A study of jointed and fractured rocks—Part II," Felsmechanik und Ingenieungeologie, V(4):7.

Cundall, P.A. 1986. Distinct element models of rock and soil structure, in E. T. Brown (ed.), Analytical and computational methods in engineering rock mechanics, p.129-163. London, Allen & Unwin.

Itasca Consulting Group, Inc. 1987. FLAC: Fast Lagrangian analysis of continua, user manual, version 2.0. Minneapolis, Itasca Consulting Group, Inc.

Johansson, Erik, Reijo Riekkola, & Loren Lorig. 1987. Design analysis of multiple parallel caverns using explicit finite difference methods, submitted for presentation at the 29th U.S. Rock Mechanics Symposium.

Lin, Ming. 1987. The analytical and numerical solutions in limit load plasticity problems. MSc. thesis, Univeristy of Minnesota.

Malvern, L.E. 1969. Introduction to the mechanics of a continuous medium. New Jersey, Prentice-Hall.

Marti, J. & P.A.Cundall. 1982. Mixed discretization procedure for accurate modelling of plastic collapse, Int. J. for Num. & Anal. Meth. in Geomech. 6:129-139.

Otter, J.R.H., A.C.Cassell & R.E.Hobbs. 1966. Dynamic relaxation, Proc. Inst. of Civil Engineers. 35:633-656.

Wilkins, M.L. 1969. Calculation of elastic-plastic flow, Report UCRL-7322, Lawrence Radiation Laboratory, Livermore.

Numerical Methods in Geomechanics (Innsbruck 1988), Swoboda (ed.)
© 1988 Balkema, Rotterdam. ISBN 90 6191 809 X

Finite layer analysis of layered pavements subjected to horizontal loading

J.R.Booker & J.C.Small
School of Civil and Mining Engineering, Sydney, Australia

ABSTRACT: A finite layer method is presented which may be used to compute stresses, strains and displacements in horizontally layered materials subjected to horizontal loading. Such results are of use in pavement design where the action of horizontal braking, turning and acceleration forces applied by pneumatic tyres cause deformation of the pavement. The finite layer method enables solutions to be obtained very easily on a microcomputer.

1. INTRODUCTION

There are many examples in the field of geotechnical engineering which involve loadings applied to soil profiles which consist of layers of different materials. Layered soils are common because man made fills and pavements are generally constructed of horizontal layers of different materials, and because natural soil profiles have often been produced through sedimentation processes and this results in different material layers which are very nearly horizontal.

It is of great interest to be able to predict the behaviour of structures built on such deposits and to be able to predict how the foundation will deform under vertical as well as horizontal loadings. One area of special interest is that of pavement design. The forces applied by the tyres of moving vehicles not only involve vertical forces due to the self weight of the vehicle, but also horizontal forces due to braking, turning and acceleration. These horizontal forces may be very large and contribute significantly to the deformation of the pavement.

Many researchers have examined the problem of horizontal loading, and analytic solutions have been obtained by Scott (1963), for a uniform horizontal loading applied over a strip on the surface of an infinitely deep elastic layer, and Barber (1963) who considered the case where the load was applied to a circular region.

For layered soils, solutions have been presented by Westmann (1963) (uniform shear loading over circular region) for the case where a surface layer of material overlies an infinitely deep layer of another material.

When several layers of material are involved analytic solutions become difficult, and presentation of results is complex because of the many different combinations of layer thicknesses and moduli. Use may be made of finite element methods, however this is not an efficient method for solving problems involving horizontal layering, especially when horizontal loading of circular or rectangular regions is required. A much more efficient method is the finite layer method which may make use of Fourier or Hankel transforms (Rowe and Booker (1981a, b)) or Fourier series (Cheung and Fan (1979), Tham and Cheung (1981)) to greatly simplify the solution process.

In this paper the finite layer method is used to obtain solutions to problems involving uniform horizontal loadings applied

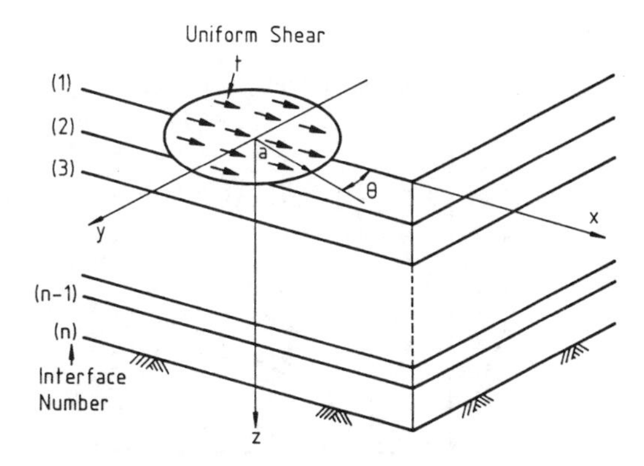

Figure 1. Section showing uniform shear loading applied to surface of layered material.

to the surface of layered anisotropic materials. The loading may be applied over a strip, circular or general region, however special emphasis is placed on the circular loaded region because of the applicability to pavement design where tyre loads may be considered to act over a circular region. The methods used in this paper are an extension of those described in Small and Booker (1984, 1986).

2. BASIC EQUATIONS

Shown in Figure 1 is a soil profile consisting of several layers of different materials. For the general case of an anisotropic elastic material we may write the following equations for each material layer:

(a) The equations of equilibrium

$$\sigma_{k\ell,\ell} = 0 \qquad (1)$$

where

σ denotes total stress increases and the summation convention for repeated indices applies. A comma denotes a partial differentiation, and the indices vary over the set x, y, z.

(b) The stress-strain relationship

$$\sigma_{ij} = R_{ijk\ell}\,\epsilon_{k\ell} \qquad (2)$$

where

$\epsilon_{k\ell}$ are the components of strain, and $R_{ijk\ell}$ are the elastic coefficients relating stress to strain.

(c) The strain-displacement relationships

$$\epsilon_{k\ell} = \tfrac{1}{2}(u_{k,\ell} + u_{\ell,k}) \qquad (3)$$

where

u_k are the components of displacement.

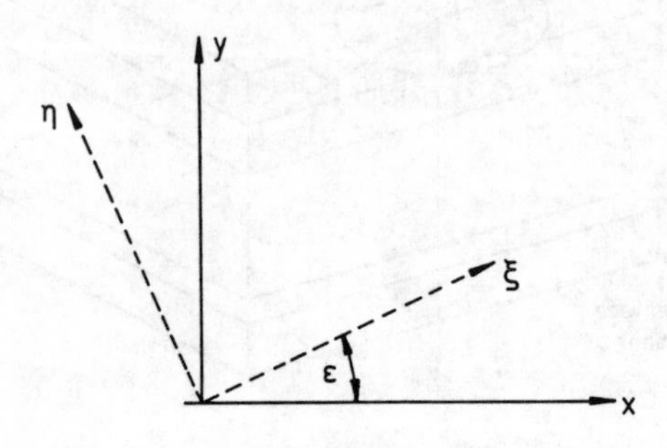

Figure 2. System of axes.

3. FOURIER TRANSFORMS

In order to simplify the governing equations we may apply a single (strip loading) or double (general region of loading) Fourier transform to the field variables. e.g.

$$U_x = \frac{1}{4\pi^2}\int_{\infty}^{+\infty}\int_{\infty}^{+\infty} i u_x \, e^{-i(\alpha x+\beta y)}\,dx\,dy$$

$$= F\{ i u_x \}$$

For the other field quantities we have:

$$(U_y,\ U_z) = F\{ i u_y,\ u_z \}$$

$$(S_{xx},\ S_{yy},\ S_{zz},\ T_{xy},\ T_{yz},\ T_{zx})$$

$$= F\{\sigma_{xx},\ \sigma_{yy},\ \sigma_{zz},\ i\sigma_{xy},\ i\sigma_{yz},\ i\sigma_{zx}\} \qquad (4)$$

It has been shown, (Small and Booker (1986)) that when such transforms are applied to the governing equations for the problem (equations 1 to 3) we may write the flexibility relationship for the layered system as

$$FP = \delta \qquad (5)$$

where $P = (S_{zz1},\ T_{\xi z2},\ S_{zz2},\ T_{\xi z2},\ \ldots\ldots\ S_{zzn},\ T_{\xi zn})^T$

$\delta = (U_{z1},\ U_{\xi 1},\ 0,\ 0,\ \ldots\ldots\ 0,\ 0,\ U_{zn},\ U_{\xi n})^T$

The assumptions made regarding continuity conditions between layers are that there is no slip at an interface and the normal stress component is continuous. In the above equation (5) we have introduced a new set of axes η and ξ as shown in Figure 2, such that $\xi = x\cos\epsilon + y\sin\epsilon$ and $\eta = -x\sin\epsilon + y\cos\epsilon$. The subscripted numbers refer to the layer interfaces; in the above equation we have n interfaces (see Figure 1). Equation 5 also applies to a general form of surface loading; i.e. one which is applied over a general shaped area and being either horizontal or vertical.

We need however, to apply the transformed boundary conditions before a solution to equation 5 may be found. Usually we specify stress or displacement boundary conditions. If the base is rough and rigid we set $U_{zn} = U_{\xi n} = 0$. If the surface is loaded with a shear traction we set $S_{zz1} = 0$, $T_{\xi z1} = T_\xi$ where T_ξ is the transform of the applied shear load.

We also obtain a set of equations for the uncoupled terms (see Small and Booker 1986) which may be written:

$$F^* P^* = \delta^* \qquad (6)$$

where

F^* is the uncoupled flexibility matrix

$$P^* = (T_{\eta z_1}, T_{\eta z_2}, \ldots\ldots, T_{\eta z n})^T$$

$$\delta^* = (U_{\eta 1}, 0, \ldots\ldots 0, U_{\eta n})^T$$

Here once again we have the shear $T_{\eta z}$ and displacement U_η transforms referred to the ξ, η axes; and solution involves specifying boundary conditions. For a surface shear load we set $T_{\eta z_1} = T_\eta$ (the transform of the shear load) and if the base is rough we set $U_{\eta n} = 0$.

4. TRANSFORM OF LOADING. (CIRCULAR REGION)

If we have a uniform shear loading t applied to the surface z = 0 over a circular area of radius a in the x-direction as shown in Figure 1, we may write the transform

$$T_{xz} = \frac{1}{4\pi^2} \int_{\infty}^{\infty} \int_{\infty}^{\infty} i t \, e^{-i(\alpha x + \beta y)} dx dy \qquad (7)$$

If we make the substitution $x = r\cos\theta$, $y = r\sin\theta$, $\alpha = \rho\cos\epsilon$, $\beta = \rho\sin\epsilon$ Equation 7 becomes:

$$T_{xz} = \frac{i}{4\pi^2} \int_0^{\infty} \int_0^{2\pi} t \, e^{-i\rho r \cos(\theta - \epsilon)} r dr d\theta \qquad (8)$$

Now since it is well known that Bessel functions of the first kind J_n have the integral representation

$$J_n(z) = \frac{1}{i^n \pi} \int_0^{\pi} e^{iz\cos\theta} \cos(n\theta) d\theta \qquad (9)$$

we may write the transform of the shear loading as:

$$T_{xz} = i t \, a J_1(\rho a)/2\pi\rho \qquad (10a)$$

and so the shear that we need to specify as boundary conditions in equations 5 and 6, are given by

$$T_\xi = i t \, a J_1(\rho a)\cos\epsilon/2\pi\rho \qquad (10b)$$
$$T_\eta = i t \, a J_1(\rho a)\sin\epsilon/2\pi\rho$$

5. SOLUTION PROCEDURE

Solving the sets of equations such as equations 5, 6 gives us the solution for the transformed stress components S_{zz}, $T_{\xi z}$, $T_{\eta z}$ at each layer interface. Once these are known all other stress and displacement components may be found. Firstly however an inverse Fourier transform is required. For example, to obtain σ_{zz} we would proceed as follows:

$$\sigma_{zz} = \int_{\infty}^{+\infty} \int_{\infty}^{+\infty} S_{zz} \, e^{i(\alpha x + \beta y)} d\alpha \, d\beta \qquad (11)$$

If we once again put $x = r\cos\theta$, $y = r\sin\theta$, $\alpha = \rho\cos\epsilon$, $\beta = \rho\sin\epsilon$ we obtain

$$\sigma_{zz} = \int_0^{\infty} \int_0^{2\pi} S_{zz} \, . \, e^{i\rho r \cos(\epsilon - \theta)} \rho d\rho d\epsilon$$

It follows from equation 10b that $S_{zz} = iA(\rho)\cos\epsilon$ and thus:

$$\sigma_{zz} = -2\pi \cos\theta \int_0^{\infty} A(\rho) \, J_1(\rho r) \, \rho d\rho \qquad (12)$$

From the above it may be seen that the vertical stress varies as the cosine of the angle θ which is the angle measured from the x axis direction. This inverse transform may be simply evaluated using numerical integration and for the results presented in this paper, Gaussian quadrature was employed. The infinite integral was approximated by making the integration range large; details are given in the paper by Small and Booker (1986). The great advantage of using the Bessel function (or Hankel transform) form of the integrals is that numerical inversion time is greatly reduced as only a single, not a double, integral need be evaluated.

6. EXAMPLES

In order to verify the foregoing theory the problem of a uniform shear loading t applied to the surface z = 0 of an infinitely deep isotropic elastic layer was examined. The loading occupies the region $0 \leqslant r \leqslant a$ and is acting in the direction of the x-axis. The solution to this problem has been presented by Barber (1963), and the results for the shear stresses at a depth z/a = 1 are shown in Figure 3 compared with the finite layer solution. The results are for the case where the Poisson's ratio ν of the soil is equal to ½. Because of the flexibility formulation used here, this value of Poisson's ratio presents no problem as it would in a stiffness formulation where the matrices become infinitely large for the incompressible case. The results are presented for shear stresses τ_{rz}, $\tau_{r\theta}$ and normal stress σ_{zz}. These stresses are presented in cylindrical coordinates to allow comparison with Barber's solution (θ is the angle turned from the x-axis as shown in Figure 1) and it may be seen from Figure 3 that excellent agreement could be obtained between the two results.

In order to investigate the effects of horizontal loading on a layered material, the problem shown schematically in the inset to Figure 4 was analysed. The problem involves a uniform shear stress t applied to the surface of a layered material. The load is applied in the x-direction over the region $0 \leqslant r \leqslant a$ (a is the radius of the loading). Layer A is much stiffer than layer B which is in turn stiffer than Layer C. Material

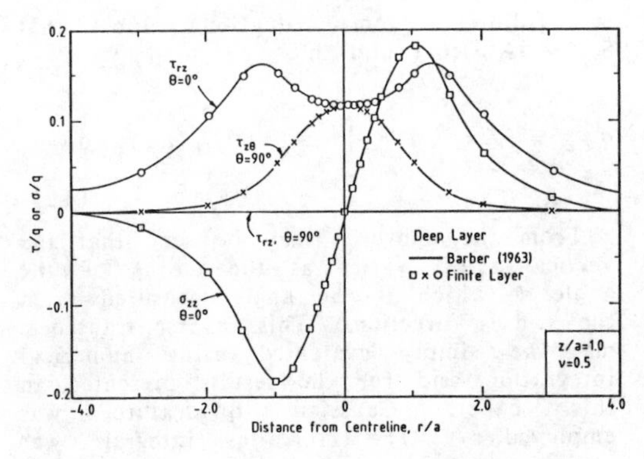

Figure 3. Shear and normal stresses computed for infinitely deep isotropic layer subjected to horizontal surface load. (Uniform shear over circular region).

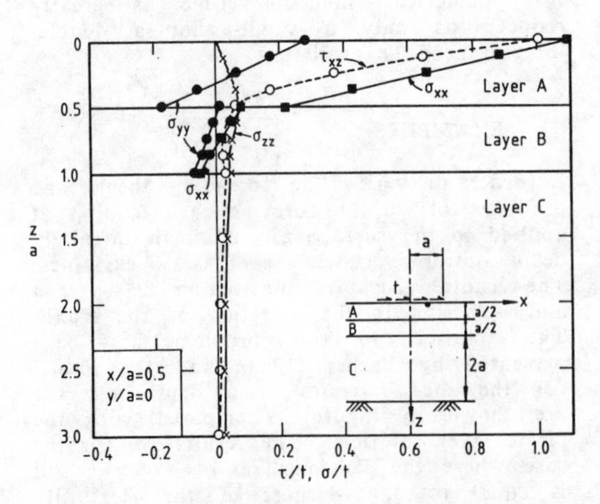

Figure 4. Shear and normal stresses computed for layered material subjected to horizontal surface load. (Uniform shear over circular region).

properties were selected such that $E_A:E_B:E_C = 25:5:1$ and $\nu_A = 0.3$, $\nu_B = 0.4$, $\nu_C = 0.5$ (E is the Young's modulus, ν is the Poisson's ratio of the material). Layers A and B are of thickness equal to half of the radius a of the loaded circular region, while layer C is of thickness 2a.

As the program is capable of computing all stress, strain and displacement components in cylindrical or Cartesian coordinates at any point within the soil mass, many plots of interest could be presented. However as an example, stress components only (plotted throughout the layer depth) are presented in Figure 4 at the position $x/a = \frac{1}{2}$, $y/a = 0$. It may be seen from the plot that at this position the stress components quickly reduce to small values with depth z, and stresses are small in the base layer C. Some negative or tensile stress is seen to exist at the bases of layers A and B.

Isotropic material properties were chosen for the previous example, however the computer program FLEA (Finite Layer Elastic Analysis) which is based on the theory presented in this paper, can easily deal with anisotropic materials. To illustrate this, the results for the displacements computed for the first example were compared with those found for a problem involving the same geometry and shear loading, but where layers A and B were anisotropic materials. Layer C remained an isotropic material as before. Material properties were chosen such that:

Layer A: $\nu_h = 0.3$ $\nu_{vh} = 0.2$
$E_h/E_v = 2$ $G_v/E_v = 0.4$

Layer B: $\nu_h = 0.4$ $\nu_{vh} = 0.2$
$E_h/E_v = 2$ $G_v/E_v = 0.4$

$(E_v)_A:(E_v)_B:(E_v)_C = 25:5:1$

The subscripts h, v refer to the horizontal and vertical directions respectively.

The results of the analysis are shown in Figure 5 where it may be seen that the effect of the anisotropy is to reduce the displacements in the pavement.

7. CONCLUSIONS

A method has been presented which may be used to compute the stresses, strains and displacements in layered anisotropic material which are subjected to horizontal shear loadings. Stresses and strains induced by a shear loading applied over a circular region are quite significant, and so horizontal tyre forces should not be neglected in pavement design.

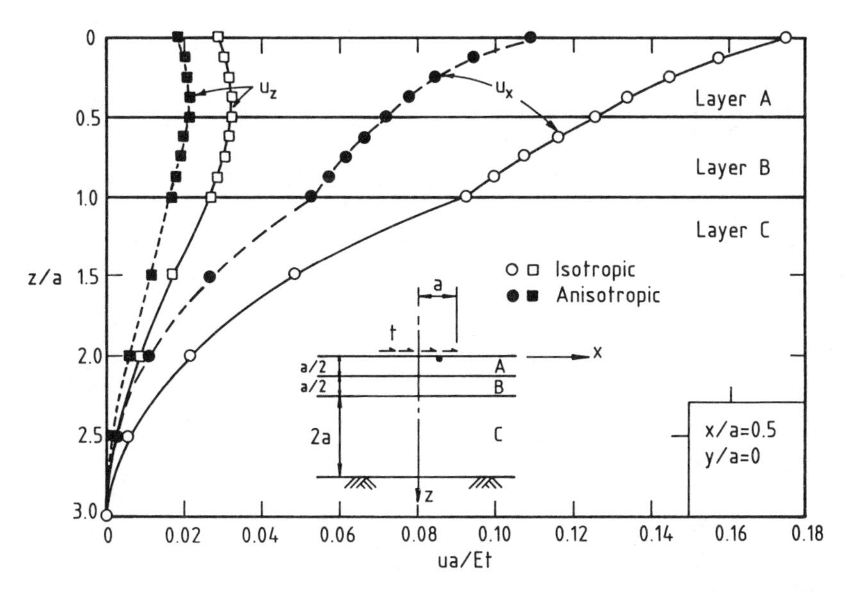

Figure 5. Horizontal and vertical displacements in layered material subjected to horizontal surface loading: effect of anisotropic materials. (Uniform shear over circular region).

8. ACKNOWLEDGEMENT

The authors gratefully acknowledge the assistance of Soon Chong Seng and Chang Si Kean in carrying out some of the programming and running test cases with the code.

REFERENCES

Barber, E.S. (1963). "Shear Loads on Pavements", Public Roads, Vol.32, No.6, pp.141-144.

Cheung, Y.K. and Fan, S.C. (1979). "Analysis of Pavements and Layered Foundations by Finite Layer Method", Proc. 3rd Int. Conf. Num. Meth. in Geomech., Aachen, Vol.3, pp.1129-1135.

Holl, D.L. (1940). "Stress Transmission in Earths", Proc. High Res. Board, Vol.20, pp.709-721.

Rowe, R.K. and Booker, J.R. (1981a). "The Behaviour of Footings Resting on a Non-Homogeneous Soil Mass with a Crust", Part I, Strip Footings", Canadian Geotech. Jl., Vol.18, No.2, pp.250-264.

Rowe, R.K. and Booker, J.R. (1981b). "The Behaviour of Footings Resting on a Non-Homogeneous Soil Mass with a Crust", Part II, Circular Footings", Canadian Geotech. Jl., Vol.18, No.2, pp.265-279.

Scott, R.F. (1963). Principles of Soil Mechanics, Addison-Wesley.

Small, J.C. and Booker, J.R. (1984). "Finite Layer Analysis of Layered Elastic Materials using a Flexibility Approach. Part 1 - Strip Loadings", Int. Jl. Num. Meth. Eng., Vol.20, pp.1025-1037.

Small, J.C. and Booker, J.R. (1986). "Finite Layer Analysis of Layered Elastic Materials Using a Flexibility Approach. Part 2 - Circular and Rectangular Loadings", Int. Jl. Num. Meth. Eng., Vol.23, pp.959-978.

Tham, L.G. and Cheung, Y.K. (1981). "Infinite Rigid Pavement on Layered Foundation", Proc. 8th Canadian Cong. Appl. Mech., Moncton, pp.875-876.

Westmann, R.A. (1963). "Layered Systems Subjected to Surface Shears", Jl. Eng. Mechs Div., ASCE, Vol.89, EM6, pp.177-191.

Numerical Methods in Geomechanics (Innsbruck 1988), Swoboda (ed.)
© 1988 Balkema, Rotterdam. ISBN 90 6191 809 X

Application of Powell's conjugate direction method to slope stability analysis

Jay S.DeNatale & Nassim R.Abifadel
University of Arizona, Tucson, USA

ABSTRACT: Much research has been directed towards the development of generalized limit equilibrium-based safety factor expressions for analyzing the stability of earth slopes. However, little attention has been paid to the implementation of advanced optimization routines to direct the search for the critical slip surface. In order to improve the efficiency of computer-based stability analyses, Powell's Conjugate Direction algorithm is integrated into an existing computer program. The new optimization-based slope stability program is shown to be accurate, efficient, and highly reliable.

1 INTRODUCTION

Evaluation of stability is a necessary consideration prior to any construction involving existing slopes, excavations, or man-made embankments. As such, slope stability analyses are an integral part of most construction projects and are therefore among the most frequent concerns of the practicing geotechnical engineer.

The objective of a given slope stability analysis is to identify the most likely failure mechanism (or critical slip surface) and the corresponding minimum factor of safety. A wide variety of analytical procedures have been developed for this purpose. However, in practice, the overwhelming majority of stability analyses are performed using limit equilibrium techniques.

A limit equilibrium-based slope stability analysis is a rather straightforward minimization problem. As such, a limit equilibrium solution involves the following two steps:

1. The development of a merit function -- in this case, the safety factor expression -- to serve as a scalar measure of the stability of a particular trial slip surface; and
2. The formulation of a search strategy to enable the minimum of this scalar function to be accurately, efficiently, and reliably found.

During the past thirty years, a great deal of research has been concerned with Step (1) above. A variety of restricted and generalized solutions are now available, including those of Bishop, Janbu, Lowe and Karafiath, Morgenstern and Price, Spencer, Taylor, and others (see, for example, Fredlund, 1984). However, little attention has been paid to the implementation of advanced optimization techniques to direct the minimization process. In fact, all stability programs described in the literature use "direct search" methods (the most simplistic class of optimization procedures) to control the search for the critical slip surface (DeNatale, 1986). The traditional grid and pattern search procedures, the alternating variable method, and the simplex routines all belong to this particular class of minimization techniques (Swann, 1972).

In order to improve the efficiency of computer-based slope analyses, a more advanced minimization algorithm -- the Conjugate Direction method of Powell (1964) -- is integrated into an existing stability code -- the STABR program of Lefebvre (1971). The relative accuracy, efficiency, and reliability of the old and new search strategies is established through extensive comparative testing involving a wide range of realistic slope problems. The role and significance of user expertise is assessed by beginning the search at different distances from the actual critical surface.

2 POWELL'S CONJUGATE DIRECTION METHOD

Nearly all minimization routines are iterative in nature; that is, given an initial estimate of the solution, they proceed by generating a sequence of improved ones. The various minimization algorithms are thus distinguished by the particular strategies employed to produce this series of improving approximations. Conjugate Direction procedures attempt to identify the most profitable search directions in the absence of any explicit gradient or curvature information.

The Conjugate Direction Method described by Powell (1964) is perhaps the most efficient member of its class, although several modifications to the basic procedure have been proposed over the years. In Powell's version, the minimum value of a general n-dimensional merit function is identified by cyclically searching along an ever-changing set of conjugate directions. A single starting point $\mathbf{P}_o$ is provided by the analyst, and the coordinate axes are used as the initial direction set $\mathbf{d}_1, \mathbf{d}_2, \ldots, \mathbf{d}_n$. A single iteration of the procedure may be outlined as follows (Powell, 1964):

1. Beginning at the current minimum $\mathbf{P}_o$, conduct a line search along direction $\mathbf{d}_1$ to identify the scalar step-length s_1 which defines the location of the function's minimum in this direction. The new current minimum $\mathbf{P}_1$ will thus be located at

$$\mathbf{P}_1 = \mathbf{P}_o + s_1\mathbf{d}_1$$

Then, repeat this process for each of the remaining n-1 directions. Hence, the position of the most current minimum at the end of the first cycle of n line searches is given by

$$\mathbf{P}_n = \mathbf{P}_{n-1} + s_n\mathbf{d}_n$$

$$\mathbf{P}_n = \mathbf{P}_o + s_1\mathbf{d}_1 + s_2\mathbf{d}_2 + s_3\mathbf{d}_3 + \ldots + s_n\mathbf{d}_n$$

$$\mathbf{P}_n = \mathbf{P}_o + \Sigma s_i\mathbf{d}_i$$

2. Define $\mathbf{d}_m$ to be the direction corresponding to the largest function value decrease, and define Δ to be the magnitude of this decrease. Let F_1 be the function value at $\mathbf{P}_o$, and let F_2 be the function value at $\mathbf{P}_n$. Evaluate the function at $(2\mathbf{P}_n - \mathbf{P}_o)$, and call this value F_3.

3. If:

$F_3 > F_1$ and/or

$2(F_1-2F_2+F_3)(F_1-F_2-\Delta)^2 > \Delta(F_1-F_3)^2$

the current direction set should be retained. Replace $\mathbf{P}_o$ with $\mathbf{P}_n$ and proceed with another iteration by returning to Step 1. If neither of the above inequalities are met, proceed to Step 4.

4. Beginning at $\mathbf{P}_n$, perform a line search in the direction $(\mathbf{P}_n - \mathbf{P}_o)$ to locate the function's minimum in this direction. Call this new minimum $\mathbf{P}_o$. Delete $\mathbf{d}_m$ from the current direction set, add $(\mathbf{P}_n - \mathbf{P}_o)$ to the current direction set, and proceed with another iteration by returning to Step 1.

The procedure is said to converge when a single iteration (or cycle of n line searches) fails to reduce the value of the function, or alter the coordinates of the projected minimum, by some preassigned amount.

Nearly all merit function evaluations are made in conjunction with the line searches. Hence, the overall success of the Conjugate Direction algorithm is directly related to the efficiency of the line search. A number of highly regarded no-derivative line search strategies are available, including those suggested by Powell (1964), Brent (1973), and Fletcher (1980). Powell's original line search algorithm is also described by Acton (1970).

3 STRUCTURE OF THE NEW PROGRAM CSLIP2

Program STABR (Lefebvre, 1971) is used extensively for stability analyses of earth slopes. It employs Bishop's Simplified Method of Slices (Bishop, 1955) to estimate the safety factors associated with circular slip surfaces. The program can examine particular trial surfaces, or it can automatically search for the critical slip circle by means of a built-in pattern search routine.

The new program CSLIP2 was created by merging STABR with the minimization program POWELL (DeNatale, 1987). The structure of CSLIP2 is illustrated in Figure 1. The program is highly modularized, with the main block consisting only of calls to various subroutines. The program begins with a call to subroutine OPEN, which in turn sends a request to the terminal for the name of the appropriate input and output files. The input data is read by making successive calls to subroutine OPDATA and subroutine SSDATA. Subroutine OPDATA reads the initial three lines of the input file which contain the parameter values needed to conduct the Conjugate Direction-based search for the minimum safety factor. Subroutine SSDATA reads the remaining lines of the input file which define the properties of the soil profile. These latter input lines are arranged in a manner nearly identical to those of the original STABR program (see Lefebvre, 1971). Inputting the data in this manner permits CSLIP2 to be easily modified to handle any alternate search procedure or safety factor formulation.

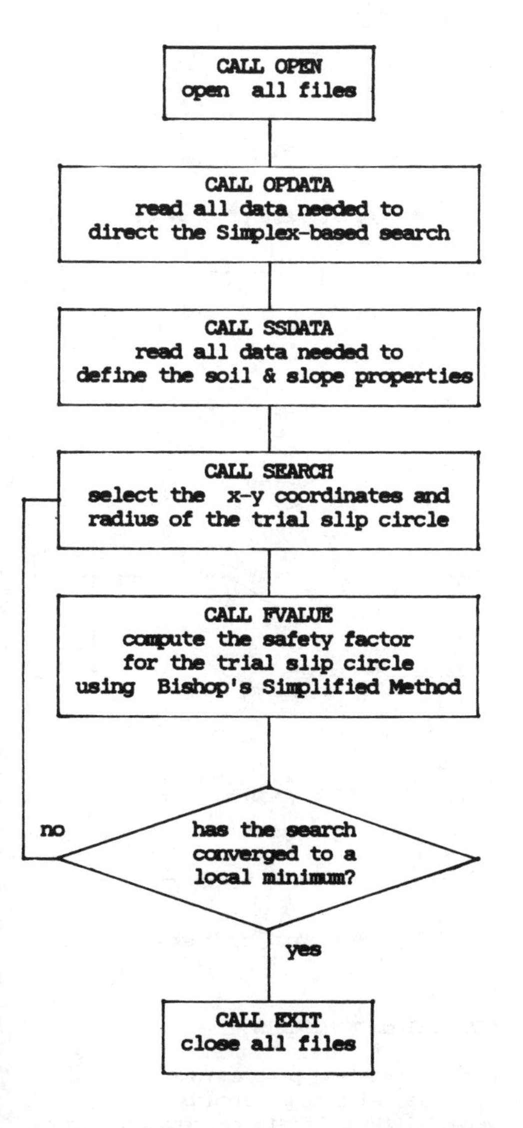

Figure 1. The structure of Program CSLIP2.

Once the input data is read, a call is made to subroutine SEARCH. This subroutine is responsible for directing the search for the minimum safety factor and the associated critical slip surface. An initial trial slip circle is specified in the input file. The x and y center coordinates and radii r of all subsequent trial circles are computed by subroutine SEARCH, based on the Conjugate Direction logic of Powell. Those segments of the original STABR program which compute the safety factor for a given (x,y,r) coordinate set are contained in subroutine FVALUE, which is called by SEARCH whenever a merit function value is required. When a locally minimum safety factor is found, control is returned to the main block, and a final call is made to subroutine EXIT to close all files and terminate the run.

It is important to recognize that any existing limit equilibrium procedure could be used in lieu of Bishop's Simplified Method. Either circular arc or piecewise linear formulations could be employed. Once the particular procedure is integrated into CSLIP2, it may be treated as simply a "black box" function subroutine.

4 THE TEST PROBLEMS

The only way to investigate the reliability of a computer program is to test it on a wide variety of representative problems. Such comprehensive testing has been performed by Abifadel (1987). However, due to space limitations, only two of these trial analyses will be discussed herein. Program CSLIP2 is examined for accuracy, efficiency, and reliability, and the following questions are asked:

1. Does the program locate the critical slip surface;

2. How many merit function evaluations (or trial slip surfaces) are required; and

3. Will the program converge to the minimum safety factor from all possible starting points.

The performance of the Conjugate Direction-based search is also compared to Lefebvre's original pattern search routine.

Problem #1 involves determination of the critical slip circle for a 60 degree slope in a purely cohesive soil (see Figure 2). According to the rules established by Taylor (1937), the critical surface passes through the toe of the slope. This restriction enables the radius of any trial circle to be treated as a dependent variable. Therefore the problem reduces to a two-dimensional search for the (x, y) center coordinates of the critical circle. This first type of slope problem is relatively easy to analyze with traditional grid search and pattern search techniques.

Problem #2 involves a relatively flat 30 degree slope in another purely cohesive deposit (see Figure 3). In this example, the soil is characterized as having a cohesion which increases linearly with depth. Taylor's (1937) research does not deal with such situations, and there is no way of identifying a priori the depth to which the critical circle will pass. Therefore, a full three-dimensional search is required to identify the (x, y) center coordinates and radius r of the critical circle. This second type of slope problem is very difficult to analyze with traditional grid search and pattern search techniques.

Any search must always begin at some user-specified starting point. In the present study, this starting point is defined in term of the (x, y, r) coordinates of the first trial slip circle. Since fewer function evaluations are normally required when the search begins closer to the true minimum, the influence of the starting error (which may be interpreted as a measure of the analyst's expertise) must also be considered. This is accomplished by defining four types of user-specified starts: "expert," "good," "fair," and "poor." An expert starting point is located at a distance of $D = 0.10\ R_c$ from the true minimum, where R_c is the radius of the critical slip surface. A good starting point is at a distance of $D = 0.25\ R_c$ from the true minimum, a fair starting point is at a distance of $D = 0.50\ R_c$ from the true minimum, and a poor starting point is at a distance of $D = 1.00\ R_c$ from the true minimum. Specific starting points for test problems #1 and #2 are indicated in Figures 2 and 3.

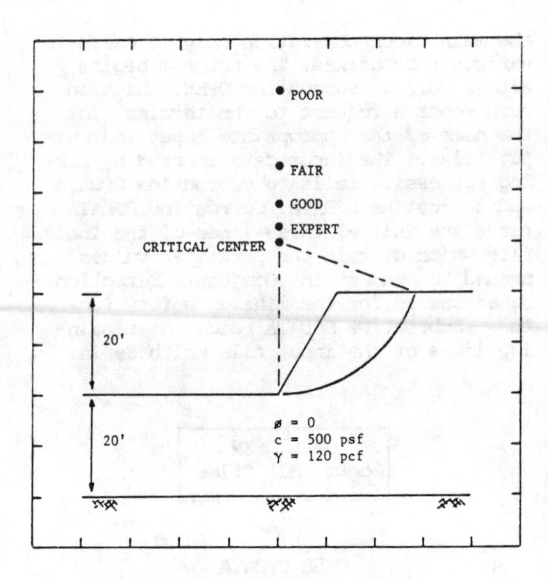

Figure 2. Test problem #1.

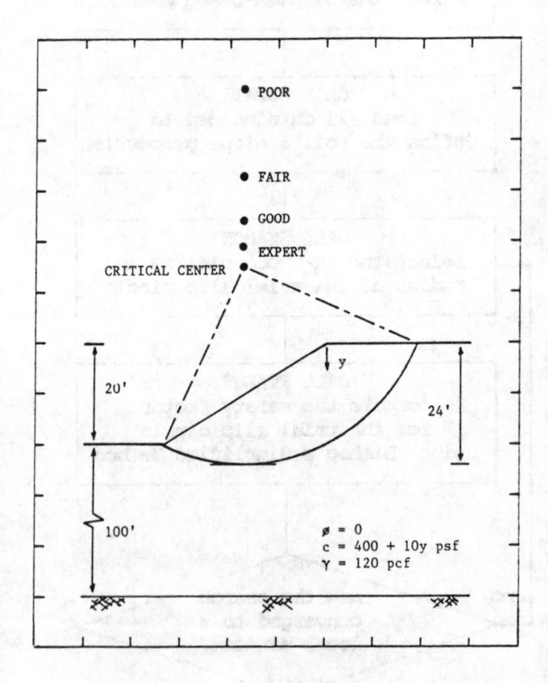

Figure 3. Test problem #2.

5 DISCUSSION OF RESULTS

Each of the test problems was analyzed many times, with the search beginning at several different distances from the true critical slip circle. The results of

Table 1. Relative Efficiency of the Conjugate Direction and Pattern Search analyses.

Problem #	Starting Point	Starting Distance Error (ft)	Number of Trial Circles Required to Locate the Minimum	
			Con-Dir	Pattern
1	Expert	3.0	17	33
1	Good	7.5	24	39
1	Fair	15.0	42	51
1	Poor	30.0	45	76
2	Expert	3.8	22	360
2	Good	9.5	26	465
2	Fair	19.0	31	-
2	Poor	38.0	56	-

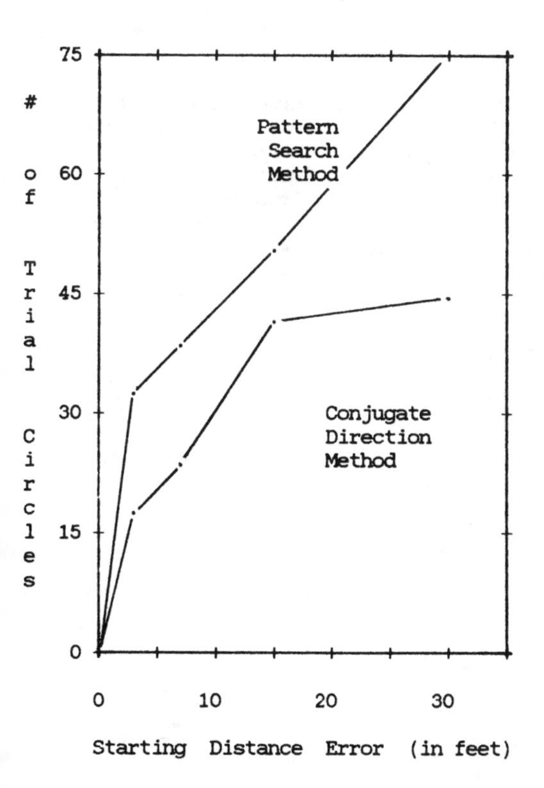

Figure 4. Effect of starting error on search efficiency for Problem #1.

these analyses are presented in Table 1 and Figure 4. As one would expect, a greater number of trial circles must be examined when the absolute starting error is increased. However, beyond a certain point, the efficiency of the Conjugate Direction-based search no longer decreases. In other words, the fair and poor starting estimates each require about the same number of trial surfaces to pinpoint the true critical circle. This occurs because the "step size" (or distance between any two consecutive evaluation points) varies continuously as the search progresses. The results demonstrate that Powell's technique is quite robust, and that a Conjugate Direction-based analysis can still be relatively efficient when performed by a novice engineer with little intuitive feeling for the correct location of the critical circle.

Problems #1 and #2 were also analyzed with the original STABR program (Lefebvre, 1971), which relies on a built-in pattern search routine. In a pattern search, the step size is always the same, regardless of the distance from the true minimum. Hence, when using such a routine, it is common to first isolate a neighborhood of the minimum by specifying a relatively large step size, and then pinpoint the minimum by means of a second search with a much smaller user-specified step length.

The number of mathematical operations required to select a new circle center and radius is always insignificant relative to the number of operations required to evaluate the safety factor for the trial surface. Therefore, the number of safety factor evaluations (or trial circles) may be used as a basis for comparing the relative efficiency of the two search routines. With this in mind, the Conjugate Direction search is clearly the more efficient of the two (see Table 1). The relative efficiency of the Conjugate Direction routine increases dramatically as the dimension of the problem increases from two (in Problem #1) to three (in Problem #2). The constant step-size restriction causes the traditional pattern search procedure to become very inefficient when the dimension of the problem exceeds two.

6 CONCLUSIONS

The new Conjugate Direction-based stability program CSLIP2 is a highly modular computer code. The generalized safety factor calculations are contained within a single function subroutine. Thus, the Conjugate Direction search logic would be very easy to integrate into any existing slope stability program. CSLIP2 is executable on both minicomputers, such as the VAX 11/750, as well as on any IBM PC-compatible machine. Since limit equilibrium-based slope stability programs are already widely used in the geotechnical community, the new procedure should be readily accepted and will therefore be of immediate benefit to the practicing engineer.

7 ACKNOWLEDGEMENT

Program POWELL (DeNatale, 1987) was developed during research sponsored by the National Science Foundation under Grant No. CEE-8404070. The senior author is grateful for this support.

8 REFERENCES

Abifadel, N.R. 1987. Application of Powell's Conjugate Direction method to slope stability analysis, MS Thesis, The University of Arizona, Tucson.

Acton, F.S. 1970. Numerical methods that work, Harper and Row, New York.

Bishop, A.W. 1955. The use of the slip circle in the stability analysis of slopes, Geotechnique, 5: 7-17.

Brent, R.P. 1973. Algorithms for minimization without derivatives, Prentice Hall, Englewood Cliffs.

DeNatale, J.S. 1986. A survey of search techniques for locating the critical slip surface in slope stability analyses, Technical Report, The University of Arizona, Tucson.

DeNatale, J.S. 1987. Program POWELL: For the minimization of general unconstrained multivariate functions by Powell's Conjugate Direction method, Technical Report, The University of Arizona, Tucson.

Fletcher, R. 1972. Conjugate Direction methods, in Numerical methods for unconstrained optimization (W. Murray, editor), Academic Press, London, pp. 73-86.

Fletcher, R. 1980. Practical methods of optimization - Volume I: unconstrained optimization, John Wiley and Sons, Chichester.

Fredlund, D.G. 1984. Analytical methods for slope stability analysis, Proceedings of the 4th International Symposium on Landslides, Vol. 1: 229-250.

Lefebvre, G. 1971. STABR user's manual, University of California, Berkeley.

Powell, M.J.D. 1964. An efficient method for finding the minimum of a function of several variables, Computer Journal, 7: 155-162.

Swann, W.H. 1972. Direct Search methods, in Numerical methods for unconstrained optimization (W. Murray, editor), Academic Press, London, pp. 13-28.

Taylor, D.W. 1937. Stability of earth slopes, Journal of the Boston Society of Civil Engineers, 24: 197-246.

Numerical Methods in Geomechanics (Innsbruck 1988), Swoboda (ed.)
© 1988 Balkema, Rotterdam. ISBN 90 6191 809 X

Z-SOIL.PC: A program for solving soil mechanics problems on a personal computer using plasticity theory

Th.Zimmermann & C.Rodriguez
Zace Services Ltd, Switzerland

B.Dendrou
ZEI Engineering Inc., USA

ABSTRACT : Plasticity theory applied to soil mechanics has gained about twenty years of research experience. Based on this experience, Z_SOIL.PC offers now an attractive tool to solve soil mechanics problems in a unified way. It is a highly interactive, user-friendly program. It includes expert knowledge which makes it possible even for the user who is unfamiliar with plasticity theory to obtain accurate and reliable results. This article describes briefly the key features of the program as well as selection of validation problems.

1. INTRODUCTION

Z_SOIL.PC offers an attractive alternative to traditional approaches to geotechnical problems. It uses recent advances in nonlinear finite element techniques and plastic modeling of soils to solve stability, load carrying capacity and deformation problems in a unified, cost-effective way.

In this paper, the constitutive model and numerical techniques are briefly described first. A more extensive presentation is given in [1]. The correspondance between the proposed approach based on plasticity and classical techniques is then addressed. The interactive program structure is described next. Selected validation problems are finally presented.

2. ON THE PLASTIC MODEL

A plastic model requires a yield criterion, a flow rule and a hardening law.

Z_SOIL.PC uses a Drucker-Prager yield criterion. A non-associated flow rule is assumed which imposes a plastic flow with no dilatancy (by default) and perfectly plastic behavior, i.e., no hardening. These default options were found to be the most adequate choices for most validation problems solved. The experienced user can however modify some of these in order to obtain special effects.

Yield criterion

The Drucker-Prager yield criterion is defined as (1) :

$$f(\underset{\sim}{\sigma}) = a\, I_1 + \quad J_2 - k = 0 \qquad (1)$$

I_1, J_2 are stress invariants and constants a, k can be defined from common geotechnical data : cohesion c and angle of friction ϕ. The characteristics of the yield surface can be adapted so that the plastic collapse matches the one obtained with the Mohr-Coulomb criterion. Under general plane-strain conditions this yields, for the default option of incompressible flow :

$$a = \sin\phi/3 \qquad k = c\cos\phi \qquad (2)$$

Flow rule

The flow rule assumes the existence of a potential surface such that :

$$\dot{\varepsilon}^p_{ij} = \lambda P_{ij} \;;\; P_{ij} = \frac{\partial g}{\partial \sigma_{ij}} = a(\psi)\,\delta_{ij} + \frac{1}{2\sqrt{J_2}}\, s_{ij} \qquad (3)$$

where λ is the plastic loading function and, g a plastic potential, s the stress deviator, and J_2 the second invariant of the stress deviator. ϕ is defined as the angle between the actual flow and purely deviatoric flow (fig. 1).

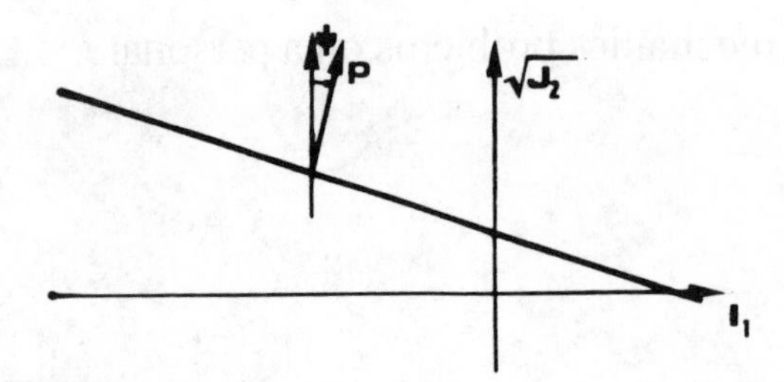

Fig. 1 : Definition of flow direction

3. NUMERICAL IMPLEMENTATION

Several aspect of the numerical modeling have a major influence on the performance of the model. Among these the tangent stiffness, the numerical integration scheme and the stress-point algorithm play an important part.

Tangent stiffness

The use of the proper tangent stiffness appears to be important if localization phenomena are to be simulated. The linearization of the discretized equilibrium equation leads to the following set of linear equations :

$$\underset{\sim}{K}_T \, \Delta \underset{\sim}{d} = \underset{\sim}{F}^{ext} - \underset{\sim}{N}(\underset{\sim}{d}) \qquad (4)$$

where K_T characterizes the global tangent stiffness obtained by assembly of elemental contributions $\underset{\sim}{K}^e_T$, $\Delta\underset{\sim}{d}$ is the displacement increment, $\underset{\sim}{F}^{ext}$ the vector of applied forces and $\underset{\sim}{N}(\underset{\sim}{d})$ a nonlinear function of the displacements :

$$\underset{\sim}{K}^e = \int_{\Omega^e} \underset{\sim}{B}^T \underset{\sim}{D}^{ep} \underset{\sim}{B} \, d\Omega^e \qquad (5)$$

where $\underset{\sim}{D}^{ep}$ is a 6 x 6 matrix obtained by compaction (using symmetries) of :

$$D^{ep}_{ijkl} - C_{ijkl} - C_{ijmn} P_{mn} Q_{op} C_{opkl} / H_o \qquad (5.1)$$

where :

$$H_o = K P_{kk} Q_{ll} + 2G P'_{mn} Q'_{mn} \qquad (5.2)$$

$\underset{\sim}{C}$ is the elasticity tensor, K and G are material constants, $\underset{\sim}{P}$ and $\underset{\sim}{Q}$ are normals to the plastic potential and yield surfaces, respectively, and $\underset{\sim}{P}'$ $\underset{\sim}{Q}'$ are their deviatoric parts.

Numerical integration

Incompressibility usually requires the volumetric contribution to $\underset{\sim}{K}_T$ to be underintegrated in order to avoid locking. An out-of-balance of the equilibrium equation (3) will result unless the same integration scheme is adopted for $\underset{\sim}{N}(\underset{\sim}{d})$.

A proper way to obviate this problem is the use of strain-projection methods [2].

Stress-point Algorithm

The stress-point algorithm computes the stress increment given the strain increment. Considerable attention has been given to this problem lately. The approach implemented in Z_SOIL.PC follows the return-algorithm proposed by Nguyen [3]. The algorithm reduces to the following steps.

Table 1 : radial-return ALGORITHM (fig. 2)

1. Compute trial stress at $t=(n+1)\,\Delta t$ (fictitious time)

$$\underset{\sim}{\sigma}^{tr}_{n+1} = \underset{\sim}{\sigma}_n + \underset{\sim}{C} \cdot \Delta\underset{\sim}{\varepsilon}$$

2. Compute σ_{n+1}

2.1 If $f(\sigma^{tr}_{n+1}) \leq 0$

$$\underset{\sim}{\sigma}_{n+1} = \underset{\sim}{\sigma}^{tr}_{n+1} \text{ (elastic process)}$$

2.2 Otherwise : $\underset{\sim}{\sigma}_{n+1} = \underset{\sim}{\sigma}^{tr} + d\underset{\sim}{\sigma}$ (plastic process) with :

$$d\bar{\sigma} = -\lambda \, 3K \, a(\psi)$$

$$d\underset{\sim}{s} = -\lambda \frac{\mu}{\sqrt{J_2^{tr}}} \underset{\sim}{s}^{tr}$$

$$\lambda = \frac{f(\sigma^{tr})}{9Ka(\phi)a(\phi) + \mu}$$

$$d\underset{\sim}{\sigma} = d\bar{\sigma}\,\underset{\sim}{\delta} + d\underset{\sim}{s}$$

N.B. : a special treatment is necessary at the vortex

σ_m = mean stress, $\underset{\sim}{s}$ = deviatoric stress

3.1 Plasticity based approach to stability problems

When no external load is applied the traditional definition of the safety factor, i.e. the ratio of the ultimate load to the service load fails and a slightly different approach needs be adopted.

The slip surface approach to soil stability problems defines the safety factor as the ratio of the sum over the slip surface of available shear strength on the sum of mobilized shear strength. That is :

$$S = \int_{\Gamma_s} (C + \bar{\sigma}_n \text{ tg } \phi) \, d\Gamma_s / \int_{\Gamma_s} \tau \, d\Gamma_s \tag{6}$$

where C is the cohesion, $\bar{\sigma}_n$ the effective normal stress, ϕ the angle of shearing resistance, S the safety factor $\tau_y = C + \bar{\sigma}_n \text{ tg } \phi$ the yield stress according to the Mohr-Coulomb criterion.

An algorithm can be deduced from (6) which fits the plasticity based approach; this is shown in table 2.

Table 2 : stability algorithm

0. $i = 1$

1. Set $S_i = 1$; $C_i = C$; $(\text{tg } \phi_i) = \text{tg } \phi$

2. Solve B.V.P.

3. Perform nonlinear iteration until $\tau \leq \tau_y$ on the problem domain or divergence occurs

4. $S_{i+1} = S_i + \Delta S$
   $C_{i+1} = C/S_{i+1}$
   $(\text{tg } \phi)_{i+1} = (\text{tg } \phi)/S_{i+1}$

5. Go to 3.

---

4. USER INTERFACE

The importance of the user interface cannot be underestimated. In order to make a nonlinear finite element program adequate for practice it is necessary that the user interface has the structure of an Expert System. This means that full proof default options must be built in which make expert knowledge accessible to the user.

In addition, the experienced user must be able to overrun these options in order to introduce his own expertise and he should be able to get advice from the program. To this extent Z_SOIL.PC works like an expert system. It does not however comment on the options chosen by the user or the obtained results.

Z_SOIL.PC gives the user access to three environments through functions keys (F1, F2, F3) (fig. 2). These three environments are : INPUT, ANALYSIS and GRAPHICS.

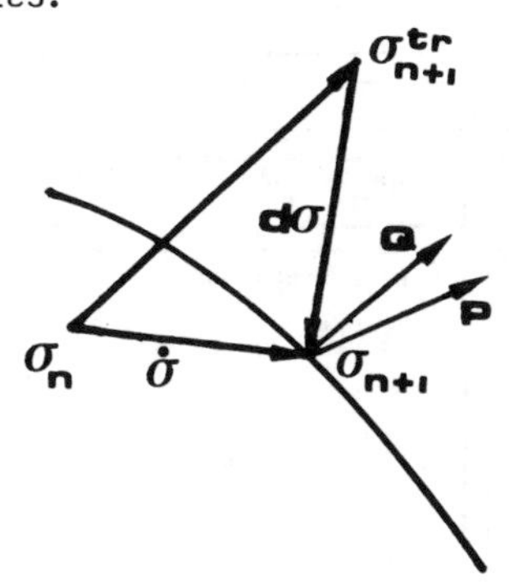

Fig. 2 : Return algorithm

In the INPUT PHASE environment the user specifies the type of action through the main menu "CREATE INPUT FILE...DENSE MESH...". For each of these options the user is prompted again to choose a particular item from the secondary menu that automatically appears on the left hand side of the screen. The user then has the flexibility to select any item on this list without prescribed order according to its own priority. A help command activated through a particular function key provides all necessary information on how to properly use that option.

The COMPUTATIONAL PHASE executes the nonlinear analysis of the soil structure. During program execution, the user has the flexibility to update the important parameters of the nonlinear algorithm such as the maximum number of iterations per loading step; to change the tolerance of the Newton-Raphson scheme; or to alter the total number of loading steps. At the execution mode a secondary menu is available on the left side of the screen allowing the user to select the type of information that will be prompted on the screen during the program execution phase. In many cases the user will be interested in intermediate results given at the end of

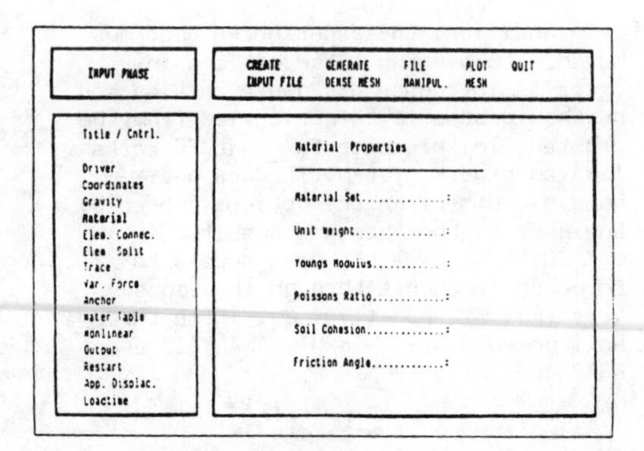

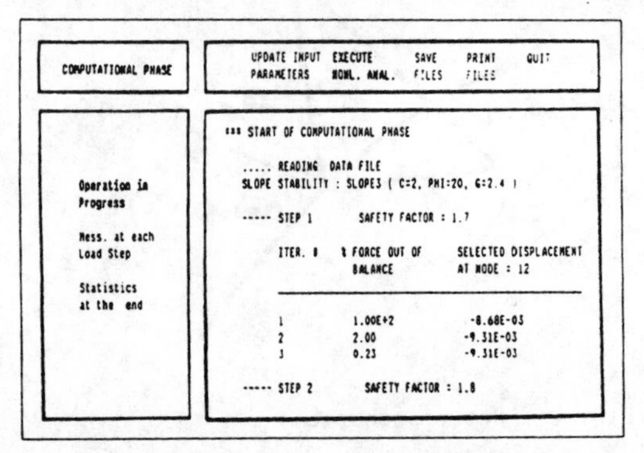

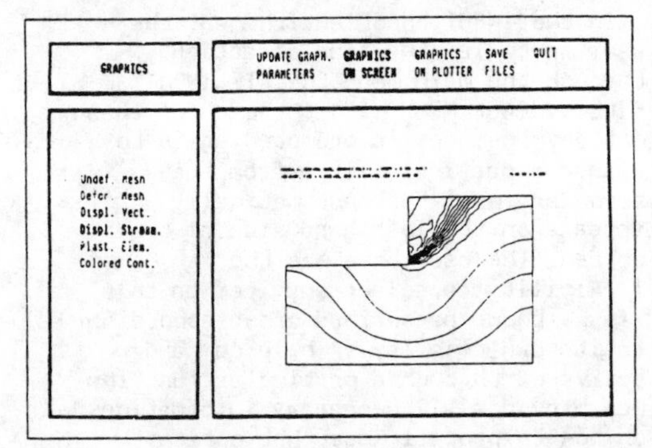

Fig. 3 : Displays of main menus

each loading step, or a pertinent statistics of the analysis at the end of the run. After a successful execution of the program two choices are available : the different output files can be either saved for later use, or printed. Such output files are for example the plotting files to be used in the graphic phase, the files containing deformations and stresses at each loading step, and the log-files.

The OUTPUT GRAPHIC PHASE provides a precise visual representation of the Z_SOIL.PC nonlinear solutions. Graphics are created first on the screen.

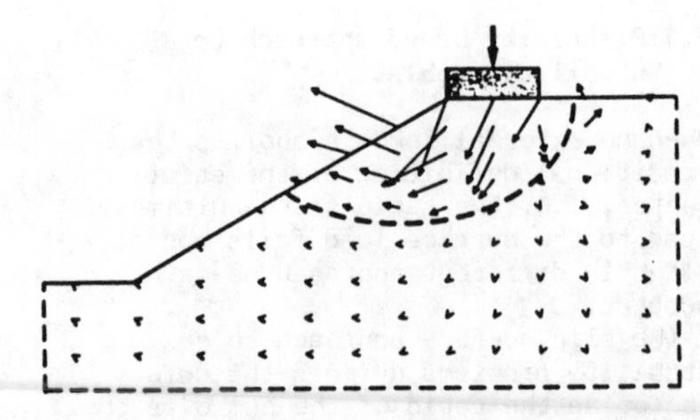

Fig. 4 : Slope stability displacement field

Selected finalized graphics are then queued to a dot-printer. The following graphics options are available on the secondary menu : deformed mesh of soil structure, displacement vectors, streamlines of displacements field. Plastified zones of the soil medium, color graphic representation of displacements and stress fields. Figure 3 illustrates a typical screen view of the input, computational and graphic phases. The top box of the display contains the main menu for each phases of the program. Secondary menus are shown on the left side of the screen. The currently "selected" menu option is easily distinguished because it appears in reverse type. Passage from one menu option to another is done easily by using the horizontal or vertical cursor arrow keys (→ and ←) or (↑ and ↓), respectively for the main menu and for the secondary menu. For each menu option a help command is available to assist the user in the implementation of that particular Z SOIL.PC function.

5. VALIDATION PROBLEMS

Two classical geotechnical problems are analyzed : the slope stability problem and the footing problem.

The slope stability problem

Results of interest for this type of problem include mainly the slope safety factor and the displacement field.

The displacement field is indeed illustrative of the failure mechanism and allows a clear identification of the slip surface as shown on figure 4.

As compared to alternative approaches, the plasticity based approach appears to yield similar results when a comparizon with classical methods is possible and it is more flexible when more general slip surfaces occur (fig. 5).

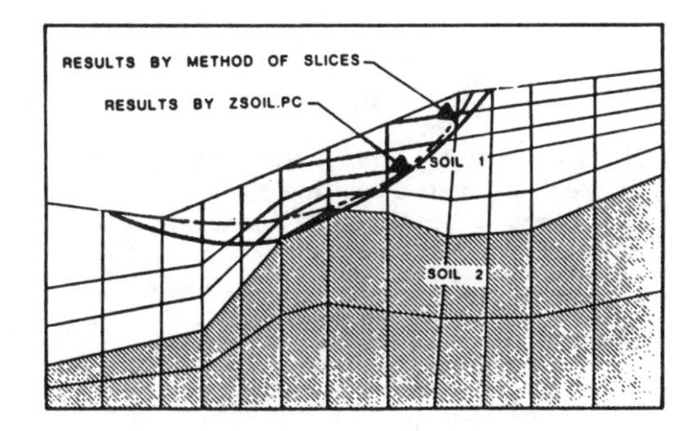

Fig. 5 : Slip surfaces computed with Z SOIL.PC

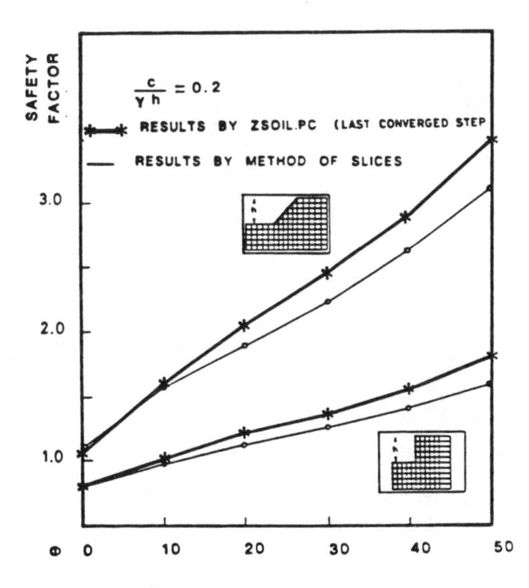

Fig. 6 : Slope safety factor predicted by Z_SOIL.PC and the method of slices

The slope safety factor as shown in the previous section, defined as the factor of reduction applied to the yield surface coefficients when failure occurs. Predictions of the slope stability factor obtained by use of Z_SOIL.PC are compared to results obtained by the conventional method of slices on figure 6.

Table 3 shows a similar comparizon with various approaches. Again slightly higher values are obtained using Z_SOIL.PC.

The footing problem

The problem of the bearing capacity of a rigid superficial footing is analyzed next. The problem definition and numerical results are shown on figure 7. Experiment and computation are in reasonably good agreement.

Table 3 : Comparizon of safety factors predicted by different models of the slope stability problem (ϕ = 20°, γh = 100)

tan Φ / c/γh	Simplified Bishop	Ord. Meth. of slices	Friction Circle	Janbu Procedure	Z_SOIL.P
2	1.17	1.12	1.14	1.10	1.20
5	1.83	1.73	1.78	1.70	2.00
8	2.48	2.30	2.36	2.26	2.60

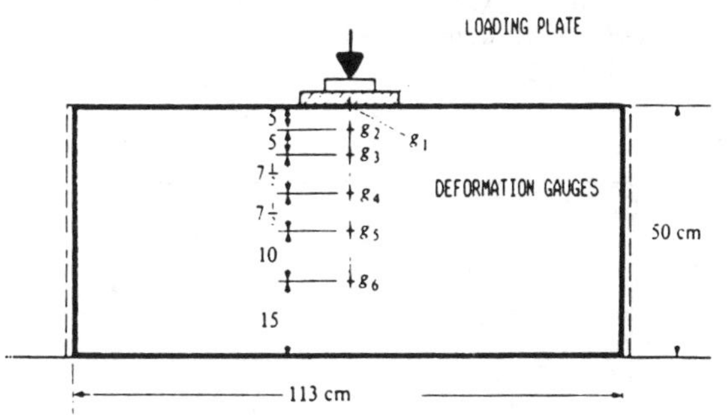

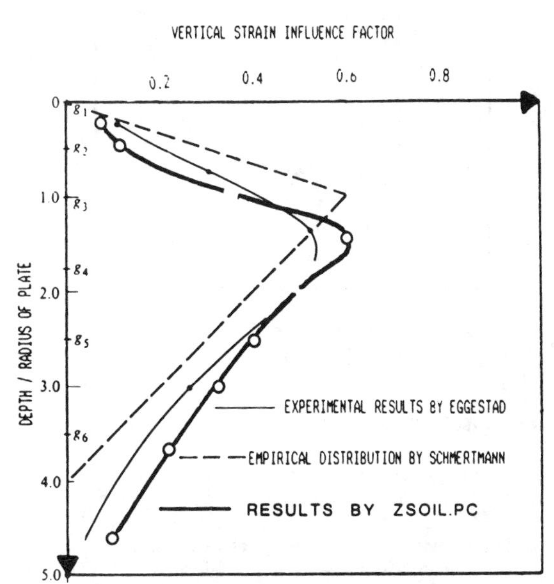

Fig. 7 : Comparizon of vertical strains predicted by Z_SOIL.PC with experimental and empirical results

Similarly, a parametric study of the ultimate load as a function of the angles of friction shows an excellent coïncidence with theoretical results obtained by Salencon for friction angles up to 40 % (fig. 8); the solution above this angle then progressively deviates from the theoretical solution and reaches and undershoot of 23 % at 50°. Results obtained for high angles of friction have been reported to be very unstable numerically by several authors. To the present state of knowledge the above results can therefore be considered to be quit encouraging.

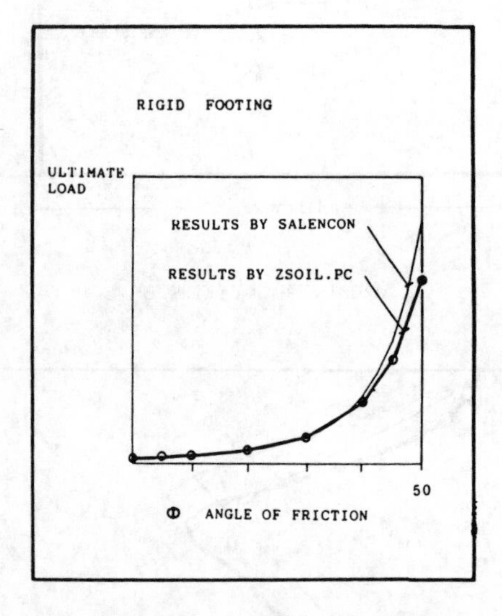

Fig. 8 : Validation results for the rigid footing

6. CONCLUSIONS

Results presented in the article show that load carrying capacity problems and slope stability problems which represent a large portion of the problems most commonly encountered in geotechnical practice can be solved in an efficient and cost effective way using plasticity theory.

The implementation in program Z-SOIL.PC is highly interactive, features expert knowledge and advanced graphic capabilities which provide the geotechnical engineers with the adequate tool for dayly practice.

REFERENCES

[1] Z_SOIL. Theory and implementation. In preparation.
[2] Hughes, T.J.R. 1987. The finite element method. Prentice Hall.
[3] Nguyen, G.S. 1977. On the elastic-plastic initial boundary value problem and its numerical integration. I.J. N.M.E, v 11, pp 817-832.

Numerical Methods in Geomechanics (Innsbruck 1988), Swoboda (ed.)
© 1988 Balkema, Rotterdam. ISBN 90 6191 809 X

Prediction of the settlement of an embankment using a micro-computer

A.S.Balasubramaniam, S.Chandra & Pisit Kuntiwattanakul
Division of Geotechnical and Transportation Engineering, Asian Institute of Technology, Bangkok, Thailand

ABSTRACT: The stress strain behavior is modelled by an approach based on the modified Cam Clay Theory and the Pender's model. Using this modified approach, expressions for the strain increment during the anisotropic consolidation of normally consolidated clay are derived. The settlement of a large embankment at the centre line was then predicted under plane strain conditions. The stress distribution was calculated using Micro-FEAP program. The SOFT-CLAY program developed on a micro-computer (using the Basic language) with the modified relationship successfully predicted the shear strain for the stress paths for which the stress ratio decreased.

1 INTRODUCTION

The ability to predict the stress-strain behavior of soft clays is very useful for geotechnical engineers, as most of the settlement problems are encountered with soft clay deposits. This study is aimed to apply the stress-strain theories to practical problems. The previous studies of the stress-strain theories were largely confined to the comparison of the results predicted from the theories with the observations of specimens prepared in the laboratory and tested under controlled conditions. However, the present study contains the applications of the theories to a practical problem of an embankment on a soft clay deposit. The theory can explain many features of the stress-strain behaviour of normally consolidated clays. However, the undrained stress paths predicted by the theory do not agree with the test results and the theory over-predicted both the shear strain and the volumetric strain and the pore water pressure.

ROSCOE & BURLAND (1968) have modified the theory by introducing a new energy balance equation and proposed a new model, called the modified Cam-clay model. This model is successful in predicting the undrained stress path and the volumetric strain of normally consolidated clays. However, the theory under-estimated the shear strain. ROSCOE & POOROOSHASB (1963) had proposed a stress-strain theory, called the incremental stress-strain theory, derived from experimental observations and empirical strain equations.

PENDER (1978) propósed a model for the behaviour of overconsolidated clays. In deriving the model, the assumptions used are similar to those used in the Cam-clay theory, and, the modified Cam-clay model. The distinct feature of this model from others is the recognition that overconsolidated soil can experience plastic strains, both volumetric and shear. The stress path method of estimation of settlement has serious limitations. An alternative approach, proposed by BURLAND (1971), is to make use of a stress-strain theory whose accuracy can be checked in the laboratory so that the limitation can be assessed. The testing of the real soil is then aimed at providing the material parameters required in the theory. BURLAND (1971) demonstrated that the modified Cam-clay model (ROSCOE & BURLAND, 1968) can be successfully used to predict the pore-water pressure and the displacement beneath embankments.

2 STRESS-STRAIN THEORIES INVOLVED IN THE COMPUTER PROGRAM

The computer program for the prediction of the stress-strain behaviour of clay is based on two models, the modified Cam-clay model of ROSCOE & BURLAND (1968) and the model proposed by PENDER (1977). Hereinafter, the model proposed by PENDER (1977) and (1978) will be called Pender's model.

PENDER (1977) extended the model for overconsolidated clay to cover the behaviour of normally consolidated clay. In this study the computation of the strain increment developed during anisotropic consolidation of normally consolidated clay will be modified from that proposed by PENDER (1977). The following paragraphs will discuss the need for a modified model.

The plastic volumetric strain increments given by PENDER (1977) are not exactly the volumetric strain increment induced during anisotropic consolidation. The expression for the plastic strain increment should be modified as follows:

Referring to Fig.1 for the meaning of subcripts a,b and c, total volumetric strain equation, along bc should be:

$$(dv)_\eta = \frac{\lambda}{(1+e)} \frac{dp_{bc}}{p_b} \qquad (1)$$

The plastic volumetric strain increment along bc should be:

$$(dv^p)_\eta = \frac{1}{(1+e)} \frac{(\lambda-\kappa)}{p_b} dp_{bc} \qquad (2)$$

Hence, the ratio of the total volumetric strain increment to the plastic volumetric strain increment for anisotropic consolidation is:

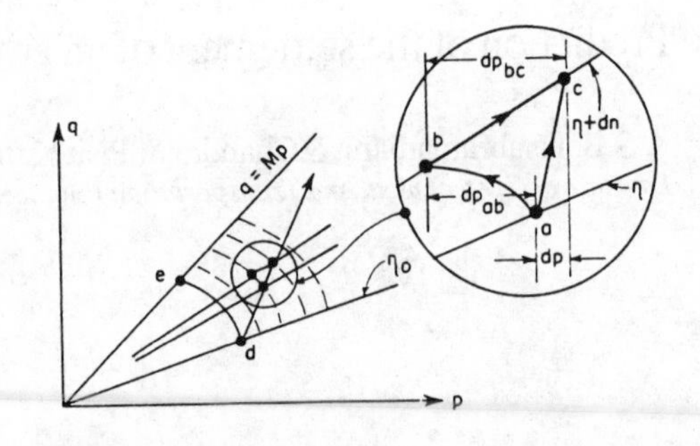

FIG. I COMPONENT OF THE GENERAL STRESS INCREMENT FOR CALCULATION OF VOLUME CHANGE IN NC. REGION.

$$\left(\frac{dv}{dv^p}\right)_\eta = \frac{\lambda}{(\lambda-\kappa)} \qquad (3)$$

The ratio of the volumetric strain increment to the shear strain increment can be written as:

$$\left(\frac{dv}{d\varepsilon}\right)_\eta = \frac{\lambda}{(\lambda-\kappa)} \left(\frac{AM-\eta}{\eta}\right) \qquad (4)$$

But, BALASUBRAMANIAM (1976) had shown that the relationship between the strain increment ratio, $dv/d\varepsilon$ and the stress ratio, η for anisotropic consolidation path is very similar to that predicted by the modified Cam-clay model, i.e.:

$$\left(\frac{dv}{d\varepsilon}\right)_\eta = \frac{\lambda}{\lambda-\kappa} \frac{M^2-\eta^2}{2\eta} \qquad (5)$$

Fig. 2 shows the relationship between the strain increment ratio and the stress ratio predicted by the Pender's Model and that predicted by the modified Cam-clay model together with laboratory test results for Kaolin. It can be seen that the values predicted by Eq.5 better fit to the tests result than those predicted by Eq.4.

From Eq.1 and 5 the plastic shear strain increment, which is assumed to be the total shear strain increment, during anisotropic consolidation can be obtained from:

$$(d\varepsilon)_\eta = \frac{(\lambda-\kappa)}{(1+e)} \frac{dp_{bc}}{Pb} \frac{2\eta}{M^2-\eta^2} \qquad (6)$$

Now the incremental shear strain from a to c is assumed to be the sum of incremental shear strains from paths along a-b and b-c as in the theory of Roscoe & Poorooshasb (1963). The shear strain contribution along path a-b ie. segment of an undrained stress path, is modelled using Pender's model. In the computer program the strain increment during anisotropic consolidation is based on Eqs. 1 and 6.

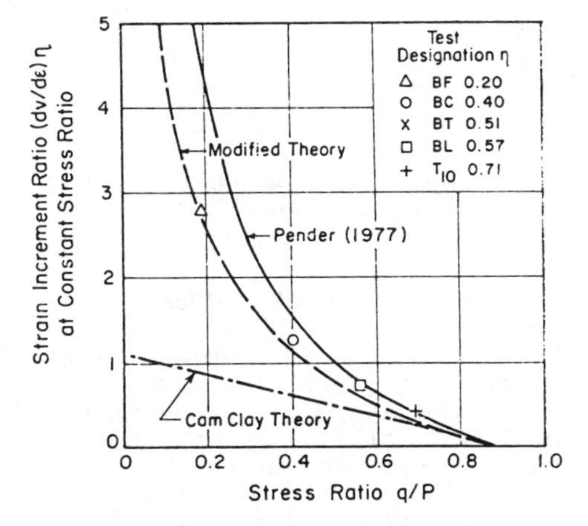

Fig.2 Observed and Predicted Relationships of Strain Increment Ratio and Stress Ratio.

The stress-strain models described above are developed to predict the results of triaxial tests, i.e. axi-symmetric conditions. But this study will be concentrated on the prediction of settlement of a long embankment under plane strain conditions. Plane strain version of the stress-strain models are therefore obtained.

For normally consolidated clay, ROSCOE & BURLAND (1968) showed that, to a close approximation, the stress-strain equations of the modified Cam-clay model for plane strain are identical in form to the corresponding axi-symmetric equations with $t = (\bar{\sigma}_1 - \bar{\sigma}_3)/2$ and $s = (\bar{\sigma}_1 + \bar{\sigma}_3)/2$ replacing q and p respectively and $M / \sqrt{3}$ replacing M. The relevant strain parameters in plane strain corresponding to the stress parameter s and t are denoted respectively by

$$dv = d\varepsilon_1 + d\varepsilon_3 \qquad (7)$$

$$d\gamma = d\varepsilon_1 - d\varepsilon_3 \qquad (8)$$

These satisfy the condition of plastic work increment:

$$dw = s.dv + t.d\gamma = \sigma_1 d\varepsilon_1 + \sigma_3 d\varepsilon_3 \qquad (9)$$

The consequence of choosing the above strain parameters, elastic strains for over consolidated clay are given by:

$$dv = \frac{2}{3}\frac{(1+\nu)}{K}.ds \qquad (10)$$

$$\text{and} \quad d\gamma = \frac{1}{G}\,.\,dt \qquad (11)$$

From the relationship between K and G, $d\gamma$ may be written as follows:

$$d\gamma = \frac{\kappa}{(1+e)}\cdot\frac{1}{(1-2\nu)}\cdot\frac{dt}{s} \qquad (12)$$

In this study it is assumed that the method proposed above is also valid for Pender's model. Though the validity of the above approach to be used in the Pender's model has not been verified yet, it will be shown subsequently that for normally consolidated clays the results of undrained stress path and stress-strain curves predicted by both models are similar. Hence replacing q and p with t and s respectively and M by $M/\sqrt{3}$ in Pender's model will also give similar results for the undrained stress path and the stress-strain curve to those predicted by the plane strain version of the modified Cam-clay.

Undrained Pore Pressure Response

SKEMPTON & BJERRUM (1957) suggested that, for saturated soil, the pore pressure, u, induced by a foundation loading can be obtained from:

$$\Delta u = \Delta\sigma_3 + A(\Delta\sigma_1 - \Delta\sigma_3) \qquad (13)$$

where A is the pore pressure parameter and $\Delta\sigma_1$ and $\Delta\sigma_3$ are change in total stresses.

However, the application of Eq. 13 to the general stress condition, other than axi-symmetric condition, is not appropriate. A more general expression for the excess pore pressure is given by

$$\Delta u = \frac{1}{3}(\Delta\sigma_1+\Delta\sigma_2+\Delta\sigma_3) + \alpha[(\Delta\sigma_1-\Delta\sigma_2)^2+(\Delta\sigma_2-\Delta\sigma_3)^2+(\Delta\sigma_3-\Delta\sigma_1)^2]^{\frac{1}{2}} \quad (14)$$

where α is a coefficient which accounts for the excess pore pressure generated during pure shear.

In this study the excess pore pressure generated by a long embankment is obtained by using Eq. 14 and is expressed as:

$$\Delta u = \frac{1}{2}(\Delta\sigma_1+\Delta\sigma_3)+\alpha\sqrt{\tfrac{3}{2}}(\Delta\sigma_1-\Delta\sigma_3) \quad (15)$$

In modifying the model, a concept similar to the threshold effects is used. The main idea of the concept is that whenever the direction of the stress increment applied to the soil differs from the previous stress, the shear strain produced by the new stress will initially be very small. Hence, the relationship between the strain increment ratio and stress ratio (Eq. 5) may be modified as given below.

For a stress path in which the stress ratio is increasing the equation is given as:

$$\frac{dv}{d\varepsilon^p} = \frac{\lambda}{(\lambda-\kappa)} \frac{(M-\eta_0)^2-(\eta-\eta_0)^2}{2(\eta-\eta_0)} \quad (16)$$

and for a path for which the stress ratio is decreasing, the equation is

$$\frac{dv}{d\varepsilon^p} = \frac{\lambda}{(\lambda-\kappa)} \frac{(M-\eta_0)^2-(\eta-\eta_0)^2}{2\sqrt{(\eta-\eta_0)\eta}} \quad (17)$$

where η_0 is the value of η from which the stress ratio changes by increasing or decreasing.

By using this new strain ratio equation, the shear strain are recalculated and plotted in Fig. 3. It can be seen that the results are noticeably improved. The shapes of the stress-strain curves are similar to those obtained from test results.

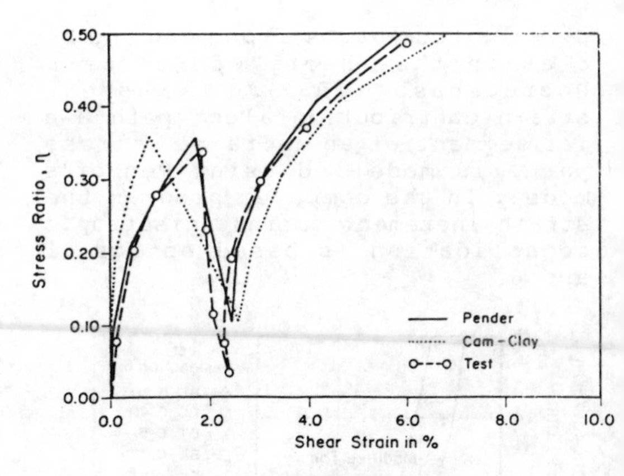

Fig. 3 Stress Ratio – Strain Relationship Using the Modified Strain Ratio Equation

3 DETAILS OF SOFTCLAY PROGRAM

The purpose of the computer program developed herein is to predict the settlement along the center line of a long embankment. In the computer program it is assumed that before yielding, the stresses in the soil mass which are induced by the embankment load can be estimated by means of the elastic theory. Once yielding occurs redistribution of the stress makes the elastic theory invalid. However, it has been shown by HOEG, et al (1968) that the vertical stress increment under a flexible loading is largely independent of yielding, but the magnitude of the horizontal stress is sensitive to yielding. The effects of local yield on the distribution of stress will be discussed in detail later.

The computer program does not take the secondary compression into account. However, if the soil is not a highly organic clay or a highly micaceous soil, the amount of secondary compression is small compared to the total settlement.

4 SUBSOIL CONDITIONS AT NONG NGOO HAO SITE

The subsoil conditions at the Nong Ngoo Hao site is typical of marine clays in South-East Asia which are soft and silty clays, light grey in colour. According to the strength and compressibility characteristics the upper clay layer has been divided into three zones, (i) the weathered zone extending to a depth of about 3-6 m below the ground surface, (ii) a highly compressible soft clay occurring to a depth of about 10 m below the bottom of the weathered zone and (iii) a layer of stiff clay extending to a depth of about 10 m below the bottom of soft clay. Below the stiff clay alternative layers of sand and clay extend to depths more than a few hundred metres (BALASUBRAMANIAM, 1979). The subsoil parameters used in the analysis for the weathered clay and soft clay are given in Table 1.

5 RESULTS OF THE SETTLEMENT ANALYSIS

The immediate settlement predicted by the computer program using finite element analysis, Poulos and Davis's charts and

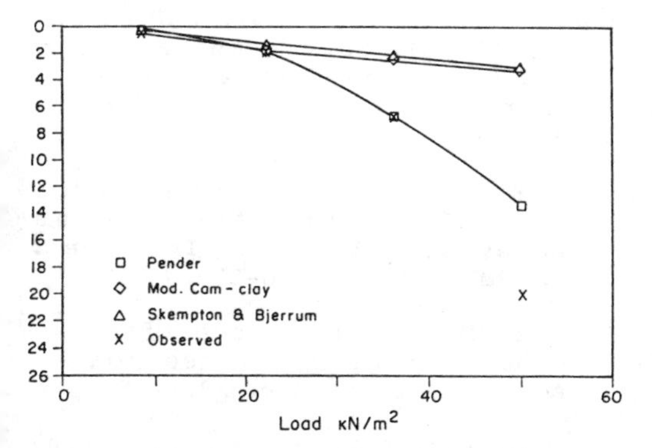

Fig.4 Predicted and Observed Immediate Settlement

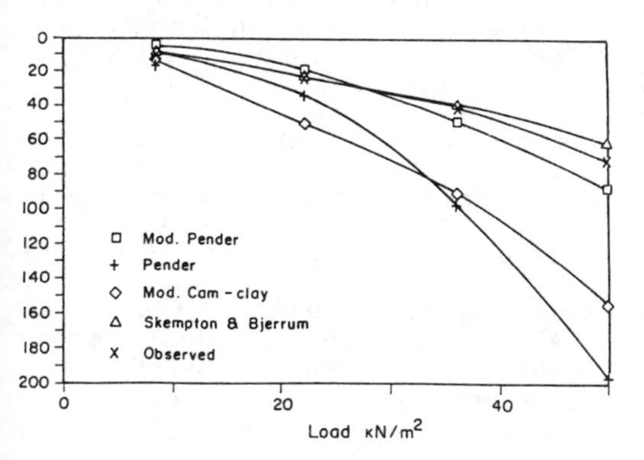

Fig. 5 Predicted and Observed Long Term Settlement

Table 1 - Subsoil Parameters

	Weathered Zone	Soft Clay
Depth	0-5 m	5-10 m
M	0.9	1.05
λ	0.52	0.51
κ	0.072	0.091
e_a	5.88	5.23
e_{cs}	5.58	4.93
γ	14 KN/m³	14 KN/m³
OCR	4-2	1-1.6

Gray's formulae were obtained for an yield stress ratio value of 0.8. The results obtained by finite element analysis are presented in Fig. 4. It was observed that the modified cam-clay underestimated the magnitude of immediate settlement for all the three approaches. The relationship between the load and the immediate settlement was observed to be linear as a consequence of the assumptionthatthe overconsolidated soil behaves as an elastic material. On the other hand, the relationship between the load and the immediate settlement predicted by using the Pender's model was found to be nonlinear.

The results of the long term settlement predicted by the computer program using the value of an yield stress ratio of 0.8 are presented in Fig. 5. This Figure shows that for high embankments, the magnitude of the long term settlement predicted by the Pender's model is always higher than those predicted by the modified Cam-clay model. This is a consequence of the overestimation of the volumetric strain in the Pender's model.

Both the modified Cam-clay and the Pender's model overestimate the magnitude of the long term settlement by an order of two. It is because both the models overestimate the shear strain for the drained stress path (the stress ratio is decreased).

The effects of the choice of the yield stress ratio on the magnitude of the immediate settlement predicted by the Pender's model for all the three methods were studied. Typical results for finite element analysis are presented in Fig. 6. It should be noted that though the construction period is short the consolidation settlement would have occurred been completed. However, in the computer program it is assumed that no drainage occurs during the construction period. Therefore, the predicted value of the immediate settlement should be equal or less than the observed values. The stress distribution computed from Poulos and Davis's charts overestimated the immediate settlement for all the yield stress values. In this study, the yield stress ratio value of 0.8 has been found to be the best value for both the settlements predicted by Gray's formulae and by finite element analysis.

The effect of the choice of the yield stress ratio value on the magnitude of the long term settlement computed by finite element analysis was studied and it was not found to be significant.

6 CONCLUSIONS

The relationship between the strain increment ratio and the stress ratio as suggested in this study can be successfully used to predict the shear strain for the stress paths for which the stress ratio decreased. The computer program developed in Basic language and run on IBM PC is found quite useful in predicting both immediate and long term settlements at the centre line of an embankment. By modifying the strain increment ratio and stress ratio relationship in accordance with this study, the predicted settlements are considerably improved.

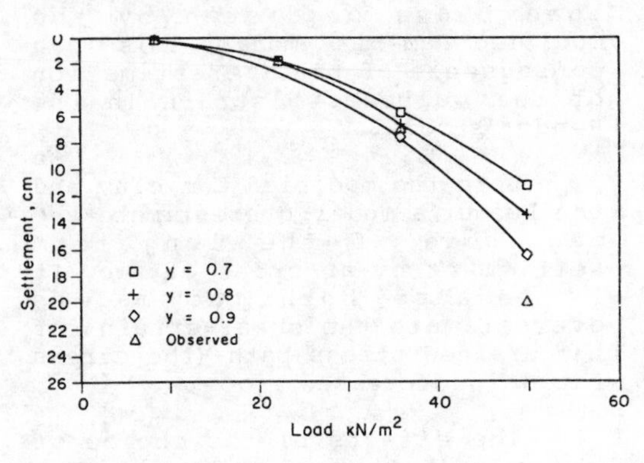

Fig. 6 Effect of the Yield Stress Ratio on Immediate Settlement

7 REFERENCES

Balasubramaniam, A.S. 1976. Stress-Strain theories for normally consolidated clays. Australian Geomechanics Journal.35-42.

Balasubramaniam, A.S. 1979. Stability and settlement of Embankment on Soft Bangkok clay. Proc. 3rd Int. Conf. on Nom. Met. Geomech., Aachen:1373-1411.

Burland, J.B. 1971. A method of estimating the pore pressure and displacement beneath embankments on soft natural clay deposits. Proc. The Roscoe Memorial Symposium , Cambridge:505-536.

Hoeg, K., Christian,J.T. & Whitman, R.Y.(1968). Settlement of strip load on Elastic-plastic soil, Jour. SMFE Divs., ASCE 94-SM2 : 431-445.

Pender, M.J. 1977. A unified model for soil stress-strains behaviour. Proc. 9th Int. Conf. on SMFE, Tokyo:213-222.

Pender, M.J. 1978. A model for the behaviour of overconsolidated soil. Geotechnique. 28:1-25

Roscoe, K.H. & Poorooshasb, H.B. 1963. A Theoretical and experimental study of strains in Triaxial tests on normally consolidated clays. Geotechnique. 13: 12-38.

Roscoe, K.H. & Burland, J.B. 1968. On the generalized stress-strain behaviour of wet clay. Engineering Plasticity, Cambridge Univ. Press. :535-609.

Skempton, A.W. & Bjerrum, L. 1957. A contribution to settlement analysis of foundations on clay. Geotechnique. 7: 168-178.

Numerical Methods in Geomechanics (Innsbruck 1988), Swoboda (ed.)
© 1988 Balkema, Rotterdam. ISBN 90 6191 809 X

Soil-structure interaction analyses on microcomputers

J.K.Jeyapalan & S.W.Ethiyajeevakaruna
Wisconsin Hazardous Waste Management Center, University of Wisconsin-Madison, USA

ABSTRACT: Full versions of soil-structure interaction finite element analyses on microcomputers have been implemented successfully. Complex soil-structure interaction projects have been analysed on the microcomputers with large meshes and full nonlinear soil stress-strain models. The graphic capabilities of these microcomputers are also found to be very useful for the validation of input data and for the representation of computational results. The use of pre- and post-processing modules are discussed. Some case studies are also presented to demonstrate the capabilities of the soil-structure interaction finite element program.

1 INTRODUCTION

Microcomputers are common in today's geotechnical engineering design offices and laboratories. These computers are generally used for performing various analyses required in day-to-day design practice. Stand-alone ability of these machines provide additional advantages over remote terminals connected to mainframes. They also provide an efficient and cost effective environment for implementing numerical techniques such as the soil-structure interaction finite element analyses. However, at present soil-structure interaction analyses are generally not performed in many design offices. This is mainly due to the presumption, that such analyses can be performed only on large computers. The purpose of the present paper is to demonstrate that such analyses can indeed be performed on microcomputers, such as the IBM PC. The IBM PC was chosen, because of the popularity of IBM PC and compatibles, mainly among the engineering community. The IBM PC with 640KB of memory and a hard disk are today's norm and best define a microcomputer. It is generally viewed by experts, in spite of the recent introduction of the powerful Personal Systems Series by IBM, and MacSE and Macintosh II computers by Apple, the IBM PC will continue to remain with a very large user base. The popularity of these machines is further endorsed by the fact, new computers from both IBM and Apple are expected to offer MS-DOS capabilty, atleast as an option.The use of microcomputers for geotechnical analyses and the capability of some programs are given by Ethiyajeevakaruna and Jeyapalan(1986). The impact of microcomputers on civil engineering is well seen by the number of short courses, and conferences. The publication of the journal "Microcomputer applications in Civil engineering", further reflects the growing interest, and awareness in this important area. This paper while presenting the use and capabilities of the sophisticated soil-structure interaction finite element analyses program, "SOILPIPE", will also attempt to draw the attention of prospective users about its limitations. The availabilty of programs such as "SOILPIPE" ready to be implemented on microcomputers does not imply a user can become an expert overnight. Reasonable results can be obtained only if the input is correct. The use of a pre-processor greatly helps a user to check the input data. The availability of a post-processor further helps a user to graphically view the response of both the soil and the structure under the design loads.

2 "SOILPIPE" PROGRAM

Finite element methods for soil-structure interaction problems have been effectively carried out on microcomputers. "SOILPIPE" is one of such programs.

The program can be implemented on an IBM or a compatible microcomputer with atleast 512kB of memory and a harddisk. This soil-structure interaction finite element program also has pre- and post-processor options. Most errors in the input phase for this type of analyses have been found to be in the input data for the mesh. The Pre-processor assists the user to check the mesh, by displaying it on the screen, the mesh and the various zones of soil materials in different colors. Soil-structure interaction FEM analyses also generate a great deal of output. The user has to seive through volumes of output to locate the numbers which are of significant interest. The use of post-processor to convert these vast amounts of numbers greatly helps the designer. The output can be displayed on a terminal, sent to a dot -matrix printer, or to a high quality plotter. The structural members in the program can either be beam elements, bar elements, or both. The program calculates stresses, strains and displacements in the soil elements and internal forces and displacements in structural elements, by means of analyses which simulate the actual sequence of construction operations in a number of steps. The nonlinear and stress-dependent stress-strain properties of the soils are approximated by varying the values of modulus and Poisson's ratio in accordance with the calculated stresses, using the Duncan model. An increment of an analysis may consist of placement of a layer of fill, placement of a structure, or application of loads to a completed structure or soil mass, such as traffic loads. The program was named "SOILPIPE", simply because it was used extensively in the analyses of underground pipes. This program can be used for many other type of installations, as well. The technical details of the program, the program organization, and the details of input data and options are well documented in the user manual. In addition, several series of analyses were performed with "SOILPIPE" using the micro version and on the mainframe and the results compared well.

The availability of a powerful program such as "SOILPIPE", on a micro computer, does not imply that anyone can perform a soil-structure interaction finite element analysis. Engineering judgement and a good understanding and appreciation of the FEM method is very essential.

The following observations by the authors are intended to aid the users:

1. The program "SOILPIPE" is essentially a non-linear FEM program, and employs an incremental approach.
2. A finer mesh as possible should be used. A mesh with about 300 soil elements and about 10 structural elements gives reasonable results. A too finer mesh will take a very long time for execution.
3. Backfill layers must be placed in very small layers, particularly when dealing with soft soils.
4. Traffic loads also must be applied in increments for better results.
5. The results should be checked whenever possible with results from installations, or other design equations.

3 CASE HISTORIES

The program has been used for a wide range of problems, including research studies with success. The use of "SOILPIPE" is demonstrated by the following of case histories of pipeline projects:

3.1 Case History 1

The Chippewa Reservior hydropower project and the Jim Falls hydropower project are located on the Chippewa River in Northern Wisconsin. Both hydropower units were constructed at existing dams owned by Northern State Power Company. In order to meet the clients' special need and constraints, siphon penstocks provided the most cost-effective engineering solutions. The project constraints and other special needs are summarized in Rajpal, Hampton, and Riley(1987) and the most significant constraint placed on the design of the projects was that no structural alterations be made to the existing embankment dams. The loading conditions on Chippewa penstock were quite complex and required special engineering design studies using the finite element method.

The geometry of the two projects are shown in Figs. 1 and 2. It should be noted that the maximum lift of water to the apex of the penstock designed for Chippewa Reservoir is 25.5 ft(7.8 m), which is about the highest lift than can be used for the siphon type system at an elevation of 1000 ft (305m) above mean sea level. Also, the penstock at this site is of a very large diameter of 108 inches

(2.75m) and a length of 280 feet (86m). Thus, proper care was taken during construction to ensure that the welds along the penstock had no leaks in order for the vaccum based priming system to work. The primary material considered for the walls of these two penstocks was A36 carbon steel. However, during the initial stages of the design analyses, steel lined precast concrete cylinder pipes were also considered as an

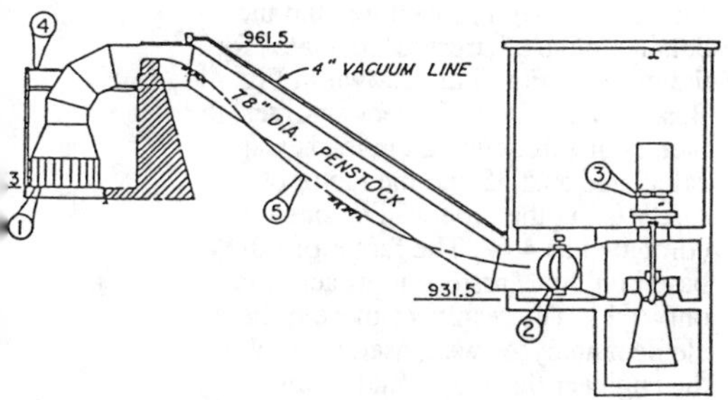

Figure 1. Jim Falls minimum flow unit.

alternate material. The construction with concrete pipe would have introduced far too many joints along the length of the Chippewa Reservoir penstock and this was of some concern for priming the system. Therefore, this proposal was eliminated in the final stage of the design analyses.

At Chippewa Reservoir, the penstock wall was designed of 3/8 inch (20mm) thick steel with 5 inch (125mm) by 1/2 inch(12.5mm) ring type stiffners at 6 foot (1.8m) centers. The wall thickness used for the penstock at Jim Walls was 1/2 inch (12.5mm) and no stiffners were provided for this penstock.

The loads on the siphon penstock designed for Chippewa Reservoir includedhydrostatic water pressures, weights of penstock and water, water hammer effects, weight of backfill over the section close to the apex, suction load from the vaccum, thermal effects, longitudinal loads, and ice loads. During the initial design stage, the road and the traffic was planned to be over the apex of the penstock. Detailed design studies and cost analyses indicated that it would be more economical to construct the road with an offset to relieve some of the stresses near the apex of the penstock as shown in Fig. 2. The thermal stresses were also relieved to some extent by insulating the exposed surfaces of the penstocks with styroform pads.

The loads on these two penstocks were of several types that the ASME code was incapable of providing enough guidance for performing the design computations. Even in the instances for

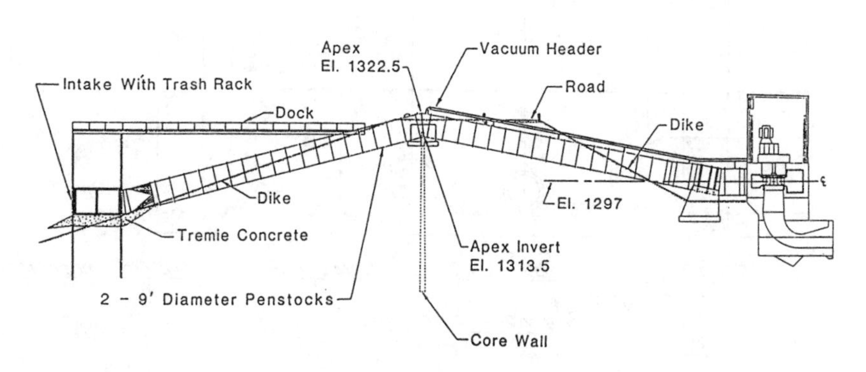

Figure 2. Chippewa reservoir hydro.

which the code provided a method of analysis, the design was somewhat conservative in comparison to the design based on the results of finite element analyses. Thus, most of the design for both projects were carried out wit appropriate finite elemnet analyses of the penstocks under all loading conditions. The finite element analyses were performed using the program "SOILPIPE" by Jeyapalan and Saleira (1986). These two dimensional plane strain finite element analyses simulated field construction sequence by placing one layer of backfill at a time and calculating the response of the penstock to that loading. The penstock wall was modelled by a series of beam elements connected at common nodes, and the surrounding soil mass was modelled by two dimensional quadrilateral elements with nonlinear stress dependent stress-strain behavior. The loads other than those due to the placement of backfill around and over the penstock were also accounted for in the finite element analyses. In addition to studying the response of the penstock in the cross sectional plane as a ring interacting with the surrounding backfill, longitudinal analyses were performed to ensure that the penstock is supported adequately at sufficient number of sections along its length.

The results from the finite element analyses included the deformed shape of the penstock in both ring behavior and longitudinal behavior, stresses, and strains in the penstock wall, and the surrounding backfill. Several series of analyses were performed until satisfactory designs for the penstock wall thickness and the support conditions were obtained for both projects.

3.2 Case History 2

A sewer poject in Phoenix, Arizona, U.S., did not include vitrified clay pipes in the specifications as one of the potential materials due to the engineer on the project relying too heavily on Marston's load theory. When the engineer computed the capacity of the clay pipes for this site, the field supporting strengths of the pipes were not adequate based on the Marston's loads. Finite element analyses were performed to demonstrate the conservativeness of the current Marston's design procedure, and the detailed plots of stresses calculated for the pipes in Fig. 3 are shown in Fig. 4. Based on the finite element analyses, the factors of safety using the loads was calculated as 2.35 and the factor of safety using the pipe wall strains was computed as 4.62. The factor of safety based on the Marston' approach was only 1.11. The results of these finite element analyses were used to convince the engineer that clay could be an economical pipe construction material at this site.

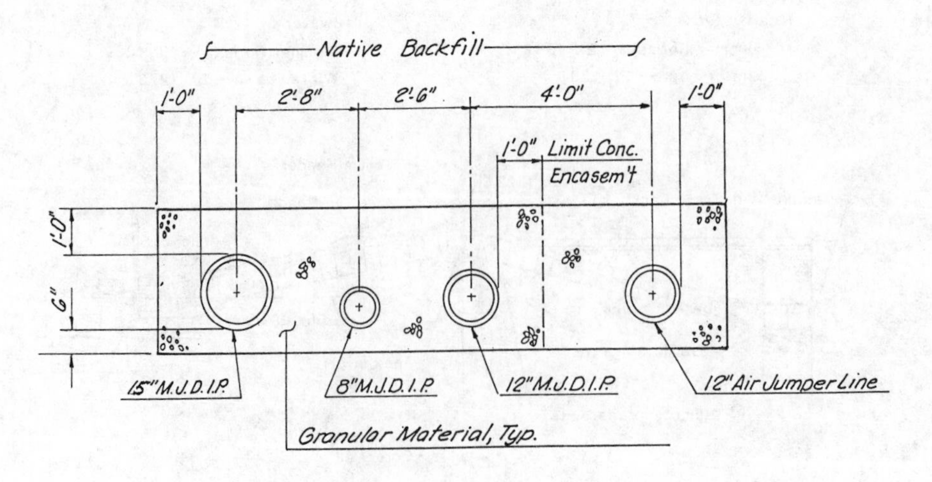

Figure 3. Cross section of the pipe installation.

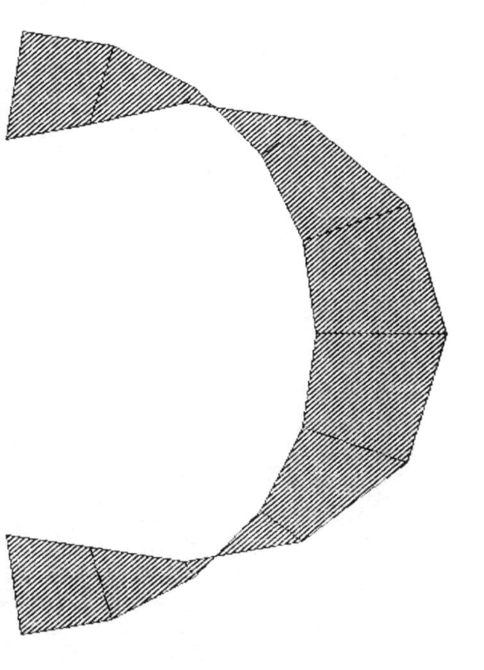

Figure 4. Fiber Stress.

3.3 Case History 3

A research program using finite element analyses showed that the loads used by the designers on clay pipes had been too high and that the field supporting strengths of these pipes are also higher than those used in the current design. The final research conclusions resulted in a recommendation to increase the bedding factors used by the underground clay pipe industry in the United States in order to constuct clay sewer pipes at a lower cost. Details of this program are given in Jeyapalan and Jiang (1986, 1987).

3.4 Case History 4

The pipeline infrastructure rehabilitation program in the Unites States are seeing many new products. One of such products is the Insituform pipe made at site inside a deteriorated pipe without having to excavate a trench. This is the only product which allows a pipeline engineer to reline a damaged pipe of any shape or size. However, the design tools used by this pipe producer were too conservative and had too many unreliable assumptions. Finite element analyses are being performed to design these Cast in Place Pipe installations for many sites in the Unites States and United Kingdom.

4 CONCLUSIONS

1. A soil-structure interaction finite element program that can be implemented on a microcomputer is presented.
2. The use of pre- and post processor modules to check input data, and interprete output data is also presented.
3. The versatility of the program is illustrated by application to actual installations. Though the examples cited refer to pipeline projects, "SOILPIPE" can be used for any type of structure that is fully or partially buried, and can be approximated by a plane strain assumption.
4. It is necessary for users to gain some expertise to use these sophisticated programs in day-to-day design. This expertise can be achieved only by using these techniques in practice.

REFERENCES

Jeyapalan, J.K., and Ethiyajeevakaruna, S.W. 1986. Geotechnical Analyses on Micro-Computers. Proceedings of the 4 th National Conference on Micro-Computers in Civil Engineering, Orlando, Florida

Jeyapalan, J. K., and Jiang, N. 1986. Load Reduction Factors for Vitrified Clay Pipes. Jounral of Transportation Engineering, ASCE, Vol. 111, No. 3: 236-249

Jeyapalan, J. K., and Jiang, N. 1987. New Bedding Factors for Buried Vitrified Clay Pipes. Paper to appear in Transportation Research Record

Jeyapalan, J. K., and Saleira, W. E. 1986. "SOILPIPE"-Finite Element Analyses of Underground Pipelines and Tanks, User Manual of FEM Code

Rajpal, A., Hampton, T. L., and Riley, T. 1987. Site, Owner Needs Impact on the Choice of Siphon Penstocks, ASCE Energy Conference, Atlantic City, New Jersey

Numerical Methods in Geomechanics (Innsbruck 1988), Swoboda (ed.)
© 1988 Balkema, Rotterdam. ISBN 90 6191 809 X

Estimation of subsurface ground layer using a data base system

M.Kawamura
Toyohashi University of Technology, Japan
A.Takada
Hokkaido Development Agency, Sapporo, Japan

ABSTRACT: This paper describes the method to estimate soil properties of subsurface ground layers quantitatively and qualitatively using a database system; especially the evaluation of the reliability of soil properties obtained from field tests and the classification of soil layers to draw soil profiles automatically. The database system is operated using a personal computer basically. The original data set obtained from a boring test at one site and stored in the computer, is consisted of thickness, soil name, color of soil, symbol for graphic display for each layers and the distribution of SPT N-value with the title of the data set.

1 INTRODUCTION

It is very important to select reliable values as soil properties for the numerical analysis on geotechnical engineering. The reliability of the input values affects the types of a model to solve and it also influences the reliability of the results of the analysis. The estimation of the soil properties based on the number of the information at a certain region is more reliable than the prediction of the properties obtained at an isolated point. It is effective if a number of data stored in database system are utilized for the estimation of the properties. In this paper the procedures to estimate the soil properties on the subsurface ground layer quantitatively and qualitatively using database system are introduced.

2 DATABASE SYSTEM

The flow chart of the database system is shown in Figure 1. This system is consisted of five portions; Data acquisition, Data classification, Data correction, Analysis and Output. Data acquisition is the arrangement of the original data concerning geographical information and so on. Data classification is the coding of the original data based on the Japan Unified

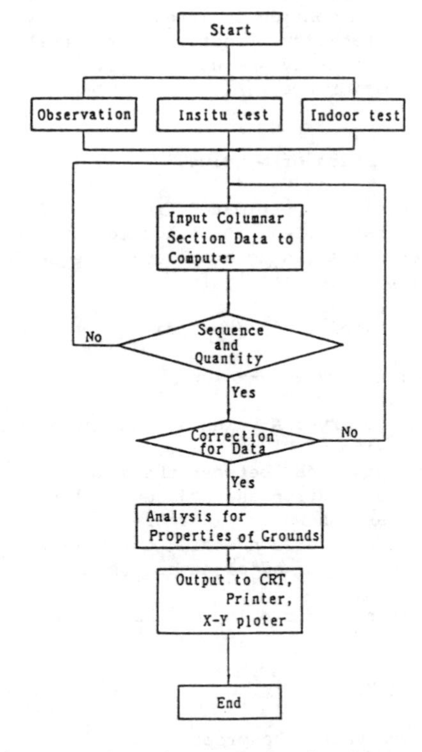

Figure 1 Flow chart of the database system

Soil Classification (Kishi, 1973). Data correction is the check of the stored data in the computer. Analysis is the calculation of the stored data such as vertical section of soil layer along arbitrary line, conter of the depth for the equi-N-value, etc. Output is the display of the stored data or results of the analysis. The details of the system was introduced in the previous paper (Kuribayashi et al, 1985). In this paper analysis are performed based on the data which were obtained at the southeast part of Aichi Prefecture in Japan.

3 RELIABILITY OF THE SOIL PROPERTIES

Reliability of the soil properties obtained from field measurements, for example, Standard Penetration Test, are evaluated as follows.

1) Comparison between the observed values of N-value and estimated values through interpolation from adjacent site data at each sampling density which corresponds to boring test number at certain area.
2) Determination of probability density function for estimation error of N-value.
3) Estimation of reliability of N-value, that is the confidence interval, applying the probability function for certain sampling density and using a given range of estimation error.

3.1 Estimation of N-value

The N-value is estimated from the values which have been investigated at the adjacent sites using a following equation as the weighted mean value,

$$Nr = \sum_{i=1}^{q} Noi(1/di)^{p} / \sum_{i=1}^{q}(1/di)^{p} \qquad (1)$$

where

Nr : Estimated N-value

Noi: Observed N-value at the adjacent site

di : Distance between the site for estimation and adjacent points as shown in Figure 2.

p : Parameter concerning the weight.

q : Number of adjacent points

Estimation error e is defined by equation 2.

$$e = (Nr - No)/ No \qquad (2)$$

To evaluate the proper value of the parameters p and q the mean values of estimation errors e are compared for several sets of the parameters as shown in Figure 3. Estimation error becomes small in the case where p=2 and q=6 -10 . To check effect of the density of observed point for the area, the relations between the mean values of estimation error and distances between the sites where the boring test were performed, were examined as shown in Figure 4. The estimated errors are small for p=2 and q=6 - 8. When the interval of boring test is 1000 m q=4 is preferable .

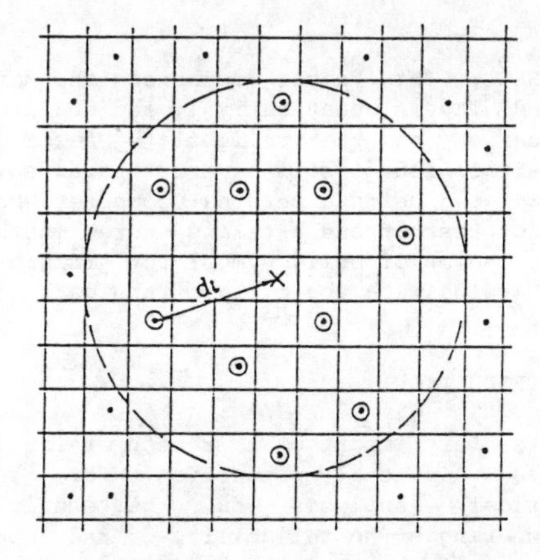

× Point where N-value is interpolated.

⊙ Adjacent point where N-value is known.

Figure 2 Distance between the estimated point and adjacent known sites

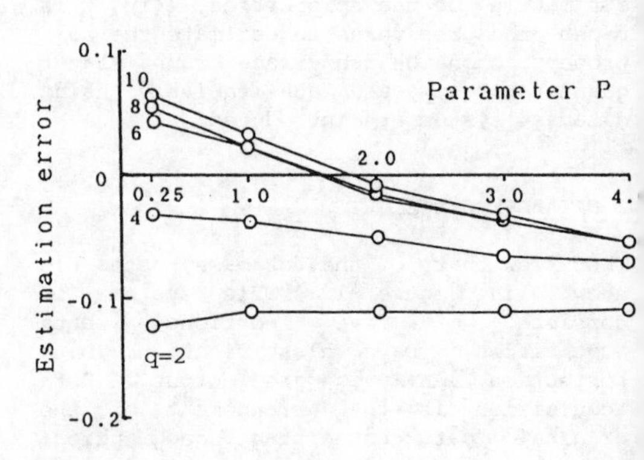

Figure 3 Effect of p and q on estimation error

Figure 5 shows autocorrelation coefficient of N-value along the horizontal direction. The autocorrelation coefficient $R(\Delta h)$ are given by following equations (Vanmarcke,1977).

$$R(\Delta h) = \sum_{l=1}^{n} ul(h) \cdot ul(h+\Delta h)/n \quad (3)$$

$$ul(h) = \{Nl(h) - Nm(h)\}/s(h) \quad (4)$$

where

Nm(h) : Mean N-value at a certain depth in the horizontal direction
S(h) : Standard deviation of Nl(h) along vertical direction
h : Distance from the original point
Δh : Distance between the boring sites

Autocorrelation coefficient of N-value in horizontal direction is an index to know continuity of the similar soil deposit from one site to another. R is 0.7 when the distances between the boring site is 700m. These results shown in Figures 4 and 5 suggest that estimated N-value is reliable if the estimation is based on the boring data with the mean distance between the sites, 700m.

3.2 Confidence interval of the estimated N-value

According to central limit theorem the value of z follows the standard normal distribution if the number of the sample is large. z is given by equation 6.

$$z = (Xm - m)/(s/\sqrt{n}) \quad (5)$$

where Xm is a mean value of the sample data, n is the number of the sample, m and s are the mean value and normal deviation of the population, respectively. Let za the value of z which corresponds to the confidence coefficient a. The confidence interval is given as follows.

$$m - za \cdot s/\sqrt{n} \leq Xm \leq m + za \cdot s/\sqrt{n} \quad (6)$$

The value of za is given by the table for standard normal distribution. For instance za is 1.96 for confidence coefficient a = 0.95. The confidence interval for several cases of distance between boring sites, that is, the sampling density for a certain region is shown in Figure 6. It is said with confidence factor of 95 % that if we have performed boring tests with 700 m interval, the estimation error of N-Value based on the test data is less than 20 % in the region. Through the procedure mentioned above, we can evaluate quantitatively the reliability of the soil property using the database system.

4 CLASSIFICATION OF SOIL LAYER

Subsurface ground layers are classified to

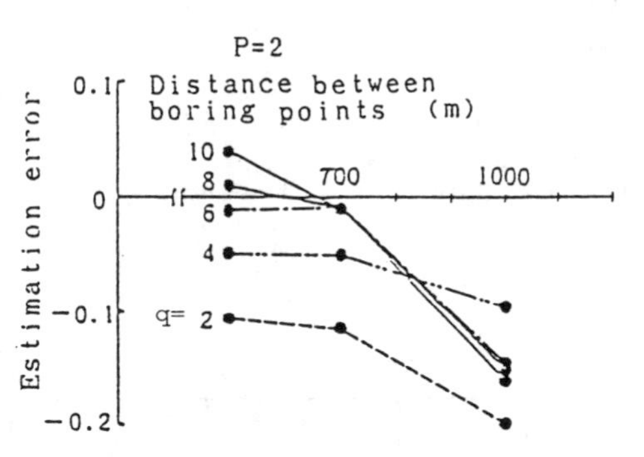

Figure 4 Relations between estimation error and interval of boring sites

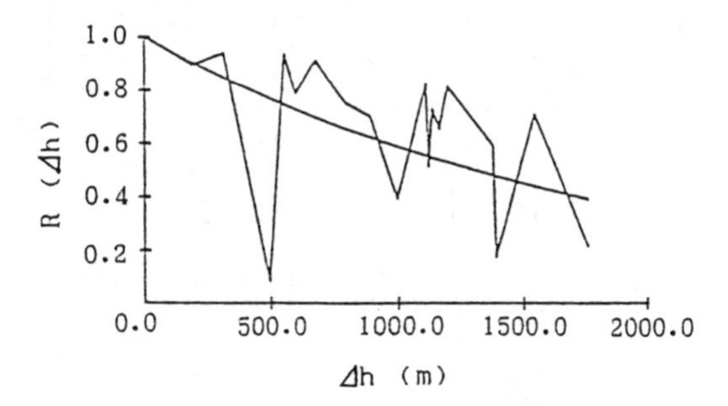

Figure 5 Autocorrelation coefficients in horizontal direction

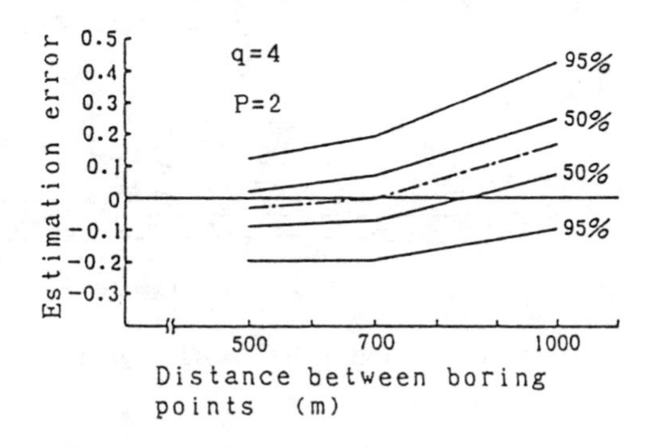

Figure 6 Relatios between confidens interval and distance of boring sites

No.43 No.44 No.45 No.46 No.47 No.48 No.49

Figure 7 Soil profile data of boring sites NO.43-No.49, which are stored in the database

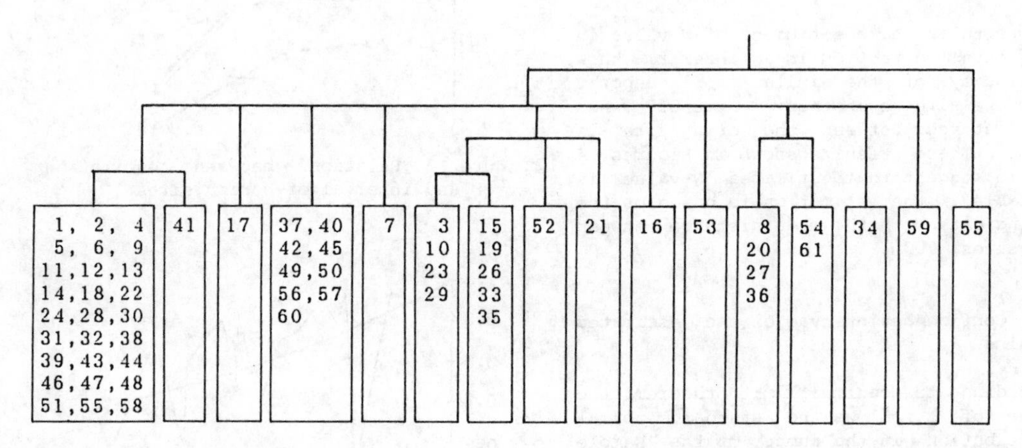

Figure 8 An example of dendrograph in cluster analysis for the soil layers in Figure 7

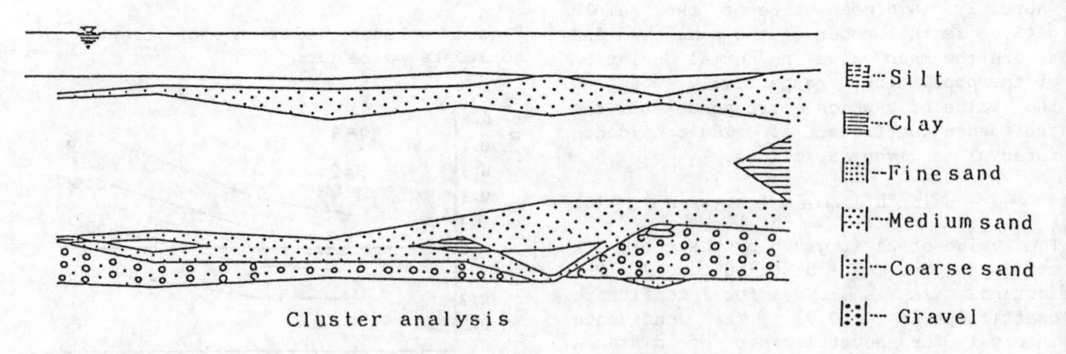

Figure 9 Cross section of the soil layers drawn based on the dendrograph in Figure 8

draw soil profile along an arbitrary section automatically and objectively as follows;

1) Thickness, soil name and N-values are selected as parameters and soil names are categorized into ten groups.

2) Dendrograph is determined through the Cluster Analysis and soil profile is drawn automatically according to the graph.

Cluster analysis is the convenient tool to classify the assemble of different kind of things into several groups(Tanaka, 1985). The main data on soil layer stored in the database system, are thickness, soil name and N-values. The position of each soil layer is expressed by the depth from the standard water level. Soil name is categorized into ten groups as shown in Table 1. A weight for soil name in the analysis is 20. N-value representing a soil layer is the mean value for ten points in the layer. Characteristics of soil layers are listed as shown in Table 2.

The example of dendrograph is shown in Figure 7 for the results of the boring tests in Figure 8. The soil layers are classified into several groups based on the dendrograph. Figure 9 is the cross section of soil profile which is drawn on the basis of the dendrograph. Through the cluster analysis we can evaluate qualitatively the soil profiles automatically using the database system.

Table 1 Categories for soil name

Category number	Soil name
1	Surface soil
2	Coarse sand
3	Medium sand
4	Fine sand
5	Cohesive soil
6	Silt
7	Volcanic ash
8	Gravel
9	Sand and gravel
10	Rock

5 CONCLUSIONS

As the results of the quantitative and qualitative analysis of the subsurface ground layer for the southeast of Aichi Prefecture in Japan, followings are made clear.

1) It is possible to know the confidence interval for estimation error of N-value of subsurface ground if the mean values and standard deviation of the population are determined using the database system.

2) Soil profile along an arbitrary section can be drawn automatically if the soil layers are classified into several groups using the database system.

BIBLIOGRAPHY

[1] S.Kishi: Retrieval by computer information,General Report of National Research Center for Disaster Prevention, No.31, Science and Technology Agency, Japan,(1973).

[2] E.Kuribayashi, M.Kawamura, Y.YUi and A.Takada: A scanning method for properties of natural soil, Proc. of the 5th ICONMIG, Vol.3,pp.1167-1670,(1985).

[3] E.H.Vanmarcke: Probabilistic modeling of soil profile, Proc. of ASCE, Vol.103,No.GT11, pp.1227-1246,(1977).

[4] Y.Tanaka Ed.: Handbook for statistical analysis applying personal computer, Kyouritsu shuppan,Tokyo,1973.

Table 2 An example of characteristics list of soil layers

Number of soil layer	Soil name										Mean N-value	Depth (m)	Weight
	1	2	3	4	5	6	7	8	9	10			
NO. 1	0	0	0	20	0	0	0	0	0	0	0.8	5.00	1
NO. 2	0	0	0	0	0	20	0	0	0	0	1.9	9.50	1
NO. 3	0	0	0	0	0	20	0	0	0	0	0.1	15.20	1
NO. 4	0	0	0	20	0	0	0	0	0	0	2.1	21.40	1
:					:						:	:	:
:					:						:	:	:
:					:						:	:	:
NO. 30	0	0	0	0	0	20	0	0	0	0	23.0	19.50	31

Numerical Methods in Geomechanics (Innsbruck 1988), Swoboda (ed.)
© 1988 Balkema, Rotterdam. ISBN 90 6191 809 X

General slope stability software package for micro-computers

M.Maksimovic
University of Belgrade, Yugoslavia

ABSTRACT: The package for different classes of personal computers performs the limit equilibrium analyses (using standard Mohr-Coulomb or proposed non-linear failure envelope) of non-homogeneous sections of soil and rock slopes with arbitrary distribution of pore pressures, external loading, and inertia forces due to earthquake acceleration, by the methods of slices considering arbitrary and circular slip surfaces. An automatic search option, using tne method of steepest descent, can be selected for determination of the critical slip circle. Programs are menu-driven; wide variety of branching in data preparation and computation can be easily carried out in an interactive work, including grafic presentation of cross sections, slip surfaces, line of thrust e.t.c. on monitor and/or on dot matrix printer. Performance of three classes of PCs is briefly evaluated in terms of program organization, storage requirements, speed and users convenience.

1. INTRODUCTION

Tne engineering methods of limit equilibrium will remain, at least for some time in future, one of the main engineering tools for assessment of the safety margins of tne soil mass. The problem of stability of soil and rock slopes is the typical case where significant participation of tne engineer is helpful in finding tne solution of the practical problem. Personal computers offer very good possibilities for interactive and comfortable work, as the presentation of grapnics on the screen permits easy evaluation of most steps of the stability analysis. Tne geotechnical engineer can actually see what he is doing, and that is one of tne advantages in the practical use of PCs in the limit equilibrium analysis.

The overall strategy in writing a slope stability computer program for personal computers differs significantly from the approach usually adopted for large systems. Most and even smaller PCs have some dialect of BASIC wnich is generaly not the fastest and most efficient programming language, particularly if tne compiler for its interpreter version is not available, but other features of tne language, make tnem rather atractive for program development and application.

Running of tne program can be performed in an interactive manner with tne appropriate use of screen graphics wnich facilitates easy checking of input data for tne geometry of cross section, position and shape of slip surfaces. Corrections and variation of data on geometry, soil parameters, pore water pressure distribution, external or eartnquake loading, functions of distribution of interslice forces, etc, can be easily performed in any stage of tne analysys.Wnen desired, tne graph snown on tne screen can be obtained on tne dot matrix printer for final presentation. Tne running of tne program is made as simple as possible in order to permit tne user to concentrate on tne geotechnical aspects of tne problem, without any need for tne use of tne programming knowledge.

2. Soil strength and the factor of safety

Methods for tne computation of tne index of relative stability called tne factor of safety are oriented towards the calculation of tne value $Fs = \tau_f / \tau_m$ wnere τ_f is tne actual snearing strengtn of tne soil available and τ_m is tne average shear stress on tne hypotnetical failure surface mobilized to maintain tne body of soil in equilibrium.

The strength of soil is traditionally described as the Coulomb's linear relationship between the effective normal stress on the failure plane with parameters of "cohesion" and "angle of the shearing resistance", both parameters being constant and independent of the stress level. The straight line failure envelope could be considered as the reasonable approximation for loose sands and normally consolidated clays. Most other soil and rock types exhibit curved failure envelopes. Some alternative possibilities for description of nonlinear failure envelope are presented by Maksimović (1979). New expression, which is found very suitable for description of failure envelope in terms of effective stresses for most soil types, describes the angle of the shearing resistance ϕ' as the function of the normal effective stress on the plane of failure, (ϕ' is the expression in square brackets), and parameters in the form:

$$\tau_f = c' + \sigma'_n \left[\tan \phi'_B + \Delta\phi'/(1+\sigma'_n/p_N)\right] \quad ..(1)$$

where c' is cohesion, ϕ'_B is the basic angle of friction and constant for a given soil, $\Delta\phi'$ is the maximum angle difference, p_N is the mean angle stress for which the secant angle of the shearing resistance equals $\phi'=\phi_B+\Delta\phi'/2$, as shown in Fig.1.

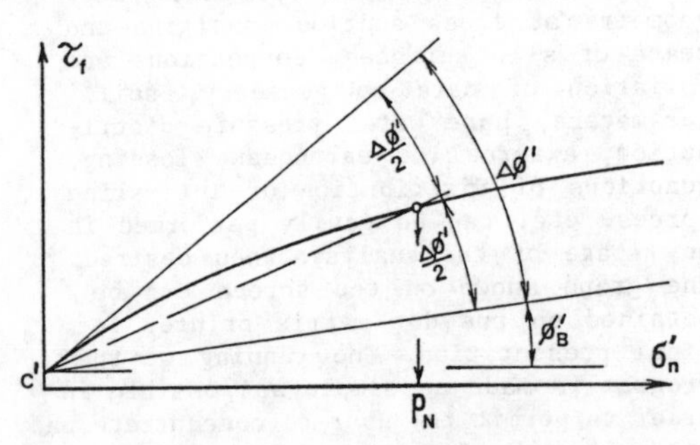

Fig.1 Parameters of the nonlinear failure envelope

In the case of noncemented coarse-grained soils, $c'=0$, $\Delta\phi'$ is the maximum angle of dilatancy and ϕ'_B is the angle of friction at constant volume when all dilatancy effects are suppressed. When describing the residual strength envelope, ϕ'_B is the angle of friction for perfect orientation of platy clay particles in the direction of the shearing plane.

The full justification of the proposed eq.(1) is prepared by Maksimović (1987).

The proposed eq.(1) reduces to the conventional Coulomb's straight line envelope in any of the following three cases:

(1) if the maximum angle difference $\Delta\phi'=0$ for any value of p_N then $\phi'=\phi'_B$
(2) if $p_N=0$ for any finite value of $\Delta\phi'$ then, just as in the case (1) $\phi'=\phi'_B$
(3) if p_N is infinite then $\phi'=\phi'_B+\Delta\phi'$.

3. Methods of the analysis

Fredlund (1984) quotes that methods which satisfy all equilibrium conditions give essentially the same results. Bishop simplified method gives virtualy the same results as methods that satisfy complete equilibrium.

General slip surfaces, as well as circular ones, are analysed by the method basically described by Maksimović (1979). It is analogous to the Morgenstern and Price (1967) method as it uses the distribution function of inclinations of interslice forces f(x), but probably simpler. The function f(x) could be determined by the application of the finite element method (Maksimovic. 1979), but as the result is rather insensitive to the choice of f(x), provided the solution is physically admissible, such an approach can be justified in some special cases, but not in practical applications. Bromhead (1986) has suggested the iteration on equilibrium of moments, instead on positions of interslice forces for the early (1970) unpublished version of the method which uses the linear failure anvelope. This modification is incorporated in the solution scheme, as it reduces the amount of computation and storage requirements. Lever arms, which define the position of the line of thrust, are computed after the convergence criteria on stresses, forces and moments are satisfied.

Large number of circular slip surfaces are ususly treated by the application of the Bishop (1955) method which is extended by the introduction of the nonlinear failure envelope and modified to include external line loads, distributed loads and inertia forces due to earthquake loading in any direction. For the general case of the partially submerged slope the expression for the factor of safety may be written as:

$$Fs = \frac{\sum\left[c'b+(W+W'-b\cdot u_s+R_v)\tan\phi'\right]m_\alpha}{\sum (W+W'+R_v)\sin\alpha + M_x} \quad ..(2)$$

where

$$R_v = R_y + p\,b + k_y(W+W_z) \quad ....(3)$$

$$M_x = R_x\, Y_r/R + k_x(W+W_z)\, Y_z/R \qquad (4)$$

R_y - resultant vertical line loads, p distributed vertical loading, k_y - vertical coefficient of seismicity, W, W' and W_z are the weights above hydrostatic level, submerged and saturated weights of parts of the slice beneath the hydrostatic level respectively, k_x-horizontal seismic coefficient, Y_r and Yz are lever arms of corresponding forces with respect to the center of the slip-circle, and R is radius. The expression (2) is fully consistent with assumptions of the Bishop's routine method. However due to introduction of the nonlinear failure envelope, (if used), the additional iterative cycle on the soil strength is required until the difference of normal effective stresses acting on the base of each slice computed in two successive iterations is less then some prescribed tollerance.

It is probably worth mentioning that the "slope rotation" technique for pseudostatic seismic analysis, though rather simple and correct in most cases, can be seriously in error on the unsafe side if applied to partially submerged slopes.

In both methods some initial value of Fs is required for iterative computation. Among the number of possibilities, it was found convenient to use basically the eq.(2) by taking that $m_\alpha=1$, for any shape of the slip surface and $Y_r/R=1$ and $Y_z/R=1$ for slip surfaces of the arbitrary shape, main reason being that the program should calculate the correct safety factor if sliding is from left to right as well as in the case if it is from right to left. The initial value computed in this way will have proper sign and its absolute value is slightly smaller than the absolute value of the final result.

4. Method of steepest descent for the search of the critical slip-circle

For the search of the most critical slip-circle three options are built in the program. Besides the single circle and the grid option, which is common in most computer programmes of this kind, the routine for the automatic search is very valuable in practical work. The problem is to find the minimum of the function Fs=Fs(x,y,R) with three variables, i.e., two center coordinates of the circle and the radius. Even if defined as the bounded problem, the time for execution of the search for the minimum of the

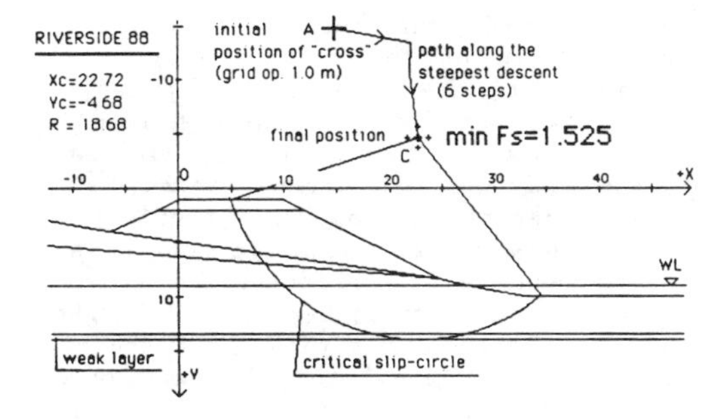

Fig.2. Dot matrix printer output of the graph with final results obtained by "BIME"

function of three variables will be excessive, at least with the present speed of most popular micros. That would leave the user in the idle state of expectation for unreasonable amount of time. The computation is for an order of magnitude shorter and significantly safer, if the problem is reduced to the search for the minimum of the function Fs(x,y,p) of two variables only, where p is a parameter treated in each search as the constant selected by the user. It specifies that each family of slip circles has either one common point (an exit point in the zone of the toe, for example) or is tangential to certain horizontal or inclined plane (like the lower boundary of some weak zone).

To reach the solution it is necessary to start from some arbitrary point A (Fig.2) and arrive to point C which is such that Fs(Xc,Yc,p)=Fs min. The paths from point A to C will follow the trajectory of steepest gradient. As the method requires calculation of first and second partial derivatives of F(x,y,p), the local finite difference grid, consisting of five points forming "the finite difference cross" as shown in Fig.2, is defined with the central point A placed initially by the user. Factors of safety for five points, and the direction of steepest descent are computed. By using Taylor's expansion to the polynomial of second degree, estimated length of the step is calculated, the finite difference cross moved to the new point and the procedure automatically repeated until the calculated step becomes some small tolerable value, like 0.1-0.2 of the opening "a" of the grid, as at the stationary point first partial derivatives vanish. Practical experience show that the value of "a", can be taken as about 1/10 of the slope height. Then the accuracy of the location of the critical center becomes 1% to 2% of the slope height.

In order to avoid excessively long jumps, its reach in one step is limited to the length of the diagonal or $2 \cdot a \cdot \sqrt{2}$.

Usual restriction on the practical significance of the automatic search is the consequence of continuity and unimodality of the function F(x,y,p). It might have more than one minima, and the procedure will reveal the minimum along the path from chosen starting point A. In order to check if some other minimum exist, user can try some different starting point, or use the grid of centers. The combined procedure of automatic search and the grid option, when used sensibly, will provide the satisfactory answer to an engineer with reasonable geotechnical training and experience. The "lattice" method, /10/, was tried in initial stages of the program development. It was found that it offers safe convergence, but it is about 20-30% less efficient then the method of steepest descent. Due to space limitations, the specifics of the type of the method could not be described here in more detail, and might be offered for publication elsewhere.

5. Software package organization

In order to make programs adaptable for most wide spread medium sized or 512kB PCs, (class of computers like IBM-PC/AT-XT or Apple Macintosh 512), it was found convenient to compose the software package (BIGEM=BIME+GEME) of two independent, but fully compatible programs, as follows:

-No.1 named "BIME"(BIshop's MEthod) uses circular slip surfaces and handles the grid or automatic search option for finding the critical slip circle.

-No.2 named "GEME" (GEneral MEthod) is used for the analysis of arbitrary slip surfaces including the circular ones, using the method basicaly described by Author with some modifications explained briefly in the previous paragraph. In order to simplify the definition of the function f(x) with minimum number of parameters, eight types are envisaged, as shown in Fig.3. Type 0 permits the definition of arbitrary function by the set of straight lines within selected intervals, but in the practical work it is rarely used. The Type 1 is constant which would yield the solution analogous to the Spencer (1967) method, and it is usually used first in most practical cases. Other types use the autnomatically generated half-wave sinus function for curved transitions between constants and are found usefull in some particular cases.

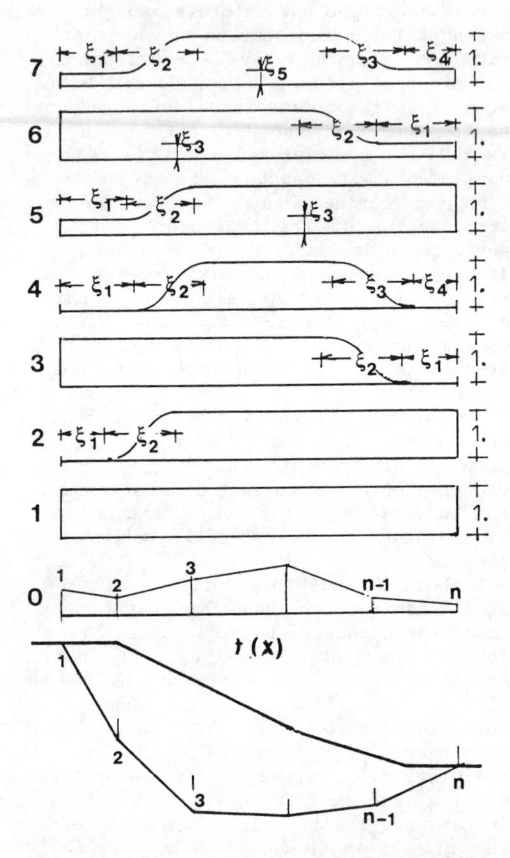

Fig.3. Choice of types of f(x) for distribution of inclinations of interslice forces. Functions from 2 to 7 defined by a few of ξ parameters only.

Particular attention is excersized that the data preparation for any of the two programs is identical (except for the definition of slip surfaces), and that they can be stored or read from files by both programs, regardles of the program which was used for input and saving of the data. As the slip circle analysis by "BIME" is usually done first, and the number of circular slip surfaces is larger by an order of magnitude when compared to the nuber of slip surfaces of arbitrary shape, after completion of the slip circle analysis by "BIME", user can read same data file by "GEME" and compare results computed by the extended Bishop's method with the result obtained by satisfying all equilibrium conditions, or check some noncircular slip surfaces.

Both programs are meny driven and each has three menues. The list of menues with options for the larger program "GEME" is given in Fig.4, while the menu for "BIME", being similar and slightly simpler, is not presented. Most of the menu options shown in Fig.4 are self-explanatory.

```
CHOICE I N I T I A L   M E N U
   1 FROM KEYBOARD MAIN DATA
   2 FROM KEYBOARD SLIP-SURF.
   3 LOAD MAIN DATA FROM FILE
   4 LOAD SLIP-SURF FROM FILE
   5 PRINT MAIN DATA - CURRENT OPTION N
   6 PRINT SLIP-SURF. - CURRENT OPTION N
   7 GRAPH
   8 PRINTING OPTION N
   9 SAVE MAIN DATA
  10 SAVE SLIP-SURFACE
  11 COMPUTE
  12 WORKING MENU
  13 CHANGES
  14 E N D
CHOICE? |

                 C H A N G E S
 1 COORDINATES    0
 2 SOIL LIM & PIEZ. LINES.  0     3 SOIL PARAMETERS   0
 4 LINE LOADS    0                5 DISTRIBUTED LOADS   0
 6 LOWER WATER LEVEL 0            7 NUMBER of SLICES   0
          8 SEIZMIC COEFICIENTS  0.00 0.00
          9 POINTS FOR SLIP SURFACES   0
10 TITLE FOR MAIN DATA
11 TITLE FOR SLIP-SURF.
12 VARIATION ON SLIP SURFACE.  0
13 FUNCTION f(x)  0
14 ORIGIN 150 150       15 SHIFT CROSS-SECTION
16 MIRROR Y             17 GRAPH              18 SCALE 0.00
                        19 INITIAL MENU
CHOICE ?

CHOICE  W O R K I N G   M E N U
   1 INITIAL MENU
   2 GRAPH N
   3 PRINT REZULTS N
   4 PRINTING OPTION N
   5 SEIZMIC COEFICIENTS  0  0
   6 FUNCTION f(x) TYPE  0
   7 EQIVALENT WEDGE
   8 NEW PROBLEM
   9 CHANGES
CHOICE ?
```

Fig. 4 List of menues for program "GEME"

Menu functions in both programs can be sumarized as follows:

-MAIN MENU permits inputing the data via keyboard or from saved file, display of cross-section and slip surface(s) on the screen, initiation of the computation, saving input data and exiting from the program.

-CHANGES is the menu which permits changes in any data or variation of parameters. It is rather handy in data input stage as well. If the user enters some wrong input, he does not have to restart the typing procedure from the beginning, but he can continue just as nothing wrong is done, and after completing the entering, he can call this menu and correct typing error(s) and check the effects on the graphic display before initiating the computation. Option 12 facilitates modifications of the geometry of the slip surface, while options 14 and 15 permit the change of the origin of the coordinate axes, or the shifting of the section with respect to the system of axes. Combined with option 18, which permits the user to choose or change the scale of the graph, full control of the screen graphics is possible during the work, or at some stage when the output of the screen graph to the dot matrix printer is asked for.

-WORKING MENU controls the branching after certain computation stage is completed, for example by selecting the new initial circle for the automatic search in "BIME" or check the solution for some different choice of the f(x) function. Graph option from this menu displays the results of computation, showing, for example, grid of centers with the critical slip circle, or the finite difference cross with smallest safety factor in program "BIME" as shown in Fig.2. The graph in "GEME" will show the result, the cross section with the slip surface analysed and line of thrust of interslice forces for used f(x) as shown in Fig.5.

The CHANGES menu for 64 kb mashines in "GEME" had to be reduced due to memory limitations, but without significant loss of efficiency, as the CHANGES menu can be used in "BIME" and then the data file used for input by "GEME".

Program "GEME" in the WORKING MENU has an EQUIVALENT WEDGE option as number 7. When chosen, the program will map all acting loads and resultant stresses on the slip surface into the system of resultant forces acting on an equivalent block on an inclined plane, give the equivalent inclination, show average normal and shear stresses on the slip surface and show the initial estimate of the critical acceleration for which the safety factor might become equal to one. If called repeatedly a few times in sequence, it will compute the critical acceleration. The main aims of this option is to provide data for the estimate of the rigid body displacements due to earhquake acceleration for which the safety factor is less then one by

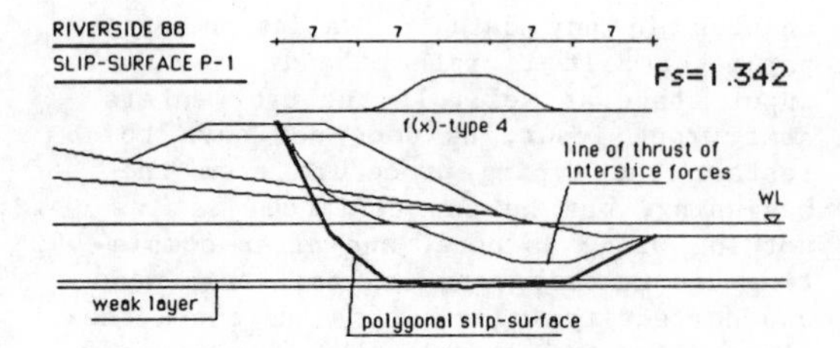

Fig.5. Dot matrix printer output of tne graph showing final results for arbitrary slip surface obtained by program "GEME"

using the Newmark's (1965) principle or, alternatively, to compute one point for evaluation of tne resistance envelope proposed by Casagrande and advocated by Janbu (1977).

6. Concluding remarks

As mentioned in tne text, tne software package is written for three classes of computers:

(a) IBM-PC-AT/XT 512 k and compatibles,
(b) Apple Mecintosn 512 k, and
(c) AMSTRAD CPC series (464,664 and 6128).

Both, (a) and (b) offer similar RAM, and with the described program organization any practical problem, including complex cross-sections of non- nomogeneous large fill dams could be handled without difficulties with the capacity of the memory.

However, for (c), the maximum number of slices had to be reduced from 100 to 50, the maximum number of soil zones from 20 to 12, and the maximum number of points and lines from 80 to 40. Even with such limitations, the smallest PC could handle most of tne typical problems occuring in practice, but using tne interpreter version of tne language only. For examples shown in tnis paper, the time for the computation of tne Fs for single slip circle with about 40 slices witn "BIME" is typically about 28 sec. in tne case of classical linear envelope, while in the case of application of nonlinear failure envelopes, it increases to 45 sec approximately. For tne arbitrary slip surface, computation with "GEME", wnich satisfies all equilibrium conditions, tne time of computation is 3-4 times longer for other parameters being same.

Speed of tne execution for (a) and (b) largely depends on available compilers and their performance, but it is generally, 10 to 40 times faster, reducing the time of computation to quite acceptable few seconds in considered cases.

The quality of the screen grapnics is rather similar, but the writer would give slight advantage to mashines the resolution wnich coresponds closer to tne proportions of tne screen.

REFERENCES

/1/ Bishop.A.W. 1955. Tne use of slip circle for stability analysis, Geotechnique 5.
/2/ Bromhead.E.N. 1986. Tne stability of slopes. Surrey University Press, US: Cnapman & Hall.New York
/3/ Fredlund.D.G. 1984. State of tne Art Lecture-Analytical metnods for slope stability analysis. IV Int.Symposium on Land- slides, Toronto
/4/ Janbu.N.1977. State of tne Art Report 9tn Int.Conf.SMFE. Vol.2, Tokyo
/5/ Maksimović.M. 1979. Limit equilibrium for nonlinear failure envelope and arbitrary slip surface. Numerical methods in geomechanics. Vol.2. Aacnen.
/6/ Maksimović,M. 1987. Nonlinear failure envelope for soils. (In preparation)
/7/ Morgenstern & Price, 1965. The analysis of tne stability of general slip surfaces, Geotechnique 15.
/8/ Newmark,N.M.1965. Effects of eartnquakes on dams and embankments. Geotechnique, 15.
/9/ Spencer.E. 1967. A method of analysis of tne stability of embankments assuming parallel interslice forces, Geotechnique 17.
/10/Wrignt, S.G. 1974. SSTAB1-A General computer program for slope stability analyses. Researcn Report No. GE-74-1, Dept of Civ.Eng. The Univ. of Texas.

Numerical Methods in Geomechanics (Innsbruck 1988), Swoboda (ed.)
© 1988 Balkema, Rotterdam. ISBN 90 6191 809 X

A comparison of four slip surface search routines

Jay S.DeNatale & Susan G.Gillett
University of Arizona, Tucson, USA

ABSTRACT: Nearly all computer programs for slope stability analysis include some routine which directs the search for the critical slip surface and the associated minimum factor of safety. The role and significance of this aspect of the program is examined by comparing the relative performance of four different slip surface search routines. A mechanism is proposed for improving the efficiency of any existing limit equilibrium-based slope stability program.

1 INTRODUCTION

A slope stability analysis is normally performed to identify the most probable failure mechanism (or critical slip surface) and the associated minimum factor of safety. Although analyses by means of empirically developed field slope charts and/or stability charts are sometimes justified, more precise solutions based on a thorough program of site exploration and laboratory testing are generally recommended. A wide variety of specific analytical methods have been developed over the years (see, for example, Fredlund, 1984). However, in practice, the overwhelming majority of stability analyses are performed with the aid of computer programs based on well-established limit equilibrium techniques.

A limit equilibrium-based slope stability analysis is a rather straightforward minimization problem. As such, a limit-equilibrium solution consists of the following two steps:

1. The development of a merit function -- in this case, the safety factor expression -- to serve as a scalar measure of the stability of a particular trial slip surface; and
2. The selection of a search strategy to enable the minimum of this scalar function to be accurately, efficiently, and reliably found.

Although considerable effort has been devoted to the development of accurate and general safety factor formulations, little attention has been paid to the implementation of advanced minimization techniques to direct the search for the critical slip surface. This article is concerned with the applicability and effectiveness of some of these alternate search procedures.

2 AVAILABLE MINIMIZATION ROUTINES

An extremely large number of optimization routines have been developed during the past 30 years, and the performance of a given approach is highly dependent on the particular type of problem to which it is applied. It is generally agreed that there is no single "best" optimization algorithm but, rather, only strategies which perform most effectively when applied to certain classes of problems.

Nearly all methods developed to locate the minimum of a general nonlinear function are iterative in nature. The iterative search may be guided by varying degrees of data, including function values as well as perhaps gradient and curvature information. Since the safety factor expression in a generalized slice procedure is not differentiable, search procedures which require only function evaluations would seem to be best suited to computer-aided slope stability analyses. Four such no-derivative minimization techniques are considered herein.

3 THE GRID SEARCH ROUTINE

The grid search routine is one of the earliest known minimization strategies. A significant number of the trial evaluation points are selected before the search even begins. Hence, the search direction is unaffected by the true geometry of the merit function, which causes the technique to be relatively inefficient. Nevertheless, this strategy is the one most often used in conjunction with computer-aided slope stability analyses (Gillett, 1987).

When used in conjunction with circular slip surface analyses, the search is usually conducted to pinpoint the center coordinates of the critical circle. The radius of the circle is turned into a dependent variable by specifying either a tangent elevation or a point through which all circles must pass. If the critical tangent elevation is unknown, a series of two-dimensional searches is conducted using two or more regularly spaced trial tangent elevations.

Initially, a rectangular area is selected which includes the center of the critical circle. The area is subdivided into a number of equal-shaped elements, and the safety factor is evaluated at each of the perimeter and interior nodes (or element corners). The critical center is assumed to be located at the node associated with the smallest safety factor value.

If the specified search area is rather large, two or three iterations may be required to pinpoint the location of the critical center. In this event, the initial subdivision should be relatively coarse. Then, those elements which surround the point associated with the smallest safety factor should be further subdivided into a second network of regular elements and nodes. A second round of safety factor evaluations is then made at each new node, and a refined estimate of the critical center is thereby acquired. If necessary, the process can be repeated a third time using a still finer node spacing.

4 THE PATTERN SEARCH ROUTINE

A variety of specific pattern search logics have been developed over the years (Swann, 1972). The procedures are similar to the basic grid search, in the sense that only function information is required. However, instead of being specified in advance, the trial evaluation points are computed internally as the search proceeds.

The pattern search algorithm employed in the STABR slope stability program (Lefebvre, 1971) is representative of this class of minimization procedures. As with the grid search, the pattern search technique works best when only the (x, y) center coordinates of the critical circle are being sought.

The safety factor is first evaluated at the user-specified starting point. This point becomes the "center of rotation," and subsequent evaluations will be made at a specified distance to the left, below, to the right, and above this current "center." This specified distance is referred to as the "step-length" and is held constant throughout the analysis. If a smaller safety factor is found at any one of these points, that point becomes the new center of rotation, and the process is repeated.

If a full four-point rotation fails to identify a smaller safety factor, the step-length is halved, and another four-point rotation is performed. The search is assumed to have converged when a full four-point rotation with this new step-length fails to reduce the value of the safety factor. A second search with a still smaller step-length may be required if additional precision is desired.

5 THE SIMPLEX METHOD

The Simplex Method of Nelder and Mead (1965) is one of the most efficient no-derivative minimization routines. It is superior to the earlier version attributed to Spendley et al (1962). The general procedure is based on the evolutionary operation technique originally introduced by Box in 1957, in which the solution space is explored by means of a geometrical configuration of points (known as a "simplex") rather than a set of directions.

A regular simplex in n dimensions is defined by n+1 mutually equidistant points. Hence, for a two-dimensional problem (n = 2) the simplex is an equilateral triangle, for a three-dimensional problem (n = 3) the simplex becomes a regular tetrahedron, and so on. The general search procedure is based on the observation that only one new point is needed to transform an existing simplex into a new one.

A single starting point is provided by the analyst, and a regular simplex in the n-dimensional solution space is then automatically constructed. The merit function is evaluated at each of the n+1 vertices of this initial simplex. The simplex is then reflected (about some existing face), expanded (in size), or contracted (in size), so that the vertex associated with the maximum function value is continually replaced by a point closer to the true minimum. The procedure is said to have converged when the standard deviation of the function values at the simplex vertices falls below some preassigned limit.

6 THE CONJUGATE DIRECTION METHOD

In Powell's (1964) Conjugate Direction algorithm, the minimum value of an n-dimensional merit function is identified by cyclically searching along an ever-changing set of conjugate directions. A single starting point $\mathbf{P}_o$ is provided by the analyst, and the coordinate axes are used as the initial direction set $\mathbf{d}_1, \mathbf{d}_2, \ldots, \mathbf{d}_n$. A detailed description of the algorithm's structure is presented by DeNatale and Abifadel (1988).

7 MODIFICATION OF EXISTING PROGRAMS

A program architecture was developed to allow each of the four search routines to be used in conjunction with an existing limit-equilibrium based stability code, In the present study, the STABR program (Lefebvre, 1971) was selected to receive the search upgrade. However, any other existing stability code could just as easily have been used. The STABR program employs Bishop's Simplified Method of Slices to estimate the safety factors associated with circular slip surfaces.

Program SSDRIV, which drives the optimization-based slope analysis, is completely modular, with the main block consisting only of calls to various subroutines. The program begins with a call to subroutine OPEN, which in turn sends a request to the terminal for the name of the appropriate input and output files. The input data is read by making successive calls to subroutine OPDATA and subroutine SSDATA. Subroutine OPDATA reads the initial lines of the input file which contain the parameter values needed to conduct the grid-, pattern-, Simplex-, or Conjugate Direction-based search for the minimum safety factor. Subroutine SSDATA reads the remaining lines of the input file which define the relevant characteristics of the actual slope.

Once the input data is read, a call is made to subroutine SEARCH. This subroutine is responsible for directing the search for the minimum factor of safety and the associated critical slip surface. An initial trial slip circle is specified in the input file. The x and y center coordinates and radii r of all subsequent trial circles are computed by subroutine SEARCH, using one of the four previously described search logics. Those segments of the original STABR program which compute the safety factor for a given (x,y,r) coordinate set are contained in subroutine FVALUE, which is called by SEARCH whenever a merit function value is required. When a locally minimum safety factor is found, control is returned to the main block, and a final call is made to subroutine EXIT to close all files and terminate the run.

The program architecture described above enables any search procedure to be readily incorporated into any existing limit equilibrium-based slope stability code. It is only necessary to place into subroutine FVALUE those portions of the existing program which carry out the safety factor calculations for a specified trial slip surface.

8 THE TEST PROBLEMS

The relative merits of the four previously described slip surface search routines were established by means of a parametric study involving a wide variety of realistic slope problems (Gillett, 1987). The results of analyses with three of these slope problems are discussed herein.

Problem #1 involves determination of the critical slip circle for a 30 degree slope in a purely cohesive soil (see Figure 1). According to the rules established by

Taylor (1937), the critical surface passes tangent to the top of the underlying firm stratum. This restriction enables the radius of any trial circle to be treated as a dependent variable. Therefore the problem reduces to a two-dimensional search for the (x, y) center coordinates of the critical circle. This first type of slope problem is relatively easy to analyze with traditional grid search and pattern search techniques.

An idealized model of the Birch Dam project (Figure 2) is used as Problem #2. This particular case history has also been examined by Celestino and Duncan (1981) and Nguyen (1985). By recognizing that the critical circle must pass tangent to the base of the relatively weak lowermost soil layer, it is once again possible to restrict the search to two dimensions.

Problem #3 involves a relatively flat 30 degree slope in another purely cohesive deposit (see Figure 3). In this example, the soil is characterized as having a cohesion which increases linearly with depth. Taylor's (1937) research does not deal with such situations, and there is no way of identifying a priori the depth to which the critical circle will pass. Therefore, a full three-dimensional search is required to identify the (x, y) center coordinates and radius r of the critical circle. This second type of slope problem is very difficult to analyze with traditional grid search and pattern search techniques.

Any search must always begin at some user-specified starting point. In the present study, this starting point is defined in terms of the (x, y, r) coordinates of the first trial slip circle. Since fewer function evaluations are normally required when the search begins closer to the true minimum, the influence of the starting error (which may be interpreted as a measure of the analyst's expertise) must also be considered. Hence, four different starting points were defined for each problem, as shown in Figures 1 through 3. Pattern, Simplex, and Conjugate Direction searches were all begun from these points. The grid searches involved a square area centered about the true minimum, as indicated in Figures 1 through 3.

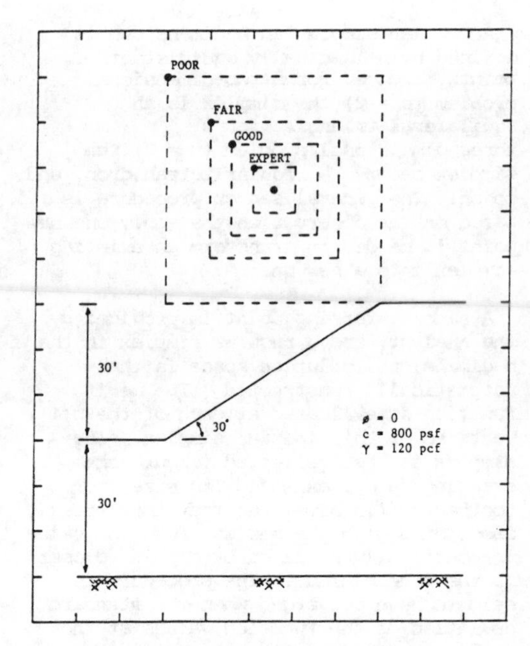

Figure 1. Test problem #1.

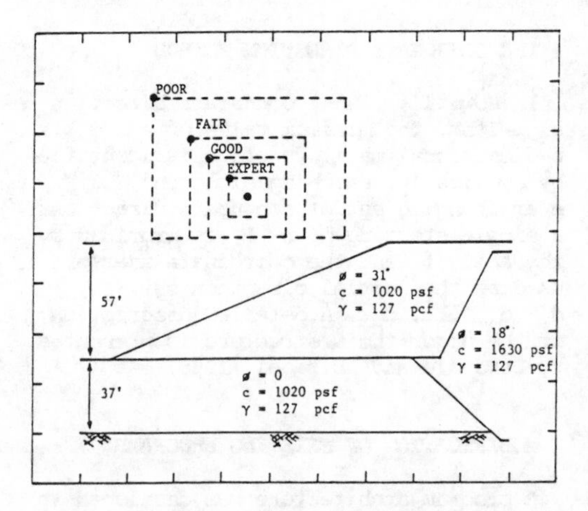

Figure 2. Test problem #2.

9 DISCUSSION OF RESULTS

Each of the test problems was analyzed many times, with the search beginning at several different distances from the true critical slip circle. The results of these analyses are presented in Table 1. As one would expect, a greater number of trial circles must be examined when the absolute starting error is increased.

The number of mathematical operations required to select a new circle center and radius is always insignificant relative to the number of operations required to evaluate the safety factor for the trial

Table 1. Relative efficiency of the four slip surface search routines.

Problem #	Starting Point	Number of Trial Circles Required to Locate the Minimum			
		Simplex	Con-Dir	Pattern	Grid
1	Expert	10	13	15	36
1	Good	14	19	23	50
1	Fair	18	22	31	76
1	Poor	29	32	47	148
2	Expert	10	18	19	36
2	Good	17	26	23	50
2	Fair	24	33	31	76
2	Poor	37	41	45	126
3	Expert	54	64	170	360
3	Good	56	82	230	500+
3	Fair	70	108	310	--
3	Poor	78	121	450	--

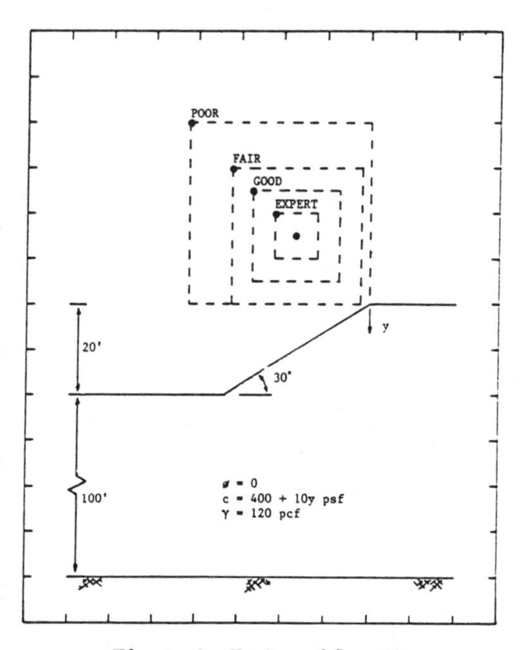

Figure 3. Test problem #3.

surface. Hence, the number of safety factor evaluations (or trial circles) may be used as a basis for comparing the relative efficiency of the four search routines. On this basis, the Simplex search strategy is clearly the most efficient (see Table 1). This occurs because the "step size" (or distance between any two consecutive evaluation points) varies continuously as the search progresses. The results demonstrate that a Simplex-based analysis would still be very efficient when performed by a novice engineer with little intuitive feeling for the correct location of the critical circle.

The relative superiority of the Simplex and Conjugate Direction routines increases dramatically when the dimension of the problem increases from two (in Problems #1 and #2) to three (in Problem #3). The constant step-size restriction causes the traditional grid search and pattern search procedures to become very inefficient when the dimension of the problem exceeds two.

10 CONCLUSIONS

The role and significance of the slip surface search algorithm has been all but ignored during the past thirty years. The very basic grid and pattern search routines have a long history of reliable performance in conjunction with computer-aided slope stability analyses. However, the Simplex procedure appears to be far more efficient and just as robust. The Simplex and Conjugate Direction routines are particularly effective when a full three-dimensional search is required. A few other optimization algorithms exist which could possibly outperform the Simplex and Conjugate Direction routines (DeNatale, 1986), and experimentation with these alternate procedures is currently underway.

11 REFERENCES

Celestino, T.B. and Duncan, J.M. (1981). Simplified search for noncircular slip surfaces, Proceedings of the 10th International Conference on Soil Mechanics and Foundation Engineering, Vol. 3: 391-394.

DeNatale, J.S. 1983. On the calibration of constitutive models by multivariate optimization. A case study: the Bounding Surface Plasticity Model. Ph.D. Thesis, The University of California, Davis.

DeNatale, J.S. and Abifadel, N.R. 1988. Application of Powell's Conjugate Direction method to slope stability analysis, Proceedings of the 6th International Conference on Numerical Methods in Geomechanics (in press).

Fredlund, D.G. 1984. Analytical methods for slope stability analysis, Proceedings of the 4th International Symposium on Landslides, Vol. 1: 229-250.

Gillett, S.G. 1987. An examination of search routines used in slope stability analyses. M.S. Thesis, The University of Arizona, Tucson.

Lefebvre, G. 1971. STABR user's manual, University of California, Berkeley.

Nelder, J.A. and Mead, R. 1965. A simplex method for function minimization, Computer Journal, 7: 308-313.

Nguyen, V.U. 1985. Determination of critical slope failure surfaces, Journal of Geotechnical Engineering, ASCE, 111: 348-352.

Powell, M.J.D. 1964. An efficient method for finding the minimum of a function of several variables, Computer Journal, 7: 155-162.

Spendley, W., Hext, G.R., and Himsworth, F.R. 1962. Sequential application of simplex designs in optimization and evolutionary design, Technometrics, 4: 441-461.

Swann, W.H. 1972. Direct search methods, in Numerical methods for unconstrained optimization (W. Murray, editor), Academic Press, London, pp. 13-28.

Taylor, D.W. 1937. Stability of earth slopes, Journal of the Boston Society of Civil Engineers, 24: 197-246.

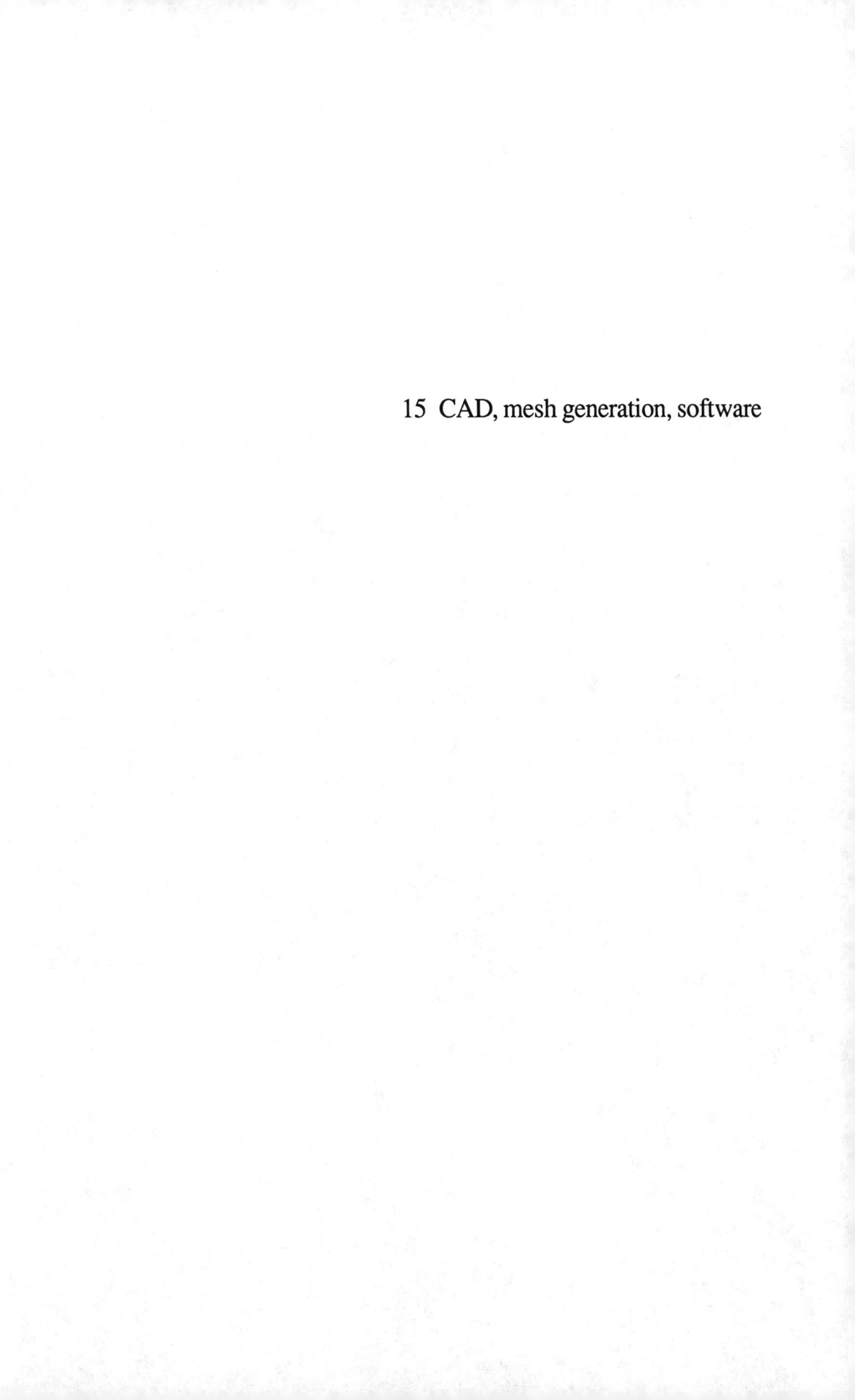

15 CAD, mesh generation, software

Numerical Methods in Geomechanics (Innsbruck 1988), Swoboda (ed.)
© 1988 Balkema, Rotterdam. ISBN 90 6191 809 X

Interactive modelling system for finite element applications in geomechanics

B.Kröplin & V.Bettzieche
Anwendung Numerischer Methoden, University of Dortmund, FR Germany
C.Bremer
Beratungszentrum CIM-Technologie, Dortmund, FR Germany

ABSTRACT: The finite element method (FEM) has been proved suitable since long time for the calculation of geotechnical problems as well as structural problems. However, an efficient use of FEM requires program systems, which allow a comfortable and fast process of modelling, investigation and optimization of structural variants and interactive design and calculation. Classical FEM programs and their pre- and postprocessors do not meet these tasks. The paper describes a software tool, which was designed in order to assist the engineer in the modelling process. It allows for parameter studies, optimization of design and use of the expert's knowledge.

1 INTRODUCTION

The finite element method (FEM) is an ideal method for engineering tasks in geotechnics (Duddeck 1986). However, the knowledge of the approximative features of FEM and the sensibility regarding discretization and assumptions, particularly in dynamic and non-linear problems requires a conscientious handling of FEM. Two conditions are important for a qualified application of FEM:

1. The calculation and above all the interpretation of the results must be handled by an experienced engineer.
2. The calculation must be carried out with various sets of input parameters which represent upper and lower bounds of possible realistic parameter combinations (Bomhard 1984).

If these two requirements are met, a correct and reliable modelling of the actual behaviour is possible. In addition the cost-optimal and construction-optimal solution can be found by investigation of variants of different geometries, materials, construction operations, construction methods etc.

For intensive use of FEM within the iterative calculation and development process an efficient and comfortable software system is necessary.

2 PROCESS OF MODELLING

The most important task in the use of FEM is the selection of a realistic finite element model. This model has to represent the essential characteristics of the structure and the foundation soil. The transformation from the construction idea to the finite element model is done in three steps (fig. 1).

The structural model is formed by classification.

The calculation model is found by idealization. In this step the engineer will primarily rely on his experience and intuition. The computer is only of limited help.

The transformation of the calculation model into the finite element model by discretization is the third step.

3 AVAILABLE FINITE ELEMENT SYSTEMS

Several FE-systems with pre-processor, FE-calculation program and post-processor are available. Mostly the modules are designed as a chain. Often the CAD-coupling is limited. The procedure chain "pre-main-post-processing" is rigid and irreversible. A feedback coupling for the adaptation of the FE-model according to an analysis of approximation errors is not possible yet. The systems available are mainly calculation aids. The process of modelling and development of variants is not sufficiently supported.

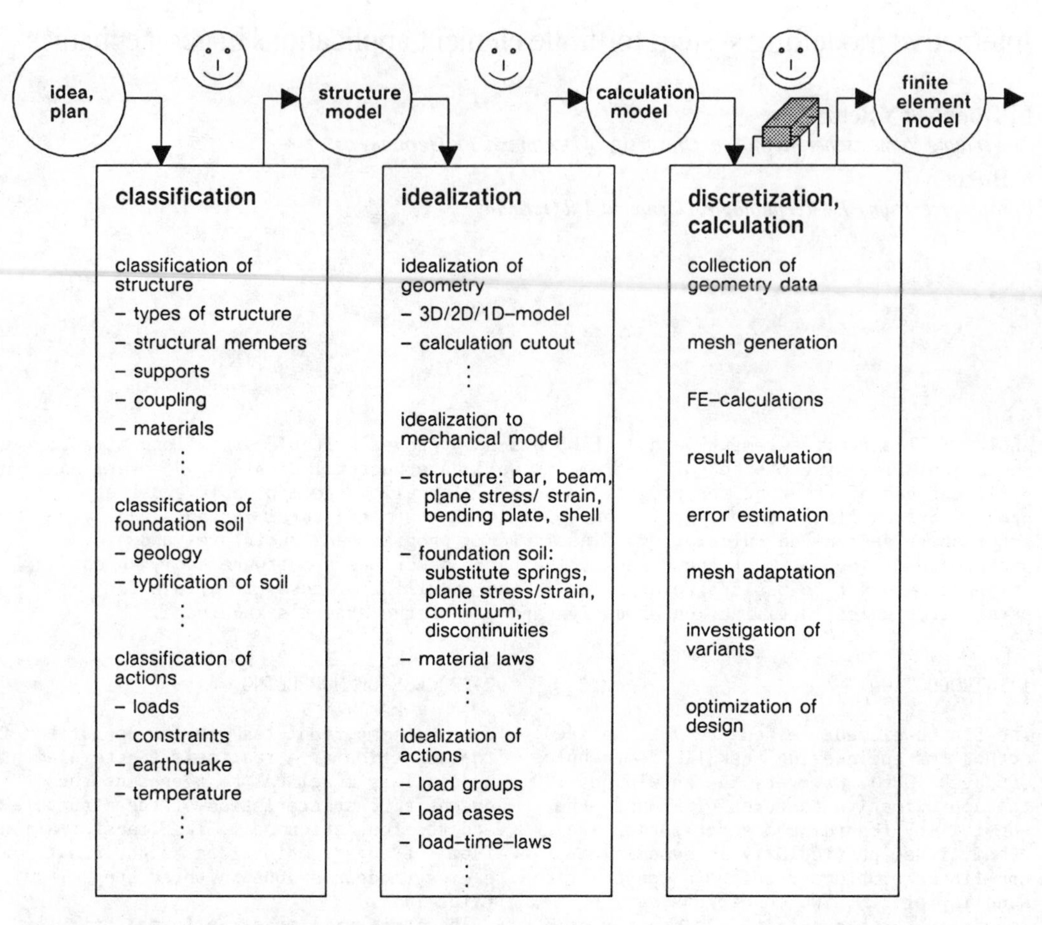

Figure 1: From the idea to the finite element model

4 TARGET CONCEPTION

The engineer is given an instrument which will enable him to ingeniously utilize the potentials of FEM for simulation and optimization. The user interface provides graphic assistance. The system is able to switch interactively among the processing steps of modelling, parameter studies and design optimizations. The engineer can almost introduce his full expert knowledge into the process.

5 LIMITS OF AUTOMATION

Present systems are designed to generate in one step the problem solution (output file) from a problem description (input file) once prepared. Even up-to-date systems which are developed as expert systems with AI-components often try to realize the solution in one-step.

In contrary to that it is the user's experience, that the solution thus obtained must still be corrected which is resulting in an iteration process outside of the program (fig. 2). The problems of automation will particularly occur in the steps of data collection, discretization and development of variants.

In data collection problem-relevant data (CAD-files) have to be filtered from a - mostly great - number of data. In this case algorithms are able to pre-sort. But valuation and final decision depend on the use of individual information and should interactively be made by the engineer. In the discretization process such a lot of information and relations are of importance, that here as well an automatic algorithm or even an expert system will rarely be able to achieve an optimal solution. This is mainly due to the fact that it is not yet possible to comprehend the overall experience of a calculation engineer, including intuition, by algorithms. This implies as well the

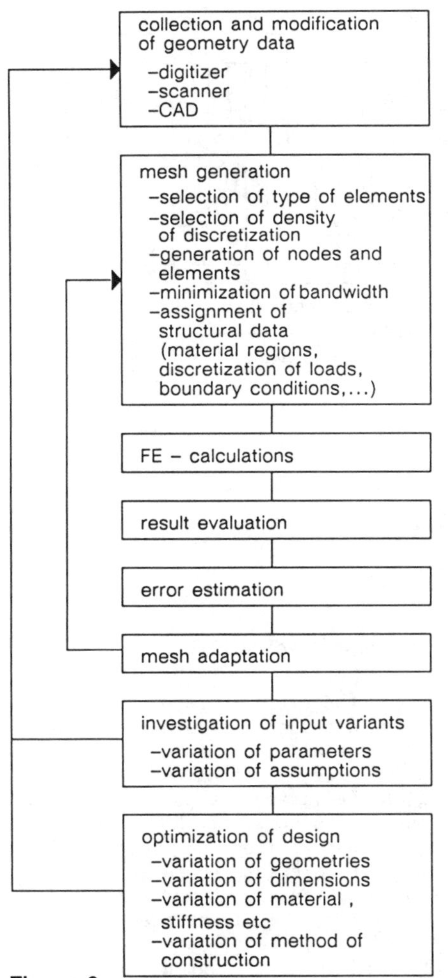

Figure 2:
Finite element modelling and calculation

knowledge of the behaviour of different element types, their approximation functions and possibilities of coupling to other element types etc. Particularly here the efficiency of an interactive expert system becomes evident by preparing decisions, making available its methods, perhaps even interpreting user decisions, but leaving the decision itself to the engineer.

In the development of variants as well the experience of the user is irreplaceable. The formulation of rules for the gradation of mesh refinements, the mutual impacts of refinements and the evaluation of the necessary grade of refinement should not be carried out by the system total automatically. The interpretation of arithmetical results by the engineer is indispensible. He decides, if further variants are to be calculated or if model or design are to be modified.

6 DATA COLLECTION

Since the data are available on different media the interface to data collection must be as variable as possible. Adequate data collection programs (digitizer programs, scanner programs, interface programs to CAD) should be individually attachable for these different media. It is obvious that in construction engineering CAD is increasingly prevailing in the area of design and construction (Kröplin 1987). The geometry data collected should possibly be described in its raw state, processed and corrected by the engineer. A sufficiently simple description, i.e. a simple data form must be chosen for this data interface.

7 PROCESS OF FINITE ELEMENT MODELLING

The computer-aided modelling should be adapted to the conventional procedure of a calculation engineer and should support it (fig. 3). This implies :

- Determination of problem-relevant data (contour):
  The user must be given the opportunity to classify and interpret the input data and perhaps exclude them as irrelevant for further processing.
- Construction of a rough division (superelement mesh):
  By the aid of the most important data the user will roughly divide the problem geometry into uncritical - i.e. areas to be roughly discretized - and refinement areas.
- Detail mesh construction (element meshes):
  In the structure areas which call for a high discretization quality the user must be able to select locally dense discretization.
- Linking of detail solutions (total mesh geometry):
  The individual detail solutions are coordinated by the aid of expansion and transition meshes and combined to a total discretizations.
- Attribute assignment:
  The attribute assignment, i.e. the assignment of material data, loads and the remaining input data relevant for the calculation program should be effected by the aid of the computer.

The procedure of a user is not merely forward-oriented but consists of forward- and backward-steps, i.e. an optimization process which is based on the valuation of various variants and variant combinations. This stepwise procedure and its iteration possibilities must not even be taken from

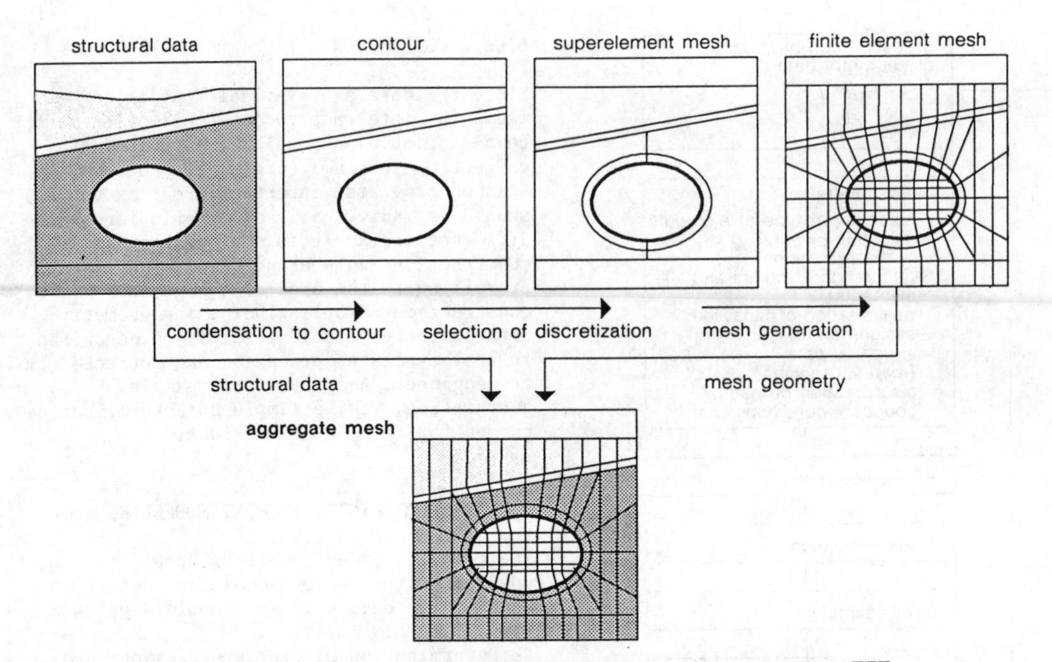

Figure 3: Generation of a FE input data file with the finite element editor FEE

the user by a computer aid. Thus the system has to store variants and keep them ready for calling at any time.

8 INTEGRATION OF VARIOUS FINITE ELEMENT PROGRAMS

The field of application of the method of finite elements in geotechnic engineering ranges from soil statics via ground-water calculation to thermal calculation. Often one problem belongs to several of these areas, e.g. in case of the static and flow calculation of a sheet piling or in case of static calculations of frost structures. At present these problems must be calculated with different programs and in addition different problem descriptions - i.e. input files - must be generated. A great part of the extra work consists of double work in discretization and assignment of problem-overlapping parameters. Here an efficient preprocessor should at first generate and manage the general data for both tasks together and only for calculation complete them by problem-specific data.

9 SOLUTION APPROACH AND REALIZATION

The program system FEE (Finite Element Editor) uses the tools of modern software technology (fig. 4) to solve these problems. Based on a standard graphic module, a data base system, the efficient operation system UNIX and the standardized data interface FEDIS (Groth et al. 1985) an user interface has been realized, which allows a simple handling of the various steps of modelling. At present the capacity of FEE comprises the areas of geometry collection, mesh generation, FE-calculation and result evaluation (see fig. 2). Some FE-meshes, generated by FEE, are illustrated in fig. 5.

Corresponding to the steps of modelling there are individual data groups in the program system which are complete in itself but are logically linked with the preceding and following groups. This leads from a hierarchic to a rational structure which was realized in an adequate relational data base.

The graphic kernel system GKS (DIN/ISO-standard) forms the interface to the graphic terminal. Additional modules enlarge and support the possibilities of GKS in graphics and in the user dialogue. A layer management is included which allows to visualize on the screen different modelling stages such as foils. Even the data at the bottom of processing can at any time be activated, corrected and faded out again. Thus the transparency of modelling will be increased without loosing the survey of the data used.

The performance of the system was increased by the modular structure. In this way time-intensive modules are working, such as mesh generators and assignment modules, controlled by a control module (master) in

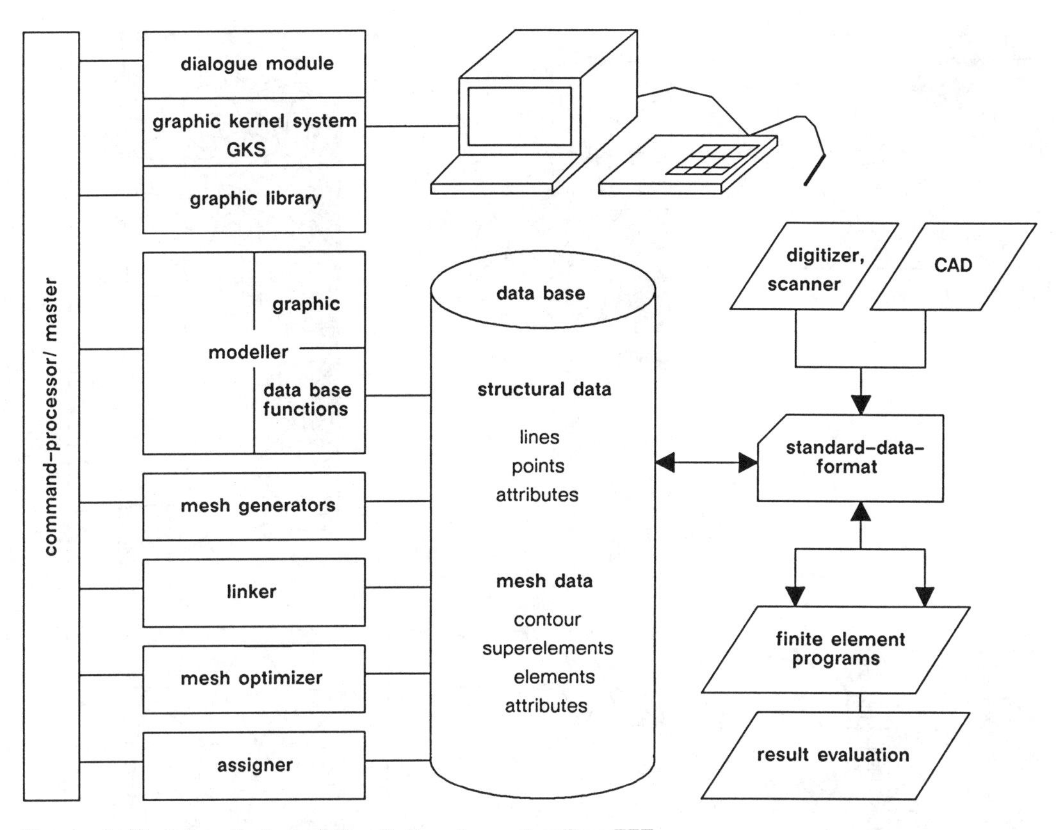

Figure 4: Modular design of the finite element editor FEE

the background (batch) and after having finished their work they supply the data to the data base. Meanwhile the user can continue his processing with the system. The mesh generators used work according to various methods (Bremer 1986). Apart from conventional generators different interpolators and triangulators are available. Their parameters can also be interactively controlled by the user.

The output of the data is effected in the way defined in FEDIS (Finite Element Data Interface Standard) which allows a connection to many available finite element program systems.

10 CONCLUSIONS AND PROSPECT

The finite element editor FEE is a tool, which will assist the engineer on his way towards a set of finite element models. The graphic and interactive program system enables him to carry out efficiently the finite element process of modelling, parameter studies and design optimizations.

Future developments deal with a further support of the individual steps on this way using knowledge based tools as well as classical programming. Capable superelement generators are provided which subdivide a structure into refinement areas and model libraries allowing to store recurrent refinement strategies and to make them available for later utilization. Last not least adaptive methods have to find access to preprocessing. This will lead to an interactive, graphic dialogue system including simple result representation as well as the complete integration of finite element modelling and calculation.

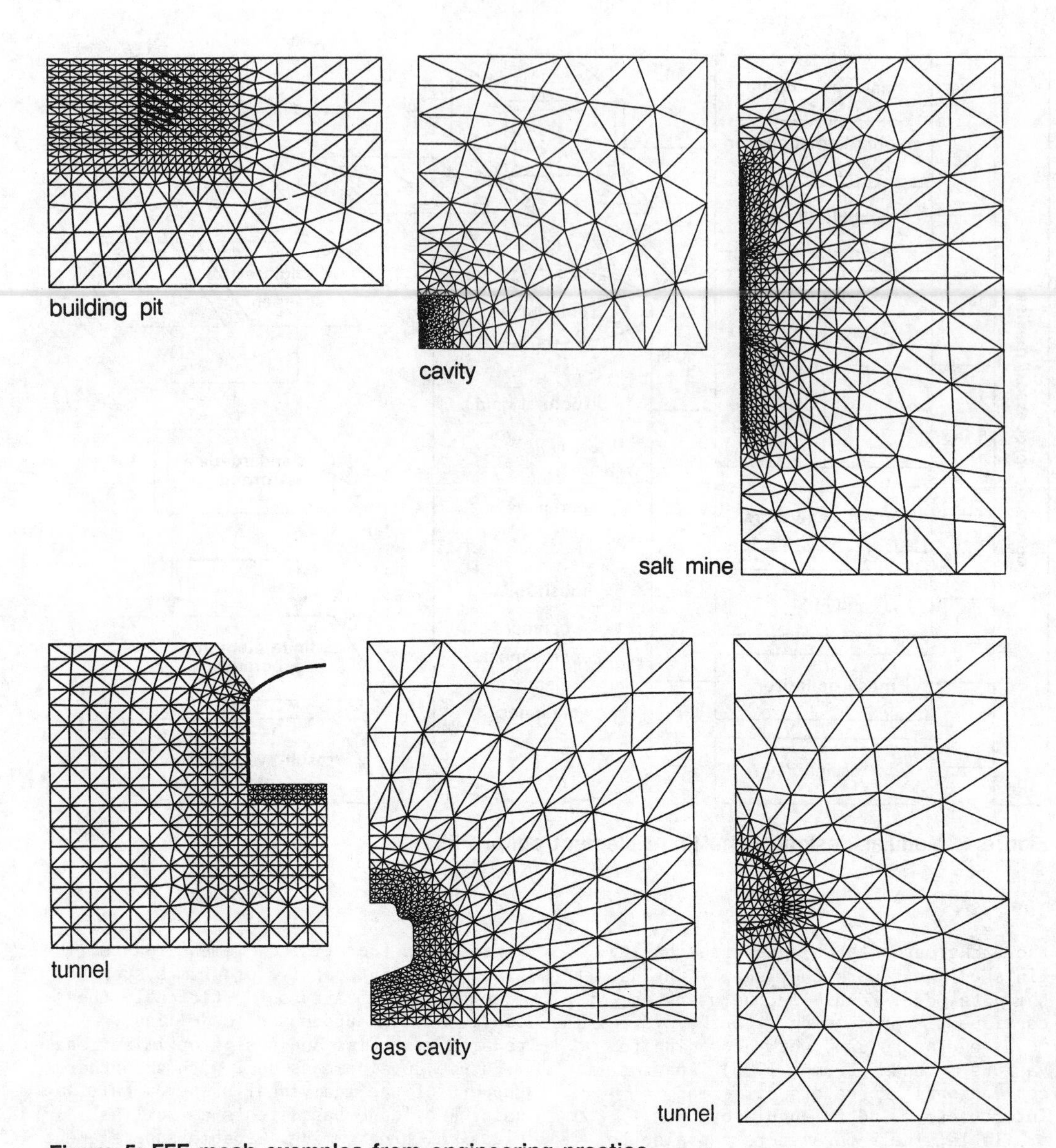

Figure 5: FEE mesh examples from engineering practice

REFERENCES

Bomhard, H. 1984. Die Finite-Element-Methode und die Baupraxis. Finite Elemente - Anwendungen in der Baupraxis. Ed.: H. Grundmann, München: 34 - 39.

Bremer, C. 1986. Algorithmen zum effizienteren Einsatz der Finite-Element-Methode. Report No. 86-48, Institut für Statik, Technische Universität Braunschweig.

DIN 66252. 1986. GKS-Graphic Kernel System.

Duddeck, H. 1986. Leistungsfähigkeit und Grenzen der Methode der Finiten Elemente in der Geotechnik. Felsbau, Vol. 4, No. 3: 126 - 133.

Groth, P. et al. 1985. FEDIS - Finite Element Data Interface Standard. Report KfK-PF 114, Kernforschungszentrum Karlsruhe.

Kröplin, B. 1987. Zur Problematik der Gesamtplanung mit CAD-Systemen. VDI-Seminar - Planen und Konstruieren im Hochbau mit CAD, München: 1 - 12.

Numerical Methods in Geomechanics (Innsbruck 1988), Swoboda (ed.)
© 1988 Balkema, Rotterdam. ISBN 90 6191 809 X

The computer-aided analytical system for the geotechnical characterization problems

J.L.Meek & S.Akutagawa
University of Queensland, Australia

ABSTRACT: This paper presents the on-going research work for the development of computer-aided design and analytical system for geotechnical characterization problems. The system is currently being evolved on the MicroVax hardware configuration using the Graphic Kernel System in FORTRAN 77 language. The main goal is to set up the fully interactive analytical system in which the unknown parameters concerning (1) geometry, (2) material constitutive law and (3) loading system are determined through the use of FEM program and appropriate optimization procedure. Such system characterization analysis is based on the concept of minimizing the discrepancy between the numerical prediction and the actual behaviour of the geotechnical system and therefore requires, as the standard procedure, the in-situ field measurement data to define the suitable error function. Previous studies have proved that such a system is not only a feasible project but also has great potential to improve the monitoring system of geotechnical engineering practices.

1 INTRODUCTION

In order to perform reliable construction monitoring it is essential that the numerically predicted behaviour of the geotechnical media coincides with field observation. However, due to difficulties in assessing the "true" mechanical properties of the rock mass and complicated geological conditions, the analytical predictions are not always guaranteed to be an exact simulation of the actual behaviour. Characterization, system identification, or back analysis approach intends to minimize such "discrepancy" between the numerical prediction and the actual behaviour of the geotechnical media from the standpoint that the field measurements data exhibit the true mechanical characteristics of the physical system under consideration. The "tuned" model which matches as closely as possible the actual movement of the rock mass is then regarded as the best achievable numerical model and parameters defined in the tuned model are used for subsequent design purposes.

Solution methods for such characterization analysis require the in-situ measurement data as the input data and determine the so-called "decision variables" (unknown variables) through either (1) closed form of inverse algorithms or (2) iteration-based mathematical optimization procedures. The inverse approach requires the actual inversion of the governing equation; however it can solve the problem basically in one analysis. On the other hand, the second approach (direct search method) is based on the combined use of standard FEM code and general form of mathematical optimization routes and reaches to the solution on an iteration basis. Applications are versatile and any required modification of code is relatively inexpensive whereas the computational cost is somewhat higher.

On the other hand, it has been widely recognized that the use of interactive computer graphics greatly enhances the efficiency of numerical analyses, especially for pre- and post-processing of large amounts of data. Construction of such an interactive system requires state-of-the-art programming techniques and overall reorganization of the program management scheme, thus leveling up the standard of the software itself. Kulhawy et al have successfully applied this concept and demonstrated the efficiency and productivity of the system designed for geotechnical analysis purposes (Ref.19-21).

This paper briefly discusses the various types of characterization analyses and describes development directions for the computer-aided design system in which the various characterization analyses can be performed in a user-friendly manner.

2 GEOTECHNICAL CHARACTERIZATION ANALYSIS

As mentioned before and discussed in detail in references 4 and 17, characterization analysis is based on either (1) the inverse method or (2) the direct search method from the mathematical point of view. However, from the engineering point of view it is worthwhile to classify the problem solution techniques developed for (1) the materially linear system and (2) the materially non-linear system.

2.1 Characterization for the materially linear system

Characterization analyses developed for the materially linear geotechnical system have various forms depending on the type of problem, type of unknown variables to be determined and the type of solution technique being used.

The closed form inverse approach was adopted by Kavanah (Ref.1) and Kirsten (Ref.2) for determination of material elastic constants with or without the use of the least square method. The least square method was also adopted by Sakurai and Takeuchi (Ref.7) to determine in-situ initial stress state, later systemized on the microcomputer as a convenient tool for underground excavation monitoring by Sakurai and Shinji (Ref.8). Boundary element formulation was also demonstrated for determination of elastic constants by Hermans (Ref. 10) and initial stress state by Sakurai and Shimizu (Ref.9).

For types of problems in which unknown variables range over more than one data category, e.g. unknown parameters concerning material properties and geometric parameters at the same time, the closed form inverse formulation is often not available. Such problems, therefore, must be solved by appropriate mathematical optimization techniques coupled with the standard FEM programs. Gioda (Ref.3) has shown the applicability of the function minimization algorithm for determination of average elastic and bulk moduli. Sakurai (Ref.11) combined the least square method and the Rosenbrock method for determining in-situ initial stress (load parameter) and the directional and stiffness parameters of anisotropic elastic material at the same time.
Cividini et al (Ref.5) employed the Bayesian statistical identification approach to identify the length and location of unknown inclusion (geometry data) and material properties (material data) within the single problem formulation.

These studies have shown that for materially linear characterization analysis, either the inverse method (if necessary conditions are satisfied) or the direct search method can be used effectively with a fairly high degree of confidence.

2.2 Characterization analysis for materially non-linear system

Characterization analyses developed for determining the unknown parameters of a materially non-linear system reflect the complexity of the material constitutive law and take more sophisticated forms than those for the linear system. However, it is still feasible to adopt generalized mathematical optimization techniques by treating the non-linear stress analysis process as a "function" of N-decision variables.

Applicability of direct search solution techniques previously applied successfully to linear system identification has been proved by several authors. Such an approach has been used to determine non-linear material properties and in-situ stress by Gioda and Maier (Ref.12), soil-structure interaction pressure by Gioda and Jurina (Ref.13) and rheological material properties by Cividini et al (Ref.16). Parameters on yield limits were identified through mathematical programming by Maier et al (Ref.14) and through the statistical identification approach by Maier et al (Ref.15).

These studies have indicated that even though the physical implication of material non-linearity is more complicated than the linear counterpart, implementation of the direct search method is quite straightforward from the mathematical standpoint. Therefore, in general terms, characterization analysis for the materially non-linear system can be accommodated by the same procedure by the proper use of the generalized function minimization technique.

2.3 Summary

To summarise previous discussion on various forms of geotechnical characterization analyses, some noteworthy indices are shown in Table 1. Most problem formulations are based on the finite element method and the appropriate selection of function minimization techniques. Inverse methods are applied to limited elastic problems. It is interesting to note that almost all approaches use displacement data as field measurement data to define the proper error function. Also it is found that most attention is paid to determination of material properties and relatively little effort spent on determination of the loading system and geometric uncertainties.

3 CAD DESCRIPTION

In order to perform various types of characterization analyses in an interactive workstation environment, correct and reliable modules must be organized systematically while maintaining the simplicity of the operation through maximum use of data visualisation. This section discusses the requirements for the characterization analysis system in general terms and refers to the fundamental methodology of data and software management which are central issues in the CAD development process.

3.1 Specification for analytical modules

The primary objective of the development is construction of the flexible characterization system. However, since problem formulation is based on the finite element method, the entire FEM code of full analytical capability must be built into the system as the primary requirement.

As can be imagined from Table 1, except for a few cases in which the inverse formulation is possible, characterization analysis rests its foundation on the coupling of FEM code and the general form of function minimization routine. This coupled formulation is regarded as a purely mathematical operation which the entire stress analysis process is treated as one "function" whose output depends explicitly on the finite number of input decision variables. Therefore, whole varieties of different types of characterization analysis can be formed simply by specifying which analytical module is to be coupled with the optimization routine and which decision variables (geometric, material property, or load system parameters) are to be determined through the iterative search algorithm.

Since the characterization analysis is a deliberately organized dynamic flow of various types of data groups, executable modules are designed and organized with respect to each different data group. For each data category, data specification and data representation modules are created. For example, the specification module for geometry data means the interactive mesh generator, and the representation module for stress data means the graphic stress contouring program. Specification of control parameters governing execution of the analytical module can be regarded in a broad sense as a special data group and can be defined independently or at the time of module execution.

Interactions necessary for those CAD operations are either in the form of "keyboard interaction" or in "graphic interaction" where the user is requested to send signals or data into the system by using the alphanumeric terminal or the graphic input device, respectively.

The proposed system is currently being evolved and not yet capable of satisfying all requirements. However, the end product of the characterization analysis system is specified to fulfil the requirements shown in Table 2.

3.2 Data management and software structure

A well-designed data base system and systematic organization principles are the fundamental requirement for successful construction of the flexible CAD system. Software used in an interactive CAD system is usually subject to constant upgrading and periodical reorganization. To minimize the necessary coding modifications in these circumstances, a data base management scheme and software organization principles must be as simple as possible while maintaining the generality and efficiency of available memory usage.

For all the coding employed in this study, the "naming-based" simple, yet fairly well designed data management scheme is employed. This method succeeds directly from the work of Hoit and Wilson (Ref.24), and all data entries registered in a single dynamic parent array are identified by the name of finite length. All data-related operations are targeted to the intended data object through their names, therefore the user need not worry about the exact location of the required data entry even during active modification of the source code. This method not only removes complicated address manipulations but also favourably affects the coding structure in such a way that data flow is uniquely centralized at limited locations in the program. This centralization is advantageous for error detection and maintenance purposes. A name-oriented data manipulation method also maximizes freedom in choosing the dynamic variables of any data category in the versatile characterization analysis process.

To establish the software in such a way that interdepencies between different modules are minimized and greater efficiency achieved for constant software evolution and possible software transfer from the main frame to a personal computer, particular care has been taken to structure the software.

To quantitatively identify modules registered in the CAD system, 3 different levels of interactability index are introduced for (1) keyboard input mode, (2) graphic input mode, and (3) graphic output mode. For all modes, interactability indices are defined as 0 (specified function is never called; 1 (specified function is occasionally called) and 2 (specified function is definitely called).

For the analytical modules which operates only on digital data, modes are set strictly as (0,0,0) in which the three digits represent keyboard input, graphic input and graphic output modes respectively. Any interactive operations related to that particular analysis module are deliberately separated from the main module and make use of full interactive capability of available hardware with its modes (2,2,2). These interaction capabilities are essential for a user-friendly operational environment and at the same time introduction of these graphical interactive functions inevitably increases the size of the source codes. However by building up these interactive functions on top of non-interactive standard coding with maximum care, the software portability is enhanced and the transfer of the system to a smaller computer becomes, at least, easier, at the risk of losing full interactability and complete access to all the different modules at one time.

4 DISCUSSION

In the extreme sense, the characterization analysis system is a hybrid software structure in which the numerical discretization technique of inverse or normal forms is optimally integrated in the mathematical function minimization scheme. Several attempts in this direction have already been reported by DeVore et al (Ref.22) and Vanderplaats et al (Ref.23). This concept has already proved valid for various practical examples. Increasing CPU time with the problem complexity should not be a major obstacle considering the rapid advancement of hardware capability. The emergence of parallel processors or multi-workstation-type hardware configuration will also reinforce the system's overall performance. The characterization system, once built up, may be applied usefully for a real time monitoring system of various geotechnical engineering observations. Extensive application will also lead to further sophistication of characterization theory.

On the other hand, there are several points to be made clear concerned with fundamental assumptions employed in the characterization analysis. First of all, as pointed out in Ref.14, the characterization analysis assumes that measurement values are correct and represent the true behaviour of the geotechnical system. Therefore, the decision variables of only the numerical model side are modified until the model can adjust itself sufficiently closely to actual measured behaviour. In this process, the possible existence of man- or machine-dependent measurement errors or local failure of the rock mass where measurement devices are installed is not considered. Also the discrepancy may accrue from the theoretical limit of modelling, for example, the hihgly discontinuous media by the continuum mechanics-based problem formulation. If these circumstances arise, "permanent" (or "residual") discrepancy, left over after the mathematically defined optimum model is obtained, remains untouched and unexplained under the current trend of the characterization analysis concept.

The proper investigation of the permanent error will have to include the new concept which treats the optimization process not only as "(1) the fixed function defined by N-dynamic variables", (the currently accepted approach), but also as "(2) the dynamic function defined by N-dynamic variables". The characterization algorithm therefore will include the additional "function identification analysis" loop in which the permanent error remaining after the type (1) "parameter identification analysis" is exposed to the further minimization which obviously follows different principles from those used for type (1) analysis.

5 CONCLUSION

This paper briefly reviewed geotechnical characterization analysis and discussed the methodology of setting up the interactive characterization analysis system in the light of data management and software organisation. Characterization analysis can be generally reiterated as the optimization problem of the fixed function subject to N-dynamic variables. It is pointed out that further interest in research lies in the reasonable quantification of the permanent error through optimization of the dynamic function subject to N-dynamic variables.

6. REFERENCES

1. Kavanagh,K.T. 1973. Experiment versus analysis: computational techniques for the description of static material response. Int Jnl Numerical Methods in Eng. Vol.5:503-515.
2. Kirsten,H.A.D., Msaimm,M. 1976. Determination of rock mass elastic moduli by back analysis of deformation measurements. Proc Symp on Exploration for Rock Eng. Johannesburg. 165-172.
3. Gioda,G. 1980. Indirect identification of the average elastic characteristics of rock masses. Proc Int Conf Structural

Foundation on Rock. Sydney. 65-73.
4. Cividini,A., Jurina,L., Gioda,G. 1981. Some aspects of characterization problems in geomechanics. Int J Rock Mech Min Sci & Geomech. Abstr. Vol.18:487-503.
5. Cividini,A., Maier,G., Nappi,A. 1983. Parameter estimation of a static geotechnical model using a Bayes' approach. Int J Rock Mech Min Sci & Geomech. Abstr. Vol.20 No.5:215-226.
6. Asaoka,A., Matsuo,M. 1979. Bayesian approach to inverse problem in consolidation and its application to settlement prediction. Proc 3rd Int Conf on Numerical Methods in Geomech. Aachen. 115-123.
7. Sakurai,S., Takeuchi,K. 1983. Back analysis of measured displacements of tunnels. Rock Mech & Rock Eng 16:173-180.
8. Sakurai,S., Shinji,M. 1984. A monitoring system of excavation of underground openings based on microcomputers. Proc Int Symp on Design & Performance of Underground Excavation. Cambridge.
9. Shimizu,N., Sakurai,S. 1983. Application of boundary element method for back analysis associated with tunnelling problems. Proc 5th Int Conf on Boundary Element Method in Eng. Hiroshima. 645-654.
10. Hermans. 1982. Boundary integral equations applied in the characterization of elastic material. Proc Int Conf on Computational Methods & Experimental Measurements. 189-199.
11. Sakurai,S. 1987. Interpretation of the results of displacement measurements in cut slopes. Proc 2nd Int Symp on Field Measurements in Geomechanics. Kobe. 528-539.
12. Gioda,G., Maier,G. 1980. Direct search solution of an inverse problem in elastoplasticity: Identification of cohesion, friction angle and in situ stress by pressure tunnel tests. Int J Numerical Methods in Eng. Vol.15:1823-1848.
13. Gioda,G., Jurina,L. 1981. Numerical identification of soil-structure interaction pressures. Int J Numerical & Analytical Methods in Geomechanics. Vol.5:33-56.
14. Maier,G., Ginnessi,F., Nappi,A. 1982. Indirect identification of yield limits by mathematical programming. Eng Struct Vol.4:86-97.
15. Maier,G., Nappi,A., Cividini,A. 1982. Statistical identification of yield limits in piece-wise linear structural models. ISCME Int Conf Comput Methods & Exper Measurements. Washington DC. 812-829.
16. Cividini,A., Gioda,G., Barla,G. 1985. Calibration of a rheological material model on the basis of field measurements. Proc 5th Int Conf Numerical Methods in Geomech. Nagoya. 1621-1628.
17. Gioda,G. 1985. Some remakrs on back analysis and characterization problems in geomechanics. Proc 5th Int Conf on Numerical Methods in Geomechanics. Nagoya. 47-61.
18. Gioda,G., Cividini,A. 1987. A comparative evaluation of some back analysis algorithms and their application to in-situ load tests. Proc 2nd Int Symp Field Measurement in Geomech. Kobe.
19. Kulhawy,F.H., Ingraffea,A.R., Han,T.Y., Huang,Y.P. 1985. Interactive computer graphics in 3-D nonlinear geotechnical FEM analysis. Proc 5th Int Conf on Numerical Methods in Geomech. Nagoya. 1673-1681.
20. Ingraffea,A.R., Kulhawy,F.H., Abel,J.F. 1981. Interactive computer graphics for analysis of geotechnical structures. Proc Symp Implementation of Computer Procedures & Stress-Strain Laws in Geotech. Eng. Chicago. Vol.1:190-202.
21. Han, T.Y. 1984. Adaptive substructuring and interactive graphics for 3-dimensional elasto-plastic finite element analysis. PhD Thesis, Cornell Univ.
22. DeVore,C.R., Briggs,H.C., DeWispelare, A.R. 1987. Application of multiple objective optimization techniques to finite element model tuning. Computers & Structures Vol.24 No.5:683-690.
23. Vanderplaatz,G.N., Sugimoto,H. 1986. A general-purpose optimization program for engineering design. Computers & Structures Vol.24 No.1:13-21.
24. Hoit,M.I. 1983. New computer programming techniques for structural engineering. PhD Thesis, Univ of California, Berkeley.

TABLE (1) Variations of geotechnical characterization analysis

Authors	Ref.	Year	Unknown data	Measurement data	Optimization method	Numerical method
MATERIALLY LINEAR GEOTECHNICAL SYSTEM						
Kavanah	1	1973	M	D,E	LSQ	Mathematical
Kirsten	2	1976	M	D	LSQ	FEM
Asaoka et.al	6	1979	M	D	STT	Mathematical
Gioda 3	3	1980	M	D	LSQ,DRS	FEM
Cividini et.al	4	1981	GML	D	LSQ,DRS	FEM
Hermans	10	1982	M	D	LSQ	BEM
Sakurai et.al	7	1983	L	D	LSQ	FEM
Shimizu et.al	9	1983	L	D	LSQ	BEM
Cividini et.al	5	1983	GM	D	STT	FEM
Sakurai	11	1987	ML	D	LSQ,DRS	FEM
MATERIALLY NON-LINEAR GEOTECHNICAL SYSTEM						
Gioda et.al	12	1980	ML	D	DRS	FEM
Gioda et.al	13	1981	L	DL	DRS	FEM
Maier et.al	14	1982	M	D	DRS	FEM
Maier et.al	15	1982	M	D	STT	FEM
Cividini et.al	16	1985	M	D	DRS	FEM

Notations : Unknown data (M = material property, G = geometry , L = load parameter)
Measurement data (D = displacement , E = strain , L = load parameter)
Optimization method (LSQ = least square , DRS = direct search , STT = statistical , INV = inverse)

TABLE (2) Executable menus to be installed in the proposed system

Data category	GENERALIZED INTERACTIVE GRAPHIC EDITOR			
	DATA SPEC.		MAIN EXEC.	DATA REPRESENTATION
	NG mode	G mode		
Geometry	A	B	C	B
Boundary condition	A	B	C	B
Material property E	A	B	C	B
EP	A	B	C	B
EV	A	B	C	B
EVP	A	B	C	B
Load vector : Nodal	A	B	C	B
Pressure	A	B	C	B
Body force	A	B	C	B
Excavation	A	B	C	B
Forced disp.	A	B	C	B
Time	A	B	C	B
Displacement	A	B	C	B
Strain	A	B	C	B
Stress	A	B	C	B

Module execution mode	A	B	C
Keyboard input	2	2	1
Graphic input	0	2	0
Graphic output	0	2	0

Unknown data category	CHARACTERIZATION ANALYSIS			
	LINEAR SYSTEM		NON LINEAR SYSTEM	
	INVERSE	DIRECT SEARCH	INVERSE	DIRECT SEARCH
G	*	Δ	*	Δ
M	Δ	Δ	Δ	Δ
L	Δ	Δ	Δ	Δ
GM	*	Δ	*	Δ
ML	*	Δ	*	Δ
LG	*	Δ	*	Δ
GML	*	Δ	*	Δ

G ... geometry * ... Not available Δ ... Available
M ... material prop.
L ... load parameter

Numerical Methods in Geomechanics (Innsbruck 1988), Swoboda (ed.)
© 1988 Balkema, Rotterdam. ISBN 90 6191 809 X

A practical application of MAKEBASE: An information retrieval system for structural mechanics – Examples of information management in geomechanics

J.Mackerle
Linköping Institute of Technology, Sweden

ABSTRACT: Computer program development based on the finite element/boundary element methods, as well as satellite programs in the form of pre- and postprocessors is now receiving considerable attention of engineering community. Output of related literature from the theoretical and practical application point of view is growing at prodigious rate in recent years.

MAKEBASE is a special purpose, menu-driven, database system which stores literature references and detailed information on the finite element and boundary element software, and software information on pre- and postprocessors interfacing the finite element and boundary element programs. Additional topics included: optimization techniques, and expert systems in connection to the structural/mechanical engineering.

EXMAKE is an expert system under development, helping to an unexperienced MAKEBASE user to choose keywords for an effective subject search.

MAKEBASE is implemented on VAX-11/780 (VMS) computer and regularly updated. The presented paper describes MAKEBASE and outlines its applications in the field of geomechanics.

1 INTRODUCTION

Information is the most valuable but least valued tool the engineer/scientist has. The volume of scientific and technical literature is growing exponentially, and most professionals are no longer able to be up-to-date with all the relevant literature. Some topic areas tend to become overloaded with information very rapidly. It is, for example, the situation which we have in the field of computerized numerical analysis in structural mechanics.

Figure 1 illustrates an output of the literature in finite element (FE) and boundary element (BE) technology between 1970-1986. Number of references between 1970-1974 is taken from the Norrie and de Vries (1976) and between 1975-1986 from MAKEBASE. The author is aware that a large amount of literature references is not yet included in MAKEBASE. In our opinion, there have been published about 2000-2500 papers on boundary elements, and 25,000-30,000 papers on finite element (excl. fluid flow problems, which are not included in the database).

FE and BE software development, as well as development of pre- and postprocessor programs is now receiving considerable attention of engineering community.

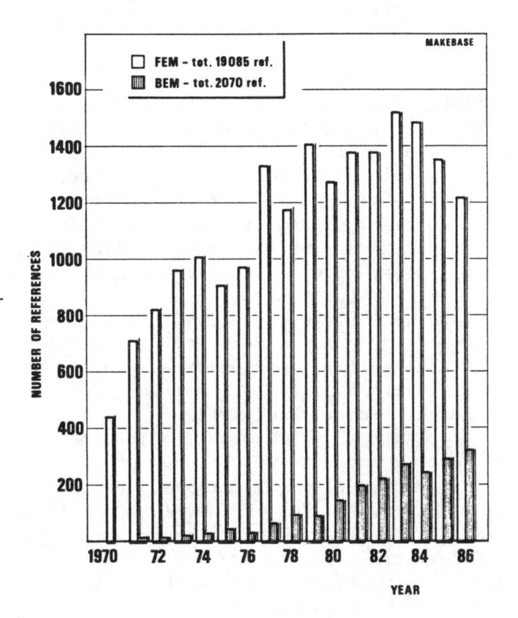

Fig.1 Output of FE/BE literature

To successfully and economically model, run, and interpret the analysis results, we need an expertise, an user with theoretical knowledge, and a quick access to different sources of information.

It has been estimated that engineers/ scientists spend as much as 30% of their working time looking for information. Even a saving of a few per cent of the total effort at present spent in "selecting out" may in fact represent a considerable saving of manpower time.

There is a dilemma, both on the literature and software, namely:

Literature:

- On the one hand scientists have to focus more narrowly than before just to stay abreast of their field of interest,
- On the other hand they have to become generalists in order to satisfy the diverse environmental and economic constraints placed on technology.

Software:

- To find the developed software, based on BE/FE methods is not easy; to the author´s knowledge no database covering this type of information has been in existence at the start of MAKEBASE development.
- To compare different programs and evaluate their quality and reliability is extremely difficult; we have not yet benchmark problems standardized and accepted by engineering community for the evaluations.
- Papers published in the literature on used software present only successful analysis and excellent results; nobody writes about the problems during the work; which program has been used is often not mentioned.

Now, how shall we store/retrieve all available information? Without the computers assistance is the effective information storage/retrieval today impossible. In our opinion, the best access to general, respective "wide-range" area of information are large information databases. In 1985 there were 2400 databases available through 345 retrieval services, and this number is growing rapidly. From these databases we are able to retrieve information in any field of human activity. Large databases are usually command-driven where a higher level of skill is required (i e knowledge of query language).

For a "narrow", high-special field of engineering activity we can use special purpose, user-friendly, databases and this was a main reason for the development of MAKEBASE. We also wanted to share both type of information, literature and software respectively, in the same database. Existing information is useless if it is not accessible in an effective way.

2 DATABASE DESCRIPTION

MAKEBASE is a highly, subject-limited, menu driven database. This type of database is very easy to use because it provides an unexperienced user with on-screen prompts.

The database is implemented on the VAX 11/780 (VMS), and new records are inserted on a daily basis. The program for data management is written in FORTRAN.

2.1 Subject areas of information stored

The database provides storage of information in the area of structural mechanics with emphasis on FE (fluid mechanics is not included), BE (all areas of application), optimization techniques, and expert systems

The following type of information is stored:

- Literature references covering FE and BE analysis techniques from the theoretical and application point of view. Retrospective to 1975.
- Literature references covering the field of pre- and postprocessing in connection to the FE/BE programs, and FE/BE software considerations in general.
- Program documentations available.
- Literature references on expert systems in connection to structural mechanics.
- Detailed FE software information. Programs included are general and special-purpose, academic and commercial. Retrospective to 1970.
- Detailed BE software information. Programs included are general and special-purpose, academic and commercial. Retrospective to 1970.
- Detailed information about pre- and postprocessors interfacing FE and/or BE programs. Retrospective to 1970.

At time of writing this paper the MAKEBASE contains information about 1400 programs and approximately 21,000 literature references.

2.2 How the information is stored

Records for literature references have the following fields:

- article title;
- author name/names;
- journal name/conference name;
- volume/place (if conference);
- number/month (if conference);
- year of issue;
- pages;
- note;
- keywords for subject and other aspects of significance description (up to 10 different keywords are permitted/record).

Records for the FE/BE software information have the following fields:
- program name;
- type of program;
- range of application;
- phenomena;
- program abstract;
- element library;
- material library;
- type of permitted loading;
- other notable program options;
- formulation;
- solution methods;
- modelling capabilities;
- related pre- and postprocessors;
- operational on (computer types);
- source for program (address).

MAKEBASE architecture is illustrated in Figure 2.

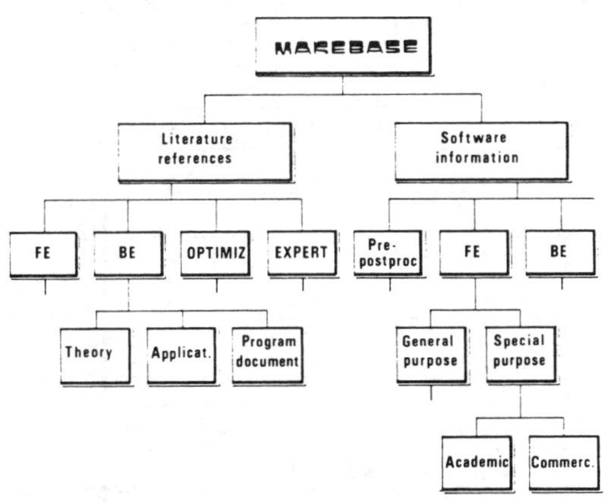

Fig.2 Database architecture

2.3 How the information is retrieved

Information retrieval is a selective recall of stored information. It can be done for literature references and software information.

Retrieval for literature references has options for:
- specific author´s name;
- specific journal´s name;
- year of issue (from, all years, or for a specific year);
- subject.

In the subject search the user describes the field of interest by means of keywords. Up to 10 different keywords in combination can be used.

As an output from the database a list of references is given. These may be sorted alphabetically according to the author and printed all or selected.

In the software retrieval a dialogue is directed by the program where the user specifies his/her requirements on requested options (in present about 150 possibilities). These options can be combined.

As an output the user obtains list of program names fulfilling the requirements together with the addresses of program developers. This list of programs will give an idea of which programs to investigate further. An another program module displays/prints detailed information about the program specified by the user by the respective program name.

MAKEBASE includes information from both the primary sources (books, periodicals, research reports, conference proceedings, theses and dissertations, correspondence, etc) and the secondary sources respectively (abstract periodicals, bibliographies, program catalogues, textbooks, etc).

More detailed MAKEBASE description has been presented by Mackerle (1986).

3 EXPERT SYSTEM DESCRIPTION

Information retrieval from a database provides the most important function for the user. To search for records which have a particular journal name, author´s name, and FE/BE program name is easy. More difficult is a selection of software according to the user´s requirements or to define the subject. There has to be information, first of all that will uniquely identify the particular object in question. To facilitate an effective search it is necessary that the user utilizes the same criteria as are used in the process of indexing (Figure 3). In other words the user has to use the same keywords as the expert providing the indexing.

Indexing techniques fail to fit the needs of the engineer/scientist. More often than not, engineers/scientists prefer to use limited and dated sources of information in their work, but they have little chance of finding what they need.

Our effort for MAKEBASE is that it would be used directly by the engineer/scientist who is problem-solver and needs an access to the instant information retrieval. To make the process of search easier and more effective, an expert system, EXMAKE, is under development. This system will advise an unexperienced MAKEBASE user what file and keywords use in the subject retrieval process.

4 SOME EXAMPLES ON THE USE OF MAKEBASE

MAKEBASE database covers FE/BE technology in many fields of engineering activity,

Table 1. The partial list of FE and coupled FE/BE programs for geomechanics.

Name	Country	Capabilities														Developer
		Consolidation	Foundations	Slopes/dams	Underground	Linear stat.	Nonl. static	Linear dyn.	Nonl. dynam.	Heat transf.	Fluid-solid	FEM/BEM ▲	3-D □	Minicomputer	Microcomp. ●	Note: □ default= 2-D ▲ default= FEM ● default= main-frame
ABLE-I	JAP							●			●		●			Ohbayashi Corp.
AFENA	AUS	●	●	●	●	●	●			●			●	●	●	University of Sydney
BEFE	AUS		●		●	●	●					●	●	●		Univ. of Queensland
CAVERN	USA				●	●						●	●	●	●	MIT, Cambridge
COMET	JAP	●	●	●	●	●	●									Taisei Corp.
CONDEP	CHI	●	●				●								●	Nanjing Hydr. Res.
DIANA-J	JAP	●	●	●	●	●	●	●	●		●		●	●		Takenaka Tech. Res.
DWTNOLIN	JAP								●		●					C.R.I. Electr.Power
ELAT	A				●	●	●					●		●		ILF, Innsbruck
ELFI	F		●				●									IMG, Saint Martin
FEAGMEC	USA		●	●	●	●	●									W Virginia Univ.
FEDYN	AUS							●								C.S.I.R.O.
FEMCON	JAP	●	●													Yamaguchi University
FESPON	USA			●			●									S.Dakota Sch.Mines
FESTAT	AUS			●	●	●										C.S.I.R.O.
FINAL	A		●	●	●	●	●	●	●				●	●		Univ. of Innsbruck
FINETIM	AUS	●				●					●			●		James Cook Univ.
FOCALS	AUS		●			●										C.S.I.R.O.
GEFDYN	F		●	●	●		●		●		●		●	●		Ecole Centr.,Paris
IESM	JAP		●	●	●	●	●						●			Disaster Prev. Res.
LAM	UK		●	●	●	●	●						●	●		Univ.Coll.Swansea
LAWPILE	UK		●			●	●					●	●			Queen Mary College
NTJTEP	AUS				●	●	●									C.S.I.R.O.
PAM-GEOM	F	●	●	●	●	●	●	●	●				●			Engng.Syst.Internat.
POLYANNA	USA		●				●		●				●			New Mexico Univ.
QUAD-4	USA		●	●				●	●							NISEE, Berkeley
RAFTS	UK		●			●	●					●	●			Queen Mary College
RHEO-STAUB	CH		●				●								●	ETH, Zurich
RHEO-2D	JAP		●	●			●									Disaster Prev. Res.
RODSIM	UK			●		●										Univ. of Surrey
ROSALIE	F	●	●	●	●	●	●	●		●	●		●			IRIGM, Grenoble
SAFE-CAN	CAN			●	●	●	●						●			Univ. of Alberta
SATURN/GEO	USA	●		●		●	●	●	●		●		●			Danes & Moore Co.
SET	D			●	●	●	●									Tech.Univ.Munchen
SLSM	JAP		●						●							Kyushu University
SSIDY	USA		●					●				●				W Virginia Univ.

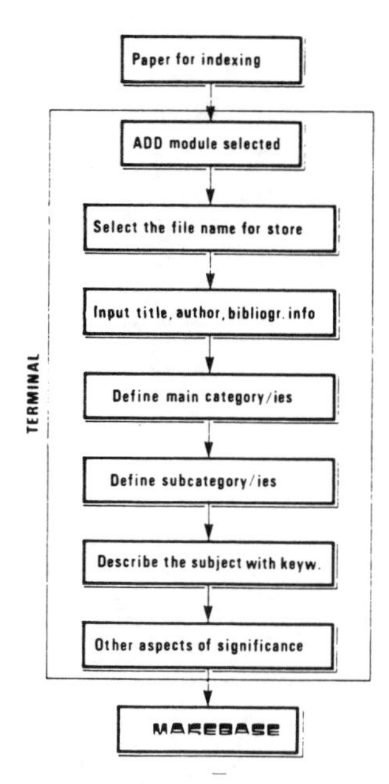

Fig.3 The process of indexing

i e aerospace engineering, automobile engineering, civil and building engineering, reactor technology, piping and pressure vessels industry, geomechanics, biomechanics, etc.

To illustrate the usage of MAKEBASE some typical problem queries are given as follow. Questions are represented by means of names, keywords or numerical codes.

Query no 1: How many papers between 1976-1986 have been published on the FE applications in geomechanics?

Answer from database: 786 references.

Query no 2: How many papers between 1976-1986 have been published on the BE applications in geomechanics?

Answer from database: 188 references.

Query no 3: How many BE programs stored in database have a possibility for analyses of geomechanics problems?

Answer from database: 18 programs (of 136 programs).

Query no 4: How many general purpose FE programs stored in database have a possibility for the soil-structure interaction analysis?

Answer from database: 27 programs (of 221 general purpose FE programs).

Query no 5: How many stored special purpose FE programs are written for geomechanics applications?

Answer from database: 62 programs (of 907 programs).

Query no 6: How many of these special purpose FE programs can be run on IBM PC?

Answer from database: 9 programs.

Query no 7: Which books have been published on the BE and geomechanics?

Answer from database: 1 book is listed.

Query no 8: Which books have been published on the FE and geomechanics?

Answer from database: 8 books are listed.

Query no 9: How many references are stored in database on ADINA application in geomechanics?

Answer from database: 17 references are listed.

Query no 10: I am interested in program SMART. Which program evaluations are known in the literature?

Answer from database: 1 reference is listed.

Query no 11: Give me a list of SMART documentation.

Answer from database: 4 references are listed.

Query no 12: Give me a list of interactive graphics pre- and postprocessor programs which have "ready-to-use" interface to the ADINA and are running on VAX.

Answer from database: 7 programs are listed (of 51 programs).

5 CONCLUDING REMARKS

Originally our intention was to review FE and BE programs used in geomechanics. From MAKEBASE and from answers on our questionnaire we have received so much material, that it was quite impossible to review these programs on a limited space of this paper.

However, we have chosen 36 programs which in few details are presented in Table 1. We had no criteria for the selection.

The author thanks all companies and individuals responding to the questionnaire. Information received has been included in the database.

REFERENCES

Norrie, D. & G.de Vries 1976. Finite element bibliography. Plenum Press, NY.

Mackerle, J.1986. MAKEBASE, an on-line information retrieval system for structural mechanics. Computers & Structures 24: 977-983.

Mackerle, J. 1986. MAKEBASE: an information system on structural mechanics software and applications. Adv. Eng. Software 8: 81-87.

Numerical Methods in Geomechanics (Innsbruck 1988), Swoboda (ed.)
© 1988 Balkema, Rotterdam. ISBN 90 6191 809 X

A case history of selecting and implementing a major CADD (computer-aided design and drafting) system

D.W.Harris
US Bureau of Reclamation, Denver, Colo.

ABSTRACT: The purchase of a major Computer-Aided Design and Drafting (CADD) system involves developing the requirements, evaluating vendor products and a decision for selection. Methods used to plan and estimate the size and cost of a major CADD system are presented. Benefit/cost considerations are given. Important requirements specified for proposals are discussed and evaluated in retrospect.

The staff support and planning required to oversee, operate, maintain and provide technical support to the system and the costs of this support are provided as guidance for other systems. Placing of equipment, training needs, and other aspects of a successful implementation are discussed. Finally, lessons learned are given and summarized from the project manager's point of view.

INTRODUCTION

In 1986, the USBR (United States Bureau of Reclamation) awarded a $19.5 million contract for the purchase of approximately 100 stations to be used in offices throughout the western United States. More than one-half of the stations will be used as an integrated system in the central design office to support the engineering of civil, mechanical, structural and electrical features. The other stations will be distributed in smaller offices which are in geographically different areas throughout the seventeen western United States. Federal government regulations require a very precise and detailed evaluation of vendor proposals. Methods which proved successful in this process are discussed in subseqent sections and offered as guidance to others. For a system of this type, a functional demonstration to verify and evaluate capabilities is very important. Examples of work to have prepared before the demonstration and real-time problems which were used are discussed and suggestions made for other such requirements.

Important considerations for the specification, selection and operation of the system include: (a) the size of the system, (b) benefit/cost analysis, (c) the final configuration of the equipment, and (d) the staff required for support and operation. These considerations are discussed in subsequent sections as well as some lessons learned in the purchase and implementation.

CONSIDERATIONS IN SELECTION

Many considerations are necessary in the selection of a CADD system; such as: (1) size of the system, (2) benefit/cost and (3) special requirements. The following discussion is not intended to provide an exhaustive reference for CADD selection. It is provided as a brief discussion of items learned during this particular acquisition process.

A. Size of System

There are several critical parameters that must be determined or assumed to size a system. These parameters include the volume of work which is applicable for CADD, the hours of production time available on the system and the production ratio which can be achieved. To determine the volume of work, a simple survey was conducted of current USBR work. CADD applicable work was defined as study drawings, specification drawings, revisions to drawings, and computation time associated with analysis of drawing production, (volumes, storage capacity of reservoirs, etc.). A yearly workload at the central design office was established as 2,200 study drawings, 3,100 published specification drawings, approximately 2 major revisions during preparation of specification drawings, and approximately 5,700 hours of computation time. Thus, in summary, 18,800 plots and 5,700 hours of computation time were defined to be accomplished yearly with the CADD system. The total labor associated with this work is approximately 43,500 staff days per year, or approximately 40 percent of the total ongoing workload was applicable to CADD implementation.

The number of hours which is available for CADD production time can become an important decision. In some cases, the cost of a major CADD system may require multiple shifts to justify the expense. At the USBR a flextime system which allows adjustable work hours readily provides a 10 hour work day. A 10 percent overtime availability and/or use is common at the USBR. For the purposes of the CADD stations, weekends and government holidays were not considered as work days; yielding 255 working days with the system.

Lastly, the production ratio possible with the CADD system must be decided to determine the number of stations and the benefit/cost of the system. A review of existing literature was completed, contacts were made with current CADD users, and a review of the existing USBR small CADD system was made. From this review a production ratio of 2.5:1 was selected.

Using all of the above parameters, the size of the system is a simple calculation. The total staff days needed with CADD (considering the increase in production) becomes 17,400 staff days. Dividing the number of station hours needed by the number of hours available in a year produces the conclusion that 50 stations are necessary at the central design facility. Similar calculations were performed for each site.

B. Benefit/Cost Analysis

To determine the cost of a system, the hardware and software costs associated with the system must obviously be considered. However, the hardware and software costs are not the only costs associated with a system. Table 1 is provided to show costs incurred through contract costs during the initial phase of the USBR CADD contract. This table is normalized to the cost of the stations to demonstrate additional costs which need to be considered. Other costs also need to be considered such as organizational costs associated with adopting standards to CADD or CADD software needed to adopt existing standards,

Table 1. - Costs of USBR CADD contract relative to hardware costs

	50 stations system	2 stations system
Station cost	1.00	1.00
Procurement award	0.11	n/a
Software	0.21	0.43
Peripheral H/W	0.24	0.36
H/W maintenance (5 years)[1]	0.10	0.62
S/W maintenance (5 years)[1]	0.12	0.63
Training[2]	0.18	0.14
	1.96	3.18

Notes:

[1] Maintenance adjusted to 5 years by multiplying monthly fee by 60.

[2] Training is estimated in order to include labor and travel costs. Costs represent only first year vendor training.

site preparation, staff associated with the CADD management and operation, quality assurance of software and data, and other items.

Benefits from a system cannot be considered to exist immediately due to many factors; for example, the need for training, the time required for operators to reach a proficency level, and time to build existing items or symbols which can be reused, thus saving time. In addition, some large projects may take time to establish before major advantages of CADD such as interference checking and rapid changes of integrated descriptions can be realized. For the purposes of the USBR cost study, a 20 percent potential savings was used per year to account for these factors (that is, 1.3 production ratio in the first year, 1.6 the second, etc., until 2.5 is realized in the fifth year).

For benefit/cost calculations, a system life of five years was used. An inflation on labor of 5 percent per year was incorporated in the calculations for both benefits and costs. The actual use of the equipment is anticipated to be longer than five years. A fund will be established from the CADD system revenue for a replacement system or stations as the need arises. It is of particular note that using these assumptions requires the recovery of dollars equivalent to the salary of a mid-level technical person for each station.

C. Special requirements in contract

To implement a complete CADD system for ongoing work, the USBR chose to specify 33 different hardware items and 53 software packages (see Appendix). Additional contract requirements included an onsite applications engineer, maintenance, training, documentation, and other miscellaneous items. This list is provided to identify specified items in the contract and provide initial guidelines to persons considering a new CADD system.

Systems that are available from vendors have a range of configurations from a central main frame with multiple graphics terminals to multiple processors with graphics and computation capability available locally and a network for communications. Either some decision must be made "a priori" as to which system configuration is desired or a selection mechanism must be defined to technically evaluate various proposals. The latter case was used during the USBR evaluations. In order to accomplish this objective a concept named the total system resources was developed. Total system resources implied that within the system some total amount of resources was required. Some minimums were established such as 2 MB (megabytes) of memory and 140 MB of disk storage available locally for stations which used a network approach. A minimum of one small CPU was required in all configurations for centralized storage of data. Table 2 shows the total system resources required for various configurations. Option A is a fully distributed system and thus requires the most resource to account for multiple copies of software, local processing, etc. Option D is fully centralized and requires the least total resources. Option C requires work done locally on a station primarily for design and Option B is the same as Option C with drafting functions also done locally.

EVALUATION TECHNIQUES

Once the system size is established and all items which are desirable are selected, some method to evaluate vendor products must be established. Federal regulations require very precise and detailed evaluations of vendor proposals. This particular contract was advertised and executed as a negotiated procurement with the weighting assigned as 65 percent technical, 30 percent cost, and 5 percent other factors. Consistent with Federal regulations, independent technical and cost evaluation teams were established.

Written proposals were required for the evaluation of the technical scores. In the preparation of the specification preferred features were given in some specified descriptions to allow vendors to optionally propose additional features in an effort to enhance their technical scores. These items were indeed preferred by the USBR but not required due to a lack of absolute need for the feature or the suspicion that the feature may be vendor specific. In some cases entire items were specified as preferred. This approach allows vendors to improve their technical score by proposing additional features, although,

Table 2. - Total System Resources
(53 station system)

Option	Station Resources: Design (21 stations) Memory	Design Disk Storage	Drafting (32 stations) Memory	Drafting Disk Storage	CPU Resources: Speed	CPU Memory	CPU Disk Storage	Total System Resources: Speed[2]	Total Memory	Total Disk Storage
A	42	2,940	64	4,480	1.0	32	4,400	22.2	138	11,820
B	42	2,940	*1	*1	6	192	7,600	14.4	234	10,540
					6.4	128		14.8	170	
					7.5	96		15.9	138	
					9.0	64		17.4	106	
					13.0	36		21.4	78	
C	42	2,940	0	0	9	192	7,600	17.4	234	10,540
					10	128		18.4	170	
					13	96		21.4	138	
					18	48		26.4	90	
D	0	0	0	0	15	192	9,700	15	192	9,700
					18	128		18	128	
					22	60		22	60	

[1] Sufficient to process specified functions

[2] Speed of processing for workstations approximately .8 MIPS each. Two stations per processor assumed. Distibution proposed by vendor.

NOTE: Memory and Storage are in megabytes, speed is in MIPS.

in most cases this will also increase the total proposed cost of the system.

Another feature of the technical scoring was that a scale of 2.0, 1.5, 1.0, and 0.5 was used. Each specified item had a point value which was published in the specification. Proposals for items which met the specified requirements received a 1.0 score, proposal for items which exceeded the requirements or which provided preferred items scored 1.5 or 2.0. Conversely, proposals for items which met the requirements but which were cumbersome or appeared marginal in their application received a score of 0.5. This approach allowed an improvement of the final technical scores for items which contained exceptional products, or allowed a lower score for marginal products.

Throughout the scoring of all items, a special team approach was utilized. Each item was independently evaluated and scored by two persons. A consensus score was then entered into the evaluation process by the two reviewers. All scores were posted in the team room as they were entered. Persons who had scored related items reviewed scores entered by other parts of the team. In many cases, vendor software products do not align exactly with specified software functions. In these cases, in particular, only portions of a software package are evaluated by various pairs and a consensus amongst everyone reviewing that particular package may be important. Once all scores were completed, the team as a whole reviewed all scores to assure consistency. Finally, relative ranking of vendors was checked to assure an objective evaluation. The required written documentation was maintained of all reviewers' comments and scores in notebooks by item for each vendor. This was not only useful as reference during the scoring process but proved to be very valuable during subsequent litigation.

A significant portion (25 percent) of the final technical score was a functional demonstration of the system capability at the vendor site. In the development of a functional demonstration, it is important to simulate work which is anticipated to be done in the system. Efforts should not focus on a functional demonstration which tests single functions or items within the requirements independently, although this may appropriate to be a limited extent. For the USBR functional demonstration, a project was conceived which involved major elements of the design process which could be completed simultaneously on multiple stations within an 8-hour period. A project of this type must have been tested before the first functional demonstration even if it is done using traditional hand methods. Some of the required work needed to be completed ahead of time by the vendor due to its time consuming nature, e.g., digitizing of topographic maps previous to siting of features. These items were identified and the necessary materials were sent to the vendors previous to the functional demonstration. As many items as possible were reserved for real time execution during the functional demonstration, e.g., excavations with cut and fill quantities, alignment of pipelines using various inputs such as surveys or coordinates, design of members, etc. It is important that enough information is given to vendors before the functional demonstration to understand the requirements. In this way, the tests will be for the system and not merely of an individuals skill level on the system.

IMPLEMENTING A CADD SYSTEM

It is clear that a CADD system is a major capital investment for any group. Therefore, the selection of the system becomes critical to the success of the project and a desirable benefit/cost ratio. Some considerations for implementing a system are as follows: (1) configuration and siting of equipment, (2) staffing required for the system, and (3) training. Each consideration is discussed separately in the following sections.

A. Configuration and siting of equipment

The system which was accepted by the USBR was designed and proposed by the GE Calma Company (note disclaimer which follows). Option A, a networked system, was used with two work stations being used with a MicroVAX computer. A VAX785 system is used as the main storage CPU in the system. The CPU, large disk drives, large line printers, and other typical computer equipment are maintained in the computer center. Two to six stations are clustered together in rooms near working areas with periphal equipment such as printers, electrostatic plotters, hard copy units, CRT terminals, etc. Each room is individually air conditioned, lighted with no-glare lighting, has nonstatic carpet, provides a power island, has manuals, etc. A disadvantage of equipment directly in the work place of individuals is that warranty problems may occur depending on 24-hour temperatures. An advantage of stations being together, at least initially, is the synergism of having people together for questions and assistance. Other advantages are the possibility of finding available stations when needed, no perception of ownership or distractions in an individual's office, more efficiency of maintenance by having necessary access space combined and by having stations together.

B. Staffing required for the system

The management, operation, and support of a large CADD system is an important consideration to assure that expected benefits are achieved. For the USBR system, the support office consists of a manager, a contract representative, two applications programmers, a system manager, and a representative from the contractor. Operations requires one contract person per shift of usage. In addition to the central group, an individual to contact for CADD needs is maintained as a part-time responsibility in each functional portion (a branch for USBR) of the organization.

C. Training

To initiate new software on a major CADD system, training in many aspects is required. It is difficult to evaluate this requirement and consistently estimates are low. It is obvious that having personnel trained in the drafting software procedures and in various software packages for functional areas is required. Not all personnel can be trained by the vendor, but some form of on-the-job training must be devised. Probably modifications, enhancements, or new applications will be required in a CADD system. Personnel need to be trained to support these areas.

What is also important is to train personnel who have knowledge of the organizational structure and ongoing work in the CADD system data structures and methods of integrating work. Some person or persons needs to set standards and define work flow and long-term storage needs acceptable to the corporate philosphy of accomplishing work. A failure to train and establish sound working principles in this area will be a downfall to any major system since benefits desired will not be achieved.

LESSONS LEARNED

In any project there are always some aspects which would have been nice to know beforehand. This final section is written to provide a summary and a checklist based on experiences from this project.

- The estimated training required is likely to be too low.
- It is important to have a well defined implementation plan for staffing and work adoption to CADD before purchase of equipment.
- The work suitable for CADD needs to be carefully defined and a straight forward method developed to use this data to size the system. Note that not all ongoing work can be used to justify the system.
- All costs associated with the system need to be considered, not just hardware and software costs.
- As many items as possible should be incorporated into the same contract.
- The allowance of various configurations with a definition of similar Total System Resources proved to be a good selection mechanism and defensible in litigation.
- Evaluation of systems should be well documented in writing.
- The evaluation should allow the different strengths of different vendors to be awarded in some way.
- A team approach with consensus mechanisms needs to be incorporated in the evaluation process.
- A functional demonstration of equipment and software should be done on work similar to ongoing work.
- Stations grouped together in an area specialized for computer use aids in implementation and efficiency of the systems.
- A staff to support the CADD system needs to be defined and implemented.
- Billing of costs will need to be approximately the same as billing for a middle level employee for the use of each system.

DISCLAIMER

The mention of product names, trademarks, or discussion of application by any vendor is for information purposes and does not constitute endorsement or recommendation by the U.S. Government.

APPENDIX
USBR Specified Items

Hardware

- Drafting Workstation
  - Operator Console Table
  - Primary Graphics Display (PF)
  - Secondary Graphics Display (P)
  - Workstation Processor (PF)
  - Function Entry System
  - Alphanumeric Keyboard (PF)
  - Function Entry System
  - Alphanumeric Keyboard (PF)
  - Color Graphics Hardcopy (PF)
  - Monochromatic Graphics Hardcopy (PF)
  - Digitizing Cursor Control
  - Digitizing Stylus Pen (P)
  - Large Digitizer
  - Small Digitizer
- Designer Workstation (PF)
  - Graphics Display (PF)
  - Workstation Processor (PF)
  - Function Entry System
  - Alphanumeric Keyboard
- CRT Terminal and Printer
- Central Processing Unit (PF)
- Operator's Console
- Magnetic Tape System (PF)
- Online Storage Devices
- Line Printers
- Pen Plotters (PF)
- Monochromatic Electrostatic Plotters (PF)
- Color Electrostatic Plotter (PF)
- Color Graphics Film Recorder
- Other Auxillary Hardware
- Synchronous Communications
- Asynchronous Communications

Additional Contract Requirements

- Functional Demonstration
- Contractor Personnel
- Maintenance
- Training
- Documentation
- Software Support

Software

- Operating System
  - Key Features
  - System Performance Monitor
  - System Accounting
- Language Processors
  - Assembly
  - Fortran (PF)
  - Fortran Library
- System Utilities
  - Sort/Merge
  - Text Editor
  - Applications Program Maintenance
  - Debugging Aids
  - File Management
  - Tape Management (P)
  - Special Provisions
- Common Data Storage System (PF)
- DBMS (Data Base Manangment System) (P)
- Drawing Management System (PF)
- Graphics Software Interface
- Interative Graphics Support
  - Two and Three Dimensional Capacity
  - Input Capability
    - Digitizer Input
    - Keyboard Entry
    - Graphics Display Input
    - Three Dimensional Input
  - Basic Geometric Construction
  - Additional Drawing Support
  - Graphics Display Capability
  - Hidden Line Removal
  - Output Capabilities
  - File Conversion
    - IGES (P)
    - Calcomp Plot Type Input
  - Bulk Test Input (P)
  - Bill of Materials (PF)
  - Properties (PF)
  - FEM Input/Output (PF)
- Civil Engineering Support
  - Survey Data Processing
  - Mapping
  - Digital Terrain Modeling
  - Plan and Profile Layout
  - Geotechnical Analysis (P)
  - Geologic Logs (P)
  - Interference Checking
- Mechanical Engineering
  - Mechanical Design & Drafting (PF)
  - Piping Detailing & Design
  - HVAC (PF)
- Electrical Engineering (PF)
  - Plant Control & Power Distribution (PF)
  - Conduit Systems
  - Grounding Systems (P)
  - Power System Analysis (P)
  - Transmission Line Design (P)
  - Switchyard Design (P)
- Architectural Support
- Structural Engineering Support
  - Steel Design System (PF)
  - Reinforcement Steel Detail (P)
- Master Planning/Facilities Management (P)
  - General Cartography (P)
  - Seismic Evaluation (P)
  - Drill-Hole Log Digitizing (PF)
  - Solids Modeling Support (PF)
  - Interference Checking (PF)
  - Design Documentation (PF)
  - Communications Requirments
    - General
    - Synchronous
    - Asynchronous
  - Security Requirements
  - System Auditing (PF)

(PF) Item contains preferred features
(P) Item is a preferred item

Numerical Methods in Geomechanics (Innsbruck 1988), Swoboda (ed.)
© 1988 Balkema, Rotterdam. ISBN 90 6191 809 X

Interactive graphic data generation and FEM analysis in geomechanics on workstations and personal computers

W.Haas
Technische Datenverarbeitung, Graz, Austria

ABSTRACT: A meshgenerator is presented, that has been developed for medium size computers. Special care has been taken in the development to support up to date devices and to provide software links to features, that may be implemented in hardware in the future. The general requirements for a mesh-generator, as well as the special requirements for the solution of rock and soil mechanics problems are discussed.

1 INTRODUCTION

The application of the Finite Element Method to geomechanics problems has almost become a standard matter.

With the advent of better constitutive models, faster and cheaper hardware and, last but not least, better educated engineers, also the demands concerning easy and error-proof input have increased. Interactive, graphic data preparation is a prerequisite to enable the analyst to concentrate on the main problem, that is definition of the numerical model and interpretation of its results.

Workstations and Personal Computers now offer high computational performance, combined with fast, high resolution colour graphics at relative low prices. Thus they are an ideal basis for running medium size problems or to prepare large size ones. For software development, it is a challenge to take full advantage of these capabilities, particularly concerning graphics.
The Finite Element Method program MISES3 and its graphic processor GAME have been specially designed for medium size computers. Already in the normal analysis part of MISES3 alphanumeric dialogue input may be selected as well as batch input from files. This feature makes the program a useful tool in education and for commercial applications. In the preprocessor GAME, all necessary data from geometry to generation of results may be specified graphically and in an interactive manner.

2 GENERAL CONCEPT

The Finite Element Program MISES3 and its application to Geomechanics is described in several publications (see e.g. /1/, /2/) and shall not be discussed further in this text.

The basic ideas, that have been persued in the design of the GAME-preprocessor, will be outlined in the following.
It has to be mentioned, that a number of existing interactive mesh generators has been investigated and analysed closely, concerning their functionality and their connection to MISES3. As they all suffered from one or more drawbacks (e.g. very high memory requirements, support of only a few terminal types, inadequate mesh-generator or simply very high price), it was decided to produce a new preprocessor, according to the following ten "commandments".

2.1 Easy definition of geometry independent of a CAD-system

It should not be necessary to buy a complete CAD-system in order to define the geometry of the part to be analyzed by FEM. All those functions, however, which are essential, must be included. Those are, for example the computation of intersection points, relative coordinate input, viewport selection, correction and move functions etc.

2.2 Link to CAD-systems

In the opposite case, when somebody already has a CAD-system, a data transfer to the meshgenerator is necessary. The user is usually more familiar with his CAD-system and may therefore prefer to prepare his geometry by using CAD rather than the meshgenerator. Even more often, the geometry is already defined by means of CAD. For complicated machinery parts, redefinition of geometry would mean the loss of many man days of ineffective work. Therefore, in GAME the standard IGES interface for exchange of data with CAD-systems is implemented (see Figure 1.). Also a specialised interface for a specific CAD-system (MEDUSA) and a neutral datafile format for exchange of data is implemented.

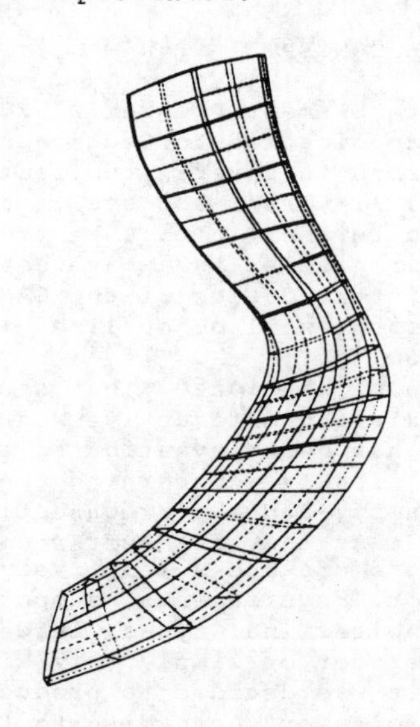

Figure 1. Complex geometry retrieved from CAD via IGES-file

2.3 Command interpreter

There are many, often contradictory demands to the input conventions of an interactive program. Commands should be as short as possible (one to four letters). At the same time the commands must be easy to keep in mind by using mnemotechnical names.

For the beginner, or in case of questions, a help function is available at every stage.

The experienced user may take advantage of features like typing ahead or concatenating commands to jump from any point to any other point in the menu hierarchy.

The various functions may be accessed by selecting from menu in a hierarchic manner.

The essential functions, like modification of view-parameters and element parameters, are accessible from any level in the hierarchic command level.

The total input dialogue is stored and may be used for full or partial reruns. This is especially useful in the case of repetitive variations for one problem.

2.4 Graphic input/output on standard hardware

One main objective in the program development was not to restrict graphic i/o to specific hardware. A set of interface subroutines has been established, to which graphic-i/o libraries (e.g. GKS, Plot10, VDI-drivers on PC) may be linked by simple means. Thus, crosshair cursor, mouse or joystick may be used to define objects or coordinates on the screen.

Also the usage of digitizers is supported. They may be used for accurate coordinate input or for input of menu-commands. The menus may be defined and plotted by the user according to his needs by an additional utility.

2.5 Definition and display of viewports

Particularly for 3-D-applications, it is essential to have not only one, but a number of viewports arbitrarily distributed on the screen. For maximum convenience, all view parameters (scale, viewpoint, orientation of axes, line types,...) can be set individually for each viewport. Depending on the hardware configuration, also the size and location of the alphanumeric input/output window may be defined in GAME.

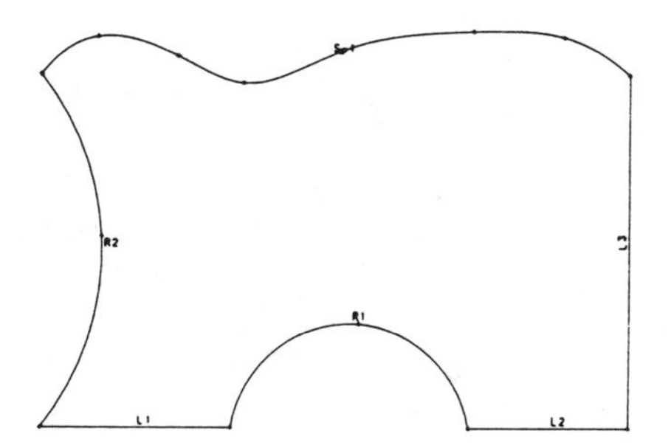

Figure 2. Example 1 for logical quadrilateral

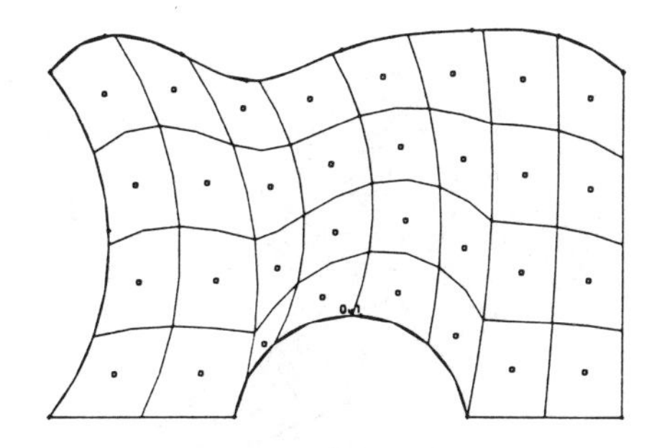

Figure 3. Mesh generated from logical quadrilateral by transfinite mappings

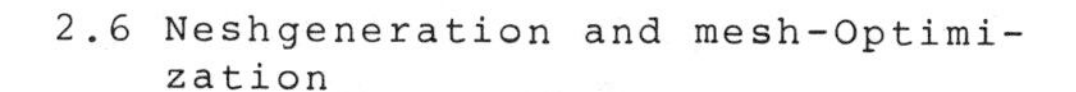

2.6 Neshgeneration and mesh-Optimization

MISES3 utilizes isoparametric 8-node and 20-node-brick elements as standard element types for 2-D and 3-D analyses. An efficient and elegant way of meshgeneration is the method of "Transfinite Mappings" /3/, /4/. This method is applied to "logical quadrilaterals" in 2-D and to "logical cubes" in 3-D- Geometry data in GAME are also hierarchically structured. At the lowest level, there are geometry points. At the next level, lines, arcs and splines are possible.

In the next higher level, boundaries may be defined. The data element ments of each level may consist of one or more elements of the next lower level.

3-D-surfaces may be either specified as rotational surfaces, by extraction from boundaries, or as "Cons-surface" /5/. The special property of these surfaces is that they are fully described by their boundaries.

The definition of the subdivision is almost trivial once the logical quadrilateral or cube is defined. The subdivision long a boundary may be equally spaced, proportional to 2^{**n} or may be arbitrarily given, provided the number of subdivisions is equal on opposite boundaries.

An example is shown in Figures 2 and 3.

Very often, the automatically generated mesh does not fulfil all the requirements for a reasonable analysis.
The elements may suffer from unfavourable aspect ratios or angles. Very often, there is also need for logical refinement.

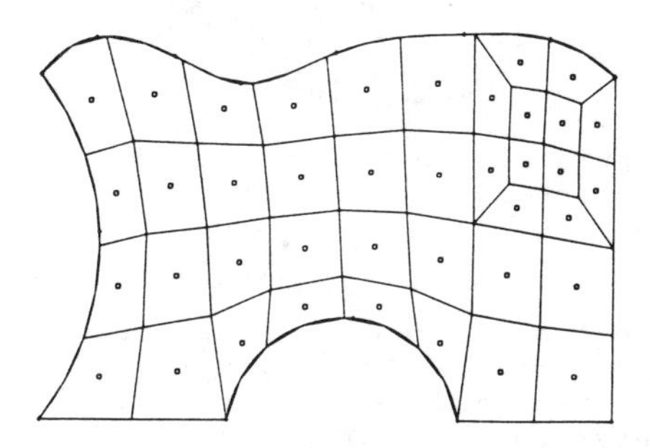

Figure 4. Optimized and refined mesh for example 1.

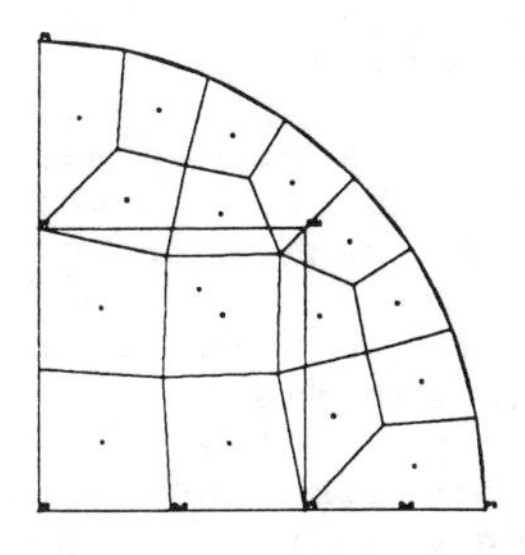

Figure 5. Plane optimized mesh for projection onto a surface

For both problems, a solution is offered in GAME. FE-nodes may be specified, and all the elements connected to these nodes will be optimized with respect to element size. The goals of optimization are an optimum angle of 90 degrees between element sides, equal length of all sides, and equal volume of adjacent elements. Wherever necessary, mesh refinement may also be specified locally (Figure 4.).
Very useful is also a projection facility, that allows to define (and optimize) a mesh in an arbitrary plane (Figure 5.) and to project that mesh onto a surface (Figure 6.).

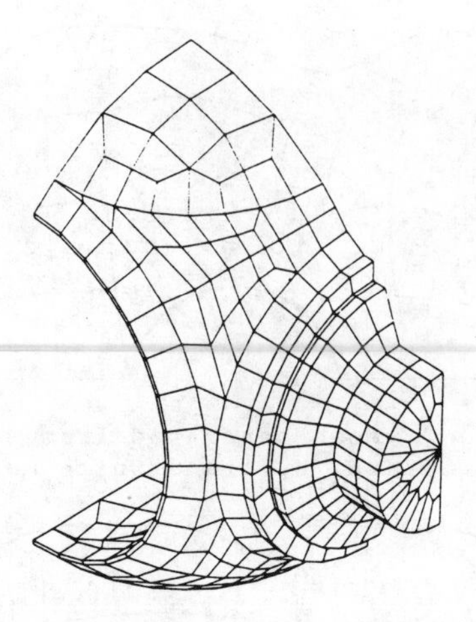

Figure 6. 3-D-mesh produced by projection of a plane mesh onto surfaces

The combination of a fast meshgeneration strategy with optional mesh-refinement and mesh-optimization offers very good performance for an interactive mesh generator.

2.7 Definition of all analysis data in the mesh generator

The basic goal of a meshgenerator is the interactive description of the geometry and the generation of a FE-mesh.

Effectivity can be highly increased, however, when also material description, loads and other data, e.g. special elements like anchor elements, can be defined within the preprocessor. In GAME, also activation of elements for different construction stages, material, loads, boundary conditions and necessary output can be defined, for static problems as well as for seepage or temperature analysis.

2.8 Interfaces to other FE-systems

Not all analysis requirements can be fulfilled by a single FE-system. Very often, specialised FE-programs need to be applied, because of special advantages they offer for a certain class of problems. GAME includes a number of interfaces to various program systems. Data can always be transferred to the FEM-program, in many cases also an exchange in the opposite direction (FEM to GAME) is possible and can be utilized for mesh changes, refinement or optimization.

The neutral datafile allows also an easy access to GAME-data for other FEM-systems, in case they provide a little program to convert the neutral file.

2.9 Portability

Due to rapid change in hardware, it is essential to take special care of program portability.

The high investment for software development is only justified, if the program can be adapted to new hardware features without difficulties.

For reasons of its standardisation and general availability, FORTRAN-77 was selected as programming language. All the functions, that are not included in FORTRAN-standard, are suitably moved into interface subroutines. This concerns disk and terminal input output and mainly the graphic i/o. Thus, it is fairly easy to interface to different graphic software standards, like Plot1O, GKS, VDI or other If a "graphic engine" with enhanced 3-D capabilities is available, also the viewing and transformation operations can be easily linked to the respective libraries. The programming strategy is explained more detailed e.g. in /6/.

2.1O Adjustable parameters

Closely correlated to the problem of portability is the question of how a mesh generator can be adjusted to different hardware-, software- and user-configurations. In GAME, this is also solved in a three-layer hierarchic way:

- Main features are predefined at installation time and cannot be changed by the user (e.g. hardware configuration, available interfaces, 2-D or 3-D option, dialogue-language, plotter-interface).
- Data base parameters may be specified at the program start, e.g. conveniently in a file specified within the start procedure.
- All parameters concerning display of data (viewport, viewplane, line types, symbols, colours) are set to reasonable default-values at program start-up, but may be modified at any time within a GAME-session. The al-

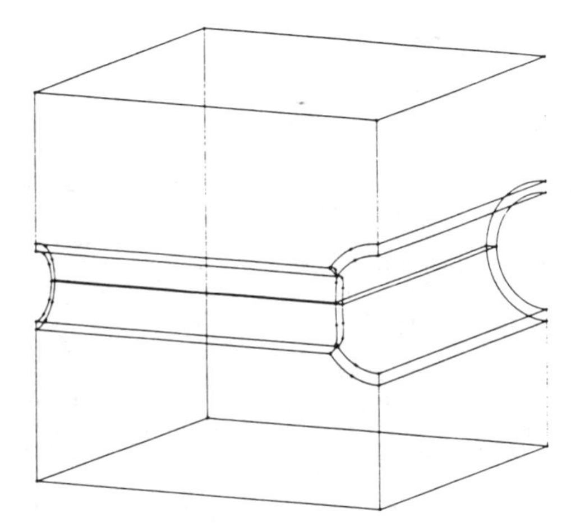

Figure 7. Geometry definition for tunnel intersection

tered parameter-values may be stored and used as further standard default values.

3 EXAMPLES

3.1 Tunnel intersection

One of the situations frequently analysed in rock and soil mechanics is the intersection of tunnels. The high effort that is necessary to prepare a FE-mesh for this situation is very often the most cost-intensive factor in the calculation, if computer capacity is anyway available.

Using GAME, it is shown how its concepts may be utilized to produce such a mesh. For reasons of clarity, symmetry has been assumed.

In a first step, the geometry is defined by lines, arcs and surfaces. The intersection of the tunnels is modelled by splines (Figure 7.). A cube of suitable size is placed around the tunnels and forms the logical cube for meshgeneration, together with the cylinder surfaces of the tunnel. It is to be noted, that only one logical cube is necessary to produce the mesh.

In Figure 8 and 9, the logical cube and the generated mesh in the surrounding soil are shown.

Finally, also the shell elements are generated. They are displayed in different views in Figure 10.

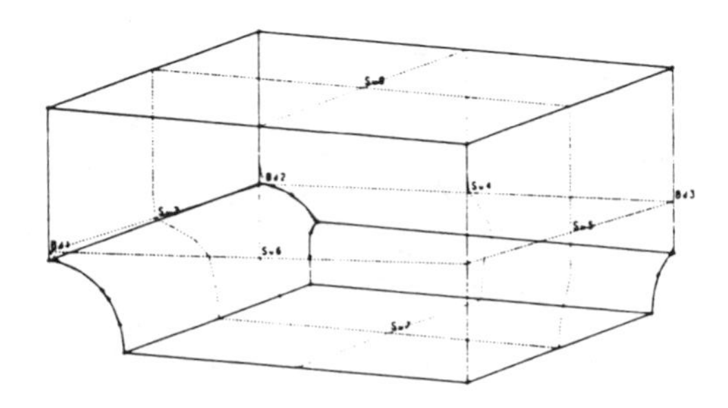

Figure 8. Logical cube for tunnel intersection problem

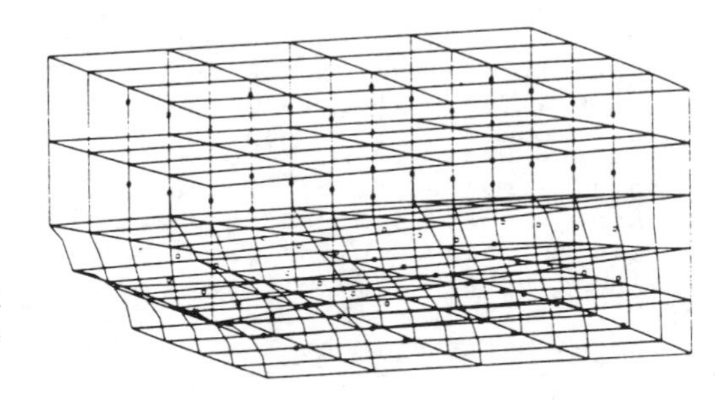

Figure 9. Mesh in the surrounding soil

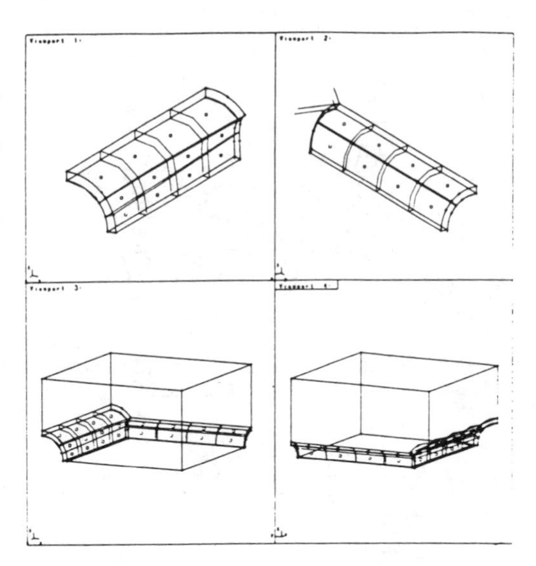

Figure 10. Shell elements for tunnel lining

For the final analysis, the model may be further extended by adding large or infinite elements to the generated mesh. By mirroring, a mesh for unsymmetric load conditions may be obtained. Manual correction and optimization of the 3-D-elements in the soil close to the tunnel was also tried, but did not produce observable improvements in the results.

All the input data for the example are prepared in such a way, that they may be used for similar situations by a few simple input modifications.

3.2 Avalanche protection tunnel

The problem of an avalanche protection tunnel was to be analysed. The loading intensity for the concrete shell was determined in a plane strain model, taking into account different excavation and backfill stages. Also interface elements were used between soil and concrete shell /2/. The mesh was generated, again using the possibilities of mesh optimization (Figure 11.). In a subsequent step, the load intensity obtained from the 2-D model was applied to a 3-D-shell analysis. The shell-mesh could be easily generated by extracting the appropriate elements normal to the x-y-plane (Figure 12.).

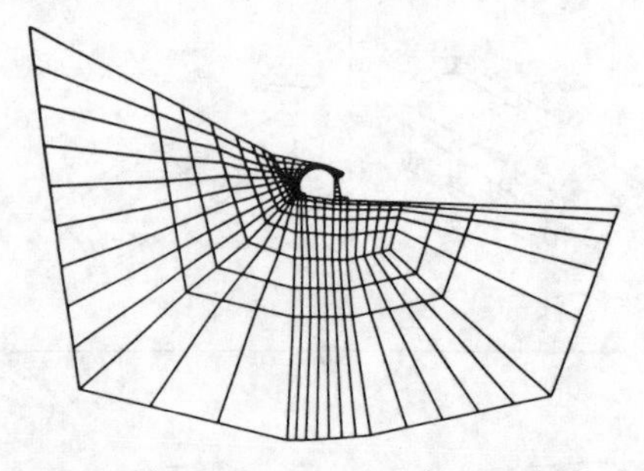

Figure 11. Plane strain model for avalanche protection tunnel

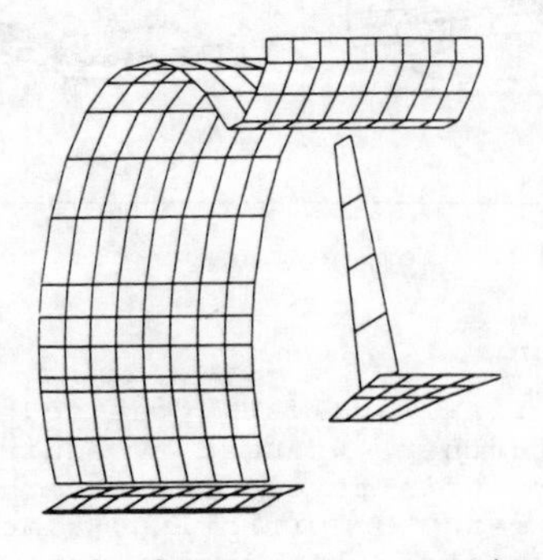

Figure 12. 3-D-shell, mesh generated by extraction

4 CONCLUSION

The overall concept of the meshgenerator GAME and its application to problems of rock and soil mechanics has been demonstrated.

Three points have been emphasized:
- Utilization of small computers (PC) and of modern graphic hardware
- Generation of all analysis data within the generator (activation, loads, seepage analysis)
- Combination of "transfinite mappings" with subsequent optimization and refinement of mesh yields "nice" meshes within acceptable response time in an interactive way.

5 REFERENCES

/1/ H.F.Schweiger, W.Aldrian, W.Haas, The influence of joint orientation and elastic anisotropy in the analysis of tunnels in jointed rock masses, NUMOG II, Ghent, 1986

/2/ H.F.Schweiger, W.Haas, Application of the thin-layer element to geotechnical problems, to be published (6th ICONMIG, Innsbruck, 1988)

/3/ R.Haber, M.S.Shepard, I.F.Abel, R.H.Gallagher, D.P.Greenberg, A general two-dimensional graphical finite element preprocessor utilizing discrete transfinite mappings, Int. Jou. Num. Meth. Engng., 17, 1015-1044, 1981

/4/ R.Haber, J.F.Abel, Discrete transfinite mappings for the description and meshing of threedimensional surfaces using interactive computer graphics, Int. Jou. Num. Meth. Engng., 18, 41-66, 1982

/5/ H.Bantli, Theorie von B-spline-Kurven and Flächen im Raum unter Berücksichtigung der Bezier-Darstellung als Spezialfall, Escher-Wyss Technischer Bericht z/H-841/83-31, Zürich, 1983

/6/ W.Haas, A general programming concept for micro- and minicomputers, 1st Conference Engineering software for microcomputers, Venice, 1984

Numerical Methods in Geomechanics (Innsbruck 1988), Swoboda (ed.)
© 1988 Balkema, Rotterdam. ISBN 90 6191 809 X

A method and a program for mesh generation in geomechanics

T.N.Hristov & V.H.Vassilev
Institute of Water Problems, Sofia, Bulgaria

ABSTRACT: A method is proposed for automatic generation of finite element meshes as imposed deformation of basic figures of standard grids to the arbitrary shape of the area of the structure subdomains. In the case of two-dimensional problems, the basic figures are rectangles and triangles (or trapezia treated as truncated triangles). The task is reduced to the solution of the problem of deformation of joint linear bar systems by FEM at imposed displacements to the contour and to some internal points, if required. A calculation program has been made for triangular element mesh generation in the case of the two-dimensional problems.

1 INTRODUCTION

In making up meshes for the solution of continuum problems by FEM, a number of rules must be observed, such as:

- comparatively precise presentation of the complex geometry of the structure area;
- compressing the meshes in the areas of the expected stress concentration;
- restricting the number of elements surrounding the node (it is required in some algorithms for the solution of the system of linear equations);
- forming elements, as regular as possible in shape;
- observing the restriction that the difference between the numbers of two adjacent nodes shouldn't go beyond certain limit;
- convient introduction of data relative to the continuum properties, boundary displacement conditions, force and temperature loading etc.

Depending on the method used and the algorithm sophistication, the programs for mesh generation satisfy more or less the above conditions [1,2,3].

Of essential importance is also the machine time spent in automation.

2 SPIDER'S WEB METHOD FOR FINITE ELEMENT MESH GENERATION

The method proposed in this paper is formally similar to the mesh generation method described by H. A. Kamel and H. K. Eisenstain [1] as far as the idea for transformation of the standard grid to the shape of the subdomains of the area is concerned (Figure 1).

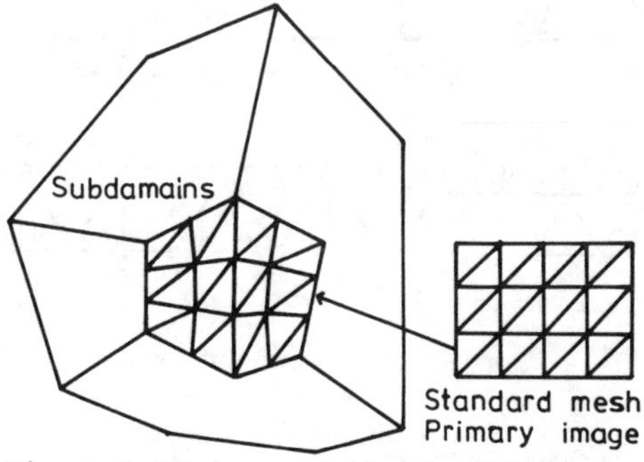

Figure 1 Mesh generation by deformation of standard grid to the shape of the subdomains

The transformation of these standard grids of finite elements, however, takes place and is based on the principle of deformation of a joint linear bar system at given imposed displacements of its contour to the one of the subdonain.

This mode of deformation is similar both in performance and visually to the deformation of a spider's web, the contour of which is drawn out by us in such a manner as to get the shape we wish. All internal cells and nodes stretch and move, respectively in proportion to the modified mesh contour.

The principle adopted here reveals possibilities for additional mesh deformation, if necessary, by "drawing out" of internal nodes or contours of it.

The idea was realized in algorithm and program for 2D meshes of triangular elements. The principles on which they are based, however, can be used successfully to three-dimensional problems as well as to any kind of polygonal elements.

3 ALGORITHM AND PROGRAM FOR MESH AND INPUT DATA GENERATION FOR TRIANGULAR ELEMENTS

In working out the algorithm and the program it was accepted that the primary image of the subdomain can be a rectangle of an uniform finite element grid (Type 1), a rectangle of a nonuniform finite element grid (Type 3) or a triangle or a trapezium, treated as truncated triangle (Type 2). They have standard grids and numerations, as shown in Figure 2.

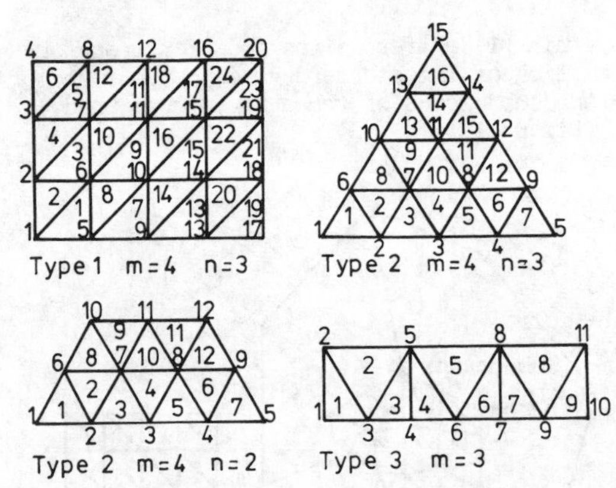

Figure 2 Primary images of subdomains

The examples shown in Figure 3 for the deformations of these primary images to the shape of different subdomains illustrate the possibilities to obtain meshes of arbitrary shape and element density.

The handiwork preparing the input data for solving a geomechanical problem by FEM, according to the method proposed, includes the following:

Taking into account the specific reguirements of the problem concerning the type of mesh, the hole area is subdivided into subdomains and the type of primary image (1,2 or 3) and the finite elements mesh density for every one of them are determined (Fi-

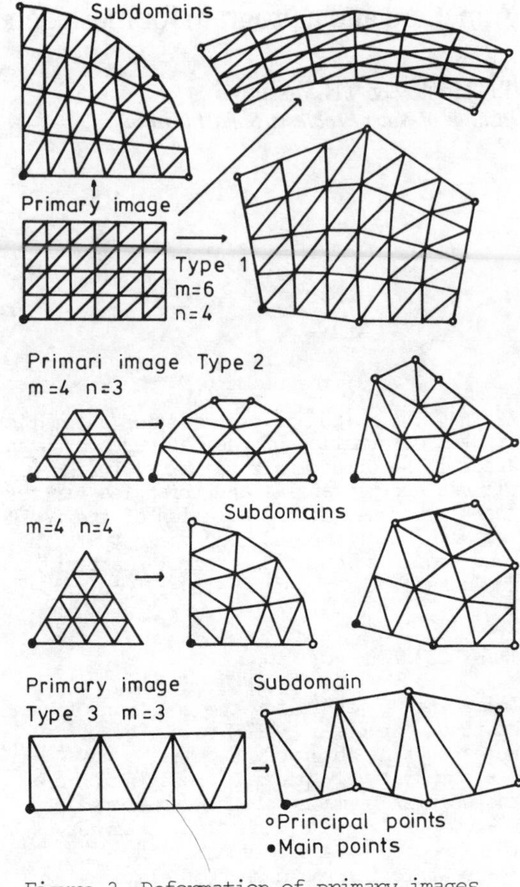

Figure 3 Deformation of primary images to the shape of subdomains

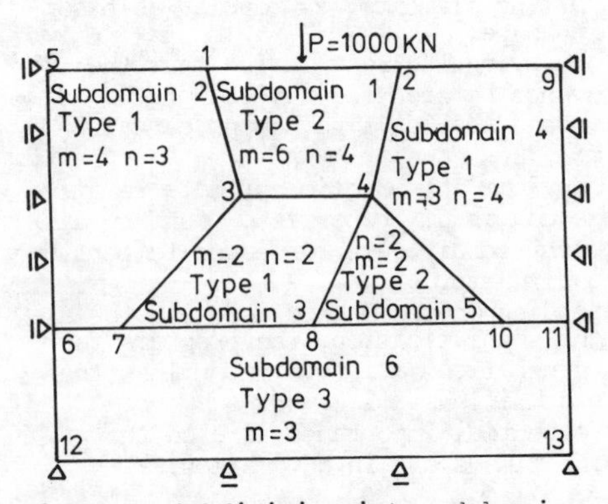

Figure 4 Subdivision into subdomains

gure 4). The density is defined by two numbers, m and n, which for primary images of type 1 and 2 correspond to the number of the sides of the triangles in the base and in the height of the primary image, respectively.

For primary images of type 3, this is the number of the sides of the triangles, lying on the opposite side of the base (Figure 3).

It is necessary to co-ordinate the number of triangles on both sides of the boundaries between the subdomains. This is practically achieved by preliminary sketching out the mesh distribution by sight.

To co-ordinate the subdomain with the primary image, one point is taken to be the main one. In the primary image this is node number 1 from the standard numeration, and in the subdomain this is the principal point which we want to coincide with the main node from the primary image (Figure 3).

The subdomains are defined with the numbers of their principal points. In addition to the points, characterizing the subdomain from geometrical point of view, they also include the points, corresponding to the apexes of the primary image.

By following the rule in defining the subdomain that the description of the principal points should begin from the main point and continue back-clockwise, orientation of the finite element grid is achieved.

At this formulation for the automatic mesh generation, the following input data are prepared:

A. For the area:
- number of subdomains;
- number of the principal points;
- coordinates of the principal points.

B. For a subdomain:
- type of the primary image (1,2 or 3);
- density (the numbers m and n);
- the principal point numbers along the contour of the subdomain;
- number of the finite elements detween two adjacent principal points along the contour of the subdomain in the order of their description.

These are the compulsory input data for each subdomain, required and sufficient for automatic mesh generation.

With a view, however, to the additional effect on the deformations of the primary image so as to obtain more suitable distribution of the finite elements or presentation of structure elements within the subdomain, it is possible, using additional information for the coordinates of some of the nodes to additionally correct the internal side of the mesh or the contour of the subdomain. This correction by input data uses the standard numeration of the primary image.

In the order to solve a problem in geomechanics some data as elements thickness, temperature difference, the zone number (relative to specific continuum properties to which the element belongs) are also reguired. For some nodes of certain subdomains information should be supplied about the imposed displacements of the system (boundary conditions) or force loading. These data are also given in the standard numeration of the primary image.

For the example, shown in Figure 4, the input data are as folows :

A. For the area:
- number of subdomains: 6;
- number of the prinsipal points: 13;
- coordinates of the principal points: they follow the coordinates of 13 points.

B. For the subdomains:

Subdomain 1:
- type of primary image: 2;
- density: m = 6, n = 4;
- principal points: 2,1,3,4,2;
- number of the finite elements along the contour: 6,4,2,4;
- zones of the continuum properties to which the triangular elements belong: 32x1, if all elements belong to zone 1. Atherwise the corresponding numbers of the zone are introduced following the standard numeration of the elements in the primary image of type 2;
- thickness of the finite elements: 32x1, if the thickness of all elements is 1;
- temperature differences: 32x0, if there is no temperature loading;
- force loading to node 4 (according to the standard numeration of type 2) with "x" and "Y" components: 0,-1000.

...

...

...

Subdomain 6:
- type of primary image: 3;
- density: m = 3;
- principal points: 11,10,8,7,6,12,13,11;
- number of finite elements along the contour: 1,2,2,1,1,3,1;
- zones: 9x2;
- temperature differences: 9x0;
- boundary conditions for nodes: 1,2,5,8, 10,11 (they follow the imposed displacements for these nodes).

This is the complete information that should be prepared by hand for automatic mesh generation of triangular elements for the given example (Figure 5). Of course, more subdomains and primary images of different type have been used in this case, so that the problem could be demonstrative in nature. The method and the algorithm offer numerous combinations of subdomains and primary images (beginning with a single one) to obtain whatever mesh and density of elements we wish to.

A peculiarity of the algorithm is that the problem of the imposed deformation of

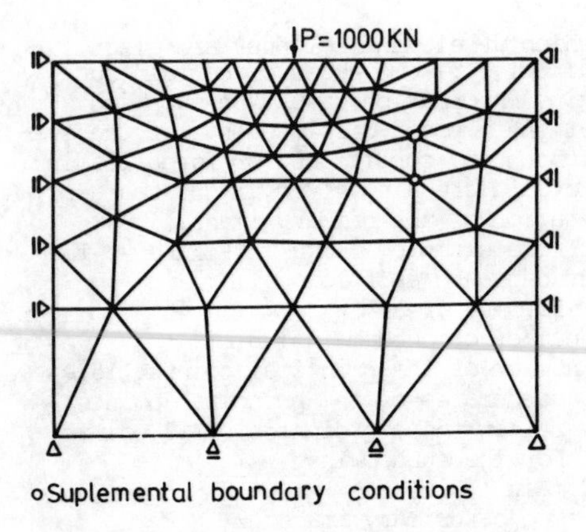

Figure 5 Automatic generated mesh

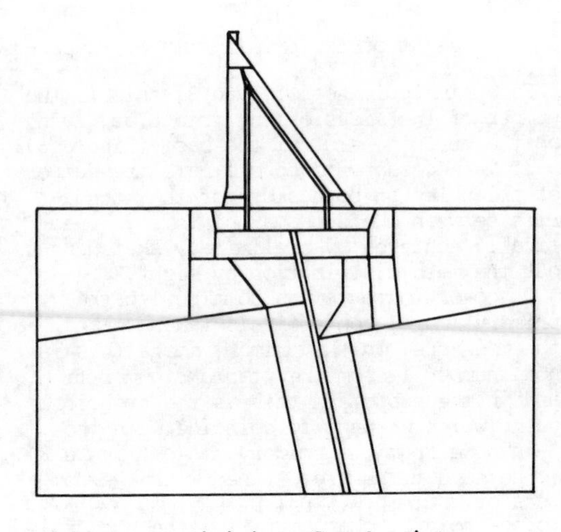

Figure 6 Subdivision of a dam into subdomains

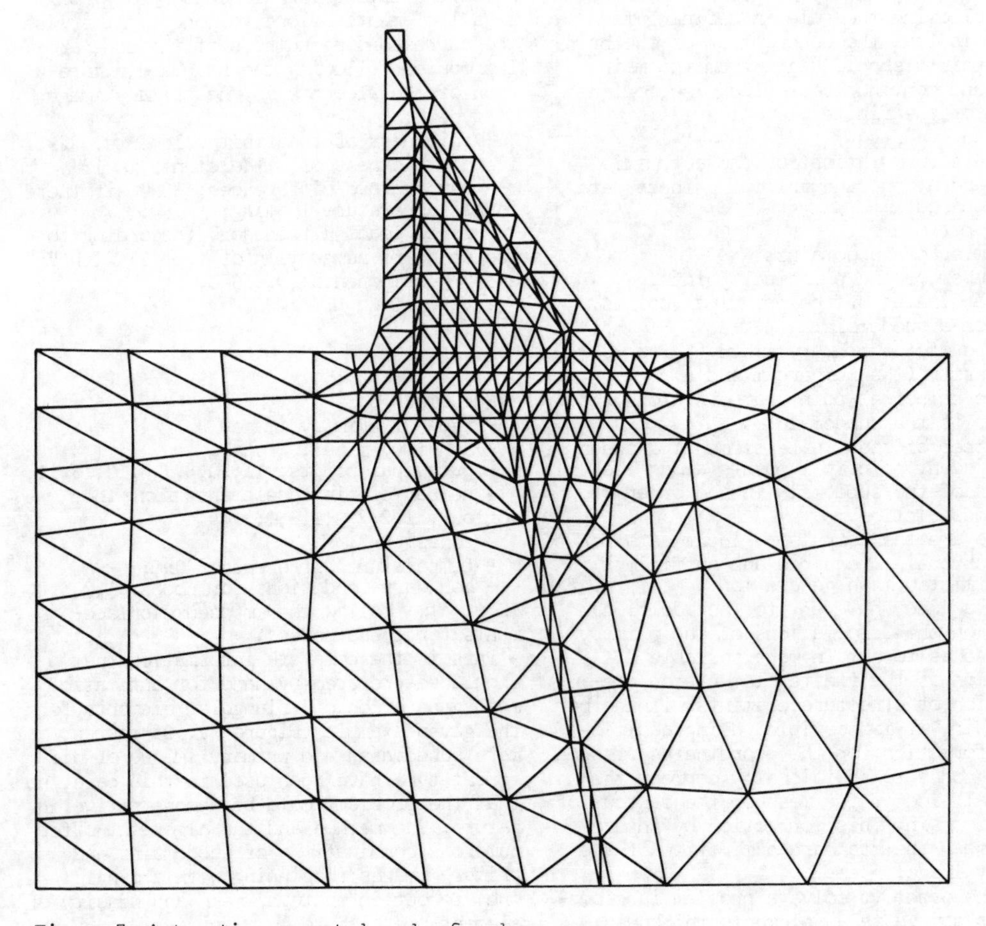

Figure 7 Automatic generated mesh of a dam

the joint bar system to the shape of the subdomain is also solved by FEM. In the case of the triangilar mesh the shape of the joint bar system is adequate to the standard grid. In order to save machine time, the program is simplified for the solution of meshes composed of linear elements only and for the solution of the system of linear equations using Gauss-Seidel method.

The data required for this purpose are generated automatically and appear tobe "transparent" for the program user.

In working out the algorithm it is accepted that the nodes along the contour settle down at equal distances in a straight line between the principal nodes. It is always possible, however, to introduce a more complex law for their distribution by sophisticating the algorithm.

The stiffness matrix of the joint bar system is calculated provided the cross section and the Young's modulus are egual for all elements. It is possible in principle to additionally influence the nature of mesh deformation by creating certain differences in these values.

The primary numeration of the nodes and elements of the whole area is made followind subdomains numeration. Parallel to the numeration of the nodes and elements, the files with coordinates and elements nodes are built.

To minimize the stiffness matrix width and with a view to applying the frontal method to obtain it, renumbering of nodes and elements is envisaged.

The subdivision of a dam and its foundation into 29 subdomains and 62 principal nodes is presented in Figure 6. The automatically generated mesh of triangular elements is given in Figure 7.

The program is realized in a PC IBM-AT where the input data are introduced by hand at interactive regime and the results are visualized. The possibility ia foreseen to introduce corrections whenever required, and to add new nodes and elements (including special ones) after automatic mesh generation.

4 CONCLUSIONS

Thanks to the experimentation of the algorithm and the program, the following basic conclusions can be drawn:

The presence of three standard primary images and the great freedom in the shape of the subdomain allow the automatic mesh generation of systems the arbitrary from geometrical point of view, without any essential difficulties.

The program has been worked aut for a triangular mesh. The principles on which the method and algorithm are grounded, however, allow the performance of both two-dimensional and three-dimensional meshes with polygonal elements of arbitrary shape.

The lack of restrictions in the primary numeration of nodes and elements makes it convenient as far as the corrections and the additions in the mesh are concerned. On the other hand, the non-compulsory principle of successive numeration of nodes and elements according to subdomains is convenient for selective printing and graphs.

The user gets an idea of the type of mesh before its automatic generation which allows to make corrections and additions parallel to the data preparation for mesh automation.

REFERENCES

[1] H.A.Kamel and H.K.Eisenstein, Automatic mesh generation in two and three dimensional inter-connected domains, Proc. Symp. Internat. Un. Theor. Appl. Mech., High Speed Comput. Elastic Struct. 455-475, Univ. Liege, 1971

[2] Y.N.Babich and A.S.Tsibenko, Methodc and algorithms for automatic triangular element mesh generation (in russian), Academy of Ukraine SSrR, Kiev, 1978

[3] J.C.Cavendish, Automatic triangulation of arbitrary planar domains for FEM, Int. J. Num. Meth. Engng 8, 679-699, (1974)

Numerical Methods in Geomechanics (Innsbruck 1988), Swoboda (ed.)
© 1988 Balkema, Rotterdam. ISBN 90 6191 809 X

SACI – A system for instrumentation control and follow-up

Luiz C.S.Domingues, Sebastião M.Burin, Tarcísio B.Celestino & Carlos T.Mitsuse
Themag Engenharia Ltda, São Paulo, Brazil
Edmundo R.Esquivel, Oladivir A.Ferrari & Alberto P.Horta Neto
Companhia do Metropolitano de São Paulo, Brazil

ABSTRACT: This paper presents the SACI System, wich was developed to speed up the treatment of instrumentation data collected in civil engineering works for the São Paulo Metro. This system allows to set up and maintain data bases that contain reading values, construction sequences and important occurences. Queries can be made yielding graphs reports allowing for efficient interpretation and quick decision making process. A case history is presented, in wich the methodology was widely testede and approved. The success in that case history led to the decision of implementing the system for the construction works of future lines.

1 INTRODUCTION

Instrumentation has several applications in engineering projects. According to Franklin [1], the most usual functions are the following:

a. to record natural variations in the environment before the start of an engineering project;

b. to ensure safety during construction by means of warnings;

c. to check the data and assumptions used in design;

d. to control the implementation of ground treatmemt and remedial works.

The purpose of any instrumentation is to complement, not to substitue visual information. So, besides the instrument readings, a number of personal observations and reports ought to be collected as well, filed and conveniently manipulated.

During the construction of urban tunnels, often involving several excavation faces, a large number of reading devices are usually installed. This is in part due to the "design-as-you-go" concept adopted for tun-

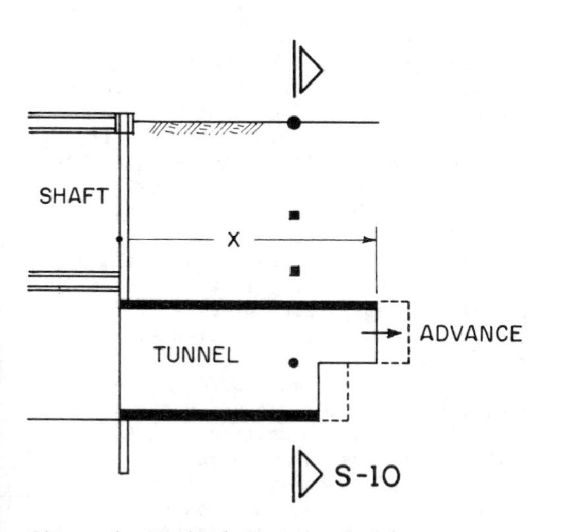

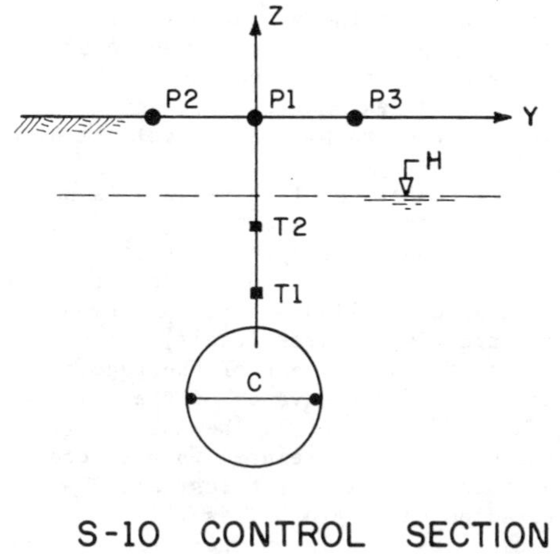

Figure 1. Typical Instrumentation

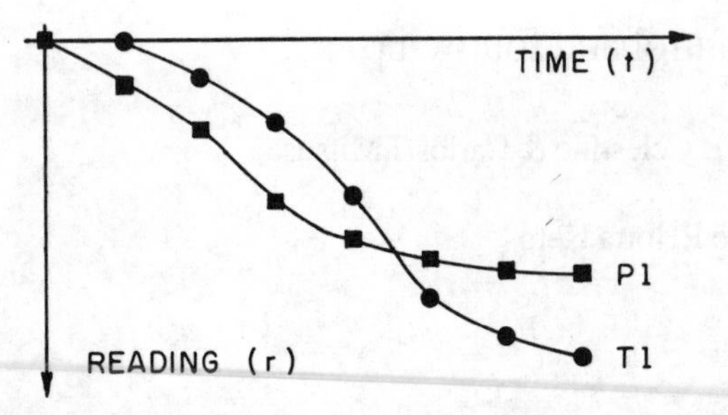

Figure 2. Typical Graph (Reading vs Time)

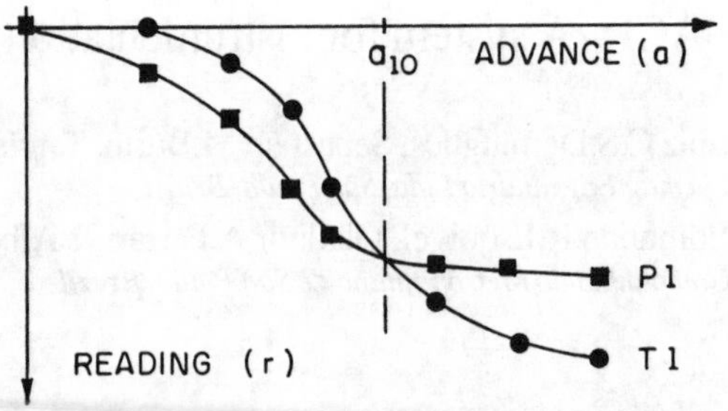

Figure 3. Typical Graph (Reading vs Advance)

nels. Information obtained during the construction phase is used to feed back the initial design, which is flexible, in order to allow for adaptations of the construction method as the work proceeds. Quick and effective decisions must be taken constantly in order to ensure high standard performance.

These requirements make it clear that a digital information system must be ideal to fulfill the purpose of following up and controlling instrumentation results of up-to-date construction works. Its advantage is clearly seen when it is compared to the usual manual procedure for graph updating, which requires much more manpower and yields much slower response.

Figure 1 shows a typical example of instrumentation. It depicts the case of an urban tunnel excavated from a shaft. It shows sets of devices for a particular control section (S-10) where the settlements are continuously monitored by means of surface (P1, P2, P3), and deep (T1, T2), settlement devices, as well as the phreatic level (H) and the convergence of the diameter of the support (C). Important pieces of information about the evolution of the construction process and observed occurrences (support cracks, changes in the materials used, excavation face conditions, construction problems, etc) are also periodicly recorded.

The interpretation of these data is usually made by means of graphs of readings (r) as functions of elapsed time (t) (Figure 2). As the most important action that can be controlled is the evolution of the construction process itself, it is also convenient to have graphs of the readings as functions of the advance of the excavation face (a) (Figure 3). The instant (t) and the position (a) related to the occurrence of any relevant fact must also be annotated on those graphs, as well as the particular position (a) where the excavation face reached the S-10 section. If the purpose of the process is to be achieved, these graphs must be constantly updated.

Besides the graphs shown in Figures 2 and 3, a number of other analyses and checks must be carried out periodicly. Both the absolute values of the readings and their variations must be compared to previously established limits. Graphs of relative values (differential settlements and distortions) are usually computed and plotted. More complex models can also be applied to the data (for example the adjustment of a Gaussian curve to the readings of the surface settlement devices P1, P2 and P3, and calculation, by integration, of the volume of the settlement trough).

2 PRESENTATION OF THE SACI SYSTEM

The SACI systems, conceived to operate in network environment, allows bases of data obtained from instrumentation readings and other observations of a construction work to be set up, updated and maintained. It also allows for simultaneous and interactive queries, producing reports and graphs that can be displayed on the screen or printed out.

There are three basic data types to be treated, genericly referred to as annotations. All of them are characterized as functions of time. They are:

a. Readings: the values obtained from the different devices installed (settlements, convergences, strains, flow rates, etc); the variation between two readings is assumed to be linear;

b. Events: construction stages that occur cyclicly. For example, in the case of a tunnel, the operations of excavation of the face can be considered as events, as they are repeated cyclicly;

c. Occurrences: description of random facts, relevant for the understanding and analysis of instrumentation results.

Those data must also be related to a cartesian reference system. In the case of tunnel construction, the x axis of the reference system corresponds to the direction of advance of the excavation, the y axis is horizontal and the z axis is vertical, upward directed.

The SACI system yields two basic types of graphs: reading versus time (Figure 2) and reading versus event (Figure 3). Other types of graphs can also be produced, relating any data available, considering even the space position of the devices. For example, the settlement trough at given location and time can be plotted.

The data basis of the system consists of the following files, whose contents are indicated:

a. CadUs: information of the users allowed to use the system;

b. LogUt: a log of the operations carried out by the system.

c. CadAn: information about the groups of annotations, for example unit system, warning values, etc.

d. ArqDat: annotations and indices for optimization of access to these data;

e. Graphs: elements necessary for the definition of graphs (options, lay-out).

The main data flow is depicted in the flow chart of Figure 4:

1. the process of maintenance takes charge of the generation and control of the data basis, including protection procedures and communication support (when applicable).

2. updating the data basis is carried out by the group of users at the construction site, using the data collected in the field; during this procedure, a number of consistence checks is done on the data and their variations.

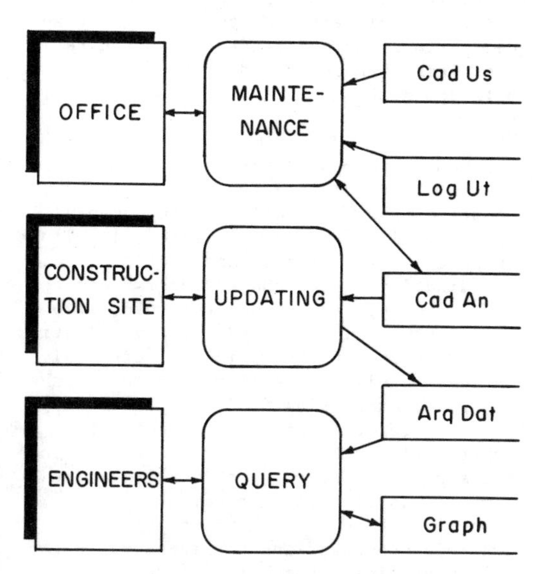

Figure 4. SACI Data Flow Chart
(main flow, control flows excluded)

3. the query procedure allows for access to the data basis for the generation of either reports or graphs, yielding quick and efficient observation of the annotations.

3 DEVELOPMENT

The SACI system was developed for IBM-PC or compatible microcomputers, using Turbo-Pascal programming language, under the DOS operational system.

3.1 Data organization

The SACI system adopts a data structure organized with the main purpose of minimizing the time for disk access and remote communication, ensuring both the integrity and the protection of the data.

The different data resulting from the instrumentation of a construction work are organized into groups of similar type annotations (devices, events, occurrences) and same general characteristics (reference system, warning values, reading variation range, etc). These pieces of information are recorded in the CadAn file.

The data file (ArqDat) is made out of blocks of fixed size. Each block is composed by a variable number of registers referring to consecutive instants. Each register contains a given instant and its correponding data.

The access to the annotations is controlled by the file containing the register of users (CadUs). Each user has a password and is assigned a different level of authorization which defines which operations he or she is allowed to carry out.

3.2 Operational structure

The operation of the SACI system is fully interactive. A number of options is made available to the user by means of menus (Figure 5). These options are briefly described in the following items.

a. Register: this is the procedure that allows the user to define new groups of annotations or to redefine existing groups (correction).

b. Updating: this procedure updates the annotation file ArqDat; a consistence check is made for each reading group; values out of the possible variation range are not accepted, and a flag is set in association to those values in excess of the warning levels; the differences of the current reading to both the previous and the initial reading are presented, as well as

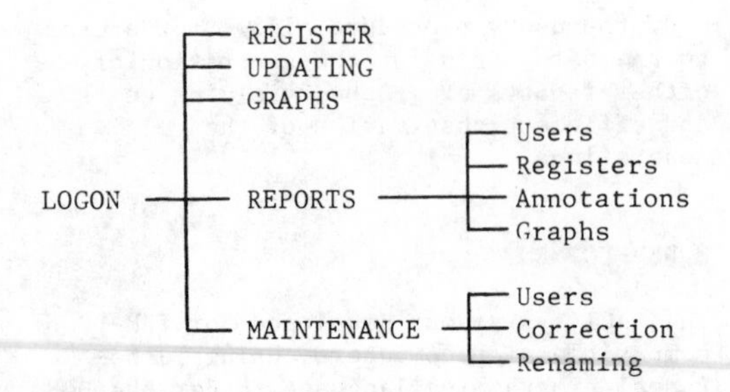

Figure 5. Options of the SACI System

the rates of variation of the readings with respect to time; SACI also accepts special conditions, such as not-read, damaged or eliminated devices; any annotation may be assigned a mathematical function for updating; in this case, the values are generated automatically upon request by the user as a function of other annotations already available.

c. Copy: this is the procedure used for copy of annotations between two data bases of two either local or remote workstations.

d. Graphs: graphs to be produced are defined with this procedure; the definition is made in two different levels; in the first level, the user selects a set of data (reading groups, events and occurrences) to be used for the graphs; in the second level, layout specifications are provided for the set of graphs to be generated; the graphs can be either displayed on the screen or hard-copied through a dot matrix printer. The user has always the two last processed graphs available, one on the screen and the other one in the memory that can be switched (very useful for comparison) by simply stroking a key.

e. Report: this option allows the user to print out the contents of a data basis (registered users, register of annotations, annotations and graphs).

f. Maintenance: critical alterations of the files CadUs and CadAn can be made with this procedure. Critical alterations are those witch may jeopardize the integrity of the data basis. These are, for example, the level of authorization of users, parameters of installation of devices, etc.

4 UTILIZATION

SACI was first implemented and widely used for the technical assistance during construction of two single-track parallel tunnels for the North Extension of the São Paulo Subway.

These tunnels, with 6 m diameter, were constructed using the New Austrian Tunnelling Method (NATM). Being an observational method, its success depends heavily on an efficient process of interpretation of instrumentation, with quick responses to allow for proper interventions in the construction process in due time.

4.1 General description of the application

Along the 65 m of extension of each tunnel, nine control sections with surface and deep settlement devices were installed involving both the tunnels, and four convergence measurement sections for each tunnel. Bench marks for settlement measurement were also installed on the nearby building, adding up to about 200 installed devices. The evolution of the construction was controlled by the annotation of the advance of the excavation faces, for both the heading and bench of each tunnel.

The annotations were organized into 41 groups of devices, two groups of events and one group of occurrences. According to the follow-up program, and considering the different instruments, 20 readings were performed daily during the excavation of the shaft (six months). This average reached 30 readings during the excavation of the tunnels (three months from the portal of the first tunnel to the end of the second tunnel), and dropped to about 10 weekly redings for the following month.

All those readings, together with the annotations of groups of events and occurrences, take 55 Kbytes of a standard floppy disk. The total data basis, including the files of registers of groups, users, graph definitions and indices takes 105 Kbytes, i.e. less than one-third of the capacity of a floppy disk.

4.2 Remote communication

Remote data communication is fundamental for a job like the one presented. It allows technical staff at the construction site, coordination office and engineers' office to have access to the updated information.

The network implemented for communication was foreseen, in the beginning, with only one data basis in the microcomputer of the central coordination office. This basis would be remotely accessed by both the construction site (updating and query) and the enginners (query) and locally by the central coordination office.

In order to optimize the use of the central microcomputer and to reduce the use of telephone lines, improving the performence of the system, one separate data basis for each office is now adopted. Technical staff in the construction site updates the data basis with reading results. The updated annotations are appended to the data basis of the central microcomputer after establishing communication. The same operation is repeated for the engineers' microcomputer.

With this new procedure, the time of remote communication has been reduced to about 30 minutes a day. Besides that, queries can be done at any time, at any place, without the need for establishing communication between the microcomputers.

5 EVOLUTION

For the job described above the first version of SACI was used. Eventhough only two types of graphs were implemented in that version, genericly presented in Figures 2 and 3, a rather significant improvement in the organization and interpretation of instrument readings was observed, in relation to the manual procedure adopted in previous similar jobs.

The other types of graphs are being implemented in a new version under development at the present time, with options for high-resolution screen and for pen plotter. Automatic calculations involving functions of the readings will also be implemented.

The remote communication will be improved in a later version. Built-in communication software will be used, in order to reduce the time for copy of annotations. Thus, manipulation of the files will be local at each one of the network points, reducing communications overhead.

6 CONCLUSIONS

The system tested for the construction jobs of the North Extension was fully approved. As a consequence, the São Paulo subway company decided to invest for the evolution of the SACI system. It has been decided to use it for all future jobs, thus contributing for use increase of microinformatic for geotechnical engineering.

7 ACKNOWLEDGMENTS

The anthors wish to acknowledge the São Paulo Subway Company for the support provided for the development and implementation of the SACI system.

The important backing was the necessary incentive to achieve what seemed to be just a futuristic idea for some time.

8 REFERENCES

[1] J.A.Franklin, Rock Mechnics Review - The monitoring of structures in rock, Int. J. Rock Mech. Min. Sci. Geotech. Abst. 14, 163-192 (1977)

[2] SACI - Sistema para Acompanhamento e Controle de Instrumentação, User's Guide, Internal Repport, Companhia do Metropolitano de São Paulo, 1987

[3] SACI - Sistema para Acompanhamento e Controle de Instrumentação, System Manual, Internal Repport, Companhia do Metropolitano de São Paulo, 1987

Numerical Methods in Geomechanics (Innsbruck 1988), Swoboda (ed.)
© 1988 Balkema, Rotterdam. ISBN 90 6191 809 X

The automatic generation of the finite element mesh

J.Bartůněk
Stavební geologie, Prague, Czechoslovakia

ABSTRACT: This paper deals with the process of the generation of a finite element mesh composed from triangular elements and joints. The contribution is concentrated on solving the problems of stability computations. The whole process is assembled from segments and enables the generation of a suitable mesh in a very short time.

1 INTRODUCTION

The main problem that we must face when solving the problems of stability computations by the finite element method is the construction of a suitable mesh. This paper describes the method, which has been successfully practised in our organisation for several years.

The process of the automatic generation consists of separate moduls that can be combined, used multiply, or combined freely in any logical order.

All the modules are included in the program system GEOSTAB. They work in an interactive way of processing and enable generating the finite element mesh even for untrained users. The conception of the modules usage is shown in the figure 1.

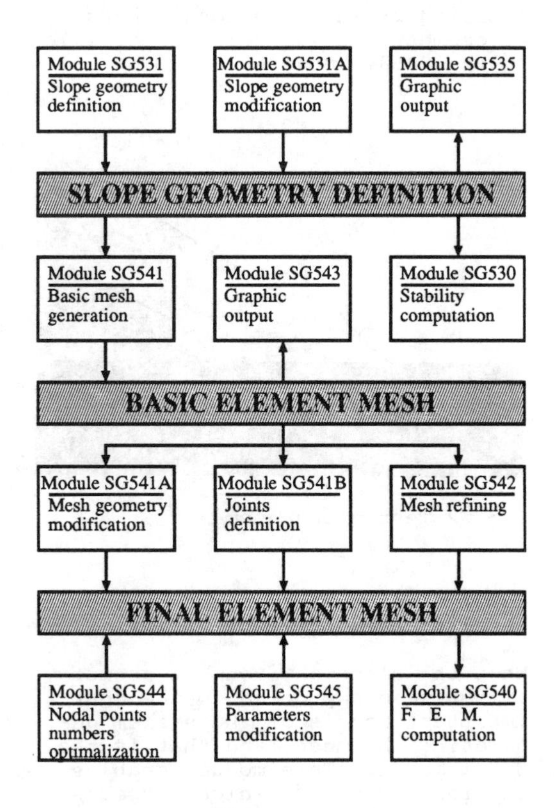

Fig. 1 Conception of segments usage

2 THE DEFINITION OF THE SLOPE ENVIRONMENT GEOMETRY

For an easy construction of the input file containing the slope geometry definition the system includes the module SG531. This module enables to define the slope geometry in a logical order of layers.

The slope surface and its division into layers with different geotechnical parameters are defined by means of group of limiting lines which form the upper boundaries of the layers. The lower boundary of the layers is formed potentially by defining the next lower layer. Vertical limits of the layers are supposed implicitly. Each line is completely defined by the serial numbers of outer points. These numbers are automatically defined by the computer as the points come to input. This method of the slope geometry definition is very clear and simple.

A file, set by this method, can be at the same time used for stability computations using any method from the set of limit equilibrium condition methods.

The example of the test slope with three layers and eleven nodal points is shoxn in the figure 2.

By means of the SG535 module we can print the geometry definition file and do basic control of its correctness. In case of a wrong definition or an intentional change of geometry we can correct the file by the SG531A module.

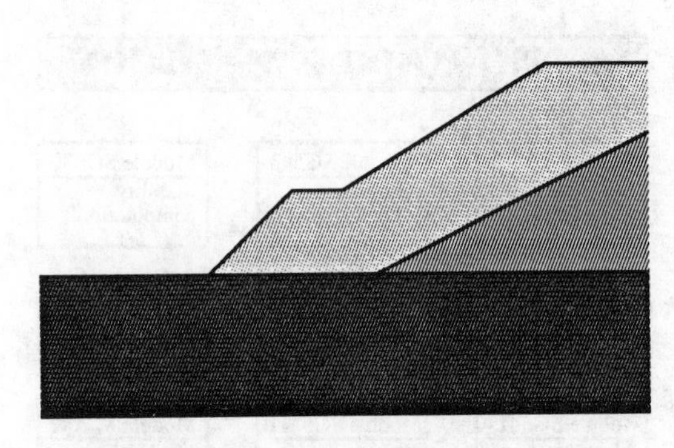

Fig. 2 Test slope geometry

2.1 The generation of the basic finite element mesh

When the slope environment geometry is set we can proceed to the most important module of the system generating the mesh, and that is the SG541 module. This module enables to create the mesh which respects the division of the slope into the layers with different geotechnical parameters depending on the material properties.

Process of the generation of the mesh is shown in the figure 3.

Primarily vertical lines are layed through all the nodal points and they divide the slope environment into stripes. Every nodal point from the definition is now projected on one or several limiting lines. As you can see in the figure 3 the stripes consist of triangels and trapezis.

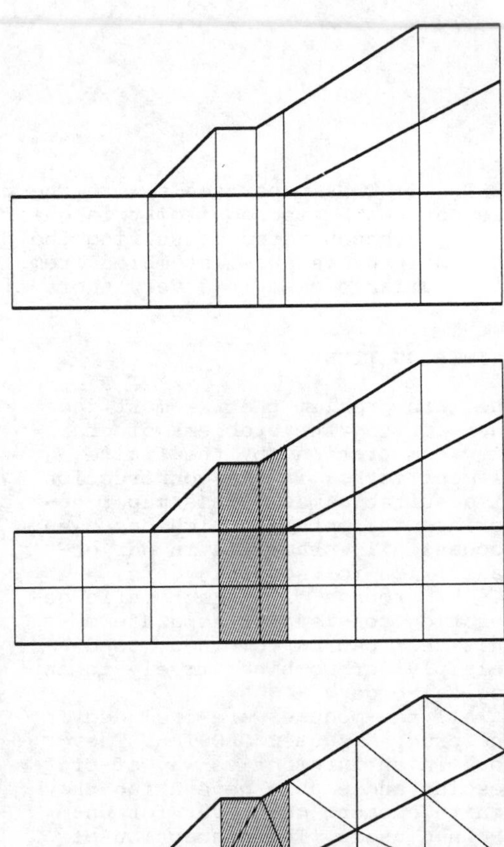

Fig. 3 Creating of the mesh

In the further phase the stripes are divided by means of inserting further vertical lines into the parts of a demanded size. These vertical lines do not run through any nodal point.

Too big areas in the layers are

corrected by means of inserting fictive layers, certainly with the same geotechnical parameters, until they have a correct size.

Eventually all the trapezium parts are divided into triangels by their diagonals. The stripes smaller than demanded are divided into triangular parts together with the next neighbour stripe. When doing this, the program differs nodal points defines from the nodes generated by a vertical projection. The projected nodes can be left out by the program. See marked areas in the figure 3.

2.2 The final element mesh

The automatically generated mesh usually does not fulfill all the criteria we had assigned, because we do not know the exact mathematical formulation of the criteria before. That is why the system includes the SG541A module. This module enables to modify the generated mesh in a very easy way.

This process is realized by the identification of specified nodes. Redundant nodal points are left out as well as the elements whose areas are equal to zero after the identification.

The next step in the mesh generation is the definition of joints. These allow a better description of the environment behaviour. The joints are placed in whole lines, defined by means of the set of nodes. The program module SG541B replaces these lines by joints. Adjoining points are separated and the gap is replaced by a joint. There is certainly no need to emphasize the point, that it is convenient to use the map of a basic mesh with numbered nodal points for this operation, which may be printed by the SG543 module.

The last thing that usually must be done is refining the whole mesh, or only a part of it. This operation is done by help of the program module SG542. This program takes all elements and sets a division flag at those elements which are to be refined. Then the program takes again only flagged elements and test the sides whether the next element is flagged or not. When it is flagged, the side division flag is set. Actually four combinations can appear, as they are shown in the figure 4.

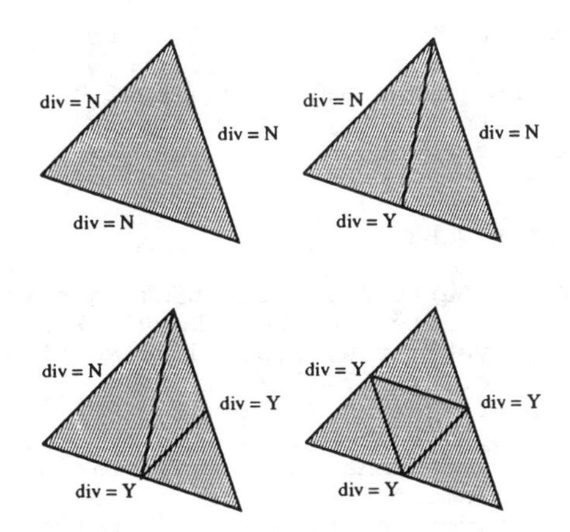

Fig. 4 Refining of the element

In case of the whole division of the element, which means that all sides are flagged, the angles of newly created elements are identical to the ones of the original element. This division does not have any negative influence on the convergence of the finite element method.

2.3 The last modifications before the finite element computation

Perfectly generated covering of the slope environment by the finite element mesh is only the base of the computation. There are two others operations that must be done. First to optimalize the nodal points numbering and second to complete necessary material parameters. The first operation which is inevitable on the computers with a lower computing speed and a lower memory capacity is done with the help of the SG544 module, the second then with the help of the SG544 module.

3 CONCLUSION

The positive quality of this program system is its orientation on a user which enables easily to overcome the difficulties in generating the mesh and allows to pay attention to other problems which arise, for example the input data for choosen type of constitutive laws etc.

REFERENCES

/1/ J.Bartůněk, Automatická generace sítě trojúhelníkových konečných prvků, Proceeding NUMEG 87, part 2, 134-137 (1987)

/2/ R.Haber, M.S.Shephard, J.F.Abel, D.P.Gallagher, A General, two dimensional, Graphical Finit element Preprocessor Utilizing Discrete Transfinite Mappings. Int. Jour. for Num Meth in Eng. Vol 17, No 7, 1015 - 1044(1981)

/3/ F.T.Tracy, Graphical pre- and post-processor for two-dimensional finite element method programs, SIGGRAPH' 77, vol 11, 8-12,(1977)

Advertisements

DESIGN AND SUPERVISION IN EUROPE AND OVERSEAS
TUNNELS · RAILWAYS · BRIDGES · ROADS

INGENIEURBÜRO
LAABMAYR

ROCK MECHANICS, TUNNELING
FOUNDATION AND STRUCTURAL
ENGINEERING

MUNICH – SALZBURG – VIENNA

NATM – NEW AUSTRIAN TUNNELING METHOD

15 YEARS IN THE DEVELOPMENT OF MODERN TUNNELING HAND IN HAND WITH FE-METHOD

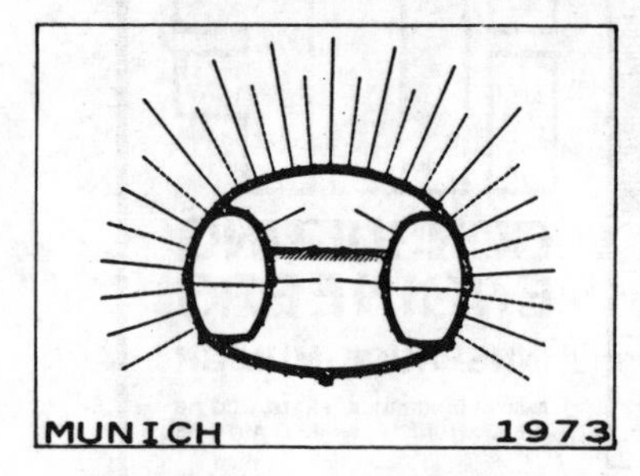
MUNICH 1973

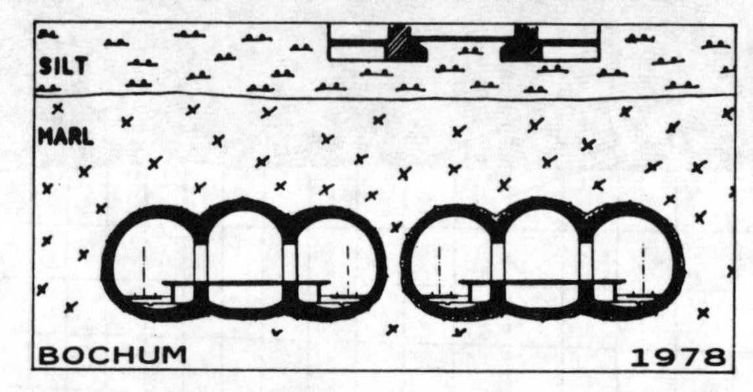

BOCHUM 1978

MUNICH 1974

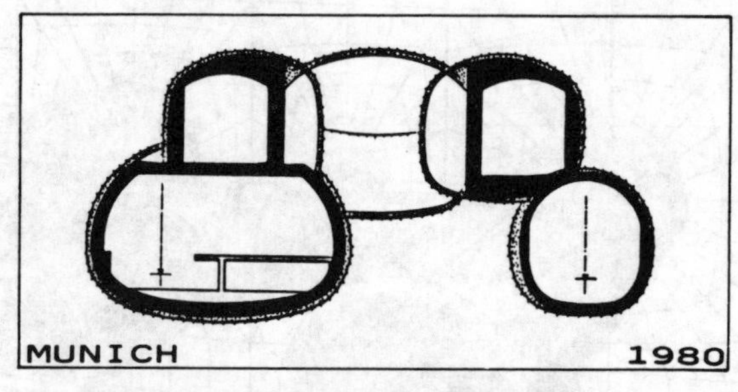
MUNICH 1980

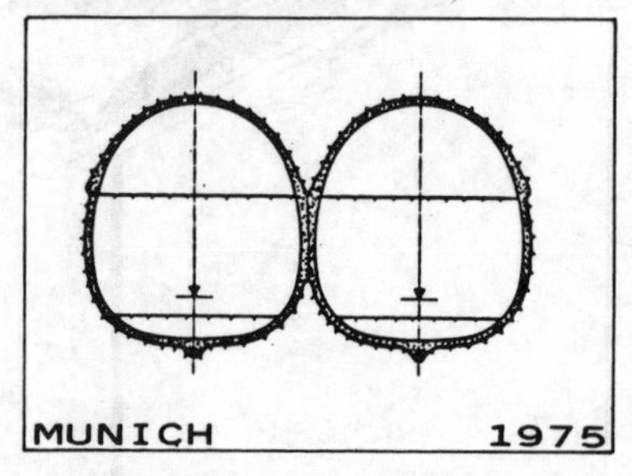
MUNICH 1975

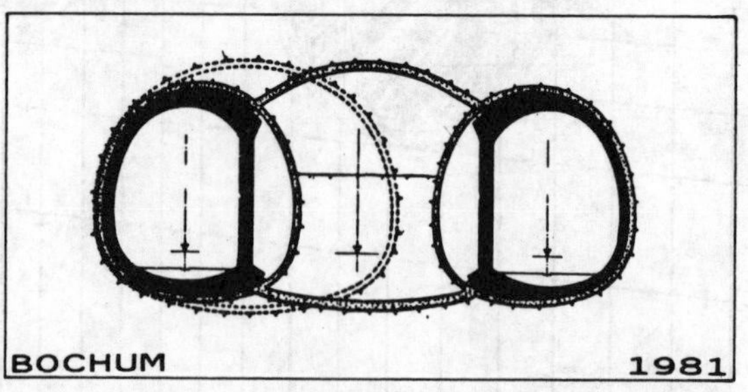
BOCHUM 1981

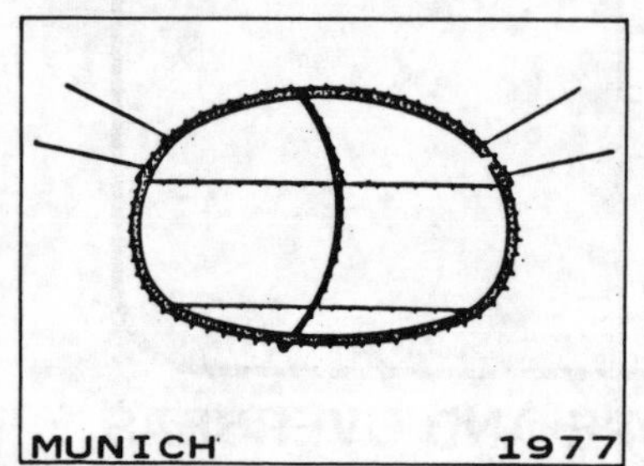
MUNICH 1977

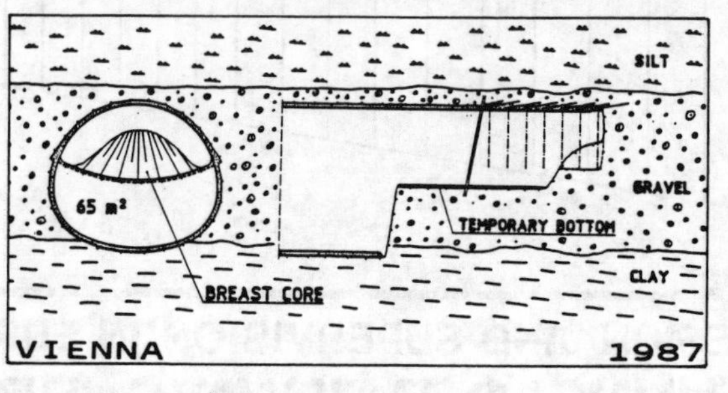

VIENNA 1987

DIPL. ING. FRANZ LAABMAYR - PREISHARTLWEG 4, A-5020 SALZBURG, AUSTRIA TEL: 0662 / 30703 OR 30704

6020 INNSBRUCK, ADAMGASSE 7A
TELEFON 0 52 22 - 23 7 68

HELMUT PASSER, DIPL.-ING.
ZIVILINGENIEUR FÜR BAUWESEN

HORST PASSER, DIPL.-ING. DR. TECHN.
ZIVILINGENIEUR FÜR BAUWESEN

OSWALD NEUNER, DIPL.-ING. DR. TECHN.
ZIVILINGENIEUR FÜR BAUWESEN

IBP
AUSTRIA

FUNCTIONAL PLANNING
WORK UP
SITE MANAGEMENT
TIME AND COST CONTROL

ALL TYPES OF BRIDGE CONSTRUCTION
TRAFFIC SYSTEMS WITH AVALANCHE PROTECTIVE SYSTEMS
DESIGN OF SPORTS FACILITIES
WIDE-SPAN HALL CONSTRUCTION
WATER SUPPLY AND WASTE DISPOSAL INSTALLATION

Beton-u Monierbau Ges.m.b.H.

A-6020 Innsbruck, Zeughausgasse 3

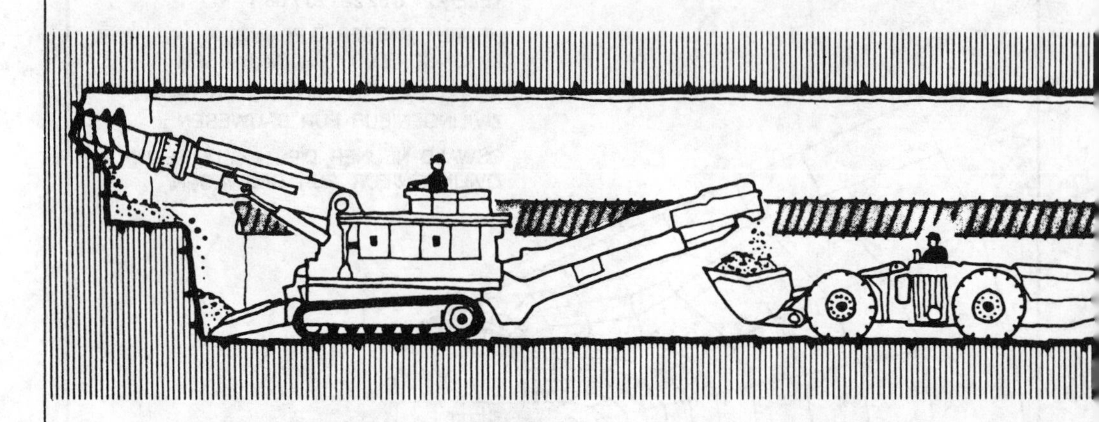

YOUR SPECIALIST IN TUNNELING

APPLIED NUMERICAL METHODS IN

VIENNA SUBWAY CONSTRUCTION

JOINT VENTURE

U 6/4 LÄNGENFELDGASSE

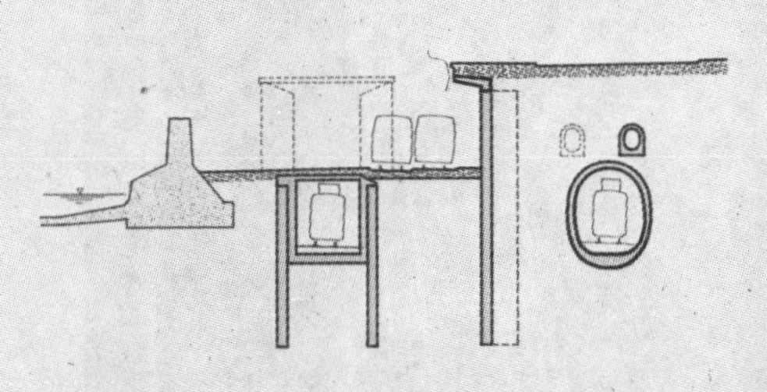

UNIVERSALE — MAYREDER

NEUE REFORMBAU, STETTIN
STUAG

JOINT VENTURE

U 3/5 LANDSTRASSE

UNIVERSALE — PORR — MAYREDER

LANG & MENHOFER, STUAG
ZUBLIN

JOINT VENTURE

U 3/10 VOLKSTHEATER

UNIVERSALE — MAYREDER

DYCKERHOFF & WIDMANN
NEUE REFORMBAU
POLENSKY & ZÖLLNER
WAYSS & FREITAG

JOINT VENTURE

U 3/11 MARIAHILFER STRASSE

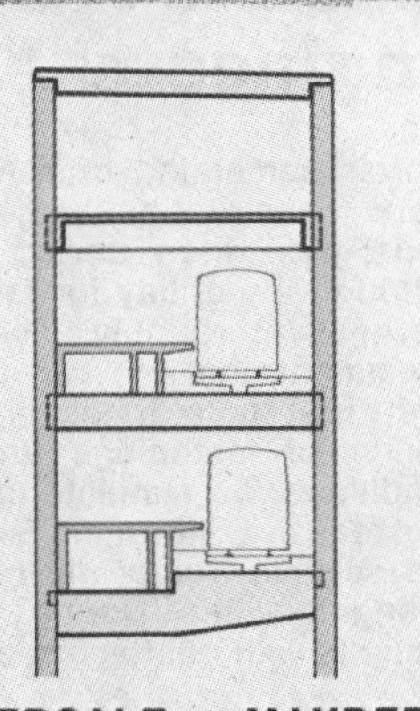

UNIVERSALE — MAYREDER

DYCKERHOFF & WIDMANN
NEUE REFORMBAU
POLENSKY & ZÖLLNER
WAYSS & FREITAG

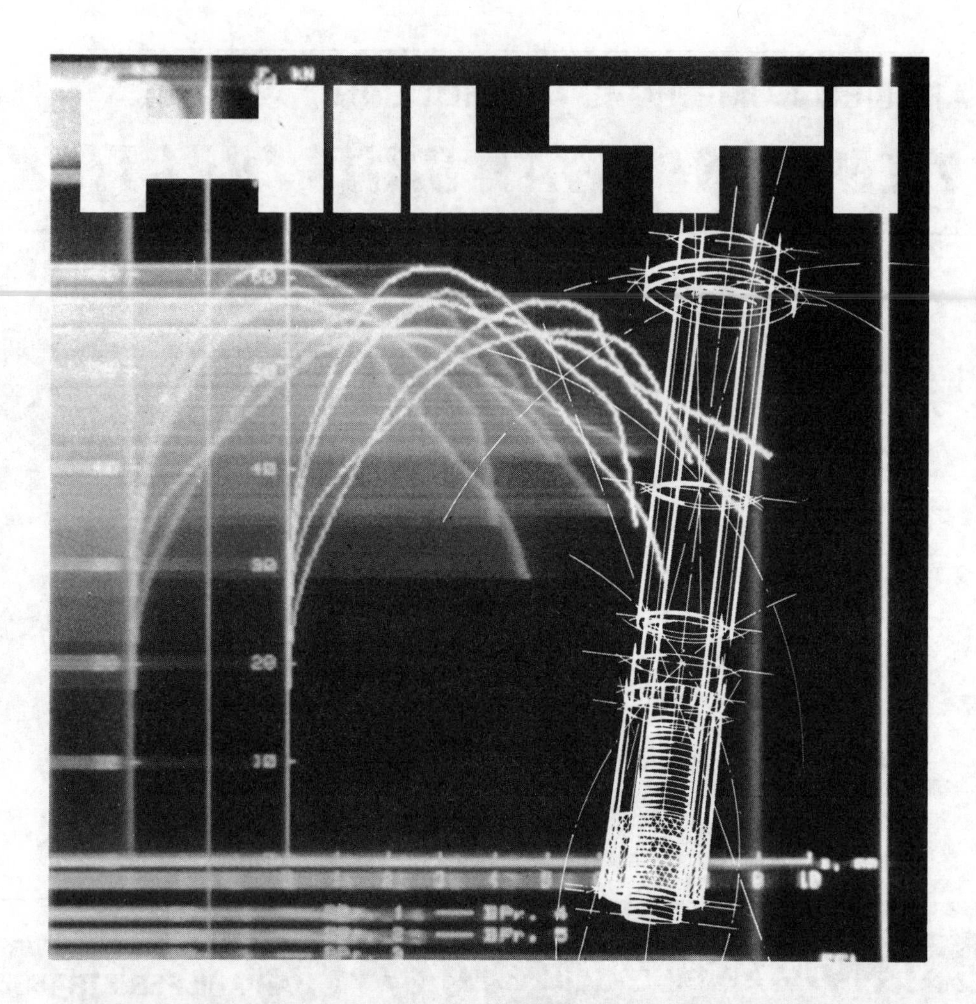

Reliable Fastenings also need Engineers' Know-how!

There's something from Hilti in almost every construction project: anchors, studs, pipe rings, ceiling hangers etc. Although the outlay for fastenings only averages at roughly 1% of a project investment, this 1% can be vital when safety is at stake. To research the performance of fastenings under extreme conditions, we simulate their exposure to excessive vibration, shock loading and seismic tremors in highly modern testing facilities. During some 30,000 contacts with customers every day, our salesmen in the markets pass on their knowledge and, at the same time, learn about new fastening problems for which the construction industry expects our solutions.

Hilti Corporation
FL-9494 Schaan

More safety. More value.

THE SUPER(COMPUTER)SOLUTION
in the
FINITE-ELEMENT-WORLD

Abaqus
Adina
Ansys
AOS/Magnum
Aska
Beasy
Dyna 2D + 3D
Final
Gifts

Marc
MSC-Nastran
Nike 2D + 3D
Permas
Symflex 2
Space/Plot
Staad 3
Stardyne
TPS 10

The 64-bit Supercomputer with integrated Vectorprocessors and the super price/performance ratio